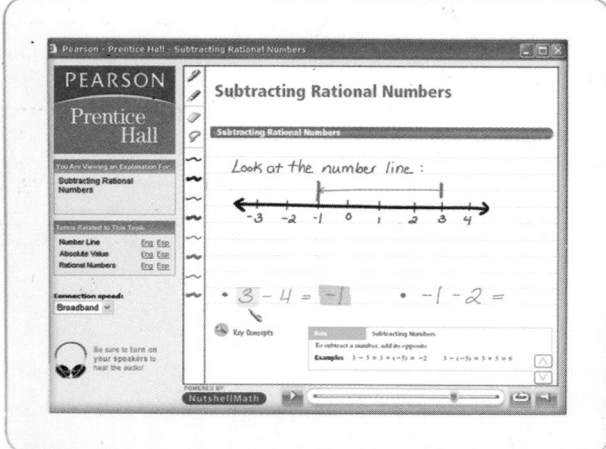

PRENTICE HALL MATHEMATICS **NEW YORK**

INTEGRATED ALGEBRA

Allan E. Bellman

Sadie Chavis Bragg

Randall I. Charles

Basia Hall

William G. Handlin, Sr.

Dan Kennedy

PEARSON

Prentice Hall

Boston, Massachusetts
Upper Saddle River, New Jersey

Authors

Series Authors

Dan Kennedy, Ph.D., is a classroom teacher and the Lupton Distinguished Professor of Mathematics at the Baylor School in Chattanooga, Tennessee. A frequent speaker at professional meetings on the subject of mathematics education reform, Dr. Kennedy has conducted more than 50 workshops and institutes for high school teachers. He is coauthor of textbooks in calculus and precalculus, and from 1990 to 1994 he chaired the College Board's AP Calculus Development Committee. He is a 1992 Tandy Technology Scholar and a 1995 Presidential Award winner.

Randall I. Charles, Ph.D., is Professor Emeritus in the Department of Mathematics and Computer Science at San Jose State University, San Jose, California. He began his career as a high school mathematics teacher, and he was a mathematics supervisor for five years. Dr. Charles has been a member of several NCTM committees and is the former Vice President of the National Council of Supervisors of Mathematics. Much of his writing and research has been in the area of problem solving. He has authored more than 75 mathematics textbooks for kindergarten through college.

Basia Hall currently serves as Manager of Instructional Programs for the Houston Independent School District. With 30 years teaching experience, Ms. Hall has served as a department chair, instructional specialist, instructional supervisor, a school improvement facilitator, and a professional development (TEXTEAMS) trainer. Ms. Hall has developed curriculum for Algebra 1, Geometry, and Algebra 2 and contributed to the development of the Texas Essential Knowledge and Skills. A recipient of the 1992 Presidential Award for Excellence in Mathematics Teaching, Ms. Hall is also a past president of the Texas Association of Supervisors of Mathematics, and she is a state representative for the National Council of Supervisors of Mathematics (NCSM).

Acknowledgments appear on pages 895–896, which constitute an extension of this copyright page.

PEARSON
Prentice
Hall

13-digit ISBN 978-0-13-365787-6
10-digit ISBN 0-13-365787-6
3 4 5 6 7 8 9 10   11 10 09 08

Integrated Algebra and Algebra 2 Authors

Allan E. Bellman is a Lecturer/Supervisor in the School of Education at the University of California, Davis. Before coming to Davis, he was a mathematics teacher for 31 years in Montgomery County, Maryland. He has been an instructor for both the Woodrow Wilson National Fellowship Foundation and the T^3 program. Mr. Bellman has a particular expertise in the use of technology in education and speaks frequently on this topic. He was a 1992 Tandy Technology Scholar.

Sadie Chavis Bragg, Ed.D., is Senior Vice President of Academic Affairs at the Borough of Manhattan Community College of the City University of New York. A former professor of mathematics, she is a past president of the American Mathematical Association of Two-Year Colleges (AMATYC), co-director of the AMATYC project to revise the standards for introductory college mathematics before calculus, and an active member of the Benjamin Banneker Association. Dr. Bragg has coauthored more than 50 mathematics textbooks for kindergarten through college.

William G. Handlin, Sr., is a classroom teacher and Department Chairman of Technology Applications at Spring Woods High School in Houston, Texas. Awarded Life Membership in the Texas Congress of Parents and Teachers for his contributions to the well-being of children, Mr. Handlin is also a frequent workshop and seminar leader in professional meetings throughout the world.

Geometry Authors

Laurie E. Bass is a classroom teacher at the 9–12 division of the Ethical Culture Fieldston School in Riverdale, New York. Ms. Bass has a wide base of teaching experience, ranging from grade 6 through Advanced Placement Calculus. She was the recipient of a 2000 Honorable Mention for the RadioShack National Teacher Awards. She has been a contributing writer of a number of publications, including software-based activities for the Algebra 1 classroom. Among her areas of special interest are cooperative learning for high school students and geometry exploration on the computer. Ms. Bass has been a presenter at a number of local, regional, and national conferences.

Art Johnson, Ed.D., is a professor of mathematics education at Boston University. He is a mathematics educator with 32 years of public school teaching experience, a frequent speaker and workshop leader, and the recipient of a number of awards. Dr. Johnson received the Tandy Prize for Teaching Excellence in 1995, a Presidential Award for Excellence in Mathematics Teaching in 1992, and New Hampshire Teacher of the Year, also in 1992. He was profiled by the Disney Corporation in the American Teacher of the Year Program.

Reviewers

Integrated Algebra Reviewers

Mary Lou Beasley
Southside Fundamental
Middle School
St. Petersburg, Florida

Blanche Smith Brownley
Washington, D.C., Public
Schools
Washington, D.C.

Joseph Caruso
Somerville High School
Somerville, Massachusetts

Belinda Craig
Highland West Junior High
School
Moore, Oklahoma

Jane E. Damaske
Lakeshore Public Schools
Stevensville, Michigan

Stacey A. Ego
Warren Central High School
Indianapolis, Indiana

Earl R. Jones
Formerly, Kansas City
Public Schools
Kansas City, Missouri

Jeanne Lorenson
James H. Blake High School
Silver Spring, Maryland

John T. Mace
Hibbett Middle School
Florence, Alabama

Ann Marie Palmieri-Monahan
Director of Mathematics
Bayonne Board of Education
Bayonne, New Jersey

Marie Schalke
Woodlawn Middle School
Long Grove, Illinois

Julie Welling
LaPorte High School
LaPorte, Indiana

Sharon Zguzenski
Naugatuck High School
Naugatuck, Connecticut

Geometry Reviewers

Marian Avery
Great Valley High School
Malvern, Pennsylvania

Mary Emma Bunch
Farragut High School
Knoxville, Tennessee

Karen A. Cannon
K–12 Mathematics
Coordinator
Rockwood School District
Eureka, Missouri

Johnnie Ebbert
Department Chairman
DeLand High School
DeLand, Florida

Russ Forrer
Math Department Chairman
East Aurora High School
Aurora, Illinois

Andrea Kopco
Midpark High School
Middleburg Heights, Ohio

Gordon E. Maroney III
Camden Fairview High
School
Camden, Arkansas

Charlotte Phillips
Math Coordinator
Wichita USD 259
Wichita, Kansas

Richard P. Strausz
Farmington Public Schools
Farmington, Michigan

Jane Tanner
Jefferson County
International
Baccalaureate School
Birmingham, Alabama

Karen D. Vaughan
Pitt County Schools
Greenville, North Carolina

Robin Washam
Math Specialist
Puget Sound Educational
Service District
Burien, Washington

Algebra 2 and Trigonometry Reviewers

Josiane Fouarge
Landry High School
New Orleans, Louisiana

Susan Hvizdos
Math Department Chair
Wheeling Park High School
Wheeling, West Virginia

Kathleen Kohler
Kearny High School
Kearny, New Jersey

Julia Kolb
Leesville Road High School
Raleigh, North Carolina

Deborah R. Kula
Sacred Hearts Academy
Honolulu, Hawaii

Betty Mayberry
Gallatin High School
Gallatin, Tennessee

John L. Pitt
Formerly, Prince William
  County Schools
Manassas, Virginia

Margaret Plouvier
Billings West High School
Billings, Montana

Sandra Sikorski
Berea High School
Berea, Ohio

Tim Visser
Grandview High School
Cherry Creek School District
Aurora, Colorado

Content Consultants

Ann Bell
Mathematics
Prentice Hall Consultant
Franklin, Tennessee

Blanche Brownley
Mathematics
Prentice Hall Consultant
Olney, Maryland

Joe Brumfield
Mathematics
Prentice Hall Consultant
Altadena, California

Linda Buckhalt
Mathematics
Prentice Hall Consultant
Derwood, Maryland

Andrea Gordon
Mathematics
Prentice Hall Consultant
Atlanta, Georgia

Eleanor Lopes
Mathematics
Prentice Hall Consultant
New Castle, Delaware

Sally Marsh
Mathematics
Prentice Hall Consultant
Baltimore, Maryland

Bob Pacyga
Mathematics
Prentice Hall Consultant
Darien, Illinois

Judy Porter
Mathematics
Prentice Hall Consultant
Fuquay-Varena, North Carolina

Rose Primiani
Mathematics
Prentice Hall Consultant
Egg Harbor City, New Jersey

Jayne Radu
Mathematics
Prentice Hall Consultant
Scottsdale, Arizona

Pam Revels
Mathematics
Prentice Hall Consultant
Sarasota, Florida

Barbara Rogers
Mathematics
Prentice Hall Consultant
Raleigh, North Carolina

Michael Seals
Mathematics
Prentice Hall Consultant
Edmond, Oklahoma

Margaret Thomas
Mathematics
Prentice Hall Consultant
Indianapolis, Indiana

New York Mathematics Student Handbook

New York Program Reviewers

John R. Chubb
Mathematics Teacher
Akron High School
Akron, NY

Carole D. Desoe
Math Department Chair
Scarsdale High School
Scarsdale, NY

Lisa Weber
K-12 Mathematics Coordinator
White Plains High School
White Plains, NY

Irene "Sam" Jovell
Senior Mathematics Specialist
Questar III
Castleton, NY
Irene "Sam" Jovell is a former President of NY State Association of Mathematics Supervisors. Sam taught Mathematics at Niskayuna, NY for 33 years. She is also a member of MST Standards writing team.

A. Rose Primiani, Ed.D.
A. Rose Primiani, Ed.D., is Prentice Hall's Program Consultant/Plannner for Prentice Hall Mathematics. Formerly a Supervisor of Mathematics for the Yonkers Public Schools and Director of Mathematics for Community School District 10 in New York City.

Table of Contents

Contents in Brief

Chapter
1

Variables, Function Patterns, and Graphs

Student Support

✓ Instant Check System

Check Your Readiness 2

Check Skills You'll Need 4, 9, 17, 27, 33, 40

Quick Check 4, 5, 6, 10, 11, 12, 17, 18, 19, 20, 27, 28, 29, 33, 34, 41, 42, 43

Checkpoint Quiz 23, 37

Vocabulary 🔊

New Vocabulary 4, 9, 17, 27, 33, 40

Vocabulary Tip 4, 18, 19, 41

Vocabulary Review 47

GO Online

Video Tutor Help 10

Active Math 6, 19, 41

Homework Video Tutor 7, 13, 21, 30, 36, 44

Lesson Quizzes 7, 15, 23, 31, 37, 45

Vocabulary Quiz 47

Chapter Test 50

Algebra at Work 32

Chapter 2

Rational Numbers

Assessment and Test Prep

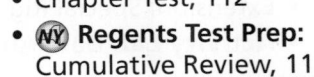

Solving Equations

Chapter 4

Solving Inequalities

Chapter 5

Graphs and Functions

Chapter 6

Linear Equations and Their Graphs

Chapter 8

Exponents and Exponential Functions

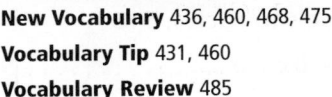

Polynomials and Factoring

Chapter 10

Quadratic Equations and Functions

Radical Expressions and Equations

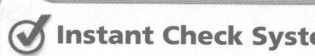

Chapter 12

Rational Expressions and Functions

Student Support

Assessment and Test Prep

Chapter NY

Assessment and Test Prep

Student Support

Connect Your Learning

through problem solving, activities, and the Web

Applications: Real-World Applications

Applications: **Algebra at Work**

Applications: **A Point in Time**

Activities

Applications: Interdisciplinary Connections

Activities: Activity Labs

Activities: Guided Problem Solving

Go Online

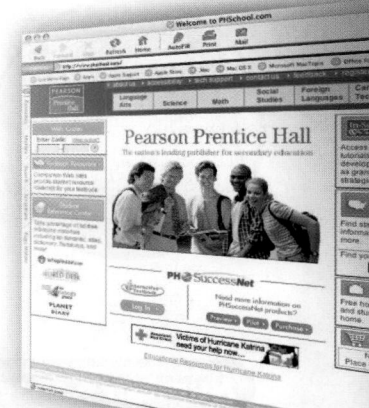

Throughout this book you will find links to the Prentice Hall Web site. Use the Web Codes provided with each link to gain direct access to online material. Here's how to *Go Online*:

1. Go to PHSchool.com
2. Enter the Web Code
3. Click Go!

Lesson Web Codes

Lesson Quiz Web Codes: There is an online quiz for every lesson. Access these quizzes with Web Codes ata-0101 through ata-1208 for Lesson 1-1 through Lesson 12-8. See page 88.

Lesson Quizzes
Web Code format: ata-0204
02 = Chapter 2   04 = Lesson 4

Homework Video Tutor Web Codes: For every lesson, there is additional support online to help students complete their homework. Access this homework help online with Web Codes ate-0101 through ate-1208 for Lesson 1-1 through Lesson 12-8. See page 341.

Homework Video Tutor
Web Code format: ate-0605
06 = Chapter 6   05 = Lesson 5

Chapter Web Codes

Chapter	Vocabulary Quizzes	Chapter Tests	Activity Labs
1	atj-0151	ata-0152	
2	atj-0251	ata-0252	ate-0253
3	atj-0351	ata-0352	
4	atj-0451	ata-0452	ate-0453
5	atj-0551	ata-0552	
6	atj-0651	ata-0652	ate-0653
7	atj-0751	ata-0752	
8	atj-0851	ata-0852	ate-0853
9	atj-0951	ata-0952	
10	atj-1051	ata-1052	ate-1053
11	atj-1151	ata-1152	
12	atj-1251	ata-1252	ate-1253
End-of-Course		ata-1254	

Additional Web Codes

Video Tutor Help:
Use Web Code ate-0775 to access engaging online instructional videos to help bring math concepts to life. See page 390.

Data Updates:
Use Web Code atg-9041 to get up-to-date government data for use in examples and exercises. See page 351.

Algebra at Work:
For information about each Algebra at Work feature, use Web Code atb-2031. See page 452.

New York State Learning Standards

To help you master essential mathematical knowledge and deepen your critical thinking skills, the New York State Board of Education established Learning Standards. Here is an overview of the strands for Number Sense and Operations, Algebra, Measurement, and Statistics and Probability.

Number Sense and Operations Strand

The standards in this strand (A.N.1 through A.N.8) cover two topics: *Number Theory (A.N.1) and Operations (A.N.2 through A.N.8)*.

When you have mastered these standards you will understand:

- Numbers, multiple ways of representing numbers, relationships among numbers, and number systems.
- Meanings of operations and procedures, and how they relate to one another.

What It Means To You

In standard A.N.1 you are asked to identify and apply different properties of real numbers. These include the closure, commutative, associative, distributive, identity, and inverse properties. Understanding these different properties will help you to justify the steps you use to solve a problem.

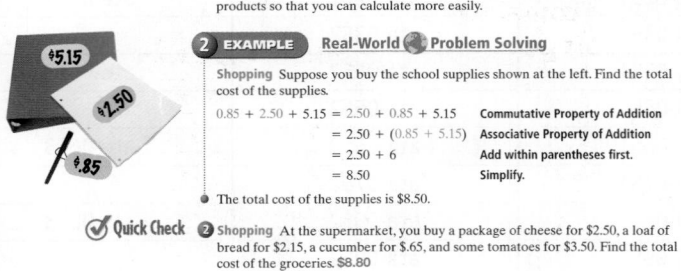

You can also use the properties to reorganize the order of numbers in sums or products so that you can calculate more easily.

2 EXAMPLE   Real-World Problem Solving

Shopping Suppose you buy the school supplies shown at the left. Find the total cost of the supplies.

$$0.85 + 2.50 + 5.15 = 2.50 + 0.85 + 5.15$$ **Commutative Property of Addition**
$$= 2.50 + (0.85 + 5.15)$$ **Associative Property of Addition**
$$= 2.50 + 6$$ **Add within parentheses first.**
$$= 8.50$$ **Simplify.**

- The total cost of the supplies is $8.50.

Quick Check **2 Shopping** At the supermarket, you buy a package of cheese for $2.50, a loaf of bread for $2.15, a cucumber for $.65, and some tomatoes for $3.50. Find the total cost of the groceries. $8.80

The Operations standards, A.N.2 through A.N.8, give you an idea of the type of operations and procedures that you will encounter throughout the book.

Algebra Strand

The standards in this strand (A.A.1 through A.A.45) cover five topics:

Variables and Expressions (A.A.1 through A.A.2 and A.A.12 through A.A.20), Equations and Inequalities, (A.A.3 through A.A.12 and A.A.21 through A.A.28), Patterns Relations, and Functions (A.A.29 through A.A.31), Coordinate Geometry (A.A.32 through A.A.41), and Trigonometric Functions (A.A.42 through A.A.45).

When you have mastered these standards you will be able to:
- Represent and analyze algebraically a wide variety of problem solving situations.
- Perform algebraic procedures accurately.
- Recognize, use, and represent algebraically patterns, relations, and functions.

What It Means To You

Algebra is a tool you can use to solve and model situations. For example you can model the height of an object moving under the influence of gravity using a quadratic function.

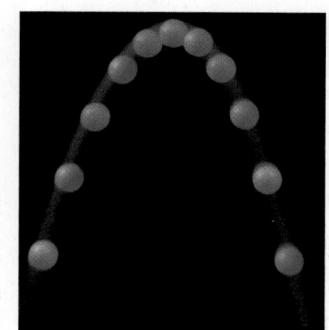

Measurement Strand

The standards in this strand (A.M.1 through A.M.3) cover two topics:

Units of Measurement (A.M.1 and A.M.2) and Error and Magnitude (A.M.3).

When you have mastered these standards you will be able to:
- Determine what can be measured and how, using appropriate methods and formulas.
- Understand that all measurement contains error and be able to determine its significance.

What It Means To You

When you work with different measurements it is important that you are using appropriate units. To find out which container of apple juice is the better buy, you need to find the cost per ounce for each container. It is important to use the appropriate units so that you can compare the costs.

Price of Apple Juice

Price	Volume
$ .72	16 oz
$1.20	32 oz
$1.60	64 oz

Statistics and Probability Strand

The 23 standards in this strand (A.S.1 through A.S.23) cover four topics:

Organization and Display of Data (A.S.1 through A.S.8), Analysis of Data (A.S.9 through A.S.14), Predictions from Data (A.S.15 through A.S.17), and Probability (A.S.18 through A.S.23).

When you have mastered these standards you will be able to:
- Collect, organize, display, and analyze data.
- Make predictions that are based upon data analysis.
- Understand and apply concepts of probability.

Workout for New York Standards Mastery

NEW YORK

Ready to go after Chapter 1

❓ **For help, go to the lesson in green.**

1. Alan built an additional room onto his house. The length of the room is 3 times the width.

$$\boxed{}\ w$$
$$3w$$

The perimeter of the room is 60 feet. What is the length of the room?
(Lesson 1-2)

(1) 7.5 feet   **(3)** 22.5 feet
(2) 15 feet   **(4)** 45 feet

2. Which expression represents "5 more than two-thirds of a number"?
(Lesson 1-1)

(1) $5 \cdot 2 + 3 \cdot n$
(2) $5 + 2 \cdot 3 + n$
(3) $5 + \frac{2n}{3}$
(4) $\frac{5 + 2}{3} \cdot n$

3. The table shows the cost of renting a van for one day, where c is the cost in dollars and m is the number of miles driven. Which equation is the function rule for the relationship?
(Lesson 1-4)

m	10	20	30
c	25	30	35

(1) $c = 2m - 10$
(2) $c = m + 15$
(3) $c = 3m - 5$
(4) $c = 0.5m + 20$

4. What is the value of the expression $-3|a| + 4|b|$ when $a = 2$ and $b = -5$?
(Lesson 1-2)

(1) -26 **(3)** 14
(2) -14 **(4)** 26

For problems 5–6, clearly indicate the necessary steps, including appropriate formula substitutions, diagrams, graphs, and charts.

5. Constructed Response The Garcias want to compare their monthly electricity bills for the summers of 2005 and 2006. Explain how they can find the difference in their mean monthly electricity bills for the two summers?
(Lesson 1-6)

Month	2005	2006
June	$167.00	$149.00
July	$258.00	$276.00
August	$301.00	$310.00

6. Constructed Response Angela decided to use a twelve-week program to help her train for a half-marathon. The following table shows how far she could run, without resting, at different times during her training.
(Lesson 1-5)

Week	Distance
0	1.25
2	2.0
4	2.8
6	4.2
8	5.9
10	8.6
12	11.5

Make a scatter plot of the data.

Does the scatter plot indicate improvement in Angela's ability to run distances? Explain your answer.

Learning Standards
NY A.A.1, NY A.A.5, NY A.N.6, NY A.S.4, NY A.S.7, NY A.S.12

Workout for New York Standards Mastery

Ready to go after Chapter 2

❓ **For help, go to the lesson in green.**

1. Which expression has a value of 13?
   (Lesson 2-3)
   (1) $(3 + 2)^2$
   (2) $3^2 + 2^2$
   (3) $3 + (2)^2$
   (4) $3^3 + 2^2$

2. Which of these expressions is equivalent to $3(b - 4) + 2b$?
   (Lesson 2-4)

   (1) $5b - 4$
   (2) $5b - 12$
   (3) $6b - 4$
   (4) $6b - 12$

3. Which of these expressions is equivalent to $rs - ts$?
   (Lesson 2-4)
   (1) $s(r - t)$
   (2) $s - (r \cdot t)$
   (3) $(r \cdot s) - t$
   (4) $r(s - t)$

4. Which of these properties justifies the statement below?

   $3(x - 2) - 2x = 3x - 6 - 2x$
   (Lesson 2-5)
   (1) Commutative Property of Multiplication
   (2) Identity Property of Addition
   (3) Distributive Property
   (4) Associative Property of Addition

5. A quality-control worker inspects 50 computers and finds 2 of them to be defective. Based on this information, what is the probability that the next computer inspected will *not* be defective?
   (Lesson 2-6)

   (1) 60%
   (2) 75%
   (3) 98%
   (4) 96%

For problems 6–7, clearly indicate the necessary steps, including appropriate formula substitutions, diagrams, graphs, and charts.

6. **Constructed Response** A quiz has 2 true-false questions and 1 multiple-choice question. The multiple-choice question has 4 possible answers. If Lori guesses at all 3 questions, what is the probability that she will answer all 3 correctly? Show your work and explain how you found your answer.
   (Lesson 2-7)

7. **Constructed Response** A spinner is divided into three equal sections numbered 1 to 3. Latrisha spins the spinner 50 times. The table below shows her results.

Outcome	1	2	3
Frequency	13	18	19

   Which is greater, the experimental probability of spinning a 2 or the theoretical probability of spinning a 2?

   Explain how to find the experimental probability and the theoretical probability and the meaning of each.
   (Lesson 2-6)

New York Workout

Learning Standards
NY A.N.1, NY A.N.6, NY A.A.1, NY A.A.13, NY A.S.19, NY A.S.20, NY A.S.21, NY A.S.22, NY A.S.23

Workout for New York Standards Mastery

Ready to go after Chapter 3

? For help, go to the lesson in green.

1. What is the solution to $5 - 2n = 17$?
(Lesson 3-1)

(1) $n = -11$ **(3)** $n = -6$
(2) $n = -10$ **(4)** $n = 7$

2. The liquid in each beaker shown has the same mass.

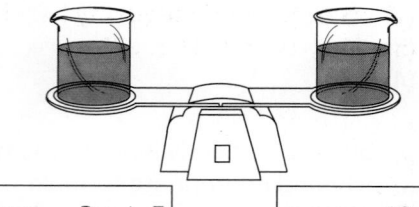

mass = $3g + 5$ mass = $10g + 12$

What is the value of g?
(Lesson 3-3)

(1) -1 **(3)** 1
(2) $\frac{17}{13}$ **(4)** $\frac{17}{7}$

3. What is the solution to $\frac{a + 3}{2} = \frac{4 - 2}{8}$?
(Lesson 3-4)

(1) $a = -1.5$ **(3)** $a = 0.5$
(2) $a = -2.5$ **(4)** $a = 2.25$

4. A tree casts a shadow that is 21 feet long at the same time as a fence post casts a shadow that is 6 feet long. The fence post is 4 feet tall.

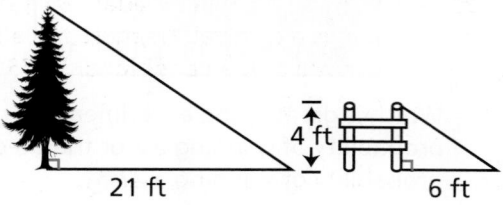

21 ft   4 ft   6 ft

What is the height of the tree?
(Lesson 3-5)

(1) 12 ft   **(3)** 25 ft
(2) 14 ft   **(4)** 30 ft

5. Mandy's store is having a sale. All sweaters are being sold at a price of $12.00. If the original price of a certain sweater was $15.00, what is the percent of decrease?
(Lesson 3-7)

(1) 18%   **(3)** 22%
(2) 20%   **(4)** 24%

For problems 6–7, clearly indicate the necessary steps, including appropriate formula substitutions, diagrams, graphs, and charts.

6. Constructed Response Kim rode her bike at a rate of 24 km/h to the post office and 16 km/h back home. If her trip to the post office and back took a total of 75 minutes, how far does Kim live from the post office?

Write an equation to represent the distance Kim traveled. Explain why you chose that equation.

Solve the equation you wrote in part A.
(Lesson 3-6)

7. Constructed Response Marian wants to determine the distance across Mirror Lake.
(Lesson 3-9)

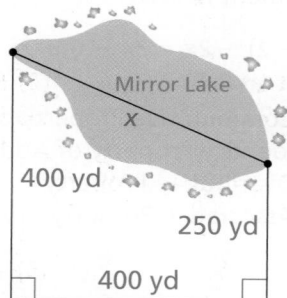

Mirror Lake
x
400 yd
250 yd
400 yd

Write an equation that can be used to find the distance x. Explain the process you used to determine your answer.

Solve the equation you wrote in A to find the distance across Mirror Lake to the nearest yard. Show your work.

Learning Standards
NY A.A.5, NY A.A.6, NY A.A.22, NY A.A.25, NY A.A.26, NY A.A.45, NY A.N.5

Workout for New York Standards Mastery

NEW YORK

Ready to go after Chapter 4

For help, go to the lesson in green.

1. At least 26 students must sign up for a painting class or it will be canceled. So far, 11 students have signed up. Which inequality **best** represents n, the number of additional students who must sign up to prevent the class from being canceled?
   (Lesson 4-1)

   (1) $n \geq 15$ (3) $n \geq 37$
   (2) $n \leq 15$ (4) $n \leq 37$

2. Which of these graphs represents the solution to $q - 8 < -3$?
   (Lesson 4-2)

   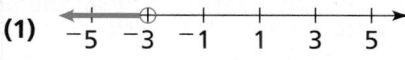

   (1) ![number line](−5 −3 −1 1 3 5, open circle at −3)

   (2) ![number line](−5 −3 −1 1 3 5, closed circle at −1 arrow left)

   (3) ![number line](−5 −3 −1 1 3 5, open circle at 5)

   (4) ![number line](−5 −3 −1 1 3 5, closed circle at 5)

3. A construction company is building new homes on 400 acres of land. If each home must have 1.5 acres, what is the greatest number of homes that can be built?
   (Lesson 4-3)

   (1) 266   (3) 599
   (2) 267   (4) 600

4. Solve the inequality $-4(s + 2) - 2s > 28$.
   (Lesson 4-4)

   (1) $s > -\frac{26}{6}$ (3) $s > -6$
   (2) $s < -\frac{26}{6}$ (4) $s < -6$

5. Solve $-3 \leq t - 4 < 7$.
   (Lesson 4-5)

   (1) $-7 \leq t < 3$
   (2) $7 \leq t < -3$
   (3) $1 \leq t < 11$
   (4) $-1 \leq t < -11$

6. Which compound inequality has the same meaning as $|x + 8| < 23$?
   (Lesson 4-6)

   (1) $-15 < x < 31$
   (2) $-31 < x < 15$
   (3) $-15 < x$ or $x > 31$
   (4) $x > -15$ or $x < 31$

For problems 7–8, clearly indicate the necessary steps, including appropriate formula substitutions, diagrams, graphs, and charts.

7. **Constructed Response** Harvey is a sales associate in an electronics store. He earns a weekly salary of $225 plus a commission equal to 5% of his sales. This week, he expects to earn at least $510. Write and solve an inequality to find the dollar amount he must sell. Explain the steps you used to find the answer.
   (Lesson 4-4)

8. **Constructed Response** Juan has $15.00 to spend at a carnival. He must pay a $5.00 entrance fee plus an additional $0.75 for each ride he goes on.
   (Lesson 4-4)

   Write an inequality that can be used to find r, the number of rides that Juan can go on. Explain how you chose the inequality.

   Does Juan have enough money to go on 12 rides? Explain the process you used to determine your answer.

Learning Standards
NY A.A.4, NY A.A.5, NY A.A.6, NY A.A.21, NY A.A.24, NY A.G.4

New York Workout

Workout for New York Standards Mastery

Ready to go after Chapter 5

? **For help, go to the lesson in green.**

1. Which best represents the equation of the line shown?
(Lesson 5-3)

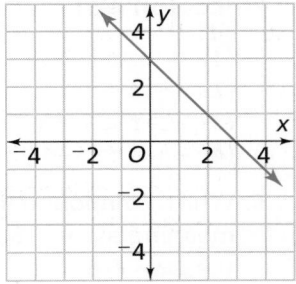

  (1) $y = x + 3$ **(3)** $y = -x + 3$
  (2) $y = x - 3$ **(4)** $y = -x - 3$

2. Which set of ordered pairs is a function?
(Lesson 5-2)

  (1) $\{(3, 5), (3, 6), (3, 7), (3,8)\}$
  (2) $\{(1, 2), (2, 3), (3, 4), (4, 5)\}$
  (3) $\{(-2, 0), (0, -2), (1, -2), (0, 1)\}$
  (4) $\{(4, -3), (-5, -3), (4, 3), (5, 3)\}$

3. Which table of values can be generated by $f(x) = 3x - 2$?
(Lesson 5-3)

(1)

x	$f(x)$
−5	−1
−2	0
1	1

(3)

x	$f(x)$
$\frac{1}{3}$	−1
$\frac{2}{3}$	0
1	1

(2)

x	$f(x)$
−1	$\frac{1}{3}$
0	$\frac{2}{3}$
1	1

(4)

x	$f(x)$
−1	−5
0	−2
1	−1

4. Which equation is a direct variation?
(Lesson 5-5)

  (1) $y = x + 7$ **(3)** $y = 3x$
  (2) $y = 4x - 3$ **(4)** $y = x^2$

5.

r	$f(r)$
−2	2
−1	1
0	0
1	−1
2	−2

Which function is satisfied by all the points in the table?
(Lesson 5-4)

  (1) $f(r) = r + 4$
  (2) $f(r) = -r$
  (3) $f(r) = r - 2$
  (4) $f(r) = r$

For problem 6, clearly indicate the necessary steps, including appropriate formula substitutions, diagrams, graphs, and charts.

6. Constructed Response The Paymore Company was founded in 2000 with 38 employees. Management plans to add 75 employees per year.

Write a function $f(n)$ to describe the number of employees after n years. Explain how you chose the function.

In what year will the company reach 1000 employees? Show your work and explain how you found your answer.
(Lesson 5-4)

NY **Learning Standards**
NY A.A.5, NY A.N.5, NY A.G.3, NY A.G.4

Workout for New York Standards Mastery

Ready to go after Chapter 6

? **For help, go to the lesson in green.**

1. What is the slope of the line that contains points (4, ⁻10) and (⁻5, 0)?
(Lesson 6-1)

(1) $-\frac{10}{9}$ (3) $\frac{9}{10}$

(2) $-\frac{9}{10}$ (4) $\frac{10}{9}$

2.

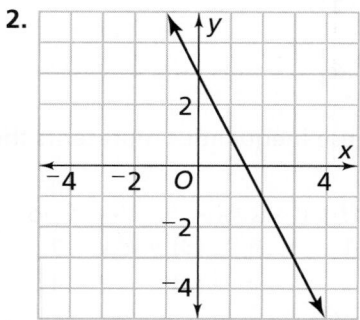

Which of these equations represents the line shown on the graph?
(Lesson 6-2)

(1) $y = -2x + 3$

(2) $y = 2x + 3$

(3) $y = -\frac{1}{2}x$

(4) $y = \frac{1}{2}x + 3$

3. Which is the equation of a line that passes through (−3, 8) with slope 4?
(Lesson 6-5)

(1) $y = 4x + 3$ (3) $y - 8 = 4(x - 3)$

(2) $y = 4x + 8$ (4) $y - 8 = 4(x + 3)$

4. What is the equation of the line that is perpendicular to the line $2x - 3y = 6$ and goes through the point (6, −17)?
(Lesson 6-6)

(1) $y = \frac{2}{3}x + 2$ (3) $y = \frac{2}{3}x - 8$

(2) $y = -\frac{3}{2}x - 8$ (4) $y = -\frac{2}{3}x - 17$

For problems 5–6, clearly indicate the necessary steps, including appropriate formula substitutions, diagrams, graphs, and charts.

5. Constructed Response The graph below shows the relationship between hours of practice and change in typing speed.

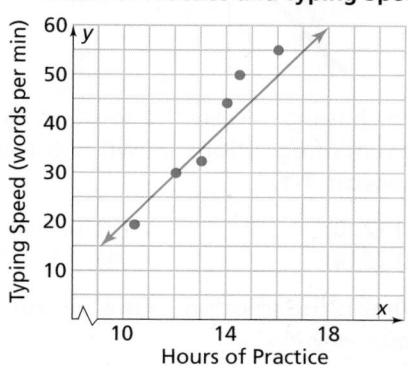

Hours of Practice and Typing Speed

Explain how to find the equation of the line of best fit.

Based on the equation you wrote in Part A, predict typing speed after 20 hours of practice. Show your work.
(Lesson 6-7)

6. Constructed Response Keisha wants to buy a scanner that costs $129 for her computer. She already has $25 and plans to save an additional $15 each week.
(Lesson 6-3)

Write an equation that relates the number of weeks and the amount saved. Explain how you chose the equation. Then, use the equation to make a graph. Be sure to title your graph and label the axes.

Determine how many weeks it will take Keisha to save enough money to buy the scanner. Explain the process you used to determine your answer.

Learning Standards
NY A.A.5, NY A.A.33, NY A.A.34, NY A.G.4, NY A.S.8, NY A.S.17

New York Workout

Workout for New York Standards Mastery

Ready to go after Chapter 7

? For help, go to the lesson in green.

1.

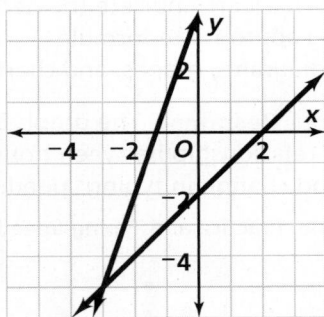

Which is most likely the solution to the system of equations shown in the graph?
(Lesson 7-1)

(1) $(2, 0)$ **(3)** $(0, -2)$
(2) $(0, 4)$ **(4)** $(-3, -5)$

2. What is the solution to this system of equations?
$$\begin{cases} y = x + 2 \\ y = 2x - 1 \end{cases}$$
(Lesson 7-2)

(1) $(-3, -1)$ **(3)** $(1, 3)$
(2) $(1, 1)$ **(4)** $(3, 5)$

3. Connie sells birdhouses at craft fairs. If x represent the number of birdhouses, her expenses can be represented by the equation $y = 5x + 30$, and her income can be represented by the equation $y = 8x$. How many birdhouses must Connie sell in order to break even?
(Lesson 7-4)

(1) 0

(2) 30

(3) 3

(4) 10

4.

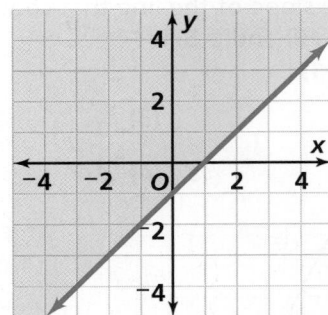

Which of these inequalities represents the graph?
(Lesson 7-5)

(1) $y > x - 1$ **(3)** $y \geq x - 1$
(2) $y < x - 1$ **(4)** $y \leq x - 1$

For problems 5–6, clearly indicate the necessary steps, including appropriate formula substitutions, diagrams, graphs, and charts.

5. Constructed Response A group of 3 adults and 25 children visited Family Fun Park and spent $486 for admission. Another group of 5 adults and 60 children spent $1,085 for admission. What is the cost of admission for adults and children? Show your work to support your answer. Explain each step.
(Lesson 7-3)

6. Constructed Response Kristin has two after school jobs and can work no more than 20 hours per week. When she baby-sits, she earns $7 per hour; she earns $9.00 per hour as a cashier. She needs to earn at least $100 per week.

Write a system of inequalities to model the problem. Explain how you chose the inequalities. Then graph the system.

What are 2 possible ways she could divide her hours between the two jobs? Explain how you found your answer.
(Lesson 7-6)

Learning Standards
NY A.A.7, NY A.A.10, NY A.G.6, NY A.G.7, NY A.G.9

Workout for New York Standards Mastery

Ready to go after Chapter 8

? For help, go to the lesson in green.

1. Which is equivalent to $\dfrac{4r^0s}{-12t^3u^{-5}}$?
(Lesson 8-1)

(1) $\dfrac{48su^5}{t^3}$ (3) $\dfrac{5rs}{-3t^3u}$

(2) $\dfrac{3t^3}{su^5}$ (4) $\dfrac{su^5}{-3t^3}$

2. A dust particle has a diameter of 6×10^{-5} meters. Which of the following shows the diameter of a dust particle in standard notation?
(Lesson 8-2)

(1) 0.0000006 m

(2) 0.000006 m

(3) 0.00006 m

(4) 0.0006 m

3.

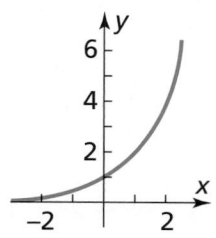

9×10^{-2} m

4.2×10^6 m

What is the area of the triangle?
(Lesson 8-3)

(1) 1.89×10^{-12} m^2

(2) 1.89×10^4 m^2

(3) 1.89×10^5 m^2

(4) 1.89×10^8 m^2

4. Which is equivalent to $\left(\dfrac{(u^{-3}v)^{-6}}{(u^{-1}v)^{-2}}\right)^{-2}$?
(Lesson 8-5)

(1) u^4v^6 (3) $\dfrac{v^8}{u^{32}}$

(2) $u^{72}v^{64}$ (4) $\dfrac{v^8}{u^4}$

5. Which function is shown in the graph?
(Lesson 8-7)

(1) $y = 2^x$ (3) $y = \left(\dfrac{1}{2}\right)^x$

(2) $y = -(2^x)$ (4) $y = -\left(\dfrac{1}{2}\right)^x$

For problem 6, clearly indicate the necessary steps, including appropriate formula substitutions, diagrams, graphs, and charts.

6. Constructed Response When Michael started a job in 2002, he earned $32,000 annually. He receives a 5% increase each year.

Write an equation to model Michael's salary. Explain how you determined the equation and what each variable represents.

Predict what Michael's salary will be in 2010 if he continues to receive a 5% increase each year. Round your answer to the nearest $10. Explain how you found your answer.
(Lesson 8-8)

Learning Standards
NY A.A.9, NY A.A.12, NY A.N.4, NY A.N.6, NY A.G.4

New York Workout

Workout for New York Standards Mastery

Ready to go after Chapter 9

❓ **For help, go to the lesson in green.**

1. Simplify: $(10a^2 - 8a + 3) + (-4a^2 + 9a - 4)$
(Lesson 9-1)

(1) $-40a^2 - 72a - 12$
(2) $2a^2 + 5a - 1$
(3) $6a^2 + a - 1$
(4) $14a^2 + 17a + 7$

2. Which is equivalent to $3x(x^2 + 2x + 1)$?
(Lesson 9-2)

(1) $x^3 + x^2 + x$
(2) $x^3 + 2x^2 + x$
(3) $3x^3 + 5x^2 + 4x$
(4) $3x^3 + 6x^2 + 3x$

3. Multiply: $(7x - 3)(-2x + 1)$
(Lesson 9-3)

(1) $5x^2 + 2x - 2$
(2) $5x^2 - 5x - 2$
(3) $-14x^2 + 6x - 3$
(4) $-14x^2 + 13x - 3$

4.

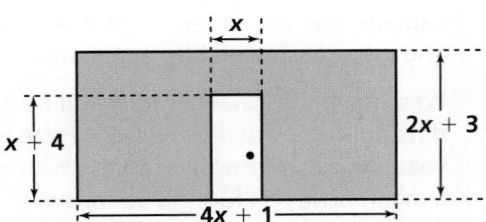

Rachel paints a wall, except for the door. Which expression represents the area that she paints?
(Lesson 9-3)

(1) $2x^2 + 11x + 12$
(2) $4x^2 + 17x + 4$
(3) $7x^2 + 10x + 3$
(4) $8x^2 + 14x + 3$

5. $(r + 9)(r - 9)$ is equivalent to —
(Lesson 9-4)

(1) $r^2 + 81$
(2) $r^2 - 81$
(3) $r^2 + 18r - 81$
(3) $r^2 - 18r - 81$

6. Which is the complete factorization of $3z^2 + 18z - 48$?
(Lesson 9-6)

(1) $(z + 8)(3z - 6)$
(2) $3(z + 8)(z - 2)$
(3) $(3z + 24)(z - 2)$
(4) $3(z - 3)(z + 16)$

7. Factor completely: $x^2 + 6x + 9$
(Lesson 9-7)

(1) $(x - 3)^2$ **(3)** $(x - 9)(x - 1)$
(2) $(x + 3)^2$ **(4)** $(x + 3)(x - 3)$

For problem 8, clearly indicate the necessary steps, including appropriate formula substitutions, diagrams, graphs, and charts.

8. Constructed Response The volume of a rectangular prism is represented by $20y^4 - 45y^2$. Find expressions for the possible dimensions of the prism. Show your work and explain.
(Lesson 9-8)

NY **Learning Standards**
NY A.A.13, NY A.A.20

Workout for New York Standards Mastery

Ready to go after Chapter 10

? For help, go to the lesson in green.

1. If $y = ax^2 + bx + c$, then $x = \frac{-b}{2a}$ is the x-coordinate of the vertex. What is the vertex of $f(x) = 3x^2 + 12x + 5$?
(Lesson 10-2)

   (1) $(0, 5)$

   (2) $(2, 41)$

   (3) $(-2, -7)$

   (4) $(-5, 20)$

2. Which is the solution to $x^2 + 5 = 69$?
(Lesson 10-3)

   (1) $x = \pm 8$

   (2) $x = \pm 9$

   (3) $x = \pm 64$

   (4) no real solution

3. Solve: $p^2 - 3p = 40$
(Lesson 10-4)

   (1) $p = -5$ or $p = 8$

   (2) $p = 5$ or $p = -8$

   (3) $p = -4$ or $p = 10$

   (4) $p = 4$ or $p = -10$

4. The equation for a projectile fired straight up at 80 ft/s is $h = -16t^2 + 80t$, where h is the height and t is the time. At what time will the height be 100 ft?
(Lesson 10-4)

   (1) 0.4 s   **(3)** 2.5 s

   (2) 4 s   **(4)** 25 s

5. What are the solutions to $d^2 + 6d = 72$?
(Lesson 10-6)

   (1) $d = -8$ or $d = 9$

   (2) $d = -6$ or $d = 12$

   (3) $d = 6$ or $d = -12$

   (4) $d = 8$ or $d = -9$

For problem 6, clearly indicate the necessary steps, including appropriate formula substitutions, diagrams, graphs, and charts.

6. Constructed Response

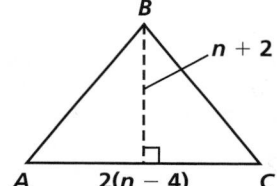

If the area of $\triangle ABC$ is 91 cm^2, find the height of the triangle. Explain your work and why you chose the method you used to solve the problem.
(Lesson 10-4)

Learning Standards
NY A.A.8, NY A.A.27, NY A.A.41

New York Workout

Workout for New York Standards Mastery

Ready to go after Chapter 11

 For help, go to the lesson in green.

1. What is equivalent to $\sqrt{32w^3}$?
(Lesson 11-1)

    (1) $2w\sqrt{2w}$ **(3)** $8w\sqrt{2w}$

    (2) $4w\sqrt{2w}$ **(4)** $16w\sqrt{2w}$

2. Simplify: $\sqrt{\dfrac{12}{x^3}}$

(Lesson 11-1)

    (1) $\dfrac{2\sqrt{3}}{x}$

    (2) $\dfrac{4\sqrt{3x}}{x}$

    (3) $\dfrac{2\sqrt{3x}}{x}$

    (4) $\dfrac{2\sqrt{3x}}{x^2}$

3. Simplify: $2\sqrt{50} - 3\sqrt{18}$
(Lesson 11-2)

    (1) $-\sqrt{32}$

    (2) $\sqrt{2}$

    (3) $10\sqrt{5} - 3\sqrt{2}$

    (4) $4\sqrt{2}$

4. Simplify: $\dfrac{8}{\sqrt{6} - \sqrt{2}}$

(Lesson 11-2)

    (1) 4

    (2) $\dfrac{1}{4}$

    (3) $\sqrt{6} - \sqrt{2}$

    (4) $2\sqrt{6} + 2\sqrt{2}$

For problem 5–6, clearly indicate the necessary steps, including appropriate formula substitutions, diagrams, graphs, and charts.

5. Constructed Response

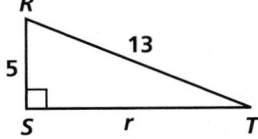

In $\triangle RST$, find the value of r, and then find sin R, cos R, and tan R. Show your work to support your answers.
(Lesson 11-5)

6. Constructed Response Kim is looking up at the top of the spire on a tower at an angle of elevation of 25°. The tower is 945 feet tall. How far from the base of the tower is Kim standing?

Draw and label a diagram to represent the problem. Then find the answer. Round your answer to the nearest foot. Explain how you found your answer.
(Lesson 11-6)

NY

Learning Standards
NY A.A.42, NY A.A.44, NY A.N.2, NY A.N.3

Workout for New York Standards Mastery

Ready to go after Chapter 12

? For help, go to the lesson in green.

1. What is the vertical asymptote of
$y = \dfrac{3}{-x + 1}$?
(Lesson 12-1)

 (1) $x = -2$ **(3)** $x = 0$
 (2) $x = -1$ **(4)** $x = 1$

2. Simplify: $\dfrac{12x + 24}{3x^2 - 6x - 24}$
(Lesson 12-3)

 (1) $\dfrac{1}{x - 1}$ **(3)** $\dfrac{x}{x - 1}$
 (2) $\dfrac{4}{x - 4}$ **(4)** $\dfrac{x}{x - 4}$

3. Divide: $(15x^4 - 5x^3 + 10x) \div (5x^3)$
(Lesson 12-4)

 (1) $3x^4 - x^3 + 2x$
 (2) $3x^7 - x^6 + 2x^4$
 (3) $10x + \dfrac{5}{x^2}$
 (4) $3x - 1 + \dfrac{2}{x^2}$

4. Which is equivalent to $\dfrac{3}{r + 3} + \dfrac{2}{r - 6}$?
(Lesson 12-5)

 (1) $\dfrac{5}{2r - 3}$ **(3)** $\dfrac{5r - 12}{(r + 3)(r - 6)}$
 (2) $\dfrac{5}{(r + 3)(r - 6)}$ **(4)** $\dfrac{5r - 12}{2r - 3}$

5. What is the solution to $\dfrac{6}{x} = \dfrac{x}{2} + \dfrac{1}{2}$?
(Lesson 12-6)

 (1) $x = -3$ or $x = 4$
 (2) $x = 3$ or $x = -4$
 (3) $x = \dfrac{4}{3}$ or $x = -1$
 (4) $x = \dfrac{1}{3}$ or $x = -\dfrac{1}{4}$

6. A restaurant's lasagna dinner comes with a choice of 2 appetizers, 4 side dishes, and 2 types of bread. How many different lasagna dinners can customers choose from if they get to pick one appetizer, one side dish, and one type of bread?
(Lesson 12-7)

 (1) 8   **(3)** 16
 (2) 9   **(4)** 32

For problem 7, clearly indicate the necessary steps, including appropriate formula substitutions, diagrams, graphs, and charts.

7. Constructed Response   Sheila can mow her lawn in 45 minutes. It takes Jamie 75 minutes to mow the same lawn. How many minutes will it take them to mow the lawn if they work together?

Write an equation to represent the problem and explain what the equation represents. Then solve the problem. Round your answer to the nearest minute.
(Lesson 12-6)

Learning Standards
NY A.G.4, NY A.A.14, NY A.A.15, NY A.A.16,
NY A.A.17, NY A.A.26, NY A.N.7, NY A.N.8

Using Your New Yor[k]

There are many features built into the daily lessons of your New York textbook that will help you learn the important skills and concepts you will need to be successful in this course... and on the Regents Exam! Look through the following pages for some study tips that you will find useful as you complete each lesson.

Mastering the New York Standards

Identifying the NY Standards

Every Chapter opens with a list of the NY Standards you already learned and an overview of the NY Standards that will be covered in the upcoming chapter.

The NY Standards are also correlated to each lesson within the chapter.

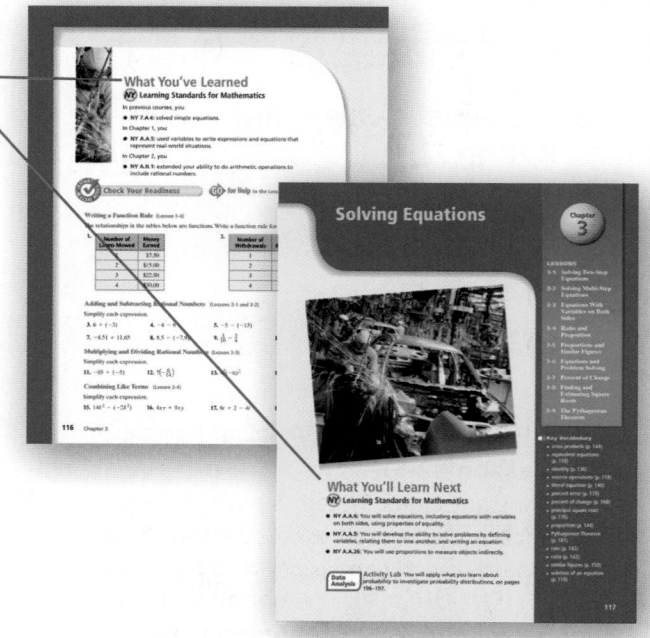

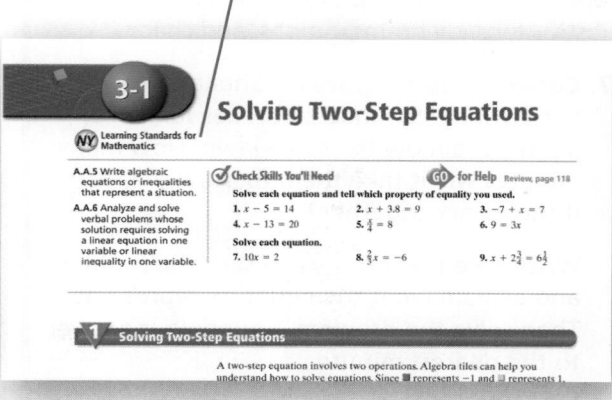

Leveled Exercise Set – including Regents Preparation

Your textbook provides comprehensive exercise sets after every lesson to help you practice what you've learned. Plus, specific New York Regents Exam preparation ensures exam success.

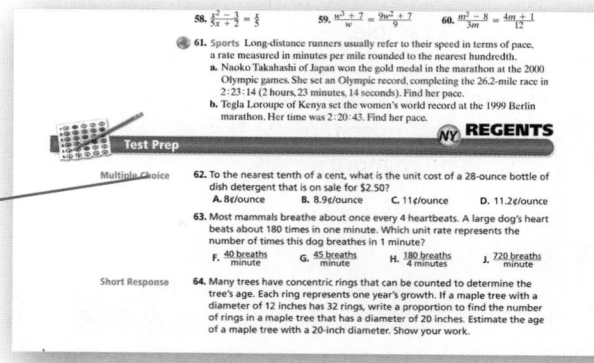

ook for Success

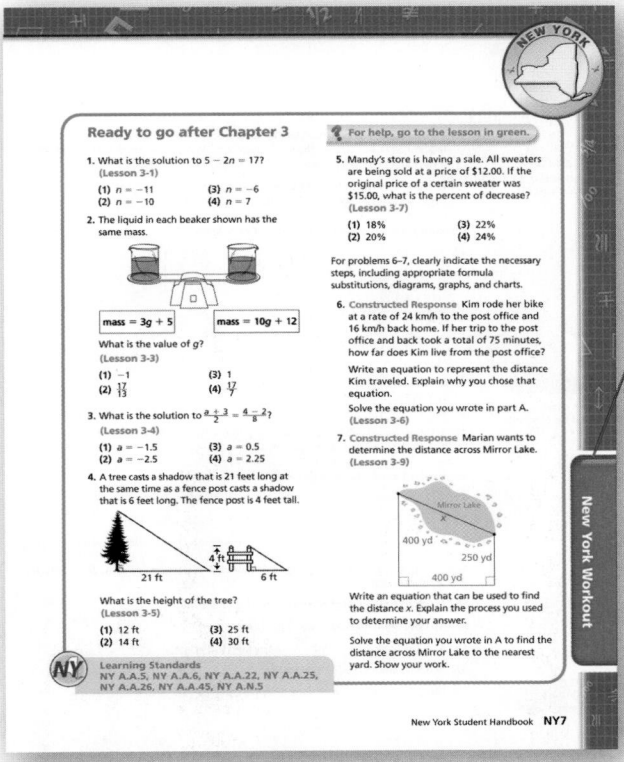

Workout for New York Standards Mastery

In addition to the Regents Exam test prep in the exercise sets, every Chapter has a Workout for New York Standards Mastery to help you succeed on the Regents Exam.

Instant Check System

The Instant Check System is a way for you to monitor your progress on the New York standards quickly and easily in every lesson.

- Check Skills You'll Need questions assess prerequisite skills before every lesson.
- Quick Check questions after every example ensure you understand before moving to the next example.
- Check Point Quizzes throughout the chapter provide a way to assess your cumulative understanding before moving on to more lessons.

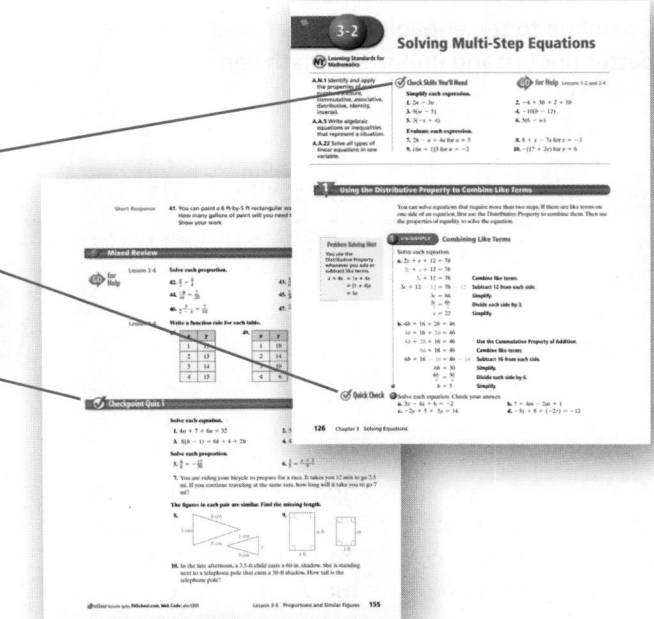

Built-In Help

There are many features built in to every lesson to help support your understanding of the mathematics.

Go for Help

Look for the green labels throughout your book that tell you where to "Go" for help. You'll see this built-in help in the lessons and in the homework exercises.

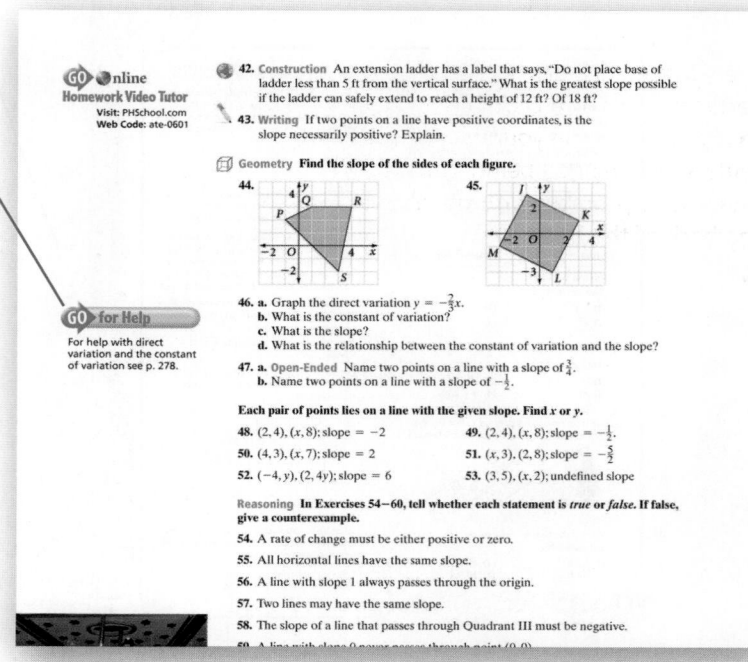

Video Tutor Help

Go online to see engaging videos to help you better understand important math concepts.

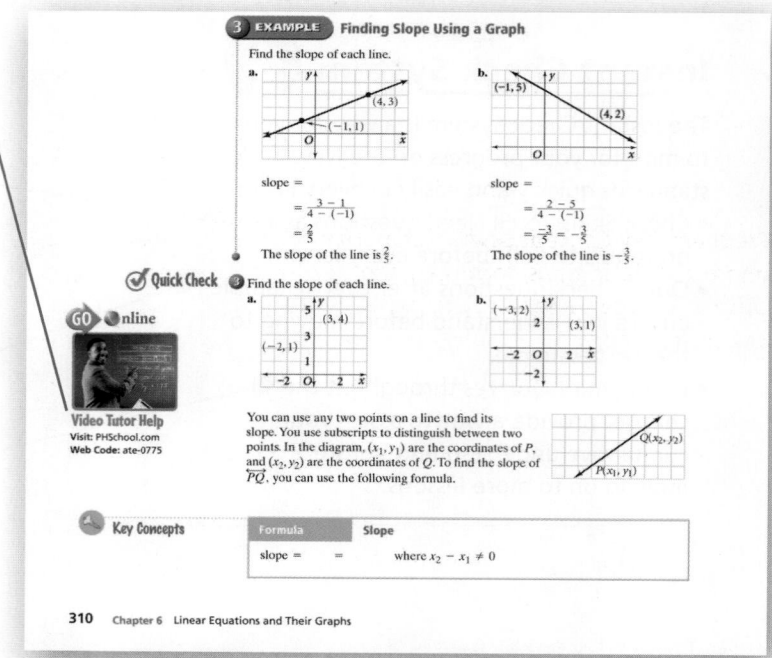

Quick Check

Every lesson includes numerous Examples, each followed by a *Quick Check* question that you can do on your own to see if you understand the skill being introduced. Check your progress with the answers at the back of the book.

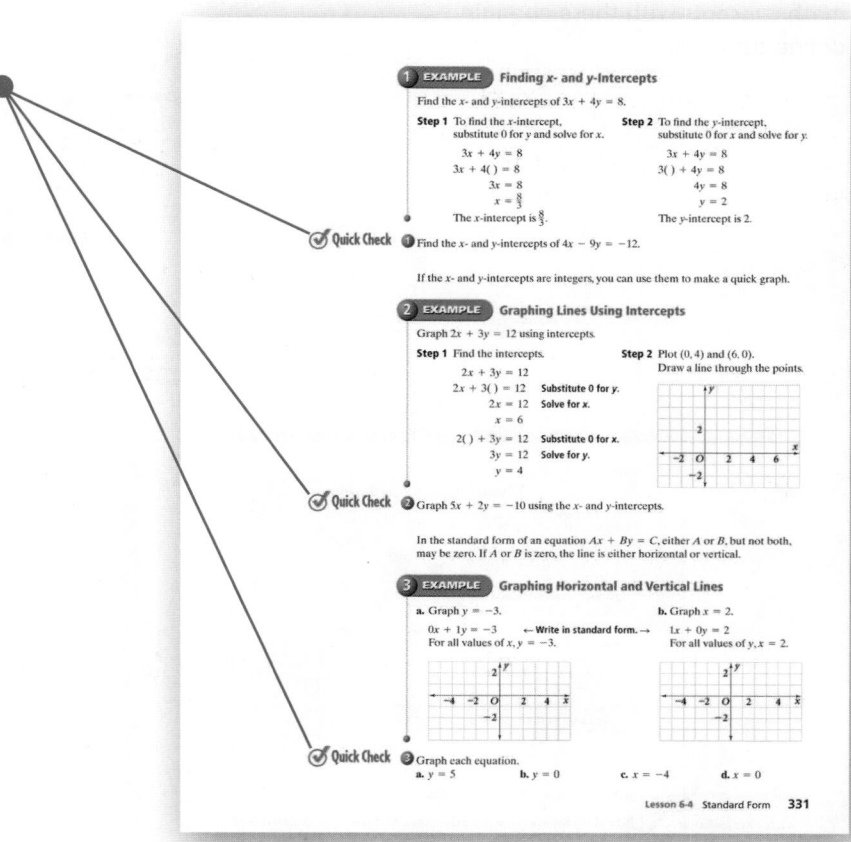

1 EXAMPLE Finding *x*- and *y*-Intercepts

Find the *x*- and *y*-intercepts of $3x + 4y = 8$.

Step 1 To find the *x*-intercept, substitute 0 for *y* and solve for *x*.

$3x + 4y = 8$
$3x + 4( ) = 8$
$3x = 8$
$x = \frac{8}{3}$
The *x*-intercept is $\frac{8}{3}$.

Step 2 To find the *y*-intercept, substitute 0 for *x* and solve for *y*.

$3x + 4y = 8$
$3( ) + 4y = 8$
$4y = 8$
$y = 2$
The *y*-intercept is 2.

Quick Check ① Find the *x*- and *y*-intercepts of $4x - 9y = -12$.

If the *x*- and *y*-intercepts are integers, you can use them to make a quick graph.

2 EXAMPLE Graphing Lines Using Intercepts

Graph $2x + 3y = 12$ using intercepts.

Step 1 Find the intercepts.

$2x + 3y = 12$
$2x + 3( ) = 12$ **Substitute 0 for *y*.**
$2x = 12$ **Solve for *x*.**
$x = 6$
$2( ) + 3y = 12$ **Substitute 0 for *x*.**
$3y = 12$ **Solve for *y*.**
$y = 4$

Step 2 Plot (0, 4) and (6, 0). Draw a line through the points.

Quick Check ② Graph $5x + 2y = -10$ using the *x*- and *y*-intercepts.

In the standard form of an equation $Ax + By = C$, either *A* or *B*, but not both, may be zero. If *A* or *B* is zero, the line is either horizontal or vertical.

3 EXAMPLE Graphing Horizontal and Vertical Lines

a. Graph $y = -3$.

$0x + 1y = -3$ ← **Write in standard form.** → $1x + 0y = 2$
For all values of *x*, $y = -3$.

b. Graph $x = 2$.

For all values of *y*, $x = 2$.

Quick Check ③ Graph each equation.
a. $y = 5$ **b.** $y = 0$ **c.** $x = -4$ **d.** $x = 0$

Lesson 6-4   Standard Form   **331**

Understanding Key Concepts

Frequent *Key Concept* boxes summarize important definitions, formulas, and properties. Use these to review what you've learned.

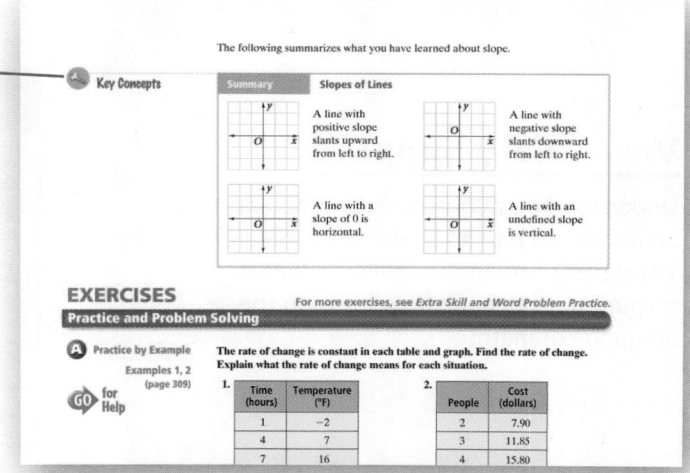

The following summarizes what you have learned about slope.

Key Concepts   **Summary**   **Slopes of Lines**

A line with positive slope slants upward from left to right.

A line with negative slope slants downward from left to right.

A line with a slope of 0 is horizontal.

A line with an undefined slope is vertical.

EXERCISES   For more exercises, see *Extra Skill and Word Problem Practice*.
Practice and Problem Solving

A Practice by Example
Examples 1, 2
(page 309)

GO for Help

The rate of change is constant in each table and graph. Find the rate of change. Explain what the rate of change means for each situation.

1.

Time (hours)	Temperature (°F)
1	-2
4	7
7	16

2.

People	Cost (dollars)
2	7.90
3	11.85
4	15.80

Built-In Help

Online Active Math

Make math come alive with these online activities. Review and practice important math concepts with these engaging online tutorials.

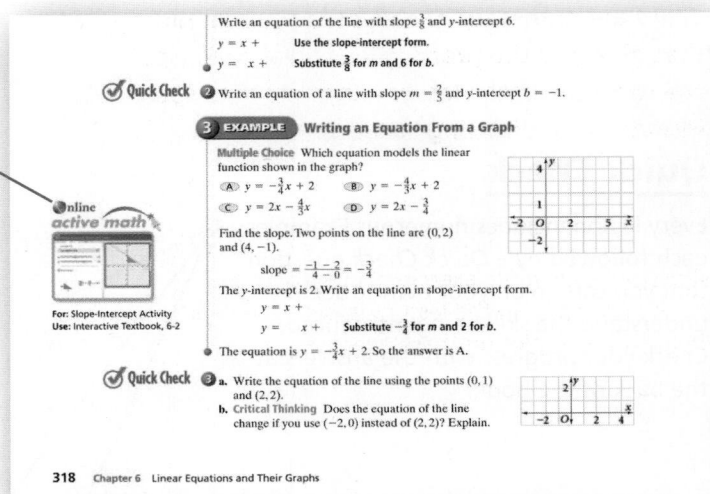

Vocabulary Support

Understanding mathematical vocabulary is an important part of studying mathematics. *Vocabulary Tips* and *Vocabulary Builders* throughout the book help focus on the language of math.

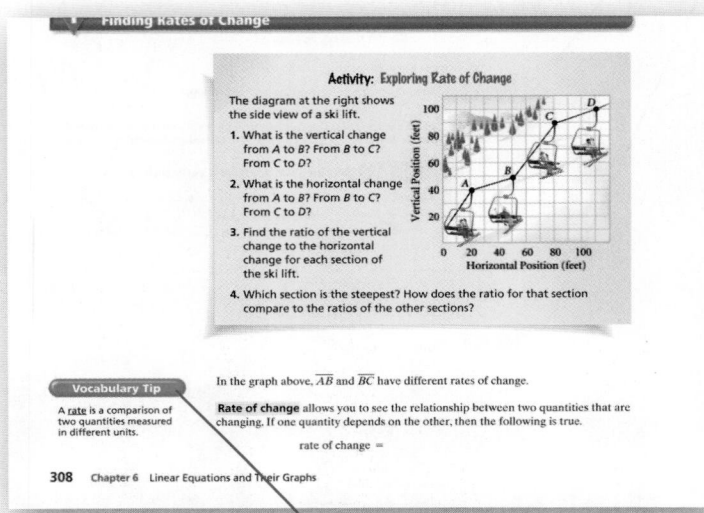

Understanding Vocabulary

There are three forms of a linear equation that you have studied in this chapter:
- slope-intercept form
- standard form
- point-slope form

To understand and remember these forms, it may help you to connect the common meaning of the words with their specialized meanings in mathematics.

Word	Common Meaning	Mathematical Meaning
Slope	An inclined surface (for example, the slope of a hill)	The rate of change that gives the steepness of a line: $\text{slope} = \dfrac{\text{vertical change}}{\text{horizontal change}} = \dfrac{\text{rise}}{\text{run}}$
Intercept	To cut off from a path (for example, to intercept a football)	The values of the points at which a line intersects (or cuts) the x-axis or y-axis
Standard	Generally accepted	A general form of an equation
Point	A dot or speck (noun)	A fixed location on a coordinate plane; every point has a unique x-value and y-value.

Slope-intercept form, standard form, and point-slope form all give clues about the lines they will graph.

- Slope-intercept form tells the slope of the line and its y-intercept.

Example $y = 3x - 1$ The graph of this equation has a slope of 3 and a y-intercept of -1.

Practice What You've Learned

There are numerous exercises in each lesson that give you the practice you need to master the concepts in the lesson. Each practice set includes the following sections.

A: Practice by Example

Practice by Example exercises refer you back to the Examples in the lesson, in case you need help with completing these exercises.

B: Apply Your Skills

Apply Your Skills exercises combine skills from earlier lessons to offer you richer skill exercises and multi-step application problems.

C: Challenge

Challenge exercises give you an opportunity to solve problems that extend and stretch your thinking.

Homework Video Tutor

Go online to get help completing your homework! For every lesson, there is a narrated, interactive tutorial to help you review the lesson and understand key concepts.

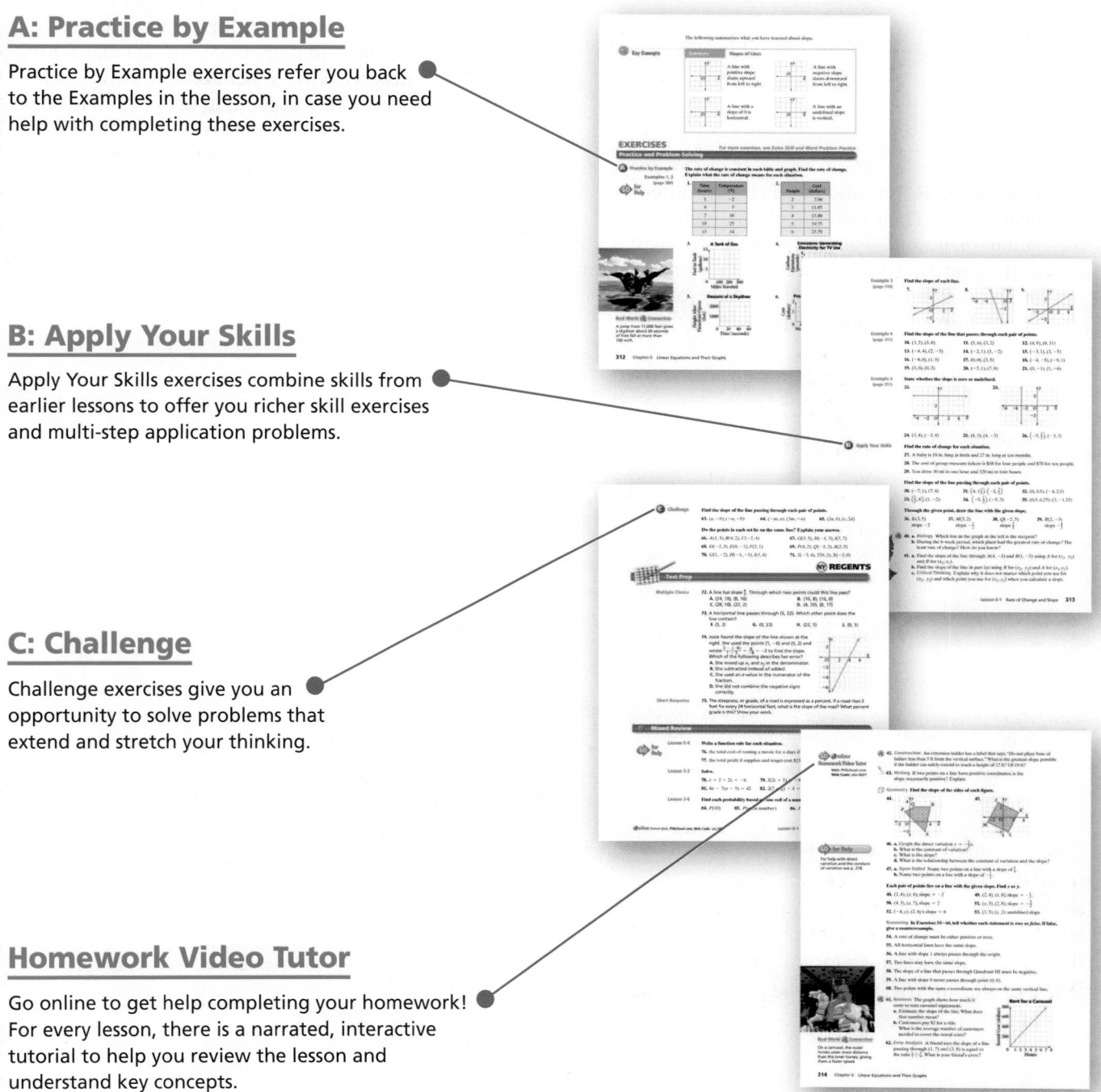

Beginning-Of-Course Diagnostic Test

1. Estimate the sum of $14.30, $143.08, and $19.74 by rounding.

2. Divide $\frac{2}{5} \div \frac{1}{8}$. Write the answer in simplest form.

3. Paul, Steve, Robin, and Ryan all play different instruments. Their instruments are guitar, bass guitar, piano, and drums. Robin's instrument is not a string instrument. Paul does not play bass guitar or piano. Ryan's instrument has only four strings. Which instrument does each play?

4. Find the perimeter.

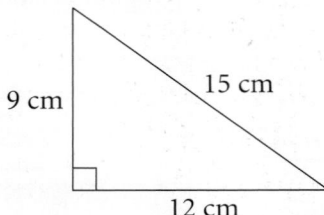

5. Write five equivalent fractions for $\frac{7}{8}$.

6. Find three consecutive even integers whose sum is 180.

7. Use a factor tree to write the prime factorization of 430.

8. Write 3.04 as a percent.

9. This week Lera withdrew $150 from her checking account. She wrote a check for $275, made a deposit of $200, and then wrote another check for $75. She now has $185 in her account. How much did Lera have in her account at the beginning of the week?

10. Find the GCF of 15 and 27.

11. Write 3.1818... as a fraction in simplest form.

12. Find a four digit number that is divisible by 3, 5, and 8.

13. Graph the image of $\triangle ABC$ after a translation of 3 units left and 2 units up.

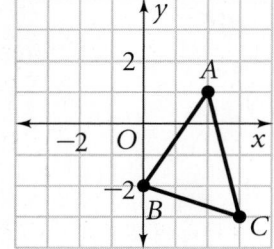

14. Subtract $6\frac{3}{4} - 4\frac{11}{12}$. Write the answer in simplest form.

15. Troy is writing a book of short stories. It is his goal to write one short story this month, two short stories next month, three short stories the following month, and so on for 13 more months. How many stories will he have written at the end of sixteen months?

16. Find the volume of the following figure. Use 3.14 for π.

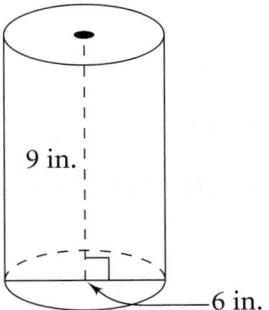

9 in.

6 in.

17. Find the LCM of 15 and 27.

18. Graph the triangle with vertices $A(-1, 3)$, $B(-3, -2)$, $C(0, -1)$. Then graph its image after a reflection across the y-axis.

19. Use a protractor to measure the angle and classify it as *acute, right, obtuse,* or *straight.*

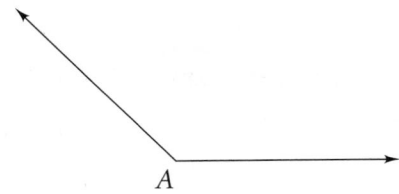

A

20. Multiply $4\frac{2}{5} \cdot 5\frac{1}{6}$. Write the answer in simplest form.

21. Draw a line plot for the frequency table.

Number	8	9	10	11	12
Frequency	3	4	6	2	1

22. When seating guests at a round table, two arrangements are considered the same if each person has the same neighbor to the left and to the right in each arrangement. Find the number of unique arrangements when seating 2, 3, and 4 guests at a round table. Use these results and the fact that 24 unique arrangements are possible when seating 5 guests to find the number of unique arrangements when seating 6 guests.

23. Write 3624 in expanded form using exponents.

24. There are 20 guests at a party. If each person shakes hands with every other person exactly once, how many total handshakes will occur?

25. Write $6^2 \cdot 3^3$ in standard form.

What You've Learned

 **Learning Standards for Mathematics**

In previous courses, you

- **NY A.N.3:** performed calculations involving addition, subtraction, multiplication, and division with whole numbers, fractions, and decimals.

- **NY 8.A.1:** compared numerical expressions using inequality and equality symbols.

- **NY 8.A.19:** read and interpreted graphs.

Check Your Readiness

 for Help to the Lesson in green.

Simplifying Fractions (Skills Handbook page 758)

Write in simplest form.

1. $\frac{12}{15}$ **2.** $\frac{20}{28}$ **3.** $\frac{33}{77}$ **4.** $\frac{8}{56}$ **5.** $\frac{48}{52}$

Adding and Subtracting Fractions (Skills Handbook page 760)

Add or subtract. Write each answer in simplest form.

6. $\frac{1}{8} + \frac{1}{6}$ **7.** $\frac{27}{33} - \frac{6}{22}$ **8.** $\frac{3}{4} + \frac{7}{10}$ **9.** $\frac{12}{13} - \frac{1}{3}$

10. $5\frac{7}{10} + 6\frac{7}{8}$ **11.** $7\frac{5}{6} - 3\frac{1}{4}$ **12.** $3\frac{5}{12} - 1\frac{7}{12}$ **13.** $4\frac{5}{7} + 8\frac{3}{4}$

Exponents (Skills Handbook page 763)

Write using exponents.

14. $9 \cdot 9 \cdot 9 \cdot 9 \cdot 9$ **15.** $5 \cdot 7 \cdot 7 \cdot 7 \cdot 7 \cdot 7 \cdot 7$ **16.** $2 \cdot 2 \cdot 3 \cdot 3 \cdot 3 \cdot 3 \cdot 3 \cdot 3$

Analyzing Graphs (Skills Handbook page 769)

Use the graph at the right.

17. What was the approximate difference in the voting-age populations of Texas and of Florida in 1988?

18. Between 1992 and 1996 which state had the greater increase in voting-age population? Estimate that increase.

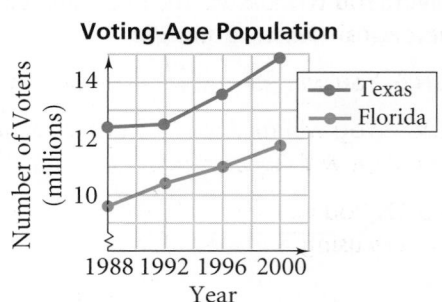

Voting-Age Population

Variables, Function Patterns, and Graphs

🔊 Key Vocabulary

- absolute value (p. 20)
- coordinates (p. 24)
- domain (p. 29)
- equation (p. 5)
- exponent (p. 9)
- function (p. 27)
- independent variable (p. 28)
- integers (p. 17)
- irrational numbers (p. 18)
- mean (p. 40)
- order of operations (p. 10)
- ordered pair (p. 24)
- origin (p. 24)
- positive correlation (p. 34)
- rational numbers (p. 17)
- real numbers (p. 18)
- scatter plot (p. 33)
- stem-and-leaf plot (p. 42)
- variable (p. 4)

What You'll Learn Next

(NY) Learning Standards for Mathematics

- **NY A.A.1:** You will use variables to transform English phrases into mathematical expressions.

- **NY A.N.6:** You will use the order of operations to simplify expressions.

- **NY A.A.5:** You will explore function rules and learn to identify relationships within functions.

- **NY A.S.12:** You will show the relationship between two sets of real-world data using a scatter plot.

Data Analysis **Activity Lab** You will apply integers to locate sunken ships, on pages 52–53.

3

1-1

Using Variables

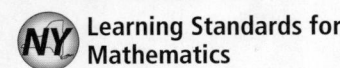 **Learning Standards for Mathematics**

A.A.1 Translate a quantitative verbal phrase into an algebraic expression.

A.A.3 Distinguish the difference between an algebraic expression and an algebraic equation.

A.A.4 Translate verbal sentences into mathematical equations or inequalities.

✓ **Check Skills You'll Need**

GO for Help Skills Handbook page 757

Estimate to find whether each answer is reasonable.

1.	$154.38	2.	478.23	3.	76.425	4.	$316.24
	22.45		−199.30		18.94		− 48.76
	276.12		378.93		182.6		$267.48
	+28.98				+54.769		
	$481.93				232.729		

🔊 **New Vocabulary** • variable • algebraic expression • equation • open sentence

Modeling Relationships With Variables

Vocabulary Tip

Each expression below means 6.50 multiplied by h.

$6.50 \times h$

$6.50 \cdot h$

$6.50(h)$

$(6.50)h$

$6.50h$

$(6.50)(h)$

If you earn an hourly wage of $6.50, your pay is the number of hours you work multiplied by 6.50.

In the table at the right, the variable h stands for the number of hours you worked. A **variable** is a symbol, usually a letter, that represents one or more numbers. The expression $6.50h$ is an algebraic expression. An **algebraic expression** is a mathematical phrase that can include numbers, variables, and operation symbols. Algebraic expressions are sometimes called variable expressions.

Hours Worked	Pay (dollars)
1	6.50×1
2	6.50×2
3	6.50×3
h	$6.50 \times h$

1 EXAMPLE Writing an Algebraic Expression

Write an algebraic expression for each phrase.

a. seven more than n

$n + 7$ "More than" indicates addition. Add the first number 7 to the second number n.

b. the difference of n and 7

$n - 7$ "Difference" indicates subtraction. Begin with the first number n. Then subtract the second number 7.

c. the product of seven and n

$7n$ "Product" indicates multiplication. Multiply the first number 7 by the second number n.

d. the quotient of n and seven

$\frac{n}{7}$ "Quotient" indicates division. Divide the first number n by the second number 7.

✓ **Quick Check** ❶ Write an algebraic expression for each phrase.
a. the quotient of 4.2 and c **b.** t minus 15

To translate an English phrase into an algebraic expression, you may need to define one or more variables first.

2 EXAMPLE Writing an Algebraic Expression

Define a variable and write an algebraic expression for each phrase.

a. two times a number plus 5

Relate	two times	a number	plus 5

Define Let n = the number.

Write	2	·	n	+	5

$2n + 5$

b. 7 less than three times a number

Relate	7	less than	three times	a number

Define Let a = the number.

Write	3	·	a	−	7

$3a − 7$

Problem Solving Hint

In math, you write the phrase "seven less than twelve" as $12 − 7$. The order of the numbers in the math phrase is different than the order of the numbers in the English phrase.

✓ Quick Check **2** Define a variable and write an algebraic expression for each phrase.
a. 9 less than a number
b. the sum of twice a number and 31

---

2 Modeling Relationships With Equations

An **equation** is a mathematical sentence that uses an equal sign. If an equation is true, then the two expressions on either side of the equal sign represent the same value. An equation that contains one or more variables is an **open sentence.** In everyday language, the word "is" suggests an equal sign in the associated equation.

Test-Taking Tip

When choosing the best equation, test your answer. In Example 3, three CDs cost $36. Only answer D gives $36 when $n = 3$.

3 EXAMPLE Writing an Equation

Multiple Choice Track One Media sells all CDs for $12 each. Which equation best represents the cost c of a given number n of CDs?

Ⓐ $12 + c = n$ Ⓑ $n + 12 = c$ Ⓒ $12c = n$ Ⓓ $c = 12n$

Relate	The total cost	is	$12	times	the number of CDs bought.

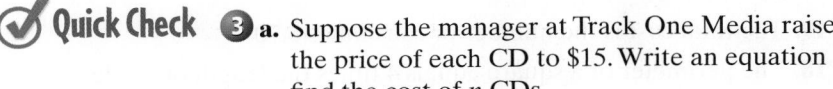

Define Let n = the number of CDs bought.

Let c = the total cost.

Write	c	=	12	·	n

The equation is $c = 12n$, so D is the correct answer.

✓ Quick Check **3 a.** Suppose the manager at Track One Media raises the price of each CD to $15. Write an equation to find the cost of n CDs.

b. Critical Thinking Suppose the manager at Track One Media uses the equation $c = 10.99n$. What could this mean?

When you write an equation for data in a table, it may help to write a short sentence describing the relationship between the data. Then translate the sentence into an equation. Be sure to tell what each variable represents.

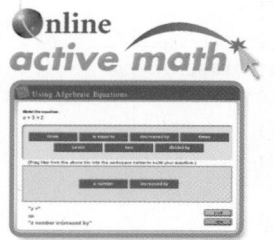

For: Equation Activity
Use: Interactive Textbook, 1-1

4 EXAMPLE **Real-World** **Problem Solving**

Sales Write an equation for the data in the table.

Relate change equals $20.00 minus cost.

Define Let c = cost of item purchased.
Let a = amount of change.

Write a = 20 − c

● $a = 20 - c$

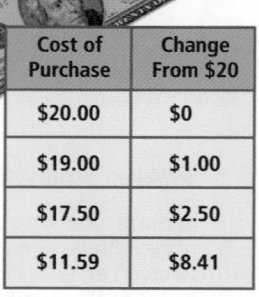

Cost of Purchase	Change From $20
$20.00	$0
$19.00	$1.00
$17.50	$2.50
$11.59	$8.41

Quick Check **4** Write an equation for the data in the table.

Amounts Earned and Saved

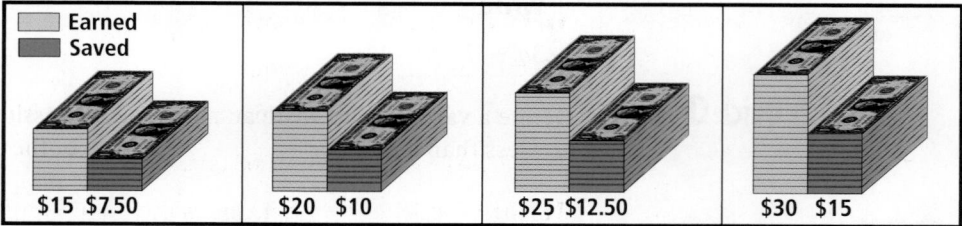

■ Earned
■ Saved

$15 $7.50 $20 $10 $25 $12.50 $30 $15

EXERCISES

For more exercises, see *Extra Skill and Word Problem Practice*.

Practice and Problem Solving

A Practice by Example

Example 1
(page 4)

 **for Help**

Write an algebraic expression for each phrase.

1. 4 more than p
2. y minus 12
3. 12 minus m
4. the product of 15 and c
5. the quotient of n and 8
6. the quotient of 17 and k
7. 23 less than x
8. the sum of v and 3

Example 2
(page 5)

Define a variable and write an expression for each phrase.

9. 2 more than twice a number
10. a number minus 11
11. 9 minus a number
12. a number divided by 82
13. the product of 5 and a number
14. the sum of 13 and twice a number
15. the quotient of a number and 6
16. the quotient of 11 and a number

Example 3
(page 5)

Define variables and write an equation to model each situation.

17. The total cost is the number of cans times $.70.
18. The perimeter of a square equals 4 times the length of a side.
19. The total length of rope, in feet, used to put up tents is 60 times the number of tents.
20. What is the number of slices of pizza left from an 8-slice pizza after you have eaten some slices?

Example 4
(page 6)

Define variables and write an equation to model the relationship in each table.

21.

Number of Workers	Number of Radios Built
1	13
2	26
3	39
4	52

22.

Number of Tapes	Cost
1	$8.50
2	$17.00
3	$25.50
4	$34.00

23.

Number of Sales	Total Earnings
5	$2.00
10	$4.00
15	$6.00
20	$8.00

24.

Number of Hours	Total Pay
4	$32
6	$48
8	$64
10	$80

 **Apply Your Skills**

Write an expression for each phrase.

25. the sum of 9 and k minus 17   **26.** 6.7 more than 5 times n

27. 9.85 less than the product of 37 and t **28.** the quotient of $3b$ and 4.5

29. 15 plus the quotient of 60 and w **30.** 7 minus the product of v and 3

31. the product of 5 and m, minus the quotient of t and 7

32. the sum of the quotient of p and 14, and the quotient of q and 3

33. 8 minus the product of 9 and r

GO Online
Homework Video Tutor
Visit: PHSchool.com
Web Code: ate-0101

Write a phrase for each expression.

34. $q + 5$ **35.** $3 - t$ **36.** $9n + 1$ **37.** $\frac{y}{5}$ **38.** $7hb$

Define variables and write an equation to model the relationship in each table.

39.

Number of Days	Change in Height (meters)
1	0.165
2	0.330
3	0.495
4	0.660

40.

Time (months)	Length (inches)
1	4.1
2	8.2
3	12.3
4	16.4

41. Use the table at the right.
  a. Does each statement fit the data in the table? Explain.
    i. hours worked = lawns mowed · 2
    ii. hours worked = lawns mowed + 3
  b. Writing Which statement in part (a) better describes the relationship between hours worked and lawns mowed? Explain.

Lawns Mowed	Hours
1	
2	
3	6

42. Multiple Choice Which equation best describes the relationship between the amount of money a in a bag of quarters and the number of quarters q?

Ⓐ $a = 0.25q$ Ⓑ $a = 0.25 + q$ Ⓒ $q = 0.25a$ Ⓓ $q = 0.25 + q$

Real-World Connection

People in the United States have spent more than $8.5 billion annually on lawn care.

🌐 Physics **The table at the left shows the height of the first bounce when a ball is dropped from different heights.**

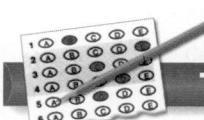

Drop Height (ft)	Height of First Bounce (ft)
1	$\frac{1}{2}$
2	1
3	$1\frac{1}{2}$
4	2
5	$2\frac{1}{2}$

43. a. Write an equation to describe the relationship between the height of the first bounce and the drop height.
   b. Suppose you drop the ball from a window 20 ft above the ground. Predict how high the ball will bounce.

44. Suppose the second bounce is $\frac{1}{4}$ of the original drop height. Write an equation to relate the height of the second bounce to the drop height.

Open-Ended **Describe a real-world situation that each equation could represent. Include a definition for each variable.**

45. $d = 5t$ **46.** $a = b + 3$ **47.** $c = \frac{40}{h}$

NY REGENTS

Test Prep

Multiple Choice

48. Which is an algebraic expression for "six less than k"?
   A. $\frac{6}{k}$ **B.** $\frac{k}{6}$ **C.** $6 - k$ **D.** $k - 6$

49. Which is an algebraic expression for "the product of 10 and a"?
   F. $a + 10$ **G.** $a - 10$ **H.** $10a$ **J.** $\frac{a}{10}$

50. Which is an algebraic expression for "9 more than v"?
   A. $v + 9$ **B.** $v - 9$ **C.** $9 - v$ **D.** $9v$

51. A container of milk contains 64 ounces. Which equation models the number n of ounces remaining after you have drunk m ounces?
   F. $m - 64 = n$ **G.** $64 - m = n$ **H.** $n - 64 = m$ **J.** $n - m = 64$

52. Which equation models the relationship in the table if r represents the row number and t represents the number of tulips?
   A. $r = 3t$ **B.** $\frac{r}{t} = 3$
   C. $t = r + 3$ **D.** $t = 3r$

Row Number	Number of Tulips
1	3
2	6
3	9
4	12

53. Which is an algebraic expression for "the quotient of $r + 5$ and b"?
   F. $\frac{r + 5}{b}$ **G.** $\frac{r}{b + 5}$
   H. $\frac{b}{r + 5}$ **J.** $\frac{b}{r} + 5$

Mixed Review

Skills Handbook
for Help

Find the area of each figure. For Exercises 54 and 55, find the perimeter of each figure.

54.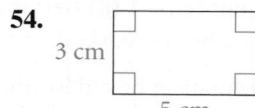
3 cm
5 cm

55.
9 in. 15 in.
12 in.

56.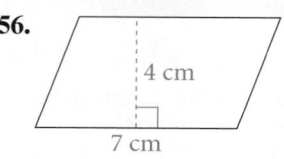
4 cm
7 cm

 **1-2**

Exponents and Order of Operations

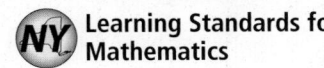 **Learning Standards for Mathematics**

A.N.6 Evaluate expressions involving factorial(s), absolute value(s), and exponential expression(s).

A.A.5 Write algebraic equations or inequalities that represent a situation.

✓ **Check Skills You'll Need**

GO for Help Skills Handbook page 755

Find the greatest common factor of each set of numbers.

1. 4 and 8   **2.** 12 and 15   **3.** 5 and 7

4. 8 and 12   **5.** 14 and 21   **6.** 12 and 20

Find the least common multiple of each set of numbers.

7. 4 and 8   **8.** 12 and 15   **9.** 5 and 7

10. 3 and 9   **11.** 6 and 9   **12.** 9 and 12

🔊 **New Vocabulary** • simplify • exponent • base • power
• order of operations • evaluate

---

1 | **Simplifying and Evaluating Expressions and Formulas**

Problem Solving Hint

Drawing a diagram may help you understand the problem.

w [rectangle diagram] ℓ

Activity: Order of Operations

📦 **Geometry** Two formulas for the perimeter of a rectangle are $P = 2\ell + 2w$ and $P = 2(\ell + w)$.

1. The length ℓ of a rectangle is 8 in. and its width w is 3 in. Find the perimeter of the rectangle using each of the formulas.

2. When you used $P = 2\ell + 2w$, did you add first or multiply first?

3. When you used $P = 2(\ell + w)$, did you add first or multiply first?

4. Which formula do you prefer to use? Why?

To **simplify** a numerical expression, you replace it with its simplest name. The simplest name for $2 \cdot 8 + 2 \cdot 3$ and for $2(8 + 3)$ is 22.

Expressions may include exponents. Using an exponent provides a shorthand way to show a product of equal factors.

$$\text{base} \rightarrow 2^{\overset{\text{exponent}}{\downarrow}4} = \underbrace{2 \cdot 2 \cdot 2 \cdot 2}_{\text{power}}$$

An **exponent** tells how many times a number, the **base,** is used as a factor. A **power** has two parts, a base and an exponent.

You read the expression 2^4 as "two to the fourth power." To simplify 2^4, you replace it with its simplest name, 16. There are special names for 2^3, "two cubed," and 2^2, "two squared."

Look at the expression below. It is simplified in two ways.

$\underbrace{3 + 5} - 6 \div 2$	$3 + 5 - \underbrace{6 \div 2}$
$\underbrace{8 - 6} \div 2$	$\underbrace{3 + 5} - 3$
$\underbrace{2 \div 2}$	$\underbrace{8 - 3}$
1 ✗	5 ✓

← Different results! →

To avoid having two different results when simplifying the same expression, mathematicians have agreed on an order for doing operations.

Key Concepts

Summary **Order of Operations**

1. Perform any operation(s) inside grouping symbols.
2. Simplify powers.
3. Multiply and divide in order from left to right.
4. Add and subtract in order from left to right.

GO Online

Video Tutor Help
Visit: PHSchool.com
Web Code: ate-0775

1 EXAMPLE Simplifying a Numerical Expression

Simplify $25 - 8 \cdot 2 + 3^2$.

$25 - 8 \cdot 2 + 3^2 = 25 - \underbrace{8 \cdot 2} + 9$ **Simplify the power: $3^2 = 3 \cdot 3 = 9$.**

$= \underbrace{25 - 16} + 9$ **Multiply 8 and 2.**

$= 9 + 9$ **Add and subtract in order from left to right.**

$= 18$ **Add.**

Quick Check 1 Simplify each expression.

a. $6 - 10 \div 5$ **b.** $3 \cdot 6 - 4^2 \div 2$ **c.** $4 \cdot 7 + 4 \div 2^2$ **d.** $5^3 + 90 \div 10$

You **evaluate** an algebraic expression by substituting a given number for each variable. Then simplify the numerical expression using the order of operations.

2 EXAMPLE Evaluating an Algebraic Expression

Evaluate $3a - 2^3 \div b$ for $a = 7$ and $b = 4$.

$3a - 2^3 \div b = 3 \cdot 7 - 2^3 \div 4$ **Substitute 7 for *a* and 4 for *b*.**

$= 3 \cdot 7 - 8 \div 4$ **Simplify the power.**

$= 21 - 2$ **Multiply and divide from left to right.**

$= 19$ **Subtract.**

Quick Check 2 Evaluate each expression for $c = 2$ and $d = 5$.

a. $4c - 2d \div c$ **b.** $d + 6c \div 4$ **c.** $c^4 - d \cdot 2$ **d.** $40 - d^2 + cd \cdot 3$

You can use expressions with variables to model many real-world situations.

3 **EXAMPLE** **Real-World** **Problem Solving**

Sales The equation $c = p + 0.06p$ represents the cost c of a pair of sneakers with price p and sales tax of 6%. Use a table to find the cost for \$20, \$30, \$45, and \$59 pairs of sneakers.

Price p	$p + 0.06p$	Cost c
\$20	$20 + 0.06(20)$	\$21.20
\$30	$30 + 0.06(30)$	\$31.80
\$45	$45 + 0.06(45)$	\$47.70
\$59	$59 + 0.06(59)$	\$62.54

 **3** The equation $c = p + 0.05p$ represents the cost c of an item with a price p and a 5% sales tax. Make a table to find the cost of items with prices \$5, \$10, \$15, and \$20.

2 Simplifying and Evaluating Expressions With Grouping Symbols

When you simplify expressions with parentheses, work within the parentheses first. Inside parentheses, use the order of operations.

4 **EXAMPLE** **Simplifying an Expression With Parentheses**

Simplify $15(13 - 7) \div (8 - 5)$.

$$15(13 - 7) \div (8 - 5) = 15(6) \div 3 \quad \textbf{Simplify within parentheses first.}$$
$$= 90 \div 3 \quad \textbf{Multiply and divide from left to right.}$$
$$= 30 \quad \textbf{Divide.}$$

 **4** Simplify each expression.
a. $(5 + 3) \div 2 + (5^2 - 3)$ **b.** $8 \div (9 - 7) + (13 \div 2)$

The base for an exponent is the number, variable, or expression directly to the left of the exponent. For $(cd)^2$, cd is the base. For cd^2, d is the base. Grouping symbols show which part of the expression is the base of the power.

5 **EXAMPLE** **Evaluating Expressions With Exponents**

Evaluate each expression for $c = 15$ and $d = 12$.
a. $(cd)^2$ **b.** cd^2
$$(cd)^2 = (15 \cdot 12)^2 \quad \leftarrow \textbf{Substitute 15 for } c \textbf{ and 12 for } d. \rightarrow \quad cd^2 = 15 \cdot 12^2$$
$$= (180)^2 \quad \leftarrow \textbf{Simplify within parentheses.} \quad = 15 \cdot 144$$
$$= 32{,}400 \quad \leftarrow \textbf{Simplify.} \rightarrow \quad = 2160$$

 **5** Evaluate each expression for $r = 9$ and $t = 14$.
a. rt^2 **b.** r^2t **c.** $(rt)^2$

You can also use brackets [ ] as grouping symbols. When an expression has several grouping symbols, simplify the innermost expression first.

Calculator Hint

To simplify a power such as 6^3, press 6 $\boxed{y^x}$ 3. On a graphing calculator, use 6 $\boxed{\wedge}$ 3. There is a special key for squares, so you can simplify 6^2 by pressing 6 $\boxed{x^2}$.

6 EXAMPLE Simplifying an Expression

Simplify $2[(13 - 7)^2 \div 3]$.

$$
\begin{aligned}
2[(13 - 7)^2 \div 3] &= 2[(6)^2 \div 3] && \text{First simplify } (13 - 7). \\
&= 2[36 \div 3] && \text{Simplify the power.} \\
&= 2[12] && \text{Divide within the brackets.} \\
&= 24 && \text{Multiply.}
\end{aligned}
$$

✓ Quick Check 6 Simplify each expression.
a. $5[4 + 3(2^2 + 1)]$ **b.** $12 + 3[18 - 5(16 - 13)]$ **c.** $5 + [(2 + 1)^3 - 3]$

A fraction bar is also a grouping symbol. For an expression like $\frac{2 + 8}{5 - 3}$, do the calculations above and below the fraction bar before simplifying the fraction.

7 EXAMPLE Real-World 🌎 Problem Solving

Urban Planning A neighborhood association turned a vacant lot into a park. The park is shaped like the trapezoid below. Use the formula $A = h\left(\frac{b_1 + b_2}{2}\right)$ to find the area of the lot.

$$
\begin{aligned}
A &= h\left(\frac{b_1 + b_2}{2}\right) \\
&= 130\left(\frac{100 + 200}{2}\right) && \text{Substitute 130 for } h, \\
& && \text{100 for } b_1, \text{ and 200 for } b_2. \\
&= 130\left(\frac{300}{2}\right) && \text{Simplify the numerator.} \\
&= 130(150) && \text{Simplify the fraction.} \\
&= 19{,}500 && \text{Multiply.}
\end{aligned}
$$

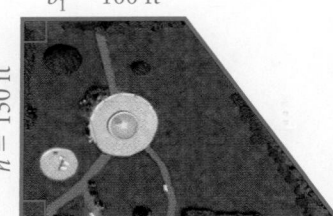

$b_1 = 100$ ft

$h = 130$ ft

$b_2 = 200$ ft

The area of the park is 19,500 ft^2.

✓ Quick Check 7 Find the area of a trapezoid with height $h = 300$ ft and bases $b_1 = 250$ ft and $b_2 = 170$ ft.

EXERCISES

For more exercises, see *Extra Skill and Word Problem Practice*.

Practice and Problem Solving

A Practice by Example

GO for Help

Example 1
(page 10)

Simplify each expression.

1. $5 + 6 \cdot 9$ **2.** $40 - 2 \cdot 3^2$ **3.** $8 + 12 \div 6 - 3$

4. $8 \cdot 4 + 9^2$ **5.** $5 \cdot 3^2 - 13$ **6.** $21 + 49 \div 7 + 1$

Example 2
(page 10)

Evaluate each expression for $a = 5$, $b = 12$, and $c = 2$.

7. $a + b + 2c$ **8.** $2b \div c + 3a$ **9.** $b^2 - 4a$

10. $ca + a$ **11.** $abc + ab$ **12.** $5a + 12b$

Example 3
(page 11)

13. The equation $s = p - 0.15p$ represents the sale price s of an item with an original price p, after a 15% discount. Make a table to find the discount prices for items with original prices of $12, $16, $20, and $25.

14. The equation $a = \frac{1}{2}(8)h$ represents the area a of a triangle with base of 8 cm and a height of h cm. Make a table to find the areas of triangles with heights of 6 cm, 7 cm, 8 cm, and 9 cm.

Example 4
(page 11)

Simplify each expression.

15. $2(5 + 9) - 6$ 16. $(17 - 7) \div 5 + 1$ 17. $(2 + 9) \cdot (8 - 4)$

18. $(7^2 - 3^2) \div 8$ 19. $17 - 5^2 \div (2^4 + 3^2)$ 20. $(10^2 - 4 \cdot 8) \div (8 + 9)$

Example 5
(page 11)

Evaluate each expression for $s = 11$ and $v = 8$.

21. sv^2 22. $(sv)^2$ 23. $s^2 + v^2$ 24. $(s + v)^2$

25. $s^2 - v^2$ 26. $(s - v)^2$ 27. $2s^2v$ 28. $(2s)^2v$

Example 6
(page 12)

Simplify each expression.

29. $6[13 - 2(4 + 1)]$ 30. $[3(7 + 4) - 2]6$ 31. $20 - [4(3 + 2)]$

32. $1^{11} + 3\left[\left(\frac{22}{11} + 8\right) \div 5\right]$ 33. $27[5^2 \div (4^2 + 3^2) + 2]$ 34. $9 + [4 - (10 - 9)^2]^3$

Example 7
(page 12)

Evaluate the formula $V = \frac{Bh}{3}$ for each pair of values.

35. $B = 4$ cm^2, $h = 6$ cm   36. $B = 21$ in.^{2}, $h = 13$ in.

37. $B = 7$ ft^2, $h = 9$ ft   38. $B = 8.4$ cm^2, $h = 10$ cm

39. $B = 500$ ft^2, $h = 90$ ft   40. $B = 118$ m^2, $h = 66$ m

 Apply Your Skills

Simplify each expression.

41. $(2 + 3)^2 - 10$ 42. $2^3 + 3^2 - 10$ 43. $(2^3 + 3)^2 - 10$

44. $(2^3 + 3^2) - 16$ 45. $3 + 6 \cdot 8$ 46. $(5.2 - 1) \cdot 12$

47. $1 + 2(3 + 4) \div (5 \cdot 6)$ 48. $4^3 \div 8 - 1 + 5 \div 8$

Visit: PHSchool.com
Web Code: ate-0102

49. A student wrote that $(a + b)^2 = a^2 + b^2$.
   a. Evaluate each side of the equation for $a = 0$ and $b = 1$.
   b. Evaluate each side of the equation for $a = 1$ and $b = 1$.
   c. Open-Ended Choose another pair of values for a and b. Evaluate each side of the equation for those values.
   d. Writing An equation is true if each side of the equation simplifies to the same value or expression. Is the equation $(a + b)^2 = a^2 + b^2$ true? Explain.

 50. **Multiple Choice** The frame at the left is 2 in. wide. The outer height h is 17 in., and the outer width w is 14 in. Use the formula $A = (h - 4)(w - 4)$ to find the area of the picture.

   Ⓐ 130 in.^{2} Ⓑ 169 in.^{2} Ⓒ 180 in.^{2} Ⓓ 238 in.^{2}

Evaluate each expression for $m = 3$, $p = 7$, and $q = 4$.

51. $mp - q$ 52. $m(p - q)$ 53. $mp^2 - q$ 54. $m(p^2 - q)$

55. $(mp^2) - q$ 56. $m(p - q)^2$ 57. $m \div q + 2p$ 58. $qp^2 + pq^2$

—2 in.

—2 in.

Lesson 1-2 Exponents and Order of Operations   **13**

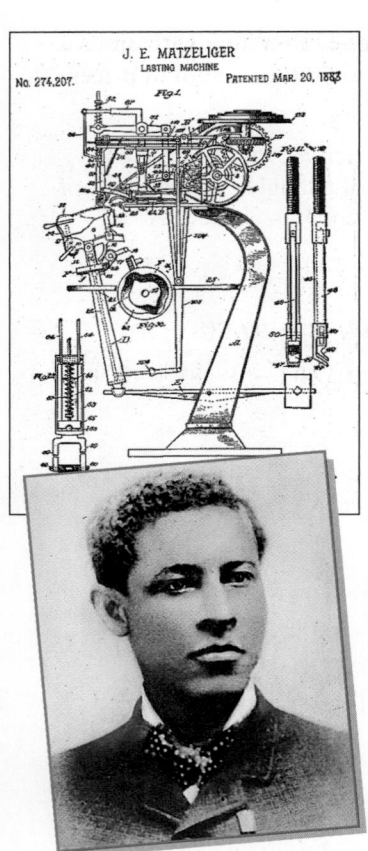

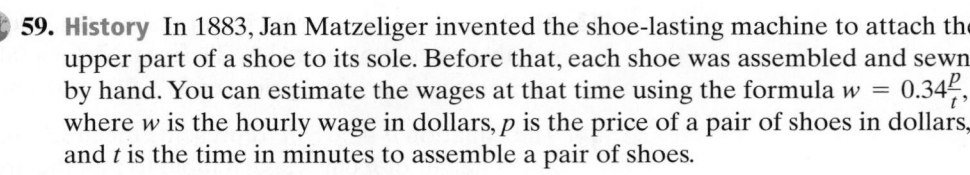 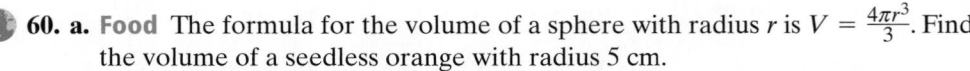

59. History In 1883, Jan Matzeliger invented the shoe-lasting machine to attach the upper part of a shoe to its sole. Before that, each shoe was assembled and sewn by hand. You can estimate the wages at that time using the formula $w = 0.34\frac{p}{t}$, where w is the hourly wage in dollars, p is the price of a pair of shoes in dollars, and t is the time in minutes to assemble a pair of shoes.

In 1891, a worker who used the shoe-lasting machine could assemble one pair of shoes in two minutes. A pair of shoes cost about $.94. Estimate the worker's hourly wage to the nearest cent.

60. a. Food The formula for the volume of a sphere with radius r is $V = \frac{4\pi r^3}{3}$. Find the volume of a seedless orange with radius 5 cm.
b. Suppose the peel of the orange is 0.5 cm thick. What volume of the orange is edible?
c. Express the edible part of the orange as a percent of the whole orange. Round your answer to the nearest percent.

Complete each table.

61.

h	$(5h^2 - 4)$
3	
7	■
8	■
10	■

62.

a	$(5 \cdot 2 - a + 4)$ 10
2	■
5	■
9	■
12	■

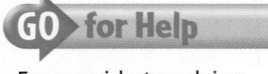

Real-World 🌐 **Connection**

Jan Matzeliger (1852–1889) immigrated to the United States and worked in a shoe factory in Lynn, Massachusetts.

63. a. Geometry The formula for the volume of a cylinder is $V = \pi r^2 h$. What is the volume of the cylinder at the right? Round your answer to the nearest hundredth of a cubic inch.

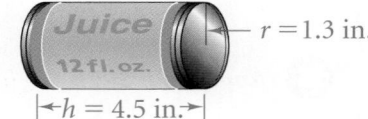

r = 1.3 in.
h = 4.5 in.

b. Critical Thinking About how many cubic inches does an ounce of juice fill? Round your answer to the nearest tenth of a cubic inch.
c. The formula for the surface area of a cylinder is $SA = 2\pi r(r + h)$. What is the surface area of the cylinder? Round your answer to the nearest hundredth of a square inch.

GO for Help

For a guide to solving Exercise 64, see p. 16.

64. Sports The formula for the volume of a sphere with radius r is $V = \frac{4\pi r^3}{3}$. Find the volume of a croquet ball that has radius 4.6 cm. Round your answer to the nearest hundredth.

65. Writing Suppose you have a numerical expression. Is there only one number that is the simplest form of the expression? Explain.

C Challenge

Use grouping symbols to make each equation true.

66. $10 + 6 \div 2 - 3 = 5$ **67.** $14 - 2 + 5 - 3 = 4$

68. $3^2 + 9 \div 9 = 2$ **69.** $6 - 4 \div 2 = 1$

70. a. Simplify $12 + (3 + 7)$ and $(12 + 3) + 7$.
b. Does it seem that the placement of the parentheses affects the value of an expression when only addition is involved? Explain.

71. a. Simplify $(12 - 3) + 7$ and $12 - (3 + 7)$.
b. Does it seem that the placement of the parentheses affects the value of the expression when both addition and subtraction are involved? Explain.

72. Open-Ended Use the numbers 1, 2, 4, and 5 in any order to write expressions to equal each integer from 1 to 20. Examples:

$$(2 \cdot 4) + 1^5 = 9 \qquad 4(5 - 2) + 1 = 13$$

73. a. Geometry A trapezoid has height $h = 8$ ft, and bases $b_1 = 4$ ft and $b_2 = 5.5$ ft. Use the formula $A = h\left(\dfrac{b_1 + b_2}{2}\right)$ to find the area of the trapezoid.

   b. Critical Thinking Does the area of the trapezoid double if the height doubles? If one base doubles? If both bases double? Explain your answers.

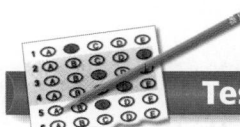

NY REGENTS

Multiple Choice

74. Simplify $5^3 - 15 \div 2 + 2$.
   A. 2   **B.** 57   **C.** 112   **D.** 119.5

75. Simplify $8(5 - 3)^3 + 9$.
   F. 73   **G.** 57   **H.** 40   **J.** 22

76. Evaluate $2ab + c$ for $a = 3.3$, $b = 4.5$ and $c = 2$.
   A. 18.4   **B.** 31.7   **C.** 32.8   **D.** 41.6

77. Evaluate $(r - s)^2$ for $r = 9$ and $s = 6.5$.
   F. 2.5   **G.** 3.5   **H.** 6.25   **J.** 12.25

78. A shirt is on sale for $25 at the local department store. There is also a 4% sales tax. What is the total cost of the shirt, including the sales tax?
   A. $25   **B.** $26   **C.** $29   **D.** $35

79. You can find the distance d an object falls in feet for time t in seconds using the formula $d = 16t^2$. Suppose a ball is dropped out of a window of a tall building. How far will the ball fall in 3 seconds?
   F. 144 ft   **G.** 96 ft   **H.** 48 ft   **J.** 16 ft

Mixed Review

GO for Help

Lesson 1-1

Write an expression for each phrase.

80. 2 more than c **81.** the product of 36 and m

82. the difference of t and 21   **83.** the quotient of y and 5

Skills Handbook

Write each decimal as a percent.

84. 0.5   **85.** 0.34   **86.** 0.95   **87.** 1.45   **88.** 0.06

Find each answer.

89. What is 50% of 86?   **90.** What is 18% of 105?

Add or subtract. Write each answer in simplest form.

91. $\dfrac{3}{10} + \dfrac{1}{5}$ **92.** $\dfrac{3}{8} - \dfrac{1}{4}$ **93.** $\dfrac{5}{12} + \dfrac{2}{3}$

94. $\dfrac{3}{4} - \dfrac{1}{3}$ **95.** $\dfrac{5}{8} + \dfrac{5}{12}$ **96.** $\dfrac{7}{8} - \dfrac{2}{3}$

Understanding Word Problems Read through the problem below.
Then follow along with what Alana thinks as she solves the problem. Check
your understanding with the exercise at the bottom of the page.

The formula for the volume of a sphere with radius r is $V = \frac{4\pi r^3}{3}$.
Find the volume of a croquet ball that has a radius of 4.6 cm.
Round your answer to the nearest hundredth.

What Alana Thinks

I need to read the problem carefully so that I can understand it. I'll write down the important information. I need to know what each variable stands for; r is radius and V is volume.

Now I should substitute 4.6 for r, so that I can find the volume.

I can use a calculator to evaluate the formula. I enter 4 $\boxed{\pi}$ $\boxed{\times}$ 4.6 $\boxed{\wedge}$ 3 $\boxed{/}$ 3 in my calculator.

I should round the calculator answer 407.7200834 to the nearest hundredth.

I need to remember to include the unit in my final answer. For volume, I must use cubic units.

What Alana Writes

Formula for volume: $V = \frac{4\pi r^3}{3}$

Radius of the croquet ball: 4.6 cm

$V = \frac{4\pi(4.6)^3}{3}$

≈ 407.72

407.72 cm³

EXERCISES

1. The formula $t = m + 0.06m + 0.15m$ gives the total cost t for a meal that has a menu price m dollars and has 6% tax and 15% tip. Use the formula to find the total cost for a meal with a menu price $14.20.

2. The formula $A = h\left(\frac{b_1 + b_2}{2}\right)$ gives the area A of trapezoid with height h and bases b_1 and b_2. Use the formula to find the area of a trapezoid with height 5 cm and bases of 7 cm and 12 cm.

3. The formula S.A. $= 4\pi r^2$ gives the surface area S.A. of a sphere with radius r. Find the surface area of a ball that has a radius of 5 in. Round to the nearest hundredth.

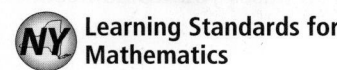

Exploring Real Numbers

NY Learning Standards for Mathematics

A.N.6 Evaluate expressions involving factorial(s), absolute value(s), and exponential expression(s).

✓ **Check Skills You'll Need**   **GO** **for Help** Skills Handbook page 759

Write each decimal as a fraction and each fraction as a decimal.

1. 0.5   **2.** 0.05   **3.** 3.25   **4.** 0.325

5. $\frac{2}{5}$ **6.** $\frac{3}{8}$ **7.** $\frac{2}{3}$ **8.** $3\frac{5}{9}$

🔊 **New Vocabulary**
- natural numbers
- whole numbers
- integers
- rational number
- irrational number
- real numbers
- counterexample
- inequality
- opposites
- absolute value

1   Classifying Numbers

The thermometer shows positive numbers, zero, and negative numbers.

Each of the graphs below shows a set of numbers on a number line. The number below a point is its coordinate on the number line.

Natural numbers	$1, 2, 3, \ldots$
Whole numbers	$0, 1, 2, 3, \ldots$
Integers	$\ldots -2, -1, 0, 1, 2, \ldots$

As you can see, the set of integers has negative numbers as well as zero and positive numbers. There are also numbers that are not integers, such as 0.37 or $\frac{1}{2}$, which are rational numbers. A **rational number** is any number that you can write in the form $\frac{a}{b}$, where a and b are integers and $b \neq 0$. A rational number in decimal form is either terminating, such as 6.27, or repeating, such as 8.222 ..., which you can write as $8.\overline{2}$.

All integers are rational numbers because you can write any integer n as $\frac{n}{1}$.

1 EXAMPLE   Classifying Numbers

Name the set(s) of numbers to which each number belongs.

a. $-\frac{17}{31}$ rational numbers

b. 23   natural numbers, whole numbers, integers, rational numbers $\left(\frac{23}{1}\right)$

c. 0   whole numbers, integers, rational numbers $\left(\frac{0}{1}\right)$

d. 4.581   rational numbers $\left(\frac{4581}{1000}\right)$

✓ **Quick Check** ❶ Name the set(s) of numbers to which each number belongs.

     a. -12 **b.** $\frac{5}{12}$ **c.** -4.67 **d.** 6

You may need to determine the set of numbers that is reasonable for a given situation. For example, suppose 110 students are going on a field trip. Each bus can hold 40 students. To find the number of buses needed, divide 110 by 40. The answer is a rational number, 2.75. However, it is not the number of buses you would need. You would need a whole number of buses, or 3 buses.

 **EXAMPLE** <u>Real-World Problem Solving</u>

Which set of numbers is most reasonable for each situation?
a. the number of students who will go on the class trip   whole numbers
b. the height of the door frame in your classroom   rational numbers

 **Quick Check** ❷ Which set of numbers is most reasonable for the cost of a scooter?

Vocabulary Tip

The <u>irrational number</u> π represents the ratio $\frac{\text{circumference}}{\text{diameter}}$ of a circle. This means that either the circumference or the diameter is not rational.

An **irrational number** cannot be expressed in the form $\frac{a}{b}$, where a and b are integers. Here are three irrational numbers.

$$0.101001000\ldots \qquad \pi \qquad \sqrt{10}$$

Decimal representations of each of these are nonrepeating and nonterminating.

Together, rational numbers and irrational numbers form the set of **real numbers.**

The Venn diagram below shows the relationships of the sets of numbers that make up the real numbers.

🔑 **Key Concepts**

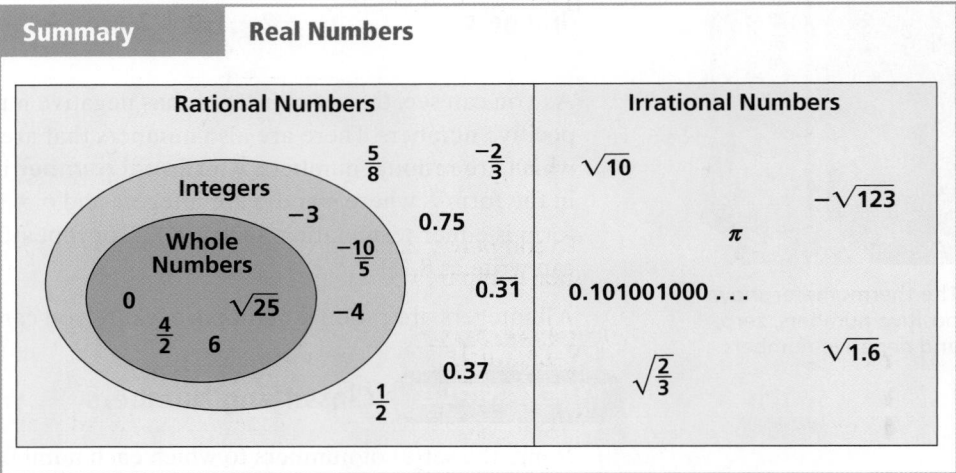

Summary	Real Numbers

Suppose a friend says that all integers are whole numbers. You respond that -3 is an integer but not a whole number. You are using a counterexample to prove that a statement is false.

Any example that proves a statement false is a **counterexample.** You need only one counterexample to prove that a statement is false, while a proof to show that a statement is true may be more complicated.

3 EXAMPLE Using Counterexamples

Is each statement *true* or *false*? If it is false, give a counterexample.

a. All whole numbers are rational numbers.

Every whole number can be written in the form $\frac{n}{1}$, so all whole numbers are rational numbers. The statement is true.

b. The square of a number is always greater than the number.

The square of 0.5 is 0.25, and 0.25 is *not* greater than 0.5. The statement is false.

Quick Check ③ **Critical Thinking** Is each statement true or false? If it is false, give a counterexample.

a. All whole numbers are integers.　　　**b.** No fractions are whole numbers.

---

2 Comparing Numbers

Vocabulary Tip

You can read an inequality from left to right or right to left. 3 < 5 means "3 is less than 5" and "5 is greater than 3." So 3 < 5 and 5 > 3 have the same mathematical meaning.

An **inequality** is a mathematical sentence that compares the value of two expressions using an inequality symbol, such as $<$ or $>$.

When you compare two real numbers, only one of these can be true:

$a < b$　　　or　　　$a = b$　　　or　　　$a > b$

is less than　　　　　**is equal to**　　　　　　**is greater than**

There are three other symbols that compare two values.

$a \le b$　　　or　　　$a \ne b$　　　or　　　$a \ge b$

is less than or equal to　　**is not equal to**　　**is greater than or equal to**

The number line below shows how values of numbers increase as you go to the right on a number line.

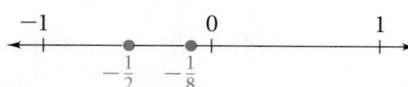

$-\frac{1}{8}$ is to the right of $-\frac{1}{2}$, so $-\frac{1}{8} > -\frac{1}{2}$ and $-\frac{1}{2} < -\frac{1}{8}$.

To compare fractions, you may find it helpful to write the fractions as decimals and then compare the decimals.

For: Number Line Activity
Use: Interactive Textbook, 1-3

4 EXAMPLE Ordering Fractions

Write $-\frac{3}{8}$, $-\frac{1}{2}$, and $-\frac{5}{12}$ in order from least to greatest.

$-\frac{3}{8} = -0.375$ 　　　　　　　**Write each fraction as a decimal.**

$-\frac{1}{2} = -0.5$

$-\frac{5}{12} = -0.4166\ldots = -0.41\overline{6}$

$-0.5 < -0.41\overline{6} < -0.375$ 　　　**Order the decimals from least to greatest.**

From least to greatest, the fractions are $-\frac{1}{2}$, $-\frac{5}{12}$, and $-\frac{3}{8}$.

Quick Check ④ Write $\frac{1}{12}$, $-\frac{2}{3}$, and $-\frac{5}{8}$ in order from least to greatest.

Two numbers that are the same distance from zero on a number line but lie in opposite directions are **opposites.**

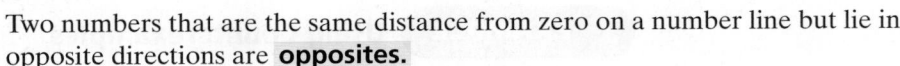

3 units   3 units

$-4 \ -3 \ -2 \ -1 \ \ 0 \ \ 1 \ \ 2 \ \ 3 \ \ 4$

−3 and 3 are the same distance from 0.
So −3 and 3 are opposites.

The **absolute value** of a number is its distance from 0 on a number line. Both −3 and 3 are 3 units from zero. Both have an absolute value of 3. You write "the absolute value of −3" as $|-3|$.

Real-World  **Connection**

The two teams pull in opposite directions. Depending on direction, a pull on the rope results in a positive or a negative direction for your team.

5 **EXAMPLE** **Finding Absolute Value**

Find each absolute value.

a. $|12|$ 12 is 12 units from 0 on a number line. $|12| = 12$

b. $\left|-\frac{2}{3}\right|$ $-\frac{2}{3}$ is $\frac{2}{3}$ units from 0 on a number line. $\left|-\frac{2}{3}\right| = \frac{2}{3}$

c. $|0|$ 0 is at 0 on a number line. $|0| = 0$

 **Quick Check** ⑤ Find each absolute value.

a. $|5|$　　　　　　**b.** $|-4|$　　　　　　**c.** $|-3.7|$　　　　　　**d.** $\left|\frac{5}{7}\right|$

EXERCISES

For more exercises, see *Extra Skill and Word Problem Practice*.

Practice and Problem Solving

Ⓐ Practice by Example

Example 1
(page 17)

GO for Help

Name the set(s) of numbers to which each number belongs.

1. -1　　　　**2.** $\frac{1}{3}$　　　　**3.** -4.8　　　　**4.** 7　　　　**5.** $-\frac{32}{95}$

6. $-\frac{20}{4}$　　　　**7.** 0　　　　**8.** -7.34　　　　**9.** $\frac{7}{1239}$　　　　**10.** $\sqrt{5}$

Open-Ended **Give an example of each kind of number.**

11. negative integer　　　**12.** whole number　　　**13.** positive real number

Example 2
(page 18)

Are *whole numbers*, *integers*, or *rational numbers* the most reasonable for each situation?

14. your shoe size

15. the number of siblings you have

16. a temperature in a news report

17. the number of quarts of paint you need to buy to paint a room

18. the number of quarts of paint you use when you paint a room

Example 3
(page 19)

Is each statement *true* or *false*? If the statement is false, give a counterexample.

19. All integers are rational numbers.

20. All negative numbers are integers.

21. Every multiple of 3 is odd.

22. No positive number is less than its absolute value.

23. No negative number is less than its absolute value.

Example 4
(page 19)

Use <, =, or > to compare.

24. $\frac{2}{3}$ ■ $\frac{1}{6}$　　　　**25.** $-\frac{2}{3}$ ■ $-\frac{1}{6}$　　　**26.** $\frac{15}{8}$ ■ $1\frac{6}{8}$　　　**27.** $\frac{3}{5}$ ■ 0.6

Order the numbers in each group from least to greatest.

28. $2.01, 2.1, 2.001$　　　**29.** $-9\frac{2}{3}, -9\frac{7}{12}, -9\frac{3}{4}$　　　**30.** $-\frac{5}{6}, -\frac{1}{2}, \frac{2}{3}$

31. $-1.01, -1.001, -1.0009$　**32.** $\frac{7}{11}, 0.63, 0.636$　　　**33.** $\frac{22}{25}, \frac{8}{9}, 0.8888$

Example 5
(page 20)

Find each absolute value.

34. $|4|$　　　　**35.** $|-9|$　　　　**36.** $\left|\frac{-9}{14}\right|$　　　**37.** $|-0.5|$

38. $\left|\frac{3}{5}\right|$　　　**39.** $|0|$　　　　**40.** $|-1295|$　　**41.** $\left|-\frac{4}{5}\right|$

B Apply Your Skills

Write each number in the form $\frac{a}{b}$ using integers to show that it is a rational number.

42. 0.2　　　**43.** 5　　　　**44.** 21.3　　　**45.** 1.034　　　**46.** -4

Name the set(s) of numbers to which each number belongs.

47. $\left|\frac{93}{3}\right|$　　　**48.** $|-782|$　　　**49.** $|-1.93|$　　　**50.** $\left|\frac{37}{59}\right|$

Use <, =, or > to compare.

51. $|19|$ ■ $|-19|$　　**52.** $|-18|$ ■ $|-17|$　　**53.** $\left|\frac{1}{2}\right|$ ■ $|-0.51|$

54. $|-3.121|$ ■ $|3.12|$　**55.** $\left|\frac{-8}{10}\right|$ ■ $\left|\frac{-16}{20}\right|$　**56.** $\left|\frac{1}{3}\right|$ ■ $|-0.333|$

Simplify each expression. (*Hint:* Absolute value symbols are grouping symbols.)

57. $4 + |3 - 1|$　　　**58.** $|41 - 38| + 6$　　　**59.** $|a - a| + a$

60. $|24| + |-4|$　　　**61.** $|12| \cdot |-4|$　　　**62.** $|-6 + 4| + |3|$

63. a. Math in the Media In the cartoon below, what type of number is pi (π)?
　　b. Will the football ever be hiked? Explain.

FoxTrot　　　　　　　　　　　　　by Bill Amend

64. Multiple Choice Use the number line. If R and T are opposites, what is the value of Q?

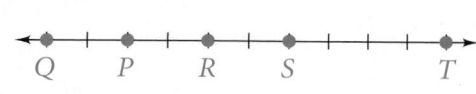

(A) 7　　　(B) 0　　　(C) -3　　　(D) -7

65. Writing Are natural numbers, whole numbers, and integers also rational numbers? Explain.

Problem Solving Hint
Test each statement using a variety of numbers.

Determine whether each statement is *sometimes*, *always*, or *never* true.

66. The difference of two rational numbers is an integer.

67. The product of two numbers is greater than either number.

68. The opposite of a number is less than the number.

69. The quotient of two nonzero integers is a rational number.

Evaluate each expression for *c* = 5, *d* = 1, and *e* = 6.

70. $-|c + d|$

71. $2e + \left|\dfrac{c}{d}\right|$

72. $\dfrac{|e - d|}{c}$

73. $|d + 2| + |-7|$

 **Challenge** 🌐 **74. a. Science** Use the formula $d = \dfrac{m}{v}$ to find the density *d* of each substance in the table below.

Densities of Some Substances

	Mass (*m*)	Volume (*v*)	Density (*d*)
Aluminum	38.5 g	14 cm³	■
Gold	38.6 g	2 cm³	■
Silver	42 g	4 cm³	■
Diamond	1.75 g	0.5 cm³	■

b. List the substances from least to greatest density.

75. a. Open-Ended Find a number between -2 and -3 on a number line.
  b. Find a number between -2.8 and -2.9.
  c. Find a number between $-2\frac{1}{16}$ and $-2\frac{3}{8}$.
  d. Make a Conjecture On a number line, is it possible to find a number between any two different given numbers? Explain.

NY REGENTS

Test Prep

Multiple Choice

76. Which number has the same value as $-\left|-\frac{3}{4}\right|$?
  A. -0.75 **B.** -0.34 **C.** 0.34 **D.** 0.75

77. Which group of numbers is ordered from least to greatest?
  F. $-0.7, -\frac{3}{4}, -1$ **G.** $-1, -\frac{3}{4}, -0.7$ **H.** $-1, -0.7, -\frac{3}{4}$ **J.** $-0.7, -1$

78. Suppose *a* is a nonzero integer. Which statement is never true?
  A. $a > -a$ **B.** $a < -a$ **C.** $|a| = -a$ **D.** $|a| = -|a|$

79. Which number is NOT an integer?
  F. -10 **G.** $-\frac{2}{3}$ **H.** 0 **J.** 5

80. Which set of numbers is most reasonable to use to describe the area of your kitchen floor?
  A. whole numbers   **B.** integers
  C. rational numbers   **D.** real numbers

81. Use the double-bar graph. In which year was there the least difference between the numbers of metal workers and of textile workers?

    F. 1970
    G. 1980
    H. 1990
    J. 2000

82. Suppose your brother says that fractions are rational numbers, but fractions are not integers. Which number is a counterexample for this statement?

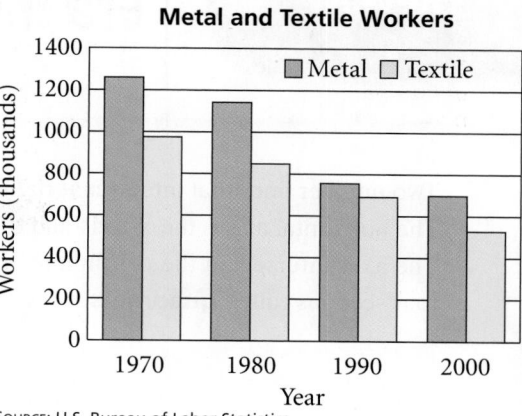

Metal and Textile Workers

Source: U.S. Bureau of Labor Statistics

    A. $\frac{10}{3}$ **B.** $\frac{5}{3}$ **C.** $-\frac{5}{3}$ **D.** $-\frac{9}{3}$

Mixed Review

Lesson 1-2

Simplify each expression.

83. $3 + 5 \cdot 6$ **84.** $(3 + 5)6$ **85.** $3 + 52 \cdot 6$

86. $(33 + 9) \div 6$ **87.** $\frac{12 - 4 \cdot 3}{7 + 13 \cdot 15}$ **88.** $8 - 3 \div 6$

Lesson 1-1

Define variables and write an equation to model the relationship in each table.

89.

Number of Hours	Distance Traveled
1	7 mi
2	14 mi
3	21 mi
4	28 mi

90.

Number of Books	Total Cost
1	$3.50
2	$7.00
3	$10.50
4	$14.00

✓ Checkpoint Quiz 1   Lessons 1-1 through 1-3

Write a variable expression for each phrase.

  1. the sum of b and 4   **2.** the quotient of c and 2

  3. the product of a and 4.3   **4.** b plus c plus twice a

Evaluate each expression for $a = 3$ and $b = 7$.

  5. $(b - a)b$ **6.** $a^2 + b^2$

  7. $ba^2 - a$ **8.** $(2a)^2 b$

  9. Is it true that a number is always greater than its opposite? Explain.

  10. What set of numbers is reasonable to use for the number of loaves of bread a bakery bakes in one day?

Graphing on the Coordinate Plane

Two number lines that intersect at right angles form a **coordinate plane.**
The horizontal axis is the **x-axis** and the vertical axis is the **y-axis.**
The axes intersect at the **origin** and divide the coordinate plane into
four sections called **quadrants.**

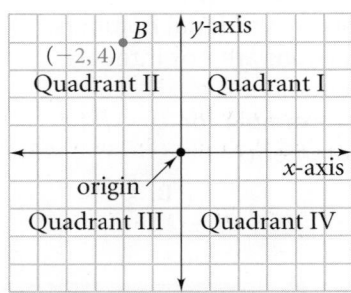

An **ordered pair** of numbers identifies the location of a point. These numbers are the **coordinates** of the
point on the graph. Point B has coordinates $(-2, 4)$.

$$(-2, 4)$$

x-coordinate   **y-coordinate**
   or abscissa   or ordinate

The x-coordinate tells you how far to move right (positive) or left (negative)
from the origin. The y-coordinate tells you how far to move up (positive) or down
(negative) from the origin.

1 EXAMPLE   Identifying Coordinates

a. Name the coordinates of point Z in the graph.

Move 2 units to the left of the origin. Then move
3 units down. The coordinates of Z are $(-2, -3)$.

b. Name the coordinates of point Q in the graph.

Since Q is directly above the origin, the
x-coordinate is 0. Move 2 units up from the origin.
The coordinates of Q are $(0, 2)$.

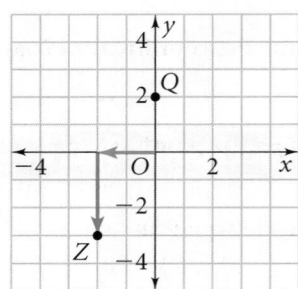

2 EXAMPLE   Graphing Points

a. Graph the point $A\left(-2\frac{1}{2}, 3\right)$ on the coordinate plane.

Move $2\frac{1}{2}$ units to the left of the origin.
Then move 3 units up.

b. Graph the point $B(3, 0)$ on the coordinate plane.

Since the y-coordinate is 0, point B is on the x-axis.
Move 3 units to the right of the origin.

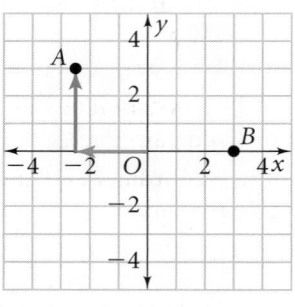

You can determine which quadrant a point is in by graphing the point or by considering the signs of the x- and y-coordinates. A point on an axis is not considered to be in a quadrant.

3 EXAMPLE Identifying Quadrants

In which quadrant or on which axis would you find each point?

a. $(-1, 5)$

Since the x-coordinate is negative and the y-coordinate is positive, the point is in Quadrant II.

b. $(0, 3)$

Since the x-coordinate is 0, the point is on the y-axis.

The **midpoint** of a segment is the point halfway between the endpoints. You can use the coordinates of the endpoints to find the midpoint of a segment.

Given a segment with endpoints $A(x_1, y_1)$ and $B(x_2, y_2)$,

the midpoint M is $\left(\dfrac{x_1 + x_2}{2}, \dfrac{y_1 + y_2}{2} \right)$.

4 EXAMPLE Midpoint of a Line Segment

Find the midpoint of $\overline{GH}$.

$\left(\dfrac{x_1 + x_2}{2}, \dfrac{y_1 + y_2}{2} \right)$ **Use the midpoint formula.**

$= \left(\dfrac{-2 + 6}{2}, \dfrac{4 + (-3)}{2} \right)$ **Replace (x_1, y_1) with $(-2, 4)$ and (x_2, y_2) with $(6, -3)$.**

$= \left(\dfrac{4}{2}, \dfrac{1}{2} \right) = \left(2, \dfrac{1}{2} \right)$ **Simplify.**

The midpoint of $\overline{GH}$ is $\left(2, \dfrac{1}{2} \right)$.

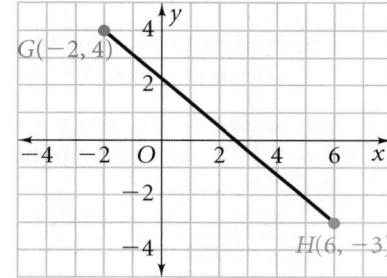

EXERCISES

Name the coordinates of each point on the graph at the right.

1. S **2.** T **3.** U **4.** V

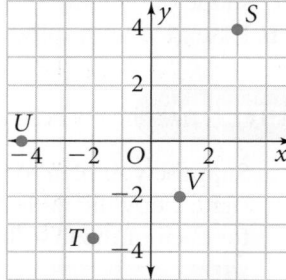

Graph the points on the same coordinate plane.

5. $(3, 0)$ **6.** $\left(-1\frac{1}{2}, 2 \right)$ **7.** $(-2, -3)$ **8.** $\left(4, -2\frac{1}{2} \right)$

In which quadrant or on which axis would you find each point?

9. $\left(-10\frac{1}{2}, 6 \right)$ **10.** $(-12, 0)$ **11.** $(8, -18)$ **12.** $(0, 5)$

Find the midpoint of the segment with the given endpoints.

13. $C(8, 1)$ and $D(-4, 3)$ **14.** $S(6, -9)$ and $T(0, 2)$

15. $A(-2, 4)$ and $B(-6, -2)$ **16.** $P(5, -6)$ and $Q(-3, 5)$

Complete each statement.

17. If the x-coordinate and the y-coordinate of an ordered pair are positive, the ordered pair is in Quadrant ? .

18. The x-coordinate is 0 and the y-coordinate is negative. Is the point in Quadrant III? Explain.

Patterns and Expressions

In Lesson 1-1, you wrote variable expressions for word phrases. You can also write variable expressions to represent patterns.

NY A.PS.3: Observe and explain patterns to formulate generalizations and conjectures.

1 ACTIVITY

Use the figures below to answer the following questions.

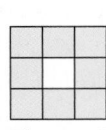   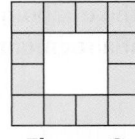   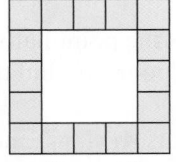

Figure 1   **Figure 2**   **Figure 3**

1. Copy the table at the right. Write the number of tiles you would need for each of the first three figures.

2. Predict How many tiles would you need to make Figure 4? Make a sketch of Figure 4 on graph paper. Write your results in the table.

3. Predict How many tiles would you need to make Figure 5? Make a sketch of Figure 5 on graph paper. Write your results in the table.

4. a. Generalize Write a sentence that describes the relationship between the figure number and the number of tiles needed to build it.
   b. Write an algebraic expression using n that tells the total number of tiles needed. Write your expression in the table.
   c. Use your expression. How many tiles would be needed for Figure 40?

5. a. Graph the data for Figures 1–5. Use the horizontal axis for the figure number and the vertical axis for the number of tiles. What do you notice about the shape of the graph?
   b. Extend your graph to find how many tiles would be needed for Figure 8.

Figure	Total Number of Tiles Needed
1	■
2	■
3	■
4	■
5	■
n	■

2 ACTIVITY

6. a. Suppose you plan to build a patio in a design like the ones shown at the right. Copy the table below. Write the total number of bricks needed for each of the first three figures.

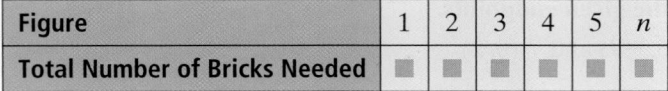

Figure	1	2	3	4	5	n
Total Number of Bricks Needed	■	■	■	■	■	■

Figure 1   Figure 2   Figure 3

   b. Predict How many bricks will you need to make Figures 4 and 5? Make sketches and write your results in the table.

7. a. Generalize Write a sentence that describes the relationship in the table.
   b. Write an algebraic expression using n that tells the total number of bricks needed. Write your expression in the table.
   c. Suppose you have at most 50 bricks to use. Can you use them all to make a figure like those above? Explain.

Patterns and Functions

NY Learning Standards for Mathematics

A.A.5 Write algebraic equations or inequalities that represent a situation.

✓ Check Skills You'll Need

Write an algebraic expression for each phrase.

1. 10 more than twice a number

2. a number divided by 4

3. 8 minus six times a number

4. twice a number subtracted from 7

GO for Help Lesson 1-1

🔊 **New Vocabulary** • function • function rule • dependent variable
• independent variable • domain • range

1 Writing a Function Rule

In Lesson 1-2, the equation $C = p + 0.06p$ shows the relationship between the original price p of a pair of sneakers and the total cost C after the 6% sales tax is included. The relationship between these quantities is a function.

A **function** is a relationship that assigns exactly one output value for each input value. For each input there is only one corresponding output. A **function rule,** such as $C = p + 0.06p$, is an equation that describes a functional relationship.

1 EXAMPLE Writing a Function Rule

Laundry Suppose you are washing and drying clothes at a self-service laundry. The relationship between the number of loads (input) and the cost (output) is a function. Use the table to write a function rule.

Number of Loads	1	2	3	4
Cost	$2.75	$5.50	$8.25	$11.00

Relate   Total cost is 2.75 · number of loads   **Describe how the quantities relate.**

Define   Let n = the number of loads.   **The number of loads is the input.**

Let c = the total cost.   **The total cost is the output.**

Write c =   2.75 · n **Translate this relationship to a function rule.**

● The function rule is $c = 2.75n$.

✓ **Quick Check** ❶ Write a function rule for the relationship between the number of hours (input) and the number of miles (output).

Hours	1	2	3	4
Total Miles	60	120	180	240

Lesson 1-4 Patterns and Functions   **27**

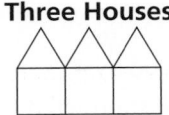

One House   Two Houses

Three Houses

2 EXAMPLE   Writing a Function Rule

The relationship between the number of houses (input) and the number of toothpicks (output) is a function. Use the table to write a function rule.

Number of Houses	1	2	3	4
Total Number of Toothpicks	6	11	16	21

Relate   Total number of toothpicks is one more than five times the number of houses.

Describe how the quantities relate.

Define   Let n = the number of houses.
Let t = the total number of toothpicks.

The number of houses is the input. The total number of toothpicks is the output.

Write $t = 5n + 1$ **Translate this relationship to a function rule.**

● The function rule is $t = 5n + 1$.

✓ Quick Check   **2 a.** How many toothpicks are needed to make seven houses?

b. Reasoning   Could a set of houses be built with exactly 12 toothpicks? Explain.

c. Write a rule for the function represented by the data in the table.

Input x	1	2	3	4
Output y	5	8	11	14

2 Relationships in a Function

In Example 2, the total number of toothpicks t depends on the number of houses n. The value of the **dependent variable** t depends on the value of the **independent variable** n. In real-world situations, you must determine which varying quantity is dependent and which is independent.

3 EXAMPLE   Identifying Independent and Dependent Quantities

Memory Stick   The table and graph model a function relating the storage capacity of a memory stick and its cost. Identify the independent and dependent quantities.

Memory (mb)	Cost ($)
32	19
64	29
128	39
256	49

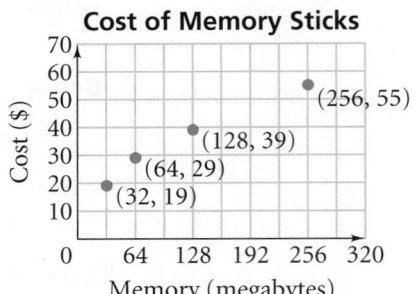

Cost of Memory Sticks

The cost is the *dependent* quantity because it *depends* on the storage capacity of
● the memory stick. Memory is the *independent* quantity.

✓ Quick Check   **3** The cooking time for an unstuffed turkey is about 20 minutes per pound. What are the independent quantity and dependent quantity for this situation?

The possible values for the input, or the independent variable, of a function are the **domain** of the function. The possible values of the output, or the dependent variable, are the **range** of the function.

4 EXAMPLE **Reasonable Domain and Range**

Maria earns $7 per hour for baby-sitting after school and on Saturday. She works no more than 16 hours a week.

a. Identify the independent and dependent quantities for this situation.

The amount Maria earns depends on the number of hours she works in a week. So the amount Maria earns is the dependent variable. The number of hours she works is the independent variable.

b. Find reasonable domain and range values for this situation.

A reasonable domain is from 0 to 16 hours. If Maria works 0 hours, she earns $0. If she works 16 hours, she earns 7 · $16, or $112. So, a reasonable range is from $0 to $112.

✓ **Quick Check** **4** Charlie downloads songs for $.75 each. He has between $3.00 and $6.00 to spend on songs. Identify the independent and dependent quantities for this situation and find reasonable domain and range values.

EXERCISES

For more exercises, see *Extra Skill and Word Problem Practice*.

Practice and Problem Solving

A Practice by Example

Example 1
(page 27)

The relationships in the tables below are functions. Write a function rule for each.

1.

Cans of Soup	Number of Servings
1	4
2	8
3	12
4	16

2.

Number of Cans of Frozen Orange Juice	Total Cost
1	$1.25
2	$2.50
3	$3.75
4	$5.00

Example 2
(page 28)

3.

Number of Hours a Plumber Works	Cost to Home Owner
1	$65
2	$90
3	$115
4	$140

4.

Time (hours)	Cost of Bike Rental
1	$10
2	$16
3	$22
4	$28

Example 3
(page 28)

Identify the independent and dependent quantity in each situation.

5. The cost of a long-distance telephone call increases, with the number of minutes of the call.

6. Water pressure increases 0.44 pounds per square inch (0.44 psi) with each increase of one foot in depth below sea level.

Example 4
(page 29)

Identify the independent and dependent quantities for each situation, and find reasonable domain and range values.

7. Tara's car travels about 25 miles on one gallon of gas. She has between 10 and 12 gallons of gas in the tank.

8. Sal and three friends plan to bowl one or two games each. Each game costs $2.50.

 **Apply Your Skills**

Copy and complete each table. Then write a function rule for each relationship.

9.

Number of Minutes Reading	Number of Words Read
1	125
2	250
3	375
4	▇
5	▇

10.

Number of Hours of Yoga Lessons	Total Cost
1	$15
2	$27
3	$39
4	▇
5	▇

11. The graph shows the average size of a school's ninth-grade class over several years.

  a. Does this graph represent a function? Explain.

  b. If the graph represents a function, identify the independent and dependent quantities.

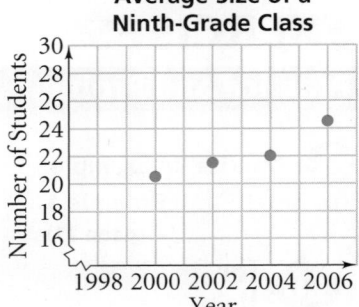

Average Size of a Ninth-Grade Class

GO Online
Homework Video Tutor
Visit: PHSchool.com
Web Code: ate-0104

12. a. The weight of the books carried in a backpack depends on the number of books in the backpack. Susan's books weigh from 1 pound up to 3.5 pounds each and she carries at most 4 books. Find a reasonable domain and range for the situation.

  b. **Critical Thinking** Can noninteger numbers be part of the domain? of the range? Explain.

13. Each side of the first figure is one unit. Copy and complete the table. Then find a function rule for the relationship between a figure's number and its perimeter.

Figure Number	Perimeter
1	▇
2	▇
3	▇
4	▇

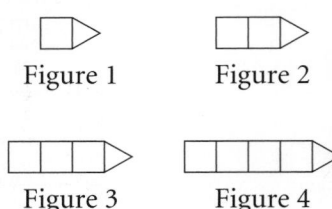

Figure 1   Figure 2

Figure 3   Figure 4

14. **Motorcycles** The table below shows the price of several motorcycles and the size of their engines. Identify the independent and dependent quantities. Explain.

Engine (cc)	249	398	645	996
Price ($)	3199	6099	6699	8599

15. a. Copy the table at the right. Evaluate the expression $x^2 + 1$ for $x = 1, 2, 3,$ and 4. Record your results in the second column.
  b. Critical Thinking Do the numbers in the table represent a function? Explain.

x	$x^2 + 1$
1	■
2	■
3	■
4	■

16. a. Look at the pattern below. Make a table that relates the figure number n to the number of hexagons h.

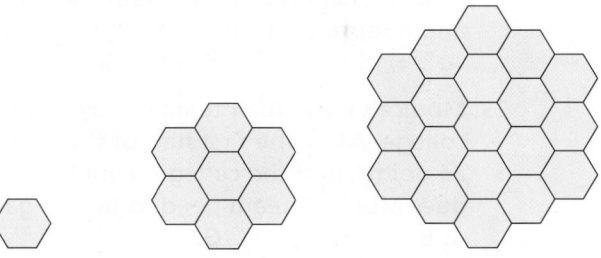

Figure 1   **Figure 2**   **Figure 3**

  b. Predict What will be the number of hexagons in Figure 4?
  c. Does your table represent a function? Explain.

17. a. Complete the table for the side lengths and areas of squares.
  b. Identify the independent and the dependent quantities.

Side length	1	1.5	2	2.5	3
Area	■	■	■	■	■

  c. Graph the data. Show the data for the independent quantity on the horizontal axis and the data for the dependent quantity on the vertical axis.

 **Test Prep**

NY REGENTS

Multiple Choice

18. The graph at the right represents the temperature of water that was boiled and then left to cool. Which statement is true?
  A. Time depends on the water temperature.
  B. The water temperature depends on time.
  C. The cooling time depends on how long the water boils.
  D. The water temperature depends on the boiling time.

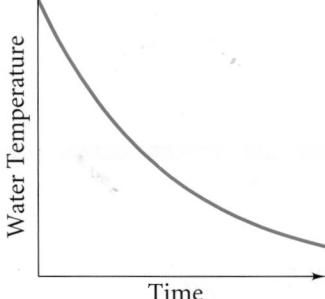

19. The function $c = 23(n - 1) + 37$ represents the cost c in cents for mailing a letter weighing n ounces. Which statement is true?
  F. The weight of a letter depends on its cost.
  G. The number of letters mailed depends on the total number of ounces of the letters.
  H. The cost to mail a letter depends on its weight.
  J. Each letter costs 37¢ an ounce.

20. Evaluate the expression $3x - 5$ for $x = -2, 0, 1,$ and 5. Which value of x produces the greatest value from the expression?
  A. −2   **B.** 0   **C.** 1   **D.** 5

21. Simplify $|-7 + 8| + |-9|$.

    F. 26   **G.** 10   **H.** –26   **J.** 8

22. Jacob recorded the height of a sunflower each week for 5 weeks.

Week	1	2	3	4	5
Height (cm)	15	30	45	60	75

Which function best describes the relationship, if w represents the week and h represents the height?

    A. $h = 15w$ **B.** $w = 15h$ **C.** $h = w + 15$ **D.** $w = h + 15$

23. Monica's basketball team is playing an exhibition game against a local college. After the first half of the game, Monica sees that her team has 24 points, and the college team has 32 points. How many more points does Monica's team need to tie the game?

    F. 6   **G.** 8   **H.** 12   **J.** 56

24. A collection of quarters and nickels is worth $3.40. There are 24 coins. How many of the coins are quarters?

    A. 9   **B.** 11   **C.** 13   **D.** 15

Mixed Review

Lesson 1-3

Are whole numbers, integers, or rational numbers the most reasonable for each situation?

25. your height

26. the number of pages in your math text

27. the overnight low temperature in degrees Celsius for Barrow, Alaska, on January 15

Lesson 1-2

Evaluate each expression for $c = 5$ and $d = 8$.

28. $d^2 - c^3$ **29.** cd^2 **30.** dc^2

31. $(3cd)^2$ **32.** $cd - c^2$ **33.** $(4c)^2d$

Algebra at Work

Cartographer

A cartographer, or mapmaker, makes measurements of an area being mapped. The cartographer uses these measurements to create the scale of the map showing the ratio of map distance to actual distance. Knowing the scale of a map means that you can use a proportion to calculate any distance on the map.

Go Online
PHSchool.com

For: Information about cartographers
Web Code: atb-2031

Scatter Plots

NY Learning Standards for Mathematics

A.S.7 Create a scatter plot of bivariate data.

A.S.12 Identify the relationship between the independent and dependent variables from a scatter plot (positive, negative, or none).

✓ **Check Skills You'll Need**

GO for Help Review page 24

Graph each point on the same coordinate grid.

1. $(6, 4)$ **2.** $(-5, 1)$ **3.** $\left(2\frac{1}{2}, -5\right)$ **4.** $(0, -1)$

🔊 **New Vocabulary** • scatter plot • positive correlation • negative correlation
• no correlation • trend line

1 Analyzing Data Using Scatter Plots

A **scatter plot** is a graph that relates two groups of data. To make a scatter plot, plot the two groups of data as ordered pairs. Most scatter plots are in the first quadrant of a coordinate plane, because the data are usually positive numbers.

1 EXAMPLE Making a Scatter Plot

Data Collection The table at the left shows data students collected on their test scores and the number of hours they watched television the previous day. Make a scatter plot of the data.

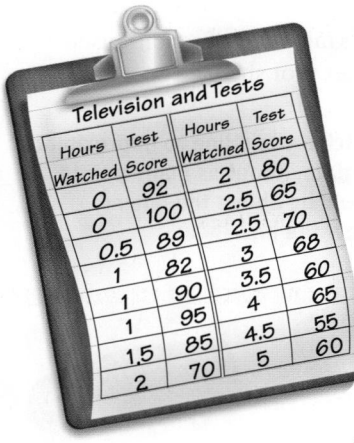

For 2 hours of television watched and a test score of 80, plot (2, 80).

The highest score is 100. So a reasonable scale on the vertical axis is 0 to 100 with every 20 points labeled.

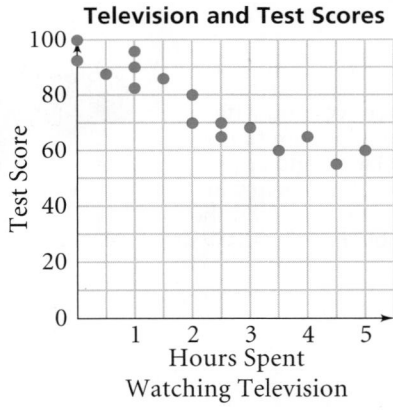

✓ **Quick Check** **1** Use the data in the table below. Make a scatter plot of newspaper circulation and the number of households with television.

Year	1950	1960	1970	1980	1990	2000
Daily Newspaper Circulation (millions)	54	59	62	62	62	55
Households With Television (millions)	4	46	59	76	92	101

You can use scatter plots to find trends in data. The scatter plots below show the three types of relationships that two sets of data may have.

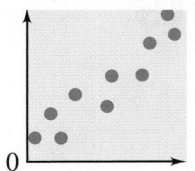

Positive correlation
In general, both sets of data increase together.

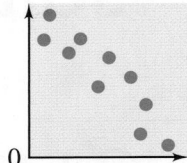

Negative correlation
In general, one set of data decreases as the other set increases.

No correlation
Sometimes data sets are not related.

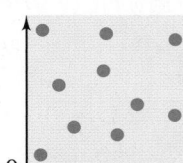

A **trend line** on a scatter plot shows a correlation more clearly.

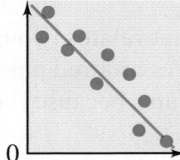

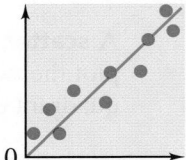

2 EXAMPLE **Real-World** **Problem Solving**

Multiple Choice The scatter plot shows the age and asking price of several used mid-sized cars. What type of relationship does the scatter plot show?

Test-Taking Tip

When working with a graph, be sure to read the labels and units carefully to help you better understand the problem.

Prices of Mid-Sized Used Cars

Sketch a trend line to approximate the data. About as many points should be above the line as below the line.

(A) a negative correlation   (B) no correlation
(C) an undefined correlation   (D) a positive correlation

As the age of a car increases, the asking price generally decreases. There is a negative correlation, so the correct answer is A.

 **Quick Check** **2 a. Critical Thinking** In the graph above, what does the data point at (4, 14,900) represent?
b. Use the graph to predict the asking price of a 7-year-old car.

Practice and Problem Solving

A Practice by Example

Example 1
(page 33)

for Help

Make a scatter plot for each set of data below.

1. **Gasoline Purchases**

Dollars Spent	10	11	9	10	13	5	8	4
Gallons Bought	6.3	6.1	5.6	5.5	8.3	2.9	5.2	2.7

2. **Jeans Sales**

Average Price	$21	$28	$36	$40
Number Sold	130	112	82	65

Example 2
(page 34)

Describe the trend in each scatter plot below.

3. 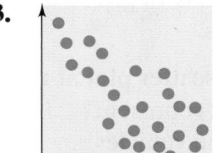 **4.** 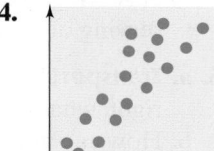 **5.**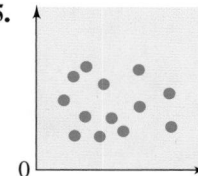

B Apply Your Skills

Critical Thinking Would you expect a *positive correlation*, a *negative correlation*, or *no correlation* between the two data sets? Explain why.

6. the amount of free time you have and the number of classes you take

7. the sales of snow shovels and the amount of snowfall

8. the air pollution levels for a city and the number of cars registered in that city

9. length of a baby at birth and the month in which the baby was born

10. the number of calories burned and the time spent exercising

11. a high-school sprinter's age and the time it takes to finish a 100-meter race

12. **Open-Ended** Describe three situations: one that shows a positive correlation, one that shows a negative correlation, and one that shows no correlation.

13. During one month at a local deli, the number of pounds of ham sold decreased as the number of pounds of turkey sold increased.
 a. Is this an example of a *positive correlation, negative correlation,* or *no correlation*?
 b. **Reasoning** Is it reasonable to conclude that the change in turkey sales caused the decrease in ham sales? Explain.

14. a. Think about the weather and its effect on voters. What correlation would you expect between the amount of precipitation and voter turnout? Explain.
 b. **Reasoning** Should candidates in an election be concerned about the weather forecast? Explain.

supplement with complete guide to ballot questions.

Daily Tribune

RAIN, HEAVY AT TIMES
CHANCE OF SNOW
HIGH: 38 LOW: 30
FULL REPORT ON PAGE 20

Tuesday, November 3, 2005

Election process

15. a. Nutrition Draw a scatter plot of the data below. Graph the grams of fat on the *x*-axis and the number of Calories on the *y*-axis.

Calories Per Serving of Some Common Foods

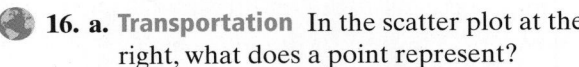

Food	Grams of Fat	Number of Calories	Food	Grams of Fat	Number of Calories
Whole Milk	8	150	Eggs	6	80
Chicken	4	90	Ham	19	245
Corn	1	70	Broccoli	1	45
Ground Beef	10	185	Cheese	9	115

b. Draw a trend line on the scatter plot. Describe the correlation, if any, between Calories and grams of fat.

c. Writing Write a statement describing the relationship between calories and grams of fat.

d. A serving of ice cream has 14 grams of fat. Predict the number of calories a serving of ice cream has.

16. a. Transportation In the scatter plot at the right, what does a point represent?

b. How can you tell if some vehicles traveled the same distance?

c. How can you tell which vehicles paid the same toll charge?

d. Is there a correlation between distance traveled and toll charges? Explain.

17. Open-Ended Describe a situation in which the correlation of two sets of data is undefined.

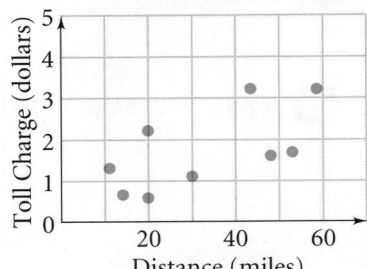

Toll Charges for 9 Vehicles on the Indiana Toll Road

Source: Indiana Department of Highways

nline
Homework Video Tutor
Visit: PHSchool.com
Web Code: ate-0105

C Challenge

If change in one quantity *causes* change in a second quantity, then the quantities have a *causal relationship*. Quantities can be correlated but not have a causal relationship. Is there a causal relationship in the following situations? Explain.

18. the number of cavities and the size of an elementary student's vocabulary

19. the depth of a snowfall and the amount of time spent clearing the driveway

20. the time of sunrise and the daily high temperature

REGENTS

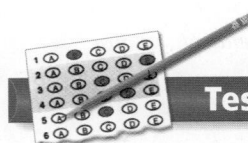

Test Prep

Multiple Choice

21. Suppose you take a survey of all the schools in your state. What would you expect the relationship between the number of students and the number of teachers in each school to be?
   A. positive correlation   **B.** negative correlation
   C. no correlation   **D.** none of the above

22. When 18 gallons of water are pumped into an empty tank, the tank is filled to three fourths of its capacity. How many gallons of water will the tank hold?
   F. 12   **G.** 13.5   **H.** 18.5   **J.** 24

23. Which sets of data would most likely have a negative correlation?
   A. the population of Detroit over a 10-year period and the population of Kansas City over the same 10-year period
   B. the height of a person and that person's shoe size
   C. the number of times a car stops to fill its gas tank and the amount of gas the tank can hold
   D. the size of an animal and the amount of food it needs each day

24. Evaluate $(2m^2 + 5) - (m + 1)$ for $m = 3$.
   F. 13   **G.** 19   **H.** 22   **J.** 37

25. Evaluate $\frac{(p + 3t)}{6} \div (p \div 6)$ for $p = 6$ and $t = 4$.
   A. $\frac{1}{4}$ **B.** $\frac{1}{2}$ **C.** 3   **D.** 4

Mixed Review

Lesson 1-4

26. Jillian's scooter can travel about 60 miles on one gallon of gas. The gas tank holds as much as 1.5 gallons. Identify the independent and dependent quantities for this situation, and reasonable domain and range values.

Lesson 1-3

Decide whether each statement is *true* or *false*. If the statement is false, give a counterexample.

27. All positive integers are natural numbers.

28. A number cannot have a value equal to its square.

29. All integers are rational numbers.

✓ Checkpoint Quiz 2   Lessons 1-4 through 1-5

1. Write a function rule for the relationship between the number of minutes and the number of words.

Minutes	1	2	3	4
Words Typed	35	70	105	140

Identify the independent and the dependent quantity in each situation.

2. increase in pressure with depth below sea level

3. time in a cell phone account and amount of money in the account

4. the number of papers delivered and the amount of money earned

5. a. Make a scatter plot of the data below.

Weeks Web Site Has Been Online	1	2	5	6	8
Visits to Web Site (thousands)	5.5	5.6	4.1	3.4	3.5

   b. Does the scatter plot show a *positive correlation*, a *negative correlation*, or *no correlation*?

   c. What values are in the domain?

Activity Lab

Interpreting Graphs

> **NY A.S.13** Understand the difference between correlation and causation.

As you learned in the previous lesson, a correlation is a relationship between two sets of data. Two quantities have a *causal relationship* if the change in one quantity causes the change in the other quantity. Two quantities can correlate, but not have a causal relationship.

Examples: Height of a plant increases as the amount of sunlight it receives increases
Correlation: Yes
Causal Relationship: Yes

A survey of a class shows that the students who get lower grades spend more time sleeping each week.
Correlation: Yes
Causal Relationship: No

1 ACTIVITY

The scatter plot shows the relationship between the distance from school and the times it takes each bus to get to school.

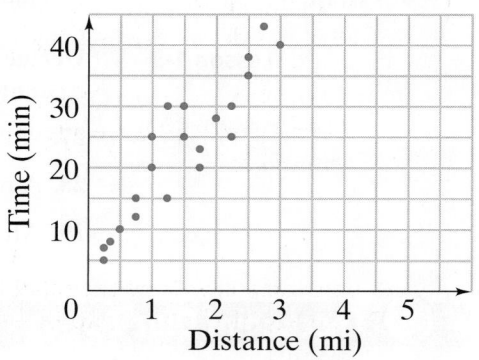

1. Is there a correlation between the distance from school and the time it takes to get to school? Explain.

2. What type of relationship does the scatter plot show?

3. Is there a causal relationship between the two quantities? Explain.

4. **Critical Thinking** Examine newspaper and magazine articles or television news programs. Find an example of a correlation that may or may not be a causal relationship.

Bar graphs, line graphs, and circle graphs are three of the most common ways to represent data. Venn diagrams are another way to show relationships among data.

Bar graphs are used to compare amounts. One axis shows the categories and the other axis shows the amounts.

2 ACTIVITY

The bar graph shows the number of hours per week that Central High School students study.

5. How many hours per week do juniors study?

6. **Writing** How can you identify which class spends the *least* amount of time studying, without determining the actual value? Explain.

7. Do seniors study twice as long as freshmen?

8. **Critical Thinking** Suppose you saw the same data in a bar graph where the *y*-axis was labeled from 0 to 20. Explain how the new bar graph might convey a different impression from the one above.

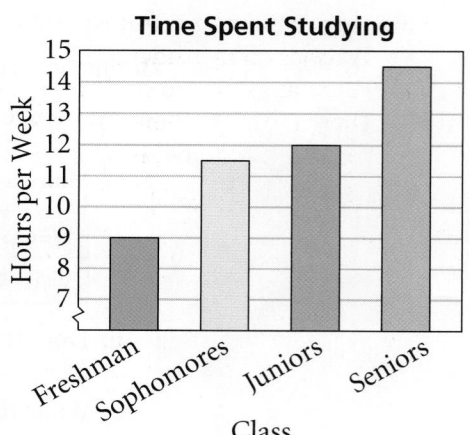

Line graphs are used to show the change in a set of data over a period of time. You can use a line graph to look for trends and make predictions.

3 ACTIVITY

The line graph shows the profits of a small company over several years.

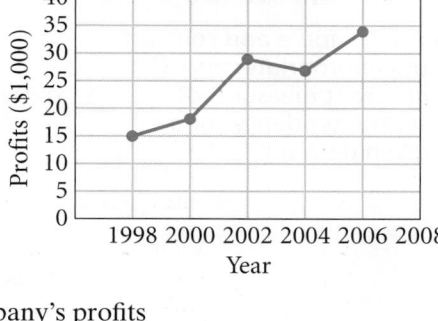

Company Profits

9. **Multiple Choice** Choose the statement that best describes the profits for this company.
   - (A) Profits generally decreased from 1998 to 2006.
   - (B) Profits showed no general trend.
   - (C) Profits generally increased from 1998 to 2006.
   - (D) Profits showed no change from 1998 to 2006.

10. What were the company's profits in 2002?

11. **Estimate** Based on the data shown in the graph, estimate the company's profits in 2005.

12. **Predict** Based on the data shown in the graph, predict the company's profits in 2008.

Interpolation is estimating a value among existing data. *Extrapolation* is predicting a value outside the existing data.

13. **Critical Thinking** In Question 7 you interpolated and in Question 8 you extrapolated. Why are interpolated values generally more reliable than extrapolated values?

A Venn diagram shows the relationships among collections of objects or numbers. The intersection, or overlap, of two circles indicates what is common to both collections.

4 ACTIVITY

The Venn diagram shows the results of a survey of students who are in either band or chorus or both.

14. **Multiple Choice** Based on the diagram, which of the following statements is true?
   - (A) Twice as many students are in band than are in chorus.
   - (B) There are 18 more students in band than in chorus.
   - (C) There are 68 students in band.
   - (D) Seventy-five students were surveyed.

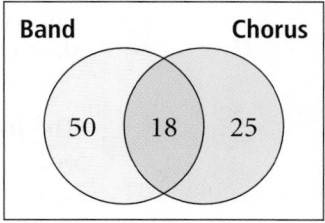

15. What does the number 18 in the intersection represent?

Mean, Median, Mode, and Range

1-6

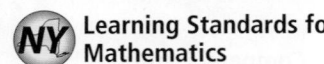
NY Learning Standards for Mathematics

A.S.4 Compare and contrast the appropriateness of different measures of central tendency for a given data set.

GO for Help Lesson 1-3

Write the numbers in each group in order from least to greatest.

1. 2.4, 9.8, 3.6, 7.5, 1.9

2. 144, 235, 98, 72, 58, 195

3. $-12, 14, -3, -8, 7, 0$

4. $2\frac{1}{2}, -3\frac{2}{3}, -4\frac{3}{8}, 6\frac{1}{4}, -2\frac{5}{8}, 4\frac{1}{2}$

Use mental math to simplify.

5. $\dfrac{3 + 4 + 5 + 6 + 7}{5}$

6. $\dfrac{5 + 6 + 8 + 9}{4}$

🔊 **New Vocabulary** • measures of central tendency • mean • outlier
• median • mode • range • stem-and-leaf plot

---

1 **Finding Mean, Median, and Mode**

To understand a set of data, you need to organize and summarize the data using a measure of central tendency. Mean, median, and mode are all **measures of central tendency.**

You must decide which measure of central tendency best describes a set of data. Below is a review of mean, median, and mode, and where you would use each as the measure of central tendency.

Key Concepts

Review	**Mean, Median, Mode**

Mean $= \dfrac{\text{sum of the data items}}{\text{total number of data items}}$

Use the mean to describe the middle of a set of data that *does not* have an outlier. An **outlier** is a data value that is much higher or lower than the other data values in the set. The mean is often referred to as the average.

The **median** is the middle value in the set when the numbers are arranged in order. For a set containing an even number of data items, the median is the mean of the two middle data values.

Use the median to describe the middle of a set of data that *does* have an outlier.

The **mode** is the data item that occurs the most times. It is possible for a set of data to have no mode, one mode, or more than one mode.

Use the mode when the data are nonnumeric or when choosing the most popular item.

---

40 Chapter 1 Variables, Function Patterns, and Graphs

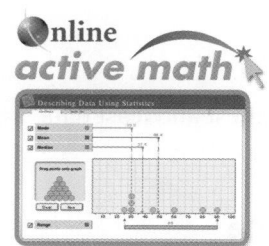
1 EXAMPLE <u>Real-World</u> Problem Solving

Wages Find the mean, median, and mode of the data in the line plot below. Which measure of central tendency best describes the data?

Hourly Wages of Employees at a Local Restaurant

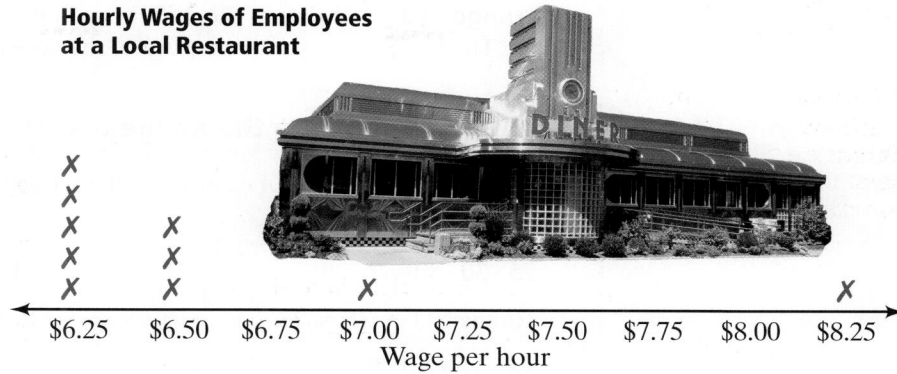

X								
X								
X		X						
X		X						
X		X		X				X
$6.25	$6.50	$6.75	$7.00	$7.25	$7.50	$7.75	$8.00	$8.25

Wage per hour

↓ **5(6.25) is a shortcut for adding 6.25 + 6.25 + 6.25 + 6.25 + 6.25.**

Mean: $\dfrac{5(6.25) + 3(6.50) + 7.00 + 8.25}{10} = 6.6$

↑ **total number of employees**

Median: 6.25 6.25 6.25 6.25 6.25 6.50 6.50 6.50 7.00 8.25   **List data in order. 6.25 and 6.50 are the two middle data values.**

$\dfrac{6.25 + 6.50}{2} = 6.375$ **The median of an even number of data items is the mean of the two middle data values.**

Mode: 6.25   **the data item that occurs most often**

The mean is $6.60, the median is about $6.38, and the mode is $6.25. The mean is greater than the salary of 8 workers. The mode is the salary of the 5 workers with the lowest salary. The median best describes the data.

 **Quick Check** **1 a. Wages** The employee who earns $8.25 per hour resigns. She is replaced by an employee earning $7.50 per hour. Find the mean, median, and mode of the data.
b. Critical Thinking Which measure best describes the data? Explain why.

Students often ask, "What grade do I need on the next test to bring up my average?" The example below shows you how to solve this kind of problem.

2 EXAMPLE Solving an Equation

Suppose your grades on three history exams are 80, 93, and 91. What grade do you need on your next exam to have a 90 average on the four exams?

$\dfrac{80 + 93 + 91 + x}{4} = 90$ **Use the formula for mean. Let x = the grade on the fourth exam.**

$\dfrac{264 + x}{4} = 90$ **Simplify the numerator.**

$4\left(\dfrac{264 + x}{4}\right) = 4(90)$ **Multiply each side by 4.**

$264 + x = 360$ **Simplify.**

$264 + x - 264 = 360 - 264$ **Subtract 264 from each side.**

$x = 96$ **Simplify.**

Your grade on the next exam must be 96 for you to have an average of 90.

✓ Quick Check ② **Critical Thinking** If 100 is the highest possible score on the fourth exam, is it possible to raise your average to 92? Explain.

The **range** of a set of data is the difference between the greatest and least data values. The range gives you a measure of the spread of the data.

③ **EXAMPLE** Finding the Range and Mean of Data

Find the range and mean of each set of data. Use the range to compare the spread of the two sets of data.

25 30 30 47 28   34 28 31 36 31

Range: $47 - 25 = 22$ Range: $36 - 28 = 8$

Mean: $\dfrac{25 + 30 + 30 + 47 + 28}{5}$ Mean: $\dfrac{34 + 28 + 31 + 36 + 31}{5}$

$\dfrac{160}{5} = 32$ $\dfrac{160}{5} = 32$

Both sets of data have a mean of 32. The range of the first set of data is 22, and the range of the second set of data is 8. The second set of data is less spread out.

✓ Quick Check ③ For the first five days in February, the low temperatures in northern Maine were 7°F, 4°F, −3°F, −6°F, and 0°F. During the same time period, the low temperatures in northern Michigan were 24°F, 15°F, −2°F, −10°F, and −5°F. Find the mean and range of each set of data. Compare the spreads of the temperature data.

---

2 ▸ Stem-and-Leaf Plots

You can use a stem-and-leaf plot to organize data. A **stem-and-leaf plot** is a display of data made by using the digits of the values. To make a stem-and-leaf plot, separate each number into a stem and a leaf. This is the stem and leaf for the number 2.39.

all digits
to the left of   last
last digit   digit
↳ 2.3 | 9
↑   ↑
stem   leaf

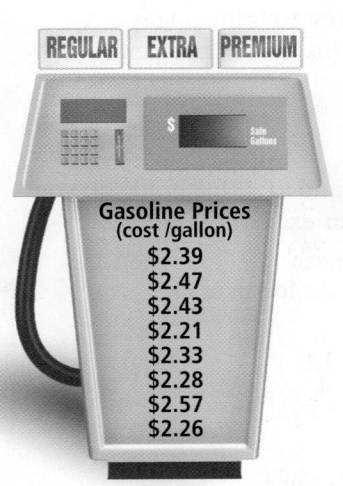

REGULAR EXTRA PREMIUM

$

Safe
Gallons

Gasoline Prices
(cost /gallon)
$2.39
$2.47
$2.43
$2.21
$2.33
$2.28
$2.57
$2.26

④ **EXAMPLE** Making a Stem-and-Leaf Plot

Make a stem-and-leaf plot for the data at the left.

Use the first two digits for the "stems."

2.2	1 6 8
2.3	3 9
2.4	3 7
2.5	7

Use the corresponding last digits for the "leaves." Arrange the numbers in order.

2.5 | 7 means 2.57

✓ Quick Check ④ Make a stem-and-leaf plot for the data below.
4.5 4.3 0.8 3.5 2.6 1.4 0.2 0.8 4.3 6.0

You can find the measures of central tendency of data displayed in a stem-and-leaf plot. The stem-and-leaf plot in the next example is a back-to-back stem-and-leaf plot. The stem is between the two bars, and the leaves are on each side. Leaves are in increasing order from the stems.

5 EXAMPLE **Using a Stem-and-Leaf Plot**

Find the mean of the city mileage and highway mileage for nine new cars.

New Car Mileage (mi/gal)

City		Highway
9	1	
9 8 3 3 0	2	7 8
4 1 1	3	0 2 2 7 8 8
	4	1

means 20 mi/gal ← 0 | 2 | 7 → means 27 mi/gal

Mean City Mileage: $\dfrac{19 + 20 + 23 + 23 + 28 + 29 + 31 + 31 + 34}{9} = 26.\overline{4}$ mi/gal

Mean Highway Mileage: $\dfrac{27 + 28 + 30 + 32 + 32 + 37 + 38 + 38 + 41}{9} = 33.\overline{6}$ mi/gal

 **Quick Check** **5** **a.** Find the median of the city mileage and of the highway mileage.
b. Find the mode(s) of the city mileage and of the highway mileage.
c. Find the range of the city mileage and of the highway mileage.

EXERCISES

For more exercises, see *Extra Skill and Word Problem Practice.*

Practice and Problem Solving

A Practice by Example

Example 1
(page 41)

  for Help

Example 2
(page 41)

Example 3
(page 42)

Find the mean, median, and mode. Which measure of central tendency best describes the data?

1. weights of textbooks in ounces
12 10 9 15 16 10

2. ages of students on math team
14 14 15 15 16 15 15 16

3. time spent on Internet in min/day
75 38 43 120 65 48 52

4. weights of channel catfish in pounds
4.8 5 2.3 4.4 4.8 5.1

Write and solve an equation to find the value of *x*.

5. 3.8, 4.2, 5.3, *x*; mean 4.8

6. 99, 86, 76, 95, *x*; mean 91

7. 100, 121, 105, 113, 108, *x*; mean 112

8. 31.7, 42.8, 26.4, *x*; mean 35

Find the range.

9. 12 15 17 28 30

10. 5.3 6.2 3.1 4.8 7.3

11. −12 −15 5 3 −2 0 −7

12. $2\frac{1}{2}$ $3\frac{1}{3}$ $-5\frac{3}{4}$ $\frac{3}{8}$ $3\frac{5}{8}$

13. For each list of data, find the range and the mean.
Use the range to compare the spread of the data.

List 1	List 2
64 43 55 28 71	48 53 61 47 52

Example 4
(page 42)

Make a stem-and-leaf plot for each set of data.

14. 18 35 28 15 36 10 25 22 15

15. 18.6 18.4 17.6 15.7 15.3 17.5

16. 785 776 788 761 768 768 785

17. 0.8 0.2 1.4 3.5 4.3 4.5 2.6 2.2

Example 5
(page 43)

Find the mean, median, mode, and range of each side of the stem-and-leaf plot.

18.

Time Spent on Homework (minutes/day)		
Class A		Class B
6 6 4 3	4	1 1 4 5 7
9 8 6 4 4 4	5	0 2 2 2 4
5 2 1 0	6	4 5 8 9
8 7 6 6 4 2	7	3 6 7 9 9 9

means 43 ← 3 | 4 | 1 → means 41

19.

Growth of Two Varieties of Tulip Plants (inches/day)		
Type A		Type B
6 3 3	2	
3 2 1 1	3	1 1 2
1	4	3 5 8
	5	2 4

means 0.33 ← 3 | 3 | 1 → means 0.31

B Apply Your Skills

Find the mean, median, mode, and range.

20. 9.8 7.2 6.3 8.7 5.8 9.4 5.1 6.2

21. 3 −12 −1 −7 −2 0 −5 −1 −4 −2

22. 42.1 46.4 58.2 67.3 49.1 40.2 22.3 46.6

23. Critical Thinking The mean of a set of data is 7.8, the mode is 6.6, and the median is 6.8. What is the least possible number of data values? Explain.

24. Wildlife Management A wildlife manager working at the Everglades National Park in Florida measured and tagged adult male crocodiles. The data he collected are at the right.
  a. What are the mean and median lengths of the crocodiles?
  b. The wildlife manager captured another crocodile. Its length was 3.3 m. What is the mean with this new piece of data? What is the median? Round to the nearest tenth.

Crocodile Lengths (meters)			
2.4	2.5	2.5	2.3
2.8	2.4	2.3	2.4
2.1	2.2	2.5	2.7

25. Manufacturing Two manufacturing plants create sheets of steel for medical instruments. The back-to-back stem-and-leaf plot at the right shows data collected from the two plants.
  a. Find the mean, median, mode, and range of each set of data.
  b. Which measure of central tendency best describes each set of data? Explain.
  c. Reasoning Which plant has the better quality control? Explain.

Width of Steel (millimeters)		
Manufacturing Plant A		Manufacturing Plant B
	4	3 5 9
8 7 4 4 2	5	2 7
4 3 1	6	3 4
	7	2

means 6.1 ← 1 | 6 | 3 → means 6.3

26. Open-Ended Give an example of a set of data for which the mode best represents the data. Explain.

27. Sports The median height of the 21 players on a girls' soccer team is 5 ft 7 in. What is the greatest possible number of girls who are less than 5 ft 7 in. tall?

28. Writing How does an outlier affect the mean of a set of data?

GO Online
Homework Video Tutor
Visit: PHSchool.com
Web Code: ate-0106

 **Challenge**

29. Make a back-to-back stem-and-leaf plot of the data below.

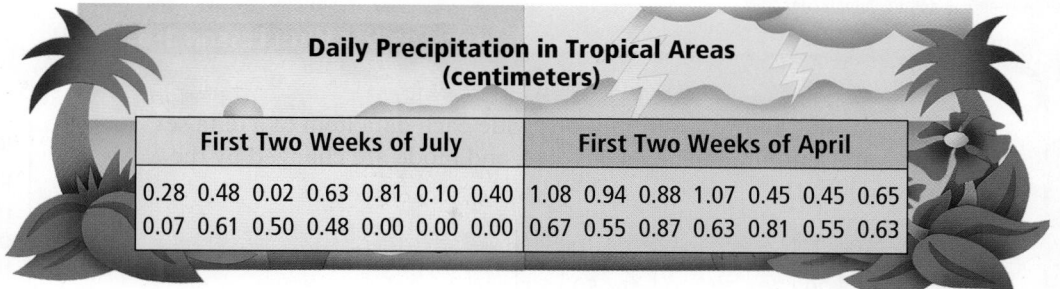

Daily Precipitation in Tropical Areas (centimeters)

First Two Weeks of July	First Two Weeks of April
0.28 0.48 0.02 0.63 0.81 0.10 0.40	1.08 0.94 0.88 1.07 0.45 0.45 0.65
0.07 0.61 0.50 0.48 0.00 0.00 0.00	0.67 0.55 0.87 0.63 0.81 0.55 0.63

30. Data Collection Record the high and low temperatures in your town for one week. Make a back-to-back stem-and-leaf plot with the data you collect.

31. During the first 6 hours of a trip, you average 44 mi/h. During the last 4 hours of your trip, you average 50 mi/h. What is your average speed for the whole trip? (*Hint:* First find the total number of miles traveled.)

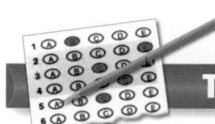

NY REGENTS

Test Prep

Multiple Choice

32. You have a mean score of 84 after taking five 100-point tests. What do you need to score on the sixth 100-point test to have a mean score of 85?
A. 85 **B.** 89 **C.** 90 **D.** 92

33. The average speeds of the winners of the Daytona 500 from 1999 to 2004 are listed at the right. What is the mean of the given speeds, rounded to the nearest tenth?
F. 158.6 mi/h **G.** 153.4 mi/h
H. 156.1 mi/h **J.** 151.9 mi/h

Daytona 500

Year	Average Speed
1999	161.6 mi/h
2000	155.9 mi/h
2001	161.7 mi/h
2002	142.0 mi/h
2003	133.9 mi/h
2004	156.3 mi/h

SOURCE: *2005 Sports Almanac*

34. Find the sum of the mean, the median, and the mode of the following data:
22, 18, 17, 18, 25, 24, 24, 18, 29, 23.
A. 58.75 **C.** 60.25
B. 59.5 **D.** 62.3

35. The average low temperature for a 4-day period in January for the city of Orlando, Florida, was 58°F. After the fifth day, the 5-day average was 59°F. What was the low temperature on the fifth day?
F. 59°F **G.** 60°F **H.** 61.5°F **J.** 63°F

Mixed Review

for Help

Lesson 1-6 **Find the mean, median, mode, and range for each set of data.**

36. 1.1, 1.4, 2.2, 1.3, 2.5 **37.** 73, 68, 79, 86, 98, 92

Lesson 1-2 **Evaluate each expression. Use $a = 4$, $b = 2$, and $c = 1$.**

38. $3a^2 + (b - c)$ **39.** $-\dfrac{4a + b}{2}$

Measures of Central Tendency and Linear Transformations

When you add, subtract, multiply, or divide each data item of a data set by the same amount, the mean, median, and mode are changed by the same magnitude.

Day	1	2	3	4	5	6	7
Minutes	20	20	20	30	41	50	50

NY A.S.16
Recognize how linear transformations of one-variable data affect the data's mean, median, mode, and range.

EXAMPLE

During the first week of training, an athlete records the time spent on a treadmill each day. The results are shown in the table above.

a. Find the mean, median, and mode of the data.

mean $= \dfrac{20 + 20 + 20 + 30 + 41 + 50 + 50}{7} = 33$ median $= 30$ mode $= 20$

b. The athlete spends 15 minutes more each day during the second week of training. Find the mean, median, and mode for the second week.

mean $= 33 + 15 = 48$
median $= 30 + 15 = 45$ **Add 15 to each measure of central tendency.**
mode $= 20 + 15 = 35$

c. During the third week of training, the athlete doubles the amount of time spent each day of the first week. Find the mean, median, and mode for the third week.

mean $= 33(2) = 66$
median $= 30(2) = 60$ **Multiply each measure of central tendency by 2.**
mode $= 20(2) = 40$

EXERCISES

For Exercises 1–5, find the mean, median, and mode of each data set.

1. A pump operator controls the flow of water as indicated in the table.

2. Each data value in the table is increased by 7 gallons per minute.

3. Each data value in the table is decreased by 5 gallons per minute.

4. Each data value in the table is tripled.

5. Each data value in the table is divided by two.

6. Critical Thinking Examine the effect of linear transformations on the *range* of a data set. How are addition and subtraction different from multiplication and division?

7. Writing in Math Explain how to use linear transformations to predict the mean, median, mode, and range of the heights of your classmates. Suppose each person in your class grows 2 inches in the next year.

8. Data Collection Survey 10 people to find the mean, median, mode, and range of the number of pencils they have in class. Suppose each person has twice as many pencils tomorrow. Use linear transformations to find the same measures.

Pipe Flow Data

Time	Gallons per Minute
12:00–12:15	20
12:15–12:30	25
12:30–12:45	25
12:45–1:00	45
1:00–1:15	55
1:15–1:30	30
1:30–1:45	10

Chapter Review

Vocabulary Review

🔊 absolute value (p. 20)
algebraic expression (p. 4)
base (p. 9)
coordinate plane (p. 24)
coordinates (p. 24)
counterexample (p. 18)
dependent variable (p. 28)
domain (p. 29)
equation (p. 5)
evaluate (p. 10)
exponent (p. 9)
function (p. 27)
function rule (p. 27)
independent variable (p. 28)
inequality (p. 19)
integers (p. 17)

irrational numbers (p. 18)
mean (p. 40)
measures of central tendency (p. 40)
median (p. 40)
midpoint (p. 25)
mode (p. 40)
natural numbers (p. 17)
negative correlation (p. 34)
no correlation (p. 34)
open sentence (p. 5)
opposites (p. 20)
order of operations (p. 10)
ordered pair (p. 24)
origin (p. 24)
outlier (p. 40)
positive correlation (p. 34)

power (p. 9)
quadrants (p. 24)
range (p. 42)
range of a function (p. 29)
rational numbers (p. 17)
real numbers (p. 18)
scatter plot (p. 33)
simplify (p. 9)
stem-and-leaf plot (p. 42)
trend line (p. 34)
variable (p. 4)
whole numbers (p. 17)
x-axis (p. 24)
x-coordinate (p. 24)
y-axis (p. 24)
y-coordinate (p. 24)

**For: Vocabulary quiz
Web Code: atj-0151**

Choose the term that correctly completes each sentence.

1. The (median, mode) is the middle value in the set when the numbers are arranged in order.

2. (Evaluate, Simplify) an algebraic expression by substituting a given number for each variable.

3. A mathematical phrase that uses numbers, variables, and operation symbols is an (algebraic expression, equation).

4. The number $-\frac{5}{8}$ belongs to the set of (irrational, rational) numbers.

5. The (absolute value, opposite) of a number is its distance from 0 on a number line.

6. You express the fraction of a pizza you have eaten by using a(n) (rational number, integer).

7. In an ordered pair, the first number is the (*x*-coordinate, *y*-coordinate), which tells how far to move to the left or right of the origin as you graph the point represented by the ordered pair.

8. A (coordinate plane, scatter plot) is a graph that relates data from two different sets.

9. When one set of data increases while another set of data decreases, there is a (positive correlation, negative correlation) between the two sets of data.

10. Simplify a (power, exponent) by multiplying the base by itself the indicated number of times.

11. The (quadrant, origin) is the point where the *x*- and *y*-axes intersect.

12. A (function, variable) is a relationship that assigns exactly one output value for each input value.

Skills and Concepts

1-1 and 1-2 Objectives

▼ To model relationships with variables (p. 4)

▼ To model relationships with equations and formulas (p. 5)

▼ To simplify and evaluate expressions and formulas (p. 9)

▼ To evaluate expressions containing grouping symbols (p. 11)

A **variable** represents one or more numbers. To **evaluate** a variable expression, you substitute a given number for each variable. Then you **simplify** the expression using the **order of operations**.

Order of Operations

1. Perform any operation(s) inside grouping symbols.
2. Simplify powers.
3. Multiply and divide in order from left to right.
4. Add and subtract in order from left to right.

Define a variable and write an expression for each phrase.

13. the sum of 5 and three times a number

14. 30 minus a number

15. the quotient of 7 and a number

16. the product of a number and 12

Evaluate each expression. Use $a = 3$, $b = 2$, and $c = 1$.

17. $2a^2 - (4b + c)$ **18.** $9(a + 2b) + c$ **19.** $\dfrac{2a + b}{2}$ **20.** $4a - b^2$

1-3 Objectives

▼ To classify numbers (p. 17)

▼ To compare numbers (p. 19)

Real numbers can be classified as either rational numbers or irrational numbers. A **rational number**, like $\frac{5}{8}$, is a ratio of two integers. An **irrational number**, like π or $\sqrt{2}$, cannot be written as a ratio of integers. Rational numbers include **natural numbers** $(1, 2, 3, \dots)$, **whole numbers** $(0, 1, 2, 3, \dots)$, and **integers** $(\dots, -2, -1, 0, 1, 2, \dots)$.

Name the set(s) of numbers to which each number belongs.

21. -3.21 **22.** $\sqrt{7}$ **23.** $-\frac{1}{2}$ **24.** 18 **25.** $\frac{35}{5}$

1-4 Objectives

▼ To write a function rule (p. 27)

▼ To recognize relationships in functions (p. 28)

A **function** is a relationship that assigns exactly one output value for each input value. A **function rule** is an equation that describes a functional relationship.

In a functional relationship, the output value depends on the input value. The variable that describes the input is the **independent variable**. The variable that describes the output is the **dependent variable**.

The possible values for the input, or independent variable, of a function are the **domain** of the function. The possible values of the output, or dependent variable, are the **range** of the function.

Identify independent and dependent quantities for each situation and find reasonable domain and range values.

26. Tom has $8.00. He wants to buy oranges which cost $.75 a piece.

27. Anna earns $6.50 per hour at her job. She can work up to 18 hours per week.

The relationships in the tables below are functions. Write a function rule for each table.

28.

x	y
1	7
2	14
3	21
4	28

29.

x	y
1	6
2	9
3	12
4	15

30.

x	y
1	20
2	16
3	12
4	8

1-5 Objective

▼ To analyze data using scatter plots (p. 33)

A **scatter plot** is a graph that relates two sets of data. There are four kinds of relationships that the data may have.

positive correlation: In general, both sets of data increase together.

negative correlation: In general, one set of data decreases as the other set increases.

no correlation: The data sets are not related.

31. a. Make a scatter plot of the data below.

Height (meters)	1.5	1.8	1.7	2.0	1.7	2.1	1.6	1.9	1.9
Arm Span (meters)	1.4	1.7	1.7	1.9	1.6	2.0	1.6	1.8	1.9

b. Is there a *positive correlation,* a *negative correlation,* or *no correlation* between the sets of data?

1-6 Objectives

▼ To use mean, median, mode, and range (p. 40)

▼ To make and use stem-and-leaf plots (p. 42)

Mean, median, and mode are three measures of central tendency. The **range** of a data set is the difference between the greatest and least items. A **stem-and-leaf plot** is a display that organizes the data by showing each item in order.

Find the mean, median, and mode for each set of data.

32. 85, 87, 81, 92, 87, 80, 83   **33.** 24, 45, 33, 27, 24

34. 2.4, 2.3, 2.1, 2.5, 2.3, 2.2   **35.** 42, 18, 55, 37, 57, 37, 48, 47, 37

The stem-and-leaf plot at the right shows kilometers walked during a benefit walk. Use it for Exercises 36–38.

36. Find the mean, median, mode, and range.

37. How many people walked more than 19 km?

38. How many people walked less than 17.1 km?

Benefit Walk
(km)

16	1 1 2 3 5 5
17	0 2 2
18	4 5 8 9
19	3 6 7 9 9 9

19 | 3 means 19.3

Chapter Test

Go Online
PHSchool.com
For: Chapter Test
Web Code: ata-0152

Define variables and write an equation to model the relationship in each table.

1.

Number	Cost
1	$2.30
2	$4.60
3	$6.90

2.

Payment	Change
$1	$9
$2	$8
$3	$7

Simplify each expression.

3. $3 + 5 - 4$

4. $8 - 2^4 \div 2$

5. $\dfrac{2 \cdot 3 - 1}{3^2}$

6. $36 - (4 + 5 \cdot 4)$

Explain why each statement is true or false.

7. All rational numbers are integers.

8. The absolute value of a number is always positive.

Write an expression for each phrase.

9. ten times the result of eleven minus two

10. the quantity p minus five eighths times the result of one fourth plus p

11. Open-Ended Write four rational numbers. Use a number line to order them from least to greatest.

12. Sales The price p of a CD is $17.95. The sales tax rate r is 7.5%. Use the formula $C = p + rp$ to find the total cost C of the CD.

Use the scatter plot below for Exercises 13–15.

13. Is there a *positive correlation, negative correlation,* or *no correlation* between daily mean temperature and latitude?

Climate Data

(scatter plot: Daily Mean Temperature (°F) vs Location (north latitude), x-axis 30°, 40°, 50°; y-axis 0, 40, 50, 60)

14. What is the daily mean temperature for the location at latitude 58° N?

15. What is the range of daily mean temperatures?

16. Writing Explain why $\left|\dfrac{a}{b}\right| = \dfrac{a}{b}$ is *not* always true.

Use the table below for Exercises 17–18.

Percent of People Who Speak a Language Other Than English at Home

State	Percent
Connecticut	15
Massachusetts	15
Maine	9
New Hampshire	9
New Jersey	20
New York	23
Pennsylvania	7
Rhode Island	17
Vermont	6

SOURCE: U.S. Census Bureau
Go to www.PHSchool.com for a data update.
Web Code: atg-9041

17. Make a stem-and-leaf plot for the data.

18. Find the mean, median, mode, and range of the data.

19. Purchasing The price of a turkey depends on its weight. This week, whole turkeys sell for $1.59 per lb.
a. Write a rule to describe the function.
b. What is the price of a 14-lb turkey?
c. If you had $10 to buy a turkey, how big a turkey could you buy?

Identify the independent and dependent variables and write a function rule to describe each situation.

20. the amount of money you earn mowing lawns at $15 per lawn

21. the profit you make by selling flowers at $1.50 each when each flower costs you $.80

22. a. Write a function rule for the relationship shown in the table.

b. Use the function to find the total savings after 7 weeks.

Number of Weeks	Total Savings
1	$52
2	$64
3	$76
4	$88

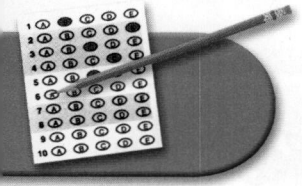

Regents Test Prep

Reading Comprehension Read the passage below, and then answer the questions on the basis of what is *stated* or *implied* in the passage.

Travel Math Often one of the biggest complications of visiting a foreign country is doing the math. Things that would cause no problem at home are suddenly challenging because the units are different. Consider a quarter-pound hamburger. At 2.2 pounds (lbs) per kilogram (kg), what is the mass of a quarter-pound hamburger in the metric system?

Once you have ordered lunch, you have to pay for it. Let's see . . . one U.S. dollar is worth 1.53 Canadian dollars, 0.70 British pounds, or 1.07 European euros. So how much does a quarter-pound hamburger cost in other countries?

Then there is figuring out the temperature. Even if you can remember the formula Fahrenheit $= \frac{9}{5} \cdot$ Celsius $+ 32$, you must do some work to find whether 20°C is a beach day or a ski day.

Fortunately, it doesn't have to be quite so hard. If you look for some simple approximations, you can usually get around foreign countries without having to pack a calculator. As with many other aspects of travel, the trick is to think ahead, anticipate what might be coming, and have a plan for dealing with it. Before you go, figure out which formulas you will be using and come up with some simple approximations. Then you can leave your calculator home.

---

1. About how many kilograms are there in 10 lb?

   (1) 25 kg   **(2)** 20 kg
   (3) 11 kg   **(4)** 5 kg

2. About how many kilograms correspond to a quarter pound?

   (1) 0.1 kg   **(2)** 0.5 kg
   (3) 5 kg   **(4)** 8.8 kg

3. Based on the article above, which could you use to estimate an exchange of Canadian dollars and United States dollars?

   (1) 1 U.S. Dollar $= \frac{1}{2}$ Canadian Dollar
   (2) 1 U.S. Dollar $= \frac{2}{3}$ Canadian Dollar
   (3) 2 U.S. Dollars $=$ 3 Canadian Dollars
   (4) 3 U.S. Dollars $=$ 2 Canadian Dollars

4. Suppose you pay $3.00 for a hamburger in Canada. How much is this in United States currency?

   (1) $1.95   **(2)** $3.00
   (3) $4.53   **(4)** $4.59

5. According to the article, about how many euros could you get for 5 U.S. dollars?

   (1) 0.20 euros   **(2)** 1.50 euros
   (3) 4.20 euros   **(4)** 5.50 euros

6. Suppose you are traveling in France. You see a T-shirt for 40 euros. Is this a reasonable price? Justify your answer.

7. A taxicab ride in London costs you 15 pounds. How much is this in United States currency?

8. The formula Fahrenheit $= 2 \cdot$ Celsius $+ 30$ gives a good estimate of the temperature in degrees Fahrenheit when you know the temperature in degrees Celsius. Is 20°C a beach day? Justify your answer.

9. A foreign exchange student could use the formula Celsius $= \frac{5}{9}$(Fahrenheit $- 32$) to find the temperature in degrees Celsius when he knows the temperature in degrees Fahrenheit. Write a formula the student could use to get a good estimate of the Celsius temperature.

Data Analysis

Activity Lab

Locating Sunken Ships

Applying Integers Marine archeologists are scientists who study sunken ships. They use scanning devices to locate objects on the ocean floor. When they find a "hot spot," divers take a closer look. If they find a sunken ship, the divers take underwater photographs and record the ship's latitude and longitude, identifying a specific point on Earth's surface.

Ancient World

This globe includes latitude (lines that run east–west) and longitude (lines that run north–south) rings.

ACTIVITY

Which part of Earth's coordinate globe corresponds to the indicated part of a coordinate plane?

1. the *x*-axis   **2.** the *y*-axis   **3.** the origin

Write and simplify an expression to show how the depth of a submersible robot changes.

4. from the *Edmund Fitzgerald* to the *Andrea Doria*

5. from the *Titanic* to the *Atocha*

6. from the water's surface to the *Atocha*

7. Estimation About how many times the depth of the *Atocha* is the depth of the *Andrea Doria*? Write an equation to model this relationship.

8. Research Pick one of the ships discussed here or another sunken ship. What factors led to its sinking? Explain.

Prime Meridian
0° Longitude

Titanic
41° N 49° W
−12,600 ft

Edmund Fitzgerald
47° N 85° W
−530 ft

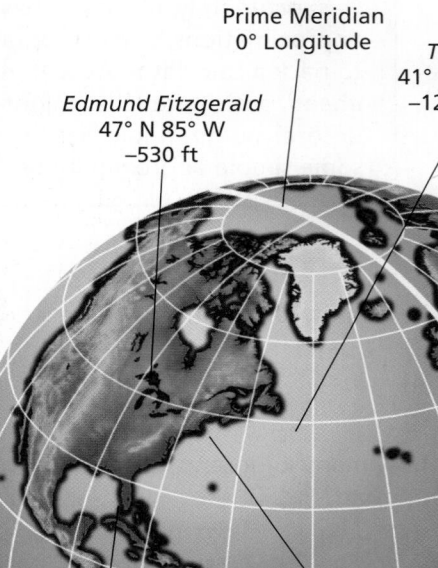

Equator
0° Latitude

Embedded Treasure

Time and water pressure pushed these coins into a piece of wood.

Atocha
24° N 82° W
−55 ft

Andrea Doria
40° N 69° W
−240 ft

Sunken World

Scientists explore shipwrecks in submersible vehicles such as Alvin (above) when the ship is too deep for scuba gear (worn by the diver).

What You've Learned

 **Learning Standards for Mathematics**

In Chapter 1, you

- **NY A.A.5:** used variables to write expressions and equations that represent real-world situations.

 **Check Your Readiness**

GO for Help to the Lesson in green.

Writing an Equation (Lesson 1-1)

Write an equation to model each situation.

1. The total cost of n cartons of milk is \$3.60. Each carton costs \$.45.

2. The perimeter of an equilateral triangle is 3 times the length of a side s. The perimeter is 124 in.

Using the Order of Operations (Lesson 1-2)

Evaluate each expression for $a = 4, b = 13,$ and $c = 2$.

3. $2a + cb$ **4.** $cb - a^2$ **5.** $5c + 2a - b$ **6.** $38 - a^2 \div c$

Simplify each expression.

7. $5 + 4 \cdot 7$ **8.** $(5 + 4)(7)$ **9.** $5(4)(7)$

10. $\dfrac{15 - 3 \cdot 2}{3^2}$ **11.** $(42 - 6) \div 2$ **12.** $42 - (6 \div 2)$

Use $<, =,$ or $>$ to compare.

13. $\dfrac{5}{8}$ ■ $\dfrac{10}{15}$ **14.** $-\dfrac{2}{5}$ ■ $-\dfrac{3}{10}$ **15.** $\dfrac{16}{9}$ ■ $\dfrac{4}{3}$ **16.** $\dfrac{4}{5}$ ■ 0.8

Writing a Function Rule (Lesson 1-4)

Write a function rule for the relationship in each table.

17.

Hours	Miles Traveled
1	55
2	110
3	165
4	220

18.

Color Copies	Total Cost ($)
1	1.20
2	2.00
3	2.80
4	3.60

19.

n	$n + (-7)$
-4	■
-1	■
3	■
■	-1

Rational Numbers

🔊 **Key Vocabulary**

- additive inverse (p. 56)
- coefficient (p. 81)
- complement of an event (p. 94)
- constant (p. 81)
- deductive reasoning (p. 88)
- dependent events (p. 103)
- element (p. 59)
- event (p. 93)
- experimental probability (p. 95)
- independent events (p. 102)
- like terms (p. 81)
- matrix (p. 59)
- multiplicative inverse (p. 72)
- outcome (p. 93)
- probability (p. 93)
- reciprocal (p. 73)
- sample space (p. 93)
- term (p. 81)
- theoretical probability (p. 93)

What You'll Learn Next

NY Learning Standards for Mathematics

- **NY A.N.1:** You will extend your ability to calculate with whole numbers, decimals, and fractions to include integers.

- **NY A.N.1:** You will use the Order of Operations and the Distributive Property to simplify expressions.

- **NY A.S.21:** You will calculate theoretical and experimental probability.

Activity Lab Applying what you learn, you will do activities related to baseball, on pages 114–115.

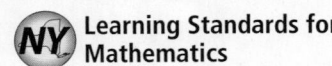

Adding Rational Numbers

NY Learning Standards for Mathematics

A.N.1 Identify and apply the properties of real numbers (closure, commutative, associative, distributive, identity, inverse).

✓ **Check Skills You'll Need**

GO for Help Skills Handbook page 760

Find each sum.

1. $4 + 2$
2. $10 + 7$
3. $9 + 5$
4. $27 + 32$
5. $0.4 + 0.9$
6. $5.2 + 0$
7. $4.1 + 6.8$
8. $7.6 + 9.5$
9. $\frac{1}{5} + \frac{3}{5}$
10. $\frac{4}{9} + \frac{7}{9}$
11. $\frac{1}{2} + \frac{3}{4}$
12. $\frac{3}{8} + \frac{1}{4}$

🔊 **New Vocabulary**
- Identity Property of Addition
- additive inverse
- Inverse Property of Addition
- matrix
- element

---

1   Adding Rational Numbers

In previous math courses, you learned that the sum of a number and 0 is the original number. This is true for any number, whether it is positive or negative.

 Key Concepts

Property	Identity Property of Addition

For every real number n, $n + 0 = n$ and $0 + n = n$

Examples $-5 + 0 = -5$ $0 + 5 = 5$

The opposite of a number is its **additive inverse.** The number line shows the sum of $4 + (-4)$.

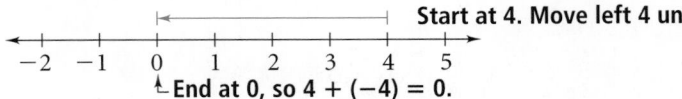

Start at 4. Move left 4 units.
End at 0, so $4 + (-4) = 0$.

The additive inverse of a negative number is a positive number. The number line below shows the sum of $-5 + 5$.

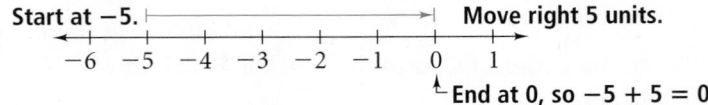

Start at −5. Move right 5 units.
End at 0, so $-5 + 5 = 0$.

 Key Concepts

Property	Inverse Property of Addition

For every real number n, there is an additive inverse $-n$ such that $n + (-n) = 0$.

Examples $17 + (-17) = 0$ $-17 + 17 = 0$

You can use number lines as models to add integers.

For: Addition Activity
Use: Interactive Textbook, 2-1

1 EXAMPLE **Using a Number Line Model**

Simplify each expression.

a. $2 + 6$ Start at 2. |⟶| Move right 6 units. $2 + 6 = 8$

$0 \quad 2 \quad 4 \quad 6 \quad 8 \quad 10$

b. $2 + (-6)$ Start at 2. Move left 6 units. $2 + (-6) = -4$

$-8 \quad -6 \quad -4 \quad -2 \quad 0 \quad 2$

c. $-2 + 6$ Start at -2. |⟶| Move right 6 units. $-2 + 6 = 4$

$-4 \quad -2 \quad 0 \quad 2 \quad 4 \quad 6$

d. $-2 + (-6)$ Start at -2. Move left 6 units. $-2 + (-6) = -8$

$-10 \quad -8 \quad -6 \quad -4 \quad -2 \quad 0$

 Quick Check **1** Use a number line to find each sum.
   a. $-6 + 4$ **b.** $4 + (-6)$ **c.** $-3 + (-8)$ **d.** $9 + (-3)$

You can also find the sums in Example 1 using rules. Recall that numbers being added are called addends.

Key Concepts

Rule	**Adding Numbers With the Same Sign**

To add two numbers with the same sign, *add* their absolute values. The sum has the same sign as the addends.

Examples $2 + 6 = 8$ $-2 + (-6) = -8$

Rule	**Adding Numbers With Different Signs**

To add two numbers with different signs, find the *difference* of their absolute values. The sum has the same sign as the addend with the greater absolute value.

Examples $-2 + 6 = 4$ $2 + (-6) = -4$

2 EXAMPLE **Adding Numbers**

Simplify each expression.

a. $-5 + (-6) = -11$ Since both addends are negative, add their absolute values. The sum is negative.

b. $13 + (-34) = -21$ The difference of the absolute values is 21. The negative addend has the greater absolute value, so the sum is negative.

c. $3.4 + 9.7 = 13.1$ Since both addends are positive, add their absolute values. The sum is positive.

d. $-1.5 + 3.4 = 1.9$ The difference of the absolute values is 1.9. The positive addend has the greater absolute value, so the sum is positive.

 **Quick Check** **2** Find each sum.
   a. $-7 + (-4)$ **b.** $-26.3 + 8.9$ **c.** $-\frac{3}{4} + \left(-\frac{1}{2}\right)$ **d.** $\frac{8}{9} + \left(-\frac{5}{6}\right)$

You can use negative numbers to model real-world situations.

3 EXAMPLE Real-World  Problem Solving

Football A football team gains 2 yd and then loses 7 yd in two plays. You express a loss of 7 yd as −7. Use addition to find the result of the two plays.

$$2 + (-7) = -5$$

The result of the two plays is a loss of 5 yd.

✓ **Quick Check** **3** **Temperature** The temperature falls 15 degrees and then rises 18 degrees. Use addition to find the change in temperature.

---

2 Applying Addition

You can evaluate expressions that involve addition. Substitute a value for the variable(s). Then simplify the expression. The expression −n means the opposite of n. The expression −n can represent a negative number, zero, or a positive number.

4 EXAMPLE Evaluating Expressions

Evaluate −n + 8.9 for n = −2.3.

$$\begin{aligned} -n + 8.9 &= -(-2.3) + 8.9 & &\textbf{Substitute −2.3 for } n. \\ &= 2.3 + 8.9 & &\textbf{− (−2.3) means the opposite of −2.3, which is 2.3.} \\ &= 11.2 & &\textbf{Simplify.} \end{aligned}$$

✓ **Quick Check** **4** Evaluate each expression for t = −7.1.
a. $t + (-4.3)$ **b.** $-2 + t$ **c.** $8.5 + (-t)$ **d.** $-t + 7.49$

You can write and evaluate expressions to model real-world situations.

5 EXAMPLE Real-World 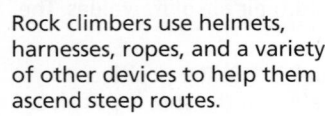 Problem Solving

Climbing A rock climber climbs a mountain. The base of the mountain is 132 ft below sea level.
a. Write an expression to represent the climber's height below or above sea level.

Relate   132 ft below sea level   plus   number of feet climbed

Define   Let h = number of feet climbed.

Write   −132   + h

−132 + h

b. Find the climber's height above sea level when he is 485 ft above the base of the mountain.

$$\begin{aligned} -132 + h &= -132 + 485 & &\textbf{Substitute 485 for } h. \\ &= 353 & &\textbf{Simplify.} \end{aligned}$$

His height is 353 ft above sea level.

Real-World  Connection

Rock climbers use helmets, harnesses, ropes, and a variety of other devices to help them ascend steep routes.

✓ **Quick Check** **5** **Temperature** The temperature one winter morning is −14°F. Define a variable and write an expression to find the temperature after it changes. Then evaluate your expression for a decrease of 11 degrees Fahrenheit.

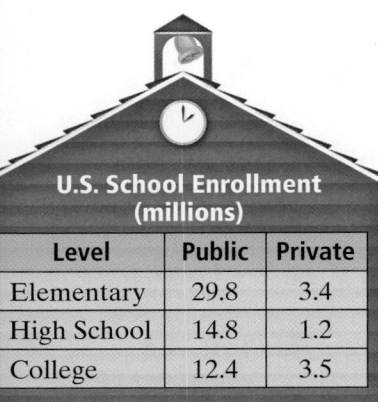

U.S. School Enrollment (millions)

Level	Public	Private
Elementary	29.8	3.4
High School	14.8	1.2
College	12.4	3.5

SOURCE: *Statistical Abstract of the United States.*
Go to **www.PHSchool.com** for a data update.
Web Code: atg-9041

You can use matrices to add rational numbers. A **matrix** is a rectangular arrangement of numbers in rows and columns. The plural of matrix is matrices (pronounced MAY-truh-seez). The matrix below shows the data in the table.

$$
\begin{array}{c}
\ \ \text{Public}\quad\text{Private} \\
\begin{array}{r}
\text{Elementary} \\
\text{High School} \\
\text{College}
\end{array}
\left[
\begin{array}{cc}
29.8 & 3.4 \\
14.8 & 1.2 \\
12.4 & 3.5
\end{array}
\right]
\leftarrow \text{row}
\end{array}
$$

↑ column

You identify the size of a matrix by the number of rows and the number of columns. The matrix above has 3 rows and 2 columns, so it is a 3×2 matrix. Each item in a matrix is an **element.**

Matrices are equal if the elements in corresponding positions are equal.

$$
\begin{bmatrix} -1 & 2 \\ 4 & 0 \end{bmatrix} = \begin{bmatrix} -1 & \frac{4}{2} \\ \frac{20}{5} & 0 \end{bmatrix}
$$

You add matrices that are the same size by adding the corresponding elements.

6 EXAMPLE **Adding Matrices**

Add $\begin{bmatrix} -5 & 2.7 \\ 7 & -3 \end{bmatrix} + \begin{bmatrix} -3 & -3.9 \\ -4 & 2 \end{bmatrix}$.

$\begin{bmatrix} -5 & 2.7 \\ 7 & -3 \end{bmatrix} + \begin{bmatrix} -3 & -3.9 \\ -4 & 2 \end{bmatrix} = \begin{bmatrix} -5 + (-3) & 2.7 + (-3.9) \\ 7 + (-4) & -3\ + 2 \end{bmatrix}$ Add corresponding elements.

$= \begin{bmatrix} -8 & -1.2 \\ 3 & -1 \end{bmatrix}$ Simplify.

✓ **Quick Check** **6** Find each sum.

a. $\begin{bmatrix} 5 \\ 3.2 \\ -4.9 \end{bmatrix} + \begin{bmatrix} -9 \\ -1.7 \\ -11.1 \end{bmatrix}$

b. $\begin{bmatrix} -4 & \frac{7}{8} \\ \frac{3}{4} & 0 \end{bmatrix} + \begin{bmatrix} -5 & -\frac{3}{4} \\ \frac{1}{2} & -1 \end{bmatrix}$

EXERCISES

For more exercises, see *Extra Skill and Word Problem Practice.*

Practice and Problem Solving

A Practice by Example

Example 1 (page 57)

Write the expression modeled by each number line. Then find the sum.

1.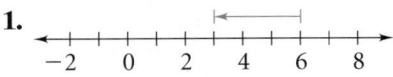

2.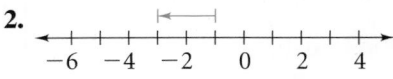

3.

4.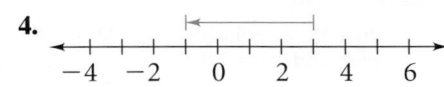

Example 2
(page 57)

Simplify.

5. $3 + 12$ **6.** $-7 + (-4)$ **7.** $-8.7 + (-10.3)$ **8.** $5.04 + 7.1$

9. $5 + (-9)$ **10.** $-8 + 13$ **11.** $-27 + 19$ **12.** $45 + (-87)$

13. $-2.3 + 4.5$ **14.** $-8.05 + 7.4$ **15.** $9.51 + (-17)$ **16.** $3.42 + (-2.09)$

17. $\frac{4}{5} + \frac{2}{15}$ **18.** $-\frac{5}{9} + \left(-\frac{1}{3}\right)$ **19.** $2\frac{1}{4} + 3\frac{15}{16}$ **20.** $-4\frac{3}{8} + \left(-1\frac{3}{4}\right)$

21. $-\frac{2}{3} + \frac{4}{6}$ **22.** $\frac{1}{9} + \left(-\frac{5}{6}\right)$ **23.** $-5\frac{7}{12} + 10\frac{3}{4}$ **24.** $\frac{9}{7} + \left(-2\frac{3}{14}\right)$

Example 3
(page 58)

25. A diver dives 47 ft below the surface of the water and then rises 12 ft. Use addition to find the diver's depth.

26. On two football plays, a team gains 8 yd, and then loses 5 yd. Use addition to find the result of the two plays.

27. One morning in Detroit, Michigan, the temperature at 6 A.M. was $-6°F$. The temperature rose 13 degrees Fahrenheit by noon. Use addition to find the temperature at noon.

Example 4
(page 58)

Evaluate each expression for $n = 3.5$.

28. $5.2 + n$ **29.** $-5.2 + n$ **30.** $-n + 5.2$ **31.** $-n + (-5.2)$

32. $9.1 + n$ **33.** $-9.1 + n$ **34.** $-n + 9.1$ **35.** $-9.1 + (-n)$

Example 5
(page 58)

36. Temperature The temperature one winter morning is $-8°F$. Define a variable and write an expression to find the temperature after each change below. Then evaluate your expression for each change.
a. a rise of 7°F   **b.** a decrease of 3°F
c. a rise of 19°F   **d.** a decrease of 12°F

37. Money You have $74 in a checking account. Define a variable and write an expression to find the balance in your account after each deposit or withdrawal below. Then evaluate your expression for each change.
a. a deposit of $18   **b.** a withdrawal of $29
c. a withdrawal of $47   **d.** a deposit of $120

Example 6
(page 59)

Simplify.

38. $\begin{bmatrix} 7 & -8 \\ -12 & 6.2 \end{bmatrix} + \begin{bmatrix} -8 & 9.4 \\ -9 & 17 \end{bmatrix}$ **39.** $\begin{bmatrix} -7.2 \\ 3.2 \\ -4.9 \end{bmatrix} + \begin{bmatrix} -11 \\ 8.4 \\ 24 \end{bmatrix}$

40. $\begin{bmatrix} \frac{1}{2} \\ 8 \\ -9 \end{bmatrix} + \begin{bmatrix} -\frac{1}{2} \\ 17 \\ -3 \end{bmatrix}$ **41.** $\begin{bmatrix} 1.3 & 26 \\ \frac{1}{8} & -2 \end{bmatrix} + \begin{bmatrix} 0.5 & -4 \\ -\frac{5}{8} & 9 \end{bmatrix}$

B Apply Your Skills

Copy and complete each table.

42.

a	$a + 5$
-7	▪
-5	▪
0	▪
▪	8

43.

x	$-3 + x$
-2	▪
0	▪
$1\frac{1}{2}$	▪
▪	6

44.

n	$n + (-7)$
-4	▪
-1	▪
3	▪
▪	-1

GO Online
Homework Video Tutor
Visit: PHSchool.com
Web Code: ate-0201

Real-World Connection

The art of photography is changing because of the development of digital cameras.

Art Use the table for Exercises 45–48.

Number of People Who Participate in Art Activities (millions)

Ages	Drawing	Pottery	Weaving	Photography	Creative Writing
18–24	9.2	5.0	5.2	6.6	7.6
25–34	7.2	6.8	10.0	7.2	5.2
35–44	6.8	8.2	13.1	8.2	5.4
45–54	4.4	6.1	9.8	6.1	3.4
55–64	1.9	2.1	6.1	2.1	1.0

SOURCE: *Statistical Abstract of the United States*

45. How many people aged 18 to 34 participate in photography?

46. How many people aged 45 to 64 draw?

47. Which of the activities is most popular? Explain how you found your answer.

48. a. Write a fraction to compare the number of people aged 25 to 34 who weave to the number of people aged 18 to 64 who weave.
   b. Write your answer from part (a) as a decimal to the nearest hundredth.
   c. What percent of the people who weave are aged 25 to 34?

Evaluate each expression for $a = -2$, $b = 3$, and $c = -4$.

49. $-a + 2 + c$ **50.** $-|a|$ **51.** $a + b$ **52.** $a + (-b)$

53. $a + 3b$ **54.** $c + 3b$ **55.** $c + a + 5$ **56.** $-(c + a + 5)$

57. Writing Without calculating, which is greater, the sum of -227 and 319 or the sum of 227 and -319? Explain.

58. Reasoning Explain what is wrong with the reasoning in the statement: *Since 20 is the opposite of -20, then $20°F$ must be very hot, because $-20°F$ is very cold.*

Evaluate each expression for $b = -3.5$.

59. $b + 3.2$ **60.** $-9 + b + (-1.2)$ **61.** $b + |-2.9|$

62. $8.5 + b + 3.7$ **63.** $|b| + (-3.4)$ **64.** $-5.6 + b + 7.2$

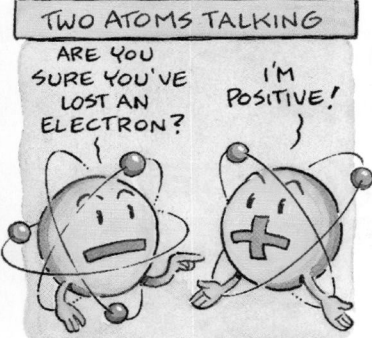

65. Multiple Choice A charged atom of magnesium has 12 protons and 10 electrons. Each proton has a charge of $+1$, and each electron has a charge of -1. What is the total charge of the atom?
  Ⓐ -2 Ⓑ -1 Ⓒ $+1$ Ⓓ $+2$

66. Open-Ended Write a 2×3 matrix.

67. Error Analysis A student added two matrices as shown. What error did the student make?

$$\begin{bmatrix} 4 & -1 \\ -3 & 2 \\ 1.5 & 6 \end{bmatrix} + \begin{bmatrix} -2 & 2.3 & 0 \\ 7 & -4 & 5.1 \end{bmatrix} = \begin{bmatrix} 11 & -3 \\ -7 & 4.3 \\ 6.6 & 6 \end{bmatrix}$$

68. Math in the Media In the cartoon below, does the total "12 27" make sense? Explain.

Frank and Ernest

© 1980 Thaves / Reprinted with permission. Newspaper dist. by NEA, Inc.

Number of Employees

Saturday Schedule

Shift	Hourly Wage			
	$6.25	$6.50	$7.00	$7.50
Day	8	3	5	1
Evening	10	2	2	1
Night	4	1	0	1

Sunday Schedule

Shift	Hourly Wage			
	$6.25	$6.50	$7.00	$7.50
Day	5	2	1	1
Evening	8	2	0	1
Night	2	1	0	1

69. Jobs Use the data in the tables at the left.
  a. Write the data in each table as a matrix.
  b. Add the matrices to find the total number of workers in each pay category for each work shift.
  c. How many weekend employees on the evening shift earn $6.50 per hour?
  d. How many weekend employees work the night shift?
  e. Critical Thinking Suppose all employees work 8-hour shifts both Saturday and Sunday. How would you use the matrix to find the total wages of the weekend employees?
  f. Find the total wages of the weekend employees.

70. Suppose you overdrew your bank account. You have a balance of $-\$34$. You then deposit checks for $17 and $49. At the same time the bank charges you a $25 fee for overdrawing your account. What is your balance?

71. a. What is the value of $-n$ when $n = -4$?
  b. What is the value of $-n$ when $n = 4$?
  c. Reasoning For what values of n will $-n$ be positive? Negative?

C Challenge

Simplify each expression.

72. $\frac{w}{5} + \left(-\frac{w}{10}\right)$

73. $-\frac{c}{4} + \left(-\frac{c}{4}\right)$

74. $3\left(\frac{a}{7}\right) + 7\left(\frac{a}{3}\right)$

75. $-1\left(\frac{b}{9}\right) + \left(-\frac{b}{9}\right)$

76. $\frac{-x}{4} + \frac{x}{3}$

77. $\frac{x}{4} + \left(-\frac{x}{3}\right)$

Tell whether each sum is *positive*, *negative*, or *zero*. Explain.

78. n is positive and m is negative. $n + (-m)$ is ___?___.

79. n is positive, and m is negative. $-n + m$ is ___?___.

80. $|n| = |m|$, n is positive, and m is negative. $n + (-m)$ is ___?___.

81. $|n| = |m|$, n is positive, and m is negative. $-n + (-m)$ is ___?___.

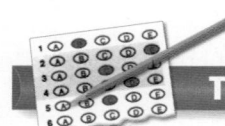

Test Prep

NY REGENTS

Multiple Choice

82. Simplify $10 + |-3| + (-3)$.
  A. 16 **B.** 10 **C.** 7 **D.** 4

83. Evaluate $(3a + b) + (-20)$ for $a = 5$ and $b = -1$.
  F. -6 **G.** -2 **H.** 20 **J.** 22

84. Which expression has a value that is not the same as the value of the other expressions?

  A. $-7 + 3$ **B.** $5 + (-9)$ **C.** $-8\frac{2}{3} + 4\frac{2}{3}$ **D.** $-9 + 13$

85. In a 12-hour period, the temperature rose from $-12°F$ to $18°F$. Find the increase in temperature in degrees.

  F. 30   **G.** 6   **H.** -6 **J.** -30

86. The value of $-(-(-27))$ is NOT the same as which of the following expressions?

  A. $-29 + 2$ **B.** $-12.8 + (-14.2)$
  C. $-42 + 17$ **D.** $8 + (-35)$

87. Suppose you have $95 in your checking account. You pay for a $34 sweater using your debit card. Then you deposit a $32 check. The next day you withdraw $16 at the supermarket. What is the balance in your account?

  F. $145   **G.** $81   **H.** $77   **J.** $13

Mixed Review

Lesson 1-3

Use $<$, $=$, or $>$ to compare.

88. $-1.23 \blacksquare -1.18$

89. $1\frac{2}{4} \blacksquare 1\frac{5}{10}$

90. $|-5| \blacksquare |-6|$

91. $|-4.1| \blacksquare |-3.9|$

92. $\left|-\frac{3}{10}\right| \blacksquare \left|\frac{2}{9}\right|$

93. $|1.2| \blacksquare \left|-\frac{6}{5}\right|$

Lesson 1-2

Simplify each expression.

94. $(5 - 2)^2$ **95.** $-4 + 3.1(2)$ **96.** $9[5 + (-3)]$ **97.** $4^2 + 3^2 - 2^2$

A Point in Time

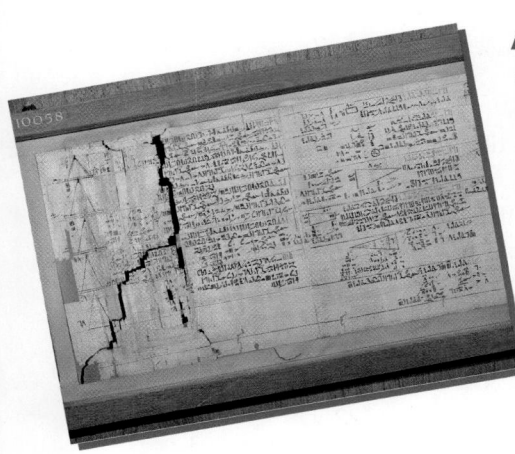

2000 B.C.   1000   0   1000   2000

A papyrus scroll discovered in Egypt shows that Egyptians were using symbols for "plus," "minus," "equals," and "unknown quantity" more than 3500 years ago. Named for the scribe who copied it, the Ahmes Papyrus is 18 ft long and 1 ft wide. It is also known as the Rhind Papyrus, after the British Egyptologist who bought the papyrus. It is a practical handbook containing 85 problems that include work with rational numbers. In the Ahmes Papyrus, rational numbers are written as sums of unit fractions. A unit fraction has 1 as its numerator. Here are some examples.

$$\frac{3}{4} = \frac{1}{2} + \frac{1}{4} \qquad \frac{3}{8} = \frac{1}{4} + \frac{1}{8} \qquad \frac{21}{30} = \frac{1}{6} + \frac{1}{5} + \frac{1}{3}$$

For: Information about the Ahmes Papyrus
Web Code: ate-2032

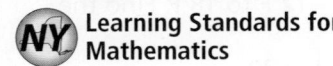

Subtracting Rational Numbers

NY Learning Standards for Mathematics

A.N.6 Evaluate expressions involving factorial(s), absolute value(s), and exponential expression(s).

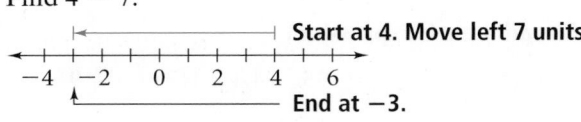

✓ **Check Skills You'll Need**

Find the opposite of each number.

1. 6   **2.** -7 **3.** 3.79   **4.** $-\frac{7}{19}$

Simplify.

5. $3 + (-2)$ **6.** $9.5 + (-3.5)$ **7.** $13 + (-8)$ **8.** $\frac{2}{3} + \left(-\frac{1}{6}\right)$

GO for Help Lessons 1-3 and 2-1

Subtracting Rational Numbers

You have learned to add rational numbers and to find the opposite of a number. You can use these two concepts to understand how to subtract rational numbers.

1 EXAMPLE Using a Number Line Model

Find $4 - 7$.

Start at 4. Move left 7 units.

$$\begin{array}{ccccccc} -4 & -2 & 0 & 2 & 4 & 6 \end{array}$$

End at -3.

• $4 - 7 = -3$

✓ **Quick Check**   **1** Use a number line to find each difference.

     a. $2 - 6$ **b.** $-1 - 4$ **c.** $-3 - 8$ **d.** $7 - 2$

You can use tiles to model subtraction. A ▢ represents $+1$, and a ■ represents -1. A negative tile and a positive tile together (▢■) represent zero. This is called a zero pair.

2 EXAMPLE Using a Tile Model

Find $3 - (-5)$.

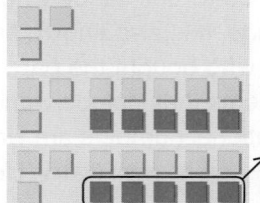

Start with 3 positive tiles.

Add zero pairs until there are 5 negative tiles.

Remove 5 negative tiles.
There are 8 positive tiles left.

• $3 - (-5) = 8$

✓ **Quick Check**   **2** Draw tiles to find each difference.

     a. $4 - 5$ **b.** $5 - (-9)$ **c.** $-6 - (-10)$ **d.** $6 - 3$

The number line below models the sum $2 + (-6)$ *and* the difference $2 - 6$.

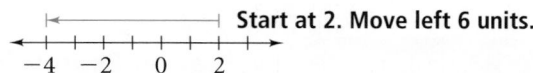

Start at 2. Move left 6 units.

Both $2 + (-6)$ and $2 - 6$ have the same value, -4. This illustrates the following rule for subtracting rational numbers.

 Key Concepts

Rule	Subtracting Numbers

To subtract a number, add its opposite.

Examples $3 - 5 = 3 + (-5) = -2$ $3 - (-5) = 3 + 5 = 8$

Video Tutor Help
Visit: PHSchool.com
Web Code: ate-0775

3 EXAMPLE   **Subtracting Rational Numbers**

Simplify each expression.

a. $-4 - (-9)$

$\qquad -4 - (-9) = -4 + 9$ **The opposite of -9 is 9.**

$\qquad\qquad\qquad\quad = 5$ **Add.**

b. $\frac{3}{4} - \left(-\frac{11}{12}\right)$

$\qquad \frac{3}{4} - \left(-\frac{11}{12}\right) = \frac{3}{4} + \frac{11}{12}$ **The opposite of $-\frac{11}{12}$ is $\frac{11}{12}$.**

$\qquad\qquad\qquad\quad = \frac{9}{12} + \frac{11}{12}$ **Use common denominators.**

$\qquad\qquad\qquad\quad = \frac{20}{12}$ **Add.**

$\qquad\qquad\qquad\quad = \frac{5}{3}$ or $1\frac{2}{3}$ **Write $\frac{20}{12}$ in simplest form.**

 **Quick Check**   **3** Find each difference.

     a. $-6 - 2$ **b.** $8 - (-4)$ **c.** $3.7 - (-4.3)$ **d.** $-\frac{8}{9} - \left(-\frac{5}{6}\right)$

2   Applying Subtraction

Recall that when you simplify an expression, you work within grouping symbols first. Absolute value symbols are grouping symbols, so find the value of an expression within the absolute value symbols before finding the absolute value.

4 EXAMPLE   **Absolute Values**

Simplify $|5 - 11|$.

$\qquad |5 - 11| = |-6|$ **Subtract within absolute value symbols.**

$\qquad\qquad\quad = 6$ **Find the absolute value.**

 **Quick Check**   **4** Simplify each expression.

     a. $|8 - 7|$ **b.** $|7 - 8|$ **c.** $|-10 - (-4)|$ **d.** $|-4 - (-10)|$

You evaluate expressions that involve subtraction by substituting for the variable. Then simplify the expression.

5 **EXAMPLE** **Evaluating Expressions**

Evaluate $-a - b$ for $a = -3$ and $b = -5$.

$$\begin{aligned} -a - b &= -(-3) - (-5) & &\text{Substitute } -3 \text{ for } a \text{ and } -5 \text{ for } b. \\ &= 3 - (-5) & &\text{The opposite of } -3 \text{ is } 3. \\ &= 3 + 5 & &\text{To subtract } -5, \text{ add its opposite, } 5. \\ &= 8 & &\text{Add.} \end{aligned}$$

 **Quick Check** **5** Evaluate each expression for $t = -2$ and $r = -7$.

a. $r - t$ **b.** $t - r$ **c.** $-t - r$ **d.** $-r - (-t)$

You can write expressions to model real-world situations.

6 **EXAMPLE** **Real-World** 🌐 **Problem Solving**

Stock Price Find the closing price of stock XYZ on Wednesday by subtracting the change in price from the closing price on Thursday.

$$\begin{aligned} 17.37 - (-0.87) &= 17.37 + 0.87 & &\text{Add the opposite.} \\ &= 18.24 & &\text{Simplify.} \end{aligned}$$

The closing share price on Wednesday was $18.24.

	Thurs	
Stock	Close	Change
ABC	32.79	0.32
PQR	14.23	-1.23
XYZ	17.37	-0.87

C3
STOCK PRICES

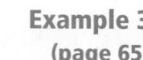 **Quick Check** **6** **Stock Price** Find the closing price of stocks ABC and PQR on Wednesday.

EXERCISES

For more exercises, see *Extra Skill and Word Problem Practice*.

Practice and Problem Solving

Ⓐ **Practice by Example**

Examples 1, 2
(page 64)

GO for Help

Draw a number line or tiles to model each difference. Then find each difference.

1. $1 - 2$ **2.** $7 - 9$ **3.** $-4 - 2$ **4.** $-2 - 3$

5. $-5 - (-6)$ **6.** $-3 - 8$ **7.** $5 - (-9)$ **8.** $-1 - (-4)$

Example 3
(page 65)

Simplify each expression.

9. $3 - 7$ **10.** $2 - (-9)$ **11.** $-4 - 6$ **12.** $-5 - (-1)$

13. $6.2 - 8.3$ **14.** $-7.4 - 1.8$ **15.** $5.3 - (-8.4)$ **16.** $-3.6 - (-7.1)$

17. $\frac{1}{3} - \frac{1}{2}$ **18.** $-\frac{2}{5} - \frac{7}{10}$ **19.** $\frac{2}{12} - \left(-\frac{3}{4}\right)$ **20.** $-\frac{5}{12} - \left(-\frac{1}{10}\right)$

Example 4
(page 65)

21. $|5 - 2|$ **22.** $|-7 - 1|$ **23.** $|4 - 10|$ **24.** $|-3 - (-5)|$

25. $|-6 - 7|$ **26.** $|8 - 6|$ **27.** $|3 - 9|$ **28.** $|-11 - (-8)|$

Example 5
(page 66)

Evaluate each expression for $x = 3$, $y = -4$, and $z = 6$.

29. $y - z$ **30.** $x - y$ **31.** $-y - x$ **32.** $-x - y$

33. $z - x$ **34.** $2x - z$ **35.** $x + y - z$ **36.** $-z + y - x$

Example 6
(page 66)

37. On Friday, the closing price of a KJL company share was $51.72. It had risen $1.08 from the previous day. Find the closing price of KJL on Thursday.

Evaluate each expression for $a = -2$, $b = 3.5$, and $c = -4$.

38. $a - b + c$ **39.** $-c - b + a$ **40.** $-|a|$ **41.** $|a| + |b|$

42. $|a + b|$ **43.** $-|3 + a|$ **44.** $4b - a$ **45.** $-4b - |a|$

46. $|c + a - 5|$ **47.** $|c + a + 5|$ **48.** $|a - c| - |c|$ **49.** $|a| + |3b|$

50. Multiple Choice Archaeologists found a 1500-year-old ship at the bottom of the Black Sea. The ship is well preserved because oxygen could not make the ship decay. The ship is at a depth of 1000 ft below the surface. This is about 350 ft below the boundary between surface water, which has oxygen, and water below, which does not have oxygen. How deep is the boundary?

    Ⓐ 650 ft   Ⓑ 750 ft   Ⓒ 1150 ft   Ⓓ 1350 ft

51. Open-Ended Write two matrices with the same dimensions. Find the difference of the two matrices.

Decide if each statement is always true. If the statement is not always true, give a counterexample.

52. The difference of two numbers is less than the sum of those two numbers.

53. The difference of two numbers is less than each of those two numbers.

54. A number minus its opposite is twice the number.

🌐 **55. a. Sports** Write the data in each table below as a matrix.

**City-Wide Participation in Sports Activities
(thousands)**

	Sport	Elementary	High School	College
	Basketball	5.5	8.2	4.9
2000	Tennis	1.4	3.2	3.9
	Soccer	4.2	3.8	1.3
	Volleyball	1.6	5.2	5.1

	Sport	Elementary	High School	College
	Basketball	6.8	7.9	4.9
2005	Tennis	1.0	1.8	1.7
	Soccer	5.6	4.1	1.3
	Volleyball	1.8	4.9	2.9

  b. Subtract the 2000 matrix from the 2005 matrix to find the changes in participation in sports activities.

✏️ **c. Writing** Suppose you invest in sporting goods. In which sport would you invest? Use elements from your matrix to explain.

Real-World 🌐 Connection

Careers Archaeologists uncover and study the remains of ancient cultures. Archaeologists may look for remains in remote locations, underground, in caves, or underwater.

GO Online
Homework Video Tutor
Visit: PHSchool.com
Web Code: ate-0202

Subtract.

56. $\begin{bmatrix} -3 & 4 \\ 0 & -1 \end{bmatrix} - \begin{bmatrix} -5 & 6 \\ 9 & -4 \end{bmatrix}$ **57.** $\begin{bmatrix} \frac{3}{8} & \frac{1}{5} & 4 \end{bmatrix} - \begin{bmatrix} \frac{5}{8} & \frac{2}{10} & 7 \end{bmatrix}$ **58.** $\begin{bmatrix} \frac{1}{4} \\ -3 \end{bmatrix} - \begin{bmatrix} \frac{2}{3} \\ -2 \end{bmatrix}$

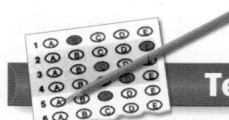

C Challenge

59. Critical Thinking Use examples to illustrate your answers.
    a. Is $|a - b|$ always equal to $|b - a|$?
    b. Is $|a + b|$ always equal to $|a| + |b|$?

Simplify each expression.

60. $1 - \frac{1}{2} - \frac{1}{3} - \frac{1}{4} - \frac{1}{5} - \frac{1}{6}$

61. $1 - \left(\frac{1}{2} - \left(\frac{1}{3} - \left(\frac{1}{4} - \left(\frac{1}{5} - \frac{1}{6} \right) \right) \right) \right)$

62. $-8t - 5m + 3m - 7t - (-3t) + m$

63. Order the expressions $|x + y|$, $|x - y|$, $|x| - |y|$, and $x - y$ from least to greatest for $x = -8$ and $y = -10$.

Test Prep

Gridded Response

64. Evaluate $-|a - b| + |2c|$ for $a = -3$, $b = 4$, and $c = -4$.

65. Find the value of $12 + (-7)$.

66. Find the next number in the pattern 14, 11, 8, 5, . . .

67. Use the bar graph at the right. In dollars, estimate the average hourly earnings of U.S. nonfarm workers in 2010.

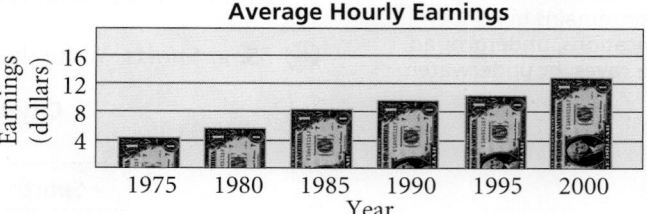

68. One January day, the temperature at dawn is −8°F. By noon, the temperature has risen 19 degrees. What is the noon temperature in degrees Fahrenheit?

Mixed Review

GO for Help

Lesson 2-1

Simplify each expression.

69. $6 + (-2)$ **70.** $-5 + (-4)$

71. $-3.4 + 2.7$ **72.** $5.9 + (-10)$

Simplify.

73. $\begin{bmatrix} 3 & -8 \\ 2 & 11 \end{bmatrix} + \begin{bmatrix} 1 & 8 \\ -2 & 5 \end{bmatrix}$ **74.** $\begin{bmatrix} 1.3 \\ -6.7 \\ 7.1 \end{bmatrix} + \begin{bmatrix} -0.1 \\ 4.2 \\ -1.9 \end{bmatrix}$ **75.** $\begin{bmatrix} \frac{1}{2} & -1 \\ 6 & \frac{2}{3} \end{bmatrix} + \begin{bmatrix} -4 & \frac{1}{3} \\ \frac{1}{3} & -5 \end{bmatrix}$

Lesson 1-1

Define variables and write an equation to model each situation.

76. The total cost equals the number of pounds of pears times $1.19/lb.

77. You have $20. Then you buy a bouquet. How much do you have left?

78. You go out to lunch with five friends and split the check equally. What is your share of the check?

2-3

Multiplying and Dividing Rational Numbers

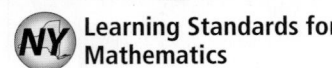

NY Learning Standards for Mathematics

A.N.1 Identify and apply the properties of real numbers (closure, commutative, associative, distributive, identity, inverse).

A.N.6 Evaluate expressions involving factorial(s), absolute value(s), and exponential expression(s).

☑ **Check Skills You'll Need**

GO **for Help** Lessons 2-1 and 2-2

Simplify each expression.

1. $-2 + (-2) + (-2) + (-2)$

2. $-5 + (-5) + (-5) + (-5) + (-5)$

3. $-6 - 6 - 6 - 6$

4. $-12 - 12 - 12 - 12 - 12 - 12$

Write the next three numbers in each pattern.

5. $2, 4, 6, \blacksquare, \blacksquare, \blacksquare$

6. $6, 4, 2, \blacksquare, \blacksquare, \blacksquare$

7. $12, 9, 6, \blacksquare, \blacksquare, \blacksquare$

8. $-18, -12, -6, \blacksquare, \blacksquare, \blacksquare$

🔊 **New Vocabulary**
- Identity Property of Multiplication
- Multiplication Property of Zero
- Multiplication Property of -1
- Inverse Property of Multiplication
- multiplicative inverse • reciprocal

---

1 Multiplying Rational Numbers

Activity: Multiplying Integers

1. Patterns Use patterns to complete each statement.

a. $2 \cdot 3 = \blacksquare$

$2 \cdot 2 = \blacksquare$

$2 \cdot 1 = \blacksquare$

$2 \cdot 0 = \blacksquare$

$2(-1) = \blacksquare$

$2(-2) = \blacksquare$

$2(-3) = \blacksquare$

b. $3(-2) = \blacksquare$

$2(-2) = \blacksquare$

$1(-2) = \blacksquare$

$0(-2) = \blacksquare$

$-1(-2) = \blacksquare$

$-2(-2) = \blacksquare$

$-3(-2) = \blacksquare$

2. Make a Conjecture From the patterns you found in Question 1, what seems to be the sign of the product of a positive number and a negative number?

3. Make a Conjecture From the patterns you found in Question 1, what seems to be the sign of the product of two negative numbers?

The product of a number and 1 is the original number. It does not matter whether the original number is positive or negative. The product of 0 and a number is 0. The product of -1 and a number is the opposite of the original number.

 **Key Concepts**

Property	Identity Property of Multiplication

For every real number n, $1 \cdot n = n$ and $n \cdot 1 = n$.

Examples $1 \cdot (-5) = -5$ $-5 \cdot 1 = -5$

Property	Multiplication Property of Zero

For every real number n, $n \cdot 0 = 0$ and $0 \cdot n = 0$.

Examples $35 \cdot 0 = 0$ $0 \cdot 35 = 0$

Property	Multiplication Property of −1

For every real number n, $-1 \cdot n = -n$ and $n \cdot -1 = -n$.

Examples $-1 \cdot 5 = -5$ $5 \cdot (-1) = -5$

From the examples for the properties above, you can see a pattern for multiplying positive and negative numbers.

Multiplying Numbers With the Same Sign

$1 \cdot 5 = 5$
positive · positive = positive

$-1 \cdot (-5) = 5$
negative · negative = positive

Multiplying Numbers With Different Signs

$1 \cdot (-5) = -5$
positive · negative = negative

$-1 \cdot 5 = -5$
negative · positive = negative

This pattern also holds true when multiplying by numbers other than 1 and −1.

 **Key Concepts**

Rule	Multiplying Numbers With the Same Sign

The product of two positive numbers or two negative numbers is positive.

Examples $5 \cdot 2 = 10$ $-5(-2) = 10$

Rule	Multiplying Numbers With Different Signs

The product of a positive number and a negative number, or a negative number and a positive number, is negative.

Examples $3(-6) = -18$ $-3 \cdot 6 = -18$

1 EXAMPLE **Multiplying Numbers**

Simplify each expression.

a. $-9(-4) = 36$ **The product of two negative numbers is positive.**

b. $5\left(-\frac{2}{3}\right) = -\frac{10}{3}$ **The product of a positive number and a negative number is negative.**

$= -3\frac{1}{3}$ **Write $-\frac{10}{3}$ as a mixed number.**

 **Quick Check** ❶ Simplify each expression.
   a. $4(-6)$ **b.** $-10(-5)$ **c.** $-4.9(-8)$ **d.** $-\frac{2}{3}\left(\frac{3}{4}\right)$

When you simplify or evaluate expressions with three or more negative numbers, you must be careful to account for all of the negative signs as you multiply.

2 EXAMPLE Evaluating Expressions

Evaluate $-2xy$ for $x = -20$ and $y = -3$.

$-2xy = -2(-20)(-3)$ **Substitute -20 for x and -3 for y.**

$\quad\quad = -120$ **$-2(-20)$ results in a positive number, 40. 40(-3) results in a negative number, -120.**

✓ Quick Check ❷ Evaluate each expression for $c = -8$ and $d = -7$.
  a. $-(cd)$ **b.** $(-2)(-3)(cd)$ **c.** $c(-d)$

You can use expressions involving multiplication to model real-world situations.

3 EXAMPLE Real-World 🌐 Problem Solving

Multiple Choice You can use the function $t = -5.5\left(\frac{a}{1000}\right)$ to calculate the change in temperature t in degrees Fahrenheit for an increase in altitude a, measured in feet. A hot-air balloon starts on the ground and then rises 8000 ft. What is the change in temperature at the altitude of the balloon?

Ⓐ 44°F   Ⓑ 40°F   Ⓒ −40°F   Ⓓ −44°F

$-5.5\left(\frac{a}{1000}\right) = -5.5\left(\frac{8000}{1000}\right)$ **Substitute 8000 for a.**

$\quad\quad\quad = -5.5(8)$ **Divide within parentheses.**

$\quad\quad\quad = -44$ **Multiply.**

● The change in temperature is -44 degrees Fahrenheit. The answer is D.

✓ Quick Check ❸ **a.** Find the change in temperature if a balloon rises 4500 ft from the ground.
  b. Suppose the temperature is 40°F at ground level. What is the approximate air temperature at the altitude of the balloon?

The expression -3^4 means the opposite of 3^4. The exponent 4 applies to the base 3. The negative sign is not part of the base. In the expression $(-3)^4$, the negative sign is part of the base -3. The exponent 4 applies to the base -3.

4 EXAMPLE Simplifying Exponential Expressions

Use the order of operations to simplify each expression.
  a. -3^4
  $-3^4 = -(3 \cdot 3 \cdot 3 \cdot 3)$ **Write as repeated multiplication.**
  $\quad\quad = -81$ **Simplify.**
  b. $(-3)^4$
  $(-3)^4 = (-3)(-3)(-3)(-3)$ **Write as repeated multiplication.**
  $\quad\quad\quad = 81$ **Simplify.**

Test-Taking Tip

Remember that an exponent tells how many times a number is used as a factor.

✓ Quick Check ❹ Simplify each expression.
  a. -4^3 **b.** $(-2)^4$ **c.** $(-0.3)^2$ **d.** $-\left(\frac{3}{4}\right)^2$

The rules for finding the sign when dividing rational numbers are the same as the rules for finding the sign when multiplying rational numbers.

Key Concepts

Rule	Dividing Numbers With the Same Sign

The quotient of two positive numbers or two negative numbers is positive.

Examples $6 \div 3 = 2$ $-6 \div (-3) = 2$

Rule	Dividing Numbers With Different Signs

The quotient of a positive number and a negative number, or a negative number and a positive number, is negative.

Examples $-6 \div 3 = -2$ $6 \div (-3) = -2$

You can use the rules for dividing numbers to simplify expressions.

5 EXAMPLE   Dividing Numbers

Simplify each expression.

a. $12 \div (-4) = -3$ **The quotient of a positive number and a negative number is negative.**

b. $-12 \div (-4) = 3$ **The quotient of a negative number and a negative number is positive.**

 **5** Simplify each expression.

a. $-42 \div 7$ **b.** $-8 \div (-2)$ **c.** $8 \div (-8)$ **d.** $-39 \div (-3)$

You can evaluate expressions that involve division.

Vocabulary Tip

A fraction bar means divide: $\frac{15}{5} = 15 \div 5$.

6 EXAMPLE   Evaluating Expressions

Evaluate $\frac{-x}{-4} + 2y \div z$ for $x = -20$, $y = 6$, and $z = -1$.

$\frac{-x}{-4} + 2y \div z = \frac{-(-20)}{-4} + 2(6) \div (-1)$ **Substitute -20 for x, 6 for y, and -1 for z.**

$= -5 + (-12)$ **Divide and multiply.**

$= -17$ **Add.**

 **6** Evaluate each expression for $x = 8$, $y = -5$, and $z = -3$.

a. $3x \div (2z) + y \div 10$ **b.** $\frac{2z + x}{2y}$ **c.** $3z^2 - 4y \div x$

A number and its multiplicative inverse have a special relationship.

 Key Concepts

Property	Inverse Property of Multiplication

For every nonzero real number a, there is a **multiplicative inverse** $\frac{1}{a}$ such that $a\left(\frac{1}{a}\right) = 1$.

Examples $5\left(\frac{1}{5}\right) = 1$ $-5\left(-\frac{1}{5}\right) = 1$

The **multiplicative inverse,** or **reciprocal,** of a nonzero rational number $\frac{a}{b}$ is $\frac{b}{a}$. Zero does not have a reciprocal. Division by zero is undefined.

GO for Help

For help with dividing fractions, see Skills Handbook p. 761.

7 EXAMPLE Division Using the Reciprocal

Evaluate $\frac{x}{y}$ for $x = -\frac{3}{4}$ and $y = -\frac{5}{2}$.

$\frac{x}{y} = x \div y$ **Rewrite the expression.**

$\quad = -\frac{3}{4} \div \left(-\frac{5}{2}\right)$ **Substitute $-\frac{3}{4}$ for x and $-\frac{5}{2}$ for y.**

$\quad = -\frac{3}{4}\left(-\frac{2}{5}\right)$ **Multiply by $-\frac{2}{5}$, the reciprocal of $-\frac{5}{2}$.**

$\quad = \frac{3}{10}$ **Simplify.**

✓ **Quick Check**   **7** Evaluate the expression in Example 7 for $x = 8$ and $y = -\frac{4}{5}$.

EXERCISES

For more exercises, see *Extra Skill and Word Problem Practice.*

Practice and Problem Solving

A Practice by Example

Example 1
(page 70)

GO for Help

Simplify each expression.

1. $3(-5)$ **2.** $5(-3)$ **3.** $3(5)$

4. $-3(-5)$ **5.** $8(-4.3)$ **6.** $9\left(-\frac{5}{18}\right)$

7. $10(-12)$ **8.** $7(-15)$ **9.** $-4(20)$

10. $-20(-4)$ **11.** $13(-6)$ **12.** $-9(-9)$

Example 2
(page 71)

Evaluate each expression for $m = -4$, $n = 3$, and $p = -1$.

13. mn **14.** $-mn$ **15.** $3m - n$

16. $-5p$ **17.** $2m$ **18.** $7p - 2n$

19. $8p \cdot (-2n)$ **20.** $p \cdot (m + n)$ **21.** mnp

22. $m \cdot (3 + p)$ **23.** $4n^3 \cdot m$ **24.** $m \cdot p + (-n)$

Example 3
(page 71)

Evaluate each expression for $x = -12$ and $y = 4$.

25. $xy - 4y$ **26.** $2xy + 9$ **27.** $x + 4y$

28. $-x + 3y$ **29.** $3y - 2x$ **30.** $6y + x$

🌐 **31. Weather** The function $w = -39 + \frac{3}{2}t$, where t is the actual air temperature, gives the approximate wind chill temperature w when the wind speed is 20 mi/h. Find the approximate wind chill temperature for the given air temperatures with a 20 mi/h wind.
     a. $10°F$ **b.** $-24°F$ **c.** $-8°F$ **d.** $5°F$

Example 4
(page 71)

Simplify each expression.

32. $(-1)^5$ **33.** $-(-2)^3$ **34.** -5^2 **35.** $(-9)^2$

36. -9^2 **37.** $3(-4)^3$ **38.** $-5(-1)^4$ **39.** $-5^2(-3)^3$

Example 5
(page 72)

Simplify each expression.

40. $\frac{6}{-3}$ **41.** $\frac{-36}{9}$ **42.** $\frac{3 - 14}{-2}$ **43.** $-18 \div (-3)$

44. $-121 \div 11$ **45.** $-64 \div (-5)$ **46.** $2^3 \div (-4)$ **47.** $-56 \div (4 + 3)$

Example 6
(page 72)

Evaluate each expression for $x = -2$, $y = 3$, and $z = 3.5$.

48. $(y + 3x) \div y$ **49.** $4z \div x$

50. $4x^3 - \frac{2z}{x}$ **51.** $(3x + 2y) \div (2x + 3y)$

52. $(2z + 7) \div y$ **53.** $8 + 6x \div (4y) - \frac{3z}{y}$

Example 7
(page 73)

Evaluate each expression.

54. $\frac{x}{y}$, for $x = \frac{2}{5}$ and $y = \frac{3}{10}$ **55.** $\frac{-3m}{t}$, for $m = \frac{5}{6}$ and $t = \frac{1}{6}$

56. $\frac{r}{-3s}$, for $r = -\frac{1}{8}$ and $s = \frac{3}{4}$ **57.** $\frac{3x}{5y}$, for $x = \frac{1}{5}$ and $y = -\frac{1}{2}$

B Apply Your Skills

Copy and complete each table.

58.

m	$-5m$
-4	▪
-1	▪
2	▪
▪	-25

59.

p	$\frac{3}{4}p - 5$
-4	▪
-3	▪
2	▪
▪	-2

60.

n	$|2n - 3|$
-5	▪
-1	▪
0	▪
4	▪

61. a. Find each product.
     i. $(-1)(-1)$ **ii.** $(-1)(-1)(-1)$
     iii. $(-1)(-2)$ **iv.** $(-1)(-2)(-3)$
     v. $(-1)(-2)(-3)(-4)$ **vi.** $(-1)(-2)(-3)(-4)(-5)$
     b. Patterns For an even number of negative factors, the product will be __?__ .
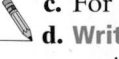   **c.** For an odd number of negative factors, the product will be __?__ .
     d. Writing For a product that includes negative and positive factors, do the positive factors affect the sign of the product? Explain.

62. Suppose a and b are integers.
     a. When is the product ab positive?
     b. When is the product ab negative?

Evaluate each expression for the given value(s).

63. $\frac{3}{4}w - 7$, for $w = 1\frac{1}{3}$ **64.** $\frac{x}{2y}$, for $x = 3.6$ and $y = -0.4$

65. $\frac{n}{m}$, for $n = -\frac{4}{5}$ and $m = 8$ **66.** $\frac{3a}{b} + c$, for $a = -2$, $b = -5$, and $c = -1$

Open-Ended Use $a = -3$, $b = 2$, and $c = -5$ to write an algebraic expression that has each value.

67. 17 **68.** 0 **69.** -1 **70.** 1 **71.** 7

72. History A toll bridge in Maine in the early 1900s charged 2¢ per person and $6\frac{1}{4}$¢ for a dozen sheep.
     a. How much would the toll for 3 people and 4 dozen sheep have been?
     b. If the toll was 56¢ and there were 8 dozen sheep, how many people were there?

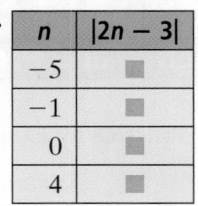

73. Reasoning Does $|ab|$ always equal $|a| \cdot |b|$? Explain.

74. a. Simplify each expression.
$(-2)^2 \quad (-2)^3 \quad (-2)^4 \quad (-2)^5$
$(-3)^2 \quad (-3)^3 \quad (-3)^4 \quad (-3)^5$
b. Make a Conjecture Do you think a negative number raised to an even power will be positive or negative? Explain.
c. What is the sign of a negative number raised to an odd power? Explain.

75. Is -10 or 0.1 the multiplicative inverse of 10? Explain.

76. Explain why the reciprocal of a nonzero number is *not* the same as the opposite of the number.

In *scalar multiplication*, you multiply the elements in a matrix by a number, called a *scalar*. Find each product.

Sample $3\begin{bmatrix} 4 & -1.5 \\ \frac{1}{2} & -6 \end{bmatrix} = \begin{bmatrix} 3 \cdot 4 & 3 \cdot (-1.5) \\ 3 \cdot \frac{1}{2} & 3 \cdot (-6) \end{bmatrix}$ Multiply each entry by the scalar.

$= \begin{bmatrix} 12 & -4.5 \\ \frac{3}{2} & -18 \end{bmatrix}$ **Simplify.**

77. $-2\begin{bmatrix} 11 & -5 \\ -9 & 6 \\ -4 & 3 \end{bmatrix}$
78. $\frac{3}{5}\begin{bmatrix} -25 & 35 \\ \frac{10}{9} & -15 \end{bmatrix}$
79. $-0.1\begin{bmatrix} -47 & 13 & -7.9 \\ 0.2 & -64 & 0 \end{bmatrix}$

80. $-4\begin{bmatrix} 3 & \frac{2}{3} \end{bmatrix}$
81. $2\begin{bmatrix} -4 & -5.3 & 2 \\ 3.1 & 0 & 6 \end{bmatrix}$
82. $\frac{1}{4}\begin{bmatrix} -1 & \frac{3}{4} \\ \frac{8}{9} & 0 \end{bmatrix}$

 83. Entertainment As riders plunge down the hill of a roller coaster, you can approximate the height h, in feet, above the ground of their roller-coaster car. Use the function $h = 155 - 16t^2$ where t is the number of seconds since the start of the descent.
a. How far is a rider from the bottom of the hill after 1 second? 2 seconds?
b. Critical Thinking Does it take more than or less than 4 seconds to reach the bottom? Explain.

84. Multiple Choice The formula $C = \frac{5}{9}(F - 32)$ changes a temperature reading from the Fahrenheit scale F to the Celsius scale C. What is the temperature measured in Celsius if the Fahrenheit temperature is $-13°$?
(A) $25°C$ (B) $15°C$ (C) $-15°C$ (D) $-25°C$

85. a. Sales The table at the left shows the monthly sales in March and October for three departments of a clothing store. Organize the data into a matrix.
b. Use a scalar to find the matrix for each month's average daily sales. Assume the store is open every day. Round to the nearest tenth.
c. Each department expects sales in March and October to increase by 10% next year. Find the matrix that shows the projected sales for these months.

	Dept. Sales		
March	180	210	200
October	170	230	190

C Challenge   **Evaluate each expression for $b = -\frac{1}{2}$.**

86. b^3 **87.** b^4 **88.** b^5 **89.** b^6 **90.** $-b^6$

91. What is the greatest integer n for which $(-n)^3$ is positive and the value of the expression has a 2 in the ones place?

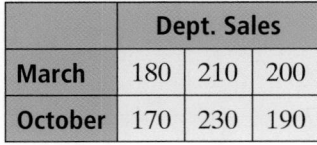
155 ft

Rewrite each expression using the symbol ÷. Then find each quotient.

Sample $\dfrac{\frac{-7}{12}}{4} = \dfrac{-7}{12} \div 4$ Rewrite as $\dfrac{-7}{12} \div 4$.

$\qquad\qquad = \dfrac{-7}{12} \cdot \left(\dfrac{1}{4}\right)$ Multiply by $\frac{1}{4}$, the reciprocal of 4.

$\qquad\qquad = -\dfrac{7}{48}$

92. $\dfrac{\frac{5}{4}}{9}$ **93.** $\dfrac{\frac{3}{8}}{\frac{-2}{3}}$ **94.** $\dfrac{\frac{-5}{6}}{8}$ **95.** $\dfrac{\frac{-2}{5}}{\frac{-4}{5}}$

Test Prep

Multiple Choice

96. Simplify $(-3)(-3)(2)(2)(-1)$.
   A. -36 **B.** -6 **C.** 6   **D.** 36

97. Evaluate $-ac + bc$ for $a = -2$, $b = 6$, and $c = -3$.
   F. -24 **G.** -6 **H.** 6   **J.** 24

98. A Mach number M indicates the speed of a supersonic airplane. You can find an airplane's speed a in miles per hour using the formula $a = Ms$ where s is the speed of sound at the altitude of the airplane. Find an airplane's speed in miles per hour if the airplane travels at Mach 2.5 at an altitude where the speed of sound is 710 mi/h.
   A. 177.5 mi/h   **B.** 284 mi/h   **C.** 1775 mi/h   **D.** 2840 mi/h

99. Which expression does NOT have the same value as $-11 + (-11) + (-11) + (-11)$?
   F. -44 **G.** $4(-11)$ **H.** $(-11)^4$ **J.** $33 - 77$

100. Use the table below. What was the average temperature for the week?

Day	Mon.	Tues.	Wed.	Thur.	Fri.	Sat.	Sun.
Temperature	$-3°F$	$4°F$	$-2°F$	$-5°F$	$-3°F$	$1°F$	$1°F$

   A. $-7°F$ **B.** $-1°F$ **C.** $1°F$ **D.** $4°F$

Mixed Review

Lesson 2-2   **Subtract.**

101. $6 - 8$ **102.** $-3 - 17$ **103.** $-2.3 - (-3.1)$

104. $7 - (-2.8)$ **105.** $1\frac{3}{4} - \left(-\frac{1}{2}\right)$ **106.** $-\frac{7}{8} - \left(-\frac{8}{9}\right)$

Lesson 1-3   **Find each absolute value.**

107. $|4.95|$ **108.** $|-56|$ **109.** $|-4.59|$ **110.** $\left|-\frac{3}{4}\right|$

Lesson 1-2   **Evaluate each expression for $a = 4$ and $b = 7$.**

111. $a^2 + b$ **112.** $(a + b)^2$ **113.** ab^2

114. A sweater costs $32.95. The sales tax rate is 6%. Find the total cost.

The Distributive Property

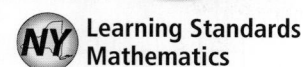

NY Learning Standards for Mathematics

A.N.1 Identify and apply the properties of real numbers (closure, commutative, associative, distributive, identity, inverse).

A.A.13 Add, subtract, and multiply monomials and polynomials.

 Check Skills You'll Need

GO for Help   Lessons 1-2 and 2-3

Use the order of operations to simplify each expression.

1. $3(4 + 7)$ **2.** $-2(5 + 6)$ **3.** $-1(-9 + 8)$

4. $-0.5(8 - 6)$ **5.** $\frac{1}{2}t(10 - 4)$ **6.** $m(-3 - 1)$

🔊 **New Vocabulary** • **Distributive Property** • **term** • **constant** • **coefficient** • **like terms**

---

1 Using the Distributive Property

You can use the Distributive Property to multiply a sum or difference by a number.

 Key Concepts

Property	**Distributive Property**

For every real number a, b, and c,

$a(b + c) = ab + ac$ $(b + c)a = ba + ca$

$a(b - c) = ab - ac$ $(b - c)a = ba - ca$

Examples $5(20 + 6) = 5(20) + 5(6)$ $(20 + 6)5 = 20(5) + 6(5)$

$9(30 - 2) = 9(30) - 9(2)$ $(30 - 2)9 = 30(9) - 2(9)$

You can use the Distributive Property to multiply some numbers using mental math. For instance, you can think of 102 as $100 + 2$ and 98 as $100 - 2$.

1 EXAMPLE   **Simplifying a Numerical Expression**

Use the Distributive Property to simplify $34(102)$.

$34(102) = 34(100 + 2)$ **Rewrite 102 as 100 + 2.**

$= 34(100) + 34(2)$ **Use the Distributive Property.**

$= 3400 + 68$ **Simplify.**

$= 3468$

 **Quick Check** ❶ Simplify each expression.

    a. $13(103)$ **b.** $21(101)$ **c.** $24(98)$ **d.** $15(99)$

You can also use the Distributive Property and mental math to calculate costs.

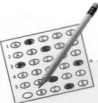

2 EXAMPLE Using Mental Math

Gridded Response At Mars' Deli, you can find the cost c of n sandwiches using the function $c = 2.95n$. Find the cost of 8 sandwiches.

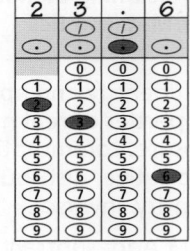

$$c = 2.95n$$
$$c = 2.95(8) \qquad \text{Substitute 8 for } n.$$
$$= (3 - 0.05)8 \qquad \text{Write 2.95 as 3 − 0.05.}$$
$$= (3)8 - (0.05)8 \qquad \text{Use the Distributive Property.}$$
$$= 24 - 0.40 \qquad \text{Simplify.}$$
$$= 23.60$$

The cost of 8 sandwiches is \$23.60. Grid 23.6.

 **Quick Check** ② Find the total cost of 6 pairs of socks that are \$2.95 per pair.

---

2 Simplifying Algebraic Expressions

You can use the Distributive Property to simplify an algebraic expression. An algebraic expression in simplest form has no grouping symbols.

3 EXAMPLE Simplifying an Expression

Simplify each expression.

a. $2(5x + 3)$

$$2(5x + 3) = 2(5x) + 2(3) \qquad \text{Use the Distributive Property.}$$
$$= 10x + 6 \qquad \text{Simplify.}$$

b. $(3b - 2)\left(\frac{1}{3}\right)$

$$(3b - 2)\left(\frac{1}{3}\right) = 3b\left(\frac{1}{3}\right) - 2\left(\frac{1}{3}\right) \qquad \text{Use the Distributive Property.}$$
$$= b - \frac{2}{3} \qquad \text{Simplify.}$$

Quick Check ③ Simplify each expression.   **a.** $2(3 - 7t)$ **b.** $(0.4 + 1.1c)(3)$

To simplify an expression like $-(6x + 4)$, rewrite the expression as $-1(6x + 4)$, using the Multiplication Property of -1.

4 EXAMPLE Using the Multiplication Property of −1

Simplify $-(6x + 4)$.

$$-(6x + 4) = -1(6x + 4) \qquad \text{Rewrite the expression using −1.}$$
$$= -1(6x) + (-1)(4) \qquad \text{Use the Distributive Property.}$$
$$= -6x - 4 \qquad \text{Simplify.}$$

 **Quick Check** ④ Simplify each expression.   **a.** $-(2x + 1)$ **b.** $(3 - 8a)(-1)$

In an algebraic expression, a **term** is a number, a variable, or the product of a number and one or more variables.

$$6a^2 - 5ab + 3b - 12 \leftarrow \text{A } \boxed{\textbf{constant}} \text{ is a term that has no variable.}$$

A **coefficient** is a numerical factor of a term.

Think of $3b - 12$ as $3b + (-12)$ to determine that the constant is -12.

Like terms have exactly the same variable factors.

Like Terms	Not Like Terms
$3x$ and $-2x$	$8x$ and $7y$
$-5x^2$ and $9x^2$	$5y$ and $2y^2$
xy and $-xy$	$4y$ and $5xy$
$-7x^2y^3$ and $15x^2y^3$	x^2y and xy^2

An algebraic expression in simplest form has no like terms. You can use the Distributive Property to combine like terms when simplifying an expression. Think of the Distributive Property as $ba + ca = (b + c)a$.

5 EXAMPLE Combining Like Terms

Simplify each expression.
a. $3x^2 + 5x^2$

$\quad 3x^2 + 5x^2 = (3 + 5)x^2$ **Use the Distributive Property.**

$\qquad\qquad\quad = 8x^2$ **Simplify.**

b. $-5c + c$

$\quad -5c + c = -5c + 1c$ **Rewrite c as $1c$.**

$\qquad\qquad = (-5 + 1)c$ **Use the Distributive Property.**

$\qquad\qquad = -4c$ **Simplify.**

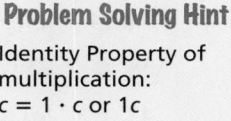

Problem Solving Hint

Identity Property of multiplication:
$c = 1 \cdot c$ or $1c$

✓ **Quick Check** ❺ Simplify each expression.
a. $7y + 6y$ **b.** $3t - t$ **c.** $-9w^3 - 3w^3$ **d.** $8d + d$

You can write an expression from a verbal phrase. The word *quantity* indicates that two or more terms are within parentheses.

6 EXAMPLE Writing an Expression

Write an expression for "3 times the quantity x minus 5."

Relate 3 times the quantity x minus 5

Write 3 $\cdot$ $(x - 5)$

$3(x - 5)$

✓ **Quick Check** ❻ Write an expression for each phrase.
a. -2 times the quantity t plus 7
b. the product of 14 and the quantity 8 plus w

EXERCISES

For more exercises, see *Extra Skill and Word Problem Practice.*

Practice and Problem Solving

A Practice by Example

Example 1
(page 79)

GO for Help

Simplify each expression using the Distributive Property.

1. 12(201)   **2.** 51(13)   **3.** 11(499)   **4.** 8(306)

5. 7(98)   **6.** 3(999)   **7.** 41(502)   **8.** 24(1020)

Example 2
(page 80)

Mental Math **Use the Distributive Property to find each price.**

9. 4($.99)   **10.** 6($1.97)   **11.** 5($5.91)   **12.** 7($29.93)

13. The school librarian got money to buy reference works on CDs. To find the cost c of n CDs, she used the function $c = 32.99n$. How much did she spend on 3 CDs?

14. You stopped on your way to basketball practice and bought four cans of fruit punch for $.69 each. How much did you spend in all?

Example 3
(page 80)

Simplify each expression.

15. $7(t - 4)$ **16.** $-2(n - 6)$ **17.** $3(m + 4)$

18. $(5b - 4)\frac{1}{5}$ **19.** $-2(x + 3)$ **20.** $\frac{2}{3}(6y + 9)$

21. $0.25(6q + 32)$ **22.** $(3n - 7)(6)$ **23.** $(8 - 3r)\frac{5}{16}$

24. $-4.5(b - 3)$ **25.** $\frac{2}{5}(5w + 10)$ **26.** $(9 - 4n)(-4)$

Example 4
(page 80)

27. $-(x + 3)$ **28.** $-(x - 3)$ **29.** $-(3 + x)$ **30.** $-(3 - x)$

31. $-(6k + 5)$ **32.** $-(7x - 2)$ **33.** $-(2 - 7x)$ **34.** $(4 - z)(-1)$

Example 5
(page 81)

35. $4t - 7t$ **36.** $12k^2 + 8k^2$ **37.** $9x - 2x$ **38.** $w + 23w$

39. $-18v^2 + 23v^2$ **40.** $7m - m$ **41.** $13q - 30q$ **42.** $x - 46x$

Example 6
(page 81)

Write an expression for each phrase.

43. 3 times the quantity m minus 7

44. -4 times the quantity 4 plus w

45. twice the quantity b plus 9

46. 2 times the quantity 3 times c plus 9

B Apply Your Skills

Simplify each expression.

47. 9(4998)   **48.** 12(7.001)   **49.** 7(2.003)

50. $144\left(\frac{15}{16}b + \frac{8}{9}\right)$ **51.** $3(6.2 + 5m)$ **52.** $\frac{7}{8}\left(\frac{10}{14}d - 48\right)$

Copy and complete each table.

53.

k	$2(k - 4)$
-10	■
-5	■
2.5	■
■	0

54.

n	$-(2n + 5)$
-3	■
-1	■
2	■
■	-15

55.

a	$-3(2 - a)$
-4	■
0	■
2	■
■	9

Write an expression for each phrase.

56. $2\frac{1}{4}$ times the quantity $5\frac{1}{2}$ minus k

57. $6\frac{7}{100}$ times the quantity 8 plus $\frac{4}{3}p$

58. the product of $\frac{11}{20}$ and the quantity b minus $\frac{13}{30}$

59. 17 divided by the quantity z minus 34

60. $4\frac{1}{3}$ times the quantity x minus $\frac{11}{12}$

61. **Multiple Choice** Which expression represents the perimeter of the figure?

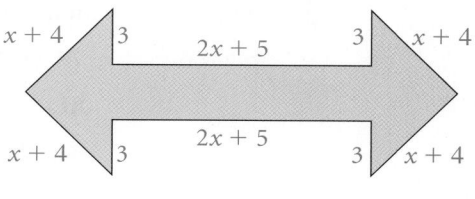

- Ⓐ $8x + 38$
- Ⓑ $46x$
- Ⓒ $4x + 38$
- Ⓓ $8x$

 62. **Writing** Does $2ab = 2a \cdot 2b$? Explain.

63. **Error Analysis** A student rewrote $4(3x + 10)$ as $12x + 10$. Explain the student's error.

64. **Open-Ended** Write a variable expression that you could simplify using the Distributive Property. Then simplify your expression.

Simplify each expression.

65. $4.78d + 0d$

66. $-21p - 76p^2 - 9 + p$

67. $3.3t^2 + 8.7t - 9.4t^2 + 5t$

68. $1.5m - 4.2m - 12.5v + 4.2m$

69. $\frac{6}{7}n + n^3 - \left(-\frac{5}{6}n\right)$

70. $-\frac{16}{15}k + \frac{3}{20}h + \frac{7}{40}k$

71. $8m^2 - 5mz + 4mz - m^2 + 4$

72. $9 - 4t + 6y - 3t + 10$

73. $1.4b - 3b^2 + 4c - 2b^2 + c$

74. $8xyz + 4xy - 12yzx + 2xy$

75. Identify the terms, the coefficients, and the constant(s) in the expression $-7t + 6v + 7 - 19y$.

76. **Sports** A high school basketball court is 84 ft long by 50 ft wide. A college basketball court is 10 ft longer than a high school court but has the same width.

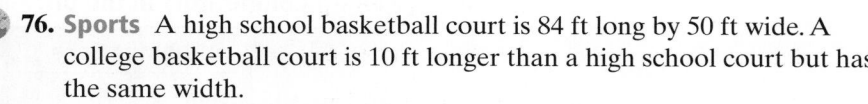

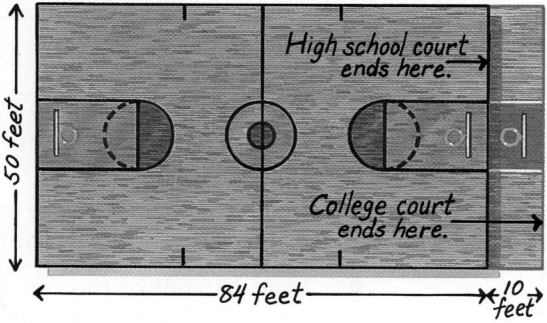

a. Write an expression using parentheses for the area of a college basketball court.

b. Simplify your expression.

77. A student searched the pockets of his jeans before doing laundry and found the following numbers of coins: 4 pennies, 2 nickels, and a quarter; 3 quarters and 6 pennies; 1 dime and 5 nickels. How many of each type of coin does the student have?

78. Shopping Suppose you buy 4 cans of tomatoes at $1.02 each, 3 cans of tuna for $.99 each, and 3 boxes of pasta at $.52 each. Write an expression to model this situation. Then use the Distributive Property to find the total cost.

 **Challenge**

Simplify each expression.

79. $9(5 + t) - 6(t + 3)$

80. $4(r + 8) - 5(2r - 1)$

81. $-(m + 3) - 2(m + 3)$

82. $a[2 + b(2 + c)]$

83. $7b[8 + 6(b - 1)]$

84. $-[-5(y + 2z) - 3z]$

85. Critical Thinking If $y = 3x - 10$, what is an expression for $\frac{y}{3}$?

86. a. Evaluate $(a + b) \div c$ and $a \div c + b \div c$ for $a = -10, b = 6,$ and $c = -2$.
  b. Evaluate each expression for $a = -9, b = -3,$ and $c = 6$.
  c. Reasoning Does it appear that $(a + b) \div c = a \div c + b \div c$? Explain.

87. a. Evaluate $a \div (b + c)$ and $a \div b + a \div c$ for $a = -60, b = 3,$ and $c = -5$.
  b. Evaluate each expression for $a = -24, b = -4,$ and $c = 2$.
  c. Reasoning Does it appear that $a \div (b + c) = a \div b + a \div c$? Explain.

NY REGENTS

Test Prep

Multiple Choice

88. Which expression is another form of $14x - 21x$?
  A. $-7x^2$ **B.** $7x^2(2 - 3x)$ **C.** $7x(2x - 3)$ **D.** $-7x$

89. Which expression is another form of $-6(k - 5)$?
  F. $-6k^2 - 30k$ **G.** $-6k + 30$ **H.** $-6k^2 - 5$ **J.** $-6k^2 + 5$

Use the table below for Exercises 90–92.

Competitors in the Ultra-Marathon 2000

Category	130-Mile Race		350-Mile Race	
	Number Started	Number Finished	Number Started	Number Finished
Bicycle	51	41	24	16
Foot	43	37	11	9
Ski	9	6	3	3

90. Which category had the most competitors in each race?
  A. bicycle   **B.** foot   **C.** ski   **D.** all of them

91. What percent of the competitors on bicycles completed the 350-mile race?
  F. 67%   **G.** 75%   **H.** 80%   **J.** 82%

92. What percent of the competitors completed the 130-mile race?
  A. 67%   **B.** 75%   **C.** 82%   **D.** 86%

93. Rocky Reifenstuhl won the 130-mile race on his bicycle. He finished in 11 hours and 45 minutes. What was his average speed?

    F. 11.8 mi/h   **G.** 11.4 mi/h   **H.** 11.1 mi/h   **J.** 10.8 mi/h

94. Simplify $7q + 8pq - 4pq - 9q$.

    A. $-2q + 12pq$ **B.** $16q + 4pq$

    C. $-2q + 4pq$ **D.** $16q + 12pq$

95. A notebook costs $1.89 at the school store. How much would notebooks for a class of 30 students cost?

    F. $59.70   **G.** $59.67   **H.** $57.30   **J.** $56.70

Mixed Review

Lessons 2-2, 2-3

GO for Help

Simplify each expression.

96. $8(-7) + 4(-3)$ **97.** $8 + (-4) \cdot 3$

98. $(-3)^2 + (-5)$ **99.** $(9^2 - 60) \div 3$

100. $\dfrac{-7 + 5}{-7 - 5}$ **101.** $\dfrac{7 - 2}{-12 + 8}$

102. $\dfrac{1 + (-4)}{21 - 6}$ **103.** $-3^4 \div 9 - 4 \cdot 2$

Lesson 2-1

Simplify each expression.

104. $6.034 + (-8.42)$ **105.** $9.73 + 2.397$

106. $-54.1 + 99.4$ **107.** $|-28.2| + 17.5$

108. $-6.45 + |-9.02|$ **109.** $3.02 + (-2.1)$

110. $-5.7 + (-3.9)$ **111.** $14.7 + |-8.3|$

Lesson 1-1

112. a. Write an expression for the phrase "4 more than the quotient of m and 3."
    b. Evaluate your expression for $m = 9, m = 3,$ and $m = 12$.

✓ Checkpoint Quiz 1   Lessons 2-1 through 2-4

Simplify each expression.

1. $\left(-\dfrac{1}{3}\right) + \left(\dfrac{2}{9}\right)$ **2.** $-3 + (-17)$

3. $|-5| + (-5) + 5$ **4.** $\begin{bmatrix} 2 & -5 \\ 4.1 & 7 \\ -10 & 8 \end{bmatrix} + \begin{bmatrix} -2 & 0 \\ -6.2 & 1 \\ -13 & 3 \end{bmatrix}$

Evaluate each expression for $a = 4$ and $b = -5$.

5. $2a \cdot (-3b)$ **6.** $-b - a$ **7.** $(2a^2) \div (4b)$

8. Temperature The temperature one winter morning is $-5°$F. Define a variable and write an expression to find the temperature after it changes. Then evaluate your expression for an increase of 13 degrees Fahrenheit.

Use the Distributive Property to simplify each expression.

9. $\dfrac{1}{3}(24x - 36)$ **10.** $6t + 43t$

2-5

Properties of Numbers

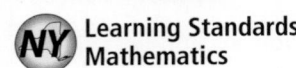 **Learning Standards for Mathematics**

A.N.1 Identify and apply the properties of real numbers (closure, commutative, associative, distributive, identity, inverse).

☑ **Check Skills You'll Need**

Simplify each expression.

1. $8 + (9 + 2)$
2. $3 \cdot (-2 \cdot 5)$
3. $7 + 16 + 3$
4. $-4(7)(-5)$
5. $-6 + 9 + (-4)$
6. $0.25 \cdot 3 \cdot 4$
7. $3 + x - 2$
8. $2t - 8 + 3t$
9. $-5m + 2m - 4m$

🔊 **New Vocabulary** ● deductive reasoning

---

1 Identifying and Using Properties

The summary below reviews properties of real numbers. Recall that real numbers include rational numbers (the subject of this chapter), as well as irrational numbers (which you will study further in Chapter 3).

 Key Concepts

Property	Properties of Real Numbers

For every real number a, b, and c,

Commutative Property of Addition
$a + b = b + a$ **Example** $3 + 7 = 7 + 3$

Commutative Property of Multiplication
$a \cdot b = b \cdot a$ **Example** $3 \cdot 7 = 7 \cdot 3$

Associative Property of Addition
$(a + b) + c = a + (b + c)$ **Example** $(6 + 4) + 5 = 6 + (4 + 5)$

Associative Property of Multiplication
$(a \cdot b) \cdot c = a \cdot (b \cdot c)$ **Example** $(6 \cdot 4) \cdot 5 = 6 \cdot (4 \cdot 5)$

Identity Property of Addition
$a + 0 = a$ **Example** $9 + 0 = 9$

Identity Property of Multiplication
$a \cdot 1 = a$ **Example** $6 \cdot 1 = 6$

Inverse Property of Addition
For every a, there is an additive inverse $-a$ such that $a + (-a) = 0$.   **Example** $5 + (-5) = 0$

Inverse Property of Multiplication
For every a ($a \neq 0$), there is a multiplicative inverse $\frac{1}{a}$ such that $a\left(\frac{1}{a}\right) = 1$.   **Example** $5 \cdot \frac{1}{5} = 1$

Symmetric Property
If $a = b$, then $b = a$ **Example** $2 \cdot 3 = 6$, so $6 = 2 \cdot 3$

The following summary reviews some additional properties of real numbers.

 **Key Concepts**

Property	Properties of Real Numbers

For every real number a, b, and c,

Distributive Property   **Examples**

$a(b + c) = ab + ac$ $5(4 + 2) = 5 \cdot 4 + 5 \cdot 2$

$a(b - c) = ab - ac$ $5(4 - 2) = 5 \cdot 4 - 5 \cdot 2$

Multiplication Property of Zero

For every real number n, $n \cdot 0 = 0$. $-35 \cdot 0 = 0$

Multiplication Property of −1

For every real number n, $-1 \cdot n = -n$. $-1 \cdot (-5) = 5$

1 EXAMPLE   **Identifying Properties**

Name the property that each equation illustrates. Explain.

a. $9 + 7 = 7 + 9$ Commutative Property of Addition, because the order of the addends changes

b. $(d \cdot 4) \cdot 3 = d \cdot (4 \cdot 3)$ Associative Property of Multiplication, because the grouping of the factors changes

c. $t + 0 = t$ Identity Property of Addition, because the sum of a number and zero is the number

d. $-q = -1q$ Multiplication Property of −1, because the opposite of a value is the same as −1 times the value

 **Quick Check**   **1** Name the property that each equation illustrates. Explain.

   a. $1m = m$ **b.** $(-3 + 4) + 5 = -3 + (4 + 5)$ **c.** $(3 \cdot 8)0 = 3(8 \cdot 0)$

   d. $2 + 0 = 2$ **e.** $np = pn$ **f.** $p + q = q + p$

You can also use the properties to reorganize the order of numbers in sums or products so that you can calculate more easily.

2 EXAMPLE   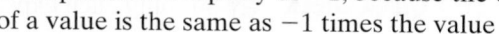**Real-World** 🌐 **Problem Solving**

Shopping Suppose you buy the school supplies shown at the left. Find the total cost of the supplies.

$0.85 + 2.50 + 5.15 = 2.50 + 0.85 + 5.15$ **Commutative Property of Addition**

$= 2.50 + (0.85 + 5.15)$ **Associative Property of Addition**

$= 2.50 + 6$ **Add within parentheses first.**

$= 8.50$ **Simplify.**

The total cost of the supplies is $8.50.

 **Quick Check**   **2 Shopping** At the supermarket, you buy a package of cheese for $2.50, a loaf of bread for $2.15, a cucumber for $.65, and some tomatoes for $3.50. Find the total cost of the groceries.

Deductive reasoning is the process of reasoning logically from given facts to a conclusion. Using deductive reasoning, you justify each step in simplifying an expression with reasons such as properties, definitions, or rules.

3 EXAMPLE Justifying Steps

Simplify each expression. Justify each step.

a. $-4b + 9 + b$

Step	Reason
$-4b + 9 + b = -4b + 9 + 1b$	Identity Property of Multiplication
$= -4b + 1b + 9$	Commutative Property of Addition
$= (-4 + 1)b + 9$	Distributive Property
$= -3b + 9$	addition

b. $7z - 5(3 + z)$

Step	Reason
$7z - 5(3 + z) = 7z - 15 - 5z$	Distributive Property
$= 7z + (-15) + (-5z)$	definition of subtraction
$= 7z + (-5z) + (-15)$	Commutative Property of Addition
$= [7 + (-5)]z + (-15)$	Distributive Property
$= 2z + (-15)$	addition
$= 2z - 15$	definition of subtraction

> **Problem Solving Hint**
>
> To use the Commutative Property of Addition, write subtraction as addition of the opposite.

 **Quick Check** ❸ Simplify each expression. Justify each step.
a. $5a + 6 + a$ **b.** $2(3t - 1) + 2$

EXERCISES

For more exercises, see *Extra Skill and Word Problem Practice.*

Practice and Problem Solving

 A **Practice by Example**

Example 1
(page 87)

 **GO for Help**

Name the property that each equation illustrates. Explain.

1. $-\frac{6}{7} + 0 = -\frac{6}{7}$ **2.** $8 + 43 = 43 + 8$ **3.** $1 \cdot \frac{21}{23} = \frac{21}{23}$

4. $(-7 + 4) + 1 = -7 + (4 + 1)$ **5.** $-0.3 + 0.3 = 0$

6. $9(7.3) = 7.3(9)$ **7.** $5(12 - 4) = 5(12) - 5(4)$

8. $8(9 \cdot 11) = (8 \cdot 9) \cdot 11$ **9.** $-0.5 \cdot (-2) = 1$

Example 2
(page 87)

Mental Math Simplify each expression.

10. $47 + 39 + 3 + 11$ **11.** $25 \cdot 74 \cdot 2 \cdot 2$ **12.** $4.75 + 2.95 + 1.25 + 6$

13. $10 \cdot 6 \cdot 7 \cdot 10$ **14.** $2(5 - 3.5) - 8$ **15.** $6\frac{1}{2} + 4\frac{1}{3} + 1\frac{1}{2} + \frac{2}{3}$

🌐 **16. Shopping** You buy 3 grapefruits for $1.50, a pound of apples for $.79, some grapes for $2.50, and some bananas for $1.21. Find the total cost of the fruit.

Example 3
(page 88)

Give a reason to justify each step.

17. **a.** $3y - 5y = 3y + (-5y)$ ___?___
   b. $= [3 + (-5)]y$ ___?___
   c. $= -2y$ ___?___

18. **a.** $3 \cdot (12 \cdot 10) = 3 \cdot (10 \cdot 12)$ ___?___
   b. $= (3 \cdot 10) \cdot 12$ ___?___
   c. $= 30 \cdot 12$ ___?___
   d. $= 360$ ___?___

Simplify each expression. Justify each step.

19. $25 \cdot 1.7 \cdot 4$

20. $-5(7y)$

21. $8 + 9m + 7$

22. $12x - 3 + 6x$

23. $29c + (-29c)$

24. $43\left(\frac{1}{43}\right) + 1$

B **Apply Your Skills**

Simplify each expression. Justify each step.

25. $2 + g\left(\frac{1}{g}\right)$

26. $36jkm - 36mjk$

27. $(3^2 - 2^3)(8759)$

28. $(7^6 - 6^5)(8 - 8)$

29. $4 + 6(8 - 3m)$

30. $5\left(w - \frac{1}{5}\right) - w(9)$

31. Shopping Suppose you are buying soccer equipment: a pair of cleats for $31.50, a soccer ball for $14.97, and shin guards for $6.50. Use mental math to find the total cost.

Tell whether the expressions in each pair are equivalent.

32. $6m + 1$ and $1 \cdot 6 + m$

33. $9y$ and $9 + y$

34. $mp + nq$ and $mq + np$

35. $-(5 - 9)$ and $9 - 5$

36. $8 - 4c$ and $4c - 8$

37. $3(5 + z)$ and $15 + z$

38. $6t - 4$ and $2[(2 + 1)t - 2]$

39. $vwx \cdot yz$ and $v \cdot w \cdot zxy$

GO for Help

For a guide to solving Exercise 38, see p. 91.

Reasoning **Explain your answer to each question.**

40. Is subtraction commutative?

41. Is subtraction associative?

42. Is division commutative?

43. Is division associative?

44. Give a reason to justify each step.
   a. $5t + 6 + 3(t + 2) = 5t + 6 + 3t + 6$ ___?___
   b. $= 5t + 3t + 6 + 6$ ___?___
   c. $= 5t + 3t + (6 + 6)$ ___?___
   d. $= 5t + 3t + 12$ ___?___
   e. $= (5 + 3)t + 12$ ___?___
   f. $= 8t + 12$ ___?___

45. Reasoning The Distributive Property states that $a(b + c) = ab + ac$. Use this and the other properties to explain why $(b + c)a = ba + ca$ is also true.

46. Multiple Choice A member of the track team records the distance she runs each day. How many miles did she run?
   Ⓐ 14.2 miles   Ⓑ 14.1 miles
   Ⓒ 13.9 miles   Ⓓ 13.8 miles

Day	Distance (miles)
Sunday	4.2
Monday	0
Tuesday	1.5
Wednesday	1.8
Thursday	2.5
Friday	0
Saturday	3.9

47. Writing Suppose you make a peanut butter and jelly sandwich. Do you think it tastes the same regardless of whether the jelly is on top or the peanut butter is on top? Relate your answer to one of the properties of real numbers.

GO Online
Homework Video Tutor
Visit: PHSchool.com
Web Code: ate-0205

Challenge **Critical Thinking** The *closure properties* for a set of numbers assure that the sum and product of two numbers in a given set of numbers are also in the set of numbers. Determine whether each set of numbers is closed for addition, for multiplication, or for both. If not, give a counterexample.

48. rational numbers   **49.** integers   **50.** whole numbers

51. negative numbers   **52.** odd numbers   **53.** even numbers

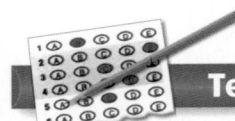

NY REGENTS

Test Prep

Multiple Choice

54. Simplify $-4 + 17 - 29 + 4 + 29 - 3$.
  A. 14   **B.** 20   **C.** 22   **D.** 72

55. Which of the following has the same result as dividing a number by $\frac{5}{2}$ and then multiplying by $\frac{1}{2}$?
  F. multiplying by 2   **G.** dividing by 2
  H. multiplying by 5   **J.** dividing by 5

56. In the formula $A = \pi r^2$, if the value of r is doubled, then what is the value of A multiplied by?
  A. $\frac{1}{4}$ **B.** $\frac{1}{2}$ **C.** 2   **D.** 4

57. The variable a is an integer. Which of the following could NOT equal a^3?
  F. -27 **G.** -16 **H.** 8   **J.** 64

58. What is the total cost if you buy 3 goldfish for $1.90 each, 3 angelfish for $6.10 each, and 12 neon tetras for $1.53 each?
  A. $39.36   **B.** $42.36   **C.** $60.36   **D.** $66.36

59. Simplify $-2h - (5 - 3h)$.
  F. $h - 5$ **G.** $-5h - 5$ **H.** h **J.** $-4h$

Mixed Review

GO for Help

Lesson 2-4   **Simplify each expression.**

60. $5(1.2 + k)$ **61.** $\frac{1}{3}\left(33 - b\right)$ **62.** $-2.5(4p + 14)$

63. $4(7 - n)$ **64.** $-(-7.4m + 0.05)$ **65.** $(3v - 5.2)(-6)$

Lesson 2-3   **Write an expression for each phrase.**

66. 7 plus the sum of m and -17

67. 8 minus the quantity 9 minus t

68. one half of the quotient of b and 4

69. one third of the sum of x and 5.1

Lesson 2-1   **Simplify.**

70. $12 + (-5)$ **71.** $9.2 + (-27.5)$ **72.** $\frac{5}{8} + \left(-\frac{7}{8}\right)$

73. $-4\frac{6}{10} + 3\frac{2}{5}$ **74.** $-11 + (-124)$ **75.** $|-2.4| + |6.8|$

Understanding Math Problems Read through the problem below. Let Franklin's thinking guide you through the solution. Check your understanding with the exercises at the bottom of the page.

Tell whether the expressions $6t - 4$ and $2[(2 + 1)t - 2]$ are equivalent.

What Franklin Thinks	What Franklin Writes
My first step is to make sure that I understand what the problem is about. Two expressions are equivalent if they simplify to the same expression.	$6t - 4 \overset{?}{=} 2[(2 + 1)t - 2]$
I'm going to look at the expression $2[(2 + 1)t - 2]$. If I can simplify this expression to $6t - 4$, then the expressions are equivalent.	$2[(2 + 1)t - 2]$
This expression looks complicated, but I know the Order of Operations. So I'll begin by working inside the innermost grouping symbols.	$2[(3)t - 2]$
I will write $(3)t$ without the parentheses.	$2[3t - 2]$
I will use the Distributive Property to simplify completely.	$6t - 4$
Now I need to write my answer.	Yes, $6t - 4$ and $2[(2 + 1)t - 2]$ are equivalent because $2[(2 + 1)t - 2]$ simplifies to $6t - 4$.

EXERCISES

Tell whether the expressions in each pair are equivalent.

1. $-(y - 4)$ and $4 - y$

2. $8 + 5m$ and $8 + 5 + m$

3. $2(t - 7)$ and $7(t - 2)$

4. $4(2b - 3)$ and $2[(2 - 1)b - 2]$

5. $3[2(a - 6)] - 4a$ and $2(a - 18)$

6. $-(b + 7) + 3(b - 7)$ and $2b$

7. $5(-8 + 3p) - 2[2(-4p + 3)]$ and $31p - 18$

8. $-2(m - 8) - 3[2(m + 5)]$ and $-2(4m + 7)$

Understanding Probability

You can use number cubes to investigate probability concepts.

NY A.S.19: Determine the number of elements in a sample space and the number of favorable events.

1 ACTIVITY

When you toss a number cube, there are a variety of possible results. You can find the probabilities of these results by first listing all the possible results.

1. List all the possible results of tossing one number cube.

2. How many possible results are there?

3. List all the possible results that are divisible by 3.

4. Use the ratio below to find the probability of tossing a number divisible by 3.

$$\frac{\text{number of possible results that are divisible by 3}}{\text{total number of possible results}}$$

2 ACTIVITY

The table shows all the possible results of tossing two number cubes.

5. How many results are there?

6. a. List the results that have a sum of 1.
   b. What is the probability of a sum of 1?

7. a. List the results that have a sum of 4.
   b. What is the probability of a sum of 4?

8. a. List the results that have a sum of 11.
   b. What is the probability of a sum of 11?

9. Find another sum that has the same probability of occurring as 11.

10. Reasoning Are the sums 2–12 equally likely? Explain.

11. Are the probabilities of getting an even sum or an odd sum equal?

Results for Two Number Cubes

(1, 1)	(1, 2)	(1, 3)	(1, 4)	(1, 5)	(1, 6)
(2, 1)	(2, 2)	(2, 3)	(2, 4)	(2, 5)	(2, 6)
(3, 1)	(3, 2)	(3, 3)	(3, 4)	(3, 5)	(3, 6)
(4, 1)	(4, 2)	(4, 3)	(4, 4)	(4, 5)	(4, 6)
(5, 1)	(5, 2)	(5, 3)	(5, 4)	(5, 5)	(5, 6)
(6, 1)	(6, 2)	(6, 3)	(6, 4)	(6, 5)	(6, 6)

3 ACTIVITY

Assume you are tossing a pair of regular tetrahedrons with the numbers 1 through 4 printed on the faces, like those at the right. The same number appears at the bottom of each face, and this number is the result of the toss.

12. Construct a table that shows all the possible results.

13. How many results are there?

14. a. List the results that have a sum of 4.
   b. What is the probability of a sum of 4?

15. Critical Thinking Why is the probability of a sum of 4 with two regular tetrahedrons not the same as the probability of a sum of 4 with two cubes?

Theoretical and Experimental Probability

Learning Standards for Mathematics

A.S.20 Calculate the probability of an event and its complement.

A.S.21 Determine empirical probabilities based on specific sample data.

 **Check Skills You'll Need**

 **for Help** Skills Handbook page 762

Rewrite each decimal or fraction as a percent.

1. 0.32   **2.** 0.09   **3.** $\frac{45}{200}$ **4.** $\frac{9}{50}$

🔊 **New Vocabulary**   • probability   • outcome   • event   • sample space
                  • theoretical probability   • complement of an event
                  • odds   • experimental probability

---

1   Theoretical Probability

Vocabulary Tip

Read *P(event)* as "the probability of an event."

The **probability** of an event, or *P*(event), tells you how likely it is that something will occur. An **outcome** is the result of a single trial, like one roll of a number cube. The **sample space** is all of the possible outcomes. An **event** is any outcome or group of outcomes.

Here is how these terms apply to finding the probability of rolling an even number on a number cube.

event	sample space	favorable outcome
↓	↓	↓
rolling an even number	1, 2, 3, 4, 5, 6	2, 4, 6

The possible outcomes of rolling a fair number cube are *equally likely* to occur. When all possible outcomes are equally likely, you can find the theoretical probability of an event using the following formula.

theoretical probability $P(\text{event}) = \dfrac{\text{number of favorable outcomes}}{\text{number of possible outcomes}}$

$$P(\text{rolling an even number}) = \frac{3}{6} = \frac{1}{2}$$

You can write the probability of an event as a fraction, a decimal, or a percent. The probability of an event ranges from 0 to 1.

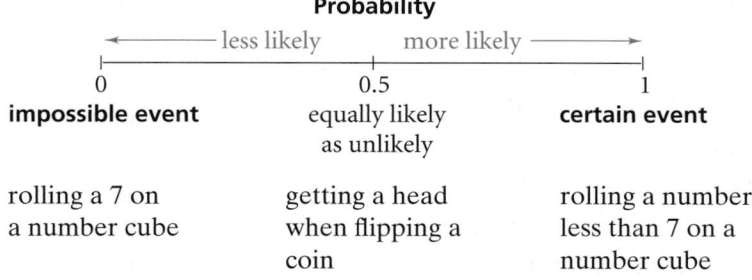

Probability

←——— less likely   more likely ———→

0   0.5   1

impossible event   equally likely   **certain event**
                   as unlikely

rolling a 7 on   getting a head   rolling a number
a number cube   when flipping a   less than 7 on a
                  coin   number cube

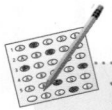

Test-Taking Tip

Ask yourself if your answer makes sense. If you find a probability greater than 1, then you have made an error.

1 EXAMPLE Finding Theoretical Probability

A bowl contains 12 slips of paper, each with a different name of a month. Find the theoretical probability that a slip selected at random from the bowl has a name of a month that starts with the letter J.

$$P(\text{event}) = \frac{\text{number of favorable outcomes}}{\text{number of possible outcomes}}$$

$$= \frac{3}{12} \quad \textbf{There are 3 months out of 12 that begin with the letter J:} \\ \textbf{January, June, and July.}$$

$$= \frac{1}{4} \quad \textbf{Simplify.}$$

The probability of picking a month that begins with the letter J is $\frac{1}{4}$.

✓ Quick Check **1** Suppose you write the names of days of the week on identical pieces of paper. Find the theoretical probability of picking a piece of paper at random that has the name of a day that starts with the letter T.

Vocabulary Tip

An event and its <u>complement</u> represent the *complete* set of possible outcomes.

The **complement of an event** consists of all the outcomes not in the event.

possible outcomes for rolling a number cube	outcomes for rolling an even number	complement of rolling an even number
↓	↓	↓
1, 2, 3, 4, 5, 6	2, 4, 6	1, 3, 5

The sum of the probabilities of an event and its complement is 1.

$$P(\text{event}) + P(\text{not event}) = 1$$

or

$$P(\text{not event}) = 1 - P(\text{event})$$

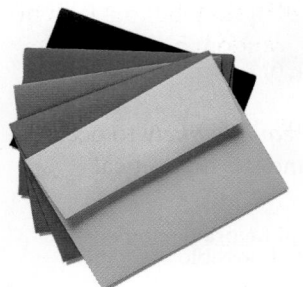

2 EXAMPLE Finding the Complement of an Event

Games On a popular television game show, a contestant must choose one of five envelopes. One envelope contains the grand prize, a car. Find the probability of not choosing the car.

$$P(\text{car}) = \frac{\text{number of favorable outcomes}}{\text{number of possible outcomes}} = \frac{1}{5}$$

$$P(\text{not choosing the car}) = 1 - P(\text{car}) \quad \textbf{Use the complement formula.}$$

$$= 1 - \frac{1}{5} = \frac{4}{5} \quad \textbf{Simplify.}$$

The probability of not choosing the car is $\frac{4}{5}$.

✓ Quick Check **2** **Critical Thinking** In Example 2, what happens to $P(\text{not choosing the car})$ as the number of envelopes increases?

Odds describes the likelihood of an event by comparing favorable and unfavorable outcomes.

odds in favor of an event → number of *favorable*   to   number of *unfavorable*
                                          outcomes   outcomes

odds against an event → number of *unfavorable*   to   number of *favorable*
                                          outcomes   outcomes

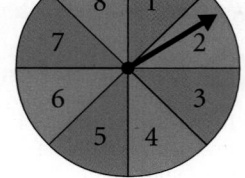

3 EXAMPLE Finding Odds

Find the odds in favor of the spinner landing on a number greater than or equal to 6.

Favorable outcomes: 6, 7, 8
Unfavorable outcomes: 1, 2, 3, 4, 5

● The odds are 3 : 5 in favor of the event.

✓ Quick Check ③ Find the odds against the spinner landing on a number less than 3.

2 Experimental Probability

Probability based on data collected from repeated trials is experimental probability.

> **experimental probability** $P(\text{event}) = \dfrac{\text{number of times an event occurs}}{\text{number of times the experiment is done}}$

4 EXAMPLE Finding Experimental Probability

Quality Control After receiving complaints, a skateboard manufacturer inspected 1000 skateboards at random. The manufacturer found no defects in 992 skateboards. What is the probability that a skateboard selected at random had no defects? Write the probability as a percent.

$P(\text{no defects}) = \dfrac{\text{number of times an event occurs}}{\text{number of times the experiment is done}}$

$\qquad\qquad\quad = \dfrac{992}{1000}$ **Substitute.**

$\qquad\qquad\quad = 0.992$ or 99.2% **Simplify and write as a percent.**

● The probability that a skateboard has no defects is 99.2%.

Real-World ● Connection

More than 100,000 skateboards are manufactured each month.

✓ Quick Check ④ The manufacturer inspects 2500 skateboards. There are 2450 skateboards with no defects. Find the probability that a skateboard selected at random has no defects.

You can use experimental probability to make a prediction. Predictions are not exact, so round your results.

●nline active math

For: Probability Activity
Use: Interactive Textbook, 2-6

5 EXAMPLE Using Experimental Probability

Quality Control The same manufacturer has 8976 skateboards in its warehouse. If the probability that a skateboard has no defect is 99.2%, predict how many skateboards are likely to have no defect.

number with no defects $= P(\text{no defects}) \cdot$ number of skateboards

$\qquad\qquad\qquad\qquad\quad = 0.992 \times 8976$ **Substitute. Use 0.992 for 99.2%.**

$\qquad\qquad\qquad\qquad\quad = 8904.192$ **Simplify.**

● Approximately 8900 boards are likely to have no defect.

✓ Quick Check ⑤ A manufacturer inspects 700 light bulbs. She finds that the probability that a light bulb works is 99.6%. There are 35,400 light bulbs in the warehouse. Predict how many light bulbs are likely to work.

EXERCISES

For more exercises, see *Extra Skill and Word Problem Practice.*

Practice and Problem Solving

A Practice by Example

Example 1
(page 94)

Example 2
(page 94)

For Exercises 1–13, use the spinner at the right. Find the theoretical probability of landing on the given section(s) of the spinner.

1. *P*(purple) **2.** *P*(green) **3.** *P*(5)

4. *P*(even) **5.** *P*(purple or white) **6.** *P*(8)

7. *P*(greater than 4) **8.** *P*(even or odd) **9.** *P*(1 or 6)

10. *P*(not white) **11.** *P*(not 2) **12.** *P*(not purple) **13.** *P*(not 8)

14. Suppose the probability that you will be picked for a committee at school is 20%. What is the probability that you will not be picked?

Example 3
(page 95)

For Exercises 15–20, use the spinner at the right above. Find each odds.

15. odds in favor of green **16.** odds against a factor of 6

17. odds against white **18.** odds in favor of purple

19. odds in favor of a multiple of 5 **20.** odds against 2

Examples 4, 5
(page 95)

The results of a survey of 100 randomly selected students at a 2000-student high school are below. Find the experimental probability that a student selected at random makes the given response.

21. *P*(community college)

22. *P*(4-year college)

23. *P*(trade school)

24. *P*(not trade school)

25. *P*(trade school or community college)

26. *P*(community or 4-year college)

Plans for After Graduation

Response	Number of Respondents
Go to community college	24
Go to 4-year college	43
Take a year off before college	12
Go to trade school	15
Do not plan to go to college	6

27. A forest contains about 500 trees. You randomly pick 67 trees and find that 27 of them are oaks.
a. What is the experimental probability that a tree in the forest is an oak?
b. Predict how many oak trees there are in the forest.

28. Suppose 12 out of 30 families on your street have a cat or a dog as a pet.
a. What is the experimental probability that a randomly selected family in your neighborhood will have a cat or a dog as a pet?
b. Based on *P*(cat or dog) from part (a), predict how many cat- or dog-owning families you can expect among 57 families in your neighborhood.

B Apply Your Skills

Suppose you roll a number cube. Find each probability.

29. *P*(5) **30.** *P*(7) **31.** *P*(3 or 4) **32.** *P*(not 5)

Real-World 🌐 **Connection**

Each day about 10,700 people celebrate their sixteenth birthday.

Suppose you select a 3-digit number at random from the set of all positive 3-digit numbers. Find each probability or odds. (*Hint:* First find how many positive 3-digit numbers there are.)

33. P(odd number)

34. P(number less than 900)

35. P(243 or 244)

36. P(number less than 100)

37. P(number is a multiple of 30)

38. P(number less than 500)

39. odds in favor of a multiple of 10

40. odds against 243 or 244

41. odds against a number greater than 400

🌐 **42. Birthdays** Each day in the United States, about 753,000 people have a birthday. Use the information at the left to find the probability that someone celebrating a birthday today will turn 16. Round to the nearest percent.

🌐 **43. Multiple Choice** The United States has a land area of about 3,536,278 mi². Illinois has a land area of about 57,918 mi². What is the probability that a location in the United States chosen at random is in Illinois?

Ⓐ 0.016%   Ⓑ 0.61%   Ⓒ 1.6%   Ⓓ 61%

44. Open-Ended Suppose your teacher chooses a student at random from your algebra class.
a. What is the probability that you are selected?
b. What is the probability that a boy is not selected?

🌐 **45. Data Analysis** The population of the United States is about 281,000,000. Use the circle graph at the right to answer the following questions. Round to the nearest percent.
a. What is the probability of selecting at random a person whose age is between 10 and 19?
b. What is the probability of selecting at random a person whose age is between 40 and 49?

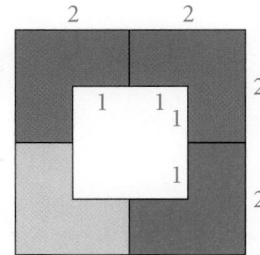

U.S. Population (millions)

1.4 — 7.7 — 16.2 — 20.3 — 31 — 42.5 — 43.2 — 38.3 — 40.7 — 39.7

Age
☐ 0–9
☐ 10–19
☐ 20–29
☐ 30–39
☐ 40–49
☐ 50–59
☐ 60–69
☐ 70–79
☐ 80–89
☐ 90+

SOURCE: U.S. Census Bureau

✏️ **46. Writing** Explain the difference between theoretical probability and experimental probability.

🔲 **Geometry** Use the diagram at the right. Assume that the white square is centered in the large square. If you choose a point inside the figure, what is the probability that it will be in each shaded region?

47. P(red)   **48.** P(purple)

49. P(yellow or white)   **50.** P(not purple)

For Exercises 51–56, use the spinner at the left.

51. odds in favor of green   **52.** odds against white

53. odds against an even number   **54.** odds against a factor of 4

55. odds in favor of green or white **56.** odds in favor of 6 or 7

57. Critical Thinking Explain how you can use odds to find probability. Include an example.

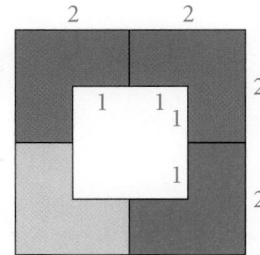

2   2
2
1 1
1
1
2

🖥️ **Homework Video Tutor**
Visit: PHSchool.com
Web Code: ate-0206

Real-World  Connection

The greatest total score of both teams in a Super Bowl was 75, in 1995.

Data Analysis The stem-and-leaf plot at the right shows the difference between the points scored by the winning and losing teams in the Super Bowl from 1981 to 2000.

58. Find the probability that the winning team won by less than 10 points.

59. Find the odds that the winning team won by 10 to 15 points.

60. Find the probability that the winning team won by more than 20 points.

Difference Between Winning and Losing Super Bowl Scores

0	1 4 5 7 7
1	0 0 3 4 5 7 7 9
2	2 3 9
3	2 5 6
4	5

Key: 1|0 means 10 points

61. Data Collection Roll a pair of number cubes and record the product of the results. Repeat 30 times.
  a. Use your data to find the following experimental probabilities: $P(5), P(6), P(12), P(36)$.
  b. Make a chart to find the sample space for the product of two cubes.
  c. Find the theoretical probabilities for $P(5), P(6), P(12), P(36)$.
  d. Are your experimental probabilities exactly the same as the theoretical probabilities?
  e. Suppose you roll the number cubes 1000 times. Would you expect the experimental probabilities and the theoretical probabilities to be about the same? Explain.

NY REGENTS

Test Prep

Multiple Choice

62. Suppose you roll a number cube with the numbers 1–6 on it. Which has the same probability as P(1 or prime)?
  A. P(factor of 6)
  B. P(1 or 2)
  C. P(less than 3)
  D. P(not odd)

63. Of 150 widgets inspected, 142 passed inspection. Out of 2855 widgets, about how many would you predict would fail an inspection?
  F. 140   **G.** 150   **H.** 2700   **J.** 2850

64. There are 28 right-handed students in a class of 31. What is the probability that a student chosen at random will be left-handed?
  A. $\frac{31}{28}$ **B.** $\frac{28}{31}$ **C.** $\frac{3}{28}$ **D.** $\frac{3}{31}$

65. A "Bag-of-Beads" for crafts contains five different colors of beads. The company that makes them claims that each bag contains the same number of each color.

Rosheeda purchased a bag to test the company's claim. She recorded the number of each color. Based on the results in the table to the right, what is the experimental probability of picking a red bead out of the bag?

Color	Number
red	55
blue	53
green	64
yellow	47
purple	61

  F. $\frac{9}{56}$ **G.** $\frac{11}{56}$
  H. $\frac{11}{45}$ **J.** $\frac{9}{45}$

66. An inspector for an office-supply company checked a batch of 360 staplers. He found that 18 of them were defective. What is the experimental probability of getting a defective stapler in this batch?

A. 18%   B. 5%   C. 0.5%   D. 0.05%

Mixed Review

Lessons 2-4 and 2-5

GO for Help

Name the property that each equation illustrates.

67. $-6g + 10 + g = -6g + 10 + 1g$

68. $-6g + 10 + 1g = -6g + 1g + 10$

69. $-6g + 1g + 10 = (-6 + 1)g + 10$

70. $-4s - 8 + 3(s + 5) = -4s - 8 + 3s + 15$

71. $-4s - 8 + 3s + 15 = -4s + 3s + (-8) + 15$

72. $-4s + 3s + (-8) + 15 = (-4 + 3)s + (-8) + 15$

Lesson 1-6

Find the mean, median, and mode of each set of numbers.

73. 3 4 5 5 8 11

74. 1 8 9 11 22 34 34

Make a stem-and-leaf plot for each set of data.

75. 34 37 39 41 49 65 71

76. 12 14 16 23 27 47 68 79

✓ Checkpoint Quiz 2   Lessons 2-5 through 2-6

Give a reason to justify each step.

1. $-7d + 3 + 5(d + 2) = -7d + 3 + 5d + 10$?

2. $= -7d + 5d + 3 + 10$?

3. $= -7d + 5d + (3 + 10)$?

4. $= -7d + 5d + 13$?

5. $= (-7 + 5)d + 13$?

6. $= -2d + 13$?

Use the spinner at the left.

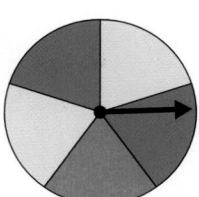

7. Find $P(\text{not yellow})$ 8. Find the odds against yellow.

9. Suppose you write the days of the week on identical pieces of paper. You mix them in a bowl and choose one at random. What is the probability that the day you select will have the letter E in it?

10. A manufacturer inspects 100 bicycles at random. She finds that 98 of them have no defects. There are 2400 bicycles in the warehouse. Predict how many bicycles are likely to be free of defects.

Activity Lab

Conducting a Simulation

FOR USE WITH LESSON 2-7

Go Online
PHSchool.com
For: Graphing calculator procedures
Web Code: ate-2115

A *simulation* is a model of a real-life situation. One way to do a simulation is to use random numbers generated by a graphing calculator or a computer program.

On a graphing calculator, the command randInt generates random integers. To create a list of random integers, press MATH ◄ 5. You will see randInt(. After the parenthesis, press 0 ⟩ 99, and press ENTER repeatedly to create 1- and 2-digit random numbers.

NY **A.S.21:** Determine empirical probabilities based on specific sample data.

ACTIVITY

About 40% of the people in the United States have type A blood. Find the experimental probability that the next two people who donate blood have type A blood.

To simulate this problem, use 2-digit numbers to represent groups of 2 people. Use your calculator to generate 40 random numbers.

25	71	47	46
66	13	63	36
01	59	27	07
83	25	72	24
73	52	59	81
14	09	40	64
81	72	02	38
21	09	92	10
93	34	36	45
53	18	23	75

Define how the simulation will be done.

Since about 40% of people have type A blood, let 40%, or 4 out of 10 digits, represent people in this group. Using numbers from the random number table, let 0, 1, 2, 3 represent people with type A blood. Let 4, 5, 6, 7, 8, 9 represent people who do not have type A blood.

Interpret the simulation.

The six numbers in red represent "the next two people have type A blood." Each of the other groups has at least one person with a different blood type.

P(next two people have type A blood) =

$$\frac{\text{number of times an event happens}}{\text{number of times the experiment is done}} = \frac{6}{40} = 0.15 = 15\%$$

● The probability that the next two people will have type A blood is about 15%.

EXERCISES

1. a. Blood Types In the United States, about 50% of people have type O blood. Design an experiment using spinners, coins, or number cubes to find the experimental probability that the next two donors have type O blood. Describe your simulation.
   b. Conduct your simulation and interpret your results.

2. A basketball player has made 12 or her last 18 free throws. Design and describe a simulation to find the experimental probability that she will make her next five free throws. Conduct your simulation and interpret your results.

3. a. Open-Ended Write an experimental probability problem that you could solve using a simulation.
   b. Design and describe your simulation.
   c. Conduct your simulation and interpret your results.

Probability of Compound Events

A.S.23 Calculate the probability of: a series of independent events; a series of dependent events; two mutually exclusive events; two events that are not mutually exclusive.

 Check Skills You'll Need

GO for Help Lesson 2-6

Find each probability for one roll of a number cube.

1. P(multiple of 3)

2. P(greater than 4)

3. P(greater than 5)

4. P(greater than 6)

Simplify.

5. $\frac{2}{14} \cdot \frac{7}{6}$

6. $\frac{15}{24} \cdot \frac{12}{30}$

7. $\frac{6}{55} \cdot \frac{44}{3}$

🔊 **New Vocabulary** • independent events • dependent events

---

1 **Finding the Probability of Independent Events**

Activity: Compound Events

Suppose you draw cards at random from the following collection.

| R | R | A | N | N | N | D | O | M | M |

1. You draw an R card and replace it. What is the probability that the next card you draw will be an R card?

2. You draw an R card and do *not* replace it. What is the probability that the next card you draw will be an R card?

3. Copy and complete each table.

Probability With Replacement		Probability Without Replacement	
First Card	**Second Card Matches**	**First Card**	**Second Card Matches**
$P(R) = $ ▪	$P(R) = $ ▪	$P(R) = $ ▪	$P(R) = $ ▪
$P(A) = $ ▪	$P(A) = $ ▪	$P(A) = $ ▪	$P(A) = $ ▪
$P(N) = $ ▪	$P(N) = $ ▪	$P(N) = $ ▪	$P(N) = $ ▪
$P(D) = $ ▪	$P(D) = $ ▪	$P(D) = $ ▪	$P(D) = $ ▪
$P(O) = $ ▪	$P(O) = $ ▪	$P(O) = $ ▪	$P(O) = $ ▪
$P(M) = $ ▪	$P(M) = $ ▪	$P(M) = $ ▪	$P(M) = $ ▪

4. For each letter, the probability of drawing the first card is the same with replacement and without replacement. Explain why the probability of drawing the second card is not the same.

The diagram at the right shows the results of randomly choosing a checker, putting it back, and choosing again. The probability of getting a red on either pick is $\frac{1}{2}$. The first pick, or first event, does not affect the second event. The events are independent.

Independent events are events that do not influence one another.

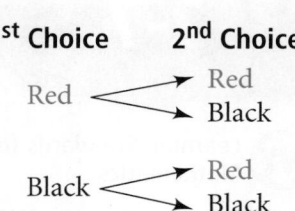

1st Choice   2nd Choice

Red   Red / Black

Black   Red / Black

 Key Concepts

Rule	Probability of Two Independent Events

If A and B are independent events, $P(A \text{ and } B) = P(A) \cdot P(B)$.

1 EXAMPLE   Independent Events

GO Online

Video Tutor Help
Visit: PHSchool.com
Web Code: ate-0775

Multiple Choice Suppose you roll a red number cube and a blue number cube. What is the probability that you will roll a 3 on the red cube and an even number on the blue cube?

   Ⓐ $\frac{1}{12}$ Ⓑ $\frac{1}{6}$ Ⓒ $\frac{1}{2}$ Ⓓ $\frac{2}{3}$

$P(\text{red } 3) = \frac{1}{6}$ **There is one way to get a 3 out of six numbers.**

$P(\text{blue even}) = \frac{3}{6} = \frac{1}{2}$ **There are three even numbers out of six numbers.**

$P(\text{red } 3 \text{ and blue even}) = P(\text{red } 3) \cdot P(\text{blue even})$

$\qquad\qquad\qquad\qquad = \frac{1}{6} \cdot \frac{1}{2}$ **Substitute.**

$\qquad\qquad\qquad\qquad = \frac{1}{12}$ **Simplify.**

The probability that you will roll a 3 on the red number cube and an even number on the blue cube is $\frac{1}{12}$. The answer is A.

✓ Quick Check ➊ Suppose you roll a red number cube and a blue number cube. What is the probability that you will roll a 5 on the red cube and a 1 or 2 on the blue cube?

2 EXAMPLE   Selecting With Replacement

Games In a word game, you choose a tile at random from a bag containing the letter tiles shown. You *replace* the first tile in the bag and then choose again. What is the probability that you will choose an A and then an E?

Since you replace the first tile, the events are independent.

$P(A) = \frac{4}{15}$ **There are 4 A's in the 15 tiles.**

$P(E) = \frac{3}{15}$ **There are 3 E's in the 15 tiles.**

$P(A \text{ and } E) = P(A) \cdot P(E)$

$\qquad\qquad = \frac{4}{15} \cdot \frac{3}{15}$ **Multiply.**

$\qquad\qquad = \frac{12}{225} = \frac{4}{75}$

The probability that you will choose an A and then an E is $\frac{4}{75}$.

✓ Quick Check ➋ Find the probability that you will choose at random a U and then an I after replacing the first tile.

2 Finding the Probability of Dependent Events

When you select one tile from a bag of 15 tiles and you do *not* replace it, there are only 14 tiles when you make your second selection. These events are dependent.

Dependent events are events that influence each other. The occurrence of one event affects the probability of a second event.

 Key Concepts

Rule	Probability of Two Dependent Events

If A and B are dependent events,

$$P(A \text{ then } B) = P(A) \cdot P(B \text{ after } A).$$

---

3 EXAMPLE   Selecting Without Replacement

Games Suppose you choose a tile at random from the letter tiles shown in Example 2. Without replacing the tile, you select a second tile. What is the probability that you will choose an A and then an E?

$P(\text{A}) = \frac{4}{15}$ **There are 4 A's in the 15 tiles.**

$P(\text{E after A}) = \frac{3}{14}$ **There are 3 E's in the 14 remaining tiles.**

$P(\text{A then E}) = P(\text{A}) \cdot P(\text{E after A})$

$\qquad = \frac{4}{15} \cdot \frac{3}{14}$ **Multiply.**

$\qquad = \frac{12}{210} = \frac{2}{35}$

The probability that you will choose an A and then an E is $\frac{2}{35}$.

Quick Check ❸ Find the probability that you will choose at random a U and then an O without replacing the first tile.

Test-Taking Tip

Before you take a test, make sure you understand the terms used in probability, such as *independent* and *dependent events*.

---

4 EXAMPLE   Real-World 🌎 Problem Solving

Selecting Representatives Suppose a teacher must select 2 high school students to represent their school at a conference. The teacher randomly picks names from a hat that contains the names of 3 freshmen, 2 sophomores, 4 juniors, and 4 seniors. What is the probability that a sophomore and then a freshman are chosen?

$P(\text{sophomore}) = \frac{2}{13}$ **There are 2 sophomores among 13 students.**

$P(\text{freshman after sophomore}) = \frac{3}{12}$ **There are 3 freshmen among the 12 remaining students.**

$P(\text{sophomore then freshman}) = P(\text{sophomore}) \cdot P(\text{freshman after sophomore})$

$\qquad = \frac{2}{13} \cdot \frac{3}{12}$ **Substitute.**

$\qquad = \frac{1}{26}$ **Simplify.**

The probability that the teacher chooses a sophomore and then a freshman is $\frac{1}{26}$.

Quick Check ❹ **a.** What is the probability that the teacher chooses a sophomore and then a junior?
b. What is the probability that the teacher chooses a junior and then a sophomore?
c. **Critical Thinking** Does the probability of choosing without replacement change if the order of the events is reversed? Explain.

EXERCISES

For more exercises, see *Extra Skill and Word Problem Practice*.

Practice and Problem Solving

A Practice by Example

Example 1
(page 102)

You roll a blue number cube and a green number cube. Find each probability.

1. P(blue 1 and green 1)

2. P(blue 1 and green 1 or 2)

3. P(blue 1 or 2 and green 1)

4. P(blue 1 or 2 and green 1 or 2)

5. P(blue even and green even)

6. P(blue and green both less than 6)

7. P(blue and green less than 7)

8. P(blue 6 and green 7)

Example 2
(page 102)

Suppose you choose a tile at random from a bag containing 2 A's, 3 B's, and 4 C's. You *replace* the first tile in the bag and then choose again. Find each probability.

9. P(A and A)

10. P(A and B)

11. P(B and B)

12. P(C and C)

13. P(B and C)

14. P(C and B)

Example 3
(page 103)

You select a card at random from those below. Without replacing the card, you choose a second card. Find each probability. Consider Y to be a vowel.

15. P(vowel then vowel)

16. P(consonant then consonant)

17. P(I then I)

18. P(consonant then vowel)

19. P(A then A)

20. P(letter then letter)

Example 4
(page 103)

21. Four girls and three boys volunteer to represent their class at a school assembly. The teacher selects one name and then another from a bag containing the seven students' names. What is the probability that both representatives will be girls?

22. A refrigerator contains 11 drinks: 5 lemon drinks, 3 apple drinks, and 3 orange drinks. Abby is first in line for drinks. Telly is second. What is the probability that Abby will get a lemon drink and Telly will get an orange drink, if they are given drinks at random?

B Apply Your Skills

You pick two marbles from the bag at the left. You pick the second one without replacing the first one. Find each probability.

23. P(red then blue)

24. P(two blues)

25. P(blue then green)

26. P(two reds)

27. P(green then yellow)

28. P(two greens)

Are the two events dependent or independent? Explain.

29. Toss a penny and a nickel.

30. Pick a name from a hat. Without replacement, pick a different name.

31. Pick a ball from a basket of both yellow and pink balls. Return the ball and pick again.

32. Writing Use your own words to explain the difference between independent and dependent events. Give an example of each.

Real-World Connection

There are roughly 430 million acres of cropland in the United States, about the same area as Texas, Arizona, California, and New Mexico combined.

33. a. Agriculture An acre of land in Indiana is chosen at random. What is the probability that it is cropland?

b. An acre of land is chosen at random from each of the three states listed. What is the probability that all three acres will be cropland?

Percent of Cropland	
Alabama	8%
Florida	7%
Indiana	58%

34. Open-Ended Find the number of left-handed students in your class. Suppose your teacher randomly picks two students to work on a problem at the board.

a. Find the probability that they are both left-handed.
b. Find the probability that they are both right-handed.
c. Find the probability that the first student is right-handed and the second student is left-handed.

35. Multiple Choice The probability that a new spark plug is defective is 0.06. You need two new spark plugs for a motorcycle. What is the probability that both spark plugs you buy are defective?

Ⓐ 0.06   Ⓑ 0.12   Ⓒ 0.0030   Ⓓ 0.0036

You have three $1 bills, two $5 bills, and a $20 bill in your pocket. You choose two bills without looking. Find each probability.

36. P($5 then $1) with replacing

37. P($1 then $20) without replacing

38. P($20 then $1) with replacing

39. P($1 then $1) without replacing

40. P($5 then $20) without replacing

41. P($20 then $5) with replacing

42. A class has 12 girls and 10 boys. A hat contains the names of all the students in the class. To select representatives of the class to attend a meeting, the teacher draws two names from the hat at random without replacing the first name.
a. Find P(two girls).   **b.** Find P(two boys).
c. Find P(boy then girl).   **d.** Find P(girl then boy).
e. Predict the sum of the probabilities in parts (a)–(d). Check to see that the sum agrees with your prediction.

43. Marketing The marketing director of a department store interviewed 50 customers who had bought appliances the previous week. They found that 20% of the customers bought a washing machine and 40% bought a dishwasher. Also 48% of the customers bought an appliance other than a washing machine or dishwasher.
a. What is the probability that a surveyed customer selected at random bought both a washing machine and a dishwasher?
b. Of the 50 customers surveyed, how many were likely to have bought both a washing machine and a dishwasher?

44. Surveys Manuel surveyed thirty students in a freshman English class. Twelve of the students take Biology, five students take Geometry, and fifteen students take neither Biology nor Geometry.
a. What is the probability that a student selected at random takes Biology? Geometry?
b. How many students take both Biology and Geometry?
c. What is the probability that a student selected at random takes both Biology and Geometry?
d. Suppose the classes taken by the students that Manuel surveyed are representative of the classes taken by the 249 freshman in the school. Predict the number of students who take both Biology and Geometry.

GO Online
Homework Video Tutor
Visit: PHSchool.com
Web Code: ate-0207

C **Challenge**

45. a. You take a five-question multiple-choice quiz. You guess on all of the questions, selecting one of five answers randomly each time. What is the probability you will get a perfect score?
b. What is the probability that you would get a perfect score if there were six questions on the quiz?
c. Find the ratio of P(perfect score on five-question quiz) to P(perfect score on six-question quiz).

46. Suppose you roll a red number cube and a yellow number cube.
a. Find P(red 1 and yellow 1).
b. Find P(red 2 and yellow 2).
c. Find the probability of rolling any matching pair of numbers. (*Hint:* Add the probabilities of each of the six matches.)

47. A two-digit number is formed by randomly selecting from the digits 1, 2, 3, and 5 without replacement.
a. How many different two-digit numbers can be formed?
b. What is the probability that a two-digit number contains a 2 or a 5?
c. What is the probability that a two-digit number is prime?

> **Problem Solving Hint**
> For Exercise 47, making an organized list can help you understand the problem.

Test Prep

Multiple Choice

48. You roll a pair of number cubes. What is the probability of getting even numbers on both cubes?
A. 1   **B.** $\frac{1}{2}$ **C.** $\frac{1}{4}$ **D.** $\frac{1}{6}$

49. You take a three-question true or false quiz. You guess on all the questions. What is the probability that you will get a perfect score?
F. $\frac{1}{8}$ **G.** $\frac{3}{8}$ **H.** $\frac{1}{2}$ **J.** $\frac{3}{2}$

50. A standard domino set has 28 dominoes. Seven of these are called "doubles" since they have the same number on both ends or are blank on both ends. The first and second player each take a domino at random. What is the probability that they will both draw a double?
A. $\frac{1}{16}$ **B.** $\frac{1}{18}$ **C.** $\frac{3}{56}$ **D.** $\frac{13}{756}$

51. You have a bag containing 3 green marbles, 4 red marbles, and 2 yellow marbles. You select 1 marble randomly. What are the odds *against* selecting a green or yellow marble?
F. 5 : 4   **G.** 9 : 4   **H.** 4 : 9   **J.** 4 : 5

Mixed Review

Lesson 2-6

Suppose you select a two-digit number at random from 10 to 30 (including 10 and 30). Find each probability.

52. P(number is even)   **53.** P(number is a multiple of 6)

54. P(number is prime)   **55.** P(number is less than 18)

Lesson 2-3

Multiply or divide.

56. $-5(-3)$ **57.** $3(-21)$ **58.** $8 \div (-4)$ **59.** $12(-5)$

60. $5\left(\frac{3}{10}\right)$ **61.** $-2 \div \left(\frac{5}{8}\right)$ **62.** $-4\left(-\frac{1}{8}\right)$ **63.** $\frac{5}{8}\left(-\frac{1}{15}\right)$

Mutually Exclusive Events

When two events cannot happen at the same time, the events are **mutually exclusive events**. If A and B are mutually exclusive events, then $P(A \text{ and } B) = 0$. You need to determine whether events A and B are mutually exclusive before you can find the probabilty of $(A \text{ or } B)$.

> If A and B are mutually exclusive events, then
> $P(A \text{ or } B) = P(A) + P(B)$.
>
> If A and B are not mutually exclusive events, then
> $P(A \text{ or } B) = P(A) + P(B) - P(A \text{ and } B)$.

NY A.S.21: Determine empirical probabilities based on specific sample data.

1 EXAMPLE Mutually Exclusive Events

Are the events mutually exclusive? Explain.

Rolling an odd number and rolling a number greater than 5 on a number cube.

● Since you cannot roll an odd number and a number greater than 5, at the same time, the events are mutually exclusive.

2 EXAMPLE Probability of Events

A spinner has twenty equal-size sections numbered from 1 to 20. If you spin the spinner, what is the probabilty that the number you spin will be a multiple of 2 or a multiple of 3?

Since a number can be a multiple of 2 and a multiple of 3, the events are not mutually exclusive. Use the $P(A \text{ or } B)$ formula for events that are not mutually exclusive.

$P(\text{multiple of 2 or multiple of 3}) = P(\text{multiple of 2}) + P(\text{multiple of 3}) - P(\text{multiple of 2 and 3})$

$$= \frac{10}{20} + \frac{6}{20} - \frac{3}{20}$$

$$= \frac{13}{20}$$

● The probability of the number you spin being a multiple of 2 and a multiple of 3 is $\frac{13}{20}$.

EXERCISES

A standard number cube is tossed. Find each probabilty.

1. $P(3 \text{ or odd})$ **2.** $P(\text{odd or greater than 2})$ **3.** $P(4 \text{ or even})$

4. $P(\text{odd or prime})$ **5.** $P(\text{even or less than 4})$ **6.** $P(4 \text{ or less than 6})$

Writing Short Responses

Short-response questions in this textbook are usually worth a maximum of 2 points. To get full credit you need to give the correct answer (including appropriate units, if applicable) and justify your reasoning or show your work.

EXAMPLE

The cost for using a phone card is 35 cents per call plus 25 cents per minute. Write an expression for the cost of a call that is n minutes long. Evaluate your expression for an eight-minute call. Show your work.

Below is a rubric that shows the number of points awarded for different types of answers.

Scoring Rubric

[2]   The expression and the cost are correct and work is shown.

[1]   There is no expression, but there is a method to show how the correct cost is found.

[1]   There is a correct expression but there is an error in finding the cost.

[0]   There is no response, it is completely incorrect, or it is a correct response, but no procedure is shown.

Three responses are below with the points each received.

2 points	1 point	0 points
For n minutes, the cost is $0.35 + 0.25n$ $0.35 + 0.25(8)$ $0.35 + 2.00$ An eight-minute call costs $2.35.	$.25 \times 8 = \$2.00$ $\$2.00 + \$.35 = \$2.35$	$2.35

EXERCISES

Use the rubric above to answer each question.

1. Explain why each response above received the indicated points.

2. Write a 1-point response that begins with a correct expression.

3. Write a 2-point response that includes the expression $35 + 25n$.

4. **Error Analysis** Suppose a student used the expression $25 + 35x$. Explain why this expression is incorrect.

Chapter Review

Vocabulary Review

additive inverse (p. 56)
coefficient (p. 81)
complement of an event (p. 94)
constant (p. 81)
deductive reasoning (p. 88)
dependent events (p. 103)
Distributive Property (p. 79)

element (p. 59)
event (p. 93)
experimental probability (p. 95)
independent events (p. 102)
like terms (p. 81)
matrix (p. 59)
multiplicative inverse (p. 73)

odds (p. 94)
outcome (p. 93)
probability (p. 93)
reciprocal (p. 73)
sample space (p. 93)
term (p. 81)
theoretical probability (p. 93)

Choose the correct term to complete each sentence.

1. A (*constant, term*) is a number, variable, or the product of a number and one or more variables.

2. A (*sample space, matrix*) is a rectangular arrangement of numbers in rows and columns.

3. Dividing by a nonzero number is the same as multiplying by its (*complement, reciprocal*).

4. Use the (*Distributive Property, Identity Property of Multiplication*) to write the expression $-4(2x - 3)$ as $-8x + 12$.

5. To change the order of addends in a sum use the (*Associative, Commutative*) Property of Addition.

6. The result of a single trial, such as one toss of a coin, is (*an outcome, a sample space*).

7. (*Theoretical probability, Complement of an event*) is all the outcomes not in an event.

8. You select a card, replace it, then select another card. The events are (*dependent, independent*).

For: Vocabulary quiz
Web Code: atj-0251

9. The (*probability, sample space*) for rolling a number cube is 1, 2, 3, 4, 5, and 6.

10. The (*reciprocal, additive inverse*) of a number is its opposite.

Skills and Concepts

2-1 and 2-2 Objectives

▼ To add rational numbers using models and rules (p. 56)

▼ To apply addition (p. 58)

▼ To subtract rational numbers (p. 64)

▼ To apply subtraction (p. 65)

To add rational numbers with the same sign, add their absolute values. The sum has the same sign as the rational numbers. To add rational numbers with different signs, find the difference of their absolute values. The sum has the sign of the rational number with the greater absolute value. To subtract a rational number, add its opposite.

Simplify each expression.

11. $(-13) + (-4)$ **12.** $-12 - (-7)$ **13.** $-12.4 + 22.3$

14. $|54.3 - 29.4|$ **15.** $5 - 17$ **16.** $-3^2 + (-3)^2$

2-3 Objectives

▼ To multiply rational numbers (p. 69)

▼ To divide rational numbers (p. 72)

To multiply or divide rational numbers, multiply or divide the absolute values of the numbers. If the numbers have the same sign, the product or quotient is positive. If the numbers have different signs, the product or quotient is negative.

Simplify each expression.

17. $4 - 3(-2)$

18. $5(4)(-2)$

19. $\left(\frac{5}{6}\right)\left(-\frac{2}{3}\right)$

20. $\frac{4 - (-2)}{3}$

21. $\frac{5}{6} \div \left(-\frac{2}{3}\right)$

22. $\frac{5}{6} \div \left(-\frac{2}{3}\right)^2$

2-4 Objectives

▼ To use the Distributive Property (p. 79)

▼ To simplify algebraic expressions (p. 80)

Terms with exactly the same variable factors are **like terms**. You can combine like terms and use the distributive property to simplify expressions.

Distributive Property   For all real numbers a, b, and c, $a(b + c) = ab + ac$ and $a(b - c) = ab - ac$.

Simplify each expression.

23. $9m - 5m + 3$

24. $2b + 8 - b + 2$

25. $-5(w - 4)$

26. $9(4 - 3j)$

27. $-(3 - 10y)$

28. $-2\left(r - \frac{1}{2}\right)$

29. $(7b + 1)(5)$

30. $7 - 16v - 9v$

31. $\frac{3}{5}(15t - 2)$

32. $(6 - 3m)(-3)$

33. $-(4 - x)$

34. $0.5(20g + 3)$

2-5 Objectives

▼ To identify properties (p. 86)

▼ To use deductive reasoning (p. 88)

Use properties of numbers to simplify expressions. Use the **Commutative Property** to change order. Use the **Associative Property** to change grouping.

Which property does each equation illustrate?

35. $62 + 15 + 38 = 62 + (15 + 38)$

36. $62 + 0 + (15 + 38) = 62 + (15 + 38)$

37. $50 \cdot 17 \cdot 2 = 50 \cdot 2 \cdot 17$

38. $9(2^3 - 4^2) = 9(2^3) - 9(4^2)$

Simplify each expression. Justify each step.

39. $19 + 56\left(\frac{1}{56}\right)$

40. $-12p + 45 - 7p$

41. $2(7 - v) - 3v$

42. $24abc - 24bac$

43. $4 \cdot 13 \cdot 25 \cdot 1$

44. $4[m - 2(2m + 3)]$

2-6 Objectives

▼ To find theoretical probability (p. 93)

▼ To find experimental probability (p. 95)

The **probability** of an event, or $P(\text{event})$, tells you how likely it is that something will occur. An **outcome** is the result of a single trial. An **event** is any outcome or group of outcomes. The **sample space** is all of the possible outcomes.

theoretical probability $P(\text{event}) = \frac{\text{number of favorable outcomes}}{\text{number of possible outcomes}}$

experimental probability $P(\text{event}) = \frac{\text{number of times an event occurs}}{\text{number of times the experiment is done}}$

Odds describe the likelihood of an event by comparing favorable and unfavorable outcomes.

$$\text{odds in favor of an event} = \frac{\text{number of } favorable \text{ outcomes}}{\text{number of } unfavorable \text{ outcomes}}$$

$$\text{odds against an event} = \frac{\text{number of } unfavorable \text{ outcomes}}{\text{number of } favorable \text{ outcomes}}$$

Find each probability for one roll of a number cube.

45. $P(\text{number} \geq 7)$ 　　　**46.** $P(\text{not } 5)$ 　　　**47.** $P(2 \text{ or } 6)$

48. $P(3)$ 　　　**49.** $P(\text{number} < 6)$ 　　　**50.** $P(\text{not odd})$

Use the spinner at the left. Find each odds.

51. odds in favor of 1 　　　**52.** odds against even

53. odds in favor of green 　　　**54.** odds against red

55. odds in favor of a multiple of 3 　　　**56.** odds against a factor of 6

57. Suppose you toss a coin 4 times.
    a. What is the probability that you toss exactly 3 heads?
    b. Explain what $P(\text{not } 3 \text{ heads})$ means.

58. **Science** An astronomer calculates that the probability that a visible meteor shower will occur in May is $\frac{3}{14}$. What is the probability that a visible meteor shower will not occur in May?

2-7 Objectives

▼ To find the probability of independent events (p. 101)

▼ To find the probability of dependent events (p. 103)

Independent events do not affect one another. When the outcome of one event affects the outcome of a second event, the events are **dependent events.**

For independent events A and B: 　　　For dependent events A and B:

$P(A \text{ and } B) = P(A) \cdot P(B)$. 　　　$P(A \text{ then } B) = P(A) \cdot P(B \text{ after } A)$.

For Exercises 59–63, suppose you choose two numbers from a box containing ten cards with the numbers 1–10. State whether the two events are independent or dependent. Then find each probability.

59. $P(6 \text{ then an even number})$ without replacing the card

60. $P(1 \text{ and an odd number})$ with replacing the card

61. $P(\text{an even number then an odd number})$ without replacing the card

62. $P(\text{an even number and an odd number})$ with replacing the card

63. $P(5 \text{ then an even number})$ without replacing the card

64. **Writing** Explain what the word *dependent* means in probability and what it means in everyday language.

Are the two events dependent or independent? Explain.

65. Roll a red and a blue number cube.

66. Randomly select a green sock and then another green sock from a drawer when you are getting ready for school.

Chapter 2

Chapter Test

Go Online
PHSchool.com

For: Online chapter test
Web Code: ata-0252

Simplify each expression.

1. $-1.8 + 12.1 + (-7.6)$ **2.** $6.3 - 3.9 + 3.7$

3. $1\frac{7}{8} + 14\frac{1}{8}$ **4.** $-7\frac{7}{8} + \left(-2\frac{1}{2}\right)$

5. $-8\frac{5}{6} + 4\frac{1}{2}$ **6.** $3 + 5 - 4$

Find each sum or difference.

7. $\begin{bmatrix} 3 & 2 \\ -1 & 5 \end{bmatrix} + \begin{bmatrix} 8 & -5 \\ 3 & 0 \end{bmatrix}$

8. $\begin{bmatrix} \frac{1}{5} & -2 \\ 0 & \frac{4}{9} \end{bmatrix} + \begin{bmatrix} 4 & -\frac{1}{2} \\ -5 & \frac{4}{9} \end{bmatrix}$

9. $\begin{bmatrix} \frac{1}{2} & -\frac{3}{4} \\ -2 & 1 \end{bmatrix} - \begin{bmatrix} \frac{1}{4} & 1 \\ -\frac{1}{2} & 5 \end{bmatrix}$

10. $\begin{bmatrix} 1 & 9 & -4 \\ 5 & 2 & -1 \\ -6 & -2 & -1 \end{bmatrix} - \begin{bmatrix} 2 & -6 & 7 \\ -8 & 3 & -3 \\ 4 & -7 & 9 \end{bmatrix}$

11. On four plays, a football team gained 22 yd, lost 18 yd, gained 8 yd, and lost 14 yd. What is the total number of yards gained or lost on the four plays?

12. **Banking** Marcus had $163 in his checking account. On Monday, he wrote a check for $315. How much does Marcus need to deposit into his account to prevent the account balance from dipping below the minimum of $25?

Evaluate each expression for $x = 3$, $y = -1$, and $z = 2$.

13. $2x + 3y + z$ **14.** $-xyz$

15. $-3x - 2z - 7$ **16.** $-z^3 - 2z + z$

17. $\frac{xy - 3z}{-5}$ **18.** $|y - x| - z$

19. $x^2 + (-x)^2$ **20.** $y^2 - (-y)^2$

Simplify each expression.

21. $9(-5)$ **22.** $8 - 2^4 \div 2$

23. $-4\left(-\frac{2}{3}\right)$ **24.** $3\frac{1}{6} \div \left(-4\frac{2}{3}\right)$

25. $\frac{2 \cdot 3 - 1}{3^2}$ **26.** $36 - (4 + 5 \cdot 4)$

27. $3^3 \cdot 11 \div 9$ **28.** $-12 \div (-4) - 3$

29. **Writing** Tell if each of the subtraction sentences would *always*, *sometimes*, or *never* be true. Support your answer with two examples.
  a. $(+) - (+) = (+)$ **b.** $(+) - (-) = (-)$
  c. $(-) - (-) = (-)$ **d.** $(-) - (+) = (+)$

Simplify each expression.

30. $m(-3 + n)$ **31.** $6(2d - 5)$

32. $\frac{y}{5}(10 - x)$ **33.** $(7 - 42a)\left(-\frac{3}{7}\right)$

Simplify each expression. Justify each step.

34. $10x + 3\left(\frac{1}{3} - x\right)$ **35.** $2(25a - 5) - 100a$

36. $(3^3 - 3^3)(1 - 2^2)$ **37.** $-36 \div (-12)$

38. $27rst + 27tsr$ **39.** $6a - 5 + 3(3 - a)$

40. $5 + h\left(\frac{6}{h}\right) - h$ **41.** $4c - 6 - (-2c)$

42. **Survey** A random survey of 60 students showed that 36 students used calculators for computation. What is the probability that a student chosen at random used a calculator for computation?

43. A softball player made a hit 34 times in the last 170 times at bat. Find the probability that the softball player will get a hit the next time at bat.

Suppose you have a bag containing 3 red, 4 blue, 5 white, and 2 black marbles. You select one marble at random. Find each probability or odds.

44. $P(\text{not white})$ **45.** $P(\text{red or blue})$

46. $P(\text{orange})$ **47.** $P(\text{not black})$

48. odds in favor of red   **49.** odds against blue

50. odds against black or white

51. odds in favor of red, white, or blue

52. **Open-Ended** Write and solve a probability problem involving dependent events.

53. You have 8 red checkers and 8 black checkers in a bag. You choose two checkers. Find each probability.
  a. $P(\text{red and red})$ with replacing
  b. $P(\text{red then black})$ without replacing
  c. $P(\text{black and red})$ with replacing

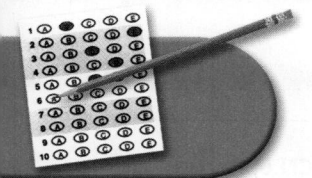

Regents Test Prep

Multiple Choice

For Exercises 1–12, choose the correct letter.

1. Simplify $2(5 - 3)^2 + 4 \div 2$.
   - **(1)** 6
   - **(2)** 8
   - **(3)** 10
   - **(4)** 18

2. Evaluate $\frac{3x - (-4)}{7}$ for $x = -6$.
   - **(1)** -3
   - **(2)** -2
   - **(3)** 2
   - **(4)** 3

3. Name the property illustrated by the equation $(8 + 2) + 7 = (2 + 8) + 7$.
   - **(1)** Commutative Property of Addition
   - **(2)** Associative Property of Addition
   - **(3)** Distributive Property
   - **(4)** Identity Property of Addition

4. Ben has a cell phone plan where he pays $12 per month plus $.10 per minute of talk time. The equation $b = 0.10m + 12$ can be used to calculate his monthly phone bill b given the number of minutes m he spends talking. Which of the following represents a reasonable range for the function?
   - **(1)** $\{0, 12\}$
   - **(2)** $0 \le b \le 12$
   - **(3)** $12 \le b$
   - **(4)** $0 \le b$

5. Evaluate $2y + 7$ for $y = 4$.
   - **(1)** 11
   - **(2)** 15
   - **(3)** 22
   - **(4)** 31

6. Which product equals $9h - 36$?
   - **(1)** $4(h - 9)$
   - **(2)** $(9 - h)4$
   - **(3)** $(4 - h)9$
   - **(4)** $9(h - 4)$

7. Tamara's teacher allows students to decide whether to use the mean, median, or mode for their test averages. Tamara will receive the highest average if she uses the mean. Which set of test scores are Tamara's?
   - **(1)** 95, 82, 76, 95, 96
   - **(2)** 79, 80, 91, 83, 80
   - **(3)** 65, 84, 75, 74, 65
   - **(4)** 100, 87, 94, 94, 81

8. Evaluate $3n - 5$ for $n = -7$.
   - **(1)** -12
   - **(2)** -16
   - **(3)** -26
   - **(4)** -35

9. A bag contains 10 red marbles and 20 white marbles. You draw a marble, keep it, and draw another. What is the probability of drawing two red marbles?
   - **(1)** $\frac{3}{29}$
   - **(2)** $\frac{1}{10}$
   - **(3)** $\frac{1}{9}$
   - **(4)** $\frac{1}{3}$

10. Evaluate $\frac{4a^2}{2b - 3}$ for $a = 3$ and $b = 6$.
    - **(1)** 3
    - **(2)** 4
    - **(3)** 6
    - **(4)** 16

11. Angie uses the equation $e = 0.03s + 25{,}000$ to find her yearly earnings e based on her total sales s. What is the independent variable?
    - **(1)** e
    - **(2)** 0.03
    - **(3)** s
    - **(4)** 25,000

12. Courtney's town studied the relationship between the number of birds and the number of mosquitoes in an area. According to the study, as the number of birds increases, the number of mosquitoes decreases. What type of relationship exists between the bird population and mosquito population?
    - **(1)** Constant correlation
    - **(2)** Positive correlation
    - **(3)** Negative correlation
    - **(4)** No correlation

Short Response

Find each answer.

13. Simplify $\frac{3}{8} + \frac{3}{4}$.

14. Simplify $\frac{3^2 + 3^2 + 3^2}{4^2 + 4^2 + 4^2}$.

15. Simplify $\frac{3}{5}(28 - 3) - (-6)$.

16. Use the formula $A = \frac{1}{2}bh$ to find the area in square feet of a triangle with $b = 3.2$ ft and $h = 7.5$ ft.

17. Sean deposited $78.35 into his savings account. The next day he withdrew $29.59. What is the difference in dollars between the deposit and the withdrawal?

18. Jenna earns $7.95 per hour as a salesclerk. Find her total earnings in dollars for 6 hours.

19. Evaluate $ac^4 + (a - b)c$ for $a = 4$, $b = 5$, and $c = -2$.

Show all of your work.

20. In 1996, an exhibit in Washington, D.C. showed 21 paintings by Johannes Vermeer. This is about $\frac{3}{5}$ of his known paintings. About how many of his paintings are known to exist?

Activity Lab

NY A.CN.6: Recognize and apply mathematics to situations in the outside world.

A Swing of the Bat

Applying Probability One of the most difficult athletic feats is also the one attempted most often in the United States—hitting a baseball with a baseball bat. A good batting average is .280 or better, which means that the batter gets a hit at least 28% of the time. The last major-league player with a season batting average above .400 was Ted Williams of the Boston Red Sox, who hit .406 in 1941. How do real batting averages compare with the probability of getting a hit?

Activity 1

Materials: baseball bat, tape measure or ruler, pencil and paper

Suppose you are standing at the plate, ready to swing at a ball. Use the assumptions below.

- Every pitch will be in the strike zone.
- The timing of your swing will be correct so that the ball and bat are over the plate at the same time.

a. Use the given information to estimate the area of your strike zone.

b. Measure the bat and estimate the area that passes through the strike zone.

c. Use your answers to parts (a) and (b) to estimate the probability of making contact with the ball during a given swing.

Borders Cracks the Barrier

In 1997, Ila Borders, a left-handed pitcher with the St. Paul Saints, became the first woman to start and win a professional baseball game since the 1940s.

Activity 2

a. Use your estimate from Activity 1 to calculate the probability of missing the ball during a given swing.

b. Use your answer to part (a) to estimate the probability of missing the ball three times in a row, or striking out.

c. Use your answer to part (b) to calculate the probability of not striking out.

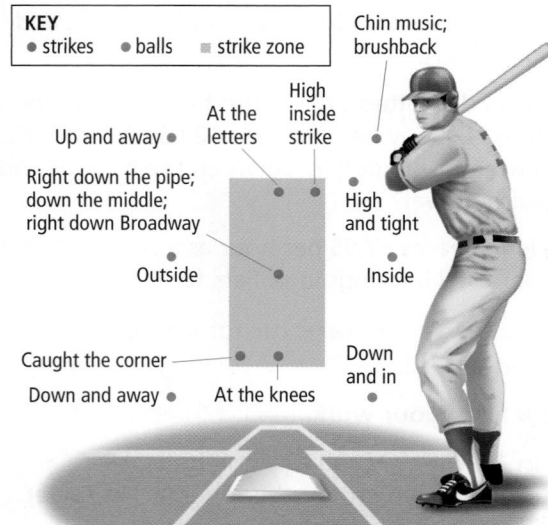

KEY
- • strikes • balls ▪ strike zone

Chin music; brushback

High inside strike

At the letters

Up and away •

Right down the pipe; down the middle; right down Broadway

High and tight

Outside

Inside

Caught the corner

Down and in

Down and away •   At the knees

The Strike Zone

Vertically, the strike zone extends from the hollow just below the batter's knee cap to a point midway between the top of the batter's belt and top of the batter's shoulders.

Horizontally, the strike zone is the width of home plate, which is 17 in., plus twice the diameter of the ball, or 5.8 in.

Power Hitter
The bat bends from the power of Mark McGwire's swing during the 1992 All-Star Game.

Anatomy of a Hit
Begin with your feet about shoulders-width apart. As you swing the bat, your hips turn and your hands follow your hips. As the bat makes contact with the ball, snap your wrists and watch the ball soar.

Activity 3

a. Having the bat make contact with the ball doesn't always mean that you get a hit. Estimate the percentage of contacts with the ball that result in a hit, either through interviewing baseball players in your school or by researching baseball statistics.

b. Use your answer to part (a) and your results from Activity 2 to estimate the batting average you could expect to have if you kept your eyes closed.

Heavy Hitter
Ted Williams of the Boston Red Sox watches the ball sail over the crowd.

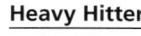
Go Online
PHSchool.com
For: Information about baseball
Web Code: ate-0253

115

What You've Learned

NY Learning Standards for Mathematics

In previous courses, you

- **NY 7.A.4:** solved simple equations.

In Chapter 1, you

- **NY A.A.5:** used variables to write expressions and equations that represent real-world situations.

In Chapter 2, you

- **NY A.N.1:** extended your ability to do arithmetic operations to include rational numbers.

 **Check Your Readiness**

 for Help to the Lesson in green.

Writing a Function Rule (Lesson 1-4)

The relationships in the tables below are functions. Write a function rule for each.

1.

Number of Lawns Mowed	Money Earned
1	$7.50
2	$15.00
3	$22.50
4	$30.00

2.

Number of Withdrawals	Money Remaining
1	$187
2	$174
3	$161
4	$148

Adding and Subtracting Rational Numbers (Lessons 2-1 and 2-2)

Simplify each expression.

3. $6 + (-3)$

4. $-4 - 6$

5. $-5 - (-13)$

6. $-7 + (-1)$

7. $-4.51 + 11.65$

8. $8.5 - (-7.9)$

9. $\frac{3}{10} - \frac{3}{4}$

10. $\frac{1}{5} + \left(-\frac{2}{3}\right)$

Multiplying and Dividing Rational Numbers (Lesson 2-3)

Simplify each expression.

11. $-85 \div (-5)$

12. $7\left(-\frac{6}{14}\right)$

13. $4^2(-6)^2$

14. $22 \div (-8)$

Combining Like Terms (Lesson 2-4)

Simplify each expression.

15. $14k^2 - (-2k^2)$

16. $4xy + 9xy$

17. $6t + 2 - 4t$

18. $9x - 4 + 3x$

Solving Equations

What You'll Learn Next

(NY) **Learning Standards for Mathematics**

- **NY A.A.6:** You will solve equations, including equations with variables on both sides, using properties of equality.

- **NY A.A.5:** You will develop the ability to solve problems by defining variables, relating them to one another, and writing an equation.

- **NY A.A.26:** You will use proportions to measure objects indirectly.

Data Analysis **Activity Lab** You will apply what you learn about probability to investigate probability distributions, on pages 196–197.

Solving One-Step Equations

A **solution of an equation** is the value (or values) of the variable that makes the equation true. To find a solution, you can use properties of equality to form equivalent equations. **Equivalent equations** are equations that have the same solution (or solutions).

Addition Property of Equality

For all real numbers a, b, and c, if $a = b$, then $a + c = b + c$.

Example $8 = 5 + 3$, so $8 + 4 = 5 + 3 + 4$.

Subtraction Property of Equality

For all real numbers a, b, and c, if $a = b$ then $a - c = b - c$.

Example $8 = 5 + 3$, so $8 - 2 = 5 + 3 - 2$.

Multiplication Property of Equality

For all real numbers a, b, and c, if $a = b$, then $a \cdot c = b \cdot c$.

Example $\frac{6}{2} = 3$, so $\frac{6}{2} \cdot 2 = 3 \cdot 2$.

Division Property of Equality

For all real numbers a, b, and c, with $c \neq 0$, if $a = b$ then $\frac{a}{c} = \frac{b}{c}$.

Example $3 + 1 = 4$, so $\frac{3 + 1}{2} = \frac{4}{2}$.

One way to find the solution of an equation is to get the variable alone on one side of the equal sign. You can do this using **inverse operations,** which are operations that undo one another. Addition and subtraction are inverse operations. Multiplication and division are also inverse operations.

1 EXAMPLE   Solving Using Addition or Subtraction

a. Solve $x - 3 = -8$.

$x - 3 + 3 = -8 + 3$ **Add 3 to each side of the equation.**

$x = -5$ **Simplify.**

b. Solve $g + 7 = 11$.

$g + 7 - 7 = 11 - 7$ **Subtract 7 from each side of the equation.**

$g = 4$ **Simplify.**

2 EXAMPLE   Solving Using Multiplication or Division

a. Solve $\frac{3}{4}x = 9$.

$\frac{4}{3}\left(\frac{3}{4}x\right) = \frac{4}{3}(9)$ **Multiply each side by $\frac{4}{3}$, the reciprocal of $\frac{3}{4}$.**

$x = 12$ **Simplify.**

b. Solve $-96 = 4c$.

$\frac{-96}{4} = \frac{4c}{4}$ **Divide each side by 4.**

$-24 = c$ **Simplify.**

Solve each equation.

1. $x - 8 = 0$

2. $c - 4 = 9$

3. $-4 = \frac{2}{5}a$

4. $-8n = -64$

5. $b + 5 = -13$

6. $6 = x + 2$

7. $-7y = 28$

8. $-101 = -\frac{r}{3}$

9. $67 = w - 65$

10. $5b = 145$

11. $\frac{m}{7} = 12$

12. $-4 = k + 19$

Solving Two-Step Equations

NY Learning Standards for Mathematics

A.A.5 Write algebraic equations or inequalities that represent a situation.

A.A.6 Analyze and solve verbal problems whose solution requires solving a linear equation in one variable or linear inequality in one variable.

 **Check Skills You'll Need**

GO for Help   Review, page 118

Solve each equation and tell which property of equality you used.

1. $x - 5 = 14$

2. $x + 3.8 = 9$

3. $-7 + x = 7$

4. $x - 13 = 20$

5. $\frac{x}{4} = 8$

6. $9 = 3x$

Solve each equation.

7. $10x = 2$

8. $\frac{2}{3}x = -6$

9. $x + 2\frac{3}{4} = 6\frac{1}{2}$

 ## 1   Solving Two-Step Equations

A two-step equation involves two operations. Algebra tiles can help you understand how to solve equations. Since ■ represents -1 and □ represents 1, ■ □ is a zero pair representing zero. A green tile represents a variable.

The model below relates solving an equation algebraically and with algebra tiles.

$3x - 2 = 4$
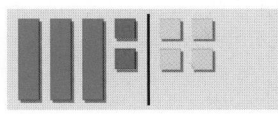
The tiles model the equation.

$3x - 2 + 2 = 4 + 2$
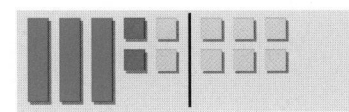
Add 2 to each side.

$3x = 6$
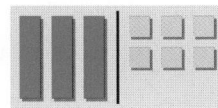
Simplify by removing zero pairs.

$\frac{3x}{3} = \frac{6}{3}$

Divide each side into three equal groups.

$x = 2$

Each green tile equals two yellow tiles, so $x = 2$.

Using the addition, subtraction, multiplication, and division properties of equality, you can form equivalent equations. The equations $3x - 2 = 4$, $3x = 6$, and $x = 2$ are all equivalent equations.

To solve equations, you use the properties of equality repeatedly to get the term with a variable alone on one side of the equation and with a coefficient of 1.

1 EXAMPLE Solving a Two-Step Equation

Solve $10 = \frac{m}{4} + 2$.

$10 - 2 = \frac{m}{4} + 2 - 2$ **Subtract 2 from each side.**

$8 = \frac{m}{4}$ **Simplify.**

$4 \cdot 8 = 4 \cdot \frac{m}{4}$ **Multiply each side by 4.**

$32 = m$ **Simplify.**

Check $10 = \frac{m}{4} + 2$

$10 \stackrel{?}{=} \frac{32}{4} + 2$ **Substitute 32 for m.**

$10 \stackrel{?}{=} 8 + 2$

$10 = 10 \checkmark$

✓ Quick Check ❶ Solve each equation. Check your answer.
    a. $7 = 2y - 3$ **b.** $\frac{x}{9} - 15 = 12$ **c.** $-x + 15 = 12$

You can write two-step equations to model real-world situations.

2 EXAMPLE <u>Real-World 🌐 Problem Solving</u>

A music store sells a copy of a deluxe electric guitar for \$295. This is \$30 more than $\frac{1}{3}$ the cost of the deluxe electric guitar it is modeled after. What is the cost of the deluxe electric guitar?

Relate   cost of the copy   is   \$30   more than $\frac{1}{3}$ the cost of the deluxe electric guitar

Define   Let c = the cost of the deluxe electric guitar.

Write   295   =   30   + $\frac{1}{3}$ · c

$295 = 30 + \frac{1}{3}c$

$295 - 30 = 30 - 30 + \frac{1}{3}c$ **Subtract 30 from each side.**

$265 = \frac{1}{3}c$ **Simplify.**

$3 \cdot 265 = 3 \cdot \frac{1}{3}c$ **Multiply each side by 3.**

$795 = c$ **Simplify.**

The deluxe electric guitar costs \$795.

✓ Quick Check ❷ Suppose your high school's baseball team scored 5 runs in Saturday's game. This was four more than $\frac{1}{2}$ the runs scored by the opposing team. How many runs did the opposing team score?

Functions can model many real-world situations. If you have enough information, you can write a one-variable equation based on a function. Also, some real-world situations require whole numbers. Make sure your answer is reasonable for the given situation.

3 EXAMPLE Writing a Function

Retailing In a catalog, tulips cost $0.75 each and shipping costs are $3.00. Write a rule that describes the amount spent as a function of the number of bulbs ordered. Then determine the greatest number of bulbs that you can order for $14.

Relate   amount spent   equals   cost per tulip bulb   times   number of tulip bulbs   plus   shipping

Define Let a = amount spent.

Let b = number of bulbs you can order.

Write a = 0.75 · b + 3

$14 = 0.75b + 3$	Substitute 14 for a.
$14 - 3 = 0.75b + 3 - 3$	Subtract 3 from each side.
$11 = 0.75b$	Simplify.
$\dfrac{11}{0.75} = \dfrac{0.75b}{0.75}$	Divide each side by 0.75.
$14.\overline{6} = b$	Simplify.

You can order 14 bulbs.

Check Is the solution reasonable? You can only order whole tulip bulbs. Since 15 bulbs would cost $15 \cdot \$0.75 = \$11.25$ plus $3 for handling, which is more than $14, you can only order 14 tulip bulbs.

 **Quick Check** ③ **a.** Suppose you have $25 to spend on tulips. What is the greatest number of bulbs that you can order?

b. Mrs. Simmons works at a furniture store. Her base salary is $125 a week plus $\frac{1}{12}$ of her sales. Write a rule that describes her total weekly salary as a function of her sales. Then find the amount of her sales if her total weekly salary is $255.

---

2 Using Deductive Reasoning

You can use deductive reasoning to justify steps as you solve an equation.

4 EXAMPLE Using Deductive Reasoning

Solve $1 = \frac{k}{12} + 5$. Justify each step.

Steps	Reasons
$1 = \frac{k}{12} + 5$	original equation
$1 - 5 = \frac{k}{12} + 5 - 5$	Subtraction Property of Equality
$-4 = \frac{k}{12}$	Simplify.
$(12)(-4) = (12)\frac{k}{12}$	Multiplication Property of Equality
$-48 = k$	Simplify.

Quick Check ④ Solve $-9 - 4m = 3$. Justify each step.

GO for Help

For help with deductive reasoning see p. 88.

EXERCISES

For more exercises, see *Extra Skill and Word Problem Practice*.

Practice and Problem Solving

 Practice by Example

Example 1
(page 120)

 **for Help**

Solve each equation. Check your answer.

1. $1 + \frac{a}{5} = -1$ **2.** $2n - 5 = 7$ **3.** $-1 = 3 + 4x$ **4.** $\frac{y}{2} + 5 = -12$

5. $3b + 7 = -2$ **6.** $\frac{x}{3} - 9 = 0$ **7.** $14 + \frac{h}{5} = 2$ **8.** $-10 = -6 + 2c$

9. $\frac{m}{8} + 4 = 16$ **10.** $\frac{a}{4} - 21 = 7$ **11.** $3x - 1 = 8$ **12.** $10 = 2n + 1$

13. $35 = 3 + 5x$ **14.** $41 = \frac{2}{5}x - 7$ **15.** $-3 + \frac{m}{3} = 12$ **16.** $9 + \frac{n}{5} = 19$

17. $-x - 4 = -20$ **18.** $-y + 10 = 25$ **19.** $5 = -z - 3$ **20.** $9 = -x + 8$

Example 2
(page 120)

Define a variable and write an equation for each situation. Then solve.

21. Donations A library receives a large cash donation and uses the funds to double the number of books it owns. Then a book collector gives the library 4028 books. After this, the library has 51,514 books. How many books did the library have before the cash donation and the gift of books?

22. Cooking Suppose you are helping to prepare a large meal. You can peel 2 carrots per minute. You need 60 peeled carrots. How long will it take you to finish if you have already peeled 18 carrots?

Example 3
(page 121)

23. Cell Phones One cell phone plan costs $39.95 per month. The first 500 minutes of usage are free. Each minute thereafter costs $.35. Write a rule that describes the total monthly cost as a function of the number of minutes of usage (over 500 minutes). Then find the number of minutes of usage over 500 minutes for a bill of $69.70.

Example 4
(page 121)

Justify each step.

24.
$$\frac{x}{5} + 9 = 11$$
$$\frac{x}{5} + 9 - 9 = 11 - 9$$
$$\frac{x}{5} = 2$$
$$5\left(\frac{x}{5}\right) = 5(2)$$
$$x = 10$$

25.
$$-y - 5 = 11$$
$$-y - 5 + 5 = 11 + 5$$
$$-y = 16$$
$$-1(-y) = -1(16)$$
$$y = -16$$

26.
$$18 - n = 21$$
$$18 - n - 18 = 21 - 18$$
$$-n = 3$$
$$-1(-n) = -1(3)$$
$$n = -3$$

27.
$$12 - 2h = 8$$
$$12 - 2h - 12 = 8 - 12$$
$$-2h = -4$$
$$\frac{-2h}{-2} = \frac{-4}{-2}$$
$$h = 2$$

B **Apply Your Skills**

Solve each equation.

28. $\frac{5}{7}x + \frac{1}{7} = 3$ **29.** $\frac{a}{5} + 15 = 30$ **30.** $-\frac{1}{5}t - 2 = 4$ **31.** $-6 + 6z = 0$

32. $3.5 + 10m = 7.32$ **33.** $7 = -2x + 7$ **34.** $\frac{1}{2} = \frac{2}{5}c - 3$ **35.** $10.7 = -d + 4.3$

36. $0.4x + 9.2 = 10$ **37.** $4x + 92 = 100$ **38.** $-t - 0.4 = -3$ **39.** $-10t - 4 = -30$

Solve each equation. Justify each step.

40. $8 + \frac{c}{-4} = -6$ **41.** $7 - 3k = -14$ **42.** $14 = 6 - 2p$ **43.** $\frac{-y}{2} + 14 = -1$

44. Multiple Choice Beneath Earth's surface, the temperature increases 10°C every kilometer. Suppose that the surface temperature is 22°C, and the temperature at the bottom of a coal mine is 45°C. Which equation could be used to find the depth d of the coal mine?

   Ⓐ $10d + 22 = 45$ Ⓑ $45d - 10 = 22$

   Ⓒ $d = 22 + 10$ Ⓓ $22 = 10d + 45$

45. Insurance One health insurance policy pays people for claims by multiplying the claim amount by 0.8 and then subtracting $500. Write a rule that describes the insurance payment as a function of the claim amount. Then find the claim amount for an insurance payment of $4650.

46. Library The Library of Congress in Washington, D.C., is the largest library in the world. It contains nearly 128 million items. The library adds about 10,000 items to its collection daily. Write a rule that describes the number of items in the Library of Congress as a function of the number of items added daily. Then find the number of days it will take the library to reach about 150 million items.

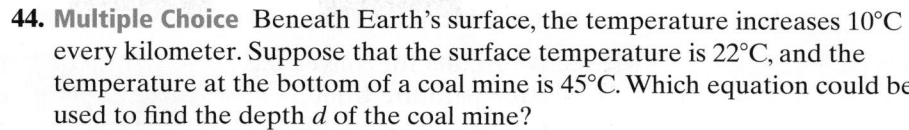

Real-World Connection

Since 1950, the size of the collections and the size of the staff of the Library of Congress have tripled.

Solve each equation. (*Hint:* As your first step, multiply each side by the denominator of the fraction.)

47. $\dfrac{x + 2}{9} = 5$ **48.** $\dfrac{y + 1}{3} = 2$ **49.** $\dfrac{a - 10}{-4} = 2$ **50.** $\dfrac{b - 7}{2} = 6$

51. $\dfrac{x - 5}{2} = 10$ **52.** $\dfrac{x - 3}{7} = 12$ **53.** $\dfrac{x + 4}{3} = -8$ **54.** $\dfrac{x + 6}{4} = -7$

Geometry In each triangle, the measure of $\angle A =$ the measure of $\angle B$. Find the value of x.

55. 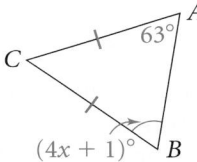   **56.** 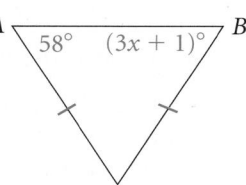   **57.**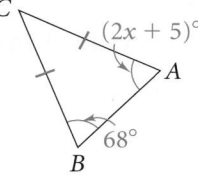

58. Writing Miles has saved $40. He wants to buy a CD player for $129 in about four months. To find how much he should save each week, he wrote $40 + 16x = 129$. Explain his equation.

Error Analysis What is the error in the work? Solve each equation correctly.

59.

$$12 - 3y = 15$$
$$3y = 3$$
$$y = 1$$

60.

$$\frac{m}{3} - 9 = -21$$
$$\frac{m}{3} - 9 + 9 = -21 + 9$$
$$\frac{m}{3} = -12$$
$$m = -4$$

61. Open-Ended Write a problem that you can model with a two-step equation. Write an equation and solve the problem.

62. You can find the value of each variable in the matrices below by writing and solving equations. For example, to find the value of a, you solve the equation $2a + 1 = 11$. Find the values of a, x, y, and k.

$$\begin{bmatrix} 2a + 1 & -6 \\ -7 & -3k \end{bmatrix} = \begin{bmatrix} 11 & x - 5 \\ 5 - 2y & 27 \end{bmatrix}$$

GO **Ⓝnline**
Homework Video Tutor

Visit: PHSchool.com
Web Code: ate-0301

Use the table at the right for Exercises 63 and 64.

 **63.** A formula for converting a temperature from Celsius C to Fahrenheit F is $F = 1.8C + 32$. Copy and complete the table. Round to the nearest degree.

Fahrenheit	Celsius	Description of Temperature
212°	■	boiling point of water
■	37°	human body temperature
68°	■	room temperature
■	0°	freezing point of water
■	−40°	

64. a. Estimation Use the formula $F = 2C + 30$ to estimate the Fahrenheit temperatures not shown in the table.

  b. Critical Thinking Compare your estimated values with the actual values. How good is your formula at estimating the actual temperatures? Explain.

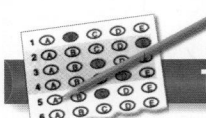 **Challenge**

65. Critical Thinking If you multiply each side of $0.24r + 5.25 = -7.23$ by 100, the result is an equivalent equation. Explain why it might be helpful to do this.

Solve the first equation for x. Then substitute your result into the second equation and solve for y.

66. $x + y = 8$
$2x + 3y = 21$

67. $x + 2y = 1$
$6x - y = -20$

68. $x + y = 12$
$2x + y = 17$

Test Prep

Gridded Response

69. What is the value of the expression $-\frac{3}{5}x - 2$ when $x = -6$?

70. Dana has a photograph that is 4 in. wide and 5 in. long. She has the photo enlarged so that both dimensions are tripled. How many times larger than the original area will the area of the enlargement be?

71. What is the value of the following expression? $\dfrac{3^2 \times 4 - (-5)^2}{(-2)^3 + 3 \times 4}$

72. Last season, Everett scored 48 points. This is 6 less than twice the number of points Max scored. How many points did Max score?

73. A cable television company charges $24.95 a month for basic cable service and $6.95 a month for each additional premium channel. If Sami's monthly bill is $45.80, how many premium channels is he receiving?

Mixed Review

 **Lesson 2-3**

Simplify each expression.

74. $-6(-4)$

75. $2(-5.5)$

76. $\dfrac{45}{-3}$

77. $\dfrac{-5 - 31}{-9}$

78. $-72 \div (4 - 12)$

79. $(18 - 10) \div (-2)$

Lessons 1-3, 2-1

Simplify.

80. $-2 + 6$

81. $9 + (-3)$

82. $-7 + (-4)$

83. $16 + (-4)$

Tables, Graphs, and Equations

FOR USE WITH LESSON 3-1

You can solve some equations by using a table.

> **NY** **A.PS.2:** Recognize and understand equivalent representations of a problem situation or a mathematical concept.

1 ACTIVITY

Solve $3x + 7 = 25$ using a table.

Make a table and evaluate the expression with a variable until you find the value that matches the constant side of the equation.

The value of x is the solution of the original equation. Since $3(6) + 7 = 25$, the solution of the equation is 6.

Equation: Variable side   Constant side
$$3x + 7 = 25$$

x	$3x + 7$
0	$3(0) + 7 = 7$
2	$3(2) + 7 = 13$
4	$3(4) + 7 = 19$
6	$3(6) + 7 = 25$ ✓

You can solve equations by using a graphing calculator to graph each side of the equation. The x-coordinate of the point where the graphs intersect is the solution of the equation.

2 ACTIVITY

Solve $9 = -\frac{3}{2}n + 4$ using a graphing calculator.

Step 1   Enter the equations using the **Y=** feature. First clear any equations. For $Y_1 =$ enter 9. For $Y_2 =$ enter $-\frac{3}{2}x + 4$ by pressing **((−) 3 ÷ 2) X,T,θ,n + 4.**

Step 2   Graph the equations. Use a standard graphing window, which you can find using the **ZOOM** feature.

Step 3   Use the **CALC** feature, select **intersect** to find the point where the lines intersect.

The calculator value for the x-coordinate of the point of intersection is −3.333333. The actual point of intersection is $-3\frac{1}{3}$.

The solution of the equation $9 = -\frac{3}{2}n + 4$ is $-3\frac{1}{3}$.

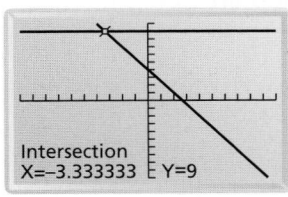

Intersection
X=−3.333333   Y=9

EXERCISES

Use a table to solve each equation.

1. $5c + 4 = 19$

2. $23 - 3x = 11$

3. $33 = 8t - 7$

4. $13 = -2x + 21$

5. $\frac{1}{2}n + 41 = 51$

6. $-9 = 3 + 6k$

Use a graphing calculator to solve each equation.

7. $8a - 12 = 6$

8. $-4 = -3t + 2$

9. $4 + \frac{3}{2}n = -7$

10. $-5 = -0.5g - 2$

11. $\frac{5}{4}d - \frac{1}{2} = 6$

12. $-3w - 1 = 3.5$

3-2

Solving Multi-Step Equations

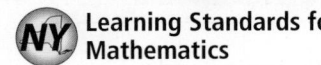 **Learning Standards for Mathematics**

A.N.1 Identify and apply the properties of real numbers (closure, commutative, associative, distributive, identity, inverse).

A.A.5 Write algebraic equations or inequalities that represent a situation.

A.A.22 Solve all types of linear equations in one variable.

✓ **Check Skills You'll Need**

Simplify each expression.

1. $2n - 3n$

2. $-4 + 3b + 2 + 5b$

3. $9(w - 5)$

4. $-10(b - 12)$

5. $3(-x + 4)$

6. $5(6 - w)$

Evaluate each expression.

7. $28 - a + 4a$ for $a = 5$

8. $8 + x - 7x$ for $x = -3$

9. $(8n + 1)3$ for $n = -2$

10. $-(17 + 3y)$ for $y = 6$

 **for Help** Lessons 1-2 and 2-4

---

1 **Using the Distributive Property to Combine Like Terms**

You can solve equations that require more than two steps. If there are like terms on one side of an equation, first use the Distributive Property to combine them. Then use the properties of equality to solve the equation.

Problem Solving Hint

You use the Distributive Property whenever you add or subtract like terms.

$x + 4x = 1x + 4x$
$= (1 + 4)x$
$= 5x$

1 **EXAMPLE** **Combining Like Terms**

Solve each equation.

a. $2c + c + 12 = 78$
$2c + c + 12 = 78$
$3c + 12 = 78$ Combine like terms.
$3c + 12 - 12 = 78 - 12$ Subtract 12 from each side.
$3c = 66$ Simplify.
$\frac{3c}{3} = \frac{66}{3}$ Divide each side by 3.
$c = 22$ Simplify.

b. $4b + 16 + 2b = 46$
$4b + 16 + 2b = 46$
$4b + 2b + 16 = 46$ Use the Commutative Property of Addition.
$6b + 16 = 46$ Combine like terms.
$6b + 16 - 16 = 46 - 16$ Subtract 16 from each side.
$6b = 30$ Simplify.
$\frac{6b}{6} = \frac{30}{6}$ Divide each side by 6.
$b = 5$ Simplify.

✓ **Quick Check** **1** Solve each equation. Check your answer.
a. $3x - 4x + 6 = -2$ **b.** $7 = 4m - 2m + 1$
c. $-2y + 5 + 5y = 14$ **d.** $-3z + 8 + (-2z) = -12$

You can model real-world situations using multi-step equations.

Real-World Connection

In the United States, there are approximately 10,000 cities with community gardens.

2 EXAMPLE **Real-World Problem Solving**

Multiple Choice A gardener is planning a rectangular garden area in a community garden. His garden will be next to an existing 12-ft fence. The gardener has a total of 44 ft of fencing to build the other three sides of his garden. How long will the garden be if the width is 12 ft?

- A) 32 ft
- B) 24 ft
- C) 22 ft
- D) 16 ft

Relate | length of side | plus | 12 ft | plus | length of side | equals | amount of fencing

Define Let x = length of a side adjacent to the fence.

Write | x | + | 12 | + | x | = | 44

$$x + 12 + x = 44$$

$$2x + 12 = 44$$ Combine like terms on the left side of the equation.

$$2x + 12 - 12 = 44 - 12$$ Subtract 12 from each side.

$$2x = 32$$ Simplify.

$$\frac{2x}{2} = \frac{32}{2}$$ Divide each side by 2.

$$x = 16$$ Simplify.

The garden will be 16 ft long. So D is the correct answer.

 **Quick Check** **2** A carpenter is building a rectangular fence for a playground. One side of the playground is the wall of a building 70 ft wide. He plans to use 340 ft of fencing material. What is the length of the playground if the width is 70 ft?

---

2 Using the Distributive Property to Solve Equations

In the equation $-2(b - 4) = 12$, the parentheses indicate multiplication. Use the Distributive Property to multiply each term within the parentheses by -2. Then use the properties of equality to solve the equation.

3 EXAMPLE **Solving an Equation With Grouping Symbols**

Solve $-2(b - 4) = 12$.

Test-Taking Tip

Remember to multiply the number outside the parentheses by both terms inside the parentheses.

$$-2b + 8 = 12$$ Use the Distributive Property.

$$-2b + 8 - 8 = 12 - 8$$ Subtract 8 from each side.

$$-2b = 4$$ Simplify.

$$\frac{-2b}{-2} = \frac{4}{-2}$$ Divide each side by -2.

$$b = -2$$ Simplify.

 **Quick Check** **3** Solve each equation.
  a. $3(k + 8) = 21$ **b.** $15 = -3(x - 1) + 9$

Following are two ways you can solve an equation like $\frac{2x}{3} + \frac{x}{2} = 7$.

4 EXAMPLE Solving an Equation That Contains Fractions

Vocabulary Tip

$\frac{x}{2}$ and $\frac{1}{2}x$ both represent $x \div 2$.

$\frac{2x}{3}$ and $\frac{2}{3}x$ both represent $2x \div 3$.

Solve $\frac{2x}{3} + \frac{x}{2} = 7$.

Method 1 Adding fractions

$$\frac{2x}{3} + \frac{x}{2} = 7$$

$\frac{2}{3}x + \frac{1}{2}x = 7$ Rewrite the equation with fractions as coefficients.

$\frac{4}{6}x + \frac{3}{6}x = 7$ Write the fractions with a denominator of 6.

$\frac{7}{6}x = 7$ Combine like terms.

$\frac{6}{7}\left(\frac{7}{6}x\right) = \frac{6}{7}(7)$ Multiply each side by $\frac{6}{7}$, the reciprocal of $\frac{7}{6}$.

$x = 6$ Simplify.

Method 2 Multiplying to clear fractions

$$\frac{2x}{3} + \frac{x}{2} = 7$$

$6\left(\frac{2x}{3} + \frac{x}{2}\right) = 6(7)$ Multiply each side by 6, a common multiple of 3 and 2.

$6\left(\frac{2x}{3}\right) + 6\left(\frac{x}{2}\right) = 6(7)$ Use the Distributive Property.

$4x + 3x = 42$ Multiply.

$7x = 42$ Combine like terms.

$\frac{7x}{7} = \frac{42}{7}$ Divide each side by 7.

$x = 6$ Simplify.

 **Quick Check** **4** Solve each equation. Explain why you chose the method you used.

a. $\frac{m}{4} + \frac{m}{2} = \frac{5}{8}$ **b.** $\frac{2}{3}x - \frac{5}{8}x = 26$

You can clear an equation of decimals by multiplying by a power of 10. In the equation $0.5a + 8.75 = 13.25$, the greatest number of digits to the right of a decimal point is 2. To clear the equation of decimals, multiply each side of the equation by 10^2, or 100.

Problem Solving Hint

Powers of 10:
$10^1 = 10$
$10^2 = 100$
$10^3 = 1000$
$10^4 = 10,000$

5 EXAMPLE Solving an Equation That Contains Decimals

Solve $0.5a + 8.75 = 13.25$.

$100(0.5a + 8.75) = 100(13.25)$ Multiply each side by 10^2, or 100.

$100(0.5a) + 100(8.75) = 100(13.25)$ Use the Distributive Property.

$50a + 875 = 1325$ Simplify.

$50a + 875 - 875 = 1325 - 875$ Subtract 875 from each side.

$50a = 450$ Simplify.

$\frac{50a}{50} = \frac{450}{50}$ Divide each side by 50.

$a = 9$ Simplify.

 **Quick Check** **5** Solve each equation.

a. $0.025x + 22.95 = 23.65$ **b.** $1.2x - 3.6 + 0.3x = 2.4$

Keep the steps in the summary below in mind as you solve equations that have variables on one side of the equation.

 Key Concepts

Summary	Steps for Solving a Multi-Step Equation
Step 1	Clear the equation of fractions and decimals.
Step 2	Use the Distributive Property to remove parentheses on each side.
Step 3	Combine like terms on each side.
Step 4	Undo addition or subtraction.
Step 5	Undo multiplication or division.

EXERCISES

For more exercises, see *Extra Skill and Word Problem Practice.*

Practice and Problem Solving

A **Practice by Example**

 GO for Help

Example 1
(page 126)

Solve each equation. Check your answer.

1. $4n - 2n = 18$

2. $y + y + 2 = 18$

3. $a + 6a - 9 = 30$

4. $5 - x - x = -1$

5. $72 + 4 - 14c = 36$

6. $13 = 5 - 13 + 3a$

7. $9 = -3 + n + 2n$

8. $7m - 3m - 6 = 6$

9. $-13 = 2b - b - 10$

Example 2
(page 127)

Write an equation to model each situation. Solve your equation.

10. Two friends are renting an apartment. They pay the landlord the first month's rent. The landlord also requires them to pay an additional half of a month's rent for a security deposit. The total amount they pay the landlord before moving in is $1725. What is the monthly rent?

11. You are fencing a rectangular puppy kennel with 25 ft of fence. The side of the kennel against your house does not need a fence. This side is 9 ft long. Find the dimensions of the kennel.

Example 3
(page 127)

Solve each equation. Check your answer.

12. $2(8 + p) = 22$

13. $5(a - 1) = 35$

14. $15 = -3(2q - 1)$

15. $26 = 6(5 - a)$

16. $m + 5(m - 1) = 7$

17. $-4(x + 6) = -40$

18. $48 = 8(x + 2)$

19. $5(y - 3) = 19$

20. $5(2 + y) = 77$

Example 4
(page 128)

21. $\frac{a}{7} - \frac{5}{7} = \frac{6}{7}$

22. $x - \frac{5}{8} = \frac{7}{8}$

23. $\frac{m}{6} - 7 = \frac{2}{3}$

24. $\frac{2}{3} + \frac{3k}{4} = \frac{71}{12}$

25. $4 + \frac{m}{8} = \frac{3}{4}$

26. $\frac{a}{2} + \frac{1}{5} = 17$

27. $\frac{1}{2} + \frac{7x}{10} = \frac{13}{20}$

28. $\frac{9y}{14} + \frac{3}{7} = \frac{9}{14}$

29. $\frac{1}{5} + \frac{3w}{15} = \frac{4}{5}$

Example 5
(page 128)

30. $3m + 4.5m = 15$

31. $7.8y + 2 = 165.8$

32. $3.5 = 12s - 5s$

33. $1.06y - 3 = 0.71$

34. $0.11p + 1.5 = 2.49$

35. $25.24 = 5y + 3.89$

36. $1.12 + 1.25y = 8.62$

37. $1.025x + 2.458 = 7.583$

38. $0.25m + 0.1m = 9.8$

Solve each equation.

39. $0.5t - 3t + 5 = 0$

40. $-(z + 5) = -14$

41. $\frac{a}{15} + \frac{4}{15} = \frac{9}{15}$

42. $0.5(x - 12) = 4$

43. $8y - (2y - 3) = 9$

44. $\frac{2}{3} + y = \frac{3}{4}$

45. $2 + \frac{a}{-4} = \frac{3}{5}$

46. $\frac{1}{4}(m - 16) = 7$

47. $x + 3x - 7 = 29$

48. $4x + 3.6 + x = 1.2$

49. $2(1.5c + 4) = -1$

50. $26.54 - p = 0.5(50 - p)$

51. Error Analysis Explain the error in the student's work at the right.

52. Critical Thinking Suppose you want to solve the equation $-3m + 4 + 5m = -6$. What would you do as your first step?

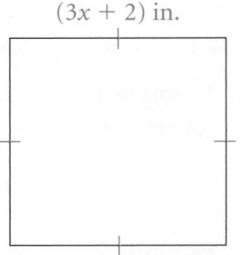

$$\frac{3}{8}x - 1 = 4$$
$$3x - 1 = 32$$
$$3x = 33$$
$$x = 11$$

53. Writing To solve $-\frac{1}{2}(3x - 5) = 7$, you can use the Distributive Property, or you can multiply each side of the equation by -2. Which method do you prefer? Explain why.

Geometry **The perimeter of each rectangle is 64 in. Find the value of x.**

54.

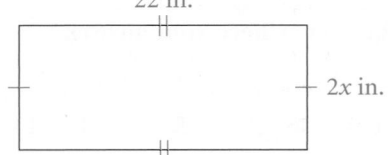

22 in.

2x in.

55.

$(3x + 2)$ in.

Use an equation to solve each problem.

56. John and two friends rent a canoe at a park. Each person must rent a life jacket. If the bill for the rental of the canoe and life jackets is $41, for how many hours did they rent the canoe?

57. Moving Costs The MacNeills rented a moving truck for $49.95 plus $.30 per mile. Before returning the truck, they filled the tank with gasoline, which cost $18.32. The total cost was $95.87. Find the number of miles the truck was driven.

58. Cell Phones Jane's cell phone plan is $40 per month plus $.15 per minute for each minute over 200 minutes of call time. If Jane's cell phone bill is $58.00, for how many extra calling minutes was she billed?

59. Open-Ended Write an expression with four terms that can be simplified to an expression with two terms.

Geometry **Find the value of x. (*Hint:* The sum of the measures of the angles of a triangle is 180°.)**

60.

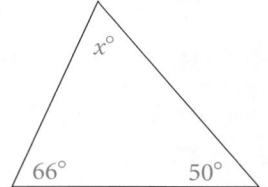

$x°$

66°   50°

61.

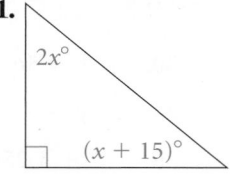

$2x°$

$(x + 15)°$

62.

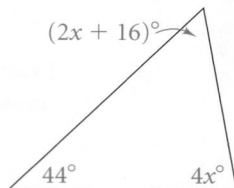

$(2x + 16)°$

44° $4x°$

63. Purchasing The function $c = p + 0.07p$ relates the price of an item p with its cost c after 7% tax is added. A student has $12.50. What is the cost of the most expensive item he can buy?

64. Finance Nadia has $70 in her bank account. Each week she deposits $3 from her allowance and $15 from babysitting. The function $b = 70 + (15 + 3)w$ relates her bank balance b to weeks w. When will she have $160 in her bank account?

65. Tyra solves an equation using her graphing calculator. Use the screens below.

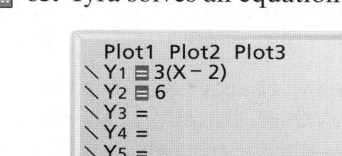
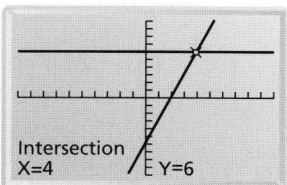

    a. What was the original equation that Tyra was solving?
    b. What is the point of intersection?
    c. Show that the x-value of the intersection solves the original equation.

C Challenge

For Exercises 66–68, use an equation to solve each problem.

66. Cars You fill your car's gas tank when it is about $\frac{1}{2}$ empty. The next week, you fill the tank a second time when it is about $\frac{3}{4}$ empty. If you buy a total of $18\frac{1}{2}$ gal of gas on these two days, about how many gallons does the tank hold?

67. A work crew has two pumps, one new and one old. The new pump can fill a tank in 5 hours. The old pump can fill the same tank in 7 hours.
    a. How much of a tank can be filled in 1 hour with the new pump? With the old pump?
    b. Write an expression for the number of tanks the new pump can fill in t hours. (*Hint:* Write the rate at which the new pump fills tanks as a fraction and then multiply by t.)
    c. Write an expression for the number of tanks the old pump can fill in t hours.
    d. Write and solve an equation for the time it will take the pumps to fill one tank if the pumps are used together.

68. Investing Mr. Fairbanks invested half his money in land, a tenth in stock, and a twentieth in bonds. He put the remaining $35,000 in a savings account. What is the total amount of money that Mr. Fairbanks saved or invested?

NY REGENTS

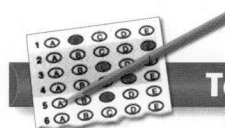
Test Prep

Multiple Choice

69. What is the value of the expression $-3r + 6 + r$ when $r = -2$?
    A. -6 **B.** -2 **C.** 10   **D.** 14

70. Solve $8n + 5 - 2n = 41$.
    F. $3\frac{1}{2}$ **G.** $4\frac{1}{2}$ **H.** 6   **J.** $7\frac{2}{3}$

71. If a number is increased by 3 and that number is doubled, the result is -8. What was the original number?
    A. -7 **B.** -5.5 **C.** 1   **D.** 6

72. The gas tank in Royston's car holds 12 gal of gasoline. The car averages 29 mi/gal. Royston filled up the tank and then drove 140 mi. About how many gallons of gasoline are left in the tank?

    F. 6 gal   **G.** 7 gal   **H.** 8 gal   **J.** 9 gal

73. Josie's goal is to run 40 miles each week. This week she has already run distances of 5.3 miles, 6.5 miles, and 6.2 miles. If she wants to spread out the remaining miles evenly over the next 4 days, which equation can you use to find how many miles (m) per day she must run?

    A. $5.3 + 6.5 + 6.2 + 40 = m$ **B.** $40 - 5.2 - 6.5 - 6.2 = m$
    C. $5.3 + 6.5 + 6.2 + 4m = 40$ **D.** $5.3 + 6.5 + 6.2 + m = \frac{40}{4}$

74. A cell phone company charges \$.35 for the first minute but only \$.10 every minute after that. Which equation can you use to find how many minutes m Eric talked if the bill for the call was \$5.45?

    F. $0.35 + 0.10(m - 1) = 5.45$ **G.** $0.35 + 0.10m = 5.45$
    H. $0.10 + 0.35(m - 1) = 5.45$ **J.** $0.10 + 0.35m = 5.45$

Mixed Review

Lesson 3-1   **Solve each equation.**

75. $2y + 4 = -6$ **76.** $3x - 15 = 33$ **77.** $-4n + 20 = 36$ **78.** $-8 - c = 11$

79. $3x + 5 = 12$ **80.** $-4y - 3 = 15$ **81.** $8m - 4 = 8$ **82.** $-p + 3 = 10$

Lesson 2-5   **Mental Math**   **Simplify each expression.**

83. $14 \cdot 4 \cdot 25$ **84.** $16 + 28 + 34 + 72$ **85.** $-8 + 15 + -9 + 2$

86. $3 \cdot 3 \cdot 10$ **87.** $2 \cdot 8 \cdot 5$ **88.** $27 + 46 - 17 - 16$

Lessons 2-1 through 2-3   **Simplify each expression.**

89. $2 - 6$ **90.** $-9 \cdot (-3)$ **91.** $-7 + (-4)$ **92.** $16 \div (-4)$

93. $-7 + (-3)$ **94.** $-5 - (-3)$ **95.** $-5 \cdot 6$ **96.** $-25 \div (-5)$

Algebra at Work

·············· Airline Pilot

Airline pilots make many calculations before, during, and after a flight. Pilots study weather conditions to determine the safest altitude, route, and speed for a flight. Pilots calculate lift, which must equal the airplane's weight in pounds. An airplane's lift capabilities are calculated using the formula $L = \frac{1}{2}dv^2sa$, where L is the lift, d is the density of the air, v is the velocity of the aircraft in feet per second, s is the wing area of the aircraft in square feet, and a is a value determined by the type of airfoil the airplane has and the pitch angle of the airplane.

Go Online
PHSchool.com   **For:** Information about a career as an airline pilot
        Web Code: atb-2031

Activity Lab
Hands-On

Modeling Equations

FOR USE WITH LESSON 3-3

Models can help you understand how to solve equations that have variables on both sides.

NY **A.R.1:** Use physical objects, diagrams, charts, tables, graphs, symbols, equations, or objects created using technology as representations of mathematical concepts.

EXAMPLE

Model and solve $3a - 2 = a + 4$.

$3a - 2 = a + 4$

The tiles model the equation.

$3a - 2 - a = a + 4 - a$
$2a - 2 = 4$

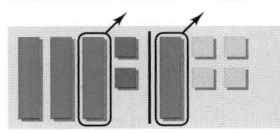

Use the Subtraction Property of Equality. Subtract a from each side to get the variable on one side of the equation.

$2a - 2 + 2 = 4 + 2$
$2a = 6$

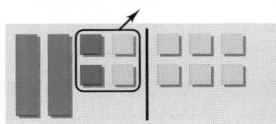

Use the Addition Property of Equality. Add 2 to each side. Remove zero pairs.

$\frac{2a}{2} = \frac{6}{2}$

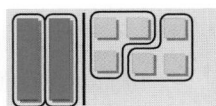

Use the Division Property of Equality. Divide each side into two identical groups.

$a = 3$

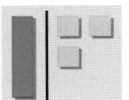

Each green tile equals three yellow tiles, so $a = 3$.

EXERCISES

Write an equation for each model. Use tiles to solve each equation.

1.
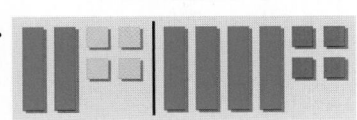

2.

3.

4.

Use tiles to model and solve each equation.

5. $4x + 2 = 2x + 6$

6. $2y - 2 = 4y + 2$

7. $2a + 2 = a + 8$

8. $5b - 4 = 2b + 5$

9. $z - 8 = 2z - 1$

10. $4(p + 1) = 2p - 2$

11. $5n - 3 = 2(n + 3)$

12. $2(k + 1) = 5(k - 2)$

3-3 Equations With Variables on Both Sides

 **Learning Standards for Mathematics**

A.A.5 Write algebraic equations or inequalities that represent a situation.

A.A.6 Analyze and solve verbal problems whose solution requires solving a linear equation in one variable or linear inequality in one variable.

✓ **Check Skills You'll Need**

Simplify.

1. $6x - 2x$ **2.** $2x - 6x$

Solve each equation.

5. $4x + 3 = -5$

7. $2t - 8t + 1 = 43$

GO for Help Lessons 2-4 and 3-2

3. $5x - 5x$ **4.** $-5x + 5x$

6. $-x + 7 = 12$

8. $0 = -7n + 4 - 5n$

◀)) **New Vocabulary** • identity

---

1 Solving Equations With Variables on Both Sides

Activity: Using a Table to Solve an Equation

Costs for a key chain business are $540 to get started plus $3 per key chain. The cost of producing k key chains is $(540 + 3k)$ dollars.

Key chains sell for $7 each. The revenue for selling k key chains is $7k$ dollars. To make a profit, revenue must be greater than costs.

1. Copy and complete the following table.

Key Chains	Cost	Revenue
k	$540 + 3k$	$7k$
100	840	700
110	■	■
120	■	■
130	■	■
140	■	■
150	■	■

2. For 110 key chains, which is greater, the cost or the revenue?

3. When will the revenue be greater than the cost?

4. Use your table to estimate the solution of $540 + 3k = 7k$.

5. Explain how solving an equation can help you decide whether a business can make a profit or not.

To solve an equation that has variables on both sides, use the Addition or Subtraction Properties of Equality to get the variables on one side of the equation.

1 **EXAMPLE** **Variables on Both Sides**

Geometry Find the value of x in the diagram below.

$$6x + 3 = 8x - 21$$ Vertical angles are congruent.
$$6x + 3 - 6x = 8x - 21 - 6x$$ Subtract $6x$ from each side.
$$3 = 2x - 21$$ Combine like terms.
$$3 + 21 = 2x - 21 + 21$$ Add 21 to each side.
$$24 = 2x$$ Simplify.
$$\frac{24}{2} = \frac{2x}{2}$$ Divide each side by 2.
$$12 = x$$ Simplify.

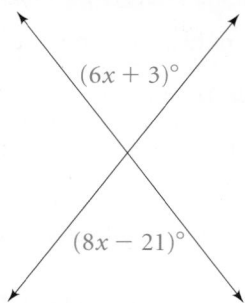

$(6x + 3)°$

$(8x - 21)°$

The value of x is 12.

✓ Quick Check **1** Solve each equation.
a. $-6d = d + 4$ **b.** $2(c - 6) = 9c + 2$
c. $m - 5 = 3m$ **d.** $7k - 4 = 5k + 16$

2 **EXAMPLE** **Real-World**  **Problem Solving**

Recreation You can buy used in-line skates from your friend for $40, or you can rent some. Either way, you must rent safety equipment. How many hours must you skate for the cost of renting and buying skates to be the same?

Relate	cost of friend's skates	plus	safety equipment rental	equals	skates plus equipment rental

Define Let $h =$ the number of hours you must skate.

Write	40	+	$1.5h$	=	$3.5h$

$$40 + 1.5h = 3.5h$$
$$40 + 1.5h - 1.5h = 3.5h - 1.5h$$ Subtract $1.5h$ from each side.
$$40 = 2h$$ Combine like terms.
$$\frac{40}{2} = \frac{2h}{2}$$ Divide each side by 2.
$$20 = h$$ Simplify.

You must skate for 20 hours for the cost to be the same.

Check Is the solution reasonable? Buying skates and renting safety equipment for 20 hours costs $40 + 1.5(20) = 70$, or $70. The cost of renting both skates and safety equipment for 20 hours is $3.5(20) = 70$, or $70. The answer is correct.

✓ Quick Check **2** **Business** A hairdresser is considering ordering a certain shampoo. Company A charges $4 per 8-oz bottle plus a $10 handling fee per order. Company B charges $3 per 8-oz bottle plus a $25 handling fee per order. How many bottles must the hairdresser buy to justify using Company B?

You can use a calculator to solve an equation. Using each side of the equation, you can graph two functions using the $\boxed{Y=}$ screen. The x-value of the point of intersection is the solution of the equation.

3 EXAMPLE Solving Using a Graphing Calculator

Solve $\frac{3}{4}m = 8 - \frac{1}{2}m$ using a graphing calculator.

Step 1 For $Y_1=$ enter $\frac{3}{4}x$. For $Y_2=$ enter $8 - \frac{1}{2}x$.

Step 2 Use the GRAPH feature to display the graph. You can adjust the window by using the ZOOM or WINDOW features.

Step 3 Use the CALC feature. Select **intersect** to find the point where the lines intersect.

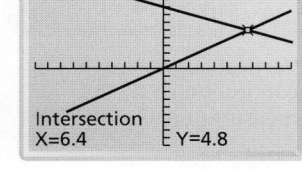
Intersection
X=6.4   Y=4.8

● The lines intersect at (6.4, 4.8). The x-value 6.4 is the solution of the equation.

 **Quick Check** ③ Solve $4 + \frac{1}{2}x = x - 1$ using a graphing calculator.

For help entering functions in the Y= screen, see p. 125.

2 Special Cases: Identities and No Solutions

An equation has **no solution** if no value of the variable makes the equation true. The equation $2x = 2x + 1$ has no solution. An equation that is true for every value of the variable is an **identity.** The equation $2x = 2x$ is an identity.

4 EXAMPLE Identities and Equations with No Solutions

a. Solve $10 - 8a = 2(5 - 4a)$.

$$10 - 8a = 10 - 8a$$
$$10 - 8a + 8a = 10 - 8a + 8a$$
$$10 = 10 \quad \textbf{Always true!}$$

This equation is true for every value of a, so the equation is an identity.

b. Solve $6m - 5 = 7m + 7 - m$.

$$6m - 5 = 7m + 7 - m$$
$$6m - 5 = 6m + 7$$
$$6m - 5 - 6m = 6m + 7 - 6m$$
$$-5 = 7 \quad \textbf{Not true.}$$

This equation has no solution.

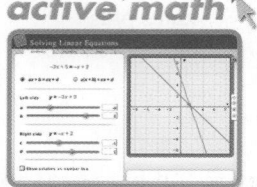
For: Solving Equations Activity
Use: Interactive Textbook, 3-3

 **Quick Check** ④ Determine whether each equation is an *identity* or whether it has *no solution*.
a. $9 + 5n = 5n - 1$ **b.** $9 + 5x = 7x + 9 - 2x$

EXERCISES

For more exercises, see *Extra Skill and Word Problem Practice.*

Practice and Problem Solving

A Practice by Example

Example 1
(page 135)

Solve each equation. Check your answer.

1. $6x - 2 = x + 13$

2. $5y - 3 = 2y + 12$

3. $4k - 3 = 3k + 4$

4. $5m + 3 = 3m + 9$

5. $8 - x = 2x - 1$

6. $2n - 5 = 8n + 7$

7. $3a + 4 = a + 18$

8. $6b + 14 = -7 - b$

Geometry Find the value of x.

9.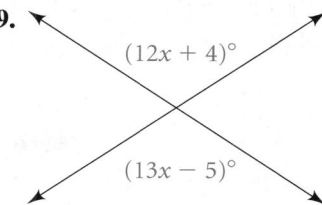
$(12x + 4)°$
$(13x - 5)°$

10.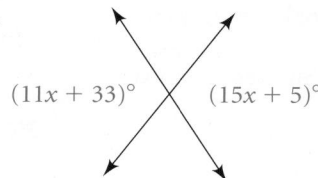
$(11x + 33)°$ $(15x + 5)°$

Example 2
(page 135)

Write and solve an equation for each situation. Check your solution.

11. Telephone Service One telephone company charges $16.95 per month and $.05 per minute for local calls. Another company charges $22.95 per month and $.02 per minute for local calls. For what number of minutes of local calls per month is the cost of the plans the same?

12. Fitness One health club charges a $44 sign-up fee and $30 per month. Another health club charges a $99 sign-up fee and $25 per month. For what number of months is the cost the same?

Example 3
(page 136)

Solve each equation using a graphing calculator.

13. $7(3 - k) = -3k + 4$

14. $a - 6 = 8 - (9 + a)$

15. $-\frac{1}{2}d + 2 = -4\left(d - \frac{1}{2}\right)$

16. $0.2n + 9 = 8(0.4n - 1)$

Example 4
(page 136)

17. a. Use the equation $9 - 6x = 3(3 - 2x)$. Substitute four different values for x and simplify.
   b. What kind of equation is $9 - 6x = 3(3 - 2x)$?

Determine whether each equation is an *identity* or whether it has *no solution*.

18. $14 - (2q + 5) = -2q + 9$

19. $6x + 1 = 6x - 8$

20. $-8x + 14 = -2(4x - 7)$

21. $y - 5 = -(5 - y)$

22. $a - 4a = 2a + 1 - 5a$

23. $9x + 3x - 10 = 3(3x + x)$

B **Apply Your Skills**

Solve each equation. If the equation is an identity, write *identity*. If it has no solution, write *no solution*.

24. $18x - 5 = 3(6x - 2)$

25. $9 + 5a = 2a + 9$

26. $3(x - 4) = 3x - 12$

27. $6x = 4(x + 5)$

28. $\frac{3}{5}k - \frac{1}{10}k = \frac{1}{2}k + 1$

29. $0 = 0.98b + 0.02b - b$

30. $5m - 2(m + 2) = -(2m + 15)$

31. $\frac{7}{8}w = \frac{4}{8}w + \frac{6}{8}w$

32. Multiple Choice A toy company spends $1500 per day for factory expenses plus $8 to make each teddy bear, like the one shown at the left. Which equation could be used to find the number of bears t the company has to sell in one day to equal its daily cost?
   Ⓐ $1500 + 8t = 12$ Ⓑ $12 + 8t = 1500$
   Ⓒ $1500 + 8t = 12t$ Ⓓ $8t = 12t + 1500$

33. Business A company manufactures tote bags. The company spends $1200 each day for overhead expenses plus $9 per tote bag for labor and materials. The tote bags sell for $25 each. How many tote bags must the company sell each day to equal its daily costs for overhead, labor, and materials? Write an equation and solve.

Find the value of each variable.

34. $\begin{bmatrix} 0.5x + 3 & w + 1.5 \\ 2.5y + 2.5 & a + 1 \end{bmatrix} = \begin{bmatrix} x + 0.5 & 2w - 1.5 \\ 5y - 2.5 & 19 - a \end{bmatrix}$

35. $\begin{bmatrix} \frac{1}{2} + a & \frac{1}{2}b + 2 \\ c - \frac{1}{3} & \frac{1}{3}d + \frac{2}{3} \end{bmatrix} = \begin{bmatrix} 6\frac{1}{2} - a & b - 1 \\ 4\frac{2}{3} & d + \frac{4}{9} \end{bmatrix}$

Error Analysis **Find the mistake in the solution of each equation. Explain the mistake and solve the equation correctly.**

36.
$$2x = 11x + 45$$
$$2x - 11x = 11x - 11x + 45$$
$$9x = 45$$
$$\frac{9x}{9} = \frac{45}{9}$$
$$x = 5$$

37.
$$4.5 - y = 2(y - 5.7)$$
$$4.5 - y = 2y - 11.4$$
$$4.5 - y - y = 2y - y - 11.4$$
$$4.5 = y - 11.4$$
$$4.5 + 11.4 = y - 11.4 + 11.4$$
$$15.9 = y$$

38. a. Solve $p + 2\left(p + \frac{1}{2}\right) = 9 - (4 - 3p)$ using a graphing calculator.
 b. Solve the same equation by hand.
 c. How do the answers in parts (a) and (b) relate to each other?

 39. Writing Is an equation that has 0 for a solution the same as an equation with no solution? Explain.

40. Spreadsheet Don set up a spreadsheet to solve $5(x - 3) = 4 - 3(x + 1)$.
 a. Does Don's spreadsheet show a solution to the equation?
 b. Between which two values of x is the solution to the equation? How do you know?
 c. For what values of x is $4 - 3(x + 1)$ less than $5(x - 3)$?

	A	B	C
	x	$5(x - 3)$	$4 - 3(x + 1)$
1			
2	−5	−40	16
3	−3	−30	10
4	−1	−20	4
5	1	−10	−2
6	3	0	−8

 **Challenge**

Open-Ended **Write an equation with a variable on each side such that you get the solution described.**

41. $x = 0$

42. x is a positive number.

43. x is a negative number.

44. All values of x are solutions.

45. No values of x are solutions.

46. $x = 1$

47. Use the equations below to find the length of the pipe.

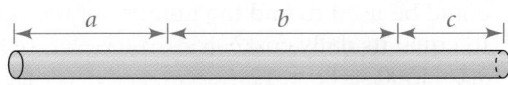

$$a + b = 15 \quad b - a = 3 \quad a + b - 12 = c$$

 48. Geometry The perimeters of the rectangles at the right are equal. Find the length and width of each rectangle.

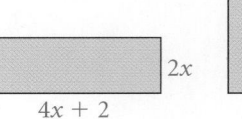

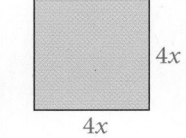

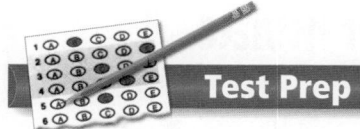

Multiple Choice

49. Solve $2y = 3y - 20$.

    A. -20 **B.** -4 **C.** 4   **D.** 20

50. Which of the following equations is NOT equivalent to the others?

    F. $-2(y - 3) = -6y$ **G.** $-2y - 6 = -6y$

    H. $y = -\frac{3}{2}$ **J.** $4y = -6$

51. Ace Truck Rental charges $54.00 a day plus 9¢ per mile. Roni's Truck Rental charges $38.00 a day plus 13¢ per mile. For how many miles will the cost of renting a truck for one day at Ace equal the cost at Roni's?

    A. 40 mi   **B.** 170 mi   **C.** 400 mi   **D.** 418 mi

52. Which equation is NOT equivalent to $3p - 2 = 6p + 4$?

    F. $3p = 6p + 6$ **G.** $-6 = 3p$

    H. $3p = 6$ **J.** $-3p - 2 = 4$

53. A record store sells CDs for $12.00 each. A music club offers 5 free CDs and charges $15.00 for each additional CD. Which equation can you use to find the number of CDs x that would cost the same under both plans?

    A. $15x - 5 = 12x$ **B.** $12x - 5 = 15x$

    C. $12x = 15(x - 5)$ **D.** $12(x - 5) = 15x$

54. Solve $2(y - 3) = 1.2 - y$.

    F. -1.6 **G.** 1.4   **H.** 1.6   **J.** 2.4

55. The perimeters of the rectangle and the triangle below are equal. Find the value of x.

    A. 6   **B.** 8   **C.** 10   **D.** 12

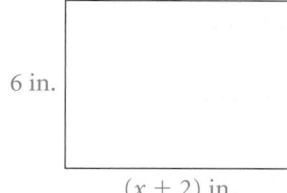 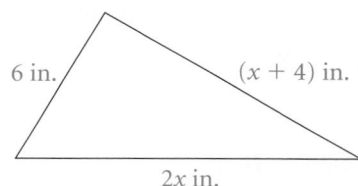

Mixed Review

Lesson 3-2   **Solve each equation.**

56. $9 = -4y + 6y - 5$ **57.** $-2(a - 3) = 14$ **58.** $0.5m + 2.8 = 3.64$

59. $\frac{1}{2}x + 4 = \frac{2}{3}$ **60.** $4.8 = 1.25(y - 17)$ **61.** $4\left(\frac{1}{4} + x\right) = 5$

Lesson 3-1   **62. Art** An art gallery owner is framing a painting. The width of the painting to be displayed is 30 in. He wants the width of the framed painting to be $38\frac{1}{2}$ in. How wide should each section of the frame be?

Lesson 1-3   **Write the numbers in each group in order from least to greatest.**

63. $-\frac{3}{5}, -\frac{5}{8}, -\frac{4}{5}$ **64.** $5.04, 5.009, 5.043$ **65.** $8.1, 8.02, 8.3$

66. $-100, 93, -87, 500$ **67.** $0.45, -1.24, 2.24, 1.23$ **68.** $9.7, -9.8, 8.6, 0.9$

Using and Transforming Formulas

A **literal equation** is an equation involving two or more variables. Formulas are special types of literal equations. To transform a literal equation, you solve for one variable in terms of the others. This is helpful when you need to use a formula numerous times.

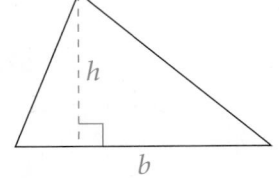

NY A.G.2: Use formulas to calculate volume and surface area of rectangular solids and cylinders.

1 EXAMPLE Transforming Geometric Formulas

Geometry Solve the formula for the area of a triangle $A = \frac{1}{2}bh$ for height h. Then find the height of the triangle if the area is 48 in.^{2} and the base is 4 in.

Step 1 Solve for h.

$$A = \frac{1}{2}bh$$

$$2A = 2\left(\frac{1}{2}\right)bh \qquad \textbf{Multiply each side by 2.}$$

$$2A = bh \qquad \textbf{Simplify.}$$

$$\frac{2A}{b} = \frac{bh}{b} \qquad \textbf{Divide each side by } \textit{b} \textbf{ to get } \textit{h} \textbf{ alone on one side of the equation.}$$

$$\frac{2A}{b} = h \qquad \textbf{Simplify.}$$

Step 2 Evaluate $\frac{2A}{b} = h$ for $A = 48$ and $b = 4$.

$$\frac{2(48)}{4} = h \qquad \textbf{Substitute 48 for } \textit{A} \textbf{ and 4 for } \textit{b.}$$

$$\frac{96}{4} = h \qquad \textbf{Simplify.}$$

$$24 = h$$

● The height is 24 in.

Formulas involving multiplication are used to calculate the volume of a cylinder.

2 EXAMPLE Using Formulas to Find Volume

Use th formula $V = \pi r^2 h$ to find the volume of the cylinder.

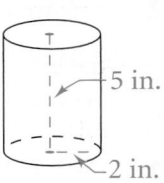

5 in.

2 in.

$$V = 3.14 r^2 h \qquad \textbf{Use 3.14 for } \pi.$$

$$V = 3.14 \cdot 2^2 \cdot 5 \qquad \textbf{Replace } \textit{r} \textbf{ with 2 and } \textit{h} \textbf{ with 5.}$$

$$V = 3.14 \cdot 4 \cdot 5 \qquad \textbf{Simplify.}$$

$$V = 62.8 \qquad \textbf{The volume is 62.8 in.}^3.$$

3 EXAMPLE Transforming Formulas for Real-World Problems

The formula $C = \frac{5}{9}(F - 32)$ gives the Celsius temperature C in terms of the Fahrenheit temperature F. Transform the formula to find Fahrenheit temperature in terms of Celsius temperature. Then find the Fahrenheit temperature when the Celsius temperature is 30°.

Step 1 Solve for F.

$$C = \frac{5}{9}(F - 32)$$

$$\frac{9}{5} \cdot C = \frac{9}{5} \cdot \frac{5}{9}(F - 32) \qquad \text{Multiply each side by } \frac{9}{5}, \text{ the reciprocal of } \frac{5}{9}.$$

$$\frac{9}{5}C = F - 32 \qquad\qquad \text{Simplify.}$$

$$\frac{9}{5}C + 32 = F - 32 + 32 \qquad \text{Add 32 to each side.}$$

$$\frac{9}{5}C + 32 = F \qquad\qquad \text{Simplify.}$$

Step 2 Find F when $C = 30$.

$$\frac{9}{5}(30) + 32 = F \qquad \text{Substitute 30 for C.}$$

$$54 + 32 = F \qquad \text{Find } \frac{9}{5}(30).$$

$$86 = F$$

30°C is equivalent to 86°F.

EXERCISES

1. a. Solve the formula $V = \ell wh$ for h.
   b. Copy and complete the table to find the height for rectangular prisms with the given volumes, lengths, and widths.

V	54 in.^{3}	64 in.^{3}	72 in.^{3}	90 in.^{3}
ℓ	3 in.	4 in.	3 in.	18 in.
w	2 in.	4 in.	8 in.	2 in.
h	■	■	■	■

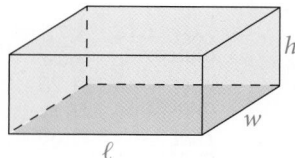

2. a. Construction Bricklayers use the formula $N = 7LH$ to estimate the number of bricks N needed to build a wall of height H given a length L in feet. Transform the formula to find the height of a wall in terms of the length and the number of bricks.

   b. What is the height of a wall that is 30 feet long and requires 2135 bricks to build?

3. You can use the number of chirps a cricket makes in one minute to estimate the outside temperature F in Fahrenheit. Transform the formula $F = \frac{n}{4} + 37$ to find the number of chirps a cricket makes in a minute in terms of a given temperature. How many chirps can you expect if the temperature is 60°F?

3-4

Ratio and Proportion

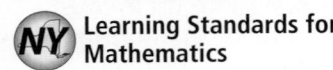

A.A.26 Solve algebraic proportions in one variable which result in linear or quadratic equations.

A.M.1 Calculate rates using appropriate units.

A.M.2 Solve problems involving conversions within systems, given the relationship between the units.

☑ **Check Skills You'll Need**   **GO for Help**   Skills Handbook pages 758 and 761

Write each fraction in simplest form.

1. $\frac{49}{84}$ 2. $\frac{24}{42}$ 3. $\frac{135}{180}$

Simplify each product.

4. $\frac{35}{25} \times \frac{40}{14}$ 5. $\frac{99}{144} \times \frac{96}{88}$ 6. $\frac{21}{81} \times \frac{108}{56}$

🔊 **New Vocabulary**   • ratio   • rate   • unit rate   • unit analysis   • proportion
• extremes of a proportion   • means of a proportion
• cross products

---

1   Ratios and Rates

A **ratio** is a comparison of two numbers by division. The ratio of a to b is $a:b$ or $\frac{a}{b}$, where $b \neq 0$. If a and b represent quantities measured in different units, then the ratio of a to b is a **rate.**

A **unit rate** is a rate with a denominator of 1. An example of a unit rate is $\frac{40 \text{ miles}}{1 \text{ hour}}$. You can write this rate as 40 miles per hour or 40 mi/h.

Price of Apple Juice

Price	Volume
$ .72	16 oz
$1.20	32 oz
$1.60	64 oz

1   EXAMPLE   **Using Unit Rates**

Comparison Shopping The table at the left gives prices for different sizes of the same brand of apple juice. Find the unit rate (cost per ounce) for each. Which has the lowest cost per ounce?

cost ⟶ $\frac{\$.72}{16 \text{ oz}}$ = $.045/oz   **Divide the numerator and denominator by 16.**
ounces →

cost ⟶ $\frac{\$1.20}{32 \text{ oz}}$ = $.0375/oz   **Divide the numerator and denominator by 32.**
ounces →

cost ⟶ $\frac{\$1.60}{64 \text{ oz}}$ = $.025/oz   **Divide the numerator and denominator by 64.**
ounces →

The unit rate for the 16 oz bottle of apple juice is 4.5¢/oz, for the 32-oz bottle is 3.75¢/oz, and for the 64-oz bottle is 2.5¢/oz. The 64-oz bottle has the lowest cost per ounce.

☑ **Quick Check**   ① Main Street Florist sells two dozen roses for $24.60. Flowers for You Florist sells six roses for $7.50. Find the unit rate for each. Which florist has the lower cost per rose?

You can use formulas like distance = rate × time ($d = rt$) to write a function to solve real-world problems. To write a function, you would use a unit rate for r.

 **EXAMPLE** Real-World Connection

Cycling In 2004, Lance Armstrong won the Tour de France, completing the 3391 km course in about 83.6 hours. Find Lance's unit rate, which is his average speed. Write a rule to describe the distance he cycles d as a function of the time t he cycles. Cycling at his average speed, about how long it would take Lance to cycle 185 km?

Step 1 Find Lance's unit rate.

$$\text{distance} \rightarrow \frac{3391 \text{ km}}{83.6 \text{ h}} \approx 40.6 \text{ kilometers per hour}$$
$$\text{time} \rightarrow$$

Step 2 Write a function that includes the unit rate.

$$d = rt$$
$$d = 40.6t$$

Step 3 Use the function to find t when $d = 185$.

$$185 = 40.6t$$
$$\frac{185}{40.6} = \frac{40.6t}{40.6}$$
$$4.56 \approx t$$

Traveling at his average speed, it would take Lance a little over $4\frac{1}{2}$ hours to cycle 185 km.

Quick Check ❷ Suppose you walk 2 miles in 35 minutes.

a. Find the average walking speed. Write a rule to describe the distance d you walk as a function of the time t you walk.

b. Use the function to find how far you would walk in an hour.

To change one unit of measure to another, you can use rates that equal 1. Since 60 min = 1 h, both $\frac{60 \text{ min}}{1 \text{ h}}$ and $\frac{1 \text{ h}}{60 \text{ min}}$ equal 1. You can use $\frac{60 \text{ min}}{1 \text{ h}}$ as a *conversion factor* to change hours into minutes and $\frac{1 \text{ h}}{60 \text{ min}}$ to change minutes into hours.

When converting from one unit to another, as in hours to minutes or minutes to hours, you must decide which conversion factor will produce the appropriate unit. This process is called **unit analysis,** or *dimensional analysis*.

❸ **EXAMPLE** Converting Rates

Speed of Cheetah A cheetah ran 300 feet in 2.92 seconds. What was the cheetah's average speed in miles per hour?

You need to convert feet to miles and seconds to hours.

$$\frac{300 \text{ ft}}{2.92 \text{ s}} \cdot \frac{1 \text{ mi}}{5280 \text{ ft}} \cdot \frac{60 \text{ s}}{1 \text{ min}} \cdot \frac{60 \text{ min}}{1 \text{ h}}$$ **Use appropriate conversion factors.**

$$= \frac{300 \text{ ft}}{2.92 \text{ s}} \cdot \frac{1 \text{ mi}}{5280 \text{ ft}} \cdot \frac{60 \text{ s}}{1 \text{ min}} \cdot \frac{60 \text{ min}}{1 \text{ h}}$$ **Divide the common units.**

$$\approx 70 \text{ mi/h}$$ **Simplify.**

The cheetah's average speed was about 70 mi/h.

Quick Check ❸ A sloth travels 0.15 miles per hour. Convert this speed to feet per minute.

A **proportion** is an equation that states that two ratios are equal.

$$\frac{a}{b} = \frac{c}{d} \text{ for } b \neq 0 \text{ and } d \neq 0$$

You read this proportion as "*a* is to *b* as *c* is to *d*." For this proportion *a* and *d* are the **extremes of the proportion,** and *b* and *c* are the **means of the proportion.** Another way you may see this proportion written is $a:b = c:d$.

You can use the Multiplication Property of Equality to solve a proportion for a variable.

Vocabulary Tip

Multiplication Property of Equality: For all numbers *a*, *b*, and *c*, if $a = b$, then $ac = bc$.

 **4 EXAMPLE** **Using the Multiplication Property of Equality**

Solve $\frac{t}{9} = \frac{5}{6}$.

$\frac{t}{9} \cdot 18 = \frac{5}{6} \cdot 18$ **Multiply each side by the least common multiple of 9 and 6, which is 18.**

$2t = 15$ **Simplify.**

$\frac{2t}{2} = \frac{15}{2}$ **Divide each side by 2.**

$t = 7.5$ **Simplify.**

 Quick Check **4** Solve each proportion.

a. $\frac{x}{8} = \frac{5}{6}$ **b.** $\frac{y}{12} = \frac{4}{7}$ **c.** $\frac{18}{50} = \frac{m}{15}$

You can use the Multiplication Property of Equality to prove an important property of proportions.

If $\frac{a}{b} = \frac{c}{d}$,

then $\frac{a}{b} \cdot bd = \frac{c}{d} \cdot bd$ **Multiplication Property of Equality**

$\frac{ab^1 d}{1\,b} = \frac{cbd^1}{1\,d}$ **Divide the common factors.**

and $ad = cb$ **Simplify.**

or $ad = bc$ **Commutative Property of Multiplication**

The products *ad* and *bc* are the **cross products** of the proportion $\frac{a}{b} = \frac{c}{d}$.

Key Concepts

Property	**Cross Products of a Proportion**
If $\frac{a}{b} = \frac{c}{d}$, then $ad = bc$.	
Example $\frac{2}{3} = \frac{8}{12}$, so $2 \cdot 12 = 3 \cdot 8$.	

Notice that for a proportion, the product of the extremes *ad* equals the product of the means *bc*. The Cross Products property is also called the means-extremes property of proportions.

Online active math

For: Proportion Activity
Use: Interactive Textbook, 3-4

5 **EXAMPLE** **Using Cross Products**

Nutrition A box of cereal weighing 354 grams contains 20 grams of fat. Find the number of grams of fat in the recommended serving size of 55 grams.

$$\begin{array}{l} \text{weight} \rightarrow \\ \text{fat} \rightarrow \end{array} \quad \frac{354}{20} = \frac{55}{x}$$

$354(x) = (55)(20)$ **Write cross products.**

$354x = 1100$ **Simplify.**

$\dfrac{354x}{354} = \dfrac{1100}{354}$ **Divide each side by 354.**

$x \approx 3.1$ **Simplify.**

A 55-gram serving has about 3.1 grams of fat.

 **Quick Check** **5** Solve each proportion by using cross products.

a. $\dfrac{x}{4} = \dfrac{25}{12}$ **b.** $\dfrac{24}{5} = \dfrac{y}{7}$ **c.** $\dfrac{54}{d} = \dfrac{72}{64}$

In Example 6, the ratios that form the proportion have variable expressions with more than one term. To solve for the variable, you will use cross products and the Distributive Property.

6 **EXAMPLE** **Solving Multi-Step Proportions**

Solve the proportion $\dfrac{x+4}{5} = \dfrac{x-2}{7}$.

$\dfrac{x+4}{5} = \dfrac{x-2}{7}$

$(x+4)(7) = 5(x-2)$ **Write cross products.**

$7x + 28 = 5x - 10$ **Use the Distributive Property.**

$2x + 28 = -10$ **Subtract 5x from each side.**

$2x = -38$ **Subtract 28 from each side.**

$x = -19$ **Divide each side by 2.**

 **Quick Check** **6** Solve each proportion.

a. $\dfrac{x+2}{14} = \dfrac{x}{10}$ **b.** $\dfrac{y-15}{y+4} = \dfrac{35}{7}$ **c.** $\dfrac{3}{w+6} = \dfrac{5}{w-4}$

EXERCISES

For more exercises, see *Extra Skill and Word Problem Practice.*

Practice and Problem Solving

A Practice by Example

Example 1
(page 142)

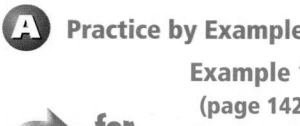

Find each unit rate.

1. $57 for 6 hours   **2.** $\dfrac{\$2}{5\text{ lb}}$ **3.** $\dfrac{524\text{ cars}}{4\text{ weeks}}$ **4.** $\dfrac{600\text{ calories}}{1.5\text{ h}}$

5. A 10-oz bottle of shampoo costs $2.40 and a 12-oz bottle costs $2.64. Find the unit rate for each. Which bottle has the lower unit cost?

6. Two students are preparing for a marathon. Hector ran 8 miles in 85 minutes. Mario ran 6 miles in 55 minutes. Who has the faster average speed?

Example 2
(page 143)

7. Mrs. Magdalino kept records on how much she spent on gasoline and the maintenance of her car. She found that it cost $485 to drive 500 mi in a month.
 a. Find the cost per mile. Write a rule to describe the cost c for gasoline and maintenance of a car as a function the number of miles m the car is driven.
 b. Use the function to find the cost for driving 1,200 miles.
 c. About how many miles are driven for a cost of $820?

8. You are riding your bicycle. It takes you 20 minutes to go 5 miles.
 a. Find your average speed. Write a rule to describe the distance d you cycle as a function of the number of minutes m you cycle.
 b. How long would it take you to cycle 12 miles?

Example 3
(page 143)

Choose A or B for the correct conversion factor for each situation.

9. quarts to gallons

 A. $\dfrac{1 \text{ gal}}{4 \text{ qt}}$ **B.** $\dfrac{4 \text{ qt}}{1 \text{ gal}}$

10. ounces to pounds

 A. $\dfrac{1 \text{ lb}}{16 \text{ oz}}$ **B.** $\dfrac{16 \text{ oz}}{1 \text{ lb}}$

11. inches to yards

 A. $\dfrac{36 \text{ in.}}{1 \text{ yd}}$ **B.** $\dfrac{1 \text{ yd}}{36 \text{ in.}}$

12. miles to feet

 A. $\dfrac{5280 \text{ ft}}{1 \text{ mi}}$ **B.** $\dfrac{1 \text{ mi}}{5280 \text{ ft}}$

Complete each statement.

13. $8 \text{ h} = \blacksquare \text{ min}$ **14.** $120 \text{ cm} = \blacksquare \text{ m}$ **15.** $3 \text{ h} = \blacksquare \text{ s}$

Example 4
(page 144)

Solve each proportion.

16. $\dfrac{5}{6} = \dfrac{c}{9}$ **17.** $\dfrac{3}{8} = \dfrac{x}{30}$ **18.** $\dfrac{2}{8} = \dfrac{n}{20}$ **19.** $\dfrac{7}{5} = \dfrac{k}{18}$

20. $\dfrac{3}{4} = \dfrac{x}{10}$ **21.** $\dfrac{4}{6} = \dfrac{m}{9}$ **22.** $\dfrac{8}{d} = -\dfrac{12}{30}$ **23.** $\dfrac{5}{9} = \dfrac{8}{w}$

Example 5
(page 145)

24. A canary's heart beats 200 times in 12 seconds. Use a proportion to find how many times its heart beats in 42 seconds.

25. Suppose you traveled 66 kilometers in 1.25 hours. Moving at the same speed, how many kilometers would you cover in 2 hours?

Example 6
(page 145)

Solve each proportion.

26. $\dfrac{x + 3}{4} = \dfrac{7}{8}$ **27.** $\dfrac{a - 6}{5} = \dfrac{7}{12}$ **28.** $\dfrac{8}{9} = \dfrac{w - 2}{6}$

29. $\dfrac{1}{c + 5} = \dfrac{2}{3}$ **30.** $\dfrac{8}{b + 10} = \dfrac{4}{2b - 7}$ **31.** $\dfrac{k + 5}{10} = \dfrac{k - 12}{9}$

B **Apply Your Skills**

Complete each statement.

32. $2/\text{lb} = \blacksquare \text{ ¢/oz}$ **33.** $3/\text{lb} = \blacksquare \text{ ¢/oz}$

34. $4¢/\text{day} = \$ \blacksquare/\text{yr}$ **35.** $5¢/\text{day} = \$ \blacksquare/\text{yr}$

36. $5 \text{ cm/min} = \blacksquare \text{ m/week}$ **37.** $1 \text{ qt/min} = \blacksquare \text{ gal/week}$

Express each rate in miles per hour.

38. 1 mi in 3 min **39.** 1 mi in 4 min **40.** 1 mi in 300 s

41. 10,560 ft in 2 h **42.** 21,120 ft in 4 h **43.** 270 ft in 10.8 min

Solve each proportion.

44. $\dfrac{m + 12}{9m} = \dfrac{5}{9}$ **45.** $\dfrac{p}{20} = \dfrac{p - 4}{5}$ **46.** $\dfrac{n + 12}{4} = \dfrac{n}{16}$

47. Hair Human hair grows at a rate of about 2.45 mm per week. Find the unit rate (millimeters per day). Write a rule that describes the amount of growth *g* as a function of *n* number of days. Use the function to find about how much hair grows in 30 days.

48. Record Speed According to the *Guinness Book of World Records*, the peregrine falcon has a record diving speed of 168 miles per hour. Write this speed in feet per second.

Data Analysis **Below are the survey results of 60 students. These results are representative of the 1250 students in the school. Use the table for Exercises 49–51.**

Question	Number Answering Yes
Do you work on weekends?	31
Do you spend 2 or more hours per night on homework?	36
Do you buy lunch in the school cafeteria?	48

49. Predict the number of students in the school who work on weekends.

50. Predict the number of students in the school who spend 2 or more hours per night on homework.

51. Predict the number of students in the school who buy lunch in the school cafeteria.

52. Writing Write an explanation telling an absent classmate how to use cross products to solve a proportion. Include an example.

53. Multiple Choice Your car averages 34 miles per gallon on the highway. If gas costs $2.10 per gallon, how much does it cost in dollars per mile, to drive your car on the highway?
  Ⓐ $.06/mi   Ⓑ 16 mi/dollar   Ⓒ $16/mi   Ⓓ 0.06 mi/dollar

54 Demographics Population density is a unit rate describing the number of individuals per unit of area. An example of population density is 5 people per square mile. Use the diagram below. Find the population densities of Mongolia, Bangladesh, and the United States. Round your answers to the nearest integer.

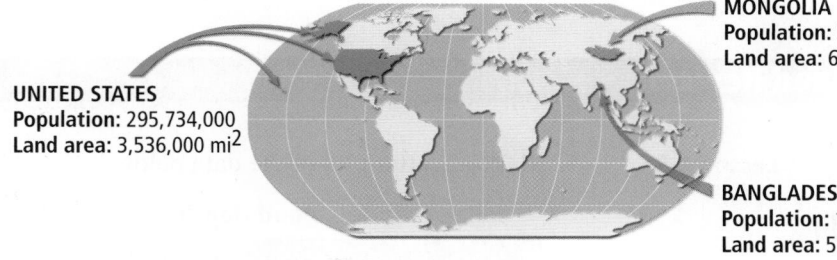

MONGOLIA
Population: 2,791,000
Land area: 604,000 mi^2

UNITED STATES
Population: 295,734,000
Land area: 3,536,000 mi^2

BANGLADESH
Population: 144,320,000
Land area: 52,000 mi^2

55. Open-Ended Estimate your walking rate in feet per second. Write this rate in miles per hour.

56. Bonnie and Tim do some yardwork for their neighbor. The ratio comparing the amount of time each one works is 7:4. The neighbor pays them $88. If Bonnie worked more, how much should each of them receive?

Real-World Connection

Enrollment in grades 9 through 12 has risen over 19% since 1990.

Homework Video Tutor
Visit: PHSchool.com
Web Code: ate-0304

57. Engineering Transformers use coils of wire to increase or decrease voltage. The following proportion relates the number of turns of wire in the coils to the voltages. (*Note:* Each semi-circle in the diagram represents a turn of wire.)

primary coil ⟩⟩ secondary coil $\dfrac{\text{primary turns}}{\text{secondary turns}} = \dfrac{\text{primary voltage}}{\text{secondary voltage}}$

Suppose the primary-coil voltage is 120 volts. Use the proportion $\dfrac{5}{2} = \dfrac{120}{v}$ to find the secondary-coil voltage.

C Challenge   **Solve each proportion.**

58. $\dfrac{x^2 - 3}{5x + 2} = \dfrac{x}{5}$ **59.** $\dfrac{w^3 + 7}{w} = \dfrac{9w^2 + 7}{9}$ **60.** $\dfrac{m^2 - 8}{3m} = \dfrac{4m + 1}{12}$

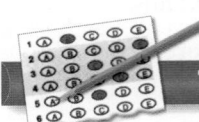 **61. Sports** Long-distance runners usually refer to their speed in terms of pace, a rate measured in minutes per mile rounded to the nearest hundredth.
 a. Naoko Takahashi of Japan won the gold medal in the marathon at the 2000 Olympic games. She set an Olympic record, completing the 26.2-mile race in 2:23:14 (2 hours, 23 minutes, 14 seconds). Find her pace.
 b. Tegla Loroupe of Kenya set the women's world record at the 1999 Berlin marathon. Her time was 2:20:43. Find her pace.

NY REGENTS

Test Prep

Multiple Choice

62. To the nearest tenth of a cent, what is the unit cost of a 28-ounce bottle of dish detergent that is on sale for $2.50?
 A. 8¢/ounce   **B.** 8.9¢/ounce   **C.** 11¢/ounce   **D.** 11.2¢/ounce

63. Most mammals breathe about once every 4 heartbeats. A large dog's heart beats about 180 times in one minute. Which unit rate represents the number of times this dog breathes in 1 minute?
 F. $\dfrac{40 \text{ breaths}}{\text{minute}}$ **G.** $\dfrac{45 \text{ breaths}}{\text{minute}}$ **H.** $\dfrac{180 \text{ breaths}}{4 \text{ minutes}}$ **J.** $\dfrac{720 \text{ breaths}}{\text{minute}}$

Short Response

64. Many trees have concentric rings that can be counted to determine the tree's age. Each ring represents one year's growth. If a maple tree with a diameter of 12 inches has 32 rings, write a proportion to find the number of rings in a maple tree that has a diameter of 20 inches. Estimate the age of a maple tree with a 20-inch diameter. Show your work.

Mixed Review

GO for Help

Lesson 1-5

65. Make a scatter plot of the data below.

Pond Depth

Annual Rainfall (inches)	50	48	64	55	68
Pond Depth (feet)	21	20	27	24	28

Lessons 3-1, 3-2   **Solve and check each equation.**

66. $15x + 30 = 90$ **67.** $34 = \dfrac{t}{4} + 5$ **68.** $-31 = 7h + 11$

69. $2b + 4.3 - b = 9.8$ **70.** $-\dfrac{3}{5}(c + 20) = 54$ **71.** $\dfrac{5v}{6} + \dfrac{7}{6} = \dfrac{1}{12}$

72. $6 = 7 - \dfrac{t}{2} + 8$ **73.** $9b + 8b - 13 = 106$ **74.** $2(p - 7.5) - 4 = 32$

Proportions and Similar Figures

Learning Standards for Mathematics

A.A.6 Analyze and solve verbal problems whose solution requires solving a linear equation in one variable or linear inequality in one variable.

A.A.25 Solve equations involving fractional expressions.

A.A.26 Solve algebraic proportions in one variable which result in linear or quadratic equations.

✓ **Check Skills You'll Need**

GO for Help Skills Handbook and Lesson 3-4

Simplify each ratio.

1. $\frac{36}{42}$

2. $\frac{81}{108}$

3. $\frac{26}{52}$

Solve each proportion.

4. $\frac{x}{12} = \frac{7}{30}$

5. $\frac{y}{12} = \frac{8}{45}$

6. $\frac{w}{15} = \frac{12}{27}$

7. $\frac{9}{a} = \frac{81}{10}$

8. $\frac{25}{75} = \frac{z}{30}$

9. $\frac{n}{9} = \frac{n+1}{24}$

◀)) **New Vocabulary** • similar figures • dilation • scale factor
• scale drawing • scale

---

1 **Similar Figures**

Real-World Connection

The triangles in the quilt are the same shape, so they are *similar*.

Activity: Proportions in Triangles

The figure below shows $\triangle ACB$ and $\triangle DCE$.

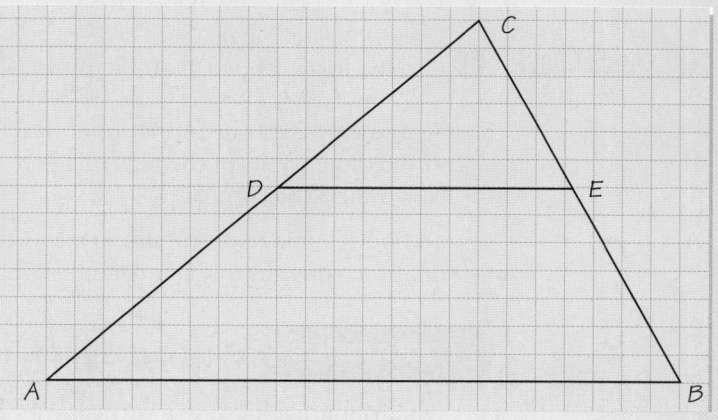

1. Measure *AB*, *CA*, *CB*, *CD*, *CE*, and *DE* using a metric ruler.

2. Find each ratio. Round to the nearest hundredth.
   a. $\frac{DE}{AB}$ **b.** $\frac{CE}{CB}$ **c.** $\frac{CD}{CA}$

3. Tell whether each statement appears to be true.
   a. $\frac{DE}{AB} = \frac{CE}{CB}$ **b.** $\frac{CD}{CA} = \frac{CE}{CB}$ **c.** $\frac{DE}{AB} = \frac{CD}{CA}$

4. Using the lengths you have measured, write two ratios that equal $\frac{CB}{CE}$.

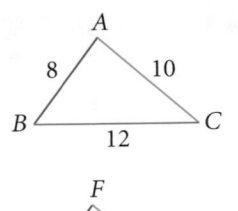

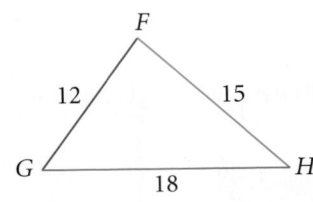

In the diagram, $\triangle ABC$ and $\triangle FGH$ are similar. **Similar figures** have the same shape but not necessarily the same size. The symbol $\sim$ means *is similar to*.

In similar triangles, corresponding angles are congruent and corresponding sides are in proportion. The order of the letters indicates the corresponding angles. If $\triangle ABC \sim \triangle FGH$, then the following is true.

$$\angle A \cong \angle F \qquad \angle B \cong \angle G \qquad \angle C \cong \angle H$$

$$\frac{AB}{FG} = \frac{AC}{FH} = \frac{BC}{GH}$$

1 EXAMPLE Finding the Length of a Side

Multiple Choice In the figure at the right, $\triangle ABC \sim \triangle DEF$. Find DE.

A 36 cm   B 28.45 cm
C 20.25 cm   D 18 cm

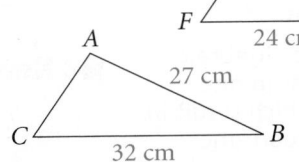

Relate $\frac{BC}{EF} = \frac{AB}{DE}$ Relate the lengths of corresponding sides.

Define Let $x = DE$.

Write $\frac{32}{24} = \frac{27}{x}$ Substitute lengths given in the diagram.

$\qquad 32x = 24(27)$ Write cross products.

$\qquad \frac{32x}{32} = \frac{648}{32}$ Divide each side by 32.

$\qquad x = 20.25$ Simplify.

● DE is 20.25 cm. The answer is C.

 **1** Use the figure above. If $AC = 14$ cm, what is DF?

A **dilation** is a transformation in which a figure and its image are similar. The ratio of the dimensions of the new image to the dimensions of the original figure is called the **scale factor.**

To find the image of a figure in a coordinate plane, with the origin as the center of dilation, you multiply the x- and y-coordinates by the scale factor.

2 EXAMPLE Dilating Figures on the Coordinate Plane

Quadrilateral $PQRS$ has vertices $P(-2, 4)$, $Q(4, 4)$, $R(4, -2)$, and $S(-4, -4)$. It is dilated by a scale factor of $\frac{1}{2}$, and the origin is the center of dilation. Graph the original figure and its dilation.

Multiply the x- and y-coordinates of each point by $\frac{1}{2}$. Then graph the figure.

$P(-2, 4) \rightarrow P'(-1, 2)$ $Q(4, 4) \rightarrow Q'(2, 2)$
$R(4, -2) \rightarrow R'(2, -1)$ $S(-4, -4) \rightarrow S'(-2, -2)$

● The coordinates of S' are $(-2, -2)$.

 **2** $\triangle MNP$ has vertices $M(3, -2)$, $N(3, 5)$, and $P(-4, -1)$. Graph the triangle and its image after a dilation with a scale factor of 2.

You can use proportions to find the dimensions of objects that are difficult to measure directly.

3 EXAMPLE Applying Similarity

Indirect Measurement A tree casts a shadow 7.5 ft long. A woman 5 ft tall casts a shadow 3 ft long. The triangle shown for the tree and its shadow is similar to the triangle shown for the woman and her shadow. How tall is the tree?

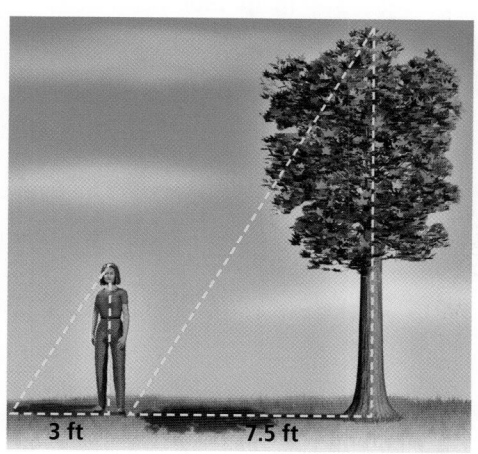

$\dfrac{3}{7.5} = \dfrac{5}{x}$ **Corresponding sides of similar figures are in proportion.**

$3x = 7.5 \cdot 5$ **Write cross products.**

$3x = 37.5$ **Simplify.**

$x = 12.5$ **Divide each side by 3.**

3 ft   7.5 ft

● The tree is 12.5 ft tall.

✓ Quick Check   **3** **a.** A tree casts a 26-ft shadow. A boy standing nearby casts a 12-ft shadow. His height is 4.5 ft. How tall is the tree?

b. A house casts a 56-ft shadow. A girl standing nearby casts a 7.2-ft shadow. Her height is 5.4 ft. What is the height of the house?

A **scale drawing** is an enlarged or reduced drawing that is similar to an actual object or place. Floor plans, blueprints, and maps are all examples of scale drawings. The ratio of a distance in the drawing to the corresponding actual distance is the **scale** of the drawing.

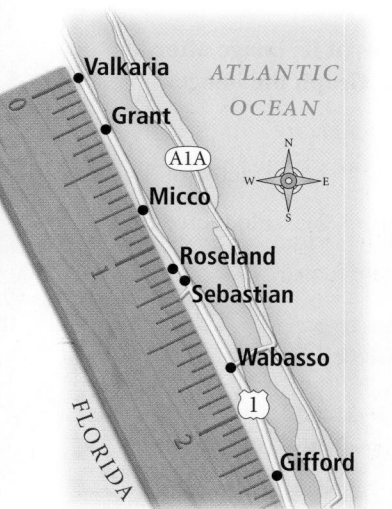

Valkaria
ATLANTIC
OCEAN
Grant
A1A
N
W   E
S
Micco
Roseland
Sebastian
Wabasso
1
FLORIDA
Gifford

4 EXAMPLE Finding Distances on Maps

The scale of the map at the left is 1 inch : 10 miles. Approximately how far is it from Valkaria to Wabasso?

Map distance = 1.75 in.   **Measure the map distance.**

$\begin{array}{cc} \text{map} \rightarrow \\ \text{actual} \rightarrow \end{array} \quad \dfrac{1}{10} = \dfrac{1.75}{d} \quad \begin{array}{c} \leftarrow \text{map} \\ \leftarrow \text{actual} \end{array}$ **Write a proportion.**

$1 \cdot d = 10 \cdot 1.75$ **Write cross products.**

$d = 17.5$ **Simplify.**

● Wabasso is about 17.5 mi from Valkaria.

✓ Quick Check   **4** **a.** On the map above, measure the map distance from Grant to Gifford. Find the actual distance.

b. Critical Thinking If another map showed the distance from Valkaria to Wabasso but had a scale of 1 inch : 5 miles, what would the map distance be between the two locations?

Practice by Example

Example 1
(page 150)

GO for Help

The figures in each pair are similar. Identify the corresponding sides and angles.

1.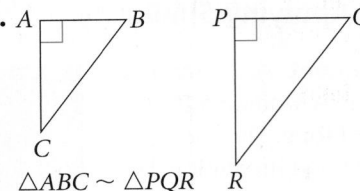
$\triangle ABC \sim \triangle PQR$

2.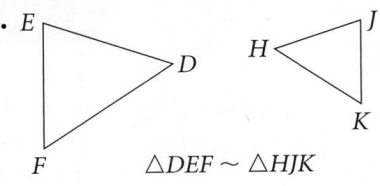
$\triangle DEF \sim \triangle HJK$

The figures in each pair are similar. Find the missing length.

3.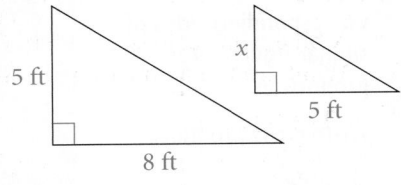

4.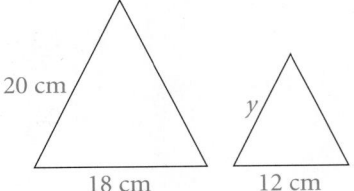

5.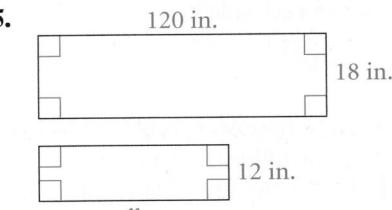

6.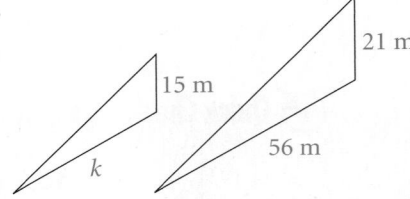

Example 2
(page 150)

Graph the coordinates of each figure. Find the coordinates of its image after a dilation centered at the origin with the given scale factor. Graph the image.

7. $\triangle ABC$; $A(3,5)$, $B(1,-2)$, $C(-6,3)$; scale factor of $\frac{1}{3}$

8. Rectangle $GHJK$; $G(-1,-2)$, $H(-1,3)$, $J(4,3)$, $K(4,-2)$; scale factor of 3

9. Quadrilateral $STUV$; $S(3,5)$, $T(4,-3)$, $U(-2,-5)$, $V(-3,2)$; scale factor of 0.5

Example 3
(page 151)

The child in the figure is 3 ft tall.

10. How tall is the tree?

11. The cat casts an 18-in. shadow. How tall is the cat?

Example 4
(page 151)

The scale of a map is 1 in. : 17.5 mi. Find the actual distance corresponding to each map distance.

12. 5 in.   **13.** 8.3 in.   **14.** 18.6 in.   **15.** 20 in.

Lincoln

San Paulo

Duncanville

SCALE
1 in. : 16 mi

16. a. Use a ruler and the map at the left. Find the distance from each town to the others.
   b. A student lives halfway between Lincoln and San Paulo and takes the shortest route to school in Duncanville. How far does the student travel each day to school?

17. The actual distance between two towns is 28 km. Suppose you measure the distance on your map and find that it is 3.5 cm. What is the scale of your map?

Using each of the following scales, find the dimensions in a blueprint of an 8 ft-by-12 ft room.

18. 1 in.:2 ft   **19.** 1 in.:3 ft   **20.** 1 in.:4 ft   **21.** 1 in.:2.5 ft

B **Apply Your Skills**

22. Two rectangles are similar. The first is 4 in. wide and 15 in. long. The second is 9 in. wide. Find the length of the second rectangle.

23. Architecture A blueprint scale is 1 in.:9 ft. On the plan, the room measures 2.5 in. by 3 in. What are the actual dimensions of the room?

24. Error Analysis The two figures are similar. Robert uses the proportion $\frac{GH}{PQ} = \frac{GK}{RQ}$ to find RQ.
   a. What is Robert's error?
   b. What proportion should he have used?

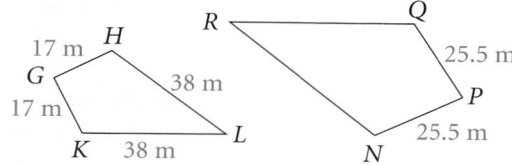

Architecture A 2-in. length in the scale drawing represents an actual length of 24 ft.

25. What is the scale of the drawing?

26. What are the actual dimensions of the kitchen?

27. Find the actual width of the doorways that lead into the kitchen and the dining room.

28. Find the actual area of the dining room.

29. Can a table 7 ft long and 4 ft wide fit into the narrower section of the dining room? Explain your answer.

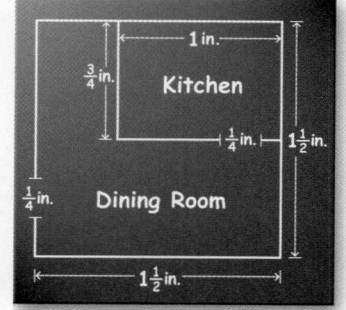

1 in.

$\frac{3}{4}$ in.   Kitchen

$\frac{1}{4}$ in. $1\frac{1}{2}$ in.

$\frac{1}{4}$ in.   Dining Room

$1\frac{1}{2}$ in.

30. Writing Are the trapezoids in the diagram similar figures? Explain your answer.

31. Open-Ended Give some examples of similar figures found in everyday life.

32. Two rectangles are similar. One is 5 cm by 12 cm. The longer side of the second rectangle is 8 cm greater than twice its shorter side. Find its length and width.

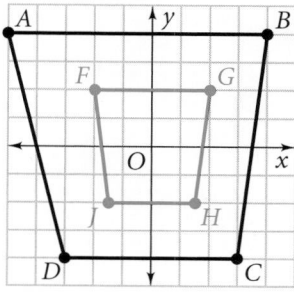

33. Multiple Choice $\triangle ABC$ and $\triangle TUV$ are similar figures. Which scale factor was used to transform $\triangle ABC$ to $\triangle TUV$?
   Ⓐ 20   Ⓑ $\frac{3}{7}$
   Ⓒ $\frac{7}{3}$ Ⓓ $\frac{7}{5}$

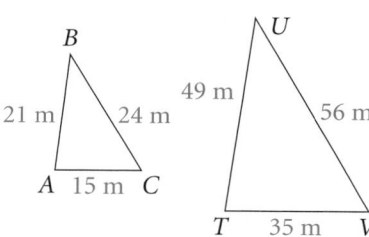

34. Geometry The perimeter of a triangle with sides a, b, and c is 24 cm. Side a is 2 cm longer than side b. The ratio of the lengths of sides b and c is $3 : 5$. What are the lengths of the three sides of the triangle?

35. The state of Alabama is about 335 mi long and 210 mi wide. What scale would you use to draw a map of Alabama on an $8\frac{1}{2}$ in.-by-11 in. paper to make the map as large as possible?

C Challenge **36. Astronomy** You can block out the moon by holding a coin up at a distance from your eye that is 110 times the diameter of the coin. Using similar figures, $\frac{\text{coin diameter}}{\text{moon diameter}} = \frac{\text{coin distance}}{\text{moon distance}}$. The moon is roughly 3640 kilometers in diameter. How far away is it?

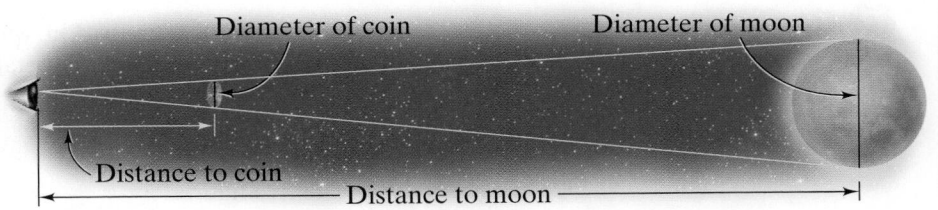

Diameter of coin   Diameter of moon

Distance to coin

Distance to moon

Not drawn to scale.

37. Geometry In the figure at the right, $\triangle ABC \sim \triangle ADE$.
  a. Substitute values from the diagram into the following proportion. $\frac{AD}{AB} = \frac{DE}{BC}$
  (*Hint:* $AB = AD + DB$.)
  b. Solve the proportion for x.
  c. Find the length of AB.
  d. What is the area of $\triangle ABC$?

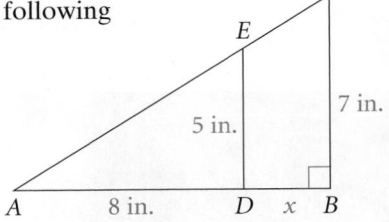

NY REGENTS

Test Prep

Multiple Choice

38. In the figure at the right, $\triangle ABC \sim \triangle XYZ$. Which proportion is incorrect?

  A. $\frac{AB}{AC} = \frac{XY}{XZ}$ **B.** $\frac{AB}{BC} = \frac{XY}{XZ}$

  C. $\frac{BC}{AC} = \frac{YZ}{XZ}$ **D.** $\frac{AC}{XZ} = \frac{BC}{YZ}$

39. A map of Kentucky is drawn with a scale of 1 cm : 11 km. The map distance between Louisville and Bowling Green is 14.5 cm. Which is the best estimate of the actual distance?
  F. 1.3 km   **G.** 14 km   **H.** 100 km   **J.** 160 km

40. Which of the following would be most helpful as the first step in solving $45t + 9 = 23 - 7t$?
  A. Add $7t$ to each side of the equation.
  B. Divide each side of the equation by 9.
  C. Add 23 to each side of the equation.
  D. Multiply each side of the equation by negative 1.

41. You can paint a 6 ft-by-5 ft rectangular wall using 0.5 gallon of paint. How many gallons of paint will you need to cover a 10 ft-by-12 ft wall? Show your work.

Mixed Review

Lesson 3-4

Solve each proportion.

42. $\frac{x}{2} = \frac{9}{4}$

43. $\frac{5}{n} = \frac{3}{10}$

44. $\frac{-8}{m} = \frac{7}{20}$

45. $\frac{12}{30} = \frac{16}{v}$

46. $\frac{5}{2-x} = \frac{7}{10}$

47. $\frac{3+x}{8} = \frac{7}{12}$

Lesson 1-4

Write a function rule for each table.

48.

x	y
1	12
2	13
3	14
4	15

49.

x	y
1	18
2	14
3	10
4	6

50.

x	y
1	35
2	38
3	41
4	44

Checkpoint Quiz 1   Lessons 3-1 through 3-5

Solve each equation.

1. $4n + 7 + 6n = 32$

2. $5(3 - d) = 2d + 1$

3. $8(h - 1) = 6h + 4 + 2h$

4. $43 = 8y - 11$

Solve each proportion.

5. $\frac{8}{k} = -\frac{12}{30}$

6. $\frac{3}{5} = \frac{y+1}{9}$

7. You are riding your bicycle to prepare for a race. It takes you 12 min to go 2.5 mi. If you continue traveling at the same rate, how long will it take you to go 7 mi?

The figures in each pair are similar. Find the missing length.

8.

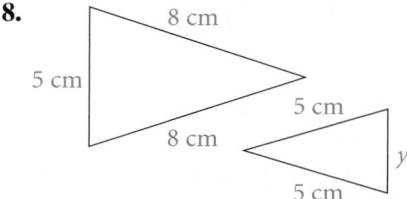

9.

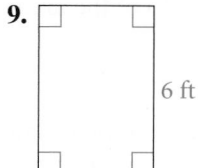
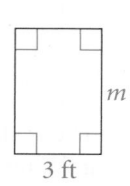

10. In the late afternoon, a 3.5-ft child casts a 60-in. shadow. She is standing next to a telephone pole that casts a 50-ft shadow. How tall is the telephone pole?

Scale Factor: Perimeter, Area, and Volume

FOR USE WITH LESSON 3-5

In this Activity Lab, you will investigate the perimeters and areas of similar figures and the surface areas and volumes of similar three-dimensional objects.

1 ACTIVITY

The graph shows four dilations of rectangle A.

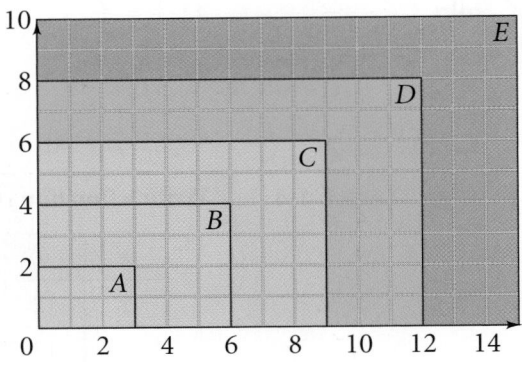

1. Copy and complete the table below.

Rectangle	Width	Length	Perimeter	Area
A	2 units	3 units	10 units	6 units^2
B	■	■	■	■
C	■	■	■	■
D	■	■	■	■
E	■	■	■	■

2. Use the information in your table from Exercise 1. Copy and complete the table below.

Rectangles	Scale Factor	Ratio of Corresponding Sides	Ratio of Perimeters	Ratio of Areas
A : B	1 : 2	1 : 2	■	■
A : C	■	■	■	■
A : D	■	■	■	■
A : E	■	■	■	■
B : C	■	■	■	■

3. a. What can you conclude about the scale factor and the ratios of perimeters?
   b. Complete this statement: If the scale factor of two similar figures is $a : b$, then the ratio of the perimeters is __?__.

4. a. What can you conclude about the scale factor and the ratios of areas?
   b. Complete this statement: If the scale factor of two similar figures is $a : b$, then the ratio of the areas is __?__.

EXERCISES

5. The scale factor of two triangles is 3 : 5.
   a. What is the ratio of the perimeters of the triangles?
   b. What is the ratio of the areas of the triangles?

6. The ratio of the areas of two similar triangles is $\frac{25}{16}$. The measure of one side of the smaller triangle is 6 in. What is the measure of the corresponding side of the larger triangle?

Use the rectangular prisms at the right to answer the following questions.

7. Copy and complete the table.

Prism	Surface Area	Volume
A	22 units^2	6 units^3
B	■	■
C	■	■

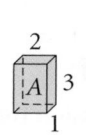
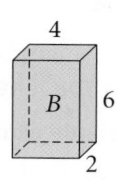
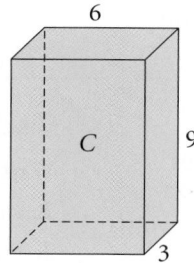

8. Use the information in your table from Exercise 7. Copy and complete the table below.

Prisms	Scale Factor	Ratio of Corresponding Sides	Ratio of Surface Areas	Ratio of Volumes
A : B	1 : 2	1 : 2	■	■
A : C	■	■	■	■
B : C	■	■	■	■

9. a. What can you conclude about the scale factor and the ratios of surface areas of similar three-dimensional objects?
   b. Complete this statement: If the scale factor of the lengths of corresponding sides of three-dimensional objects is $a : b$, then the ratio of the surface areas is __?__.

10. a. What can you conclude about the scale factor and the ratios of volumes of similar three-dimensional objects?
   b. Complete this statement: If the scale factor of the lengths of corresponding sides of three-dimensional objects is $a : b$, then the ratio of the volumes is __?__.

EXERCISES

11. The scale factor of two rectangular prisms is 5 : 8.
   a. What is the ratio of the surface areas of the rectangular prisms?
   b. What is the ratio of the volumes of the rectangular prisms?

12. The formula for the volume of a sphere is $V = \frac{4\pi r^3}{3}$. One sphere has a radius of 6 in. The second sphere has a radius of 10 in. Find the ratio of their volumes.

13. The ratio of the volumes of two similar pyramids is 27 : 125. What is the ratio of the heights of the two pyramids?

14. The ratio of the surface areas of two similar rectangular prisms is $\frac{36}{25}$. One side of the smaller prism is 4 cm long. What is the length of the corresponding side of the larger prism?

15. The ratio of the volumes of two similar cylinders is $\frac{27}{8}$. The radius of the larger cylinder is 12 cm. What is the length of the radius of the smaller cylinder?

Equations and Problem Solving

Learning Standards for Mathematics

A.A.5 Write algebraic equations or inequalities that represent a situation.

✓ Check Skills You'll Need

GO for Help Lesson 1-1

Write a variable expression for each situation.

1. value in cents of q quarters

2. twice the length ℓ

3. number of miles traveled at 34 mi/h in h hours

4. weight of 5 crates if each crate weighs x kilograms

5. cost of n items at $3.99 per item

🔊 **New Vocabulary** • consecutive integers • uniform motion

---

1 **Defining Variables**

Some problems contain two or more unknown quantities. To solve such problems, first decide which unknown quantity the variable will represent. Then express the other unknown quantity or quantities in terms of that variable.

1 **EXAMPLE** Defining One Variable in Terms of Another

Geometry The length of a rectangle is 6 in. more than its width. The perimeter of the rectangle is 24 in. What is the length of the rectangle?

Relate The length is 6 in. more than the width.

Define Let w = the width.

Then $w + 6$ = the length.

The length is described in terms of the width. So define a variable for the width first.

Write	$P = 2\ell + 2w$	Use the perimeter formula.
	$24 = 2(w + 6) + 2w$	Substitute 24 for P and $w + 6$ for ℓ.
	$24 = 2w + 12 + 2w$	Use the Distributive Property.
	$24 = 4w + 12$	Combine like terms.
	$24 - 12 = 4w + 12 - 12$	Subtract 12 from each side.
	$12 = 4w$	Simplify.
	$\dfrac{12}{4} = \dfrac{4w}{4}$	Divide each side by 4.
	$3 = w$	Simplify.

The width of the rectangle is 3 in. The length of the rectangle is 6 in. more than the width. So the length of the rectangle is 9 in.

Problem Solving Hint

For Example 1, drawing a diagram will help you understand the problem.

$w + 6$

w [ ]

✓ Quick Check

1 The width of a rectangle is 2 cm less than its length. The perimeter of the rectangle is 16 cm. What is the length of the rectangle?

Consecutive integers differ by 1. The integers 50 and 51 are consecutive integers, and so are -10, -9, and -8. For consecutive integer problems, it may help to define a variable before describing the problem in words. Let a variable represent one of the unknown integers. Then define the other unknown integers in terms of the first one.

2 EXAMPLE Consecutive Integer Problem

The sum of three consecutive integers is 147. Find the integers.

Define Let n = the first integer.
Then $n + 1$ = the second integer,
and $n + 2$ = the third integer.

Relate

first integer	plus	second integer	plus	third integer	is	147

Write

n	+	$n + 1$	+	$n + 2$	=	147

$$n + n + 1 + n + 2 = 147$$
$$3n + 3 = 147 \qquad \text{Combine like terms.}$$
$$3n + 3 - 3 = 147 - 3 \qquad \text{Subtract 3 from each side.}$$
$$3n = 144 \qquad \text{Simplify.}$$
$$\frac{3n}{3} = \frac{144}{3} \qquad \text{Divide each side by 3.}$$
$$n = 48 \qquad \text{Simplify.}$$

If $n = 48$, then $n + 1 = 49$, and $n + 2 = 50$. The three integers are 48, 49, and 50.

Check Is the solution correct? Yes; $48 + 49 + 50 = 147$.

✓ Quick Check **2** The sum of three consecutive integers is 48.
  a. Define a variable for one of the integers.
  b. Write expressions for the other two integers.
  c. Write and solve an equation to find the three integers.

---

2 Distance-Rate-Time Problems

Video Tutor Help
Visit: PHSchool.com
Web Code: ate-0775

An object that moves at a constant rate is said to be in **uniform motion.** The formula $d = rt$ gives the relationship between distance d, rate r, and time t. Uniform motion problems may involve objects going the same direction, opposite directions, or round trips.

In the diagram below, the two vehicles are traveling the same direction at different rates. The distances the vehicles travel are the same.

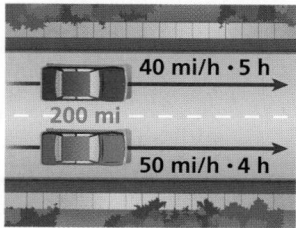

40 mi/h · 5 h

200 mi

50 mi/h · 4 h

Since the distances are equal, the products of rate and time for the two cars are equal. For the vehicles shown, $40 \cdot 5 = 50 \cdot 4$.

A table can also help you understand relationships in distance-rate-time problems.

3 EXAMPLE Same-Direction Travel

Engineering A train leaves a train station at 1 P.M. It travels at an average rate of 72 mi/h. A high-speed train leaves the same station an hour later. It travels at an average rate of 90 mi/h. The second train follows the same route as the first train on a track parallel to the first. In how many hours will the second train catch up with the first train?

Define Let t = the time the first train travels.

Then $t - 1$ = the time the second train travels.

Relate

Train	Rate	Time	Distance Traveled
1	72	t	$72t$
2	90	$t - 1$	$90(t - 1)$

Write $72t = 90(t - 1)$ The distances traveled by the trains are equal.

Method 1   Solve by using a table.

Use a table to evaluate each side of the equation. Look for matching values.

Time	$72t$	$90(t - 1)$
1	$72(1) = 72$	$90(1 - 1) = 0$
2	$72(2) = 144$	$90(2 - 1) = 90$
3	$72(3) = 216$	$90(3 - 1) = 180$
4	$72(4) = 288$	$90(4 - 1) = 270$
5	$72(5) = 360$	$90(5 - 1) = 360$ ✓

When the first train travels 5 hours, the second train travels 4 hours $(t - 1)$. The second train will catch up with the first train in 4 hours.

Method 2   Solve the equation.

$$72t = 90t - 90$$ Use the Distributive Property to simplify $90 (t - 1)$.

$$72t - 72t = 90t - 90 - 72t$$ Subtract $72t$ from each side.

$$0 = 18t - 90$$ Combine like terms.

$$0 + 90 = 18t - 90 + 90$$ Add 90 to each side.

$$90 = 18t$$ Simplify.

$$\frac{90}{18} = \frac{18t}{18}$$ Divide each side by 18.

$$t = 5$$ Simplify.

$$t - 1 = 4$$ Find the time the second train travels.

The second train will catch up with the first train in 4 h.

Real-World Connection

High-speed trains that go from Boston to New York in less than 4 hours can reach a speed of 150 mi/h.

 **Quick Check** ❸ A group of campers and one group leader left a campsite in a canoe. They traveled at an average rate of 10 km/h. Two hours later, the other group leader left the campsite in a motorboat. He traveled at an average rate of 22 km/h.
a. How long after the canoe left the campsite did the motorboat catch up with it?
b. How long did the motorboat travel?

For uniform motion problems that involve a round trip, it is important to remember that the distance going is equal to the distance returning.

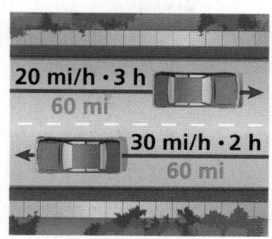

Since the distances are equal, the products of rate and time for traveling in both directions are equal. That is, $20 \cdot 3 = 30 \cdot 2$.

4 EXAMPLE Round-Trip Travel

Noya drives into the city to buy a software program at a computer store. Because of traffic conditions, she averages only 15 mi/h. On her drive home she averages 35 mi/h. If the total travel time is 2 hours, how long does it take her to drive to the computer store?

Define Let t = time of Noya's drive to the computer store.

$2 - t$ = the time of Noya's drive home.

Relate

Part of Noya's Travel	Rate	Time	Distance
To the computer store	15	t	$15t$
Return home	35	$2 - t$	$35(2 - t)$

Noya drives $15t$ miles to the computer store and $35(2 - t)$ miles back.

Problem Solving Hint

The total travel time is for a round trip. If it takes t out of a 2-hour round trip to get to the store, then $2 - t$ is the time it will take for the drive home.

Write

$15t = 35(2 - t)$	The distances traveled to and from the store are equal.
$15t = 70 - 35t$	Use the Distributive Property.
$15t + 35t = 70 - 35t + 35t$	Add $35t$ to each side.
$50t = 70$	Combine like terms.
$\dfrac{50t}{50} = \dfrac{70}{50}$	Divide each side by 50.
$t = 1.4$	Simplify.

● It took Noya 1.4 h to drive to the computer store.

✓ Quick Check ❹ On his way to work from home, your uncle averaged only 20 miles per hour. On his drive home, he averaged 40 miles per hour. If the total travel time was $1\frac{1}{2}$ hours, how long did it take him to drive to work?

For uniform motion problems involving two objects moving in opposite directions, you can write equations using the fact that the sum of their distances is the total distance.

5 EXAMPLE **Opposite-Direction Travel**

Jane and Peter leave their home traveling in opposite directions on a straight road. Peter drives 15 mi/h faster than Jane. After 3 hours, they are 225 miles apart. Find Peter's rate and Jane's rate.

Define Let r = Jane's rate.
Then $r + 15$ = Peter's rate.

Relate

Person	Rate	Time	Distance
Jane	r	3	$3r$
Peter	$r + 15$	3	$3(r + 15)$

Jane's distance is $3r$. Peter's distance is $3(r + 15)$.

Write $3r + 3(r + 15) = 225$ The sum of Jane's and Peter's distances is the total distance, 225 miles.

$$3r + 3(r + 15) = 225$$
$$3r + 3r + 45 = 225 \qquad \text{Use the Distributive Property.}$$
$$6r + 45 = 225 \qquad \text{Combine like terms.}$$
$$6r + 45 - 45 = 225 - 45 \qquad \text{Subtract 45 from each side.}$$
$$6r = 180 \qquad \text{Simplify.}$$
$$\frac{6r}{6} = \frac{180}{6} \qquad \text{Divide each side by 6.}$$
$$r = 30 \qquad \text{Simplify.}$$

Jane's rate is 30 mi/h, and Peter's rate is 15 mi/h faster, which is 45 mi/h.

 Quick Check **5** Sarah and John leave Perryville traveling in opposite directions on a straight road. Sarah drives 12 miles per hour faster than John. After 2 hours, they are 176 miles apart. Find Sarah's speed and John's speed.

EXERCISES

For more exercises, see *Extra Skill and Word Problem Practice.*

Practice and Problem Solving

A Practice by Example
Example 1
(page 158)

 for Help

1. The length of a rectangle is 3 in. more than its width. The perimeter of the rectangle is 30 in.
   a. Define a variable for the width.
   b. Write an expression for the length in terms of the width.
   c. Write an equation to find the width of the rectangle. Solve your equation.
   d. What is the length of the rectangle?

2. The length of a rectangle is 8 in. more than its width. The perimeter of the rectangle is 24 in. What are the width and length of the rectangle?

3. The width of a rectangle is one half its length. The perimeter of the rectangle is 54 cm. What are the width and length of the rectangle?

4. The length of a rectangular garden is 3 yd more than twice its width. The perimeter of the garden is 36 yd. What are the width and length of the garden?

Example 2
(page 159)

5. The sum of three consecutive integers is 915. What are the integers?

6. The sum of two consecutive *even* integers is 118.
  a. Define a variable for the smaller integer.
  b. What must you add to an even integer to get the next greater even integer?
  c. Write an expression for the second integer.
  d. Write and solve an equation to find the two even integers.

7. The sum of two consecutive *even* integers is −298. What are the integers?

8. The sum of two consecutive *odd* integers is 56.
  a. Define a variable for the smaller integer.
  b. What must you add to an odd integer to get the next greater odd integer?
  c. Write an expression for the second integer.
  d. Write and solve an equation to find the two odd integers.

Example 3
(page 160)

9. A moving van leaves a house traveling at an average rate of 40 mi/h. The family leaves the house $\frac{1}{2}$ hour later following the same route in a car. They travel at an average rate of 60 mi/h.
  a. Define a variable for the time traveled by the moving van.
  b. Write an expression for the time traveled by the car.
  c. Copy and complete the table.

Vehicle	Rate	Time	Distance Traveled
Moving van	▓	▓	▓
Car	▓	▓	▓

  d. Make a table comparing the distance traveled by the moving van and car for each hour. Use the table to find out how long it will take the car to catch up with the van.

10. Air Travel A jet leaves the Charlotte, North Carolina, airport traveling at an average rate of 564 km/h. Another jet leaves the airport one half hour later traveling at 744 km/h in the same direction. Use an equation to find how long the second jet will take to overtake the first.

Example 4
(page 161)

11. Juan drives to work. Because of traffic conditions, he averages 22 miles per hour. He returns home averaging 32 miles per hour. The total travel time is $2\frac{1}{4}$ hours.
  a. Define a variable for the time Juan takes to travel to work. Write an expression for the time Juan takes to return home.
  b. Write and solve an equation to find the time Juan spends driving to work.

12. Air Travel An airplane flies from New Orleans, Louisiana, to Atlanta, Georgia, at an average rate of 320 miles per hour. The airplane then returns at an average rate of 280 miles per hour. The total travel time is 3 hours.
  a. Define a variable for the flying time from New Orleans to Atlanta. Write an expression for the travel time from Atlanta to New Orleans.
  b. Write and solve an equation to find the flying time from New Orleans to Atlanta.

Example 5
(page 162)

13. John and William leave their home traveling in opposite directions on a straight road. John drives 20 miles per hour faster than William. After 4 hours they are 250 miles apart.
  a. Define a variable for John's rate. Write an expression for William's rate.
  b. Write and solve an equation to find John's rate. Then find William's rate.

14. Two bicyclists ride in opposite directions. The speed of the first bicyclist is 5 miles per hour faster than the second. After 2 hours they are 70 miles apart. Find their rates.

x ft

(1.5 + 2*x*) ft

B **Apply Your Skills**

15. The sum of three consecutive *odd* integers is −87. What are the integers?

16. The tail of a kite is 1.5 ft plus twice the length of the kite. Together, the kite and tail are 15 ft 6 in. long.
   a. Write an expression for the length of the kite and tail together.
   b. Write 15 ft 6 in. in terms of feet.
   c. Write and solve an equation to find the length of the tail.

17. Travel A bus traveling at an average rate of 30 miles per hour left the city at 11:45 A.M. A car following the bus at 45 miles per hour left the city at noon. At what time did the car catch up with the bus?

18. Ellen and Kate raced on their bicycles to the library after school. They both left school at 3:00 P.M. and bicycled along the same path. Ellen rode at a speed of 12 miles per hour and Kate rode at 9 miles per hour. Ellen got to the library 15 minutes before Kate. At what time did Ellen get to the library?

19. a. Which of the following numbers is not the sum of three consecutive integers?
   I. 51   **II.** 61   **III.** 72   **IV.** 81
   b. Critical Thinking What common trait do the other numbers share?

20. At 1:30 P.M., Tom leaves in his boat from a dock and heads south. He travels at a rate of 25 miles per hour. Ten minutes later, Mary leaves the same dock in her speedboat and heads after Tom. If she travels at a rate of 30 miles per hour, when will she catch up with Tom?

21. Air Travel Two airplanes depart from an airport traveling in opposite directions. The second airplane is 200 miles per hour faster than the first. After 2 hours they are 1100 miles apart. Find the speeds of the airplanes.

22. At 1:00 P.M. a truck leaves Centerville traveling 45 mi/h. One hour later a train leaves Centerville traveling 60 mi/h. They arrive in Smithfield at the same time.
   a. Use the table to find when the train and truck arrive in Smithfield.
   b. Critical Thinking What piece of information can you get from the table that you would NOT get by solving the equation $45t = 60(t - 1)$?

Time	Truck $45t$	Train $60(t-1)$
1 P.M.	45 mi	0 mi
2 P.M.	90 mi	60 mi
3 P.M.	135 mi	120 mi
4 P.M.	180 mi	180 mi
5 P.M.	225 mi	240 mi

23. Three friends were born in consecutive years. The sum of their birth years is 5982. Find the year in which each person was born.

24. Writing Describe the steps you would use to solve consecutive integer problems.

25. Open-Ended Write a word problem that could be solved using the equation $35(t - 1) = 20t$.

26. a. Write and solve an equation to find three consecutive integers with a sum of 126. Let $n =$ the first integer.
   b. Critical Thinking In part (a), could you solve the problem by letting $n =$ the middle integer, $n - 1 =$ the smallest integer, and $n + 1 =$ the largest integer?

27. Electricity A group of ten 6- and 12-volt batteries are wired in series as shown at the right. The sum of their voltages is 84 volts. How many of each type of battery are used?

Batteries in Series

GO **Online**
Homework Video Tutor
Visit: PHSchool.com
Web Code: ate-0306

28. Geometry A triangle has a perimeter of 165 cm. The first side is 65 cm less than twice the second side. The third side is 10 cm less than the second side. Write and solve an equation to find the length of each side of the triangle.

29. At 9:00 A.M., your friends begin hiking at 2 mi/h. You begin from the same place at 9:25 A.M. You hike at 3 mi/h.
 a. How long will you have hiked when you catch up with your friends?
 b. At what time will you catch up with your friends?

30. Find five consecutive *odd* integers such that the sum of the first and the fifth is one less than three times the fourth.

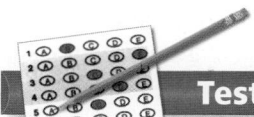

Test Prep

Multiple Choice

31. Solve $3n - 7 + 2n = 8n + 11$.
 A. -6 **B.** $1\frac{1}{3}$ **C.** $3\frac{3}{5}$ **D.** 9

32. Which expression represents the sum of 3 odd integers of which n is the least integer?
 F. $n + 3$ **G.** $3n + 3$
 H. $3n + 6$ **J.** $3n + 7$

33. Which equation does NOT have -2 as its solution?
 A. $2x + 5 = 5x + 11$ **B.** $7n + 9 = 3 - 9n$
 C. $3k + 6 - 4k = k + 10$ **D.** $4 + 3q = 7q + 12$

34. A truck traveling at an average rate of 45 miles per hour leaves a rest stop. Fifteen minutes later a car traveling at an average rate of 60 miles per hour leaves the same rest stop traveling the same route. How long will it take for the car to catch up with the truck?
 F. 15 minutes **G.** 45 minutes
 H. 1 hour 15 minutes **J.** 3 hours

35. The perimeter of the triangle at the right is 22.6 in. What is the value of n?
 A. 3.5 **B.** 4.6
 C. 7.8 **D.** 9.4

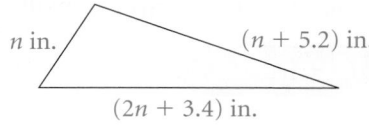

n in. $(n + 5.2)$ in. $(2n + 3.4)$ in.

Mixed Review

Lesson 3-3

Solve each equation. If the equation is an identity, write *identity*. If it has no solution, write *no solution*.

36. $2x = 7x + 10$ **37.** $2q + 4 = 4 - 2q$

38. $0.5t + 3.6 = 4.2 - 1.5t$ **39.** $2x + 5 + x = 2(3x + 3)$

40. $4 + x + 3x = 2(2x + 5)$ **41.** $8z + 2 = 2(z - 5) - z$

Lesson 3-2

42. Brendan earns $8.25 per hour at his job. He also makes $12.38 per hour for any number of hours over 40 that he works in one week. He worked 40 hours last week, plus some overtime, and made $385.71. How many overtime hours did he work?

Proportions and Percents

FOR USE WITH LESSON 3-7

To solve a percent problem such as "Find 38% of 320," you change the percent to a decimal and multiply.

$$38\% \text{ of } 320 \rightarrow 0.38 \cdot 320 = 121.6$$

For more involved percent problems, you can think of a percent as a ratio that compares a number to 100. You can solve problems involving percents using this proportion. Diagrams will help you visualize relationships.

$$\text{percent} \left\{ \frac{n}{100} = \frac{\text{part}}{\text{whole}} \right.$$

1 EXAMPLE

At Summerville High School there are 553 freshmen, which is 35% of the total number of students. How many students attend the high school?

Relate 35% of the total number of students is 553 students.

Define Let n = the total number of students.

Write $\frac{35}{100} = \frac{553}{n}$ ← part
← whole

$35n = 55{,}300$ **Find the cross products.**

$n = 1580$ **Divide each side by 35.**

There are a total of 1580 students at Summerville High School.

When you use proportions to solve real-world problems involving percents, first establish what is the part and what is the whole. In the next example, notice that the part is actually greater than the whole.

2 EXAMPLE

Band members had a fundraising goal of $2500 as they earned money to go to the state competition. The fundraiser was more successful than they expected, and they raised $3275. What percent of their goal did they raise?

Relate What percent of $2500 is $3275?

Define Let n = the percent.

Write $\frac{n}{100} = \frac{3275}{2500}$ ← part
← whole

$2500n = 327{,}500$ **Find the cross products.**

$n = 131$ **Divide each side by 2500.**

The band members raised 131% of their goal.

You can estimate some percents using fractions. These are some of the percent-fraction equivalents that you should remember.

$20\% = \frac{1}{5}$ $25\% = \frac{1}{4}$ $33.\overline{3}\% = \frac{1}{3}$ $50\% = \frac{1}{2}$ $66.\overline{6}\% = \frac{2}{3}$ $75\% = \frac{3}{4}$

166   Review   Proportions and Percents

You can also solve a percent problem by translating from words to an equation.

Water Supply   According to the United States Geological Survey, surface water accounts for 77.6% of our country's total daily water supply. The daily surface water supply is about 264 billion gallons. Estimate the total daily water supply.

Relate   77.6%   of   total daily water supply   is   264 billion gallons

Define   Let w = the total water supply.

Write   77.6%   · w =   264

$\frac{3}{4}w = 264$ **75% is close to 77.6%. Use $\frac{3}{4}$ to estimate.**

$\left(\frac{4}{3}\right)\frac{3}{4}w = \left(\frac{4}{3}\right)264$ **Multiply by $\frac{4}{3}$, the reciprocal of $\frac{3}{4}$.**

$w = 352$ **Simplify.**

The total daily water supply is about 352 billion gallons.

EXERCISES

Solve each problem.

1. What percent of 40 is 20?

2. What percent of 20 is 40?

3. 20% of what number is 40?

4. 40% of what number is 20?

5. 8% of 125 is what number?

6. What percent of 125 is 8?

7. 15% of what number is 24?

8. 120% of what number is 48?

9. Carlos worked 31.5 hours at a hospital as a volunteer. This represents 87.5% of his school's requirement for community service. How many hours does his school require for community service?

10. **Sales** Suppose you work in an electronics store. You earn a 6% commission on every item that you sell. How much commission do you earn if you sell a $545 sound system?

11. **Sales** Juan earns a 5.5% commission on his bicycle sales. In September, he earned $214.28 in commissions. What were his sales for the month?

12. **Sales Tax** Jane paid $1185 in sales tax on her new car. The car cost $15,800 before the tax was added. What percent was the sales tax rate?

13. **Finance** The formula for simple interest is $I = prt$, where I is the interest, p is the principal, r is the interest rate per year, and t is the time in years.
    a. You invest $550 for three years. Find the amount of simple interest you earn with an annual interest rate of 4.5%.
    b. Suppose you invested $900 for two years. You earned $67.50 in simple interest. What was the annual rate of interest?
    c. You invest $812 with an annual interest rate of 6.5%. You earned $316.68 in simple interest. For how many years was the money invested?

14. **Sales** A store advertises sneakers on sale for 33% off. The original price of sneakers is $56.
    a. Estimate the amount the sneakers have been marked down.
    b. Estimate the sale price of the sneakers.

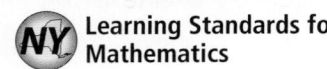

Percent of Change

NY Learning Standards for Mathematics

A.N.5 Solve algebraic problems arising from situations that involve fractions, decimals, percents (decrease/increase and discount), and proportionality/direct variation.

A.M.3 Calculate the relative error in measuring square and cubic units, when there is an error in the linear measure.

✓ Check Skills You'll Need

GO for Help Review page 166

Write an equation for each problem and solve.

1. What is 20% of 20?

2. 8 is what percent of 20?

3. 18 is 90% of what number?

4. 27 is 90% of what number?

Estimate each answer.

5. 67.3% of 24

6. 65% of 48

🔊 New Vocabulary
- percent of change
- percent of increase
- percent of decrease
- greatest possible error
- percent error

---

1 Percent of Change

Suppose the price of a $20 sweatshirt increases by $2. You can express the increase as a percent.

$$\text{increase in price} \longrightarrow \frac{2}{20} = \frac{1}{10} = 10\%$$
$$\text{original price} \longrightarrow$$

There was 10% increase in the price. This is an example of percent of change. **Percent of change** is the ratio $\frac{\text{amount of change}}{\text{original amount}}$ expressed as a percent. When a value increases from its original amount, it is the **percent of increase.** When a value decreases from its original amount, it is the **percent of decrease.**

1 EXAMPLE Finding Percent of Change

The price of a sweater decreased from $29.99 to $24.49. Find the percent of decrease.

$$\text{percent of decrease} = \frac{\text{amount of change}}{\text{original amount}}$$

$$= \frac{29.99 - 24.49}{29.99} \qquad \text{Subtract to find the amount of change. Substitute the original amount.}$$

$$= \frac{5.50}{29.99} \qquad \text{Simplify the numerator.}$$

$$\approx 0.18 \text{ or } 18\% \qquad \text{Write as a decimal and then as a percent.}$$

● The price of the sweater decreased by about 18%.

 **Quick Check** **1** **a.** Find the percent of change if the price of a CD increases from $12.99 to $13.99. Round to the nearest percent.

b. Find the percent of change if the CD is on sale, and its price decreases from $13.99 to $12.99. Round to the nearest percent.

2 EXAMPLE Real-World Problem Solving

Farming In 1990, there were 1330 registered alpacas in the United States. By the summer of 2000, there were 29,856. What was the percent of increase in registered alpacas?

$$\text{percent of increase} = \frac{\text{amount of change}}{\text{original amount}}$$

$$= \frac{29{,}856 - 1330}{1330} \qquad \textbf{Substitute.}$$

$$= \frac{28{,}526}{1330} \qquad \textbf{Simplify the numerator.}$$

$$\approx 21.448 \text{ or } 2145\% \qquad \textbf{Write as a decimal and then as a percent.}$$

● The number of registered alpacas increased by nearly 2145%.

 **2** The number of alpaca owners increased from 146 in 1991 to 2919 in 2000. Find the percent of increase. Round to the nearest percent.

2 Percent Error

Think about the last time you used a ruler. You probably measured to the nearest inch, half-inch, centimeter, or millimeter. Because no measurement is exact, you always measure to the nearest "something." The **greatest possible error** in a measurement is one half of that measuring unit.

3 EXAMPLE Finding the Greatest Possible Error

You use a beam balance to find the mass of a rock sample for a science lab. You read the scale as 3.8 g. What is your greatest possible error?

The rock's mass was measured to the nearest 0.1 g, so the greatest possible error is
● one half of 0.1 g, or 0.05 g.

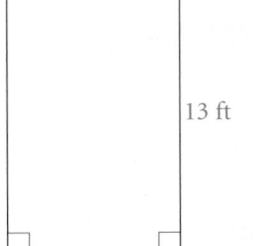 **3** You measure a picture for the yearbook and record its height as 9 cm. What is your greatest possible error?

4 EXAMPLE Finding Maximum and Minimum Areas

You measure a room and make the diagram shown at the left. Use the greatest possible error to find the maximum and minimum possible areas.

Both measurements were made to the nearest whole foot, so the greatest possible error is 0.5 ft. The length could be as little as 12.5 ft or as great as 13.5 ft. The width could be as little as 6.5 ft or as great as 7.5 ft. Find the minimum and maximum areas.

Minimum Area	**Maximum Area**
12.5 ft × 6.5 ft = 81.25 ft²	13.5 ft × 7.5 ft = 101.25 ft²

● The minimum area is 81.25 ft², and the maximum area is 101.25 ft².

13 ft

7 ft

 **4** You measure a wall of your room as 8 ft high and 12 ft wide. Find the minimum and maximum possible areas of the wall.

Percent error is another useful way to think of the error in a measurement. It is the ratio of the greater possible error and the measurement.

$$\textbf{percent error} = \frac{\text{greatest possible error}}{\text{measurement}}$$

5 EXAMPLE **Finding Percent Error**

Suppose you measure a CD and record its diameter as 12.1 cm. Find the percent error in your measurement.

Since the measurement is to the nearest 0.1 cm, the greatest possible error is 0.05 cm.

percent error $= \dfrac{\text{greatest possible error}}{\text{measurement}}$	**Use the percent error formula.**
$= \dfrac{0.05}{12.1}$	**Substitute.**
≈ 0.0041322314	**Divide.**
$\approx 0.4\%$	**Round and write as a percent.**

● The percent error is about 0.4%.

✓ Quick Check **5** **a.** You measure the length of a table as 168 inches. Find the percent error in this measurement.
b. You measure the length of a table as 168.0 inches. Find the percent error in this measurement.

6 EXAMPLE **Finding Percent Error in Calculating Volume**

The diagram at the left shows the dimensions of a cassette case. Find the percent error in calculating its volume.

The measurements are to the nearest 0.1 cm. The greatest possible error is 0.05 cm.

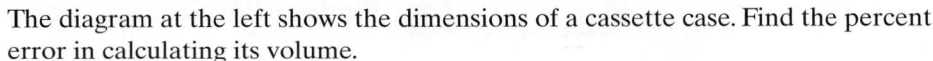
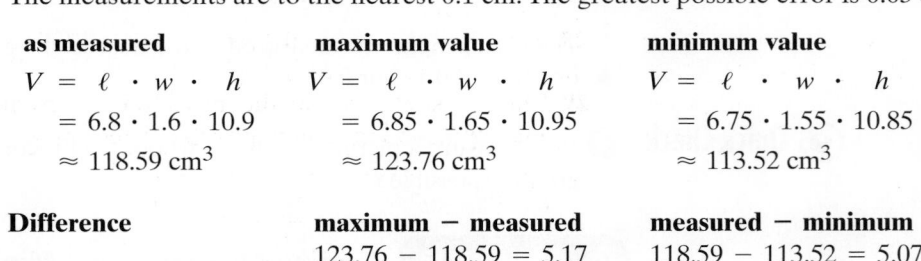

as measured	**maximum value**	**minimum value**
$V = \ell \cdot w \cdot h$	$V = \ell \cdot w \cdot h$	$V = \ell \cdot w \cdot h$
$= 6.8 \cdot 1.6 \cdot 10.9$	$= 6.85 \cdot 1.65 \cdot 10.95$	$= 6.75 \cdot 1.55 \cdot 10.85$
$\approx 118.59 \text{ cm}^3$	$\approx 123.76 \text{ cm}^3$	$\approx 113.52 \text{ cm}^3$

Difference	**maximum − measured**	**measured − minimum**
	$123.76 - 118.59 = 5.17$	$118.59 - 113.52 = 5.07$

Use the greater of the differences to find the percent error.

percent error $= \dfrac{\text{greater difference}}{\text{measurement}}$	**Use the percent error formula.**
$= \dfrac{123.76 - 118.59}{118.59}$	**Substitute.**
$= \dfrac{5.17}{118.59}$	**Simplify the numerator.**
≈ 0.0435955814	**Write as a decimal.**
$\approx 4\%$	**Round and write as a percent.**

● The percent error is about 4%.

✓ Quick Check **6** Suppose you measured your math book and recorded the dimensions as 1 in. × 9 in. × 10 in. Find the percent error in calculating its volume.

Practice by Example

Example 1
(page 168)

 GO for Help

Find each percent of change. Describe the percent of change as an increase or decrease. If necessary, round to the nearest tenth.

1. $2 to $3   **2.** $3 to $2   **3.** 4 ft to 5 ft   **4.** 5 ft to 4 ft

5. 9 m to 12 m   **6.** 12 cm to 9 cm   **7.** 12 in. to 15 in.   **8.** 15 lb to 18 lb

9. 4.5 cm to 8.3 cm   **10.** $38 to $65   **11.** $12.20 to $4.80   **12.** 125 lb to 143 lb

Example 2
(page 169)

13. Physical Therapy Physical therapists measure strength on a dynamometer, which uses a unit called a foot-pound. Suppose you increase the strength in your elbow joint from 90 foot-pounds to 125 foot-pounds. Find the percent of increase to the nearest percent.

14. Environment From 1999 to 2000, the number of days of unhealthy air quality in Charlotte, North Carolina, dropped from 5 to 2. Find the percent of decrease in the number of days of unhealthy air.

Example 3
(page 169)

Find the greatest possible error for each measurement.

15. 14 ft   **16.** 3.5 cm   **17.** 56.38 g   **18.** 17 in.

Example 4
(page 169)

Find the minimum and maximum possible areas for rectangles with the following measured areas.

19. 4 cm × 6 cm   **20.** 7 mi × 8 mi   **21.** 6 in. × 9 in.

22. 12 km × 5 km   **23.** 18 in. × 15 in.   **24.** 23 km × 14 km

Example 5
(page 170)

Find the percent error of each measurement.

25. 2 cm   **26.** 0.2 cm   **27.** 4 cm   **28.** 0.4 cm

Example 6
(page 170)

29. The table below shows the measured dimensions of the prism and the maximum and minimum possible values based on the greatest possible error.

Dimensions	ℓ	w	h
Measured	8	3	2
Maximum	8.5	3.5	2.5
Minimum	7.5	2.5	1.5

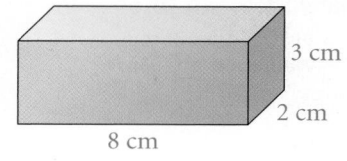

a. Find the measured volume.   **b.** Find the maximum volume.
c. Find the minimum volume.   **d.** Find the greater difference.
e. What is the percent error? Round to the nearest percent.

B **Apply Your Skills**

Find each percent of change. Describe the percent of change as an increase or decrease. Round to the nearest percent.

30. 26 to 20   **31.** $4.95 to $3.87   **32.** 21 in. to 54 in.

33. 2 ft to $5\frac{1}{2}$ ft   **34.** $24,000 to $25,000   **35.** 18 to $17\frac{1}{2}$

36. 8.99 to 3.99   **37.** 132 lb to 120 lb   **38.** $42.69 to $49.95

39. Sports In the 1988 Olympics, Florence Griffith-Joyner of the United States won the women's 100-meter run in 10.54 seconds. In 2000, Marion Jones, also of the United States, won with a time of 10.75 seconds. Find the percent of change in the winning times. Round to the nearest percent.

40. Meteorology In 1999, the National Oceanographic and Atmospheric Administration reported a total of 16 Atlantic cyclones. In 2000, there were 19 Atlantic cyclones. Find the percent of change in the number of cyclones from 1999 to 2000. Round to the nearest percent.

41. If you want accuracy of 0.5 mm, what measuring unit should you use?

42. Critical Thinking An item costs $64. The price is increased by $10, then reduced by $10. Is the percent of increase equal to the percent of decrease? Explain your answer.

43. Critical Thinking An item costs $64. The price is increased by 10%, then reduced by 10%. Is the final price equal to the original price? Explain.

44. Open-Ended Write a word problem involving percent of change. Include your solution.

Find the minimum and maximum possible areas for rectangles with the following measured dimensions. Round to the nearest tenth.

45. 4.1 cm × 6.1 cm   **46.** 7.0 mi × 8.4 mi   **47.** 6.01 in. × 9.02 in.

48. Writing Explain how to find the percent error when calculating the area of a rectangle.

49. Error Analysis Jorge found the percent of change from $15 to $10 to be 50%. What error did he make?

50. Sales Suppose that you are selling sweatshirts for a class fund-raiser. The wholesaler charges you $8 for each sweatshirt.
   a. You charge $16 for each sweatshirt. Find the percent of increase.
   b. Generalize your answer to part (a). Doubling a price is the same as a ? percent of increase.
   c. After the fund-raiser is over, you reduce the price on the remaining sweatshirts to $8. Find the percent of decrease.
   d. Generalize your answer to part (c). Cutting a price in half is the same as a ? percent of decrease.

Find the percent error in calculating the volume of each rectangular prism. Round to the nearest percent.

51.

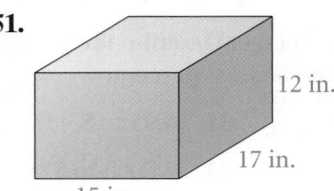

15 in.   17 in.   12 in.

52.

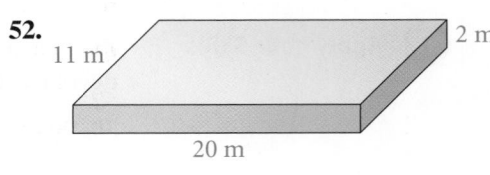

11 m   20 m   2 m

GO Online
Homework Video Tutor

Visit: PHSchool.com
Web Code: ate-0307

GO for Help

For a guide to solving Exercise 51, see p. 174.

C Challenge

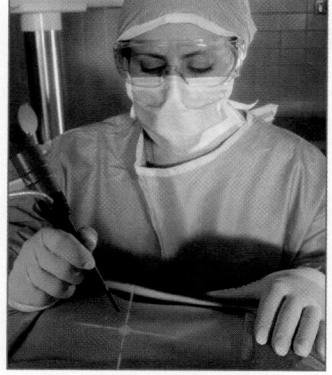

Real-World 🌐 **Connection**

Careers Students who want to become physicians must take calculus, physics, biology, and chemistry before going to medical school.

53. Suppose you measure two cubes. The smaller one measures 18 cm on each side. The larger one measures 45 cm on each side.
  a. Find the percent error of the volume of each cube. Round to the nearest percent.
  b. Critical Thinking Explain why using the same measuring unit did not yield the same percent error for the two cubes.

54. Data Analysis A reporter states, "From 1980 to 1996, the number of female physicians more than tripled." A second reporter states, "From 1980 to 1996, the number of female physicians increased about 205%." Can both reports be correct? Explain.

Year	U.S. Female Physicians
1980	48,700
1990	96,100
1996	148,300

55. a. The sides of a 12 cm × 12 cm square are all increased in length by 10%. Find the percent of increase in the area.
  b. The sides of a 14 cm × 14 cm square are all increased in length by 10%. Find the percent of increase in the area.
  c. Predict the percent of increase in the area if the sides of a 16 cm × 16 cm square are all increased by 10%. Explain your prediction and check.

Test Prep

Multiple Choice

56. Which percent of change best reflects a price increase from $32 to $36?
  A. 4%   **B.** 8%   **C.** 12%   **D.** 89%

57. A carpenter measured a rectangle as 5 in. by 8 in. Which number is the maximum possible area?
  F. 40 in.^{2} **G.** 42.5 in.^{2} **H.** 44 in.^{2} **J.** 46.75 in.^{2}

58. A student records the measured length of an object as 24.7 cm. What is the greatest possible error in this measurement?
  A. 0.05 cm   **B.** 0.2 cm   **C.** 0.5 cm   **D.** 1.0 cm

Short Response

59. A softball diamond is a 60 ft-by-60 ft square. The base lines of a baseball diamond are 50% longer than those of a softball diamond. What is the percent of increase from a softball diamond to a baseball diamond of the trip around all four bases? What is the percent of increase in the area from a softball diamond to a baseball diamond?

Mixed Review

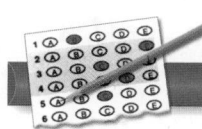

Lesson 3-6

Write an equation and solve.

60. The sum of two consecutive *odd* integers is 56. Find the two odd integers.

61. The sum of four consecutive *even* integers is 308. Find the four integers.

62. The length of a rectangle is 8 cm more than twice the width. The perimeter of the rectangle is 34 cm. What is the length of the rectangle?

Lesson 3-1

Solve each equation.

63. $5t - 2 = 14$ **64.** $4 = 3 - 8h$ **65.** $3w + 11 = -10$

Understanding Math Problems   Read through the problem below. Then follow along with what Rasia thinks as she solves the problem. Check your understanding with the exercise at the bottom of the page.

Find the percent error in calculating the volume of the rectangular prism. Round to the nearest percent.

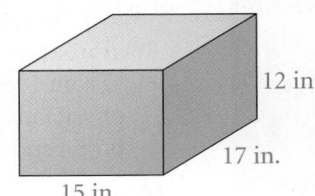

12 in.

17 in.

15 in.

What Rasia Thinks

I need to write down the important information.

I need to find the greatest possible error. Since all the measurements are to the nearest inch, the greatest possible error is 0.5 inch.

I should find the volume as measured and then find the volume using the maximum and minimum values.

I need to find which is the greater difference, maximum−measured or measured−minimum.

Maximum−measured gives the greater difference. I'll use it in the percent error formula.

My calculator shows the answer as 0.1080473856. I need to round my answer to the nearest hundredth and then write it as a percent.

What Rasia Writes

Volume of prism: V = LWH

Dimensions: L = 17 in., W = 15 in., H = 12 in.

measured: 17 · 15 · 12 = 3060
maximum: 17.5 · 15.5 · 12.5 = 3390.625
minimum: 16.5 · 14.5 · 11.5 = 2751.375

3390.625 − 3060 = 330.625

3060 − 2751.375 = 308.625

percent error = $\dfrac{330.625}{3060}$

≈ 11%

EXERCISE

A rectangular prism has dimensions of 8 cm, 12 cm, and 20 cm. Find the percent error in calculating its volume. Round to the nearest percent.

Learning Vocabulary

FOR USE WITH LESSON 3-7

As you study mathematics this year, make your own mathematics dictionary of new vocabulary terms. Use the following guide for each new term.

- Write the vocabulary term and its definition. Include any symbols for the term.
- If possible, draw a diagram. Include details using other related terms you know.
- Give one or more examples of the term.
- Give one or more "nonexamples" and explain how they are different.

EXAMPLE

Percent of change: $\dfrac{\text{amount of change}}{\text{original amount}}$

Percent of decrease: $\dfrac{\text{amount of decrease}}{\text{original amount}}$

Example: The price of an item goes from \$20 to \$18.

$$\text{Percent of decrease} = \frac{20 - 18}{20}$$

$$= \frac{2}{20}$$

$$= 0.10$$

$$= 10\%$$

Percent of increase: $\dfrac{\text{amount of increase}}{\text{original amount}}$

Example: The price of an item goes from \$18 to \$20.

$$\text{Percent of increase} = \frac{20 - 18}{18}$$

$$= \frac{2}{18}$$

$$= 0.\overline{11}$$

$$\approx 11\%$$

Non-example: "45% of students passed the test" is a percent but not a percent of change because there are no changing values.

EXERCISES

Write an entry for your dictionary for each term.

1. greatest possible error
2. percent error
3. similar figures
4. scale factor
5. dilation
6. proportion

Write an entry for the first term. Your entry should include the second term.

7. ratio, division
8. rate, ratio
9. unit rate, rate
10. cross products, proportion
11. identity, equation
12. literal equation, variables

3-8 Finding and Estimating Square Roots

 **Learning Standards for Mathematics**

A.N.2 Simplify radical terms (no variable in the radicand).

✓ Check Skills You'll Need

GO for Help Lesson 1-2

Simplify each expression.

1. 11^2 **2.** $(-12)^2$ **3.** $-(12)^2$ **4.** 1.5^2

5. 0.6^2 **6.** $\left(\frac{1}{2}\right)^2$ **7.** $\left(-\frac{2}{3}\right)^2$ **8.** $\left(\frac{4}{5}\right)^2$

🔊 **New Vocabulary**   • square root   • principal square root
                   • negative square root   • radicand   • perfect squares

---

1   Finding Square Roots

The diagram at the right shows the relationship between squares and square roots. Every positive number has *two* square roots.

4^2 ⟶ 16 ⟵ $(-4)^2$ Square root of 16 ⟶ 4, ⟶ −4

 Key Concepts

Definition	**Square Root**

The number a is a **square root** of b if $a^2 = b$.

Example $4^2 = 16$ and $(-4)^2 = 16$, so 4 and −4 are square roots of 16.

Vocabulary Tip

Read $\sqrt{16}$ as "the square root of 16." Read $\pm$ as "plus or minus."

A radical symbol $\sqrt{}$ indicates a square root. The expression $\sqrt{16}$ means the positive, or **principal square root** of 16. The expression $-\sqrt{16}$ means the **negative square root** of 16. The expression under the radical sign is a **radicand.** You can use the symbol $\pm$ to indicate both square roots.

1 EXAMPLE   Simplifying Square Root Expressions

Simplify each expression.

a. $\sqrt{64} = 8$ positive square root

b. $-\sqrt{100} = -10$ negative square root

c. $\pm\sqrt{\frac{9}{16}} = \pm\frac{3}{4}$ The square roots are $\frac{3}{4}$ and $-\frac{3}{4}$.

d. $\pm\sqrt{0} = 0$ There is only one square root of 0.

e. $\sqrt{-16}$ is undefined.   For real numbers, the square root of a negative number is undefined.

✓ **Quick Check**   ❶ Simplify each expression.
     a. $\sqrt{49}$ **b.** $\pm\sqrt{36}$ **c.** $-\sqrt{121}$ **d.** $\sqrt{\frac{1}{25}}$

Vocabulary Tip

In decimal form, a <u>rational</u> number terminates or repeats. In decimal form, an <u>irrational</u> number continues without repeating.

Some square roots are rational numbers and some are irrational numbers.

Rational: $\sqrt{100} = 10$ $\pm\sqrt{0.36} = \pm 0.6$ $\sqrt{\frac{16}{121}} = \frac{4}{11}$

Irrational: $\sqrt{10} \approx 3.16227766$ $\sqrt{\frac{1}{7}} \approx 0.377964473$

2 EXAMPLE   Rational and Irrational Square Roots

Tell whether each expression is *rational* or *irrational*.

a. $\pm\sqrt{81} = \pm 9$ rational
b. $-\sqrt{1.44} = -1.2$ rational
c. $-\sqrt{5} \approx -2.23606798$ irrational
d. $\sqrt{\frac{4}{9}} = \frac{2}{3}$ rational
e. $\sqrt{\frac{1}{3}} \approx 0.57735026$ irrational

✓ Quick Check **2** Tell whether each expression is *rational* or *irrational*.

    a. $\sqrt{8}$ **b.** $\pm\sqrt{225}$ **c.** $-\sqrt{75}$ **d.** $\sqrt{\frac{1}{4}}$

2   Estimating and Using Square Roots

The squares of integers are called **perfect squares.**

consecutive integers:	1	2	3	4	5	6
	↓	↓	↓	↓	↓	↓
consecutive perfect squares:	1	4	9	16	25	36

You can estimate square roots by using perfect squares.

Online active math

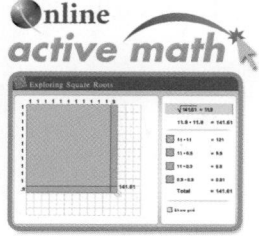

For: Square Root Activity
Use: Interactive Textbook, 3-8

3 EXAMPLE   Estimating Square Roots

Estimation Between what two consecutive integers is $\sqrt{14.52}$?

$\sqrt{9} < \sqrt{14.52} < \sqrt{16}$ **14.52 is between the two consecutive perfect squares 9 and 16.**

    ↓   ↓   ↓

$3 < \sqrt{14.52} < 4$ **The square roots of 9 and 16 are 3 and 4, respectively.**

$\sqrt{14.52}$ is between 3 and 4.

✓ Quick Check **3** Between what two consecutive integers is $-\sqrt{105}$?

You can find the approximate value of a square root using a calculator.

4 EXAMPLE   Approximating Square Roots With a Calculator

Calculator Find $\sqrt{14.52}$ to the nearest hundredth.

$\sqrt{14.52} \approx 3.810511777$ **Use the calculator sequence** √ **14.52** ENTER **.**

       ≈ 3.81 **Round to the nearest hundredth.**

✓ Quick Check **4** Find $\sqrt{17.81}$ to the nearest hundredth.

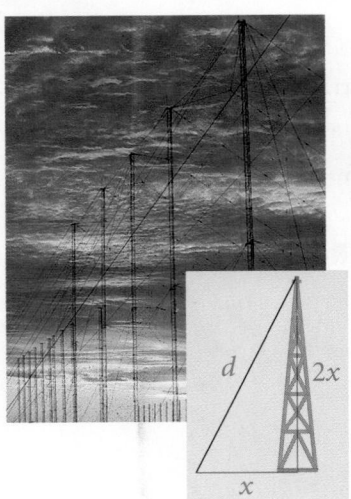

Many real-world formulas involve square roots.

5 EXAMPLE **Real-World Problem Solving**

Construction The formula $d = \sqrt{x^2 + (2x)^2}$ gives the length d of each wire for the tower at the left. Find the length of the wire if $x = 12$ ft.

$d = \sqrt{x^2 + (2x)^2}$

$d = \sqrt{12^2 + (2 \cdot 12)^2}$ **Substitute 12 for x.**

$d = \sqrt{144 + 576}$ **Simplify.**

$d = \sqrt{720}$

$d \approx 26.8$ **Use a calculator. Round to the nearest tenth.**

The wire is about 26.8 ft long.

 Quick Check **5** Suppose the tower is 140 ft tall. How long is the supporting wire? Round to the nearest tenth of a foot.

EXERCISES

For more exercises, see *Extra Skill and Word Problem Practice.*

Practice and Problem Solving

A **Practice by Example**

Example 1
(page 176)

 **for Help**

Simplify each expression.

1. $\sqrt{169}$ **2.** $\sqrt{400}$ **3.** $\sqrt{\frac{1}{9}}$ **4.** $\sqrt{900}$

5. $\sqrt{0.25}$ **6.** $\sqrt{\frac{36}{49}}$ **7.** $-\sqrt{1.21}$ **8.** $\sqrt{1.96}$

9. $\sqrt{0.36}$ **10.** $-\sqrt{144}$ **11.** $\sqrt{\frac{25}{16}}$ **12.** $\pm\sqrt{0.01}$

Example 2
(page 177)

Tell whether each expression is *rational* or *irrational*.

13. $\sqrt{37}$ **14.** $-\sqrt{0.04}$ **15.** $\pm\sqrt{\frac{1}{5}}$ **16.** $-\sqrt{\frac{16}{121}}$

Example 3
(page 177)

Between what two consecutive integers is each square root?

17. $\sqrt{35}$ **18.** $\sqrt{27}$ **19.** $-\sqrt{130}$ **20.** $\sqrt{170}$

Example 4
(page 177)

Use a calculator to find each square root to the nearest hundredth.

21. $\sqrt{12}$ **22.** $-\sqrt{203}$ **23.** $\sqrt{11,550}$ **24.** $-\sqrt{150}$

Example 5
(page 178)

25. Sports The elasticity coefficient e of a ball relates the height r of its rebound to the height h from which it is dropped. You can use the function $e = \sqrt{\frac{r}{h}}$ to find the elasticity coefficient. What is the elasticity coefficient of a tennis ball that rebounds 3 ft after it is dropped from a height of 3.5 ft? Round to the nearest hundredth.

B **Apply Your Skills**

Find the square root(s) of each number.

26. 400   **27.** 0   **28.** 625   **29.** $\frac{9}{49}$

30. 1.69   **31.** $\frac{1}{81}$ **32.** 729   **33.** 2.25

34. 256   **35.** 0.01   **36.** $\frac{64}{121}$ **37.** 40,804

38. Critical Thinking What number other than 0 is its own square root?

39. Multiple Choice The formula $d = \sqrt{12{,}800h + h^2}$ gives the distance d in kilometers to the horizon from a satellite h kilometers above Earth. Find the distance to the horizon from a satellite 4200 km above Earth. Round to the nearest kilometer.

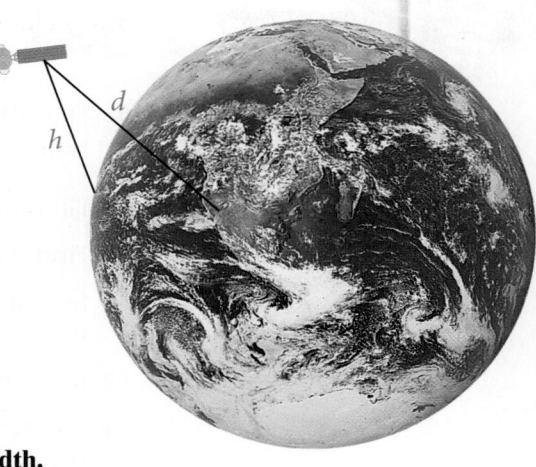

   Ⓐ 130 km   Ⓑ 7333 km

   Ⓒ 8450 km   Ⓓ 11,532 km

Find the value of each expression. If necessary, round to the nearest hundredth.

40. $\sqrt{441}$ **41.** $-\sqrt{\dfrac{4}{25}}$ **42.** $\sqrt{2}$

43. $\sqrt{1.6}$ **44.** $-\sqrt{30}$ **45.** $-\sqrt{1089}$

46. $-\sqrt{0.64}$ **47.** $\sqrt{41}$ **48.** $\sqrt{75}$

 49. Writing Explain the difference between $-\sqrt{1}$ and $\sqrt{1}$.

50. Open-Ended Find two integers a and b between 1 and 20 such that $a^2 + b^2$ is a perfect square.

51. Math in the Media In the cartoon, to what number is the golfer referring?

 52. Physics If you drop an object, the time t in seconds that it takes to fall d feet is given by the formula $t = \sqrt{\dfrac{d}{16}}$.
  a. Find the time it takes an object to fall 400 ft.
  b. Find the time it takes an object to fall 1600 ft.
  c. Critical Thinking In part (b), the object falls four times as far as in part (a). Does it take four times as long to fall? Explain.

Ⓒ Challenge

Critical Thinking For Exercises 53–56, tell whether each statement is *true* or *false*. If the statement is false, give a counterexample.

53. Every nonnegative number has two square roots.

54. The square root of a positive number is always less than the number.

55. The square root of an even perfect square is always an even number.

56. $\sqrt{p} + \sqrt{q} = \sqrt{p + q}$

57. a. What is the total area of the large square shown at the right?
  b. What is the area of each shaded triangle?
  c. What is the area of the shaded square?
  d. What is the length of the diagonal of each 1×1 square?

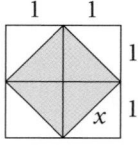

GO **nline**
Homework Video Tutor
Visit: PHSchool.com
Web Code: ate-0308

Gridded Response

58. Find the area of the shaded region.

59. Find the value of $\sqrt{1.69}$.

60. Find the value of $\sqrt{\frac{4}{81}}$.

61. Round $\sqrt{80}$ to the nearest whole number.

62. Simplify $\sqrt{4^2 + 3^2 + 11}$.

63. The formula $\ell = \sqrt{\frac{A}{6}}$ relates the surface area A of a cube to the length of its edge ℓ. A cube has a surface area of 726 cm². How many centimeters is the length of the edge?

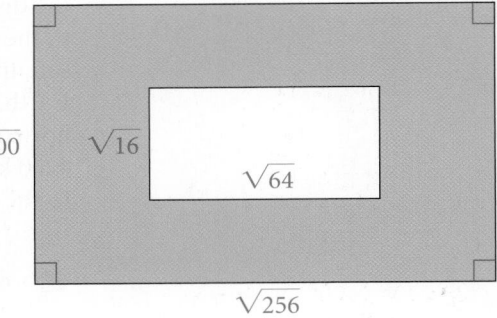

$\sqrt{100}$ $\sqrt{16}$ $\sqrt{64}$ $\sqrt{256}$

Mixed Review

Lesson 3-7

Find each percent of change. Describe the percent of change as an increase or decrease. If necessary, round to the nearest tenth.

64. $5 to $6   **65.** 40 in. to 36 in.   **66.** 10 m to 11 m   **67.** 18° to 24°

Lesson 2-5

Simplify each expression. Justify each step.

68. $8 \cdot 7 \cdot 5$ **69.** $14 + 23 + 56$ **70.** $16\left(\frac{1}{16}\right) + 5$

71. $(6^8 - 6^9)(11 - 11)$ **72.** $-7(3w)$ **73.** $8m + 145 + 4m$

Checkpoint Quiz 2   Lessons 3-6 to 3-8

1. The length of a rectangle is 5 cm more than its width. The perimeter of the rectangle is 46 cm. What is the length of the rectangle?

2. The sum of three consecutive odd integers is 465. Find the integers.

3. Marcus and Beth leave college traveling in opposite directions on a straight road. Beth drives 13 mi/h faster than Marcus. After 4 hours, they are 452 miles apart. Find their rates.

Find the percent error for each measurement.

4. 210 cm   **5.** 25 cm

6. A student measures a rectangle as 14 in. × 23 in. Find the minimum and maximum possible areas for the rectangle.

Simplify each expression.

7. $-\sqrt{256}$ **8.** $\sqrt{\frac{64}{121}}$ **9.** $\pm\sqrt{0.0025}$

10. Between what two consecutive integers is $\sqrt{52}$?

The Pythagorean Theorem

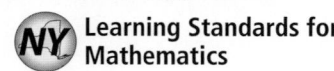
Learning Standards for Mathematics

A.A.45 Determine the measure of a third side of a right triangle using the Pythagorean Theorem, given the lengths of a two sides.

 Check Skills You'll Need

Simplify each expression.

1. $5^2 + 6^2$ **2.** $9^2 - 4^2$ **3.** $(3t)^2 + (4t)^2$

4. $\sqrt{196}$ **5.** $\sqrt{\frac{25}{49}}$ **6.** $\sqrt{1.44}$

GO for Help Lessons 1-2 and 3-8

◄)) **New Vocabulary** • hypotenuse • leg • Pythagorean Theorem
• conditional • hypothesis • conclusion • converse

---

1   Solving Problems Using the Pythagorean Theorem

Activity: The Pythagorean Theorem

1. The values in the chart represent the lengths of the sides of a right triangle. Copy and complete the chart.

a	b	c	a²	b²	a² + b²	c²
3	4	5	■	■	■	■
5	12	13	■	■	■	■
$\frac{3}{5}$	$\frac{4}{5}$	1	■	■	■	■
0.9	1.2	1.5	■	■	■	■

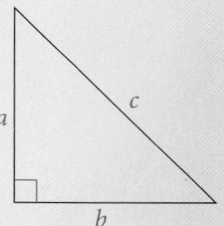

2. Compare the values of $a^2 + b^2$ for each row to the values of c^2.

3. Complete the following statement: For a right triangle, the square of the longest side __?__ the sum of the squares of the other two sides.

Vocabulary Tip

The <u>Pythagorean Theorem</u> is named after Pythagoras, a Greek philosopher and mathematician who taught about 530 B.C.

In a right triangle, the side opposite the right angle is the **hypotenuse.** It is the longest side. Each of the sides forming the right angle is a **leg.**

The **Pythagorean Theorem** describes the relationship of the lengths of the sides of a right triangle.

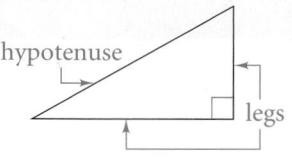

 Key Concepts

Theorem	**The Pythagorean Theorem**

In any right triangle, the sum of the squares of the lengths of the legs is equal to the square of the length of the hypotenuse.

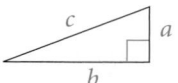

$$a^2 + b^2 = c^2$$

Sometimes you will need to find square roots to determine the length of a side or of a hypotenuse of a triangle. In real-world situations, $x^2 = a^2$, you find the principal square root of each side. So $\sqrt{x^2} = \sqrt{a^2}$, and $x = a$.

1 EXAMPLE Using the Pythagorean Theorem

What is the length of the hypotenuse of the triangle at the right?

$a^2 + b^2 = c^2$	Use the Pythagorean Theorem.
$9^2 + 12^2 = c^2$	Substitute 9 for a and 12 for b.
$81 + 144 = c^2$	Simplify.
$\sqrt{225} = \sqrt{c^2}$	Find the principal square root of each side.
$15 = c$	Simplify.

● The length of the hypotenuse is 15 cm.

✓ Quick Check ❶ What is the length of the hypotenuse of a right triangle with legs of lengths 7 cm and 24 cm?

You can also use the Pythagorean Theorem to find the length of a leg of a right triangle when you know the lengths of the hypotenuse and the other leg.

2 EXAMPLE Real-World 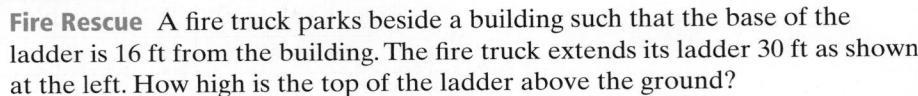 Problem Solving

Fire Rescue A fire truck parks beside a building such that the base of the ladder is 16 ft from the building. The fire truck extends its ladder 30 ft as shown at the left. How high is the top of the ladder above the ground?

Define Let b = height (in feet) of the ladder from a point 10 ft above the ground.

Relate The triangle formed is a right triangle. Use the Pythagorean Theorem.

Write		
	$a^2 + b^2 = c^2$	
	$16^2 + b^2 = 30^2$	Substitute.
	$256 + b^2 = 900$	Simplify.
	$b^2 = 644$	Subtract 256 from each side.
	$\sqrt{b^2} = \sqrt{644}$	Find the principal square root of each side.
	$b \approx 25.4$	Use a calculator and round to the nearest tenth.

The height to the top of the ladder is 10 feet higher than 25.4 ft, so it is about 35.4 ft from the ground.

✓ Quick Check ❷ Use the figure at the right. About how many miles is it from downtown to the harbor? Round to the nearest tenth of a mile.

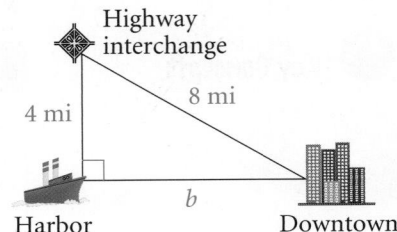

An *if-then* statement like "If an animal is a horse, then it has four legs" is called a **conditional.** Every conditional has two parts. The part following *if* is the **hypothesis,** and the part following *then* is the **conclusion.**

The **converse** of a conditional switches the hypothesis and the conclusion. Somtimes converses of conditionals are not true. For example, "If an animal has four legs, then the animal is a horse" is not a true statement.

You can rewrite the Pythagorean Theorem as an *if-then* statement, "If a triangle is a right triangle with legs of lengths a and b and hypotenuse of length c, then $a^2 + b^2 = c^2$." The Pythagorean Theorem has a converse that is always true.

 Key Concepts

Property	The Converse of the Pythagorean Theorem

If a triangle has sides of lengths a, b, and c, and $a^2 + b^2 = c^2$, then the triangle is a right triangle with hypotenuse of length c.

You can use the converse of the Pythagorean Theorem to determine whether a triangle is a right triangle. Since the Pythagorean Theorem and its converse are always true, you can also determine whether a triangle is *not* a right triangle.

3 EXAMPLE Using the Converse of the Pythagorean Theorem

Determine whether the given lengths can be sides of a right triangle.

a. 5 in., 12 in., and 13 in.
$$5^2 + 12^2 \stackrel{?}{=} 13^2$$
$$25 + 144 \stackrel{?}{=} 169$$
$$169 = 169 \checkmark$$
The triangle is a right triangle.

Determine whether $a^2 + b^2 = c^2$**, where** c **is the longest side.**

b. 7 m, 9 m, and 12 m
$$7^2 + 9^2 \stackrel{?}{=} 12^2$$
$$49 + 81 \stackrel{?}{=} 144$$
$$130 \neq 144$$
The triangle is not a right triangle.

✓ **Quick Check** **3** A triangle has sides of lengths 10 m, 24 m, and 26 m. Is the triangle a right triangle?

You can use the converse of the Pythagorean Theorem to solve a physics problem involving force.

4 EXAMPLE Real-World 🌐 Problem Solving

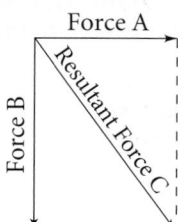

Force A

Force B

Resultant Force C

Physics If two forces pull at right angles to each other, the resultant force is represented as the diagonal of a rectangle, as shown at the left. The diagonal forms a right triangle with two of the perpendicular sides of the rectangle. For a 30-lb force and a 40-lb force, the resultant force is 50 lb. Are the forces pulling at right angles to each other?

$$30^2 + 40^2 \stackrel{?}{=} 50^2$$ **Determine whether** $a^2 + b^2 = c^2$**, where** c **is the greatest force.**
$$900 + 1600 \stackrel{?}{=} 2500$$
$$2500 = 2500 \checkmark$$

Yes, the 30-lb and 40-lb forces are pulling at right angles to each other.

 **Quick Check** **4** For a 70-lb force and a 60-lb force, the resultant force is 100 lb. Are the forces pulling at right angles to each other?

EXERCISES

For more exercises, see *Extra Skill and Word Problem Practice.*

Practice and Problem Solving

A **Practice by Example**

Example 1
(page 182)

 **GO** for Help

Use the triangle at the right. Find the length of the missing side. If necessary, round to the nearest tenth.

1. $a = 6, b = 8$ **2.** $a = 15, b = 20$

3. $a = 8, b = 15$ **4.** $a = 10, b = 24$

5. $a = 1.5, b = 2$ **6.** $a = \frac{3}{5}, b = \frac{4}{5}$

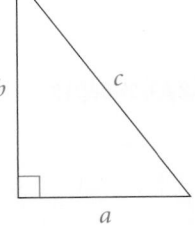

Example 2
(page 182)

7. $a = 3, c = 5$ **8.** $b = 12, c = 13$

9. $a = 9, c = 15$ **10.** $b = 7, c = 10$

11. $a = 5, c = 9$ **12.** $a = 0.8, c = 1$

13. Packaging Use the diagram at the right. Find the width w that the box needs to be for the fishing rod to fit flat inside of it.

14. A 16-ft ladder is placed 4 ft from the base of a building. How high on the building will the ladder reach?

15. A pigeon leaves its nest in New York City and flies 5 km due east. The pigeon then flies 3 km due north. How far is the pigeon from its nest?

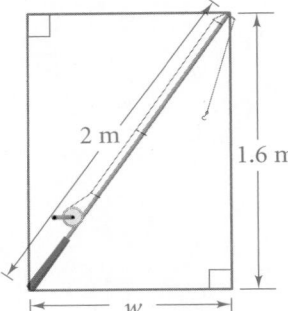

Example 3
(page 183)

Determine whether the given lengths can be sides of a right triangle.

16. 9 ft, 12 ft, 15 ft **17.** 1 in., 2 in., 3 in. **18.** 2 m, 4 m, 5 m

19. 16 cm, 30 cm, 34 cm **20.** 4 m, 4 m, 8 m **21.** 10 in., 24 in., 26 in.

Example 4
(page 183)

Physics **Determine whether the forces in each pair are pulling at right angles to each other.**

22. 45 lb, 24 lb, resultant force 51 lb **23.** 3.5 lb, 6.2 lb, resultant force 9.1 lb

24. 20 lb, 10 lb, resultant force 30 lb **25.** 1.25 lb, 3 lb, resultant force 3.25 lb

B **Apply Your Skills**

For the values given, a and b are legs of a right triangle, and c is the hypotenuse. Find the length of the missing side of each right triangle. If necessary, round to the nearest tenth.

26. $a = 1.2, b = 0.9$ **27.** $a = \frac{1}{5}, c = \frac{1}{3}$ **28.** $a = 1, c = 2$

29. $a = \frac{3}{4}, b = 1$ **30.** $a = 2.4, b = 1.0$ **31.** $a = 2\frac{1}{2}, b = 6\frac{1}{2}$

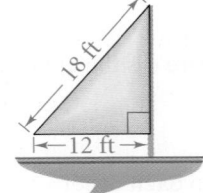

32. Sailing The diagram at the left shows a sailboat.
  a. Use the Pythagorean Theorem to find the height of the sail. Round to the nearest tenth.
  b. Use the result of part (a) and the formula for the area of a triangle to find the area of the sail. Round to the nearest tenth.

33. Multiple Choice What is the diameter of the smallest circular opening through which the rectangular rod shown at the right will fit? Round to the nearest tenth.

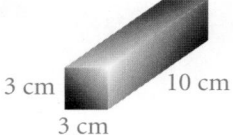

3 cm   10 cm   3 cm

   Ⓐ 2.12 cm   Ⓑ 3 cm   Ⓒ 4.3 cm   Ⓓ 9 cm

34. Physics Two utility vehicles at a 90° angle to each other try to pull a third vehicle out of the snow. If one utility vehicle exerts a force of 600 lb, and the other exerts a force of 800 lb, what is the resulting force on the vehicle stuck in the snow?

Find the missing length to the nearest tenth.

35.

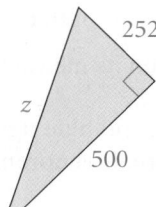

252
z
500

36.

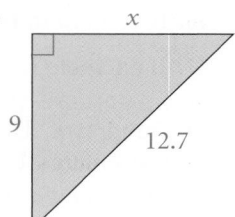

x
9
12.7

37.

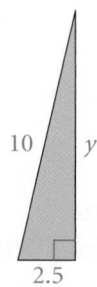

10
y
2.5

38. You know that two sides of a right triangle measure 10 in. and 8 in.
  a. Writing Explain why this is not enough information to be sure of finding the length of the third side.
  b. Give two possible values for the length of the third side.

39. Any set of three positive integers that satisfies the relationship $a^2 + b^2 = c^2$ is called a *Pythagorean triple*.
  a. Verify that the numbers 6, 8, and 10 form a Pythagorean triple.
  b. Copy the table at the right. Complete the table so that the values in each row form a Pythagorean triple.
  c. Open-Ended Find a Pythagorean triple that does not appear in the table.

a	b	c
3	4	▨
5	▨	13
▨	24	25
9	40	▨

40. Solar Power Solar cars use panels built out of photovoltaic cells, which convert sunlight into electricity. Consider a car like the one shown. Not counting the driver's "bubble," the panels form a rectangle.
  a. The length of the rectangle is 13 ft and the diagonal is 14.7 ft. Find the width. Round to the nearest tenth of a foot.
  b. Find the area of the rectangle.
  c. The panels produce a maximum power of about 11 watts/ft^2. Find the maximum power produced by the panels on the car. Round to the nearest watt.

41. Construction A carpenter braces an 8 ft × 10 ft wall by nailing a board diagonally across the wall. How long is the bracing board?

42. a. Open-Ended Find a right triangle that has legs with irrational lengths and a hypotenuse with a rational length.
  b. Use a calculator to find the area of your triangle. Round to the nearest tenth.

Real-World Connection

This solar-powered car only weighs 110 lb. It can reach a top speed of 80 mi/h.

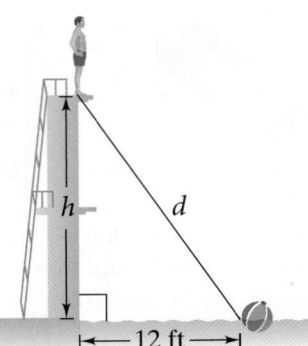

43. Diving Suppose you are standing at the top of a diving platform *h* feet tall. Looking down, you see a ball on the water 12 feet from the bottom of the diving platform as in the diagram at the left.
   a. Find the distance *d* to the ball if *h* = 6 feet; if *h* = 12 feet.
   b. Suppose you know the distance *d* to the ball is 16 feet. About how tall is the diving platform?
   c. Critical Thinking Could the distance *d* to the ball be 7 feet? Explain.

State the hypothesis and the conclusion of each conditional. Then write the converse. Tell whether the converse is true or false.

44. If an integer has 2 as a factor, then the integer is even.

45. If a figure is a square, then the figure is a rectangle.

46. If you are in Brazil, then you are south of the equator.

47. If an angle is a right angle, then its measure is 90°.

GO Online
Homework Video Tutor
Visit: PHSchool.com
Web Code: ate-0309

48. Geometry The yellow, green, and blue figures at the right are squares. Use the Pythagorean Theorem to find the area of the blue square.

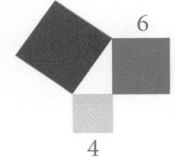

49. Geometry The diagonal of a square measures $6\sqrt{2}$ in. Find the length of the side of the square.

For any two points $P(x_1, y_1)$ and $Q(x_2, y_2)$ not on the same horizontal or vertical line, you can graph the points and form a right triangle as shown at the right. You can then use the Pythagorean Theorem to find the distance between the points. The result is called the Distance Formula.

$$PQ = \sqrt{(x_2 - x_1)^2 + (y_2 - y_1)^2}$$

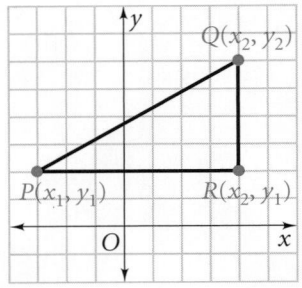

Use the Distance Formula to find the distance between each pair of points. Where necessary, round to the nearest tenth.

50. $(7, -3), (-8, -3)$ **51.** $(-2, 7), (-3, -7)$ **52.** $(0, 0), (6, -8)$

53. $(-4, -4), (4, 4)$ **54.** $(9, 10), (11, 12)$ **55.** $(3, -2), (-1, 5)$

C Challenge

Use the Pythagorean Theorem to find *s*.

56. 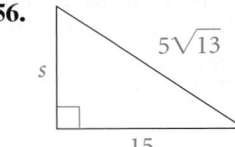   **57.** 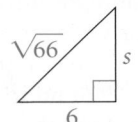   **58.**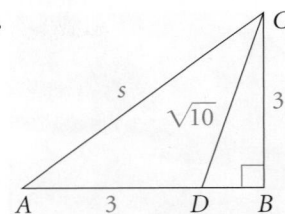

59. Geometry The lengths of the sides of a right triangle are three consecutive integers. Write and solve an equation to find the three integers.

60. a. Critical Thinking The vertex of the right angle of a right triangle is at the origin of coordinate axes. The length of the horizontal side is 5 units. The length of the vertical side is 7 units. The triangle is located in Quadrant II. Sketch the graph.
   b. Find the length of the hypotenuse. Round to the nearest tenth.

61. On graph paper, draw a right triangle like the one at the right. Then draw the square by drawing four right triangles as shown.

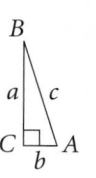
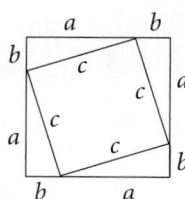

   a. Find the area of the larger square. Write your answer as a trinomial.
   b. Find the area for the smaller square. Write your answer as a monomial.
   c. Find the area of each triangle in terms of a and b.
   d. The area of the larger square equals the sum of the area of the smaller square and the areas of the four triangles. Write this equation and simplify.
   e. What do you notice about the equation you wrote for part (d)?

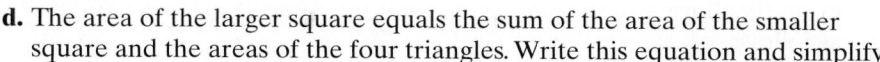

Test Prep

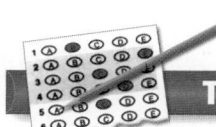

Multiple Choice

62. Find the approximate length of the hypotenuse of a right triangle with leg lengths 8.4 cm and 7.6 cm.
   A. 4.00 cm   **B.** 5.66 cm   **C.** 7.99 cm   **D.** 11.33 cm

63. Find the approximate length of the leg of a right triangle with one leg length 8 and hypotenuse length 19.
   F. 20.3   **G.** 20.6   **H.** 17.7   **J.** 17.2

64. Find the approximate perimeter of the right triangle.
   A. 29 cm   **B.** 25 cm
   C. 32 cm   **D.** 70 cm

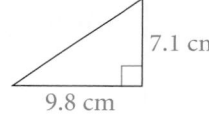

65. Find the approximate area of the right triangle.
   F. 32 in.^{2} **G.** 24 in.^{2}
   H. 36 in.^{2} **J.** 18 in.^{2}

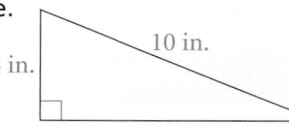

Short Response

66. The sides of a triangular garden are 8 ft, 17 ft, and 15 ft. Is the garden in the shape of a right triangle? Justify your answer.

Mixed Review

Lesson 3-8

Between which two consecutive integers is each square root?

67. $\sqrt{11}$ **68.** $\sqrt{80}$ **69.** $-\sqrt{51}$ **70.** $\sqrt{125}$

Tell whether each expression is rational or irrational.

71. $-\sqrt{1.44}$ **72.** $\sqrt{130}$ **73.** $\sqrt{\frac{2}{3}}$ **74.** $\sqrt{\frac{1}{36}}$

Lesson 2-7

Suppose you choose a tile from a bag containing 4 A's, 5 B's, and 3 C's. You replace the first tile in the bag and then choose again. Find each probability.

75. $P(\text{B and B})$ **76.** $P(\text{A and C})$ **77.** $P(\text{C and C})$

78. $P(\text{A and B})$ **79.** $P(\text{C and A})$ **80.** $P(\text{B and C})$

Special Right Triangles

FOR USE WITH LESSON 3-9

You can use the Pythagorean Theorem to explore properties of some special right triangles. You will also need the *Multiplication Property of Square Roots* to work with square roots efficiently. The property allows you to combine or separate products in square roots as follows.

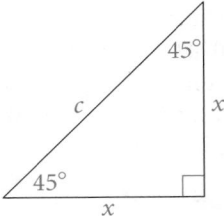

If $a \geq 0$ and $b \geq 0$ then $\sqrt{a} \cdot \sqrt{b} = \sqrt{a \cdot b}$.

Consider the isosceles right triangle. It is called a 45°-45°-90° triangle.

$c^2 = a^2 + b^2$	Use the Pythagorean Theorem.
$c^2 = x^2 + x^2$	Substitute x for b and for a.
$c^2 = 2x^2$	Simplify.
$\sqrt{c^2} = \sqrt{2x^2}$	Find the principal square root of each side.
$c = \sqrt{2} \cdot \sqrt{x^2}$	Use the Multiplication Property of Square Roots.
$c = \sqrt{2} \cdot x$, or $x\sqrt{2}$	Simplify.

Theorem   **45°-45°-90° Triangle Theorem**

In a 45°-45°-90° triangle, the length of the hypotenuse is the length of the leg times $\sqrt{2}$.

$$\text{hypotenuse} = \text{leg} \cdot \sqrt{2}$$

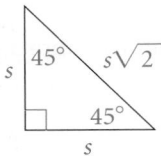

---

1   **EXAMPLE**   **Finding Lengths in a 45°-45°-90° Triangle**

Find the length of the hypotenuse in the triangle at the right.

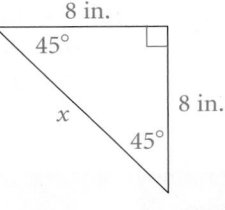

$$\text{hypotenuse} = \text{leg} \cdot \sqrt{2}$$

$x = 8 \cdot \sqrt{2}$ **The length of either leg is 8 in.**

≈ 11.3 **Round to the nearest tenth.**

● The length of the hypotenuse is about 11.3 in.

EXERCISES

Find the hypotenuse of each 45°-45°-90° triangle. Round to the nearest tenth.

1.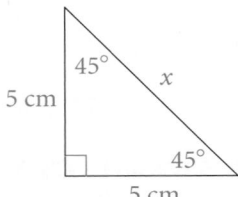

2.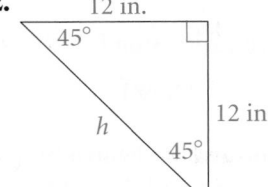

3.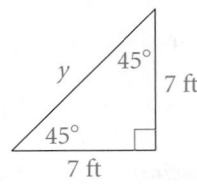

4. A baseball diamond is a square. The distance from any base to the next base is 90 ft. How far is it from home plate to second base? (*Hint:* Draw a diagram.)

5. The hypotenuse of a 45°-45°-90° triangle is 40.2 ft long. How long is each leg? Round to the nearest tenth.

Another special right triangle is the 30°-60°-90° triangle. You can form two congruent 30°-60°-90° triangles by bisecting an angle of an equilateral triangle. As the diagram at the right shows, the length of the hypotenuse is twice the length of the shorter leg. You can use the Pythagorean Theorem to find the length of the longer leg b shown in the diagram at the right below.

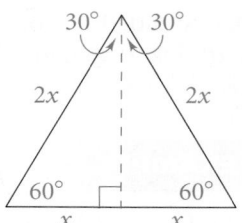

$$(2x)^2 = x^2 + b^2 \quad \text{Use the Pythagorean Theorem.}$$
$$4x^2 = x^2 + b^2 \quad \text{Simplify.}$$
$$3x^2 = b^2 \quad \text{Subtract } x^2 \text{ from each side.}$$
$$\sqrt{3x^2} = \sqrt{b^2} \quad \text{Find the principal square root of each side.}$$
$$\sqrt{3} \cdot \sqrt{x^2} = b \quad \text{Use the Multiplication Property of Square Roots.}$$
$$b = \sqrt{3} \cdot x \quad \text{Simplify.}$$

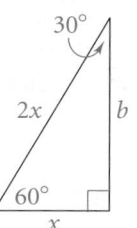

The length of the longer leg is $\sqrt{3} \cdot x$, or $x\sqrt{3}$.

Theorem	30°-60°-90° Triangle Theorem

In a 30°-60°-90° triangle, the length of the hypotenuse is twice the length of the shorter leg. The length of the longer leg is $\sqrt{3}$ times the length of the shorter leg.

$$\text{hypotenuse} = 2 \cdot \text{shorter leg}$$
$$\text{longer leg} = \sqrt{3} \cdot \text{shorter leg}$$

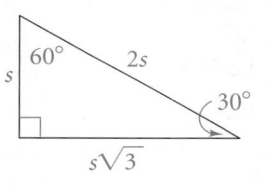

2 EXAMPLE Finding Lengths in a 30°-60°-90° Triangle

Find the missing lengths in the triangle at the right.

$$\text{hypotenuse} = 2 \cdot \text{shorter leg}$$
$$x = 2 \cdot 9 \quad \text{The length of the shorter leg is 9.}$$
$$x = 18 \quad \text{Simplify.}$$
$$\text{longer leg} = \sqrt{3} \cdot \text{shorter leg}$$
$$y = 9 \cdot \sqrt{3} \quad \text{The length of the shorter leg is 9.}$$
$$y \approx 15.6 \quad \text{Simplify. Round to the nearest tenth.}$$

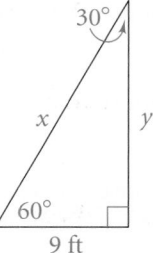

The length of the hypotenuse is 18 ft, and the length of the longer leg is about 15.6 ft.

EXERCISES

Find the missing lengths in each triangle. Round to the nearest tenth.

6.

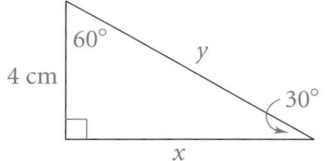

7.

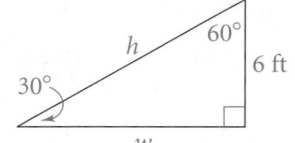

8.

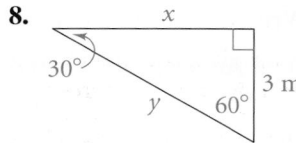

An extended-response question is usually worth a maximum of 4 points in this textbook. It sometimes has multiple parts. To get full credit, you need to answer each part and show all your work or justify your reasoning.

EXAMPLE

The Theatre Club needs to raise $440 to cover the cost of its children's play. The ticket prices are $14 for an adult and $2 for a child. The club expects that three times as many children as adults will attend the play. Write and solve an equation to find how many adults and children have to buy tickets in order for the club to cover its costs.

Three responses are below with the points each received.

4 points	3 points	1 point
x = number of adults $3x$ = number of children $14(x) + 2(3x) = 440$ $14x + 6x = 440$ $20x = 440$ $x = 22$ $3x = 66$ 22 adults and 66 children must attend the play.	x = number of adults $14(x) + 2(3x) = 440$ $14x + 5x = 440$ $19x = 440$ $x = 23.2$ 24 adults and $3(24) = 72$ children must attend the play.	x = number of adults $x + 3x = 440$ $4x = 440$ $x = 110$

The 4-point response shows the work and gives a written answer to the problem. Note that it begins by identifying the variable before writing the equation.

The 3-point response contains a computational error, but the student completed both parts.

The 1-point response shows an incorrect equation, and it does not give the number of children who must attend to cover costs.

EXERCISES

Use the Example above to answer each question.

1. Read the 3-point response. What error did the student make?

2. Write a 2-point response that begins by defining variables.

3. **Error Analysis** Why is the equation in the 1-point response incorrect?

Chapter Review

Vocabulary Review

Addition Property of Equality (p. 118)
conclusion (p. 183)
conditional (p. 183)
consecutive integers (p. 159)
converse (p. 183)
cross products (p. 144)
dilation (p. 150)
Division Property of Equality (p. 118)
equivalent equations (p. 118)
extremes of a proportion (p. 144)
greatest possible error (p. 169)
hypotenuse (p. 181)
hypothesis (p. 183)
identity (p. 136)

inverse operations (p. 118)
leg (p. 181)
literal equation (p. 140)
means of a proportion (p. 144)
Multiplication Property of Equality (p. 118)
negative square root (p. 176)
percent error (p. 170)
percent of change (p. 168)
percent of decrease (p. 168)
percent of increase (p. 168)
perfect squares (p. 177)
principal square root (p. 176)
proportion (p. 144)

Pythagorean Theorem (p. 181)
radicand (p. 176)
rate (p. 142)
ratio (p. 142)
scale (p. 151)
scale drawing (p. 151)
scale factor (p. 150)
similar figures (p. 150)
solution of an equation (p. 118)
square root (p. 176)
Subtraction Property of Equality (p. 118)
uniform motion (p. 159)
unit analysis (p. 143)
unit rate (p. 142)

For: Vocabulary quiz
Web Code: atj-0351

Choose the correct item to complete each sentence.

1. Addition and subtraction are examples of (*inverse operations, extremes of a proportion*).

2. (*Consecutive integers, Solutions of equivalent equations*) have the same value.

3. To change one unit of measure to another you can use a (*proportion, rate*) that is equal to 1.

4. You can use (*cross products, unit analysis*) to solve a proportion that involves one variable.

5. The (*greatest possible error, percent error*) in a measurement is one half of the measuring unit.

Skills and Concepts

3-1 Objectives

▼ To solve two-step equations (p. 119)

▼ To use deductive reasoning (p. 121)

A two-step equation is an equation that has two operations. You can use tiles to model and solve a two-step equation. To solve a two-step equation, first add or subtract. Then multiply or divide.

Solve each equation. Then check your solution.

6. $5x - 8 = 12$ 7. $7t - 3 = 18$ 8. $\frac{c}{5} - 4 = -3$

9. $-2q - 5 = -11$ 10. $-3m + 8 = 2$ 11. $11y + 9 = 130$

12. A state park charges admission of $6 per person plus $3 for parking. Jo paid $27 when her car entered the park. Write and solve an equation to find the number of people in Jo's car. Be sure to explain what your variable represents.

Solve each equation. Justify each step.

13. $314 = -n + 576$ 14. $-\frac{1}{4}w - 1 = 6$ 15. $3h - 4 = 5$

3-2 and 3-3 Objectives

▼ To use the Distributive Property when combining like terms (p. 126)

▼ To use the Distributive Property when solving equations (p. 127)

▼ To solve equations with variables on both sides (p. 134)

▼ To identify equations that are identities or have no solution (p. 136)

You can combine like terms and use the Distributive Property to simplify expressions and solve equations. You can also use the properties of equality to solve an equation.

An equation has no solution if no value of the variable makes the equation true. An equation is an **identity** if every value of the variable makes the equation true.

Solve each equation. If the equation is an identity or if it has no solution, write *identity* **or** *no solution.*

16. $b + 4b = -90$ **17.** $-x + 7x = 24$ **18.** $2(t + 5) = 9$

19. $-(3 - 10y) = 12$ **20.** $x - (4 - x) = 0$ **21.** $4n - 6n = 2n$

22. $3(5x - 2) - 6x = 3(3x + 2)$ **23.** $3(2t - 6) = 2(3t - 9)$

24. $\frac{3y}{4} - \frac{y}{2} = 5$ **25.** $0.36p + 0.26 = 3.86$

26. **Geometry** The width of a rectangle is 6 cm less than the length. The perimeter is 72 cm. Write and solve an equation to find the dimensions of the rectangle.

3-4 Objectives

▼ To find ratios and rates (p. 142)

▼ To solve proportions (p. 144)

A **ratio** is a comparison of two numbers by division. A **rate** is a ratio that compares quantities measured in different units. A **proportion** is a statement that two ratios are equal. You can solve a proportion involving a single variable by finding the **cross products.**

Write in miles per hour. Round to the nearest tenth where necessary.

27. 2.5 mi/min   **28.** 300 ft/min   **29.** 4 in./s

Solve each proportion.

30. $\frac{4}{12} = \frac{c}{6}$ **31.** $\frac{t}{5} = \frac{23}{50}$ **32.** $\frac{-9}{m} = \frac{3}{2}$

33. $\frac{x}{8} = \frac{x - 5}{6}$ **34.** $\frac{12}{r} = \frac{4}{0.5r - 1}$ **35.** $\frac{d - 2}{d + 9} = \frac{3}{14}$

3-5 Objectives

▼ To find missing measures of similar figures (p. 149)

▼ To use similar figures when measuring indirectly (p. 151)

Similar figures have the same shape but not necessarily the same size. If two figures are **similar,** then corresponding angles are congruent and corresponding sides are in proportion.

A scale drawing is an enlarged or reduced drawing of an object. The ratio of the length of drawing to the actual length is the **scale** of the drawing. You can use proportions to solve problems involving scale drawings. A map is an example of a scale drawing.

In the figure at the right $\triangle DEF \sim \triangle QRS$.

36. Find RS.   **37.** Find QR.

38. **Hobbies** A certain model airplane is $\frac{1}{48}$ of the airplane's actual size. The length of the model airplane's wing is $\frac{3}{4}$ ft. How long is the airplane's wing?

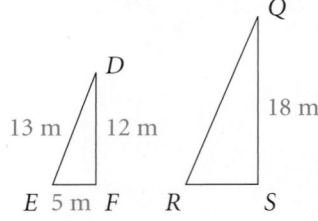

3-6 Objectives

▼ To define a variable in terms of another variable (p. 158)

▼ To model distance-rate-time problems (p. 159)

Many types of real-world problems can be solved using equations. You can also use tables and diagrams to help organize information or solve the problem.

Write and solve an equation for each situation.

39. Travel Time The Great Seto Bridge in Japan is about 7.6 mi long. How long would it take you to cross the bridge if you were walking at 4 mi/h?

40. The sum of three consecutive integers is 582. Find the three integers.

41. Ocean Travel A supertanker left port traveling north at an average speed of 10 knots. Two hours later a cruise ship leaves the same port, heading south at an average speed of 18 knots. How many hours after the cruise ship sails will the two ships be 209 nautical miles apart? (1 knot = 1 nautical mile per hour)

3-7 Objectives

▼ To find percent of change (p. 168)

▼ To find percent error (p. 169)

The **percent of change** $= \dfrac{\text{amount of change}}{\text{original amount}}$. If a value increases, the percent of change is the **percent of increase.** If a value decreases, the change is the **percent of decrease.** The greatest possible error in a measurement is one half of the measuring unit. The percent error is $\dfrac{\text{greater difference}}{\text{measurement}}$.

For Exercises 42–44, find each percent of change. Where necessary, round to the nearest percent. Describe the percent of change as a percent of increase or decrease.

42. $75,000 to $85,000   **43.** 20 ft to 15 ft   **44.** 60 h to 40 h

45. Suppose you measure a box. Its dimensions are 32 in. × 28 in. × 25 in. Find the percent of error in calculating its volume to the nearest tenth of a percent.

3-8 Objectives

▼ To find square roots (p. 176)

▼ To estimate and use square roots (p. 177)

If $a^2 = b$, then a is a **square root** of b. The positive or **principal square root** of b is indicated by $\sqrt{b}$. The **negative square root** is indicated by $-\sqrt{b}$. The squares of integers are called **perfect squares.**

Tell whether each expression is rational or irrational. Then find the value of each expression. If necessary, round to the nearest hundredth.

46. $\sqrt{86}$ **47.** $-\sqrt{121}$ **48.** $\pm\sqrt{\frac{1}{2}}$ **49.** $\sqrt{2.55}$ **50.** $-\sqrt{\frac{4}{25}}$

3-9 Objectives

▼ To solve problems using the Pythagorean Theorem (p. 181)

▼ To identify right triangles (p. 183)

For a right triangle with **legs** a and b and **hypotenuse** c, the **Pythagorean Theorem** states that $a^2 + b^2 = c^2$. The converse of the Pythagorean Theorem states that if a triangle has sides of lengths a, b, and c, and if $a^2 + b^2 = c^2$, then it is a right triangle with hypotenuse c.

Find the length of the hypotenuse with the given leg lengths. Round to the nearest tenth.

51. $a = 3, b = 5$ **52.** $a = 11, b = 14$ **53.** $a = 7, b = 13$ **54.** $a = 4, b = 9$

Determine whether the given lengths can be sides of a right triangle.

55. $XY = 16, YZ = 34, XZ = 30$ **56.** $XY = 2.5, YZ = 2.4, XZ = 0.7$

Solve each equation. Then check.

1. $3w + 2 - w = -4$ **2.** $\frac{1}{4}(k - 1) = 10$

3. $6(y + 3) = 24$ **4.** $\frac{5n + 1}{8} = \frac{1}{2}$

5. If $2t + 3 = -9$ what is the value of $-3t - 7$?

6. Solve $2x - 4 = -7$. Justify each step.

Define a variable and write an equation to model each situation. Then solve.

7. Your chorus holds a car wash. They have $25.00 for making change. At the end of the car wash, they have $453.50. How much money did they make?

8. Truck Rental The rate to rent a certain truck is $55 per day and $0.20 per mile. Your family pays $80 to rent this truck for one day. How many miles did your family drive?

9. Entertainment Movie tickets for an adult and three children cost $20. An adult's ticket costs $2 more than a child's ticket. Find the cost of an adult's ticket.

Solve each proportion.

10. $\frac{3}{4} = \frac{c}{20}$ **11.** $\frac{8}{15} = \frac{4}{w}$

12. $\frac{w}{6} = \frac{6}{15}$ **13.** $\frac{5}{t} = \frac{25}{100}$

Solve. If the equation is an identity, write *identity*. If it has no solution, write *no solution*.

14. $9j + 3 = 3(3j + 1)$ **15.** $4v - 9 = 6v + 7$

16. $2(1 - 2y) = 4y + 18$

17. $4p - 5 + p = 7 + 5p + 2$

Calculate the percent of change. Describe each as a percent of increase or a percent of decrease.

18. $4.50/h to $5/h   **19.** 60 km/h to 45 km/h

20. 150 lb to 135 lb   **21.** $18 to $24

Complete each statement.

22. 14¢/oz = $■/lb   **23.** 7 gal/wk = ■ qt/h

24. 35 mi/h = ■ ft/min   **25.** 120 ft/day = ■ in./min

26. A scale on a map is 1 in. : 25 mi. You measure 6.5 inches. How many miles is the actual distance?

27. A taxicab company charges a flat fee of $1.85, plus an additional $.40 per quarter-mile.
  a. Write a formula to find the total cost for each fare.
  b. Use this formula to find the cost for 1 person to travel 8 mi.

Find the principal and negative square root of each number. If necessary, round to the nearest hundredth.

28. 1.44   **29.** 1600   **30.** $\frac{4}{9}$ **31.** 0.4

Define a variable and write an equation to model each situation. Then solve.

32. Jan is one year younger than her brother Bill and one year older than her sister Sue. The sum of the three children's ages is 57. How old is each child?

33. Travel At noon, your family starts out from Louisville to go to Memphis driving at 40 mi/h. Your uncle leaves Memphis to come to Louisville two hours later. He is taking the same route and driving at 60 mi/h. The two cities are 380 miles apart. At what time do the cars meet?

Between what two consecutive integers is each square root?

34. $\sqrt{28}$ **35.** $\sqrt{136}$

36. $\sqrt{332}$ **37.** $-\sqrt{8.99}$

Determine whether the given lengths can be sides of a right triangle.

38. 6, 8, 10   **39.** 6, 7, 9

40. 4, 8, 11   **41.** 10, 24, 26

42. The length of each leg of an isosceles right triangle is 40.9 cm. Find the length of the hypotenuse to the nearest tenth.

43. One house is 12 mi east of a school. Another house is 9 mi north of the school. How far apart are the houses?

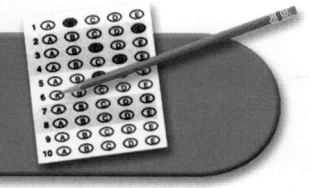

Regents Test Prep

Reading Comprehension **Read the passage below. Then answer the questions on the basis of what is *stated* or *implied* in the passage.**

> **Geometry of Earth** Long before Columbus, a Greek scholar named Eratosthenes (274–194 B.C.) calculated the circumference of Earth by making some simple observations about shadows at noon and applying some insights from geometry.
>
> The size of Earth has been estimated and measured many times since then. The French Academy of Sciences tried to make an accurate measurement in the late 1700s in order to create a new unit of measure, the meter. They proposed to define the meter as one ten-millionth of the distance from the North Pole to the equator.
>
> Today we can measure Earth's circumference very accurately using data from satellites and making calculations on computers. The circumference at the equator is 24,901.55 miles. Earth is not a perfect sphere, however, because its rotation causes it to bulge a little at the equator. So the circumference from pole to pole is a bit smaller, about 24,859.82 miles.

1. How many centuries before the French Academy of Sciences measured the circumference of Earth did Eratosthenes live?

   (1) 22 centuries   **(2)** 20 centuries

   (3) 18 centuries   **(4)** 16 centuries

2. How much greater is the circumference of Earth at the equator than the circumference at the poles?

   (1) about 20 miles
   (2) about 40 miles
   (3) about 80 miles
   (4) about 200 miles

3. Earth is divided into 24 time zones. About how wide is each time zone at the equator?
   (1) 1031 miles
   (2) 1036 miles
   (3) 1038 miles
   (4) 1042 miles

4. Which is the best estimate of the diameter of Earth measured from pole to pole?
   (1) 3960 miles
   (2) 4530 miles
   (3) 6300 miles
   (4) 7920 miles

5. A geostationary satellite moves in an orbit 22,300 miles above the equator. The satellite moves at a rate such that it stays at the same point above Earth as Earth rotates on its axis. What is the approximate speed of the satellite?
   (1) 700 miles per hour
   (2) 7000 miles per hour
   (3) 10,000 miles per hour
   (4) 14,000 miles per hour

6. **a.** One mile is equal to about 1610 meters. Write an equation expressing this relationship. Use the equation to calculate the approximate circumference of Earth from pole to pole in meters.
   b. Is the meter unit of measure you found close to the Academy of Sciences' original definition? Explain.

7. **a.** As Earth rotates, the line separating night from day moves west across Earth's surface. How fast is this line moving in miles per minute at the equator? Show your work.
   b. Does the day-night dividing line move more quickly or more slowly across your state than it does at the equator? Explain.

Data Analysis

Activity Lab

Probability Distributions

A probability distribution is a function that gives the probability for each outcome in a sample space. You can show a probability distribution using a table or a graph.

Forty people are surveyed about the number of siblings they have. The results are below, including the probability of a person making each response. The graph below shows the probability distribution.

Number of Siblings	Response	Probability of Response
0	卌 |||	$\frac{1}{5} = 20\%$
1	卌 卌 ||	$\frac{3}{10} = 30\%$
2	卌 卌	$\frac{1}{4} = 25\%$
3	卌		$\frac{3}{20} = 15\%$
4 or more	||||	$\frac{1}{10} = 10\%$

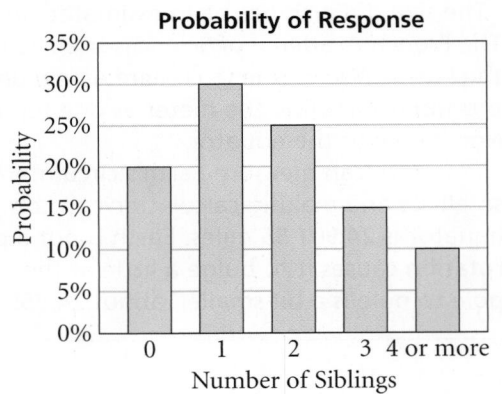

One kind of probability distribution is called a **discrete probability distribution**. In this case, the sample space contains a countable number of values. For example, there are only three possible outcomes when two coins are tossed, heads-heads, heads-tails, and tails-tails.

1 ACTIVITY

1. **Data Collection** Roll a pair of number cubes 20 times. Record the sum of each roll. Write the frequency of each sum in a table like the one at the right.

2. Find the experimental probability for each outcome and enter it in the table.

3. Make a graph of the probability distribution.

4. **Critical Thinking** Find the sum of the experimental probabilities. Does your answer make sense? Explain.

Sum of Cubes	Frequency	Experimental Probability
1	■	■
2	■	■
3	■	■
4	■	■
5	■	■
⋮	■	■

For this activity you will need 4 coins.

5. **Data Collection** Toss all of the coins at once. Record the number of tails. Record the results of 30 tosses in a table like the one at the right.

6. Find the experimental probability for each outcome (0 tails, 1 tail, 2 tails, and so on.) Round the probabilities to the nearest hundredth.

7. Make a graph of the probability distribution.

8. **Critical Thinking** Do the probabilities add to 1? If they don't, explain why.

9. **Critical Thinking** If you were to repeat this activity would you expect to see the same probability distribution? Explain.

Number of Tails	Frequency
0	■
1	■
2	■
3	■
4	■

EXERCISES

10. A bag contains tiles with letters A, B, C, D, E, and F. The probability distribution shows the probabilities of selecting each letter.
   a. If there are 80 tiles in bag, how many are B's?
   b. How many more E's than D's are in the bag?
   c. Suppose you pick a tile at random, replace it and pick a second tile. Find P(C then A).

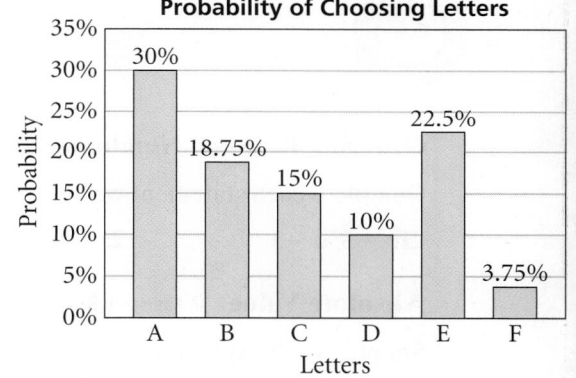

Probability of Choosing Letters

11. **Odds** Recall that the odds in favor of an event equal the ratio of the number of times the event occurs to the number of times the event does not occur.

   The odds in favor of event M are 1 : 3.

   The odds in favor of event N are also 1 : 3.

   The odds in favor of event P are 1 : 5.

   The odds in favor of event Q are 1 : 2.

   Assume the events above are independent and complete a sample space. Graph the probability distribution of events M, N, P, and Q.

12. **Data Collection** Use the color palette. Ask 20 people which of the colors shown is their favorite. Make a table and graph of the probability distribution.

What You've Learned

NY Learning Standards for Mathematics

In Chapters 1 and 3, you

● **NY A.A.22:** evaluated variable expressions and solved equations.

 **Check Your Readiness**    **for Help** to the Lesson in green.

Ordering Rational Numbers (Lesson 1-3)

Complete each statement with $<$, $=$, or $>$.

1. -3 ■ -5 **2.** 7 ■ $\frac{14}{2}$ **3.** -8 ■ -8.4 **4.** $-\frac{3}{2}$ ■ -1

Absolute Value (Lesson 1-3)

Simplify each expression.

5. $5 + |4 - 6|$ **6.** $|30 - 28| - 6$ **7.** $|-7 + 2| - 4$

Solving One-Step Equations (Review page 118)

Solve each equation. Check your solution.

8. $x - 4 = -2$ **9.** $b + 4 = 7$ **10.** $-\frac{3}{4}y = 9$ **11.** $\frac{m}{12} = 2.7$

12. $-8 + x = 15$ **13.** $n - 7 = 22.5$ **14.** $-\frac{12}{7}z = 48$ **15.** $\frac{5y}{4} = -15$

Solving Two-Step Equations (Lesson 3-1)

Solve each equation. Check your solution.

16. $-5 + \frac{b}{4} = 7$ **17.** $4.2m + 4 = 25$ **18.** $-12 = 6 + \frac{3}{4}x$ **19.** $6 = -z - 4$

20. $4m + 2.3 = 9.7$ **21.** $\frac{5}{8}t - 7 = -22$ **22.** $-4.7 = 3y + 1.3$ **23.** $12.2 = 5.3x - 3.7$

Solving Multi-Step Equations (Lesson 3-2)

Solve each equation. Check your solution.

24. $4t + 7 + 6t = -33$ **25.** $2a + 5 = 9a - 16$ **26.** $\frac{1}{3} + \frac{4y}{6} = \frac{2}{3}$

27. $6(y - 2) = 8 - 2y$ **28.** $n + 3(n - 2) = 10.4$ **29.** $\frac{1}{2}w + 3 = \frac{2}{3}w - 5$

Solving Inequalities

Key Vocabulary

- compound inequalities
(p. 227)
- equivalent inequalities
(p. 206)
- solution of an inequality
(p. 200)

What You'll Learn Next

NY Learning Standards for Mathematics

- **NY A.A.4:** You will graph inequalities.

- **NY A.A.6:** You will solve inequalities, noting the differences from the methods used for solving equations.

- **NY A.A.6:** You will write and solve compound inequalities by interpreting phrases that use *and* or *or*.

Activity Lab You will apply your knowledge of area and perimeter to composite figures, on pages 248–249.

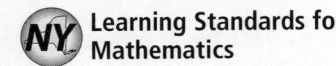

Inequalities and Their Graphs

NY Learning Standards for Mathematics

A.A.4 Translate verbal sentences into mathematical equations or inequalities.

A.A.21 Determine whether a given value is a solution to a given linear equation in one variable or linear inequality in one variable.

✓ **Check Skills You'll Need**

Graph the numbers on the same number line.

1. 4   **2.** -3 **3.** $\frac{9}{3}$ **4.** 0   **5.** 1.5

Complete each statement with <, =, or >.

6. $-3 \; \blacksquare \; -5$ **7.** $4.29 \; \blacksquare \; 4.8$ **8.** $(-3)(-4) \; \blacksquare \; 12$

9. $-1 - 2 \; \blacksquare \; 6 - 9$ **10.** $-\frac{3}{4} \; \blacksquare \; -\frac{4}{5}$ **11.** $\frac{1}{3} + \frac{1}{3} \; \blacksquare \; \frac{1}{2} + \frac{1}{2}$

GO **for Help** Lesson 1-3

🔊 **New Vocabulary** • solution of an inequality

---

1   Identifying Solutions of Inequalities

Vocabulary Tip

Less than and is less than have different meanings. For example, *x* less than 3 means $3 - x$; *x* is less than 3 means $x < 3$.

A **solution of an inequality** is any number that makes the inequality true. For example, the solutions of the inequality $x < 3$ are all numbers that are less than 3.

1 EXAMPLE   **Identifying Solutions by Mental Math**

Is each number a solution of $x \le 7$?

   a. 9   No, $9 \le 7$ is not true.

   b. -1 Yes, $-1 \le 7$ is true.

   c. $\frac{14}{2}$ $\frac{14}{2} = 7$; yes, $\frac{14}{2} \le 7$ is true.

✓ **Quick Check** ① Is each number a solution of $x \ge -4.1$?

   a. -5 **b.** -4.1 **c.** 8   **d.** 0

You can determine whether a value is a solution of an inequality by evaluating an expression.

2 EXAMPLE   **Identifying Solutions by Evaluating**

Is each number a solution of $2 - 5x > 13$?

   a. 3   **b.** -4

$2 - 5x > 13$		$2 - 5x > 13$
$2 - 5(3) > 13$	← Substitute for *x*.→	$2 - 5(-4) > 13$
$2 - 15 > 13$	← Simplify.→	$2 + 20 > 13$
$-13 \not> 13$	← Compare.→	$22 > 13$

3 does not make the original inequality true, so 3 is not a solution. -4 does make the original inequality true, so -4 is a solution.

    a. 1   **b.** 2   **c.** 3   **d.** 4

2   Graphing and Writing Inequalities in One Variable

You can use a graph to indicate all of the solutions of an inequality.

Inequality	**Graph**	

Vocabulary Tip

You normally read $-1 \geq a$ as "-1 is greater than or equal to a." The inequality symbol also indicates that a is less than or equal to -1.

$x < 3$ (number line: -1 0 1 2 3 4 5, open dot at 3, shaded left)   **The open dot shows that 3 is *not* a solution. Shade to the left of 3.**

$m \geq -2$ (number line: -3 -2 -1 0 1 2 3, closed dot at -2, shaded right)   **The closed dot shows that -2 is a solution. Shade to the right of -2.**

$-1 \geq a$ (number line: -3 -2 -1 0 1 2 3, closed dot at -1, shaded left)   **The closed dot shows that -1 is a solution. Shade to the left of -1.**

You can also write $-1 \geq a$ as $a \leq -1$.

3 EXAMPLE   **Graphing Inequalities**

a. Graph $c > -2$.

(number line: -3 -2 -1 0 1 2 3, open dot at -2, shaded right)   **The solutions of $c > -2$ are all the points to the right of -2.**

b. Graph $4 \leq m$.

(number line: -1 0 1 2 3 4 5, closed dot at 4, shaded right)   **The solutions of $4 \leq m$ are 4 and all the points to the right of 4.**

 **Quick Check** **3** Graph each inequality.

    a. $a < 1$ **b.** $n \geq -3$ **c.** $2 > p$

You can write an inequality for a graph.

Video Tutor Help
Visit: PhSchool.com
Web Code: ate-0775

4 EXAMPLE   **Writing an Inequality From a Graph**

Write an inequality for each graph.

a. (number line: -5 -4 -3 -2 -1 0 1, open dot at -4, shaded left) $x < -4$ **Numbers less than -4 are graphed.**

b. (number line: -1 0 1 2 3 4 5, closed dot at 5, shaded left) $x \leq 5$ **Numbers less than or equal to 5 are graphed.**

c. (number line: -3 -2 -1 0 1 2 3, open dot at $\frac{1}{2}$, shaded right) $x > \frac{1}{2}$ **Numbers greater than $\frac{1}{2}$ are graphed.**

d. (number line: -3 -2 -1 0 1 2 3, closed dot at -1, shaded right) $x \geq -1$ **Numbers greater than or equal to -1 are graphed.**

 **Quick Check** **4** Write an inequality for each graph.

    a. (number line: -3 -2 -1 0 1 2 3, closed dot at 2, shaded left)   **b.** (number line: -3 -2 -1 0 1 2 3, open dot at 1, shaded right)

You can describe real-world situations using an inequality.

5 EXAMPLE **Real-World Problem Solving**

Define a variable and write an inequality for each situation.

a.

Let s = a legal speed.
The sign indicates that $s \leq 65$.

b.

Let p = pay per hour (in dollars).
The sign indicates that $p \geq 6.15$.

✓ Quick Check **5 a.** **Critical Thinking** In part (a) of Example 5, can the speed be *all* real numbers less than or equal to 65? Explain.
b. In part (b) of Example 5, are all real numbers greater than or equal to $6.15 reasonable solutions of the inequality? Explain.

EXERCISES

For more exercises, see *Extra Skill and Word Problem Practice.*

Practice and Problem Solving

 Practice by Example

Example 1
(page 200)

Mental Math **Is each number following the inequality a solution of the given inequality?**

1. $v \geq -5; 4$ **2.** $0.5 > c; 2$ **3.** $b < 4; -0.5$ **4.** $d \leq \frac{17}{3}; 5$

5. $g \leq \frac{12}{5}; 3$ **6.** $k < 0; -1$ **7.** $a > 3.2; 3$ **8.** $x \geq -2.5; -2.5$

Example 2
(page 200)

Is each number a solution of the given inequality?

9. $3x - 7 > -1$ **a.** 2   **b.** 0   **c.** 5

10. $4n - 3 \leq 5$ **a.** 2   **b.** 3   **c.** −1

11. $2y + 1 < -3$ **a.** 0   **b.** −2   **c.** 1

12. $\frac{4 - m}{m} \geq 5$ **a.** 0.5   **b.** 2   **c.** −4

13. $n(n - 3) < 54$ **a.** 9   **b.** 3   **c.** 10

14. $5(2q - 8) \geq 7$ **a.** −2   **b.** $\frac{9}{2}$ **c.** 6

Example 3
(page 201)

Match each inequality with its graph.

15. $x < 4$ **16.** $x \geq 4$ **17.** $x > 4$ **18.** $x \leq 4$

A. ![number line from -3 to 5 with filled dot at 4, shaded left]
   $-3\ -2\ -1\ \ 0\ \ 1\ \ 2\ \ 3\ \ 4\ \ 5$

B. ![number line from -3 to 5 with filled dot at 4, shaded right]
   $-3\ -2\ -1\ \ 0\ \ 1\ \ 2\ \ 3\ \ 4\ \ 5$

C. ![number line from -3 to 5 with open dot at 4, shaded left]
   $-3\ -2\ -1\ \ 0\ \ 1\ \ 2\ \ 3\ \ 4\ \ 5$

D. ![number line from -3 to 5 with open dot at 4, shaded right]
   $-3\ -2\ -1\ \ 0\ \ 1\ \ 2\ \ 3\ \ 4\ \ 5$

Graph each inequality.

19. $x > 1$ **20.** $s < -3$ **21.** $y \le -4$ **22.** $t \ge -1$

23. $-2 < d$ **24.** $-\frac{3}{2} \le b$ **25.** $7 \ge a$ **26.** $4.25 > c$

Example 4
(page 201)

Write an inequality for each graph.

27. ← | ⊕ | | | | | | | | →
　　　−4 −3 −2 −1　0　1　2　3　4

28. ← | | | | | | | | ● →
　　　　−1　0　1　2　3　4　5　6　7

29. ← | | | | | ● | | | →
　　　−4 −3 −2 −1　0　1　2　3　4

30. ← | ⊕ | | | | | | | →
　　　−7 −6 −5 −4 −3 −2 −1　0　1

31. ← | | | | | | ● | | →
　　　−2 −1　0　1　2　3　4　5　6

32. ← | | | | | ○ | | | →
　　　−4 −3 −2 −1　0　1　2　3　4

Example 5
(page 202)

Define a variable and write an inequality to model each situation.

33. A bus can seat at most 48 students.

34. In many states, you must be at least 16 years old to obtain a driver's license.

35. It is not safe to use a light bulb of more than 60 watts in this light fixture.

36. At least 350 students attended the band concert Friday night.

37. Aviation The Navy's flying squad, the Blue Angels, makes more than 75 appearances each year.

B **Apply Your Skills**

Write each inequality in words.

38. $n < 5$ **39.** $b > 0$ **40.** $7 \ge x$ **41.** $z \ge -5.6$

42. $4 > q$ **43.** $-1 \ge m$ **44.** $35 \ge w$ **45.** $g - 2 < 7$

46. $a \le 3$ **47.** $6 + r > -2$ **48.** $8 \le h$ **49.** $1.2 > k$

50. Writing Explain how you choose whether to draw an open or a closed dot when you graph an inequality.

GO for Help

For help with counterexamples see p. 18.

51. Open-Ended Describe a situation that you can represent using the inequality $x \ge 18$.

52. Multiple Choice Suppose your school plans a musical. The director's goal is ticket sales of at least $4000. Adult tickets are $5.00 and student tickets are $4.00. Let a represent the number of adult tickets and s represent the number of student tickets. Which inequality represents the director's goal?

　Ⓐ $4a + 5s < 4000$　　　　Ⓑ $5a + 4s > 4000$

　Ⓒ $5a + 4s \le 4000$　　　　Ⓓ $5a + 4s \ge 4000$

Rewrite each inequality so that the variable is on the left. Then graph the solutions.

53. $2 < x$ **54.** $-5 \ge b$ **55.** $0 \le r$ **56.** $5 > a$

Graph each inequality from the given description.

57. t is nonnegative.　　　　　**58.** x is positive.

59. k is no more than 3.　　　　**60.** r is at least 2.

61. s is at most 4.　　　　　　**62.** v is no less than 7.

63. Writing Explain how you interpret the phrases "at least" and "at most" in an inequality that models a real-world situation.

Use the map below for Exercises 64–65.

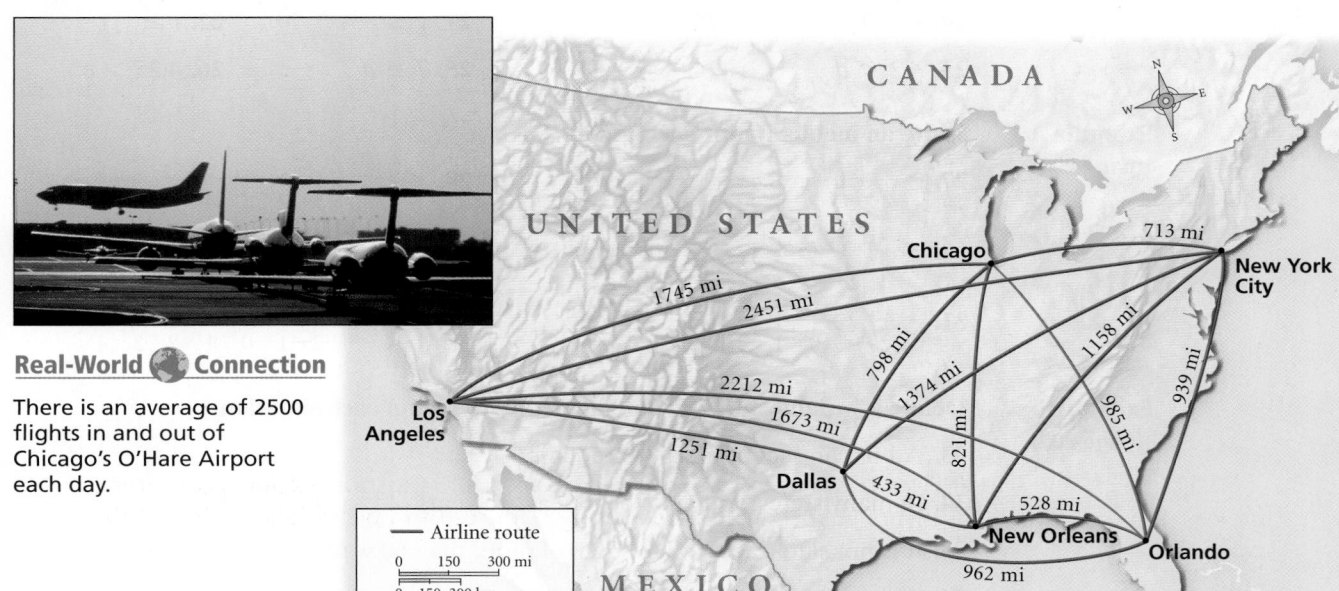

Real-World 🌐 **Connection**

There is an average of 2500 flights in and out of Chicago's O'Hare Airport each day.

GO ◯**nline**

Homework Video Tutor

Visit: PHSchool.com
Web Code: ate-0401

🌐 **64. Air Travel** You plan to go from New York City to Los Angeles. Let x be the distance in miles of any air-route between New York City and Los Angeles. The shortest route is a direct flight. Using the map, write a true statement about the mileage of any route from New York City to Los Angeles.

🌐 **65. Air Travel** Your travel agent is making plans for you to go from Chicago to New Orleans. A direct flight costs too much. Option A consists of flights from Chicago to Dallas to New Orleans. Option B consists of flights from Chicago to Orlando to New Orleans. Write an inequality comparing the mileage of these two options.

66. Error Analysis A student claims that the inequality $3x + 1 > 0$ is always true because multiplying a number by three and then adding one to it makes the number greater than zero. Use a counterexample to show why the student is not correct.

67. Critical Thinking Describe how you can display the solutions of the inequality $x \neq 3$ on a number line.

68. Critical Thinking Explain the difference between "4 greater than x" and "$4 > x$."

C Challenge

69. Critical Thinking Which is the correct graph of $-4 < -x$? Explain.

Problem Solving Hint

In Exercise 71, make logical choices for values of a and b. Use positive and negative numbers as well as zero.

70. Reasoning Give a counterexample for this statement. If $a < b$, then $a^2 < b^2$.

71. Reasoning Describe the numbers a and b for which the following statement is true. If $a < b$, then $a^2 = b^2$.

Graph on a number line.

72. all values of x such that $x > -2$ and $x \leq 2$

73. all values of x such that $x < -1$ or $x > 3$

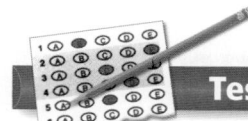

Multiple Choice

74. Which inequality has the same solutions as $n > 5$?

    A. $n < -5$ **B.** $n < 5$ **C.** $5 < n$ **D.** $-n > -5$

75. What is the least whole-number solution of $k \geq -5$?

    F. -5 **G.** -4 **H.** 0 **J.** 1

76. Employees must work at least 20 years in a company in order to receive full benefits upon retirement. Which inequality or graph does NOT describe this situation?

    A. $y \geq 20$ **B.** $y > 20$

    C. $20 \leq y$ **D.** $\begin{array}{cccccc} \vdash & + & + & + & \bullet & + \\ 0 & 5 & 10 & 15 & 20 & 25 \end{array}$

77. Which value makes the inequality $x^2 \geq x$ false?

    F. $-\frac{1}{4}$ **G.** 0 **H.** $\frac{1}{4}$ **J.** 1

78. Fire codes require that no more than 150 persons occupy a conference room. Which graph includes a room count in possible violation of the fire codes?

    A. $\begin{array}{ccccc} \bullet & \bullet & \bullet & & \\ 147 & 148 & 149 & 150 & 151 \end{array}$

    B. $\begin{array}{ccccc} & & \bullet & \bullet & \bullet \\ 147 & 148 & 149 & 150 & 151 \end{array}$

    C. $\begin{array}{ccccc} & & & \bullet & \\ 147 & 148 & 149 & 150 & 151 \end{array}$

    D. $\begin{array}{ccccc} \bullet & \bullet & \bullet & & \\ 147 & 148 & 149 & 150 & 151 \end{array}$

Short Response

79. A stretch of $12\frac{3}{5}$ mi of highway is being repaired. The project foreman reported that less than half of the job is complete. Draw a diagram to show the remaining miles to be repaired. Then write an inequality for the number of miles m that still need repair.

Mixed Review

Lesson 3-9

A right triangle has hypotenuse c and legs a and b. Find the missing side. If necessary, round to the nearest tenth.

80. $a = 6, b = 12$ **81.** $b = 5, c = 8$ **82.** $a = 1.2, b = 0.5$

83. $c = 16, b = 7$ **84.** $a = 10, c = 30$ **85.** $a = \frac{6}{7}, b = \frac{8}{7}$

Determine whether the given lengths can be sides of a right triangle.

86. $10, 13, 27$ **87.** $9, 40, 41$ **88.** $28, 96, 100$

Lesson 2-5

Name the property that each equation demonstrates.

89. $3(2 \cdot 7) = (3 \cdot 2)7$ **90.** $5 \times 1 = 1 \times 5$ **91.** $3 + 4 = 4 + 3$

Lesson 1-6

Find each measure for the following data.

    1   3   4   4   5   5   7   9   9   9   13

92. mean   **93.** median   **94.** mode

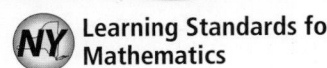

4-2

Solving Inequalities Using Addition and Subtraction

NY Learning Standards for Mathematics

A.A.6 Analyze and solve verbal problems whose solution requires solving a linear equation in one variable or linear inequality in one variable.

A.A.24 Solve linear inequalities in one variable.

☑ **Check Skills You'll Need**    **for Help**   Lesson 2-1 and Review page 118

Complete each statement with <, =, or >.

1. $-3 + 4$ ■ $-5 + 4$ **2.** $-3 + 6$ ■ $4 + 6$ **3.** $-3.4 + 2$ ■ $-3.45 + 2$

Solve each equation.

4. $x - 4 = 5$ **5.** $n - 3 = -5$ **6.** $t + 4 = -5$ **7.** $k + \frac{2}{3} = \frac{5}{6}$

◀)) **New Vocabulary** • equivalent inequalities

---

1 Using Addition to Solve Inequalities

Equivalent inequalities are inequalities with the same solutions. For example, $x + 4 < 7$ and $x < 3$ are equivalent inequalities.

$x + 4 < 7$

$x < 3$

You can add the same value to each side of an inequality, just as you did with equations.

 Key Concepts

Property	**Addition Property of Inequality**

For every real number a, b, and c,
   if $a > b$, then $a + c > b + c$;   if $a < b$, then $a + c < b + c$.

Examples $3 > 1$, so $3 + 2 > 1 + 2$. $-5 < 4$, so $-5 + 2 < 4 + 2$.

This property is also true for $\geq$ and $\leq$.

1 EXAMPLE Using the Addition Property of Inequality

Solve $x - 3 < 5$. Graph the solution.

$x - 3 + 3 < 5 + 3$ **Add 3 to each side.**

$\quad\quad x < 8$ **Simplify.**

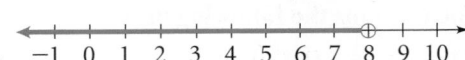

☑ **Quick Check** ❶ Solve $m - 6 > -4$. Graph your solution.

An inequality has an infinite number of solutions, so it is not possible to check all the solutions. You can check your computations and the direction of the inequality symbol. The steps below show how to check that $x < 8$ describes the solutions to Example 1.

Step 1 Check the computation. See if 8 is the solution to the equation $x - 3 = 5$.

$$x - 3 = 5$$
$$8 - 3 \stackrel{?}{=} 5 \qquad \textbf{Substitute 8 for } x.$$
$$5 = 5 \checkmark$$

Step 2 Check the inequality symbol. Choose any number less than 8 and substitute it into $x - 3 < 5$. In this case, use 7.

$$x - 3 < 5$$
$$7 - 3 < 5 \qquad \textbf{Substitute 7 for } x.$$
$$4 < 5 \checkmark$$

Since the computation and the direction of the inequality symbol are correct, $x - 3 < 5$ and $x < 8$ are equivalent inequalities. So the solution of $x - 3 < 5$ is $x < 8$.

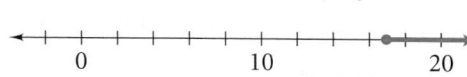

2 EXAMPLE Solving and Checking Solutions

Solve $12 \leq x - 5$. Graph and check your solution.

$$12 + 5 \leq x - 5 + 5 \qquad \textbf{Add 5 to each side.}$$
$$17 \leq x \qquad \textbf{Simplify.}$$

For: Linear Inequalities Activity
Use: Interactive Textbook, 4-2

Check $12 = x - 5$ **Check the computation.**
$12 \stackrel{?}{=} 17 - 5$ **Substitute 17 for x.**
$12 = 12 \checkmark$
$12 \leq x - 5$ **Check the direction of the inequality.**
$12 \leq 18 - 5$ **Substitute 18 for x.**
$12 \leq 13 \checkmark$

✔ Quick Check ❷ Solve $n - 7 \leq -2$. Graph and check your solution.

2 Using Subtraction to Solve Inequalities

You can subtract the same number from each side of an inequality to create an equivalent inequality.

Key Concepts

Property	Subtraction Property of Inequality
For every real number a, b, and c, $\quad$ if $a > b$, then $a - c > b - c$; $\qquad$ if $a < b$, then $a - c < b - c$.	
Examples $3 > -1$, so $3 - 2 > -1 - 2$ $\quad$ $-5 < 4$, so $-5 - 2 < 4 - 2$	
This property is also true for $\geq$ and $\leq$.	

3 EXAMPLE Using the Subtraction Property of Inequality

Solve $y + 5 < -7$. Graph the solution.

$y + 5 - 5 < -7 - 5$ **Subtract 5 from each side.**
$y < -12$ **Simplify.**

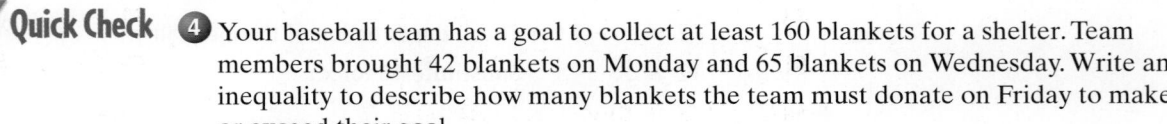

Quick Check ③ Solve $t + 3 \geq 8$. Graph and check your solution.

You can use inequalities to model real-world situations.

4 EXAMPLE Real-World 🌐 Problem Solving

Multiple Choice The maximum safe load of a chairlift is 680 lb. In the spring, a cyclist and bicycle go to the top of the slope using the chairlift. The weight of the person is 124 lb, and the weight of the bicycle is 32 lb. Which inequality best describes how much additional weight w the chairlift could safely carry?

  Ⓐ $124 + w \leq 680 + 32$ Ⓑ $124 + 32 + w \leq 680$
  Ⓒ $32 + w \geq 680 + 124$ Ⓓ $124 + 32 + w \geq 680$

Relate | weight of a person and a bicycle | plus | additional weight | is at most | safe load |

Define Let w = the amount of weight that can be added to the chairlift.

Write | $124 + 32$ | + | w | $\leq$ | 680 |

The inequality $124 + 32 + w \leq 680$ models the situation. So B is the correct answer.

Quick Check ④ Your baseball team has a goal to collect at least 160 blankets for a shelter. Team members brought 42 blankets on Monday and 65 blankets on Wednesday. Write an inequality to describe how many blankets the team must donate on Friday to make or exceed their goal.

Test-Taking Tip

You can test inequality choices by substituting a reasonable value for the variable. If the value is NOT a solution of the inequality you can eliminate that choice as the correct answer.

EXERCISES

For more exercises, see *Extra Skill and Word Problem Practice.*

Practice and Problem Solving

A Practice by Example

Examples 1, 2
(pages 206, 207)

GO for Help

State what number you would add to each side of the inequality to solve the inequality.

1. $d - 5 \geq -4$ **2.** $0 < c - 8$ **3.** $z - 4.3 \geq 1.6$

Solve each inequality. Graph and check your solution.

4. $x - 1 > 10$ **5.** $t - 3 < -2$ **6.** $-5 > b - 1$ **7.** $7 \leq d - 3$

8. $s - 2 \geq -6$ **9.** $r - 9 \leq 0$ **10.** $8 < n - 2$ **11.** $-4 \geq w - 2$

12. $-1 < -4 + d$ **13.** $y - \frac{1}{2} \leq -5$ **14.** $-\frac{2}{3} > q - 4$ **15.** $x - 2 \geq 0.5$

16. $3.2 > -1.3 + r$ **17.** $-3.4 > m - 1.8$ **18.** $b - \frac{3}{8} < \frac{1}{8}$ **19.** $n - 2\frac{1}{2} > \frac{1}{2}$

Example 3
(page 208)

State what number you would subtract from each side of the inequality to solve the inequality.

20. $w + 2 > -1$ **21.** $8 < \frac{5}{3} + r$ **22.** $5.7 \geq k + 3.1$

Solve each inequality. Graph and check your solution.

23. $w + 4 \leq 9$ **24.** $m + 5 > -3$ **25.** $1 < 8 + b$ **26.** $-2 \geq 4 + a$

27. $r + 1 \geq -5$ **28.** $k + 3 \leq 4$ **29.** $3 > 4 + x$ **30.** $-5 < 1 + p$

31. $\frac{3}{5} + z \geq -\frac{2}{5}$ **32.** $7.5 + y < 13$ **33.** $\frac{1}{2} < m + 2$ **34.** $2.7 \geq a + 3$

35. $-2.9 < 4.1 + p$ **36.** $\frac{1}{4} \geq h + \frac{3}{4}$ **37.** $5.3 + d > 3.8$ **38.** $t + \frac{3}{8} < -\frac{1}{8}$

Example 4
(page 208)

39. Vacation Budget Your brother has $2000 saved for a vacation. His airplane ticket is $637. Write and solve an inequality to find how much he can spend for everything else.

40. Weekly Budget You have an allowance of $15.00 per week. You are in a bowling league that costs $6.50 each week, and you save at least $5.00 each week. Write and solve an inequality to show how much you have left to spend each week.

41. Fund-Raising A school club is selling reflectors for Bicycle Safety Day. Each member is encouraged to sell at least 50 reflectors. You sell 17 on Monday and 12 on Tuesday. How many reflectors do you need to sell on Wednesday to meet your goal?

B **Apply Your Skills**

State what you must do to the first inequality in order to get the second.

42. $36 \leq -4 + y$; $40 \leq y$ **43.** $9 + b > 24$; $b > 15$ **44.** $m - \frac{1}{2} < \frac{3}{8}$; $m < \frac{7}{8}$

Solve each inequality.

45. $w - 3 + 1 \geq 9$ **46.** $\frac{1}{2} + c \leq 3\frac{1}{2}$ **47.** $y - 0.3 < 2.8$

48. $-6 > n - \frac{1}{5}$ **49.** $z + 4.1 < -5.6$ **50.** $-4.1 > y - 0.9$

51. $\frac{2}{3} + t - \frac{5}{6} > 0$ **52.** $5 \leq v - 4 - 7$ **53.** $3.6 + k \geq -4.5$

54. $6 + b - 7 < 5$ **55.** $m + 2.3 \leq -1.2$ **56.** $4 \geq k - \frac{3}{4}$

57. $h - \frac{1}{2} \geq -1$ **58.** $-7.7 \geq x - 2$ **59.** $-2 > 9 + 3 + w$

60. Banking Your local bank offers free checking for accounts with a balance of at least $500. Suppose you have a balance of $516.46 and you write a check for $31.96. How much must you deposit to avoid being charged a service fee?

61. a. If $45 + 47 = t$, does $t = 45 + 47$?
b. If $45 + 47 < r$, is $r < 45 + 47$?
c. Discuss the differences between these two examples.

62. Gymnastics Suppose your sister wants to qualify for a regional gymnastics competition. At today's competition she must score at least 34.0 points. She scored 8.8 on the vault, 7.9 on the balance beam, and 8.2 on the uneven parallel bars. The event that remains is the floor exercise.
a. Write and solve an inequality that models the information.
b. Explain what the solution means in terms of the original situation.
c. Open-Ended Write three scores your sister could make that would allow her to qualify for the regional gymnastics competition.

Real-World **Connection**

More than 71,000 athletes compete in gymnastic programs in the United States.

63. Computers Suppose your computer has nearly 64 megabytes (MB) of memory. Its basic systems require 12.8 MB. How much memory is available for other programs and functions?

64. To earn an A in Ms. Orlando's math class, students must score a total of at least 135 points on the three tests. On the first two tests, Amy's scores were 47 and 48. What is the minimum score she must get on the third test to earn an A?

65. a. Open-Ended Use each of the inequality symbols $<$, $\leq$, $>$, and $\geq$ to write four addition or subtraction inequalities.
   b. Solve each of the inequalities in part (a) and graph your solution.

66. a. Sam says that he can solve $z - 8.6 \geq 5.2$ by replacing z with 13, 14, and 15. When $z = 13$, the inequality is false. When $z = 14$ and $z = 15$, the inequality is true. So Sam says that the solution is $z \geq 14$. Is his reasoning correct? Justify your answer.
   b. Critical Thinking Explain why substituting values into the inequality does not guarantee that your solution is correct.

Solve each inequality.

67. $4x + 4 - 3x \geq 5$ **68.** $-5n - 3 + 6n < 2$

69. $7t - (6t - 2) \leq -1$ **70.** $5k - 2(2k + 1) > 8$

71. $-6(a + 2) + 7a \leq 12$ **72.** $-2(a - 3) + 3(a + 2) < 4$

73. Geometry The Triangle Inequality Theorem states that the sum of the lengths of any two sides of a triangle is greater than the length of the third side. Following are inequalities for sides of the triangle shown.

$$a + b > c \qquad b + c > a \qquad a + c > b$$

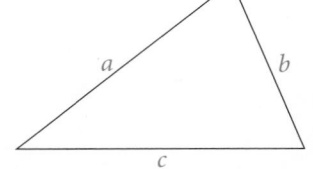

   a. Write an inequality using $c - b$ and a.
   b. Write an inequality using $a - c$ and b.
   c. Write an inequality using $b - a$ and c.
   d. Writing Write a generalization about the length of the third side and the difference of the lengths of the other two sides.

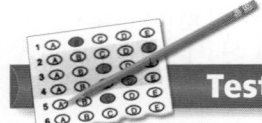 **Challenge**

Reasoning Decide if each inequality is true for all real numbers. If the inequality is *not* true, give a counterexample.

74. $a - b < a + b$ **75.** If $a \geq b$, then $a + c \geq b + c$.

76. If $c > d$, then $a - c < a - d$.   **77.** If $a < b$, then $a < b + c$.

78. Reasoning Find real numbers x, y, z, and w for which it is true that $x > y$ and $z > w$, but it is not true that $x - z > y - w$.

 **REGENTS**

Test Prep

Multiple Choice

79. Solve $x + 5 < 13$.
    A. $x > 8$ **B.** $x < 8$ **C.** $x > 18$ **D.** $x < 18$

80. Solve $-12 + n > 20$.
    F. $n < 32$ **G.** $n > 32$ **H.** $n < 8$ **J.** $n > 8$

81. Which graph represents all real number solutions of $x + 4 \geq 8$?

A.
-1 0 1 2 3 4 5

B.
-5 -4 -3 -2 -1 0 1

C.
-1 0 1 2 3 4 5

D.
-3 -2 -1 0 1 2 3

82. Which of the following is a solution for $5 < n - 0.1$?
    F. 4.99   **G.** 5.01   **H.** 5.10   **J.** 5.11

83. Hector is flying his plane. To avoid a storm, he climbs 5500 ft without going above his plane's maximum safe altitude of 35,000 ft. The inequality $a + 5500 \leq 35,000$ represents his original altitude a in feet. Which of the following could have been the original altitude?
    A. 40,500 ft   **B.** 30,000 ft   **C.** 29,750 ft   **D.** 27,750 ft

Short Response

84. The leading scorer in your high school's basketball division finished the season with a game average of 20 points for 25 games. As the division's second leading scorer, you have a 19.5 point per game average for 24 games. You have your last game yet to play.

How many points must you score in the final game of the season to overtake the division's leading scorer? Show your work.

Mixed Review

GO for Help

Lesson 4-1

Define a variable and write an inequality to model each situation.

85. An octopus can be up to 10 ft long.

86. A hummingbird migrates more than 1850 mi.

87. Your average in algebra class must be 90 or greater for you to receive an A for the term.

88. You must read at least 25 pages this weekend.

Lesson 1-2

Simplify.

89. $9^2 + 17$
    90. $4(5 - 3)^2 - 3^2$
    91. $0.2(4.2 - 3.4) + 0.4$

92. $3 \cdot 2 + 5^2$
    93. $3 + 7^2 - 4$
    94. $4^3 + 3^2$

95. $6(5 - 2)^2 + 4$
    96. $2 + 8(6 + 2^2)$
    97. $3^3 - 2^3 + 7$

Algebra at Work

·······················Marketing Director

Marketing directors rely on equations and inequalities to predict the actions their companies must take to stay competitive. For example, the marketing director of a manufacturing company determines how much the cost of raw materials can increase before the company must raise the price of its finished goods or services. The director also predicts the effect of price changes on the quantity of goods and services sold by his company.

Go Online
PHSchool.com
For: Information about a career in marketing
Web Code: atb-2031

Solving Inequalities Using Multiplication and Division

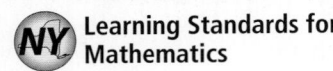 **Learning Standards for Mathematics**

A.A.6 Analyze and solve verbal problems whose solution requires solving a linear equation in one variable or linear inequality in one variable.

A.A.24 Solve linear inequalities in one variable.

✓ **Check Skills You'll Need**

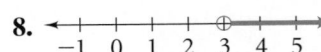 **for Help** Review page 118 and Lesson 3-1

Solve each equation.

1. $8 = \frac{1}{2}t$

2. $14 = -21x$

3. $\frac{x}{6} = -1$

4. $5d = 32$

5. $\frac{2}{3}x = -12$

6. $0.5n = 9$

Write an inequality for each graph.

7.
-3 -2 -1 0 1 2 3

8.
-1 0 1 2 3 4 5

---

1 **Using Multiplication to Solve Inequalities**

Activity: Multiplying Each Side of an Inequality

Consider the inequality $4 > 1$.

1. Copy and complete each statement at the right by replacing each ■ with $<$, $>$, or $=$.

2. What happens to the inequality symbol when you multiply each side by a positive number?

3. What happens to the inequality symbol when you multiply each side by zero?

4. What happens to the inequality symbol when you multiply each side by a negative number?

$4 \cdot 3$ ■ $1 \cdot 3$	
$4 \cdot 2$ ■ $1 \cdot 2$	
$4 \cdot 1$ ■ $1 \cdot 1$	
$4 \cdot 0$ ■ $1 \cdot 0$	
$4 \cdot -1$ ■ $1 \cdot -1$	
$4 \cdot -2$ ■ $1 \cdot -2$	
$4 \cdot -3$ ■ $1 \cdot -3$	

You can multiply each side of an inequality by the same number, just as you did with equations. When you multiply each side of an inequality by a positive number, the direction of the inequality symbol stays the same. When you multiply each side by a negative number, the direction of the inequality symbol reverses.

 Key Concepts

Property	**Multiplication Property of Inequality for $c > 0$**

For every real number a and b, and for $c > 0$,

if $a > b$, then $ac > bc$;   if $a < b$, then $ac < bc$.

Examples $4 > -1$, so $4(5) > -1(5)$. $-6 < 3$, so $-6(5) < 3(5)$.

This property is also true for $\geq$ and $\leq$.

You can use the Multiplication Property of Inequality to solve inequalities that involve division.

 **EXAMPLE** **Multiplying by a Positive Number**

Solve $\frac{x}{2} < -1$. Graph and check the solution.

$2\left(\frac{x}{2}\right) < 2(-1)$ **Multiply each side by 2. Do not reverse the inequality symbol.**

$x < -2$ **Simplify each side.**

$$\overset{\oplus}{\underset{-5\ -4\ -3\ -2\ -1\ \ 0\ \ 1\ \ 2\ \ 3\ \ 4\ \ 5}{\longleftrightarrow}}$$

Check $\quad \frac{x}{2} = -1 \qquad$ **Check the computation.**

$\qquad\quad \frac{-2}{2} \overset{?}{=} -1 \qquad$ **Substitute -2 for x.**

$\qquad\quad -1 = -1 \checkmark \quad$ **Simplify.**

$\qquad\quad \frac{x}{2} < -1 \qquad$ **Check the direction of the inequality.**

$\qquad\quad \frac{-3}{2} < -1 \checkmark \quad$ **Substitute -3 for x.**

 Quick Check ❶ Solve each inequality. Graph and check your solution.

a. $\frac{b}{4} > \frac{1}{2}$ **b.** $\frac{d}{3} \geq \frac{5}{6}$ **c.** $\frac{y}{0.5} \leq -3$

Key Concepts

Property	**Multiplication Property of Inequality for $c < 0$**

For every real number a and b, and for $c < 0$,

$\qquad\qquad$ if $a > b$, then $ac < bc$; $\qquad\qquad$ if $a < b$, then $ac > bc$.

Examples $\;4 > -1$, so $4(-2) < -1(-2)$. $\qquad -6 < 3$, so $-6(-2) > 3(-2)$.

This property is also true for $\geq$ and $\leq$.

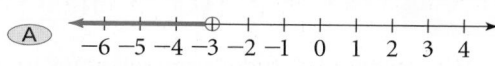 **EXAMPLE** **Multiplying by a Negative Number**

Multiple Choice Which graph shows the solution of $-\frac{2}{3}n \leq 2$?

Ⓐ $\overset{\oplus}{\underset{-6\ -5\ -4\ -3\ -2\ -1\ \ 0\ \ 1\ \ 2\ \ 3\ \ 4}{\longleftrightarrow}}$

Ⓑ $\underset{-6\ -5\ -4\ -3\ -2\ -1\ \ 0\ \ 1\ \ 2\ \ 3\ \ 4}{\longleftrightarrow}$

Ⓒ $\underset{-6\ -5\ -4\ -3\ -2\ -1\ \ 0\ \ 1\ \ 2\ \ 3\ \ 4}{\longleftrightarrow}$

Ⓓ $\underset{-4\ -3\ -2\ -1\ \ 0\ \ 1\ \ 2\ \ 3\ \ 4\ \ 5\ \ 6}{\longleftrightarrow}$

$\left(-\frac{3}{2}\right)\left(-\frac{2}{3}n\right) \geq \left(-\frac{3}{2}\right)2 \qquad$ **Multiply each side by $-\frac{3}{2}$, the reciprocal of $-\frac{2}{3}$. Reverse the inequality symbol.**

$n \geq -3 \qquad$ **Simplify.**

$\underset{-6\ -5\ -4\ -3\ -2\ -1\ \ 0\ \ 1\ \ 2\ \ 3\ \ 4}{\longleftrightarrow}$

B is the correct answer.

Test-Taking Tip

You can use the endpoint and another point on a graph to determine if the graph shows the solution of an inequality. Use the endpoint to check for equality. For the answer B, -3 is the endpoint and 0 is another point.

$-\frac{2}{3}(-3) = 2 \checkmark$

$-\frac{2}{3}(0) \leq 2 \checkmark$

This confirms that B is the graph of the solution of $-\frac{2}{3}(n) \leq 2$.

 **Quick Check** ❷ Solve each inequality. Graph and check the solution.

a. $-\frac{k}{4} > -1$ **b.** $-t < \frac{1}{2}$ **c.** $6 \leq -\frac{3}{5}w$

Solving inequalities using division is similar to solving inequalities using multiplication. Remember that division by zero is undefined.

Key Concepts

Online active math

For: Solving Inequalities Activity
Use: Interactive Textbook, 4-3

Property	Division Property of Inequality

For every real number a and b, and for $c > 0$,

$$\text{if } a > b, \text{ then } \frac{a}{c} > \frac{b}{c}; \qquad\qquad \text{if } a < b, \text{ then } \frac{a}{c} < \frac{b}{c}.$$

Examples $6 > 4$, so $\frac{6}{2} > \frac{4}{2}$. $\qquad\qquad 2 < 8$, so $\frac{2}{2} < \frac{8}{2}$.

For every real number a and b, and for $c < 0$,

$$\text{if } a > b, \text{ then } \frac{a}{c} < \frac{b}{c}; \qquad\qquad \text{if } a < b, \text{ then } \frac{a}{c} > \frac{b}{c}.$$

Examples $6 > 4$, so $\frac{6}{-2} < \frac{4}{-2}$. $\qquad\qquad 2 < 8$, so $\frac{2}{-2} > \frac{8}{-2}$.

This property also applies to $\geq$ and $\leq$.

3 EXAMPLE **Dividing to Solve an Inequality**

Solve $-5z \geq 25$. Graph the solution.

$$\frac{-5}{-5}z \leq \frac{25}{-5} \qquad \textbf{Divide each side by } -5. \textbf{ Reverse the inequality symbol.}$$
$$z \leq -5 \qquad \textbf{Simplify.}$$

$$\xleftarrow{\hspace{2cm}} \quad -9\ -8\ -7\ -6\ -5\ -4\ -3\ -2\ -1\ \ 0\ \ 1$$

✓ Quick Check **3** Solve the inequality. Graph and check your solution.
   a. $-2t < -8$ **b.** $-3w \geq 12$ **c.** $0.6 > -0.2n$

There are times when you must think about which types of numbers are acceptable as solutions of inequalities that represent real-world situations.

4 EXAMPLE **Real-World 🌎 Problem Solving**

Community Service The student council votes to buy food for a local food bank. A case of 12 jars of spaghetti sauce costs $13.75. What is the greatest number of cases of sauce the student council can buy if they use at most $216 for this project?

Relate	cost per case	times	the number of cases	is at most	total cost

Define Let c = the number of cases of spaghetti sauce.

Write	13.75	·	c	$\leq$	216

$$13.75c \leq 216$$
$$\frac{13.75c}{13.75} \leq \frac{216}{13.75} \qquad \textbf{Divide each side by 13.75.}$$
$$c \leq 15.71 \qquad \textbf{Simplify and round to the nearest hundredth.}$$

The student council does not have enough money to buy 16 cases, so they can buy at most 15 cases of sauce for the food bank.

Real-World 🌎 Connection

Careers The duties of a manager of a nonprofit organization, such as a food bank, include organizing and supervising volunteers.

 **Quick Check** ④ Students in the school band are selling calendars. They earn $.40 on each calendar they sell. Their goal is to earn more than $327. Write and solve an inequality to find the fewest number of calendars they can sell and still reach their goal.

EXERCISES

Practice and Problem Solving

For more exercises, see *Extra Skill and Word Problem Practice.*

A Practice by Example

Examples 1, 2
(page 213)

 **for Help**

Solve each inequality. Graph and check your solution.

1. $\frac{t}{4} \geq -1$ **2.** $\frac{s}{6} < 1$ **3.** $1 \leq -\frac{w}{2}$ **4.** $2 < -\frac{p}{4}$

5. $-2 < \frac{y}{2}$ **6.** $-\frac{v}{3} \geq 0.5$ **7.** $4 > \frac{2}{3}x$ **8.** $-5 \leq \frac{5}{2}k$

9. $0 < -\frac{7}{8}x$ **10.** $\frac{4}{3}y \geq 0$ **11.** $-\frac{5}{7}x > -5$ **12.** $6 \geq -\frac{3}{2}d$

13. $-\frac{4}{9} < \frac{2}{3}c$ **14.** $\frac{3}{4}b \geq -\frac{9}{8}$ **15.** $-\frac{5}{3}u > \frac{5}{6}$ **16.** $-\frac{5}{8} > -\frac{5}{6}n$

Example 3
(page 214)

17. $3t < -9$ **18.** $4m \geq 8$ **19.** $10 \leq -2w$ **20.** $-20 > -5c$

21. $-27 \geq 3z$ **22.** $-7b > 42$ **23.** $18d < -12$ **24.** $-3x \leq 16$

25. $-7 < 2q$ **26.** $16 > 3.2h$ **27.** $-1.5d < -6$ **28.** $3.6 \leq -0.8m$

Example 4
(page 214)

29. Fund-Raising The science club charges $4.50 per car at their car wash. Write and solve an inequality to find how many cars they have to wash to earn at least $300.

30. Earnings Suppose you earn $6.15 per hour working part time at a dry cleaner. Write and solve an inequality to find how many full hours you must work to earn at least $100.

B Apply Your Skills

Write four solutions to each inequality.

31. $\frac{x}{2} \leq -1$ **32.** $\frac{r}{3} \geq -4$ **33.** $-1 \geq \frac{t}{3}$ **34.** $0.5 > \frac{1}{2}c$

35. $-\frac{3}{4}q > 4$ **36.** $1 < -\frac{5}{7}s$ **37.** $-4.5 \leq -0.9p$ **38.** $-2.7w \geq 28$

Tell what you must do to the first inequality in order to get the second.

39. $-\frac{c}{4} > 3$; $c < -12$ **40.** $\frac{n}{5} \leq -2$; $n \leq -10$

41. $5z > -25$; $z > -5$ **42.** $\frac{3}{4}b \leq 3$; $b \leq 4$

43. $-12 < 4a$; $-3 < a$ **44.** $-b \geq 3.4$; $b \leq -3.4$

Replace each ▪ with the number that makes the inequalities equivalent.

45. $▪s > 14$; $s < -7$ **46.** $▪x \geq 25$; $x \leq -5$

47. $-8u \leq ▪$; $u \geq -0.5$ **48.** $-2a > ▪$; $a < -9$

49. $36 < ▪r$; $r < -3.6$ **50.** $-k \leq ▪$; $k \geq -7.5$

 **Homework Video Tutor**

Visit: PHSchool.com
Web Code: ate-0403

51. Critical Thinking If $x \geq y$ and $-x \geq -y$, what can you conclude about x and y?

Estimation **Estimate the solution of each inequality.**

52. $-2.099r < 4$ **53.** $3.87j > -24$ **54.** $20.95 \geq \frac{1}{2}p$ **55.** $-\frac{20}{39}s \leq -14$

56. Safe Load An elevator like the one at the left can safely lift at most 4400 lb. A concrete block has an average weight of 42 lb. What is the maximum number of concrete blocks that the elevator can lift?

57. Writing Explain how solving the equation $-\frac{x}{3} = 4$ is similar to and different from solving the inequality $-\frac{x}{3} > 4$.

58. Open-Ended Write four different inequalities with $x > 3$ as their solution that you can solve using multiplication or division.

Solve each inequality.

59. $4d \le -28$ **60.** $\frac{u}{7} > 5$ **61.** $2 < -8s$ **62.** $\frac{3}{2}k \ge -45$

63. $0.3y < 2.7$ **64.** $9.4 \le -4t$ **65.** $-h \ge 4$ **66.** $\frac{5}{2}x > 5$

67. $24 < -\frac{8}{3}x$ **68.** $0 < -\frac{1}{6}b$ **69.** $\frac{5}{6} > -\frac{1}{3}p$ **70.** $-0.2m \ge 9.4$

71. $6 < -9g$ **72.** $4n \ge 9$ **73.** $-3.5 < -m$ **74.** $\frac{2}{5}z \ge -1$

Real-World  **Connection**

Depending on its size, an elevator at a construction site can have a maximum load from 900 lb to 20,000 lb.

75. Michael solved the inequality $-2 > \frac{y}{-3}$ and got $6 < y$. Erica solved the same inequality and got $y > 6$. Are they both correct? Explain.

76. A friend calls you and asks you to meet at a location 3 miles from your home in 20 minutes. You set off on your bicycle after the telephone call. Write and solve an inequality to find the average rate in miles per minute you could ride to be at your meeting place within 20 minutes.

77. a. Error Analysis Kia solved $-15q \le 135$ by adding 15 to each side of the inequality. What mistake did she make?

b. Kia's solution was $q \le 150$. She checked her work by substituting 150 for q in the original inequality. Why didn't her check let her know that she had made a mistake?

c. Open-Ended Find a number that satisfies Kia's solution but does not satisfy the original inequality.

C **Challenge**

Reasoning **If a, b, and c are real numbers, for which values of a is each statement true?**

78. If $c < 0$, then $ac < a$.   **79.** If $b > c$, then $ab > ac$.

80. If $b > c$, then $a^2b > a^2c$.   **81.** If $b > c$, then $\frac{b}{a} < \frac{c}{a}$.

82. Packaging Suppose you have a plastic globe that you wish to put into a gift box. The circumference of the globe is 15 in. The edges of cube-shaped boxes are either 3 in., 4 in., 5 in., or 6 in. Write and solve an inequality to find the boxes that will hold the globe. (*Hint:* circumference $= \pi \cdot$ diameter)

83. Tiling a Floor The Sumaris' den floor measures 18 ft by 15 ft. They want to cover the floor with square tiles that are $\frac{9}{16}$ ft². Write and solve an inequality to find the least number of tiles they need to cover the floor.

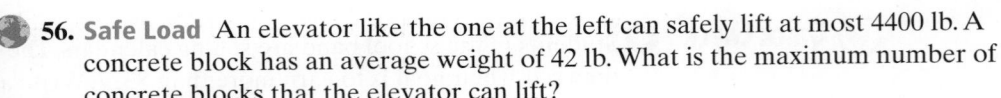

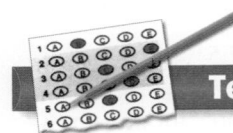

Gridded Response

84. Solve $\frac{2}{5}x = 16$.

85. Paul expects to pay $1680 in income taxes. This is no more than $\frac{1}{5}$ of his salary. What is his least possible earned income?

86. Five boxes of tiles will cover 30 ft². What is the least number of boxes of tiles needed to tile 320 ft²?

87. What is the greatest number of 37¢ stamps you can buy for $5.00?

88. The length of a rectangle is 90 in. Its area is less than 380 in². What is the greatest possible width of the rectangle, to the nearest inch?

89. City Middle School set a goal of collecting "one mile of plastic bottles." If the height of a typical plastic bottle is 11.25 in., what is the fewest number of bottles needed to reach their goal?

Mixed Review

Lesson 4-2

Solve each inequality.

90. $x + 3 \leq -4$ **91.** $\frac{5}{8} > \frac{3}{4} + w$ **92.** $t - 3.4 \geq 5.8$ **93.** $0 < d - 4$

94. $3.2 \geq m + 7.1$ **95.** $y - 2 < 6$ **96.** $-3 > a - 4$ **97.** $k + 12 \leq 15$

Lesson 3-2

Solve each equation.

98. $3t + 5 - t = 9$ **99.** $-3(2n + 1) = 9$ **100.** $\frac{3x}{4} - \frac{1}{2} = \frac{1}{4}$

Lesson 2-5

Name the property that each exercise illustrates.

101. $(-7) + 7 = 0$ **102.** $-3\left(-\frac{1}{3}\right) = 1$ **103.** $2 + 0 = 2$

104. $1x = x$ **105.** $4(2 \cdot 5) = (4 \cdot 2)5$ **106.** $4(2 \cdot 5) = 4(5 \cdot 2)$

✓ Checkpoint Quiz 1   Lessons 4-1 through 4-3

Solve each inequality. Graph your solution.

1. $6 < c + 1$ **2.** $5x < -30$ **3.** $\frac{p}{3} \leq -2$

4. $y - 4 \geq -2$ **5.** $12 + g < 4$ **6.** $-3b \geq 15$

7. Determine whether each of the following is a solution of $x + 7 \leq 3$.
   a. -4 **b.** 0 **c.** $-\frac{17}{4}$ **d.** -3.9

8. Determine whether each of the following is a solution of $-4x < -12$.
   a. 3 **b.** 0 **c.** $\frac{7}{3}$ **d.** π

Write and solve an inequality that models each situation.

9. You plan to buy a bicycle that will cost at least $180. You have saved $38 and your parents have given you $50.
   a. Write an inequality to find how much more money m you need to save.
   b. Solve your inequality.

10. Your local garden shop has plants on sale for $1.50 each. You are planning a vegetable garden. You have $20 to spend on tomato plants.
   a. Write an inequality to find the greatest number of plants p you can buy.
   b. What is the greatest number of plants you can buy?

Activity Lab

Modeling Inequalities

Technology

FOR USE WITH LESSON 4-4

Go Online
PHSchool.com

For: Graphing calculator procedures
Web Code: ate-2100

You can use a table or graphing calculator to solve inequalities.

1 ACTIVITY

Solve $4x + 8 < 20 - 2x$ using a table.

Make a table and evaluate both expressions. Find the x-value that makes the inequality change from true to false or from false to true.

The inequality changed from true to false when $x = 2$. Since the expressions were equal when $x = 2$, $x < 2$ is the solution to $4x + 8 < 20 - 2x$.

x	$4x + 8$	$<$	$20 - 2x$
0	$4(0) + 8 = 8$	true	$20 - 2(0) = 20$
1	$4(1) + 8 = 12$	**true**	$20 - 2(1) = 18$
2	$4(2) + 8 = 16$	**false**	$20 - 2(2) = 16$
3	$4(3) + 8 = 20$	false	$20 - 2(3) = 14$

2 ACTIVITY

Use a graphing calculator to solve $3(m - 2) < 5(1 - m)$.

Step 1 Enter two equations using the [Y=] feature.

For **Y₁ =** enter $3(x - 2)$. For **Y₂ =** enter $5(1 - x)$.

Step 2 Use the [GRAPH] screen to display the graphs.

Step 3 Use the **CALC** feature, then select **intersect** to find the point where the lines intersect.

The x-value of the point of intersection is 1.375, so the solution of the inequality will either be $m < 1.375$ or $m > 1.375$.

Step 4 Choose a second value and determine if it is a solution of the inequality. Is zero a solution of $3(m - 2) < 5(1 - m)$?

$$3(0 - 2) < 5(1 - 0)$$
$$-6 < 5 \checkmark$$

Since 0 is a solution, then 0 and all other numbers *also* less than 1.375 are solutions. So, $m < 1.375$ describes all solutions. (If zero had not been a solution then $m > 1.375$ would have described all solutions.)

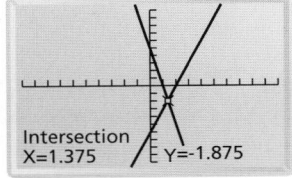

Intersection
X=1.375   Y=-1.875

EXERCISES

Solve each inequality using a table.

1. $2(5 + 3p) > 9p - 2$

2. $t + 2.5 < -0.5t + 5.5$

3. $7y - 3 \le -5(y + 3)$

Solve each inequality using a graphing calculator.

4. $2d - 6 < -4\left(\frac{1}{2} + d\right)$

5. $8n - 3 - 6n \ge 1 - 3n - 5n$

6. $-2.3 + 2.4x > -0.3x + 6.7$

Solving Multi-Step Inequalities

NY Learning Standards for Mathematics

A.A.5 Write algebraic equations or inequalities that represent a situation.

A.A.6 Analyze and solve verbal problems whose solution requires solving a linear equation in one variable or linear inequality in one variable.

A.A.24 Solve linear inequalities in one variable.

✓ Check Skills You'll Need

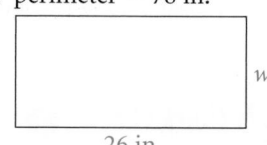 **for Help** Lessons 3-2 and 3-3

Solve each equation, if possible. If the equation is an identity or if it has no solution, write *identity* or *no solution*.

1. $3(c + 4) = 6$ **2.** $3t + 6 = 3(t - 2)$

3. $5p + 9 = 2p - 1$ **4.** $7n + 4 - 5n = 2(n + 2)$

5. $\frac{1}{2}k - \frac{2}{3} + k = \frac{7}{6}$ **6.** $2t - 32 = 5t + 1$

Find the missing dimension of each rectangle.

7. perimeter = 110 cm

15 cm

ℓ

8. perimeter = 78 in.

w

26 in.

---

1   Solving Inequalities With Variables on One Side

Sometimes you need to perform two or more steps to solve an inequality. Models can help you understand how to solve multi-step inequalities.

$2x - 3 < 1$    The tiles model the inequality.

$2x - 3 + 3 < 1 + 3$ 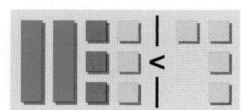   Add 3 to each side.

$2x < 4$ 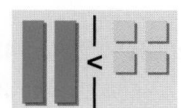   Simplify by removing the zero pairs.

$\frac{2x}{2} < \frac{4}{2}$ 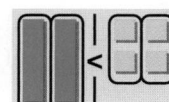   Divide each side into two equal groups.

$x < 2$ 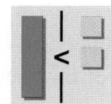   Each green tile is less than two yellow tiles, so $x < 2$.

To solve inequalities, you undo addition and subtraction first. Then undo multiplication and division.

1 EXAMPLE **Using More Than One Step**

Solve $7 + 6a > 19$. Check the solution.

$7 + 6a - 7 > 19 - 7$	**Subtract 7 from each side.**
$6a > 12$	**Simplify.**
$\frac{6a}{6} > \frac{12}{6}$	**Divide each side by 6.**
$a > 2$	**Simplify.**

Check $\quad 7 + 6a = 19 \qquad$ **Check the computation.**

$\qquad 7 + 6(2) \stackrel{?}{=} 19 \qquad$ **Substitute 2 for a.**

$\qquad\qquad\quad 19 = 19 ✓$

$\qquad 7 + 6a > 19 \qquad$ **Check the direction of the inequality.**

$\qquad 7 + 6(3) > 19 \qquad$ **Substitute 3 for a.**

$\qquad\qquad\quad 25 > 19 ✓$

✓ Quick Check ❶ Solve each inequality. Check your solution.

   a. $-3x - 4 \le 14$ **b.** $5 < 7 - 2t$ **c.** $-8 < 5n - 23$

You can adapt familiar formulas like the formula for the perimeter of a rectangle to write inequalities. You determine which inequality symbol to use from the real-world situation.

2 EXAMPLE **Real-World  Problem Solving**

Geometry The school band needs a banner to carry in a parade. The banner committee decides that the length of the banner should be 18 feet. A committee member drew the diagram at the left to help understand the problem. What are the possible widths of the banner if they can use no more than 48 feet of trim?

Relate Since the border goes around the edges of a rectangular banner, you can adapt the perimeter formula $P = 2\ell + 2w$.

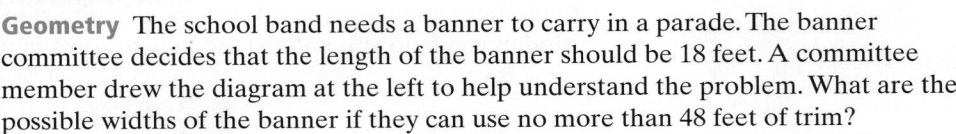

twice the length	plus	twice the width	can be no more than	the length of trim

Write $\qquad 2(18) \qquad + \qquad 2w \qquad\qquad \le \qquad\qquad 48$

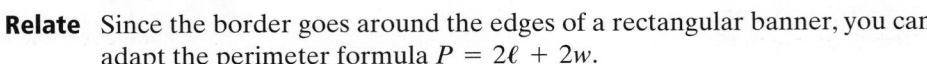

$2(18) + 2w \le 48$	
$36 + 2w \le 48$	**Simplify 2(18).**
$36 + 2w - 36 \le 48 - 36$	**Subtract 36 from each side.**
$2w \le 12$	**Simplify.**
$\frac{2w}{2} \le \frac{12}{2}$	**Divide each side by 2.**
$w \le 6$	**Simplify.**

The banner's width must be 6 feet or less.

✓ Quick Check ❷ To make a second banner, the committee decided to make the length 12 feet. They have 40 feet of a second type of trim. Write and solve an inequality to find the possible widths of the second banner.

3 **EXAMPLE** **Using the Distributive Property**

Solve $2(t + 2) - 3t \geq -1$.

$2t + 4 - 3t \geq -1$	**Use the Distributive Property.**
$-t + 4 \geq -1$	**Combine like terms.**
$-t + 4 - 4 \geq -1 - 4$	**Subtract 4 from each side.**
$-t \geq -5$	**Simplify.**
$\dfrac{-t}{-1} \leq \dfrac{-5}{-1}$	**Divide each side by -1. Reverse the inequality symbol.**
$t \leq 5$	**Simplify.**

☑ **Quick Check** **3** Solve each inequality. Check your solution.
    a. $4p + 2(p + 7) < 8$ **b.** $15 \leq 5 - 2(4m + 7)$ **c.** $8 > 3(5 - b) + 2$

---

2 **Solving Inequalities With Variables on Both Sides**

Many inequalities have variables on both sides of the inequality symbol. You need to gather the variable terms on one side of the inequality and the constant terms on the other side.

Video Tutor Help
Visit: PHSchool.com
Web Code: ate-0775

4 **EXAMPLE** **Gathering Variables on One Side of an Inequality**

Solve $6z - 15 < 4z + 11$.

$6z - 15 - 4z < 4z + 11 - 4z$	**To gather variables on the left, subtract $4z$ from each side.**
$2z - 15 < 11$	**Combine like terms.**
$2z - 15 + 15 < 11 + 15$	**To gather the constants on the right, add 15 to each side.**
$2z < 26$	**Simplify.**
$\dfrac{2z}{2} < \dfrac{26}{2}$	**Divide each side by 2.**
$z < 13$	**Simplify.**

☑ **Quick Check** **4** Solve $3b + 12 > 27 - 2b$. Check your solution.

Test-Taking Tip

To grid a mixed number, write it as an improper fraction:
$3\frac{1}{5} = \frac{16}{5}$.

5 **EXAMPLE** **Multi-Step Inequalities**

Gridded Response Solve $-3(4 - m) \geq 2(4m - 14)$.

$-12 + 3m \geq 8m - 28$	**Use the Distributive Property.**
$-12 + 3m - 8m \geq 8m - 28 - 8m$	**Subtract $8m$ from each side.**
$-12 - 5m \geq -28$	**Combine like terms.**
$-12 - 5m + 12 \geq -28 + 12$	**Add 12 to each side.**
$-5m \geq -16$	**Simplify.**
$\dfrac{-5m}{-5} \leq \dfrac{-16}{-5}$	**Divide each side by -5. Reverse the inequality symbol.**
$m \leq 3\frac{1}{5}$	**Simplify.**

☑ **Quick Check** **5** Solve $-6(x - 4) \geq 7(2x - 3)$. Check your solution.

EXERCISES

For more exercises, see *Extra Skill and Word Problem Practice*.

Practice and Problem Solving

 Practice by Example

Example 1
(page 220)

Solve each inequality. Check your solution.

1. $4d + 7 \leq 23$
2. $5m - 3 > -18$
3. $-4x - 2 < 8$

4. $5 - 3n \geq -4$
5. $8 \leq -12 + 5q$
6. $5 \leq 11 + 3h$

7. $-7 \leq 5 - 4a$
8. $10 > 29 - 3b$
9. $5 - 9c > -13$

Example 2
(page 220)

Write and solve an inequality.

10. On a trip from Virginia to Florida, the Sampson family wants to travel at least 420 miles in 8 hours of driving. What must be their average rate of speed?

 11. Geometry The perimeter of an isosceles triangle is at most 27 cm. The base is 8 cm long. Find the possible lengths of the two congruent sides.

Example 3
(page 221)

12. You want to solve an inequality containing the expression $-3(2x - 3)$. The next line in your solution would rewrite this expression as __?__.

Solve each inequality.

13. $2(j - 4) \geq -6$
14. $-(6b - 2) > 0$
15. $-2(h + 2) < -14$

16. $-3 \leq 3(5x - 16)$
17. $25 > -(4y + 7)$
18. $4(w - 2) \leq 10$

19. $-3(c + 4) - 2 > 7$
20. $-2(r - 3) + 7 \geq 8$
21. $16 \leq 4 - 3(n - 13)$

Example 4
(page 221)

22. $3w + 2 < 2w + 5$
23. $3t + 7 \geq 5t + 9$
24. $4d + 7 \geq 1 + 5d$

25. $5 - 2n \leq 3 - n$
26. $2k - 3 \leq 5k + 9$
27. $3s + 16 > 6 + 4s$

28. $6p - 1 > 3p + 8$
29. $3x + 2 > -4x + 16$
30. $2 - 3m < 4 + 5m$

Example 5
(page 221)

31. $-3(v - 3) \geq 5 - 4v$
32. $3q + 6 \leq -5(q + 2)$
33. $3(2 + r) \geq 15 - 2r$

34. $9 + x < 7 - 2(x - 3)$
35. $2(m - 8) < -8 + 3m$
36. $2v - 4 \leq 2(3v - 6)$

 Apply Your Skills

Tell what you must do to the first inequality in order to get the second.

37. $8 - 4s > 16; -4s > 8$
38. $\frac{2}{3}g + 7 \geq 9; \frac{2}{3}g \geq 2$

39. $2y - 5 > 9 + y; y > 14$
40. $-8 > \frac{z}{-5} - 2; 30 < z$

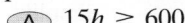 **41. Writing** Suppose a friend is having difficulty solving $2.5(p - 4) > 3(p + 2)$. Explain how to solve the inequality, showing all necessary steps and identifying the properties you would use.

42. Multiple Choice Mandela is starting a part-time word-processing business out of his home. He plans to charge $15 per hour. The table at the right shows his expected monthly business expenses. Which inequality describes the number of hours h he must work in a month to make a profit of at least $600?

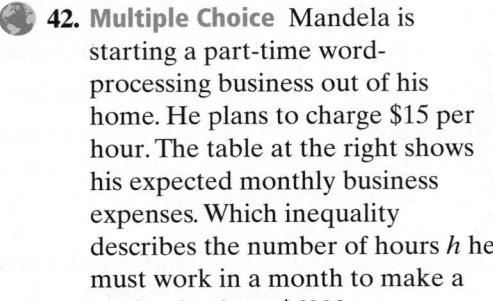

Expense	Cost
Equipment rental	$490
Materials	$45
Business phone	$65

   Ⓐ $15h \geq 600$
   Ⓑ $15h + 600 \geq 490 + 45 + 65$
   Ⓒ $15h \geq 490 + 45 + 65 + 600$
   Ⓓ $15h + 600 \leq 490 + 45 + 65$

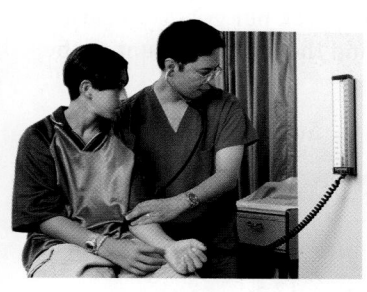

Real-World Connection

Normal blood pressure for teens is about 110/70.

43. Expenses The sophomore class is planning a picnic. The cost of a permit to use a city park is $250. To pay for the permit, there is a fee of $.75 for each sophomore and $1.25 for each guest who is not a sophomore. Two hundred sophomores plan to attend. Write and solve an inequality to find how many guests must attend for the sophomores to pay for the permit.

44. Health Care Systolic blood pressure is the higher number in a blood pressure reading. It is measured as your heart muscle contracts. The formula $P \leq \frac{1}{2}a + 110$ gives the normal systolic blood pressure P based on age a.
  a. At age 20, does 120 represent a maximum or a minimum normal systolic pressure?
  b. Find the normal systolic blood pressure for a 50-year-old person.

Match each inequality with its graph below.

45. $-2x - 2 > 4$ **46.** $2 - 2x > 4$ **47.** $2x + 2 > 4$

48. $2x + 2 > 4x$ **49.** $2x - 2 > 4$ **50.** $-2(x - 2) > 4$

A. ◁——————⊕——————▷
  −5 −4 −3 −2 −1 0 1 2 3

B. ◁————⊕——————————▷
  −5 −4 −3 −2 −1 0 1 2 3

C. ◁——————————⊕————▷
  −5 −4 −3 −2 −1 0 1 2 3

D. ◁——————————⊕————▷
  −3 −2 −1 0 1 2 3 4 5

E. ◁————⊕——————————▷
  −5 −4 −3 −2 −1 0 1 2 3

F. ◁——————————⊕————▷
  −3 −2 −1 0 1 2 3 4 5

51. Open-Ended Write two different inequalities that you can solve by adding 5 and multiplying by -3. Solve each inequality.

Solve each inequality.

52. $\frac{4}{3}r - 3 < r + \frac{2}{3} - \frac{1}{3}r$ **53.** $4 - 2m \leq 5 - m + 1$

54. $-2(0.5 - 4s) \geq -3(4 - 3.5s)$ **55.** $\frac{1}{2}n - \frac{1}{8} \geq \frac{3}{4} + \frac{5}{6}n$

56. $-(8 - s) < 0$ **57.** $3.8 - k \leq 5.2 - 2k$

58. $10 > 3(2n - 1) - 5(4n + 3)$ **59.** $3(3r + 1) - (r + 4) \leq 13$

60. $2(3x + 7) > 4(7 - 2x)$ **61.** $4(a - 2) - 6a \leq -9$

62. $4(3m - 1) \geq 2(m + 3)$ **63.** $17 - (4k - 2) \geq 2(k + 3)$

64. $2n - 3(n + 3) \leq 14$ **65.** $5x - \frac{1}{2}(3x + 8) \leq -4 + 3x$

66. $5a - 2(a - 15) < 10$ **67.** $5c + 4(c - 1) \geq 2 + 5(2 + c)$

68. a. Solve $5t + 4 \leq 8t - 5$ by gathering the variable terms on the left side and the constant terms on the right side of the inequality.
  b. Solve $5t + 4 \leq 8t - 5$ by gathering the constant terms on the left side and the variable terms on the right side of the inequality.
  c. Compare the results of parts (a) and (b).

69. a. Mental Math Like equations, some inequalities are true for all values of the variable, and some inequalities are not true for any values of the variable. Determine whether each inequality is *always* true or *never* true.
  i. $4s + 6 \geq 6 + 4s$ **ii.** $3r + 5 > 3r - 2$ **iii.** $4(n + 1) < 4n - 3$
  b. Critical Thinking How can you tell whether an inequality is always true or never true without solving?

Vocabulary Tip

Inequalities or equations that are always true are called <u>identities</u>. (See p. 136)

GO nline
Homework Video Tutor
Visit: PHSchool.com
Web Code: ate-0404

70. Commission Joleen is a sales associate in a clothing store. Each week she earns $250 plus a commission equal to 3% of her sales. This week her goal is to earn no less than $460. Write and solve an inequality to find the dollar amount of the sales she must have to reach her goal.

71. A student uses the table below to help solve $6x + 1 < 5(3 - x)$.

x	$6x + 1$	$<$	$5(3 - x)$
0	$6(0) + 1 = 1$	true	$5(3 - 0) = 15$
0.5	$6(0.5) + 1 = 4$	true	$5(3 - 0.5) = 12.5$
1	$6(1) + 1 = 7$	true	$5(3 - 1) = 10$
1.5	$6(1.5) + 1 = 10$	false	$5(3 - 1.5) = 7.5$

a. Critical Thinking Based on the table, would you expect the solution of $6x + 1 < 5(3 - x)$ to be of the form $x < n$ or $x > n$? Explain.
b. Estimate Based on the table, estimate the value of n.
c. Solve the inequality. Compare the actual solution to your estimated solution.

Error Analysis **Find and correct the mistake in each student's work.**

For a guide to solving
Exercise 72, see p. 226.

72.
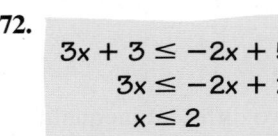
$$3x + 3 \leq -2x + 5$$
$$3x \leq -2x + 2$$
$$x \leq 2$$

73.
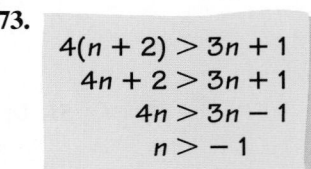
$$4(n + 2) > 3n + 1$$
$$4n + 2 > 3n + 1$$
$$4n > 3n - 1$$
$$n > -1$$

 **74. a.** Solve $ax + b > c$ for x, where a is positive.
b. Reasoning Solve $ax + b > c$ for x, where a is negative.

75. Geometry The base of a triangle is 10 in. Its height is $(x + 4)$ in. Its area is no more than 56 in.^{2}. What are the possible integer values of x?

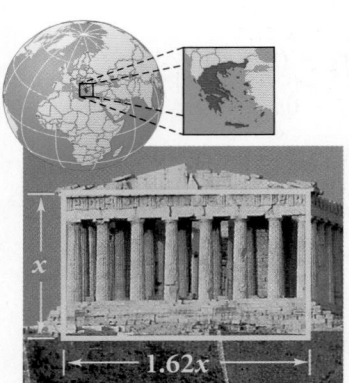

76. Architecture The rectangle shown on the building at the left is a golden rectangle. Artists often use the golden rectangle because they consider it to be pleasing to the eye. The ratio of two sides of a golden rectangle is approximately 1 : 1.62. Suppose you are making a picture frame in the shape of a golden rectangle. You have a 46-in. length of wood to use for a frame. What are the dimensions of the largest frame you can make? Round to the nearest tenth of an inch.

77. Critical Thinking Find a value of a such that the number line below shows all the solutions of $ax + 4 \leq -12$.

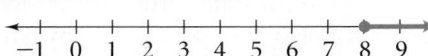

Real-World Connection

The Parthenon, an ancient Greek temple, has dimensions that form a golden rectangle.

78. Earning You can earn money by handing out flyers in the afternoon for $6.50 an hour and by typing a newsletter in the evening for $8 an hour. You have 20 hours available to work. What are the greatest number of hours you can spend handing out flyers and still make at least $145?

 79. Freight Handling The freight elevator of a building can safely carry a load of at most 4000 lb. A worker needs to move supplies in 50-lb boxes from the loading dock to the fourth floor of the building. The worker weighs 160 lb. The cart she uses weighs 95 lb.
a. What is the greatest number of boxes she can move in one trip?
b. The worker must deliver 310 boxes to the fourth floor. How many trips must she make?

Multiple Choice

80. The Science Club hopes to collect at least 200 kg of aluminum cans for recycling this semester (21 weeks). The graph at the right shows the first week's results.

Let x represent the average mass of cans required per week for the remainder of the semester. Which inequality would you use to find x?

A. $x \geq \frac{200}{21}$ **B.** $x \geq \frac{(200-8)}{21}$

C. $x \geq \frac{(200-8)}{20}$ **D.** $x > \left(\frac{200}{20}\right) - 8$

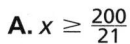

Aluminum Cans Collected in Week 1

81. Solve $2x - 8 > 4x + 2$.

   F. $x < -5$ **G.** $x > -5$ **H.** $x < 5$ **J.** $x > 5$

82. Solve $-5n + 16 \leq -7n$.

   A. $n \leq -8$ **B.** $n \geq -8$ **C.** $n \leq 8$ **D.** $n \geq 8$

83. Great Gifts pays its supplier $65 for each box of 12 bells. The owner wants to determine the least amount x he can charge his customers per bell in order to make at least a 50% profit per box. Which inequality should he use?

   F. $12x \geq 1.50(65)$ **G.** $65x \leq 1.50(12)$

   H. $0.50(12x) \geq 65$ **J.** $0.50(12x) \leq 65$

Short Response

84. Maxwell orders at least 30 bottles of flea shampoo per month for his pet-grooming business. His supplier charges $3 per quart bottle plus a $25 handling fee per order. A competing supplier offers a similar product for $4 per quart bottle plus a $5 handling fee per order. The salesman for the competitor shows Maxwell that 10 bottles from his company would cost only $45 compared to $55 from Maxwell's current supplier.

Which supplier would you advise Maxwell to use? Explain or show work to support your advice to Maxwell.

Mixed Review

Lesson 4-3

Solve each inequality.

85. $-9m \geq 36$ **86.** $-24 \leq 3y$ **87.** $\frac{x}{3} > -4$

88. $-\frac{t}{3} \leq 1$ **89.** $\frac{2}{3}b < 18$ **90.** $42 > -\frac{3}{7}w$

91. $56 < 42p$ **92.** $0.5d \geq 3.5$ **93.** $\frac{x}{5} > 10$

Lesson 3-6

94. Your family leaves your town traveling at an average rate of 45 mi/h. Two hours later, your neighbor leaves your town along the same road at an average rate of 60 mi/h. How many hours will it take your neighbor to overtake you?

Lesson 2-3

Simplify each expression.

95. -4^2 **96.** $(-4)^2$ **97.** $(-2)^3(-3)$ **98.** -2^4

Guided Problem Solving

FOR USE WITH PAGE 224, EXERCISE 72

Analyzing Errors   Read the exercise below and then follow along with what Gina thinks and writes. Check your understanding with the exercise at the bottom of the page.

Error Analysis Find and correct the mistake in the student's work at the right.

$$3x + 3 \leq -2x + 5$$
$$3x \leq -2x + 2$$
$$x \leq 2$$

What Gina Thinks

I am going to solve the inequality and see where my work is different from the student's work.

I'll subtract 3 from each side, and then simplify.

Our work agrees. There is no mistake so far.

I'll add $2x$ to each side so I can get the x's on the left. Then I'll divide each side by 5 to get x alone. My solution is different from the student's work!

I think the student *subtracted* $2x$ from each side. That's the mistake, because $-2x - 2x$ equals $-4x$, not zero! If the student thought it was zero, the equation would have x on the left side and 2 on the right side. That's what they wrote: $x \leq 2$.

I'll write my answer in a sentence.

What Gina Writes

$$3x + 3 \leq -2x + 5$$

$$3x + 3 - 3 \leq -2x + 5 - 3$$

$$3x \leq -2x + 2$$

$$3x + 2x \leq -2x + 2x + 2$$

$$5x \leq 2$$

$$\frac{5x}{5} \leq \frac{2}{5}$$

$$x \leq \frac{2}{5}$$

The student tried to subtract $2x$ from each side instead of adding it to each side.

EXERCISE

Find and correct the mistake in the work at the right.

$$1 - 3x < 7$$
$$1 - 3x - 1 < 7 - 1$$
$$-3x < 6$$
$$\frac{-3x}{-3} < \frac{6}{-3}$$
$$x < -2$$

226   **Guided Problem Solving**   Analyzing Errors

4-5

Compound Inequalities

 Learning Standards for Mathematics

A.A.5 Write algebraic equations or inequalities that represent a situation.

A.A.6 Analyze and solve verbal problems whose solution requires solving a linear equation in one variable or linear inequality in one variable.

A.A.24 Solve linear inequalities in one variable.

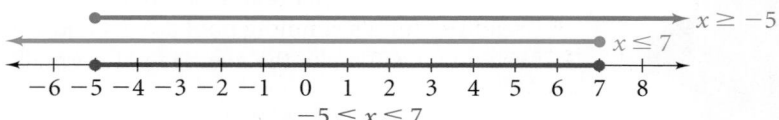

Check Skills You'll Need

GO for Help Lessons 1-2 and 4-1

Graph each pair of inequalities on one number line.

1. $c < 8; c \geq 10$ **2.** $t \geq -2; t \leq -5$ **3.** $m \leq 7; m > 12$

Use the given value of the variable to evaluate each expression.

4. $3n - 6; 4$ **5.** $7 - 2b; 5$

6. $\dfrac{12 + 13 + y}{3}; 17$ **7.** $\dfrac{2d - 3}{5}; 9$

New Vocabulary • compound inequality

---

1 Solving Compound Inequalities Containing *And*

Two inequalities that are joined by the word *and* or the word *or* form a **compound inequality.**

You can write the compound inequality $x \geq -5$ and $x \leq 7$ as $-5 \leq x \leq 7$.

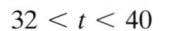

Vocabulary Tip

The word <u>inclusive</u> is related to the word <u>included</u>.

The graph above shows that a solution of $-5 \leq x \leq 7$ is in the overlap of the solutions of the inequality $x \geq -5$ and the inequality $x \leq 7$.

You can read $-5 \leq x \leq 7$ as "x is greater than or equal to -5 and less than or equal to 7." Another way to read it is "x is between -5 and 7, inclusive."

1 EXAMPLE Writing a Compound Inequality

Write a compound inequality that represents each situation. Graph the solutions.

a. all real numbers that are at least -2 and at most 4
$n \geq -2$ and $n \leq 4$
$-2 \leq n \leq 4$

b. Today's temperatures will be above 32°F, but not as high as 40°F.
$32 < t$ and $t < 40$
$32 < t < 40$

Quick Check ❶ Write a compound inequality that represents each situation. Graph your solution.
a. all real numbers greater than -2 but less than 9
b. The books were priced between \$3.50 and \$6.00, inclusive.

A solution of a compound inequality joined by *and* is any number that makes both inequalities true. One way you can solve a compound inequality is by writing two inequalities.

2 EXAMPLE Solving a Compound Inequality Containing *And*

Solve $-4 < r - 5 \le -1$. Graph your solution.

Write the compound inequality as two inequalities joined by *and*.

$-4 < r - 5$	and	$r - 5 \le -1$	
$-4 + 5 < r - 5 + 5$		$r - 5 + 5 \le -1 + 5$	**Solve each inequality.**
$1 < r$	and	$r \le 4$	**Simplify.**

$$1 < r \le 4$$

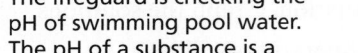

Quick Check ② Solve each inequality. Graph your solution.
   a. $-6 \le 3x < 15$ **b.** $-3 < 2x - 1 < 7$ **c.** $7 < -3n + 1 \le 13$

You could also solve an inequality like $-4 < r - 5 \le -1$ by working on all three parts of the inequality at the same time. You work to get the variable alone between the inequality symbols.

3 EXAMPLE Real-World 🌐 Problem Solving

Chemistry The acidity of the water in a swimming pool is considered normal if the average of three pH readings is between 7.2 and 7.8, inclusive. The first two readings for a swimming pool are 7.4 and 7.9. What possible values for the third reading p will make the average pH normal?

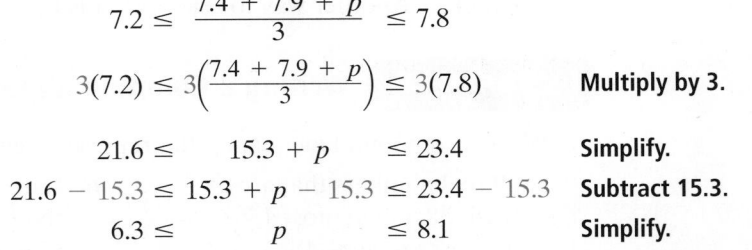

$$7.2 \le \frac{7.4 + 7.9 + p}{3} \le 7.8$$

$3(7.2) \le 3\left(\dfrac{7.4 + 7.9 + p}{3}\right) \le 3(7.8)$	**Multiply by 3.**
$21.6 \le \quad 15.3 + p \quad \le 23.4$	**Simplify.**
$21.6 - 15.3 \le 15.3 + p - 15.3 \le 23.4 - 15.3$	**Subtract 15.3.**
$6.3 \le \qquad p \qquad \le 8.1$	**Simplify.**

The value for the third reading must be between 6.3 and 8.1, inclusive.

Real-World 🌐 Connection

The lifeguard is checking the pH of swimming pool water. The pH of a substance is a measure of how acidic or basic it is. pH is measured on a scale from 0 to 14. Pure water is neutral, with a pH of 7.

Quick Check ③ **a.** Suppose the first two readings for the acidity of water in a swimming pool are 7.0 and 7.9. What possible values for the third reading will make the average pH normal?
   b. **Critical Thinking** If two readings are 8.0 and 8.4, what possible values for the third reading will make the average pH normal? Are these third readings likely? Explain.

A solution of a compound inequality joined by *or* is any number that makes either inequality true.

4 EXAMPLE Writing Compound Inequalities

Write a compound inequality that represents each situation. Graph the solution.

a. all real numbers that are less than -3 or greater than 7

$x < -3$ or $x > 7$

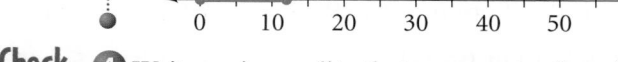

b. Discounted fares are available to children 12 and under or to adults at least 60 years of age.

$n \leq 12$ or $n \geq 60$; $n \geq 0$ because age cannot be negative.

Quick Check **4** Write an inequality that represents all real numbers that are at most -5 or at least 3. Graph your solution.

For a compound inequality joined by *or*, you must solve each of the two inequalities separately.

Online
active math

For: Compound Inequality Activity
Use: Interactive Textbook, 4-5

5 EXAMPLE Solving a Compound Inequality Containing *Or*

Solve the compound inequality $4v + 3 < -5$ or $-2v + 7 < 1$. Graph the solution.

$$4v + 3 < -5 \qquad \text{or} \qquad -2v + 7 < 1$$
$$4v + 3 - 3 < -5 - 3 \qquad\qquad -2v + 7 - 7 < 1 - 7$$
$$4v < -8 \qquad\qquad\qquad\qquad -2v < -6$$
$$\frac{4v}{4} < \frac{-8}{4} \qquad\qquad\qquad \frac{-2v}{-2} > \frac{-6}{-2}$$
$$v < -2 \qquad \text{or} \qquad\qquad v > 3$$

Quick Check **5** Solve the compound inequality $-2x + 7 > 3$ or $3x - 4 \geq 5$. Graph your solution.

EXERCISES

For more exercises, see Extra Skill and Word Problem Practice.

Practice and Problem Solving

A Practice by Example

Example 1
(page 227)

GO for
Help

Write a compound inequality that represents each situation. Graph your solution.

1. all real numbers that are between -4 and 6

2. all real numbers that are at least 2 and at most 9

3. The circumference of a baseball is between 23 cm and 23.5 cm.

4. Tropical Storm The wind speeds of a tropical storm are at least 40 mi/h but no more than 74 mi/h.

Solve each compound inequality. Graph your solution.

5. $-3 < j + 2 < 7$ **6.** $3 \le w + 2 \le 7$ **7.** $2 < 3n - 4 \le 14$

8. $7 \le 3 - 2p < 11$ **9.** $-2 < -3x + 7 < 4$ **10.** $1.5 < w + 3 \le 6.5$

11. $-16 < -3x + 8 < -7$ **12.** $-1 < 4m + 7 \le 11$ **13.** $-9 < -2s - 1 \le -7$

14. $12 \le \frac{14 + 17 + a}{3} \le 16$ **15.** $\frac{1}{2} < \frac{3x - 1}{4} < 5$ **16.** $-2 \le \frac{5 - x}{3} \le 2$

Example 4
(page 229)

For each situation write and graph an inequality.

17. all real numbers n that are at most -3 or at least 5

18. all real numbers x that are less than 3 or greater than 7

19. all real numbers h less than 1 or greater than 3

20. all real numbers b less than 100 or greater than 300

Example 5
(page 229)

Solve each compound inequality. Graph your solution.

21. $3b - 1 < -7$ or $4b + 1 > 9$ **22.** $4 + k > 3$ or $6k < -30$

23. $3c + 4 \ge 13$ or $6c - 1 < 11$ **24.** $6 - a < 1$ or $3a \le 12$

25. $7 - 3c \ge 1$ or $5c + 2 \ge 17$ **26.** $5y + 7 \le -3$ or $3y - 2 \ge 13$

27. $2d + 5 \le -1$ or $-2d + 5 \le 5$ **28.** $5z - 3 > 7$ or $4z - 6 < -10$

B **Apply Your Skills**

Write a compound inequality that each graph could represent.

29.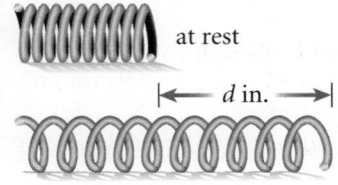

30.

31.

32.

Solve each compound inequality.

33. $3q - 2 > 10$ or $3q - 2 \le -10$ **34.** $3 - 2h > 17$ or $5h - 3 > 17$

35. $1 \le 0.25t \le 3.5$ **36.** $25r < 400$ or $100 < 4r$

37. $-20 \le 3t - 2 < 1$ **38.** $\frac{3x + 1}{4} - 4 > 3$ or $\frac{3 - 2x}{5} > 3$

39. Multiple Choice The force exerted on a spring is proportional to the distance the spring stretches from its relaxed position. Suppose you stretch a spring distance d in inches by applying force F in pounds. For a certain spring, $\frac{d}{F} = 0.8$. You apply forces between 25 and 40 pounds, inclusive. Which inequality describes the stretch of the spring?

Ⓐ $25 \le d \le 40$ Ⓑ $20 < d < 32$

Ⓒ $31.25 \le d \le 40$ Ⓓ $20 \le d \le 32$

at rest

d in.

40. Reasoning Describe the solutions of $3x - 8 < 7$ or $2x - 9 > 1$.

41. Writing Explain the difference between the words *and* and *or* in a compound inequality.

Homework Video Tutor
Visit: PHSchool.com
Web Code: ate-0405

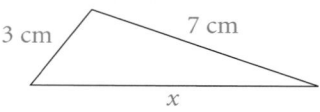 **Geometry** The sum of the lengths of any two sides of a triangle is greater than the length of the third side. The lengths of two sides of a triangle are given. Find the range of values for the possible lengths of the third side.

Sample 3 cm, 7 cm

Write inequalities for x as the longest side and for 7 cm as the longest side. The length 3 cm cannot be the longest side.

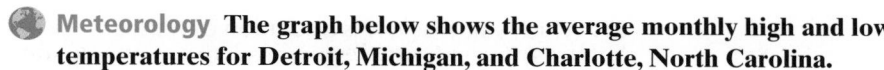

$x + 3 > 7$ and $3 + 7 > x$

$x > 4$ and $10 > x$ **Solve each inequality.**

$4 < x < 10$

The length of the third side is greater than 4 cm and less than 10 cm.

42. 2.5 in., 5 in.   **43.** 12 ft, 18 ft   **44.** 28 mm, 21 mm   **45.** 5 m, 16 m

Meteorology The graph below shows the average monthly high and low temperatures for Detroit, Michigan, and Charlotte, North Carolina.

46. Write a compound inequality for Charlotte's average temperature in June.

47. Write a compound inequality for Detroit's average temperature in January.

48. Write a compound inequality for the yearly temperature range for each city.

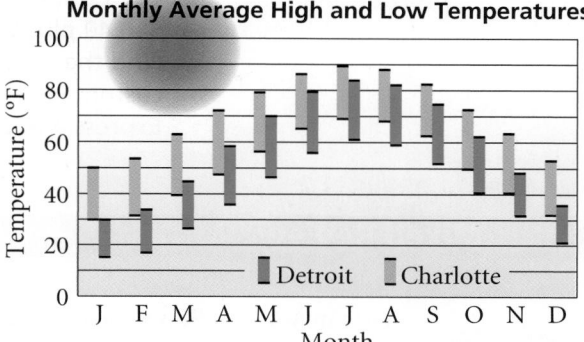

Monthly Average High and Low Temperatures

SOURCE: Statistical Abstract of the United States

49. Open-Ended Describe a real-life situation that you could represent with the inequality $-2 < x < 8$.

C Challenge 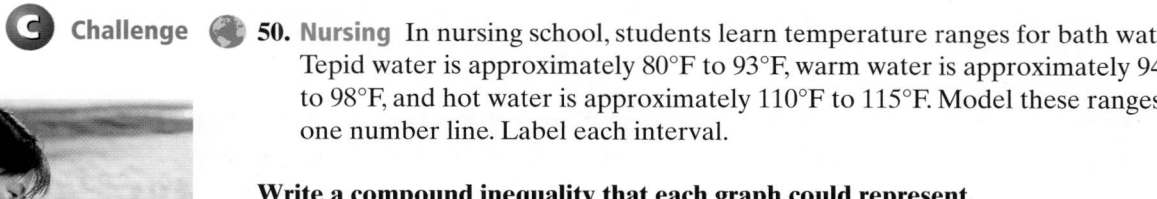 **50. Nursing** In nursing school, students learn temperature ranges for bath water. Tepid water is approximately 80°F to 93°F, warm water is approximately 94°F to 98°F, and hot water is approximately 110°F to 115°F. Model these ranges on one number line. Label each interval.

Write a compound inequality that each graph could represent.

51.
```
<-+--⊕--⊕--+--○-+->
  -4  -2   0   2   4
```

52.
```
<-+--+--●--+--●->
     -2   0   2   4
```

53. Pulse Rates When you exercise, your pulse rate rises. Recommended pulse rates vary with age and physical condition. For vigorous exercise, such as jogging, the inequality $0.7(220 - a) \leq R \leq 0.85(220 - a)$ gives a target range for pulse rate R (in beats per minute), based on age a (in years).
   a. What is the target range for pulse rates for a person 35 years old? Round to the nearest whole number.
   b. Your cousin's target pulse rate is in the range between 140 and 170 beats per minute. What is your cousin's age?

54. Find three consecutive even integers whose sum is between 48 and 60.

55. Find three consecutive even integers such that one half of their sum is between 15 and 21.

Real-World Connection

To estimate your pulse rate, count the number of beats you feel in 15 seconds at a pressure point. Multiply this number by 4.

Multiple Choice

56. An emergency vehicle responding to a 911 call for a heart attack victim traveled 5 miles to the patient's home and then delivered him to the hospital 10 miles away. Which graph below represents the possible distances the emergency vehicle was from the hospital when the call was received?

A.
```
 ◄──┼──┼──┼──●──┼──┼──◆──┼──┼──┼──►
    0  2  4  6  8  10 12 14 16 18
```
B.
```
 ◄──┼──┼──●──┼──┼──┼──┼──●──┼──┼──►
    0  2  4  6  8  10 12 14 16 18
```
C.
```
 ◄──◆━━━━━━┼──┼──┼──┼──┼──┼──┼──►
    0  2  4  6  8  10 12 14 16 18
```
D.
```
 ◄──◆━━━━━━━━━━◆──┼──┼──┼──┼──┼──►
    0  2  4  6  8  10 12 14 16 18
```

57. Which value below is a solution of neither $-3x - 7 \geq 8$ nor $-2x - 11 \leq -31$?

    F. –6   **G.** 0   **H.** 10   **J.** 16

Short Response

58. The County Water Department charges a monthly administration fee of $10.40 plus $.0059 for each gallon g of water used, up to 7,500 gallons. Find the minimum and maximum water consumption (in gallons) for customers whose monthly charge is at least $35 but no more than $50. Express amounts to the nearest gallon. Show your work.

Mixed Review

Lesson 4-4

Solve each inequality.

59. $5 < 6b + 3$ **60.** $12n \leq 3n + 27$ **61.** $2 + 4r \geq 5(r - 1)$

Lesson 3-3

Solve. If the equation is an identity or if it has no solution, write *identity* or *no solution*.

62. $x - 3 = 5x + 1$ **63.** $4(w + 3) = 10w$ **64.** $8p - 4 = 4(2p - 1)$

✓ Checkpoint Quiz 2   Lessons 4-4 through 4-5

Solve each inequality. Graph the solution.

  1. $8d + 2 < 5d - 7$ **2.** $2n + 1 \geq -3$ **3.** $-1 \leq 4m + 7 \leq 11$

  4. $5s - 3 + 1 < 8$ **5.** $5(3p - 2) > 50$ **6.** $3 - x \geq 7$ or $2x - 3 > 5$

Write an inequality that represents each situation.

  7. A cat weighs less than 8 pounds.

  8. We expect today's temperature to be between 65°F and 75°F, inclusive.

  9. **Geometry** The length of each side of a rectangular picture frame needs to be 15 in. You have only one 48 in. piece of wood to use for this frame. Write and solve an inequality that describes the possible widths for this frame.

  10. Solve $-2x + 7 \leq 45$.

Algebraic Reasoning

You can use the properties you have studied along with the four properties below to prove algebraic relationships.

NY **A.RP.4:** Develop, verify, and explain an argument, using appropriate mathematical ideas and language.

Reflexive, Symmetric, and Transitive Properties of Equality

For every real number a, b, and c:

Reflexive Property: $a = a$ **Example:** $5x = 5x$

Symmetric Property: If $a = b$, then $b = a$.   **Example:** If $15 = 3t$, then $3t = 15$.

Transitive Property: If $a = b$ and $b = c$, then $a = c$.   **Example:** If $d = 3y$ and $3y = 6$, then $d = 6$.

Transitive Property of Inequality

For all real numbers a, b, and c, if $a < b$ and $b < c$, then $a < c$.   **Example:** If $8x < 7$ and $7 < y^2$, then $8x < y^2$.

EXAMPLE

Prove each statement for all real numbers a, b, and c.

a. If $a = b$, then $ac = bc$.

$a = b$	Given
$ac = ac$	Reflexive Property
$ac = bc$	Substitute b for a.

b. If $c < 0$ and $a < b$, then $c < b - a$.

$c < 0$	Given
$a < b$	Given
$a - a < b - a$	Subtraction Property of Inequality
$a + (-a) < b - a$	Definition of subtraction
$0 < b - a$	Inverse Property of Addition
$c < b - a$	Transitive Property of Inequality

EXERCISES

Name the property that each exercise illustrates.

1. If $3.8 = z$, then $z = 3.8$.   **2.** If $x = \frac{1}{2}y$ and $\frac{1}{2}y = -2$, then $x = -2$.   **3.** $-r = -r$

4. If $k < m^2$ and $m^2 < 4$, then $k < 4$.   **5.** If $x = w^2$, then $w^2 = x$.

Supply the missing reasons to prove each statement.

6. $(a + b) + (-a) = b$

$(a + b) + (-a) = (b + a) + (-a)$	?
$= b + [a + (-a)]$	?
$= b + 0$	?
$= b$	?
$(a + b) + (-a) = b$	?

7. If $a < b$ and $c < d$, then $a + c < b + d$.

$a < b$	Given
$a + c < b + c$	?
$c < d$	Given
$b + c < b + d$	?
$a + c < b + d$	?

High-Use Academic Words

High-use academic words are words that you see often in textbooks and on tests. These words are not math vocabulary terms, but they are important for you to know to be successful in mathematics.

Words to Learn: Direction Words

Some words tell you what to do in a problem. You need to understand what these words are asking so that you give the correct answer.

Word(s)	Meaning
Describe	To write about a topic so the reader can understand or visualize it
Explain	To give the reasoning for an answer
Define	To give the meaning of a word or expressions or to tell what a variable stands for in order to model a real-world situation

EXERCISES

1. Describe how you walk from your classroom to the cafeteria.

2. Explain why you participate in your favorite school activity.

3. Define the term *extracurricular*.

Describe each of the following.

4. data that can show a negative correlation

5. a real-world situation that can be modeled by the equation $3x - 2 = 5$

Explain why each statement is *true* or *false*.

6. All integers are positive.

7. The median of a set of data may better represent the data than the mean.

Define a variable and write an inequality to model each situation.

8. The elevator can hold no more than 14 people.

9. Miguel needs to earn at least 86 on his test to maintain his A average.

10. Word Knowledge Think about the word *identify*.
    a. Choose the letter for how well you know the word.
      A. I know its meaning.
      B. I have seen it, but I do not know its meaning.
      C. I don't know it.
    b. Research Look up *identify* in a dictionary or online. Write any definition that might apply to its use in mathematics.
    c. Write a sentence involving mathematics that uses the word *identify*.

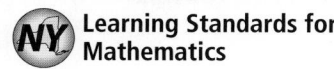

Absolute Value Equations and Inequalities

NY Learning Standards for Mathematics

A.G.4 Identify and graph linear, quadratic (parabolic), absolute value, and exponential functions.

✓ **Check Skills You'll Need**

GO for **Help** Lessons 1-3 and 2-1

Simplify.

1. $|15|$
2. $|-3|$
3. $|18 - 12|$
4. $-|-7|$
5. $|12 - (-12)|$
6. $|-10 + 8|$

Complete each statement with <, =, or >.

7. $|3 - 7|$ ▆ 4
8. $|-5| + 2$ ▆ 6
9. $|7| - 1$ ▆ 8
10. $\left|6 - 2\frac{1}{4}\right|$ ▆ $3\frac{5}{8}$
11. $\left|-4\frac{2}{3}\right| + 2\frac{1}{3}$ ▆ $2\frac{1}{2}$
12. $\left|-3\frac{1}{8} - 4\frac{1}{2}\right|$ ▆ $7\frac{5}{8}$

---

1 Solving Absolute Value Equations

Recall that the absolute value of a number is its distance from zero on a number line. Since absolute value represents distance, it can never be negative.

The graph of $|x| = 3$ is below.

Problem Solving Hint

$|3| = 3$

$|-3| = 3$

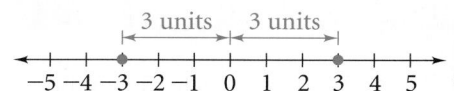

3 units   3 units

Find the numbers that are 3 units from 0.

The two solutions of the equation $|x| = 3$ are -3 and 3.

You can use the properties of equality to solve an absolute value equation.

1 EXAMPLE **Solving an Absolute Value Equation**

Solve $|x| + 5 = 11$.

$|x| + 5 - 5 = 11 - 5$ **Subtract 5 from each side.**

$|x| = 6$ **Simplify.**

$x = 6$ or $x = -6$ **Definition of absolute value.**

Check $|x| + 5 = 11$

$|6| + 5 \stackrel{?}{=} 11$ ← Substitute 6 and −6 for x. → $|-6| + 5 \stackrel{?}{=} 11$

$6 + 5 = 11$ ✓ $6 + 5 = 11$ ✓

✓ **Quick Check** **1** Solve each equation. Check your solution.
a. $|t| - 2 = -1$
b. $3|n| = 15$
c. $4 = 3|w| - 2$
d. Critical Thinking Is there a solution of $2|n| = -15$? Explain.

Some absolute value equations such as $|2p + 5| = 11$ have variable expressions within the absolute value symbols. The expression inside the absolute value symbols can be either positive or negative.

 Key Concepts

Rule	**Solving Absolute Value Equations**

To solve an equation in the form $|A| = b$, where A represents a variable expression and $b > 0$, solve $A = b$ and $A = -b$.

 **EXAMPLE** **Solving an Absolute Value Equation**

Solve $|2p + 5| = 11$.

$2p + 5 = 11$	← **Write two equations.** →	$2p + 5 = -11$
$2p + 5 - 5 = 11 - 5$	← **Subtract 5 from each side.** →	$2p + 5 - 5 = -11 - 5$
$2p = 6$		$2p = -16$
$\dfrac{2p}{2} = \dfrac{6}{2}$	← **Divide each side by 2.** →	$\dfrac{2p}{2} = \dfrac{-16}{2}$
$p = 3$		$p = -8$

● The value of p is 3 or -8.

✓ **Quick Check** ❷ Solve each equation. Check your solution.
a. $|c - 2| = 6$ **b.** $-5.5 = |t + 2|$ **c.** $|7d| = 14$

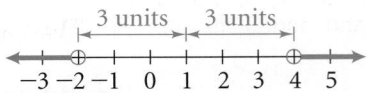
2 **Solving Absolute Value Inequalities**

You can write absolute value inequalities as compound inequalities.

The graphs below show two absolute value inequalities.

$|n - 1| < 3$ $|n - 1| > 3$

3 units   3 units   3 units   3 units

$-3\ -2\ -1\ \ 0\ \ 1\ \ 2\ \ 3\ \ 4\ \ 5$ $-3\ -2\ -1\ \ 0\ \ 1\ \ 2\ \ 3\ \ 4\ \ 5$

$|n - 1| < 3$ represents all numbers whose distance from 1 is less than 3 units. So $-3 < n - 1 < 3$.

$|n - 1| > 3$ represents all numbers whose distance from 1 is greater than 3 units. So $n - 1 < -3$ or $n - 1 > 3$.

 Key Concepts

Rule	**Solving Absolute Value Inequalities**

To solve an inequality in the form $|A| < b$, where A is a variable expression and $b > 0$, solve $-b < A < b$.

To solve an inequality in the form $|A| > b$, where A is a variable expression and $b > 0$, solve $A < -b$ or $A > b$.

Similar rules are true for $|A| \le b$ or $|A| \ge b$.

3 EXAMPLE **Solving an Absolute Value Inequality**

Solve $|v - 3| \geq 4$. Graph the solutions.

$$v - 3 \leq -4 \quad \text{or} \quad v - 3 \geq 4 \qquad \text{Write a compound inequality.}$$
$$v - 3 + 3 \leq -4 + 3 \quad \Big| \quad v - 3 + 3 \geq 4 + 3 \qquad \text{Add 3.}$$
$$v \leq -1 \quad \text{or} \quad v \geq 7 \qquad \text{Simplify.}$$

Quick Check **3 a.** Solve and graph $|w + 2| > 5$.
  b. Critical Thinking What are the solutions of $|w + 2| > -5$?

To maintain quality, a manufacturer sets limits for how much an item can vary from its specifications. You can use an absolute value equation to model a quality-control situation.

4 EXAMPLE **Real-World 🌐 Problem Solving**

Manufacturing The ideal diameter of a piston for one type of car engine is 90.000 mm. The actual diameter can vary from the ideal by at most 0.008 mm. Find the range of acceptable diameters for the piston.

Relate | difference between actual and ideal | is at most | 0.008 mm |

Define Let d = actual diameter in millimeters of the cylindrical part.

Write | $|d - 90.000|$ | $\leq$ | 0.008 mm |

$$|d - 90.000| \leq 0.008$$
$$-0.008 \leq \quad d - 90.000 \quad \leq 0.008 \qquad \text{Write a compound inequality.}$$
$$-0.008 + 90.000 \leq d - 90.000 + 90.000 \leq 0.008 + 90.000 \qquad \text{Add 90.000.}$$
$$89.992 \leq \quad d \quad \leq 90.008 \qquad \text{Simplify.}$$

The actual diameter must be between 89.992 mm and 90.008 mm, inclusive.

Real-World 🌐 Connection

Careers A quality-control inspector inspects products to maintain quality. For engines produced on an assembly line, an inspector selects engines at random to check quality of materials and manufacturing.

Quick Check **4** The ideal weight of one type of model airplane engine is 33.86 ounces. The actual weight may vary from the ideal by at most 0.05 ounce. Find the range of acceptable weights for this engine.

EXERCISES

For more exercises, see *Extra Skill and Word Problem Practice.*

Practice and Problem Solving

A Practice by Example

Example 1 (page 235)

Solve each equation. If there is no solution, write *no solution.*

1. $|b| = 2$ **2.** $4 = |y|$ **3.** $|w| = \frac{1}{2}$

4. $|n| + 2 = 8$ **5.** $7 = |s| + 4$ **6.** $|x| - 10 = -3$

7. $4|d| = 20$ **8.** $-3|m| = -6$ **9.** $|y| + 3 = 3$

10. $12 = -4|k|$ **11.** $2|z| - 5 = 1$ **12.** $16 = 5|p| - 4$

Example 2
(page 236)

Solve each equation. If there is no solution, write *no solution*.

13. $|r - 8| = 5$ **14.** $|c + 2| = 6$ **15.** $2 = |g + 1|$

16. $3 = |m + 2|$ **17.** $|v - 2| = 7$ **18.** $-3|y - 3| = 9$

19. $2|d + 3| = 8$ **20.** $-2|7d| = -14$ **21.** $1.2|5p| = 3.6$

Example 3
(page 237)

22. Complete each statement with *less than* or *greater than*.
   a. For $|x| < 5$, the graph includes all points whose distance is __?__ 5 units from 0.
   b. For $|x| > 5$, the graph includes all points whose distance is __?__ 5 units from 0.

Solve each inequality. Graph your solution.

23. $|k| > 2.5$ **24.** $|w| < 2$ **25.** $|x + 3| < 5$

26. $|n + 8| \geq 3$ **27.** $|y - 2| \leq 1$ **28.** $|p - 4| \leq 3$

29. $|2c - 5| < 9$ **30.** $|2y - 3| \geq 7$ **31.** $|3t + 1| > 8$

32. $|4x + 1| > 11$ **33.** $|5t - 4| \geq 16$ **34.** $|3 - r| < 5$

Example 4
(page 237)

35. Manufacturing The ideal diameter of a gear for a certain type of clock is 12.24 mm. An actual diameter can vary by 0.06 mm. Find the range of acceptable diameters.

36. Manufacturing The ideal width of a certain conveyor belt for a manufacturing plant is 50 in. An actual conveyor belt can vary from the ideal by at most $\frac{7}{32}$ in. Find the acceptable widths for this conveyor belt.

B   Apply Your Skills

Solve each equation or inequality.

37. $|2d| + 3 = 21$ **38.** $|-3n| - 2 = 7$ **39.** $|p| - \frac{2}{3} = \frac{5}{6}$

40. $|t| + 2.7 = 4.5$ **41.** $4|k + 1| = 16$ **42.** $-2|c - 4| = -8$

43. $|3d| \geq 6$ **44.** $|n| - 3 > 7$ **45.** $9 < |c + 7|$

46. $\frac{|v|}{-3} = -4.2$ **47.** $|6.5x| < 39$ **48.** $4|n| = 32$

49. $\left|\frac{1}{2}a\right| + 1 = 5$ **50.** $|a| + \frac{1}{2} = 3\frac{1}{2}$ **51.** $4 - 3|m + 2| > -14$

Write an absolute value inequality that represents each situation.

52. all numbers less than 3 units from 0

53. all numbers more than 7.5 units from 0

54. all numbers more than 2 units from 6

55. all numbers at least 3 units from −1

56. Manufacturing A pasta manufacturer makes 16-ounce boxes of macaroni. The manufacturer knows that not every box weighs exactly 16 ounces. The allowable difference is 0.05 ounce. Write and solve an absolute value inequality that represents this situation.

57. Elections In a poll for the upcoming mayoral election, 42% of likely voters said they planned to vote for Lucy Jones. This poll has a margin of error of ± 3 percentage points. Use the inequality $|v - 42| \leq 3$ to find the least and greatest percent of voters v likely to vote for Lucy Jones according to this poll.

58. Quality Control A box of one brand of crackers should weigh 454 g. The quality-control inspector randomly selects boxes to weigh. The inspector sends back any box that is not within 5 g of the ideal weight.
  a. Write an absolute value inequality for this situation.
  b. What is the range of allowable weights for a box of crackers?

59. Gears Acceptable diameters for one type of gear are from 6.25 mm to 6.29 mm. Write an absolute value inequality for the acceptable diameters for the gear.

60. Writing Explain why the absolute value inequality $|2c - 5| + 9 < 4$ has no solution.

61. Open-Ended Write an absolute value equation using the numbers $5, 3, -12$. Then solve your equation.

Write an absolute value equation that has the given values as solutions.

Sample $8, 2$

$\qquad |x - 5| = 3$ Since 8 and 2 are both 3 units from 5, write $|x - 5| = 3$.

62. $2, 6$ **63.** $-2, 6$ **64.** $-3, 9$ **65.** $9, 16$

66. $-1, 7$ **67.** $3, 8$ **68.** $-15, -3$ **69.** $2, 10$

70. Banking The ideal weight of a nickel is 0.176 ounce. To check that there are 40 nickels in a roll, a bank weighs the roll and allows for an error of 0.015 ounce in the total weight.
  a. What is the range of acceptable weights if the wrapper weighs 0.05 ounce?
  b. Critical Thinking For any given roll of nickels, can you be certain that all the coins are acceptable? Explain.

71. a. Meteorology A meteorologist reported that the previous day's temperatures varied 14 degrees from the normal temperature of 25°F. What were the maximum and minimum temperatures possible on the previous day?
  b. Write an absolute value equation for the temperature.

C Challenge

Solve each equation. Check your solution.

72. $|x + 4| = 3x$ **73.** $|4x - 5| = 2x + 1$ **74.** $\frac{4}{3}|2x + 3| = 4x$

Replace the ▦ with ≤, ≥, or =.

75. $|a + b| \ ▦ \ |a| + |b|$ **76.** $|a - b| \ ▦ \ |a| - |b|$

77. $|ab| \ ▦ \ |a| \cdot |b|$ **78.** $\left|\frac{a}{b}\right| \ ▦ \ \frac{|a|}{|b|}, b \neq 0$

Write an absolute value inequality that each graph could represent.

79.

80.
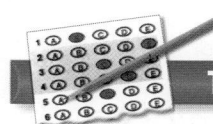

Test Prep

Multiple Choice

81. Which compound inequality has the same meaning as $|x + 4| < 8$?
  A. $-12 < x < 4$ **B.** $-12 > x > 4$
  C. $x < -12$ or $x > 4$ **D.** $x > -12$ or $x < 4$

82. Which of the following values is a solution of $|2 - x| < 4$?
  F. -2 **G.** -1 **H.** 6 **J.** 7

83. The ideal diameter of a metal rod for a lamp is 1.25 inches with an allowable error of at most 0.005 inch. Which rod below would not be suitable?
   A. a rod with diameter 1.249 inches
   B. a rod with diameter 1.251 inches
   C. a rod with diameter 1.253 inches
   D. a rod with diameter 1.355 inches

84. A delivery driver receives a bonus if he delivers pizza to a customer in 30 minutes plus or minus 5 minutes. Which inequality or equation represents the driver's allotted time to receive a bonus?
   F. $|x - 30| < 5$ **G.** $|x - 30| > 5$
   H. $|x - 30| = 5$ **J.** $|x - 30| \le 5$

85. Water is in a liquid state if its temperature t, in degrees Fahrenheit, satisfies the inequality $|t - 122| < 90$. Which graph represents the temperatures described by this inequality?

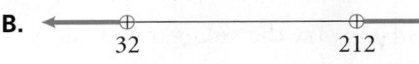

Short Response

86. A bicycling club is planning a trip. The graphs below show the number of miles three people want to cycle per day.

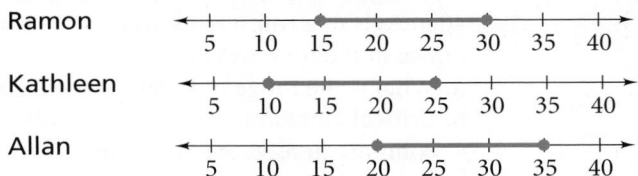

   a. Draw a graph showing a trip length that would be acceptable to all three bikers.
   b. Explain how your graph relates to the graphs above.

Mixed Review

Lesson 4-5

Write a compound inequality to model each situation.

87. Elevation in North America is between the highest elevation of 20,320 ft above sea level at Mount McKinley, Alaska, and the lowest elevation of 282 ft below sea level at Death Valley, California.

88. Normal body temperature t is within 0.6 degrees of 36.6°C.

Lesson 3-2

Solve each equation.

89. $3t + 4t = -21$ **90.** $9(-2n + 3) = -27$ **91.** $k + 5 - 4k = -10$

92. $5x + 3 - 2x = -21$ **93.** $5.4m - 2.3 = -0.5$ **94.** $3(y - 4) = 9$

Lesson 1-3

Write each group of numbers from least to greatest.

95. $3, -2, 0, -2.5, \pi$ **96.** $\frac{15}{2}, -1.5, -\frac{4}{3}, 7, -2$

97. $0.001, 0.01, 0.009, 0.011$ **98.** $-\pi, 2\pi, -2.5, -3, 3$

Sometimes, Always, Never

An equation or inequality is *sometimes* true, *always* true, or *never* true. To test the truth of an equation or inequality, you should try substituting different numerical values for the variables. Be sure to use a variety of numbers including 0, 1, fractions, and negative numbers.

> **NY** **A.RP.5** Construct logical arguments that verify claims or counterexamples that refute them.

EXAMPLE

Determine whether each statement is *sometimes*, *always*, or *never* true. Justify your reasoning.

a. $|x + y| = |x| + |y|$

Try the values $x = 3$ and $y = 5$.

$|3 + 5| \overset{?}{=} |3| + |5|$

$\quad |8| \overset{?}{=} 3 + 5$

$\quad\quad 8 = 8$ true ✓

Try the values $x = -3$ and $y = 5$.

$|-3 + 5| \overset{?}{=} |-3| + |5|$

$\quad\quad |2| \overset{?}{=} 3 + 5$

$\quad\quad\quad 2 \neq 8$ false ✗

The statement $|x + y| = |x| + |y|$ is *sometimes* true.

b. $|x| < x$

$|x| < x$, so the statement becomes $x < x$. A number cannot be less than itself. So the statement is *never* true.

EXERCISES

Determine whether each statement is *sometimes*, *always*, or *never* true. Justify your reasoning.

1. $x + x = x$

2. $|x| + |x| = 2x$

3. $x + 1 = 1 + x$

4. $3(x + 2) = 3x + 2$

5. $|xy| = |x \parallel y|$

6. $|x - y| = |x| - |y|$

7. $2 + x = x$

8. $2x = x$

9. $|x - 1| = 4$

10. $|x - 1| = -4$

11. $\dfrac{|x|}{x} > 0$

12. $|x - y| < 0$

13. $xy + 1 \geq 1$

14. $|xy| + 1 \geq 1$

15. $\left|\dfrac{x}{2}\right| \leq |x|$

16. $|x - 1| \leq |x|$

Some test items may ask you to interpret a situation displayed in a graph. To answer the questions correctly, you must read the data shown in the graph carefully.

EXAMPLE

The graph shows the number of hours Gloria worked over a seven-day period.

Based on the graph, which statement is NOT true?
- Ⓐ Gloria worked more than 21.5 hours over the course of the week.
- Ⓑ Gloria worked fewer hours on Wednesday than she did on Monday.
- Ⓒ Gloria worked more hours on Friday than she did on Monday, Tuesday, and Wednesday combined.
- Ⓓ Gloria never worked more than 8 hours on any single day.
- Ⓔ Gloria averaged more than 3 hours per day over the seven days.

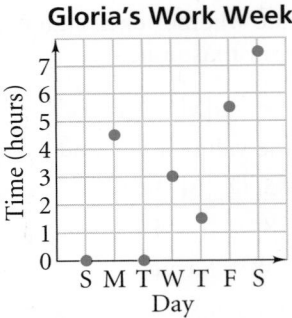

Gloria's Work Week

Read the values off the graph to test the statements.

A. Find the total number of hours Gloria worked.
Total hours = 0 + 4.5 + 0 + 3 + 1.5 + 5.5 + 7.5 = 22 ✓ Statement A is true.

B. Compare the number of hours Gloria worked on Monday and Wednesday.
3 < 4.5 ✓ Statement B is true.

C. Find the total hours Gloria worked on Monday, Tuesday, and Wednesday. Then compare with the hours she worked on Friday.
4.5 + 0 + 3 = 7.5, 5.5 > 7.5 ✗ Statement C is false.

D. The greatest number of hours Gloria worked on a single day was 7.5 hours on Saturday. 7.5 ≤ 8 ✓ Statement D is true.

E. Divide the total hours by the number of days.
$\frac{22}{7} = 3\frac{1}{7} > 3$ ✓ Statement E is true.

● Statement C is NOT true so C is the correct answer.

EXERCISE

1. Based on the graph shown above, which statement is true?
   - Ⓐ Gloria worked 3 hours more on Monday than she did on Wednesday.
   - Ⓑ If Gloria earns $7 per hour, she would have earned $140.50 over the course of the week.
   - Ⓒ Gloria worked all 7 days that week.
   - Ⓓ Gloria worked a total of 13 hours on her two longest workdays.
   - Ⓔ Gloria did not work on Saturday.

Chapter Review

🔊 compound inequalities (p. 227)   equivalent inequalities (p. 206)   solution of an inequality (p. 200)

Write the letter of the choice that correctly completes each sentence.

1. A solution of an inequality is any number that makes the inequality __?__ .
   A. complete   **B.** reverse direction
   C. true   **D.** false

2. An inequality is equivalent to another inequality if the two inequalities have __?__ .
   A. the same number of terms   **B.** the same graphs
   C. real-number solutions   **D.** no solutions

3. Compound inequalities are joined by __?__ .
   A. either the word *and* or the word *or*   **B.** the word *or*
   C. the word *and*   **D.** equations

4. Write the expression "absolute value of *x*" as __?__ .
   A. $[x]$ **B.** $-x$ **C.** $|x|$ **D.** $a = x$

5. A number's distance from 0 on a number line is the number's __?__ .
   A. solution of an inequality   **B.** compound form
   C. equivalent form   **D.** absolute value

Go Online
PHSchool.com
For: Vocabulary quiz
Web Code: atj-0451

Skills and Concepts

4-1 Objectives

▼ To identify solutions of inequalities (p. 200)

▼ To graph and write inequalities (p. 201)

A **solution of an inequality** is any number that makes the inequality true. A graph can indicate all the solutions of an inequality. On the graph, a closed dot indicates that the number is a solution. An open dot indicates that the number is *not* a solution.

Graph each inequality.

   6. $x > 3$ **7.** $m \leq -5$ **8.** $10 \geq p$ **9.** $r < 2.5$

Write an inequality for each graph.

10. ◄—+—+—⊕—+—+—+—►
       $-4 \ -3 \ -2 \ -1 \quad 0 \quad 1$

11. ◄—+—+—●—+—+—+—+—►
       $-5 \ -4 \ -3 \ -2 \ -1 \quad 0 \quad 1$

12. ◄—+—+—⊕—+—+—+—►
       $-10 \ -8 \ -6 \ -4 \ -2 \quad 0 \quad 2$

13. ◄—+—+—+—+—●—+—+—►
       $-2 \ -1 \quad 0 \quad 1 \quad 2 \quad 3 \quad 4$

Define a variable and write an inequality to model each situation.

14. At least 600 people attended a school play.

15. An elevator can carry at most 15 people.

16. The temperature was less than 32°F.

4-2 and 4-3 Objectives

▼ To use addition to solve inequalities (p. 206)
▼ To use subtraction to solve inequalities (p. 207)
▼ To use multiplication to solve inequalities (p. 212)
▼ To use division to solve inequalities (p. 214)

To solve inequalities, you may need to find a simpler, equivalent inequality. **Equivalent inequalities** have the same solution. You can add, subtract, multiply, or divide both sides of an inequality by the same number to find a simpler equivalent inequality. Multiplying or dividing by a negative number causes the direction of the inequality symbol to be *reversed*.

Properties of Inequality

For all real numbers a, b, and c:

- If $a > b$, then $a + c > b + c$ and $a - c > b - c$.
- If $a < b$, then $a + c < b + c$ and $a - c < b - c$.
- If $a > b$ and $c > 0$, then $ac > bc$ and $\frac{a}{c} > \frac{b}{c}$.
- If $a < b$ and $c > 0$, then $ac < bc$ and $\frac{a}{c} < \frac{b}{c}$.
- If $a > b$ and $c < 0$, then $ac < bc$ and $\frac{a}{c} < \frac{b}{c}$.
- If $a < b$ and $c < 0$, then $ac > bc$ and $\frac{a}{c} > \frac{b}{c}$.

These properties are also true for inequalities involving $\leq$ and $\geq$.

Solve each inequality. Graph and check the solution.

17. $h + 3 > 2$ **18.** $t - 4 < -9$ **19.** $8m \geq -24$ **20.** $-6w \leq 12$

21. $q + 0.5 > -2$ **22.** $y - 8 > -22$ **23.** $-\frac{3}{5}n \geq -9$ **24.** $\frac{5}{8}d \geq \frac{5}{2}$

25. $0 \leq 2 + t$ **26.** $0 < 4c$ **27.** $-0.3 \geq u - 2.8$ **28.** $3 > -\frac{1}{3}p$

29. Weekly Budget You have an allowance of $12.00. You buy a discount movie ticket that costs at least $3.50 and popcorn that costs $2.75. Write and solve an inequality to find how much you have for other spending.

30. Jobs Suppose you earn $7.25 per hour working part-time as a florist. Write and solve an inequality to find how many full hours you must work to earn at least $200.

4-4 Objectives

▼ To solve multi-step inequalities with variables on one side (p. 219)
▼ To solve multi-step inequalities with variables on both sides (p. 221)

When you solve equations, sometimes you need to use more than one step. The same is true for inequalities. Many inequalities have variables on both sides of the inequality symbol. You need to gather the variable terms on one side of the inequality and the constant terms on the other side.

Solve each inequality. Check your solution.

31. $3n + 5 > -1$ **32.** $4k - 1 \leq -3$ **33.** $\frac{5}{8}b < 25$

34. $6(c - 1) \leq -18$ **35.** $3m > 5m + 12$ **36.** $t - 4t < -9$

37. $0.5x - 2 \geq -4x + 7$ **38.** $-\frac{6}{7}y - 6 \geq 42$ **39.** $4 + \frac{x}{2} > 2x$

40. Commission Trenton sells electronic supplies. Each week he earns $190 plus a commission equal to 4% of his sales. This week his goal is to earn no less than $500. Write and solve an inequality to find the amount of sales he must have to reach his goal.

4-5 Objectives

▼ To solve and graph inequalities containing *and* (p. 227)

▼ To solve and graph inequalities containing *or* (p. 229)

Two inequalities that are joined by the word *and* or the word *or* are called **compound inequalities.** A solution of a compound inequality joined by *and* makes both inequalities true. A solution of a compound inequality joined by *or* makes either inequality true. A number sentence with two inequality symbols, such as $a < x < b$ represents the compound inequality $a < x$ and $x < b$.

Graph each compound inequality.

41. $x > -3$ and $x < 2$ **42.** $m < -2$ or $m \geq 1$ **43.** $-3 \leq k < 4$

Solve each compound inequality and graph the solutions.

44. $-3 \leq z - 1 < 3$ **45.** $-2 \leq d + \frac{1}{2} < 4\frac{1}{2}$ **46.** $0 < -8b \leq 12$

47. $2t \leq -4$ or $7t \geq 49$ **48.** $-1 \leq a - 3 < 2$ **49.** $-2 \leq 3a - 8 < 4$

50. Climate In Miami, Florida, July's average high temperature is 89°F. July's average low temperature is 75°F. Write a compound inequality to represent Miami's average temperature in July.

4-6 Objectives

▼ To solve equations that involve absolute value (p. 235)

▼ To solve inequalities that involve absolute value (p. 236)

Recall that the absolute value of a number is its distance from 0 on a number line. Since absolute value represents distance, it can never be negative.

Solving Absolute Value Equations and Inequalities

- To solve an equation in the form $|A| = b$, where A represents a variable expression and $b > 0$, solve the equations $A = b$ or $A = -b$.
- To solve an inequality in the form $|A| < b$, where A represents a variable expression and $b > 0$, solve $-b < A < b$.
- To solve an inequality in the form $|A| > b$, where A represents a variable expression and $b > 0$, solve $A < -b$ or $A > b$.

Similar rules are true for $|A| \leq b$ and $|A| \geq b$.

Write an absolute value inequality that represents each set of numbers.

51. all numbers n that are more than 3 units from -2

52. all numbers n that are within 5 units of 12

Solve each equation or inequality.

53. $|y| = 5$ **54.** $|n + 2| \geq 4$ **55.** $|-5x| \leq 15$

56. $|\frac{1}{2}m| < 4.8$ **57.** $|2x - 7| - 1 > 0$ **58.** $|p + 3| = 9.5$

59. $|k - 8| = 0$ **60.** $|3x + 5| > -2$ **61.** $|6 - b| = -1$

62. $4|k + 5| > 8$ **63.** $4 + |r + 2| = 7$ **64.** $-2 + |3.6z| \geq -1.1$

65. Manufacturing The ideal diameter of a steel reinforcement rod is 2.8 cm. The actual diameter may vary from the ideal by at most 0.06 cm. Find the range of acceptable diameters for this steel rod.

66. Manufacturing The ideal length of a certain nail is 20 mm. The actual length can vary from the ideal by at most 0.4 mm. Find the range of acceptable lengths of the nail.

Chapter Test

Determine whether each number is a solution of the given inequality.

1. $4z + 7 \geq 15$ **a.** -2 **b.** 2   **c.** 5

2. $-2g + 3 > 5$ **a.** -3 **b.** -1 **c.** 4

Define a variable and write an inequality to model each situation.

3. A student can take at most 7 classes.

4. The school track team needs at least 5 runners to compete at Saturday's meet.

5. Elephants can drink up to 40 gallons of water at a time.

6. Your cousin's early-morning paper route has more than 32 homes.

Write an inequality for each graph.

7.

8.

9.

10.

Solve each inequality. Graph the solution.

11. $z + 7 \leq 9$

12. $-16 \geq 4y$

13. $-\frac{1}{3}x < 2$

14. $8 - u > 4$

15. $-5 + 4t \leq 3$

16. $5w \geq -6w + 11$

17. $-\frac{7}{2}m < 14$

18. $6y - 7 < -2y + 13$

19. $|x - 5| \geq 3$

20. $|2h + 1| < 5$

21. $9 \leq 6 - b < 12$

22. $-10 < 4q < 12$

23. $4 + 3n \geq 1$ or $-5n > 25$

24. $10k < 75$ and $4 - k \leq 0$

Solve each inequality. Check your solution.

25. $3(d - 1) > -4$

26. $5(-2 + b) < 3b + 2$

27. $3(m + 3) + 4 \leq 15$

28. $0.5(x + 3) - 2.1 \geq -1$

Write a compound inequality that each graph could represent.

29.

30.

Solve each equation. Check your solution.

31. $|4k - 2| = 11$

32. $23 = |n + 10|$

33. $|3c + 1| - 4 = 13$

34. $4|5 - t| = 20$

35. Writing Explain why the solution to $ax - 1 < 3$ is not $x < \frac{4}{a}$. Use solutions of the inequality with different values of a to support your explanation.

36. Open-Ended Write an absolute value inequality that has 3 and -5 as two of its solutions.

37. Community Service The chart below shows the number of cans of food collected by a club during the first four weeks of a food drive.

Food Drive

Week	Number of Cans
1	702
2	470
3	492
4	547

The goal is to collect at least 3000 cans in 5 weeks. Write and solve an inequality to find how many cans should be collected during Week 5 to meet or exceed the goal.

38. Safe Load A freight elevator can safely hold no more than 2000 pounds. An elevator operator must take 55-pound boxes to a storage area. If he weighs 165 pounds, how many boxes can he safely move at one time?

39. Manufacturing A manufacturer is cutting plastic sheets to make rectangles that are 11.125 in. by 7.625 in. Each rectangle's length and width must be within 0.005 in. of the desired size. Write and solve inequalities to find the acceptable range for the length ℓ and for the width w.

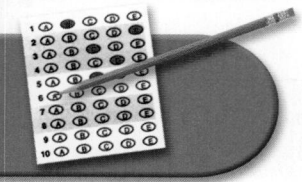

Regents Test Prep

Multiple Choice

For Exercises 1–14, choose the correct letter.

1. A store owner has a bicycle priced at $100. She raises the price 10%. During a sale, she then lowers the price 10%. What is the new price of the bicycle? **3**
   - **(1)** $101
   - **(2)** $100
   - **(3)** $99
   - **(4)** $98

2. A bag contains 8 green marbles and 12 blue marbles. You draw a marble, return it to the bag, and draw another. What is the probability of drawing two green marbles?
   - **(1)** $\frac{4}{25}$
   - **(2)** $\frac{4}{9}$
   - **(3)** $\frac{7}{50}$
   - **(4)** $\frac{4}{5}$

3. Which equation does *not* have the same solution as $\frac{7}{y} = \frac{31}{36}$?
   - **(1)** $\frac{7}{31} = \frac{y}{36}$
   - **(2)** $7 \cdot 36 = 31y$
   - **(3)** $\frac{y}{36} = \frac{7}{31}$
   - **(4)** $\frac{36}{31} = \frac{7}{y}$

4. The number of subscribers to a magazine fell from 210,000 to 190,000. Find the approximate percent of decrease.
   - **(1)** 5%
   - **(2)** 10%
   - **(3)** 20%
   - **(4)** 90%

5. Which are solutions of $3(x - 4) \le 18$ and $2(x - 1) \ge 6$?
   - I. 9
   - II. 12
   - III. 15
   - **(1)** I only
   - **(2)** II only
   - **(3)** I and III
   - **(4)** II and III

6. In February, rent increased from $875 per month to $915 per month. What was the percent of increase to the nearest percent?
   - **(1)** 4%
   - **(2)** 5%
   - **(3)** 40%
   - **(4)** 50%

7. The solutions of which inequality are $x > 9$ or $x < -1$?
   - **(1)** $|x + 4| < 5$
   - **(2)** $|x - 4| < 5$
   - **(3)** $|x + 4| > 5$
   - **(4)** $|x - 4| > 5$

8. Which equation has no solution?
   - **(1)** $4\left(\frac{1}{2}x - 1\right) = \frac{1}{2}$
   - **(2)** $5 - 6x = 2(1 - 3x)$
   - **(3)** $6(2x - 3) = 2(6x - 9)$
   - **(4)** $7(x + 8) - x = 0$

9. Simplify $15 - 5 \times 2 + 4^2$.
   - **(1)** 13
   - **(2)** 21
   - **(3)** 28
   - **(4)** 36

10. Solve $8x - 2y = 3$ for y.
    - **(1)** $y = -\frac{5x}{2}$
    - **(2)** $y = \frac{3 - 8x}{2}$
    - **(3)** $y = \frac{8x - 3}{2}$
    - **(4)** $y = \frac{8x + 3}{2}$

11. What percent of 280 is 350?
    - **(1)** 125%
    - **(2)** 80%
    - **(3)** 25%
    - **(4)** 20%

12. Between which two integers is $\sqrt{45}$?
    - **(1)** 22 and 23
    - **(2)** 45 and 46
    - **(3)** 7 and 8
    - **(4)** 6 and 7

13. Simplify $-6(2t - 8)$.
    - **(1)** $-12t + 48$
    - **(2)** $-6t + 48$
    - **(3)** $-4t - 8$
    - **(4)** $-12t - 48$

14. Which expression is equivalent to $3a + 2 + (3 - a)$?
    - **(1)** $2a + 5$
    - **(2)** $4a + 5$
    - **(3)** $2a + 6$
    - **(4)** $4a + 6$

Short Response

Find each answer.

15. A CD player that normally costs $225 would cost an employee $180. What is the percent of the employee's discount?

16. The two rectangles are similar. Find the perimeter of the smaller rectangle in centimeters.

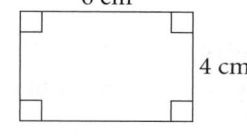

17. A right triangle has legs which are 7 in. and 9 in. Find the length of the hypotenuse to the nearest tenth of an inch.

Show your work.

18. Suppose you earn $80 a week at your summer job. Your employer offers you a $20 raise or a 20% raise. Which should you take? Explain.

19. Two boats leave a ramp traveling in opposite directions. The second boat is 10 miles per hour faster than the first. After 3 hours they are 150 miles apart. Find the speeds of the boats.

Area and Perimeter of Composite Figures

You can sometimes find the area of a composite figure by separating it into figures with areas you know how to find.

NY NY A.G.1 Find the area and/or perimeter of figures composed of polygons and circles or sectors of a circle.

1 EXAMPLE Area of Perimeter and Composite Figures

Multiple Choice A pound of grass seed covers approximately 675 ft². Grass seed comes in 3-lb bags. Find the area of the lawn at right. How many bags of grass seed do you need to buy to cover the lawn?

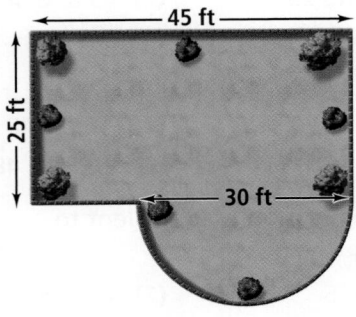

(1) 4 bags   (2) 3 bags   (3) 2 bags   (4) 1 bag

Area of region that is one half of a circle

area of circle $= \pi r^2$

area of half circle $= \frac{1}{2}\pi r^2$

$A \approx \frac{1}{2}(3.14)(15)^2$ **Replace π with 3.14 and r with 15.**

$A = 353.25$

Area of region that is a rectangle

area of rectangle $= bh$

$A = 45 \cdot 25$ **Replace b with 45 and h with 25.**

$A = 1,125$

The area of the lawn is about 353 ft² + 1,125 ft² = 1,478 ft².

$1,478 \div 675 \approx 2.19$ **Divide to find the amount of seed.**

● You need to buy one 3-lb bag of grass seed. The answer is 4.

When finding the perimeter of a composite figure you need to decide which sides or sections of the figure to include and which not to include.

② EXAMPLE **Perimeter of Composite Figures**

Find the perimeter of the figure to the nearest centimeter

Perimeter of region that is a rectangle

perimeter of rectangle = $2l + 2w$

$P = 2(40) + 2(20)$ **Find the perimeter of the entire rectangle**

$P = 120 - 20$ **Subtract the side that is composed of the half circle.**

$P = 100$

Circumference of region that is one half of a circle

circumference of a circle = $2\pi r$

circumference of half circle = πr

$C = (3.14)(10)$ **Replace π with 3.14 and r with 10.**

$C = 31.4$

● The perimeter of the composite figure is 131 cm.

EXERCISES

Find the perimeter of each figure to the nearest unit.

1.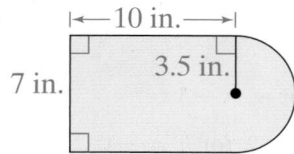

2.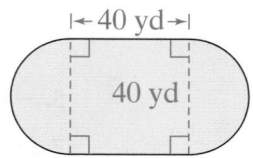

Find the area of each shaded region to the nearest square unit.

3.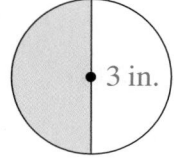

4.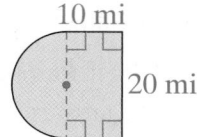

5.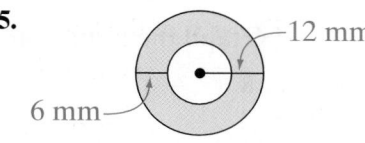

6. Lawn Care A groundskeeper wants to use sod to cover the lawn in green in the diagram. Each piece of sod covers 3 ft². About how many pieces of sod are needed?

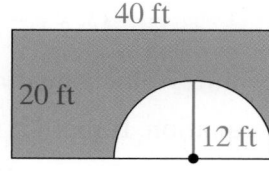

What You've Learned

NY Learning Standards for Mathematics

In Chapter 1, you

- **NY A.A.5:** represented relationships using variables, and explored function patterns.

In Chapter 3, you

- **NY A.A.26:** used methods for solving equations in one variable and applied those methods to solving proportions.

In Chapter 4, you

- **NY A.A.24:** solved inequalities.

 **Check Your Readiness**

 for Help to the Lesson in green.

Writing Equations (Lesson 1-1)

Define a variable and write an equation to model each situation.

1. The total price is the number of pens times $.59.

2. The tower is 200 feet taller than the house.

3. What is the perimeter of an equilateral triangle?

Evaluating Expressions (Lesson 1-2)

Evaluate each expression.

4. $3x - 2y$, for $x = -1$ and $y = 2$

5. $-w^2 + 3w$, for $w = -3$

6. $\frac{3 + k}{k}$, for $k = 3$

7. $h - (h^2 - 1) \div 2$, for $h = -1$

Graphing on the Coordinate Plane (Review page 24)

Graph the points on the same coordinate plane.

8. $(3, -3)$ **9.** $(0, -5)$ **10.** $(-2, 2)$ **11.** $(-2, 0)$

Using Cross Products (Lesson 3-4)

Solve the following proportions.

12. $\frac{4}{w} = \frac{5}{8}$ **13.** $\frac{c}{2.2} = \frac{3}{11}$ **14.** $\frac{4}{0.5} = \frac{36}{p}$ **15.** $-\frac{29}{2} = \frac{d}{4}$

Solving Absolute Value Equations (Lesson 4-6)

Solve each equation. If there is no solution, write *no solution*.

16. $|r + 2| = 2$ **17.** $-3|d - 5| = -6$ **18.** $-3.2 = |8p|$

Graphs and Functions

What You'll Learn Next

NY Learning Standards for Mathematics

● **NY A.A.5:** You will move from the specific case of equations in one variable to the study of functions in two variables.

● **NY A.G.4:** You will study function rules, and model data using equations, tables, and graphs.

Data Analysis **Activity Lab** Extending your knowledge and understanding of data analysis, you will interpret and construct histograms, on pages 304–305.

251

Relating Graphs to Events

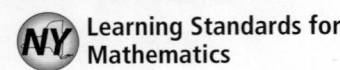
Learning Standards for Mathematics

A.CN.3 Model situations mathematically, using representations to draw conclusions and formulate new situations.

✓ **Check Skills You'll Need**

GO for Help Review page 24

Use the graph at the right.

Name the point with the given coordinates.

1. $(4, -2)$ **2.** $(4, 3)$

3. $(2, -4)$ **4.** $(-2, 1)$

Name the coordinates of each given point.

5. B **6.** F **7.** G

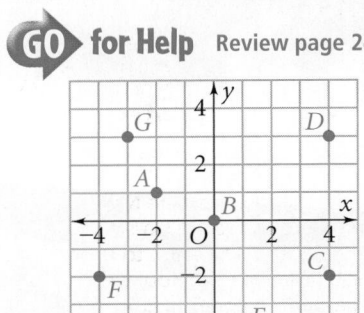

---

1   Interpreting, Sketching, and Analyzing Graphs

You can use an equation, an inequality, or a proportion to make a statement about a variable. You can use a graph to show the relationship between two variables. For example, you can use a graph to show how a quantity changes over time.

1 EXAMPLE   Interpreting Graphs

Commute One student walks and takes a bus to get from school to home each day. The graph at the right shows the student's commute by relating the time the student spends commuting and the distance he travels.

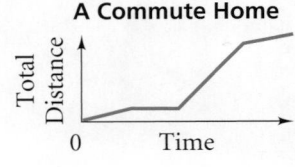

Describe what the graph shows by labeling each part.

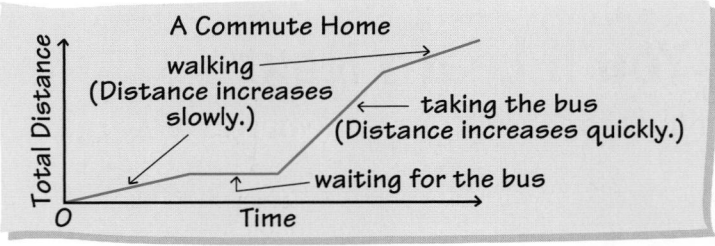

✓ **Quick Check**

1 The graph at the right shows a trip from home to school and back. The trip involves walking and getting a ride from a neighbor. Copy the graph and label each section.

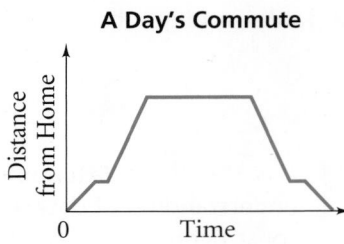

In Example 1, distance, which is on the vertical axis, depends on time, which is on the horizontal axis. When one quantity depends on another, show the dependent quantity on the vertical axis.

2 EXAMPLE Sketching a Graph

Travel A plane is flying from New York to London. Sketch a graph of the plane's altitude during the flight. Label each section.

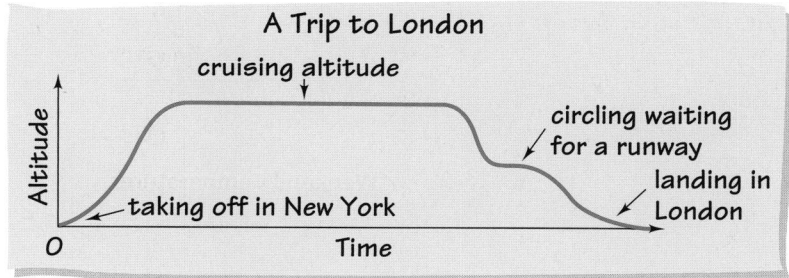

Real-World Connection

The Statue of Liberty, in New York harbor, is 151 ft from base to torch. The clock tower, which is part of the Houses of Parliament in London, is 320 ft tall.

Quick Check ② Sketch a graph of the distance from a child's feet to the ground as the child jumps rope. Label each section.

Most of the graphs in this lesson do not have numbers along the axes. You can analyze a graph based on the shape of the graph alone.

3 EXAMPLE Relating Graphs to Situations

Multiple Choice Suppose you pour water into the container at a steady rate. Which graph shows the change in the height of the liquid in the container over time?

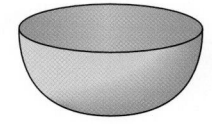

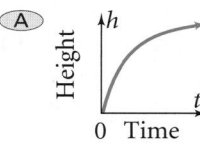
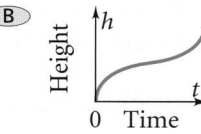
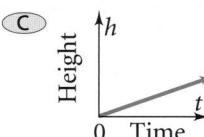
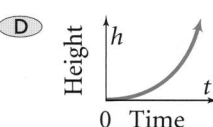

Test-Taking Tip

Even if the first choice appears to be the correct answer, review the other choices so that you can be certain of your answer.

The rate that water rises will decrease steadily because the container gets wider from the bottom to the top. So, A is the correct answer.

Quick Check ③ The graph at the right shows the time and distance of a moving object. Which of the following situations could be described by the graph?

F A car travels at a steady speed.

G A cyclist slows down as she rides up a hill and speeds up as she peddles over the top.

H A train slows down as it arrives at the station.

J A plane accelerates steadily down the runway until it takes off.

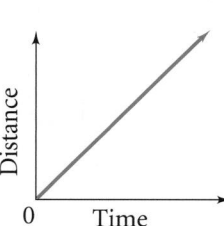

EXERCISES

For more exercises, see *Extra Skill and Word Problem Practice.*

Practice and Problem Solving

A Practice by Example

Example 1
(page 252)

Copy each graph. Label each section of the graph.

1. **Exercising**

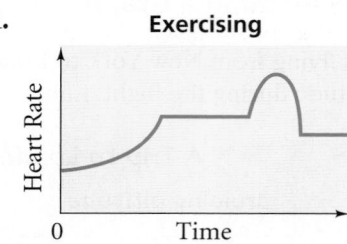

2. **Checking Account**

3. **Weekend Temperatures**

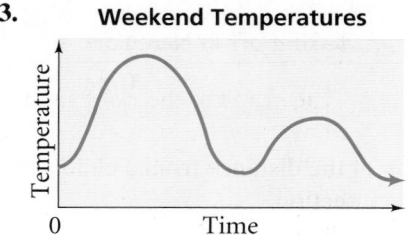

4. **Hair Length**

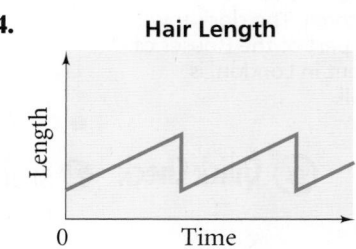

Example 2
(page 253)

Sketch a graph of each situation. Label each section.

5. hours of daylight over the course of one year

6. your distance from the ground as you ride a Ferris wheel for five minutes

7. your pulse rate as you watch a scary movie

8. your walking speed during five minutes between classes

Example 3
(page 253)

9. **Cooking** You turn on your oven to bake a casserole. Which graph best represents the oven temperature over time? Explain your choice.

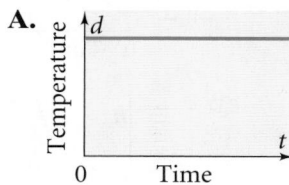
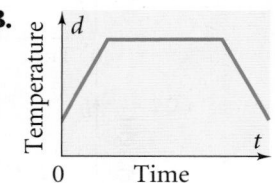
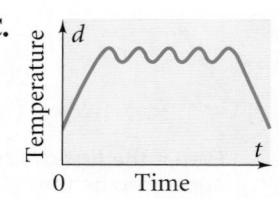

B Apply Your Skills

10. **Weather** The graph shows the barometric pressure in Pittsburgh, Pennsylvania, during a blizzard. Describe what happened to the pressure during the storm.

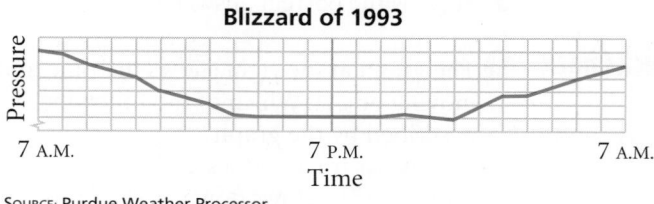

Blizzard of 1993

SOURCE: Purdue Weather Processor

11. Sketch graphs of each situation. Are the graphs the same? Explain.
   a. Your speed as you travel from the bottom of a ski slope to the top.
   b. Your speed as you travel from the top of a ski slope to the bottom.

Weight Gains

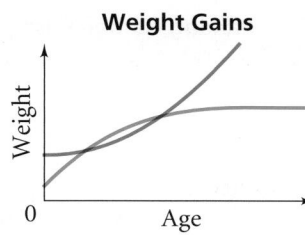

Weight

0   Age

12. The graph at the left shows the weight of a baby and the weight of a puppy for their first two years.
  a. Which curve represents the puppy's weight? The baby's weight?
  b. Writing Describe the growth patterns of the baby and the puppy.

13. You pour juice into a pitcher like the one shown in the photographs below. You pour the juice at a constant rate. Make a sketch to show the height of juice in the pitcher as you fill it.

14. Error Analysis The graph at the right shows a person's speed over the course of a bike ride. Your friend said that this graph describes a person bicycling up and then down a hill. Explain your friend's error.

Speed on a Bike Ride

Speed

0   Time

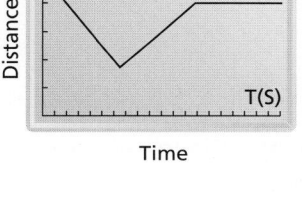

DIST(H)

Distance

T(S)

Time

15. A student used a graphing calculator, a data collector, and a motion detector to make the graph at the left, which shows a classmate's distance from the motion detector.
  a. Copy the graph and label each section.
  b. During which section was the student walking toward the motion detector?
  c. During which section(s) was the student walking at a constant speed?

16. Multiple Choice Which graph best represents a person's height from birth to age 80?

Ⓐ   h
Height
0   Age   a

Ⓑ   h
Height
0   Age   a

Ⓒ   h
Height
0   Age   a

Ⓓ   h
Height
0   Age   a

17. a. Data Collection Sketch a graph of the daily high temperatures over the course of one year for your town.
  b. Critical Thinking How would your graph be different if you lived at the equator?

GO Online
Homework Video Tutor
Visit: PHSchool.com
Web Code: ate-0501

18. a. In-Line Skating Describe what the graph at the right shows about a student's in-line skating experience.
  b. Label each section.

In-Line Skating After School

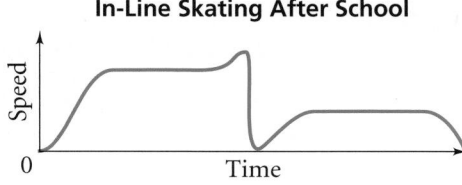

Speed

0   Time

C Challenge

Use the graph at the right for Exercises 19–22.

19. How much does it cost to park for 2 hours?

20. How much does it cost to park for 121 minutes?

21. Suppose your mother pays $6 for parking. About how long was her car parked in the garage?

22. Vocabulary This graph is a *step graph*. Does this name make sense? Explain.

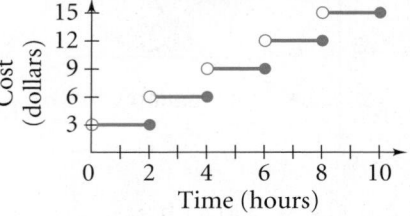

Parking Garage Costs

NY REGENTS

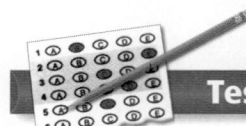

Test Prep

Multiple Choice

The graph at the right shows the distance Molly was from home throughout Tuesday. She spent about six hours at school, one hour at a friend's house, and about 30 minutes waiting for a bus. She also walked and rode the bus part of the day. Use the graph for Exercises 23–24.

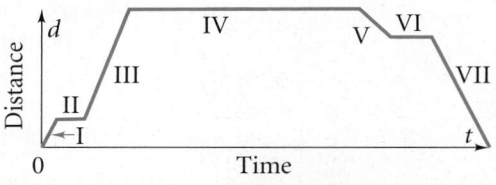

23. What section most likely represents walking to her friend's house?
  A. I   **B.** II   **C.** IV   **D.** V

24. What section most likely represents spending time at a friend's house?
  F. II   **G.** IV   **H.** VI   **J.** VII

Short Response

25. An hourglass has two compartments that hold sand. The graph at the right shows the height of the sand in the bottom container as it fills. Which hourglass does the graph represent? Explain your choice.

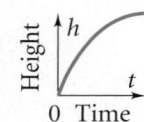

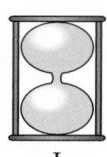
I

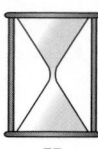
II

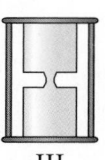
III

IV

Mixed Review

GO for Help

Lesson 4-4

Solve each inequality.

26. $5x + 2 < 37$ **27.** $x + 4 > 2x - 4$ **28.** $8x + 4 - 3x \geq 3x$

29. $7 > -4x - 9$ **30.** $7(x + 1) \leq 6(x - 1)$ **31.** $-2 + 5x < 8 - 10x$

Lesson 2-7

You roll a red number cube and a blue number cube. Find each probability.

32. $P(\text{red 1 and blue 6})$ **33.** $P(\text{red 3 and blue 5})$ **34.** $P(\text{red} > 4 \text{ and blue 5})$

35. $P(\text{red odd and blue 3})$ **36.** $P(\text{red and blue even})$ **37.** $P(\text{red and blue equal})$

5-2

Relations and Functions

NY Learning Standards for Mathematics

A.G.3 Determine when a relation is a function, by examining ordered pairs and inspecting graphs of relations.

✓ Check Skills You'll Need

GO for Help Review page 24 and Lesson 1-2

Graph each point on a coordinate plane.

1. $(2, -4)$ **2.** $(0, 3)$ **3.** $(-1, -2)$ **4.** $(-3, 0)$

Evaluate each expression.

5. $3a - 2$ for $a = -5$ **6.** $9(x - 9)$ for $x = 3$ **7.** $3x^2$ for $x = 6$

🔊 **New Vocabulary** • relation • vertical-line test • function notation

---

1 Identifying Relations and Functions

A **relation** is a set of ordered pairs. The (age, height) ordered pairs below form a relation.

Giraffes

Age (years)	18	20	21	14	18
Height (meters)	4.25	4.40	5.25	5.00	4.85

You can list the set of ordered pairs in a relation using braces.

$$\{(18, 4.25), (20, 4.40), (21, 5.25), (14, 5.00), (18, 4.85)\}$$

Recall from Lesson 1-4 that a function is a relation that assigns exactly one output (range) value for each input (domain) value.

One way you can tell if a relation is a function is by making a *mapping diagram*. List the domain values and the range values in order. Draw arrows from the domain values to their range values.

Online active math

For: Function Activity
Use: Interactive Textbook, 5-2

1 EXAMPLE Using a Mapping Diagram

Determine whether each relation is a function.

a. $\{(11, -2), (12, -1), (13, -2), (20, 7)\}$

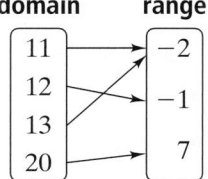

There is no value in the domain that corresponds to more than one value of the range.

b. $\{(-2, -1), (-1, 0), (6, 3), (-2, 1)\}$

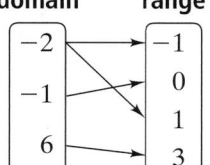

The domain value -2 corresponds to two range values, -1 and 1.

● The relation is a function.   The relation is not a function.

✓ Quick Check

1 Use a mapping diagram to determine whether each relation is a function.

a. $\{(3, -2), (8, 1), (9, 2), (3, 3), (-4, 0)\}$ **b.** $\{(6.5, 0), (7, -1), (6, 2), (2, 6), (5, -1)\}$

Another way you can tell whether a relation is a function is to analyze the graph of the relation using the **vertical-line test.** If any vertical line passes through more than one point of the graph then for some value of x there is more than one value of y. Therefore, the relation is not a function.

② EXAMPLE Using the Vertical-Line Test

Determine whether the relation $\{(3,0),(-2,1),(0,-1),(-3,2),(3,2)\}$ is a function.

Step 1 Graph the ordered pairs on a coordinate plane.

Step 2 Pass a pencil across the graph as shown.

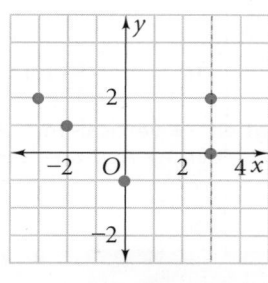
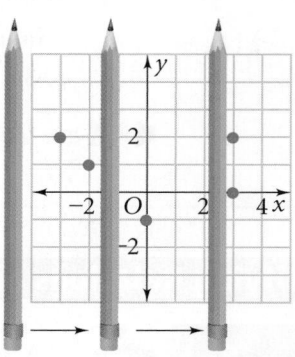

● A vertical line would pass through $(3,0)$ and $(3,2)$. The relation is not a function.

✓ **Quick Check** ② Use the vertical-line test to determine whether each relation is a function.
  a. $\{(4,-2),(1,2),(0,1),(-2,2)\}$ **b.** $\{(0,2),(1,-1),(-1,4),(0,-3),(2,1)\}$

② Evaluating Functions

Recall from Lesson 1-4 that a function rule is an equation that describes a function. You can think of a function rule as an input-output machine.

Words and Notations Used With a Function

Domain	Range
input	output
x	$f(x)$
x	y

The domain is the set of input values.

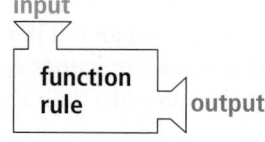

The range is the set of output values.

If you know the input values, you can use a function rule to find the output values. The output values depend on the input values.

$$\underset{\uparrow \text{ output}}{y} = \underset{\uparrow \text{ input}}{3x} + 4$$

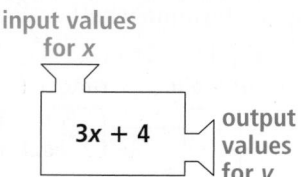

Input	Output
x	y
1	7
2	10
3	13

Another way to write the function $y = 3x + 4$ is $f(x) = 3x + 4$. A function is in **function notation** when you use $f(x)$ to indicate the outputs. You read $f(x)$ as "f of x" or "f is a function of x." The notations $g(x)$ and $h(x)$ also indicate functions of x.

In Lesson 1-4 you wrote function rules from tables. You can also make a table using values from a function rule.

3 EXAMPLE **Making a Table From a Function Rule**

Make a table for $f(n) = -2n^2 + 7$. Use 1, 2, 3, and 4 as domain values.

Vocabulary Tip

You can think of the notation $f(6)$ as "Replace n with 6 to find the value of $f(6)$."

n	$-2n^2 + 7$	$f(n)$
1	$-2(1)^2 + 7$	5
2	$-2(2)^2 + 7$	-1
3	$-2(3)^2 + 7$	-11
4	$-2(4)^2 + 7$	-25

✓ **Quick Check** ❸ Make a table for $y = 8 - 3x$. Use 1, 2, 3, and 4 as domain values.

You can use a function rule and a given domain to find the range of the function. After computing the range values, write the values in order from least to greatest.

4 EXAMPLE **Finding the Range**

Evaluate the function rule $f(a) = -3a + 5$ to find the range of the function for the domain $\{-3, 1, 4\}$.

$f(a) = -3a + 5$	$f(a) = -3a + 5$	$f(a) = -3a + 5$
$f(-3) = -3(-3) + 5$	$f(1) = -3(1) + 5$	$f(4) = -3(4) + 5$
$f(-3) = 14$	$f(1) = 2$	$f(4) = -7$

The range is $\{-7, 2, 14\}$.

✓ **Quick Check** ❹ Find the range of each function for the domain $\{-2, 0, 5\}$.
   a. $f(x) = x - 6$ **b.** $y = -4x$ **c.** $g(t) = t^2 + 1$

EXERCISES

For more exercises, see *Extra Skill and Word Problem Practice.*

Practice and Problem Solving

 **Practice by Example**

Example 1
(page 257)

 **GO for Help**

Use a mapping diagram to determine whether each relation is a function.

1. $\{(3, 7), (3, 8), (3, -2), (3, 4), (3, 1)\}$ **2.** $\{(6, -7), (5, -8), (1, 4), (5, 5)\}$
3. $\{(0.04, 0.2), (0.2, 1), (1, 5), (5, 25)\}$ **4.** $\{(4, 2), (1, 1), (0, 0), (1, -1), (4, -2)\}$

Example 2
(page 258)

Use the vertical-line test to determine whether each relation is a function.

5. $\{(2, 5), (3, -5), (4, 5), (5, -5)\}$ **6.** $\{(5, 0), (0, 5), (5, 1), (1, 5)\}$
7. $\{(3, -1), (-2, 3), (-1, -5), (3, 2)\}$ **8.** $\{(-2, 9), (3, 9), (-0.5, 9), (4, 9)\}$

Example 3
(page 259)

Make a table for each function. Use 1, 2, 3, and 4 for the domain.

9. $f(x) = x + 7$ **10.** $y = 11x - 1$ **11.** $f(x) = x^2$ **12.** $f(x) = -4x$

13. $f(x) = 15 - x$ **14.** $y = 3x + 2$ **15.** $y = \frac{1}{4}x$ **16.** $f(x) = -x + 2$

Example 4
(page 259)

Find the range of the function rule $y = 5x - 2$ for each domain.

17. $\{0.5, 11\}$ **18.** $\{-1.2, 0, 4\}$ **19.** $\{-5, -1, 0, 2, 10\}$ **20.** $\left\{-\frac{1}{2}, \frac{1}{4}, \frac{2}{5}\right\}$

B **Apply Your Skills**

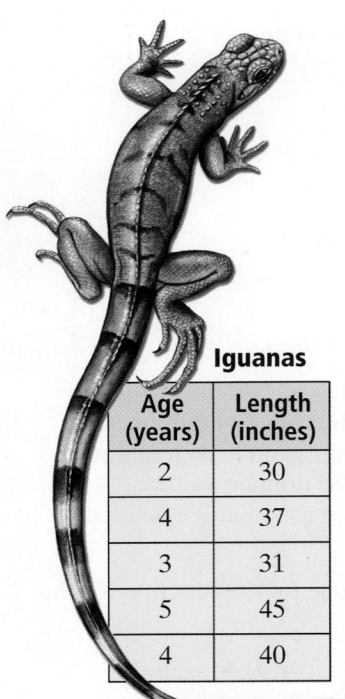

Iguanas

Age (years)	Length (inches)
2	30
4	37
3	31
5	45
4	40

Determine whether each relation is a function. If the relation is a function, state the domain and range.

21.

x	y
1	−3
6	−2
9	−1
1	3

22.

x	y
0	2
3	1
3	−1
5	3

23.

x	y
−4	−4
−1	−4
0	−4
3	−4

24. Error Analysis A student thinks that the relation $\{(2, 1), (3, −2), (4, 5), (5, −2)\}$ is not a function because two values in the domain have the same range value. What is the student's error?

25. Iguanas Use the data in the table at the left. Is an iguana's length a function of its age? Explain.

26. Open-Ended Create a data table for a relation that is *not* a function. Describe what your data might represent.

Find the range of each function for the domain $\{−1, 0.5, 3.7\}$.

27. $f(x) = 4x + 1$　**28.** $g(x) = −4x + 1$　**29.** $y = |x| − 1$　**30.** $s(t) = t^2 − 1$

31. a. Profit A store bought a case of disposable cameras for $300. The store's profit p on the cameras is a function of the number c of cameras sold. Find the range of the function $p = 6c − 300$ when the domain is $\{0, 15, 50, 62\}$.

b. Critical Thinking In this situation, what do the domain and range represent?

Determine whether each graph is the graph of a function.

32.
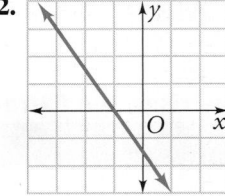

33.
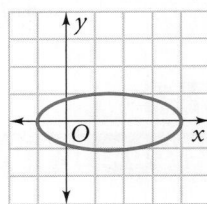

34.
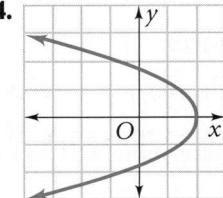

35.
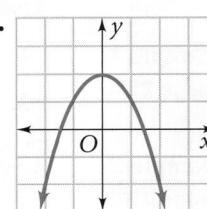

36. Physics Light travels about 186,000 miles per second. The rule $d = 186,000t$ describes the relationship between distance d in miles and time t in seconds.
a. How far does light travel in 20 seconds?
b. How far does light travel in 1 minute?

GO **Online**
Homework Video Tutor
Visit: PHSchool.com
Web Code: ate-0502

For Exercises 37–40 assume that each variable has a different value. Determine whether each relation is a function.

37. $\{(a, b), (b, a), (c, c), (e, d)\}$

38. $\{(b, b), (c, d), (d, c), (c, a)\}$

39. $\{(c, e), (c, d), (c, b)\}$

40. $\{(a, b), (b, c), (c, d), (d, e)\}$

Real-World  **Connection**

A telecommunications device for the deaf (TDD) includes a keyboard and a visual display of the conversation. This lets a hearing-impaired person use a telephone.

41. Telephone Bill The cost of a long-distance telephone call c is a function of the time t spent talking in minutes. The rule $c(t) = 0.09t$ describes the function for one service provider. At the right, a student has calculated how much a 2-hour phone call would cost.

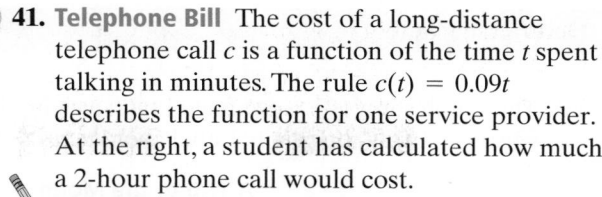

$$c = 0.09 \times 2$$
$$= 0.18$$
$$\$.18 \text{ for 2 hours}$$

  a. Writing Why does the student's answer seem unreasonable?
  b. Error Analysis What mistake(s) did the student make?
  c. How much would it cost to make a 2-hour phone call?
  d. Critical Thinking What set of numbers is reasonable for the domain values? For the range values?

42. Travel Suppose your family is driving home from vacation. The car averages 25 miles per gallon, and you are 180 miles from home. The function $d = 180 - 25g$ relates the number of gallons of gas g the car will use to your distance from home d.
  a. Make a table for $d = 180 - 25g$. Use 2, 4, 6, and 8 as domain values.
  b. Estimation Based on the table, how many gallons of gasoline are needed to get home?
  c. The gas tank holds 15 gallons when it is full. Describe a reasonable domain and range for this situation. Explain your answer.

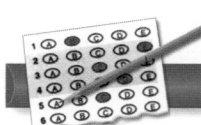 **Challenge**

Use the functions $f(x) = 2x$ and $g(x) = x^2 + 1$ to find the value of each expression.

43. $f(3) + g(4)$ **44.** $g(3) + f(4)$ **45.** $f(5) - 2g(1)$ **46.** $f(g(3))$

47. Critical Thinking Can the graph of a function be a horizontal line? A vertical line? Explain why or why not.

48. The function $y = [x]$ is called the *greatest-integer function.* $[x]$ is the greatest integer less than or equal to x. For example, $[2.99] = 2$ and $[-2.3] = -3$.
  a. Evaluate the function for 0.5, -0.1, -1.99, and -5.2.
  b. The domain of $y = [x]$ is all real numbers. What is the range of $y = [x]$?

Test Prep

Gridded Response

49. Evaluate the function rule $f(x) = 7x$ for $x = 0.75$.

50. Evaluate the function rule $f(x) = 9 - 0.2x$ for $x = 1.5$.

51. What is the greatest value in the range of $y = x^2 - 7$ for the domain $\{-2, 0, 1\}$?

Short Response

52. Determine whether the data below are a function. Show your work.

Mount Rushmore Temperatures (°F)

At Base of Mountain	At Top of Mountain
80	72
65	58
93	84
98	91
74	69

Lesson 5-1

53. The graph shows distance from home as a family drives to the mountains for a vacation. Copy the graph. Label each section of the graph.

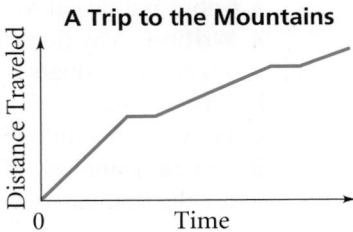

Lesson 3-5

The scale of a map is 1 in. : 15 mi. Find the actual distance corresponding to each map distance.

54. 2 in.   **55.** 1.5 in.   **56.** 0.5 in.

57. 3.25 in.   **58.** 5.5 in.   **59.** 7.25 in.

Lesson 1-6

Find the mean, median, mode, and range.

60. 34 33 35 33 32 35 34 32

61. 1 −2 0 −1 1 −2 2 0 1 −2

62. 4 5 3 7 1 12 6 9 5

63. 15 13 19 20 9 13 15 13

Checkpoint Quiz 1   **Lessons 5-1 through 5-2**

Sketch a graph of each situation. Label each section.

1. the height of a plant that grows at a steady rate

2. the temperature in a classroom after the heater is turned on

3. a child's height above the ground while on a swing

4. Is the graph at the right the graph of a function? Explain.

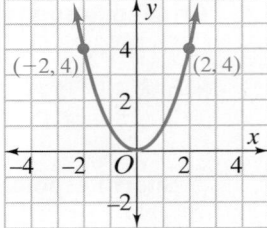

Make a table for each function. Use 1, 2, 3, and 4 for the domain.

5. $f(x) = -5x$ 6. $g(x) = x + 1.4$

7. $f(n) = 3n^2$ 8. $y = 2 - 0.5x$

Determine whether each relation is a function.

9.

x	y
0	6
1	7
5	8
8	9

10.

x	y
5	1
−6	8
5	3
6	7

5-3

Function Rules, Tables, and Graphs

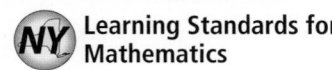
Learning Standards for Mathematics

A.G.4 Identify and graph linear, quadratic (parabolic), absolute value, and exponential functions.

✓ Check Skills You'll Need

GO> for Help Lesson 1-4

Identify the independent and dependent quantities in each situation.

1. A runner averages 7 miles per hour.

2. A customer can buy a dozen apples for $2.40 or two dozen apples for $4.50.

🔊 **New Vocabulary** • discrete data • continuous data

---

1 Modeling Functions

You can model functions using rules, tables, and graphs. A function rule shows how the variables are related. A table identifies specific input and output values of the function. A graph gives a visual picture of the function.

Recall from Lesson 1-4 that the inputs are values of the independent variable. The outputs are the corresponding values of the dependent variable.

Graph the independent variable on the horizontal axis.

Graph the dependent variable on the vertical axis.

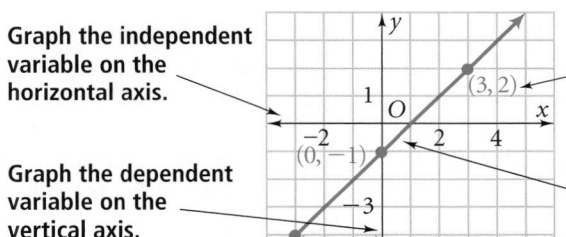

Use the input and output values as ordered pairs to plot points.

Join the points with a line or smooth curve to give a general picture of the function.

1 EXAMPLE Three Views of a Function

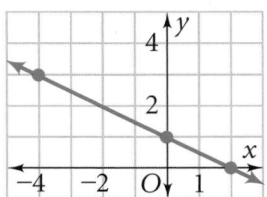

GO ●nline

Video Tutor Help
Visit: PHSchool.com
Web Code: ate-0775

Model the function rule $y = -\frac{1}{2}x + 1$ using a table of values and a graph.

Step 1 Choose input values for x. Evaluate to find y.

Step 2 Plot points for the ordered pairs.

Step 3 Join the points to form a line.

x	$y = -\frac{1}{2}x + 1$	(x, y)
-4	$y = -\frac{1}{2}(-4) + 1 = 3$	$(-4, 3)$
0	$y = -\frac{1}{2}(0) + 1 = 1$	$(0, 1)$
2	$y = -\frac{1}{2}(2) + 1 = 0$	$(2, 0)$

✓ Quick Check ❶ **a.** How would the table and graph of $f(x) = -\frac{1}{2}x + 1$ compare to the table and graph in Example 1?
b. Model the rule $f(x) = 3x + 4$ with a table of values and a graph.

When you draw a graph for a real-world situation, choose appropriate intervals for the units on the axes. Be sure the intervals are equal. Also, if the data are positive numbers, use only the first quadrant.

2 EXAMPLE Real-World 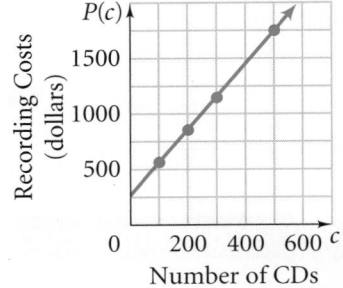 Problem Solving

Recording Costs Suppose your group recorded a CD. Now you want to copy and sell it. One company charges $250 for making a master CD and designing the art for the cover. There is also a cost of $3 to burn each CD. The total cost $P(c)$ depends on the number of CDs c burned. Use the function rule $P(c) = 250 + 3c$ to make a table of values and a graph.

c	P(c) = 250 + 3c	(c, P(c))
100	250 + 3(100) = 550	(100, 550)
200	250 + 3(200) = 850	(200, 850)
300	250 + 3(300) = 1150	(300, 1150)
500	250 + 3(500) = 1750	(500, 1750)

Real-World 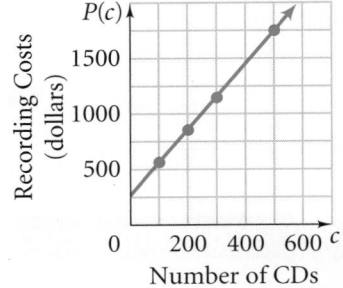 Connection

Careers Recording engineers record singers and instruments separately. The sounds are mixed later to achieve the desired effect.

✓ **Quick Check** ❷ **a.** Another company charges $300 for making a master and designing the art. It charges $2.50 for burning each CD. Use the function rule $P(c) = 300 + 2.5c$. Make a table of values and a graph.
  b. Your band decides to use the second company. You plan on making between 100 and 300 CDs. Find a reasonable range for this situation.

Continuous data are data where numbers between any two data values have meaning. Examples of continuous data include measurement of temperature, length, or weight. Use a solid line to indicate continuous data.

Discrete data are data that involve a count of items, such as number of people or number of cars. For discrete data, indicate each data item with a point.

In some cases, like Example 2, the scale of the graph makes it impractical to use a dot for each value in the domain. So a solid line is used. Be aware of the real-world situation to determine if data graphed is continuous or discrete.

3 EXAMPLE Discrete and Continuous Data

Determine whether the function rule models discrete or continuous data. Then make a table and graph each function.
a. Photography An automatic photo booth charges $3 for a sheet of photos. The function $C(p) = 3p$ describes the cost of p sheets of photos.

You cannot buy part of a sheet in a photo booth, so the data are discrete. Use points for each input value.

p	C(p) = 3p	(p, C(p))
1	3(1) = 3	(1, 3)
2	3(2) = 6	(2, 6)
3	3(3) = 9	(3, 9)
4	3(4) = 12	(4, 12)

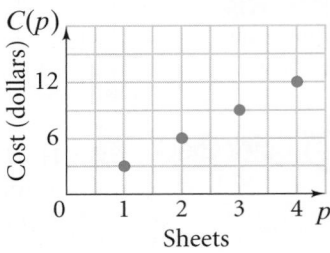

b. Physics The function $M(w) = \frac{1}{6}w$ describes your weight on the moon as a function of your weight on Earth w, in pounds.

Your weight can be measured in parts of a pound, so the data are continuous. Connect the points with a solid line.

w	$M(w) = \frac{1}{6}w$	$(w, M(w))$
60	$\frac{1}{6}(60) = 10$	$(60, 10)$
90	$\frac{1}{6}(90) = 15$	$(90, 15)$
120	$\frac{1}{6}(120) = 20$	$(120, 20)$
150	$\frac{1}{6}(150) = 25$	$(150, 25)$

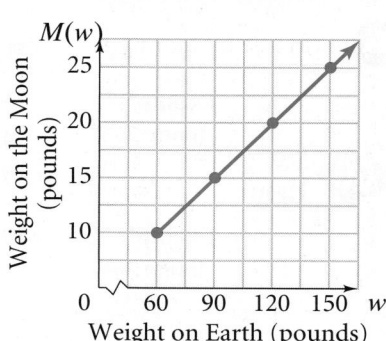

✓ Quick Check ③ Determine whether the function rule models discrete or continuous data. Then make a table and graph each function.

a. Your family is hiking uphill from a campsite 4000 feet above sea level. You climb an average of 500 vertical feet each hour. The function $h = 4000 + 500t$ relates time t to your height above sea level h.

b. The function $c = 6.50t$ represents the cost of t tickets to a baseball game.

Some functions have graphs that are not straight lines. You can graph a function as long as you know its rule. After you have graphed the ordered pairs that you have calculated from a rule, join the points with a smooth line or curve.

④ EXAMPLE **Graphing Functions**

a. Graph the function $y = |x| + 1$.

Make a table of values.   Then graph the data.

x	$y = |x| + 1$	(x, y)
-3	$|-3| + 1 = 4$	$(-3, 4)$
-1	$|-1| + 1 = 2$	$(-1, 2)$
0	$|0| + 1 = 1$	$(0, 1)$
1	$|1| + 1 = 2$	$(1, 2)$
3	$|3| + 1 = 4$	$(3, 4)$

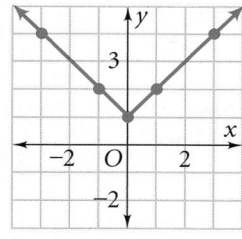

b. Graph the function $f(x) = x^2 + 1$.

Make a table of values.   Then graph the data.

x	$f(x) = x^2 + 1$	$(x, f(x))$
-2	$4 + 1 = 5$	$(-2, 5)$
-1	$1 + 1 = 2$	$(-1, 2)$
0	$0 + 1 = 1$	$(0, 1)$
1	$1 + 1 = 2$	$(1, 2)$
2	$4 + 1 = 5$	$(2, 5)$

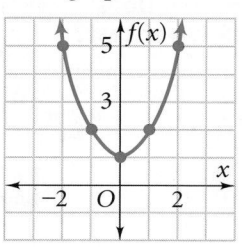

✓ Quick Check ④ Make a table of values and graph each function.
    a. $f(x) = |x| - 1$ **b.** $y = x^2 - 1$

EXERCISES

For more exercises, see *Extra Skill and Word Problem Practice*.

Practice and Problem Solving

 Practice by Example

Example 1
(page 263)

Model each rule with a table of values and a graph.

1. $f(x) = -3x$ **2.** $f(x) = -3x + 1$ **3.** $f(x) = -3x - 2$

4. $y = 2x - 7$ **5.** $f(x) = 8 - x$ **6.** $y = 5 + 4x$

7. $f(x) = \frac{1}{4}x$ **8.** $y = 4x$ **9.** $y = x + 4$

Example 2
(page 264)

10. Earnings Juan charges $3.50 per hour for baby-sitting.
  a. Write a rule to describe how the amount of money M earned is a function of the number of hours h spent baby-sitting.
  b. Make a table of values.
  c. Graph the values and join the points with a line.
  d. Estimation Use the graph to estimate how long it will take Juan to earn $15.

 11. Geometry The figure at the right is a regular pentagon. The function $P(\ell) = 5\ell$ describes the perimeter of a regular pentagon with side length ℓ.
  a. Make a table of values for $\ell = 1, 2, 3,$ and 4.
  b. Graph the function.

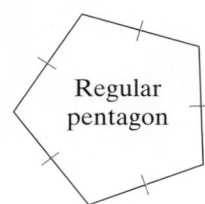

Regular pentagon

Example 3
(page 264)

Determine whether the function rule models discrete or continuous data. Then make a table and graph each function.

12. A hardware store sells bolts for $0.35 apiece. The function $C(p) = 0.35p$ relates the total cost of the bolts to the number p purchased.

13. A supermarket sells string beans for $2 a pound. The function $A(n) = 2n$ relates the total cost of string beans to the number of pounds n bought.

14. Students sell lemonade at a school fundraiser. It costs them $0.42 to make each lemonade which they sell for $0.75. The function $P(c) = 0.75c - 0.42c$ relates the number of cups of lemonade sold c to the students' profit.

Example 4
(page 265)

Graph each function.

15. $y = |x|$ **16.** $y = |x| + 2$ **17.** $y = x^2$

18. $f(x) = x^2 - 1$ **19.** $f(x) = |x| + 3$ **20.** $y = x^2 + 3$

21. $y = |x| - 4$ **22.** $f(x) = -x^2 - 1$ **23.** $f(x) = -x^2 + 2$

Apply Your Skills   **24. Conserving Water** The equation $w = 6m$ models the gallons of water w used by a standard shower head for a shower that takes m minutes. The function $w = 3m$ models the water-saving shower head.

How Long Does Your Shower Last?

- average shower: 12.2 min
- recommended shower: 6 min
- standard head uses 6 gal/min
- water-saving head cuts water flow in half

SOURCE: Opinion Research Corp.

  a. Suppose you take a 6-minute shower using a water-saving shower head. How much water do you save compared to an average shower with a standard shower head?
  b. Graph both functions on the same coordinate plane.
  c. Open-Ended How much water did you use during your last shower?
  d. How did you find your answer?

25. Open-Ended Describe a situation which involves discrete data. Write a function to model the situation.

26. a. Geometry The function $A(\ell) = \frac{1}{2}\ell^2$ describes the area of an isosceles right triangle with leg ℓ. Make a table of values for $\ell = 1, 2, 3,$ and 4.

    b. Does the function describe discrete or continuous data? Explain.

    c. Graph the function.

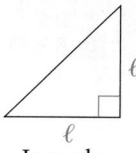
Isosceles
right triangle

Graph each function.

27. $f(x) = \frac{3}{4}x + 7$ **28.** $y = x^2 - 4x + 4$ **29.** $y = |2x|$

30. $y = x + \frac{1}{2}$ **31.** $f(x) = 7 - 5x$ **32.** $f(x) = \left|\frac{1}{2}x\right|$

33. $f(x) = \left|\frac{1}{2}x\right| + 1$ **34.** $y = 1 - x^2$ **35.** $f(x) = -5x^2$

36. Multiple Choice Which function rule best models the data in the table?

    Ⓐ $y = x + 3$ Ⓑ $y = 4x$

    Ⓒ $y = 3x + 1$ Ⓓ $y = x^2 + 3$

x	y
0	3
1	4
2	7
3	12

37. a. Graph $y = |x|$ and $y = -|x|$ on the same coordinate plane.

    b. The graph of $y = -|x|$ is the reflection of the graph of $y = |x|$. Over which axis is the graph of $y = |x|$ reflected?

    c. Write an equation of the reflection of the graph of $y = |x| + 1$ over the same axis.

GO for Help

To review reflections, see
Skills Handbook p. 767.

38. a. Make a table for the perimeters of the rectangles formed by each set of blue tiles.

    b. The perimeter $P(t)$ is a function of the number of tiles t. Write a rule for the data in your table and graph the function.

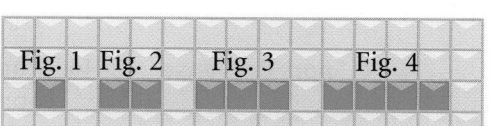
Fig. 1   Fig. 2   Fig. 3   Fig. 4

39. a. Graph each function on the same coordinate plane.

    i. $f(x) = |x| + 2$ **ii.** $f(x) = |x| + 4$ **iii.** $f(x) = |x| - 3$

    b. Critical Thinking In the function $y = |x| + b$, how does changing the value of b change the graph of the function?

40. a. Graph each function on the same coordinate plane.

    i. $f(x) = |2x|$ **ii.** $f(x) = |0.5x|$ **iii.** $f(x) = |3x|$

    b. Critical Thinking In the function $y = |ax|$, how does changing the value of a change the graph of the function?

Ⓒ Challenge

41. The function $s(x)$, sometimes called the signum function, is defined as

$$s(x) = \begin{cases} 1 \text{ if } x > 0 \\ 0 \text{ if } x = 0 \\ -1 \text{ if } x < 0 \end{cases}$$

For example, $s(17) = 1, s(0) = 0,$ and $s(-32) = -1$.

    a. Evaluate $s(3.77), s(0.003), s(-1.5),$ and $s(-2300)$.

    b. The domain of the function is all real numbers. What is the range?

    c. Make a table of values and graph the function.

    d. Make a Conjecture Do you think $s(a + b) = s(a) + s(b)$? First, test some values of a and b. If your answer is *yes*, justify your answer. If your answer is *no*, give a counterexample.

Multiple Choice

42. Suppose you hire an electrician to install several electrical outlets in your home. The electrician charges $68 for materials plus $40 per hour (or fraction of an hour). How much will the electrician charge you if the job takes $2\frac{1}{4}$ hours?

A. $148   **B.** $158   **C.** $188   **D.** $208

43. Which is the graph of the function rule $f(x) = \frac{1}{2}x - 2$?

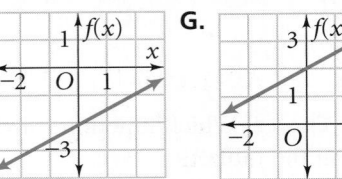
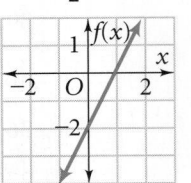
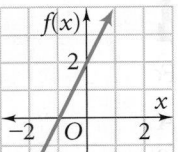

44. Which function is modeled by the table at the right?
   A. $f(x) = x - 2$ **B.** $f(x) = 2x + 1$
   C. $f(x) = -x + 1$ **D.** $f(x) = \frac{1}{2}x - 1$

x	f(x)
−3	−5
0	1
2	5
3	7

45. Which points are on the graph of the function rule $f(x) = 10 - 4x$?
   F. (18, −2), (10, 0), (2, 2)
   G. (−18, 2), (−10, 0), (−2, −2)
   H. (2, −18), (0, −10), (−2, −2)
   J. (−2, 18), (0, 10), (2, 2)

Short Response

46. Graph the equations $y = x + 3$, $y = x^2 + 3$, and $y = |x| + 3$ on the same coordinate plane. Describe the similarities and differences in the graphs.

Extended Response

47. a. Make a table of values for the function rule $f(x) = |x + 1| - 2$.
   b. Graph $f(x) = |x + 1| - 2$.

Mixed Review

Lesson 5-2

Find the range of each function for the domain {−2, 0, 3.5}.

48. $f(x) = 3x + 1$ **49.** $g(x) = 3x - 5$ **50.** $f(s) = -3s + 4$

51. $g(v) = |v| - 5$ **52.** $h(n) = 12 - n$ **53.** $g(w) = 5(w - 2)$

54. $p(n) = 6n + 1$ **55.** $f(x) = 0.5x - 8$ **56.** $k(n) = -11n + 9$

Lesson 3-5

The scale of a map is 1 cm : 16 km. Find the actual distance corresponding to each map distance.

57. 3 cm   **58.** 2.5 cm   **59.** 6.3 cm   **60.** 8.5 cm   **61.** 10.2 cm

 62. Architecture The Lyndon Johnson Presidential Library has a model of the Oval Office in the White House. The model in the Johnson Library is $\frac{7}{8}$ the size of the original. Write and solve a proportion to find each dimension in the Johnson Library given the following actual Oval Office dimensions. Round to the nearest tenth of a foot.
   a. greatest width: 29 ft
   b. greatest length: 35 ft 10 in.
   c. height: 18 ft 6 in.

Function Rules, Tables, and Graphs

You can use a graphing calculator to explore the relationship among a function rule, a table, and a graph. When you use the table feature, the calculator computes the values for y based on the values of x that you enter.

NY **A.CN.1:** Understand and make connections among multiple representations of the same mathematical idea.

1 EXAMPLE

For the function $y = -2x + 5$, find the range when the domain is $\{-12, 2, 0, 3, 8\}$.

Access the **TBLSET** feature. Use the arrow key to shade the **Ask** to the right of **Indpnt**.

Press **ENTER**.

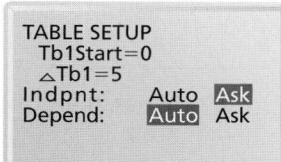

Press **Y=** . Enter the function. Access the **TABLE** feature. Enter values for x.

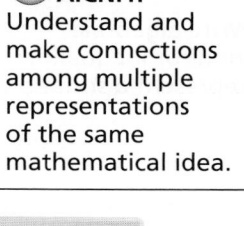

• The range is $\{29, 1, 5, -1, -11\}$.

To graph an equation, use the **GRAPH** feature. You can use the **TRACE** feature to find x- and y-values. If you graph and trace the equation in Example 1, you will see that the x- and y-values are generally given as 8-digit numbers. To see values for x that are given in tenths, press **ZOOM** 4 and then **TRACE** .

Go Online
PHSchool.com
For: Graphing calculator procedures
Web Code: ate-2104

2 EXAMPLE

Graph $y = -0.5x - 2$. Where does the graph cross each axis?

Press **Y=** . Enter the function rule. Then press **GRAPH** .

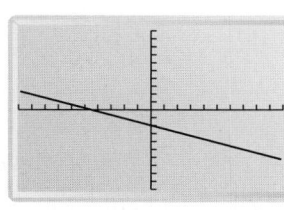

Press **ZOOM** 4 and then **TRACE** to find where the graph crosses the axes.

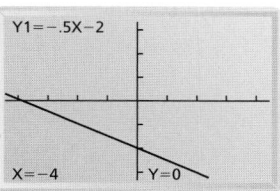

• The graph crosses the y-axis at -2 and the x-axis at -4.

EXERCISES

Find the range of each function for the given domain.

1. $y = 3x + 6; \{-5, 0, 3, 7\}$

2. $y = 0.4x - 5.1; \{-2.1, 1.35, 5.7\}$

Determine where each graph crosses the y-axis and the x-axis.

3. $y = -2x + 3$

4. $y = -0.25x - 1$

5. $y = 1.2x + 2.16$

6. Open-Ended Graph $y = -0.2x + 6$. Using the **WINDOW** screen, experiment with values for **Xmin**, **Xmax**, **Ymin**, and **Ymax** until you can see the graph crossing both axes. What values did you use for **Xmin**, **Xmax**, **Ymin**, and **Ymax**?

Writing a Function Rule

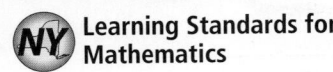

Learning Standards for Mathematics

A.A.5 Write algebraic equations or inequalities that represent a situation.

✓ Check Skills You'll Need

GO **for Help** Lesson 5-3

Model each rule with a table of values.

1. $f(x) = 5x - 1$ **2.** $y = -3x + 4$ **3.** $g(t) = 0.2t - 7$

4. $y = 4x + 1$ **5.** $f(x) = 6 - x$ **6.** $c(d) = d + 0.9$

Evaluate each function rule for $n = 2$.

7. $A(n) = 2n - 1$ **8.** $f(n) = -3 + n - 1$ **9.** $g(n) = 6 - n$

1 Writing Function Rules

You can write a rule for a function by analyzing a table of values. Look for a pattern relating the independent and dependent variables.

1 EXAMPLE Writing a Rule From a Table

Write a function rule for each table.

a.

x	f(x)
1	5
2	6
3	7
4	8

Ask yourself, "What can I do to 1 to get 5, to 2 to get 6, . . . ?"

You add 4 to each x-value to get the $f(x)$ value.

Relate $f(x)$ equals x plus   4

Write $f(x)$ = x +   4

A rule for the function is $f(x) = x + 4$.

b.

x	y
1	1
3	9
6	36
9	81

Ask yourself, "What can I do to 3 to get 9, to 6 to get 36, . . . ?"

You multiply each x-value times itself to get the $f(x)$ value.

Relate y equals x times itself

Write y = x^2

A rule for the function is $y = x^2$.

✓ Quick Check ① Write a function rule for each table.

a.

x	f(x)
1	−1
2	0
3	1
4	2

b.

x	y
1	2
2	4
3	6
4	8

c.

x	y
1	3
2	4
3	5
4	6

2 EXAMPLE Real-World Problem Solving

Scale Model The exhibit at the left is a scale model of a desktop computer. It is about 20 times the size of a normal-sized desktop computer.

a. Write a function rule to describe this relationship.

Relate	larger	is	20	times	normal

Define Let n = length of normal-sized computer.

Let $L(n)$ = length of larger size shown in museum exhibit.

Write	$L(n)$	=	20	·	n

The function rule $L(n) = 20n$ describes the relationship between the size of the computer in the exhibit and a normal-sized computer.

b. A space bar on a normal-sized computer is $4\frac{3}{8}$ in long. About how long is the space bar in the exhibit?

$L(n) = 20 \cdot n$

$L(n) = 20 \cdot 4\frac{3}{8}$ **Substitute $4\frac{3}{8}$ for n.**

$L(n) = 87\frac{1}{2}$ **Simplify.**

The space bar in the exhibit is about $87\frac{1}{2}$ in. long.

✓ Quick Check **2 a. Carpentry** A carpenter buys finishing nails by the pound. Each pound of nails costs $1.19. Write a function rule to describe this relationship.
b. How much do 12 lb of finishing nails cost?

When you write a function, the dependent variable is defined in terms of the independent variable. In Example 3 below, profit depends on the number of lawns mowed, so profit is a function of the number of lawns mowed.

3 EXAMPLE Real-World Problem Solving

Multiple Choice Suppose you borrow money from a relative to buy a lawn mower that costs $245. You charge $18 to mow a lawn. Write a rule to describe your profit $P(n)$ as a function of the number of lawns mowed n. Which equation best describes the situation?

Ⓐ $P(n) = 18n$ Ⓑ $P(n) = 18 + 245$

Ⓒ $P(n) = 18n - 245$ Ⓓ $P(n) = 245 + 18n$

Relate	total profit	is	$18	times	lawns mowed	minus	cost of mower

Write	$P(n)$	=	18	·	n	−	245

The function rule $P(n) = 18n - 245$ describes your profit as a function of the number of lawns mowed. So C is the correct answer.

✓ Quick Check **3 Earnings** Suppose you buy a word-processing software package for $199. You charge $15 per hour for word processing. Write a rule to describe your profit as a function of the number of hours you work.

EXERCISES

For more exercises, see *Extra Skill and Word Problem Practice*.

Practice and Problem Solving

A Practice by Example

Example 1
(page 270)

GO for Help

Match each table with its rule.

1. $y = 4x$

2. $y = x - 4$

3. $y = -4 - x$

A.

x	y
-2	-6
-1	-5
0	-4
1	-3

B.

x	y
-1	-4
-2	-8
-3	-12
-4	-16

C.

x	y
-1	-3
0	-4
1	-5
2	-6

Write a function rule for each table.

4.

x	f(x)
1	3
2	6
3	9
4	12

5.

x	f(x)
1	0.5
2	1.5
3	2.5
4	3.5

6.

x	f(x)
1	0.5
2	1
3	1.5
4	2

7.

x	f(x)
1	-3
2	-6
3	-9
4	-12

8.

x	y
-2	-8
-1	-4
0	0
1	4

9.

x	y
-8	64
-4	16
0	0
4	16
8	64

Example 2
(page 271)

Write a function rule for each situation.

10. the total cost $t(c)$ of c ounces of cinnamon if each ounce costs $.79

11. the total distance $d(n)$ traveled after n hours at a constant speed of 45 miles per hour

12. the height $f(h)$ of an object in feet when you know the height h in inches

13. a worker's earnings $e(n)$ for n hours when the worker's hourly wage is $6.37

14. the area $A(n)$ of a square when you know the length n of a side

15. the volume $V(n)$ of a cube when you know the length n of a side

16. the area $A(r)$ of a circle with radius r

Example 3
(page 271)

17. Food Costs At a supermarket salad bar, the price of a salad depends on its weight. Salad costs $.19 per ounce.
a. Write a rule to describe the function.
b. How much would an 8-ounce salad cost?

18. Postage In 2002, the price of mailing a letter was $.34 for the first ounce or part of an ounce and $.21 for each ounce or part of an ounce after the first ounce.
a. Write a rule to describe the function.
b. How much did it cost to mail a 4-ounce letter?

Write a function rule for each table.

19.

Distance (km)	Distance (m)
0.5	500
1.0	1000
1.5	1500
2.0	2000

20.

Inches	Centimeters
1	2.54
2	5.08
3	7.62
4	10.16

Buy additional books at our regular low Club price of $10.00 per book. To become a Book Express member, just buy 2 additional books within the first year. You may resign your membership at any time.

🌐 **Math in the Media** **Use the advertisement at the left for Exercises 21–22.**

21. **a.** Write a rule to find the total cost $C(a)$ for all the books a person buys through Book Express. Let a represent the number of additional books bought (after the first 6 books).
    b. Suppose a person buys 9 books in all. Find the total cost.
    c. Evaluate the function for $a = 6$. What does the output represent?
    d. Does the function rule model discrete or continuous data? Explain.

22. A bookstore sells the same books for an average price of $6 each.
    a. Write a function rule to model the total cost $C(b)$ of books bought at the bookstore. Let b represent the number of books bought.
    b. Evaluate your function for $b = 12$. What does the output represent?
    c. You plan to buy 12 books. What is your average cost per book as a member of Book Express?
    d. Is it less expensive to buy 12 books through the club or at the bookstore? Explain.

✏️ 23. **Writing** What advantage(s) can you see of having a function rule instead of a table of values for a function?

🌐 24. **Water Usage** Use the function in the table at the right.
    a. Identify the dependent and independent variables.
    b. Write a rule to describe the function.
    c. How many gallons of water would you use for 7 loads of laundry?
    d. **Critical Thinking** In one month, you used 442 gallons of water for laundry. How many loads did you wash?

Water Used for Laundry

1 load	34 gallons
2 loads	68 gallons
3 loads	102 gallons
4 loads	136 gallons

25. **Open-Ended** Write a function rule that models a real-world situation. Evaluate your function for an input value and explain what the output represents.

26. **Multiple Choice** The graph of $f(x) = \frac{1}{2}x$ is shown at the right. Which statement about the function is always true?

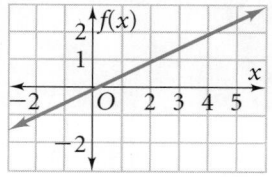

Ⓐ The value of y is always less than the value of x.
Ⓑ When the value of x is positive, the value of y is negative.
Ⓒ As the value of x increases, the value of y increases.
Ⓓ The value of y is always different from the value of x.

Make a table of values for each graph. Use the table to write a function rule.

27.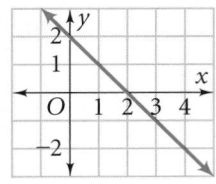

28.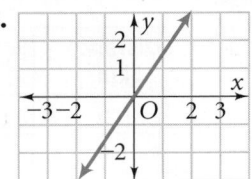

29.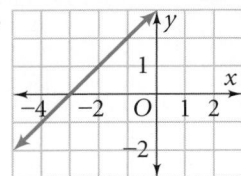

GO for Help

For a guide to solving Exercise 30, see p. 276.

30. **Tipping** You go to dinner and decide to leave a 15% tip for the server. You had $45 when you came to the restaurant.
   a. Make a table that shows how much money you would have left after buying a $12, $18, $24, or $28 meal.
   b. Write a function that relates the cost of the meal c to the amount of money you have left $A(c)$.
   c. Graph the function.

31. a. **Geometry** Make a table for the area of the largest shaded triangle in each figure.
   b. Write a function rule that relates the height of each figure h to the area of the triangle $A(h)$.

   Fig. 1   Fig. 2   Fig. 3

   c. **Critical Thinking** Does the function model discrete or continuous data? Explain.
   d. Graph the function.

C Challenge

Write a function rule for each table.

32.

x	f(x)
1	1
2	8
3	27
4	64

33.

x	f(x)
−1	1
−2	8
−3	27
−4	64

34.

x	f(x)
−1	0
−2	7
−3	26
−4	63

35. **Truck Rental** A truck rental company charges $44 per day for renting a medium-sized truck. There is also a charge of $.38 per mile.
   a. Write a function rule $c(m)$ to model the cost of renting a truck for a day and driving m miles.
   b. Evaluate your function rule for $m = 70$ and $m = 120$.
   c. You return the truck to the rental company and pay $58.44 (excluding tax). How far did you drive?
   d. Suppose you need to rent a truck for two days. You plan to drive 150 miles each day. How much will this cost?

36. **Making Pickles** The table at the right shows the relationship between the amount of pickling salt added to a gallon of water and the brine concentration, which is the percent of salt by weight.
   a. Write a function rule to describe the relationship between salt volume and brine concentration.
   b. Write a function rule to describe the relationship between salt weight and brine concentration.

Brine Strength

Salt Volume (cup)	Salt Weight (oz)	Brine Concentration (percent salt)
$\frac{1}{3}$	3.3	2.31
$\frac{1}{2}$	4.95	3.465
$\frac{2}{3}$	6.6	4.62
$\frac{3}{4}$	7.425	5.1975
1	9.9	6.93

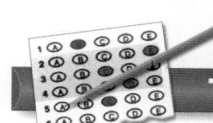
Multiple Choice

37. What is the function rule for the total cost $T(b)$ of b books, if each book costs $11.95?
 A. $T(b) = 11.95b$
 B. $T(b) = b + 11.95$
 C. $T(b) = 11.95 - b$
 D. $T(b) = b - 11.95$

38. What is the function rule for the amount of change $C(x)$ you receive from a $50 bill if you buy x pounds of dog food at $1.60 a pound?
 F. $C(x) = 1.6x - 50$
 G. $C(x) = 50x - 1.6$
 H. $C(x) = 50 - 1.6x$
 J. $C(x) = 160 - 50x$

39. What is the function rule for the table at the right?
 A. $f(x) = x - 5$
 B. $f(x) = -5x - 4$
 C. $f(x) = 5x - 1$
 D. $f(x) = -5x + 1$

x	$f(x)$
0	1
1	-4
2	-9
3	-14
4	-19

Short Response

40. The recommended dosage D in milligrams of a certain medicine depends on a person's body mass w in kilograms. The function rule $D = 0.1w^2 + 5w$ describes the relationship of the dosage to body mass. Evaluate the function for a person who has a mass of 60 kilograms. Show your work.

Mixed Review

GO for Help

Lesson 5-3

Model each rule with a table of values and a graph.

41. $f(x) = x - 3$ **42.** $y = 5 - x$ **43.** $g(x) = -x + 3$

44. $f(x) = 2x - 3$ **45.** $y = |2x| - 3$ **46.** $y = 2x^2 - 3$

Lesson 3-7

Find each percent of change. Describe the percent of change as an increase or decrease. Round to the nearest percent.

47. 12 cm to 14 cm   **48.** 98 oz to 100 oz   **49.** 65 ml to 60 ml

50. 6 ft to 1 ft   **51.** 1.4 m to 1.8 m   **52.** $1\frac{1}{2}$ in. to $\frac{7}{8}$ in.

Lesson 3-4

53. Measurement The figure at the right shows how much juice you can get from some fruits.
 a. What is the minimum number of oranges needed to make a cup of orange juice? What is the maximum number of oranges needed? (*Hint:* 16 tablespoons = 1 cup)
 b. Suppose you buy a bag of 6 lemons and a bag of 5 limes. What is the most juice you can expect to get from these bags of fruit?

How Much Juice in an Average Fruit?

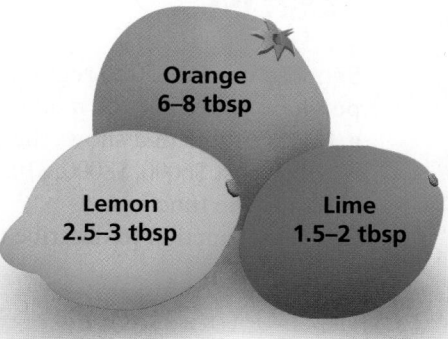

Orange
6–8 tbsp

Lemon
2.5–3 tbsp

Lime
1.5–2 tbsp

Guided Problem Solving

Understanding Math Problems   Read the exercise below and then follow along with Raymond as he uses the four-corner method to solve the problem. Check your understanding with the exercise at the bottom of the page.

Tipping You go to dinner and decide to leave a 15% tip for the server. You had $45 when you came to the restaurant.

a. Make a table that shows how much money you would have left after buying a $12, $18, $24, or $28 meal.

b. Graph the function.

c. Write a function to describe the situation.

---

1. I'll use the four-corner method to solve the problem. First, I'll write the original word problem.

You go to dinner and decide to leave a 15% tip for the server. You had $45 when you came to the restaurant.

4. To write a function, I need to define the variables.

Let d = cost of meal.

Let A(d) = amount of money remaining after paying the bill and the tip.

$$A(d) = 45 - (d + 0.15d)$$

---

2. Now I'll make a table of values.

Cost of Meal	Process	Amount Left
$12	45 − [12 + 0.15(12)]	$31.20
$18	45 − [18 + 0.15(18)]	$24.30
$24	45 − [24 + 0.15(24)]	$17.40
$28	45 − [28 + 0.15(28)]	$12.80

3. Next, I graph the function.

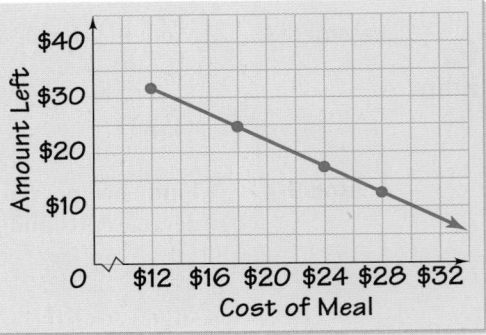

EXERCISE

Suppose it costs $50 to register your car, and you pay 7% tax on the purchase price. Use the method shown above to organize your work.

a. Make a table that shows the total registry and tax costs for cars which cost $6000, $8000, $10,000, and $12,000.

b. Graph the function.

c. Write a function to describe the situation.

5-5

Direct Variation

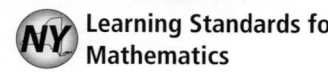
NY Learning Standards for Mathematics

A.A.5 Write algebraic equations or inequalities that represent a situation.

A.N.5 Solve algebraic problems arising from situations that involve fractions, decimals, percents (decrease/increase and discount), and proportionality/direct variation.

✓ **Check Skills You'll Need**

Solve each proportion.

1. $\frac{5}{8} = \frac{x}{12}$

2. $\frac{4}{9} = \frac{n}{45}$

3. $\frac{25}{15} = \frac{y}{3}$

4. $\frac{7}{n} = \frac{35}{50}$

5. $\frac{8}{d} = \frac{20}{36}$

6. $\frac{14}{18} = \frac{63}{n}$

GO for Help Lesson 3-4

🔊 **New Vocabulary** • direct variation • constant of variation for direct variation

---

1 ▸ **Writing the Equation of a Direct Variation**

Activity: Direct Variation

As you watch a movie, 24 individual pictures, or frames, flash on the screen each second. Here are three ways you can model the relationship between the number of frames $f(s)$ and the number of seconds s.

| **Table** | | **Graph** | **Function Rule** |

Function Rule

$f(s) = 24s$

s number of seconds	$f(s)$ number of frames
1	24
2	48
3	72
4	96
5	120

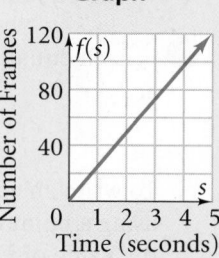

1. As the number of seconds doubles, what happens to the number of frames?

2. Find the ratio $\frac{\text{number of frames}}{\text{number of seconds}}$ for each pair of data in the table.

3. For every increase of 1 second on the horizontal axis of the graph, what is the increase on the vertical axis?

4. What do you notice about your answers to Questions 2 and 3 and the coefficient of s in the function rule?

5. a. What number of frames corresponds to $s = 0$?
   b. What is the ordered pair on the graph for the seconds and number of frames when $s = 0$?

The table in the activity on page 277 shows the number of frames of a movie that are projected over various lengths of time. The ratio $\frac{\text{number of frames}}{\text{time}} = 24$, which is constant. The number of frames is proportional to the time and 24 is the constant of variation.

 Key Concepts

Definition	Direct Variation

A function in the form $y = kx$, where $k \neq 0$, is a **direct variation.** The **constant of variation for direct variation** k is the coefficient of x. The variables y and x are said to vary directly with each other.

Vocabulary Tip

Constant means *remaining the same.* Constant of variation means changing at the *same* rate.

For $y = kx$, y is a function of x. If $x = 0$, then $y = 0$, so the graph of a direct variation is a line that passes through $(0, 0)$. To tell whether an equation represents a direct variation, solve for y. If the equation can be written in the form $y = kx$, where $k \neq 0$, it represents a direct variation.

1 EXAMPLE **Is an Equation a Direct Variation?**

Is each equation a direct variation? If it is, find the constant of variation.

a. $5x + 2y = 0$

$2y = -5x$ **Subtract 5x from each side.**

$y = -\frac{5}{2}x$ **Divide each side by 2.**

The equation has the form $y = kx$, so the equation is a direct variation. The constant of variation is $-\frac{5}{2}$.

b. $5x + 2y = 9$

$2y = 9 - 5x$ **Subtract 5x from each side.**

$y = \frac{9}{2} - \frac{5}{2}x$ **Divide each side by 2.**

The equation cannot be written in the form $y = kx$. It is not a direct variation.

 **1** Is each equation a direct variation? If it is, find the constant of variation.
a. $7y = 2x$ **b.** $3y + 4x = 8$ **c.** $y - 7.5x = 0$

To write an equation for a direct variation, you first find the constant of variation k using a point other than the origin that lies on the graph of the equation. Then use the value of k to write an equation.

GO **n**line

Video Tutor Help
Visit: PHSchool.com
Web Code: ate-0775

2 EXAMPLE **Writing an Equation Given a Point**

Write an equation of the direct variation that includes the point $(4, -3)$.

$y = kx$ **Start with the function form of a direct variation.**

$-3 = k(4)$ **Substitute 4 for x and −3 for y.**

$-\frac{3}{4} = k$ **Divide each side by 4 to solve for k.**

$y = -\frac{3}{4}x$ **Write an equation. Substitute $-\frac{3}{4}$ for k in y = kx.**

An equation of the direct variation is $y = -\frac{3}{4}x$.

 **2** Write an equation of the direct variation that includes the point $(-3, -6)$.

278   Chapter 5   Graphs and Functions

You can use a direct variation to describe a real-world situation in which the dependent variable varies directly with the independent variable.

3 EXAMPLE   Real-World  Problem Solving

Weather Your distance from lightning varies directly with the time it takes you to hear thunder. If you hear thunder 10 seconds after you see lightning, you are about 2 miles from the lightning. Write an equation for the relationship between time and distance.

Relate   The distance varies directly with the time. When $x = 10$, $y = 2$.

Define   Let x = the number of seconds between your seeing lightning and your hearing thunder.

Let y = your distance in miles from the lightning.

Write
$y = kx$	Use the general form of a direct variation.
$2 = k(10)$	Substitute 10 for *x* and 2 for *y*.
$\frac{1}{5} = k$	Divide each side by 10 to solve for *k*.
$y = \frac{1}{5}x$	Write an equation. Substitute $\frac{1}{5}$ for *k* in *y = kx*.

The equation $y = \frac{1}{5}x$ relates the time *x* in seconds it takes you to hear the thunder to the distance *y* in miles you are from the lightning.

Real-World  Connection

The total energy released by a single flash of lightning could power an ordinary light bulb for a few months.

✓ **Quick Check**   ③ A recipe for a dozen corn muffins calls for 1 cup of flour. The number of muffins varies directly with the amount of flour you use. Write a direct variation for the relationship between the number of cups of flour and the number of muffins.

---

2   Proportions and Equations of Direct Variations

You can rewrite a direct variation $y = kx$ as $\frac{y}{x} = k$. When two sets of data vary directly, the ratio $\frac{y}{x}$ is the constant of variation. It is the same for each data pair.

4 EXAMPLE   Direct Variations and Tables

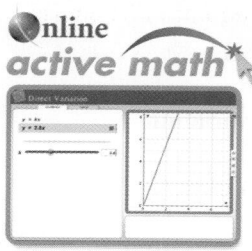

For: Direct Variation Activity
Use: Interactive Textbook, 5-5

For each table, use the ratio $\frac{y}{x}$ to tell whether *y* varies directly with *x*. If it does, write an equation for the direct variation.

a.
x	y	$\frac{y}{x}$
−3	2.25	$\frac{2.25}{-3} = -0.75$
1	−0.75	$\frac{-0.75}{1} = -0.75$
4	−3	$\frac{-3}{4} = -0.75$
6	−4.5	$\frac{-4.5}{6} = -0.75$

b.
x	y	$\frac{y}{x}$
2	−1	$\frac{-1}{2} = -0.5$
4	1	$\frac{1}{4} = 0.25$
6	3	$\frac{3}{6} = 0.5$
9	4.5	$\frac{4.5}{9} = 0.5$

Yes, the constant of variation is −0.75. The equation is $y = -0.75x$.

No, the ratio $\frac{y}{x}$ is not the same for all pairs of data.

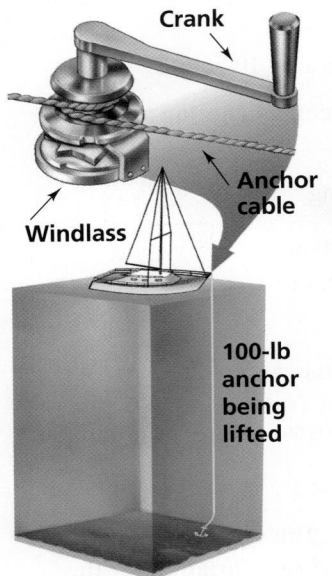

Crank

Anchor cable

Windlass

100-lb anchor being lifted

A windlass is a winch turned by a crank. It is used in a water well and to raise an anchor on a boat.

 **Quick Check** ④ For the data in each table, tell whether y varies directly with x. If it does, write an equation for the direct variation.

a.

x	y
−2	3.2
1	2.4
4	1.6

b.

x	y
4	6
8	12
10	15

In a direct variation, the ratio $\frac{y}{x}$ is the same for all pairs of data where $x \neq 0$. So the proportion $\frac{y_1}{x_1} = \frac{y_2}{x_2}$ is true for the ordered pairs (x_1, y_1) and (x_2, y_2), where neither x_1 nor x_2 are zero. You can rewrite a proportion as an equation of direct variation.

5 EXAMPLE **Real-World 🌐 Problem Solving**

Physics The force you must apply to lift an object is proportional to the object's weight. You would need to apply 0.625 lb of force to a windlass to lift a 28-lb weight. How much force would you need to lift 100 lb?

Relate $\frac{\text{force}}{\text{weight}} = \frac{0.625}{28}$, which is about 0.0223.

Define Let $n =$ the force you need to lift 100 lb.

Write Let $w =$ the weight and $f =$ the force.

$f = 0.0223w$ **Write an equation.**

$f = 0.0223(100)$ **Substitute 100 for w.**

$f = 2.23$ **Simplify.**

● You need about 2.2 lb of force to lift 100 lb.

 **Quick Check** ⑤ **Physics** Suppose a second windlass requires 0.5 lb of force to lift an object that weighs 32 lb. How much force would you need to lift 160 lb?

EXERCISES

For more exercises, see *Extra Skill and Word Problem Practice.*

Practice and Problem Solving

Ⓐ Practice by Example

Example 1
(page 278)

GO for Help

Is each equation a direct variation? If it is, find the constant of variation.

1. $2y = 5x + 1$ **2.** $8x + 9y = 10$ **3.** $-12x = 6y$

4. $y + 8 = -x$ **5.** $5x - 6y = 0$ **6.** $-4 + 7x + 4 = 3y$

7. $-x = 10y$ **8.** $0.7x - 1.4y = 0$ **9.** $\frac{1}{2}x + \frac{1}{3}y = 0$

Example 2
(page 278)

Write an equation of the direct variation that includes the given point.

10. $(1, 5)$ **11.** $(5, 1)$ **12.** $(-8, 10)$ **13.** $(-5, -9)$

14. $(-2, 3)$ **15.** $(-6, 1)$ **16.** $(3, -4)$ **17.** $(6, -8)$

18. $(-6, 8)$ **19.** $(-5, -10)$ **20.** $(12, -8)$ **21.** $(35, 7)$

Example 3
(page 279)

Define the variables. Then write a direct variation to model each relationship.

 22. Geometry The perimeter p of a regular octagon varies directly with the length ℓ of one side of the octagon.

23. Earnings When you have a job that pays an hourly wage, the amount you earn varies directly with the number of hours you work. Suppose you earn $7.10/hour working at the library.

Example 4
(page 279)

For the data in each table, tell whether y varies directly with x. If it does, write an equation for the direct variation.

24.

x	y
3	5.4
7	12.6
12	21.6

25.

x	y
−2	1
3	6
8	11

26.

x	y
−6	9
1	−1.5
8	−12

Example 5
(page 280)

27. Physics The force you apply to a lever is proportional to the weight you can lift. Suppose you can lift a 50-lb weight by applying 20 lb of force to a certain lever.
   a. What is the ratio of force to weight for the lever?
   b. Write an equation and find the force you need to lift a friend weighing 130 lb.

28. Bicycling A bicyclist traveled at a constant speed during a timed practice period. Write an equation and find the distance the cyclist traveled in 30 min.

A Bicyclist's Practices

Elapsed Time	Distance
10 min	3 mi
25 min	7.5 mi

B Apply Your Skills

Write an equation of the direct variation that includes the given point.

29. $\left(3, \frac{1}{2}\right)$ **30.** $\left(\frac{1}{4}, -5\right)$ **31.** $\left(\frac{-5}{6}, \frac{6}{5}\right)$ **32.** $(1.2, 7.2)$

33. $(0.5, 4.5)$ **34.** $\left(-2, \frac{1}{16}\right)$ **35.** $(5.2, -1.5)$ **36.** $\left(-\frac{8}{3}, -\frac{9}{8}\right)$

37. a. Writing How can you tell whether two sets of data vary directly?
   b. How can you tell if a line is the graph of a direct variation?

Critical Thinking **Is each statement true or false? Explain.**

38. The graph of a direct variation may pass through $(-2, 4)$.

39. The graph of a direct variation may pass through $(0, 3)$.

40. If you triple an x-value of a direct variation, the y-value also triples.

Graph the direct variation that includes the given point. Write an equation of the line.

41. $(2, 5)$ **42.** $(-2, 5)$ **43.** $(2, -5)$ **44.** $(-2, -5)$

45. Biology The amount of blood in a person's body varies directly with body weight. A person who weighs 160 lb has about 5 qt of blood.
   a. Find the constant of variation.
   b. Write an equation relating quarts of blood to weight.
   c. Open-Ended Estimate the number of quarts of blood in your body.

Real-World **Connection**

You must be at least 17 years old and weigh at least 110 pounds to give blood.

46. Electricity Ohm's Law $V = I \times R$ relates the voltage, current, and resistance of a circuit. V is the voltage measured in volts. I is the current measured in amperes. R is the resistance measured in ohms.

   a. Find the voltage of a circuit that has a current of 24 amperes and resistance of 2 ohms.

   b. Find the resistance of a circuit that has a current of 24 amperes and a voltage of 18 volts.

47. Graph each direct variation on the same coordinate plane.

    i. $y = x$ **ii.** $y = 2x$ **iii.** $y = 3x$ **iv.** $y = 4x$

   a. Describe how the graphs change as the constant of variation increases.

   b. Predict how the graph of $y = \frac{1}{2}x$ would appear.

 Challenge

The ordered pairs in each exercise are for the same direct variation. Find each missing value.

48. $(3, 4)$ and $(9, y)$ **49.** $(-1, 2)$ and $(4, y)$ **50.** $(-5, 3)$ and $(x, -4.8)$

51. $(1, y)$ and $\left(\frac{3}{2}, -9\right)$ **52.** $(2, 5)$ and $(x, 12.5)$ **53.** $(-2, 5)$ and $(x, -5)$

Problem Solving Hint

For Exercise 54, start with the relationship of miles and gallons: $\frac{m}{g} = 24$.

54. Gas Mileage A car gets 24 miles per gallon. The number of gallons g of gas used varies directly with the number of miles m traveled.

   a. Suppose the price of gas is \$1.83 per gallon. Write a function relating the cost c for g gallons of gas. Is this a direct variation?

   b. Write a direct variation relating the cost of gas to the miles traveled.

 **REGENTS**

Test Prep

Multiple Choice

55. Which equation is a direct variation?

   A. $y = -0.7x$ **B.** $y = \frac{21}{x}$ **C.** $y - x = 4$ **D.** $y = 3x + 2$

56. A direct variation includes the point $(-8, 2)$. Which is an equation of the direct variation?

   F. $-8y = x + 2$ **G.** $2y = -8x$ **H.** $y = \frac{x}{-4}$ **J.** $y = -4x$

57. Each point is included in a different direct variation. In which variation is the constant of variation $\frac{3}{5}$?

   A. $(-3, -5)$ **B.** $(-3, 5)$ **C.** $(-2, -3)$ **D.** $(15, 9)$

58. Each point is included in a different direct variation. Which has the greatest constant of variation?

   F. $(-1, 4)$ **G.** $(10, -5)$ **H.** $(6, 2)$ **J.** $(-3, -3)$

Short Response

59. The table at the right shows the number of hours a clerk works per week and the amount of money she earns before taxes. Write an equation for the direct variation. Then use it to find how much money the clerk would earn if she works for 34 hours per week.

Hours Worked	Dollars Earned
12	\$99.00
17	\$140.25
21	\$173.25
32	\$264.00

60. Write an equation of the direct variation that includes the point $(-1, -4)$. Show your work.

Lesson 5-4

Write a function rule for each table.

61.

Number of People	Total Bill
1	$3.00
2	$6.00
3	$9.00
4	$12.00

62.

Amount Earned	Amount Spent
$15	$5
$30	$10
$45	$15
$60	$20

63.

Number of Days	Supplies Remaining
0	12 lb
2	10 lb
4	8 lb
6	6 lb

64.

Weight on Earth (lb)	Weight on Moon (lb)
96	16
123	20.5
144	24
171	28.5

Lessons 4-2, 4-3

Solve each inequality.

65. $r + 6 > -12$ **66.** $5 + c \leq 3.2$ **67.** $7m < -21$ **68.** $a - 4.5 \geq 12.1$

69. $\frac{n}{4} < -20$ **70.** $3t \geq 9.12$ **71.** $\frac{v}{-5} \leq \frac{1}{2}$ **72.** $b + 4\frac{2}{3} > 5\frac{1}{6}$

Lesson 2-3

73. Shipping For the ships that pass through the Panama Canal, the average toll is $45,000 per ship. The canal authority earned about $700 million in the year 2000. About how many ships passed through the canal that year? Round to the nearest hundred.

✓ Checkpoint Quiz 2   Lessons 5-3 through 5-5

Model each rule with a table of values and a graph. If the rule describes a direct variation, state the constant of variation.

1. $y = 4x + 1$ **2.** $y = \frac{1}{2}x$ **3.** $f(x) = -3x$ **4.** $y = -3x + 2$

Write a function rule for each situation.

5. the total cost $t(p)$ of p pounds of potatoes at $.79 per pound

6. the total distance $d(n)$ traveled in n hours at a constant speed of 60 mi/h

Write an equation for the direct variation that includes the given point.

7. $(7, -2)$ **8.** $(-3, -6)$ **9.** $(-4, -5)$

 10. a. Bicycling The distance a wheel moves forward varies directly with the number of rotations. Suppose the distance d the wheel moves is 56 ft when the number of rotations n is 8. Find the constant of variation and write a direct variation equation to model this situation.
   b. Use the direct variation you wrote for part (a) to find the distance the wheel moves in 20 rotations.

5-6

Inverse Variation

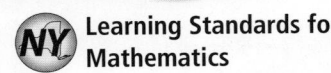
Learning Standards for Mathematics

A.CM.12 Understand and use appropriate language, representations, and terminology when describing objects, relationships, mathematical solutions, and rationale.

Suppose y varies directly with x. Find each constant of variation.

1. $y = 5x$ **2.** $y = -7x$ **3.** $3y = x$ **4.** $0.25y = x$

Write an equation of the direct variation that includes the given point.

5. $(2, 4)$ **6.** $(3, 1.5)$ **7.** $(-4, 1)$ **8.** $(-5, -2)$

 New Vocabulary   • inverse variation
• constant of variation for inverse variation

1 Solving Inverse Variations

Real-World Connection

With volunteer labor, Habitat for Humanity helped build over 115,000 homes for families around the world in its first 25 years.

SOURCE: *Habitat for Humanity*

Activity: Inverse Variation

Suppose you are part of a volunteer crew constructing affordable housing. Building a house requires a total of 160 workdays. For example, a crew of 20 people can complete a house in 8 days.

1. How long should it take a crew of 40 people?

2. Copy and complete the table.

Crew size (x)	Construction Days (y)	Total Workdays
2	80	160
5	▪	160
8	▪	▪
▪	16	▪
20	8	160
40	▪	▪

3. Graph the (x, y) data in the table above.

4. Describe what happens to construction time as the crew size increases.

In the table, the total number of workdays remains the same. The number of construction days decreases as the number of people on the crew increases. The relationship of construction days and crew size is an inverse variation.

 **Key Concepts**

Definition	Inverse Variation

An equation in the form $xy = k$ or $y = \frac{k}{x}$, where $k \neq 0$, is an **inverse variation.**

The **constant of variation for inverse variation** is k, the product $x \cdot y$ for an ordered pair (x, y).

Inverse variations have graphs with the same general shape. You can see from the graph at the right how the constant of variation k affects the graph of $xy = k$.

If you know the values of x and y for one point on the graph of an inverse variation, you can use the point to find the constant of variation k and the equation of the inverse variation.

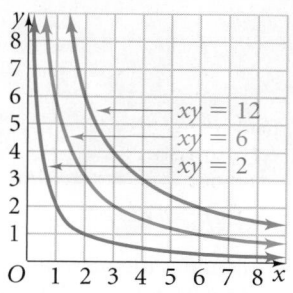

Vocabulary Tip

There are several ways to describe an inverse variation:

- y varies inversely with x.
- y varies inversely as x.
- y is inversely proportional to x.

1 EXAMPLE **Writing an Equation Given a Point**

Suppose y varies inversely with x and $y = 7$ when $x = 5$. Write an equation for the inverse variation.

$xy = k$ **Use the general form of an inverse variation.**

$5(7) = k$ **Substitute 5 for x and 7 for y.**

$35 = k$ **Multiply to solve for k.**

$xy = 35$ **Write an equation. Substitute 35 for k in $xy = k$.**

● The equation of the inverse variation is $xy = 35$, or $y = \frac{35}{x}$.

 **Quick Check** ❶ Suppose y varies inversely with x and $y = 9$ when $x = 2$. Write an equation for the inverse variation.

Suppose (x_1, y_1) and (x_2, y_2) are two ordered pairs of an inverse variation. Each ordered pair of an inverse variation has the same product k, that is $x_1 \cdot y_1 = k$ and $x_2 \cdot y_2 = k$. So $x_1 \cdot y_1 = x_2 \cdot y_2$.

2 EXAMPLE **Finding the Missing Coordinate**

The points $(3, 8)$ and $(2, y)$ are two points on the graph of an inverse variation. Find the missing value.

$x_1 \cdot y_1 = x_2 \cdot y_2$ **Use the equation $x_1 \cdot y_1 = x_2 \cdot y_2$ since you know coordinates but not the constant of variation.**

$3(8) = 2(y_2)$ **Substitute 3 for x_1, 8 for y_1, and 2 for x_2.**

$24 = 2(y_2)$ **Simplify.**

$12 = y_2$ **Solve for y_2.**

The missing value is 12. The point $(2, 12)$ is on the graph of the inverse variation that includes the point $(3, 8)$.

 **Quick Check** ❷ Each pair of points is on the graph of an inverse variation. Find the missing value.
  a. $(3, y)$ and $(5, 9)$ **b.** $(75, 0.2)$ and $(x, 3)$

3 **EXAMPLE** Real-World  **Problem Solving**

Physics The weight needed to balance a lever varies inversely with the distance from the fulcrum to the weight. Where should Julio, who weighs 150 lb, sit to balance the lever?

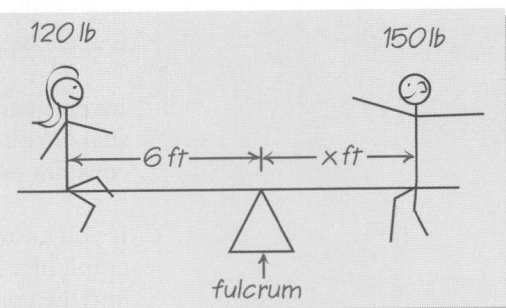

Relate A weight of 120 lb is 6 ft from the fulcrum. A weight of 150 lb is x ft from the fulcrum.
Weight and distance vary inversely.

Define Let weight$_1$ = 120 lb.
Let weight$_2$ = 150 lb.
Let distance$_1$ = 6 ft.
Let distance$_2$ = x ft.

Write weight$_1$ · distance$_1$ = weight$_2$ · distance$_2$

$$120 \cdot 6 = 150 \cdot x \qquad \text{Substitute.}$$
$$720 = 150x \qquad \text{Simplify.}$$
$$\frac{720}{150} = x \qquad \text{Solve for } x.$$
$$4.8 = x \qquad \text{Simplify.}$$

● Julio should sit 4.8 feet from the fulcrum to balance the lever.

 **Quick Check** **3** **a. Physics** A 100-lb weight is placed 4 ft from a fulcrum. How far from the fulcrum should a 75-lb weight be placed to balance the lever?
b. An 80-lb weight is placed 9 ft from a fulcrum. What weight should you put 6 ft from the fulcrum to balance the lever?

Real-World  Connection

A fulcrum is the point at which a lever pivots. Students can use levers in science labs to investigate physical properties.

---

2 **Comparing Direct and Inverse Variation**

Recall that a direct variation is an equation in the form $y = kx$. This summary will help you recognize and use direct and inverse variations.

 Key Concepts

Summary	**Direct and Inverse Variation**

Direct Variation

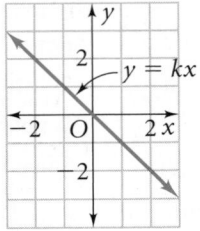

Inverse Variation

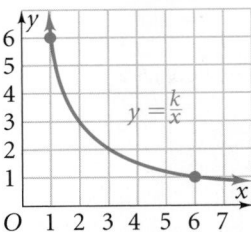

y varies directly with x.

y is directly proportional to x.

The ratio $\frac{y}{x}$ is constant.

y varies inversely with x.

y is inversely proportional to x.

The product xy is constant.

4 EXAMPLE Determining Direct or Inverse Variation

Do the data in each table represent a *direct variation* or an *inverse variation*? For each table, write an equation to model the data.

a.

x	y
2	5
4	10
10	25

The values of y seem to vary directly with the values of x. Check each ratio $\frac{y}{x}$.

$\begin{array}{l} y \to \\ x \to \end{array} \frac{5}{2} = 2.5 \qquad \frac{10}{4} = 2.5 \qquad \frac{25}{10} = 2.5$

The ratio $\frac{y}{x}$ is the same for all pairs of data. So this is a direct variation, and $k = 2.5$.

The equation is $y = 2.5x$.

b.

x	y
5	20
10	10
25	4

The values of y seem to vary inversely with the values of x. Check each product xy.

xy: $5(20) = 100$ $\qquad$ $10(10) = 100$ $\qquad$ $25(4) = 100$

The product xy is the same for all pairs of data. So this is an inverse variation, and $k = 100$.

The equation is $xy = 100$.

✓ Quick Check **4** Determine whether the data in each table represent a direct variation or an inverse variation. Write an equation to model the data in each table.

a.

x	y
3	12
6	6
9	4

b.

x	y
3	12
5	20
8	32

Many real-world situations involve variation. You can look for a constant ratio or a constant product to determine whether the relationship is a direct variation or an inverse variation.

5 EXAMPLE Real-World Problem Solving

Explain whether each situation represents a direct variation or an inverse variation.

a. **Carpooling** The cost of $20 worth of gasoline is split among several people.

The cost per person times the number of people equals the total cost of the gasoline. Since the total cost is a constant product of $20, this is an inverse variation.

b. **School Supplies** You buy several markers for 70¢ each.

The cost per marker times the number of markers equals the total cost of the markers. Since the ratio $\frac{\text{cost}}{\text{marker}}$ is constant at 70¢ each, this is a direct variation.

✓ Quick Check **5** Explain whether each situation represents a direct variation or an inverse variation.
 a. You are in a discount store. All sweaters are on sale for $15 each.
 b. You walk 5 miles each day. Your speed and time vary from day to day.

Online
active math

For: Variation Activity
Use: Interactive Textbook, 5-6

EXERCISES

For more exercises, see *Extra Skill and Word Problem Practice.*

Practice and Problem Solving

 Practice by Example

Example 1
(page 285)

 **for Help**

Suppose *y* varies inversely with *x*. Write an equation for the inverse variation.

1. $y = 6$ when $x = 3$ **2.** $y = 1$ when $x = 2$ **3.** $y = 7$ when $x = 8$

4. $y = 3$ when $x = 0.5$ **5.** $y = 10$ when $x = 2.4$ **6.** $y = 3.5$ when $x = 2.2$

7. $y = 6$ when $x = \frac{1}{3}$ **8.** $y = \frac{1}{16}$ when $x = 8$ **9.** $y = \frac{1}{10}$ when $x = \frac{3}{5}$

Example 2
(page 285)

Each pair of points is on the graph of an inverse variation. Find the missing value.

10. $(6, 12)$ and $(9, y)$ **11.** $(3, 5)$ and $(1, n)$ **12.** $(x, 11)$ and $(1, 66)$

13. $(x, 55)$ and $(5, 77)$ **14.** $(9.4, b)$ and $(6, 4.7)$ **15.** $(50, 13)$ and $(t, 5)$

16. $(4, 3.6)$ and $(1.2, g)$ **17.** $(24, 1.6)$ and $(c, 0.4)$ **18.** $(500, 25)$ and $(4, n)$

19. $\left(\frac{1}{2}, 24\right)$ and $(6, y)$ **20.** $\left(x, \frac{1}{2}\right)$ and $\left(\frac{1}{3}, \frac{1}{4}\right)$ **21.** $\left(\frac{1}{2}, 5\right)$ and $\left(b, \frac{1}{8}\right)$

Example 3
(page 286)

22. Travel Suppose you take $2\frac{1}{2}$ h to drive from your house to the lake at 48 mi/h. How long will your return trip take at 40 mi/h?

23. Bicycling Suppose a camper took 2 h to ride around a reservoir at 10 mi/h at the beginning of the summer. By the end of the summer, she can ride around the reservoir in $1\frac{1}{2}$ h. What is her rate at the end of the summer?

Example 4
(page 287)

Do the data in each table represent a direct variation or an inverse variation? Write an equation to model the data in each table.

24.

x	y
2	1
5	2.5
8	4

25.

x	y
4	15
6	10
10	6

26.

x	y
3	24
9	8
12	6

Example 5
(page 287)

Explain whether each situation represents a direct variation or an inverse variation.

27. You buy some chicken for $1.79/lb.

28. An 8-slice pizza is shared equally by a group of friends.

29. You find the length and width of several rectangles. Each has an area of 24 square units.

B **Apply Your Skills**

Find the constant of variation *k* for each inverse variation. Then write an equation for the inverse variation.

30. $y = 8$ when $x = 4$ **31.** $r = 3.3$ when $t = \frac{1}{3}$ **32.** $x = \frac{1}{2}$ when $y = 5$

33. $a = 25$ when $b = 0.04$ **34.** $p = 10.4$ when $q = 1.5$ **35.** $x = 5$ when $y = 75$

Geometry **Does each formula represent a direct or an inverse variation? Explain.**

36. the perimeter of an equilateral triangle: $P = 3s$

37. the time t to travel 150 mi at r mi/h: $t = \frac{150}{r}$

38. the circumference of a circle with radius r: $C = 2\pi r$

39. Surveying Each of two rectangular building lots is one quarter acre in size. One lot measures 99 ft by 110 ft. The other lot is 90 ft wide. What is the second lot's length?

40. Construction Suppose 4 people can paint a house if they work 3 days each. How long would it take a crew of 5 people to paint the house?

Do the data in each table represent a direct or an inverse variation? Write an equation to model the data. Then complete the table.

41.

x	y
10	4
20	■
8	3.2

42.

x	y
0.4	28
1.2	84
■	63

43.

x	y
1.6	30
4.8	10
■	96

44. Math in the Media According to the First Law of Air Travel, for each situation below, will the distance to your gate be *greater* or *less* for this trip than for your last trip?
   a. You have more luggage.
   b. You have less time to make your flight.
   c. You have less luggage.

45. a. Earnings Suppose you want to earn $80. How long will it take you if you are paid $5/h; $8/h; $10/h; $20/h?
   b. What are the two variable quantities in part (a)?
   c. Write an equation to represent this situation.

46. Open-Ended Write and graph a direct variation and an inverse variation that have the same constant of variation.

CLOSE TO HOME by John McPherson

The First Law of Air Travel
The distance to your connecting gate is directly proportional to the amount of luggage you are carrying and inversely proportional to the amount of time you have.

47. Multiple Choice Boyle's Law states that volume V varies inversely with pressure P for any gas at a constant temperature in an enclosed space. Suppose a gas at constant temperature occupies 15.3 liters at a pressure of 40 millimeters of mercury. Which equation models this situation?
   Ⓐ $612 = PV$ Ⓑ $P = 15.3V$ Ⓒ $V = 40P$ Ⓓ $P + V = 40$

48. Critical Thinking The graphs p and q represent a direct variation and an inverse variation. Write an equation for each graph.

49. Writing Explain how the variable y changes in each situation.
   a. y varies directly with x. The value of x is doubled.
   b. y varies inversely with x. The value of x is doubled.

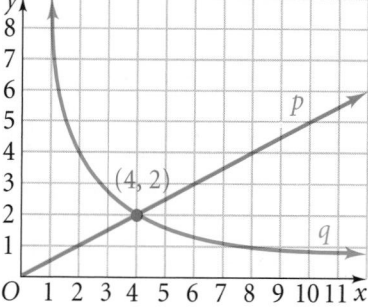

 **Challenge** **50. Physics** The intensity of a sound s varies inversely with the square of the distance d from the sound. This can be modeled by the equation $sd^2 = k$. If you move half the distance closer to the source of a sound, by what factor will the intensity of the sound increase? Explain your reasoning.

51. Write an equation to model each situation.
a. y varies inversely with the fourth power of x.
b. y varies inversely with the fourth power of x and directly with z.

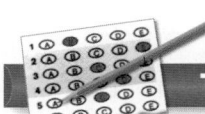

Test Prep

Multiple Choice

52. Suppose y varies inversely with x and $x = 12$ when $y = 3$. What is the equation of the inverse variation?

A. $\frac{y}{x} = 36$ **B.** $y =$ **C.** $y =$ **D.** $12 = \frac{3}{y}$

53. The volume V of a gas varies inversely with the pressure P. When the volume is 75 in.^{3}, the pressure is 30 lb/in.^{2}. What is the volume when the pressure is 25 lb/in.^{2}?

F. 90 in.^{3} **G.** $58\frac{1}{3}$ in.^{3} **H.** 60 in.^{3} **J.** 30 in.^{3}

Short Response

54. Use the table at the right. Find a value such that the y-values vary directly with the x-values. Then find a value such that the y-values vary inversely with the x-values. Show your work.

x	y
5	10
8	■

Extended Response

55. You are traveling to visit your best friend who moved 100 miles away.
a. Copy and complete the table below to find the time the trip takes at different speeds.
b. Describe the relationship of the variables.
c. How long would the round trip take if you could travel at 80 mi/h?

Distance (*d*)	100	100	100	100
Speed (*r*)	30	40	50	60
Time (*t*)	■	■	■	■

Mixed Review

Lesson 5-5

Write an equation of the direct variation that includes the given point.

56. $(1, 6)$ **57.** $(-2, 8)$ **58.** $(10, -1)$
59. $(-5, -4)$ **60.** $(7, 11)$ **61.** $(-28, 4)$

Lesson 3-9

Use the Pythagorean Theorem to find the hypotenuse or the missing leg. Round to the nearest hundredth if necessary.

62. $FH = 25$ in., $HG = 10$ in.
63. $FH = 8$ cm, $FG = 15$ cm
64. $GH = 17$ ft, $FG = 20$ ft
65. $GH = 1.4$ in., $FH = 3.2$ in.

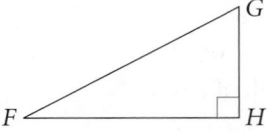

Lesson 2-4

Simplify each expression.

66. $-(x + 1)$ **67.** $6(8 - p)$ **68.** $(4n + 7)(1.5)$
69. $5k + 11k$ **70.** $0.25(24 - 16t)$ **71.** $99m - 36m$

290 Chapter 5 Graphs and Functions

Inverse Variation

In this activity you will use a paper "telescope." You will need two meter sticks, tape, two sheets of 8.5 × 11 in. paper (approximately 21.6 × 27.9 cm). Follow the steps below.

Step 1 Take each sheet of paper and draw dashed lines on them, as shown below.

Step 2 Take one sheet and add marks and labels at 4 cm, 8 cm, 12 cm, 16 cm, 20 cm, and 24 cm.

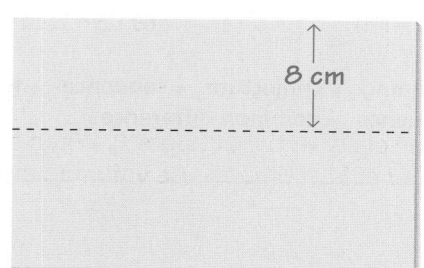

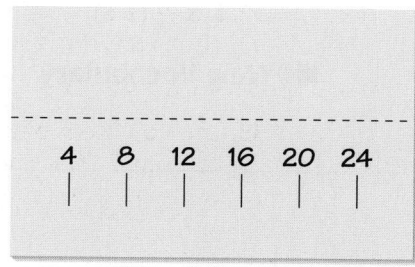

Step 3 Roll each sheet into a tube so that the bottom edge meets the dashed line. Tape it in place.

Step 4 Slide the measured tube inside the other tube.

ACTIVITY

Tape a meter stick to a wall at eye level. Stand between 1 m and 3 m from the target meter stick and mark your spot on the floor. You will take all measurements from the same distance.

Collapse the telescope so the measured tube is completely hidden. Draw a mark on the top of the telescope. Hold the telescope so the mark remains on the top. Also, do not bend the telescope.

Length of Telescope	Width of View
28 cm	■
32 cm	■
36 cm	■
40 cm	■
44 cm	■
48 cm	■
52 cm	■

1. **Data Collection** Copy the table at the right. When the telescope is completely collapsed, it will be nearly 28-cm long. Look through the telescope at the target and record the width of your view.

2. Extend the telescope so the 4-cm mark is just visible. Now the telescope is 32-cm long. Record the width of your view from the same spot on the floor. Repeat and complete the table. Graph your data.

3. **Predict** Use your graph to predict your width of view if the telescope was 24-cm long.

4. Based on the data, do you think this is a direct or inverse variation?

5. **a.** Find (length of telescope) × (width of view) for each pair of points.

 b. Find the mean of the products and use this for your value of k.

 c. Predict Use $w = \frac{k}{\ell}$ to predict the width of view w when the length of telescope ℓ is 40 cm.

 d. How does your prediction compare with the measured value?

5-7

Describing Number Patterns

 Learning Standards for Mathematics

A.PS.3 Observe and explain patterns to formulate generalizations and conjectures.

Check Skills You'll Need

GO for Help Lesson 1-2 and 2-2

Evaluate each expression for $x = 2, 3, 4$.

1. $9 + 3(x - 1)$ **2.** $8 + 7(x - 1)$ **3.** $0.4 - 3(x - 1)$

Subtract.

4. $8 - (-6)$ **5.** $-7 - 10$ **6.** $1.5 - 3.4$

New Vocabulary • inductive reasoning • conjecture • sequence • term • arithmetic sequence • common difference

---

1 Inductive Reasoning and Number Patterns

Suppose you are in a city and notice that the first three streets you pass are 10th Street, 11th Street, and 12th Street. You would probably conclude that the next street would be 13th Street. You would be basing your conclusion on inductive reasoning.

Inductive reasoning is making conclusions based on patterns you observe. A conclusion you reach by inductive reasoning is a **conjecture.**

1 EXAMPLE Extending Number Patterns

Use inductive reasoning to describe each pattern. Then find the next two numbers in each pattern.

a. 2,   5,   8,   11
     + 3   + 3   + 3

The pattern is "add 3 to the previous term." To find the next two numbers, you add 3 to each previous term: $11 + 3 = 14$ and $14 + 3 = 17$.

b. 2,   4,   8,   16
     × 2   × 2   × 2

The pattern is "multiply the previous term by 2." To find the next two numbers, you multiply each previous term by 2: $16 × 2 = 32$ and $32 × 2 = 64$.

c. $1, 4, 9, 16, \ldots$

The pattern is "square consecutive integers": $1^2, 2^2, 3^2, 4^2$. To find the next two numbers, square the next two consecutive integers: $5^2 = 25$ and $6^2 = 36$.

Vocabulary Tip

You read "..." at the end of a sequence as "and so on."

Quick Check ❶ Use inductive reasoning to describe each pattern. Then find the next two numbers in each pattern.
a. $3, 9, 27, 81, \ldots$ **b.** $9, 15, 21, 27, \ldots$ **c.** $2, -4, 8, -16, \ldots$

A number pattern is also called a **sequence.** Each number in a sequence is a **term** of the sequence.

One kind of number sequence is an arithmetic sequence. You form an **arithmetic sequence** by adding a fixed number to each previous term. This fixed number is the **common difference.**

$$-4, \quad 5, \quad 14, \quad 23$$
$$+9 \qquad +9 \qquad +9$$

2 EXAMPLE Finding the Common Difference

Find the common difference of each arithmetic sequence.

a. $-7, \quad -3, \quad 1, \quad 5$ **b.** 17,   13,   9,   5

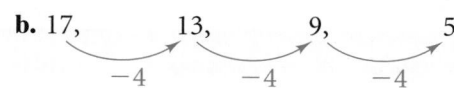

    $+4 \quad +4 \quad +4$ $-4 \quad -4 \quad -4$

The common difference is 4.   The common difference is -4.

✓ Quick Check   **2** Find the common difference of each sequence.
a. $11, 23, 35, 47, \ldots$ **b.** $8, 3, -2, -7, \ldots$

Consider the sequence $7, 11, 15, 19, \ldots$ Think of each term as the output of a function. Think of the term number as the input.

term number	1	2	3	4	← input
term	7	11	15	19	← output

You can use the common difference of the terms of an arithmetic sequence to write a function rule for the sequence. For the sequence $7, 11, 15, 19, \ldots$, the common difference is 4.

Let $n =$ the term number in the sequence.

Let $A(n) =$ the value of the nth term of the sequence.

$$A(1) = 7$$
$$A(2) = 7 + 4 = 7 + 1 \cdot 4$$
$$A(3) = 7 + 4 + 4 = 7 + 2 \cdot 4$$
$$A(4) = 7 + 4 + 4 + 4 = 7 + 3 \cdot 4$$
$$A(n) = 7 + 4 + 4 + 4 + \ldots + 4 = 7 + (n-1)4$$

4 is the common difference.
Note that the number in red is one less than the term number, which is in blue.

For an arithmetic sequence, you can use the first term, the term number, and the common difference to find the value of any given term.

Key Concepts

Rule	Arithmetic Sequence

$$A(n) = a + (n-1)d$$

nth term   first term   term number   common difference

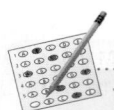

3 **EXAMPLE** **Finding Terms of a Sequence**

Gridded Response Find the tenth term of the sequence that has the rule $A(n) = 32 + (n - 1)(-2)$.

$$A(10) = 32 + (10 - 1)(-2)$$
$$= 32 + 9(-2)$$
$$= 14$$

- Fill in the grid with 14.

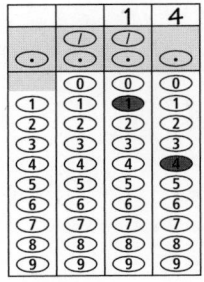

Quick Check **3** Find the first, sixth, and twelfth terms of each sequence.
  a. $A(n) = -5 + (n - 1)(3)$ **b.** $A(n) = 6.3 + (n - 1)(5)$

EXERCISES

For more exercises, see *Extra Skill and Word Problem Practice.*

Practice and Problem Solving

 **Practice by Example**

Example 1
(page 292)

Use inductive reasoning to describe each pattern. Then find the next two numbers in each pattern.

1. $4, 6, 8, 10, \ldots$ **2.** $4, 6, 9, 13\frac{1}{2}, \ldots$ **3.** $4, 6, 9, 13, \ldots$

4. $3, 3.04, 3.08, 3.12, \ldots$ **5.** $3, 3.3, 3.63, 3.993, \ldots$ **6.** $3, 1, -1, -3, \ldots$

7. $1.1, 2.2, 3.3, 4.4, \ldots$ **8.** $0.001, 0.01, 0.1, 1, \ldots$ **9.** $2, 8, 32, 128, \ldots$

10. $1, \frac{1}{4}, \frac{1}{9}, \frac{1}{16}, \ldots$ **11.** $9, -5, -19, -33, \ldots$ **12.** $1.5, 7.5, 37.5, 187.5, \ldots$

Example 2
(page 293)

Find the common difference of each arithmetic sequence.

13. $-5, -2, 1, 4, \ldots$ **14.** $-6, -10, -14, -18, \ldots$ **15.** $18, 7, -4, -15, \ldots$

16. $8, 21, 34, 47, \ldots$ **17.** $\frac{1}{2}, \frac{1}{3}, \frac{1}{6}, 0, \ldots$ **18.** $0.7, 1.5, 2.3, 3.1, \ldots$

19. $8, 6, 4, 2, \ldots$ **20.** $10, 22, 34, 46, \ldots$ **21.** $-9, -4, 1, 6, \ldots$

Example 3
(page 294)

Find the second, fifth, and ninth terms of each sequence.

22. $A(n) = 2 + (n - 1)(3)$ **23.** $A(n) = -9 + (n - 1)(6)$

24. $A(n) = -7 + (n - 1)(4)$ **25.** $A(n) = 8 + (n - 1)(9)$

26. $A(n) = 0.5 + (n - 1)(3)$ **27.** $A(n) = -5 + (n - 1)(7)$

28. $A(n) = 9 + (n - 1)(-6)$ **29.** $A(n) = -2.1 + (n - 1)(-5)$

30. $A(n) = 65 + (n - 1)(-7)$ **31.** $A(n) = 21 + (n - 1)(-4)$

32. $A(n) = -5 + (n - 1)(-3)$ **33.** $A(n) = 0.2 + (n - 1)(-1)$

B **Apply Your Skills**

Find the next two terms in each sequence.

34. $20, 14, 8, 2, \ldots$ **35.** $2, 2\frac{1}{4}, 2\frac{1}{2}, 2\frac{3}{4}, 3, \ldots$ **36.** $2, 5, 10, 17, \ldots$

37. $12, 4, 1\frac{1}{3}, \frac{4}{9}, \ldots$ **38.** $0, 3, 8, 15, 24, \ldots$ **39.** $-5, 4, 13, 22, \ldots$

40. $40, 20, 10, 5, \ldots$ **41.** $7, 7\frac{1}{4}, 7\frac{1}{2}, 7\frac{3}{4}, \ldots$ **42.** $12, -4, \frac{4}{3}, -\frac{4}{9}, \ldots$

 43. a. Writing Explain the difference between inductive and deductive reasoning.
    b. Open-Ended Give an example of inductive reasoning and of deductive reasoning.

Real-World Connection

About 15% of all trips on mass transit are students going to or from school or college.

44. Transportation Buses on your route run every 7 minutes from 6:30 A.M. to 10:00 A.M. You get to the bus stop at 7:56 A.M. How long will you have to wait for a bus?

45. Open-Ended Write a function rule for a sequence that has −30 as the eighth term.

For Exercises 46 and 47, write the first five terms in each sequence. Explain what the fifth term means in the context of the situation.

46. A baby's birth weight is 7 lb 4 oz. The baby gains 5 oz each week.

47. The balance of a car loan starts at $4,500 and decreases $150 each month.

48. Use the sequence $1, 2, 4, \ldots$
  a. Find the difference between consecutive terms in the sequence. Use inductive reasoning to make a conjecture about the next term in the sequence.
  b. Find the quotient of consecutive terms in the sequence. Use inductive reasoning to make a conjecture about the next term in the sequence.
  c. Critical Thinking Explain why having more than three terms in a sequence can help you make a conjecture that is more likely to be correct.

Is each given sequence arithmetic? Justify your answer.

49. $0.3, 3, 30, 300, \ldots$ **50.** $-3, -7, -11, -15, \ldots$ **51.** $1, 8, 27, 64, \ldots$

52. $2, 4, 8, 16, 32, \ldots$ **53.** $46, 31, 16, 1, \ldots$ **54.** $0.2, -0.6, -1.4, -2.2, \ldots$

55. The first five rows of Pascal's Triangle are at the right.
  a. Predict the numbers in the sixth row.
  b. Find the sum of the numbers in each of the first five rows. Predict the sum of the numbers in the sixth row.

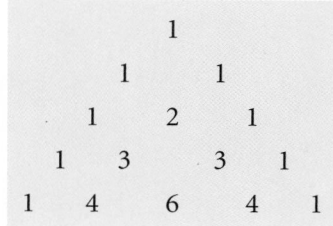

Find the second, fourth, and eighth terms of each sequence.

56. $A(n) = 11 + (n-1)\left(\frac{1}{3}\right)$ **57.** $A(n) = 9 + (n-1)(-4.5)$

58. $A(n) = -2 + (n-2)(-1.6)$ **59.** $A(n) = \frac{1}{5} + (n-1)\left(\frac{4}{5}\right)$

60. a. Complete the table at the right for an arithmetic sequence.
  b. Graph the ordered pairs (term number, term) on a coordinate plane.
  c. What do you notice about the points on your graph?

x	y
1	5
2	8
3	■
4	■

61. Music There are 52 white keys on a piano. The frequency produced when a key is struck is the number of vibrations per second the key's string makes.
  a. Reasoning Is this relation a function? Explain.
  b. Writing Describe the pattern in the relation.

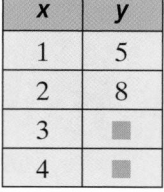

62. Number Theory The Fibonacci sequence is $1, 1, 2, 3, 5, 8, 13, \ldots$ After the first two numbers, each number is the sum of the two previous numbers.
   a. What is the next term of the sequence?
   b. What is the eleventh term of the sequence?
   c. Open-Ended Choose two other numbers to start a Fibonacci-like sequence. Write the first seven terms of your sequence.

A *recursive formula* **relates a new term of a sequence to the previous term of the sequence. Describe each of the sequences using a recursive formula.**

Sample $3, 7, 11, 15, \ldots$

$$\text{value of new term} = \text{value of previous term} + 4$$

63. $12, 18, 24, 30, \ldots$ **64.** $12, 18, 27, 40.5, \ldots$ **65.** $54, 51.5, 49, 46.5, \ldots$

66. $1.1, 5.1, 9.1, 13.1, \ldots$ **67.** $98, 14, 2, \frac{2}{7}, \ldots$ **68.** $-8, 20, -50, 125, \ldots$

C Challenge   **Find the common difference of each sequence. Then find the next term.**

69. $4, x + 4, 2x + 4, 3x + 4, \ldots$

70. $a + b + c, 4a + 3b + c, 7a + 5b + c, \ldots$

71. Use the sequence $10, 4, -2, -8, \ldots$
   a. What is the first term of the sequence?
   b. What is the common difference of the sequence?
   c. Write a function rule $A(n)$ for the sequence.

72. a. Draw the next figure in the pattern.

   b. Reasoning What is the color of the 20th figure? Explain.
   c. How many sides does the 28th figure have? Explain.

73. Use the arithmetic sequence $-5, 1, 7, 13, \ldots$
   a. What is the first term?
   b. What is the common difference?
   c. Use your answers from parts (a) and (b) to write a rule for the sequence.

 **NY REGENTS**

 **Test Prep**

Multiple Choice

74. What is the seventh term of the sequence $24, 12, 6, 3, \ldots$?
   A. 0   **B.** 0.25   **C.** 0.375   **D.** 1.5

75. What is the common difference of the arithmetic sequence $9, -1, -11, -21, \ldots$?
   F. -10 **G.** -9 **H.** 9   **J.** 10

76. What is the common difference of the arithmetic sequence $\frac{1}{5}, \frac{6}{5}, \frac{11}{5}, \frac{16}{5}, \ldots$?
   A. 1   **B.** $1\frac{1}{5}$ **C.**   **D.** 5

77. What is the seventh term of the sequence $A(n) = -9 + (n - 1)0.5$?
   F. -7 **G.** -6.5 **H.** -6 **J.** -5.5

78. What is the first term of the sequence $A(n) = (n - 1)(-3)$?

    A. -3 **B.** -2 **C.** 0 **D.** 1

79. What is the next term in the sequence $x - 4, x - 2, x, x + 2, \ldots$?

    F. $2x$ **G.** $x + 3$ **H.** $x + 4$ **J.** $2x + 2$

Short Response

80. Explain how to find the seventh term of the sequence 24, 21, 18, 15, . . .

Extended Response

81. Marta started to work at a company in the year 2001. Her yearly salary was $26,500. At the beginning of the next year she received a $2,880 raise. Assume that she receives the same raise each year.

    a. Write a function $f(n)$ to find Marta's salary n years after 2001.

    b. Find Marta's salary in 2008. Show your work.

Mixed Review

Lesson 5-6

Suppose y varies inversely with x. Write an equation for the inverse variation.

82. $y = 4$ when $x = 5$ **83.** $y = 1.2$ when $x = 8$

84. $y = \frac{1}{2}$ when $x = 24$ **85.** $y = 7$ when $x = 6.1$

Each pair of points is on the graph of an inverse variation. Find the missing value.

86. $(8, 9)$ and $(p, 6)$ **87.** $(x, 12)$ and $\left(\frac{1}{3}, 18\right)$

88. $(2.5, 16)$ and $(5, m)$ **89.** $(30, 2)$ and $(t, 6)$

Lesson 5-5

Write an equation of the direct variation that includes the given point.

90. $(4, -5)$ **91.** $(0.5, 12)$ **92.** $(-1, 14)$ **93.** $(10, 1.4)$

94. $(1.1, -3.1)$ **95.** $(11, -3.1)$ **96.** $(2, -3)$ **97.** $\left(\frac{1}{2}, \frac{1}{3}\right)$

Lesson 5-2

Find the range of each function for the domain $\{-2, 1, 5\}$.

98. $f(x) = -4x$ **99.** $g(x) = 1 - 4x$ **100.** $y = 3x + 4$

101. $y = 2|x|$ **102.** $h(x) = |2x|$ **103.** $f(x) = \frac{3}{4}x - 5$

A P●int in Time

1500   1600   1700   1800   1900   2000

In 1971, Romana Acosta Bañuelos became the first Mexican American woman to hold the office of United States Treasurer. Before her appointment to this post by President Nixon, she founded and managed her own multimillion-dollar food enterprise and established the Pan American National Bank of East Los Angeles. As a highly successful businesswoman, she had to work on a daily basis with interest rates, balance sheets, investments, and other activities required in the corporate world.

For: Information about the office of United States Treasurer

PHSchool.com   **Web Code:** ate-2032

Using a Variable

You can solve many problems by using a variable to represent an unknown quantity. Try to let the variable be the quantity that you are looking for. Then use the variable to write an equation or inequality.

1 EXAMPLE

A brand of cereal comes in two sizes. The 12-oz size costs $4.35. At that rate, how much should the 20-oz box cost?

The problem is asking for the cost of a 20-oz box. Let the variable x be the cost of the 20-oz box. Write and solve a proportion to answer the question.

$\frac{12}{20} = \frac{4.35}{x}$ **Write a proportion.**

$12x = 20(4.35)$ **Find the cross products.**

$12x = 87.00$ **Simplify.**

$x = 7.25$ **Divide each side by 12.**

● The 20-oz box should cost about $7.25.

2 EXAMPLE

One house painter charges an initial fee of $25, plus $15 per hour. A second painter charges $25 per hour. Find out how many hours a job takes for the charge of the second painter to be the same as the charge of the first painter.

Let h = number of hours each painter must work for the charges to be the same. Then write an equation that expresses the charges for each painter.

First painter   Second painter

$25 + 15h \quad = \quad 25h$

$\qquad\qquad 25 = 10h$ **Subtract 15h from each side.**

$\qquad\qquad 2.5 = h$ **Divide each side by 10.**

● The charges are the same when both painters have worked 2.5 hours.

EXERCISES

1. Another way to solve the problem in Example 2 is to try values and test them until you find the correct answer. What is the advantage of using a variable?

2. The pressure of water varies directly with the depth. At 98 meters, the pressure is 10.21 atmospheres. Round to the nearest tenth.
   a. Let x be the depth where the pressure is 5 atmospheres. Use this variable to write and solve an equation to find that depth.
   b. Let x be the pressure at a depth of 150 meters. Use this variable to write and solve an equation to find the pressure.

Chapter Review

Vocabulary Review

 arithmetic sequence (p. 293)
common difference (p. 293)
conjecture (p. 292)
constant of variation for direct
   variation (p. 278)

constant of variation for inverse
   variation (p. 285)
continuous data (p. 264)
direct variation (p. 278)
discrete data (p. 264)
function notation (p. 258)

inductive reasoning (p. 292)
inverse variation (p. 285)
relation (p. 257)
sequence (p. 293)
term (p. 293)
vertical-line test (p. 258)

Match the vocabulary term in the column on the left with the most specific description in the column on the right.

1. direct variation

2. inductive reasoning

3. independent variable

4. function

5. range

6. sequence

7. conjecture

A. *x*-coordinate

B. *y*-coordinate

C. a function that can be expressed in the form $y = kx$, where $k \neq 0$

D. drawing conclusions based on observed patterns

E. a relation with exactly one value of the dependent variable for each value of the independent variable

F. a conclusion based on inductive reasoning

G. a number pattern

Go Online
PHSchool.com
For: Vocabulary quiz
Web Code: atj-0551

Skills and Concepts

5-1 Objective

▼ To interpret, sketch, and analyze graphs from situations (p. 252)

A graph shows a visual representation of the relationship between two sets of data.

Describe a situation for each graph.

8.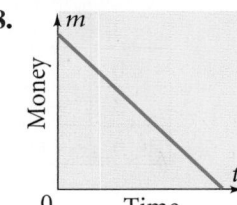

9.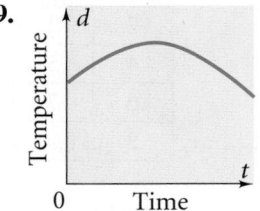

10.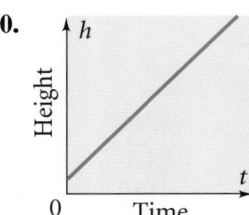

Sketch a graph of each situation. Label each section.

11. the height of a sunflower over a summer

12. the number of customers in a restaurant each hour of one day

13. the number of vehicles that enter a school parking lot during one day

14. the number of bags of peanuts sold during a 2-hour baseball game

▼ To identify relations and functions (p. 257)

▼ To evaluate functions (p. 258)

A **relation** is a set of ordered pairs. A function is a relation that assigns exactly one value in the range to each value in the domain. A function rule is an equation that describes a function. A function is in **function notation** when it uses $f(x)$ for the outputs.

Find the range of each function when the domain is {−4, 0, 1, 5}.

15. $y = 4x - 7$ **16.** $m = 0.5n + 3$ **17.** $p = q^2 + 1$ **18.** $w = 5 - 3z$

Determine whether each relation is a function.

19.

x	y
0	1
1	2
2	3
1	4

20.

x	y
0	−2
2	0
−2	−4
4	2

21.

x	y
2	−3
−1	−3
0	−3
5	−3

22. Use the vertical-line test to determine if the graph at the right is a function.

23. Writing When is a relation also a function?

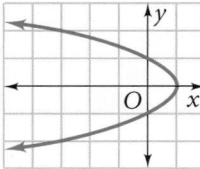

▼ To model functions using rules, tables, and graphs (p. 263)

▼ To write a function rule given a table or a real-world situation (p. 270)

When you graph data, put the independent variable on the horizontal axis and the dependent variable on the vertical axis.

Discrete data are data that involve a count of items. For discrete data, indicate each data item with a point and connect the points with a dashed line. **Continuous data** are data where numbers between any two data values have meaning. Use a solid line to indicate continuous data.

Model each rule with a table of values and a graph.

24. $f(x) = x^2 - 3$ **25.** $f(x) = -\frac{1}{2}x - 3$ **26.** $y = |x| - 7$ **27.** $y = 2x + 1$

Write a function rule for each table of values.

28.

x	f(x)
2	3
4	5
6	7
8	9

29.

x	f(x)
−3	3
0	0
3	−3
6	−6

30.

x	f(x)
3.0	6.5
3.5	7.0
4.0	7.5
4.5	8.0

 31. Weather The table at the right compares inches of snow to the corresponding amounts of rain.

    a. Write a function rule that models the data.

    b. Does the function represent discrete or continuous data? Explain.

Precipitation

Snow (in.)	Rain (in.)
3	0.3
5	0.5
10	1.0
7.5	0.75

A function is a **direct variation** if it has the form $y = kx$, where $k \neq 0$. The coefficient k is the **constant of variation**.

Is each equation a direct variation? If it is, find the constant of variation.

32. $f(x) = -3x$

33. $y = x - 3$

34. $y = 2x + 5$

35. $y = \frac{2}{5}x$

Write an equation of the direct variation that includes the given point.

36. $(5, 1)$

37. $(-2, -2)$

38. $(1, 2)$

39. $(-2, 6)$

When two quantities are related so that their product is a nonzero constant, they form an **inverse variation**. An inverse variation can be written $xy = k$, where k is the **constant of variation**.

Suppose y varies inversely with x. Write an equation for each inverse variation.

40. $x = 6$ when $y = 1$

41. $x = 90$ when $y = 0.1$

42. $x = 88$ when $y = 0.05$

Each pair of points is on the graph of an inverse variation. Find the missing value.

43. $(9, x)$ and $(3, 12)$

44. $(4, 2.65)$ and $(y, 4.24)$

45. $(r, 100)$ and $(75, 25)$

Do the data in each table represent a direct variation or an inverse variation? Write an equation to model the data in each table.

46.

x	y
2	35
5	14
10	7

47.

x	y
3	24.6
5	41
10	82

48.

x	y
1	3
4	$\frac{3}{4}$
9	$\frac{1}{3}$

Inductive reasoning is the process of making conclusions or **conjectures** based on patterns you observe. A number pattern is called a **sequence**, and each number in the sequence is a **term**.

An **arithmetic sequence** is formed by adding a fixed number, the **common difference**, to each previous term.

Use inductive reasoning to describe each pattern. Then find the next three numbers in each pattern.

49. $99, 90, 81, 72, \ldots$

50. $5, 8, 11, 14, \ldots$

51. $12, 23, 34, 45, \ldots$

Find the third, eighth, and tenth terms of each sequence.

52. $A(n) = -1 + (n - 1)\,2$

53. $A(n) = 4 + (n - 1)\,3$

54. $A(n) = 1.5 + (n - 1)\,1.5$

55. $A(n) = 4 + (n - 1)(-3)$

Determine whether each sequence is arithmetic. If it is, find the next three terms.

56. $14, 21, 28, 35, \ldots$

57. $16, -8, 4, -2, \ldots$

Chapter Test

Go Online
PHSchool.com
For: Chapter Test
Web Code: ata-0552

Sketch a graph of each situation. Label each section.

1. the speed of a bicycle during an afternoon ride

2. the amount of milk in your container over one lunch period

Determine whether each relation is a function. If the relation is a function, state the domain and range.

3.

x	y
−2	5
8	6
3	12
5	6

4.

x	y
9	6
3	8
4	9.5
9	2

5. **Writing** Explain how to use the vertical-line test to determine whether a graph is a graph of a function.

Find the range of each function when the domain is {−3, −1.5, 0, 1, 4}.

6. $r = 4t^2 + 5$

7. $m = -3n - 2$

Model each rule with a table of values and a graph.

8. $f(x) = 1.5x - 3$

9. $f(x) = -x^2 + 4$

Write a function rule to describe each statement.

10. the cost in dollars of printing dollar bills when it costs 3.8¢ to print a dollar bill

11. the amount of money you earn mowing lawns at $15 per lawn

12. the profit you make selling flowers at $1.50 each when each flower costs you $.80

Write a function rule for each table of values.

13.

x	y
0	1
1	3
2	5
−3	−5

14.

x	f(x)
0	0
1	−4.5
−1	4.5
2	−9

15. **Open-Ended** Describe a situation that could be modeled by the equation $y = 5x$.

16. **Purchasing** The price of turkey depends on its weight. Suppose turkeys sell for $.59 per lb.
    a. Write a rule to describe the function.
    b. What is the price of a 14-lb turkey?
    c. If you had $10 to buy a turkey, how big a turkey could you buy?

Write an equation of the direct variation that includes the given point.

17. $(2, 2)$ 18. $(-8, -4)$ 19. $(3, -1)$ 20. $(-5, 3)$

Determine whether each of the following graphs shows a direct variation. Write an equation for each direct variation.

21.
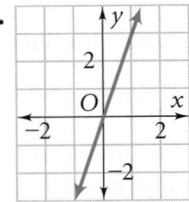

22.
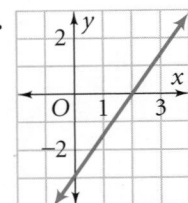

23. **Plumbing** The total amount of water dripping from a leaky faucet varies directly with time. If water drips at the rate of 5 mL/min, how much water drips in 30 min?

Find the common difference for each arithmetic sequence. Then find the next three terms.

24. $-55, -50, -45, -40, \ldots$ 25. $1.7, 2.7, 3.7, 4.7, \ldots$

Find the fifth term of each arithmetic sequence.

26. $A(n) = 2 + (n - 1)(-2.5)$
27. $A(n) = -9 + (n - 1)\,3$

Find the constant of variation k for each inverse variation.

28. $y = 5$ when $x = 6$ 29. $y = 78$ when $x = 0.1$

30. Write a function rule for the cost of catfish shown in the table below.

Weight (lb)	1	2	3	4	5
Cost (dollars)	3	6	9	12	15

Is each sequence arithmetic? Justify your answer.

31. $128, 64, 32, 16, \ldots$ 32. $3, 3.25, 3.5, 3.75, \ldots$

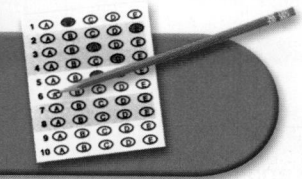

Regents Test Prep

Reading Comprehension Read the passage below. Then answer the questions on the basis of what is *stated* or *implied* in the passage.

Train Math Amtrak's Acela regional train has taken more than an hour off the old five-hour train trip from Boston to New York City. The faster Acela Express makes the 231-mile run in about 3.5 hours. The Express goes from New York to Washington, D.C., in about 2.75 hours.

A train's speed depends on how secure the track is and how well banked the curves are. On the best stretches, the Express can go as fast as 150 miles an hour.

For the New York–Washington, D.C., run, Amtrak carries 70% of the passengers traveling by either train or air. For the Boston–New York run, Amtrak carries only 30% of the passengers. The average number of riders that Amtrak carries in one month on the Boston–New York run is 100,404. The average number of riders in a month on the New York–Washington, D.C., run is 771,900. Amtrak hopes the new, faster train will increase ridership between Boston and New York.

1. Which is closest to the average speed for the old five-hour Boston–New York run?

   (1) 30 miles per hour
   (2) 40 miles per hour
   (3) 80 miles per hour
   (4) 1000 miles per hour

2. How much longer was the old five-hour Boston–New York run than the same trip on the Acela Express?

   (1) 0.75 hour
   (2) 1.5 hours
   (3) 2.25 hours
   (4) 2.5 hours

3. Which is closest to the average speed of the Boston–New York run for the Acela Express?

   (1) 57 miles per hour
   (2) 70 miles per hour
   (3) 85 miles per hour
   (4) 114 miles per hour

4. What is the percent of change between the Acela and Acela Express Boston–New York trip times?

   (1) 12.5%   **(2)** 20%
   (3) 30%   **(4)** 70%

5. If the Acela Express could make the entire trip from Boston to New York at 150 miles an hour, about how long would it take?

   (1) 1 hour
   (2) $1\frac{1}{2}$ hours
   (3) $2\frac{1}{5}$ hours
   (4) 3 hours

6. If you took the Acela Express train from Boston to Washington, D.C., what portion of your travel time would be spent on the part of the trip between Boston and New York?

   (1) $\frac{11}{14}$ **(2)** $\frac{14}{25}$
   (3) $\frac{14}{11}$ **(4)** $\frac{25}{14}$

7. Providence, Rhode Island, is on the Boston–New York run. It is about 180 miles from New York City. If the Acela travels at a constant speed, what portion of its Boston–New York run is spent on the Boston–Providence section?

8. If the number of passengers on the Boston–New York run doubled, would the percent of passengers on the Boston–New York run also double? Justify your answer.

Data Analysis

Activity Lab

NY A.S.9: Analyze and interpret a frequency distribution table or histogram, a cumulative frequency distribution table or histogram, or a box-and-whisker plot.

Histograms

A histogram is a special type of bar graph that shows the frequency a data item occurs. Histograms often combine data into intervals of equal size. The intervals do not overlap.

ACTIVITY

The histogram shows the amount of money that 50 customers spent in a supermarket.

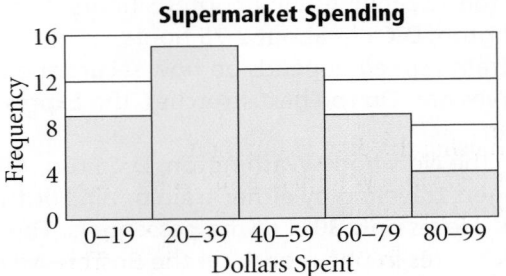

1. Based on the intervals, what is the greatest amount of money that any customer spent?

2. Which interval represents the greatest number of customers?

3. How many customers spent less than $20?

4. **Writing** Summarize the spending of the 50 customers represented in the histogram.

2 ACTIVITY

The data below shows the number of winning points scored at 15 NCAA Division I Women's Basketball Championship games from 1991 to 2005.

Winning Points Scored

84	70	73	82	68
71	62	93	68	83
70	60	84	78	70

Number of Winning Points

Winning	Tally	Frequency
59–64	▪	▪
65–70	▪	▪
71–76	▪	▪
77–82	▪	▪
83–88	▪	▪
89–94	▪	▪

5. **a.** Copy the frequency table at the right.
   b. Make a tally mark for each score in the appropriate interval.
   c. After you have tallied all the scores, record the frequency in the third column.

6. Use the intervals and the frequencies to construct a histogram.

7. What interval do most of the scores fall into?

8. **Critical Thinking** How would a histogram with 5-point intervals differ from the one you made with 6-point intervals?

You can describe histograms by their shape. Three types are shown below.

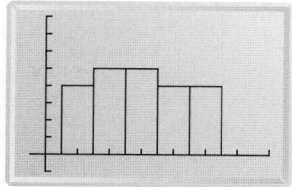

If the bars are roughly the same height, the histogram has a uniform shape.

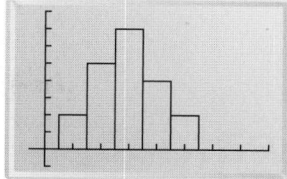

A histogram with a central peak has a symmetric shape.

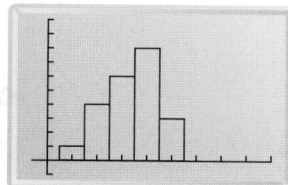

If the peak is not in the center of the distribution, the histogram has a skewed shape.

3 ACTIVITY

9. a. Make a histogram using the data at the right.
   b. Which shape best describes your histogram: *uniform, symmetric,* or *skewed*? Explain.

You can also construct a cumulative frequency histogram. This displays the *total* frequency of all data up to and including each interval.

10. Copy and complete the table below.

Interval (pounds)	0–3	0–7	0–11	0–15	0–19
Cumulative Frequency	3	13	■	■	■

Weight of Cats

Interval (pounds)	Frequency
0–3	3
4–7	10
8–11	7
12–15	4
16–19	1

11. a. Draw vertical and horizontal axes.
   b. Use the interval labels from the table to mark the horizontal axis. Label it: "Weights."
   c. Mark the vertical axis up to 25. Label it: "Cumulative Frequency."
   d. Use the cumulative frequencies you found to construct a cumulative frequency histogram.

12. Writing The bars that represent the last three intervals are all about the same height. Explain what this means.

EXERCISES

13. a. Data Collection Roll three number cubes at least 40 times. Record the sum of each roll.
   b. Make a frequency table of the sums using an interval of 3. The first interval will be 3–5. Then construct a histogram.
   c. Which shape best describes your histogram: *uniform, symmetric,* or *skewed*? Explain.

14. a. Data Collection Gather 30 or more pennies. Record their dates. Make a frequency table of the dates and construct a histogram.
   b. Which shape best describes your histogram: *uniform, symmetric* or *skewed*? Explain.

What You've Learned

NY Learning Standards for Mathematics

In Chapter 1, you

● **NY A.A.5:** explored patterns and functions.

In Chapters 3 and 4, you

● **NY A.A.6:** solved multi-step problems, including equations, inequalities, and proportions.

In Chapter 5, you

● **NY A.G.4:** graphed functions by making a table of values.

 **Check Your Readiness**

 for Help to the Lesson in green.

Adding and Subtracting Real Numbers (Lessons 2-1 and 2-2)

Simplify each expression.

1. $-5 + 7$ **2.** $2 - (-3)$ **3.** $-\frac{3}{4} + \frac{5}{6}$ **4.** $11 + (-4)$ **5.** $|1 - 8|$

Analyzing Data Using Scatter Plots (Lesson 1-5)

Make a scatter plot of the data below.

6.

Average Life Span of American Currency

Value of Currency ($)	1	5	10	20	50	100
Time (years)	1.5	1.25	1.5	2	5	8.5

Solving Equations (Lesson 3-3)

Solve each equation. Check your solution.

7. $3x + 4x = 8 - x$ **8.** $12 - 3d = d$ **9.** $6x - 8 = 7 + x$

Transforming Equations (Review page 140)

Solve each equation for y.

10. $2y - x = 4$ **11.** $3x = y + 2$ **12.** $-2y - 2x = 4$

Graphing Functions (Lesson 5-3)

Make a table of values and graph each function.

13. $y = -\frac{2}{3}x$ **14.** $y = 2x + 1$ **15.** $y = x - 5$

Linear Equations and Their Graphs

What You'll Learn Next

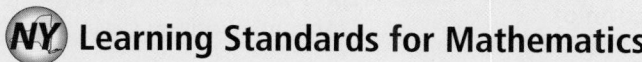 **Learning Standards for Mathematics**

● **NY A.G.4:** You will write linear equations and recognize their different forms.

● **NY A.A.32:** By working with the rate of change, you will understand how the slope of a line can be interpreted in real-world situations.

● **NY A.A.38:** You will determine whether the graphs of two linear equations are parallel or perpendicular.

 **Activity Lab** Applying what you learn, you will do activities involving pyramids, on pages 370–371.

Rate of Change and Slope

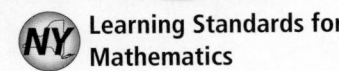
Learning Standards for Mathematics

A.A.32 Explain slope as a rate of change between dependent and independent variables.

A.A.33 Determine the slope of a line, given the coordinates of two points on the line.

✓ **Check Skills You'll Need**

Evaluate each function rule for $x = -5$.

1. $y = x - 7$

2. $y = 7 - x$

3. $y = 2x + 5$

4. $y = -\frac{2}{5}x + 3$

Write in simplest form.

5. $\frac{7 - 3}{3 - 1}$

6. $\frac{3 - 5}{6 - 0}$

7. $\frac{8 - (-4)}{3 - 7}$

8. $\frac{-1 - 2}{0 - 5}$

9. $\frac{-6 - (-4)}{-2 - 6}$

10. $\frac{0 - 1}{1 - 0}$

GO for Help Lessons 5-2 and 2-2

🔊 **New Vocabulary** • rate of change • slope

---

1 Finding Rates of Change

Activity: Exploring Rate of Change

The diagram at the right shows the side view of a ski lift.

1. What is the vertical change from *A* to *B*? From *B* to *C*? From *C* to *D*?

2. What is the horizontal change from *A* to *B*? From *B* to *C*? From *C* to *D*?

3. Find the ratio of the vertical change to the horizontal change for each section of the ski lift.

4. Which section is the steepest? How does the ratio for that section compare to the ratios of the other sections?

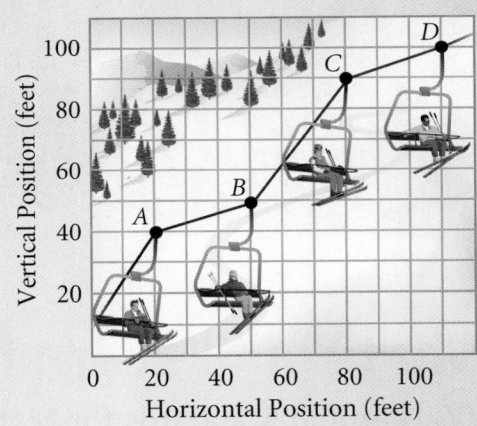

In the graph above, $\overline{AB}$ and $\overline{BC}$ have different rates of change.

Vocabulary Tip

A <u>rate</u> is a comparison of two quantities measured in different units.

Rate of change allows you to see the relationship between two quantities that are changing. If one quantity depends on the other, then the following is true.

$$\text{rate of change} = \frac{\text{change in the dependent variable}}{\text{change in the independent variable}}$$

Cost of Renting a Computer

Number of Days	Rental Charge
1	$60
2	$75
3	$90
4	$105
5	$120

1 EXAMPLE Finding Rate of Change Using a Table

Business For the data at the left, is the rate of change for each pair of consecutive days the same? What does the rate of change represent?

$$\text{rate of change} = \frac{\text{change in cost}}{\text{change in number of days}}$$ **Cost depends on the number of days.**

$$\frac{75 - 60}{2 - 1} = \frac{15}{1} \qquad\qquad \frac{90 - 75}{3 - 2} = \frac{15}{1}$$

$$\frac{105 - 90}{4 - 3} = \frac{15}{1} \qquad\qquad \frac{120 - 105}{5 - 4} = \frac{15}{1}$$

The rate of change for each consecutive pair of days is $\frac{15}{1}$. The rate of change is the same for all the data. It costs $15 for each day a computer is rented after the first day.

✓ **Quick Check** **1 a.** Find the rate of change using Days 5 and 2.
   b. Critical Thinking Does finding the rate of change for just one pair of days mean that the rate of change is the same for all the data? Explain.

The graphs of all the ordered pairs (number of days, cost) in Example 1 lie on a line as shown at the right. So, the data are linear.

You can use a graph to find a rate of change. Recall that the independent variable is plotted on the horizontal axis and the dependent variable is plotted on the vertical axis.

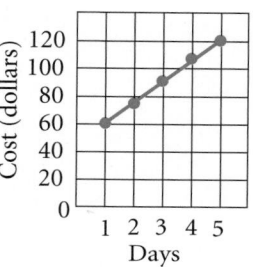

$$\text{rate of change} = \frac{\text{vertical change}}{\text{horizontal change}} = \frac{\text{change in the dependent variable}}{\text{change in the independent variable}}$$

2 EXAMPLE Finding Rate of Change Using a Graph

Airplane Altitude The graph shows the altitude of an airplane as it comes in for a landing. Find the rate of change. Explain what this rate of change means.

$$\begin{aligned}\text{rate of} \atop \text{change} &= \frac{\text{vertical change}}{\text{horizontal change}} \qquad \begin{array}{l}\leftarrow \textbf{change in altitude}\\ \leftarrow \textbf{change in time}\end{array}\\[6pt] &= \frac{1000 - 0}{60 - 180} \qquad \textbf{Use two points.}\\[6pt] &= \frac{1000}{-120} \qquad \begin{array}{l}\textbf{Divide the vertical change}\\ \textbf{by the horizontal change.}\end{array}\\[6pt] &= -8\tfrac{1}{3} \qquad \textbf{Simplify.}\end{aligned}$$

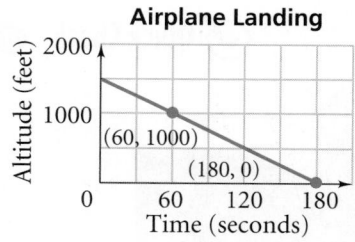

The rate of change is $-8\frac{1}{3}$. The airplane descends $8\frac{1}{3}$ feet each second.

✓ **Quick Check** **2** Find the rate of change of the data in the graph.

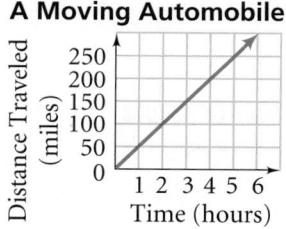

Some roads are steeper than others. A steeper road has a greater rate of change.

Real-World Connection

The grade of a road is the ratio of rise to run expressed as a percent. For example, a road with 100% grade is at a 45° angle with level ground.

The slope of a line is its rate of change.

$$\text{slope} = \frac{\text{vertical change}}{\text{horizontal change}} = \frac{\text{rise}}{\text{run}}$$

3 EXAMPLE   Finding Slope Using a Graph

Find the slope of each line.

a.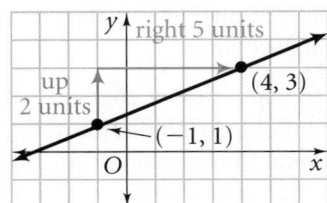

$$\text{slope} = \frac{\text{rise}}{\text{run}}$$

$$= \frac{3 - 1}{4 - (-1)}$$

$$= \frac{2}{5}$$

The slope of the line is $\frac{2}{5}$.

b.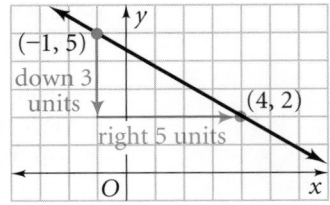

$$\text{slope} = \frac{\text{rise}}{\text{run}}$$

$$= \frac{2 - 5}{4 - (-1)}$$

$$= \frac{-3}{5} = -\frac{3}{5}$$

The slope of the line is $-\frac{3}{5}$.

✓ Quick Check   3 Find the slope of each line.

a.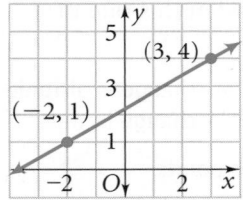

b.

Video Tutor Help

Visit: PHSchool.com
Web Code: ate-0775

You can use any two points on a line to find its slope. You use subscripts to distinguish between two points. In the diagram, (x_1, y_1) are the coordinates of P, and (x_2, y_2) are the coordinates of Q. To find the slope of $\overleftrightarrow{PQ}$, you can use the following formula.

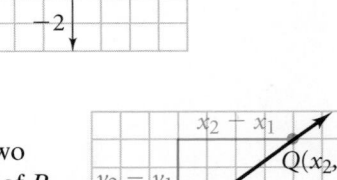

Key Concepts

Formula	Slope
$\text{slope} = \dfrac{\text{rise}}{\text{run}} = \dfrac{y_2 - y_1}{x_2 - x_1}$, where $x_2 - x_1 \neq 0$	

Keep in mind that when calculating slope, the x-coordinate you use first in the denominator must belong to the same ordered pair as the y-coordinate you use first in the numerator.

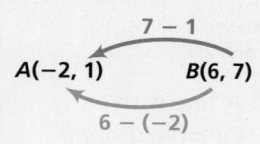
4 **EXAMPLE** **Finding Slope Using Points**

Gridded Response Find the slope of the line through $A(-2, 1)$ and $B(6, 7)$.

$$\text{slope} = \frac{y_2 - y_1}{x_2 - x_1}$$

$$= \frac{7 - 1}{6 - (-2)} \quad \begin{array}{l}\textbf{Substitute (6, 7) for } (x_2, y_2) \\ \textbf{and } (-2, 1) \textbf{ for } (x_1, y_1).\end{array}$$

$$= \frac{6}{8} \quad \textbf{Simplify.}$$

• The slope of $\overleftrightarrow{AB}$ is $\frac{6}{8}$ or $\frac{3}{4}$. Fill in the grid with 3/4.

✓ **Quick Check** **4** Find the slope of the line through each pair of points.
   a. $C(2, 5)$ and $D(4, 7)$ **b.** $P(-1, 4)$ and $Q(3, -2)$ **c.** $M(a, b)$ and $N(c, d)$

You can also analyze the graphs of horizontal and vertical lines. The next example shows why the slope of a horizontal line is 0, and the slope of a vertical line is undefined.

5 **EXAMPLE** **Horizontal and Vertical Lines**

Find the slope of each line.

a.

$$\text{slope} = \frac{y_2 - y_1}{x_2 - x_1}$$

$$= \frac{2 - 2}{4 - 1} \quad \begin{array}{l}\textbf{Substitute (4, 2) for } (x_2, y_2) \\ \textbf{and (1, 2) for } (x_1, y_1).\end{array}$$

$$= \frac{0}{3} \quad \textbf{Simplify.}$$

$$= 0$$

The slope of the horizontal line is 0.

b.

$$\text{slope} = \frac{y_2 - y_1}{x_2 - x_1}$$

$$= \frac{2 - (-1)}{4 - 4} \quad \begin{array}{l}\textbf{Substitute (4, 2) for } (x_2, y_2) \\ \textbf{and (4, -1) for } (x_1, y_1).\end{array}$$

$$= \frac{3}{0} \quad \textbf{Simplify.}$$

Division by zero is undefined. So, the slope of the vertical line is undefined.

✓ **Quick Check** **5** Find the slope of each line.

a.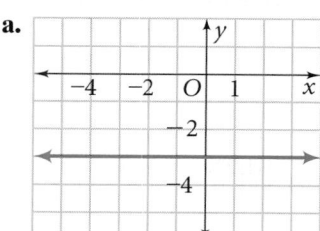

b.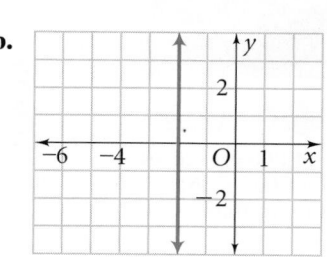

The following summarizes what you have learned about slope.

Key Concepts

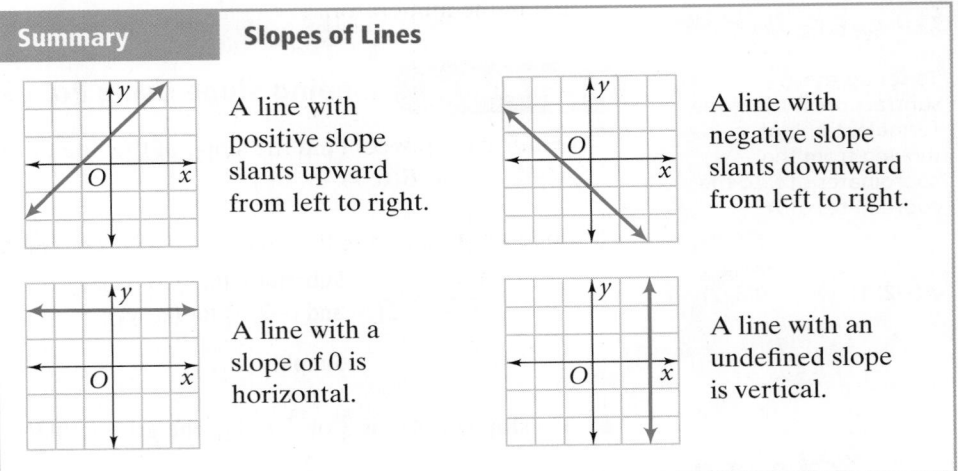

Summary	Slopes of Lines

A line with positive slope slants upward from left to right.

A line with negative slope slants downward from left to right.

A line with a slope of 0 is horizontal.

A line with an undefined slope is vertical.

EXERCISES

For more exercises, see *Extra Skill and Word Problem Practice.*

Practice and Problem Solving

A Practice by Example

Examples 1, 2
(page 309)

GO for Help

The rate of change is constant in each table and graph. Find the rate of change. Explain what the rate of change means for each situation.

1.

Time (hours)	Temperature (°F)
1	−2
4	7
7	16
10	25
13	34

2.

People	Cost (dollars)
2	7.90
3	11.85
4	15.80
5	19.75
6	23.70

3. A Tank of Gas

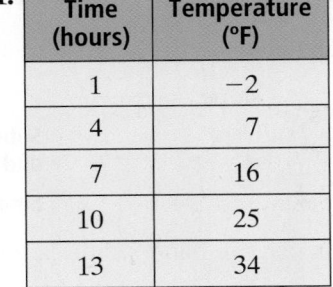

4. Emissions: Generating Electricity for TV Use

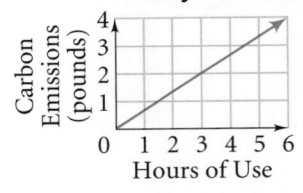

5. Descent of a Skydiver

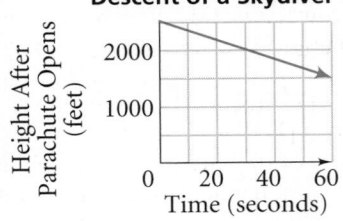

6. Price of Oregano

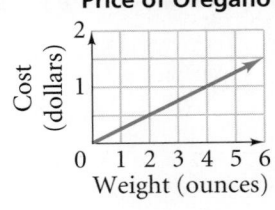

Real-World Connection

A jump from 11,000 feet gives a skydiver about 60 seconds of free fall at more than 100 mi/h.

Example 3
(page 310)

Find the slope of each line.

7.
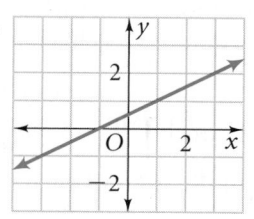

8.
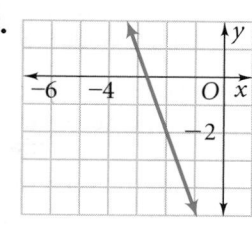

9.
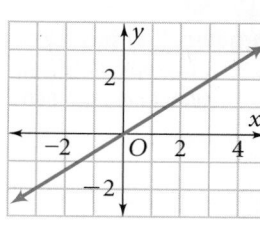

Example 4
(page 311)

Find the slope of the line that passes through each pair of points.

10. $(3, 2), (5, 6)$

11. $(5, 6), (3, 2)$

12. $(4, 8), (8, 11)$

13. $(-4, 4), (2, -5)$

14. $(-2, 1), (1, -2)$

15. $(-3, 1), (3, -5)$

16. $(-8, 0), (1, 5)$

17. $(0, 0), (3, 5)$

18. $(-4, -5), (-9, 1)$

19. $(5, 0), (0, 2)$

20. $(-7, 1), (7, 8)$

21. $(0, -1), (1, -6)$

Example 5
(page 311)

State whether the slope is zero or undefined.

22.
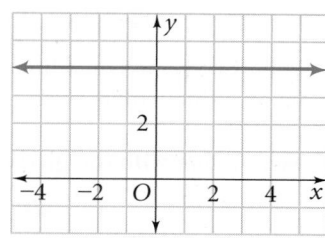

23.
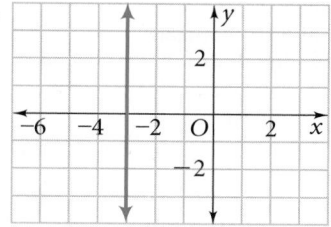

24. $(3, 4), (-3, 4)$

25. $(4, 3), (4, -3)$

26. $\left(-5, \frac{1}{2}\right), (-5, 3)$

B **Apply Your Skills**

Find the rate of change for each situation.

27. A baby is 18 in. long at birth and 27 in. long at ten months.

28. The cost of group museum tickets is $48 for four people and $78 for ten people.

29. You drive 30 mi in one hour and 120 mi in four hours.

Find the slope of the line passing through each pair of points.

30. $(-7, 1), (7, 8)$

31. $\left(4, 1\frac{2}{3}\right), \left(-2, \frac{2}{3}\right)$

32. $(0, 3.5), (-4, 2.5)$

33. $\left(\frac{1}{2}, 8\right), (1, -2)$

34. $\left(-5, \frac{1}{2}\right), (-5, 3)$

35. $(0.5, 6.25), (3, -1.25)$

Through the given point, draw the line with the given slope.

36. $K(3, 5)$
slope -2

37. $M(5, 2)$
slope $-\frac{1}{2}$

38. $Q(-2, 3)$
slope $\frac{3}{5}$

39. $R(2, -3)$
slope $-\frac{4}{3}$

40. a. Biology Which line in the graph at the left is the steepest?
   b. During the 6-week period, which plant had the greatest rate of change? The least rate of change? How do you know?

41. a. Find the slope of the line through $A(4, -3)$ and $B(1, -5)$ using A for (x_2, y_2) and B for (x_1, y_1).
   b. Find the slope of the line in part (a) using B for (x_2, y_2) and A for (x_1, y_1).
   c. Critical Thinking Explain why it does not matter which point you use for (x_2, y_2) and which point you use for (x_1, y_1) when you calculate a slope.

Plant Growth

Height (in.)

4
3
2
1

A
B
C

O 1 2 3 4 5 6
Time (weeks)

 42. Construction An extension ladder has a label that says, "Do not place base of ladder less than 5 ft from the vertical surface." What is the greatest slope possible if the ladder can safely extend to reach a height of 12 ft? Of 18 ft?

43. Writing If two points on a line have positive coordinates, is the slope necessarily positive? Explain.

Geometry Find the slope of the sides of each figure.

44.

45.

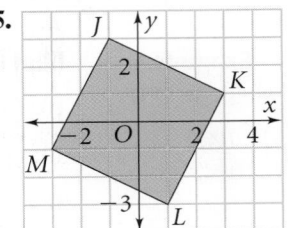

46. a. Graph the direct variation $y = -\frac{2}{3}x$.
   b. What is the constant of variation?
   c. What is the slope?
   d. What is the relationship between the constant of variation and the slope?

47. a. Open-Ended Name two points on a line with a slope of $\frac{3}{4}$.
   b. Name two points on a line with a slope of $-\frac{1}{2}$.

GO for Help

For help with direct variation and the constant of variation see p. 278.

Each pair of points lies on a line with the given slope. Find x or y.

48. $(2, 4), (x, 8)$; slope $= -2$

49. $(2, 4), (x, 8)$; slope $= -\frac{1}{2}$.

50. $(4, 3), (x, 7)$; slope $= 2$

51. $(x, 3), (2, 8)$; slope $= -\frac{5}{2}$

52. $(-4, y), (2, 4y)$; slope $= 6$

53. $(3, 5), (x, 2)$; undefined slope

Reasoning In Exercises 54–60, tell whether each statement is *true* or *false*. If false, give a counterexample.

54. A rate of change must be either positive or zero.

55. All horizontal lines have the same slope.

56. A line with slope 1 always passes through the origin.

57. Two lines may have the same slope.

58. The slope of a line that passes through Quadrant III must be negative.

59. A line with slope 0 never passes through point $(0, 0)$.

60. Two points with the same x-coordinate are always on the same vertical line.

Real-World Connection

On a carousel, the outer horses cover more distance than the inner horses, giving them a faster speed.

61. Business The graph shows how much it costs to rent carousel equipment.
   a. Estimate the slope of the line. What does that number mean?
   b. Customers pay $2 for a ride. What is the average number of customers needed to cover the rental costs?

62. Error Analysis A friend says the slope of a line passing through $(1, 7)$ and $(3, 9)$ is equal to the ratio $\frac{1 - 3}{7 - 9}$. What is your friend's error?

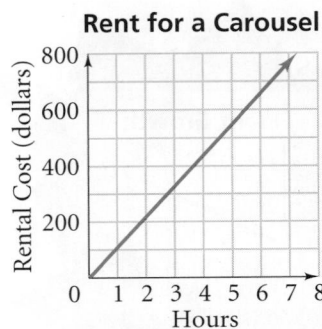

Rent for a Carousel

 **Challenge**

Find the slope of the line passing through each pair of points.

63. $(a, -b), (-a, -b)$ **64.** $(-m, n), (3m, -n)$ **65.** $(2a, b), (c, 2d)$

Do the points in each set lie on the same line? Explain your answer.

66. $A(1, 3), B(4, 2), C(-2, 4)$ **67.** $G(3, 5), H(-1, 3), I(7, 7)$

68. $D(-2, 3), E(0, -1), F(2, 1)$ **69.** $P(4, 2), Q(-3, 2), R(2, 5)$

70. $G(1, -2), H(-1, -5), I(5, 4)$ **71.** $S(-3, 4), T(0, 2), X(-3, 0)$

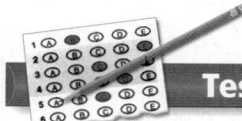 **NY REGENTS**

Test Prep

Multiple Choice

72. A line has slope $\frac{4}{3}$. Through which two points could this line pass?
- **A.** (24, 19), (8, 10)
- **B.** (10, 8), (16, 0)
- **C.** (28, 10), (22, 2)
- **D.** (4, 20), (0, 17)

73. A horizontal line passes through (5, 22). Which other point does the line contain?
- **F.** (5, 2)
- **G.** (0, 22)
- **H.** (22, 5)
- **J.** (0, 5)

74. Josie found the slope of the line shown at the right. She used the points (1, −6) and (5, 2) and wrote $\frac{2 - (-6)}{1 - 5} = \frac{8}{-4} = -2$ to find the slope. Which of the following describes her error?
- **A.** She mixed up x_1 and x_2 in the denominator.
- **B.** She subtracted instead of added.
- **C.** She used an x-value in the numerator of the fraction.
- **D.** She did not combine the negative signs correctly.

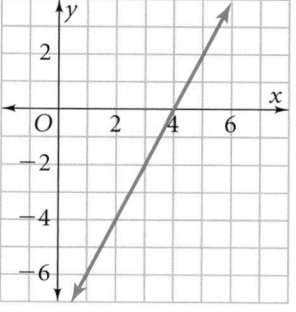

Short Response

75. The steepness, or grade, of a road is expressed as a percent. If a road rises 3 feet for every 24 horizontal feet, what is the slope of the road? What percent grade is this? Show your work.

Mixed Review

 **GO for Help**

Lesson 5-4

Write a function rule for each situation.

76. the total cost of renting a movie for n days if it costs $3.50/day

77. the total profit if supplies and wages cost $232, and each item q sells for $4.95

Lesson 3-2

Solve.

78. $x + 3 + 2x = -6$ **79.** $3(2t + 5) = -9$ **80.** $9 = y + 2(4y - 5)$

81. $4n - 7(n - 9) = 42$ **82.** $2(7 - q) - 4 = 0$ **83.** $\frac{2}{5}(p + 10) = 0$

Lesson 2-6

Find each probability based on one roll of a number cube.

84. $P(10)$ **85.** $P(\text{even number})$ **86.** $P(3 \text{ or } 5)$ **87.** $P(\text{integer})$

Investigating $y = mx + b$

You can use a graphing calculator to explore the graph of an equation in the form $y = mx + b$. For this activity, use a standard screen by pressing ZOOM 6.

Go Online
PHSchool.com
For: Graphing calculator procedures
Web Code: ate-2104

NY A.G.5: Investigate and generalize how changing the coefficients of a function affects its graph.

1. Graph these equations on the same screen. Then complete each statement.

$$y = x + 1 \qquad y = 2x + 1 \qquad y = \tfrac{1}{2}x + 1$$

 a. The graph of __?__ is closest to the y-axis.
 b. The graph of __?__ is closest to the x-axis.

2. Match each equation with the best choice for its graph.

 A. $y = \tfrac{1}{5}x - 1$ **B.** $y = 5x - 1$ **C.** $y = x - 1$

 I. 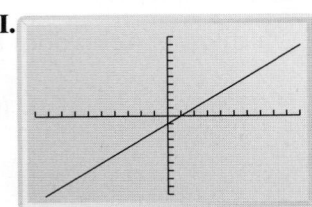   II. 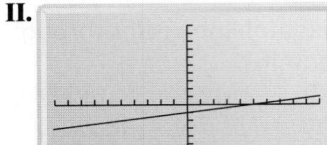   III.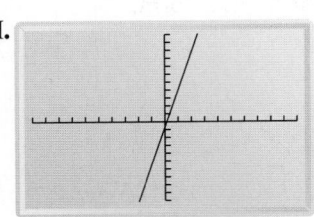

3. How does changing m affect the graph of an equation in the form $y = mx + b$?

4. Graph these equations on the same screen.

$$y = 2x + 1 \qquad y = -2x + 1$$

 How does the sign of m affect the graph of an equation?

5. Graph these equations on the same screen.

$$y = 2x + 1 \qquad y = 2x - 2 \qquad y = 2x + 2$$

 Where does the graph of each equation cross the y-axis? (*Hint:* Use the ZOOM feature to better see the points of intersection.)

6. Match each equation with the best choice for its graph.

 A. $y = \tfrac{1}{2}x - 5$ **B.** $y = \tfrac{1}{2}x$ **C.** $y = \tfrac{1}{2}x + 3$

 I.   II. 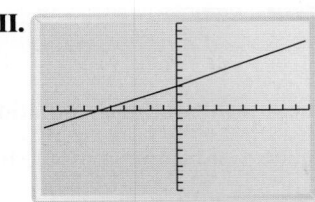   III.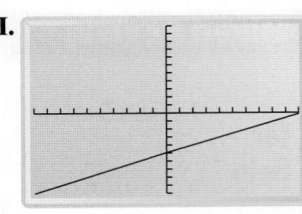

7. How does changing the value of b affect the graph of an equation in the form $y = mx + b$?

8. You can change the appearance of a graph by changing its scale in the WINDOW screen. Describe how the graph of $y = 2x + 1$ changes from its appearance on a standard screen using the following values for Xmin, Xmax, Ymin, and Ymax.

 a. Xmin = –5   Ymin = –10
       Xmax = 5   Ymax = 10

 b. Xmin = –10   Ymin = –5
       Xmax = 10   Ymax = 5

Slope-Intercept Form

 Learning Standards for Mathematics

A.A.34 Write the equation of a line, given its slope and the coordinates of a point on the line.

A.A.35 Write the equation of a line, given the coordinates of two points on the line.

A.A.37 Determine the slope of a line, given its equation in a form.

 Check Skills You'll Need

Evaluate each expression.

1. $6a + 3$ for $a = 2$

2. $-2x - 5$ for $x = 3$

3. $\frac{1}{4}x + 2$ for $x = 16$

4. $0.2x + 2$ for $x = 15$

5. $8 - 5n$ for $n = 3$

6. $-4p + 9$ for $p = 2$

GO **for Help** Lessons 1-2 and 2-3

New Vocabulary • linear function • parent function • linear parent function • linear equation • y-intercept • slope-intercept form

---

1 Writing Linear Equations

Vocabulary Tip

The word <u>linear</u> contains the word "line."

In lesson 5-5, you studied direct variations such as $y = 3x$. The graph of a direct variation is a straight line. All direct variations are linear functions. A **linear function** is a function that graphs a line. Direct variations are only part of the family of linear functions. For example, $y = -\frac{1}{2}x + 1$ is a linear function but not a direct variation because it does not go through $(0, 0)$.

A **parent function** is the simplest equation of a function. The equation $y = x$ or $f(x) = x$ is the **linear parent function**.

A **linear equation** is an equation that models a linear function. In a linear equation, the variable cannot be raised to a power other than 1. So $y = 2x$ is the equation of a linear function, but $y = x^2$ or $y = 2^x$ are not.

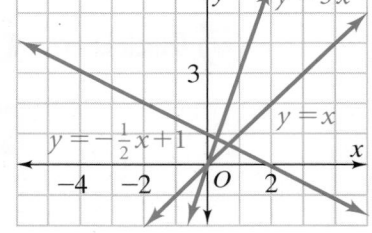

The equation of a line gives important information about its graph. Consider the table and graph of the equation $y = -2x + 1$.

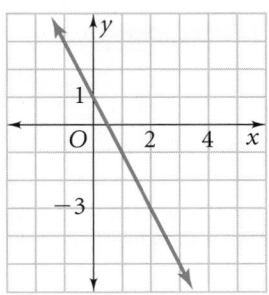

x	−2x + 1	y
0	−2(0) + 1	1
1	−2(1) + 1	−1
2	−2(2) + 1	−3

Two points on the line are $(0, 1)$ and $(2, -3)$. The slope is $\frac{1 - (-3)}{0 - 2} = -\frac{4}{2}$ or -2. The **y-intercept** is the y-coordinate of the point where a line crosses the y-axis. Since $y = -2x + 1$ crosses the y-axis at $(0, 1)$, the y-intercept is 1.

If you know the slope of a line and its y-intercept, you can write the equation of the line. The letter m refers to the *slope*.

Key Concepts

Definition	Slope-Intercept Form of a Linear Equation

The **slope-intercept form** of a linear equation is $y = mx + b$.

$$\underset{\text{slope}}{\uparrow} \quad \underset{y\text{-intercept}}{\uparrow}$$

Test-Taking Tip

As you do homework, make a list of formulas you can use later to study for quizzes and tests.

1 EXAMPLE **Identifying Slope and y-Intercept**

What are the slope and y-intercept of $y = 3x - 5$?

$y = mx + b$ **Use the slope-intercept form.**

$y = 3x + (-5)$ **Think of $y = 3x - 5$ as $y = 3x + (-5)$.**

● The slope is 3; the y-intercept is -5.

✓ Quick Check **1** **a.** Find the slope and y-intercept of $y = \frac{7}{6}x - \frac{3}{4}$.

    b. Critical Thinking For the equation in Example 1, what happens to the graph of the line and to the equation if the y-intercept is moved down 3 units?

2 EXAMPLE **Writing an Equation**

Write an equation of the line with slope $\frac{3}{8}$ and y-intercept 6.

$y = mx + b$ **Use the slope-intercept form.**

$y = \frac{3}{8}x + 6$ **Substitute $\frac{3}{8}$ for m and 6 for b.**

✓ Quick Check **2** Write an equation of a line with slope $m = \frac{2}{5}$ and y-intercept $b = -1$.

3 EXAMPLE **Writing an Equation From a Graph**

Multiple Choice Which equation models the linear function shown in the graph?

   Ⓐ $y = -\frac{3}{4}x + 2$ Ⓑ $y = -\frac{4}{3}x + 2$

   Ⓒ $y = 2x - \frac{4}{3}x$ Ⓓ $y = 2x - \frac{3}{4}$

Find the slope. Two points on the line are $(0, 2)$ and $(4, -1)$.

$$\text{slope} = \frac{-1 - 2}{4 - 0} = -\frac{3}{4}$$

The y-intercept is 2. Write an equation in slope-intercept form.

$y = mx + b$

$y = -\frac{3}{4}x + 2$ **Substitute $-\frac{3}{4}$ for m and 2 for b.**

● The equation is $y = -\frac{3}{4}x + 2$. So the answer is A.

Online active math

For: Slope-Intercept Activity
Use: Interactive Textbook, 6-2

✓ Quick Check **3** **a.** Write the equation of the line using the points $(0, 1)$ and $(2, 2)$.

    b. Critical Thinking Does the equation of the line change if you use $(-2, 0)$ instead of $(2, 2)$? Explain.

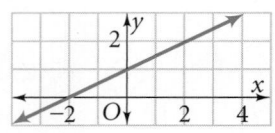

Each point on the graph of an equation is an ordered pair that makes the equation true. The graph of a linear equation is a line that indicates all the solutions of the equation. You can use the slope and y-intercept to graph a line.

4 EXAMPLE Graphing Equations

Graph $y = 3x - 1$.

Step 1
The y-intercept is –1. So plot a point at $(0, -1)$.

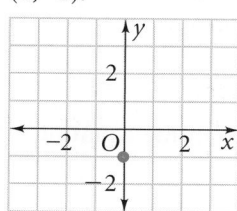

Step 2
The slope is 3, or $\frac{3}{1}$. Use the slope to plot a second point.

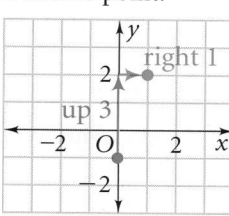

Step 3
Draw a line through the two points.

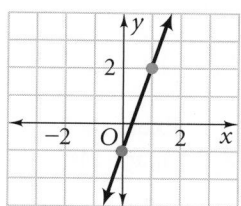

✓ Quick Check **4** Graph $y = \frac{3}{2}x - 2$.

When you graph equations for real-world situations, use scales on the x- and y-axes that are reasonable for the situation. Recall that you can avoid having a large blank space in a graph by using a zigzag line to show a break in a scale.

5 EXAMPLE Real-World 🌐 Problem Solving

Commission The base pay of a water-delivery person is $210 per week. He also earns 20% commission on any sale he makes. The equation $t = 210 + 0.2s$ relates total earnings t to sales s. Graph the equation.

Step 1 Identify the slope and y-intercept.

$t = 210 + 0.2s$

$t = 0.2s + 210$ **Rewrite the equation in slope-intercept form.**

slope y-intercept

Step 2 Plot two points. First plot $(0, 210)$, the y-intercept. Then use the slope to plot a second point.

The slope is 0.2, which equals $\frac{2}{10}$, or $\frac{20}{100}$. Plot a second point 20 units above and 100 units to the right of the y-intercept.

Step 3 Draw a line through the points.

Weekly Earnings for a Water-Delivery Person

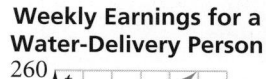
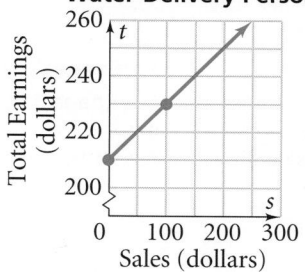

✓ Quick Check **5** Suppose the base pay of the delivery person is $150, and his commission on each sale is 30%. The equation relating his total earnings t to sales s is $t = 150 + 0.3s$. Graph the equation.

Real-World 🌐 Connection

Between 1990 and 1999, the sales of bottled water in the United States increased 107.6%, which means that sales more than doubled.

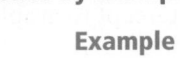 **Practice and Problem Solving**

A **Practice by Example**

Example 1
(page 318)

Find the slope and *y*-intercept of each equation.

1. $y = -2x + 1$ **2.** $y = -\frac{1}{2}x + 2$ **3.** $y = x - \frac{5}{4}$

4. $y = 5x + 8$ **5.** $y = \frac{2}{3}x + 1$ **6.** $y = -4x$

7. $y = -x - 7$ **8.** $y = -0.7x - 9$ **9.** $y = -\frac{3}{4}x - 5$

Example 2
(page 318)

Write an equation of a line with the given slope and *y*-intercept.

10. $m = \frac{2}{9}, b = 3$ **11.** $m = 3, b = \frac{2}{9}$ **12.** $m = \frac{9}{2}, b = 3$

13. $m = 0, b = 1$ **14.** $m = -1, b = -6$ **15.** $m = -\frac{2}{3}, b = 5$

16. $m = 0.3, b = 4$ **17.** $m = 0.4, b = 0.6$ **18.** $m = -7, b = \frac{1}{3}$

19. $m = -\frac{1}{5}, b = -\frac{2}{5}$ **20.** $m = -\frac{1}{4}, b = \frac{5}{4}$ **21.** $m = \frac{8}{3}, b = \frac{2}{3}$

Example 3
(page 318)

Write the slope-intercept form of the equation for each line.

22. 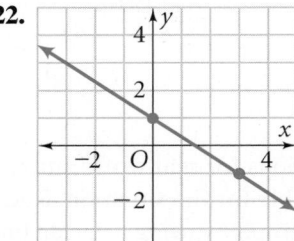 **23.** 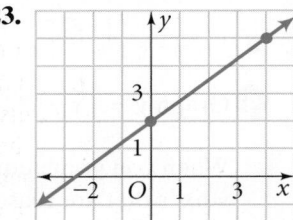 **24.**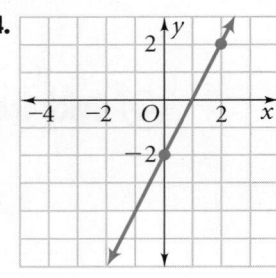

25. 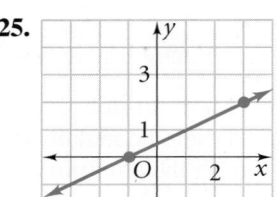 **26.** 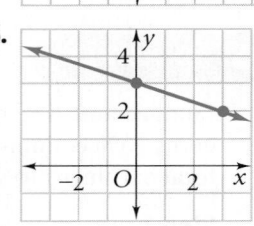 **27.**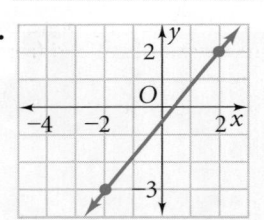

Example 4
(page 319)

Use the slope and *y*-intercept to graph each equation.

28. $y = \frac{1}{2}x + 4$ **29.** $y = \frac{2}{3}x - 1$ **30.** $y = -5x + 2$ **31.** $y = 2x + 5$

32. $y = x + 4$ **33.** $y = -x + 2$ **34.** $y = 4x - 3$ **35.** $y = -\frac{3}{2}x$

36. $y = \frac{2}{5}x - 3$ **37.** $y = -\frac{2}{3}x + 2$ **38.** $y = -\frac{4}{5}x + 4$ **39.** $y = -0.5x + 2$

Example 5
(page 319)

40. Retail Sales A music store is offering a coupon promotion on its CDs. The regular price for CDs is $14. With the coupon, customers are given $4 off the total purchase. The equation $t = 14c - 4$, where *c* is the number of CDs and *t* is the total cost of the purchase, models this situation.
a. Graph the equation.
b. Find the total cost for a sale of 6 CDs.

B **Apply Your Skills**

Find the slope and *y*-intercept of each equation.

41. $y - 2 = -3x$ **42.** $y + \frac{1}{2}x = 0$ **43.** $y - 9x = \frac{1}{2}$

44. $y = 3x - 9$ **45.** $2y - 6 = 3x$ **46.** $-2y = 6(5 - 3x)$

47. $y - d = cx$ **48.** $y = (2 - a)x + a$ **49.** $2y + 4n = -6x$

Use the slope and *y*-intercept to graph each equation.

50. $y = 7 - 3x$ **51.** $2y + 4x = 0$ **52.** $3y + 6 = -2x$

53. $y + 2 = 5x - 4$ **54.** $4x + 3y = 2x - 1$ **55.** $-2(3x - 4) + y = 0$

56. Error Analysis Fred drew the graph at the right for the equation $y = -2x + 1$. What error did he make?

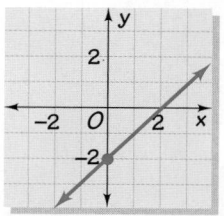

57. a. A candle begins burning at time $t = 0$. Its original height is 12 in. After 30 min the height of the candle is 8 in. Draw a graph showing the change in the height of the candle.
  b. Write an equation that relates the height of the candle to the time it has been burning.
  c. How many minutes after the candle is lit will it burn out?

Real-World **Connection**

Careers Airport ground crews direct airplanes to and from their gates.

58. Airplane Fuel The graph shows the relationship between the number of gallons of fuel in the tank of an airplane and the weight of the airplane. The equation $y = 6x + 2512$, where x is the number of gallons of fuel and y is the weight of the airplane, models this situation.
  a. What does the slope represent?
  b. Use the equation to predict the weight of the plane when the tank contains 25 gallons of fuel.

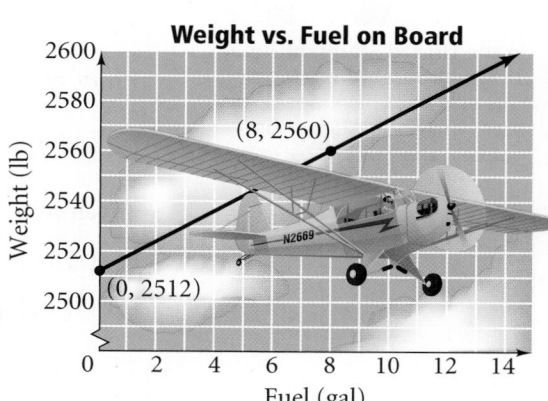

Weight vs. Fuel on Board

Is the ordered pair on the graph of the given equation?

59. $(-3, 4); y = -2x + 1$ **60.** $(-6, 5); y = -\frac{1}{2}x + 2$ **61.** $(0, -1); y = x - \frac{5}{4}$

62. Multiple Choice At the right is the graph of $y = \frac{1}{4}x - 2$. Which of the graphs below represents the linear function if the slope is doubled and the y-intercept stays the same?

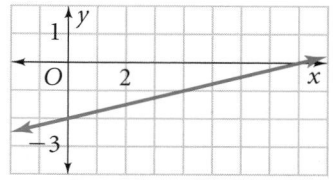

A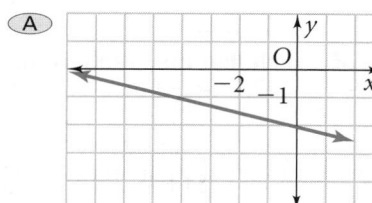

B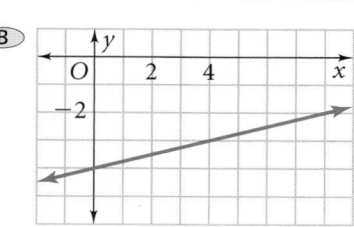

C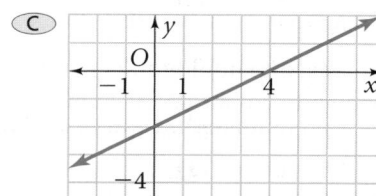

D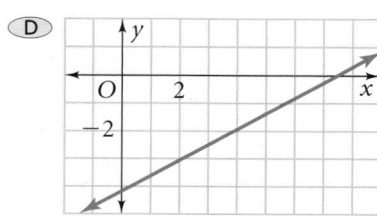

GO nline
Homework Video Tutor
Visit: PHSchool.com
Web Code: ate-0602

Real-World 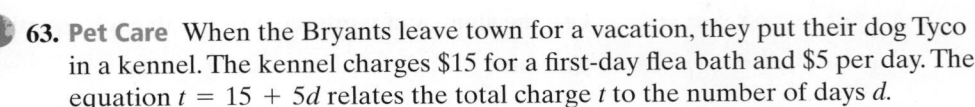 **Connection**

In the United States, although 35% of households have pet cats and 37% have pet dogs, there are about 25% more pet cats than pet dogs.

63. Pet Care When the Bryants leave town for a vacation, they put their dog Tyco in a kennel. The kennel charges $15 for a first-day flea bath and $5 per day. The equation $t = 15 + 5d$ relates the total charge t to the number of days d.
  a. Rewrite the equation in slope-intercept form.
  b. Graph the equation.
  c. Explain why the line you graph should lie only in Quadrant I.

64. Writing Explain the steps you would use to graph $y = \frac{3}{4}x + 5$.

65. Critical Thinking Which graphed line has the greater slope? Explain.

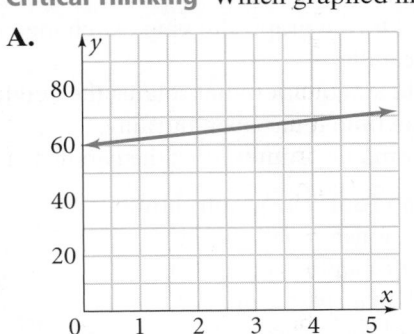
A.

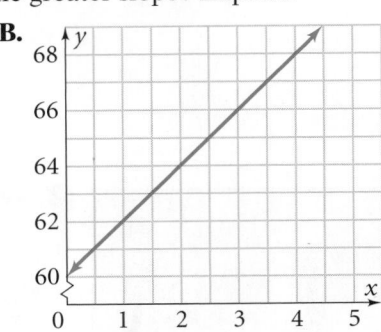
B.

Given two points on a line, write the equation of the line in slope-intercept form.

66. $(3, 5), (5, 9)$ **67.** $(5, -13), (2, -1)$ **68.** $(-4, 10), (6, 5)$

69. $(8, 7), (-12, 2)$ **70.** $(-7, 4), (11, -14)$ **71.** $(-1, -9), (2, 0)$

72. Graphing Calculator Suppose you want to graph the equation $y = \frac{5}{4}x - 3$. Enter each key sequence and display the graph.
  a. $\boxed{\text{Y=}}\ 5\ \boxed{\div}\ 4\ \boxed{\text{X,T,}\theta,n}\ \boxed{-}\ 3$ **b.** $\boxed{\text{Y=}}\ \boxed{(}\ 5\ \boxed{\div}\ 4\ \boxed{)}\ \boxed{\text{X,T,}\theta,n}\ \boxed{-}\ 3$
  c. Which equation gives you the graph of $y = \frac{5}{4}x - 3$? Explain.

73. a. What is the slope of each line?
  b. What is the y-intercept of each line?
  c. Geometry The lines in the graph are parallel. What appears to be true about the slopes of parallel lines?

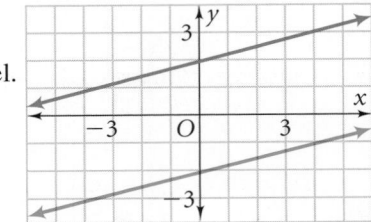

74. Open-Ended Write a linear equation. Identify the slope and y-intercept. Then graph your equation.

C **Challenge**

Find the value of a such that the graph of the equation has the given slope.

75. $y = 2ax + 4; m = -1$ **76.** $y = -\frac{1}{2}ax - 5; m = \frac{5}{2}$ **77.** $y = \frac{3}{4}ax + 3; m = \frac{9}{16}$

78. a. Geometry Graph these equations on the same grid.
  $y = 3$ $y = -3$ $x = 2$ $x = -2$
  b. What geometric figure did you draw? Justify your answer.
  c. Draw a diagonal of the figure. What is the equation of this line? Explain.

79. Recreation A group of mountain climbers begin an expedition with 265 lb of food. They plan to eat a total of 15 lb of food per day.
  a. Write an equation in slope-intercept form relating the remaining food supply r to the number of days d.
  b. Graph your equation.
  c. The group plans to eat the last of their food the day their expedition ends. Use your graph to find how many days they expect the expedition to last.

Multiple Choice

80. Which equation has the same y-intercept as $y = 4x - 3$?

   A. $y - 3 = x$ **B.** $y = 8x + 3$ **C.** $3 - y = 4x$ **D.** $y = -3 + 8x$

81. Which of the following is the equation of the line that has the same slope as $y = -\frac{3}{2}x + 2$ and the same y-intercept as $y = 3x - 2$?

   F. $y - 2 = -\frac{3}{2}x$ **G.** $-\frac{3}{2}x = y + 2$

   H. $y + 2 = -\frac{3}{2}$ **J.** $-\frac{3}{2}x = y + 3$

82. A software company started with 2 employees. In 6 months, the company had 7 employees. The number of employees increased at a steady rate. Which equation models the relationship between the number of employees n and the number of months m since the company started?

   A. $n = \frac{5}{6}m + 2$ **B.** $m = 2n + \frac{5}{6}$

   C. $n = \frac{6}{5}m + 2$ **D.** $m = \frac{5}{6}n + 2$

Short Response

83. A line passes through the points $(0, 3)$ and $(1, 5)$. Graph this line and find an equation for the line in slope-intercept form. Show your work.

Mixed Review

Lesson 6-1

GO for Help

Find the slope of the line that passes through each pair of points.

84. $(-2, 8), (5, -1)$ **85.** $(0, 0), (-6, 5)$ **86.** $(4, 6), (2, -3)$ **87.** $(1, 2), (2, 1)$

Review page 166

88. The greeting card industry sells over 6 billion cards annually. Women purchase 80% of all greeting cards sold. How many cards do women purchase annually?

A Point in Time

1500   1600   1700   1800   1900   2000

On August 30, 1984, Astronaut Judith A. Resnik became the second American woman in space, on the shuttle *Discovery*'s first voyage. Resnik was an electrical engineer with a Ph.D. from the University of Maryland. Prior to her mission, she helped to design and develop a remote manipulator system. This required skill in writing linear equations. Her job during *Discovery*'s six-day voyage was to manipulate a robotic arm and to extend and retract the shuttle's solar power array. Resnik died tragically in the *Challenger* disaster in 1986.

Go Online
PHSchool.com   **For:** Information about astronauts
                      Web Code: ate-2032

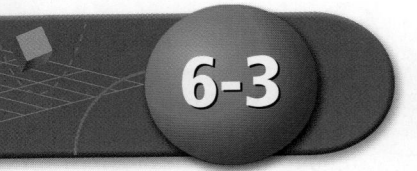

Applying Linear Functions

 **Learning Standards for Mathematics**

A.A.5 Write algebraic equations or inequalities that represent a situation.

A.G.4 Identify and graph linear, quadratic (parabolic), absolute value, and exponential functions.

✓ **Check Skills You'll Need**

GO for Help   Lesson 6-2

Find the slope and *y*-intercept of each equation.

1. $y = 4x - 3$

2. $y = \frac{4}{5}x + 7$

3. $y = -\frac{1}{3}x - 8$

4. $y = 6x$

Write an equation of a line with the given slope and *y*-intercept.

5. $m = 5, b = -1$

6. $m = \frac{2}{5}, b = 4$

7. $m = 6, b = \frac{1}{4}$

---

1 Interpreting Linear Graphs

You can model many real world situations with linear equations. Recall that the graph of an equation shows the solutions of the equation. However, for a discrete real-world situation not every point may represent a reasonable value.

1 EXAMPLE   **Real-World Connection**

Automobiles A car dealership has 40 cars in stock. The auto manufacturer will deliver new cars to the dealership by car carrier. Each carrier holds 6 cars. Write a linear function that relates the number of carriers used to the total number of cars at the dealership. Graph the function that models the situation.

Relate   total number of cars at dealership   equals   40   plus   6   times   number of car carriers

Define   The total number of cars depends on the number of car carriers. So total number of cars is the dependent variable and number of car carriers is the independent variable.

Let   *x*   = the number of car carriers.

Let   *y*   = the total number of cars at the dealership.

Write   *y*   =   40   +   6   ·   *x*

Transporting Cars

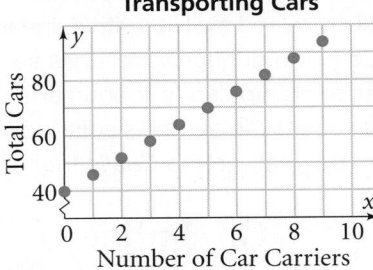

The number of car carriers is the domain. A reasonable domain is whole numbers.

✓ Quick Check **1** A sporting goods store sells cans of tennis balls. There are 3 tennis balls in each can. The coach of the tennis team is buying supplies. Write a linear function that relates the number of cans to the total number of tennis balls. Graph the function that models the situation.

2 EXAMPLE **Analyzing Linear Graphs**

Students in a ninth-grade class drew the following graph to represent how much money would be in the class fund after washing cars at a fundraiser.

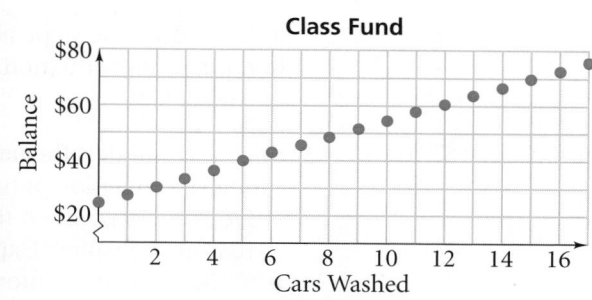

a. What does the slope and *y*-intercept of the graph mean for the given situation?

The slope is 3, which indicates the class charged $3 per car. The *y*-intercept is 25, which means they had $25 in the class fund before the fundraiser.

b. If the graph had the same slope but a *y*-intercept of 15, what could you conclude about the fundraiser?

You could conclude that there were $15 in the class fund instead of $25.

c. If the graph had a slope of 5, what could you conclude about the fundraiser?

You could conclude that the students decided to charge $5 per car.

✓ Quick Check **2 a.** Suppose you drew the following graph to represent how far you ride your bike at a steady rate. What is your rate?

b. If the graph had a slope of 10, what could you conclude about your bike ride?

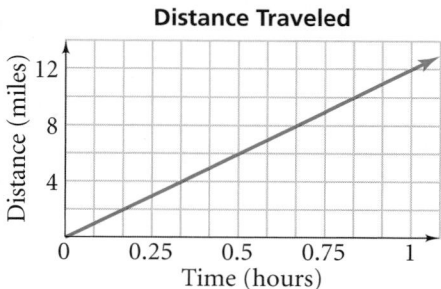

EXERCISES

For more exercises, see *Extra Skill and Word Problem Practice.*

Practice and Problem Solving

A Practice by Example

Example 1 (page 324)

GO for Help

Model each situation with a linear function and graph.

1. A lumber yard sells pre-cut 6-, 8-, 10-, and 12-foot beams for $1.20 per foot.

2. A market sells chicken for $2.99 a pound.

3. Suppose your family buys a "movie card" from a DVD store that entitles you to rent 25 movies. You rent 3 movies per week.

Example 2
(page 325)

4. A helicopter takes off from the roof of a building. The graph shows the altitude of the helicopter as it rises steadily for several minutes.

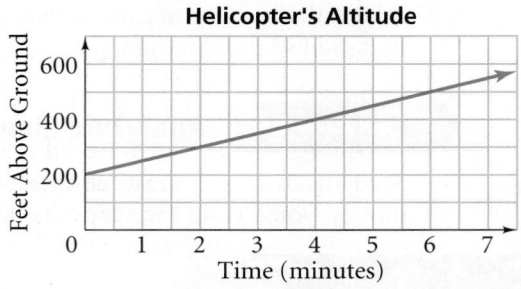

Helicopter's Altitude

a. What does the slope and *y*-intercept reveal about the original situation?
b. For a similar situation, the slope is 85 and *y*-intercept is 250. What can you conclude?

B **Apply Your Skills**

5. The graph models the height in inches of a burning candle as a function of time in hours.
   a. Does every point on the line represent a reasonable value? Explain.
   b. Write a linear function for the graph.
   c. A second 8-in. candle burns at a rate of 3 in./hr. How would the graph for that candle compare with the one shown here?

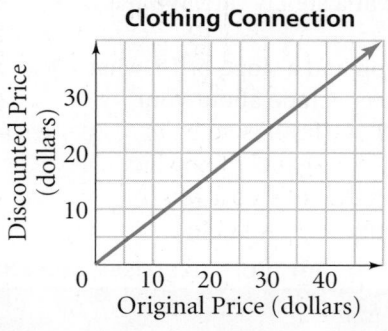

Burning Candle

6. Suppose an elevator is 400 feet above the ground. It descends at a steady rate. After 15 seconds it is 250 feet above the ground.
   a. Write a linear function for the height of the elevator as a function of time.
   b. Graph the function.
   c. **Critical Thinking** Is it reasonable to include negative numbers in the range?

7. **Shopping** Clothing Connection and Teen World are having sales. The graphs show the discounted price as a function of the original price for merchandise from each shop.

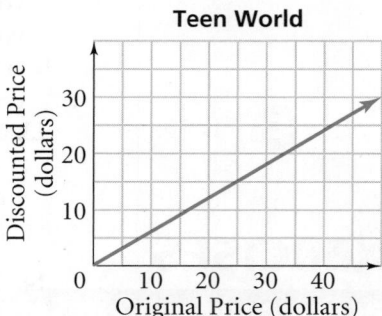

Clothing Connection

Teen World

a. Which store has a greater discount? Explain.
b. Write a linear function to model the situation in each store.
c. A customer buys merchandise originally priced at $16 from each store. What is the discounted price at each store?

8. a. **Open-Ended** Describe a real-world situation that you could model with a linear function.
   b. Describe a reasonable domain and range for the situation.

GO **Online**
Homework Video Tutor
Visit: PHSchool.com
Web Code: ate-0603

For a guide to solving Exercise 9, see p. 329.

9. Two friends are paddling a kayak downstream (with the current). The graph shows their distance from camp as a function of time.
   a. Is every point on the line a reasonable value? Explain.
   b. **Critical Thinking** Suppose they were paddling back toward the camp, upstream (against the current) rather than downstream. How would the graph compare with the downstream graph? Explain.

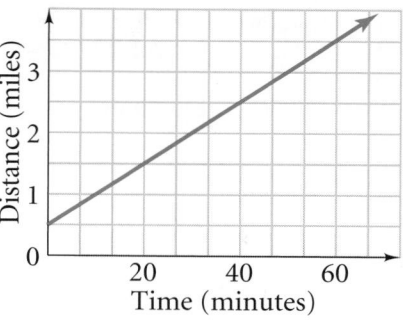

10. **a.** Direct variations model some linear relationships. How is the equation of a direct variation similar to the linear equation $y = mx + b$? How are they different?
    b. **Writing** Describe a situation that could be modeled by a linear function but NOT by a direct variation.

11. **Multiple Choice** Which situation could best be modeled by the graph at the right?
    Ⓐ the cost of buying muffins
    Ⓑ the distance between a train and the station as the train travels towards the station
    Ⓒ the amount of money left in a roll of quarters after paying a toll each day
    Ⓓ the distance a runner covers, traveling at a steady pace

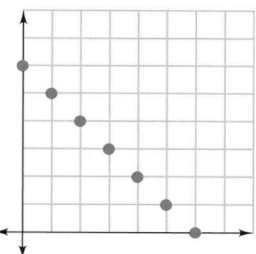

12. **a.** **Math in the Media** Write an equation relating the data in the cartoon.
    b. How many dog years are 12 human years?

Mother Goose and Grimm

13. **Sports** Four friends go bowling. For each person, it costs $2.50 to rent shoes and $2.00 per game of bowling. They all bowl the same number of games.
    a. Model the total cost with a linear function. Graph the function.
    b. Describe a reasonable domain and range.

14. **Biology** The table shows the average height of female infants.
    a. Write a linear function using the data for 3- and 9-month-old girls.
    b. Use your model to find the height of a 15-year-old girl, to the nearest foot.
    c. **Critical Thinking** Based on your answer, what can you conclude about the linear model?

Age (months)	Height (inches)
3	23.5
6	25.6
9	27.5

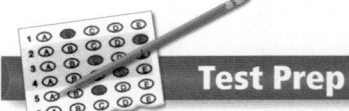
Multiple Choice

15. Which graph best models the amount of money in a bank account after the same amount is withdrawn each week?

A.
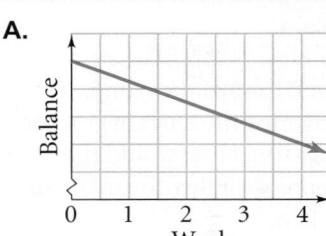

B.
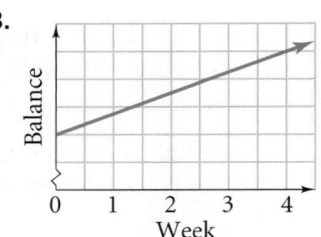

C.
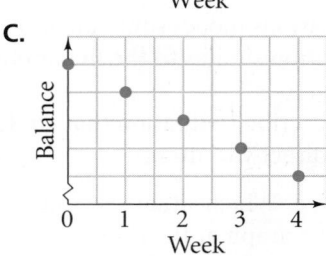

D.
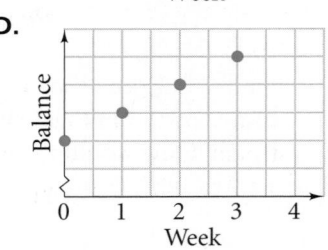

16. Find the slope of the line through the points $(-3, 1)$ and $(6, -7)$.

F. 2 **G.** $\frac{1}{2}$ **H.** $-\frac{8}{9}$ **J.** $-\frac{9}{8}$

17. Which equation models a line with the same y-intercept but half the slope of the line $y = 4 - 10x$?

A. $y = -5x + 4$ **B.** $y = 2 - 10x$ **C.** $y = 2 - 5x$ **D.** $y = -5x - 2$

Short Answer

18. The drama club plans to attend a professional production. From 10 to 35 students will go. There is a one-time handling fee of $3. Each ticket costs $25 plus a $2 surcharge. Write a linear function that models this situation. Describe a reasonable domain and range.

Mixed Review

Lesson 6-2

for Help

Write the slope-intercept form of the equation for each line.

19.
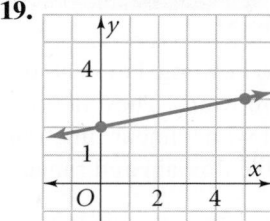

20.
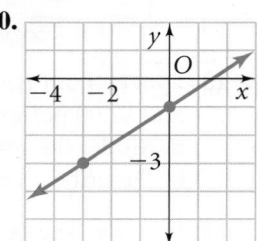

21.
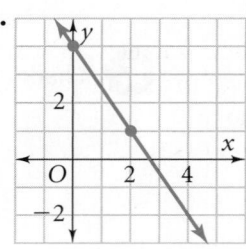

Lesson 3-8

Tell whether each number is *rational* or *irrational*.

22. $\sqrt{\frac{1}{4}}$ **23.** $\sqrt{\frac{8}{5}}$ **24.** $\sqrt{11}$ **25.** $\sqrt{\frac{121}{100}}$ **26.** $-\sqrt{0.81}$

Lesson 3-2

Solve each equation.

27. $5(x + 1) = 8$ **28.** $-33 = 6h + 7 + 2h$ **29.** $-2(3 - k) = 15$

30. $17 = 4t - 2t + 1$ **31.** $4(n + 1) - 3 = 16$ **32.** $40 = 8p + 5 - p$

Guided Problem Solving

Understanding Word Problems Read the exercise below and then follow along with what Vera thinks and writes. Check your understanding with the exercise at the bottom of the page.

Two friends are paddling a kayak downstream (with the current). The graph shows their distance from camp as a function of time.

a. Is every point on the line a reasonable value for this situation? Explain.

b. **Critical Thinking** Suppose they were paddling back toward the camp, upstream (against the current) rather than downstream. How would the graph compare with the downstream graph? Explain.

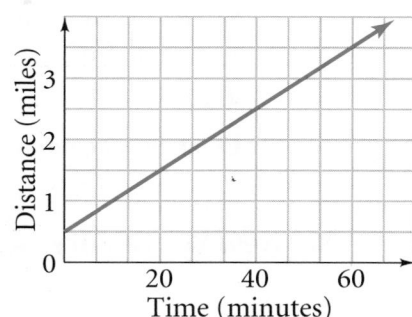

What Vera Thinks

The graph of the line is rising, so the kayakers' distance from camp is steadily increasing. After 20 minutes they are 1.5 miles from camp. After 40 minutes they are 2.5 miles away.

I can look at any point on the line between (20, 1.5) and (40, 2.5) and see a meaningful time and distance.

When you paddle upstream, the current slows you down. The slope of the line is the kayakers' rate, so the new line would be less steep than the downstream line.

Since the kayakers are traveling toward camp, their distance from the camp would decrease over time. The new line would have a negative slope.

What Vera Writes

a. Yes. Both time and distance are continuous, so every point on the line is a reasonable value.

b. The upstream graph would be less steep than the downstream graph because the kayakers would move more slowly against the current.

The graph for paddling upstream would have a negative slope because they are getting closer to camp.

EXERCISE

The graph shows the cost of buying bagels.
 a. What is the cost per bagel?
 b. Is every point on the graph a reasonable value? Explain.
 c. How would the graph be different if the bagels sold for $.75 each?

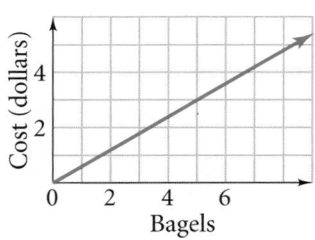

6-4

Standard Form

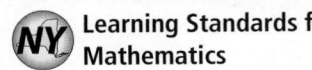 **Learning Standards for Mathematics**

A.A.36 Write the equation of a line parallel to the x- or y-axis.

A.G.4 Identify and graph linear, quadratic (parabolic), absolute value, and exponential functions.

✓ **Check Skills You'll Need**

GO for Help Review page 140 and Lesson 3-2

Solve each equation for y.

1. $3x + y = 5$ **2.** $y - 2x = 10$ **3.** $x - y = 6$

4. $20x + 4y = 8$ **5.** $9y + 3x = 1$ **6.** $5y - 2x = 4$

Clear each equation of decimals.

7. $6.25x + 8.5 = 7.75$ **8.** $0.4 = 0.2x - 5$ **9.** $0.9 - 0.222x = 1$

🔊 **New Vocabulary** • standard form of a linear equation • x-intercept

---

1   Graphing Equations Using Intercepts

Activity: Intercepts

1. Make a table of values for the equation $3y - 2x = 12$.

2. Use the table of values to graph $3y - 2x = 12$.

3. What is the y-intercept?

4. What is the value of x when the line crosses the x-axis?

5. In the equation $3y - 2x = 12$, what is the value of y when x = 0? What is the value of x when y = 0?

6. Using your answers to 3, 4, and 5, explain how you can make a graph of $3y - 2x = 12$ without making a table.

The slope-intercept form is just one form of a linear equation. Another form is standard form, which is useful in making quick graphs.

 Key Concepts

Definition	**Standard Form of a Linear Equation**

The **standard form of a linear equation** is $Ax + By = C$, where A, B, and C are real numbers, and A and B are not both zero.

You can use the x- and y-intercepts to make a graph. The **x-intercept** is the x-coordinate of the point where a line crosses the x-axis. To graph a linear equation in standard form, you can find the x-intercept by substituting 0 for y and solving for x. Similarly, to find the y-intercept, substitute 0 for x and solve for y.

1 EXAMPLE Finding *x*- and *y*-Intercepts

Find the *x*- and *y*-intercepts of $3x + 4y = 8$.

Step 1 To find the *x*-intercept, substitute 0 for *y* and solve for *x*.

$$3x + 4y = 8$$
$$3x + 4(0) = 8$$
$$3x = 8$$
$$x = \frac{8}{3}$$

The *x*-intercept is $\frac{8}{3}$.

Step 2 To find the *y*-intercept, substitute 0 for *x* and solve for *y*.

$$3x + 4y = 8$$
$$3(0) + 4y = 8$$
$$4y = 8$$
$$y = 2$$

The *y*-intercept is 2.

✓ Quick Check ❶ Find the *x*- and *y*-intercepts of $4x - 9y = -12$.

If the *x*- and *y*-intercepts are integers, you can use them to make a quick graph.

2 EXAMPLE Graphing Lines Using Intercepts

Graph $2x + 3y = 12$ using intercepts.

Step 1 Find the intercepts.

$$2x + 3y = 12$$
$$2x + 3(0) = 12 \quad \textbf{Substitute 0 for } \textit{y}.$$
$$2x = 12 \quad \textbf{Solve for } \textit{x}.$$
$$x = 6$$
$$2(0) + 3y = 12 \quad \textbf{Substitute 0 for } \textit{x}.$$
$$3y = 12 \quad \textbf{Solve for } \textit{y}.$$
$$y = 4$$

Step 2 Plot $(0, 4)$ and $(6, 0)$. Draw a line through the points.

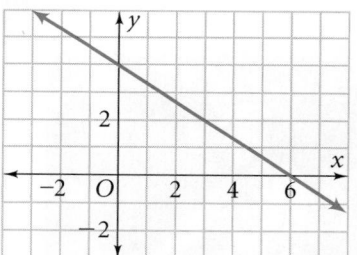

✓ Quick Check ❷ Graph $5x + 2y = -10$ using the *x*- and *y*-intercepts.

In the standard form of an equation $Ax + By = C$, either *A* or *B*, but not both, may be zero. If *A* or *B* is zero, the line is either horizontal or vertical.

3 EXAMPLE Graphing Horizontal and Vertical Lines

a. Graph $y = -3$.

$0x + 1y = -3$ ← **Write in standard form.** →
For all values of *x*, $y = -3$.

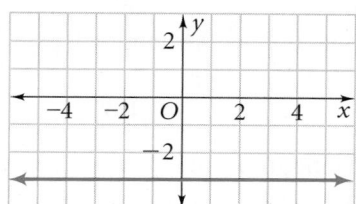

b. Graph $x = 2$.

$1x + 0y = 2$
For all values of *y*, $x = 2$.

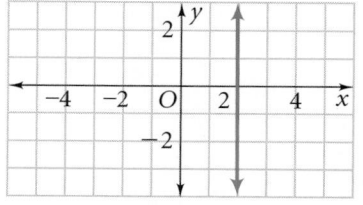

✓ Quick Check ❸ Graph each equation.

a. $y = 5$ **b.** $y = 0$ **c.** $x = -4$ **d.** $x = 0$

You can change an equation from slope-intercept form to standard form. If the equation contains fractions or decimals, multiply to write the equation using integers.

4   EXAMPLE   Transforming to Standard Form

Write $y = \frac{3}{4}x + 2$ in standard form using integers.

$$y = \frac{3}{4}x + 2$$

$$4y = 4\left(\frac{3}{4}x + 2\right) \qquad \textbf{Multiply each side by 4.}$$

$$4y = 3x + 8 \qquad \textbf{Use the Distributive Property.}$$

$$-3x + 4y = 8 \qquad \textbf{Subtract 3x from each side.}$$

The standard form of $y = \frac{3}{4}x + 2$ is $-3x + 4y = 8$.

✓ Quick Check   **4** Write $y = -\frac{2}{5}x + 1$ in standard form using integers.

You can write equations for real-world situations using standard form.

5   EXAMPLE   Real-World 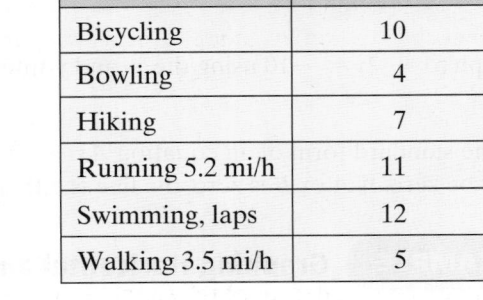 Problem Solving

Data Analysis Write an equation in standard form to find the minutes someone who weighs 150 lb would need to bicycle and swim laps in order to burn 300 calories. Use the data below.

Activity by a 150-lb Person	Calories Burned per Minute
Bicycling	10
Bowling	4
Hiking	7
Running 5.2 mi/h	11
Swimming, laps	12
Walking 3.5 mi/h	5

Real-World Connection

Doctors recommend 30 minutes of exercise each day.

Define   Let x = the minutes spent bicycling.

            Let y = the minutes spent swimming laps.

Relate   10 · minutes bicycling   plus   12 · minutes swimming laps   equals   300 calories

Write $10x$ + $12y$ =   300

The equation in standard form is $10x + 12y = 300$.

✓ Quick Check   **5** **Data Analysis** Write an equation in standard form to find the minutes someone who weighs 150 lb would need to bowl and walk to burn 250 calories.

EXERCISES

For more exercises, see *Extra Skill and Word Problem Practice.*

Practice and Problem Solving

 **Practice by Example**

Example 1
(page 331)

 for Help

Find the *x*- and *y*-intercepts of each equation.

1. $x + 2y = 18$

2. $3x - y = 9$

3. $-5x + y = 30$

4. $-6x + 3y = -9$

5. $4x + 12y = -18$

6. $9x - 6y = -72$

7. $-2x - 3y = -12$

8. $7x - 2y = 4$

9. $-8x + 10y = 40$

Example 2
(page 331)

Match each equation with its graph.

10. $2x - 5y = 10$

11. $-2x + 5y = 10$

12. $2x + 5y = 10$

A.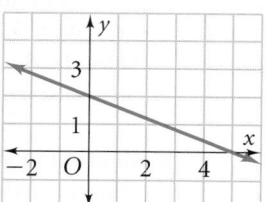

B.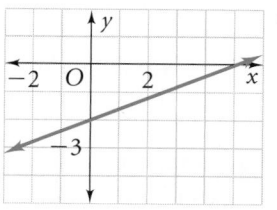

C.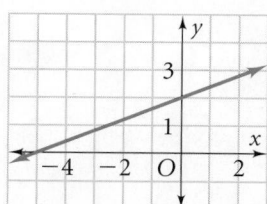

Graph each equation using *x*- and *y*-intercepts.

13. $x + y = 2$

14. $x + y = -5$

15. $x - y = -7$

16. $-3x + y = 6$

17. $-2x + y = -6$

18. $5x - 3y = 15$

Example 3
(page 331)

For each equation, tell whether its graph is a horizontal or a vertical line.

19. $y = -1$

20. $x = 4$

21. $y = 2\frac{1}{2}$

22. $x = -3.75$

Graph each equation.

23. $y = 3$

24. $x = -7$

25. $y = -1.5$

26. $x = 4.5$

Example 4
(page 332)

Write each equation in standard form using integers.

27. $y = 3x + 1$

28. $y = 4x - 7$

29. $y = \frac{1}{2}x - 3$

30. $y = \frac{2}{3}x + 5$

31. $y = -\frac{3}{4}x - 4$

32. $y = -\frac{4}{5}x - 7$

33. $y = \frac{7}{2}x + \frac{1}{4}$

34. $y = -\frac{2}{5}x + \frac{1}{10}$

35. $y = -3x$

Example 5
(page 332)

36. Fund-Raising The sophomore class holds a car wash to raise money. A local merchant donates all of the supplies. A wash costs $5 per car and $6.50 per van or truck.
  a. Define a variable for the number of cars. Define a different variable for the number of vans or trucks.
  b. Write an equation in standard form to relate the number of cars and vans or trucks the students must wash to raise $800.

37. Fitness Larry runs at an average rate of 8 mi/h. He walks at an average rate of 3 mi/h.
  a. Define a variable for time spent walking. Define a different variable for time spent running.
  b. Write an equation in standard form to relate the times he could spend running and walking if he travels a distance of 15 mi.

Graph each equation.

38. $-3x + 2y = -6$ **39.** $x + y = 1$ **40.** $2x - 3y = 18$

41. $y - x = -4$ **42.** $y = 2x + 5$ **43.** $y = -3x - 1$

44. $2 - y = x - 6$ **45.** $9 + y = 8 - x$ **46.** $6x = y$

47. Nutrition Suppose you are preparing a snack mix. You want the total protein from peanuts and granola to equal 28 grams. Peanuts have 7 grams of protein per ounce, and granola has 3 grams of protein per ounce.
  a. Write an equation for the protein content of your mix.
  b. Graph your equation. Use your graph to find how many ounces of granola you should use if you use 1 ounce of peanuts.

48. You are sent to the store to buy sliced meat for a party. You are told to get roast beef and turkey, and you are given $30. Roast beef is $4.29/lb and turkey is $3.99/lb. Write an equation in standard form to relate the pounds of each kind of meat you could buy at the store with $30.

Real-World Connection

A peanut contains about 0.24 gram of protein.

Graphing Calculator Write each equation in slope-intercept form. Then use a graphing calculator to graph each equation. Make a sketch of the graph. Include the x- and y-intercepts.

49. $8x - 10y = -100$ **50.** $-6x + 7y = 21$ **51.** $12x + 15y = -45$

52. $-5x + 9y = -15$ **53.** $16x + 11y = -88$ **54.** $3x - 27y = 18$

 55. Writing Two of the forms of a linear equation are slope-intercept form and standard form. Explain when each is the more useful.

56. Critical Thinking The definition of standard form states that A and B can't both be zero. Explain why.

57. Error Analysis A student says that the equation $3x + 2y = 6$ is a standard form of the equation $y = \frac{3}{2}x + 3$. What is the student's error?

Write an equation for each line on the graph.

58. a **59.** b **60.** c **61.** d

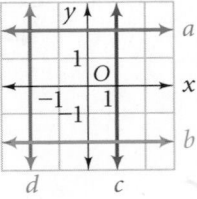

GO Online
Homework Video Tutor
Visit: PHSchool.com
Web Code: ate-0604

62. a. Fund-Raising Suppose your school is having a talent show to raise money for new music supplies. You estimate that 200 students and 150 adults will attend. You estimate $200 in expenses. Write an equation to find what ticket prices you should set to raise $1000.
  b. Open-Ended Graph your equation. Choose three possible prices you could set for students' and adults' tickets. Which is the best choice? Explain.

C Challenge

63. Write an equation of a line that has the same slope as the line $3x - 5y = 7$ and the same y-intercept as the line $2y - 9x = 8$.

64. Geometry Graph each of the four lines below on the same graph. What figure do the four lines appear to form?

$-2x + 3y = 10$ $3x + 2y = -2$ $-2x + 3y = -3$ $3x + 2y = 11$

65. a. Graph $2x + 3y = 6$ and $2x + 3y = 18$.
  b. What is the slope of each line?
  c. Compare the x-intercepts of the two lines. How are they related? How are the y-intercepts related?

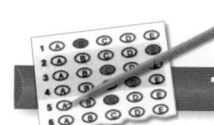
Test Prep

Multiple Choice

66. Which of the following is the standard form of $y = -\frac{2}{3}x + 6$ written using integers?

A. $\frac{2}{3}x + y = 6$ **B.** $-6 = -\frac{2}{3}x - y$ **C.** $2x + 3y = 18$ **D.** $-2x - 3y = 18$

67. If $A \neq 0$ and $B \neq 0$, which is the slope of $Ax + By = C$?

F. $-\frac{B}{A}$ **G.** $\frac{C}{A}$ **H.** $-\frac{A}{B}$ **J.** $\frac{C}{B}$

Short Response

68. A basket with 4 apples weighs 2 pounds. The same basket with 12 apples weighs 4 pounds. Write an equation in slope-intercept form for the weight y in terms of the number of apples x. Write an equation in standard form with integer coefficients that shows the relationship of the weight y and the number of apples x.

Extended Response

69. A tire dealer sells Supreme tires for $48 each and Prestige tires for $56 each. During one week, the sales for both tires totaled $2008.
  a. Write an equation that you can use to determine the possible combinations of Supreme tires x and Prestige tires y sold.
  b. Graph your equation on a coordinate plane.
  c. Use your graph to list 3 possible combinations of Supreme and Prestige tires sold.

Mixed Review

GO for Help

Lesson 6-2

Determine whether the ordered pair is a solution of the equation.

70. $(2, -3)$; $y = -x - 1$ **71.** $(6, -1)$; $y = 2x - 15$ **72.** $(-5, -7)$; $y = -3x - 8$

Lesson 3-4

Solve each proportion.

73. $\frac{a}{5} = \frac{12}{15}$ **74.** $\frac{2}{8} = \frac{w}{9}$ **75.** $\frac{x + 2}{4} = \frac{3}{8}$ **76.** $\frac{14}{4m} = \frac{16}{5m + 9}$

Lesson 2-7

Find each probability for rolling a number cube.

77. P(rolling a 2, then a 4) **78.** P(rolling a 5, then an even number)

Checkpoint Quiz 1

Lessons 6-1 through 6-4

Find the slope of the line passing through each pair of points.

1. $(-1, 3), (6, -2)$ **2.** $(4, 5), (0, 2)$ **3.** $(-2, -3), (-1, -7)$ **4.** $(4, -4), (-5, 5)$

5. Credit Cards In 1990, people charged $534 billion on the two most-used types of credit cards. In 1994, people charged $1.021 trillion on these same two types of credit cards. What was the rate of change?

Graph each equation.

6. $y = 4x - 1$ **7.** $y = -\frac{2}{5}x + 6$ **8.** $5x + 3y = -30$ **9.** $2x - 7y = 15$

10. Writing How are the graphs of $y = 3x + 5$, $y = \frac{2}{3}x + 5$, and $y = \frac{3}{5}x + 5$ alike? How are they different?

6-5

Point-Slope Form and Writing Linear Equations

 **Learning Standards for Mathematics**

A.A.34 Write the equation of a line, given its slope and the coordinates of a point on the line.

A.A.35 Write the equation of a line, given the coordinates of two points on the line.

A.G.4 Identify and graph linear, quadratic (parabolic), absolute value, and exponential functions.

✓ **Check Skills You'll Need**

GO for Help   Lessons 6-1 and 2-4

Find the rate of change of the data in each table.

1.

x	y
2	4
5	−2
8	−8
11	−14

2.

x	y
−3	−5
−1	−4
1	−3
3	−2

3.

x	y
10	4
7.5	−1
5	−6
2.5	−11

Simplify each expression.

4. $-3(x - 5)$

5. $5(x + 2)$

6. $-\frac{4}{9}(x - 6)$

◀)) **New Vocabulary**   • point-slope form

1   Using Point-Slope Form

You can use the definition of slope to find another form of a linear equation called the point-slope form.

$$\frac{y_2 - y_1}{x_2 - x_1} = m \qquad \text{Use the definition of slope.}$$

Suppose you know that a line passes through the point $(3, 4)$ and has slope 2.

$$\frac{y - 4}{x - 3} = 2 \qquad \text{Substitute } (3, 4) \text{ for } (x_1, y_1) \text{ and substitute } (x, y) \text{ for } (x_2, y_2).$$
$$\text{Substitute 2 for } m.$$

$$\frac{y - 4}{x - 3}(x - 3) = 2(x - 3) \qquad \text{Multiply each side by } x - 3.$$

$$y - 4 = 2(x - 3) \qquad \text{Simplify the left side of the equation.}$$

The equation $y - 4 = 2(x - 3)$ is in point-slope form.

$$y - 4 = 2(x - 3)$$
$$\text{y-coordinate} \quad \text{slope} \quad \text{x-coordinate}$$

 Key Concepts

Definition	Point-Slope Form of a Linear Equation

The **point-slope form** of the equation of a nonvertical line that passes through the point (x_1, y_1) and has slope m is

$$y - y_1 = m(x - x_1)$$

1 EXAMPLE Graphing Using Point-Slope Form

Graph the equation $y - 5 = \frac{1}{2}(x - 2)$.

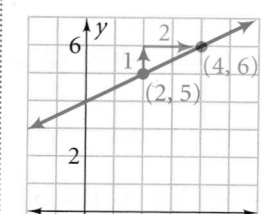

The equation shows that the line passes through (2, 5) and has a slope $\frac{1}{2}$.

Start at (2, 5). Using the slope, go up 1 unit and right 2 units to (4, 6). Draw a line through the two points.

Quick Check ❶ Graph the equation $y - 5 = -\frac{2}{3}(x + 2)$.

2 EXAMPLE Writing an Equation in Point-Slope Form

Vocabulary Tip

Square brackets, [ ], are grouping symbols commonly used when parentheses, ( ), are inside.

Write the equation of the line that has slope -3 that passes through the point $(-1, 7)$.

$y - y_1 = m(x - x_1)$	Use the point-slope form.
$y - 7 = -3[x - (-1)]$	Substitute $(-1, 7)$ for (x_1, y_1) and -3 for m.
$y - 7 = -3(x + 1)$	Simplify inside the grouping symbols.

Quick Check ❷ Write the equation of the line that has slope $\frac{2}{5}$ that passes through the point $(10, -8)$.

If you know two points on a line, first use them to find the slope. Then you can write an equation using either point.

3 EXAMPLE Using Two Points to Write an Equation

Online active math

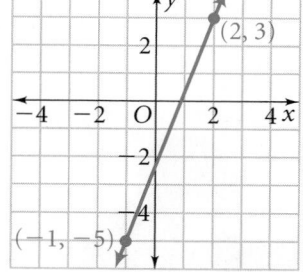

For: Point-Slope Activity
Use: Interactive Textbook, 6-5

Write equations for the line in point-slope form and in slope-intercept form.

Step 1 Find the slope.

$$\frac{y_2 - y_1}{x_2 - x_1} = m$$

$$\frac{-5 - 3}{-1 - 2} = \frac{8}{3}$$

The slope is $\frac{8}{3}$.

Step 2 Use either point to write the equation in point-slope form. Use (2, 3).

$$y - y_1 = m(x - x_1)$$

$$y - 3 = \frac{8}{3}(x - 2)$$

Step 3 Rewrite the equation from Step 2 in slope-intercept form.

$$y - 3 = \frac{8}{3}(x - 2)$$

$$y - 3 = \frac{8}{3}x - 5\frac{1}{3}$$

$$y = \frac{8}{3}x - 2\frac{1}{3}$$

Quick Check ❸ **a.** Write an equation for the line in Example 3 in point-slope form using the point $(-1, -5)$.
   b. Write the equation you found in part (a) in slope-intercept form.
   c. What is true about the equation you wrote in part (b) and the equation in Step 3 of Example 3?

You can write a linear equation to model data in tables. Two sets of data have a linear relationship if the rate of change between consecutive pairs of data is the same. For data that have a linear relationship, the rate of change is the slope.

4 **EXAMPLE** **Writing an Equation Using a Table**

Is the relationship shown by the data linear? If so, model the data with an equation.

Step 1 Find the rate of change for consecutive ordered pairs.

x	y
−1	4
3	6
5	7
11	10

4   2 $\frac{2}{4} = \frac{1}{2}$
2   1 $\frac{1}{2} = \frac{1}{2}$
6   3 $\frac{3}{6} = \frac{1}{2}$

Step 2 Use the slope and a point to write an equation.

$$y - y_1 = m(x - x_1)$$

Substitute (5, 7) for (x_1, y_1) and $\frac{1}{2}$ for m.

$$y - 7 = \frac{1}{2}(x - 5)$$

✓ **Quick Check** **4** Is the relationship shown by the data at the right linear? If so, model the data with an equation.

x	y
−11	−7
−1	−3
4	−1
19	5

5 **EXAMPLE** **Real-World** **Problem Solving**

Is the relationship shown by the data linear? If so, model the data with an equation.

Boiling Point of Water

Altitude (1000 ft)	Temperature (°F)
8	197.6
4.5	203.9
3	206.6
2.5	207.5

−3.5   6.3
−1.5   2.7
−0.5   0.9

Step 1 Find the rates of change for consecutive ordered pairs.

$\frac{6.3}{-3.5} = -1.8$ $\frac{2.7}{-1.5} = -1.8$ $\frac{0.9}{-0.5} = -1.8$

The relationship is linear. The rate of change is −1.8 degrees Fahrenheit per 1000 ft of altitude.

Step 2 Use the slope and a point to write an equation.

$y - y_1 = m(x - x_1)$ **Use the point-slope form.**
$y - 206.6 = -1.8(x - 3)$ **Substitute (3, 206.6) for (x_1, y_1) and −1.8 for m.**

The equation $y - 206.6 = -1.8(x - 3)$ relates altitude in thousands of feet x to the boiling point temperature in degrees Fahrenheit.

Real-World **Connection**

At 5280 feet above sea level it takes 17 minutes to hard-boil an egg. This is more than 40% longer than it takes the same egg to cook at sea level.

 **Quick Check** ⑤ Is the relationship shown by the data in the table linear? If it is, model the data with an equation.

Working Outdoors

Temperature	Calories Burned per Day
68°F	3030
62°F	3130
56°F	3230
50°F	3330

In Example 5 you could rewrite $y - 206.6 = -1.8(x - 3)$ as $y = -1.8x + 212$. This form gives you useful information about the y-intercept. For instance, 212°F is the boiling point of water at sea level.

Here are the three forms of linear equations you have studied.

 Key Concepts

 **for Help**

For more help with the three forms of a linear equation, see page 342.

Summary	**Linear Equations**

Slope-Intercept Form

$y = mx + b$

m is the slope and b is the y-intercept.

Standard Form

$Ax + By = C$

A and B are not both 0.

Point-Slope Form

$(y - y_1) = m(x - x_1)$

(x_1, y_1) lies on the graph of the equation, and m is the slope.

Examples

$y = -\frac{2}{3}x + \frac{5}{3}$

$2x + 3y = 5$

$y - 1 = -\frac{2}{3}(x - 1)$

EXERCISES

For more exercises, see *Extra Skill and Word Problem Practice*.

Practice and Problem Solving

Ⓐ Practice by Example

Example 1
(page 337)

 for Help

Graph each equation.

1. $y - 2 = (x - 3)$
2. $y - 2 = 2(x - 3)$
3. $y - 2 = -\frac{3}{2}(x - 3)$
4. $y + 5 = -(x - 2)$
5. $y + 1 = \frac{2}{3}(x + 4)$
6. $y - 1 = -3(x + 2)$
7. $y + 3 = -2(x - 1)$
8. $y - 4 = (x - 5)$
9. $y - 2 = 3(x + 2)$

Example 2
(page 337)

Write an equation in point-slope form for the line through the given point that has the given slope.

10. $(3, -4); m = 6$
11. $(4, 2); m = -\frac{5}{3}$
12. $(0, 2); m = \frac{4}{5}$
13. $(-2, -7); m = -\frac{3}{2}$
14. $(4, 0); m = 1$
15. $(5, -8); m = -3$
16. $(-5, 2); m = 0$
17. $(1, -8); m = -\frac{1}{5}$
18. $(-6, 1); m = \frac{2}{3}$

Example 3
(page 337)

A line passes through the given points. Write an equation for the line in point-slope form. Then rewrite the equation in slope-intercept form.

19. $(-1, 0), (1, 2)$
20. $(3, 5), (0, 0)$
21. $(4, -2), (9, -8)$
22. $(6, -4), (-3, 5)$
23. $(-1, -5), (-7, -6)$
24. $(-3, -4), (3, -2)$
25. $(2, 7), (1, -4)$
26. $(-2, 6), (5, 1)$
27. $(3, -8), (-2, 5)$
28. $\left(1, \frac{1}{2}\right), (3, 2)$
29. $\left(\frac{1}{2}, 2\right), \left(-\frac{3}{2}, 4\right)$
30. $(0.2, 1.1), (7, 3)$

Example 4
(page 338)

Is the relationship shown by the data linear? If so, model the data with an equation.

31.

x	y
−4	9
2	−3
5	−9
9	−17

32.

x	y
−10	−5
−2	19
5	40
11	58

33.

x	y
3	1
6	4
9	13
15	49

Example 5
(page 338)

34.

Speed Over Posted Speed Limit (mi/h)	Fine ($)
10	75
12	95
15	125
19	165

35.

Volume (gal)	Weight (lb)
0	0
2	16
4	33
6	50

B **Apply Your Skills**

Write an equation of each line in point-slope form.

36.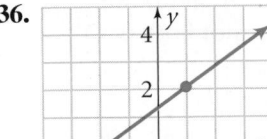

37.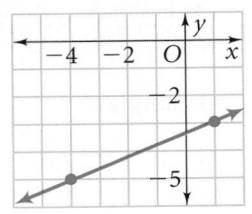

38.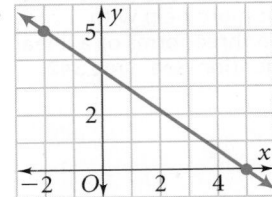

Write one equation of the line through the given points in point-slope form and one in standard form using integers.

39. $(1, 4), (−1, 1)$

40. $(6, −3), (−2, −3)$

41. $(0, 0), (−1, −2)$

42. $(0, 2), (−4, 2)$

43. $(−6, 6), (3, 3)$

44. $(2, 3), (−1, 5)$

45. $(5, −3), (3, 4)$

46. $(2, 2), (−1, 7)$

47. $(−7, 1), (5, −1)$

48. $(−8, 4), (−4, −2)$

49. $(2, 4), (−3, −6)$

50. $(5, 3), (4, 5)$

51. $(0, 1), (−3, 0)$

52. $(−2, 4), (0, −5)$

53. $(6, 2), (1, −1)$

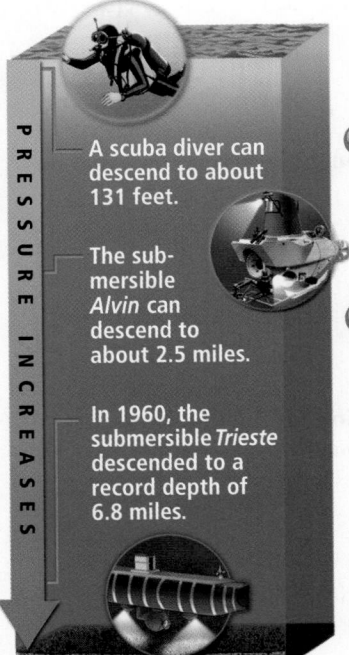

P
R
E
S
S
U
R
E

I
N
C
R
E
A
S
E
S

A scuba diver can descend to about 131 feet.

The submersible *Alvin* can descend to about 2.5 miles.

In 1960, the submersible *Trieste* descended to a record depth of 6.8 miles.

54. Science At the surface of the ocean, pressure is 1 atmosphere. At 66 ft below sea level, the pressure is 3 atmospheres. The relationship of pressure and depth is linear.
  a. Write an equation for the data.
  b. Predict the pressure at 100 ft below sea level.

55. Environment Worldwide carbon monoxide emissions are decreasing about 2.6 million metric tons each year. In 1991, carbon monoxide emissions were 79 million metric tons. Use a linear equation to model the relationship between carbon monoxide emissions and time. Let $x = 91$ correspond to 1991.

56. a. Open-Ended Write an equation in point-slope form that contains the point $(−4, −6)$. Explain your steps.
  b. How many equations could you write in part(a)? Explain.

57. Critical Thinking How would the graph of $y − 12 = 8(x − 2)$ change if all of the subtraction signs were changed to addition signs?

58. Reasoning Is $y − 5 = 2(x − 1)$ an equation of a line through $(4, 11)$? Explain.

Homework Video Tutor
Visit: PHSchool.com
Web Code: ate-0605

59. Open-Ended Write an equation in each of the following forms.
  a. slope-intercept form
  b. standard form
  c. point-slope form

 60. Science Use the scatter plot.
  a. Write an equation to model the data.
  b. What is the speed of sound at 15°C?
  c. Predict the speed of sound at 60°C.

 **Challenge** **Write an equation in slope-intercept form of each line described below.**

61. The line contains the point $(-3, -5)$ and has the same slope as $y + 2 = 7(x + 3)$.

62. The line contains the point $(1, 3)$ and has the same y-intercept as $y - 5 = 2(x - 1)$.

63. The line contains the point $(2, -2)$ and has the same x-intercept as $y + 9 = 3(x - 4)$.

64. The table shows data that you can model using a linear function.
  a. Find the value of y when $x = 6$.
  b. Find the value of y when $x = 120$.
  c. Find the value of x when $y = 11$.
  d. Find the value of x when $y = 50$.

Effect of Air Temperature on Speed of Sound

x	y
4	14
8	15.5
12	17
16	18.5

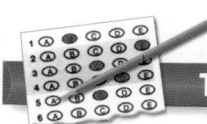
Test Prep

Gridded Response

65. What is the slope of the graph of $y - 8 = \frac{1}{2}(x + 2)$?

66. Find the y-intercept of the line $y + 3 = 4(x + 3)$.

67. What is the x-intercept of the line $y = 3x - 7$?

68. When $y - 1 = -\frac{4}{5}(x - 3)$ is written in standard form using positive integers, what is the smallest possible coefficient of x?

69. When $y = -\frac{5}{2}x + \frac{2}{3}$ is written in standard form using positive integers, what is the smallest possible coefficient of y?

Mixed Review

GO for Help

Lesson 6-4

Graph each line.

70. $6x + 7y = 14$ **71.** $-2x + 9y = -9$ **72.** $5x - 4y = 24$

73. $3x - 8y = 4$ **74.** $5x + 18y = 6$ **75.** $-7x + 4y = -21$

Lesson 5-7

Find the common difference of each sequence. Then write the next two terms.

76. $-12, -7, -2, \ldots$ **77.** $\frac{1}{2}, \frac{5}{6}, \frac{7}{6}, \ldots$ **78.** $2.45, 2.52, 2.59, \ldots$

79. $-3.2, -3.25, -3.3, \ldots$ **80.** $18, 35, 52, \ldots$ **81.** $-7, -3, 1, \ldots$

Understanding Vocabulary

There are three forms of a linear equation that you have studied in this chapter:

- slope-intercept form
- standard form
- point-slope form

To understand and remember these forms, it may help you to connect the common meaning of the words with their specialized meanings in mathematics.

Word	Common Meaning	Mathematical Meaning
Slope	An inclined surface (for example, the slope of a hill)	The rate of change that gives the steepness of a line: $\text{slope} = \dfrac{\text{vertical change}}{\text{horizontal change}} = \dfrac{\text{rise}}{\text{run}}$
Intercept	To cut off from a path (for example, to intercept a football)	The values of the points at which a line intersects (or cuts) the x-axis or y-axis
Standard	Generally accepted	A general form of an equation
Point	A dot or speck (noun)	A fixed location on a coordinate plane; every point has a unique x-value and y-value.

Slope-intercept form, standard form, and point-slope form all give clues about the lines they will graph.

- Slope-intercept form tells the slope of the line and its y-intercept.

  Example $y = 3x - 1$ **The graph of this equation has a slope of 3 and a y-intercept of −1.**

- Standard form is the general form for linear equations. You can model many real-world situations using the standard form. Also, it is easy to find the x- and y-intercepts of the graph of an equation in this form.

  Example $5x - 2y = 100$ **The graph of this equation intersects the y-axis at $\frac{100}{-2} = -50$ and intersects the x-axis at $\frac{100}{5} = 20$.**

- Point-slope form tells a point on the line and the slope of the line.

  Example $y - 2 = 3(x - 5)$ **The graph of this equation passes through the point (5, 2), and its slope is 3.**

EXERCISES

For each situation, which form of a linear equation is easiest to write?

1. A submarine started at sea level and submerged at a steady rate of 8 feet per minute.

2. A plant growing at a steady rate measured 2 in. on day 5 and 13 in. on day 21.

Find the common meaning and the mathematical meaning of each word.

3. function   **4.** relation   **5.** range   **6.** domain

7. variable   **8.** base   **9.** element   **10.** term

6-6 Parallel and Perpendicular Lines

 Learning Standards for Mathematics

A.A.34 Write the equation of a line, given its slope and the coordinates of a point on the line.

A.A.37 Determine the slope of a line, given its equation in a form.

A.A.38 Determine if two lines are parallel, given their equations in a form.

✓ **Check Skills You'll Need**

GO for Help Lessons 2-3 and 6-2

What is the reciprocal of each fraction?

1. $\frac{1}{2}$　　2. $\frac{4}{3}$　　3. $-\frac{2}{5}$　　4. $-\frac{7}{5}$

What are the slope and *y*-intercept of each equation?

5. $y = \frac{5}{3}x + 4$　　6. $y = \frac{5}{3}x - 8$　　7. $y = 6x$　　8. $y = 6x + 2$

🔊 **New Vocabulary**　• parallel lines　• perpendicular lines　• negative reciprocal

---

1 Parallel Lines

In the graph at the right, the red and blue lines are parallel. **Parallel lines** are lines in the same plane that never intersect. The equation of the red line is $y = \frac{1}{2}x + \frac{3}{2}$. The equation of the blue line is $y = \frac{1}{2}x - 1$.

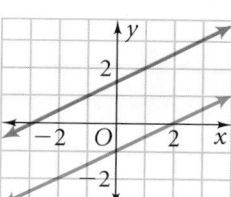

🔑 **Key Concepts**

Property　**Slopes of Parallel Lines**

Nonvertical lines are parallel if they have the same slope and different *y*-intercepts. *Any* two vertical lines are parallel.

Example　The equations $y = \frac{2}{3}x + 1$ and $y = \frac{2}{3}x - 3$ have the same slope, $\frac{2}{3}$, and different *y*-intercepts. The graphs of the two equations are parallel.

You can use slope-intercept form to determine whether the lines are parallel.

Real-World 🌐 Connection

The lanes for competitive swimming are parallel.

1 EXAMPLE　**Determining Whether Lines Are Parallel**

Are the graphs of $y = -\frac{1}{3}x + 5$ and $2x + 6y = 12$ parallel? Explain.

Write $2x + 6y = 12$ in slope-intercept form. Then compare with $y = -\frac{1}{3}x + 5$.

$6y = -2x + 12$　**Subtract 2x from each side.**

$\frac{6y}{6} = \frac{-2x + 12}{6}$　**Divide each side by 6.**

$y = -\frac{1}{3}x + 2$　**Simplify.**

The lines are parallel. The equations have the same slope, $-\frac{1}{3}$, and different *y*-intercepts.

✓ **Quick Check**　**1** Are the graphs of $-6x + 8y = -24$ and $y = \frac{3}{4}x - 7$ parallel? Explain.

Video Tutor Help
Visit: PHSchool.com
Web Code: ate-0775

You can use the fact that the slopes of parallel lines are the same to write the equation of a line parallel to a given line. To write the equation, you use the slope of the given line and the point-slope form of a linear equation.

② EXAMPLE **Writing Equations of Parallel Lines**

Write an equation for the line that contains $(5, 1)$ and is parallel to $y = \frac{3}{5}x - 4$.

Step 1 Identify the slope of the given line.

$$y = \frac{3}{5}x - 4$$

↑
slope

Step 2 Write the equation of the line through $(5, 1)$ using slope-intercept form.

$y - y_1 = m(x - x_1)$	**point-slope form**
$y - 1 = \frac{3}{5}(x - 5)$	**Substitute (5, 1) for (x_1, y_1) and $\frac{3}{5}$ for m.**
$y - 1 = \frac{3}{5}x - \frac{3}{5}(5)$	**Use the Distributive Property.**
$y - 1 = \frac{3}{5}x - 3$	**Simplify.**
$y = \frac{3}{5}x - 2$	**Add 1 to each side.**

 **② Quick Check** Write an equation for the line that contains $(2, -6)$ and is parallel to $y = 3x + 9$.

2 Perpendicular Lines

The lines at the right are perpendicular. **Perpendicular lines** are lines that intersect to form right angles. The equation of the red line is $y = -\frac{1}{4}x - 1$. The equation of the blue line is $y = 4x + 2$.

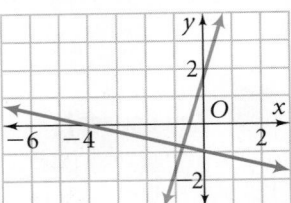

🔑 Key Concepts

Property **Slopes of Perpendicular Lines**

Two lines are perpendicular if the product of their slopes is -1. A vertical and a horizontal line are also perpendicular.

Example The slope of $y = -\frac{1}{4}x - 1$ is $-\frac{1}{4}$. The slope of $y = 4x + 2$ is 4. Since $-\frac{1}{4} \cdot 4 = -1$, the graphs of the two equations are perpendicular.

The product of two numbers is -1 if one number is the **negative reciprocal** of the other. Here is how to find the negative reciprocal of a number.

Start with a fraction: $-\frac{3}{5}$.	→	Find its reciprocal: $-\frac{5}{3}$.	→	Write the negative reciprocal: $\frac{5}{3}$.

Since $-\frac{3}{5} \cdot \frac{5}{3} = -1$, $\frac{5}{3}$ is the negative reciprocal of $-\frac{3}{5}$.

Start with an integer: 4.	→	Find its reciprocal: $\frac{1}{4}$.	→	Write its negative reciprocal: $-\frac{1}{4}$.

Since $4\left(-\frac{1}{4}\right) = -1$, $-\frac{1}{4}$ is the negative reciprocal of 4.

You can use the negative reciprocal of the slope of a given line to write an equation of a line perpendicular to that line.

3 EXAMPLE Writing Equations for Perpendicular Lines

Multiple Choice Which equation describes a line that contains $(0, -2)$ and is perpendicular to $y = 5x + 3$?

  Ⓐ $y = 5x - 2$ Ⓑ $y = \frac{1}{5}x - 2$ Ⓒ $y = \frac{1}{5}x + 2$ Ⓓ $y = -\frac{1}{5}x - 2$

Step 1   Identify the slope of the given line.

$$y = \underset{\uparrow}{5}x + 3$$
$$\text{slope}$$

Step 2   Find the negative reciprocal of the slope.

The negative reciprocal of 5 is $-\frac{1}{5}$.

Step 3   Use the slope-intercept form to write an equation.

$y = mx + b$

$y = -\frac{1}{5}x + (-2)$ **Substitute $-\frac{1}{5}$ for m, and -2 for b.**

$y = -\frac{1}{5}x - 2$ **Simplify.**

● The equation is $y = -\frac{1}{5}x - 2$. So D is the correct answer.

✓ Quick Check   **3** Write an equation of the line that contains $(1, 8)$ and is perpendicular to $y = \frac{3}{4}x + 1$.

Test-Taking Tip

After finding the slopes of the perpendicular lines, multiply the two slopes as a check.

$5\left(-\frac{1}{5}\right) = -1$

You can use equations of parallel and perpendicular lines to solve some real-world problems.

4 EXAMPLE Real-World  Problem Solving

Urban Planning A bike path for a new city park will connect the park entrance to Park Road. The path will be perpendicular to Park Road. Write an equation for the line representing the bike path.

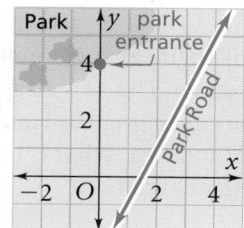

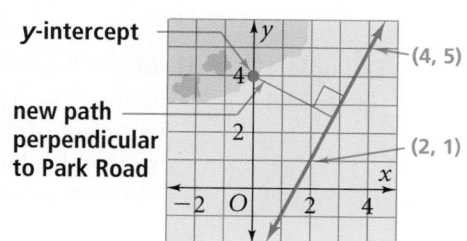

Step 1   Find the slope m of Park Road.

$m = \dfrac{y_2 - y_1}{x_2 - x_1} = \dfrac{5 - 1}{4 - 2} = \dfrac{4}{2} = 2$ **Points (2, 1) and (4, 5) are on Park Road.**

Step 2   Find the negative reciprocal of the slope.

The negative reciprocal of 2 is $-\frac{1}{2}$. So the slope of the bike path is $-\frac{1}{2}$. The y-intercept is 4.

● The equation for the bike path is $y = -\frac{1}{2}x + 4$.

✓ Quick Check   **4** A second bike path is planned. It will be parallel to Park Road and will also contain the park entrance. Write an equation for the line representing this bike path.

Real-World  Connection

Careers Urban planners use mathematics to make decisions on community problems such as traffic congestion and air pollution.

EXERCISES

For more exercises, see *Extra Skill and Word Problem Practice.*

Practice and Problem Solving

A Practice by Example

Example 1
(page 343)

GO for Help

Find the slope of a line parallel to the graph of each equation.

1. $y = \frac{1}{2}x + 2.3$

2. $y = -\frac{2}{3}x - 1$

3. $y = x$

4. $y = 6$

5. $3x + 4y = 12$

6. $7x - y = 5$

Are the graphs of the lines in each pair parallel? Explain.

7. $y = 4x + 12$
$-4x + 3y = 21$

8. $y = -\frac{3}{2}x + 2$
$3x + 2y = 8$

9. $y = \frac{1}{3}x + 3$
$x - 3y = 6$

10. $y = -\frac{1}{2}x + \frac{3}{2}$
$5x - 10y = 15$

11. $y = -3x$
$21x + 7y = 14$

12. $y = \frac{3}{4}x - 2$
$-3x + 4y = 8$

Example 2
(page 344)

Write an equation for the line that is parallel to the given line and that passes through the given point.

13. $y = 6x - 2; (0,0)$

14. $y = -3x; (3,0)$

15. $y = -2x + 3; (-3,5)$

16. $y = -\frac{7}{2}x + 6; (-4,-6)$

17. $y = 0.5x - 8; (8,-5)$

18. $y = -\frac{2}{3}x + 12; (5,-3)$

Example 3
(page 345)

Find the slope of a line perpendicular to the graph of each equation.

19. $y = 2x$

20. $y = -3x$

21. $y = \frac{7}{5}x - 2$

22. $y = -\frac{x}{5} - 7$

23. $2x + 3y = 5$

24. $y = -8$

Write an equation for the line that is perpendicular to the given line and that passes through the given point.

25. $y = 2x + 7; (0,0)$

26. $y = x - 3; (4,6)$

27. $y = -\frac{1}{3}x + 2; (4,2)$

28. $3x + 5y = 7; (-1,2)$

29. $-10x + 8y = 3; (15,12)$

30. $4x - 2y = 9; (8,-2)$

Example 4
(page 345)

31. Maps A city's civil engineer is planning a new parking garage and a new street. The new street will go from the entrance of the parking garage to Handel St. It will be perpendicular to Handel St. What is the equation of the line representing the new street?

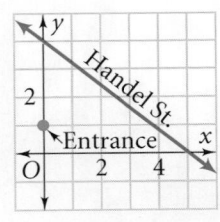

B Apply Your Skills

Tell whether the lines for each pair of equations are *parallel*, *perpendicular*, or *neither*.

32. $y = 4x + \frac{3}{4}, y = -\frac{1}{4}x + 4$

33. $y = \frac{2}{3}x - 6, y = \frac{2}{3}x + 6$

34. $y = -x + 5, y = x + 5$

35. $y = 5x, y = -5x + 7$

36. $y = \frac{x}{3} - 4, y = \frac{1}{3}x + 2$

37. $x = 2, y = 9$

38. $2x + y = 2, 2x + y = 5$

39. $3x - 5y = 3, -5x + 3y = 8$

40. Multiple Choice Which pair of equations graph as perpendicular lines?

Ⓐ $y = 4x + \frac{1}{2}$
  $y = -4x + 3$

Ⓑ $y = \frac{2}{3}x + 1$
  $y = 3x - 1$

Ⓒ $y = -\frac{3}{5}x - 9$
  $y = \frac{5}{3}x + 8$

Ⓓ $y = -\frac{3}{5}x - 9$
  $y = \frac{3}{5}x + 4$

Find the equation for each line.

41.

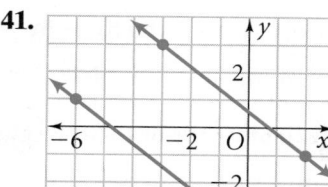

42.

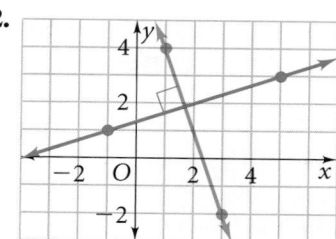

43.

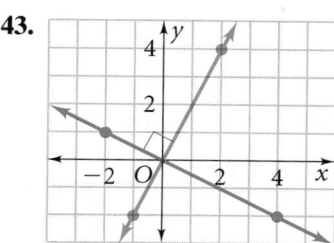

44.

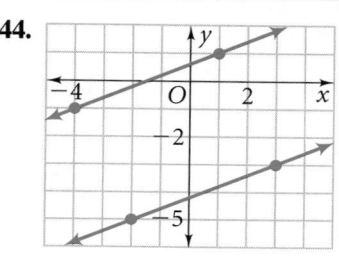

45.

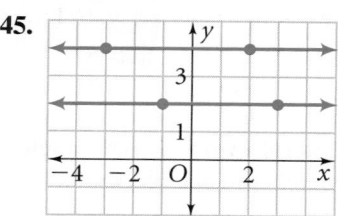

46.

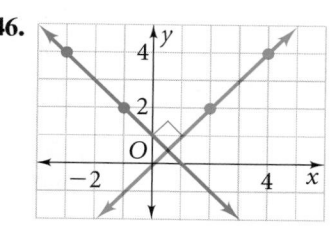

🌐 **Maps** Use the map below for Exercises 47–49.

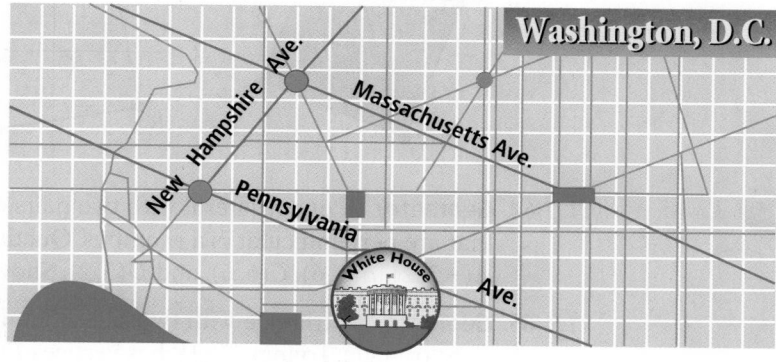

47. What is the slope of New Hampshire Avenue?

48. Show that the parts of Pennsylvania Avenue and Massachusetts Avenue near New Hampshire Avenue are parallel.

49. Show that New Hampshire Avenue is not perpendicular to Pennsylvania Avenue.

50. a. The graphs of $y = x$ and $y = -x$ are shown on the standard screen at the right. The product of the slopes is -1. Explain why the lines do not appear to be perpendicular.

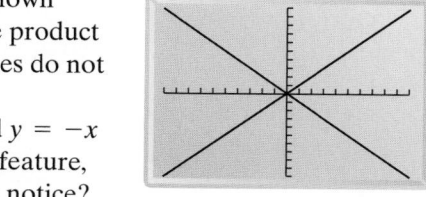

🖩 **b. Graphing Calculator** Graph $y = x$ and $y = -x$ on a graphing calculator. In the ZOOM feature, choose the square screen. What do you notice?

51. Open-Ended Write an equation for a line parallel to the graph of $4x - y = 1$.

52. Are the graphs of $2x + 7y = 6$ and $7y = 2x + 6$ parallel? Explain.

53. Are the graphs of $8x + 3y = 6$ and $8x - 3y = 6$ perpendicular? Explain.

54. Writing Are all horizontal lines parallel? Explain.

Tell whether each statement is *true* or *false*. Explain your choice.

55. Two lines with positive slopes can be perpendicular.

56. Two lines with positive slopes can be parallel.

57. The graphs of two different direct variations can be parallel.

Problem Solving Hint

For Exercises 55–57, sketch a graph to help you understand the statement in each exercise.

Geometry A quadrilateral with both pairs of opposite sides parallel is a parallelogram. Use slopes to determine whether each figure is a parallelogram.

58. 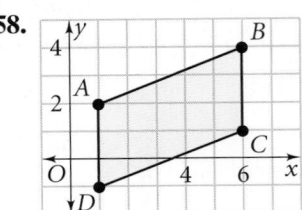 **59.** 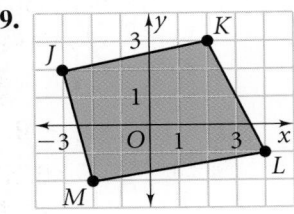 **60.**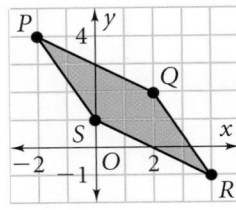

Geometry A quadrilateral with four right angles is a rectangle. Use slopes to determine whether each figure is a rectangle.

61. 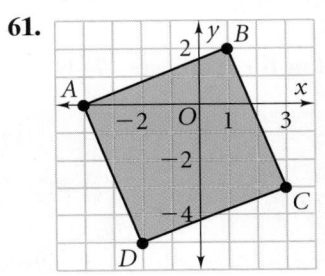 **62.** 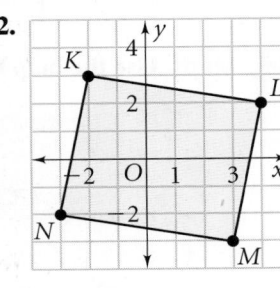 **63.**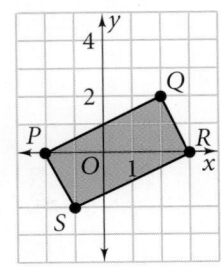

C Challenge

64. Geometry A quadrilateral with two pairs of parallel sides and with diagonals that are perpendicular is a rhombus. Quadrilateral $ABCD$ has vertices $A(-2, 2)$, $B(1, 6)$, $C(6, 6)$, and $D(3, 2)$. Show that $ABCD$ is a rhombus.

65. Geometry A triangle with two sides that are perpendicular to each other is a right triangle. Triangle PQR has vertices $P(3, 3)$, $Q(2, -2)$, and $R(0, 1)$. Determine whether PQR is a right triangle. Explain.

Tell whether the lines in each pair are *parallel*, *perpendicular*, or *neither*.

66. $ax - by = 5$; $-ax + by = 2$ **67.** $ax + by = 8$; $bx - ay = 1$

Assume the two lines are perpendicular. Find an equation for each line.

68. 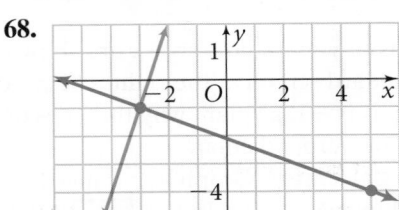 **69.**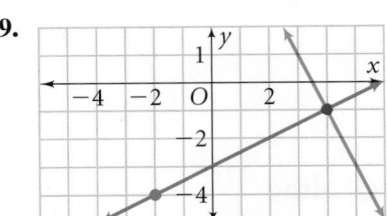

70. For what value of k are the graphs of $3x + 12y = 8$ and $6y = kx - 5$ parallel? Perpendicular?

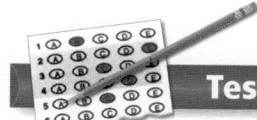
Multiple Choice

71. Which equation has as its graph a line perpendicular to a line that has a slope of $\frac{2}{3}$?

A. $y = \frac{2}{3}x + 5$ **B.** $y = \frac{3}{2}x - 1$ **C.** $y = 3x - 2$ **D.** $3x + 2y = 8$

72. Which equation has as its graph a line parallel to the graph of $-2x - 4y = 3$?

F. $y = -\frac{1}{2}x + 5$ **G.** $y = 2x - 6$ **H.** $y = -2x + 4$ **J.** $y = \frac{1}{2}x - 2$

73. A parallelogram has vertices $A(0, 2)$, $B(2, -1)$, $C(6, 3)$, and $D(p, q)$. Which of the following ordered pairs has possible values for (p, q)?

A. $(0, 6)$ **B.** $(6, 1)$ **C.** $(4, 6)$ **D.** $(6, 4)$

74. Which equation has as its graph a line perpendicular to the graph of the line that passes through the points $(0, -1)$ and $(8, 9)$?

F. $(y - 2) = -\frac{4}{5}(x + 1)$ **G.** $y = \frac{4}{5}x + 10$

H. $4x - 5y = 0$ **J.** $y = -x + 7$

75. Which equation has as its graph a line parallel to the graph of the line that passes through the points $(-8, 2)$ and $(-3, 3)$?

A. $(y - 2) = -(x + 8)$ **B.** $y = -5x + 10$

C. $x - 5y = -6$ **D.** $-y = -\frac{1}{5}x + 7$

Short Response

76. Suppose the line through points $(x, 6)$ and $(1, 2)$ is parallel to the graph of $2x + y = 3$. Find the value of x. Show your work.

Mixed Review

GO for Help

Lesson 6-4

Write an equation for the line through the given point with the given slope.

77. $(0, 4); m = 3$ **78.** $(-2, 0); m = -4$

79. $(5, -3); m = \frac{3}{4}$ **80.** $(-1, -9); m = -\frac{2}{3}$

81. $(-6, 4); m = -\frac{3}{5}$ **82.** $(7, 11); m = \frac{1}{2}$

Lesson 6-1

Find the slope of each line.

83. 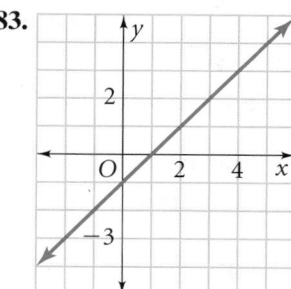   **84.** 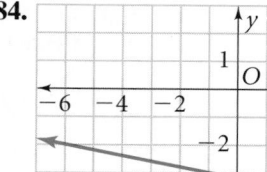   **85.**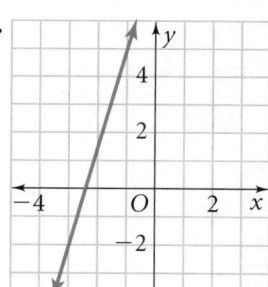

Lesson 5-2

Determine whether each relation is a function.

86. $\{(1, 1), (2, 2), (3, 3)\}$ **87.** $\{(1, 3), (2, 5), (3, 5)\}$

88. $\{(5, 1), (5, 2), (4, 3)\}$ **89.** $\{(1, 3), (2, 2), (3, 1)\}$

6-7

Scatter Plots and Equations of Lines

Learning Standards for Mathematics

A.S.7 Create a scatter plot of bivariate data.

A.S.8 Construct manually a reasonable line of best fit for a scatter plot and determine the equation of that line.

A.S.17 Use a reasonable line of best fit to make a prediction involving interpolation or extrapolation.

✓ **Check Skills You'll Need**

GO **for Help** Lesson 1-5

Use the data in each table to draw a scatter plot.

1.

x	y
1	2
2	−3
3	8
4	9
5	−25

2.

x	y
1	21
2	15
3	12
4	9
5	7

🔊 **New Vocabulary** • line of best fit • correlation coefficient

---

1 Writing an Equation for a Trend Line

In Chapter 1 you used scatter plots to determine how two sets of data are related. You can now write an equation for a trend line.

1 EXAMPLE Trend Line

Birds Make a scatter plot of the data at the left. Draw a trend line and write its equation. Use the equation to predict the wingspan of a hawk that is 28 in. long.

Length and Wingspan of Hawks

Type of Hawk	Length (in.)	Wing-span (in.)
Cooper's	21	36
Crane	21	41
Gray	18	38
Harris's	24	46
Roadside	16	31
Broad-winged	19	39
Short-tailed	17	35
Swainson's	19	46

SOURCE: *Birds of North America*

Step 1 Make a scatter plot and draw a trend line. Estimate two points on the line.

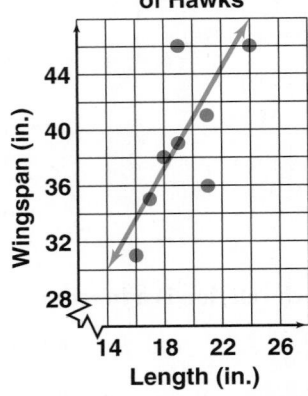

Length and Wingspan of Hawks

Two points on the trend line are $(14, 30)$ and $(19, 39)$.

Step 2 Write an equation of the trend line.

$$m = \frac{y_2 - y_1}{x_2 - x_1} = \frac{39 - 30}{19 - 14} = \frac{9}{5}$$

$y - y_1 = m(x - x_1)$ **Use point-slope form.**

$y - 30 = \frac{9}{5}(x - 14)$ **Substitute $\frac{9}{5}$ for m and $(14, 30)$ for (x_1, y_1).**

Step 3 Predict the wingspan of a hawk that is 28 in. long.

$y - 30 = \frac{9}{5}(28 - 14)$ **Substitute 28 for x.**

$y - 30 = \frac{9}{5}(14)$ **Simplify within the parentheses.**

$y - 30 = 25.2$ **Multiply.**

$y = 55.2$ **Add 30 to each side.**

The wingspan of a hawk 28 in. long is about 55 in.

 **1** Graph the data below and draw a trend line. Find an equation for the trend line. Estimate the number of Calories in a fast-food that has 14g of fat.

Calories and Fat in Selected Fast-Food Meals

Fat (g)	6	7	10	19	20	27	36
Calories	276	260	220	388	430	550	633

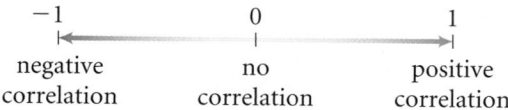

2 Writing an Equation for a Line of Best Fit

The trend line that shows the relationship between two sets of data most accurately is called the **line of best fit.** A graphing calculator computes the equation of a line of best fit using a method called linear regression.

The graphing calculator also gives you the **correlation coefficient** r, which tells you how closely the equation models the data.

$$\overset{-1}{\underset{\substack{\text{negative} \\ \text{correlation}}}{\vdash}} \quad \overset{0}{\underset{\substack{\text{no} \\ \text{correlation}}}{\vert}} \quad \overset{1}{\underset{\substack{\text{positive} \\ \text{correlation}}}{\dashv}}$$

When the data points cluster around a line, there is a strong correlation between the line and the data. So the nearer r is to 1 or −1, the more closely the data cluster around the line of best fit. In later chapters, you will learn how to find non-linear models which may better describe some data.

2 EXAMPLE Line of Best Fit

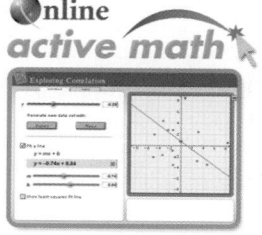

Online active math

For: Correlation Activity
Use: Interactive Textbook, 6-7

Recreation Use a graphing calculator to find the equation of the line of best fit for the data at the right. What is the correlation coefficient to three decimal places?

Step 1 Use the **EDIT** feature of the
[STAT] screen on your graphing calculator.
Let 93 correspond to 1993 and 100 correspond to 2000. Enter the data for years and then the data for expenditures.

Step 2 Use the **CALC** feature in the
[STAT] screen. Find the equation for the line of best fit.

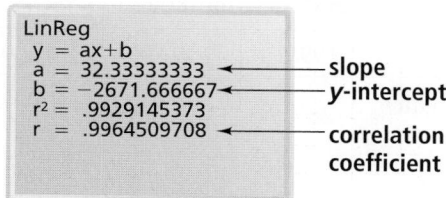

```
LinReg
y = ax+b
a = 32.33333333      ←— slope
b = −2671.666667 ←— y-intercept
r² = .9929145373
r = .9964509708  ←— correlation
                     coefficient
```

Recreation□ Expenditures

Year	Dollars (billions)
1993□	340
1994	369
1995	402
1996	430
1997	457
1998	489
1999	527
2000	574

SOURCE: *Statistical Abstract of the United States.* Go to **www.PHSchool.com** for a data update. Web Code: atg-9041

Go Online
PHSchool.com
For: Graphing calculator procedures
Web Code: ate-2122

The equation for the line of best fit is $y = 32.33x - 2671.67$ for values of a and b rounded to the nearest hundredth. The correlation coefficient is about 0.996.

 **Quick Check** ➋ Find the equation of the line of best fit. Let 95 correspond to 1995 and 103 correspond to 2003. What is the correlation coefficient to three decimal places?

Yearly Box Office Gross for Movies (billions)

1995	1996	1997	1998	1999	2000	2001	2002	2003
$5.5	$6.0	$6.4	$7.0	$7.4	$7.7	$8.4	$9.5	$9.5

EXERCISES

For more exercises, see *Extra Skill and Word Problem Practice*.

Practice and Problem Solving

A Practice by Example

Example 1
(page 350)

 GO for Help

Find an equation of a reasonable trend line for each scatter plot.

1.
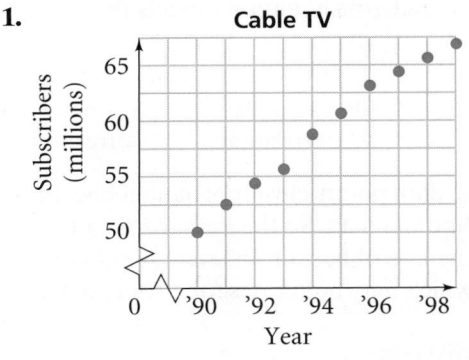

2.
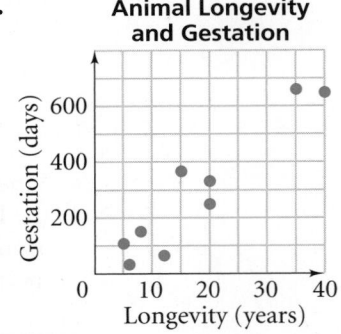

3.
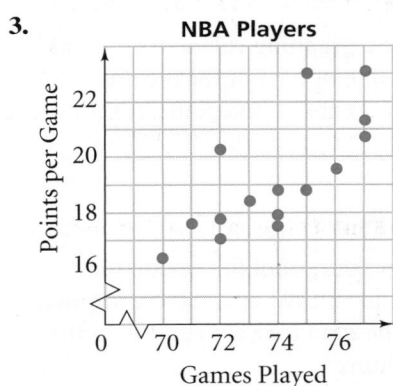

4.
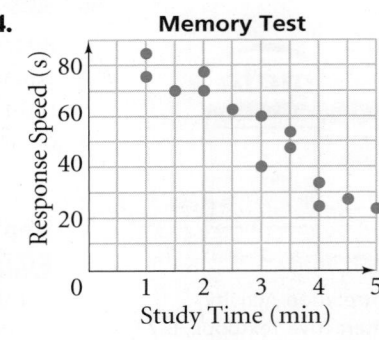

5. Graph the data in the table below for the attendance and revenue at theme parks. Find an equation for a trend line of the data.

Attendance and Revenue at U.S. Theme Parks

Year	1995	1996	1997	1998	1999	2000	2001	2002	2003
Attendance (millions)	280	290	300	300	309	317	319	324	322
Revenue (billions of dollars)	7.4	7.9	8.4	8.7	9.1	9.6	9.6	9.9	10.3

SOURCE: International Association of Amusement Parks and Attractions.
Go to **www.PHSchool.com** for a data update.   Web Code: atg-9041

6. Graph the data for the average July temperature and the annual precipitation of the cities in the table below. Find an equation for the line of best fit of the data. Estimate the average rainfall for a city with average July temperature of 75° F.

Precipitation and Temperature in Selected Eastern Cities

City	Average July Temperature (°F)	Average Annual Precipitation (in.)
New York	76.4	42.82
Baltimore	76.8	41.84
Atlanta	78.6	48.61
Jacksonville	81.3	52.76
Washington, D.C.	78.9	39.00
Boston	73.5	43.81
Miami	82.5	57.55

Source: Time *Almanac*

Example 2
(page 351)

Graphing Calculator Use a graphing calculator to find the equation of the line of best fit for the data. Find the value of the correlation coefficient r to three decimal places.

7. **Average Temperatures in Northern Latitudes**

Latitude (° N)	0	10	20	30	40	50	60	70	80
Temp. (°F)	79.2	80.1	77.5	68.7	57.4	42.4	30.0	12.7	1.0

8. **Retail Department Store Sales (billions of dollars)**

Year	1980	1985	1990	1994	1995	1996	1997	1998
Sales	86	126	166	217	231	245	261	279

Source: *Statistical Abstract of the United States.*
Go to **www.PHSchool.com** for an update. Web Code: atg-9041

9. Olympic 5000-Meter Men's Gold Medal Speed Skating Times

Year	1980	1984	1988	1992	1994	1998	2002
Time (seconds)	422	432	404	420	395	382	402

Source: International Skating Union

10. **Average Male Lung Power**

Respiration (breaths/min)	50	30	25	20	18	16	14
Heart Rate (beats/min)	200	150	140	130	120	110	100

Source: *Encyclopeadia Britannica*

Real-World Connection

The 500-meter men's speed skating race has been an Olympic event since 1924.

11. **Wind Chill Temperature for 15 mi/h Wind**

Air Temp. (°F)	35	30	25	20	15	10	5	0
Wind-Chill Temp. (°F)	16	9	2	−5	−11	−18	−25	−31

12. Geometry Students measured the diameters and circumferences of the tops of a variety of cylinders. Below is the data that they collected.

Cylinder Tops

Diameter (cm)	3	3	5	6	8	8	9.5	10	10	12
Circumference (cm)	9.3	9.5	16	18.8	25	25.6	29.5	31.5	30.9	39.5

a. Graph the data. **b.** Find the equation of a trend line.
c. What does the slope of the equation mean?
d. Find the diameter of a cylinder with a circumference of 45 cm.

Population Growth

1790

1860

Today

■ More than 2 persons per square mile

13. Population Use the data at the right.
a. Graph the data for the male and female populations of the United States.
b. Find the equation of a trend line.
c. Use your equation to predict the number of females if the number of males were to increase to 150,000,000.
d. Critical Thinking Would it be reasonable to predict the population in 2025 from these data? Explain.

14. a. Open-Ended Make a table of data for a linear function. Use a graphing calculator to find the equation of the line of best fit.
b. What is the correlation coefficient for your linear data?

15. Writing What kind of trend line do you think data for the following comparison would be likely to show? Explain.

temperature and the number of students absent from school

Estimated Population of the United States (thousands)

Year	Male	Female
1991	122,956	129,197
1992	124,424	130,606
1993	125,788	131,995
1994	127,049	133,278
1995	128,294	134,510
1996	129,504	135,724
1997	130,783	137,001
1998	132,030	138,218
1999	133,277	139,414
2000	138,054	143,368

Source: U.S. Census Bureau. Go to **www.PHSchool.com** for a data update. Web Code: atg-9041

16. Graphing Calculator A school collected data on math and science grades of nine randomly selected students.

Student	1	2	3	4	5	6	7	8	9
Math	76	89	84	79	94	71	79	91	84
Science	82	94	89	89	94	84	68	89	84

a. Use a graphing calculator to find the equation of the line of best fit for the data.
b. Critical Thinking Should the equation for the line of best fit be used to predict grades? Explain.

17. Graphing Calculator Use a graphing calculator to find the equation of the line of best fit for the data below. Predict sales of greeting cards in the year 2010.

Greeting Card Sales

Year	1989	1990	1991	1992	1993	1994	1995	1996	1997	1998
Sales (billions)	$4.2	$4.6	$5.0	$5.3	$5.6	$5.9	$6.3	$6.8	$7.3	$7.5

Source: Greeting Card Association

18. a. Data Collection Find two sets of data that you could display in a scatter plot, such as the number of boys and girls in each class in your school, or the population and the number of airports in some states. Then graph the data.
   b. Find the equation of a trend line.
   c. Use the equation to predict another value that could be on your scatter plot.
   d. What is the correlation coefficient?

19. Another way you can find a line of best fit is the *median-median method*. The graph below shows how this method works. The points in red indicate the original data.

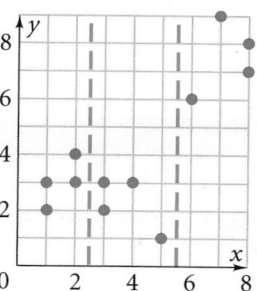

Divide the data into three groups of equal size.

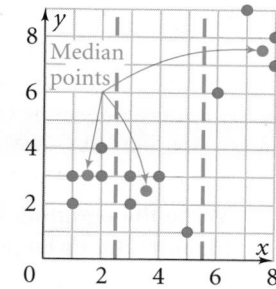

Find and plot the median points, (*x*-median, *y*-median).

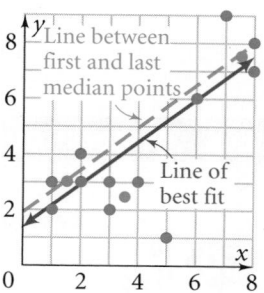

Find the line parallel to the line between the first and last median points and $\frac{1}{3}$ of the way to the middle median point.

   a. Estimate two coordinates on the purple line in the graph at the right above. Find the equation of the line of best fit.
    b. Graphing Calculator You can use a graphing calculator to find the line of best fit with the median-median method. Below are the coordinates of the points graphed in red. Use the **EDIT** feature of the STAT screen on your graphing calculator. Use the **Med-Med** feature to find a line of best fit.
   $(1, 2), (1, 3), (2, 3), (2, 4), (3, 2), (3, 3), (4, 3), (5, 1), (6, 6), (7, 9), (8, 8), (8, 7)$

 **Challenge**

20. a. Make a scatter plot of the data below. Then find the equation of the line of best fit.

Car Stopping Distances

Speed (mi/h)	10	15	20	25	30	35	40	45
Stopping Distances (ft)	27	44	63	85	109	136	164	196

   b. Use your equation to predict the stopping distance at 90 mi/h.
   c. Critical Thinking The actual stopping distance at 90 mi/h is close to 584 ft. Why do you think this is not close to your prediction?

NY REGENTS

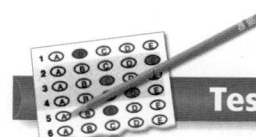

Test Prep

Multiple Choice

21. A horizontal line passes through $(5, -2)$. Which other point does it also pass through?
   A. $(5, 2)$ **B.** $(-5, -2)$ **C.** $(-5, 2)$ **D.** $(5, 0)$

22. Which of the following equations has a graph that contains the ordered pairs $(-3, 4)$ and $(1, -4)$?
   F. $x + 2y = 8$ **G.** $2x - y = 4$ **H.** $2x + y = -2$ **J.** $x - 2y = -6$

Extended Response

23. The table at the right shows the number of elderly in the United States from 1960 through 2000.
   a. Graph the data and draw a trend line.
   b. Write an equation for the trend line you drew.
   c. Predict the elderly population in the United States in 2005. Show your work.

U.S. Elderly Population

Year	Elderly (millions)
1960	16.560
1970	19.980
1980	25.550
1990	31.235
2000	34.709

SOURCE: *Statistical Abstract of the United States.*
Go to **www.PHSchool.com** for a data update.
Web Code: atg-9041

Mixed Review

GO for Help

Lesson 6-6

Write the equation for the line that is parallel to the given line and that passes through the given point.

24. $y = 5x + 1; (2, -3)$ **25.** $y = -x - 9; (0, 5)$ **26.** $2x + 3y = 9; (-1, 4)$

27. $y = -\frac{1}{2}x; (3, -4)$ **28.** $y = -2x + 3; (-2, -1)$ **29.** $y = \frac{2}{3}x + 7; (-1, 2)$

Lesson 4-4

Solve each inequality.

30. $1 + 5x + 1 > x + 9$ **31.** $7x + 3 < 2x + 28$ **32.** $4x + 4 > 2 + 2x$

33. $4x + 3 \leq 2x - 7$ **34.** $-x + 5 < 3x - 1$ **35.** $2x > 7x - 3 - 4x$

✓ Checkpoint Quiz 2   **Lessons 6-5 through 6-7**

Write an equation for the line through the given point that has the given slope.

1. $(3, 4); m = -\frac{1}{4}$ **2.** $(0, -3); m = 18$ **3.** $(-7, -5); m = 0$

4. Write an equation for the line through the points $(2, -6)$ and $(-1, -4)$.

Write an equation of the line that is parallel to the given line and that passes through the given point.

5. $x + y = 3; (5, 4)$ **6.** $3x + 2y = 1; (-2, 6)$

Write an equation of the line that is perpendicular to the given line and that passes through the given point.

7. $y = -4x + 2; (0, 2)$ **8.** $y = \frac{2}{3}x + 6; (-6, 2)$

9. Find the equation for a trend line for the data at the right.

x	1	2	3	4	5	6	7
y	7	12	19	20	28	33	40

 10. Graphing Calculator Use a graphing calculator to find the equation for the line of best fit for the data at the right.

x	1	2	3	4	5	6	7
y	54	52	45	40	33	27	18

In this activity, you will release a ball from various heights and record the maximum height after its first bounce. Complete this activity over a hard surface. Measure all heights from the bottom of the ball.

ACTIVITY

Place one end of a meter stick on the floor and tape it to the wall. Tape a second meter stick to the wall starting with the top of the first meter stick.

1. **Data Collection** Drop the ball from 50 cm. Carefully record its maximum height after the first bounce. Repeat.

2. Copy and complete the table shown below. You may make additional measurements using different starting heights.

Initial Height	Maximum Height After First Bounce	
	Trial 1	Trial 2
50 cm	■	■
100 cm	■	■
150 cm	■	■
200 cm	■	■

3. Graph the data from both trials on the same coordinate plane.

4. **Critical Thinking** Why is it reasonable to use $(0, 0)$ as a data point?

5. Draw a trend line which includes $(0, 0)$.

6. **a. Predict** Use your line to predict the maximum height of the first bounce after the ball was dropped from 175 cm.
   b. From what height would you have to drop the ball for it to reach 2 meters after the first bounce?

7. Write an equation for your trend line.

EXERCISES

8. **Multiple Choice** The graph at the right shows the bounce data for a ball. Which equation best models the data?

   (A) $y = 0.4x$ (B) $y = 0.5x$
   (C) $y = 0.6x$ (D) $y = 0.7x$

9. Suppose students used several different types of balls and found that the slopes of the trend lines were not the same. What is the significance of the slope?

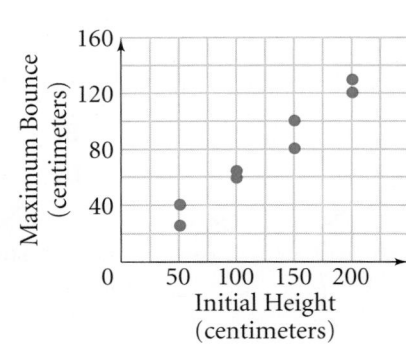

In Chapter 2, you solved absolute value equations. Here you will explore graphs of absolute value functions.

NY A.G.4: Identify and graph linear, quadratic (parabolic), absolute value, and exponential functions.

ACTIVITY

Graph the absolute value *parent function* $y = |x|$.

Press [Y=]. Select **abs(** by pressing [MATH] [▶] 1. Complete the equation by pressing [X,T,θ,n] and [ )].

To graph, press [ZOOM] 5 and then [GRAPH].

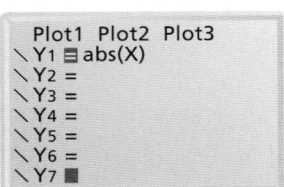

1. Graph the equations $y = |x| - 3$ and $y = |x| + 3$.
   a. The equation __?__ translates the parent graph $y = |x|$ __?__ units up.
   b. The equation __?__ translates the parent graph $y = |x|$ __?__ units down.

2. **Critical Thinking** Explain how the translations of $y = |x| + k$ as k changes are similar to translations of $y = mx + b$ as b changes.

3. Graph the equations $y = |x + 2|$ and $y = |x - 2|$. The first equation should appear as follows in the **Y=** screen.

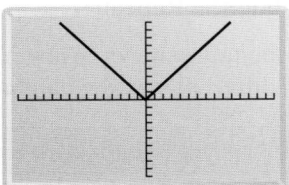

   a. The equation __?__ translates the parent graph __?__ units to the left.
   b. The equation __?__ translates the parent graph __?__ units to the right.

4. **Predict** Based on your results from questions 1 and 3, predict how the equation $y = |x - 2| + 3$ would translate the graph of $y = |x|$.

EXERCISES

5. Match each equation with the best choice for its graph.

   A. $y = |x + 6| - 2$ **B.** $y = |x - 6| - 2$ **C.** $y = |x + 6| + 2$

   I. 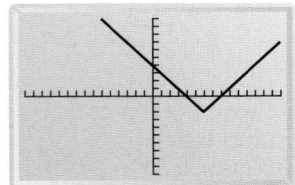   **II.** 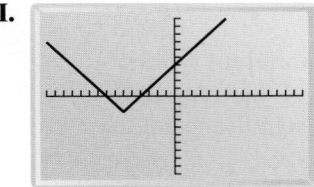   **III.**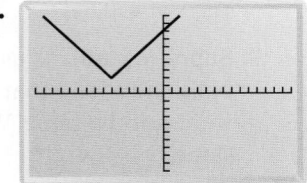

6-8

Graphing Absolute Value Equations

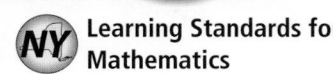

NY Learning Standards for Mathematics

A.G.4 Identify and graph linear, quadratic (parabolic), absolute value, and exponential functions.

✓ **Check Skills You'll Need**

GO for Help Lessons 2-1 and 5-3

Simplify each expression.

1. $|2 - 7|$ **2.** $|7 - 12|$ **3.** $|38 - 56|$ **4.** $|-24 + 12|$

Model each rule using a table of values.

5. $y = 6 - x$ **6.** $y = |x| + 1$ **7.** $y = |x + 1|$

🔊 **New Vocabulary** • absolute value equation • translation

---

1 Translating Graphs of Absolute Value Equations

Vocabulary Tip

The <u>absolute value</u> of a number is its distance from 0 on a number line.

A V-shaped graph that points upward or downward is the graph of an **absolute value equation.** In Lesson 5-3, you graphed absolute value equations by making tables of values.

In this lesson you will graph by translating the graph of $y = |x|$. A **translation** is a shift of a graph horizontally, vertically, or both. The result is a graph of the same shape and size, but in a different position.

1 EXAMPLE **Vertical Translations**

Below are the graphs of $y = |x|$ and $y = |x| + 2$. Describe how the graphs are the same and how they are different.

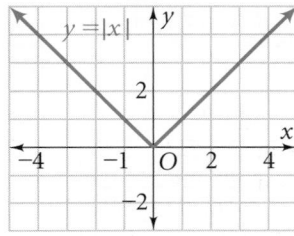   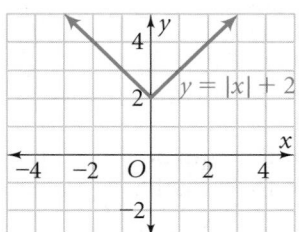

The graphs are the same shape. The y-intercept of the first graph is 0.
● The y-intercept of the second graph is 2.

✓ **Quick Check** ❶ Describe how each graph below is like $y = |x|$ and how it is different.

a. 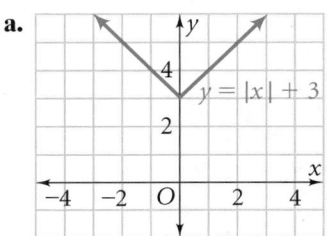   **b.**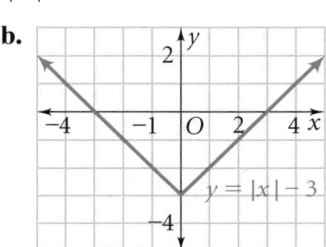

The graph of $y = |x| + k$ is a translation of $y = |x|$. Let k be a positive number. Then $y = |x| + k$ translates the graph of $y = |x|$ up k units, while $y = |x| - k$ translates the graph of $y = |x|$ down k units.

2 EXAMPLE Graphing a Vertical Translation

Graph $y = |x| - 1$.

Start with the graph of $y = |x|$.
Translate the graph *down* 1 unit.

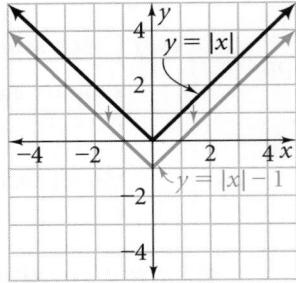

Moving this chess piece is a horizontal and vertical translation.

 **Quick Check** ❷ Graph each function by translating $y = |x|$.
  a. $y = |x| + 4$ **b.** $y = |x| - 5$

You can write an equation to describe a vertical translation.

3 EXAMPLE Writing an Absolute Value Equation

Write an equation for each translation of $y = |x|$.
  a. 8 units down
    The equation is $y = |x| - 8$.
  b. 6 units up
    The equation is $y = |x| + 6$.

Quick Check ❸ For each translation of $y = |x|$, write an equation.
  a. 2 units up **b.** 5 units down

The following tables of values and graphs show what happens when you graph $y = |x - 3|$ and $y = |x + 3|$.

x	y = |x|	y = |x - 3|
−3	$|-3| = 3$	$|-3 - 3| = 6$
−2	$|-2| = 2$	$|-2 - 3| = 5$
−1	$|-1| = 1$	$|-1 - 3| = 4$
0	$|0| = 0$	$|0 - 3| = 3$
1	$|1| = 1$	$|1 - 3| = 2$
2	$|2| = 2$	$|2 - 3| = 1$
3	$|3| = 3$	$|3 - 3| = 0$

x	y = |x|	y = |x + 3|
−3	$|-3| = 3$	$|-3 + 3| = 0$
−2	$|-2| = 2$	$|-2 + 3| = 1$
−1	$|-1| = 1$	$|-1 + 3| = 2$
0	$|0| = 0$	$|0 + 3| = 3$
1	$|1| = 1$	$|1 + 3| = 4$
2	$|2| = 2$	$|2 + 3| = 5$
3	$|3| = 3$	$|3 + 3| = 6$

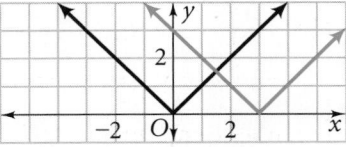

For the graph of $y = |x - 3|$, $y = |x|$ is translated 3 units to the right.

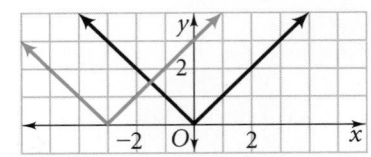

For the graph of $y = |x + 3|$, $y = |x|$ is translated 3 units to the left.

So for a positive number h, $y = |x + h|$ translates the graph of $y = |x|$ by h units to the left, and $y = |x - h|$ translates the graph of $y = |x|$ by h units to the right.

4 EXAMPLE Graphing a Horizontal Translation

Graph each equation by translating $y = |x|$.

a. $y = |x + 2|$

b. $y = |x - 2|$

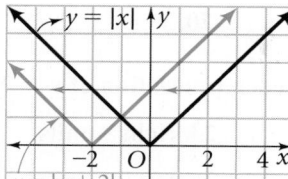

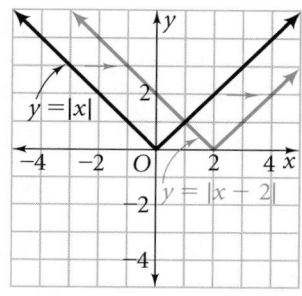

✓ **Quick Check** ❹ Graph each equation by translating $y = |x|$.
a. $y = |x - 4|$

b. $y = |x + 1|$

If you know the number of units that a function is to be translated and the direction of the translation, you can write an equation to describe the horizontal translation.

5 EXAMPLE Writing an Absolute Value Equation

Write an equation for each translation.

a. $y = |x|$, 8 units left
   The equation is $y = |x + 8|$.

b. $y = |x|$, 6 units right
   The equation is $y = |x - 6|$.

✓ **Quick Check** ❺ Write an equation for each translation of $y = |x|$.
a. 5 units right

b. 7 units left

EXERCISES

For more exercises, see *Extra Skill and Word Problem Practice*.

Practice and Problem Solving

Describe how each graph is like the graph of $y = |x|$ and how it is different.

1.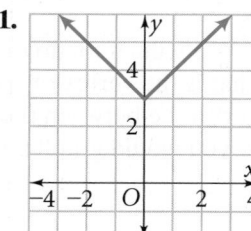

2.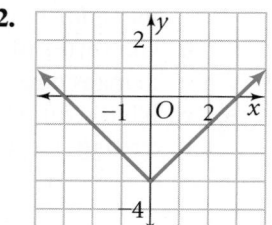

3.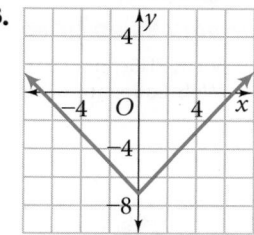

Graph each function by translating $y = |x|$.

4. $y = |x| + 2$

5. $y = |x| - 4$

6. $y = |x| + 8$

7. $y = |x| + 1$

8. $y = |x| - 6$

9. $y = |x| - 2.5$

Example 3
(page 360)

Write an equation for each translation of $y = |x|$.

10. 9 units up   **11.** 6 units down   **12.** 0.25 units up

13. $\frac{5}{2}$ units up   **14.** 5.90 units up   **15.** 1 unit down

Example 4
(page 361)

Graph each function by translating $y = |x|$.

16. $y = |x - 3|$ **17.** $y = |x + 3|$ **18.** $y = |x - 1|$

19. $y = |x + 5|$ **20.** $y = |x - 7|$ **21.** $y = |x + 2.5|$

Example 5
(page 361)

Write an equation for each translation of $y = |x|$.

22. left 9 units   **23.** right 9 units   **24.** right $\frac{5}{2}$ units

25. left $\frac{3}{2}$ units   **26.** left 0.5 unit   **27.** right 8.2 units

B   **Apply Your Skills**

At the right is the graph of $y = -|x|$.
Graph each function by translating $y = -|x|$.

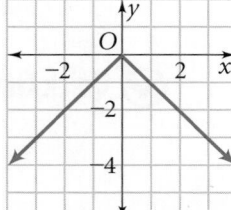

> **Problem Solving Hint**
>
> For Exercises 28–31, you can check your work by substituting ordered pairs from the graph into the corresponding equation.

28. $y = -|x| + 3$ **29.** $y = -|x| - 3$

30. $y = -|x + 3|$ **31.** $y = -|x - 3|$

Write an equation for each translation of $y = -|x|$.

32. 2 units up   **33.** 2.25 units left

34. $\frac{3}{2}$ units down   **35.** 4 units right

36. The graph at the right shows a translation of $y = |x|$ where there is both a vertical and a horizontal change. Which equation below is an equation for this graph?
     A. $y = |x + 2| - 1$ **B.** $y = |x - 2| + 1$
     C. $y = |x - 2| - 1$ **D.** $y = |x + 2| + 1$

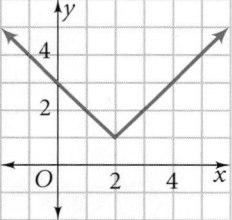

Graph each translation of $y = |x|$.

Sample For $y = |x + 3| - 2$, add 3 indicates the translation of the graph 3 units left. Subtract 2 indicates the translation of the graph 2 units down.

37. $y = |x - 1| + 2$ **38.** $y = |x + 2| - 1$

39. $y = |x - 3| - 4$ **40.** $y = |x + 3| + 4$

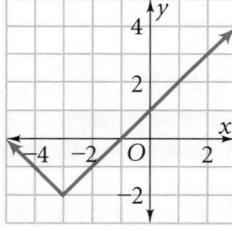

41. a. Graph $y = |x - 2| + 3$. (*Hint:* Read the sample above for Exercises 37–40.)
     b. The vertex of an absolute value function is the point at which the function changes direction. What is the vertex of $y = |x - 2| + 3$?
     c. What relationship do you see between the vertex and the equation?
     d. Writing Explain how you would graph any equation of the form $y = |x - a| + b$.

Online
Homework Video Tutor
Visit: PHSchool.com
Web Code: ate-0608

C   **Challenge**

42. a. Graph $y = |2x|$ by making a table of values.
     b. Translate $y = |2x|$ to graph $y = |2x| + 3$.
     c. Translate $y = |2x|$ to graph $y = |2(x - 1)|$.
     d. Translate $y = |2x|$ to graph $y = |2(x - 1)| + 3$.

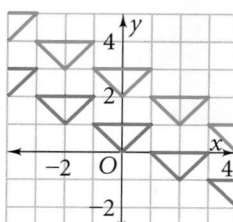

43. Programming A computer programmer is plotting the triangles at the left. The pattern extends infinitely in both directions. She will use two equations for each triangle: some translation of $y = |x|$ and $y = c$ (c is a constant).
a. What equations will the programmer use to plot the red triangle with vertices at $(0, 0)$, $(-1, 1)$, and $(1, 1)$?
b. What are the least and greatest values in the domain for the equations the programmer would use to plot the triangle with vertices $(0, 0)$, $(-1, 1)$ and $(1, 1)$?
c. What linear equation can the programmer use to find the lowest vertex on each red triangle? Each blue triangle?

NY REGENTS

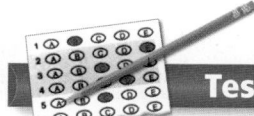

Test Prep

Multiple Choice

44. Which equation translates $y = |x|$ by 8 units to the left?
A. $y = |x| + 8$ **B.** $y = |x + 8|$ **C.** $y = |x| - 8$ **D.** $y = |x - 8|$

45. What is the lowest point of the graph of $y = |x - 9|$?
F. $(0, -9)$ **G.** $(-9, 0)$ **H.** $(9, 0)$ **J.** $(0, 9)$

46. What point do the graphs of $y = |x - 3|$ and $y = |x + 5|$ have in common?
A. $(-1, 4)$ **B.** $(1, 4)$ **C.** $(4, 1)$ **D.** $(4, -1)$

47. The graph of which equation contains the point $(3, 5)$?
F. $y = |x + 3| + 5$ **G.** $y = |x - 3| + 5$
H. $y = |x + 3| - 5$ **J.** $y = |x - 3| - 5$

Extended Response

48. a. Graph the equation $y = |x| - 4$ on a coordinate plane.
b. Graph the equation $y = |x| + 4$ on the same coordinate plane.
c. Describe the relationship of the ordered pairs in the graphs of $y = |x| - 4$ to the graph of $y = |x| + 4$.

Mixed Review

Lesson 6-7 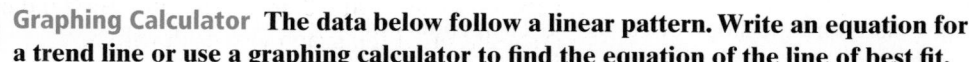 **Graphing Calculator** The data below follow a linear pattern. Write an equation for a trend line or use a graphing calculator to find the equation of the line of best fit.

49.

Year	Sales
1988	$27,000
1989	$32,000
1990	$37,000
1991	$42,000
1992	$47,000
1993	$52,000
1994	$57,000

50.

Year	Sales
1990	$47,000
1991	$51,000
1992	$55,000
1993	$59,000
1994	$63,000
1995	$67,000
1996	$71,000

Lesson 2-1   **Add the matrices.**

51. $\begin{bmatrix} 5 & 3 \\ 1 & 2 \end{bmatrix} + \begin{bmatrix} 7 & 2 \\ 1 & 4 \end{bmatrix}$ **52.** $\begin{bmatrix} -3 & 2 \\ -7 & 4 \end{bmatrix} + \begin{bmatrix} 7 & -1 \\ 8 & 0 \end{bmatrix}$ **53.** $\begin{bmatrix} -5.6 & 9.8 \\ -4.2 & 3.2 \end{bmatrix} + \begin{bmatrix} 8.1 & 4.2 \\ 2.2 & 7.5 \end{bmatrix}$

For some problems, it may help to draw a diagram of the given information if one is not provided.

1 EXAMPLE

The points R, S, and T lie on a line in order such that the length of $\overline{ST}$ is twice the length of $\overline{RS}$. The length of $\overline{RT}$ is 5 cm more than the length of $\overline{ST}$. Find the length of $\overline{RS}$ and $\overline{ST}$.

Draw $\overline{RT}$. Since $\overline{ST}$ is twice as long as $\overline{RS}$, let $RS = x$ and $ST = 2x$. Since the length of $\overline{RT}$ is 5 cm more than the length of $\overline{ST}$, let $RT = 5 + 2x$.

You can see from your diagram that $2x + x = 5 + 2x$. Solve this equation, and you find that $x = 5$. The length of $\overline{RS}$ is 5 cm and the length of $\overline{ST}$ is 10 cm.

2 EXAMPLE

The points $A(-8, 1)$, $B(-2, 7)$ and $C(4, -11)$ are the vertices of a triangle. Is Triangle ABC a right triangle?

Two lines form a right angle if the product of their slopes is -1.

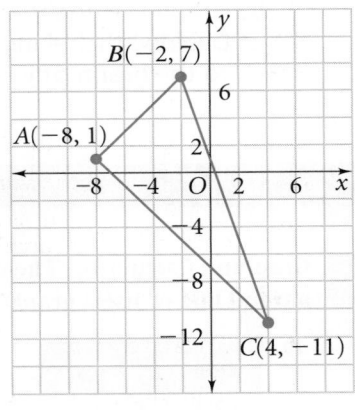

When you draw a diagram, you can see that the right angle cannot be at point C. You need to see if the product of the slopes of $\overline{AC}$ and $\overline{AB}$ or of $\overline{BC}$ and $\overline{AB}$ is -1. The slope of $\overline{AB}$ is 1, and the slope of $\overline{AC}$ is -1. The product of 1 and -1 is -1. So Triangle ABC is a right triangle.

EXERCISES

Draw a diagram to solve each exercise.

1. The points $L(4, 0)$, $M(10, 0)$, and $N(7, 5)$ form $\triangle LMN$. What is the sum of the slopes of the three sides of the triangle?

2. Three towns A, B, and C lie on a straight road in that order. The distance from B to C is 6 miles more than twice the distance from A to B. The distance from A to C is 2 miles more than four times the distance from A to B. What is the distance from A to B?

3. $P(1, 3)$ and $R(5, 5)$ are the endpoints of a diagonal of the rectangle $PQRS$. Point Q has the same x-coordinate as point P. Which sides of the rectangle are parallel to the x-axis? To the y-axis? What is the perimeter of the rectangle?

Chapter Review

Vocabulary Review

🔊 absolute value equation (p. 359)
correlation coefficient (p. 351)
line of best fit (p. 351)
linear equation (p. 317)
linear function (p. 317)
linear parent function (p. 317)
negative reciprocal (p. 344)

parallel lines (p. 343)
parent function (p. 317)
perpendicular lines (p. 344)
point-slope form (p. 336)
rate of change (p. 308)
slope (p. 310)

slope-intercept form (p. 318)
standard form of a
    linear equation (p. 330)
translation (p. 359)
x-intercept (p. 330)
y-intercept (p. 317)

Go Online
PHSchool.com
For: Vocabulary quiz
Web Code: atj-0651

Choose the vocabulary term that correctly completes the sentence.

1. Two lines are _?_ if the product of their slopes is −1.

2. Two lines in the same plane that never intersect are _?_.

3. A(n) _?_ shifts a graph horizontally, vertically, or both.

4. The ratio of the vertical change to the horizontal change is called the _?_.

5. The *y*-coordinate of the point at which the graph of a line crosses the vertical axis is called the _?_.

Skills and Concepts

6-1 Objectives

▼ To find rates of change from tables and graphs (p. 308)

▼ To find slope (p. 310)

Rate of change allows you to look at how two quantities change relative to each other.

$$\text{rate of change} = \frac{\text{change in the dependent variable}}{\text{change in the independent variable}}$$

Slope is the ratio of the vertical change to the horizontal change.

$$\text{slope} = \frac{\text{vertical change}}{\text{horizontal change}} = \frac{\text{rise}}{\text{run}}$$

Find the rate of change for each situation.

6. A kitten grows from 5 oz at birth to 3 lb 5 oz at 6 months. (*Hint:* 1 lb = 16 oz)

7. A plant measures 0.5 in. at the end of Week 1 and 14 in. at the end of Week 5.

Find each rate of change. Explain what *rate of change* means in each situation.

8.
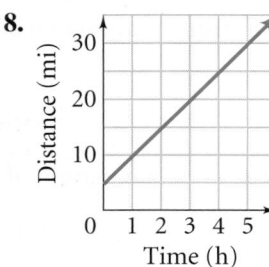

9.
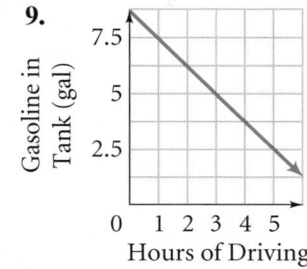

10.
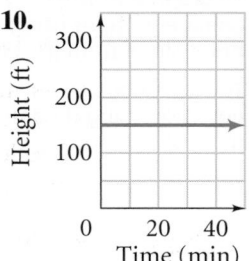

Find the slope of the line that passes through each pair of points.

11. (3, −2) and (−5, −4)   **12.** (4.5, −1) and (4.5, 2.6)   **13.** (2, 5) and (−5, −2)

The graph of a **linear equation** is a line. The **x-intercept** of a line is the x-coordinate of the point where the line crosses the x-axis, and the **y-intercept** is the y-coordinate of the point where the line crosses the y-axis.

The **slope-intercept form of a linear equation** is $y = mx + b$, where m is the slope and b is the y-intercept.

Write an equation of a line with the given slope and y-intercept. Then graph the equation.

14. $m = 0, b = -3$ **15.** $m = -7, b = \frac{1}{2}$ **16.** $m = \frac{2}{5}, b = 0$

Write the slope-intercept form of the equation for each line.

17. 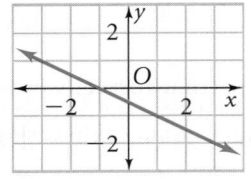   **18.**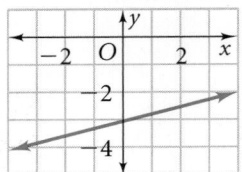

🌐 **19. Earnings** A job at a retail store pays $75 each week plus 25% commission on total weekly sales.
   a. Write an equation for the total weekly pay p for total weekly sales s.
   b. Use p as the vertical axis and s as the horizontal axis. Graph your equation.
   c. What is the total weekly pay if total weekly sales are $800?
   d. What is the p intercept? What does it mean in this situation?

The **standard form of a linear equation** is $Ax + By = C$, where A, B, and C are real numbers, and A and B are not both zero.

Find the x- and y-intercepts. Then graph each equation.

20. $5x + 2y = 10$ **21.** $6.5x - 4y = 52$ **22.** $x + 3y = -1$

Write each equation in standard form.

23. $y = \frac{3}{5}x + 7$ **24.** $y = -\frac{1}{3}x + 2$ **25.** $y = \frac{4}{3}x - 5$

The **point-slope form of a linear equation** is $y - y_1 = m(x - x_1)$, which passes through the point (x_1, y_1) and has slope m.

Use point-slope form to write an equation of a line that passes through the point (1, −2) with slope m.

26. $m = 2$ **27.** $m = \frac{3}{4}$ **28.** $m = -3$ **29.** $m = 0$

Use the point-slope form to write an equation of a line through the given points.

30. $(4, 3), (-2, 1)$ **31.** $(5, -4), (0, 2)$ **32.** $(-1, 0), (-3, -1)$

6-6 Objectives

▼ To determine whether lines are parallel (p. 343)

▼ To determine whether lines are perpendicular (p. 344)

Parallel lines are lines in the same plane that never intersect. Nonvertical lines are parallel if they have the same slope. Two lines are **perpendicular lines** if they intersect to form right angles. For perpendicular lines that are not horizontal and vertical, the product of their slopes is -1.

Write an equation for each of the following conditions.

33. parallel to $y = 5x - 2$, through $(2, -1)$

34. perpendicular to $y = -3x + 7$, through $(3, 5)$

35. parallel to $y = 9x$, through $(0, -5)$

36. perpendicular to $y = 8x - 1$, through $(4, 10)$

6-7 Objectives

▼ To write an equation for a trend line and use it to make predictions (p. 350)

▼ To write the equation for a line of best fit and use it to make predictions (p. 351)

You can find an equation to model the relationship between two sets of data in a scatter plot by sketching a trend line and using two points on the line to write an equation.

The **line of best fit** of a scatter plot is the most accurate trend line for the data. You can find the equation of a line of best fit using a graphing calculator. The **correlation coefficient** tells how well the equation of the line of best fit models the data.

37. Write an equation of a reasonable trend line for the scatter plot.

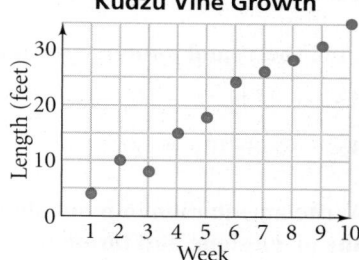

Kudzu Vine Growth

38. Graphing Calculator The table shows the average consumption of poultry in the United States in pounds per person from 1970 to 2000.
   a. Find the equation of a trend line or use a graphing calculator to find the equation of the line of best fit.
   b. Use your equation to **predict** how much poultry the average person will eat in 2010.

Years	1970	1975	1980	1985	1990	1995	2000
Pounds	33.8	32.9	40.8	45.5	56.3	62.9	68.4

6-8 Objective

▼ To translate the graph of an absolute value equation (p. 359)

The graph of an **absolute value equation** is a V-shaped graph that points upward or downward.

A **translation** shifts a graph either vertically, horizontally, or both. It results in a graph of the same shape and size in a different position.

Graph each equation by translating $y = |x|$.

39. $y = |x - 2|$

40. $y = |x| - 3$

Chapter Test

Go Online
PHSchool.com
For: Chapter Test
Web Code: ata-0652

Tell whether each statement is *true* or *false*. Explain.

1. A rate of change must be positive.

2. The rate of change for a vertical line is 0.

Find the slope of the line that passes through each pair of points.

3. $(4, 3), (3, 8)$ **4.** $(-2, 1), (6, -1)$

Graph each equation.

5. $x - 4y = 8$ **6.** $2x + 4y = -4$

7. $y = \frac{1}{3}x + 2$ **8.** $y - 1 = -3(x - 3)$

Write each equation in slope-intercept form.

9. $-7y = 8x - 3$ **10.** $x - 3y = -18$

11. $5x + 4y = 100$ **12.** $9x = 2y + 13$

Find the *x*- and *y*-intercepts of each line.

13. $3x + 4y = -24$ **14.** $-6x + 2y = -8$

15. $-5x + 10y = 60$ **16.** $x + y = 1$

Write an equation in point-slope form for the line with the given slope and through the given point.

17. slope $= \frac{8}{3}, (-2, -7)$ **18.** slope $= 3, (4, -8)$

19. slope $= \frac{-1}{2}, (0, 3)$ **20.** slope $= -5, (9, 0)$

Write an equation in point-slope form for the line through the given points.

21. $(4, 9), (-2, -6)$ **22.** $(-1, 0), (3, 10)$

23. $(5, -8), (-9, -8)$ **24.** $(0, 7), (1, 5)$

25. Which of the following lines is *not* perpendicular to $y = -2.5x + 13$?
    A. $y = 0.4x - 7$ **B.** $-2x + 5y = 8$
    C. $y = \frac{2}{5}x + 4$ **D.** $2y = 5x + 1.5$

Write an equation in slope-intercept form for a line that passes through the given point and is parallel to the given line.

26. $y = 5x; (2, -1)$ **27.** $y = 5; (-3, 6)$

Write an equation in slope-intercept form for a line that passes through the given point and that is perpendicular to the given line.

28. $y = -2x; (4, 0)$ **29.** $x = -7; (0, 2)$

30. Open-Ended Write the equation of a line parallel to $y = 0.5x - 10$.

31. You start a pet-washing service. You spend $30 on supplies. You plan to charge $5 to wash each pet.
    a. Write an equation to relate your profit y to the number of pets x you wash.
    b. Graph the equation. What are the x- and y-intercepts?

Write an equation for each translation of $y = |x|$.

32. 2 units down   **33.** right $\frac{3}{4}$ unit

Graph each function by translating $y = |x|$.

34. $y = |x - 4|$ **35.** $y = |x| + 2$

Use the data below for Exercises 36 and 37.

Local Governments in the United States (thousands)

Year	Municipalities	School Districts
1967	18.0	21.8
1972	18.5	15.8
1977	18.9	15.2
1982	19.1	14.9
1987	19.2	14.7
1992	19.3	14.4
1997	19.4	13.7

SOURCE: *Statistical Abstract of the United States.*
Go to **www.PHSchool.com** for a data update.
Web Code: atg-2041

36. a. Graphing Calculator Find an equation of a trend line or the line of best fit for the number of municipalities and the year.
    b. Predict the number of municipalities in 2010.

37. a. Graphing Calculator Find the equation of a trend line or the line of best fit for the number of school districts and the year.
    b. Predict the number of school districts in the year 2010.

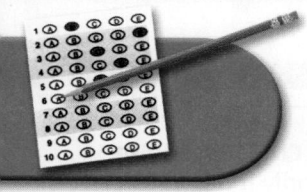

Regents Test Prep

Multiple Choice

For Exercises 1–11, choose the correct letter.

1. Suppose you earn $74.25 for working 9 hours. How much will you earn for working 15 hours?

(1) $120   **(2)** $123.75
(3) $124.50   **(4)** $127.25

2. Which is *not* a solution of $5x - 4 < 12$?

(1) −2   **(2)** 0   **(3)** 3   **(4)** 4

3. A line perpendicular to $y = 3x - 2$ passes through the point (0, 6). Which other point lies on the line?

(1) (9, 3)   **(2)** (−9, 3)
(3) (9, −3)   **(4)** (−9, −3)

4. If *a*, *b*, and *c* are three consecutive positive integers, which of the following is true?

  I. $a + c < 2b$ **II.** $a + b < c$
III. $a + c > 2b$ **IV.** $b + c > a$

(1) I only   **(2)** IV only
(3) I and II   **(4)** III and IV

5. A scatter plot shows a positive correlation. Which of the following could be an equation of the line of best fit?

(1) $y = -5x + 1$ **(2)** $2x + 3y = 6$
(3) $x = 16$ **(4)** $y = 2x - 1$

6. Which of the following is the solution of $6(4x - 3) = -54$?

(1) −3   **(2)** −1.5   **(3)** 1.5   **(4)** 3

7. Find $f(-2)$ when $f(x) = -3x + 4$.

(1) −10   **(2)** −2   **(3)** 2   **(4)** 10

8. Mariko runs 800 ft in one minute. What is her approximate speed in miles per hour? (*Hint:* 5280 ft = 1 mi)

(1) 6   **(2)** 8   **(3)** 9   **(4)** 12

9. Which of the following formulas correctly represent(s) the perimeter of the rectangle?

  I. $p = c + c + d + d$
 II. $p = cd$
III. $p = 2c + 2d$

(1) I and III   **(2)** II and III
(3) I only   **(4)** II only

10. A local club has 15 girls and 21 boys who are members. A representative from the club is randomly selected to meet with the town mayor. To the nearest percent, what is the probability that the representative is a girl?

(1) 40%   **(2)** 42%   **(3)** 58%   **(4)** 71%

11. Which situation is best modeled by the graph shown below?

(1) the amount of fuel remaining in a plane as it flies

(2) the cost of buying blank DVDs

(3) the amount of money remaining in a bank account after monthly withdrawals

(4) the cost of buying hamburger meat in the supermarket

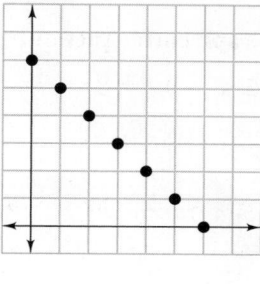

Short Response

Find each answer.

12. A car rental company charges $19.95 per day plus $.15 per mile. Calculate the cost in dollars to travel 250 miles over a 2-day period.

13. The ratio of crocus bulbs to tulip bulbs at a nursery is 5 to 2. The nursery has 175 crocus bulbs. How many crocus and tulip bulbs does the nursery have altogether?

14. A train moving at a constant speed travels 260 miles in 5 hours. At this rate, how many miles does the train travel in 9 hours?

15. Francesca made a rectangular tablecloth that has an area of 5 square feet. Her mother made a similar tablecloth with three times the length and three times the width. What is the area of her mother's tablecloth?

Show all of your work.

16. Write an equation in slope-intercept form of the line through (2, −3) that is perpendicular to the line $y = \frac{2}{5}x - \frac{7}{8}$.

17. Solve $-3 \le 2x + 1 < 7$. Graph the solutions.

Activity Lab

A.CN.6: Recognize and apply mathematics to situations in the outside world.

Mathematically Inclined

Measuring Force Although there are more than 80 pyramids in Egypt, the most famous are the three at Giza, near Cairo. Archaeologists think that the ancient Egyptians used ramps, either similar to the one below, or in a spiral around the perimeter of the pyramid, to lift the stone blocks into place. Moving heavy objects up a ramp requires a certain amount of effort, or force. The heavier the object, the more force is required.

Made of mud bricks, the ramp grew in height as layers were added to the pyramid.

Stone blocks were dragged on sleds with wooden rollers underneath.

Solidly Built

The Great Pyramid, around the Pharaoh's Chamber, is almost entirely solid. It contains more than 2,300,000 blocks, or 90 million ft^3 of stone. The same volume of brick and stone would build 40 Empire State Buildings.

Activity

Materials: paper and pencil, yardstick, wooden board at least 30 in. long, toy truck with a rubber band attached to the front axle

a. Raise one end of the plank to a height of 10 in. Hold the rubber band and pull the truck up the ramp. When your hand reaches the top, stop pulling, keeping the rubber band stretched. Hold the truck still while your partner records the height of the ramp (x) and the length of the rubber band (y).

b. Data Collection Raise the end of the ramp in 2-in. increments. Repeat part (a) until you have at least 10 (x, y) pairs of data. Record your data.

c. Graph your data and draw a line of best fit. Determine the equation of the line.

d. Writing The variable y represents the force required to move the toy truck. How does the height of the plank affect the force required?

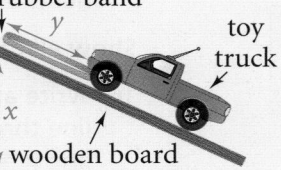

rubber band

y

toy truck

x

wooden board

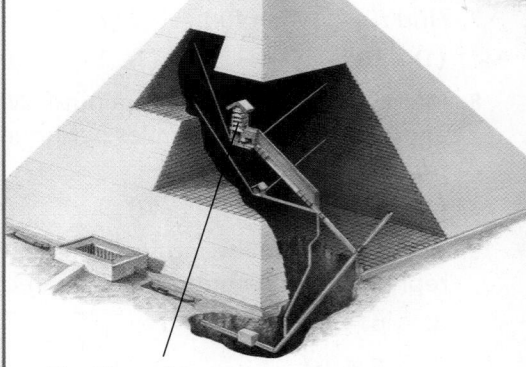

The Pharaoh's Chamber lies almost in the center of the pyramid.

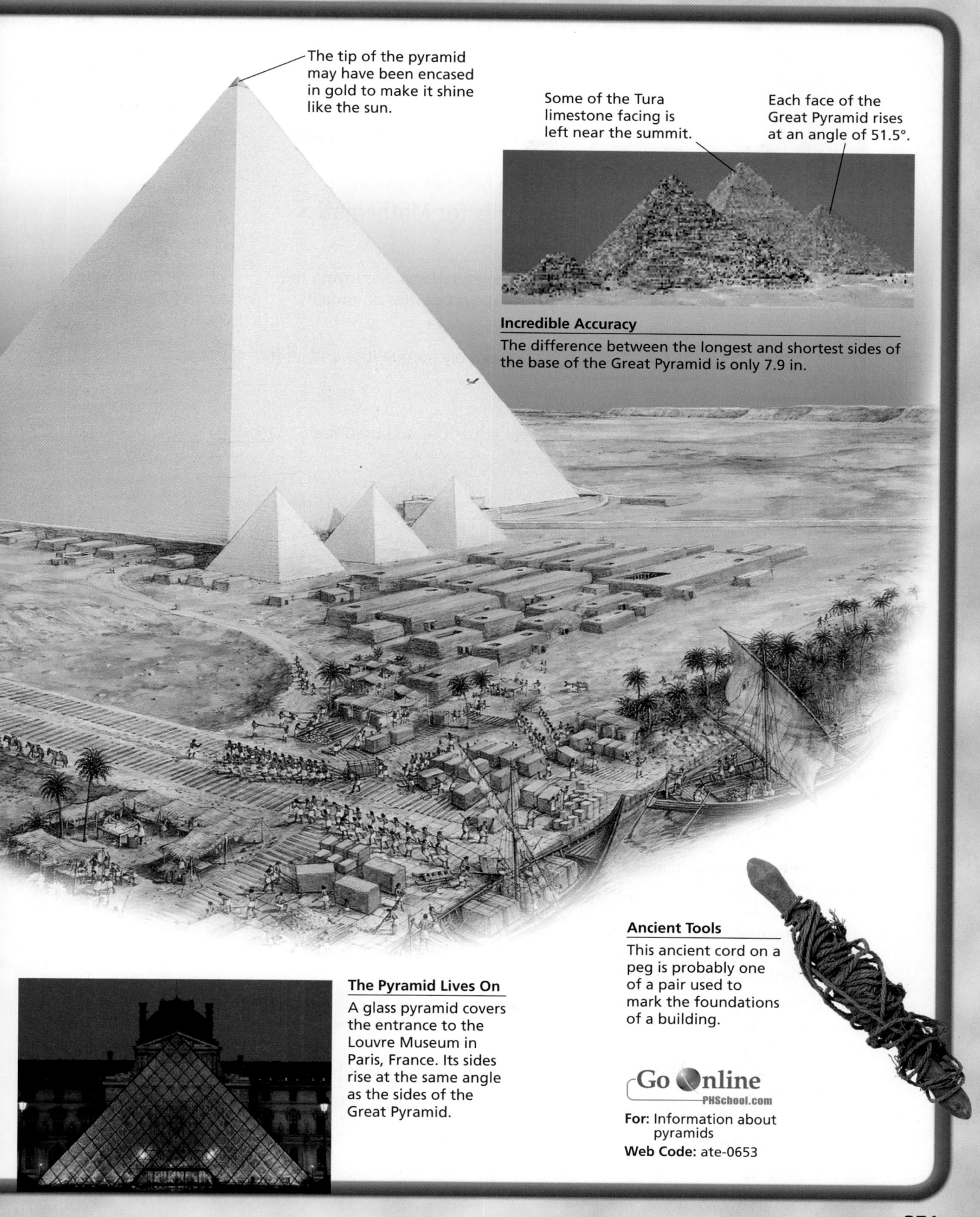

The tip of the pyramid may have been encased in gold to make it shine like the sun.

Some of the Tura limestone facing is left near the summit.

Each face of the Great Pyramid rises at an angle of 51.5°.

Incredible Accuracy

The difference between the longest and shortest sides of the base of the Great Pyramid is only 7.9 in.

Ancient Tools

This ancient cord on a peg is probably one of a pair used to mark the foundations of a building.

The Pyramid Lives On

A glass pyramid covers the entrance to the Louvre Museum in Paris, France. Its sides rise at the same angle as the sides of the Great Pyramid.

Go Online
PHSchool.com

For: Information about pyramids
Web Code: ate-0653

What You've Learned

NY Learning Standards for Mathematics

In Chapter 3, you

- **NY A.A.6:** solved multi-step equations, employing the Distributive Property and the properties of equality.

In Chapter 4, you

- **NY A.A.21:** graphed one-variable inequalities and interpreted their solutions.

In Chapters 5 and 6, you

- **NY A.G.4:** graphed linear equations and used them to model real-world situations.

 **Check Your Readiness**

 for Help to the Lesson in green.

Solving Equations (Lesson 3-3)

Solve each equation. If the equation is an identity, write *identity*. If it has no solution, write *no solution*.

1. $3(2 - 2x) = -6(x - 1)$ **2.** $3m + 1 = -m + 5$ **3.** $4x - 1 = 3(x + 1) + x$

4. $\frac{1}{2}(6x - 4) = 4 + x$ **5.** $5x = 2 - (x - 7)$ **6.** $x + 5 = x - 5$

Solving for a Variable (Review page 140)

Solve for y in terms of x.

7. $3x - 2y = -2$ **8.** $10 = x + 5y$ **9.** $2y = -2x - 8$

Writing Compound Inequalities (Lesson 4-5)

Write an inequality that represents each situation. Graph the solutions.

10. all real numbers that are between -10 and 3

11. Discounts are given to children under 12 and seniors over 60.

Writing a Function Rule (Lesson 5-4)

12. For every $35 ticket, the box office charges a fee of $4.50.
    a. Write a function rule that relates the total cost $C(t)$ to t, the number of tickets.
    b. What is the total cost for 3 tickets?
    c. How many tickets were purchased if the total cost was $237?

Graphing Linear Equations (Lessons 6-2, 6-4, and 6-5)

Graph each line.

13. $2x + 4y = -8$ **14.** $y = -\frac{2}{3}x + 3$ **15.** $2x = y - 4$

Systems of Equations and Inequalities

Chapter 7

🔊 Key Vocabulary

- elimination method (p. 387)
- infinitely many solutions (p. 376)
- linear inequality (p. 405)
- no solution (p. 376)
- solution of a system (p. 374)
- solution of a system of linear inequalities (p. 411)
- solutions of an inequality (p. 405)
- substitution method (p. 382)
- system of linear equations (p. 374)
- system of linear inequalities (p. 411)

What You'll Learn Next

NY Learning Standards for Mathematics

- **NY A.A.10:** You will extend your ability to solve equations to include solving a system of two equations with two variables.

- **NY A.A.10:** You will use methods of solving a linear system, including graphing, substitution, and elimination, and determine which method is best for a given situation.

Data Analysis **Activity Lab** You will apply equations to given data, on pages 426–427.

7-1

Solving Systems by Graphing

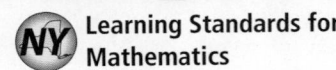 Learning Standards for Mathematics

A.A.7 Analyze and solve verbal problems whose solution requires solving systems of linear equations in two variables.

A.G.7 Graph and solve systems of linear equations and inequalities with rational coefficients in two variables.

A.G.9 Solve systems of linear and quadratic equations graphically.

✓ Check Skills You'll Need

GO for Help Lessons 3-3 and 6-2

Solve each equation.

1. $2n + 3 = 5n - 2$ **2.** $8 - 4z = 2z - 13$ **3.** $8q - 12 = 3q + 23$

Graph each pair of equations on the same coordinate plane.

4. $y = 3x - 6$
$y = -x + 2$

5. $y = 6x + 1$
$y = 6x - 4$

6. $y = 2x - 5$
$6x - 3y = 15$

7. $y = x + 5$
$y = -3x + 5$

🔊 **New Vocabulary** • system of linear equations • solution of a system of linear equations • no solution • infinitely many solutions

---

1 Solving Systems by Graphing

Two or more linear equations together form a **system of linear equations.** One way to solve a system of linear equations is by graphing each equation. Look for any point common to all the lines. Any ordered pair in a system that makes *all* the equations true is a **solution of the system of linear equations.**

1 EXAMPLE Solving a System of Equations

Solve by graphing. $y = 2x - 3$
$y = x - 1$

Graph both equations on the same coordinate plane.

$y = 2x - 3$ **The slope is 2. The y-intercept is −3.**
$y = x - 1$ **The slope is 1. The y-intercept is −1.**

Find the point of intersection.

The lines intersect at $(2, 1)$, so $(2, 1)$ is the solution of the system.

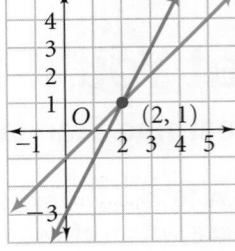

Check See if $(2, 1)$ makes both equations true.

$y = 2x - 3$ $y = x - 1$
$1 \stackrel{?}{=} 2(2) - 3$ ← **Substitute (2, 1)** → $1 \stackrel{?}{=} 2 - 1$
$1 \stackrel{?}{=} 4 - 3$ **for (x, y).** $1 = 1$ ✓
$1 = 1$ ✓

✓ **Quick Check** ❶ Solve by graphing. Check your solution.

a. $y = x + 5$
$y = -4x$

b. $y = -\frac{1}{2}x + 2$
$y = -3x - 3$

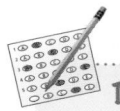

2 **EXAMPLE** Real-World Connection

Multiple Choice Suppose you are testing two fertilizers on bamboo plants A and B, which are growing under identical conditions. Plant A is 6 cm tall and growing at a rate of 4 cm/day. Plant B is 10 cm tall and growing at a rate of 2 cm/day. Which system of equations models the height of each plant $H(d)$ as a function of days d?

Ⓐ $H(d) = 6d + 4$
  $H(d) = 10d + 2$

Ⓑ $H(d) = 4d + 6$
  $H(d) = 10d + 2$

Ⓒ $H(d) = 6d + 4$
  $H(d) = 2d + 10$

Ⓓ $H(d) = 4d + 6$
  $H(d) = 2d + 10$

Relate plant height   is   initial height   plus   daily growth

Write Plant A: $H(d)$ =   6   + $4d$

  Plant B: $H(d)$ =   10   + $2d$

The system is $H(d) = 4d + 6$
  $H(d) = 2d + 10$.

● So, D is the correct answer.

✓ **Quick Check** ❷ You are testing two fertilizers on bamboo plants C and D. Plant C is 5 cm tall and growing at a rate of 3 cm/day. Plant D is 1 cm tall and growing at a rate of 4 cm/day. Write a system of equations that models the height $H(d)$ of each plant as a function of days d.

3 **EXAMPLE** Interpreting Solutions

The system below models the heights of the bamboo plants in Example 2. Find the solution of the system by graphing. What does the solution mean in terms of the original situation?

Part 1 Find the solution by graphing.

$H(d) = 4d + 6$ **The slope is 4. The y-intercept is 6.**
$H(d) = 2d + 10$ **The slope is 2. The y-intercept is 10.**

Graph the equations.

$H(d) = 4d + 6$

$H(d) = 2d + 10$

The lines intersect at $(2, 14)$.

Part 2 Interpret the solution.

After 2 days, both plants will be the same height, 14 cm tall.

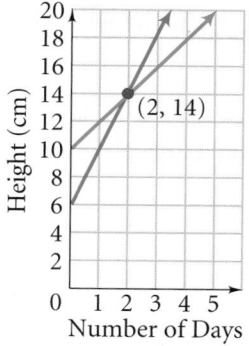

✓ **Quick Check** ❸ Two friends are walking around a quarter-mile track. One person has completed six laps before the second one starts. The system below models the distance $d(t)$ in miles each walker covers as a function of time t in hours.

  $d(t) = 3t + 1.5$ $d(t) = 4t$

a. Find the solution of the system by graphing. Use units of 0.5 on your graph.

b. What does the solution mean in terms of the original situation?

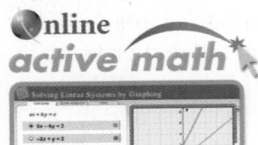

For: Linear System Activity
Use: Interactive Textbook, 7-1

When two lines are parallel, there are no points of intersection. So a system of linear equations has **no solution** when the graphs of the equations are parallel.

4 **EXAMPLE** **Systems With No Solution**

Solve by graphing. $y = -2x + 1$
$\qquad\qquad\qquad y = -2x - 1$

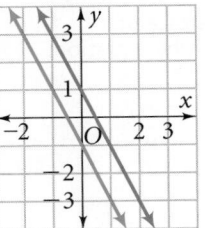

Graph both equations on the same coordinate plane.

$y = -2x + 1$ **The slope is −2. The y-intercept is 1.**
$y = -2x - 1$ **The slope is −2. The y-intercept is −1.**

The lines are parallel. There is no solution.

✓ **Quick Check** **4** **Critical Thinking** Without graphing, how can you tell if a system has no solution? Give an example.

A system of linear equations has **infinitely many solutions** when the graphs of the equations are the same line. The coordinates of the points on the common line are all solutions of the system.

5 **EXAMPLE** **Systems With Infinitely Many Solutions**

Vocabulary Tip

Infinitely many solutions is another way of saying that there is an infinite number of solutions of a system.

Solve by graphing. $2x + 4y = 8$
$\qquad\qquad\qquad y = -\frac{1}{2}x + 2$

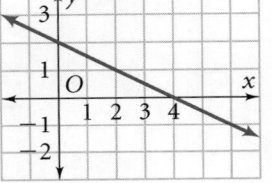

Graph both equations on the same coordinate plane.

$2x + 4y = 8$ **The y-intercept is 2. The x-intercept is 4.**
$y = -\frac{1}{2}x + 2$ **The slope is −$\frac{1}{2}$. The y-intercept is 2.**

The graphs are the same line. The solutions are an infinite number of ordered pairs (x, y) such that $y = -\frac{1}{2}x + 2$.

✓ **Quick Check** **5** Solve by graphing. $y = \frac{1}{5}x + 9$
$\qquad\qquad\qquad\qquad 5y = x + 45$

 Key Concepts

Summary	**Numbers of Solutions of Systems of Linear Equations**	
different slopes	same slope different y-intercepts	same slope same y-intercept
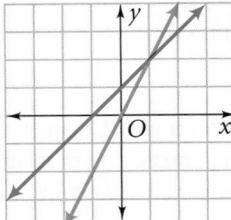	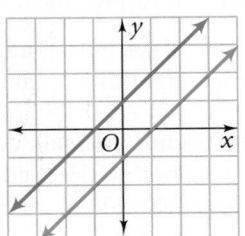	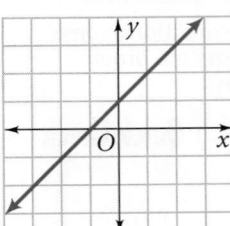
The lines intersect so there is one solution.	The lines are parallel so there are no solutions.	The lines are the same so there are infinitely many solutions.

EXERCISES

For more exercises, see *Extra Skill and Word Problem Practice*.

Practice and Problem Solving

 Practice by Example

Example 1
(page 374)

 for Help

Is $(-1, 5)$ a solution of each system? Explain.

1. $x + y = 4$
$\quad x = -1$

2. $y = -x + 4$
$\quad y = -\frac{1}{5}x$

3. $y = 5$
$\quad x = y - 6$

4. $y = 2x + 7$
$\quad y = x + 6$

Solve by graphing. Check your solution.

5. $y = x + 2$
$\quad y = -2x + 2$

6. $y = x$
$\quad y = 5x$

7. $y = 1$
$\quad y = x$

8. $y = x + 4$
$\quad y = 4x + 1$

9. $y = -\frac{1}{3}x + 1$
$\quad y = \frac{1}{3}x - 3$

10. $y = \frac{1}{2}x + 1$
$\quad y = -3x + 8$

11. $3x + 4y = 12$
$\quad 2x + 4y = 8$

12. $y = \frac{1}{2}x + 2$
$\quad y = -x + 5$

Examples 2, 3
(page 375)

13. Suppose you have $20 in your bank account. You start saving $5 each week. Your friend has $5 in his account and is saving $10 each week. Assume that neither you nor your friend makes any withdrawals.
  a. After how many weeks will you and your friend have the same amount of money in your accounts?
  b. How much money will each of you have?

14. Suppose you have $55 in your bank account. You start saving $10 each week. Your friend has $20 in her account and is saving $15 each week. When will you and your friend have the same amount of money in your accounts?

Examples 4, 5
(page 376)

Graph each system. Tell whether the system has *no solution* or *infinitely many solutions*.

15. $y = -2x + 1$
$\quad y = -2x - 3$

16. $x + 2y = 10$
$\quad 2x + 4y = 10$

17. $y = 3x + 4$
$\quad -12x + 4y = 16$

18. $y = 2x + 6$
$\quad 4x - 2y = 8$

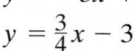 **Apply Your Skills**

Without graphing, decide whether each system has *one solution*, *no solution*, or *infinitely many solutions*. Explain.

19. $y = 2x$
$\quad y = 2x - 5$

20. $x + y = 4$
$\quad 2x + 2y = 8$

21. $y = -3x + 1$
$\quad y = 3x + 7$

22. $3x - 5y = 0$
$\quad y = \frac{3}{5}x$

23. Which graphing calculator screen shows the solution of the system below?
$y = -5x + 4$
$y = \frac{3}{4}x - 3$

A.

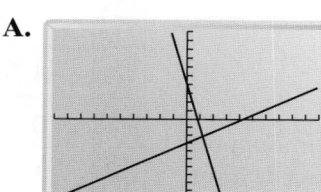

B.

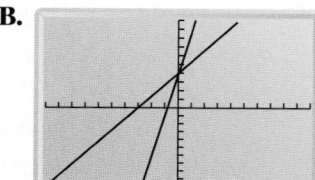

 24. Communications A communications company offers a variety of calling card options. Card A has a 30¢ connection fee and then costs 2¢ per minute. Card B has a 10¢ connection fee and then costs 6¢ per minute. Find the length of the call that would cost the same with both cards.

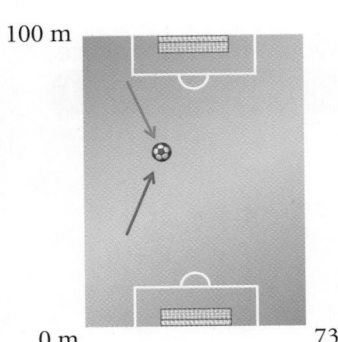

100 m

0 m   73 m

25. Soccer Jim and Tony are on opposing teams in a soccer match. They are running after the same ball. Jim's path is the line $y = 3x$. Tony's path is the line $y = -2x + 100$. Solve by graphing to find the coordinates of the ball.

Open-Ended **Write a system of two linear equations with the given characteristics.**

26. One solution; perpendicular lines

27. No solution; one equation is $y = 2x + 5$.

28. Infinitely many solutions; the graph of one equation has a y-intercept of 3.

Solve by graphing. Check your solution.

29. $y = 4x + 12$	**30.** $y = 3x - 5$	**31.** $y = x + 18$	**32.** $y = 4x + 80$
$y = -2x + 24$	$y = 2x + 10$	$y = -\frac{1}{2}x + 36$	$y = \frac{1}{2}x + 10$

33. Below is a retelling of one of Aesop's fables. Read it and use the story to answer the questions below.

*O*ne day, the tortoise challenged the hare to a race. The hare laughed while bragging about how fast a runner he was. On the day of the race, the hare was so confident that he took a nap during the race. When he awoke, he ran as hard as he could, but he could not beat the slow-but-sure tortoise across the finish line.

a. The graph at the right shows the race of the tortoise and the hare. Which label should be on each axis?
b. Writing Which color indicates the tortoise? Which indicates the hare? Explain your answers.
c. What does the point of intersection mean?

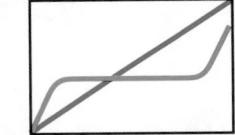

Graphing Calculator **Find the solution of each system. If necessary, round answers to the nearest tenth.**

34. $y = 1.5x + 2$	**35.** $y = -\frac{7}{3}x + \frac{16}{3}$	**36.** $y = 0.2x + 3.5$	**37.** $y = 3.2x + 4.5$
$y = 2.5x + 14$	$y = \frac{4}{3}x + \frac{38}{3}$	$y = 0.4x + 9.5$	$y = -8.7x - 6.1$

38. Use the spreadsheet to find the solution of the following system.
$y = -4x + 11$
$y = 3x - 3$

	A	B	C
1	x	$y = -4x + 11$	$y = 3x - 3$
2	-1	15	-6
3	0	11	-3
4	1	7	0
5	2	3	3
6	3	-1	6

39. Recording Music Suppose you and your friends form a band. You want to record a demo. Studio A rents for $100 plus $50/hour. Studio B rents for $50 plus $75/hour.
a. Write a system that models the situation and solve by graphing.
b. Explain what the solution of the system means in terms of renting a studio.

Online
Homework Video Tutor
Visit: PHSchool.com
Web Code: ate-0701

40. a. Critical Thinking For what values of w and v does the system have exactly one solution?
$y = -5x + w$
$y = -5x + v$
   b. For what values of w and v does the system have no solution?
   c. For what values of w and v does the system have infinitely many solutions?

41. a. If $g \geq h$, the system at the right has no solution *always, sometimes,* or *never*?
$y = gx + 3$
$y = hx + 7$
   b. If $g \leq h$, the system has infinitely many solutions *always, sometimes,* or *never*?

42. The slope of the line joining point P to the origin is $\frac{2}{9}$. The slope of the line joining point P to $(-4, 3)$ is 1. Find the coordinates of point P.

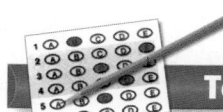

NY REGENTS

Test Prep

Multiple Choice

43. Which ordered pair is the solution of the system?
$6x - 6y = 2$
$3x + 9y = -7$
   A. $\left(\frac{2}{3}, -\frac{1}{3}\right)$ **B.** $\left(\frac{1}{3}, \frac{2}{3}\right)$
   C. $\left(-\frac{2}{3}, \frac{1}{3}\right)$ **D.** $\left(-\frac{1}{3}, -\frac{2}{3}\right)$

44. Which value of b will make the graphs of $y = 2x + 3$ and $y = 2.5x + b$ intersect at (2, 7)?
   F. 2   **G.** 3   **H.** 5   **J.** 7

Short Response

45. The first equation in a system of two equations is $x - 2y = 10$. The graph of the second equation does not intersect the first.
   a. Write a possible second equation for the system.
   b. Explain your answer to part (a).

Extended Response

46. The advertisements at the right are for two jobs you are considering.
   a. Write a system of equations that relates the amount of sales x to the money y earned in a week at each job.
   b. How much would you need to sell in a week at each job to earn the same amount of money at both?
   c. After talking with salespeople, you estimate weekly sales of about $600 at either job. At which job would you earn more money?

> **Sales Position**
> Salesperson Wanted
> Knowledge of Cellular Phones
> On-Site Sales
> $150/week + 20% commission

> **CAREER OPPORTUNITY**
> Sell Stereo Equipment in
> National Electronics Retail Chain!
> $200/week + 10% commission

Mixed Review

 **GO for Help**

Lesson 6-8

Graph each equation and describe its translation from $y = |x|$.

47. $y = |x| + 2$ **48.** $y = |x + 3|$ **49.** $y = |x - 2| + 5$

Lesson 3-7

Find each percent of change. Describe the percent of change as an increase or decrease.

50. 4 cm to 5 cm   **51.** 12 in. to 8 in.   **52.** $20 to $24   **53.** 10 ft to 25 ft
54. $9 to $6   **55.** 12 cm to 15 cm   **56.** 50 m to 55 m   **57.** $48 to $42

Solving Systems Using Tables and Graphs

FOR USE WITH LESSON 7-1

1 ACTIVITY

Solve the system by using a table.

$$y = 3x - 6$$
$$y = -4x + 29$$

> **NY** A.G.7: Graph and solve systems of linear equations and inequalities with rational coefficients in two variables.

Step 1

Enter the equations in the **Y=** screen.

```
 Plot1  Plot2  Plot3
\Y1 ▣3X–6
\Y2 ▣-4X+29
\Y3 =
\Y4 =
\Y5 =
\Y6 =
\Y7 =▮
```

Step 2

Use the **TBLSET** function. Set **TblStart** to 0 and ΔTbl to 1.

```
TABLE SETUP
  TblStart = 0
  △Tbl = 1
Indpnt: Auto  Ask
Depend: Auto  Ask
```

Step 3

Press 2nd GRAPH to access the **TABLE** screen.

X	Y1	Y2
0	-6	29
1	-3	25
2	0	21
3	3	17
4	6	13
5	9	9
6	12	5
X=0		

1. Which *x*-value gives the same value for **Y1** and **Y2**?

2. Complete the following statement:

The point (__, __) is a solution of each linear equation, so it is the solution of the system.

2 ACTIVITY

Solve the system by using a graph.

$$y = 4x + 14$$
$$y = -3x - 3.5$$

Step 1 Enter the equations in the **Y=** screen.

Step 2 Graph the equations. Use a standard graphing window.

Step 3 Use the **CALC** feature, select **intersect** to find the point where the lines intersect.

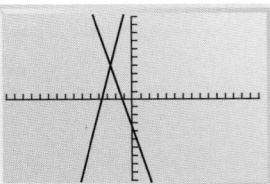

3. Complete the following statement:

The lines intersect at (__, __), so it is the solution of the system.

EXERCISES

Choose Use a table or a graph to solve each system.

4. $y = 3x - 3$
   $y = 2x - 1$

5. $y = -4.5x - 8.5$
   $y = 4.5x + 18.5$

6. $y = -x - 2.5$
   $y = 2.5x - 11.25$

7. $y = -\frac{1}{2}x - 1$
   $y = 3x - 15$

8. $y = -2x - 7$
   $y = 2x + 17$

9. $y = \frac{1}{2}x + \frac{9}{2}$
   $y = 2x - 6$

Solving Systems Using Algebra Tiles

You can model and solve some linear systems using algebra tiles.

Solve the following system.

$$2x + y = 5$$
$$y = x - 1$$

NY **A.G.7:** Graph and solve systems of linear equations and inequalities with rational coefficients in two variables.

Model the expression for y, which is $x - 1$.

To model the equation $2x + y = 5$, substitute the expression for y, which is $x - 1$.

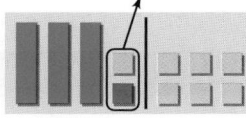

$$2x + y = 5$$
$$2x + x - 1 = 5$$
$$3x - 1 = 5$$

Using the Addition Property of Equality, add 1 to each side. Simplify the model by removing the zero pair.

$$3x - 1 + 1 = 5 + 1$$
$$3x = 6$$

Divide each side into three identical groups.

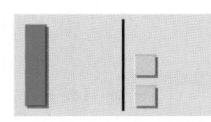

$$\frac{3x}{3} = \frac{6}{3}$$

Solve for x.

$$x = 2$$

To find y, substitute the value of x into the equation $y = x - 1$.

$$y = x - 1$$
$$= 2 - 1$$
$$= 1$$

The solution of the system is $(2, 1)$.

EXERCISES

Model and solve each system.

1. $y = x + 1$
$2x + y = 10$

2. $x + 4y = 1$
$x = y - 4$

3. $y = 2x - 1$
$y = x + 5$

4. $x = 3y + 2$
$2x = y + 9$

5. $x - 4y = 2$
$x = y + 1$

6. $y = x + 3$
$y = 2x + 6$

7. Open-Ended Let the equation $y = x + 2$ be part of a system. Write the second equation of the system such that the system could be solved using algebra tiles.

7-2 Solving Systems Using Substitution

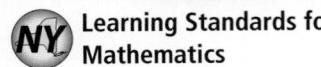
Learning Standards for Mathematics

A.A.7 Analyze and solve verbal problems whose solution requires solving systems of linear equations in two variables.

A.A.10 Solve systems of two linear equations in two variables algebraically.

✓ Check Skills You'll Need

GO ▶ for Help Lessons 3-3 and 7-1

Solve each equation.

1. $m - 6 = 4m + 8$ **2.** $4n = 9 - 2n$ **3.** $\frac{1}{3}t + 5 = 10$

For each system, is the ordered pair a solution of both equations?

4. $(5, 1)$ $y = -x + 4$
                 $y = x - 6$

5. $(2, 2.4)$ $4x + 5y = 20$
                     $2x + 6y = 10$

◀))) **New Vocabulary** • substitution method

---

1 Using Substitution

Vocabulary Tip

Substitution means one value or expression is used in place of another.

You can solve a system of equations by graphing when the solution contains integers or when you have a graphing calculator. Another method for solving systems of equations is the **substitution method.** By replacing one variable with an equivalent expression containing the other variable, you can create a one-variable equation that you can solve using methods shown in Chapter 3.

1 EXAMPLE Using Substitution

Solve using substitution. $y = -4x + 8$
                            $y = x + 7$

Step 1 Write an equation containing only one variable, and solve it.

$y = -4x + 8$	**Start with one equation.**
$x + 7 = -4x + 8$	**Substitute x + 7 for y.**
$5x + 7 = 8$	**Add 4x to each side.**
$5x = 1$	**Subtract 7 from each side.**
$x = 0.2$	**Divide each side by 5.**

Step 2 Solve for the other variable in either equation.

$y = 0.2 + 7$	**Substitute 0.2 for x in y = x + 7.**
$y = 7.2$	**Simplify.**

Since $x = 0.2$ and $y = 7.2$, the solution is $(0.2, 7.2)$.

Check $7.2 \overset{?}{=} -4(0.2) + 8$ Since $y = x + 7$ was used in step 2, see if $(0.2, 7.2)$ solves $y = -4x + 8$.

$7.2 = 7.2$ ✓   **Simplify.**

✓ **Quick Check** ❶ Solve using substitution. Check your solution. $y = 2x$
                                                                       $7x - y = 15$

To use the substitution method, you must have an equation that has already been solved for one of the variables.

2 EXAMPLE **Using Substitution and the Distributive Property**

Solve using the substitution method. $3y + 2x = 4$
$$-6x + y = -7$$

Step 1 Solve the second equation for y because it has a coefficient of 1.
$$-6x + y = -7$$
$$y = 6x - 7 \qquad \textbf{Add 6\textit{x} to each side.}$$

Step 2 Write an equation containing only one variable and solve.

$3y + 2x = 4$	**Start with the other equation.**
$3(6x - 7) + 2x = 4$	**Substitute 6\textit{x} − 7 for \textit{y}. Use parentheses.**
$18x - 21 + 2x = 4$	**Use the Distributive Property.**
$20x = 25$	**Combine like terms and add 21 to each side.**
$x = 1.25$	**Divide each side by 20.**

Step 3 Solve for the other variable in either equation.

$-6(1.25) + y = -7$	**Substitute 1.25 for \textit{x} in −6\textit{x} + \textit{y} = −7.**
$-7.5 + y = -7$	**Simplify.**
$y = 0.5$	**Add 7.5 to each side.**

Since $x = 1.25$ and $y = 0.5$, the solution is $(1.25, 0.5)$.

☑ **Quick Check** **2** Solve using substitution. Check your solution. $6y + 8x = 28$
$$3 = 2x - y$$

3 EXAMPLE **Real-World** 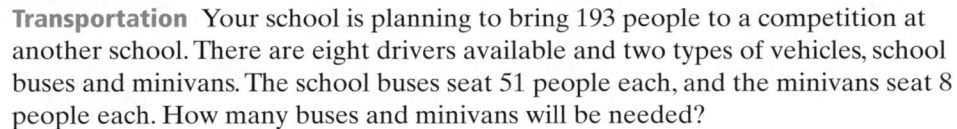 **Problem Solving**

Transportation Your school is planning to bring 193 people to a competition at another school. There are eight drivers available and two types of vehicles, school buses and minivans. The school buses seat 51 people each, and the minivans seat 8 people each. How many buses and minivans will be needed?

Let b = number of school buses.   Let m = number of minivans.
   drivers $b + m = 8$ people $51b + 8m = 193$

Step 1	$b + m = 8$	**Solve the first equation for \textit{m}.**
	$m = -b + 8$	**Subtract \textit{b} from each side.**
Step 2	$51b + 8(-b + 8) = 193$	**Substitute −\textit{b} + 8 for \textit{m} in the second equation.**
	$51b - 8b + 64 = 193$	**Solve for \textit{b}.**
	$43b + 64 = 193$	
	$43b = 129$	
	$b = 3$	
Step 3	$(3) + m = 8$	**Substitute 3 for \textit{b} in \textit{b} + \textit{m} = 8.**
	$m = 5$	

Three school buses and five minivans will be needed to transport 193 people.

☑ **Quick Check** **3** **Geometry** A rectangle is 4 times longer than it is wide. The perimeter of the rectangle is 30 cm. Find the dimensions of the rectangle.

EXERCISES

For more exercises, see *Extra Skill and Word Problem Practice*.

Practice and Problem Solving

A Practice by Example

Example 1
(page 382)

Mental Math Match each system with its solution at the right.

1. $y = x + 1$
   $y = 2x - 1$

2. $y = \frac{1}{2}x + 4$
   $2y + 2x = 2$

A. $(3, 2)$

3. $2y = x + 3$
   $x = y$

4. $x - y = 1$
   $x = \frac{1}{2}y + 2$

B. $(3, 3)$

C. $(-2, 3)$

D. $(2, 3)$

Solve each system using substitution. Check your solution.

5. $y = 4x - 8$
   $y = 2x + 10$

6. $C(n) = -3n - 6$
   $C(n) = n - 4$

7. $m = 5p + 8$
   $m = -10p + 3$

8. $y = -4x + 12\frac{1}{2}$
   $y = \frac{1}{4}x + 4$

9. $h = 6g - 4$
   $h = -2g + 28$

10. $a = \frac{2}{5}b - 3$
   $a = 2b - 18$

Example 2
(page 383)

11. $y = x - 2$
   $2x + 2y = 4$

12. $c = 3d - 27$
   $4d + 10c = 120$

13. $3x - 6y = 30$
   $y = -6x + 34$

14. $m = 4n + 11$
   $-6n + 8m = 36$

15. $7x - 8y = 112$
   $y = -2x + 9$

16. $t = 0.2s + 10$
   $4s + 5t = 35$

Example 3
(page 383)

17. Geometry The length of a rectangle is 5 cm more than twice the width. The perimeter of the rectangle is 34 cm. Find the dimensions of the rectangle.

18. Suppose you have $28.00 in your bank account and start saving $18.25 every week. Your friend has $161.00 in his account and is withdrawing $15 every week. When will your account balances be the same?

B Apply Your Skills

Solve each system by substitution. Check your solution.

19. $a - 1.2b = -3$
   $0.2b + 0.6a = 12$

20. $0.5x + 0.25y = 36$
   $y + 18 = 16x$

21. $y = 0.8x + 7.2$
   $20x + 32y = 48$

For Exercises 22–24, define variables and write a system of equations for each situation. Solve using substitution.

22. Agriculture A farmer grows only sunflowers and flax on his 240-acre farm. This year he wants to plant 80 more acres of sunflowers than of flax. How many acres of each crop does the farmer need to plant?

23. Renting Videos Suppose you want to join a video store. Big Video offers a special discount card that costs $9.99 for one year. With the discount card, each video rental costs $2.49. A discount card from Main Street Video costs $20.49 for one year. With the Main Street Video discount card, each video rental costs $1.79. After how many video rentals is the cost the same?

Real-World Connection

Sunflower seeds are sold as snacks and as bird food, and they are a source of cooking oil.

24. Buying a Car Suppose you are thinking about buying one of two cars. Car A will cost $17,655. You can expect to pay an average of $1230 per year for fuel, maintenance, and repairs. Car B will cost about $15,900. Fuel, maintenance, and repairs for it will average about $1425 per year. After how many years are the total costs for the cars the same?

25. Multiple Choice You have 28 coins that are all nickels n and dimes d. The value of the coins is \$2.05. Which system of equations can be used to find the number of nickels and the number of dimes?

Ⓐ $n + d = 28$
   $10n + 5d = 2.05$

Ⓑ $n + d = 205$
   $n + d = 28$

Ⓒ $10n + 5d = 205$
   $n + d = 28$

Ⓓ $n + d = 28$
   $5n + 10d = 205$

Estimation **Graph each system to estimate the solution. Then use substitution to find the exact solution of the system.**

26. $y = 2x$
     $y = -6x + 4$

27. $y = \frac{1}{2}x + 4$
     $y = -4x - 5$

28. $x + y = 0$
     $5x + 2y = -3$

29. $y = 2x + 3$
     $y = 0.5x - 2$

30. $y = -x + 4$
     $y = 2x + 6$

31. $y = 0.7x + 3$
     $y = -1.5x - 7$

32. Open-Ended Write a system of linear equations with exactly one solution. Use substitution to solve your system.

33. a. Solve each system below using substitution.

   $y = 0.5x + 4$ $6x - 2y = 10$
   $-x + 2y = 8$ $y = 3x + 1$

   b. Solve each system by graphing.

   c. Critical Thinking Make a general statement about the solutions you get when solving by graphing and the results you get when solving by substitution.

Solve each system using substitution.

34. $y = 2x$
     $6x - y = 8$

35. $y = 3x + 1$
     $x = 3y + 1$

36. $x - 3y = 14$
     $x - 2 = 0$

37. $2x + 2y = 5$
     $y = \frac{1}{4}x$

38. $4x + y = -2$
     $-2x - 3y = 1$

39. $3x + 5y = 2$
     $x + 4y = -4$

Ⓒ **Challenge**

40. There are 1170 students in a school. The ratio of girls to boys is $23 : 22$. The system below describes relationships between the number of girls and the number of boys.

   $g + b = 1170$ $\frac{g}{b} = \frac{23}{22}$

   a. Solve the proportion for g.
   b. Solve the system.
   c. How many more girls are there than boys?

🌐 **41. Sprinting** The graph at the left represents the start of a 100-meter race between Joetta and Gail. The red line and blue line represent Joetta's and Gail's time and distance. Joetta averages 8.8 m/s. Gail averages 9 m/s but started 0.2 s after Joetta. At time 0.2 s, Gail's distance is 0 m. You can use point-slope form to write an equation that relates Gail's time t to her distance d.

   $y - y_1 = m(x - x_1)$
   $d - 0 = 9(t - 0.2)$
   $d = 9t - 1.8$

Since Joetta started at $t = 0$, the equation $d = 8.8t$ relates her time and distance.
   a. Solve the system using substitution.
   b. Will Gail overtake Joetta before the finish line?

Distance
(meters)

Time
(seconds)

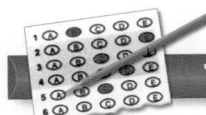

Test Prep

Gridded Response

42. Find the value of the *y*-coordinate of the solution to the given system.

$5x + 5y = 179$
$x = 5y - 143$

43. Find the value of the *y*-coordinate of the solution to the given system.

$y = 9x + 3480$
$y = 81x - 7104$

44. Tina has $220 in her account. Cliff has $100 in his account. Starting in July, Tina adds $25 to her account on the first of each month, while Cliff adds $35 to his. How many dollars will they have in their accounts when the amounts are the same?

Short Response

45. Is $(-2, -7)$ the solution of the following system? Justify your answer.

$7y - 4x = 29$
$x = y - 5$

Mixed Review

Lesson 7-1

Solve each system by graphing.

46. $y = x - 2$
$ y = \frac{1}{2}x + 4$

47. $y = -2x + 5$
$ y = -x + 3$

48. $y = \frac{3}{4}x - 1$
$ y = \frac{1}{4}x + 1$

Lesson 5-3

Graph each function.

49. $y = 3x - 2$

50. $f(x) = x + 1$

51. $f(x) = -2x$

52. $y = |x| + 5$

53. $y = -2|x|$

54. $y = |x + 3| - 1$

✓ Checkpoint Quiz 1   Lessons 7-1 through 7-2

Solve each system by graphing.

1. $y = 3x - 4$
$ y = -2x + 1$

2. $y = \frac{4}{3}x - 2$
$ y = \frac{2}{3}x$

3. $y = \frac{1}{4}x - 1$
$ y = -2x - 10$

Solve each system using substitution.

4. $y = 3x - 14$
$ y = x - 10$

5. $y = 2x + 5$
$ y = 6x + 1$

6. $x = y + 7$
$ y = 8 + 2x$

7. $3x + 4y = 12$
$ y = -2x + 10$

8. $4x + 9y = 24$
$ y = -\frac{1}{3}x + 2$

In Exercises 9 and 10, write and solve a system of equations for each situation.

9. A rectangle is 3 times longer than it is wide. The perimeter is 44 cm. Find the dimensions of the rectangle.

10. A farmer grows only pumpkins and corn on her 420-acre farm. This year she wants to plant 250 more acres of corn than of pumpkins. How many acres of each crop does the farmer need to plant?

GO for Help

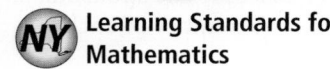 **7-3**

Solving Systems Using Elimination

NY Learning Standards for Mathematics

A.A.7 Analyze and solve verbal problems whose solution requires solving systems of linear equations in two variables.

A.G.7 Graph and solve systems of linear equations and inequalities with rational coefficients in two variables.

✓ **Check Skills You'll Need**

Solve each system using substitution.

1. $y = 4x - 3$
$y = 2x + 13$

2. $y + 5x = 4$
$y = 7x - 20$

3. $y = -2x + 2$
$3x - 17 = 2y$

 **for Help** Lesson 7-2

🔊 **New Vocabulary** • elimination method

---

1 Adding or Subtracting to Solve Systems

Vocabulary Tip

Addition Property of Equality
If $a = b$,
then $a + c = b + c$.

Subtraction Property of Equality
If $a = b$,
then $a - c = b - c$.

The Addition and Subtraction Properties of Equality can be extended to state,

If $a = b$ and $c = d$, then $a + c = b + d$.   If $a = b$ and $c = d$, then $a - c = b - d$.

You can use the Addition and Subtraction Properties of Equality to solve a system by the **elimination method.** You can add or subtract equations to eliminate a variable.

1 EXAMPLE Adding Equations

Solve by elimination. $5x - 6y = -32$
$3x + 6y = 48$

Step 1 Eliminate y because the sum of the coefficients of y is zero.

$5x - 6y = -32$
$\underline{3x + 6y = 48}$ **Add the two equations.**
$8x + 0 = 16$ **Addition Property of Equality**
$x = 2$ **Solve for x.**

Step 2 Solve for the eliminated variable y using either of the original equations.

$3x + 6y = 48$ **Choose the second equation.**
$3(2) + 6y = 48$ **Substitute 2 for x.**
$6 + 6y = 48$ **Simplify. Then solve for y.**
$y = 7$

Since $x = 2$ and $y = 7$, the solution is $(2, 7)$.

Check $5(2) - 6(7) \overset{?}{=} -32$ **See if (2, 7) solves $5x - 6y = -32$.**
$10 - 42 \overset{?}{=} -32$
$-32 = -32$ ✓

✓ **Quick Check** ❶ Solve by elimination. $6x - 3y = 3$
$-6x + 5y = 3$

Real-World Connection

There are 188 basketball teams in 26 conferences in the National Wheelchair Basketball Association.

2 EXAMPLE Real-World Problem Solving

Ticket Sales Suppose your community center sells a total of 292 tickets for a basketball game. An adult ticket costs $3. A student ticket costs $1. The sponsors collect $470 in ticket sales. Write and solve a system to find the number of each type of ticket sold.

Define Let a = number of adult tickets.
Let s = number of student tickets.

Relate total number of tickets   total amount of sales
Write $a + s = 292$ $3a + 1s = 470$

Solve by elimination.

Step 1 Eliminate s because the difference of the coefficients of s is zero.

$$\begin{array}{rl} a + s = & 292 \\ 3a + s = & 470 \\ \hline -2a + 0 = & -178 \\ a = & 89 \end{array}$$

Subtract the two equations.

Subtraction Property of Equality

Solve for a.

Step 2 Solve for the eliminated variable using either of the original equations.

$a + s = 292$ **Choose the first equation.**
$89 + s = 292$ **Substitute 89 for a.**
$s = 203$ **Solve for s.**

There were 89 adult tickets sold and 203 student tickets sold.

Check Is the solution reasonable? The answers 89 and 203 are close to 90 and 200. The total number of tickets is about $90 + 200 = 290$, close to 292. The total sales is about $3(90) + \$1(200)$ or $470. The solution is reasonable.

✔ **Quick Check** ② Your class sells a total of 64 tickets to a play. A student ticket costs $1, and an adult ticket costs $2.50. Your class collects $109 in total ticket sales. How many adult tickets did you sell? How many student tickets did you sell?

Multiplying First to Solve Systems

From Examples 1 and 2 you can see that to eliminate a variable its coefficients must have a sum or difference of zero. Sometimes you may need to multiply one or both of the equations by a nonzero number first.

3 EXAMPLE Multiplying One Equation

Solve by the elimination method. $2x + 5y = -22$
                                     $10x + 3y = 22$

Step 1 Eliminate one variable.

Start with the given system.	To prepare for eliminating x, multiply the first equation by 5.	Subtract the equations to eliminate x.
$2x + 5y = -22 \quad \rightarrow$	$5(2x + 5y = -22) \quad \rightarrow$	$10x + 25y = -110$
$10x + 3y = 22 \quad \rightarrow$	$\underline{10x + 3y = 22}$ $\rightarrow$	$\underline{10x + 3y = 22}$
		$0 + 22y = -132$

Step 2 Solve for y.
$$22y = -132$$
$$y = -6$$

Step 3 Solve for the eliminated variable using either of the original equations.

$2x + 5y = -22$ **Choose the first equation.**

$2x + 5(-6) = -22$ **Substitute −6 for y.**

$2x - 30 = -22$ **Solve for x.**

$$2x = 8$$
$$x = 4$$

● The solution is $(4, -6)$.

 **Quick Check** ❸ Solve by elimination. $-2x + 15y = -32$
$7x - 5y = 17$

To solve problems that arise from real-world situations, you can also use the elimination method.

4 **EXAMPLE**   <u>Real-World 🌐 Problem Solving</u>

Gridded Response Suppose your class sells gift wrap for $4 per package and greeting cards for $10 per package. Your class sells 205 packages in all and receives a total of $1084. Find the number of packages of gift wrap and the number of packages of greeting cards sold.

Define Let w = number of packages of gift wrap sold.
Let c = number of packages of greeting cards sold.

Relate   total number of packages   total amount of sales
Write $w + c = 205$ $4w + 10c = 1084$

Step 1 Eliminate one variable.

Start with the given system.	To prepare for eliminating w, multiply the first equation by 4.	Subtract the equations to eliminate w.
$w + \quad c = 205$ →	$4(w + \quad c = 205)$ →	→ $\quad 4w + \quad 4c = \quad 820$
$4w + 10c = 1084$ →	$\underline{4w + 10c = 1084}$	→ $\underline{4w + 10c = 1084}$
		$0 \ - \ 6c = -264$

Step 2 Solve for c.
$$-6c = -264$$
$$c = 44$$

Step 3 Solve for the eliminated variable using either of the original equations.

$w + c = 205$ **Use the first equation.**

$w + 44 = 205$ **Substitute 44 for c.**

$w = 161$ **Solve for w.**

● The class sold 161 packages of gift wrap and 44 packages of greeting cards.

Quick Check ❹ Suppose your younger brother's elementary school class sells a different brand of gift wrap, which costs $2 per package, and cards, which cost $5 per package. His class sells 220 packages in all and earns a total of $695. Find the number of each type of package sold.

Test-Taking Tip

If you are asked to grid one of the solutions of a system, such as the number of packages of gift wrap, be sure to select the correct variable.

Video Tutor Help

Visit: PHSchool.com
Web Code: ate-0775

To eliminate a variable, you may need to multiply both equations in a system by a nonzero number. Multiply each equation by values such that when you write equivalent equations, you can then add or subtract to eliminate a variable.

5 EXAMPLE **Multiplying Both Equations**

Solve by elimination. $4x + 2y = 14$
$7x - 3y = -8$

Step 1 Eliminate one variable.

Start with the given system.	To prepare for eliminating y, multiply one equation by 3 and the other equation by 2.	Add the equations to eliminate y.
$4x + 2y = 14$ →	$3(4x + 2y = 14)$ →	$12x + 6y = 42$
$7x - 3y = -8$ →	$2(7x - 3y = -8)$ →	$\underline{14x - 6y = -16}$
		$26x + 0 = 26$

Step 2 Solve for x.
$$26x = 26$$
$$x = 1$$

Step 3 Solve for the eliminated variable y using either of the original equations.

$4x + 2y = 14$ **Use the first equation.**

$4(1) + 2y = 14$ **Substitute 1 for x.**

$2y = 10$

$y = 5$

● The solution is $(1, 5)$.

✓ Quick Check **5** Solve by elimination. $15x + 3y = 9$
$10x + 7y = -4$

When you solve systems using elimination, plan a strategy. A flowchart like the one below can help you to decide how to eliminate a variable.

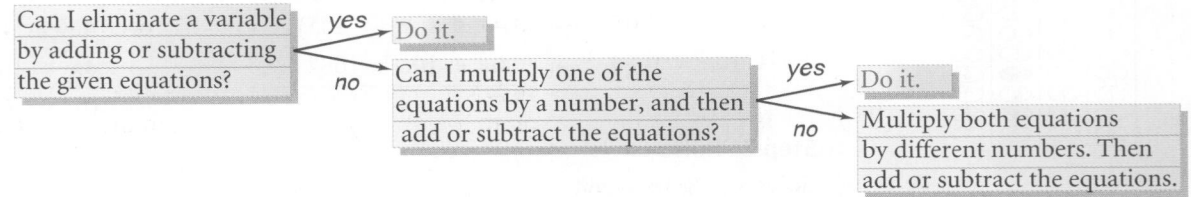

EXERCISES

For more exercises, see *Extra Skill and Word Problem Practice.*

Practice and Problem Solving

A Practice by Example

Solve by elimination.

Example 1
(page 387)

 for Help

1. $2x + 5y = 17$
$6x - 5y = -9$

2. $7x + 2y = 10$
$-7x + y = -16$

3. $2x - 3y = 61$
$2x + y = -7$

4. $8x + 11y = 20$
$5x - 11y = -59$

5. $2x + 18y = -9$
$4x + 18y = -27$

6. $20x + 3y = 20$
$-20x + 5y = 60$

Example 2
(page 388)

7. The sum of two numbers is 20. Their difference is 4.
  a. Write a system of equations that describes this situation.
  b. Solve by elimination to find the two numbers.

8. Ticket Sales Your school sold 456 tickets for a high school play. An adult ticket cost \$3.50. A student ticket cost \$1. Total ticket sales equaled \$1131. Let a equal the number of adult tickets sold, and let s equal the number of student tickets sold.
  a. Write a system of equations that relates the number of adult and student tickets sold to the total number of tickets sold and to the total ticket sales.
  b. Solve by elimination to find the number of each type of ticket sold.

Example 3
(page 388)

Solve by elimination.

9. $3x - 10y = -25$
$4x + 40y = 20$

10. $7x + 15y = 32$
$x - 3y = 20$

11. $x - 8y = 18$
$-16x + 16y = -8$

12. $24x + 2y = 52$
$6x - 3y = -36$

13. $88x - 5y = 39$
$-8x + 3y = -1$

14. $2x + 4y = 8$
$5x + y = -7$

Example 4
(page 389)

15. Sales A photo studio that takes school pictures offers several different packages. Let w equal the cost of a wallet-sized portrait, and let ℓ equal the cost of an 8×10 portrait.

Basic Package
30 wallet-sized photos
1 8" x 10" portrait
\$17.65

Deluxe Package
20 wallet-sized photos
3 8" x 10" portraits
\$25.65

  a. Write a system of equations that relates the cost of wallet-sized portraits and 8×10 portraits to the cost of the basic and deluxe packages.
  b. Find the cost of each type of portrait.

16. Two groups of students order burritos and tacos at a local restaurant. One order of 3 burritos and 4 tacos costs \$11.33. The other order of 9 burritos and 5 tacos costs \$23.56.
  a. Write a system of equations that describes this situation.
  b. Solve by elimination to find the cost of a burrito and the cost of a taco.

Example 5
(page 390)

Solve by elimination.

17. $3x + 2y = -9$
$-10x + 5y = -5$

18. $4x + 5y = 15$
$6x - 4y = 11$

19. $3x - 2y = 10$
$2x + 3y = -2$

20. $-2x + 5y = 20$
$3x - 7y = -26$

21. $10x + 8y = 2$
$8x + 6y = 1$

22. $9x + 5y = 34$
$8x - 2y = -2$

B **Apply Your Skills**

Solve each system using any method. Tell why you chose the method you used.

23. $y = 2x$
$y = x - 1$

24. $7x + 8y = 25$
$9x + 10y = 35$

25. $x = 12y - 14$
$3y + 2x = 26$

26. $-20x + 7y = 137$
$4x + 5y = 43$

27. $5y = x$
$2x - 3y = 7$

28. $y = x + 2$
$y = -2x + 3$

29. Vacation A weekend at the Beach Bay Hotel in Florida includes 2 nights and 4 meals. A week includes 7 nights and 10 meals. Let n = the cost of 1 night and m = the cost of 1 meal. Find the cost of 1 night and the cost of 1 meal.

30. a. Business A company sells brass and steel machine parts. One shipment contains 3 brass and 10 steel parts and costs $48. A second shipment contains 7 brass and 4 steel parts and costs $54. Find the cost of each type of machine part.

    b. How much would a shipment containing 10 brass and 13 steel machine parts cost?

31. Error Analysis Beth is solving a system by elimination. Her work is shown below. What error did she make?

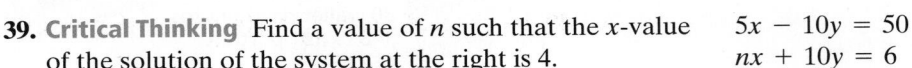

$$4x - 6y = 1 \longrightarrow 20x - 30y = 5$$
$$3x + 5y = -8 \longrightarrow 18x + 30y = -8$$

32. Open-Ended Write a system of equations that can be solved by elimination. Solve your system.

Solve by elimination.

33. $\frac{1}{2}x + y = -1$
   $16x - \frac{1}{2}y = 163$

34. $\frac{1}{4}x - 6y = -70$
   $5x + \frac{3}{4}y = 49$

35. $-0.2x + 4y = -1$
   $x + 0.5y = -15.5$

36. $y = 0.5x + 2$
   $1.5x + y = 42$

37. $\frac{1}{4}x + \frac{33}{2} = y$
   $y - 12 = -2x$

38. $\frac{2}{3}x - y = 70$
   $\frac{1}{3}x - \frac{2}{3}y = 43$

39. Critical Thinking Find a value of n such that the x-value of the solution of the system at the right is 4.
   $5x - 10y = 50$
   $nx + 10y = 6$

40. Writing Explain how to solve a system using elimination. Give examples of when you use addition, subtraction, and multiplication.

41. Electricity Two batteries produce a total voltage of 4.5 volts ($B_1 + B_2 = 4.5$). The difference in their voltages is 1.5 volts ($B_1 - B_2 = 1.5$). Find the voltages of the two batteries.

C **Challenge**

Solve by elimination.

42. $\frac{6}{x} - \frac{4}{y} = -4$
   $\frac{3}{x} + \frac{8}{y} = 3$

43. $ax + y = c$
   $ax + by = c$

44. $x + y + z = 41$
   $x - y + z = 15$
   $3x - z = 4$

45. Music Suppose your band wants to sell CDs and cassette tapes of your music. You use a production company that offers two different production packages.

	CDs	Tapes	Mastering	Artwork	Total Cost
Package #1	300	400	✓	✓	$2080
Package #2	500	600	✓	✓	$3120

Both companies charge $100 to master your original recording and $240 to create cover artwork. Find the average production cost of each CD and cassette tape.

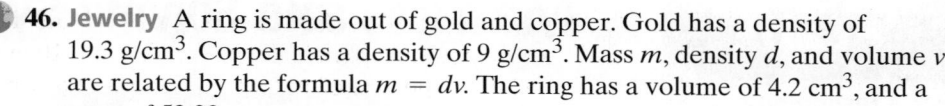

46. Jewelry A ring is made out of gold and copper. Gold has a density of 19.3 g/cm³. Copper has a density of 9 g/cm³. Mass m, density d, and volume v are related by the formula $m = dv$. The ring has a volume of 4.2 cm³, and a mass of 52.22 g.

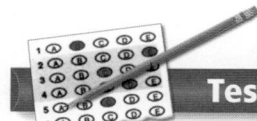

Vocabulary Tip

Density is the ratio of mass to volume.
Since $d = \frac{m}{v}$, $m = dv$.

Let a = volume of gold.   mass of gold = $dv = 19.3a$
Let c = volume of copper.   mass of copper = $dv = 9c$

a. Solve the following system by elimination to find out how many grams of gold are in the ring.
$$a + c = 4.2$$
$$19.3a + 9c = 52.22$$

b. What is the percent of gold by mass?

Test Prep

NY REGENTS

Multiple Choice

47. Which of the following systems does NOT have the same solution as the system at the right?

$$7x - 4y = 5$$
$$6x + 7y = -11$$

A. $49x - 28y = 35$
$24x + 28y = -44$

B. $42x - 24y = 30$
$42x + 49y = -77$

C. $-14x + 8y = -10$
$12x + 14y = -22$

D. $21x + 12y = 15$
$-24x - 28y = 44$

48. Use the solution of the system below to find $x - y$.
$$4x - 2y = 11$$
$$3x - 4y = -6$$

F. 11.3   **G.** 0.1   **H.** −0.1   **J.** −11.3

Short Response

49. Solve the following system by elimination. Show your work.
$$y - x = 13$$
$$7y + x = 11$$

Extended Response

50. A trapezoid is formed by lines with the following equations.
$$2x + 4y = 16 \qquad x = 4 \qquad x = 0 \qquad y = 0$$
Find the area of the trapezoid.

Mixed Review

GO for Help

Lesson 7-2

Solve using substitution. Give the solutions in alphabetical order.

51. $y = 4x + 2$
$y = 6x - 10$

52. $p = q - 5$
$3p + q = 1$

53. $w + a = 4$
$w + 2a = 13$

Lesson 2-7

You have a bag with two red marbles, three blue marbles, and five green marbles. You choose a marble at random. Without replacing the marble, you choose a second marble. Find each probability.

54. $P(\text{red then green})$ **55.** $P(\text{two greens})$ **56.** $P(\text{blue then red})$

Lesson 1-6

Write and solve an equation to find the value of x.

57. $5, 7, 3, 4, 8, x$; mean 5   **58.** $1.2, 1.4, 1.5, 1.1, x$; mean 1.2

59. $10, 15, 9, 11, 8, x$; mean 10.5   **60.** $4, 1, 3, x$; mean 3.25

In Chapter 2, you learned how to add and subtract two matrices. You can also multiply matrices. You multiply the elements in a row of the first matrix by the corresponding elements in a column of the second matrix. Then you add the products.

Go Online
PHSchool.com
For: Graphing calculator procedures
Web Code: ate-2013

3 elements in a row → $\begin{bmatrix} 2 & 5 & -3 \\ 3 & 1 & 6 \end{bmatrix}\begin{bmatrix} 7 \\ 2 \\ 1 \end{bmatrix} = \begin{bmatrix} 2 \cdot 7 + 5 \cdot 2 + (-3 \cdot 1) \\ 3 \cdot 7 + 1 \cdot 2 + 6 \cdot 1 \end{bmatrix} = \begin{bmatrix} 21 \\ 29 \end{bmatrix}$

↑
3 elements in a column

The second matrix can have more than one column.

$$\begin{bmatrix} 2 & 5 & -3 \\ 3 & 1 & 6 \end{bmatrix}\begin{bmatrix} 7 & 3 \\ 2 & 9 \\ 1 & 2 \end{bmatrix} = \begin{bmatrix} 2 \cdot 7 + 5 \cdot 2 + (-3 \cdot 1) & 2 \cdot 3 + 5 \cdot 9 + (-3 \cdot 2) \\ 3 \cdot 7 + 1 \cdot 2 + 6 \cdot 1 & 3 \cdot 3 + 1 \cdot 9 + 6 \cdot 2 \end{bmatrix} = \begin{bmatrix} 21 & 45 \\ 29 & 30 \end{bmatrix}$$

first row and first column **first row and second column**

second row and first column **second row and second column**

You can use a graphing calculator to multiply matrices. To enter matrices, you must know their dimensions. A matrix with two rows and three columns is a 2×3 matrix.

1 ACTIVITY

Use a graphing calculator to find $A \times B$.

$$A = \begin{bmatrix} 2 & 5 & -3 \\ 3 & 1 & 6 \end{bmatrix} \qquad B = \begin{bmatrix} 7 & 3 \\ 2 & 9 \\ 1 & 2 \end{bmatrix}$$

Step 1 Use the **MATRX** feature. Edit the dimensions and enter the values of the elements. You have to quit the first matrix screen before using the **MATRX** feature to enter the second matrix.

Matrix A

```
Matrix [A]   2×3
[2          5         −3]
[3          1          6]



2,3=6
```

Matrix B

```
Matrix [B]   3×2
[7          3          ]
[2          9          ]
[1          2          ]



3, 2=2
```

Step 2 Use the **NAMES** list on the matrix screen. Select [A]. Then use the name list on the matrix screen to select [B]. Press ENTER. The product matrix will appear on the main screen.

```
[A] [B]

            [[21     45]
             [29     30]]
```

You can use matrices to solve systems of equations. Start with equations in standard form.

System of Equations   Matrices for the System

$$2x + 6y = 80$$
$$4x + 5y = -1$$

$$\begin{bmatrix} 2 & 6 \\ 4 & 5 \end{bmatrix}\begin{bmatrix} x \\ y \end{bmatrix} = \begin{bmatrix} 80 \\ -1 \end{bmatrix}$$

$$A \quad \cdot X \quad = \quad B$$

A is the matrix for the coefficients of the variables, X is a matrix for the variables, and B is a matrix for the constants. To solve the system you must use the inverse of A, which is A^{-1}. The product $A^{-1} \times B$ gives you X.

2 ACTIVITY

Solve the system at the right by using matrix multiplication.

$$A = \begin{bmatrix} 2 & 6 \\ 4 & 5 \end{bmatrix} \text{ and } B = \begin{bmatrix} 80 \\ -1 \end{bmatrix}$$

$$2x + 6y = 80$$
$$4x + 5y = -1$$

Step 1

Use the matrix feature. Edit the dimensions and enter the values of the elements for each matrix.

Step 3

Use the matrix feature. Select [B]. Then press [ENTER].

The values of matrix X will appear as shown at the right.

$\begin{bmatrix} -29 \\ [23] \end{bmatrix}$ corresponds to $\begin{bmatrix} x \\ y \end{bmatrix}$, so $x = -29$ and $y = 23$.

● The solution of the system is $(-29, 23)$.

Step 2

Use the matrix feature. Select [A]. Then press [x⁻¹]. [A]⁻¹ will appear on the main screen.

$[A]^{-1} [B]$

$\begin{bmatrix} -29 \\ [23] \end{bmatrix}$

EXERCISES

Find each product.

1. $[4 \quad 2 \quad 9]\begin{bmatrix} 3 \\ 1 \\ 7 \end{bmatrix}$

2. $\begin{bmatrix} 12 & 10 \\ 8 & -11 \end{bmatrix}\begin{bmatrix} 0 & 4 \\ 9 & -1 \end{bmatrix}$

3. $\begin{bmatrix} 44 & -12 \\ 27 & 35 \\ 25 & -16 \end{bmatrix}\begin{bmatrix} 21 & -41 \\ 25 & 17 \end{bmatrix}$

4. The table at the near right shows the number of three sizes of widgets made at a manufacturing plant on Monday and Tuesday. The table at the far right shows the production costs of each widget. Write each table as a matrix. Then multiply to find the total production cost each day.

Number of Widgets

Day	Size A	Size B	Size C
Monday	212	318	175
Tuesday	185	292	221

Cost of Widgets

Cost A	$.96
Cost B	$1.23
Cost C	$1.51

Solve each system using matrix multiplication.

5. $1x + 8y = 16$
$9x + 12y = 66$

6. $29x + 7y = 1012$
$8x - 25y = 737$

7. $76x + 18y = 86$
$189x + 47y = 132$

7-4

Applications of Linear Systems

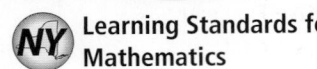
Learning Standards for Mathematics

A.A.7 Analyze and solve verbal problems whose solution requires solving systems of linear equations in two variables.

A.G.7 Graph and solve systems of linear equations and inequalities with rational coefficients in two variables.

☑ **Check Skills You'll Need**

 **for Help** Lesson 3-6

1. Two trains run on parallel tracks. The first train leaves a city $\frac{1}{2}$ hour before the second train. The first train travels at 55 mi/h. The second train travels at 65 mi/h. How long does it take for the second train to pass the first train?

2. Carl drives to the beach at an average speed of 50 mi/h. He returns home on the same road at an average speed of 55 mi/h. The trip home takes 30 min less. What is the distance from his home to the beach?

---

1   Writing Systems of Linear Equations

Below is a summary of the methods you have used to solve systems of equations. You must choose a method before you solve a word problem.

 Key Concepts

Summary	Methods for Solving Systems of Linear Equations
Graphing	Use graphing for solving systems that are easily graphed. If the point of intersection does not have integers for coordinates, find the exact solution by using one of the methods below or by using a graphing calculator.
Substitution	Use substitution for solving systems when one variable has a coefficient of 1 or −1.
Elimination	Use elimination for solving any system.

1 EXAMPLE   Real-World 🌐 Problem Solving

Metallurgy A metalworker has some ingots of metal alloy that are 20% copper and others that are 60% copper. How many kilograms of each type of ingot should the metalworker combine to create 80 kg of a 52% copper alloy?

Define   Let g = the mass of the 20% alloy.
          Let h = the mass of the 60% alloy.

Relate   mass of alloys   mass of copper

Write $g + h = 80$ $0.2g + 0.6h = 0.52(80)$

Solve using substitution.

Step 1   Choose one of the equations and solve for a variable.

     $g + h = 80$ **Solve for g.**

         $g = 80 - h$ **Subtract h from each side.**

Real-World 🌐 Connection

The melting point of copper is 1083°C.

Step 2 Find h.

$$0.2g + 0.6h = 0.52(80)$$
$$0.2(80 - h) + 0.6h = 0.52(80) \quad \textbf{Substitute } 80 - h \textbf{ for } g. \textbf{ Use parentheses.}$$
$$16 - 0.2h + 0.6h = 0.52(80) \quad \textbf{Use the Distributive Property.}$$
$$16 + 0.4h = 41.6 \quad \textbf{Simplify. Then solve for } h.$$
$$0.4h = 25.6$$
$$h = 64$$

Step 3 Find g. Substitute 64 for h in either equation.

$$g = 80 - 64$$
$$g = 16$$

To make 80 kg of 52% copper alloy, you need 16 kg of 20% copper alloy and 64 kg of 60% copper alloy.

✓ Quick Check

1 Suppose you combine ingots of 25% copper alloy and 50% copper alloy to create 40 kg of 45% copper alloy. How many kilograms of each do you need?

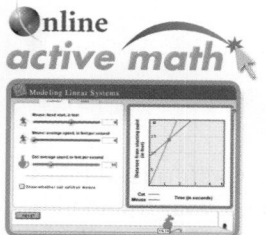

Online active math

For: Linear System Activity
Use: Interactive Textbook, 7-4

When starting a business, people want to know the *break-even point*, the point at which their income equals their expenses. The graph at the right shows the break-even point for one business.

☐ Lose money ▨ Make money

Notice that the values of y on the red line represent dollars spent on expenses, and the values of y on the blue line represent dollars received as income. So y is used to represent both expenses and income.

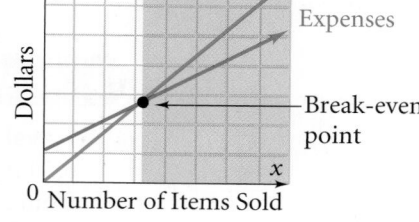

2 EXAMPLE **Finding a Break-Even Point**

Publishing Suppose a model airplane club publishes a newsletter. Expenses are $.90 for printing and mailing each copy, plus $600 total for research and writing. The price of the newsletter is $1.50 per copy. How many copies of the newsletter must the club sell to break even?

Define Let $x =$ the number of copies.
Let $y =$ the amount of dollars of expenses or income.

Relate Expenses are printing costs   Income is price
plus research and writing.   times copies sold.

Write $y = 0.9x + 600$ $y = 1.5x$

Choose a method to solve this system. Use substitution since it is easy to substitute for y with these equations.

$$y = 0.9x + 600 \quad \textbf{Start with one equation.}$$
$$1.5x = 0.9x + 600 \quad \textbf{Substitute } 1.5x \textbf{ for } y.$$
$$0.6x = 600 \quad \textbf{Solve for } x.$$
$$x = 1000$$

To break even, the model airplane club must sell 1000 copies.

❷ Suppose an antique car club publishes a newsletter. Expenses are $.35 for printing and mailing each copy, plus $770 total for research and writing. The price of the newsletter is $.55 per copy. How many copies of the newsletter must the club sell to break even?

In Lesson 3-6, you modeled rate-time-distance problems using one variable. You can also model rate-time-distance problems using two variables. The steady west-to-east winds across the United States act as tail winds for planes traveling from west to east. The tail winds increase a plane's groundspeed. For planes traveling east to west, the head winds decrease a plane's groundspeed.

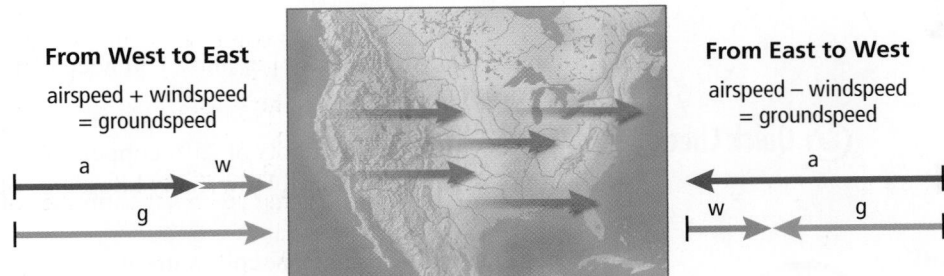

From West to East

airspeed + windspeed = groundspeed

a   w

g

From East to West

airspeed − windspeed = groundspeed

a

w   g

3 EXAMPLE Real-World 🌐 Problem Solving

Travel Suppose you fly from Miami, Florida, to San Francisco, California. It takes 6.5 hours to fly 2600 miles against a head wind. At the same time, your friend flies from San Francisco to Miami. Her plane travels at the same average airspeed, but her flight only takes 5.2 hours. Find the average airspeed of the planes. Find the average wind speed.

Define Let A = the airspeed. Let W = the wind speed.

Relate with tail wind   with head wind
(rate)(time) = distance   (rate)(time) = distance
$(A + W)$(time) = distance $(A - W)$(time) = distance

Write $(A + W)5.2 = 2600$ $(A - W)6.5 = 2600$

Step 1 Divide to get the variables of each equation with coefficients of 1 or −1.
$(A + W)5.2 = 2600 \rightarrow A + W = 500$ **Divide each side by 5.2.**
$(A - W)6.5 = 2600 \rightarrow A - W = 400$ **Divide each side by 6.5.**

Step 2 Eliminate W.
$A + W = 500$
$\underline{A - W = 400}$ **Add the equations to eliminate W.**
$2A + 0 = 900$

Step 3 Solve for A.
$A = 450$ **Divide each side by 2.**

Step 4 Solve for W using either of the original equations.
$A + W = 500$ **Use the first equation.**
$450 + W = 500$ **Substitute 450 for A.**
$W = 50$ **Solve for W.**

● The average airspeed of the planes is 450 mi/h. The average wind speed is 50 mi/h.

 **Quick Check** **3** A plane takes about 6 hours to fly 2400 miles from New York City to Seattle, Washington. At the same time, your friend flies from Seattle to New York City. His plane travels with the same average airspeed, but his flight takes 5 hours. Find the average airspeed of the planes. Find the average wind speed.

EXERCISES

For more exercises, see *Extra Skill and Word Problem Practice.*

Practice and Problem Solving

 Practice by Example

Example 1
(page 396)

 **for Help**

1. Tyrel and Dalia bought some pens and pencils. Tyrel bought 4 pens and 5 pencils, which cost him $6.71. Dalia bought 5 pens and 3 pencils, which cost her $7.12. Let a equal the price of a pen. Let b equal the price of a pencil.
   a. Write an equation that relates the number of pens and pencils Tyrel bought to the amount he paid for them.
   b. Write an equation that relates the number of pens and pencils Dalia bought to the amount she paid for them.
   c. Solve the system you wrote for parts (a) and (b) to find the price of a pen and the price of a pencil.

2. Suppose you have just enough money, in coins, to pay for a loaf of bread priced at $1.95. You have 12 coins, all quarter and dimes. Let q equal the number of quarters and d equal the number of dimes. Which system models the given information?
   A. $q + d = 12$
   $\quad\;\, q + d = 1.95$
   B. $25q + 10d = 195$
   $\quad\;\, q + 12 = d$
   C. $10q + 25d = 12$
   $\quad\;\;\, q + d = 1.95$
   D. $q + d = 12$
   $\quad\;\, 25q + 10d = 195$

3. Suppose you want to combine two types of fruit drink to create 24 kilograms of a drink that will be 5% sugar by weight. Fruit drink A is 4% sugar by weight, and fruit drink B is 8% sugar by weight.
   a. Copy and complete the table below.

	Fruit Drink A 4% Sugar	Fruit Drink B 8% Sugar	Mixed Fruit Drink 5% Sugar
Fruit Drink (kg)	■	■	■
Sugar (kg)	■	■	■

   b. Write a system of equations that relates the amounts of fruit drink A and fruit drink B to the total amount of drink needed and to the total amount of sugar needed.
   c. Solve the system to find how much of each type of fruit drink you need to use.

4. You have $22 in your bank account and deposit $11.50 each week. At the same time your cousin has $218 but is withdrawing $13 each week.
   a. When will your accounts have the same balance?
   b. How much money will each of you have after 12 weeks?

Example 2
(page 397)

5. **Business** Suppose you invest $10,410 in equipment to manufacture a new board game. Each game costs $2.65 to manufacture and sells for $20. How many games must you make and sell before your business breaks even?

Lesson 7-4 Applications of Linear Systems **399**

6. Business Several students decide to start a T-shirt company. After initial expenses of $280, they purchase each T-shirt wholesale for $3.99. They sell each T-shirt for $10.99. How many must they sell to break even?

Example 3
(page 398)

7. Travel A family is canoeing downstream (with the current). Their speed relative to the banks of the river averages 2.75 mi/h. During the return trip, they paddle upstream (against the current), averaging 1.5 mi/h relative to the riverbank.
   a. Write an equation for the rate of the canoe downstream.
   b. Write an equation for the rate of the canoe upstream.
   c. Solve the system to find the family's paddling speed in still water.
   d. Find the speed of the current of the river.

8. Travel John flies from Atlanta, Georgia, to San Francisco, California. It takes 5.6 hours to travel 2100 miles against the head wind. At the same time Debby flies from San Francisco to Atlanta. Her plane travels with the same average airspeed but, with a tail wind, her flight takes only 4.8 hours.
   a. Write a system of equations that relates time, airspeed, and wind speed to distance for each traveler.
   b. Solve the system to find the airspeed.
   c. Find the wind speed.

B **Apply Your Skills**

Open-Ended Without solving, what method would you choose to solve each system: *graphing*, *substitution*, or *elimination*? Explain your reasoning.

9. $4s - 3t = 8$
   $t = -2s - 1$

10. $y = 3x - 1$
   $y = 4x$

11. $3m - 4n = 1$
   $3m - 2n = -1$

12. $y = -2x$
   $y = -\frac{1}{2}x + 3$

13. $2x - y = 4$
   $x + 3y = 16$

14. $u = 4v$
   $3u - 2v = 7$

15. Chemistry A piece of glass with an initial temperature of 99°C is cooled at a rate of 3.5 degrees Celsius per minute (°C/min). At the same time, a piece of copper with an initial temperature of 0°C is heated at a rate of 2.5°C/min. Let m = the number of minutes, and t = the temperature in degrees Celsius after m minutes.
   a. Write a system of equations that relates the temperature t of each material to the time m. Solve the system.
   b. Writing Explain what the solution means in this situation.

16. Geometry The perimeter of the rectangle is 34 cm. The perimeter of the triangle is 30 cm. Find the values of m and n.

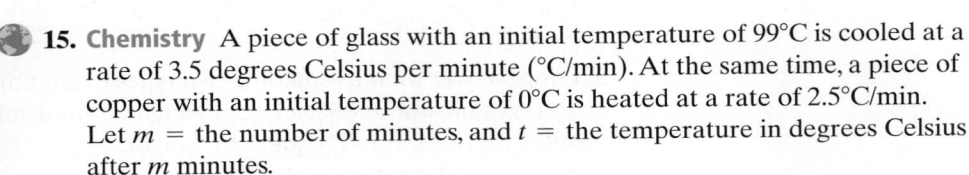

Real-World **Connection**

Glass can be drawn into optical fibers 16 km long. One fiber can carry 20 times as many phone calls as 500 copper wires.

17. Open-Ended Write a problem for the total of two types of coins. Then solve the problem.

18. Sales A garden supply store sells two types of lawn mowers. Total sales of mowers for the year were $8379.70. The total number of mowers sold was 30. The small mower costs $249.99. The large mower costs $329.99. Find the number sold of each type of mower.

For a guide to solving Exercise 18, see p. 403.

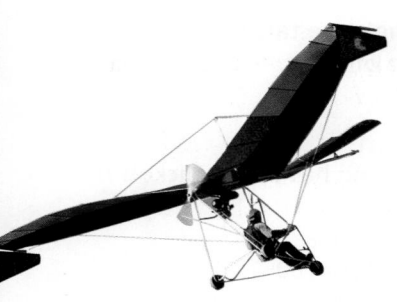

19. Aviation Suppose you are flying an ultralight aircraft like the one pictured at the left. You fly to a nearby town, 18 miles away. With a tail wind, the trip takes $\frac{1}{3}$ hour. Your return flight with a head wind takes $\frac{3}{5}$ hour.
   a. Find the average airspeed of the ultralight aircraft.
   b. Find the average wind speed.

20. Suppose the ratio of girls to boys in your school is 19 : 17. There are 1908 students altogether.
   a. Solve the proportion $\frac{g}{b} = \frac{19}{17}$ for g.
   b. Write and solve the system of equations to find the total number of boys b and girls g.

Real-World Connection

Ultralight aircraft like the one pictured above can weigh less than 400 lb.

21. Consumer Decisions Suppose you are trying to decide whether to buy ski equipment. Typically, it costs you $60 a day to rent ski equipment and buy a lift ticket. You can buy ski equipment for about $400. A lift ticket alone costs $35 for one day.
   a. Find the break-even point.
   b. Critical Thinking If you expect to ski five days a year, should you buy the ski equipment? Explain.

GO Online
Homework Video Tutor
Visit: PHSchool.com
Web Code: ate-0704

22. Geometry Find the values of x and y.

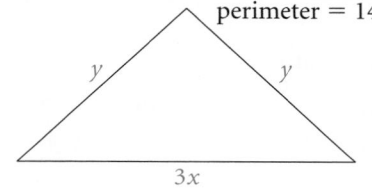

perimeter = 14

y y

$3x$

$\frac{5}{4}y$

x

$\frac{5}{4}y$

perimeter = 12

C Challenge

23. You can represent the value of any two-digit number with the expression $10a + b$, where a is the tens' place digit and b is the ones' place digit. If a is 5 and b is 7, then the value of the number is $10(5) + 7$, or 57.
   Use a system of equations to find the two-digit number described below.
   • The ones' place digit is one more than twice the tens' place digit.
   • The value of the number is two more than five times the ones' place digit.

24. Sales An artist sells original hand-painted greeting cards. He makes $2.50 profit on a small card and $4.00 profit on a large card. He generally sells 5 large cards for every 2 small cards. He wants a profit of $10,000 from large and small cards this year.
   a. Find the quantity of each card the artist needs to sell to reach his goal.
   b. The artist can create a card every 12 minutes. How many hours will he need to make enough to reach his profit target if he sells them all?
   c. What is the artist's hourly rate of pay?

NY REGENTS

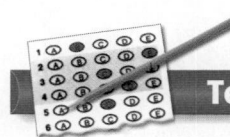

Test Prep

Multiple Choice

25. Which system describes the following situation: The sum of two numbers is 20. The difference between three times the larger and twice the smaller is 40.
   A. $x + y = 20$
       $3x + 2y = 40$
   B. $x - y = 20$
       $3x - 2y = 40$
   C. $x + y = 20$
       $3x - 2y = 40$
   D. $x - y = 20$
       $3x + 2y = 40$

26. The federal tax on a \$12,000 salary was 8 times the state tax. If the combined taxes were \$2700, find the state's share of taxes.
   F. \$400   G. \$150
   H. \$300   J. \$350

27. Which system describes the following situation? Craig has 80¢ in nickels n and dimes d. He has four more nickels than dimes.
   A. $d + n = 4$ B. $n - d = 4$
      $10d + 5n = 80$ $10d + 5n = 80$
   C. $d - n = 4$ D. $d + n = 4$
      $10d + 5n = 80$ $10d - 5n = 80$

Short Response

28. A large group of students wants to go to the movies. If the students take 3 vans and 1 car, they can transport 22 people. If they take 2 vans and 4 cars, they can transport 28 people. Write and solve a system of equations to find the number of people that can be transported in a van. Show your work.

Mixed Review

Lesson 7-3   **Solve by elimination.**

29. $2x + 5y = 13$ **30.** $4x + 2y = -10$ **31.** $7x + 6y = 30$
      $3x - 5y = 7$ $-2x + 3y = 33$ $9x - 8y = 15$

Lesson 6-1   **Find the slope of the line that passes through each pair of points.**

32. $(2, 4), (6, 10)$ **33.** $(-3, 1), (10, 14)$ **34.** $(8, -11), (5, -12)$

35. $(1.2, 7), (4.6, 0.2)$ **36.** $\left(5, -\frac{1}{2}\right), \left(-6, 3\frac{1}{2}\right)$ **37.** $(8, 0), (8, 5)$

Lesson 4-5   **Solve each inequality and graph the solutions.**

38. $6 < y < 10$ **39.** $-8 < n \le 3$ **40.** $2 < k + 1 < 7$

41. $4 \le 4p \le 16$ **42.** $-13 < 3c + 2 \le 17$ **43.** $21 > 5w - 4 > 1$

Algebra at Work

......................**Businessperson**

Restriction Polygon

— Advertising
— Raw Materials
— Transportation
— Packaging
— Equipment
— Labor

Some of the goals of a business are to minimize costs and maximize profits. People in business use systems of linear inequalities to analyze data in order to achieve these goals.

The illustration lists some of the variables involved in operating a small manufacturing company. To solve a problem, a businessperson must identify the variables and restrictions, and then search for the best of many possible solutions.

Go Online
PHSchool.com   **For:** Information about a career in business
Web Code: atb-2031

Understanding Word Problems Read the exercise below and then follow along with what Bill thinks and writes. Check your understanding by solving the exercise at the bottom of the page.

A garden supply store sells two types of lawn mowers. Total sales of mowers for the year were $8379.70. The total number of mowers sold was 30. The small mowers cost $249.99. The large mowers cost $329.99. Find the number of each type of mower sold.

What Bill Thinks

I'll read the problem and write down the important information.

Where should I start? Well, it's always helpful to write sentences based on the information I'm given. Total sales include the sales for both the small mowers and the large mowers.

Total number of mowers is the number of small mowers *plus* the number of large mowers.

Now I'll define some variables. The problem asks for the number of small mowers and the number of large mowers. I'll use 2 variables.

Now I can write 2 equations.

Since the first equation has large numbers, it's probably easier to rewrite the second equation and substitute into the first equation. I'll then solve for *w* and *s*.

I'll write my answer in a sentence.

What Bill Writes

Total sales = $8379.70
Total number of mowers = 30
Small mowers cost $249.99.
Large mowers cost $329.99.

Total sales = sales from small mowers + sales from large mowers

Total number of mowers = number of small mowers + number of large mowers

Number of small mowers = s
Number of large mowers = w

Total sales: $8379.70 = 249.99s + 329.99w$
Total number: $30 = s + w$

$s = 30 - w$
$8379.70 = (249.99)(30 - w) + 329.99w$
$8379.70 = 7499.70 + 80w$
$880 = 80w$
$11 = w; s = 30 - 11 = 19$

The store sold 19 small mowers and 11 large mowers.

EXERCISE

A nursery sells small apple trees for $19.99 and large apple trees for $35.99. Total sales for the year were $1907.27. The total number of apple trees sold was 73. Find the number of each type of apple tree sold.

Linear Inequalities

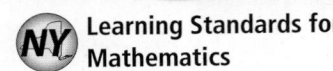
NY Learning Standards for Mathematics

A.A.6 Analyze and solve verbal problems whose solution requires solving a linear equation in one variable or linear inequality in one variable.

A.A.21 Determine whether a given value is a solution to a given linear equation in one variable or linear inequality in one variable.

A.G.6 Graph linear inequalities.

✓ **Check Skills You'll Need**

GO **for Help** Lessons 4-1 and 6-2

Describe each statement as *always*, *sometimes*, or *never* true.

1. $-3 > -2$ **2.** $8 \leq 8$ **3.** $4n \geq n$

Write each equation in slope-intercept form.

4. $2x - 3y = 9$ **5.** $y + 3x = 6$ **6.** $4y - 3x = 1$

◀)) **New Vocabulary**   • linear inequality   • solutions of an inequality

---

1 Graphing Linear Inequalities

Activity: Graphing Inequalities

1. Graph $y = x + 4$ on a coordinate plane.

2. Test three points that lie above the graph of $y = x + 4$. Substitute the coordinates of each of the points for (x, y) in the inequality $y > x + 4$. If the results are true statements, mark the points on your graph.

3. Test three points that lie below the graph of $y = x + 4$. Substitute the coordinates of each of the points for (x, y) in the inequality $y > x + 4$. If the results are true statements, mark the points on your graph.

4. Critical Thinking To graph $y > x + 4$, would you choose points above or below $y = x + 4$?

5. Determine whether you would graph points above or below the graph of $y = x - 2$ to graph the inequality $y < x - 2$.

Just as you have used inequalities to describe graphs on a number line, you can use inequalities to describe regions of a coordinate plane.

GO **for Help**

To review graphing inequalities in one variable see p. 20.

Number line
$x < 1$

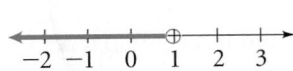

Coordinate plane
$x < 1$

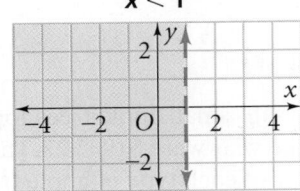

A **linear inequality** describes a region of the coordinate plane that has a boundary line. The **solutions of an inequality** are the coordinates of the points that make the inequality true.

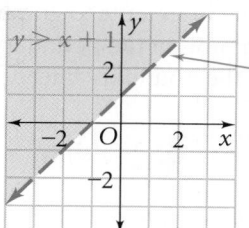

Each point on a *dashed* boundary line is not a solution.

Each point on a *solid* boundary line is a solution.

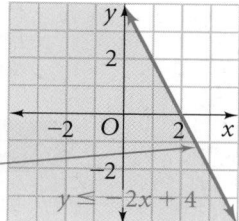

As you can see in the graphs above, you can tell from an inequality whether to shade above or below the boundary line. For an inequality written in the form of $y <$ or $y \leq$, shade below the boundary line. For an inequality written in the form of $y >$ or $y \geq$, shade above the boundary line.

Online active math

For: Linear Inequality Activity
Use: Interactive Textbook, 7-5

1 EXAMPLE Graphing an Inequality

Graph $y < 2x + 3$.

First graph the boundary line $y = 2x + 3$.

The coordinates of points on the boundary line do not make the inequality true. So, use a dashed line.

Shade below the boundary line.

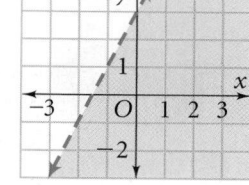

Check   The point $(0, 0)$ is in the region of the graph of the inequality. See if $(0, 0)$ satisfies the inequality.

$y < 2x + 3$

$0 < 2(0) + 3$ **Substitute (0, 0) for (x, y).**

$0 < 3$ ✓

✓ Quick Check ❶ Graph $y \geq 3x - 1$.

In order to tell whether you shade above or below a boundary line, you may need to write the inequality in slope-intercept form.

2 EXAMPLE Rewriting to Graph an Inequality

Graph $3x - 5y \leq 10$.

Solve $3x - 5y \leq 10$ for y.

$3x - 5y \leq 10$

$\quad -5y \leq -3x + 10$ **Subtract 3x from each side.**

$\quad\quad y \geq \frac{3}{5}x - 2$ **Divide each side by −5. Reverse the inequality symbol.**

Graph $y = \frac{3}{5}x - 2$.

The coordinates of points on the boundary line make the inequality true. So, use a solid line.

Since $y \geq \frac{3}{5}x - 2$, shade above the boundary line.

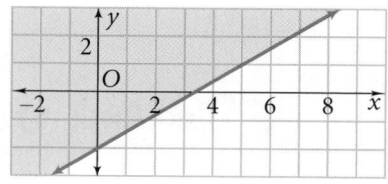

✓ Quick Check ❷ Graph $6x + 8y \geq 12$.

Many situations are modeled by inequalities that have a boundary line of the form $Ax + By = C$. You can use the intercepts to graph the boundary line of the inequality. Choose a test point not on the boundary line to determine whether the solutions are above or below the boundary line.

3 EXAMPLE   Real-World 🌐 Problem Solving

Budget Suppose your budget for a party allows you to spend no more than $12 on peanuts and cashews. Peanuts cost $2/lb and cashews cost $4/lb. Find three possible combinations of peanuts and cashews you can buy.

Relate | cost of peanuts | plus | cost of cashews | is less than or equal to | total budget |

Define Let x = the number of pounds of peanuts.

Let y = the number of pounds of cashews.

Write $2x$ + $4y$ $\leq$ 12

Graph $2x + 4y = 12$ by graphing the intercepts, $(6, 0)$ and $(0, 3)$.

The coordinates of points on the boundary line make the inequality true. So, use a solid line.

Graph only in Quadrant I, since you cannot buy a negative amount of peanuts or cashews.

Test the point $(1, 1)$.

$$2x + 4y \leq 12$$
$2(1) + 4(1) \leq 12$ **Substitute (1, 1) for (x, y).**
$6 \leq 12$ **Since the inequality is true, (1, 1) is a solution.**

Shade the region containing $(1, 1)$. The graph below shows all the possible solutions of the problem.

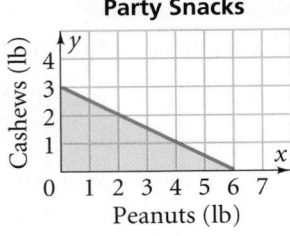

Party Snacks

Since the boundary line is included in the graph, the intercepts are also solutions of the inequality. The solution $(2, 2)$ means that if you buy 2 lb of peanuts, you can buy 2 lb of cashews. Three solutions are $(2, 2)$, $(3, 1)$, and $(1, 2)$.

✓ **Quick Check** ③ **Cooking** Suppose you plan to spend no more than $24 on meat for a cookout. At your local market, hamburger costs $3.00/lb and chicken wings cost $2.40/lb. Find three possible combinations of hamburger and chicken wings you can buy.

Real-World 🌐 Connection

One ounce of peanuts has 9 g of protein. One ounce of cashews has 5.4 g of protein.

Practice and Problem Solving

 Practice by Example

Example 1
(page 405)

GO for Help

Determine whether point P is a solution of the linear inequality.

1. $y \le -2x + 1$; $P(2, 2)$ **2.** $x < 2$; $P(1, 0)$ **3.** $y \ge 3x - 2$; $P(0, 0)$

4. $y > x - 1$; $P(0, 1)$ **5.** $y \ge -\frac{2}{5}x + 4$; $P(0, 0)$ **6.** $y > \frac{5}{3}x - 4$; $P(0, 1)$

Choose the linear inequality that describes each graph.

7.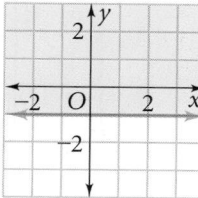
   A. $y \ge -1$
   B. $y \le -1$

8.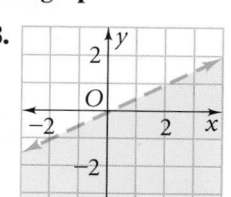
   A. $y > \frac{1}{2}x$
   B. $y < \frac{1}{2}x$

9.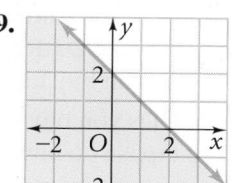
   A. $y \ge -x + 2$
   B. $y \le -x + 2$

10.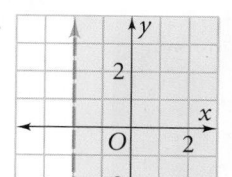
   A. $x > -2$
   B. $x < -2$

Graph each linear inequality.

11. $y \le \frac{1}{4}x - 1$ **12.** $y \ge \frac{1}{4}x - 1$ **13.** $y < -4x - 1$ **14.** $y \ge 4x - 1$

15. $y < 5x - 5$ **16.** $y \le \frac{2}{5}x - 3$ **17.** $y \le -3x$ **18.** $y \ge -\frac{1}{2}x$

Example 2
(page 405)

Write each linear inequality in slope-intercept form. Then graph the inequality.

19. $2x - 3y \ge 7$ **20.** $5x - 3y \le 6$ **21.** $4x - 6y \ge 16$ **22.** $-4y - 6x > 8$

Example 3
(page 406)

23. Budget Suppose you are shopping for crepe paper to decorate the school gym for a dance. Gold crepe paper costs $5 per roll, and blue crepe paper costs $3 per roll. Your budget allows you to spend at most $48 for crepe paper. How many rolls of gold and blue crepe paper can you buy without exceeding your budget?

     Let x = the number of rolls of blue crepe paper.
     Let y = the number of rolls of gold crepe paper.

   a. Write a linear inequality that describes the situation.
   b. Graph the linear inequality.
   c. Write three possible solutions to the problem.
   d. Critical Thinking The point $(-2, 5)$ is a solution of the inequality. Is it a solution of the problem? Explain.

 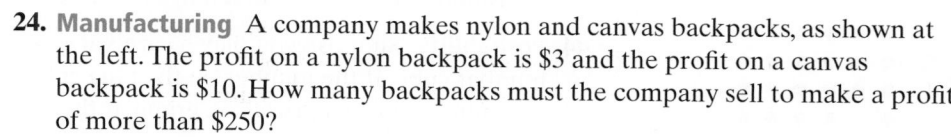
24. Manufacturing A company makes nylon and canvas backpacks, as shown at the left. The profit on a nylon backpack is $3 and the profit on a canvas backpack is $10. How many backpacks must the company sell to make a profit of more than $250?

   a. Write a linear inequality that describes the situation.
   b. Graph the linear inequality.
   c. Write three possible solutions to the problem.
   d. Critical Thinking Which values are reasonable for the domain and for the range? Explain.

Apply Your Skills

Graph each linear inequality.

25. $y \le \frac{2}{5}x + 2$ **26.** $y \ge -\frac{2}{5}x + 2$ **27.** $4x - 5y \le 10$ **28.** $4x + 5y \le 10$

29. $4y < 6x + 2$ **30.** $2x + 3y \le 6$ **31.** $4x - 4y \le 8$ **32.** $y - 2x < 2$

33. Writing Explain how you can tell from a linear inequality whether you will shade above or below the graph of the boundary line.

Write the linear inequality shown in each graph.

34. 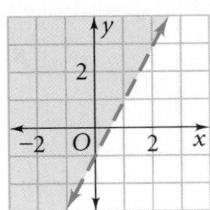   **35.** 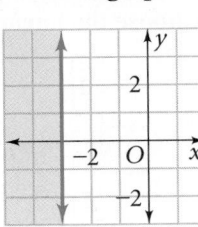   **36.**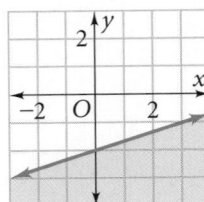

37. Budget Suppose you work at a local radio station. You are in charge of a $180 budget for new tapes and CDs. Record companies will give you 21 promotional (free) CDs. You can buy tapes for $8 and CDs for $12.

     Let x = the number of CDs you buy.
     Let y = the number of tapes you buy.

     a. Write an inequality that shows the number of tapes and CDs you can buy.
     b. Graph the inequality.
     c. Is $(8, 9)$ a solution of the inequality? Explain what the solution means.
     d. If you buy only tapes and you buy as many as possible, how many new recordings will the station get?

Write the linear inequality described. Then graph the inequality.

38. x is positive.

39. y is negative.

40. y is not negative.

41. x is less than y.

42. Error Analysis Jan's graph of the inequality $4x + 6y > 12$ is shown below. What is wrong with the graph?

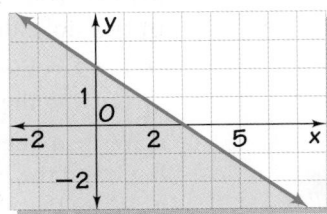

43. Critical Thinking Write an inequality that describes the entire part of the coordinate plane *not* included in the solution of $y \ge x + 2$.

44. Probability Suppose you play a carnival game. You toss one blue and one red number cube. If the number on the blue cube is greater than the number on the red cube, you win a prize. The graph at the left shows all the possible outcomes of tossing the cubes.
     a. Copy and shade the graph to show the winning outcomes.
     b. Write an inequality that describes the shaded region.
     c. What is the probability that you will win a prize?

Real-World Connection

In 2000, there were about 12,900 licensed radio stations in the United States.

Comparing Cubes

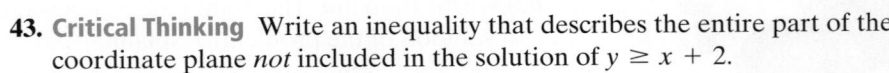
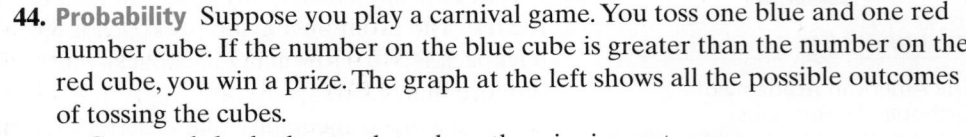

45. Geometry You want to fence a rectangular area of your yard for a garden. You plan to use no more than 50 ft of fencing.
    a. Write and graph a linear inequality that describes this situation.
    b. Open-Ended What are two possible sizes for a square garden?
    c. Can you make the garden 12 ft by 15 ft? Justify your answer.

 Challenge

For Exercises 46–47, write the inequality that has the solution described.

46. The points $(0, -3)$ and $(8, 5)$ lie on the boundary line, but neither point is a solution. The point $(1, 1)$ is not a solution.

47. The points $(7, 12)$ and $(-3, -8)$ lie on the boundary line, and each point is a solution. The point $(1, 1)$ is also a solution.

48. a. Open-Ended Write and graph an inequality in the form $Ax + By > C$, where A, B, and C are all positive.
    b. Write and graph an inequality in the form $Ax + By < C$, where A, B, and C are all positive.
    c. Reasoning Both inequalities are in standard form. Make a conjecture about the inequality symbol and the region shaded.
    d. Would your conjecture in part (c) be different if B were negative?

49. a. Is the point $(4, 5)$ a solution of the inequality $y > x - 1$?
    b. Is the point $(4, 5)$ a solution of the inequality $y < 3x$?
    c. Find one other point that is a solution of both inequalities.
    d. Draw a graph that shows all the points that are solutions of both inequalities.

 **NY REGENTS**

Test Prep

Multiple Choice

50. Which of the following is true of the graph of $y \geq -x + 1$?
    A. The line is solid, and the shading is above the line.
    B. The line is dashed, and the shading is above the line.
    C. The line is solid, and the shading is below the line.
    D. The line is dashed, and the shading is below the line.

51. Which linear inequality describes the graph at the right?
    F. $y < -\frac{2}{3}x - 4$ **G.** $y > -\frac{2}{3}x - 4$
    H. $y \leq -\frac{2}{3}x - 4$ **J.** $y \geq -\frac{2}{3}x - 4$

52. The graph of which of the following is shaded above the line?
    A. $x + y < 9$ **B.** $x + y < -9$
    C. $y - x < 9$ **D.** $x - y < 9$

53. Which inequality below models the following situation?

You want to spend less than \$20 on asparagus and bananas. Asparagus is \$3.00 per pound and bananas are \$.50 per pound. Let a represent the weight of the asparagus and b represent the weight of the bananas.

    F. $3a + 0.5b < 20$ **G.** $3a + 0.5b > 20$
    H. $3a + 0.5b \leq 20$ **J.** $3a + 0.5b \geq 20$

Short Response

54. Explain how to graph $y \leq 3x - 4$. Then graph the inequality.

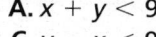

Lesson 7-4

 For Exercises 55–56, define the variables and write a system of equations for each situation. Solve by any method.

55. Business Suppose you invest $12,000 in equipment to manufacture a new board game. Each game costs $2.50 to manufacture and sells for $18. How many games must you make and sell for your business to break even?

56. Suppose you are canoeing along a river with a steady current. Your average speed upstream is 2.5 mi/h. On the return trip you paddle with the current, and your average speed is 4 mi/h. Find the average speed of the current and your average speed if you were paddling in still water.

Lesson 5-7

Find the common difference of each arithmetic sequence.

57. $-8, -3, 2, \ldots$ **58.** $4, 11, 18, \ldots$ **59.** $13, 24, 35, \ldots$ **60.** $11, 5, -1, \ldots$

Find the second and fourth terms of each sequence.

61. $A(n) = 3 + (n - 1)(5)$ **62.** $A(n) = -9 + (n - 1)(2.3)$

Lesson 3-4

Solve each proportion.

63. $\frac{3}{4} = \frac{m}{16}$ **64.** $\frac{6}{7} = \frac{24}{g}$ **65.** $\frac{4}{w} = \frac{8}{22}$ **66.** $\frac{9}{10} = \frac{15}{a}$

67. $\frac{x + 1}{3} = \frac{2}{9}$ **68.** $\frac{n - 2}{5} = \frac{6}{15}$ **69.** $\frac{8}{r + 1} = \frac{4}{7}$ **70.** $\frac{9}{x + 3} = \frac{18}{19}$

 **Checkpoint Quiz 2**   **Lessons 7-3 through 7-5**

For Exercises 1–5, solve each system using elimination.

1. $2x + 5y = 2$
$ 3x - 5y = 53$

2. $-8x - 3y = 69$
$ 8x + 7y = -65$

3. $4x + 2y = 34$
$ 10x - 4y = -5$

4. $11x - 13y = 89$
$ -11x + 13y = 107$

5. $3x + 6y = 42$
$ -7x + 8y = -109$

6. You have a total of 21 coins, all nickels and dimes. The total value is $1.70. Write and solve a system of equations to find the number of dimes d and the number of nickels n that you have.

7. Business Suppose you start an ice cream business. You buy a freezer for $200. It costs you $.35 to make each single-scoop ice cream cone. You sell each cone for $1.20. Write and solve a system of equations to find the break-even point for your business.

8. To go to a campsite 12 miles away, you paddle a canoe against the current of a river for 4 hours. During your return trip you paddle with the current, and you travel the same distance in 3 hours. Write and solve a system of equations to find your paddling speed in still water. Find the speed of the current of the river.

Graph each inequality.

9. $y \geq 2x - 4$ **10.** $3x + 4y < 18$

Systems of Linear Inequalities

NY Learning Standards for Mathematics

A.A.40 Determine whether a given point is in the solution set of a system of linear inequalities.

A.G.7 Graph and solve systems of linear equations and inequalities with rational coefficients in two variables.

✓ **Check Skills You'll Need**

GO **for Help** Lessons 7-1 and 7-5

Solve each system by graphing.

1. $y = 3x - 6$
$y = -x + 2$

2. $y = -\frac{1}{2}x + 4$
$y = -\frac{1}{2}x + 3$

3. $x + y = 4$
$2x - y = 8$

Graph each inequality.

4. $y > 5$

5. $y \leq \frac{2}{3}x - 1$

6. $4x - 8y \geq 4$

🔊 **New Vocabulary** • system of linear inequalities
• solution of a system of linear inequalities

---

1 **Solving Systems of Linear Inequalities by Graphing**

Two or more linear inequalities together form a **system of linear inequalities.** The system below describes the lavender-shaded region of the graph. Notice that there are two boundary lines.

System of Linear Inequalities

$x \geq 3$
$y < -2$

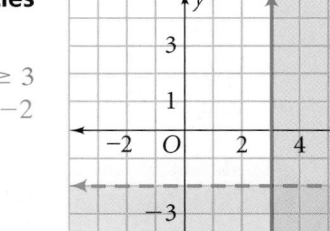

You can describe all the points of a quadrant with a system of linear inequalities.

$x < 0$
$y > 0$

$x > 0$
$y > 0$

$x < 0$
$y < 0$

$x > 0$
$y < 0$

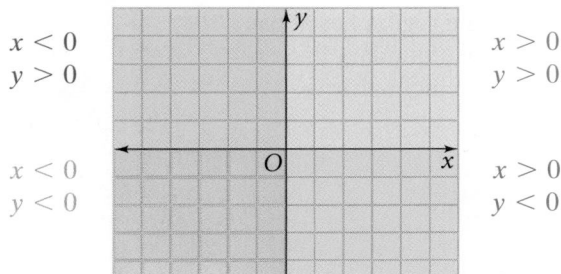

A **solution of a system of linear inequalities** makes each inequality in the system true. The graph of a system shows all of its solutions.

1 EXAMPLE Graphing a System of Inequalities

Solve by graphing. $y > 2x - 5$
$3x + 4y < 12$

Graph $y > 2x - 5$ and $3x + 4y < 12$.

For: Systems of Inequalities
  Activity
Use: Interactive Textbook, 7-6

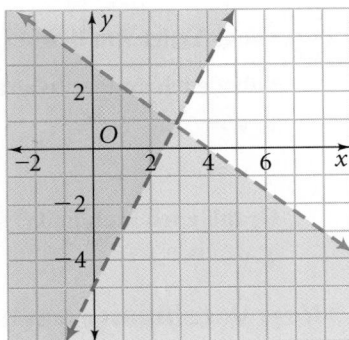

The coordinates of the points in the region where the graphs of the two inequalities overlap (shaded in lavender) are solutions of the system.

Check The point $(0, 0)$ is in the region graphed by both inequalities. See if $(0, 0)$ satisfies both inequalities.

$y > 2x - 5$ $3x + 4y < 12$
$0 > 2(0) - 5$ ⟵ **Substitute (0, 0) for (x, y).** ⟶ $3(0) + 4(0) < 12$
$0 > -5$ ✔ $0 < 12$ ✔

☑ Quick Check ❶ Solve by graphing. $y \geq -x + 2$
$2x + 4y < 4$

You can combine your knowledge of linear equations with your knowledge of inequalities to describe a graph using a system of inequalities.

2 EXAMPLE Writing a System of Inequalities From a Graph

Write a system of linear inequalities from each shaded region below.

GO Online

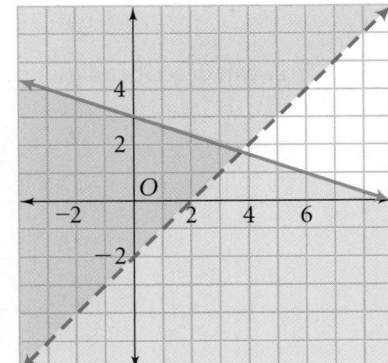

Video Tutor Help
Visit: PHSchool.com
Web Code: ate-0775

red region
boundary: $y = x - 2$
The region lies above the boundary line, so the inequality is $y > x - 2$.

blue region
boundary: $y = -\frac{1}{3}x + 3$
The region includes the boundary line and the points lying below the boundary line, so the inequality is $y \leq -\frac{1}{3}x + 3$.

system for the lavender region: $y > x - 2$
$y \leq -\frac{1}{3}x + 3$

❷ Write a system of inequalities for the lavender region in each of the following graphs.

a.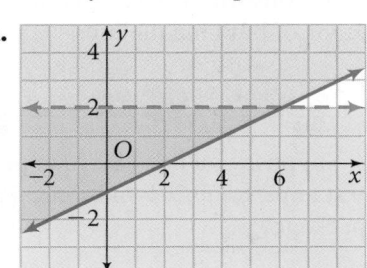

b.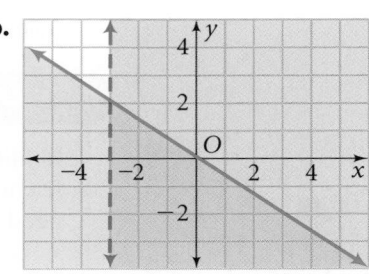

---

❷ Writing and Using Systems of Linear Inequalities

You can model some real-world situations by graphing linear inequalities. When you graph real-world situations, you often need to plan how you will scale each axis. Use the values for the *x*- and *y*-intercepts to determine your scale.

❸ EXAMPLE   Real-World 🌐 Problem Solving

Animal Habitat A zoo keeper wants to fence a rectangular habitat for goats. The length of the habitat should be at least 80 ft, and the distance around it should be no more than 310 ft. What are the possible dimensions of the habitat?

Relate | the length | is at least | 80 ft | | the perimeter | is no more than | 310 ft

Define Let x = width of the habitat.
Let y = length of the habitat.

Write | y | $\geq$ | 80 | | $2x + 2y$ | $\leq$ | 310

Solve by graphing. $y \geq 80$
$2x + 2y \leq 310$.

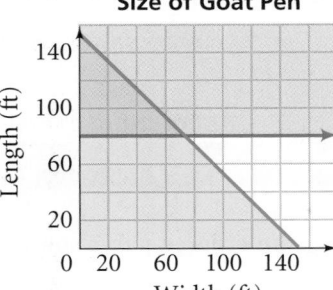

Size of Goat Pen

$y \geq 80$
$m = 0$
$b = 80$

Shade above
$y = 80$.

$2x + 2y \leq 310$

Graph the intercepts
$(155, 0)$ and $(0, 155)$.

Test $(0, 0)$.
$2(0) + 2(0) \leq 310$
$0 \leq 310$

Shade below
$2x + 2y = 310$.

The solutions are the coordinates of the points that lie in the region shaded lavender and on the parts of the lines $y = 80$ and $2x + 2y = 310$ that border the lavender region.

✓ **Quick Check** ❸ Suppose you want to fence a rectangular garden plot. You want the length of the garden to be at least 50 ft and the perimeter to be no more than 140 ft. Solve by graphing to show all of the possible dimensions of the garden.

Real-World 🌐 Connection

Careers A zoologist studies individual animals and the processes that sustain an animal within its group and its environment. To adapt an animal to a zoo habitat, a zoologist must research ways to help an animal adapt to a restricted environment.

Some real-world situations have a domain and range that include only integers. In such cases, the solutions will be some, but not all, of the points in the region included in the graphs of both inequalities.

4 EXAMPLE Real-World Problem Solving

Mailing Packages Suppose you need \$2.40 in postage to mail a package to a friend. You have 9 stamps, some 20¢ and some 34¢. How many of each do you need to mail the package?

Relate

the number of 20¢ and 34¢ stamps	is less than or equal to	9	the value of 20¢ and 34¢ stamps	is at least	240¢

Define Let a = the number of 20¢ stamps.
Let b = the number of 34¢ stamps.

Write

$a + b$	$\leq$	9	$20a + 34b$	$\geq$	240

Solve by graphing.
$a + b \leq 9$
$20a + 34b \geq 240$

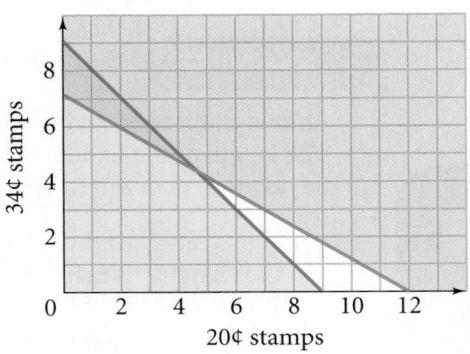

The solutions are all of the coordinates of points that are nonnegative integers lying in the region shaded lavender and on its boundary lines.

4
a. Give two solutions from the graph in Example 4.
b. Does either solution give you the exact postage needed to mail the package?
c. Critical Thinking Why are the solutions to the problem only nonnegative integers?

EXERCISES

For more exercises, see *Extra Skill and Word Problem Practice.*

Practice and Problem Solving

 Practice by Example

Example 1
(page 412)

Is the given ordered pair a solution of the system of inequalities?

1. $(1, 19)$
$y \leq 7x - 13$
$y > 3x + 6$

2. $(4, 10)$
$9x - y \geq 23$
$5x + 0.2y \geq 20$

3. $(-2, 40)$
$y > -13x + 29$
$y \leq 9x + 11$

Solve each system of inequalities by graphing.

4. $y < 2x + 4$
$-3x - 2y \geq 6$

5. $y < 2x + 4$
$2x - y \leq 4$

6. $y > 2x + 4$
$2x - y \leq 4$

7. $y > \frac{1}{4}x$
$y \leq -x + 4$

8. $y < 2x - 3$
$y > 5$

9. $y \leq -\frac{1}{3}x + 7$
$y \geq -x + 1$

10. $x + 2y \leq 10$
$x + 2y \geq 9$

11. $y \geq -x + 5$
$y \leq 3x - 4$

12. $y \leq 0.75x - 2$
$y > 0.75x - 3$

13. $8x + 4y \geq 10$
$3x - 6y > 12$

14. $2x - \frac{1}{4}y < 1$
$4x + 8y > 4$

15. $6x - 5y < 15$
$x + 2y \geq 7$

Example 2
(page 412)

Write a system of inequalities for each graph.

16.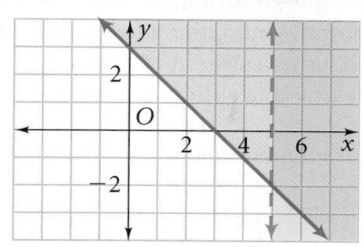

17.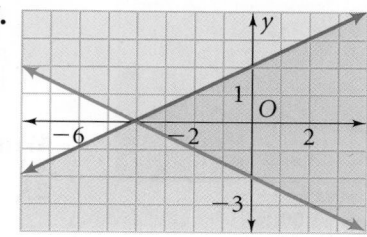

18.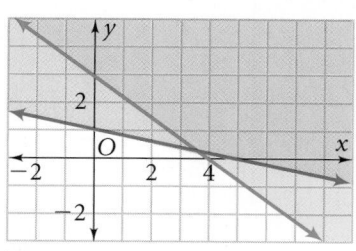

19.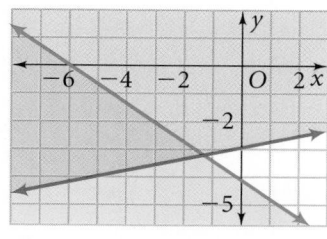

Example 3
(page 413)

20. Budget Suppose you buy flour and cornmeal in bulk to make flour tortillas and corn tortillas. Flour costs $1.50/lb. Cornmeal costs $2.50/lb. You want to spend less than $9.50 on flour and cornmeal, and you need at least 4 lb altogether.
a. Write a system of inequalities that describes this situation.
b. Graph the system to show all possible solutions.

21. Suppose you want to fence a rectangular area for your dog. You will use the house as one of the four sides. Since the house is 40 ft wide, the length ℓ needs to be no more than 40 ft. You plan to use at least 150 ft of fencing. Graph the following system to find possible dimensions for the rectangle.

$$\ell \leq 40$$
$$\ell + 2w \geq 150$$

Example 4
(page 414)

22. Suppose you receive a $50 gift certificate to the Cityside Music and Books store. All CDs at the store cost $9.99, and all books cost $5.99. You want to buy some books and at least one CD.
a. Write a system of inequalities for x books and y CDs that describes this situation.
b. Graph the system to show all possible solutions.
c. What purchase does the ordered pair $(2, 6)$ represent? Is it a solution to your system? Explain.
d. Find a solution in which you spend almost all of the gift certificate.

23. Business A seafood restaurant owner orders perch and salmon. He wants to buy at least 50 pounds of fish but cannot spend more than $180. Write and graph a system of inequalities to show the possible combinations of perch and salmon he could buy.

24. Earnings Suppose you have a job in an ice cream shop that pays $6 per hour. You also have a babysitting job that pays $4 per hour. You want to earn at least $60 per week but would like to work no more than 12 hours per week.
a. Graph and write a system of linear inequalities that describes this situation.
b. Give three possible solutions to the system.

Perch $4.00/lb

Salmon $3.00/lb

Write a system of inequalities for each of the following graphs.

25.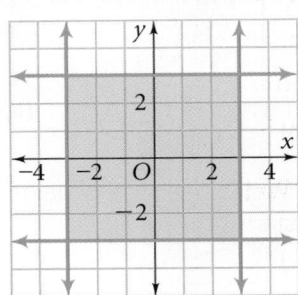

26.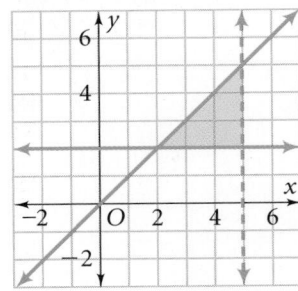

27.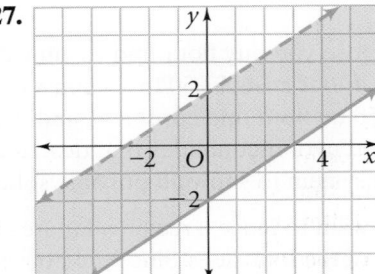

28.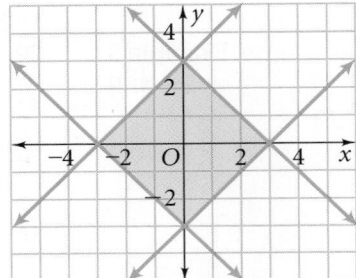

29. Open-Ended Write a system of four inequalities that describes a rectangle. Graph the system.

30. Geometry The following system of inequalities describes a right isosceles triangle.
a. Find m.
b. Find the area of the triangle.

$$x > 0$$
$$y > 0$$
$$y < mx + 4$$

Geometry The solution region of each system of linear inequalities below forms a figure. **(a) Describe the shape. (b) Find the vertices. (c) Find the area.**

31. $y \geq \frac{1}{2}x + 1$
$y \leq 2$
$x \geq -4$

32. $x \geq 1$
$x \leq 5$
$y \geq -1$
$y \leq 3$

33. $x \geq 0$
$x \leq 2$
$y \geq -4$
$y \leq -x + 2$

34. $x \geq 2$
$y \geq -3$
$x + y \leq 4$

35. a. Business A clothing store has a going-out-of-business sale. They are selling pants for $10.99 and shirts for $4.99. You can spend as much as $45 and want to buy at least one pair of pants. Write and graph a system of inequalities that describes this situation.
b. Suppose you need to buy at least three pairs of pants. From your graph, find all the ordered pairs that are possible solutions.

36. a. Graph each inequality. $y > 4x + 1$
$y < 4x - 2$

 **b. Writing** Will the boundary lines $y = 4x + 1$ and $y = 4x - 2$ ever intersect? Explain.

c. Will the shaded regions you drew in part (a) overlap?

d. Does the system of inequalities have any solutions?

37. a. Graph the system of inequalities. $y > 3x - 5$
$y < 3x + 4$

b. Will the boundary lines $y = 3x - 5$ and $y = 3x + 4$ ever intersect? Explain.

c. Describe the shape of the overlapping region.

Open-Ended **Write a system of linear inequalities with the given characteristics.**

38. $(0, 0)$ is a solution.

39. Solutions are only in Quadrant II.

40. There is no solution.

41. $(3, 7)$ is not a solution.

42. Solutions are only in Quadrant IV.

43. Multiple Choice Which region represents the solution to the system?

$y \geq \frac{1}{2}x + 1$
$4x + 2y \leq 8$

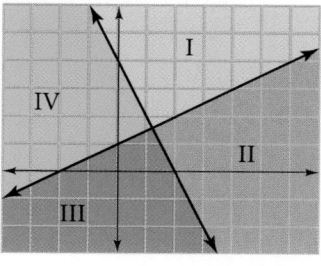

Ⓐ I   Ⓑ II

Ⓒ III   Ⓓ IV

Challenge **44. Business** A jeweler plans to produce a ring made of silver and gold. The price of gold is approximately \$10/g. The price of silver is approximately \$.15/g. She considers the following in deciding how much gold and silver to use in the ring.

• The total mass must be more than 10 g but less than 20 g.
• The ring must contain at least 3 g of gold.
• The total cost of the gold and silver must be less than \$60.

Let s = the mass of silver in grams and d = the mass of gold in grams.
a. Write and graph the four inequalities that describe this situation.
b. For one solution (s, d), find the mass of the ring and the cost of the gold and silver.

45. Solve $|y| \geq x$. (*Hint:* Write two inequalities; then graph them.)

Write a system of linear inequalities with the given characteristics.

46. $(2, 5)$ and $(5, 2)$ are not solutions; $(5, 5)$ is a solution.

47. $(-3, 2)$ and $(3, 2)$ are not solutions; $(-2, 6)$ is a solution.

 **48. Business** A drum maker sells two sizes of frame drums like the ones at the left. A 14-in. drum sells for \$180 and an 18-in. drum sells for \$240. He is trying to decide how many drums to build and considers the following:

• He wants to produce and sell at least \$2700 worth of drums.
• He has materials to make no more than 17 drums.
• He plans to make more 14-in. drums than 18-in. drums.
• He wants to make at least four 18-in. drums.

a. Write and graph the four inequalities that describe this situation.
b. Give one possible solution to the system.

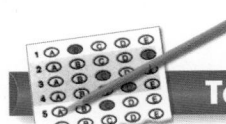

Multiple Choice

49. Which point is a solution of the following system? $y > x$
$y < 3x - 4$

I. (1, 2)   **II.** (3, 4)   **III.** (3, 9)
A. I only   **B.** I and II   **C.** I and III   **D.** II only

50. There are at most 12 bicycles and tricycles in a school playground. There are at least 17 wheels altogether. Let b equal the number of bicycles and t equal the number of tricycles. Which system describes this situation?

F. $b + t < 12$
$2b + 3t \geq 17$

G. $b + t \leq 12$
$2b + 3t \geq 17$

H. $b + t \leq 12$
$2b + 3t > 17$

J. $b + t \leq 12$
$2b + 3t \leq 17$

Short Response

51. Describe the solution to the following system. $3x + 4y \geq 12$
$3x + 4y \leq 12$

Extended Response

52. Suppose you and your friends are going out for pizza.
  a. Write a system of equations for the cost of a large pizza at each restaurant, based on the information at the right.
  b. Solve the system. Interpret your results.
  c. Where will you go for pizza? Explain your reasons.

Tony's Pizza	Maria's Pizza
Large cheese $7	Large cheese $8
Each topping $.75	Each topping $.50

Mixed Review

Lesson 7-5

Graph each linear inequality.

53. $y > x - 5$ **54.** $y \leq -2x + 4$ **55.** $y > -3$

56. $y + x \leq 7$ **57.** $3y - x \geq 6$ **58.** $4y + 2x < 8$

Lesson 6-6

Find the slope of a line parallel to the graph of each equation.

59. $5x - 2y = 8$ **60.** $y - 17 = -3x$ **61.** $0.5y - 10 + 4x = 0$

Find the slope of a line perpendicular to the graph of each equation.

62. $y = 4x$ **63.** $y = 5x - 7$ **64.** $y = \frac{3}{8}x + 19$

65. $y = -\frac{9}{10}x - 3$ **66.** $6y + 13x = 22$ **67.** $-4x - 15y = 74$

Lesson 5-4

Write a function rule for each table.

68.

x	f(x)
1	7
2	14
3	21
4	28
5	35

69.

x	f(x)
1	7
2	8
3	9
4	10
5	11

70.

x	f(x)
-2	4
-1	1
0	0
1	1
2	4

Graphing Linear Inequalities

You can use a graphing calculator to show the solutions of an inequality or a system of inequalities. The symbol before each Y in the [Y=] window indicates the graph style. You can use the graph style to shade above or below a line. The standard style, indicated by \, shows only the line.

NY A.G.6: Graph linear inequalities.

To change the graph style, select \ and press [ENTER] to rotate through the seven styles available. You can use ▼ to shade above the line and ▲ to shade below the line.

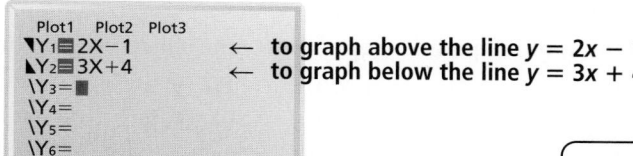

← to graph above the line $y = 2x - 1$
← to graph below the line $y = 3x + 4$

Go Online
PHSchool.com
For: Graphing calculator procedures
Web Code: ate-2108

The graphing calculator does not make a distinction between a boundary line that is dotted ($y < 2x - 1$) and a boundary line that is solid ($y \le 2x - 1$). You must decide whether a boundary line should be solid or dotted when you sketch the inequality.

1 ACTIVITY

Graph $y > -4x + 1$.

Enter the equation for the boundary line $y = -4x + 1$.
● Select ▼ to shade above the boundary line.

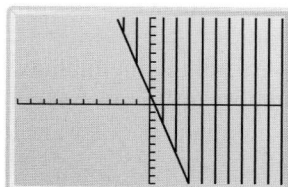

2 ACTIVITY

Graph the system: $y < -x + 4$
$\qquad\qquad y > 2x + 3$

Enter the equation of the first boundary line as Y_1.
Enter the equation of the second boundary line as Y_2.

Select ▲ to shade below $y = -x + 4$.
● Select ▼ to shade above $y = 2x + 3$.

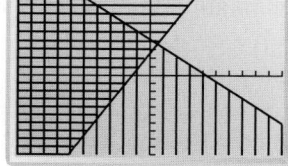

EXERCISES

Use a graphing calculator to graph each inequality. Sketch your graph.

1. $y < x$ **2.** $y > 2x - 3$ **3.** $y \ge -x + 3$ **4.** $y \le 5$

Use a graphing calculator to graph each system of inequalities. Sketch your graph.

5. $y \ge -1$ **6.** $y \ge 0.5x - 2$ **7.** $y < x$ **8.** $y \ge -4x + 6$
    $y \ge 2x$ $y \le x + 2$ $y \ge 1$ $y \ge -2x + 5$

In multiple correct answer questions, you have to determine the truth or falsehood of a number of statements. As you test each statement, mark it as true or false. Then choose the option with all those that are true.

1 EXAMPLE

Which system(s) of inequalities represent(s) the shaded region below?

I. $y \geq x$
$y \geq -x$
$y \leq 1$

II. $x + y \geq 0$
$x - y \geq 0$
$y \leq 1$

III. $y \geq |x|$
$y \leq 1$

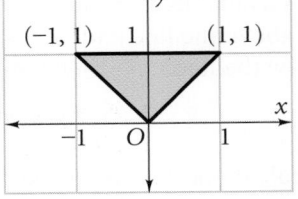

- Ⓐ I only
- Ⓒ I and II only
- Ⓔ II and III only
- Ⓑ I and III only
- Ⓓ I, II, and III

Method 1 Graph each system in I, II, and III to see which systems match the shaded region. The graphs of the systems in I and III match the shaded region. The correct answer is B.

Method 2 Choose a point, such as $(0, 0.5)$, inside the given shaded region. Test each of the three statements with the values $x = 0$ and $y = 0.5$. When $x = 0$ and $y = 0.5$, all the inequalities in I are true. The second inequality in II is $x - y = 0 - 0.5 \geq 0$, which is false. All the inequalities in III are true. The correct answer is B.

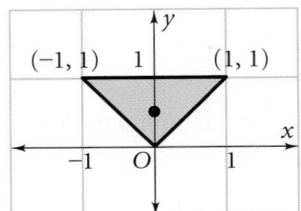

When you use Method 2, you should choose more than one point to test each system. This is because the point you choose may show a system to be true when it is really false. For example, if you choose to test the point $(1, 1)$, all the inequalities in II will be true, so you might think that II is a correct system.

EXERCISES

1. Choose another point in the shaded region and test each system.

2. **Writing** The point $(0, 0)$ is in the shaded region. Test each of the three statements with its coordinates to determine which are true. Explain why this point is not a good choice with which to answer this question.

3. Louis is selling lemonade for $.25 per cup. He bought the lemonade mix for $8.40 and the cups for $.05 each. Which statement(s) must be true?
   I. His break-even point is 42 cups.
   II. When he sells 20 cups, his income will be $13.40.
   III. His income is greater than his expenses when he sells 20 cups.
   - Ⓐ I and II only
   - Ⓒ I only
   - Ⓔ II and III only
   - Ⓑ I and III only
   - Ⓓ III only

Chapter Review

Vocabulary Review

🔊 elimination method (p. 387)
infinitely many solutions (p. 376)
linear inequality (p. 405)
no solution (p. 376)

solution of a system of linear
   equations (p. 374)
solution of a system of linear
   inequalities (p. 411)

solutions of an inequality (p. 405)
substitution method (p. 382)
system of linear equations (p. 374)
system of linear inequalities (p. 411)

Choose the vocabulary term that correctly completes each sentence.

1. _____?_____ is a method for solving a system of linear equations in which you multiply one or both equations by a nonzero number to get a variable term with coefficients that have a sum or difference of zero.

2. Any ordered pair that makes all equations in a system of equations true is a(n)_____?_____.

Go Online
PHSchool.com
For: Vocabulary quiz
Web Code: atj-0751

3. A(n)_____?_____ is formed by two or more linear inequalities.

4. Each point whose coordinates make an inequality true is a(n)_____?_____.

5. _____?_____ is a method for solving a system of linear equations in which at least one equation must first be solved for a single variable.

Skills and Concepts

7-1 Objectives

▼ To solve systems by graphing (p. 374)

▼ To analyze special types of systems (p. 376)

Two or more linear equations form a **system of linear equations.** You can solve a system of linear equations by graphing. A point where all the lines intersect is a **solution of the system.**

6. Which graph shows the solution of the following system?
$y = x - 1$
$y = -x + 3$

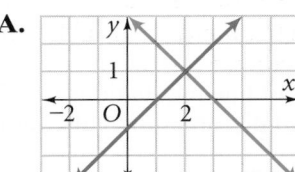

A.

B.

7. Is $(2, 5)$ a solution of the following system? Explain.
$y = 2x + 1$
$2x - y = 8$

8. How many solutions does the following system have? Explain.
$y = -\frac{1}{2}x + 2$
$3x + 6y = 12$

9. Critical Thinking What kinds of systems would be hard to solve by graphing?

Solve each system by graphing.

10. $y = 3x - 1$
    $y = -x + 3$

11. $x - y = -3$
    $3x + y = -1$

12. $-x + 2y = -2$
    $y = \frac{1}{2}x + 3$

13. $y = -2x + 1$
    $y = 2x - 3$

You can also solve a system of linear equations using the **substitution method.** By replacing one variable with an equivalent expression containing the other variable, you create a one-variable equation to solve.

Solve each system using substitution.

14. $y = 3x + 11$
$y = -2x + 1$

15. $4x - y = -12$
$-6x + 5y = -3$

16. $8x = -2y - 10$
$2x = 4y$

17. $y = 5x - 8$
$5y = 2x + 6$

18. Writing Explain how you determine if a system has no solution or infinitely many solutions when you solve a system using substitution.

19. There are 24 questions on a test. Each question is worth either 4 points or 5 points. The total is 100 points.
a. Write a system of equations to find the number of each type of question.
b. Solve the system by substitution.
c. How many questions of each type are on the test?

You can solve a system of linear equations using the **elimination method.** You add or subtract the equations to eliminate one variable. You can multiply one or both of the equations by a nonzero number before adding or subtracting.

Solve each system using elimination. Check your solution.

20. $y = -3x + 5$
$y = -4x - 1$

21. $2x - 3y = 5$
$x + 2y = -1$

22. $x + y = 10$
$x - y = 2$

23. $x + 4y = 12$
$2x - 3y = 6$

24. Farming A farmer raises chickens and cows. There are 34 animals in all. The farmer counts 110 legs on these animals. Write a system of equations to find the number of each type of animal. Solve the system by elimination. How many of each animal does the farmer have?

You can use systems of linear equations to solve word problems. First, define variables. Then model the situation with a system of linear equations.

25. A furniture finish consists of turpentine and linseed oil. It contains twice as much turpentine as linseed oil. If you plan to make 16 fluid ounces of furniture finish, how much turpentine do you need?

26. Geometry The difference between the measures of two complementary angles is 36°. Find both angle measures. (*Hint:* Two angles are complementary if the sum of their measures is 90°.)

27. Geometry The perimeter of a rectangle is 114 feet. Its length is three more than twice its width. Find the dimensions of the rectangle.

28. Supplies Marcella and Rupert bought some party supplies. Marcella bought 3 packages of balloons and 4 packages of favors for $14.63. Rupert bought 2 packages of balloons and 5 packages of favors for $16.03. Find the price of a package of balloons.

29. An airplane flew for 6 hours with a 22-km/h tail wind. The return flight against the same wind took 8 hours. Find the speed of the plane in still air.

7-5 Objectives

▼ To graph linear inequalities (p. 404)

▼ To use linear inequalities when modeling real-world situations (p. 406)

A **linear inequality** describes a region of the coordinate plane. The **solutions of the inequality** are the coordinates of the points that make the inequality true.

Graph each linear inequality.

30. $y < -3x + 8$ **31.** $y \geq 2x - 1$ **32.** $y \leq 0.5x + 6$ **33.** $y > -\frac{1}{4}x - 2$

Write the linear inequality shown in each graph.

34. 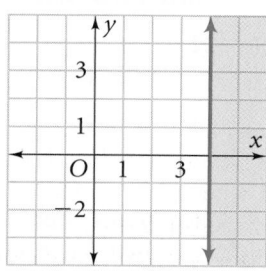   **35.** 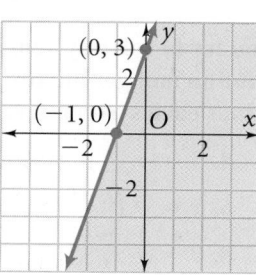   **36.**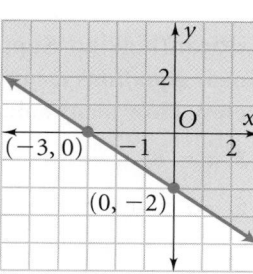

7-6 Objectives

▼ To solve systems of linear inequalities by graphing (p. 411)

▼ To model real-world situations using systems of linear inequalities (p. 413)

Two or more linear inequalities form a **system of linear inequalities.** To find the **solution of a system of linear inequalities,** graph each linear inequality. The solution region is where all the inequalities are true.

Solve each system of linear inequalities by graphing.

37. $y \geq -4x + 1$ **38.** $x - y < 10$ **39.** $y \leq x - 3$ **40.** $y < 5x$
     $y \leq \frac{5}{2}x - \frac{9}{2}$ $x + y \leq 8$ $y > x - 7$ $y \geq 0$

Write the system of inequalities shown in each graph.

41. 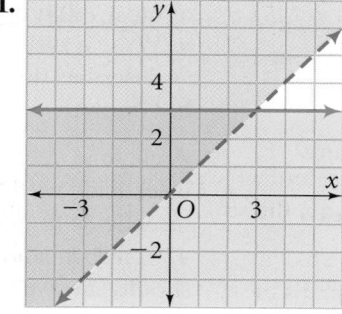   **42.**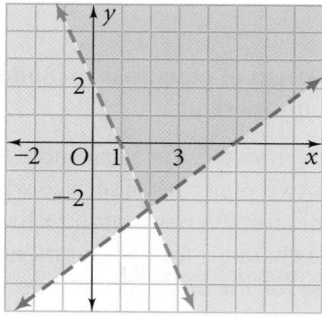

43. 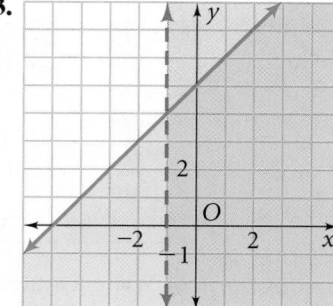   **44.**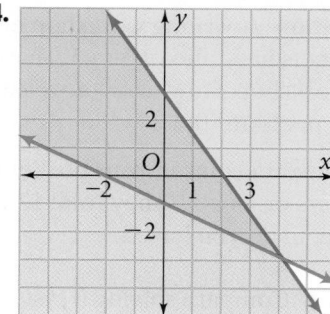

Chapter Test

Go Online
PHSchool.com

For: Chapter Test
Web Code: ata-0752

Solve each system by graphing.

1. $y = 3x - 7$
$y = -x + 1$

2. $4x + 3y = 12$
$2x - 5y = -20$

Critical Thinking Suppose you try to solve systems of linear equations using substitution and get the results below. How many solutions does each system have?

3. $x = 8$

4. $5 = y$

5. $-7 = 4$

6. $x = -1$

7. $2 = y$

8. $9 = 9$

Solve each system using substitution.

9. $y = 4x - 7$
$y = 2x + 9$

10. $y = -2x - 1$
$y = 3x - 16$

11. $8x + 2y = -2$
$y = -5x + 1$

12. $y + 6 = 2x$
$4x - 10y = 4$

Solve each system using elimination.

13. $4x + y = 8$
$-3x - y = 0$

14. $2x + 5y = 20$
$3x - 10y = 37$

15. $x + y = 10$
$-x - 2y = -14$

16. $3x + 2y = -19$
$x - 12y = 19$

Write a system of equations to model each situation. Solve by any method.

17. Cable Service Your local cable television company offers two plans: basic service with one movie channel for $35 per month or basic service with two movie channels for $45 per month. What is the charge for the basic service and the charge for each movie channel?

18. Education A writing workshop enrolls novelists and poets in a ratio of 5 to 3. There are 24 people at the workshop. How many novelists are there? How many poets are there?

19. You have 15 coins in your pocket that are either quarters or nickels. They total $2.75. How many of each coin do you have?

 20. Writing Compare solving a system of linear equations with solving a system of linear inequalities. What are the similarities? What are the differences?

21. Which point is *not* a solution of $y < 3x - 1$?
 A. $(2, -4)$ **B.** $(5, 7)$ **C.** $(0, -1)$ **D.** $(-2, -9)$

Solve each system by graphing.

22. $y > 4x - 1$
$y \le -x + 4$

23. $y \ge 3x + 5$
$y > x - 2$

24. $x > -3$
$-3x + y \ge 6$

25. $2x - y \le 2$
$y \ge 4$

26. Open-Ended Write a system of two linear equations. Solve by any method.

27. Garage Sale Leo held a garage sale. He priced all the items at a dime or a quarter. His sales totaled less than $5.
 a. Write a linear inequality that describes the situation. Graph the linear inequality.
 b. What is the maximum possible number of items that could have been sold for a dime?
 c. What is the maximum possible number of items that could have been sold for a quarter?

28. Gardening Mrs. Paulson bought chicken wire to enclose a rectangular garden. She is restricted to a width of no more than 30 ft. She would like to use at most 180 ft of chicken wire.
 a. Write a system of linear inequalities that describes this situation.
 b. Graph the system to show all possible solutions.

29. A chemist needs to mix a solution containing 30% insecticide with a solution containing 50% insecticide to make 200 L of a solution that is 42% insecticide. How much of each solution should she use?
 a. Complete the table below.

	30% Insecticide	50% Insecticide	42% Insecticide
Liters of Solution	■	■	■
Liters of Insecticide	■	■	■

 b. Write a system of equations that describes the situation. Solve the system.

Regents Test Prep

Reading Comprehension Read the passage below. Then answer the questions on the basis of what is *stated* or *implied* in the passage.

> **Music to Our Ears** The way in which the music industry delivers music has changed dramatically since 1985. In that year, according to industry sources, there were 22.6 million CDs shipped. By 1990, the number of CDs shipped increased to 286.5 million. In 1999, shipments swelled to 938.9 million CDs, an increase of over 4000% from the number in 1985.
>
> From 1985 to 1999, cassette shipments decreased from 339.1 million cassettes to 123.6 million, and record album shipments went from 167.0 million albums to 2.9 million. Clearly, CDs have replaced both cassettes and record albums as listeners' favorites.

1. What was the total number of CDs, cassettes, and record albums shipped in 1985?

 (1) 22.6 million   **(2)** 528.7 million

 (3) 1065.4 million   **(4)** 1445 million

2. Which graph correctly illustrates data in the article above?

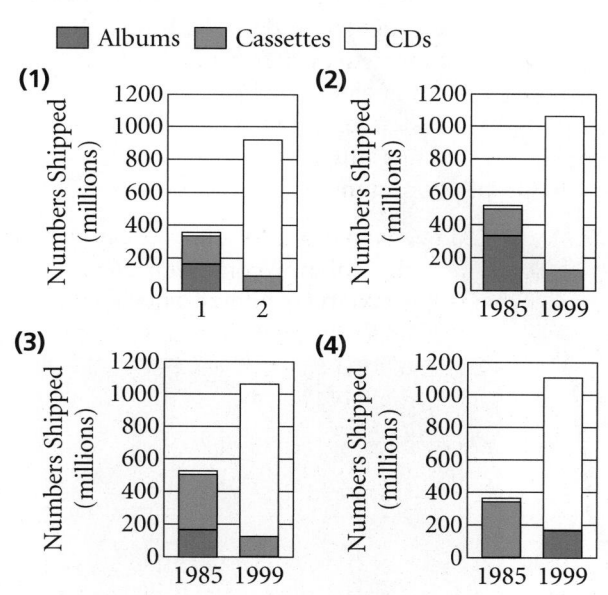

3. Which expression could you use to find the percent of change in the number of cassettes shipped from 1985 to 1999?

 (1) $\dfrac{339.1 + 123.6}{123.6}$ **(2)** $\dfrac{339.1 - 123.6}{123.6}$

 (3) $\dfrac{339.1 + 123.6}{339.1}$ **(4)** $\dfrac{339.1 - 123.6}{339.1}$

4. Suppose a linear equation models record album shipments from 1985 to 1999. What is a correct interpretation of the slope of the model?

 (1) about 12 million fewer shipped each year
 (2) 164.1 million fewer shipped in all
 (3) about 12 million more shipped each year
 (4) 164.1 million more shipped in all

5. Which of the following statements is true?

 (1) There is a positive correlation between the numbers of CDs and record albums shipped.
 (2) There is a negative correlation between the numbers of CDs and record albums shipped.
 (3) There is a positive correlation between the numbers of cassettes and CDs shipped.
 (4) There is a negative correlation between the numbers of cassettes and record albums shipped.

6. Use percent of change to describe the change in the number of cassettes shipped between 1985 and 1999.

7. According to the article, the increase in CD shipments from 1985 to 1999 was over 4000%. Do you agree? Explain.

8. Write a linear equation to model the number of CDs shipped from 1985 to 1999. Use the equation to predict the number of CDs shipped in 2010.

9. Do you agree with the main conclusion stated in the article? Explain.

Activity Lab

Wireless Style

Applying Equations Cell-phone use has increased dramatically since the mid-1990s. Millions of people worldwide own cell phones. If you are one of them, you probably purchased a calling plan from a service provider. These providers charge different fees for a variety of services.

Throw It Away!
A credit-card-sized disposable cell phone offers approximately one hour of talk time.

The circuits are printed metallic ink instead of tiny wires.

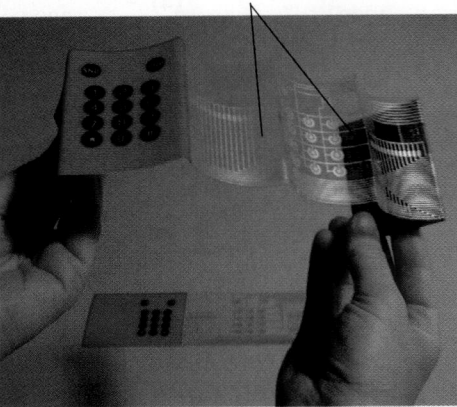

Activity

1. Suppose you are shopping for a calling plan. You expect to use 10 long-distance minutes per month.
   a. Use the table below and the total-cost equation at the right to find out how much you will pay for the first month of each calling plan.
   b. **Writing in Math** Which plan would you choose? Explain.

2. Suppose a friend is also shopping for a calling plan. Your friend expects to use 60 long-distance minutes each month.
   a. Use the table below and the total-cost equation to find out how much your friend will pay for the first month of each calling plan.
   b. Which plan do you think your friend would choose? Explain.

3. **Number Sense** Without calculating, which plan would be the least expensive to use in the second month? Explain.

Total-Cost Equation

$$c = m + d\ell + a$$

c = total cost
m = monthly fee
d = long-distance minutes
ℓ = long-distance rate
a = activation fee

Monthly Fee The amount a customer pays each month for basic service

Long-Distance Rate The amount a customer pays for each minute of a call made outside the local calling area

Activation Fee A one-time fee paid to start phone service

Calling Plan	A	B	C	D
Monthly Fee	$19.99	$34.99	$19.99	$29.99
Long-Distance Rate	$.15	$.15	$.00	$.20
Activation Fee	$36.00	$24.00	$30.00	$35.00

Where's the Cell-Phone Tower?

Cell phone companies often camouflage their towers to make them blend in with the surrounding landscape.

Antenna

Antenna

What You've Learned

 **Learning Standards for Mathematics**

In Chapter 1, you

● **NY A.N.6:** applied the order of operations to simplify expressions containing exponents.

In Chapter 5, you

● **NY A.A.5:** studied number patterns and practiced recognizing an arithmetic sequence. You also wrote equations for function rules.

In Chapter 6, you

● **NY A.S.8:** found a line of best fit for a set of data.

 Check Your Readiness   **GO for Help** to the Lesson in green.

Converting Fractions to Decimals (Skills Handbook page 759)

Write as a decimal.

1. $\frac{7}{10}$ **2.** $6\frac{2}{5}$ **3.** $\frac{8}{1000}$ **4.** $\frac{7}{2}$ **5.** $\frac{3}{11}$

Using the Order of Operations (Lesson 1-2)

Simplify each expression.

6. $(9 \div 3 + 4)^2$ **7.** $5 + (0.3)^3$ **8.** $3 - (1.5)^2$ **9.** $64 \div 2^4$

Evaluating Expressions (Lessons 2-1 to 2-3)

Evaluate each expression for $a = -2$ and $b = 5$.

10. $(ab)^2$ **11.** $(a - b)^2$ **12.** $a^3 + b^2$ **13.** $b - (3a)^2$

Finding Percent of Change (Lesson 3-7)

Find the percent of change.

14. $15.00 to $20.00   **15.** $20.00 to $15.00

Understanding Domain and Range (Lesson 5-2)

Find the range of each function with domain $\{-2, 0, 3.5\}$.

16. $f(x) = -2x^2$ **17.** $g(x) = 10 - x^3$ **18.** $y = 5x - 1$

Finding Terms of a Sequence (Lesson 5-7)

Find the next two terms of each sequence.

19. $1, 3, 5, 7, \ldots$ **20.** $-1, 0, 2, 5, 9, \ldots$ **21.** $7, 13, 19, 25, \ldots$

Exponents and Exponential Functions

🔊 **Key Vocabulary**

- common ratio (p. 460)
- compound interest (p. 476)
- decay factor (p. 478)
- exponential decay (p. 478)
- exponential function (p. 468)
- exponential growth (p. 475)
- geometric sequence (p. 460)
- growth factor (p. 475)
- interest period (p. 476)
- scientific notation (p. 436)

What You'll Learn Next

NY Learning Standards for Mathematics

- **NY A.N.6:** You will extend your knowledge about exponents to include zero and negative exponents.

- **NY A.A.12:** You will apply the properties of exponents, and see how exponents are used to write a geometric sequence.

- **NY A.G.4:** You will graph exponential functions by making a table of values.

Zero and Negative Exponents

NY Learning Standards for Mathematics

A.N.6 Evaluate expressions involving factorial(s), absolute value(s), and exponential expression(s).

✓ **Check Skills You'll Need**

GO for Help   Lessons 1-2 and 2-3

Simplify each expression.

1. 2^3

2. $\frac{1}{4^2}$

3. $4^2 \div 2^2$

4. $(-3)^3$

5. -3^3

6. $6^2 \div 12$

Evaluate each expression for $a = 2$, $b = -1$, and $c = 0.5$.

7. $\frac{a}{2a}$

8. $\frac{bc}{c}$

9. $\frac{ab}{bc}$

1 Zero and Negative Exponents

Activity: Exponents

1. a. Copy the table below. Replace each blank with the value of the power in simplest form.

2^x	5^x	10^x
$2^4 = $ ■	$5^4 = $ ■	$10^4 = $ ■
$2^3 = $ ■	$5^3 = $ ■	$10^3 = $ ■
$2^2 = $ ■	$5^2 = $ ■	$10^2 = $ ■

b. Look at the values that you used to replace the blanks. What pattern do you see as you go down each column?

2. Copy the table below. Use the pattern you described in Question 1 to complete the table.

2^x	5^x	10^x
$2^1 = $ ■	$5^1 = $ ■	$10^1 = $ ■
$2^0 = $ ■	$5^0 = $ ■	$10^0 = $ ■
$2^{-1} = $ ■	$5^{-1} = $ ■	$10^{-1} = $ ■
$2^{-2} = $ ■	$5^{-2} = $ ■	$10^{-2} = $ ■

3. Critical Thinking What pattern do you notice in the row with 0 as an exponent?

4. Copy and complete each expression.

a. $2^{-1} = \frac{1}{2^■}$

b. $2^{-2} = \frac{1}{2^■}$

c. $2^{-3} = \frac{1}{2^■}$

Consider 3^3, 3^2, and 3^1. Decreasing the exponent by one is the same as dividing by 3. Continuing the pattern, 3^0 equals 1 and 3^{-1} equals $\frac{1}{3}$.

 Key Concepts

Property	Zero as an Exponent

For every nonzero number a, $a^0 = 1$.

Examples $\qquad 5^0 = 1 \qquad\qquad (-2)^0 = 1 \qquad\qquad (1.02)^0 = 1 \qquad\qquad \left(\frac{1}{3}\right)^0 = 1$

Property	Negative Exponent

For every nonzero number a and integer n, $a^{-n} = \frac{1}{a^n}$.

Examples $\qquad 6^{-4} = \frac{1}{6^4} \qquad\qquad (-8)^{-1} = \frac{1}{(-8)^1}$

Why can't you use 0 as a base? By the first property, $3^0 = 1$, $2^0 = 1$, and $1^0 = 1$, which implies $0^0 = 1$. However, the pattern $0^3 = 0$, $0^2 = 0$, and $0^1 = 0$ implies $0^0 = 0$. Since both 1 and 0 cannot be the answer, 0^0 is undefined. In the second property, using 0 as a base results in division by zero, which you know is undefined.

1 EXAMPLE **Simplifying a Power**

Vocabulary Tip

Read 4^{-3} as "four to the negative three".

Simplify.

a. $4^{-3} = \frac{1}{4^3}$ $\qquad$ **Use the definition of negative exponent.**

$\qquad = \frac{1}{64}$ $\qquad$ **Simplify.**

b. $(-1.23)^0 = 1$ $\qquad$ **Use the definition of zero as an exponent.**

 **Quick Check** **1** Simplify each expression.

$\qquad$ **a.** 3^{-4} $\qquad\qquad$ **b.** $(-7)^0$ $\qquad\qquad$ **c.** $(-4)^{-3}$ $\qquad\qquad$ **d.** 7^{-1} $\qquad\qquad$ **e.** -3^{-2}

An algebraic expression is in simplest form when it is written with only positive exponents. If the expression is a fraction in simplest form, the only common factor of the numerator and denominator is 1.

2 EXAMPLE **Simplifying an Exponential Expression**

Simplify each expression.

a. $4yx^{-3} = 4y\left(\frac{1}{x^3}\right)$ $\qquad$ **Use the definition of negative exponent.**

$\qquad = \frac{4y}{x^3}$ $\qquad$ **Simplify.**

b. $\frac{1}{w^{-4}} = 1 \div w^{-4}$ $\qquad$ **Rewrite using a division symbol.**

$\qquad = 1 \div \frac{1}{w^4}$ $\qquad$ **Use the definition of negative exponent.**

$\qquad = 1 \cdot w^4$ $\qquad$ **Multiply by the reciprocal of $\frac{1}{w^4}$, which is w^4.**

$\qquad = w^4$ $\qquad$ **Identity Property of Multiplication**

 **Quick Check** **2** Simplify each expression.

$\qquad$ **a.** $11m^{-5}$ $\qquad\qquad$ **b.** $7s^{-4}t^2$ $\qquad\qquad$ **c.** $\frac{2}{a^{-3}}$ $\qquad\qquad$ **d.** $\frac{n^{-5}}{v^2}$

When you evaluate an exponential expression, you can write the expression with positive exponents before substituting values.

3 EXAMPLE **Evaluating an Exponential Expression**

Evaluate $3m^2t^{-2}$ for $m = 2$ and $t = -3$.

Method 1 Write with positive exponents first.

$$3m^2t^{-2} = \frac{3m^2}{t^2} \qquad \text{Use the definition of negative exponent.}$$

$$= \frac{3(2)^2}{(-3)^2} \qquad \text{Substitute 2 for } m \text{ and } -3 \text{ for } t.$$

$$= \frac{12}{9} = 1\frac{1}{3} \qquad \text{Simplify.}$$

Method 2 Substitute first.

$$3m^2t^{-2} = 3(2)^2(-3)^{-2} \qquad \text{Substitute 2 for } m \text{ and } -3 \text{ for } t.$$

$$= \frac{3(2)^2}{(-3)^2} \qquad \text{Use the definition of negative exponent.}$$

$$= \frac{12}{9} = 1\frac{1}{3} \qquad \text{Simplify.}$$

 **Quick Check** ❸ Evaluate each expression for $n = -2$ and $w = 5$.

a. $n^{-3}w^0$ **b.** $\dfrac{n^{-1}}{w^2}$ **c.** $\dfrac{w^0}{n^4}$ **d.** $\dfrac{1}{nw^{-2}}$

You can also evaluate exponential expressions that model real-world situations.

4 EXAMPLE **Real-World 🌐 Problem Solving**

Population Growth A biologist is studying green peach aphids, like the one shown at the left. In the lab, the population doubles every week. The expression $1000 \cdot 2^w$ models an initial population of 1000 insects after w weeks of growth.

a. Evaluate the expression for $w = 0$. Then describe what the value of the expression represents in the situation.

$$1000 \cdot 2^w = 1000 \cdot 2^0 \quad \text{Substitute 0 for } w.$$

$$= 1000 \cdot 1 \quad \text{Simplify.}$$

$$= 1000$$

The value of the expression represents the initial population of insects. This makes sense because when $w = 0$, no time has passed.

b. Evaluate the expression for $w = -3$. Then describe what the value of the expression represents in the situation.

$$1000 \cdot 2^w = 1000 \cdot 2^{-3} \quad \text{Substitute } -3 \text{ for } w.$$

$$= 1000 \cdot \frac{1}{8} \quad \text{Simplify.}$$

$$= 125$$

There were 125 aphids 3 weeks before the present population of 1000 insects.

Real-World 🌐 Connection

During the months of June and July, green peach aphids in a field of potato plants can double in population every three days.

Quick Check ❹ A sample of bacteria triples each month. The expression $5400 \cdot 3^m$ models a population of 5400 bacteria after m months of growth. Evaluate the expression for $m = -2$ and $m = 0$. Describe what each value of the expression represents in the situation.

EXERCISES

For more exercises, see *Extra Skill and Word Problem Practice.*

Practice and Problem Solving

 Practice by Example

Example 1
(page 431)

Simplify each expression.

1. $-(2.57)^0$ **2.** 4^{-2} **3.** $(-5)^{-2}$ **4.** -5^{-2}

5. $(-4)^{-2}$ **6.** -3^{-4} **7.** 2^{-6} **8.** -12^{-1}

9. $\frac{1}{2^0}$ **10.** 78^{-1} **11.** $(-4)^{-3}$ **12.** -4^{-3}

Example 2
(page 431)

Copy and complete each equation.

13. $4n^{\blacksquare} = \frac{4}{n^2}$ **14.** $\frac{x^{\blacksquare}}{2y^{\blacksquare}} = \frac{1}{2x^{-3}y^4}$ **15.** $\frac{a^{\blacksquare}}{3b^{\blacksquare}} = \frac{b^3}{3}$ **16.** $3xy^{\blacksquare} = \frac{3x}{y^5}$

Simplify each expression.

17. $3ab^0$ **18.** $5x^{-4}$ **19.** $\frac{1}{x^{-7}}$ **20.** $\frac{1}{c^{-1}}$

21. $\frac{5^{-2}}{p}$ **22.** $a^{-4}c^0$ **23.** $\frac{3x^{-2}}{y}$ **24.** $\frac{7ab^{-2}}{3w}$

25. $x^{-5}y^{-7}$ **26.** $x^{-5}y^7$ **27.** $\frac{8}{2c^{-3}}$ **28.** $\frac{7s}{5t^{-3}}$

29. $\frac{6a^{-1}c^{-3}}{d^0}$ **30.** $2^{-3}x^2z^{-7}$ **31.** $9^0y^7t^{-11}$ **32.** $\frac{7s^0t^{-5}}{2^{-1}m^2}$

Example 3
(page 432)

Evaluate each expression for $r = -3$ and $s = 5$.

33. s^{-2} **34.** r^{-2} **35.** $-r^{-2}$ **36.** s^0

37. $3s^{-2}$ **38.** $(2s)^{-2}$ **39.** $r^{-4}s^2$ **40.** $\frac{1}{r^{-4}s^2}$

41. s^2r^{-3} **42.** r^0s^{-2} **43.** $5r^3s^{-1}$ **44.** $2^{-4}r^3s^{-2}$

Example 4
(page 432)

45. a. Suppose your allowance doubles every week. This week you receive $2.56. How much will your allowance be three weeks from now? How much was your allowance three weeks ago?

     b. Critical Thinking From a parent's point of view, is doubling your allowance each week a good plan? Explain.

 Apply Your Skills

Mental Math Is the value of each expression *positive* or *negative*?

46. -2^2 **47.** $(-2)^2$ **48.** 2^{-2} **49.** $(-2)^3$ **50.** $(-2)^{-3}$

Write each number as a power of 10 using negative exponents.

51. $\frac{1}{10}$ **52.** $\frac{1}{100}$ **53.** $\frac{1}{1000}$ **54.** $\frac{1}{10,000}$ **55.** $\frac{1}{100,000}$

Write each expression as a decimal.

56. 10^{-3} **57.** 10^{-6} **58.** $7 \cdot 10^{-1}$ **59.** $3 \cdot 10^{-2}$ **60.** $5 \cdot 10^{-4}$

61. a. Patterns Complete the pattern using powers of 5.

     $\frac{1}{5^2} = \blacksquare$ $\frac{1}{5^1} = \blacksquare$ $\frac{1}{5^0} = \blacksquare$ $\frac{1}{5^{-1}} = \blacksquare$ $\frac{1}{5^{-2}} = \blacksquare$

     b. Write $\frac{1}{5^{-4}}$ using a positive exponent.

     c. Rewrite $\frac{1}{a^{-n}}$ so that the power of a is in the numerator.

62. Multiple Choice Which expression is equivalent to $\frac{3x^{-2}y^3}{9x^3y^{-5}}$?

     Ⓐ $\frac{3x^{-5}}{y^8}$ Ⓑ $\frac{xy^2}{3}$ Ⓒ $3xy^2$ Ⓓ $\frac{y^8}{3x^5}$

Simplify each expression.

63. $45 \cdot (0.5)^0$ **64.** $54 \cdot 3^{-2}$ **65.** $\dfrac{5^{-2}}{10^{-3}}$ **66.** $\dfrac{4^{-1}}{9^0}$ **67.** $\dfrac{(-3)^{-4}}{-3}$

Evaluate each expression for $a = 3$, $b = 2$, and $c = -4$.

68. c^b **69.** $a^{-b}b$ **70.** b^{-a} **71.** b^c **72.** $c^{-a}b^{ab}$

73. Copy and complete the table below.

a	4	■	■	$\frac{7}{8}$	■
a^{-1}	■	3	$\frac{1}{6}$	■	0.5

74. a. Critical Thinking Simplify $a^n \cdot a^{-n}$.
   b. What is the mathematical relationship of a^n and a^{-n}? Justify your answer.

75. Which expressions equal $\frac{1}{4}$?
   A. 4^{-1} **B.** 2^{-2} **C.** -4^1 **D.** $\dfrac{1}{2^2}$ **E.** 1^4 **F.** -2^{-2}

76. Open-Ended Choose a fraction to use as a value for the variable a. Find the values of a^{-1}, a^2, and a^{-2}.

77. Critical Thinking Are $3x^{-2}$ and $3x^2$ reciprocals? Explain.

78. Error Analysis A student simplified an expression as shown at the right. What error did the student make?

$$\frac{x^n}{a^{-n}b^0} = \frac{a^n x^n}{b^0}$$
$$= \frac{a^n x^n}{0} \text{ undefined}$$

79. Probability Suppose your history teacher gives a multiple-choice quiz. There are four questions, each with five answer choices. The probability p of guessing the answer to a question correctly is $\frac{1}{5}$. The probability q of guessing the answer to each question incorrectly is $\frac{4}{5}$.
   a. The table has expressions to find the probability of correctly guessing a certain number of answers on this quiz. Copy and complete the table.

Multiple-Choice Quiz

Number Correct	Expression	Probability
0	$p^0 q^4$	$\left(\frac{1}{5}\right)^0 \left(\frac{4}{5}\right)^4 = 0.4096$
1	$4p^1 q^3$	■
2	$6p^2 q^2$	■
3	$4p^3 q^1$	■
4	$p^4 q^0$	■

   b. Which number of correct answers is most likely?

80. Communication Suppose you are the only person in your class who knows a certain story. After a minute you tell a classmate. Every minute after that, every student who knows the story tells another student (sometimes the person being told already will have heard it). In a class of 30 students, the expression $\frac{30}{1 + 29 \cdot 2^{-t}}$ predicts the approximate number of people who will have heard the story after t minutes. About how many students will have heard your story after 2 min? After 5 min? After 10 min?

C Challenge

Simplify each expression.

81. $2^3(5^0 - 6m^2)$
82. $(-5)^2 - (0.5)^{-2}$
83. $\frac{6}{m^2} + \frac{5m^{-2}}{3^{-3}}$

84. $(0.8)^{-3} + 19^0 - 2^{-6}$
85. $\frac{2r^{-5}y^3}{n^2} \div \frac{r^2y^5}{2n}$
86. $2^{-1} - \frac{1}{3^{-2}} + 5\left(\frac{1}{2^2}\right)$

87. For what values of n is $n^{-3} = \left(\frac{1}{n}\right)^5$?

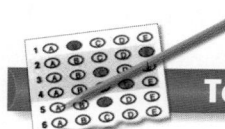

Test Prep

Gridded Response

88. Evaluate the expression xy^{-1} for $x = 2$ and $y = 3$.

89. Simplify $\frac{3^{-2}b^2}{a^0b^2}$.

90. Evaluate the expression $(4cd)^{-2}$ for $c = 2$ and $d = 1$.

91. Simplify $-6(-6)^{-1}$.

92. Write $26 \cdot 10^{-2}$ as a decimal.

93. Write $0.258 \cdot 10^2$ as a decimal.

Mixed Review

Lesson 7-6

Solve each system by graphing.

94. $y > 3x + 4$
$y \le -3x + 1$

95. $y \le -2x + 1$
$y < 2x - 1$

96. $y \ge 0.5x$
$y \le x + 2$

Lesson 6-7

97. Hat Sales Use the data in the table at the right.
   a. Make a scatter plot of the data. Use 87 for 1987.
   b. Draw a trend line.
   c. Write an equation for the trend line.
   d. Use your trend line to predict the retail sales of women's hats in 2005.

Estimated Women's Retail Hat Sales

Year	Sales (millions of dollars)
1987	300
1988	345
1989	397
1990	457
1991	510
1992	587
1993	664
1994	700
1995	770
1996	792
1997	830
1998	872
1999	915

Source: Headwear Information Bureau

Lesson 6-2

Write an equation of the line with the given slope and y-intercept.

98. $m = -1, b = 4$

99. $m = 5, b = -2$

100. $m = \frac{2}{5}, b = -3$

101. $m = -\frac{3}{11}, b = -17$

102. $m = \frac{5}{9}, b = \frac{1}{3}$

103. $m = 1.25, b = -3.79$

Scientific Notation

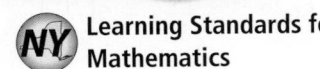

A.N.4 Understand and use scientific notation to compute products and quotients of numbers.

✓ **Check Skills You'll Need**

Simplify each expression.

1. $6 \cdot 10^4$

2. $7 \cdot 10^{-2}$

3. $8.2 \cdot 10^5$

4. $3 \cdot 10^{-3}$

5. $3.4 \cdot 10^1$

6. $5.24 \cdot 10^2$

7. Simplify $3 \times 10^2 + 6 \times 10^1 + 7 \times 10^0 + 8 \times 10^{-1}$.

GO **for Help** Lesson 8-1

🔊 **New Vocabulary** • scientific notation

1 Writing Numbers in Scientific and Standard Notation

Calculator Hint

The E on a calculator readout means exponentiation. The **EE** or **EXP** keys let you input an exponent for a power of 10. So to enter 4×10^6, you can enter 4 **EE** 6.

The planet Jupiter has an average radius of 69,111 km. What is Jupiter's volume?

Since Jupiter is a sphere, to answer this question you use the formula for the volume of a sphere.

$V = \frac{4}{3}\pi r^3$

$= \frac{4}{3}\pi(69,111)^3$ **Substitute 69,111 for r.**

$\approx 1.382706933E15$ **Use a calculator.**

In standard notation, you write the number above as 1,382,706,933,000,000. In scientific notation, you write the number as $1.382706933 \times 10^{15}$. Scientific notation is a shorthand way to write very large or very small numbers.

🔑 **Key Concepts**

Definition	**Scientific Notation**

A number in **scientific notation** is written as the product of two factors in the form $a \times 10^n$, where n is an integer and $1 \le a < 10$.

Examples 3.4×10^6 $\qquad$ 5.43×10^{13} $\qquad$ 2.1×10^{-10}

1 EXAMPLE **Recognizing Scientific Notation**

Is each number written in scientific notation? If not, explain.

a. 56.29×10^{12} $\quad$ No; 56.29 is greater than 10.

b. 0.84×10^{-3} $\quad$ No; 0.84 is less than 1.

c. 6.11×10^5 $\quad$ yes

✓ **Quick Check**  Is each number written in scientific notation? If not, explain.

a. 3.42×10^{-7} $\qquad$ **b.** 52×10^4 $\qquad$ **c.** 0.04×10^{-5}

In scientific notation, you use positive exponents to write a number greater than 1. You use negative exponents to write a number between 0 and 1.

2 EXAMPLE Writing a Number in Scientific Notation

Write each number in scientific notation.

a. 56,900,000

$56,900,000. = 5.69 \times 10^7$ **Move the decimal point 7 places to the left and use 7 as an exponent. Drop the zeros after the 9.**

b. 0.00985

$0.00985 = 9.85 \times 10^{-3}$ **Move the decimal point 3 places to the right and use −3 as an exponent. Drop the zeros before the 9.**

✓ Quick Check **2** Write each number in scientific notation.

a. 267,000   **b.** 46,205,000   **c.** 0.0000325   **d.** 0.000000009

e. **Critical Thinking** You express 1 billion as 10^9. Explain why you express 436 billion as 4.36×10^{11} in scientific notation.

3 EXAMPLE Writing a Number in Standard Notation

Physical Science Write each number in standard notation.

a. temperature at the sun's core: 1.55×10^6 kelvins

$1.55 \times 10^6 = 1\,550000.$ **A positive exponent indicates a number greater than 10. Move the decimal point 6 places to the right.**

$= 1,550,000$

b. lowest temperature recorded in a lab: 2×10^{-11} kelvin

$2 \times 10^{-11} = 0.00000000002$ **A negative exponent indicates a number between 0 and 1. Move the decimal point 11 places to the left.**

$= 0.00000000002$

✓ Quick Check **3** Write each number in standard notation.

a. 3.2×10^{12} **b.** 5.07×10^4 **c.** 5.6×10^{-4} **d.** 8.3×10^{-2}

2 Using Scientific Notation

Masses of Planets (kilograms)

Jupiter 3.7×10^{27}

Uranus 8.7×10^{25}

Neptune 1.0×10^{26}

Saturn 5.7×10^{26}

You can compare and order numbers in scientific notation. First compare the powers of 10, and then compare the decimals.

4 EXAMPLE Real-World Problem Solving

Astronomy List the planets in order from least to greatest mass.

Order the powers of 10. Arrange the decimals with the same power of 10 in order.

8.7×10^{25}	1.0×10^{26}	5.7×10^{26}	3.7×10^{27}
Uranus	Neptune	Saturn	Jupiter

From least to greatest mass, the order of the planets is Uranus, Neptune, Saturn, and Jupiter.

✓ Quick Check **4** The following masses of parts of an atom are measured in grams. Order the parts of an atom from least to greatest mass.

neutron: 1.6749×10^{-24}, electron: 9.1096×10^{-28}, proton: 1.6726×10^{-24}

You can write numbers like 815×10^5 and 0.078×10^{-2} in scientific notation.

$$815 \times 10^5 = 81,500,000 = 8.15 \times 10^7 \qquad 0.078 \times 10^{-2} = 0.00078 = 7.8 \times 10^{-4}$$

The examples above show this pattern: When you move a decimal n places left, the exponent of 10 increases by n; when you move a decimal point n places right, the exponent of 10 decreases by n.

5 EXAMPLE **Using Scientific Notation to Order Numbers**

Order 0.052×10^7, 5.12×10^5, 53.2×10, and 534 from least to greatest.

Write each number in scientific notation.

0.052×10^7	5.12×10^5	53.2×10	534
↓	↓	↓	↓
5.2×10^5	5.12×10^5	5.32×10^2	5.34×10^2

Problem Solving Hint

Remember that
53.2×10 is 53.2×10^1.

Order the powers of 10. Arrange the decimals with the same power of 10 in order.

$$5.32 \times 10^2 \quad 5.34 \times 10^2 \quad 5.12 \times 10^5 \quad 5.2 \times 10^5$$

Write the original numbers in order.

$$53.2 \times 10 \quad 534 \quad 5.12 \times 10^5 \quad 0.052 \times 10^7$$

 **Quick Check** **5** Order 60.2×10^{-5}, 63×10^4, 0.067×10^3, and 61×10^{-2} from least to greatest.

You can multiply a number that is in scientific notation by another number. If the product is less than one or greater than 10, rewrite the product in scientific notation.

6 EXAMPLE **Multiplying a Number in Scientific Notation**

Simplify. Write each answer using scientific notation.

a. $7(4 \times 10^5) = (7 \cdot 4) \times 10^5$	Use the Associative Property of Multiplication.
$= 28 \times 10^5$	Simplify inside the parentheses.
$= 2.8 \times 10^6$	Write the product in scientific notation.
b. $0.5(1.2 \times 10^{-3}) = (0.5 \cdot 1.2) \times 10^{-3}$	Use the Associative Property of Multiplication.
$= 0.6 \times 10^{-3}$	Simplify inside the parentheses.
$= 6 \times 10^{-4}$	Write the product in scientific notation.

 **Quick Check** **6** Simplify. Write each answer using scientific notation.
   a. $2.5(6 \times 10^3)$ **b.** $0.4(2 \times 10^{-9})$

EXERCISES

For more exercises, see *Extra Skill and Word Problem Practice*.

Practice and Problem Solving

A **Practice by Example**

Example 1
(page 436)

GO for Help

Is each number written in scientific notation? If not, explain.

1. 55×10^4 **2.** 3.2×10^5 **3.** 0.9×10^{-2}

4. 7.3×10^{-5} **5.** 1.12×10^1 **6.** 46×10^7

Example 2 (page 437)	**Write each number in scientific notation.**	

Example 2
(page 437)

Write each number in scientific notation.

7. 9,040,000,000   **8.** 0.02   **9.** 9.3 million   **10.** 21,700

11. 0.00325   **12.** 8,003,000   **13.** 0.00092   **14.** 0.0156

Example 3
(page 437)

Write each number in standard notation.

15. 5×10^2 **16.** 5×10^{-2} **17.** 2.04×10^3 **18.** 7.2×10^5

19. 8.97×10^{-1} **20.** 1.3×10^0 **21.** 2.74×10^{-5} **22.** 4.8×10^{-3}

Examples 4, 5
(pages 437, 438)

Order the numbers in each list from least to greatest.

23. $10^5, 10^{-3}, 10^0, 10^{-1}, 10^1$

24. $9 \times 10^{-7}, 8 \times 10^{-8}, 7 \times 10^{-6}, 6 \times 10^{-10}$

25. $50.1 \times 10^{-3}, 4.8 \times 10^{-1}, 0.52 \times 10^{-3}, 56 \times 10^{-2}$

26. $0.53 \times 10^7, 5300 \times 10^{-1}, 5.3 \times 10^5, 530 \times 10^8$

27. Measuring instruments may have different degrees of precision. Instrument A is precise to 10^{-2} cm, Instrument B is precise to 5×10^{-2} cm, and Instrument C is precise to 8×10^{-3} cm. Order the instruments from most precise (least possible error) to least precise (greatest possible error).

Example 6
(page 438)

Simplify. Write each answer using scientific notation.

28. $8(7 \times 10^{-3})$ **29.** $8(3 \times 10^{14})$ **30.** $0.2(3 \times 10^2)$

31. $6(5.3 \times 10^{-4})$ **32.** $0.3(8.2 \times 10^{-3})$ **33.** $0.5(6.8 \times 10^5)$

 Apply Your Skills

For Exercises 34–39, find the missing value.

Selected Masses (kilograms)

		Standard Notation	Scientific Notation
34.	Elephant	■	5.4×10^3
35.	Adult human	70	■
36.	Dog	10	■
37.	Golf ball	0.046	■
38.	Paper clip	■	5×10^{-4}
39.	Oxygen atom	0.000000000000000000000000003	■

Real-World 🌐 Connection

Elephant calves weigh from 100 to 145 kilograms.

40. Critical Thinking Is the number 10^5 in scientific notation? Explain.

41. Writing Explain how to write 48 million and 48 millionths in scientific notation.

42. Health Care In 2010, the population in the United States will be about 3.09×10^8. Spending for health care will be about $8754 per person. About how much will the United States spend on health care in 2010? Use scientific notation.

43. Computers A computer can perform 4.66×10^8 instructions per second. How many instructions is that per minute? Per hour? Use scientific notation.

44. Multiple Choice The national debt at the beginning of 2005 was about $7,600,000,000,000. Which represents the debt in scientific notation?

   Ⓐ 76×10^{11} Ⓑ 7.6×10^{12} Ⓒ 76×10^{12} Ⓓ 7.6×10^{13}

45. Math in the Media Use the cartoon below.

FOX TROT by Bill Amend

a. Write 500 trillion in scientific notation.

b. Since the 10-second length of the movie is off by a factor of 500 trillion, what time span does the movie actually represent?

C Challenge **46. World Population** The world population in 2025 may reach 7.84×10^9 persons. This is about 3 times the world population in 1950. What was the world population in 1950?

47. Astronomy Use a calculator to find the volume of each planet with the given radius.
  a. Mercury: 2439 km   b. Earth: 6378 km   c. Saturn: 60,268 km

48. Write $\frac{1}{300}$ using scientific notation.

NY REGENTS

Test Prep

Multiple Choice **49.** Simplify $90(1.2 \times 10^{-5})$. Write the answer in scientific notation.
  A. 1.08×10^{-7} **B.** 108.0×10^{-5} **C.** 108.0×10^{-3} **D.** 1.08×10^{-3}

50. Which answer has the states in the table at the right ordered from least to greatest projected population?
  F. New York, Florida, Virginia, Vermont
  G. Vermont, Virginia, New York, Florida
  H. Vermont, Virginia, Florida, New York
  J. Florida, New York, Virginia, Vermont

Projected Population in 2025

State	Population
Florida	2.07×10^7
Virginia	8.47×10^6
Vermont	6.78×10^5
New York	1.98×10^7

51. Which equals 275 million?
  A. 275×10^5 **B.** 2.75×10^6
  C. 2.75×10^8 **D.** 275×10^9

Short Response **52.** A microscope set on 1000X makes an object appear 1000 times its actual size. If a bacterium is 8×10^{-4} millimeters in diameter, how large will it appear under this microscope? Use scientific notation. Show your work.

Mixed Review

GO for Help

Lesson 8-1   **Simplify each expression.**
  53. $4(1.8)^0$ **54.** $12 \cdot 2^{-2}$ **55.** $6 \cdot 3^{-2}$ **56.** $\frac{4^3}{7^2}$ **57.** $\frac{3^{-2}}{9^0}$

Lesson 7-5   **Graph each linear inequality.**
  58. $y < -\frac{1}{4}x + 2$ **59.** $y \geq \frac{2}{3}x$ **60.** $y < 3x - 4$

8-3

Multiplication Properties of Exponents

A.N.4 Understand and use scientific notation to compute products and quotients of numbers.

A.A.12 Multiply and divide monomial expressions with a common base, using the properties of exponents.

✓ **Check Skills You'll Need**

 **GO for Help** Lesson 2-3

Rewrite each expression using exponents.

1. $t \cdot t \cdot t \cdot t \cdot t \cdot t \cdot t \cdot t$

2. $(6 - m)(6 - m)(6 - m)$

3. $(r + 5)(r + 5)(r + 5)(r + 5)(r + 5)$

4. $5 \cdot 5 \cdot 5 \cdot s \cdot s \cdot s$

Simplify.

5. -5^4 **6.** $(-5)^4$ **7.** $(-5)^0$ **8.** $(-5)^{-4}$

1 Multiplying

Activity: Exponents With the Same Base and Multiplication

1. Copy and complete the table.

2. What is true for $2^2 \cdot 2^2$ and $2^3 \cdot 2^1$?

3. **Patterns** What relationship do you see between the sum of the exponents of 2 in the first column and the exponent of 2 in the third column?

Factors	Product Using Repeated Factors	Power of 2
$2^1 \cdot 2^1$	$2 \cdot 2$	2^2
$2^2 \cdot 2^2$	■	■
$2^3 \cdot 2^1$	■	■
$2^4 \cdot 2^2$	■	■
$2^3 \cdot 2^3$	■	■

You can write the product of powers with the same base, like $2^4 \cdot 2^2$, using one exponent.

$$2^4 \cdot 2^2 = (2 \cdot 2 \cdot 2 \cdot 2) \cdot (2 \cdot 2) = 2^6$$

Notice that the sum of the exponents of the expression $2^4 \cdot 2^2$ equals the exponent of 2^6.

 Key Concepts

Property	**Multiplying Powers With the Same Base**

For every nonzero number a and integers m and n, $a^m \cdot a^n = a^{m + n}$.

Examples $3^5 \cdot 3^4 = 3^{5 + 4} = 3^9$ $h^2 \cdot h^9 = h^{2 + 9} = h^{11}$

1 EXAMPLE Multiplying Powers

Rewrite each expression using each base only once.

a. $11^4 \cdot 11^3 = 11^{4\,+\,3}$ Add exponents of powers with the same base.

$= 11^7$ Simplify the sum of the exponents.

b. $5^{-2} \cdot 5^2 = 5^{-2\,+\,2}$ Add exponents of powers with the same base.

$= 5^0$ Simplify the sum of the exponents.

$= 1$ Use the definition of zero as an exponent.

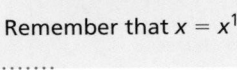 **❶** Rewrite each expression using each base only once.

a. $5^3 \cdot 5^6$ **b.** $2^4 \cdot 2^{-3}$ **c.** $7^{-3} \cdot 7^2 \cdot 7^6$

When variable factors have more than one base, be careful to combine only those powers with the same base.

2 EXAMPLE Multiplying Powers in an Algebraic Expression

Simplify each expression.

a. $2n^5 \cdot 3n^{-2} = (2 \cdot 3)(n^5 \cdot n^{-2})$ Commutative Property of Multiplication

$= 6(n^{5\,+\,(-2)})$ Add exponents of powers with the same base.

$= 6n^3$ Simplify.

> **Problem Solving Hint**
>
> Remember that $x = x^1$.

b. $5x \cdot 2y^4 \cdot 3x^8 = (5 \cdot 2 \cdot 3)(x \cdot x^8)(y^4)$ Commutative and Associative Properties of Multiplication

$= 30(x^1 \cdot x^8)(y^4)$ Multiply the coefficients. Write x as x^1.

$= 30(x^{1\,+\,8})(y^4)$ Add exponents of powers with the same base.

$= 30x^9y^4$ Simplify.

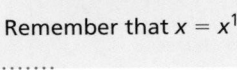 **❷** Simplify each expression.

a. $n^2 \cdot n^3 \cdot 7n$ **b.** $2y^3 \cdot 7x^2 \cdot 2y^4$ **c.** $m^2 \cdot n^{-2} \cdot 7m$

---

2 | Working With Scientific Notation

In Lesson 8-2, you wrote numbers in scientific notation using patterns to move decimal points. You can now use the property for multiplying powers with the same base to write numbers and to multiply numbers in scientific notation.

3 EXAMPLE Multiplying Numbers in Scientific Notation

Simplify $(7 \times 10^2)(4 \times 10^5)$. Write the answer in scientific notation.

$(7 \times 10^2)(4 \times 10^5) = (7 \cdot 4)(10^2 \cdot 10^5)$ Commutative and Associative Properties of Multiplication

$= 28 \times 10^7$ Simplify.

$= 2.8 \times 10^1 \cdot 10^7$ Write 28 in scientific notation.

$= 2.8 \times 10^{1\,+\,7}$ Add exponents of powers with the same base.

$= 2.8 \times 10^8$ Simplify the sum of the exponents.

 **❸** Simplify each expression. Write each answer in scientific notation.

a. $(2.5 \times 10^8)(6 \times 10^3)$ **b.** $(1.5 \times 10^{-2})(3 \times 10^4)$ **c.** $(9 \times 10^{-6})(7 \times 10^{-9})$

4 **EXAMPLE** <u>Real-World</u> Problem Solving

Biology A human body contains about $3.2 \times 10^4 \ \mu L$ (microliters) of blood for each pound of body weight. Each microliter of blood contains about 5×10^6 red blood cells. Find the approximate number of red blood cells in the body of a 125-lb person.

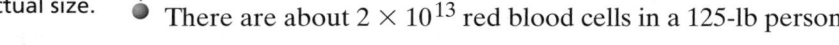

$$\text{red blood cells} = \text{pounds} \cdot \frac{\text{microliters}}{\text{pound}} \cdot \frac{\text{cells}}{\text{microliter}} \qquad \textbf{Use dimensional analysis.}$$

$$= 125 \text{ lb} \cdot (3.2 \times 10^4)\frac{\mu L}{\text{lb}} \cdot (5 \times 10^6)\frac{\text{cells}}{\mu L} \qquad \textbf{Substitute.}$$

$$= (125 \cdot 3.2 \cdot 5) \times (10^4 \cdot 10^6) \qquad \begin{array}{l}\textbf{Commutative and}\\ \textbf{Associative Properties of}\\ \textbf{Multiplication}\end{array}$$

$$= (2000) \times (10^{4\ +\ 6}) \qquad \textbf{Simplify.}$$

$$= 2000 \times 10^{10} \qquad \textbf{Add exponents.}$$

$$= 2 \times 10^3 \cdot 10^{10} \qquad \begin{array}{l}\textbf{Write 2000 in scientific}\\ \textbf{notation.}\end{array}$$

$$= 2 \times 10^{13} \qquad \textbf{Add the exponents.}$$

There are about 2×10^{13} red blood cells in a 125-lb person.

<u>**Real-World**</u> **Connection**

The blood cells shown in the photo above are magnified 5×10^3 times their actual size.

✓ **Quick Check** ❹ About how many red blood cells are in the body of a 160-lb soccer player?

EXERCISES

For more exercises, see *Extra Skill, Word Problem, and Proof Practice.*

Practice and Problem Solving

 **Practice by Example**

Example 1
(page 442)

 GO for Help

Rewrite each expression using each base only once.

1. $2^6 \cdot 2^4$

2. $5^{-13} \cdot 5^5 \cdot 2^5$

3. $10^{-6} \cdot 10^5 \cdot 10^1$

4. $(0.99)^3 \cdot (0.99)^0$

5. $6^6 \cdot 6^{-2} \cdot 6^5$

6. $(1.025)^2(1.025)^{-2}$

Example 2
(page 442)

Simplify each expression.

7. $c^{-2}c^7$

8. $3r \cdot r^4$

9. $5t^{-2} \cdot 2t^{-5}$

10. $(7x^5)(8x)$

11. $3x^2 \cdot x^2$

12. $(-2.4n^4)(2n^{-1})$

13. $b^{-2} \cdot b^4 \cdot b$

14. $(-2m^3)(3.5m^{-3})$

15. $(15a^3)(-3a)$

16. $(x^5y^2)(x^{-6}y)$

17. $(5x^5)(3y^6)(3x^2)$

18. $(4c^4)(ac^3)(3a^5c)$

19. $x^6 \cdot y^2 \cdot x^4$

20. $a^6b^3 \cdot a^2b^{-2}$

21. $-m^2 \cdot 4r^3 \cdot 12r^{-4} \cdot 5m$

Example 3
(page 442)

Simplify each expression. Write each answer in scientific notation.

22. $(2 \times 10^3)(3 \times 10^2)$

23. $(2 \times 10^6)(3 \times 10^3)$

24. $(4 \times 10^6) \cdot 10^{-3}$

25. $(1 \times 10^3)(3.4 \times 10^{-8})$

26. $(8 \times 10^{-5})(7 \times 10^{-3})$

27. $(5 \times 10^7)(3 \times 10^{14})$

Example 4
(page 443)

Write each answer in scientific notation.

28. **Astronomy** The distance light travels in one year (one light-year) is about 5.88×10^{12} miles. The closest star to Earth (other than the sun) is Alpha Centauri, which is 4.35 light-years from Earth. About how many miles from Earth is Alpha Centauri?

Write each answer in scientific notation.

29. **Geology** Earth's crust contains approximately 120 trillion metric tons of gold. One metric ton of gold is worth about $9 million. What is the approximate value of the gold in Earth's crust?

30. **Astronomy** Light travels through space at a constant speed of about 3×10^5 km/s. Sunlight reflecting from the moon takes about 1.28×10^0 s to reach Earth. Find the distance from the moon to Earth.

B Apply Your Skills **Complete each equation.**

31. $5^2 \cdot 5^\blacksquare = 5^{11}$ 32. $5^7 \cdot 5^\blacksquare = 5^3$ 33. $2^\blacksquare \cdot 2^4 = 2^1$

34. $c^{-5} \cdot c^\blacksquare = c^6$ 35. $m^\blacksquare \cdot m^{-4} = m^{-9}$ 36. $a \cdot a \cdot a^3 = a^\blacksquare$

37. $a^\blacksquare \cdot a^4 = 1$ 38. $a^{12} \cdot a^\blacksquare = a^{12}$ 39. $x^3 y^\blacksquare \cdot x^\blacksquare = y^2$

Geometry **Find the area of each figure.**

40.
$2x$
$3x^2 + x$

41.
$2x^2$

GO Online
Homework Video Tutor
Visit: PHSchool.com
Web Code: ate-0803

42.
$4y^2$
$y^3 + 2$

43.
$4c$
$2c^3$

Error Analysis **Correct each error.**

44.
$$(3x^2)(-2x^4) = 3(-2)x^{2 \cdot 4}$$
$$= -6x^8$$

45.
$$4a^2 \cdot 3a^5 = (4 + 3)a^{2 + 5}$$
$$= 7a^7$$

46.
$$x^6 \cdot x \cdot x^3 = x^{6 + 3}$$
$$= x^9$$

47.
$$3^4 \cdot 2^2 = 6^{4 + 2}$$

Simplify each expression. Write each answer in scientific notation.

48. $(9 \times 10^7)(3 \times 10^{-16})$ 49. $(8 \times 10^{-3})(0.1 \times 10^9)$

50. $(0.7 \times 10^{-12})(0.3 \times 10^8)$ 51. $(0.4 \times 10^0)(3 \times 10^{-4})$

52. $(0.2 \times 10^5)(4 \times 10^{-12})$ 53. $(0.5 \times 10^{13})(0.3 \times 10^{-4})$

Calculator Hint
To simplify $(6.12 \times 10^5) \cdot (12.5 \times 10^8)$, press
6.12 [EE] 5 [ENTER]
[×] 12.5 [EE] 8 [ENTER].

54. **Chemistry** The term *mole* can be used in chemistry to refer to 6.02×10^{23} atoms of a substance. The mass of a single hydrogen atom is approximately 1.67×10^{-24} gram. What is the mass of 1 mole of hydrogen atoms?

55. a. **Open-Ended** Write y^8 as a product of two powers with the same base in four different ways. Use only positive exponents.
   b. Write y^8 as a product of two powers with the same base in four different ways using negative or zero exponents in each.
   c. **Reasoning** How many ways are there to write y^8 as the product of two powers? Explain your reasoning.

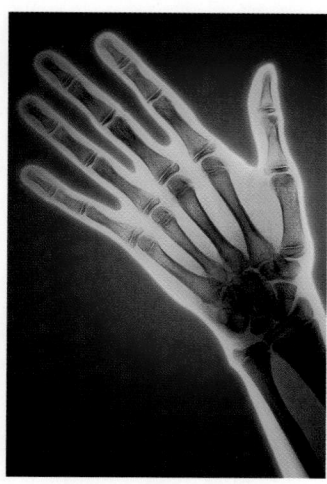

56. Medicine Medical X-rays, with a wavelength of about 10^{-10} meter, can penetrate your skin.

   a. Ultraviolet rays, which cause sunburn by penetrating only the top layers of skin, have a wavelength about 1000 times the wavelength of an X-ray. Find the wavelength of ultraviolet rays.

   b. Critical Thinking The wavelengths of visible light are between 4×10^{-7} meters and 7.5×10^{-7} meters. Are these wavelengths longer or shorter than those of ultraviolet rays? Explain.

57. Writing Explain why $x^3 \cdot y^5$ cannot be written with fewer bases.

58. Technology A CD-ROM stores about 650 megabytes (6.5×10^8 bytes) of information along a spiral track. Each byte uses about 9 micrometers (9×10^{-6} m) of space along the track. Find the length of the track.

Use a calculator. Simplify each expression. Write each answer in scientific notation.

59. $(6.12 \times 10^5)(12.5 \times 10^8)$ **60.** $(1.98 \times 10^{-3})(2.04 \times 10^{11})$

61. $(9.55 \times 10^7)(7.371 \times 10^{-15})$ **62.** $(6.934 \times 10^{-9})(2.579 \times 10^{-4})$

63. Measurement There are about 3.35×10^{25} molecules in a liter of water. Pine Lake, in New York, has about 2×10^8 liters of water. About how many molecules of water are in Pine Lake?

64. Measurement About 8.4×10^{11} drops of water flow over Niagara Falls each minute. Each drop of water contains about 1.7×10^{21} molecules of water. About how many molecules of water flow over the falls each minute?

Simplify each expression.

65. $\dfrac{1}{x^2 \cdot x^{-5}}$ **66.** $\dfrac{1}{a^3 \cdot a^{-2}}$ **67.** $\dfrac{5}{c \cdot c^{-4}}$

68. $2a^2(3a + 5)$ **69.** $8m^3(m^2 + 7)$ **70.** $-4x^3(2x^2 - 9x)$

 **Challenge**

Simplify.

71. $3^x \cdot 3^{2-x} \cdot 3^2$ **72.** $2^n \cdot 2^{n+2} \cdot 2$ **73.** $3^x \cdot 2^y \cdot 3^2 \cdot 2^x$

74. $(a + b)^2(a + b)^{-3}$ **75.** $(t + 3)^7(t + 3)^{-5}$ **76.** $5^{x+1} \cdot 5^{1-x}$

77. a. Geometry Find the volume of a rectangular prism with length 1.3×10^{-3} km, width 1.5×10^{-3} km, and height 9.4×10^{-4} km. Write your answer in scientific notation.

   b. What is the volume of the prism in cubic meters?

78. Science An illustrator plans to draw a diagram of a protozoan for a science book. A protozoan is 1.1×10^{-4} meters long. The illustrator wants the diagram to be 7.7 centimeters long. The diagram will be how many times greater than the protozoan in length?

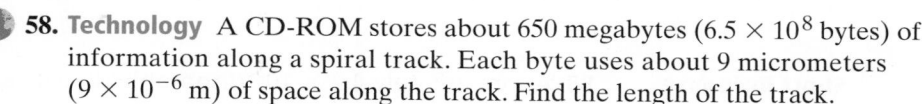

NY REGENTS

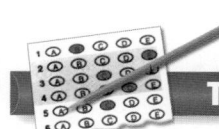

Test Prep

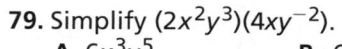

Multiple Choice

79. Simplify $(2x^2y^3)(4xy^{-2})$.

   A. $6x^3y^5$ **B.** $6x^2y^6$ **C.** $8x^2y$ **D.** $8x^3y$

80. In the 2000 Olympics, the winning time for the women's 100-meter race was 1.79×10^{-1} min. Which is another way of expressing this time in minutes?

   F. 0.179   **G.** 17.9   **H.** 179×10^1 **J.** 179×10^{-2}

Real-World Connection

X-rays are absorbed more by dense objects such as bones, and less by soft tissue. So bones show up as darker than soft tissues.

81. Which is the product of (4.5×10^7) and (2.4×10^{-1}) in scientific notation?

    A. 1.08×10^7 **B.** 10.8×10^6

    C. 10.8×10^{-7} **D.** 1.08×10^{-8}

82. Which is an expression for the area of the triangle shown at the right?

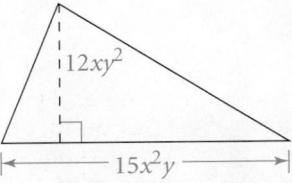

    F. $60x^2y^2$ **G.** $90x^3y^3$

    H. $180x^2y^2$ **J.** $180x^3y^3$

Short Response

83. Approximately 4.7×10^7 disposable diapers are thrown away each day in the United States. About how many are thrown away in one year? Write your answer in scientific notation. Show your work.

Extended Response

84. Sophie's Desserts packages its cheesecake in boxes with square bottoms, as shown below. Answer each of the following, showing all of your work.

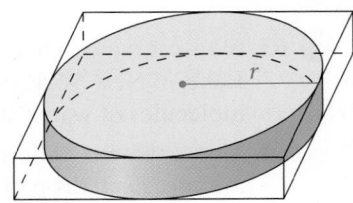

    a. Write an expression for the area of the bottom of the box.

    b. If the cheesecake has a radius of 5 in., what is the area of the bottom of the box?

    c. The area of the bottom of a second box is 144 in.^{2}. What is the diameter of the largest cheesecake the box can hold?

Mixed Review

Lesson 8-2

Write each number in scientific notation.

85. 1,280,000   **86.** 0.0035   **87.** 0.00009   **88.** 6.2 million

Write each number in standard form.

89. 8.76×10^8 **90.** 1.052×10^{-3} **91.** 9.1×10^{11} **92.** 2.9×10^{-4}

Lesson 7-6

Solve each system by graphing.

93. $y < 3x + 2$ **94.** $y < x + 6$ **95.** $y > x + 4$

      $2x + y \geq 4$ $x - 3y \leq 6$ $x + 2y \leq 6$

Lesson 5-7

Find the third, seventh, and tenth terms of each sequence.

96. $A(n) = 10 + (n - 1)(4)$ **97.** $A(n) = -5 + (n - 1)(2)$

98. $A(n) = 12 + (n - 1)(-4)$ **99.** $A(n) = 1.2 + (n - 1)(-4)$

More Multiplication Properties of Exponents

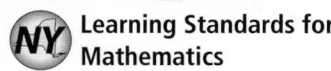

Learning Standards for Mathematics

A.N.4 Understand and use scientific notation to compute products and quotients of numbers.

A.A.12 Multiply and divide monomial expressions with a common base, using the properties of exponents.

✓ **Check Skills You'll Need**

GO for Help Lesson 8-3

Rewrite each expression using each base only once.

1. $3^2 \cdot 3^2 \cdot 3^2$

2. $2^3 \cdot 2^3 \cdot 2^3 \cdot 2^3$

3. $5^7 \cdot 5^7 \cdot 5^7 \cdot 5^7$

4. $7 \cdot 7 \cdot 7$

Simplify.

5. $x^3 \cdot x^3$

6. $a^2 \cdot a^2 \cdot a^2$

7. $y^{-2} \cdot y^{-2} \cdot y^{-2}$

8. $n^{-3} \cdot n^{-3}$

---

1 Raising a Power to a Power

Activity: Powers of Powers

You can use what you learned in the previous lesson to find a shortcut for simplifying expressions with powers. Copy and complete each statement.

1. $(3^6)^2 = 3^6 \cdot 3^6 = 3^{\blacksquare + \blacksquare} = 3^{6 \cdot \blacksquare} = 3^{\blacksquare}$

2. $(5^4)^3 = 5^4 \cdot 5^4 \cdot 5^4 = 5^{\blacksquare + \blacksquare + \blacksquare} = 5^{4 \cdot \blacksquare} = 5^{\blacksquare}$

3. $(2^7)^4 = 2^7 \cdot 2^7 \cdot 2^7 \cdot 2^7 = 2^{\blacksquare + \blacksquare + \blacksquare + \blacksquare} = 2^{7 \cdot \blacksquare} = 2^{\blacksquare}$

4. $(a^3)^2 = a^3 \cdot a^3 = a^{\blacksquare + \blacksquare} = a^{3 \cdot \blacksquare} = a^{\blacksquare}$

5. $(g^4)^3 = g^4 \cdot g^4 \cdot g^4 = g^{\blacksquare + \blacksquare + \blacksquare} = g^{4 \cdot \blacksquare} = g^{\blacksquare}$

6. $(c^3)^4 = c^3 \cdot c^3 \cdot c^3 \cdot c^3 = c^{\blacksquare + \blacksquare + \blacksquare + \blacksquare} = c^{3 \cdot \blacksquare} = c^{\blacksquare}$

7. a. Make a Conjecture What pattern do you see in your answers to Questions 1–6?
   b. Use your pattern to simplify $(8^6)^3$.

Raising a power to a power is the same as raising the base to the product of the exponents.

 Key Concepts

Property	**Raising a Power to a Power**

For every nonzero number a and integers m and n, $(a^m)^n = a^{mn}$.

Examples $(5^4)^2 = 5^{4 \cdot 2} = 5^8$ $(x^2)^5 = x^{2 \cdot 5} = x^{10}$

1 EXAMPLE Simplifying a Power Raised to a Power

Simplify $(x^3)^6$.

$(x^3)^6 = x^{3 \cdot 6}$ **Multiply exponents when raising a power to a power.**

$= x^{18}$ **Simplify.**

✓ Quick Check ❶ Simplify $(a^4)^7$ and $(a^{-4})^7$.

Be sure to use the order of operations. Simplify expressions in parentheses first.

GO **⊙nline**

Video Tutor Help
Visit: PHSchool.com
Web Code: ate-0775

2 EXAMPLE Simplifying an Expression With Powers

Simplify $c^5(c^3)^{-2}$.

$c^5(c^3)^{-2} = c^5 \cdot c^{3 \cdot (-2)}$ **Multiply exponents in $(c^3)^{-2}$.**

$= c^5 \cdot c^{-6}$ **Simplify.**

$= c^{5 + (-6)}$ **Add exponents when multiplying powers with the same base.**

$= c^{-1}$ **Simplify.**

$= \frac{1}{c}$ **Write using only positive exponents.**

✓ Quick Check ❷ Simplify each expression. **a.** $t^2(t^7)^{-2}$ **b.** $(a^4)^2 \cdot (a^2)^5$

---

2 Raising a Product to a Power

You can use repeated multiplication to simplify expressions like $(5y)^3$.

$$(5y)^3 = 5y \cdot 5y \cdot 5y$$
$$= 5 \cdot 5 \cdot 5 \cdot y \cdot y \cdot y$$
$$= 5^3 y^3$$
$$= 125y^3$$

Notice that $(5y)^3 = 5^3 y^3$. This illustrates another property of exponents.

 Key Concepts

Property	**Raising a Product to a Power**
For every nonzero number a and b and integer n, $(ab)^n = a^n b^n$.	
Example $(3x)^4 = 3^4 x^4 = 81x^4$	

3 EXAMPLE Simplifying a Product Raised to a Power

 **Test-Taking Tip**

When raising a product to a power, make sure each factor of the product is raised to the power.

Multiple Choice Which expression represents the area of the square?

Ⓐ $8x^4$ Ⓑ $2x^8$ Ⓒ $4x^6$ Ⓓ $4x^8$

$(2x^4)^2 = 2^2(x^4)^2$ **Raise each factor to the 2nd power.**

$= 2^2 x^8$ **Multiply exponents of a power raised to a power.**

$= 4x^8$ **Simplify.**

The correct answer is D.

$2x^4$

✓ Quick Check ❸ Simplify each expression. **a.** $(2z)^4$ **b.** $(4g^5)^{-2}$

Some expressions have more than one power raised to a power.

4 EXAMPLE **Simplifying a Product Raised to a Power**

Simplify $(x^{-2})^2(3xy^2)^4$.

$$(x^{-2})^2(3xy^2)^4 = (x^{-2})^2 \cdot 3^4x^4(y^2)^4 \qquad \text{Raise the three factors to the 4th power.}$$
$$= x^{-4} \cdot 3^4x^4y^8 \qquad \text{Multiply the exponents of a power raised to a power.}$$
$$= 3^4 \cdot x^{-4} \cdot x^4 \cdot y^8 \qquad \text{Use the Commutative Property of Multiplication.}$$
$$= 3^4x^0y^8 \qquad \text{Add exponents of powers with the same base.}$$
$$= 81y^8 \qquad \text{Simplify.}$$

✓ **Quick Check** ❹ Simplify each expression.
   a. $(c^2)^3(3c^5)^4$ **b.** $(2a^3)^5(3ab^2)^3$ **c.** $(6mn)^3(5m^{-3})^2$

You can use the property of raising a product to a power to solve problems involving scientific notation. For an expression like $(3 \times 10^8)^2$, raise both 3 and 10^8 to the second power.

5 EXAMPLE **Real-World** 🌐 **Problem Solving**

Physical Science All objects, even resting ones, contain energy. A raisin has a mass of 10^{-3} kg. The expression $10^{-3} \cdot (3 \times 10^8)^2$ describes the amount of resting energy in joules the raisin contains. Simplify the expression.

$$10^{-3} \cdot (3 \times 10^8)^2 = 10^{-3} \cdot 3^2 \cdot (10^8)^2 \qquad \text{Raise each factor within parentheses to the second power.}$$
$$= 10^{-3} \cdot 3^2 \cdot 10^{16} \qquad \text{Simplify } (10^8)^2.$$
$$= 3^2 \cdot 10^{-3} \cdot 10^{16} \qquad \text{Use the Commutative Property of Multiplication.}$$
$$= 3^2 \cdot 10^{-3 + 16} \qquad \text{Add exponents of powers with the same base.}$$
$$= 9 \times 10^{13} \qquad \text{Simplify. Write in scientific notation.}$$

Real-World 🌐 **Connection**

Albert Einstein is famous for discovering the relationship $E = mc^2$, where E is energy (in joules), m is mass (in kg), and c is the speed of light (about 3×10^8 meters per second).

✓ **Quick Check** ❺ **Energy** An hour of television use consumes 1.45×10^{-1} kWh (kilowatt-hour) of electricity. Each kilowatt-hour of electric use is equivalent to 3.6×10^6 joules of energy.
   a. Simplify the expression $(1.45 \times 10^{-1})(3.6 \times 10^6)$ to find how many joules a television uses in 1 hour.
   b. **Critical Thinking** Suppose you could release the resting energy in a raisin. About how many hours of television use could be powered by that energy?

EXERCISES

For more exercises, see *Extra Skill and Word Problem Practice.*

Practice and Problem Solving

A **Practice by Example**

Examples 1, 2
(page 448)

Simplify each expression.

1. $(c^5)^2$ **2.** $(c^2)^5$ **3.** $(n^8)^4$ **4.** $(q^{10})^{10}$

5. $(c^5)^3c^4$ **6.** $(d^3)^5(d^3)^0$ **7.** $(t^2)^{-2}(t^2)^{-5}$ **8.** $(x^3)^{-1}(x^2)^5$

Example 3
(page 448)

Simplify each expression.

9. $(5y)^4$ **10.** $(4m)^5$ **11.** $(7a)^2$ **12.** $(12g^4)^{-1}$

13. $(6y^2)^2$ **14.** $(3n^6)^4$ **15.** $(2y^4)^{-3}$ **16.** $(2p^6)^0$

Example 4
(page 449)

17. $(x^2)^5(x^3)^2$ **18.** $(2xy)^3x^2$ **19.** $(mg^4)^{-1}(mg^4)$

20. $(c^{-2})^3c^{-12}$ **21.** $(3b^{-2})^2(a^2b^4)^3$ **22.** $(2a^2c^4)^{-5}(c^{-1}a^7)^6$

Example 5
(page 449)

Simplify. Write each answer in scientific notation.

23. $(4 \times 10^5)^2$ **24.** $(3 \times 10^5)^2$ **25.** $(2 \times 10^{-10})^3$ **26.** $(2 \times 10^{-3})^3$

27. $(7 \times 10^4)^2$ **28.** $(6 \times 10^{12})^2$ **29.** $(4 \times 10^8)^{-2}$ **30.** $(3.5 \times 10^{-4})^3$

31. Geometry The length of one side of a cube is 9.5×10^{-4} m. What is the volume of the cube?

B **Apply Your Skills**

Complete each equation.

32. $(x^2)^\blacksquare = x^6$ **33.** $(m^\blacksquare)^3 = m^{-12}$ **34.** $(b^2)^\blacksquare = b^8$

35. $(y^{-4})^\blacksquare = y^{12}$ **36.** $(n^9)^\blacksquare = 1$ **37.** $7(c^1)^\blacksquare = 7c^8$

38. $(5x^\blacksquare)^2 = 25x^{-4}$ **39.** $(3x^3y^\blacksquare)^3 = 27x^9$ **40.** $(m^2n^3)^\blacksquare = \frac{1}{m^6n^9}$

41. Error Analysis One student simplified $x^5 + x^5$ to x^{10}. A second student simplified $x^5 + x^5$ to $2x^5$. Which student is correct? Explain.

Simplify each expression.

42. $(4.1)^5 \cdot (4.1)^{-5}$ **43.** $3^2(3x)^3$ **44.** $(b^5)^3b^2$

45. $(-5x)^2 + 5x^2$ **46.** $(2x^{-3})^2 \cdot (0.2x)^2$ **47.** $(-2a^2b)^3(ab)^3$

48. $(3^7)^2 \cdot (3^{-4})^3$ **49.** $(10^3)^4(4.3 \times 10^{-8})$ **50.** $(4xy^2)^4(-y)^{-3}$

51. a. Geometry Write an expression for the surface area of each cube.
  b. How many times greater than the surface area of the small cube is the surface area of the large cube?
  c. Write an expression for the volume of each cube.
  d. How many times greater than the volume of the small cube is the volume of the large cube?

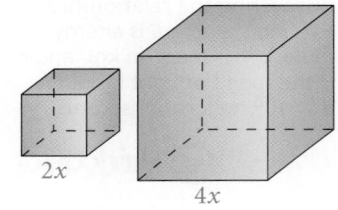

$2x$

$4x$

Write each expression with only one exponent. Use parentheses.

52. $m^4 \cdot n^4$ **53.** $(a^5)(b^5)(a^0)$ **54.** $49x^2y^2z^2$ **55.** $\frac{12x^2}{3y^2}$

56. Open-Ended Choose a value of n for the expression a^n. Express the power you wrote as a product of the form $(a^c)^d$ in four different ways.

57. Measurement Write each answer as a power of 10.
  a. How many cubic centimeters are in a cubic meter?
  b. How many cubic millimeters are in a cubic meter?
  c. How many cubic meters are in a cubic kilometer?
  d. How many cubic millimeters are in a cubic kilometer?

GO **Online**
Homework Video Tutor
Visit: PHSchool.com
Web Code: ate-0804

58. Computers Write each answer as a power of 2.
    a. Computer capacity is often measured in bits and bytes. A bit is the smallest unit, a 1 or 0 in the computer's memory. A byte is 2^3 bits. A megabyte (MB) is 2^{20} bytes. How many bits are in a megabyte?
    b. A gigabyte (GB) is 2^{10} megabytes. How many bytes are there in a gigabyte? How many bits are there in a gigabyte?

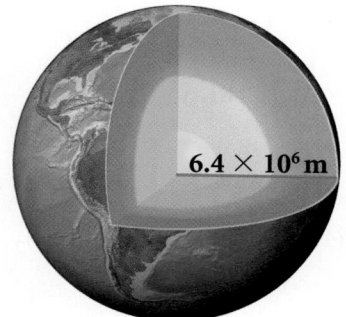

 59. a. Geography Earth has a radius of about 6.4×10^6 m. Approximate the surface area of Earth using the formula for the surface area of a sphere, $S = 4\pi r^2$.
    b. Earth's surface is about 70% water, almost all of it in oceans. About how many square meters of Earth's surface are covered with water?
    c. The oceans have an average depth of 3795 m. Estimate the volume of water on Earth.

6.4×10^6 m

60. Which expression or expressions do *not* equal 64?
    A. $2^5 \cdot 2$ **B.** 2^6 **C.** $2^2 \cdot 2^3$ **D.** $(2^3)^2$ **E.** $(2^2)(2^2)^2$

61. Writing Explain how you know when to add the exponents of powers and when to multiply the exponents.

C **Challenge**

Solve each equation.

Sample $\quad 25^3 = 5^x$
$\quad\quad\quad (5^2)^3 = 5^x$ **Write 25 as a power of 5.**
$\quad\quad\quad\quad\quad 5^6 = 5^x$ **Simplify $(5^2)^3$.**
$\quad\quad\quad\quad\quad\quad 6 = x$ **Since the bases are the same, the exponents are equal.**

62. $5^6 = 25^x$ **63.** $8^2 = 2^x$ **64.** $3^x = 27^4$

65. $4^x = 2^6$ **66.** $3^{2x} = 9^4$ **67.** $2^x = \frac{1}{32}$

68. Critical Thinking Simplify $(x^3)^4$ and x^{3^4}. Are the expressions equivalent?

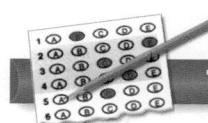  **REGENTS**

Test Prep

Multiple Choice **69.** Which expression could you use for the area of the triangle at the right?
    A. $3x$ **B.** $4.5x^2$
    C. $9x^2$ **D.** $22.5x^2$

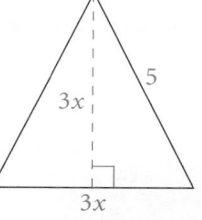
5
3x
3x

70. Evaluate $3a^2$ for $a = 5.1 \times 10^{-5}$.
    F. 1.53×10^{-10} **G.** 7.803×10^{-10}
    H. 1.53×10^{-9} **J.** 7.803×10^{-9}

71. If 7^{-2} is raised to the power of 3, which of the following describes the result?
    A. a number less than -7 **B.** a number between -1 and 0
    C. a number between 0 and 1   **D.** a number greater than 1

72. Which expression does NOT equal $25n^{12}$?
    F. $(5n^6)^2$ **G.** $(5n^3)(5n^9)$ **H.** $25(n^3)^9$ **J.** $5^2(n^2)^6$

Short Response **73.** Does $(x^2 + 3y)^2$ equal $x^4 + 9y^2$? Substitute values for x and y to justify your answer.

Lesson 8-3   Simplify each expression.

74. $bc^{-6} \cdot b$ **75.** $(a^2 b^3)(a^6)$ **76.** $9m^3(6m^2 n^4)$ **77.** $2t(-2t^4)$

Lesson 7-2   Solve each system using substitution.

78. $y = 3x + 5$ **79.** $y = 0.5x - 1$ **80.** $y = 5x - 9$ **81.** $y = x + 4$
     $y = -4x + 12$ $y = 0.2x + 0.4$ $y = 3x + 5$ $y = -5$

Lesson 6-1   Find the slope of the line that passes through each pair of points.

82. $(0, 3), (4, 0)$ **83.** $(2, -5), (3, 1)$ **84.** $(-3, 6), (1, 0)$ **85.** $(0, 0), (11, -9)$

 Checkpoint Quiz 1   **Lessons 8-1 through 8-4**

Simplify each expression.

1. $5^{-1}(3^{-2})$ **2.** $(r^{-5})^{-4}$ **3.** $(2x^5)(3x^{12})$ **4.** $\dfrac{mn^{-4}}{p^0 q^{-2}}$

5. $a^2 b^0 (a^{-3})$ **6.** $(3^2)^{-1}(4m^2)^3$ **7.** $(2m^3)(3m^6)$ **8.** $(3t^2)^3(2t^0)^{-3}$

9. A certain bacteria population doubles in size every day. Suppose a sample
   starts with 500 bacteria. The expression $500 \cdot 2^x$ models the number of bacteria
   in the sample after x days. Evaluate the expression for $x = 0, 2, 5$.

10. Astronomy The diameter of Mars is about 6800 km.
   a. Write this number in scientific notation.
   b. Approximate the surface area of Mars using the formula for the surface area
      of a sphere, $S = 4\pi r^2$. Write your answer in scientific notation.
   c. Write your answer from part (b) in standard form.

Algebra at Work

.... **Dr. Jewel Plummer Cobb**

Dr. Jewel Plummer Cobb was born in 1924 and obtained her master's and
doctor's degrees in cell physiology from New York University. Dr. Cobb has
concentrated on the study of normal and malignant skin cells and has
published nearly 50 books, articles, and reports. Because the number of
cancer cells grows exponentially, cell biologists often write cancer cell data
in scientific notation.

For: Information about a career in cancer research
PHSchool.com **Web Code:** atb-2031

8-5

Division Properties of Exponents

Learning Standards for Mathematics

A.N.4 Understand and use scientific notation to compute products and quotients of numbers.

A.A.12 Multiply and divide monomial expressions with a common base, using the properties of exponents.

 Check Skills You'll Need

GO for Help Skills Handbook page 758

Write each fraction in simplest form.

1. $\frac{5}{20}$

2. $\frac{125}{25}$

3. $\frac{60}{100}$

4. $\frac{124}{4}$

5. $\frac{6}{15}$

6. $\frac{8}{30}$

7. $\frac{10}{35}$

8. $\frac{18}{63}$

9. $\frac{5xy}{15x}$

10. $\frac{6y^2}{3x}$

11. $\frac{3ac}{12a}$

12. $\frac{24m}{6mn^2}$

---

1 Dividing Powers With the Same Base

You can use repeated multiplication to simplify fractions. Expand the numerator and the denominator using repeated multiplication. Then cancel like terms.

$$\frac{5^6}{5^2} = \frac{5 \cdot 5 \cdot 5 \cdot 5 \cdot 5 \cdot 5}{5 \cdot 5} = 5^4$$

This illustrates the following property of exponents.

 Key Concepts

Property	**Dividing Powers With the Same Base**

For every nonzero number a and integers m and n, $\frac{a^m}{a^n} = a^{m-n}$.

Example $\frac{3^7}{3^3} = 3^{7-3} = 3^4$

Since division by zero is undefined, assume that no base is equal to zero.

GO Online

Video Tutor Help
Visit: PHSchool.com
Web Code: ate-0775

1 EXAMPLE Simplifying an Algebraic Expression

Simplify each expression.

a. $\frac{a^6}{a^{14}} = a^{6-14}$ **Subtract exponents when dividing powers with the same base.**

$= a^{-8}$ **Simplify the exponents.**

$= \frac{1}{a^8}$ **Rewrite using positive exponents.**

b. $\frac{c^{-1}d^3}{c^5d^{-4}} = c^{-1-5}d^{3-(-4)}$ **Subtract exponents when dividing powers with the same base.**

$= c^{-6}d^7$ **Simplify.**

$= \frac{d^7}{c^6}$ **Rewrite using positive exponents.**

 **Quick Check** ❶ Simplify each expression.

a. $\dfrac{b^4}{b^9}$ **b.** $\dfrac{z^{10}}{z^5}$ **c.** $\dfrac{a^2b}{a^4b^3}$ **d.** $\dfrac{m^{-1}n^2}{m^3n}$ **e.** $\dfrac{x^2y^{-1}z^4}{xy^4z^{-3}}$

When you divide numbers that are in scientific notation, you can use the property of dividing powers with the same base. In real-world situations, decide whether to write the result in standard or scientific notation.

❷ **EXAMPLE**   **Real-World** 🌎 **Problem Solving**

Recycling In 2000, the total amount of paper and paperboard recycled in the United States was 37 million tons. The population of the United States in 2000 was 281.4 million. On average, how much paper and paperboard did each person recycle?

$$\frac{37 \text{ million tons}}{281.4 \text{ million people}} = \frac{3.7 \times 10^7 \text{ tons}}{2.814 \times 10^8 \text{ people}} \qquad \text{Write in scientific notation.}$$

$$= \frac{3.7}{2.814} \times 10^{7-8} \qquad \text{Subtract exponents when dividing powers with the same base.}$$

$$= \frac{3.7}{2.814} \times 10^{-1} \qquad \text{Simplify the exponent.}$$

$$\approx 1.3 \times 10^{-1} \qquad \text{Divide. Round to the nearest tenth.}$$

$$= 0.13 \qquad \text{Write in standard notation.}$$

Real-World 🌎 **Connection**

Worldwide, about 43% of the paper that is discarded is recovered for recycling.

● There was about 0.13 ton of paper and paperboard recycled per person in 2000.

 **Quick Check** ❷ Find each quotient. Write each answer in scientific notation.

a. $\dfrac{2 \times 10^3}{8 \times 10^8}$ **b.** $\dfrac{7.5 \times 10^{12}}{2.5 \times 10^{-4}}$ **c.** $\dfrac{4.2 \times 10^5}{12.6 \times 10^2}$

d. In 2000 the total amount of glass recycled in the United States was 2.7 million tons. The population of the United States in 2000 was 281.4 million people. On average, about how many tons of glass were recycled per person?

2 Raising a Quotient to a Power

You can use repeated multiplication to simplify the expression $\left(\dfrac{x}{y}\right)^3$.

$$\left(\frac{x}{y}\right)^3 = \frac{x}{y} \cdot \frac{x}{y} \cdot \frac{x}{y}$$

$$= \frac{x \cdot x \cdot x}{y \cdot y \cdot y}$$

$$= \frac{x^3}{y^3}$$

This illustrates another property of exponents.

 Key Concepts

Property	**Raising a Quotient to a Power**
For every nonzero number a and b and integer n, $\left(\dfrac{a}{b}\right)^n = \dfrac{a^n}{b^n}$.	
Example $\left(\dfrac{4}{5}\right)^3 = \dfrac{4^3}{5^3} = \dfrac{64}{125}$	

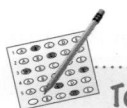

Test-Taking Tip

Check your answer when simplifying by substituting a value for the variable.
For $x = 2$,
$\left(\frac{4}{2^2}\right)^3 = \left(\frac{4}{4}\right)^3 = 1^3 = 1$
$\frac{64}{2^6} = \frac{64}{64} = 1$

3 EXAMPLE Raising a Quotient to a Power

Multiple Choice Which expression is equivalent to $\left(\frac{4}{x^2}\right)^3$?

Ⓐ $\dfrac{12}{x^5}$ Ⓑ $\dfrac{12}{x^6}$ Ⓒ $\dfrac{64}{x^5}$ Ⓓ $\dfrac{64}{x^6}$

$\left(\dfrac{4}{x^2}\right)^3 = \dfrac{4^3}{(x^2)^3}$ **Raise the numerator and the denominator to the third power.**

$\qquad\quad = \dfrac{4^3}{x^6}$ **Multiply the exponents in the denominator.**

$\qquad\quad = \dfrac{64}{x^6}$ **Simplify.**

● The correct answer is D.

✓ **Quick Check** ③ Simplify each expression.   **a.** $\left(\dfrac{3}{x^2}\right)^2$ **b.** $\left(\dfrac{x}{y^2}\right)^3$

You can use what you know about exponents to rewrite an expression in the form $\left(\frac{a}{b}\right)^{-n}$ using positive exponents.

$\left(\dfrac{a}{b}\right)^{-n} = \dfrac{1}{\left(\frac{a}{b}\right)^n}$ Use the definition of negative exponent.

$\qquad\quad = \dfrac{1}{\frac{a^n}{b^n}}$ Raise the quotient to a power.

$\qquad\quad = \dfrac{1}{\frac{a^n}{b^n}} \cdot \dfrac{b^n}{b^n}$ Use the Identity Property of Multiplication to multiply by $\frac{b^n}{b^n}$.

$\qquad\quad = \dfrac{b^n}{a^n}$ Simplify.

$\qquad\quad = \left(\dfrac{b}{a}\right)^n$ Write the quotient using one exponent.

So, $\left(\dfrac{a}{b}\right)^{-n} = \left(\dfrac{b}{a}\right)^n$.

4 EXAMPLE Simplifying an Exponential Expression

Simplify each expression.

a. $\left(\dfrac{3}{5}\right)^{-2} = \left(\dfrac{5}{3}\right)^2$ **Rewrite using the reciprocal of $\frac{3}{5}$.**

$\qquad\quad = \dfrac{5^2}{3^2}$ **Raise the numerator and denominator to the second power.**

$\qquad\quad = \dfrac{25}{9}$ or $2\frac{7}{9}$ **Simplify.**

b. $\left(-\dfrac{2x}{y}\right)^{-4} = \left(-\dfrac{y}{2x}\right)^4$ **Rewrite using the reciprocal of $-\frac{2x}{y}$.**

$\qquad\quad = \left(\dfrac{-y}{2x}\right)^4$ **Write the fraction with a negative numerator.**

$\qquad\quad = \dfrac{(-y)^4}{(2x)^4}$ **Raise the numerator and denominator to the fourth power.**

$\qquad\quad = \dfrac{y^4}{16x^4}$ **Simplify.**

✓ **Quick Check** ④ Simplify each expression.

a. $\left(\dfrac{3}{4}\right)^{-3}$ **b.** $\left(\dfrac{-1}{2}\right)^{-5}$ **c.** $\left(\dfrac{2r}{s}\right)^{-1}$ **d.** $\left(\dfrac{7a}{m}\right)^{-2}$

EXERCISES

For more exercises, see *Extra Skill and Word Problem Practice*.

Practice and Problem Solving

A Practice by Example

Example 1
(page 453)

GO for Help

Copy and complete each equation.

1. $\dfrac{5^9}{5^2} = 5^\blacksquare$

2. $\dfrac{2^4}{2^3} = 2^\blacksquare$

3. $\dfrac{3^2}{3^5} = 3^\blacksquare$

4. $\dfrac{5^2 5^3}{5^3 5^2} = 5^\blacksquare$

Simplify each expression.

5. $\dfrac{2^5}{2^7}$

6. $\dfrac{2^7}{2^5}$

7. $\dfrac{c^{12}}{c^{15}}$

8. $\dfrac{m^{-2}}{m^{-5}}$

9. $\dfrac{3s^{-9}}{6s^{-11}}$

10. $\dfrac{x^{13}y^2}{x^{13}y}$

11. $\dfrac{c^2 d^{-3}}{c^3 d^{-1}}$

12. $\dfrac{3^2 m^3 t^6}{3^5 m^7 t^{-5}}$

Example 2
(page 454)

Simplify each quotient. Write each answer in scientific notation.

13. $\dfrac{6.5 \times 10^{15}}{1.3 \times 10^8}$

14. $\dfrac{2.7 \times 10^{-8}}{9 \times 10^{-4}}$

15. $\dfrac{4.2 \times 10^8}{7 \times 10^5}$

16. $\dfrac{8.4 \times 10^{-5}}{2 \times 10^{-8}}$

17. $\dfrac{4.65 \times 10^{-4}}{3.1 \times 10^2}$

18. $\dfrac{3.5 \times 10^6}{5 \times 10^8}$

19. **Television** In 2000, people in the United States over age 2 watched television a total of 386 billion hours. The population of the United States over age 2 was about 265 million people.
   a. Write each number in scientific notation.
   b. Find the average number of hours of TV viewing per person older than age 2 for 2000.
   c. On average, how many hours per day did each person older than age 2 watch television in 2000?

20. **Computers** The speed of computers is measured in number of calculations per picosecond. There are 3.6×10^{15} picoseconds per hour. What fraction of a second is a picosecond?

Example 3
(page 455)

Simplify each expression.

21. $\left(\dfrac{3}{5}\right)^2$

22. $\left(\dfrac{1}{x}\right)^3$

23. $\left(\dfrac{2x}{y}\right)^5$

24. $\left(\dfrac{3a}{2b}\right)^4$

25. $\left(\dfrac{2^2}{5}\right)^3$

26. $\left(\dfrac{3^3}{3^4}\right)^2$

27. $\left(\dfrac{6}{n^6}\right)^2$

28. $\left(\dfrac{2p}{5}\right)^3$

Example 4
(page 455)

29. $\left(\dfrac{2}{3}\right)^{-1}$

30. $\left(\dfrac{2}{3}\right)^{-2}$

31. $\left(-\dfrac{2}{3}\right)^{-2}$

32. $\left(-\dfrac{2}{3}\right)^{-3}$

33. $\left(\dfrac{3x^4}{15}\right)^2$

34. $\left(\dfrac{4n}{2n^2}\right)^3$

35. $\left(\dfrac{c^5}{c^9}\right)^3$

36. $\left(\dfrac{3b^2}{5}\right)^0$

B Apply Your Skills

Explain why each expression is *not* in simplest form.

37. $5^3 m^3$

38. $x^5 y^{-2}$

39. $(2c)^4$

40. $x^0 y$

41. $\dfrac{d^7}{d}$

Simplify each expression.

42. $\dfrac{3^2 \cdot 5^0}{2^3}$

43. $\left(\dfrac{2m^5}{m^2}\right)^{-4}$

44. $\dfrac{5x^3}{(5x)^3}$

45. $\dfrac{(2a^7)(3a^2)}{6a^3}$

46. $\left(\dfrac{7t^3}{21t}\right)^3$

47. $\left(\dfrac{n^4 n}{n^{-2}}\right)^{-4}$

48. $\left(\dfrac{2k^3}{3k^{-2}}\right)^{-2}$

49. $\dfrac{7^9 \cdot (10)^2}{7^7}$

1920's

1950's

Today

50. **Telecommunications** In 2000, there were 97.4 million households with telephones. The people in these households made 544 billion local calls and 97 billion long distance calls.
   a. Write each number in scientific notation.
   b. What was the average number of local calls placed per household? Round to the nearest whole number.
   c. What was the average number of long distance calls placed per household? Round to the nearest whole number.

51. a. **Writing** While simplifying the expression $\frac{c^4}{c^6}$, Kneale said, "I've found a property of exponents that's not in my algebra book!" Write an explanation of why Kneale's method works.
   b. **Open-Ended** Apply Kneale's method to an example you create.

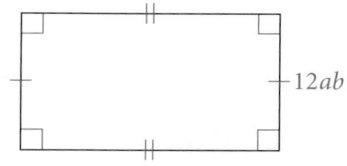

Kneale
$$\frac{c^4}{c^6} = \frac{1}{c^{6-4}} = \frac{1}{c^2}$$

Simplify each expression.

52. $\left(\frac{2ab^6}{a^3b}\right)^{-2}$ 53. $\frac{a^3b^2c^{-4}}{a^{-2}b^5c^{-9}}$ 54. $\frac{\left(\frac{1}{3}\right)^{-3}}{\left(\frac{1}{6}\right)^{-2}}$ 55. $\frac{0.2^2 \cdot 0.2^3}{0.2^6}$

56. $\left(\frac{p^{-2}q^4r}{p^3q^5}\right)^5$ 57. $\left(\frac{(-3)^2}{(-2)^{-4}}\right)^2$ 58. $\left(\frac{(3x)^2y}{x^2y^4}\right)^{-2}$ 59. $\frac{(5a^2)(6b^3)}{(2a^3)(25b^{-2})}$

60. **Multiple Choice** The area of the rectangle is $60a^2b^5$. What is the width of the rectangle?

   F $\frac{a^2b^5}{5}$ G $\frac{5}{a^2b^5}$
   H $5ab^4$ J $5a^3b^6$

$-12ab$

61. **Critical Thinking** Lena and Jared used different methods to simplify $\left(\frac{b^7}{b^3}\right)^2$. Why are both methods correct?

Lena
$$\left(\frac{b^7}{b^3}\right)^2 = \frac{b^{14}}{b^6}$$
$$= b^8$$

Jared
$$\left(\frac{b^7}{b^3}\right)^2 = (b^4)^2$$
$$= b^8$$

62. a. **Finance** In 1990, the United States government owed $3.23 trillion to its creditors. The population of the United States was 248.7 million people. How much did the government owe per person in 1990? Round to the nearest dollar.
   b. In 1999 the debt had grown to $5.66 trillion, with a population of 273 million. How much did the government owe per person? Round to the nearest dollar.
   c. What was the percent of increase in the average amount owed per person from 1990 to 1999?

GO **Online**
Homework Video Tutor
Visit: PHSchool.com
Web Code: ate-0805

63. a. **Error Analysis** What error did the student make in simplifying the expression at the right?
   b. What is the correct answer?

$$5^4 \div 5 = \frac{5^4}{5}$$
$$= 1^4$$
$$= 1$$

Write each expression with only one exponent. You may need to use parentheses.

64. $\dfrac{3^5}{5^5}$ **65.** $\dfrac{m^7}{n^7}$ **66.** $\dfrac{d^8}{d^5}$ **67.** $\dfrac{10^7 \cdot 10^0}{10^{-3}}$

68. $\dfrac{27x^3}{8y^3}$ **69.** $\dfrac{4m^2}{169m^4}$ **70.** $\dfrac{49m^2}{25n^2}$ **71.** $\dfrac{125c^7}{216c^4}$

 72. Medicine If you donate blood regularly, the American Red Cross recommends a 56-day waiting period between donations. One pint of blood contains about 2.4×10^{12} red blood cells. Your body normally produces about 2×10^6 red blood cells per second.
   a. At its normal rate, in how many seconds will your body replace the red blood cells lost by giving one pint of blood?
   b. Convert your answer from part (a) to days.

73. a. Open Ended Write three numbers in scientific notation.
   b. Divide each number by 2.
   c. Critical Thinking Is the power of 10 divided by 2 when you divide a number in scientific notation by 2? Explain.

Which property or properties of exponents would you use to simplify each expression?

74. 2^{-3} **75.** $\dfrac{2^2}{2^5}$ **76.** $\left(\dfrac{1}{2}\right)^3$ **77.** $\dfrac{1}{2^{-4}2^7}$ **78.** $\dfrac{(2^4)^3}{2^{15}}$

 **Challenge**

Simplify each expression.

79. $n^{x+2} \div n^x$ **80.** $n^{5x} \div n^x$ **81.** $\left(\dfrac{x^m}{x^{m-2}}\right)^2$ **82.** $\dfrac{\left(\dfrac{n^5}{n^4}\right)}{n^3}$

83. Astronomy The ratio of a planet's maximum to minimum distance from the sun is related to how circular its orbit is.
   a. Copy and complete the table below. Round decimals to the nearest hundredth.
   b. Reasoning How can you use the ratio maximum : minimum to determine whether a planet's orbit is close to circular?
   c. Which planet has the least circular orbit? The most circular orbit?

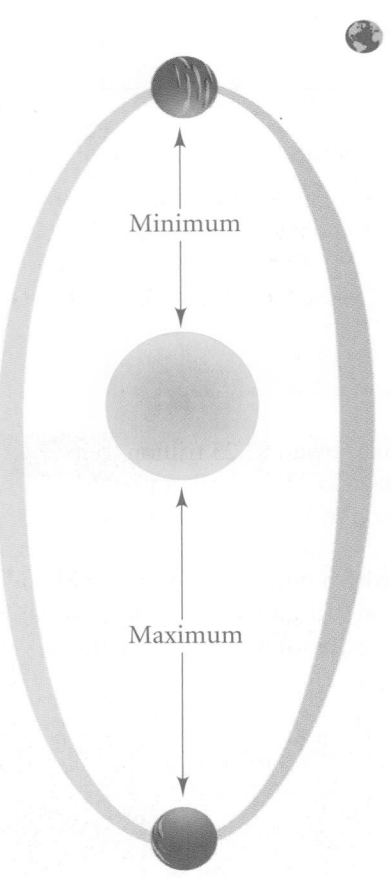

Minimum

Maximum

Distance From the Sun (kilometers)

Planet	Maximum	Minimum	Maximum : Minimum
Mercury	6.97×10^7	4.59×10^7	$\blacksquare : \blacksquare = \dfrac{6.97 \times 10^7}{4.59 \times 10^7} = \dfrac{6.97}{4.59} \approx 1.52$
Venus	1.089×10^8	1.075×10^8	$1.089 \times 10^8 : \blacksquare \approx \blacksquare$
Earth	1.521×10^8	1.471×10^8	$\blacksquare : 1.471 \times 10^8 \approx \blacksquare$
Mars	2.491×10^8	2.067×10^8	$\blacksquare : \blacksquare \approx \blacksquare$
Jupiter	8.157×10^8	7.409×10^8	$\blacksquare : \blacksquare \approx \blacksquare$
Saturn	1.507×10^9	1.347×10^9	$\blacksquare : \blacksquare \approx \blacksquare$
Uranus	3.004×10^9	2.735×10^9	$\blacksquare : \blacksquare \approx \blacksquare$
Neptune	4.537×10^9	4.457×10^9	$\blacksquare : \blacksquare \approx \blacksquare$
Pluto	7.375×10^9	4.425×10^9	$\blacksquare : \blacksquare \approx \blacksquare$

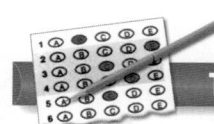

Multiple Choice

84. Simplify the expression $\frac{(-6)^5}{6^5}$.

   A. -6^5 **B.** -1 **C.** 1 **D.** 6^7

85. Evaluate $\frac{-5x^3y^5}{15x^{-7}y^5z^{-2}}$ for $x = -1$, $y = 5$, and $z = 3$.

   F. -9 **G.** -3 **H.** -1 **J** 0

86. Which point on the number line below could be the graph of 2^n if n is a negative integer?

$$\begin{array}{cccccc} Y & W\ Z & & X & \\ \hline -2 & -1 & 0 & 1 & 2 \end{array}$$

   A. W **B.** X **C.** Y **D.** Z

87. Which of the following statements is true?

   F. Any number raised to the zero power equals zero.

   G. Any number raised to a negative power equals a negative number.

   H. Negative numbers can be written in scientific notation.

   J. The product of two powers with the same base equals the base raised to the sum of the exponents.

88. Find the quotient $\frac{1.8 \times 10^{-4}}{3.6 \times 10^3}$.

   A. 5×10^{-8} **B.** 5×10^{-6} **C.** 5×10^{-2} **D.** 5×10^{-1}

89. Which expression is NOT equivalent to $\left(\frac{4n}{3m^5}\right)^{-2}$?

   F. $\left(\frac{3m^5}{4n}\right)^2$ **G.** $\frac{4n^{-2}}{3m^3}$ **H.** $\frac{9m^{10}}{16n^2}$ **J** $\left(\frac{16n^2}{9m^{10}}\right)^{-1}$

Extended Response

90. At its closest, Saturn is about 743,000,000 miles from Earth. A deep-space probe travels from Earth to Saturn at an average speed of 25,000 miles per hour. Assume that the probe can go straight from Earth to Saturn. How many hours will it take the probe to get from Earth to Saturn? About how many years will it take? Show your work.

Mixed Review

GO for Help

Lesson 8-4

Simplify each expression.

91. $(3y^2)^3$ **92.** $(2m^{-7})^3$ **93.** $(r^2t^{-5})^{-4}$ **94.** $2(3s^{-2})^{-3}$

95. $(2^3c^2)^{-1}$ **96.** $(-3)^2(-r^3)^2$ **97.** $(7^0n^{-3})^2(n^5)^2$ **98.** $(7^2y^{12})^0$

Lesson 7-1

Solve each system by graphing.

99. $y = 3x$ **100.** $y = 2x + 1$ **101.** $y = 5$ **102.** $y = 7$
      $y = -2x$ $y = x - 3$ $x = 3$ $y = 8$

Lesson 6-8

103. Graph $y = |x|$ and its translation $y = |x| + 3$.

Geometric Sequences

NY Learning Standards for Mathematics

A.PS.3 Observe and explain patterns to formulate generalizations and conjectures.

✓ **Check Skills You'll Need**

GO **for Help** Lesson 5-7

Find the common difference of each sequence.

1. $1, 3, 5, 7, \ldots$ **2.** $19, 17, 15, 13, \ldots$

3. $1.3, 0.1, -1.1, -2.3, \ldots$ **4.** $18, 21.5, 25, 28.5, \ldots$

Use inductive reasoning to find the next two numbers in each pattern.

5. $2, 4, 8, 16, \ldots$ **6.** $4, 12, 36, \ldots$

7. $0.2, 0.4, 0.8, 1.6, \ldots$ **8.** $200, 100, 50, 25, \ldots$

🔊 **New Vocabulary**   • geometric sequence   • common ratio

---

1   Geometric Sequences

Recall that a number pattern is also called a sequence, and each number in a sequence is a term of the sequence.

Vocabulary Tip

When you write a ratio of one term to the previous term in a geometric sequence, the ratios are equal. Thus the name is <u>common ratio</u>.

In Chapter 5 you studied arithmetic sequences, where you found each new term by adding the same amount to each previous term. Another kind of number sequence is a geometric sequence. In a **geometric sequence,** the ratio between consecutive terms is constant. This ratio is called the **common ratio.**

Term   2,   10,   50,   250

Common Ratio $\times 5$ $\times 5$ $\times 5$

1 EXAMPLE   **Finding the Common Ratio**

Find the common ratio of each sequence.

a. $3, 12, 48, 192, \ldots$

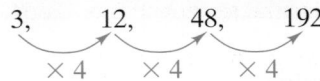

3,   12,   48,   192

   $\times 4$ $\times 4$ $\times 4$

The common ratio is 4.

b. $80, 20, 5, \frac{5}{4}, \ldots$

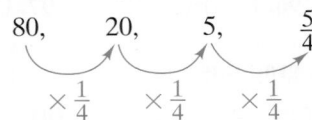

80,   20,   5, $\frac{5}{4}$

   $\times \frac{1}{4}$ $\times \frac{1}{4}$ $\times \frac{1}{4}$

The common ratio is $\frac{1}{4}$.

✓ **Quick Check**   ❶ Find the common ratio of each sequence.

     a. $750, 150, 30, 6, \ldots$ **b.** $-3, -6, -12, -24, \ldots$ **c.** $4, 6, 9, 13.5, \ldots$

2 EXAMPLE Finding the Next Terms in a Sequence

Find the next three terms of the sequence $2, -6, 18, -54, \ldots$

$$2, \quad -6, \quad 18, \quad -54$$
$$\times(-3) \quad \times(-3) \quad \times(-3)$$

The common ratio is -3. The next three terms are $-54(-3) = 162$, $162(-3) = -486$, and $-486(-3) = 1458$.

✓ **Quick Check** ❷ Find the next three terms of each sequence.
  a. $1, 3, 9, 27, \ldots$ **b.** $120, -60, 30, -15, \ldots$ **c.** $1.1, 2.2, 4.4, 8.8, \ldots$

You can look for a common difference or common ratio to determine whether a sequence is arithmetic or geometric. If there is no common difference or common ratio, the sequence is neither arithmetic nor geometric.

3 EXAMPLE Arithmetic or Geometric Sequence

Determine whether each sequence is arithmetic or geometric.
  a. $-7, -5, -3, -1, \ldots$

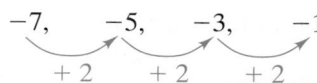

$$-7, \quad -5, \quad -3, \quad -1$$
$$+2 \quad +2 \quad +2$$

The sequence has a common difference. The sequence is arithmetic.

  b. $56, 28, 14, 7, \ldots$

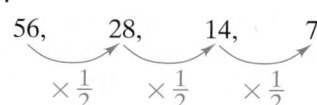

$$56, \quad 28, \quad 14, \quad 7$$
$$\times\frac{1}{2} \quad \times\frac{1}{2} \quad \times\frac{1}{2}$$

The sequence has a common ratio. The sequence is geometric.

✓ **Quick Check** ❸ Determine whether each sequence is arithmetic or geometric.
  a. $2, 4, 6, 8, \ldots$ **b.** $2, 4, 8, 16, \ldots$ **c.** $1, 3, 5, 7, \ldots$

2 Using a Formula

You can use the common ratio of a geometric sequence to write a function rule for the sequence. Consider the sequence $2, 6, 18, 54, \ldots$ Its common ratio is 3.

Let n = the term number in a sequence.

Let $A(n)$ = the value of the nth term of the sequence.

$A(1) = 2$

$A(2) = 2 \cdot 3 = 2 \cdot 3^1$

$A(3) = 2 \cdot 3 \cdot 3 = 2 \cdot 3^2$

$A(4) = 2 \cdot 3 \cdot 3 \cdot 3 = 2 \cdot 3^3$

$\vdots$

$A(n) = 2 \cdot 3 \cdot 3 \cdot 3 \cdot 3 \ldots \cdot 3 = 2 \cdot 3^{n-1}$

Note that each exponent is one less than its term number.

In general, you can write a function rule using the first term, the term number, and the common ratio. For the sequence above, the rule is $A(n) = 2 \cdot 3^{n-1}$.

Rule	Geometric Sequence

$$A(n) = a \cdot r^{n-1}$$

nth term · first term · common ratio · term number

---

4 EXAMPLE **Finding Terms of a Sequence**

Find the first, fifth, and tenth terms of the sequence that has the rule $A(n) = 5(-2)^{n-1}$.

first term: $A(1) = 5$

fifth term: $A(5) = 5(-2)^{5-1} = 5(-2)^4 = 5(16) = 80$

tenth term: $A(10) = 5(-2)^{10-1} = 5(-2)^9 = 5(-512) = -2560$

 **Quick Check** **4** Find the first, sixth, and twelfth terms of each sequence.

a. $A(n) = 4 \cdot 3^{n-1}$ **b.** $A(n) = -2 \cdot 5^{n-1}$

You can write and evaluate a rule for a geometric sequence that models a real-world situation.

5 EXAMPLE **Real-World Problem Solving**

Sports You drop a rubber ball from a height of 1 meter and it bounces back to lower and lower heights. Each curved path has 80% the height of the previous path. Write a rule for the height of each successive path. What height will the ball reach at the top of the fifth path?

Draw a diagram to help understand the problem.

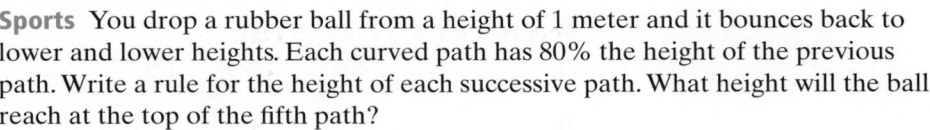

First   Second   Third   Fourth   Fifth

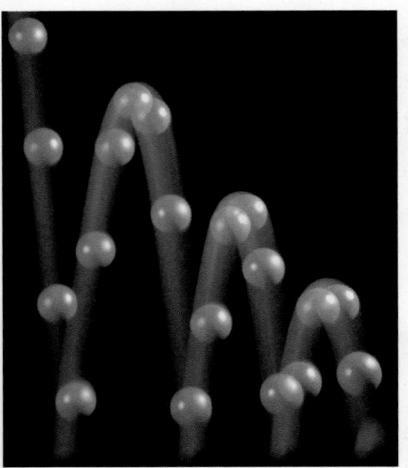

Real-World Connection

Strobe-light photography is often used to highlight details of the motion of objects.

The height of the first path is 100 cm. So the height is 100 cm for the first term, with $n = 1$. The height of the fifth path is given by the term $n = 5$. The common ratio is 80%, or 0.8.

A rule for the sequence is $A(n) = 100 \cdot 0.8^{n-1}$.

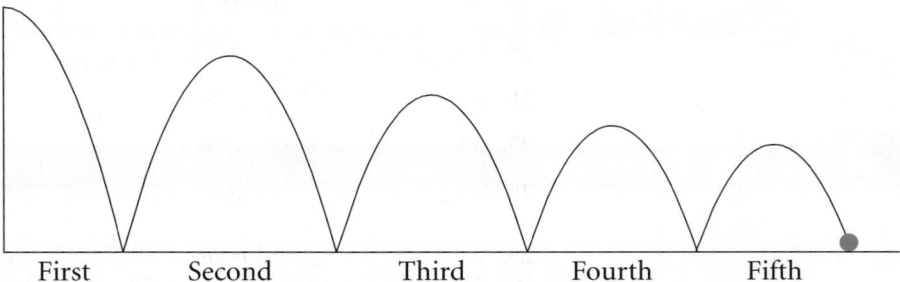

$A(n) = 100 \cdot 0.8^{n-1}$ **Use the sequence to find the height of the fifth path.**

$A(5) = 100 \cdot 0.8^{5-1}$ **Substitute 5 for n.**

$\quad = 100 \cdot 0.8^4$ **Simplify exponents.**

$\quad = 100 \cdot 0.4096$ **Evaluate powers.**

$\quad = 40.96$ **Simplify.**

The height of the fourth bounce will be 40.96 cm.

 **Quick Check** 5 **Basketball** You drop a basketball from a height of 2 meters. Each curved path has 56% of the height of the previous path. Using the height in centimeters, write a rule for the sequence. What height will the basketball reach at the top of the fourth path (when $n = 4$)? Round to the nearest tenth of a centimeter.

EXERCISES

For more exercises, see *Extra Skill and Word Problem Practice.*

Practice and Problem Solving

 A Practice by Example

 **GO for Help**

Example 1
(page 460)

Find the common ratio of each sequence.

1. $2, 8, 32, 128, \ldots$ **2.** $-3, -12, -48, -192, \ldots$ **3.** $70, 7, 0.7, 0.07, \ldots$

4. $8, 20, 50, 125, \ldots$ **5.** $-80, 20, -5, 1.25, \ldots$ **6.** $0.45, 0.9, 1.8, 3.6, \ldots$

Example 2
(page 461)

Find the next three terms of each sequence.

7. $2.5, 5, 10, 20, \ldots$ **8.** $3, 6, 12, 24, \ldots$ **9.** $4, 6, 9, 13.5, \ldots$

10. $-8, 4, -2, 1, \ldots$ **11.** $225, 45, 9, 1.8 \ldots$ **12.** $-3, 6, -12, 24, \ldots$

Example 3
(page 461)

Determine whether each sequence is *arithmetic* or *geometric*.

13. $2, 14, 98, 686, \ldots$ **14.** $12, 8, 4, 0, \ldots$ **15.** $9, -36, 144, -576, \ldots$

16. $-5, -10, -15, -20, \ldots$ **17.** $0.6, 1.3, 2, 2.7, \ldots$ **18.** $9, 12, 16, 21\frac{1}{3}, \ldots$

Example 4
(page 462)

Find the first, fourth, and eighth terms of each sequence.

19. $A(n) = 5 \cdot 3^{n-1}$ **20.** $A(n) = -5 \cdot 3^{n-1}$ **21.** $A(n) = 5 \cdot (-3)^{n-1}$

22. $A(n) = 0.5 \cdot 3^{n-1}$ **23.** $A(n) = -2 \cdot 5^{n-1}$ **24.** $A(n) = -1.1 \cdot (-4)^{n-1}$

Example 5
(page 462)

Write a rule and find the given term in each geometric sequence described below.

25. What is the fifth term when the first term is 6 and the common ratio is 0.5?

26. What is the tenth term when the first term is -6 and the common ratio is 2?

27. What is the fourth term when the first term is 7 and the common ratio is 1.1?

28. What is the seventh term when the first term is 1 and the common ratio is -4?

29. You drop a handball from a height of 1 meter. Each curved path has 64% of the height of the previous path.
a. Write a rule for the sequence using centimeters. The initial height is when $n = 1$.
b. What height will the ball reach at the top of the sixth path?

 B Apply Your Skills

Find the next three terms of each sequence. Then write a rule for each sequence.

30. $216, 72, 24, 8, \ldots$ **31.** $625, 125, 25, 5, \ldots$

32. $0.1, 0.9, 8.1, 72.9, \ldots$ **33.** $16, -8, 4, -2, \ldots$

Problem Solving Hint

For Exercises 30 and 31, making a list of the terms of a sequence, and their factors, can help you write a rule.

34. Open-Ended Write four terms of a geometric sequence. Then write a rule for your sequence.

35. Writing How can you determine whether a sequence is arithmetic or geometric?

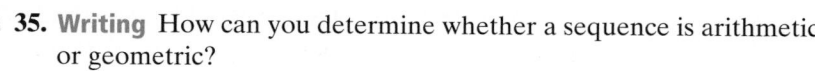

36. Internet An internet site for driving directions increases the magnification of a map 20% for each level of zoom.
  a. Copy and complete the table to show the magnification at each level.
  b. Write a rule for this situation using the decimal equivalents of the percent values.

Zoom Level	Magnification
0	100%
1	120%
2	■
3	■
4	■

Determine whether each sequence is *arithmetic*, *geometric*, or *neither*. Find the next three terms of each sequence.

37. $11, 9, 7, 5, \ldots$

38. $7, 6, 4, 1, \ldots$

39. $18, 9, 4.5, 2.25, \ldots$

40. $12, 14, 16, 18, \ldots$

41. Physics On the first swing, a pendulum swings through an arc of length 36 centimeters. On each successive swing, the length of the arc is 90% of the length of the previous swing.
  a. Write a rule to model this situation.
  b. Critical Thinking What value of n would you use to find the length of the arc on the sixth swing? Explain.
  c. Find the length of the arc on the sixth swing, to the nearest tenth of a centimeter.

42. a. Geometry What fraction of each figure is shaded?
  b. Rewrite each fraction from part (a) in the form $2^{■}$.
  c. Write a rule that relates the figure number n to the shaded rectangle r.
  d. What portion of the square would be shaded in Figure 10?

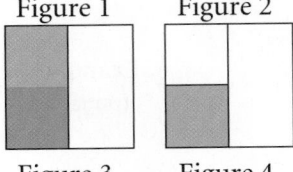
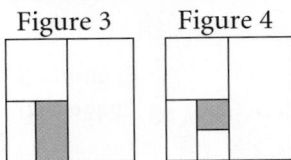

43. Reasoning Can zero be a term of a geometric sequence that has terms that are not zero? Explain.

C Challenge **44. Fractal Geometry** The figures below show the first four steps in making Sierpinski's Triangle.

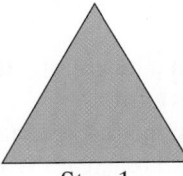 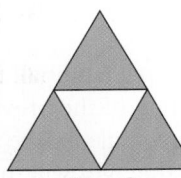 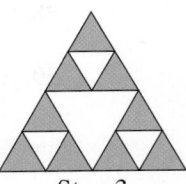 
Step 1 Step 2 Step 3 Step 4

  a. Patterns What fraction of each step is shaded?
  b. Use your answer from part (a) to write a rule that relates the step number n to the fraction r of the figure that is shaded.
  c. What fraction of Step 6 would be shaded?
  d. Patterns What fraction of each step is *not* shaded?
  e. Write a rule that relates the step number n to the fraction r of the figure that is *not* shaded.
  f. What fraction of Step 8 would *not* be shaded?

Find each common ratio. Then find the next three terms in each sequence.

45. $x, x^2, x^3, x^4, \ldots$ **46.** $\frac{1}{3}, x, 3x^2, 9x^3, \ldots$ **47.** $xy, x^2y^3, x^3y^5, x^4y^7, \ldots$

48. $\frac{2}{b^2}, \frac{2a}{b}, 2a^2, 2a^3b, \ldots$ **49.** $2 \times 10^7, 1.2 \times 10^6, 7.2 \times 10^4, 4.32 \times 10^3, \ldots$

50. What term is 512 in the geometric sequence with the first term 2 and the common ratio 4?

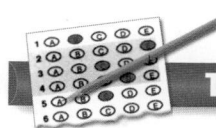

Multiple Choice

51. Which set of numbers continues the pattern 27, 9, 3, 1, . . . ?

   A. $-3, -9, -27$ **B.** $-\frac{1}{3}, -\frac{1}{6}, -\frac{1}{9}$

   C. $\frac{1}{3}, \frac{1}{6}, \frac{1}{9}$ **D.** $\frac{1}{3}, \frac{1}{9}, \frac{1}{27}$

52. Charlie is stacking cans at a grocery store. The picture shows the first four layers. How many cans are there in 6 layers?

   F. 36 cans   **G.** 61 cans

   H. 91 cans   **J.** 729 cans

53. Which equation could you use to find the next term in the pattern 3, 6, 12, 24, 48, . . . ?

   A. $A(n) = 3^{n-1}$ **B.** $A(n) = 3(2)^{n-1}$

   C. $A(n) = 3 \cdot 2n$ **D.** $A(n) = 3n^2$

Short Response

54. In a research laboratory, bacteria of a certain species double in number each day. If the number of bacteria at the beginning of a day is 350, how many bacteria will there be at the beginning of the 5th day? Show your work.

Mixed Review

Lesson 8-5

Simplify each expression.

55. $\left(\frac{a^2}{a^3}\right)^{-4}$ **56.** $\left(\frac{1}{2}\right)^{-4}$ **57.** $\left(\frac{x^2z}{z^{-3}}\right)^{-5}$ **58.** $\left(\frac{m^{-3}}{n^4}\right)^0$

59. $\left(\frac{8}{9}\right)^{-2}$ **60.** $\left(\frac{m^4}{m^2}\right)^{-7}$ **61.** $\left(\frac{pq^0}{p^4}\right)^5$ **62.** $\left(\frac{c^2d^{-2}}{d^3}\right)^{-1}$

Lesson 8-2

63. Write 0.002467 in scientific notation.

64. Water Conservation The Folsom Dam in California holds 1 million acre-feet of water in a reservoir. An acre-foot of water is the amount of water that covers an acre to the depth of one foot, or 326,000 gal. How many gallons are in the reservoir? Write your answer in scientific notation.

Lesson 5-5

Write an equation of the direct variation that includes the given point.

65. $(3, 8)$ **66.** $(-5, 2)$ **67.** $(6, -7)$ **68.** $(-3, -5)$

69. $(4, 7)$ **70.** $(-16, 4)$ **71.** $(9, 5)$ **72.** $(4, -2)$

In this Activity Lab, you will collect data from geometric situations and develop functions to model the situations.

1 ACTIVITY

1. Fold a piece of paper in half. How many layers are there?

2. Continue folding the paper. Copy and complete the table.

Number of Folds	Number of Layers
0	1
1	▪
2	▪
3	▪
4	▪
5	▪
6	▪

3. In the graphs below, the scale of the horizontal axis is from 0 to 6 and the scale of the vertical axis is from 0 to 10. Which graph best models the data in the table? Explain.

A.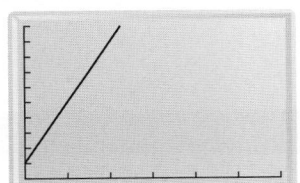

B.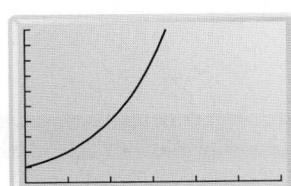

4. What kind of sequence is formed by the values in the column labeled "Number of Layers"?

5. Write a rule for this sequence.

6. Revise your sequence rule to write a function rule so that the number of layers $f(x)$ is a function of the number of folds x. The function rule should give $f(0) = 1$ and $f(1) = 2$.

7. Find the number of layers in 12 folds.

8. If there were 256 layers, how many folds would there be?

9. Is it physically possible to fold an $8\frac{1}{2}$ in. $\times$ 11 in. piece of paper in halves and get 144 layers? Explain.

10. Describe a reasonable domain for the function in Exercise 6.

11. Suppose the paper you were folding in Activity 1 measured 8 in. by 10 in. Complete the table for the area of a layer after each fold.

Number of Folds	Area of Layer
0	80 in.^{2}
1	▦
2	▦
3	▦
4	▦
5	▦
6	▦

12. In the graphs below, the scale of the horizontal axis is from 0 to 6 and the scale of the vertical axis is from 0 to 80. Which graph best models the data in the table? Explain.

A.

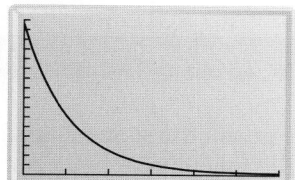

B.

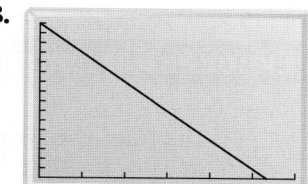

13. How is the graph you chose for Question 12 like the graph you chose for Question 3? How is it different?

14. What kind of sequence is formed by the values in the column labeled "Area of Layer"?

15. Write a rule for this sequence.

16. Revise your sequence rule to write a function rule so that the area of layer $g(x)$ is a function of the number of folds x. The function rule should give $f(0) = 80$.

17. Find the area of a layer after 8 folds.

EXERCISES

18. The rules you wrote for Questions 6 and 16 are examples of exponential functions. Why is "exponent" appropriate as part of the name for this type of function?

19. a. Copy the table below. Extend the table to $x = 5$.

x	f(x)
1	18
2	6
3	2

  b. Graph the data in your table.
  c. Write a function rule for the data in your table.
  d. Use the function rule to find $f(7)$.

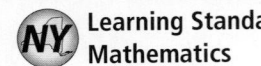

Exponential Functions

NY Learning Standards for Mathematics

A.N.6 Evaluate expressions involving factorial(s), absolute value(s), and exponential expression(s).

A.G.4 Identify and graph linear, quadratic (parabolic), absolute value, and exponential functions.

✓ **Check Skills You'll Need**

Graph each function.

1. $y = 3x$ **2.** $y = 4x$ **3.** $y = -2x$

Simplify each expression.

4. 3^2 **5.** 5^{-3} **6.** $2 \cdot 3^4$

7. $2 \cdot 3^{-2}$ **8.** $3 \cdot 2^{-1}$ **9.** $10 \cdot 3^2$

GO **for Help**   Lessons 6-2 and 8-1

◀)) **New Vocabulary**   • exponential function

---

1   Evaluating Exponential Functions

The rules you wrote in Lesson 8-6 to describe geometric sequences, such as $A(n) = 3 \cdot 4^{n-1}$, are examples of exponential functions.

🔑 **Key Concepts**

Definition	**Exponential Function**

An **exponential function** is a function in the form $y = a \cdot b^x$, where a is a nonzero constant, b is greater than 0 and not equal to 1, and x is a real number.

Examples $y = 0.5 \cdot 2^x$ $f(x) = -2 \cdot 0.5^x$

You can evaluate an exponential function for given values of the domain to find the corresponding values of the range.

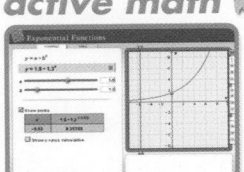

For: Exponential Functions Activity
Use: Interactive Textbook, 8-7

1 **EXAMPLE**   **Evaluating an Exponential Function**

Evaluate each exponential function.

a. $y = 5^x$ for $x = 2, 3, 4$

x	5^x	y
2	$5^2 = 25$	25
3	$5^3 = 125$	125
4	$5^4 = 625$	625

b. $t(n) = 4 \cdot 3^n$ for the domain $\{-3, 6\}$

n	$4 \cdot 3^n$	$t(n)$
-3	$4 \cdot 3^{-3} = 4 \cdot \frac{1}{27} = \frac{4}{27}$	$\frac{4}{27}$
6	$4 \cdot 3^6 = 4 \cdot 729 = 2916$	2916

✓ **Quick Check** **1** Evaluate each exponential function for the domain $\{-2, 0, 3\}$.

a. $y = 4^x$ **b.** $f(x) = 10 \cdot 5^x$ **c.** $g(x) = -2 \cdot 3^x$

You can evaluate exponential functions to solve real-world problems.

② EXAMPLE — Real-World 🌐 Problem Solving

Gridded Response Suppose 20 rabbits are taken to an island. The rabbit population then triples every half year. The function $f(x) = 20 \cdot 3^x$, where x is the number of half-year periods, models this situation. How many rabbits would there be after 2 years?

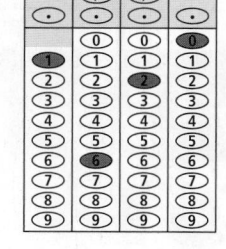

$$f(x) = 20 \cdot 3^x$$
$$= 20 \cdot 3^4 \quad \textbf{In 2 years, there are 4 half years. Evaluate the function for } x = 4.$$
$$= 20 \cdot 81 \quad \textbf{Simplify powers.}$$
$$= 1620 \quad \textbf{Simplify.}$$

After two years, there would be 1620 rabbits.

✓ Quick Check ② Suppose 10 animals are taken to an island, and then the population of these animals quadruples every year. Use the function $f(x) = 10 \cdot 4^x$. How many animals would there be after 6 years?

2 Graphing Exponential Functions

Here are two graphs that show what exponential functions generally look like.

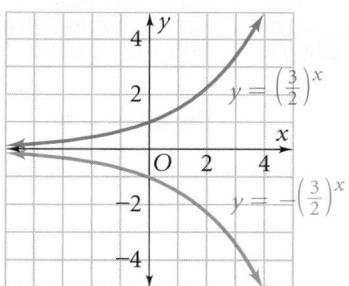
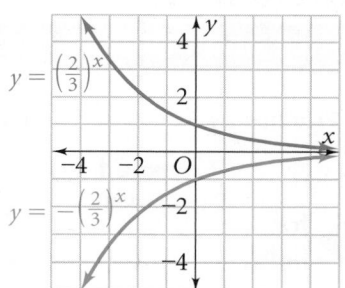

To graph an exponential function, make a table of values. Plot the points. Then join the points to form a smooth curve.

③ EXAMPLE — Graphs of Exponential Functions

Graph $y = 3 \cdot 2^x$.

x	$3 \cdot 2^x$	(x, y)
-2	$3 \cdot 2^{-2} = \frac{3}{2^2} = \frac{3}{4}$	$\left(-2, \frac{3}{4}\right)$
-1	$3 \cdot 2^{-1} = \frac{3}{2^1} = 1\frac{1}{2}$	$\left(-1, 1\frac{1}{2}\right)$
0	$3 \cdot 2^0 = 3 \cdot 1 = 3$	$(0, 3)$
1	$3 \cdot 2^1 = 3 \cdot 2 = 6$	$(1, 6)$
2	$3 \cdot 2^2 = 3 \cdot 4 = 12$	$(2, 12)$

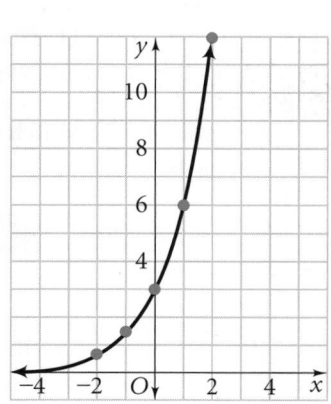

✓ Quick Check ③ Graph each exponential function. **a.** $y = 0.5 \cdot 2^x$ **b.** $y = -0.5 \cdot 2^x$

You can graph exponential functions to model real-world situations.

4 EXAMPLE Real-World 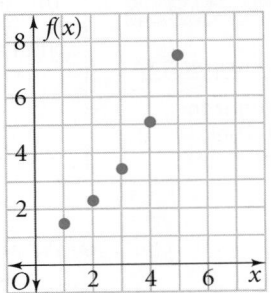 Problem Solving

Photocopying Many photocopiers allow you to choose how large you want an image to be. The function $f(x) = 1.5^x$ models the new size of an image being copied over and over at 150%, where x is the number of enlargements. Graph the function.

x	1.5^x	$(x, f(x))$
1	$1.5^1 = 1.5$	$(1, 1.5)$
2	$1.5^2 = 2.25 \approx 2.3$	$(2, 2.3)$
3	$1.5^3 = 3.375 \approx 3.4$	$(3, 3.4)$
4	$1.5^4 = 5.0625 \approx 5.1$	$(4, 5.1)$
5	$1.5^5 = 7.59375 \approx 7.6$	$(5, 7.6)$

✓ Quick Check **4** **a.** You can also make images that are smaller than the original on a photocopier. The function $f(x) = 0.9^x$ models the new size of an image being copied over and over at 90%. Graph the function.

b. Critical Thinking Explain why the function in Example 4 models discrete data.

EXERCISES

For more exercises, see *Extra Skill and Word Problem Practice.*

Practice and Problem Solving

A Practice by Example

Example 1
(page 468)

GO for Help

Evaluate each function rule for the given value.

1. $f(x) = 6^x$ for $x = 3$

2. $g(t) = 2 \cdot 3^t$ for $t = -2$

3. $y = 20 \cdot (0.5)^x$ for $x = 3$

4. $h(w) = 0.5 \cdot 4^w$ for $w = 3$

5. $y = 50 \cdot (0.3)^x$ for $x = 2$

6. $f(x) = 1.8 \cdot 2^x$ for $x = 6$

7. $y = 100 \cdot \left(\frac{1}{2}\right)^x$ for $x = -4$

8. $y = 9 \cdot \left(\frac{5}{2}\right)^x$ for $x = -3$

Example 2
(page 469)

9. Finance Suppose an investment of $10,000 doubles in value every 13 years. How much is the investment worth after 52 years? After 65 years?

10. Finance Suppose an investment of $500 doubles in value every 15 years. How much is the investment worth after 30 years? After 45 years?

11. Finance Suppose an investment of $2000 doubles in value every 8 years. How much is the investment worth after 24 years? After 32 years?

Example 3
(page 469)

Match each table with the function that models the data.

12. $y = 3x$

13. $y = x^3$

14. $y = 3^x$

A.

x	y
1	3
2	6
3	9
4	12

B.

x	y
1	3
2	9
3	27
4	81

C.

x	y
1	1
2	8
3	27
4	64

Match each function rule with the graph of the function.

15. $y = 2^x$ **16.** $y = -(2^x)$ **17.** $y = \left(\frac{1}{2}\right)^x$ **18.** $y = -\left(\frac{1}{2}\right)^x$

A.
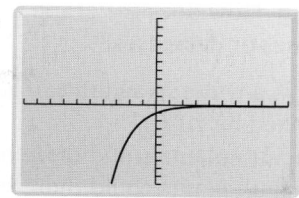

B.
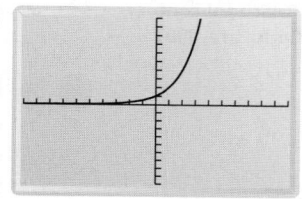

C.
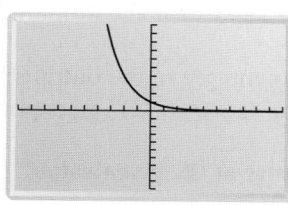

D.
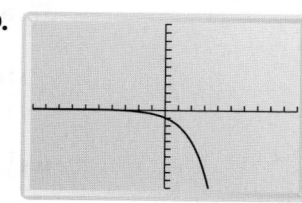

Graph each function.

19. $y = 10 \cdot 2^x$ **20.** $y = 0.1 \cdot 2^x$ **21.** $y = \frac{1}{4} \cdot 2^x$ **22.** $y = 4^x$

Example 4
(page 470)

23. Photocopying Suppose you are photocopying an image, reducing it to 85% its original size. The function $y = 0.85^x$ models the size of an image after x number of times it is reduced. Graph the function.

24. Science A population of 100 insects triples in size every month. The function $y = 100 \cdot 3^x$ models the population after x months. Graph the function.

B   **Apply Your Skills**

Evaluate each function for the domain {−2, −1, 0, 1, 2, 3}. As the values of the domain increase, do the values of the range *increase* or *decrease*?

25. $f(x) = 5^x$ **26.** $y = 2.5^x$ **27.** $h(x) = 0.1^x$ **28.** $f(x) = 5 \cdot 4^x$

29. $y = 0.5^x$ **30.** $y = \left(\frac{2}{3}\right)^x$ **31.** $g(x) = 4 \cdot 10^x$ **32.** $y = 100 \cdot 0.3^x$

33. Multiple Choice The population of Texas in 2000 was about 20.852 million people. The function $p(n) = 20.852(1.02071)^n$ estimates the population where $n = 0$ corresponds to the year 2000. Which is a reasonable estimate in millions of the population of Texas in 2020?

     Ⓐ 21.284   Ⓑ 31.420   Ⓒ 41.740   Ⓓ 49.842

 34. Biology A certain species of bacteria in a laboratory culture begins with 75 cells and doubles in number every 20 min.

     a. Copy, complete, and extend the table to find when there will be more than 5,000 bacteria cells.

Time (min)	Number of 20-min Time Periods	Pattern	Number of Bacteria Cells
Initial	0	75	75
20	1	$75 \cdot 2$	$75 \cdot 2^\blacksquare = \blacksquare$
40	■	$75 \cdot 2 \cdot 2$	$75 \cdot 2^\blacksquare = \blacksquare$
60	■	■	$75 \cdot 2^\blacksquare = \blacksquare$
■	■	■	■

     b. Write a function rule to model the situation.

35. a. Graph $y = 2^x$, $y = 4^x$, and $y = (0.25)^x$.
    b. What point is on each graph?
    c. Does the graph of an exponential function intersect the x-axis? Explain.
    d. Critical Thinking How does the graph of an exponential function change as the base increases or decreases?

 36. Ecology In 50 days, a water hyacinth can generate 1000 offspring (the number of plants is multiplied by 1000).
    a. How many hyacinth plants could there be after 150 days?
    b. How many hyacinth plants could there be after 200 days?

37. a. Make a table of values for the domain $\{1, 2, 3, 4, 5\}$ of the function $y = (-2)^x$.
    b. What pattern do you see in the outputs?
    c. Critical Thinking Is $y = (-2)^x$ an exponential function? Justify your answer.

Which function is greater at the given value?

38. $y = 5^x$ or $y = x^5$ at $x = 3$ **39.** $f(t) = 10 \cdot 2^t$ or $f(t) = 200 \cdot t^2$ at $t = 7$

40. $y = 3^x$ or $y = x^3$ at $x = 4$ **41.** $f(x) = 2^x$ or $f(x) = 100x^2$ at $x = 10$

 42. Writing Analyze the range of the function $f(x) = 500 \cdot 1^x$ using the domain $\{1, 2, 3, 4, 5\}$. Explain why the definition of *exponential function* includes the restriction that $b \neq 1$.

 **43. a. Graphing Calculator** Graph the functions $y = x^2$ and $y = 2^x$.
    b. What happens to the graphs between $x = 1$ and $x = 3$?
    c. Critical Thinking How do you think the graph of $y = 6^x$ would compare to the graphs of $y = x^2$ and $y = 2^x$?

C Challenge

Solve each equation.

44. $3^x = 9$ **45.** $3^x = \frac{1}{27}$ **46.** $2^x = 64$

47. $3 \cdot 2^x = 24$ **48.** $2 \cdot 3^x = 162$ **49.** $5 \cdot 2^x - 152 = 8$

50. Suppose $(0, 4)$ and $(2, 36)$ are on the graph of an exponential function.
    a. Use $(0, 4)$ in the general form of an exponential function $y = a \cdot b^x$ to find the value of the constant a.
    b. Use your answer from part (a) along with $(2, 36)$ to find the value of the constant b.
    c. Write a rule for the function.
    d. Evaluate the function for $x = -2$ and $x = 4$.

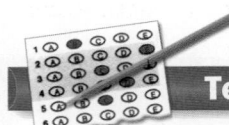

NY REGENTS

Test Prep

Multiple Choice

51. For the function $y = -3^x$, what is the value of y when $x = -2$?
    A. -9 **B.** $-\frac{1}{9}$ **C.** $\frac{1}{9}$ **D.** 9

52. Which function contains the points $(1, 3)$ and $(3, 6.75)$?
    F. $y = 1.675x + 1.325$ **G.** $y = 2 \cdot 1.5^x$
    H. $y = 1.5 \cdot 2^x$ **J.** $y = 1.325x + 1.675$

53. Which function has the same y-intercept as $y = 2^x$?
    A. $y = x + 1$ **B.** $y = 2x$ **C.** $y = x$ **D.** $y = 2(x + 1)$

54. A population of 6000 doubles in size every 10 years. Which equation relates the size of the population y to the number of 10-year periods x?

 F. $y = 6000 \cdot 10^x$ **G.** $y = 10 \cdot 2^x$

 H. $y = 6000 \cdot 2^x$ **J.** $y = 2 \cdot 100^x$

Short Response   **55.** Between what two integer values of x do the graphs of $y = 20(0.5)^x$ and $y = 0.5 \cdot 4^x$ intersect? Show your work.

Mixed Review

Lesson 8-6

Find each common ratio. Then find the next three terms in each sequence.

56. $2, 10, 50, 250, \ldots$ **57.** $7, -21, 63, -189, \ldots$

58. $-0.2, -0.4, -0.8, -1.6, \ldots$ **59.** $27, -9, 3, -1, \ldots$

60. $450, 45, 4.5, 0.45, \ldots$ **61.** $7168, 1792, 448, 112, \ldots$

Lesson 6-6

Write an equation for the line that passes through the given point and is parallel to the given line.

62. $y = 5x + 1; (0, 0)$ **63.** $y = 3x - 2; (0, 1)$

64. $y = -2x + 5; (4, 0)$ **65.** $y = 0.4x + 5; (2, -3)$

✓ Checkpoint Quiz 2   Lessons 8-5 through 8-7

Simplify each expression.

1. $\left(\dfrac{3^2}{3^{-1}}\right)^4$ **2.** $\left(\dfrac{x^2}{y^3}\right)^{-5}$ **3.** $\left(\dfrac{10m^{-3}}{25n^{-6}}\right)^2$ **4.** $\left(\dfrac{6^2 t^{-3}}{6^2 r^0 t^2}\right)^2$

Determine whether each sequence is *arithmetic* or *geometric*.

5. $22, 11, 5.5, 2.75, \ldots$ **6.** $5, 10, 20, 40, 80, \ldots$ **7.** $5, 10, 15, 20, 25, \ldots$

8. Use the sequence $-100, 20, -4, \ldots$.
 a. What is the first term?
 b. What is the common ratio?
 c. Write a rule for the sequence.
 d. Use your rule to find the fifth and seventh terms in the sequence.

9. Physics On the first swing, a pendulum swings through an arc of length 40 cm. On each successive swing, the length of the arc is 85% of the length of the previous swing.
 a. Write a rule to model this situation.
 b. Find the length of the arc on the fifth swing. Round your answer to the nearest millimeter.

10. Commuting Refer to the information at the left.
 a. Write the number of vehicles that crossed the George Washington Bridge in scientific notation.
 b. The Port Authority collected about $249 million in tolls from this bridge. Write this number in scientific notation.
 c. What was the average toll per vehicle?

Real-World 🌐 Connection

About 108 million vehicles cross the George Washington Bridge between New York and New Jersey in a year.

Fitting Exponential Curves to Data

Go Online
PHSchool.com

For: Graphing calculator procedures
Web Code: ate-2122

In Chapter 6 you learned how to find a line of best fit for a set of data. You can model some data better using an exponential function. To graph an exponential function, you may need to adjust your viewing window. Use your data to choose appropriate **Xmax** and **Ymax** values.

EXAMPLE

The table at the right shows the estimated number of customers downloading music files. Use a graphing calculator to find the best-fitting exponential function for the data. Then graph the function.

Step 1 Use your calculator's STAT feature. Enter the data. Let $x = 0$ correspond to 2000.

Step 2 To get the equation of the best-fitting exponential function, press STAT ▶ 0 ENTER.

```
ExpReg
 y=a*b^x
 a=.4236368467
 b=2.221570366
 r²=.9960912309
 r=.9980437019
```

Digital Download

Year	Customers (millions)
2000	0.4
2001	1.0
2002	2.2
2003	4.4

Step 3 To view the graph of the function press Y= CLEAR VARS 5 ▶ ▶ 1 GRAPH.

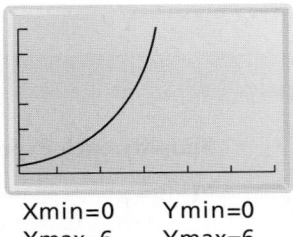

Xmin=0   Ymin=0
Xmax=6   Ymax=6

EXERCISES

Use a graphing calculator to find the exponential function that fits each set of data. Then (a) write a function for the data rounding decimals to the nearest hundredth, (b) sketch a graph of the function, and (c) use your answer from part (a) to predict the value of the function in the year 2010.

1. Shipment of Record Singles Let $x = 0$ correspond to 1990.

Year	1990	1995	2000	2001	2002	2003	2004
Record Singles (millions)	27.6	10.2	4.8	5.5	4.4	3.8	3.5

Source: Record Industry Association of America

2. U.S. Energy Consumption Let $x = 0$ correspond to 1990.

Year	1990	1995	2000	2001	2002	2003	2004
Btu (quadrillions)	84.7	91.2	98.9	96.4	97.8	98.2	99.7

Source: U.S. Department of Energy

3. U.S. Homes Heated by Coal Let $x = 0$ correspond to 1950.

Year	1950	1960	1970	1980	1991	2001
Percent of Homes	34.6	12.2	2.9	0.6	0.3	0.1

Source: U.S. Department of Energy

Exponential Growth and Decay

NY Learning Standards for Mathematics

A.A.9 Analyze and solve verbal problems that involve exponential growth and decay.

 Check Skills You'll Need

 for Help Lesson 3-7

Find each percent of change. Describe the percent of change as an increase or decrease. If necessary, round to the nearest percent.

1. The original cost of a shirt is $25. On sale the shirt costs $22.

2. In one week, a plant's height went from 15 cm to 18 cm.

3. The population of a town went from 38,356 in 1990 to 40,481 in 2000.

4. A computer that cost $1450 last year costs $999 this year.

5. An elephant that weighed 220 lb at birth weighed 300 lb at one month.

New Vocabulary • **exponential growth** • **growth factor** • **compound interest** • **interest period** • **exponential decay** • **decay factor**

---

1 Exponential Growth

Real-World Connection

In 2000, Florida's population was about 16 million. Roughly 23% of the population was under the age of 18.

In 2000, Florida's population was about 16 million. Since 2000, the state's population has grown about 2% each year. This means that Florida's population is growing exponentially.

To find Florida's population in 2001, multiply the 2000 population by 2% and add this to the 2000 population. So the population in 2001 is (2% + 100%) of the 2000 population, or 102% of the 2000 population. Here is a function that models Florida's population since 2000.

population in millions
↓
$$y = 16(1.02)^x \quad \longleftarrow \text{number of years since 2000}$$
↑
102% as a decimal

The following is a general rule for modeling exponential growth.

Key Concepts

Rule	Exponential Growth

Exponential growth can be modeled with the function
$y = a \cdot b^x$ for $a > 0$ and $b > 1$.

starting amount (when $x = 0$)
↓
$$y = a \cdot b^x \quad \longleftarrow \text{exponent}$$
↑
The base, which is greater than 1, is the **growth factor.**

1 EXAMPLE Modeling Exponential Growth

Medical Care Since 1995, the daily cost of patient care in community hospitals in the United States has increased about 4% per year. In 1995, such hospital costs were an average of $968 per day.

a. Write an equation to model the cost of hospital care since 1995.

Relate $y = a \cdot b^x$ **Use an exponential function.**

Define Let $x =$ the number of years since 1995.
Let $y =$ the cost of community hospital care at various times.
Let $a =$ the initial cost in 1995, $968.
Let $b =$ the growth factor, which is $100\% + 4\% = 104\% = 1.04$.

Write $y = 968 \cdot 1.04^x$

b. Use your equation to estimate the approximate cost per day in 2010.

$y = 968 \cdot 1.04^x$

$y = 968 \cdot 1.04^{15}$ **2010 is 15 years after 1995, so substitute 15 for x.**

≈ 1743 **Use a calculator. Round to the nearest dollar.**

The average cost per day in 2010 will be about $1743.

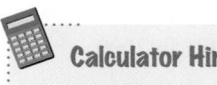

Calculator Hint

To evaluate
$968 \cdot 1.04^{15}$, press
968 ✕ 1.04 ∧
15 ENTER .

 **Quick Check** ❶ **a.** Suppose your community has 4512 students this year. The student population is growing 2.5% each year. Write an equation to model the student population.
b. What will the student population be in 3 years?

When a bank pays interest on both the principal *and* the interest an account has already earned, the bank is paying **compound interest.** An **interest period** is the length of time over which interest is calculated.

2 EXAMPLE Compound Interest

Savings Suppose your parents deposited $1500 in an account paying 3.5% interest compounded annually (once a year) when you were born. Find the account balance after 18 years.

Relate $y = a \cdot b^x$ **Use an exponential function.**

Define Let $x =$ the number of interest periods.
Let $y =$ the balance.
Let $a =$ the initial deposit, $1500.
Let $b = 100\% + 3.5\% = 103.5\% = 1.035$.

Write $y = 1500 \cdot 1.035^x$

$= 1500 \cdot 1.035^{18}$ **Once a year for 18 years is 18 interest periods. Substitute 18 for x.**

≈ 2786.23 **Use a calculator. Round to the nearest cent.**

The balance after 18 years will be $2786.23.

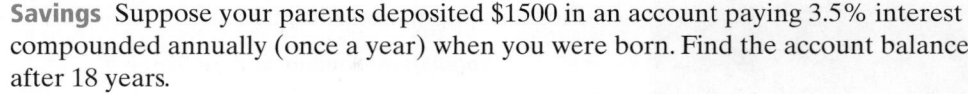

BANK 1

Deposit **$1500**

Interest compounded annually **3.5%**

Balance after 18 years **$2786.23**

 **Quick Check** ❷ **a.** Suppose the interest rate on the account in Example 2 was 4%. How much would be in the account after 18 years?
b. Another formula for compound interest is $B = p(1 + r)^x$, where B is the balance, p is the principal, and r is the annual interest rate in decimal form. Use this formula to find the balance in the account in part (a).
c. Critical Thinking Explain why the two formulas for finding compound interest are actually the same.

When interest is compounded quarterly (four times per year), you divide the interest rate by 4, the number of interest periods per year. To find the number of payment periods, you multiply the number of years by the number of interest periods per year.

Annual Interest Rate of 8%

Compounded	Periods per Year	Interest Rate per Period
annually	1	8% every year
semi-annually	2	$\frac{8\%}{2} = 4\%$ every 6 months
quarterly	4	$\frac{8\%}{4} = 2\%$ every 3 months
monthly	12	$\frac{8\%}{12} = 0.\overline{6}\%$ every month

3 EXAMPLE Compound Interest

Savings Suppose the account in Example 2 paid interest compounded quarterly instead of annually. Find the account balance after 18 years.

Relate $y = a \cdot b^x$ Use an exponential function.

Define Let x = the number of interest periods.
Let y = the balance.
Let a = the initial deposit, $1500.

Let $b = 100\% + \frac{3.5\%}{4}$ **There are 4 interest periods in 1 year, so divide the interest into 4 parts.**

$= 1 + 0.00875 = 1.00875$

Write $y = 1500 \cdot 1.00875^x$

$= 1500 \cdot 1.00875^{72}$ **Four interest periods a year for 18 years is 72 interest periods. Substitute 72 for x.**

≈ 2808.71 **Use a calculator. Round to the nearest cent.**

● The balance after 18 years will be $2808.71.

 **Quick Check** ❸ **a.** Suppose the account in Example 3 paid interest compounded monthly. How much money would be in the account after 18 years?

b. You deposit $200 into an account earning 5%, compounded monthly. How much will be in the account after 1 year? After 2 years? After 5 years?

BANK 2

Deposit **$1500**

Interest compounded
quarterly **3.5%**

Balance after
18 years
$2808.71

2 Exponential Decay

The graphs at the right show exponential growth and exponential decay. For exponential growth, as x increases, y increases exponentially. For exponential decay, as x increases, y decreases exponentially.

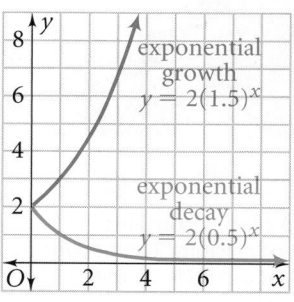

exponential
growth
$y = 2(1.5)^x$

exponential
decay
$y = 2(0.5)^x$

A real-world example of exponential decay is radioactive decay, in which radioactive elements break down by releasing particles and energy.

For: Exponential Functions
Activity
Use: Interactive Textbook, 8-8

4 EXAMPLE Real-World Problem Solving

Medicine The half-life of a radioactive substance is the length of time it takes for one half of the substance to decay into another substance. To treat some forms of cancer, doctors use radioactive iodine. The half-life of iodine-131 is 8 days. A patient receives a 12-mCi (millicuries, a measure of radiation) treatment. How much iodine-131 is left in the patient 16 days later?

In 16 days, there are two 8-day half-lives.

After one half-life, there are 6 mCi left in the patient.

After two half-lives, there are 3 mCi left in the patient.

 Quick Check 4 **a.** How many half-lives of iodine-131 occur in 32 days?
b. Suppose you start with a 50-mCi sample of iodine-131. How much iodine-131 is left after one half-life? After two half-lives?
c. Chemistry Cesium-137 has a half-life of 30 years. Suppose a lab stored a 30-mCi sample in 1973. How much of the sample will be left in 2003? In 2063?

The function $y = a \cdot b^x$ can model exponential decay as well as exponential growth.

Key Concepts

Rule	Exponential Decay

The function $y = a \cdot b^x$ models exponential decay for $a > 0$ and $0 < b < 1$.

starting amount (when $x = 0$)

$$y = a \cdot b^x \longleftarrow \text{exponent}$$

The base, which is between 0 and 1, is the **decay factor.**

When a number is decreased by 5%, the result is 95% of the original number. So when you find the decay factor, think 100% minus the percent a number is decreasing.

5 EXAMPLE Modeling Exponential Decay

Milk Consumption Since 1980, the number of gallons of whole milk each person in the United States drinks each year has decreased 4.1% each year. In 1980, each person drank an average of 16.5 gallons of whole milk per year.

a. Write an equation to model the gallons of whole milk drunk per person.

Relate $y = a \cdot b^x$ **Use an exponential function.**

Define Let x = the number of years since 1980.
Let y = the consumption of whole milk, in gallons.
Let a = 16.5, the initial number of gallons in 1980.
Let b = the decay factor, which is 100% − 4.1% = 95.9% = 0.959.

Write $y = 16.5 \cdot 0.959^x$

Real-World Connection

One cup of milk contains 300 mg of calcium. The body absorbs about 32% of the calcium in milk.

b. Use your equation to find the approximate consumption per person of whole milk in 2000.

$y = 16.5 \cdot 0.959^x$

$y = 16.5 \cdot 0.959^{20}$ **2000 is 20 years after 1980, so substitute 20 for *x*.**

≈ 7.1 **Use a calculator. Round to the nearest tenth of a gallon.**

The average annual consumption of whole milk in 2000 was about 7 gal/person.

 **5 Statistics** In 1990, the population of Washington, D.C., was about 604,000 people. Since then the population has decreased about 1.8% per year.
  a. What is the initial number of people?
  b. What is the decay factor?
  c. Write an equation to model the population of Washington, D.C., since 1990.
  d. Suppose the current trend in population change continues. Predict the population of Washington, D.C., in 2010.

EXERCISES

For more exercises, see *Extra Skill and Word Problem Practice.*

Practice and Problem Solving

 Practice by Example

Example 1
(page 476)

Identify the initial amount *a* and the growth factor *b* in each exponential function.

1. $g(x) = 20 \cdot 2^x$ **2.** $y = 200 \cdot 1.0875^x$ **3.** $y = 10{,}000 \cdot 1.01^x$ **4.** $f(t) = 1.5^t$

5. Suppose the population of a city is 50,000 and is growing 3% each year.
  a. The initial amount *a* is ■.
  b. The growth factor *b* is 100% + 3%, which is 1 + ■ = ■.
  c. To find the population after one year, you multiply 50,000 · ■.
  d. Complete the equation $y = ■ \cdot ■^{■}$ to find the population after *x* years.
  e. Use your equation to predict the population after 25 years.

Examples 2, 3
(pages 476, 477)

Each percent is an annual interest rate. In the formula $y = a \cdot b^x$, what value would you use for *b*?

6. 4%   **7.** 5%   **8.** 3.7%   **9.** 8.75%   **10.** 0.5%

Assume each interest rate below is an annual interest rate. Find the interest rate for an account that is compounded quarterly and monthly.

11. 3%   **12.** 4%   **13.** 4.5%   **14.** 7.6%   **15.** 6.25%

Find the balance in each account.

16. $4000 principal earning 6% compounded annually, after 5 years

17. $12,000 principal earning 4.8% compounded annually, after 7 years

18. $500 principal earning 4% compounded quarterly, after 6 years

19. $20,000 deposit earning 3.5% compounded quarterly, after 10 years

Example 4
(page 478)

20. Chemistry The half-life of iodine-124 is 4 days. A technician measures a 40-mCi sample of iodine-124.
  a. How many half-lives of iodine-124 occur in 16 days?
  b. How much iodine-124 is in the sample 16 days after the technician measures the original sample?

21. Chemistry The half-life of carbon-11 is 20 min. A sample of carbon-11 has 25 mCi.
  a. How many half-lives of carbon-11 occur in 1 hour?
  b. How much carbon-11 is in the sample 1 hour after the original sample is measured?

Example 5
(page 478)

Identify the decay factor in each function.

22. $y = 5 \cdot 0.5^x$ **23.** $f(x) = 10 \cdot 0.1^x$

24. $g(x) = 100 \cdot \left(\frac{2}{3}\right)^x$ **25.** $y = 0.1 \cdot 0.9^x$

Identify each function as *exponential growth* or *exponential decay*.

26. $y = 0.68 \cdot 2^x$ **27.** $y = 2 \cdot 0.68^x$ **28.** $y = 68 \cdot 2^x$ **29.** $y = 68 \cdot 0.2^x$

30. Cars The value of a new car decreases exponentially. Suppose your mother buys a new car for $22,000. The value of the car decreases by 20% each year.
  a. What is the initial price of the car? The decay factor?
  b. Write an equation to model the value of the car x years after she buys it.
  c. Find the value of the car after 6 years.

B   **Apply Your Skills**

Write an exponential function to model each situation. Find each amount after the specified time.

31. A population of 130,000 grows 1% per year for 9 years.

32. A population of 3,000,000 decreases 1.5% annually for 10 years.

33. A $2400 principal earns 7% compounded annually for 10 years.

34. A $2400 principal earns 7% compounded monthly for 10 years.

35. Education Since 1985, the average annual cost y (in dollars) for tuition and fees at public two-year colleges in the United States has increased about 6.5% per year. In 1985, tuition and fees were an average of $584 per year.
  a. Write an equation to model the cost of two-year colleges. Predict the average annual cost for 2015.
  b. Open-Ended Predict the average annual cost for the year you plan to graduate from high school.

Tell whether each graph shows a *linear function*, an *exponential function*, or *neither*. Justify your reasoning.

36.

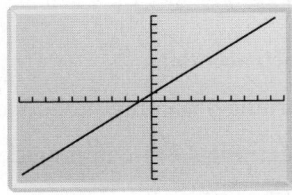

37.

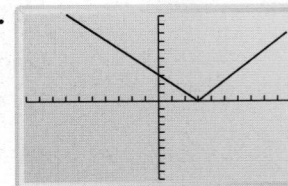

38.

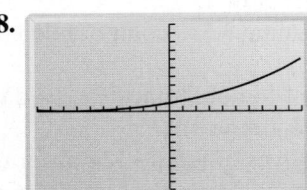

39.

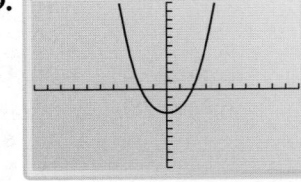

GO **O**nline
Homework Video Tutor
Visit: PHSchool.com
Web Code: ate-0808

Graph the function represented in each table. Then tell whether the table represents a *linear function* or an *exponential function*.

40.

x	y
1	20
2	40
3	60
4	80

41.

x	y
1	3
2	9
3	27
4	81

42.

x	y
1	3
2	9
3	15
4	21

43. Writing Would you rather have $500 in an account paying 6% interest compounded quarterly or $600 in an account paying 5% compounded annually? Summarize your reasoning.

GO for Help

For a guide to solving Exercise 46 see p. 483.

How many half-lives occur in each period of time?

44. 2 days (1 half-life = 8 h)

45. 300 years (1 half-life = 75 yr)

46. Medicine The function $y = 15 \cdot 0.84^x$ models the amount y of a 15-mg dose of antibiotic remaining in the bloodstream after x hours.
  a. Estimation Use the graphing calculator screen to estimate the half-life of this antibiotic in the bloodstream.
  b. Use your estimate to predict the amount of antibiotic that will remain in the bloodstream after 8 hours.
  c. Verify your prediction by using the function to find the amount of antibiotic remaining after 8 hours.

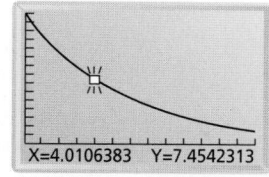

Antibiotic Decay in the Bloodstream

X=4.0106383   Y=7.4542313
Xmin=0   Ymin=0
Xmax=13   Ymax=15

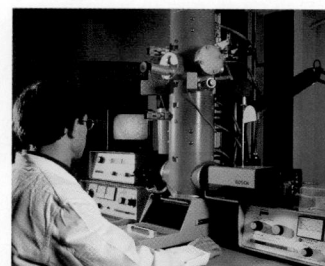

Real-World Connection

Careers A medical researcher may use equipment such as a scanning electron microscope.

47. Population Growth Since 1990, the population of Virginia has grown at an average annual rate of about 1%. In 1990, the population was about 6,284,000.
  a. Write an equation to model the population growth in Virginia since 1990.
  b. Suppose this rate of growth continues. Predict Virginia's population in 2010.

By which percent would you multiply a number to decrease it by the given amount?

48. 6%   **49.** 12%   **50.** 3.5%   **51.** 53.9%

52. Multiple Choice Use the graph at the right. What is a reasonable estimate of the half-life of cesium-134?
  Ⓐ $\frac{1}{2}$ year   Ⓑ 2 years
  Ⓒ $3\frac{1}{2}$ years   Ⓓ 4 years

Cesium-134 Decay

(graph: Level (mCi) vs Time (years); y-axis 100–500, x-axis 0–14)

Ⓒ Challenge

53. Credit Card Balances Suppose you charge $250 for a new suit. If you do not pay the whole amount the first month, you are charged 1.8% monthly interest on your account balance. Suppose you can make a $30 payment each month.
  a. What is your balance after your first payment?
  b. How much interest are you charged after your first payment?
  c. What is your balance just before you make your second payment?
  d. What is your balance after your second payment?
  e. How many months will it take for you to pay off the entire bill?
  f. How much interest will you have paid in all?

54. Data Collection Complete the table at the right using any ball. The height 0 is the starting height. Record the maximum height after the first, second, and third bounce.

a. Graph your data.

b. Write an exponential decay function that models your data.

Bounce	Height (centimeters)
0	■
1	■
2	■
3	■

55. On January 1, 2000, Chessville had a population of 40,000 people. Its population increases 7% each year. On the same day, Checkersville had a population of 60,000 people. Its population decreases 4% each year. During what year will the population of Chessville exceed that of Checkersville?

Multiple Choice

56. For which function will values of y decrease as values of x increase?
   A. $y = 12.5(1.325)^x$ **B.** $y = 300(1.06)^x$
   C. $y = 5000(0.98)^x$ **D.** $y = 1.02^x$

57. Suppose you deposit $1000 in an account earning 6% interest. You make no further deposits to the account and interest is compounded semi-annually. What is the balance after 5 years?
   F. $538.62   **G.** $1006.00   **H.** $1343.92   **J.** $1790.85

58. Read the passage below and answer the following problem.

Manhattan, Then and Now

In 1626, the Dutch landed on the island we now call Manhattan. They bought the island for $24 worth of merchandise. Today Manhattan is one of the most expensive places in the world to live. Rent for a one-bedroom apartment averages $2000 a month.

Suppose $24 had been invested in 1626 in an account paying 4.5% interest compounded annually. Which amount is closest to the balance in 2000?
   A. $339 million   **B.** $89 million   **C.** $9400   **D.** $8900

Short Response

59. Which is greater, the amount in an account that pays 5% interest compounded quarterly for 5 years or the amount in an account that pays 5.5% compounded annually for 5 years? Assume the accounts start with the same amount. Show your work.

Mixed Review

Lesson 8-7   **Graph each function.**

60. $y = 2 \cdot 10^x$ **61.** $f(x) = 100 \cdot 0.9^x$ **62.** $g(x) = \frac{1}{10} \cdot 0.1^x$

Lesson 8-2   **63. Geography** In 2000, about 1.4×10^4 ships passed through the Panama Canal. About 5.2×10^7 gallons of water flow out of the canal with each ship. About how many gallons of water flowed out of the canal with ships in 2000? Write your answer in scientific notation.

Understanding Math Problems Read the exercise below, and then learn how to use a graphing calculator to solve it. Check your understanding by solving the exercise at the bottom of the page.

Medicine The function $y = 15 \cdot 0.84^x$ models the amount y of a 15-mg dose of antibiotic remaining in the bloodstream after x hours.

a. Estimation Use the graphing calculator screen to estimate the half-life of this antibiotic in the bloodstream.

b. Use your estimate to predict the amount of antibiotic that will remain in the bloodstream after 8 hours.

c. Verify your prediction by using the function to find the amount of antibiotic remaining after 8 hours.

Antibiotic Decay in the Bloodstream

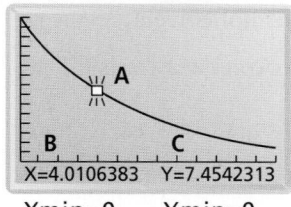

X=4.0106383 Y=7.4542313

D Xmin=0
   Xmax=13

E Ymin=0
   Ymax=15

To solve this problem, use the graphing calculator screen shown at the left. Parts of this screen are explained below.

A   The point shown on the screen is a point (x, y) on the graph.

B and C   These are the x-coordinates and y-coordinates of the highlighted point. The coordinates of the point are (4.0106383, 7.4542313).

D and E   These give the viewable range for the x- and y-values. The screen displays the graph for x-values from 0 to 13 and y-values from 0 to 15.

a. *Half-life* is the time required for the body to eliminate half of the initial dose. How do the coordinates (4.0106383, 7.4542313) relate to the half-life of the antibiotic? At time $x = 0$, the amount in the bloodstream is $y = 15$. When $y = \frac{15}{2} = 7.5$, the corresponding value of x represents the half-life. When $y \approx 7.5, x \approx 4$, so the half-life is about 4 hours.

b. If 7.5 mg remain after 4 hours, then $\frac{7.5}{2} = 3.75$ mg will remain after 8 hours.

c. $y = 15 \cdot 0.84^x$
  $y = 15 \cdot 0.84^8$ **Substitute 8 for *x*.**
  $y \approx 3.72$ **Use a calculator.**

The amount of antibiotic remaining after 8 hours will be about 3.72 mg.

EXERCISE

Memory Suppose the function $y = 40 \cdot 0.75^x$ models the number y of foreign-language words recalled from a list of 40 words after x weeks (without additional practice or study).

a. Estimation Use the graphing calculator to estimate the number of vocabulary words recalled after 5 weeks.

b. Verify your prediction by using the function to find the number of vocabulary words recalled after 5 weeks.

c. Critical Thinking What does 0.75 represent in the function?

Testing Multiple Choices

One advantage of multiple-choice tests is that the correct answer is among the choices. A strategy is to work backward by taking answers and testing them in the original problem.

1 EXAMPLE

Find the value of x if $x, x + 2,$ and $4x$ are three consecutive terms of a geometric sequence.

   (A) 0   (B) 1   (C) 2   (D) 3   (E) 4

The ratios of consecutive terms of a geometric sequence are the same. So $\frac{x + 2}{x}$ must equal $\frac{4x}{x + 2}$. Find the number for which this is true by substituting each answer choice for x.

Let $x = 0$. The sequence $0, 2, 0$ is not a geometric sequence because $\frac{2}{0} \neq \frac{0}{2}$. So A is not correct.

Let $x = 1$. The sequence $1, 3, 4$ is not a geometric sequence because $\frac{3}{1} \neq \frac{4}{3}$. So B is not correct.

Let $x = 2$. The sequence $2, 4, 8$ is a geometric sequence because $\frac{4}{2} = \frac{8}{4}$.   The correct answer is C.

● You do not need to try choice D or E.

2 EXAMPLE

Find the value of x if $2x^{-3} = \frac{1}{4}$.

   (A) 1   (B) 2   (C) 3   (D) 4   (E) 5

Solve by substituting each answer choice into the original equation.

Let $x = 1$. $\longrightarrow$ $2(1)^{-3} = 2(1) = 2$ $\longrightarrow$ $2 \neq \frac{1}{4}$.   A is not the answer.

Let $x = 2$. $\longrightarrow$ $2(2)^{-3} = 2\left(\frac{1}{8}\right) = \frac{2}{8}$ $\longrightarrow$ $\frac{1}{4} = \frac{1}{4}$.   B is the answer.

● You do not have to test answer choices C, D, and E.

EXERCISES

Solve each of the following by working backward.

1. Find the value of x if $x - 2, x,$ and $x + 3$ are three consecutive terms of a geometric sequence.
   (A) 4   (B) 6   (C) 8   (D) 10   (E) 12

2. Find the value of x if $x^{-1} + x^{-2} = 0$.
   (A) -2 (B) -1 (C) 1   (D) 2   (E) 4

3. The area of a square is 1.21×10^{-4}. What is its perimeter?
   (A) 0.00044   (B) 0.0044   (C) 0.044   (D) 0.44   (E) 4.4

4. Find the value of x if $x, 3x + 1,$ and $6x - 1$ are three consecutive terms of an arithmetic sequence.
   (A) -3 (B) -1 (C) 1   (D) 3   (E) 4

Chapter Review

Vocabulary Review

common ratio (p. 460)
compound interest (p. 476)
decay factor (p. 478)
exponential decay (p. 478)

exponential function (p. 468)
exponential growth (p. 475)
geometric sequence (p. 460)
growth factor (p. 475)

interest period (p. 476)
scientific notation (p. 436)

Go Online
PHSchool.com
For: Vocabulary quiz
Web Code: atj-0851

Choose the correct term to complete each sentence.

1. The function $y = a \cdot b^x$ models __?__ for $a > 0$ and $b > 1$.

2. For the function $y = a \cdot b^x$, where $a > 0$ and $b > 1$, b is the __?__.

3. __?__ is a shorthand way to write very large and very small numbers.

4. The function rule $y = a \cdot b^x$ models __?__ for $a > 0$ and $0 < b < 1$.

5. For the function $y = a \cdot b^x$, where $a > 0$ and $0 < b < 1$, b is the __?__.

6. __?__ is calculated using both the principal and the interest that an account has already earned.

7. Each term of a geometric sequence is found by multiplying the previous term by a fixed number called the __?__.

8. The length of time over which interest is calculated is the __?__.

9. When a sequence has a common ratio, it is a(n) __?__.

10. The rule $y = 7^x$ is a(n) __?__.

Skills and Concepts

8-1 Objectives

▼ To simplify expressions with zero and negative exponents (p. 430)

▼ To evaluate exponential expressions (p. 432)

You can use zero and negative integers as exponents. For every nonzero number a, $a^0 = 1$. For every nonzero number a and any integer n, $a^{-n} = \frac{1}{a^n}$.

Simplify each expression.

11. $b^{-4}c^0d^6$

12. $\frac{x^{-2}}{y^{-8}}$

13. $7k^{-8}h^3$

14. $\frac{1}{p^2q^{-4}r^0}$

15. $\left(\frac{2}{5}\right)^{-4}$

16. $(-2)^{-3}$

17. -2^{-3}

18. $7^{-2}y^{-4}$

19. $\frac{9w^{-4}}{x^{-2}y^7}$

Evaluate each expression for $p = 2$, $q = -3$, and $r = 0$.

20. p^2q^2

21. $(-p)^2q^{-2}$

22. p^qq^p

23. p^rq^r

24. $-p^2q^3$

25. Which expression has the greatest value for $a = 4$, $b = -3$, and $c = 0$?

   A. a^b **B.** b^c **C.** $\frac{1}{b^{-a}}$ **D.** $\frac{a^c}{b^c}$ **E.** $\frac{c}{a^{-b}}$

26. **Critical Thinking** Is $(-3b)^4 = -12b^4$? Explain why or why not.

You can use **scientific notation** to express very large or very small numbers. A number is in scientific notation if it is in the form $a \times 10^n$, where $1 \le a < 10$, and n is an integer.

Is each number written in scientific notation? If not, explain.

27. 950×10^5 **28.** 72.35×10^8 **29.** 1.6×10^{-6} **30.** 0.84×10^{-5}

31. The space probe Voyager 2 traveled 2,793,000 miles. Write the number of miles in scientific notation.

32. There are 189 million passenger cars and trucks in use in the United States. Write the number of passenger cars and trucks using scientific notation.

To multiply powers with the same base, add the exponents.
$$a^m \cdot a^n = a^{m+n}$$

To raise a power to a power, multiply the exponents.
$$(a^m)^n = a^{mn}$$

To raise a product to a power, raise each factor in the product to the power.
$$(ab)^n = a^n b^n$$

Simplify each expression.

33. $2d^2 d^3$ **34.** $(q^3 r)^4$ **35.** $(5c^{-4})(-4m^2 c^8)$

36. $(1.34^2)^5 (1.34)^{-8}$ **37.** $(12x^2 y^{-2})^5 (4xy^{-3})^{-8}$ **38.** $(-2r^{-4})^2 (-3r^2 z^8)^{-1}$

39. Estimation Each square inch of your body has about 6.5×10^2 pores. Suppose the back of your hand has an area of about 0.12×10^2 in.^{2}. About how many pores are on the back of your hand?

40. Open-Ended Write and solve a problem that involves multiplying exponents.

To divide powers with the same base, subtract the exponents.
$$\frac{a^m}{a^n} = a^{m-n}$$

To raise a quotient to a power, raise the dividend and the divisor to the power.
$$\left(\frac{a}{b}\right)^n = \frac{a^n}{b^n}$$

Simplify each expression.

41. $\frac{w^2}{w^5}$ **42.** $(8^3) \cdot 8^{-5}$ **43.** $\left(\frac{21x^3}{3x}\right)$ **44.** $\left(\frac{n^5}{v^3}\right)^7$ **45.** $\frac{e^{-6}c^3}{e^5}$

Simplify each quotient. Give your answer in scientific notation.

46. $\frac{4.2 \times 10^8}{2.1 \times 10^{11}}$ **47.** $\frac{3.1 \times 10^4}{12.4 \times 10^2}$ **48.** $\frac{4.5 \times 10^3}{9 \times 10^7}$ **49.** $\frac{5.1 \times 10^5}{1.7 \times 10^2}$

50. Writing List the steps that you would use to simplify $\left(\frac{5a^8}{10a^6}\right)^{-3}$.

You find each term of a **geometric sequence** by multiplying the previous term by a fixed number called the common ratio.

Find the common ratio in each geometric sequence.

51. $750, 75, 7.5, 0.75, \ldots$ **52.** $0.04, 0.12, 0.36, 1.08, \ldots$ **53.** $20, -10, 5, -\frac{5}{2}, \ldots$

Determine whether each sequence is *arithmetic*, *geometric*, or *neither*. Find the next three terms.

54. $1600, 400, 100, 25, \ldots$ **55.** $-40, -39, -37, -34, \ldots$ **56.** $14, 21, 28, 35, \ldots$

You can use exponents to show repeated multiplication. An **exponential function** involves repeated multiplication of an initial amount by the same positive number.

Evaluate each function for the given values.

57. $f(x) = 3 \cdot 2^x$ for the domain $\{1, 2, 3, 4\}$

58. $y = 10 \cdot (0.75)^x$ for the domain $\{1, 2, 3\}$

59. a. One kind of bacteria in a laboratory culture triples in number every 30 minutes. Suppose a culture is started with 30 bacteria cells. How many bacteria will there be after 2 hours?

   b. After how many minutes will there be more than 20,000 bacteria cells?

The general form of an exponential function is $y = a \cdot b^x$.

When $a > 0$ and $b > 1$, the function increases, and the function shows **exponential growth.** The base of the exponent, b, is called the **growth factor.** An example of exponential growth is **compound interest.**

When $a > 0$ and $0 < b < 1$, the function decreases, and the function shows **exponential decay.** Then the base of the exponent b is called the **decay factor.** An example of exponential decay is the half-life model.

Identify the initial amount a and the growth or decay factor b in each exponential function.

60. $y = 100 \cdot 1.025^x$ **61.** $y = 32 \cdot 0.75^x$ **62.** $y = 0.4 \cdot 2^x$

Identify each function as *exponential growth* or *exponential decay*. Then identify the growth or decay factor.

63. $y = 5.2 \cdot 3^x$ **64.** $y = 0.15 \cdot \left(\frac{3}{2}\right)^x$ **65.** $y = 7 \cdot 0.32^x$ **66.** $y = 1.3 \cdot \left(\frac{1}{4}\right)^x$

Graph each function.

67. $f(x) = 2.5^x$ **68.** $y = 0.5 \cdot (0.5)^x$ **69.** $f(x) = \left(\frac{1}{2}\right) \cdot 3^x$ **70.** $y = 0.1^x$

71. The function $y = 25 \cdot 0.80^x$ models the amount y of a 25-mg dose of medicine remaining in the bloodstream after x hours. How many milligrams of medicine remain in the bloodstream after 5 hours?

Chapter
8

Chapter Test

Go Online
PHSchool.com
For: Chapter Test
Web Code: ata-0852

Simplify each expression.

1. $\dfrac{r^3 t^{-7}}{t^5}$

2. $\left(\dfrac{a^3}{m}\right)^{-4}$

3. $\dfrac{t^{-8} m^2}{m^{-3}}$

4. $c^3 v^9 c^{-1} c^0$

5. $h^2 k^{-5} d^3 k^2$

6. $9 y^4 j^2 y^{-9}$

7. $(w^2 k^0 p^{-5})^{-7}$

8. $2y^{-9} h^2 (2y^0 h^{-4})^{-6}$

9. $(1.2)^5 (1.2)^{-2}$

10. $(-3q^{-1})^3 q^2$

11. If $n = -3$, which expression has the least value?

  A. $n^2 n^0$

  B. n^n

  C. $n^8 n^{-5}$

  D. $-n^n n^{-4}$

Write each number in scientific notation.

12. **History** There were about 62,041,000 votes cast for George Bush in the 2004 presidential election.

13. **Pets** More than 450,000 households in the United States have reptiles as pets.

Is each number written in scientific notation? If not, explain.

14. 76×10^{-9}

15. 7.3×10^5

16. $4.05 \times 10 \times 10^{-8}$

17. 32.5×10^{13}

18. **a. Astronomy** The speed of light in a vacuum is about 186,300 mi/s. Use scientific notation to express how far light travels in one hour.

  b. At its farthest, Saturn is about 1.03×10^9 mi from Earth. About how many hours does it take for light to travel from Earth to Saturn?

19. Use the sequence $-32, 16, -8, 4, \ldots$.

  a. What is the common ratio?

  b. What are the next three terms?

  c. Write a rule for the sequence.

  d. What is the ninth term of the sequence?

20. You drop a ball from a height of 12 ft. Each bounce has $\frac{3}{5}$ the height of the previous bounce.

  a. Write a rule for the sequence. The initial height is given by the term $n = 1$.

  b. What height will the ball reach at the top of the fourth path ($n = 4$)?

21. Find the fifth term of the sequence $A(n) = -3(-2)^{n+1}$.

Evaluate each function for $x = 1, 2,$ and 3.

22. $y = 3 \cdot 5^x$

23. $f(x) = \frac{1}{2} \cdot 4^x$

24. $f(x) = 4(0.95)^x$

25. $g(x) = 5\left(\frac{3}{4}\right)^x$

Graph each function.

26. $y = \frac{1}{2} \cdot 2^x$

27. $y = 2 \cdot \left(\frac{1}{2}\right)^x$

28. $f(x) = 3^x$

29. **Open-Ended** Write and solve a problem involving exponential decay.

30. **Writing** Explain when the function $y = a \cdot b^x$ shows exponential growth and when it shows exponential decay.

31. **Banking** A customer deposits $1000 in a savings account that pays 4% interest compounded quarterly. How much money will the customer have in the account after 2 years? After 5 years?

32. The function $y = 1.3 \cdot (1.07)^x$ models a city's annual electrical consumption for x years since 1985, where y is kilowatt-hours.

  a. Determine whether the function models exponential growth or decay, and find the growth or decay factor.

  b. According to the model, what will be the annual electrical usage in 2010?

  c. According to the model, what was the annual electrical usage in 1975?

  d. What value of x should you substitute to find the value of y now? Use this value for x to find y.

33. **Automobiles** Suppose a new car is worth $20,000. You can use the function $y = 20{,}000(0.85)^x$ to estimate the car's value after x years.

  a. What is the decay factor? What does it mean?

  b. Estimate the car's value after one year.

  c. Estimate the car's value after four years.

34. The function $y = 10 \cdot 1.08^x$ models the cost of annual tuition (in thousands of dollars) at a local college x years after 1997.

  a. What is the annual percent increase?

  b. How much was tuition in 1997? In 2000?

  c. How much will the tuition be the year you plan to graduate from high school?

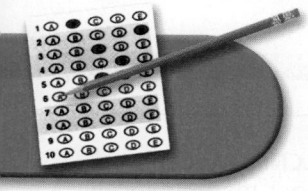

Regents Test Prep

Multiple Choice

For Exercises 1–12, choose the correct letter.

1. If a is positive and b is negative, which of the following is negative?
   - **(1)** $a + |b|$
   - **(2)** $a|b|$
   - **(3)** $|a|b$
   - **(4)** $|a| - b$

2. The scores on your first five algebra tests are 88, 78, 81, 83, and 90. What score must you get on your next test to raise the mean to 85?
   - **(1)** 90
   - **(2)** 87
   - **(3)** 86
   - **(4)** 85

3. Which value of x is NOT a solution to the inequality $5 - 6x < -x + 2$?
   - **(1)** -1
   - **(2)** 1
   - **(3)** 3
   - **(4)** 5

4. You earn a commission of 6% on your first $500 of sales and 10% on all sales above $500. If you earn $130 in commission, what are your total sales?
   - **(1)** $800
   - **(2)** $1000
   - **(3)** $1300
   - **(4)** $1500

5. Find the solution of the system of equations.
   $$\frac{1}{3}x - y = 4$$
   $$x + 3y = 0$$
   - **(1)** $(9, -1)$
   - **(2)** $(-6, 2)$
   - **(3)** $(6, -2)$
   - **(4)** $(-9, 1)$

6. You flip a coin and roll a number cube. What is the probability of getting a head and a multiple of three?
   - **(1)** $\frac{1}{12}$
   - **(2)** $\frac{1}{6}$
   - **(3)** $\frac{1}{4}$
   - **(4)** $\frac{3}{2}$

7. Which number has the least value?
   - **(1)** 2.8×10^{-5}
   - **(2)** 5.3×10^{-4}
   - **(3)** 8.3×10^{-7}
   - **(4)** 1.6×10^{-8}

8. Potassium-42 has a half-life of 12.5 h. How many half-lives are in 75 h?
   - **(1)** 6
   - **(2)** 8
   - **(3)** 25
   - **(4)** 150

9. Simplify $-3a^8 \cdot cb^{-3} \cdot b^{12} \cdot 9c^5$.
   - **(1)** $6a^9b^5c^6$
   - **(2)** $-27a^8b^9c^6$
   - **(3)** $-27a^8b^{15}c^6$
   - **(4)** $-3abc$

10. Which statement is true for every solution of the following system?
    $$y > x + 4$$
    $$y + x > 4$$
    - **(1)** $x \le -3$
    - **(2)** $y < 5$
    - **(3)** $x > 4$
    - **(4)** $y > 4$

11. Doug's teacher put 5 red markers, 5 blue markers, and 5 green markers in a box. Each student will pick a marker without looking and pass the box to the next student, until all the markers are gone. What is the probability that the first 2 students both pick red?
    - **(1)** $\frac{1}{25}$
    - **(2)** $\frac{2}{21}$
    - **(3)** $\frac{4}{35}$
    - **(4)** $\frac{2}{5}$

12. Which system of equations is shown in the graph below?

    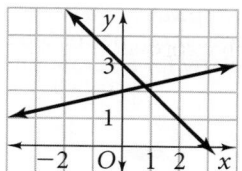

    - **(1)** $3x + 2y = 6$
        $y = \frac{1}{4}x + 2$
    - **(2)** $2x + 3y = 12$
        $y = -\frac{1}{4}x + 2$
    - **(3)** $3x + 3y = 9$
        $y = \frac{1}{4}x + 2$
    - **(4)** $3x - 2y = 12$
        $y = 4x + 2$

Short Response

13. A cafeteria charges $.21/oz for frozen yogurt. How many dollars would a 9-oz serving cost?

14. For a spinner numbered from 1 to 6, the outcomes are equally likely. What is the probability of getting an odd number?

Show all work.

15. On April 1, 2000, the day of the 2000 national census, the population of the United States was 281,421,906 people. This was a 13.2% increase from the 1990 census. What was the 1990 population of the United States?

16. Graph the inequality $|x - 2| \le 9$ on a number line.

Extended Response

17. The slopes of four different lines are $\frac{3}{5}, -\frac{10}{6}, -\frac{5}{3}$, and $\frac{9}{15}$. Do these lines determine a rectangle? Explain why or why not.

Activity Lab

NY **A.CN.3:** Model situations mathematically, using representations to draw conclusions and formulate new situations..

How Fast Can You Run?

Applying Linear Equations Animals run to escape predators and to chase prey. The display below compares animals as if they were able to sprint along at their top velocities or speeds for a whole hour. Linear equations and graphs are good tools for describing and comparing motion at constant velocities.

Instant Records

A stopwatch is a watch used to time races. It can be started and stopped quickly for accurate timing.

Giraffe, 32 mi/h

Tiger, 35 mi/h

Kangaroo, 30 mi/h

Mongolian wild ass, 40 mi/h

Mule deer, 35 mi/h

Elephant, 25 mi/h

Camel, 20 mi/h

Rhinoceros, 32 mi/h

Zebra, 40 mi/h

Grizzly Bear, 30 mi/h

Reindeer, 32 mi/h

Cape hunting dog, 45 mi/h

Hyena, 40 mi/h

Jackal, 35 mi/h

White-tailed deer, 30 mi/h

Whippet, 36 mi/h

Coyote, 43 mi/h

Human, 21 mi/h

Wart hog, 30 mi/h

Cat, 30 mi/h

Greyhound, 40 mi/h

Fox, 42 mi/h

Car Racing

Formula 1 cars compete at speeds as high as 200 mi/h. The speed limit on most U.S. highways and interstates is 65 mi/h.

Activity 1

Materials: graph paper, pencil

a. Suppose two animals are in a race. Choose the two animals and calculate the speed of each animal in yards per second.

b. Decide how much of a head start (in yards) the faster animal offers the slower animal. For each animal, write an equation relating distance from the starting line to time.

c. Graph the two equations on the same coordinate plane.

d. How long will it take the faster animal to overtake the slower animal? How many yards from the starting line are the animals when the faster animal overtakes the slower animal?

e. Reduce the head start by half and repeat parts (c) and (d).

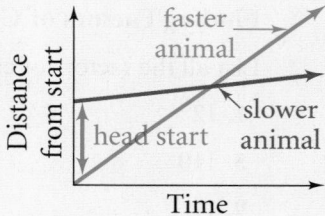

Activity 2

a. Choose a third animal. Calculate its speed in yards per second. Compare its speed with the speed of the two animals you chose in Activity 1.

b. Decide which two animals should get head starts and how much of a head start each should get. Write three equations relating distance from the starting line to time.

c. Graph all three equations on the same coordinate plane.

Ostrich,
45 mi/h

Moose,
45 mi/h

Horse,
43 mi/h

Wildebeest,
50 mi/h

Lion,
50 mi/h

Thomson's
gazelle, 50 mi/h

Cheetah,
70 mi/h

Go Online
PHSchool.com

For: Information about speeds
Web Code: ate-0853

491

What You've Learned

 Learning Standards for Mathematics

In Chapters 1 and 2, you

- **NY A.N.1:** used variables and applied the Distributive Property to variable expressions.

In Chapter 3, you

- **NY A.A.22:** combined like terms to solve equations.

In Chapter 8, you

- **NY A.A.12:** simplified variable expressions with exponents by using the multiplication and division properties of exponents.

 **Check Your Readiness**

GO for Help to the Lesson in green.

Finding Factors of Composite Numbers (Skills Handbook page 754)

List all the factors of each number.

1. 12
2. 56
3. 31
4. 27

5. 110
6. 65
7. 50
8. 200

9. 11
10. 42
11. 66
12. 73

Simplifying Expressions (Lesson 2-4)

Simplify each expression.

13. $2x^2 - x + x^2 - 3x$

14. $-b + 2 + 3b + 4$

15. $-5y - y^2 + 4y^2 - 6y$

16. $(3w - 2w^2 + 4w - 2w^2)\frac{1}{6}$

17. $-8(z + 2) + 5(3z - 10)$

18. $2(x + 4x^2 - 2x - 2x^2)$

19. $12t - 5t^2 - 2t - t^2$

20. $p - 3 - (p^2 - 3) - 3p$

Multiplying Expressions With Exponents (Lessons 8-3 and 8-4)

Simplify each expression.

21. $(7w)^2$
22. $(-5n^2)(-5n)$
23. $(3z^2)^2$
24. $(2t^3)(5t^4)$

25. $(4y^3)^2$
26. $(-9ab)^2$
27. $4(x^2)^2$
28. $(-6p^4)^2$

Dividing Expressions With Exponents (Lesson 8-5)

Simplify each expression.

29. $\dfrac{x^5 y^8}{x^3 y^4}$

30. $\dfrac{(3c)^2}{(3c)}$

31. $\dfrac{-5t}{(10t^3)(2t)}$

32. $\dfrac{(3a)(4a^3)}{6a^2}$

Polynomials and Factoring

Key Vocabulary

- binomial (p. 495)
- degree of a monomial (p. 495)
- degree of a polynomial (p. 495)
- difference of squares (p. 514)
- factor by grouping (p. 534)
- monomial (p. 494)
- perfect-square trinomial (p. 528)
- polynomial (p. 495)
- standard form of a polynomial (p. 495)
- trinomial (p. 495)

What You'll Learn Next

NY Learning Standards for Mathematics

- **NY A.A.13:** You will add, subtract, and multiply polynomials.

Activity Lab You will interpret statistical results for one- and two-variable data, on pages 546–547.

9-1

Adding and Subtracting Polynomials

A.A.13 Add, subtract, and multiply monomials and polynomials.

 **Check Skills You'll Need**

GO for Help Lesson 2-4

Simplify each expression.

1. $6t + 13t$　　　　**2.** $5g + 34g$　　　　**3.** $7k - 15k$

4. $2b - 6 + 9b$　　**5.** $4n^2 - 7n^2$　　　**6.** $8x^2 - x^2$

New Vocabulary • monomial • degree of a monomial • polynomial • standard form of a polynomial • degree of a polynomial • binomial • trinomial

---

1　Describing Polynomials

Activity: Using Polynomials

Business Suppose you work at a pet store. The spreadsheet below shows the details of several customers' orders.

	A	B	C	D	E	F
1	Customer	Seed	Cuttlebone	Millet	G. Paper	Perches
2	Davis			✔		
3	Brooks	✔	✔			
4	Casic	✔		✔		
5	Martino	✔			✔	✔

The following variables represent the number of each item ordered.

s = bags of birdseed　　　　m = bags of millet
c = packages of cuttlebone　g = packages of gravel paper
p = packages of perches

1. Which expression represents the cost of Casic's order?
A. $27.99(s + m)$　　　**B.** $3.99s + 24m$　　　**C.** $27.99sm$

2. Write expressions to represent each of the other customers' orders.

3. Martino buys 10 bags of birdseed, 4 packages of gravel paper, and 2 packages of perches. What is the total cost of his order?

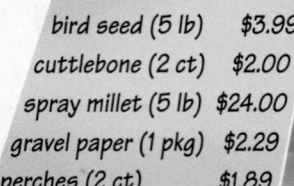

Rocky's Friends Bird Supplies

bird seed (5 lb)	$3.99
cuttlebone (2 ct)	$2.00
spray millet (5 lb)	$24.00
gravel paper (1 pkg)	$2.29
perches (2 ct)	$1.89

A **monomial** is an expression that is a number, a variable, or a product of a number and one or more variables. Each of the following is a monomial.

$$12 \qquad\qquad y \qquad\qquad -5x^2y \qquad\qquad \frac{c}{3}$$

The fraction $\frac{c}{3}$ is a monomial, but the expression $\frac{c}{x}$ is *not* a monomial because there is a variable in the denominator.

The **degree of a monomial** is the sum of the exponents of its variables. For a nonzero constant, the degree is 0. Zero has no degree.

1 EXAMPLE Degree of a Monomial

Find the degree of each monomial.

a. $\frac{2}{3}x$ Degree: 1 $\frac{2}{3}x = \frac{2}{3}x^1$. The exponent is 1.

b. $7x^2y^3$ Degree: 5   The exponents are 2 and 3. Their sum is 5.

c. -4 Degree: 0   The degree of a nonzero constant is 0.

✓ **Quick Check** ❶ **Critical Thinking** What is the degree of $9x^0$? Explain.

A **polynomial** is a monomial or the sum or difference of two or more monomials.

$$3x^4 + 5x^2 - 7x + 1$$
$$\uparrow \quad \uparrow \quad \uparrow \quad \uparrow$$
degree → $\quad$ 4 $\quad$ 2 $\quad$ 1 $\quad$ 0

The polynomial shown above is in standard form. **Standard form of a polynomial** means that the degrees of its monomial terms decrease from left to right. The **degree of a polynomial** in one variable is the same as the degree of the monomial with the greatest exponent. The degree of $3x^4 + 5x^2 - 7x + 1$ is 4.

After you simplify a polynomial by combining like terms, you can name the polynomial based on its degree or the number of monomials it contains.

Polynomial	Degree	Name Using Degree	Number of Terms	Name Using Number of Terms
$7x + 4$	1	linear	2	**binomial**
$3x^2 + 2x + 1$	2	quadratic	3	**trinomial**
$4x^3$	3	cubic	1	monomial
$9x^4 + 11x$	4	fourth degree	2	binomial
5	0	constant	1	monomial

2 EXAMPLE Classifying Polynomials

Write each polynomial in standard form. Then name each polynomial based on its degree and the number of its terms.

a. $5 - 2x$

$-2x + 5$ Place terms in order.

linear binomial

b. $3x^4 - 4 + 2x^2 + 5x^4$

$3x^4 + 5x^4 + 2x^2 - 4$ Place terms in order.

$8x^4 + 2x^2 - 4$ Combine like terms.

fourth degree trinomial

✓ **Quick Check** ❷ Write each polynomial in standard form. Then name each polynomial based on its degree and the number of its terms.

a. $6x^2 + 7 - 9x^4$ **b.** $3y - 4 - y^3$ **c.** $8 + 7v - 11v$

You can add polynomials by adding like terms.

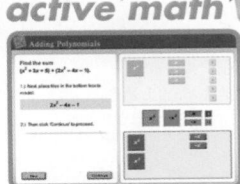

Online active math

For: Polynomial Addition Activity
Use: Interactive Textbook, 9-1

3 EXAMPLE **Adding Polynomials**

Simplify $(4x^2 + 6x + 7) + (2x^2 - 9x + 1)$.

Method 1 Add vertically.

Line up like terms. Then add the coefficients.

$$
\begin{array}{r}
4x^2 + 6x + 7 \\
+\ 2x^2 - 9x + 1 \\
\hline
6x^2 - 3x + 8
\end{array}
$$

Method 2 Add horizontally.

Group like terms. Then add the coefficients.

$(4x^2 + 6x + 7) + (2x^2 - 9x + 1) = (4x^2 + 2x^2) + (6x - 9x) + (7 + 1)$
$$= 6x^2 - 3x + 8$$

✓ **Quick Check** ❸ Simplify each sum.
   a. $(12m^2 + 4) + (8m^2 + 5)$ **b.** $(t^2 - 6) + (3t^2 + 11)$
   c. $(9w^3 + 8w^2) + (7w^3 + 4)$ **d.** $(2p^3 + 6p^2 + 10p) + (9p^3 + 11p^2 + 3p)$

In Chapter 1, you learned that subtraction means to add the opposite. So when you subtract a polynomial, change each of the terms to its opposite. Then add the coefficients.

4 EXAMPLE **Subtracting Polynomials**

Simplify $(2x^3 + 5x^2 - 3x) - (x^3 - 8x^2 + 11)$.

Method 1 Subtract vertically.

$$
\begin{array}{r}
2x^3 + 5x^2 - 3x \\
-\ (x^3 - 8x^2\ \ \ \ \ + 11)
\end{array}
$$ **Line up like terms.**

$$
\begin{array}{r}
2x^3 + 5x^2 - 3x \\
-x^3 + 8x^2\ \ \ \ \ \ - 11 \\
\hline
x^3 + 13x^2 - 3x\ - 11
\end{array}
$$ **Then add the opposite of each term in the polynomial being subtracted.**

Method 2 Subtract horizontally.

$(2x^3 + 5x^2 - 3x) - (x^3 - 8x^2 + 11)$
$= 2x^3 + 5x^2 - 3x - x^3 + 8x^2 - 11$ **Write the opposite of each term in the polynomial being subtracted.**

$= (2x^3 - x^3) + (5x^2 + 8x^2) - 3x - 11$ **Group like terms.**
$= x^3 + 13x^2 - 3x - 11$ **Simplify.**

✓ **Quick Check** ❹ Simplify each difference.
   a. $(v^3 + 6v^2 - v) - (9v^3 - 7v^2 + 3v)$ **b.** $(30d^3 - 29d^2 - 3d) - (2d^3 + d^2)$
   c. $(4x^2 + 5x + 1) - (6x^2 + x + 8)$

EXERCISES

For more exercises, see *Extra Skill and Word Problem Practice*.

Practice and Problem Solving

A Practice by Example

Example 1
(page 495)

GO for Help

Find the degree of each monomial.

1. $4x$ **2.** $7c^3$ **3.** -16 **4.** $6y^2w^8$

5. $8ab^3$ **6.** 6 **7.** $-9x^4$ **8.** 11

Example 2
(page 495)

Name each expression based on its degree and number of terms.

9. $5x^2 - 2x + 3$ **10.** $\frac{3}{4}z + 5$ **11.** $7a^3 + 4a - 12$

12. $\frac{3}{x} + 5$ **13.** -15 **14.** $w^2 + 2$

Write each polynomial in standard form. Then name each polynomial based on its degree and number of terms.

15. $4x - 3x^2$ **16.** $4x + 9$ **17.** $c^2 - 2 + 4c$

18. $9z^2 - 11z^2 + 5z - 5$ **19.** $y - 7y^3 + 15y^8$ **20.** $-10 + 4q^4 - 8q + 3q^2$

Example 3
(page 496)

Simplify each sum.

21. $\begin{array}{r} 5m^2 + 9 \\ + \ 3m^2 + 6 \\ \hline \end{array}$ **22.** $\begin{array}{r} 3k - 8 \\ + \ 7k + 12 \\ \hline \end{array}$ **23.** $\begin{array}{r} w^2 + w - 4 \\ + \ 7w^2 - 4w + 8 \\ \hline \end{array}$

24. $(8x^2 + 1) + (12x^2 + 6)$ **25.** $(g^4 + 4g) + (9g^4 + 7g)$

26. $(a^2 + a + 1) + (5a^2 - 8a + 20)$ **27.** $(7y^3 - 3y^2 + 4y) + (8y^4 + 3y^2)$

Example 4
(page 496)

Simplify each difference.

28. $\begin{array}{r} 6c - 5 \\ - \ (4c + 9) \\ \hline \end{array}$ **29.** $\begin{array}{r} 2b + 6 \\ -(b + 5) \\ \hline \end{array}$ **30.** $\begin{array}{r} 7h^2 + 4h - 8 \\ - \ (3h^2 - 2h + 10) \\ \hline \end{array}$

31. $(17n^4 + 2n^3) - (10n^4 + n^3)$ **32.** $(24x^5 + 12x) - (9x^5 + 11x)$

33. $(6w^2 - 3w + 1) - (w^2 + w - 9)$ **34.** $(-5x^4 + x^2) - (x^3 + 8x^2 - x)$

B Apply Your Skills

Simplify. Write each answer in standard form.

35. $(7y^2 - 3y + 4y) + (8y^2 + 3y^2 + 4y)$ **36.** $(2x^3 - 5x^2 - 1) - (8x^3 + 3 - 8x^2)$

37. $(-7z^3 + 3z - 1) - (-6z^2 + z + 4)$ **38.** $(7a^3 - a + 3a^2) + (8a^2 - 3a - 4)$

Geometry Find an expression for the perimeter of each figure.

39.

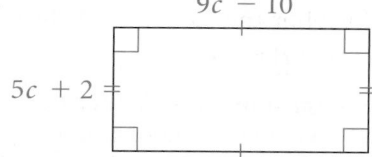

40.

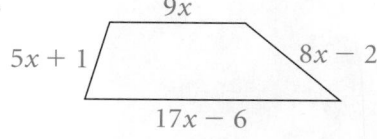

41. Error Analysis Kwan's work is shown below. What mistake did he make?

$$(5x^2 - 3x + 1) - (2x^2 - 4x - 2) = 5x^2 - 3x + 1 - 2x^2 - 4x - 2$$
$$= 5x^2 - 2x^2 - 3x - 4x + 1 - 2$$
$$= 3x^2 - 7x - 1$$

Lesson 9-1 Adding and Subtracting Polynomials   **497**

42. a. Writing Write the definition of each word. Use a dictionary if necessary.

monogram   binocular   tricuspid   polyglot

b. Open-Ended Find other words that begin with *mono*, *bi*, *tri*, or *poly*.

c. Do these prefixes have meanings similar to those in mathematics?

Simplify. Write each answer in standard form.

43. $(x^3 + 3x) + (12x - x^4)$

44. $(6g - 7g^8) - (4g + 2g^3 + 11g^2)$

45. $(2h^4 - 5h^9) - (-8h^5 + h^{10})$

46. $(-4t^4 - 9t + 6) + (13t + 5t^4)$

47. $(8b - 6b^7 + 3b^8) + (2b^7 - 5b^9)$

48. $(11 + k^3 - 6k^4) - (k^2 - k^4)$

📦 **Geometry** **Find an expression for each missing length.**

49. Perimeter = $25x + 8$

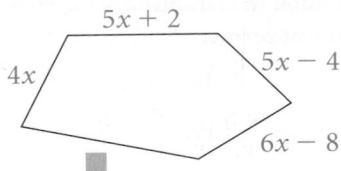

50. Perimeter = $23a - 7$

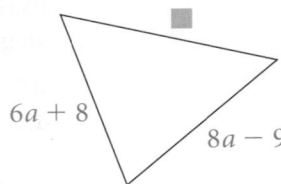

51. Critical Thinking Is it possible to write a binomial with degree 0? Explain.

C **Challenge**

52. a. Write the equations for line P and line Q. Use slope-intercept form.

b. Use the expressions on the right side of each equation to write a function for the vertical distance $D(x)$ between points on lines P and Q with the same x-value.

c. For what value of x does $D(x)$ equal zero?

d. Critical Thinking How does the x-value in part (c) relate to the graph?

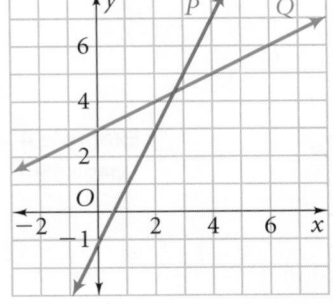

Simplify each expression.

53. $(ab^2 + ba^3) + (4a^3b - ab^2 - 5ab)$

54. $(9pq^6 - 11p^4q) - (-5pq^6 + p^4q^4)$

Real-World 🌐 **Connection**

There were about 12 million students enrolled in college in 1980, 13.8 million in 1990, and 15 million in 2000.

🌐 **55. Graduation** You can model the number of men and women in the United States who enrolled in college within a year of graduating from high school with the linear equations shown below. Let t equal the year of enrollment, with $t = 0$ corresponding to 1990. Let $m(t)$ equal the number of men in thousands, and let $w(t)$ equal the number of women in thousands.

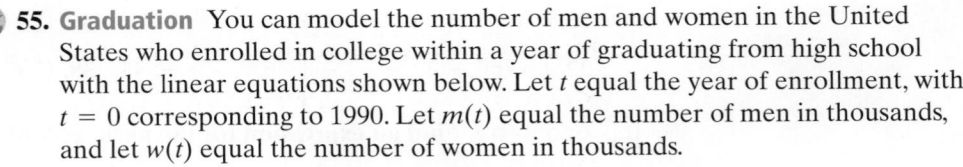

$m(t) = 35.4t + 1146.8$ men enrolled in college

$w(t) = 21.6t + 1185.5$ women enrolled in college

a. Add the expressions on the right side of each equation to model the total number of recent high school graduates $p(t)$ who enrolled in college between 1990 and 1998.

b. Use the equation you created in part (a) to find the number of high school graduates who enrolled in college in 1995.

c. Critical Thinking If you had subtracted the expressions on the right side of each equation above, what information would the resulting expression model?

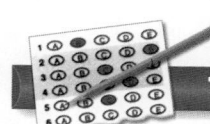
Multiple Choice

56. Which expression represents the sum of an odd integer n and the next three odd integers?

A. $n + 6$ **B.** $n + 12$ **C.** $4n + 6$ **D.** $4n + 12$

57. Simplify $(8x^2 + 3) + (7x^2 + 10)$.

F. $x^2 - 7$ **G.** $15x^2 + 13$ **H.** $15x^4 + 13$ **J.** $56x^4 + 30$

58. Which sum is equivalent to $(11k^2 + 3k + 1)$?

A. $(4k^2 + k - 5) + (7k^2 + 2k + 6)$
B. $(k^2 - 2) + (11k^2 + 3k + 3)$
C. $(8k^2 + k) + (3k + 1)$
D. $(5k + 1) + (6k - 3)$

59. How many terms does the following sum have when it is written in standard form? $(4x^9 - 3x^2 + 10x - 4) + (4x^9 + x^3 + 2x)$

F. 7   **G.** 5   **H.** 3   **J.** 1

60. Simplify $(x^3 + x^2 + 4x) - (x^2 - 5x + 6)$.

A. $4x + 10$ **B.** $4x^2 - 2x$
C. $x^3 + 9x - 6$ **D.** $x^3 + 2x^2 - x + 6$

Short Response

61. Write and simplify a polynomial expression that represents the perimeter of the rectangle. Show your work.

62. Simplify $(9x^3 - 4x^2 + 1) - (x^2 + 2)$. Show your work.

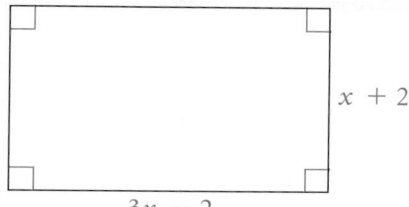
$x + 2$
$3x - 2$

Mixed Review

Lesson 8-8

Identify the growth factor in each function.

63. $y = 6 \cdot 2^x$ **64.** $y = 0.6 \cdot 1.4^x$ **65.** $y = 2 \cdot 5^x$ **66.** $y = 0.3 \cdot 5^x$

Identify each function as *exponential growth* or *exponential decay*.

67. $y = 10 \cdot 3^x$ **68.** $y = 1.8 \cdot 0.4^x$ **69.** $y = 0.3 \cdot 7^x$ **70.** $y = 0.3 \cdot 0.5^x$

Lesson 8-3

Simplify each expression.

71. $7^8 \cdot 7^{10}$ **72.** $2^6 \cdot 2^{-7}$ **73.** $(4x^2)(9x^3)$ **74.** $(3ab)(a^2b)$

75. $(-5t^2)(6t^{-9})$ **76.** $(-3)^6 \cdot (-3)^{-4}$ **77.** $(6h^2)(-2h^8)$ **78.** $(2q^5)(5q^2)$

Lesson 6-8

Write an equation for each translation of $y = |x|$.

79. 5 units up   **80.** right 6 units   **81.** 12 units down   **82.** 7 units up

83. left 10 units   **84.** left 0.4 units   **85.** 5.2 units up   **86.** 2.3 units down

9-2

Multiplying and Factoring

Learning Standards for Mathematics

A.A.13 Add, subtract, and multiply monomials and polynomials.

A.A.20 Factor algebraic expressions completely, including trinomials with a lead coefficient of one (after factoring a GCF).

1   Distributing a Monomial

In Chapter 2 you used the Distributive Property to multiply a number by a sum or difference.

$$5(a + 3) = 5a + 15 \quad (x - 2)(3) = 3x - 6 \quad -2(2y + 7) = -4y - 14$$

You can also use the Distributive Property or an area model to multiply polynomials. Consider the product $2x(3x + 1)$.

$$2x(3x + 1) = 2x(3x) + 2x(1)$$
$$= 6x^2 + 2x$$

	$3x$	$+$	1
x^2	x^2	x^2	x
x^2	x^2	x^2	x

$2x$

You can use the Distributive Property for multiplying powers with the same base when multiplying by a monomial.

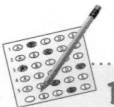

Test-Taking Tip

When you distribute a negative expression such as $-4y^2$, be sure to apply the negative sign to each term within the parentheses.

1 EXAMPLE   Multiplying a Monomial and a Trinomial

Multiple Choice Simplify $-4y^2(5y^4 - 3y^2 + 2)$.

Ⓐ $-20y^6 + 12y^4 - 8y^2$ Ⓑ $-20y^8 + 12y^4 - 8y^2$

Ⓒ $-20y^6 - 12y^4 - 8y^2$ Ⓓ $-20y^6 + 12y^4 + 8y^2$

$-4y^2(5y^4 - 3y^2 + 2)$

$= -4y^2(5y^4) - 4y^2(-3y^2) - 4y^2(2)$ **Use the Distributive Property.**

$= -20y^{2 + 4} + 12y^{2 + 2} - 8y^2$ **Multiply the coefficients and add the exponents of powers with the same base.**

$= -20y^6 + 12y^4 - 8y^2$ **Simplify.**

● So A is the correct answer.

 **Quick Check** ❶ Simplify each product.

a. $4b(5b^2 + b + 6)$ **b.** $-7h(3h^2 - 8h - 1)$ **c.** $2x(x^2 - 6x + 5)$

2 Factoring a Monomial From a Polynomial

Factoring a polynomial reverses the multiplication process. To factor a monomial from a polynomial, first find the greatest common factor (GCF) of its terms.

2 EXAMPLE Finding the Greatest Common Factor

Find the GCF of the terms of $4x^3 + 12x^2 - 8x$.

List the prime factors of each term. Identify the factors common to all terms.

$$4x^3 = 2 \cdot 2 \cdot x \cdot x \cdot x$$
$$12x^2 = 2 \cdot 2 \cdot 3 \cdot x \cdot x$$
$$8x = 2 \cdot 2 \cdot 2 \cdot x$$

● The GCF is $2 \cdot 2 \cdot x$ or $4x$.

✓ Quick Check **2** Find the GCF of the terms of each polynomial.
   a. $5v^5 + 10v^3$ **b.** $3t^2 - 18$ **c.** $4b^3 - 2b^2 - 6b$

To factor a polynomial completely, you must factor until there are no common factors other than 1.

3 EXAMPLE Factoring Out a Monomial

Factor $3x^3 - 12x^2 + 15x$.

Step 1 Find the GCF.

$$3x^3 = 3 \cdot x \cdot x \cdot x$$
$$12x^2 = 2 \cdot 2 \cdot 3 \cdot x \cdot x$$
$$15x = 3 \cdot 5 \cdot x$$

Step 2 Factor out the GCF.

$$3x^3 - 12x^2 + 15x$$
$$= 3x(x^2) + 3x(-4x) + 3x(5)$$
$$= 3x(x^2 - 4x + 5)$$

● The GCF is $3 \cdot x$ or $3x$.

✓ Quick Check **3** Use the GCF to factor each polynomial.
   a. $8x^2 - 12x$ **b.** $5d^3 + 10d$ **c.** $6m^3 - 12m^2 - 24m$

EXERCISES

For more exercises, see *Extra Skill and Word Problem Practice*.

Practice and Problem Solving

Simplify each product.

1. $8m(m + 6)$ **2.** $(x + 10)3x$ **3.** $9k(7k + 4)$

4. $-5a(a - 1)$ **5.** $2x^2(9 + x)$ **6.** $-p^2(p - 11)$

7. $2x(6x^3 - x^2 + 5x)$ **8.** $4y^2(9y^3 + 8y^2 - 11)$ **9.** $-5c^3(9c^2 - 8c - 5)$

10. $-7q^2(6q^5 - 2q - 7)$ **11.** $-3g^7(g^4 - 6g^2 + 5)$ **12.** $-4x^6(10x^3 + 3x^2 - 7)$

Find the GCF of the terms of each polynomial.

13. $15w + 21$ **14.** $6a^2 - 8a$ **15.** $36v + 24$

16. $x^3 + 7x^2 - 5x$ **17.** $5b^3 + 15b - 30$ **18.** $9x^3 - 6x^2 + 12x$

Example 3
(page 501)

Factor each polynomial.

19. $6x - 4$

20. $v^2 + 4v$

21. $10x^3 - 25x^2 + 20$

22. $2t^2 - 10t^4$

23. $15n^3 - 3n^2 + 12n$

24. $6p^6 + 24p^5 + 18p^3$

 **Apply Your Skills**

25. Error Analysis Kevin said that $-2x(4x - 3) = -8x^2 - 6x$. Karla said that $-2x(4x - 3) = -8x^2 + 6x$. Who is correct? Explain.

26. Open-Ended Write a polynomial that has a common factor in each term. Factor your polynomial.

Simplify. Write in standard form.

27. $-3a(4a^2 - 5a + 9)$

28. $-7p^2(-2p^3 + 5p)$

29. $12c(-5c^2 + 3c - 4)$

30. $y(y + 3) - 5y(y - 2)$

31. $x^2(x + 1) - x(x^2 - 1)$

32. $4t(3t^2 - 4t) - t(7t)$

33. Building Models Suppose you are building a model of the square castle shown at the left. The moat of the model castle is made of blue paper.
  a. Find the area of the moat using the diagram with the photo.
  b. Write your answer in factored form.

Factor each polynomial.

34. $9m^{12} - 36m^7 + 81m^5$

35. $24x^3 - 96x^2 + 48x$

36. $16n^3 + 48n^2 - 80n$

37. $5x^4 + 4x^3 + 3x^2$

38. $13ab^3 + 39a^2b^4$

39. $7g^2k^3 - 35g^5k^2$

40. Critical Thinking The GCF of two numbers p and q is 5. What is the GCF of p^2 and q^2? Explain your answer.

41. a. Factor $n^2 - n$.
  b. Writing Suppose n is an integer. Is $n^2 - n$ *always*, *sometimes*, or *never* even? Justify your answer.

42. A triangular number is a number you can represent with a triangular arrangement of objects. A triangular number can also be written as a product of two factors, as in the table.
  a. Find the values of $a, b, c,$ and d, and then write an expression in factored form for the nth triangular number.

	1	2	3	4
Triangular Number	1	3	6	10
Factored Form	$\frac{a}{2}(a + 1)$	$\frac{b}{2}(b + 1)$	$\frac{c}{2}(c + 1)$	$\frac{d}{2}(d + 1)$

GO Online
Homework Video Tutor
Visit: PHSchool.com
Web Code: ate-0902

  b. Use the expression you wrote to find the 100th triangular number.

C Challenge **43. a. Geometry** How many sides does the polygon have? How many of its diagonals come from one vertex?
  b. Suppose a polygon has n sides. How many diagonals will it have from one vertex?
  c. The number of diagonals from all the vertices is $\frac{n}{2}(n - 3)$. Multiply the two factors.
  d. For a polygon with 8 sides, what is the total number of diagonals that can be drawn from the vertices?

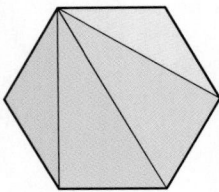

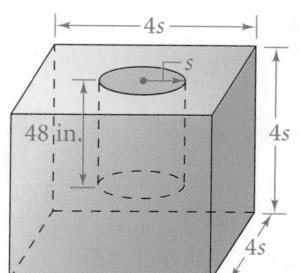

44. Manufacturing The diagram shows a cube of metal with a cylinder cut out of it. The formula for the volume of a cylinder is $V = \pi r^2 h$, where r is the radius and h is the height.
 a. Write a formula for the volume of the cube in terms of s. **$V = 64s^3$**
 b. Write a formula for the volume of the cylinder in terms of s. **$V = 48\pi s^2$**
 c. Write a formula in terms of s for the volume V of the metal left after the cylinder has been removed. **$V = 64s^3 - 48\pi s^2$**
 d. Factor your formula from part (c). **$V = 16s^2(4s - 3\pi)$**
 e. Find V in cubic inches for $s = 15$ in. **about 182,071 in.^{3}**

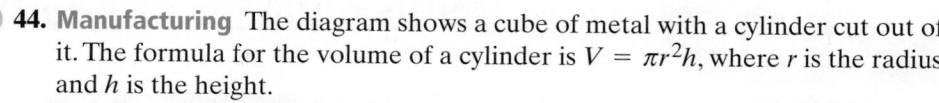

Multiple Choice

45. $x(6x^2 - 4x - 2)$ equals which of the following expressions?
 A. $6x^3 - 4x - 2$ **B.** $6x^3 - 4x^2 - 2x$
 C. $6x^3 - 4x^2 - 2$ **D.** $7x^3 - 5x^2 - 3x$

46. Simplify $(p^2 - 3) - (5 - p + 2p^2) - (4p + 5 - 2p^2)$. What is the coefficient of p^2?
 F. 1   **G.** 2   **H.** 4   **J.** 5

47. Let $\boxed{n}$ represent the number of different pairs of integers whose product is n. For example, -1×10, -2×5, $1 \times (-10)$, and $2 \times (-5)$ give -10. So $\boxed{-10} = 4$. What does $\boxed{-24}$ equal?
 A. 4   **B.** 6   **C.** 8   **D.** 10

48. Which of the following represents an odd number for any integer n?
 F. $n + 1$ **G.** $2n + 1$ **H.** $3n$ **J.** $3n + 1$

49. Factor $6g^8 - 3g^4 + 9g^2$ completely.
 A. $g^2(6g^4 - 3g^2 + 9)$ **B.** $3g^2(2g^6 - g^2 + 3)$
 C. $g^2(6g^6 - 3g^2 + 9g)$ **D.** $3g^2(2g^6 - g^2 + 3g)$

Short Response

50. How do you know if you've factored out the GCF of a polynomial? Illustrate your explanation by using the GCF to factor $10x^4 + 6x^3 + 2x^2$.
See margin.

Lesson 9-1

Simplify. Write each answer in standard form. 52–54. See margin.

51. $(x^2 + 3) - (4x^2 - 7)$ **$-3x^2 + 10$**   **52.** $(m^3 + 8m + 6) + (-5m^2 + 4m)$
53. $(g^2 + 6g - 2) + (4g^2 - 7g + 2)$ **54.** $(3r^2 - 8r + 7) - (2r^2 + 8r - 9)$
55. $(t^4 - t^3 + 1) + (t^3 + 5t^2 - 10)$ **56.** $(3b^3 - b^2) - (5b^2 + 12)$
   $t^4 + 5t^2 - 9$ **$3b^3 - 6b^2 - 12$**

Lesson 8-1

Simplify each expression.

57. 5^{-1} **$\frac{1}{5}$**   **58.** 5^{-2} **$\frac{1}{25}$**   **59.** $(-2)^{-3}$ **$-\frac{1}{8}$**   **60.** 8^0 **1**
61. $n^{-3}m^2$ **$\frac{m^2}{n^3}$**   **62.** v^3w^{-5} **$\frac{v^3}{w^5}$**   **63.** $\frac{4}{c^{-3}}$ **$4c^3$**   **64.** $\frac{ab^{-8}}{c^5}$ **$\frac{a}{b^8c^5}$**

Lesson 7-3

Solve by elimination.

65. $7x + 6y = 33$ **66.** $8x + 4y = 28$ **67.** $4x + 2y = 16$ **(1, 6)**
  $2x - 6y = -6$ **(3, 2)** $3x - 2y = 21$ **(5, −3)** $11x - 3y = -7$

Using Models to Multiply

You can use algebra tiles to multiply two binomials.

> **NY** **A.A.13:** Add, subtract, and multiply monomials and polynomials.

1 EXAMPLE Multiplying Binomials

Find the product $(2x + 1)(x + 5)$.

$(x + 5)$

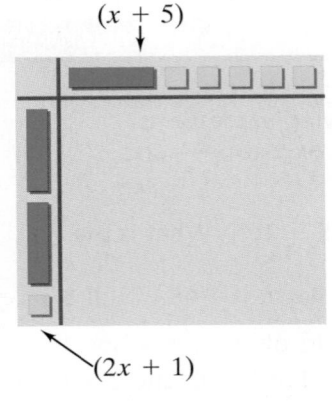

$(2x + 1)$

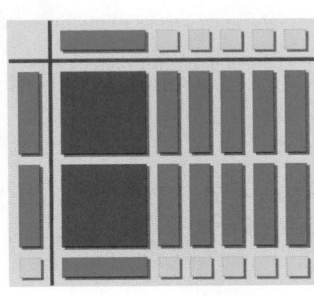

$2x^2 + 10x + x + 5$
$2x^2 + 11x + 5$ **Add coefficients of like terms.**

● The product is $2x^2 + 11x + 5$.

You can also model products that involve subtraction. Red tiles indicate negative variables and negative numbers.

2 EXAMPLE Multiplying With Negative Tiles

Find the product $(x - 2)(3x + 1)$.

$(3x + 1)$

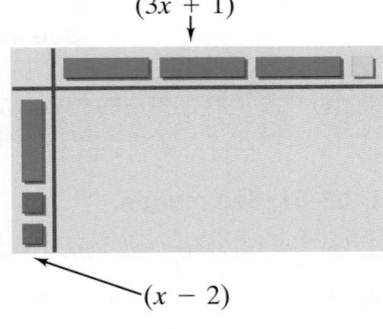

$(x - 2)$

$3x^2 - 6x + x - 2$
$3x^2 - 5x - 2$ **Add coefficients of like terms.**

● The product is $3x^2 - 5x - 2$.

EXERCISES

Use algebra tiles to find each product.

1. $(x + 1)(x + 6)$

2. $(x + 1)(x - 2)$

3. $(x + 1)(4x - 1)$

4. $(x + 4)(2x + 1)$

5. $(x - 3)(3x + 5)$

6. $(2x + 3)(3x + 5)$

Multiplying Binomials

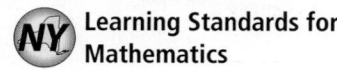
Learning Standards for
Mathematics

A.A.13 Add, subtract, and multiply monomials and polynomials.

Check Skills You'll Need

GO **for Help**   Lessons 9-1 and 9-2

Find each product.

1. $4r(r - 1)$

2. $6h(h^2 + 8h - 3)$

3. $y^2(2y^3 - 7)$

Simplify. Write each answer in standard form.

4. $(x^3 + 3x^2 + x) + (5x^2 + x + 1)$

5. $(3t^3 - 6t + 8) + (5t^3 + 7t - 2)$

6. $w(w + 1) + 4w(w - 7)$

7. $6b(b - 2) - b(8b + 3)$

8. $m(4m^2 - 6) + 3m^2(m + 9)$

9. $3d^2(d^3 - 6) - d^3(2d^2 + 4)$

1   Multiplying Two Binomials

You can use an area model to multiply two binomials. The diagram below shows $(2x + 3)(x + 4)$.

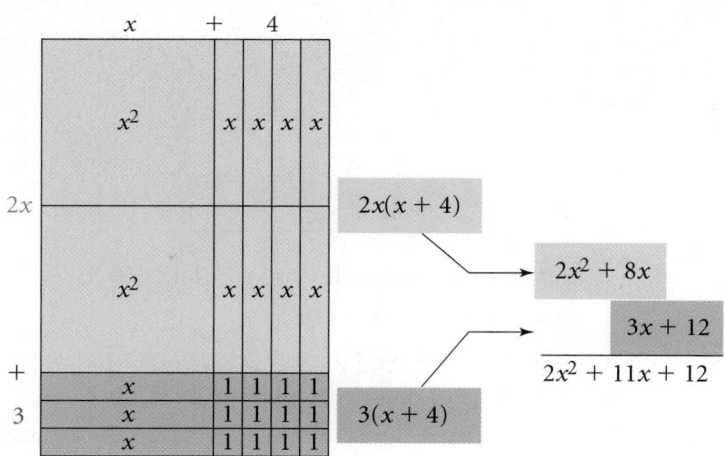

You can also use the Distributive Property to find the product of two binomials.

1 EXAMPLE   Using the Distributive Property

Simplify $(2x + 3)(x + 4)$.

$(2x + 3)(x + 4) = 2x(x + 4) + 3(x + 4)$ **Distribute x + 4.**

$\qquad\qquad\qquad = 2x^2 + 8x + 3x + 12$ **Now distribute 2x and 3.**

$\qquad\qquad\qquad = 2x^2 + 11x + 12$ **Simplify.**

Quick Check **1** Simplify each product.

a. $(6h - 7)(2h + 3)$

b. $(5m + 2)(8m - 1)$

c. $(9a - 8)(7a + 4)$

One way to organize multiplying two binomials is to use FOIL, which stands for "First, Outer, Inner, Last." The term FOIL is a memory device for applying the Distributive Property to the product of two binomials.

2 EXAMPLE **Multiplying Using FOIL**

Simplify $(3x - 5)(2x + 7)$.

	First	Outer	Inner	Last
$(3x - 5)(2x + 7)$ =	$(3x)(2x)$ +	$(3x)(7)$ −	$(5)(2x)$ −	$(5)(7)$
=	$6x^2$ +	$21x$ −	$10x$ −	35
=	$6x^2$ +		$11x$	− 35

● The product is $6x^2 + 11x - 35$.

✓ **Quick Check** **2** Simplify each product using FOIL.
 a. $(3x + 4)(2x + 5)$ b. $(3x - 4)(2x + 5)$
 c. $(3x + 4)(2x - 5)$ d. $(3x - 4)(2x - 5)$

3 EXAMPLE **Applying Multiplication of Polynomials**

Multiple Choice Which expression best describes the area of the shaded region?
 A $x^2 + 2x$ B $6x^2 + 17x + 5$
 C $5x^2 + 15x + 5$ D $7x^2 + 9x + 5$

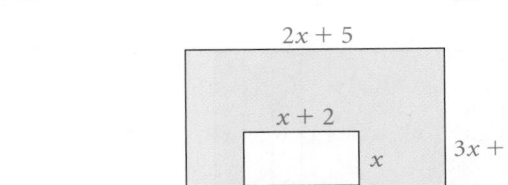

area of outer rectangle = $(3x + 1)(2x + 5)$

area of hole = $x(x + 2)$

area of shaded region

 = area of outer rectangle − area of hole

 = $(3x + 1)(2x + 5) - x(x + 2)$ **Substitute.**

 = $6x^2 + 15x + 2x + 5 - x^2 - 2x$ **Use FOIL to simplify $(3x + 1)(2x + 5)$ and the Distributive Property to simplify $-x(x + 2)$.**

 = $6x^2 - x^2 + 15x + 2x - 2x + 5$ **Group like terms.**

 = $5x^2 + 15x + 5$ **Simplify.**

● So C is the correct answer.

✓ **Quick Check** **3** Find an expression for the area of each shaded region. Simplify.
 a. 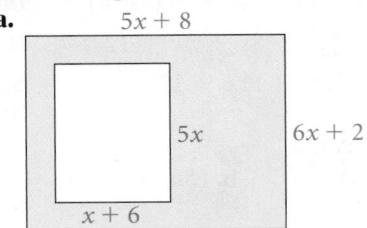 b.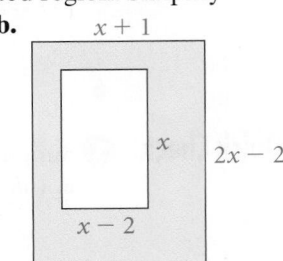

2 Multiplying a Trinomial and a Binomial

FOIL works when you multiply two binomials, but it is not helpful when multiplying a trinomial and a binomial. You can use the vertical method or the horizontal method to distribute each term in such factors.

4 EXAMPLE Multiplying a Trinomial and a Binomial

Simplify the product $(4x^2 + x - 6)(2x - 3)$.

Method 1 Multiply using the vertical method.

$$
\begin{array}{r}
4x^2 + x - 6 \\
2x - 3 \\
\hline
-12x^2 - 3x + 18 \\
8x^3 + 2x^2 - 12x \\
\hline
8x^3 - 10x^2 - 15x + 18 \\
\end{array}
$$

Multiply by −3.
Multiply by 2x.
Add like terms.

Method 2 Multiply using the horizontal method.

$(2x - 3)(4x^2 + x - 6)$

$$= 2x(4x^2) + 2x(x) + 2x(-6) - 3(4x^2) - 3(x) - 3(-6)$$
$$= 8x^3 + 2x^2 - 12x - 12x^2 - 3x + 18$$
$$= 8x^3 - 10x^2 - 15x + 18 \qquad \textbf{Add like terms.}$$

The product is $8x^3 - 10x^2 - 15x + 18$.

 Quick Check ④ Simplify $(6n - 8)(2n^2 + n + 7)$ using both methods shown in Example 4.

EXERCISES

For more exercises, see *Extra Skill and Word Problem Practice.*

Practice and Problem Solving

A Practice by Example

Example 1
(page 505)

 for Help

Copy and fill in each blank.

1. $(5a + 2)(6a - 1) = \blacksquare a^2 + 7a - 2$ **2.** $(3c - 7)(2c - 5) = 6c^2 - 29c + \blacksquare$

3. $(z - 4)(2z + 1) = 2z^2 - \blacksquare z - 4$ **4.** $(2x + 9)(x + 2) = 2x^2 + \blacksquare x + 18$

Simplify each product using the Distributive Property.

5. $(x + 2)(x + 5)$ **6.** $(h + 3)(h + 4)$ **7.** $(k + 7)(k - 6)$

8. $(a - 8)(a - 9)$ **9.** $(2x - 1)(x + 2)$ **10.** $(2y + 5)(y - 3)$

Example 2
(page 506)

Simplify each product using FOIL.

11. $(r + 6)(r - 4)$ **12.** $(y + 4)(5y - 8)$ **13.** $(x + 6)(x - 7)$

14. $(m - 6)(m - 9)$ **15.** $(4b - 2)(b + 3)$ **16.** $(8w + 2)(w + 5)$

17. $(x - 7)(x + 9)$ **18.** $(a + 11)(a + 5)$ **19.** $(p - 1)(p + 10)$

Example 3
(page 506)

Geometry Find an expression for the area of each shaded region. Simplify.

20.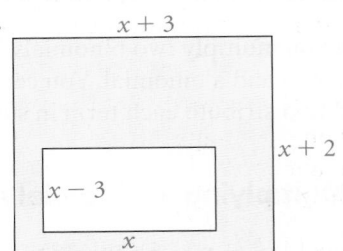

21.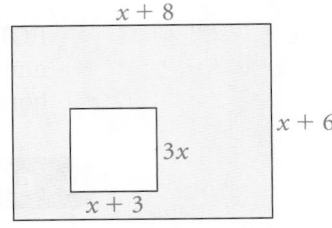

Example 4
(page 507)

Simplify. Use the vertical method.

22. $(x + 9)(x^2 - 4x + 1)$ **23.** $(a - 4)(a^2 - 2a + 1)$

24. $(g - 3)(2g^2 + 3g + 3)$ **25.** $(k + 8)(3k^2 - 5k + 7)$

Simplify. Use the horizontal method.

26. $(x^2 + 2x + 1)(9x - 3)$ **27.** $(t^2 - 6t + 3)(2t - 5)$

28. $(7p^2 + 5p - 1)(8p + 9)$ **29.** $(12w^2 - w - 1)(4w - 2)$

 **Apply Your Skills**

Simplify each product. Write in standard form.

30. $(p - 7)(p + 8)$ **31.** $(-7 + p)(8 + p)$ **32.** $(p^2 - 7)(p + 8)$

33. $(5c - 9)(5c + 1)$ **34.** $(n^2 + 3)(n + 11)$ **35.** $(3k^2 + 2)(k + 5k^2)$

36. $(6h - 1)(4h^2 + h + 3)$ **37.** $(9y^2 + 2)(y^2 - y - 1)$ **38.** $(8q - 4)(6q^2 + q + 1)$

GO Online
Homework Video Tutor
Visit: PHSchool.com
Web Code: ate-0903

 39. Construction You are planning a rectangular garden. Its length is twice its width x. You want a walkway 2 ft wide around the garden.
   a. Write an expression for the area of the garden and walk.
   b. Write an expression for the area of the walk only.
   c. You have enough gravel to cover 76 ft² and want to use it all on the walk. How big should you make the garden?

40. Open-Ended Write a binomial and a trinomial. Find their product.

41. Writing Which method do you prefer for multiplying a binomial and a trinomial? Explain.

Geometry Write an expression for the area of each shaded region. Write your answer in simplest form.

For a guide to solving Exercise 42, see p. 511.

42.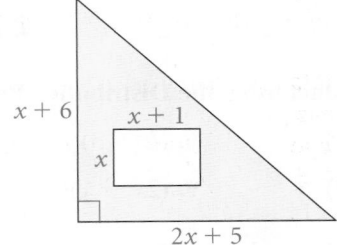

43.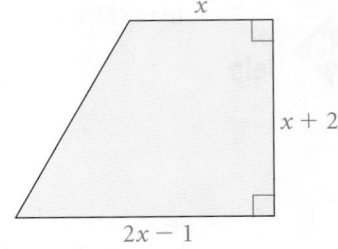

44. a. Simplify each pair of products.

   i. $(x + 1)(x + 1)$ **ii.** $(x + 1)(x + 2)$ **iii.** $(x + 1)(x + 3)$
        $11 \cdot 11$ $11 \cdot 12$ $11 \cdot 13$

   b. Critical Thinking What are the similarities between the two answers in each pair of products?

45. Geometry Use the formula $V = \ell wh$ to write a polynomial in standard form for the volume of the box.

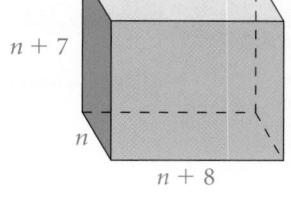
$n + 7$
n
$n + 8$

46. Multiple Choice If n represents an even number, which expression represents the product of the next two even numbers?

(A) $n^2 + 3n + 2$ (B) $2n + 3$ (C) $2n + 6$ (D) $n^2 + 6n + 8$

C Challenge

For Exercises 47–49, each expression represents the side length of a cube. Write an expression in standard form for the surface area of each cube.

47. $x + 3$ **48.** $4t + 1$ **49.** $2w^2 + 7$

50. a. Vegetable Consumption Multiply the expressions on the right side of each equation to create a model for the total number of pounds of fresh vegetables $V(t)$ consumed in a year in the United States.

$C(t) = 3.2t + 157$ the U.S. annual per capita consumption of fresh vegetables, in pounds, from 1990 to 1997

$P(t) = 3.3t + 250$ the U.S. population, in millions, from 1990 to 1997

b. Evaluate the equation you found in part (a) with $t = 5$ to find the total vegetable consumption for 1995. ($t = 0$ corresponds to the year 1990.)

51. Financial Planning Suppose you deposit $2000 for college in a savings account that has an annual interest rate r. At the end of three years, the value of your account will be $2000(1 + r)^3$ dollars.
  a. Rewrite the expression $2000(1 + r)^3$ by finding the product $2000(1 + r)(1 + r)(1 + r)$. Write your answer in standard form.
  b. Find the amount of money in the account if the interest rate is 3%.

Real-World Connection

In 2003, the U.S. consumption of fresh tomatoes was 15.4 lb per person.

For Exercises 52–55, find each product using lattice multiplication, which is explained below.

Lattice multiplication probably originated in India in the twelfth century. It came into use in Italy in the fourteenth century.

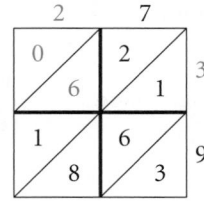

This example shows $27 \cdot 39$. Each number is treated as a binomial. The four products are placed in the small, diagonally split squares. The product of 2 and 3, shown in red, is 6. The first square shows 0/6, which indicates 6. The product of 7 and 3 is 21. The second square shows 2/1.

The products are totaled diagonally. For the diagonal shaded blue, the tens place of the sum $1 + 6 + 8$ is carried into the diagonal above and added into that diagonal: $1 + (2 + 6 + 1)$. The product 1053 appears down the left side of the lattice and across the bottom.

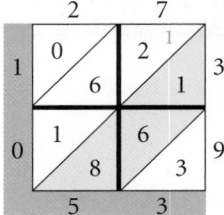

52. $14 \cdot 72$ **53.** $53 \cdot 87$

54. $91 \cdot 64$ **55.** $38 \cdot 64$

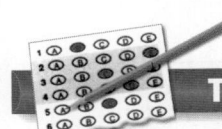

Test Prep

Multiple Choice

56. $(n - 1)(n - 4)$ is equivalent to which expression?
    A. $n^2 - 5n + 4$ **B.** $n^2 - 3n + 4$
    C. $n^2 + 3n + 4$ **D.** $n^2 - 5n - 5$

57. $(8k - 3)(k^2 - k + 1)$ is equivalent to which expression?
    F. $8k^3 + 11k^2 - 11k - 3$ **G.** $9k^3 - 8k^2 + 8k - 2$
    H. $8k^3 - 11k^2 + 11k - 3$ **J.** $9k^3 - 3k^2 + 3k - 3$

58. Which of the following products is always odd for integer values of n?
    A. $(n + 1)(n + 1)$ **B.** $(2n - 1)(2n + 1)$
    C. $(2n - 1)(n + 1)$ **D.** $(2n + 1)(n - 1)$

Short Response

59. Explain how to find the product of $(4v - 1)(2v^2 + v + 1)$, and simplify.

Extended Response

60. Find an expression for the area of the shaded region. Show your work.

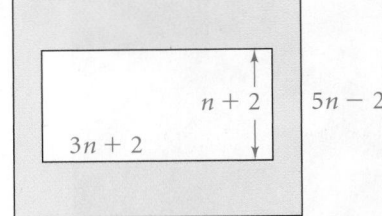

Mixed Review

Lesson 9-2

Simplify each product.

61. $4v(5v - 7)$ **62.** $(c - 9)3c$ **63.** $8t^2(t + 6)$ **64.** $y(3y - 10)$

65. $5x^2(11 - x)$ **66.** $-t^3(6t - 1)$ **67.** $4r(3 - r^5)$ **68.** $9b^2(b^3 + 2b)$

Factor.

69. $5w + 45$ **70.** $3x^2 - 11x$ **71.** $4a^2 + 12a$ **72.** $9n^2 - n^3$

73. $34t - 51$ **74.** $63v^2 + 45v$ **75.** $25m - 60m^3$ **76.** $11k + 77k^6$

Lesson 8-5

Simplify each expression.

77. $\frac{3^5}{3^2}$ **78.** $\frac{3^2}{3^5}$ **79.** $\frac{y^{12}}{y^8}$ **80.** $\frac{2w^{-3}}{6w^2}$ **81.** $\frac{x^{-8}}{2x^3}$

82. $\left(\frac{5}{3}\right)^{-1}$ **83.** $\left(\frac{5}{3}\right)^{-2}$ **84.** $\left(\frac{5}{3}\right)^{0}$ **85.** $\left(\frac{4x}{7}\right)^{-2}$ **86.** $\left(\frac{y^{-2}}{8}\right)^{-2}$

Checkpoint Quiz 1   Lessons 9-1 through 9-3

Simplify each expression.

1. $(4x^2 + x + 3) + (5x^2 + 9x - 2)$ **2.** $(7b^2 - 5b + 3) - (b^2 + 8b - 6)$

3. $3w(12w - 1) - 8w$ **4.** $6k(4k + k^2) + 9k(2k - 6k^2)$

5. $(x + 3)(x - 5)$ **6.** $(2n^3 - 5)(6n^2 + n)$ **7.** $(g^2 + 4)(4g^2 + 8g - 9)$

Factor each polynomial.

8. $12y^2 - 10$ **9.** $5t^6 + 25t^3 - 10t$ **10.** $18v^4 + 27v^3 + 36v^2$

Guided Problem Solving

Understanding Math Problems Read the exercise below and then follow along with what Terry thinks and writes. Check your understanding with the exercise at the bottom of the page.

Geometry Write an expression for the area of the shaded region. Write your answer in simplest form.

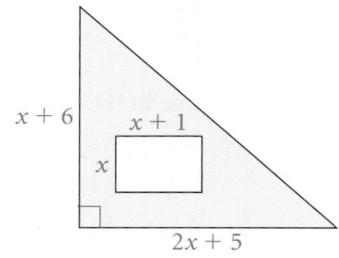

What Terry Thinks

There are two figures—a right triangle and a rectangle inside it. I'll write down the dimensions of each shape.

The area of the shaded region is the difference between the area of the triangle and the area of the rectangle.

I can rewrite my equation using the area formulas for each shape.

Next, I'll substitute the expressions for the dimensions of the triangle and rectangle into the formula.

I can use FOIL and the Distributive Property to simplify the expressions. Lastly, I combine like terms and write my answer in simplest form.

What Terry Writes

Triangle: base = $2x + 5$
height = $x + 6$

Rectangle: length = x
width = $x + 1$

$$\text{Area of shaded region} = \text{Area of triangle} - \text{Area of rectangle}$$

$$= \frac{1}{2}(\text{base})(\text{height}) - (\text{length})(\text{width})$$

$$= \frac{1}{2}(2x + 5)(x + 6) - x(x + 1)$$

$$= \frac{1}{2}(2x^2 + 12x + 5x + 30) - (x^2 + x)$$

$$= x^2 + \frac{17}{2}x + 15 - x^2 - x$$

$$= 7.5x + 15$$

EXERCISE

Geometry Write an expression for the area of the colored region. Write your answer in simplest form.

Multiplying Special Cases

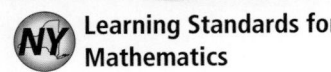
A.A.13 Add, subtract, and multiply monomials and polynomials.

 **Check Skills You'll Need**

GO **for Help**   Lessons 8-4 and 9-3

Simplify.

1. $(7x)^2$ **2.** $(3v)^2$ **3.** $(-4c)^2$ **4.** $(5g^3)^2$

Use FOIL to find each product.

5. $(j + 5)(j + 7)$ **6.** $(2b - 6)(3b - 8)$

7. $(4y + 1)(5y - 2)$ **8.** $(x + 3)(x - 4)$

9. $(8c^2 + 2)(c^2 - 10)$ **10.** $(6y^2 - 3)(9y^2 + 1)$

🔊 **New Vocabulary**   • **difference of squares**

---

1   **Finding the Square of a Binomial**

Activity: Exploring Special Products

1. Find each product.

Row 1: $(x + 8)(x + 8)$ $(y + 5)(y + 5)$ $(2p + 3)(2p + 3)$

Row 2: $(d - 3)(d - 3)$ $(t - 1)(t - 1)$ $(9r - 2)(9r - 2)$

Row 3: $(x + 4)(x - 4)$ $(k + 9)(k - 9)$ $(3c + 7)(3c - 7)$

2. Describe the pattern or patterns you found in each row.

3. Based on the patterns you found, predict each product.

$(p + 6)(p + 6)$ $(v - 5)(v - 5)$ $(x + 8)(x - 8)$

4. Use FOIL to find each product in Question 2. Were your predictions correct?

You can write the expression $(a + b)^2$ as $(a + b)(a + b)$. You can find $(a + b)^2$ using the methods you learned in Lesson 9-3.

Area Model

	a	$+\ b$
a	a^2	ab
$+$ b	ab	b^2

$a^2 + 2ab + b^2$

FOIL

$(a + b)(a + b)$
$= a^2 + ab + ba + b^2$ **Use FOIL.**
$= a^2 + 2ab + b^2$ **Simplify.**

The expressions $(a - b)^2$ and $(a + b)^2$ are squares of binomials. To square a binomial, you can use FOIL or the following rule.

Key Concepts

Rule	The Square of a Binomial

$(a + b)^2 = a^2 + 2ab + b^2$

$(a - b)^2 = a^2 - 2ab + b^2$

The square of a binomial is the square of the first term plus twice the product of the two terms plus the square of the last term.

1 EXAMPLE **Squaring a Binomial**

a. Find $(x + 7)^2$.

$(x + 7)^2 = x^2 + 2x(7) + 7^2$ **Square the binomial.**

$\quad\quad\quad = x^2 + 14x + 49$ **Simplify.**

b. Find $(4k - 3)^2$.

$(4k - 3)^2 = (4k)^2 - 2(4k)(3) + 3^2$ **Square the binomial.**

$\quad\quad\quad = 16k^2 - 24k + 9$ **Simplify.**

✓ **Quick Check** ❶ Find each square.

  a. $(t + 6)^2$ **b.** $(5y + 1)^2$ **c.** $(7m - 2p)^2$ **d.** $(9c - 8)^2$

You can square binomials to find probabilities that apply to real-world situations.

2 EXAMPLE **Real-World 🌐 Problem Solving**

Real-World 🌐 Connection

The color of a Labrador retriever is determined by a pair of genes. The offspring inherits a single gene at random from each of its parents.

Among Labrador retrievers, the dark-fur gene D is dominant, and the yellow-fur gene Y is recessive. This means that a dog with at least one dominant gene (DD or DY) will have dark fur. A dog with two recessive genes (YY) will have yellow fur.

The Punnett square at the right models the possible combinations of color genes that parents who carry both genes can pass on to their offspring. Since YY is $\frac{1}{4}$ of the outcomes, the probability that a puppy has yellow fur is $\frac{1}{4}$.

	D	Y
D	DD	DY
Y	DY	YY

You can model the probabilities found in the Punnett square with the expression $\left(\frac{1}{2}D + \frac{1}{2}Y\right)^2$. Show that this product gives the same result as the Punnett square.

$\left(\frac{1}{2}D + \frac{1}{2}Y\right)^2 = \left(\frac{1}{2}D\right)^2 + 2\left(\frac{1}{2}D\right)\left(\frac{1}{2}Y\right) + \left(\frac{1}{2}Y\right)^2$ **Square the binomial.**

$\quad\quad\quad = \frac{1}{4}D^2 + \frac{1}{2}DY + \frac{1}{4}Y^2$ **Simplify.**

The expressions $\frac{1}{4}D^2$ and $\frac{1}{4}Y^2$ indicate that the probability offspring will have either two dominant genes or two recessive genes is $\frac{1}{4}$. The expression $\frac{1}{2}DY$ indicates that there is $\frac{1}{2}$ chance that the offspring will inherit both genes. These are the same probabilities shown in the Punnett square.

 **Quick Check** **2** **Games** When you play a game with two number cubes, you can find probabilities by squaring a binomial. Let A represent rolling 1 or 2 and B represent rolling 3, 4, 5, or 6. The probability of A is $\frac{1}{3}$, and the probability of B is $\frac{2}{3}$.

    a. Find $\left(\frac{1}{3}A + \frac{2}{3}B\right)^2$.

    b. What is the probability that both number cubes you roll show 1 or 2?

    c. What is the probability that one number cube shows a 1 or 2 and the other shows 3, 4, 5, or 6?

    d. What is the probability that both number cubes show 3, 4, 5, or 6?

Using mental math, you can square a binomial to find the square of a number.

3 **EXAMPLE** **Mental Math**

    a. Find 51^2 using mental math.

      $51^2 = (50 + 1)^2$

          $= 50^2 + 2(50 \cdot 1) + 1^2$ ⟵ **Square the binomial.** ⟶

          $= 2500 + 100 + 1 = 2601$ ⟵ **Simplify.** ⟶

    b. Find 49^2 using mental math.

      $49^2 = (50 - 1)^2$

          $= 50^2 - 2(50 \cdot 1) + 1^2$

          $= 2500 - 100 + 1 = 2401$

Quick Check **3** Find each square using mental math.

    a. 31^2 **b.** 29^2 **c.** 98^2 **d.** 203^2

2   Difference of Squares

The product of the sum and difference of the same two terms also produces a pattern.

$$(a + b)(a - b) = a^2 - ab + ba - b^2$$
$$= a^2 - b^2$$

Notice that the sum of $-ab$ and ba is 0, leaving $a^2 - b^2$. This product is called the **difference of squares**.

 Key Concepts

Rule	**The Difference of Squares**

$(a + b)(a - b) = a^2 - b^2$

The product of the sum and difference of the same two terms is the difference of their squares.

4 **EXAMPLE** **Finding the Difference of Squares**

Find $(t^3 - 6)(t^3 + 6)$.

$(t^3 - 6)(t^3 + 6) = (t^3)^2 - (6)^2$ **Find the difference of squares.**

                $= t^6 - 36$ **Simplify.**

Quick Check **4** Find each product.

    a. $(d + 11)(d - 11)$ **b.** $(c^2 + 8)(c^2 - 8)$ **c.** $(9v^3 + w^4)(9v^3 - w^4)$

You can use the difference of squares to calculate products using mental math.

5 EXAMPLE Mental Math

Find $82 \cdot 78$.

$82 \cdot 78 = (80 + 2)(80 - 2)$ **Express each factor using 80 and 2.**

$\qquad\quad = 80^2 - 2^2$ **Find the difference of squares.**

$\qquad\quad = 6400 - 4 = 6396$ **Simplify.**

✓ **Quick Check** ❺ Find each product.

a. $18 \cdot 22$ **b.** $19 \cdot 21$ **c.** $59 \cdot 61$ **d.** $87 \cdot 93$

EXERCISES

For more exercises, see *Extra Skill and Word Problem Practice.*

Practice and Problem Solving

A Practice by Example

Examples 1, 2
(page 513)

Find each square.

1. $(c + 1)^2$ **2.** $(x + 4)^2$ **3.** $(2v + 11)^2$ **4.** $(3m + 7)^2$

5. $(w - 12)^2$ **6.** $(b - 5)^2$ **7.** $(6x - 8)^2$ **8.** $(9j - 2)^2$

9. Games Suppose you play a game with two spinners like the one shown at the right. Let C represent spinning an even number. Let D represent spinning an odd number. The probability of C is $\frac{1}{4}$. The probability of D is $\frac{3}{4}$.
a. Simplify $\left(\frac{1}{4}C + \frac{3}{4}D\right)^2$.
b. Find $P(C \text{ and } C)$.
c. How does the answer in part (b) relate to the polynomial in part (a)?

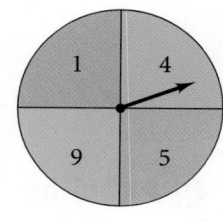

Example 3
(page 514)

Mental Math **Find each square.**

10. 61^2 **11.** 99^2 **12.** 48^2 **13.** 302^2 **14.** 499^2

Example 4
(page 514)

Find each product.

15. $(x + 4)(x - 4)$ **16.** $(a + 8)(a - 8)$ **17.** $(d + 7)(d - 7)$

18. $(h + 15)(h - 15)$ **19.** $(y + 12)(y - 12)$ **20.** $(k + 5)(k - 5)$

Example 5
(page 515)

Mental Math **Find each product.**

21. $31 \cdot 29$ **22.** $89 \cdot 91$ **23.** $52 \cdot 48$ **24.** $197 \cdot 203$ **25.** $299 \cdot 301$

B Apply Your Skills   ⬡ **Geometry** **Find an expression for the area of each shaded region. Write your answers in standard form.**

26.

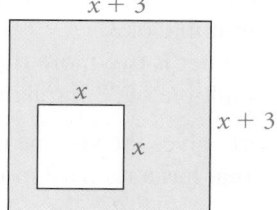

27.

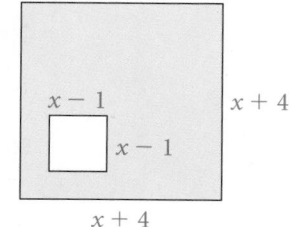

Find each square.

28. $(x + 3y)^2$ **29.** $(5p - q)^2$ **30.** $(6m + n)^2$ **31.** $(x - 7y)^2$

32. $(4k + 7j)^2$ **33.** $(2y - 9x)^2$ **34.** $(3w + 10t)^2$ **35.** $(6a + 11b)^2$

36. $(5p - 6q)^2$ **37.** $(6h - 8p)^2$ **38.** $(y^5 - 9x^4)^2$ **39.** $(8k + 4h)^2$

40. Biology The coat color of shorthorn cattle is determined by two genes, Red R and White W. RR produces red, WW produces white, and RW produces a third type of coat color called roan.

	R	W
R	RR	RW
W	RW	WW

  a. Model the Punnett square with the square of a binomial.
  b. If both parents have RW, what is the probability the offspring will also be RW?
  c. Write an expression to model a situation where one parent is RW while the other is RR.
  d. What is the probability that the offspring of the parents in step (c) will have a white coat?

Real-World Connection

The cow in the photo shows a typical roan coat.

41. a. Copy and complete the table.
  b. Describe any patterns you see.
  c. Writing How does the difference of squares account for the pattern in the table?

$4^2 = 16$	$3 \cdot 5 = 15$
$5^2 = \blacksquare$	$4 \cdot 6 = 24$
$6^2 = \blacksquare$	$5 \cdot 7 = \blacksquare$
$7^2 = \blacksquare$	$6 \cdot 8 = \blacksquare$

42. Open-Ended Give a counterexample to show that $(x + y)^2 = x^2 + y^2$ is false.

43. Critical Thinking Does $\left(3\frac{1}{2}\right)^2 = 9\frac{1}{4}$? Explain.

Find each product.

Homework Video Tutor

Visit: PHSchool.com
Web Code: ate-0904

44. $(3y + 5w)(3y - 5w)$ **45.** $(p + 9q)(p - 9q)$ **46.** $(2d + 7g)(2d - 7g)$

47. $(7b - 8c)(7b + 8c)$ **48.** $(g + 7h)(g - 7h)$ **49.** $(g^3 + 7h^2)(g^3 - 7h^2)$

50. $(2a^2 + b)(2a^2 - b)$ **51.** $(11x - y^3)(11x + y^3)$ **52.** $(4k - 3h^2)(4k + 3h^2)$

 Challenge

53. Write the expression $(a + b + c)^2$ in standard form.

54. Games Suppose you play a game by tossing 3 coins. You can find the probabilities by simplifying $\left(\frac{1}{2}H + \frac{1}{2}T\right)^3$.
  a. Simplify the expression.
  b. Use the answer you found in part (a) to find the probability of getting a head and two tails $\left(HT^2\right)$.

55. Number Theory You can use factoring to show that the sum of two multiples of 3 is also a multiple of 3.

  If m and n are integers, then $3n$ and $3m$ are multiples of three.
  $3m + 3n = 3(m + n)$
  Since $(m + n)$ is an integer, $3(m + n)$ is a multiple of three.

  a. Show that if a number is one more than a multiple of 3, then its square is also one more than a multiple of 3.
  b. Reasoning If a number is two more than a multiple of 3, is its square also two more than a multiple of 3? Explain.

56. The formula $V = \frac{4}{3}\pi r^3$ gives the volume of a sphere. Find the formula for the volume of a sphere that has a radius 3 more than r. Write your answer in standard form.

57. The area of the shaded region in the diagram is $9^2 - 2^2$.

   a. Copy the figure. Make a single cut across the shaded region and reassemble it to show that $9^2 - 2^2 = (9 - 2)(9 + 2)$.

   b. Draw your reassembled figure. Include its dimensions.

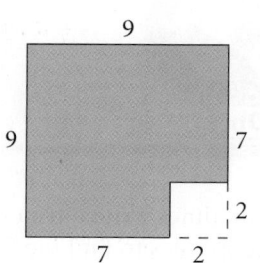

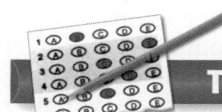
Multiple Choice

58. Which value of a makes $(9x - 1)^2 = ax^2 - 18x + 1$ true?
   A. 9   **B.** 18   **C.** 64   **D.** 81

59. Which value of n makes $(b^7 + 2)^2 = b^n + 4b^7 + 4$ true?
   F. 14   **G.** 28   **H.** 42   **J.** 49

60. Simplify $(x - 1)^2 + (x + 1)^2$.
   A. $2x$ **B.** $-2x$ **C.** $2x^2 + 2$ **D.** $2x^2$

61. Find the product of $(2x - 3)$, $(4x^2 + 9)$, and $(2x + 3)$.
   F. $16x^2 + 18$ **G.** $16x^4 + 18$
   H. $16x^4 - 81$ **J.** $64x^4 - 81$

62. Which of the following correctly shows how to use the Difference of Squares rule to multiply 17 and 23?

   A. $17 \cdot 23 = (16 + 1)(16 + 7)$
               $= 16^2 + (1 \cdot 7)$
               $= 256 + 7$
               $= 263$

   B. $17 \cdot 23 = (20 - 3)(20 + 3)$
               $= 20^2 - 9$
               $= 400 - 9$
               $= 391$

   C. $17 \cdot 23 = (19 - 2)(19 + 4)$
               $= 19^2 - (2 \cdot 4)$
               $= 361 - 8$
               $= 353$

   D. $17 \cdot 23 = (18 - 1)(22 + 1)$
               $= 18 \cdot 22 - 1^2$
               $= 396 - 1$
               $= 395$

Short Response

63. Explain how to compute the xy term of the product $(3x - 4y)^2$.

Mixed Review

Lesson 9-3

Find each product.

64. $(k + 7)(k - 9)$ **65.** $(2x - 11)(x - 6)$ **66.** $(5p + 4)(3p - 1)$

67. $(3y + 1)(y + 1)$ **68.** $(4h - 2)(6h + 1)$ **69.** $(9b + 7)(8b + 2)$

70. $(2w^2 + 5)(w + 8)$ **71.** $(r - 7)(r^2 + 3r - 9)$ **72.** $(5m^2 - 2)(6m^3 + 4m)$

Lesson 8-2

Write each number in scientific notation.

73. 8713   **74.** 0.031   **75.** 68,952   **76.** 1.2 million

77. 11   **78.** 523   **79.** 6 billion   **80.** 0.72

Using Models to Factor

You can sometimes write a trinomial as the product of two binomial factors. You can use algebra tiles to find the factors by arranging all of the tiles to form a rectangle. The lengths of the sides of the rectangle are the factors of the trinomial.

NY A.A.20: Factor algebraic expressions completely, including trinomials with a lead coefficient of one (after factoring a GCF).

ACTIVITY

Write $2x^2 + 7x + 6$ as the product of two binomial factors.

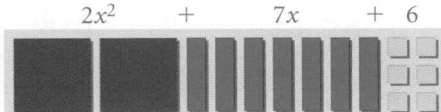

$2x^2 \quad + \quad 7x \quad + \quad 6$

Model of polynomial

Use the tiles to form a rectangle.

First try:

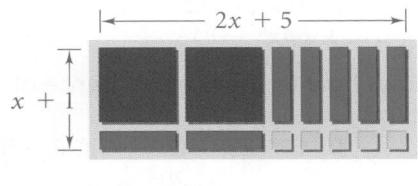

$2x + 5$

$x + 1$

Left over: ▢

Second try:

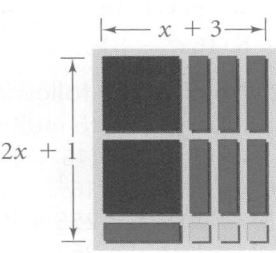

$x + 3$

$2x + 1$

Left over: ▢ ▢ ▢

Third try:

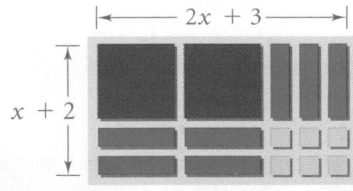

$2x + 3$

$x + 2$

Correct! No tiles are left over.
$2x^2 + 7x + 6 = (2x + 3)(x + 2)$

EXERCISES

Use tiles to find binomial factors of each trinomial.

1. $x^2 + 8x + 15$

2. $x^2 + 4x + 4$

3. $x^2 + 8x + 7$

4. $2x^2 + 7x + 3$

5. $4x^2 + 12x + 5$

6. $6x^2 + 7x + 2$

7. Critical Thinking Explain why the trinomial $x^2 + 3x + 5$ cannot be represented as a rectangle using algebra tiles.

8. Critical Thinking Complete $2x^2 + \blacksquare x + 6$ with three different integers so that each trinomial has two binomial factors. Write each trinomial as the product of binomial factors.

Factoring Trinomials of the Type $x^2 + bx + c$

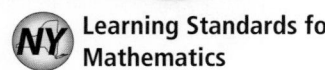

NY Learning Standards for Mathematics

A.A.20 Factor algebraic expressions completely, including trinomials with a lead coefficient of one (after factoring a GCF).

✓ Check Skills You'll Need

GO for Help   Skills Handbook p. 754

List all of the factors of each number.

1. 24   **2.** 12   **3.** 54   **4.** 15

5. 36   **6.** 56   **7.** 64   **8.** 96

---

1 ▸ Factoring Trinomials

In earlier courses, you learned how to find the factors of whole numbers like 15. Since $3 \times 5 = 15$, 3 and 5 are factors of 15. You can also find the factors of some trinomials. Consider the product below.

$$(x + 3)(x + 5) = x^2 + 5x + 3x + 3 \cdot 5$$
$$(5 + 3)x$$
$$= x^2 + 8x + 15$$

Notice that the coefficient of the middle term $8x$ is the sum of 3 and 5. Also the constant term 15 is the product of 3 and 5. To factor a trinomial of the form $x^2 + bx + c$, you must find two numbers that have a sum of b and a product of c.

The next example shows how to use a table to list the factors of the constant term c and how to add the factors until the sum is the middle term b.

1 EXAMPLE Factoring $x^2 + bx + c$

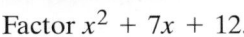

Online active math

For: Factoring Activity
Use: Interactive Textbook, 9-5

Factor $x^2 + 7x + 12$.

Find the factors of 12. Identify the pair that has a sum of 7.

Factors of 12	Sum of Factors
1 and 12	13
2 and 6	8
3 and 4	7 ✓

$x^2 + 7x + 12 = (x + 3)(x + 4)$

Check $x^2 + 7x + 12 \stackrel{?}{=} (x + 3)(x + 4)$
$= x^2 + 4x + 3x + 12$
$= x^2 + 7x + 12$ ✓

✓ Quick Check **1** Factor each expression. Check your answer.

a. $g^2 + 7g + 10$ **b.** $v^2 + 21v + 20$ **c.** $a^2 + 13a + 30$

Some factorable trinomials have a negative middle term and a positive constant term. If the middle term is negative, you need to inspect the negative factors of c to find the factors of the trinomial.

2 EXAMPLE Factoring $x^2 - bx + c$

Factor $d^2 - 17d + 42$.

Since the middle term is negative, find the negative factors of 42 until you find the pair that has a sum of -17.

Factors of 42	Sum of Factors
-1 and -42	-43
-2 and -21	-23
-3 and -14	-17 ✓

$d^2 - 17d + 42 = (d - 3)(d - 14)$

 **Quick Check** **2** Factor each expression.
   a. $k^2 - 10k + 25$ **b.** $x^2 - 11x + 18$ **c.** $q^2 - 15q + 36$

When you factor trinomials with a negative constant, you will need to inspect pairs of positive and negative factors of c.

3 EXAMPLE Factoring Trinomials With a Negative c

a. Factor $m^2 + 6m - 27$.

Identify the pair of factors of -27 that has a sum of 6.

Factors of -27	Sum of Factors
1 and -27	-26
27 and -1	26
3 and -9	-6
9 and -3	6 ✓

$m^2 + 6m - 27 = (m - 3)(m + 9)$

b. Factor $p^2 - 3p - 18$.

Identify the pair of factors of -18 that has a sum of -3.

Factors of -18	Sum of Factors
1 and -18	-17
18 and -1	17
-6 and 3	-3 ✓

$p^2 - 3p - 18 = (p + 3)(p - 6)$

Quick Check **3** Factor each expression.
   a. $m^2 + 8m - 20$ **b.** $p^2 - 3p - 40$ **c.** $y^2 - y - 56$

You can also factor some trinomials that have more than one variable. Consider the product $(p + 10q)(p + 3q)$.

$$(p + 10q)(p + 3q) = p^2 + 3pq + 10pq + 10q \cdot 3q$$
$$(3 + 10)pq$$
$$= p^2 + 13pq + 30q^2$$

You can see that the first term is the square of the first variable, the middle term includes both variables, and the last term includes the square of the second variable.

4 **EXAMPLE**   **Factoring Trinomials With Two Variables**

Factor $h^2 - 4hk - 77k^2$.

Find the factors of -77. Identify the pair that has a sum of -4.

Factors of -77	Sum of Factors
1 and -77	-76
77 and -1	76
7 and -11	-4 ✓

● $h^2 - 4hk - 77k^2 = (h + 7k)(h - 11k)$

✓ Quick Check **4** Factor each expression.
a. $x^2 + 11xy + 24y^2$ **b.** $v^2 + 2vw - 48w^2$ **c.** $m^2 - 17mn - 60n^2$

EXERCISES

For more exercises, see *Extra Skill and Word Problem Practice*.

Practice and Problem Solving

 **Practice by Example**

Examples 1, 2
(pages 519, 520)

 **for Help**

Complete.

1. $t^2 + 7t + 10 = (t + 2)(t + \blacksquare)$ **2.** $y^2 - 13y + 36 = (y - 4)(y - \blacksquare)$
3. $x^2 - 8x + 7 = (x - 1)(x - \blacksquare)$ **4.** $x^2 + 9x + 18 = (x + 3)(x + \blacksquare)$

Factor each expression. Check your answer.

5. $r^2 + 4r + 3$ **6.** $n^2 - 3n + 2$ **7.** $k^2 + 5k + 6$
8. $y^2 + 6y + 8$ **9.** $x^2 - 2x + 1$ **10.** $p^2 + 19p + 18$
11. $k^2 - 16k + 28$ **12.** $w^2 + 6w + 5$ **13.** $m^2 - 9m + 8$
14. $d^2 + 21d + 38$ **15.** $t^2 - 13t + 42$ **16.** $q^2 - 18q + 45$

Example 3
(page 520)

Complete.

17. $m^2 + 3m - 10 = (m - 2)(m + \blacksquare)$ **18.** $v^2 - 2v - 24 = (v + 4)(v - \blacksquare)$
19. $k^2 - 8k - 9 = (k + 1)(k - \blacksquare)$ **20.** $q^2 + 3q - 18 = (q - 3)(q + \blacksquare)$

Factor each expression.

21. $x^2 + 3x - 4$ **22.** $q^2 - 2q - 8$ **23.** $y^2 + y - 20$
24. $h^2 + 16h - 17$ **25.** $x^2 - 14x - 32$ **26.** $d^2 + 6d - 40$
27. $m^2 - 13m - 30$ **28.** $p^2 + 3p - 54$ **29.** $p^2 - 15p - 54$

Example 4
(page 521)

Choose the correct factoring for each expression.

30. $p^2 + 10pq + 9q^2$ **A.** $(p + 9q)(p + q)$ **B.** $(p + 9)(p + q^2)$

31. $m^2 + 4mn + 3n^2$ **A.** $(m + n)(3m + n)$ **B.** $(m + 3n)(m + n)$

32. $x^2 + 8xy + 15y^2$ **A.** $(x + 15y^2)(x + 1)$ **B.** $(x + 5y)(x + 3y)$

Factor each expression.

33. $t^2 + 7tv - 18v^2$ **34.** $x^2 + 12xy + 35y^2$ **35.** $p^2 - 10pq + 16q^2$

36. $m^2 - 3mn - 54n^2$ **37.** $h^2 + 18hj + 17j^2$ **38.** $x^2 - 10xy - 39y^2$

B Apply Your Skills

Open-Ended Find three different values to complete each expression so that it can be factored into the product of two binomials. Show each factorization.

39. $x^2 - 3x - \blacksquare$ **40.** $x^2 + x - \blacksquare$ **41.** $x^2 + \blacksquare x + 12$

42. Writing Suppose you can factor $x^2 + bx + c$ into the product of two binomials.
   a. Explain what you know about the factors if $c > 0$.
   b. Explain what you know about the factors if $c < 0$.

Factor each expression.

43. $k^2 + 10k + 16$ **44.** $m^2 + 10m - 24$ **45.** $n^2 + 10n - 56$

46. $g^2 + 20g + 96$ **47.** $x^2 + 8x - 65$ **48.** $t^2 + 28t + 75$

49. $x^2 - 11x - 42$ **50.** $k^2 + 23k + 42$ **51.** $m^2 + 14m - 51$

52. $x^2 + 29xy + 100y^2$ **53.** $t^2 - 10t - 75$ **54.** $d^2 - 19de + 48e^2$

Write the standard form for each of the polynomials modeled below. Then factor each expression.

55.
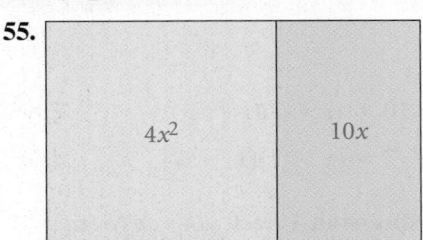

56.
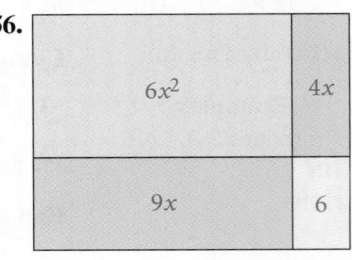

**nline**
Homework Video Tutor
Visit: PHSchool.com
Web Code: ate-0905

57. Critical Thinking Let $x^2 - 12x - 28 = (x + a)(x + b)$.
   a. What do you know about the signs of a and b?
   b. Suppose $|a| > |b|$. Which number, a or b, is a negative integer? Explain.

58. Critical Thinking Let $x^2 + 12x - 28 = (x + a)(x + b)$.
   a. What do you know about the signs of a and b?
   b. Suppose $|a| > |b|$. Which number, a or b, is a negative integer? Explain.

C Challenge

Factor each trinomial.

Sample $n^6 + n^3 - 56 = n^{3+3} + n^3 - 56$
$$= (n^3 + 8)(n^3 - 7)$$

59. $x^{12} + 12x^6 + 35$ **60.** $t^8 + 5t^4 - 24$ **61.** $r^6 - 21r^3 + 80$

62. $m^{10} + 18m^5 + 17$ **63.** $x^{12} - 19x^6 - 120$ **64.** $p^6 + 14p^3 - 72$

In the next example, c is negative. In this case, you need to consider combinations that equal -8, like $(-8)(1)$. You must also consider $(-1)(8)$.

2 EXAMPLE *c* **Is Negative**

Factor $7x^2 - 26x - 8$.

$7x^2$	$-26x$	-8
$1 \cdot 7$	$(1)(-8) + (1)(7) = -1$	$(1)(-8)$
	$(1)(1) + (-8)(7) = -55$	$(-8)(1)$
	$(1)(-4) + (2)(7) = 10$	$(2)(-4)$
	$(1)(2) + (-4)(7) = -26$ ✓	$(-4)(2)$

$7x^2 - 26x - 8 = (1x + -4)(7x + 2)$

 **Quick Check** ❷ Factor each expression.

 a. $5d^2 - 14d - 3$ **b.** $2n^2 + n - 3$ **c.** $20p^2 - 31p - 9$

Some polynomials can be factored repeatedly. Continue the process of factoring until there are no common factors other than 1. If a trinomial has a common monomial factor, factor it out before trying to find binomial factors.

3 EXAMPLE **Factoring Out a Monomial First**

Factor $20x^2 + 80x + 35$ completely.

$20x^2 + 80x + 35 = 5(4x^2 + 16x + 7)$ **Factor out the GCF.**

Factor $4x^2 + 16x + 7$.

$4x^2$	$16x$	7
$1 \cdot 4$	$1 \cdot 7 + 1 \cdot 4 = 11$	$1 \cdot 7$
	$1 \cdot 1 + 7 \cdot 4 = 29$	$7 \cdot 1$
$2 \cdot 2$	$2 \cdot 7 + 1 \cdot 2 = 16$ ✓	$1 \cdot 7$

$4x^2 + 16x + 7 = (2x + 1)(2x + 7)$

$20x^2 + 80x + 35 = 5(2x + 1)(2x + 7)$ **Include the GCF in your final answer.**

 **Quick Check** ❸ Factor each expression.

 a. $2v^2 - 12v + 10$ **b.** $4y^2 + 14y + 6$ **c.** $18k^2 - 12k - 6$

EXERCISES

For more exercises, see *Extra Skill and Word Problem Practice.*

Practice and Problem Solving

A Practice by Example

Example 1
(page 524)

 **for Help**

Factor each expression.

1. $2n^2 + 15n + 7$ **2.** $7d^2 + 50d + 7$ **3.** $11w^2 - 14w + 3$

4. $3x^2 - 17x + 10$ **5.** $6t^2 + 25t + 11$ **6.** $3d^2 - 17d + 20$

7. $16m^2 + 26m + 9$ **8.** $15p^2 - 26p + 11$ **9.** $8y^2 + 30y + 13$

10. $2y^2 + 35y + 17$ **11.** $7x^2 - 30x + 27$ **12.** $8x^2 + 18x + 9$

Example 2
(page 525)

Factor each expression.

13. $2t^2 - t - 3$ **14.** $8y^2 - 10y - 3$ **15.** $2q^2 - 11q - 21$

16. $7x^2 - 20x - 3$ **17.** $13p^2 + 8p - 5$ **18.** $5k^2 - 2k - 7$

19. $10w^2 + 11w - 8$ **20.** $12d^2 - d - 20$ **21.** $14n^2 + 23n - 15$

Example 3
(page 525)

22. $24m^2 - 32m + 8$ **23.** $21v^2 - 70v + 49$ **24.** $6t^2 + 26t + 24$

25. $25x^2 - 10x - 15$ **26.** $11p^2 + 77p + 66$ **27.** $24v^2 + 10v - 6$

B **Apply Your Skills**

Open-Ended Find three different values that complete each expression so that the trinomial can be factored into the product of two binomials. Factor your trinomials.

28. $4g^2 + \blacksquare g + 10$ **29.** $15m^2 + \blacksquare m - 24$ **30.** $35g^2 + \blacksquare g - 16$

31. a. Write each area as a product of two binomials.

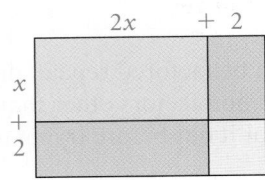
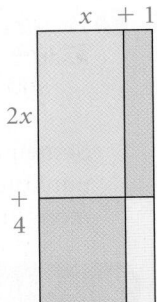

b. Are the products equal?

c. **Critical Thinking** Explain how the two products you found in part (a) can equal the same trinomial.

 32. **Writing** Explain how you would factor the expression $50x^2 - 90x + 16$.

Homework Video Tutor
Visit: PHSchool.com
Web Code: ate-0906

Factor each expression.

33. $54p^2 + 87p + 28$ **34.** $66r^2 + 57r + 12$ **35.** $14x^2 - 53x + 14$

36. $28m^2 + 28m - 56$ **37.** $21h^2 + 72h - 48$ **38.** $55n^2 - 52n + 12$

39. $36y^2 + 114y - 20$ **40.** $63w^2 - 89w + 30$ **41.** $99q^2 - 92q + 9$

42. **Critical Thinking** If a and c in $ax^2 + bx + c$ are prime numbers, and the trinomial is factorable, how many positive values are possible for b?

43. **Open-Ended** Write three different factorable trinomials that are of the form $\blacksquare x^2 - 12x + \blacksquare$. Factor your trinomials.

C **Challenge**

Factor each expression.

44. $56x^3 + 43x^2 + 5x$ **45.** $49p^2 + 63pq - 36q^2$ **46.** $108g^2h - 162gh + 54h$

47. The graph of the function $y = x^2 + 5x + 6$ is shown at the right.

a. What are the x-intercepts?

b. Factor $x^2 + 5x + 6$.

c. **Critical Thinking** Describe the relationship between the binomial factors you found in part (b) and the x-intercepts.

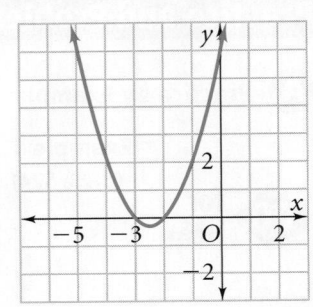

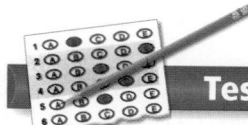

Multiple Choice

48. Which of the following expressions does NOT equal $12n^2 + 32n - 140$?
A. $4(n + 5)(3n - 7)$ **B.** $(4n + 20)(3n - 7)$
C. $(2n + 10)(6n - 14)$ **D.** $(n + 5)(12n - 7)$

49. Which value would make the expression $8p^2 + \blacksquare p + 11$ factorable?
F. 24   **G.** 46   **H.** 48   **J.** 52

50. Which binomial is one of the factors of $13x^2 + 32x - 21$?
A. $13x + 3$ **B.** $13x + 7$ **C.** $13x + 21$ **D.** $13x - 7$

51. A rectangle has dimensions that are the binomial factors of $3x^2 + 22x + 24$. Which of the following expressions describes the perimeter of the rectangle?
F. $4x + 10$ **G.** $4x + 25$ **H.** $8x + 20$ **J.** $8x + 50$

52. The table at the right shows the atomic masses, rounded to the nearest whole number, for the first five elements. Which of the following values is the median of the data?
A. 5.5   **B.** 6
C. 6.4   **D.** 7

Element	Atomic Mass
Hydrogen	1
Helium	4
Lithium	7
Beryllium	9
Boron	11

Short Response

53. What are the factors of $3x^2 + 40x - 75$? Show your work.

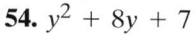

Lesson 9-5

Factor each expression.

54. $y^2 + 8y + 7$ **55.** $t^2 - 7t + 12$ **56.** $p^2 - p - 20$

57. $m^2 - 15m + 36$ **58.** $k^2 + 16k - 36$ **59.** $g^2 + 17g + 72$

60. $h^2 - 13h - 48$ **61.** $x^2 - 13x - 30$ **62.** $d^2 - 18d + 56$

Lesson 9-4

Mental Math Find each square.

63. 89^2 **64.** 401^2 **65.** 903^2 **66.** 197^2

Mental Math Find each product.

67. $39 \cdot 41$ **68.** $38 \cdot 42$ **69.** $198 \cdot 202$ **70.** $73 \cdot 67$

Lesson 8-7

Evaluate each exponential function for the domain $\{-3, 0, 2\}$.

71. $f(x) = 4 \cdot 2^x$ **72.** $h(x) = -3 \cdot 3^x$ **73.** $k(x) = \frac{1}{3} \cdot 3^x$

74. $g(x) = 5 \cdot \left(\frac{1}{10}\right)^x$ **75.** $g(x) = \frac{1}{10} \cdot 5^x$ **76.** $h(x) = 8 \cdot (0.2)^x$

Graph each function.

77. $y = 3 \cdot 2^x$ **78.** $y = -3 \cdot 2^x$ **79.** $y = \frac{1}{2} \cdot 2^x$ **80.** $y = \frac{1}{3} \cdot 3^x$

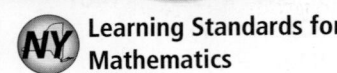

Factoring Special Cases

NY Learning Standards for Mathematics

A.A.19 Identify and factor the difference of two perfect squares.

A.A.20 Factor algebraic expressions completely, including trinomials with a lead coefficient of one (after factoring a GCF).

✓ Check Skills You'll Need

GO for Help Lessons 8-4 and 9-4

Simplify each expression.

1. $(3x)^2$ **2.** $(5y)^2$ **3.** $(15h^2)^2$ **4.** $(2ab^2)^2$

Simplify each product.

5. $(c - 6)(c + 6)$ **6.** $(p - 11)(p - 11)$ **7.** $(4d + 7)(4d + 7)$

◀)) **New Vocabulary** • perfect-square trinomial

---

1 Factoring Perfect-Square Trinomials

Activity: Perfect-Square Trinomials

1. Factor each trinomial.

$x^2 + 6x + 9$ $x^2 + 10x + 9$ $m^2 + 15m + 36$

$m^2 + 12m + 36$ $k^2 + 26k + 25$ $k^2 + 10k + 25$

2. a. Which trinomials have pairs of binomial factors that are identical?
  b. Describe the relationship between the middle and last terms of the trinomials that have identical pairs of factors.

In Lesson 9-4 you found the square of a binomial.

$$(a + b)^2 = (a + b)(a + b) = a^2 + 2ab + b^2 \text{ and}$$
$$(a - b)^2 = (a - b)(a - b) = a^2 - 2ab + b^2$$

Any trinomial of the form $a^2 + 2ab + b^2$ or $a^2 - 2ab + b^2$ is a **perfect-square trinomial.** You can factor a perfect-square trinomial into identical binomial factors.

 Key Concepts

Rule	Perfect-Square Trinomials

For every real number a and b:

$a^2 + 2ab + b^2 = (a + b)(a + b) = (a + b)^2$
$a^2 - 2ab + b^2 = (a - b)(a - b) = (a - b)^2$

Examples $x^2 + 10x + 25 = (x + 5)(x + 5) = (x + 5)^2$
$x^2 - 10x + 25 = (x - 5)(x - 5) = (x - 5)^2$

You can factor a perfect-square trinomial using the method shown in the previous lesson. Or you can recognize a perfect-square trinomial and then factor it quickly. Here is how to recognize a perfect-square trinomial.

- The first and the last terms can both be written as the product of two identical factors.

- The middle term is twice the product of one factor from the first term and one factor from the last term.

Consider the following trinomials.

$$4x^2 \quad + \quad 12x \quad + \quad 9$$
$$2x \cdot 2x \qquad\qquad\qquad 3 \cdot 3$$
$$2(2x \cdot 3) = 12x$$

This is a perfect-square trinomial. In factored form the trinomial is $(2x + 3)(2x + 3)$, or $(2x + 3)^2$.

$$4x^2 \quad + \quad 20x \quad + \quad 9$$
$$2x \cdot 2x \qquad\qquad\qquad 3 \cdot 3$$
$$2(2x \cdot 3) \neq 20x$$

This is not a perfect-square trinomial. Factor by listing factors, as shown in Lesson 9-6.

When you factor a perfect-square trinomial, it may help to write the first and last terms as the products of identical factors.

1 EXAMPLE   Factoring a Perfect-Square Trinomial With $a = 1$

Factor $x^2 - 8x + 16$.

$x^2 - 8x + 16 = x \cdot x - 8x + 4 \cdot 4$	**Rewrite first and last terms.**
$= x \cdot x - 2(x \cdot 4) + 4 \cdot 4$	**Does the middle term equal 2ab? $8x = 2(x \cdot 4)$**
$= (x - 4)^2$	**Write the factors as the square of a binomial.**

 **Quick Check** ❶ Factor each expression.
   a. $x^2 + 8x + 16$ **b.** $n^2 + 16n + 64$ **c.** $n^2 - 16n + 64$

Problem Solving Hint

For $a \neq 0$, $b \neq 0$, and all integers n,
$a^n b^n = (ab)^n$.
$9g^2 = 3^2 g^2 = (3g)^2$

When you write the identical factors of the first and last terms, you can write them as square terms. Notice in Example 2, $9g^2$ is written as $(3g)^2$ and 4 is written as 2^2.

2 EXAMPLE   Factoring a Perfect-Square Trinomial With $a \neq 1$

Geometry The area of the square shown at the right is $(9g^2 + 12g + 4)$ cm^2. Find an expression for the length of a side.

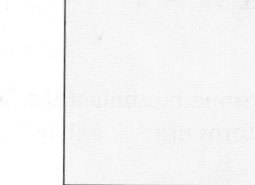

$9g^2 + 12g + 4 = (3g)^2 + 12g + 2^2$	**Rewrite $9g^2$ as $(3g)^2$ and 4 as 2^2.**
$= (3g)^2 + 2(3g)(2) + 2^2$	**Does the middle term equal 2ab? $12g = 2(3g)(2)$ ✓**
$= (3g + 2)^2$	**Write the factors as the square of a binomial.**

The side of the square has length of $(3g + 2)$ cm.

 **Quick Check** ❷ Factor each expression.
   a. $9g^2 - 12g + 4$ **b.** $4t^2 + 36t + 81$ **c.** $4t^2 - 36t + 81$

2 Factoring the Difference of Squares

Recall from Lesson 9-4 that $(a + b)(a - b) = a^2 - b^2$. So you can factor a difference of two squares as $(a + b)(a - b)$.

Key Concepts

| **Rule** | **Difference of Two Squares** |

For every real number a and b:

$a^2 - b^2 = (a + b)(a - b)$

Examples $\quad x^2 - 81 = (x + 9)(x - 9)$
$16x^2 - 49 = (4x + 7)(4x - 7)$

3 EXAMPLE The Difference of Two Squares for $a = 1$

Factor $x^2 - 64$.

$\begin{aligned} x^2 - 64 &= x^2 - 8^2 &&\textbf{Rewrite 64 as 8}^2. \\ &= (x + 8)(x - 8) &&\textbf{Factor.} \end{aligned}$

Check   Use FOIL to multiply.

$\qquad (x + 8)(x - 8)$
$\qquad x^2 - 8x + 8x - 64$
$\qquad x^2 - 64 \checkmark$

Quick Check **3** Factor each expression. Check your answer.
 a. $x^2 - 36$ **b.** $m^2 - 100$ **c.** $p^2 - 49$

4 EXAMPLE The Difference of Two Squares for $a \neq 1$

Factor $4x^2 - 121$.

$\begin{aligned} 4x^2 - 121 &= (2x)^2 - (11)^2 &&\textbf{Rewrite 4}x^2 \textbf{ as (2}x)^2 \textbf{ and 121 as 11}^2. \\ &= (2x + 11)(2x - 11) &&\textbf{Factor.} \end{aligned}$

Quick Check **4** Factor each expression.
 a. $9v^2 - 4$ **b.** $25x^2 - 64$ **c.** $4w^2 - 49$

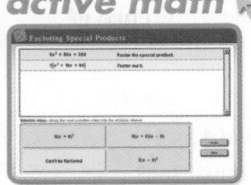

For: Common Factor Activity
Use: Interactive Textbook, 9-7

Some binomials that do not appear to be the difference of squares may have the form $n(a^2 - b^2)$ after a GCF is factored out.

5 EXAMPLE Factoring Out a Common Factor

Factor $10x^2 - 40$.

$\begin{aligned} 10x^2 - 40 &= 10(x^2 - 4) &&\textbf{Factor out the GCF of 10.} \\ &= 10(x - 2)(x + 2) &&\textbf{Factor }(x^2 - 4). \end{aligned}$

Quick Check **5** Factor each expression.
 a. $8y^2 - 50$ **b.** $3c^2 - 75$ **c.** $28k^2 - 7$

EXERCISES
For more exercises, see *Extra Skill and Word Problem Practice.*

Practice and Problem Solving

 Practice by Example

Example 1
(page 529)

 **for Help**

Example 2
(page 529)

Factor each expression.

1. $c^2 + 10c + 25$ **2.** $x^2 - 2x + 1$ **3.** $h^2 + 12h + 36$

4. $m^2 - 24m + 144$ **5.** $k^2 - 16k + 64$ **6.** $t^2 - 14t + 49$

The given expression represents the area. Find the side length of each square.

7.
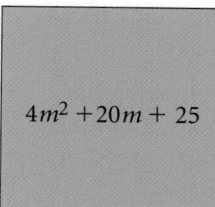
$4m^2 + 20m + 25$

8.
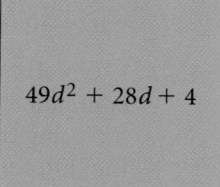
$49d^2 + 28d + 4$

9.
$25g^2 - 40g + 16$

Factor each expression. Check your answer.

10. $25g^2 - 30g + 9$ **11.** $64r^2 - 144r + 81$ **12.** $100v^2 - 220v + 121$

Example 3
(page 530)

13. $x^2 - 4$ **14.** $y^2 - 81$ **15.** $k^2 - 196$

16. $r^2 - 144$ **17.** $h^2 - 100$ **18.** $m^2 - 225$

19. $w^2 - 256$ **20.** $x^2 - 400$ **21.** $y^2 - 900$

Example 4
(page 530)

22. $25q^2 - 9$ **23.** $49y^2 - 4$ **24.** $9c^2 - 64$

25. $4m^2 - 81$ **26.** $16k^2 - 49$ **27.** $144p^2 - 1$

28. $81v^2 - 100$ **29.** $400n^2 - 121$ **30.** $25w^2 - 196$

Example 5
(page 530)

31. $3m^2 - 12$ **32.** $5k^2 - 245$ **33.** $3x^2 + 48x + 192$

34. $2t^2 - 36t + 162$ **35.** $6r^3 - 150r$ **36.** $7h^2 - 56h + 112$

B **Apply Your Skills**

37. Writing Summarize the procedure for factoring a perfect-square trinomial. Give at least two examples.

38. Error Analysis Suppose a classmate factored the binomial at the right. What error did your classmate make?

$$4x^2 - 121 = (4x - 11)(4x - 11)$$
$$= (4x - 11)^2$$

Mental Math **Find a pair of factors for each number by using the difference of two squares.**

Sample $143 = 144 - 1$ Write 143 as the difference of two squares.

$= 12^2 - 1^2$ Rewrite 144 as 12^2 and 1 as 1^2.

$= (12 - 1)(12 + 1)$ Factor.

$= (11)(13)$ Simplify.

39. 99   **40.** 91   **41.** 75   **42.** 117   **43.** 224

44. a. Open-Ended Write an expression that is a perfect-square trinomial.
   b. Explain how you know your trinomial is a perfect-square trinomial.

Factor each expression.

45. $100v^2 - 25w^2$

46. $16p^2 - 48pq + 36q^2$

47. $28c^2 + 140cd + 175d^2$

48. $\frac{1}{4}m^2 - \frac{1}{9}$

49. $x^2 + x + \frac{1}{4}$

50. $64g^2 - 192gh + 144h^2$

51. $\frac{1}{4}p^2 - 2p + 4$

52. $\frac{1}{9}n^2 - \frac{1}{25}$

53. $\frac{1}{25}k^2 + \frac{6}{5}k + 9$

54. a. Geometry Write an expression in terms of n and m for the area of the top of the block that was drilled at the right. Use 3.14 for π. Factor your expression.

   b. Find the area of the top of the block if $n = 10$ in. and $m = 3$ in.

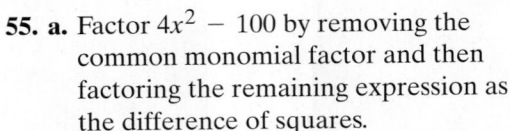

55. a. Factor $4x^2 - 100$ by removing the common monomial factor and then factoring the remaining expression as the difference of squares.

   b. Factor $4x^2 - 100$ as the difference of squares, and then remove the common monomial factors.

   c. Critical Thinking Why can $4x^2 - 100$ be factored in two different ways?

   d. Can you factor $3x^2 - 75$ in the two ways you factored $4x^2 - 100$ in parts (a) and (b)? Explain your answer.

 **Challenge**

Factor each expression.

56. $64r^6 - 144r^3 + 81$

57. $p^6 + 40p^3q + 400q^2$

58. $36m^4 + 84m^2 + 49$

59. $81p^{10} + 198p^5 + 121$

60. $108m^6 - 147$

61. $x^{20} - 4x^{10}y^5 + 4y^{10}$

62. $256g^4 - 100h^6$

63. $45x^4 - 60x^2y + 20y^2$

64. $37g^8 - 37h^8$

65. a. The expression $(t - 3)^2 - 16$ is a difference of two squares. Identify a and b.

   b. Factor $(t - 3)^2 - 16$ and simplify.

66. The binomial $16 - 81n^4$ can be factored twice as the difference of squares.

   a. Factor $16 - 81n^4$ completely.

   b. Critical Thinking What characteristics do 16 and $81n^4$ share that make this possible?

   c. Open-Ended Write a binomial that can be factored twice as the difference of squares.

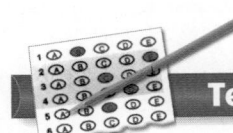 **Test Prep**

Gridded Response

67. The area of the square shown at the right is $4x^2 + 28x + 49$. What is the sum of a and b?

68. For what value of p would $(x + p)(x + p)$ be the factored form of $x^2 - 24x + 144$?

69. For what positive value of k are the factors of $x^2 - kx + 225$ the same?

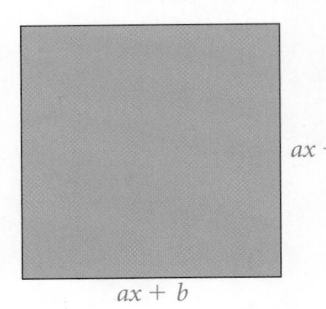

$ax + b$

$ax + b$

70. The diagram shows two squares. The area of the region shaded green is $4x^2 + 16x + 16$. The area of the region shaded purple is $5x^2 + 14x + 9$. What is the value of b?

71. For what value of a does $144x^4 - 121 = (ax^2 + 11)(ax^2 - 11)$?

72. The expression $81x^2 - 36$ can factored as $9(ax + b)(ax - b)$. What is the mean of a and b?

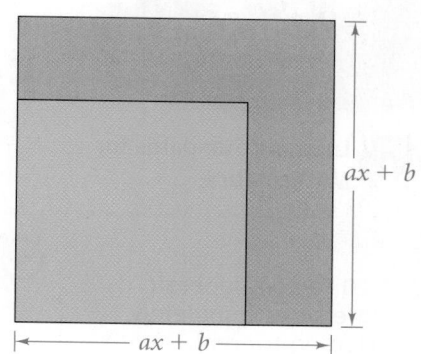
$ax + b$
$ax + b$

Mixed Review

for
Help

Lesson 9-6

Factor each expression.

73. $2d^2 + 11d + 5$

74. $2x^2 - 11x + 12$

75. $4t^2 + 16t + 7$

76. $5w^2 - 44w - 9$

77. $6t^2 + 19t + 8$

78. $21m^2 - 20m - 9$

79. $14x^2 - 11x - 9$

80. $4y^2 + 32y + 55$

81. $12k^2 - 5k - 2$

Lesson 8-6

Find the next three terms of each sequence. Then write a rule for each sequence.

82. $3, 12, 48, 192, \ldots$

83. $-3, 5, 13, 21, \ldots$

84. $25, 16, 7, -2, \ldots$

85. $200, 20, 2, 0.2, \ldots$

86. $-2, 4, -8, 16, \ldots$

87. $0.1, 0.6, 3.6, 21.6, \ldots$

88. $10, 4, \frac{8}{5}, \frac{16}{25}, \ldots$

89. $\frac{1}{2}, 2, 3\frac{1}{2}, 5, \ldots$

90. $\frac{1}{32}, \frac{1}{8}, \frac{1}{2}, 2, \ldots$

Lesson 6-7

91. A teacher is comparing time her students spent studying, in hours, with their grades on a math test.

 a. Graphing Calculator Use a graphing calculator to find the equation of the line of best fit.

b. What test score would you predict for a student who studied 2.5 hours?

c. What test score would you predict for a student who studied 1.25 hours?

Student	Time Studying	Grade
1	1	82
2	2	92
3	1.5	80
4	2	88
5	1	70
6	3	97
7	0.5	70

✓ Checkpoint Quiz 2   Lessons 9-4 through 9-7

Simplify each expression.

1. $(k - 7)^2$

2. $(5t + 9)^2$

3. $(h - 11)(h - 11)$

Factor each expression.

4. $v^2 + 20v + 100$

5. $p^2 - 6p - 40$

6. $k^2 - 17k + 60$

7. $2x^2 + 13x + 11$

8. $10m^2 + 19m + 7$

9. $3w^2 - 6w - 24$

10. $9t^2 - 25$

9-8

Factoring by Grouping

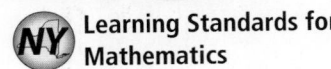
A.A.20 Factor algebraic expressions completely, including trinomials with a lead coefficient of one (after factoring a GCF).

✓ **Check Skills You'll Need**

GO **for Help** Lessons 9-2 and 9-3

Find the GCF of the terms of each polynomial.

1. $6y^2 + 12y - 4$

2. $9r^3 + 15r^2 + 21r$

3. $30h^3 - 25h^2 - 40h$

4. $16m^3 - 12m^2 - 36m$

Find each product.

5. $(v + 3)(v^2 + 5)$

6. $(2q^2 - 4)(q - 5)$

7. $(2t - 5)(3t + 4)$

8. $(4x - 1)(x^2 + 2x + 3)$

🔊 **New Vocabulary** • factor by grouping

---

1 Factoring Polynomials With Four Terms

You can use the Distributive Property to **factor by grouping** if two groups of terms have the same factor.

$$y^3 + 3y^2 \quad + \quad 4y + 12$$

$$y^2(y + 3) \quad + \quad 4(y + 3)$$

$$(y^2 + 4)(y + 3)$$

These factors are the same, so factor again.

To factor by grouping, look for a common binomial factor of two pairs of terms.

1 EXAMPLE Factoring a Four-Term Polynomial

Factor $4n^3 + 8n^2 - 5n - 10$.

$4n^3 + 8n^2 - 5n - 10 = 4n^2(n + 2) - 5(n + 2)$ Factor the GCF from each group of two terms.

$= (4n^2 - 5)(n + 2)$ Factor out $(n + 2)$.

Check $4n^3 + 8n^2 - 5n - 10 \stackrel{?}{=} (4n^2 - 5)(n + 2)$

$= 4n^3 + 8n^2 - 5n - 10$ ✓   Use FOIL.

✓ **Quick Check** ❶ Factor each expression. Check your answer.

a. $5t^4 + 20t^3 + 6t + 24$

b. $2w^3 + w^2 - 14w - 7$

Before you factor by grouping, you may need to factor the GCF of all the terms of a polynomial. Remember, a polynomial is not completely factored until there are no common factors other than 1.

2 EXAMPLE Factoring Completely

Factor $12p^4 + 10p^3 - 36p^2 - 30p$.

$$
\begin{aligned}
12p^4 + 10p^3 - 36p^2 - 30p &= 2p(6p^3 + 5p^2 - 18p - 15) && \text{Factor out the GCF, } 2p. \\
&= 2p[(p^2(6p + 5) - 3(6p + 5)] && \text{Factor by grouping.} \\
&= 2p(p^2 - 3)(6p + 5) && \text{Factor again.}
\end{aligned}
$$

✓ **Quick Check** ❷ Factor $45m^4 - 9m^3 + 30m^2 - 6m$.

Factoring Trinomials by Grouping

You can also factor by grouping to find the factors of a trinomial of the form $ax^2 + bx + c$. You may want to use this method when you cannot quickly factor a trinomial using the method you learned in Lesson 9-6.

You can use these steps to factor a trinomial such as $48x^2 + 46x + 5$.

Step 1 Find the product ac.

$$48 \cdot 5 = 240$$

Step 2 Find the two factors of ac that have sum b.

Factors	$\rightarrow$	Sum		Factors	$\rightarrow$	Sum
$1 \cdot 240$	$\rightarrow$	$1 + 240 = 241$		$4 \cdot 60$	$\rightarrow$	$4 + 60 = 64$
$2 \cdot 120$	$\rightarrow$	$2 + 120 = 122$		$5 \cdot 48$	$\rightarrow$	$5 + 48 = 53$
$3 \cdot 80$	$\rightarrow$	$3 + 80 = 83$		$6 \cdot 40$	$\rightarrow$	$6 + 40 = 46$ ✓

Step 3 Rewrite the trinomial using the sum.

$$
\begin{aligned}
48x^2 + 46x + 5 &= 48x^2 + (6 + 40)x + 5 \\
&= 48x^2 + 6x + 40x + 5
\end{aligned}
$$

Step 4 Factor by grouping.

$$48x^2 + 6x + 40x + 5$$
$$6x(8x + 1) + 5(8x + 1)$$
$$(6x + 5)(8x + 1)$$

As you look for factors of the product ac that have a sum b, you do not have to list all the possible factors. Use mental math to eliminate those factors that give sums too great or too small to be reasonable choices.

3 EXAMPLE Factoring a Trinomial by Grouping

Factor $24q^2 + 25q - 25$.

Step 1 $24(-25) = -600$ Find the product ac.

Step 2 Factors $\rightarrow$ Sum

$(-12)(50) \rightarrow -12 + 50 = 38$ Find two factors of ac that have sum b.
$(-15)(40) \rightarrow -15 + 40 = 25$ ✓   Use mental math to determine a good place to start.

Step 3 $24q^2 - 15q + 40q - 25$ Rewrite the trinomial.

Step 4 $3q(8q - 5) + 5(8q - 5)$ Factor by grouping.
$(3q + 5)(8q - 5)$ Factor again.

GO Online

Video Tutor Help
Visit: PHSchool.com
Web Code: ate-0775

✅ **Quick Check** ❸ Factor each trinomial by grouping.
　　a. $63d^2 + 44d + 5$　　　**b.** $11k^2 + 49k + 20$　　　**c.** $4y^2 + 33y - 70$

Given a polynomial expression for the volume of a rectangular prism, you can sometimes factor to find possible expressions for the length, width, and height.

4 **EXAMPLE**　**Finding the Dimensions of a Rectangular Prism**

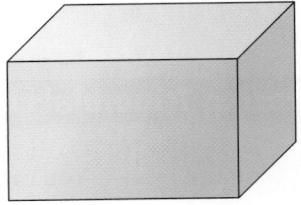

Geometry The volume (ℓwh) of the rectangular prism at the right is $80x^3 + 224x^2 + 60x$. Factor to find possible expressions for the length, width, and height of the prism.

Factor $80x^3 + 224x^2 + 60x$.

Step 1 $4x(20x^2 + 56x + 15)$　　　**Factor out the GCF, 4x.**

Step 2 $20 \cdot 15 = 300$　　　**Find the product ac.**

Step 3 Factors　→　Sum
　　　　　$5 \cdot 60$　→　$5 + 60$ = 65　　**Find two factors of ac that have**
　　　　$10 \cdot 30$　→　$10 + 30$ = 40　　**sum b. Use mental math to**
　　　　$15 \cdot 20$　→　$15 + 20$ = 35　　**determine a good place to start.**
　　　　　$6 \cdot 50$　→　$6 + 50$ = 56 ✓

Step 4 $4x(20x^2 + 50x + 6x + 15)$　　　**Rewrite the trinomial.**

Step 5 $4x[10x(2x + 5) + 3(2x + 5)]$　　**Factor by grouping.**
　　　　$4x(10x + 3)(2x + 5)$　　　　　　**Factor again.**

● The possible dimensions of the prism are $4x$, $(10x + 3)$, and $(2x + 5)$.

✅ **Quick Check** ❹ Find expressions for the possible dimensions of each rectangular prism.

a.

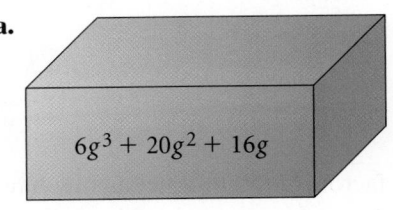

$6g^3 + 20g^2 + 16g$

b.

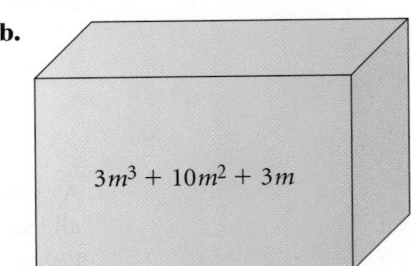

$3m^3 + 10m^2 + 3m$

Here is a summary of what to remember as you factor polynomials.

 Key Concepts

Summary	**Factoring Polynomials**

1. Factor out the greatest common factor (GCF).
2. If the polynomial has two terms or three terms, look for a difference of two squares, a product of two squares, or a pair of binomial factors.
3. If there are four or more terms, group terms and factor to find common binomial factors.
4. As a final check, make sure there are no common factors other than 1.

For more exercises, see *Extra Skill and Word Problem Practice*.

Practice and Problem Solving

A **Practice by Example**

Example 1
(page 534)

for
Help

Find the GCF of the first two terms and the GCF of the last two terms for each polynomial.

1. $2m^3 + 6m^2 + 3m + 9$

2. $10p^3 - 25p^2 + 4p - 10$

3. $2z^3 + 12z^2 - 5z - 30$

4. $6n^3 + 3n^2 + 2n + 1$

Factor each expression.

5. $6n^3 + 8n^2 + 3n + 4$

6. $14t^3 + 21t^2 + 16t + 24$

7. $27t^3 + 45t^2 - 3t - 5$

8. $13y^3 - 8y^2 + 13y - 8$

9. $45x^3 + 20x^2 + 9x + 4$

10. $10w^3 + 16w^2 - 15w - 24$

Example 2
(page 535)

Factor completely.

11. $12v^3 - 32v^2 + 6v - 16$

12. $7q^4 - 4q^3 + 28q^2 - 16q$

13. $20m^3 - 18m^2 + 40m - 36$

14. $6x^4 + 4x^3 - 6x^2 - 4x$

15. $12y^3 - 20y^2 + 30y - 50$

16. $9c^3 - 12c^2 + 18c - 24$

Example 3
(page 535)

Factor by grouping.

17. $12p^2 + 16p + 5$

18. $16t^2 + 24t + 9$

19. $18n^2 + 57n - 10$

20. $9w^2 - 27w + 20$

21. $24m^2 + 8m - 2$

22. $36v^2 - 9v - 7$

23. $6x^2 + 11x - 10$

24. $20v^2 - 41v + 9$

25. $63q^2 - 52q - 20$

Example 4
(page 536)

Find expressions for the possible dimensions of each rectangular prism.

26.

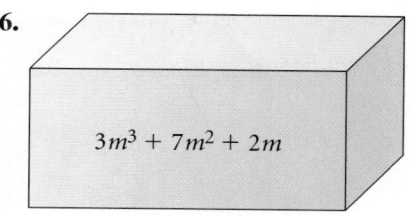

$3m^3 + 7m^2 + 2m$

27.

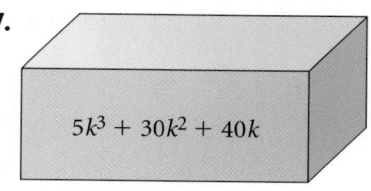

$5k^3 + 30k^2 + 40k$

B **Apply Your Skills**

Factor completely.

28. $7h^3 - 35h^2 - 42h$

29. $60t^3 - 200t^2 - 66t + 220$

30. $8d^3 + 16d^2 + 24d + 48$

31. $12x^2 - 4xy - 56y^2$

32. $54r^3 - 45r^2 + 9r$

33. $150k^3 + 350k^2 + 180k + 420$

34. a. Factor $(28x^3 - 7x^2) + (36x - 9)$.

  b. Factor $(28x^3 + 36x) + (-7x^2 - 9)$.

  c. **Critical Thinking** Why can you factor the same polynomial using different pairs of terms?

Write each expression in standard form and factor.

35. $-8w + 49w^2 + 14w^3 - 28$

36. $2m^3 + 16 - m - 32m^2$

37. $-6 + 44t^3 - 4t^2 + 66t$

38. $2 - 50x - x^2 + 25x^3$

39. Geometry The polynomial shown at the right represents the volume of the rectangular prism. Factor the polynomial to find possible expressions for the length, width, and height of the prism.

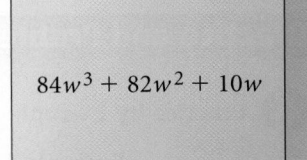

$$84w^3 + 82w^2 + 10w$$

40. Open-Ended Write a four-term polynomial that can be factored by grouping. Factor your polynomial.

41. Writing Describe how to factor the expression $10x^3 - 15x^2 + 2x - 3$ by grouping.

C Challenge

Factor by grouping.

42. $30m^5 + 24m^3n - 35m^2n^2 - 28n^3$ **43.** $x^2p + x^2q^5 + yp + yq^5$

44. $h^3 + 11h^2 - 4h - 44$ **45.** $w^6 - w^4 - 9w^2 + 9$

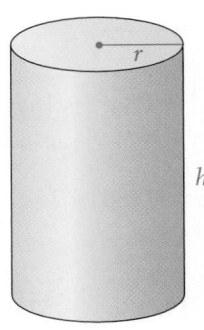

46. Geometry The polynomial $2\pi x^3 + 12\pi x^2 + 18\pi x$ represents the volume of a cylinder. The formula for the volume of a cylinder is $V = \pi r^2 h$.
  a. Factor $2\pi x^3 + 12\pi x^2 + 18\pi x$.
  b. Based on your answer to part (a), write an expression for a possible radius of the cylinder.

The number 63 can be written as $2^5 + 2^4 + 2^3 + 2^2 + 2^1 + 2^0$. For Exercises 47 and 48, factor each expression by grouping. Then simplify the powers of 2 to write 63 as the product of two numbers.

47. $(2^5 + 2^4 + 2^3) + (2^2 + 2^1 + 2^0)$

48. $(2^5 + 2^4) + (2^3 + 2^2) + (2^1 + 2^0)$

49. a. Open-Ended For the rectangular prism below, let $x = 3$. Write linear expressions for the length, width, and height of the prism.

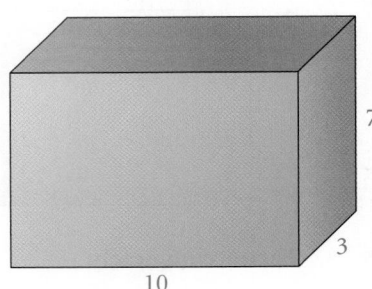

  b. Using your answers from part (a), write a polynomial that represents the volume of the prism.

NY REGENTS

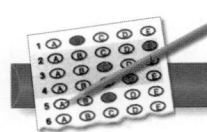

Test Prep

Multiple Choice

50. Which of the following expressions is a factor of $27x^4 + 15x^3 + 63x + 35$?
  A. $3x + 7$ **B.** $3x^2 + 7$ **C.** $3x^3 + 7$ **D.** $3x^4 + 7$

51. Which of the following expressions equals the sum of the binomial factors of $6q^3 - 5q^2 + 24q - 20$?
  F. $7q^3 - 1$ **G.** $6q^3 - 1$ **H.** $q^2 + 6q - 1$ **J.** $q^2 + 6q + 1$

Short Response

52. Factor $9a^4 - 54a^3 - 2a + 12$ completely. Show your work.

Extended Response

53. The volume of the square prism is $96x^3 + 48x^2 + 6x$. Find an expression that could describe the perimeter of one of the prism's square faces. Show your work.

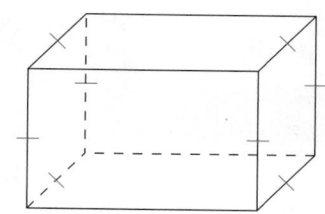

Mixed Review

GO for Help

Lesson 9-7 Factor each expression.

54. $k^2 + 14k + 49$ 55. $r^2 + 6r + 9$ 56. $y^2 - 16y + 64$

57. $2t^2 + 12t + 18$ 58. $m^2 - 64$ 59. $4g^2 + 40g + 100$

60. $4d^2 - 25$ 61. $5n^2 - 45$ 62. $25q^2 + 40q + 16$

Lesson 8-4 Simplify each expression.

63. $(b^2)^2$ 64. $x^4 \cdot x^{-2}$ 65. $(t^3)^5$ 66. $(c^5 d)^7$

67. $(2y)^3$ 68. $(9m)^0$ 69. $(x^3)(x^7)^{-2}$ 70. $(3w^2 v^3)^4$

Simplify . Write each answer in scientific notation.

71. $(2 \times 10^5)^4$ 72. $(3 \times 10^6)^2$ 73. $(7 \times 10^{-6})^2$ 74. $(2 \times 10^7)^5$

75. $(5.3 \times 10^2)^2$ 76. $(8.1 \times 10^{-3})^2$ 77. $(1.9 \times 10^8)^3$ 78. $(4 \times 10^{-3})^{-2}$

Lesson 7-2 Solve each system using substitution.

79. $y = -7x + 12$
    $y = 3x + 2$

80. $y = -3x + 4$
    $y = -5x + 12$

81. $10x + 2y = 15$
    $y = -4x + 7$

82. $x + y = -28$
    $y = -2x - 26$

83. $8x + 2y = 50$
    $y = -4x + 25$

84. $y = x - 5$
    $11x - 6y = 65$

A P●int in Time

1000 1200 1400 1600 1800 2000

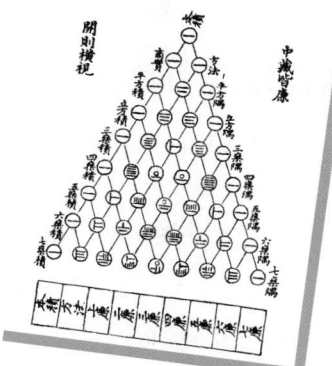

Very little is known about the life of Chu Shih-Chieh, the Chinese mathematician and teacher who had many students during the last two decades of the 1200s. In 1303, Chu wrote *Ssu-yüan-yü-chien*, or *The Precious Mirror of Four Elements*. He described what is now known as Pascal's Triangle and explained how it could be used to solve polynomial equations. He also invented "the method of celestial element" to write and solve polynomial equations and linear systems with up to four variables.

For: Information about Pascal's Triangle
PHSchool.com **Web Code:** ate-2032

Before you do all the work involved in solving a multiple-choice problem, you usually can eliminate some answer choices. This can save you time in arriving at the answer to the problem, or in making an "educated" guess if you do not actually know how to find the answer.

1 EXAMPLE

One factor of $x^3 + x^2 + x - 3$ is $x - 1$. Which of the following is also a factor?

A. $x^2 + 2x - 3$ **B.** $x^2 + 2x + 3$ **C.** $x^3 + 2x + 3$ **D.** $x^2 - 2x + 3$ **E.** $x^3 - 2x + 3$

Look at the degrees of the polynomials. The original polynomial has degree 3 and the given factor has degree 1. The answer must have degree less than or equal to $3 - 1 = 2$. Choices C and E have degree 3, so you can eliminate them.

Look at the constant terms of the polynomials. The constant term of the original polynomial is -3. Since the constant term of the factor is -1, the other constant term must be 3 for the product of the two factors to be -3. The constant term of answer choice A is -3, so you can eliminate choice A.

● The correct answer is either B or D.

2 EXAMPLE

$(3x + 5)(x - 2) = \blacksquare$
A. $3x^2 - 2x - 10$ **B.** $3x^2 - 5x - 2$ **C.** $3x^2 + 2x - 10$ **D.** $3x^2 + 7x - 10$ **E.** $3x^2 - x - 10$

Substitute a number. The original polynomial has $x - 2$ as a factor. If x equals 2, then the polynomial equals 0. The correct answer will equal 0 when x is 2.

Choice A: $3(2)^2 - 2(2) - 10 = 3(4) - 4 - 10 = 12 - 4 - 10 = -2$.   You can eliminate A.
Choice B: $3(2)^2 - 5(2) - 2 = 3(4) - 10 - 2 = 12 - 10 - 2 = 0$.   B is a possible answer.
Choice C: $3(2)^2 + 2(2) - 10 = 3(4) + 4 - 10 = 12 + 4 - 10 = 6$.   You can eliminate C.
Choice D: $3(2)^2 + 7(2) - 10 = 3(4) + 14 - 10 = 12 + 14 - 10 = 16$.   You can eliminate D.
Choice E: $3(2)^2 - 2 - 10 = 3(4) - 2 - 10 = 12 - 2 - 10 = 0$.   E is a possible answer.

● The correct answer is either B or E.

EXERCISES

1. Multiply $x - 1$ by choices B and D in Example 1 to find the correct answer.

2. Look at the constant terms in choices B and E in Example 2, and explain why you can eliminate choice B.

3. Consider the following multiple-choice problem.
One factor of $3x^3 + 2x^2 + x - 6$ is $x - 1$. What is the other factor?
A. $3x^2 - 5x - 6$ **B.** $3x^2 + 5x + 6$ **C.** $3x^3 + x^2 - 3$ **D.** $3x^2 - 5x + 6$ **E.** $3x^2 + 6$

a. Explain why you can eliminate answer choices A and C.
b. What is the correct answer to the problem?

Chapter Review

Vocabulary Review

binomial (p. 495)
degree of a monomial (p. 495)
degree of a polynomial (p. 495)
difference of squares (p. 514)

factor by grouping (p. 534)
monomial (p. 494)
perfect-square trinomial (p. 528)

polynomial (p. 495)
standard form of a polynomial (p. 495)
trinomial (p. 495)

Go Online
PHSchool.com

For: Vocabulary quiz
Web Code: atj-0951

Match the vocabulary term in the column on the left with the most accurate description in the column on the right.

1. binomial

2. degree of a monomial

3. monomial

4. perfect-square trinomial

5. standard form of a polynomial

A. a polynomial with two terms

B. a polynomial in which the terms decrease in degree from left to right and there are no like terms

C. a polynomial with two identical binomial factors

D. the sum of the exponents of the variables

E. an expression that is a number, a variable, or a product of a number and one or more variables

Skills and Concepts

9-1 Objectives

▼ To describe polynomials (p. 494)

▼ To add and subtract polynomials (p. 496)

The degree of a term with one variable is the exponent of the variable. A **polynomial** is one monomial or the sum or difference of two or more monomials. The **degree of a polynomial** is the same as the degree of the term with the highest degree. A polynomial can be named by its degree or by the number of its terms. You can simplify polynomials by adding the coefficients of like terms.

Write each polynomial in standard form. Then name each polynomial based on its degree and number of terms.

6. $5y + 2 - 6y^2 + 3y$

7. $1 + 9h^2$

8. $k^3 + 3k^5 + k - k^3$

9. $6t^3 + 9 + 8t + 7t^2 - 6t^3$

10. x^2y^2

11. $5 + x^2 + x^3$

12. **Open-Ended** Write a polynomial using the variable z. What is the degree of your polynomial?

Simplify. Write each answer in standard form.

13. $(-4b^5 + 3b^3 - b + 10) + (3b^5 - b^3 + b - 4)$

14. $(3g^4 + 5g^2 + 5) + (5g^4 - 10g^2 + 11g)$

15. $(3x^3 + 8x^2 + 2x + 9) - (-4x^3 + 5x - 3)$

16. $(2t^3 - 4t^2 + 9t - 7) - (t^3 + t^2 - 3t + 1)$

17. $(6y^2 + 3y + 5) - (2y^2 + 1)$

18. $(7w^5 - 7w^3 + 3w) - (5w^4 - w^2 + 3)$

9-2 Objectives

▼ To multiply a polynomial by a monomial (p. 500)

▼ To factor a monomial from a polynomial (p. 501)

You can multiply a monomial and a polynomial using the Distributive Property. You can factor a polynomial by finding the greatest common factor (GCF) of the terms of the polynomial.

Simplify each product. Write in standard form.

19. $8x(2 - 5x)$
20. $5g(3g + 7g^2 - 9)$
21. $8t^2(3t - 4 - 5t^2)$

22. $5m(3m + m^2)$
23. $-2w^2(4w - 10 + 3w^2)$
24. $b(10 + 5b - 3b^2)$

Find the GCF of the terms of each polynomial. Then factor the polynomial.

25. $9x^4 + 12x^3 + 6x$
26. $4t^5 - 12t^3 + 8t^2$
27. $40n^5 + 70n^4 - 30n^3$

28. $2k^4 + 4k^3 - 6k - 8$
29. $3d^2 - 6d$
30. $10m^4 - 12m^3 + 4m^2$

31. $10v - 5$
32. $12w^3 + 8w^2 + 20w$
33. $18d^5 + 6d^4 + 9d^3$

34. Critical Thinking The GCF of two numbers x and y is 3. Can you predict the GCF of $4x$ and $4y$? Explain your answer.

35. Critical Thinking Amanda says the GCF of $8m^2n$ and $4mn$ is 4. Kris says the GCF is $4n$. Kim says the GCF is $4mn$. Which student is correct? Explain your answer.

9-3 and 9-4 Objectives

▼ To multiply binomials using FOIL (p. 505)

▼ To multiply trinomials by binomials (p. 507)

▼ To find the square of a binomial (p. 512)

▼ To find the difference of squares (p. 514)

You can use tiles or the Distributive Property to multiply polynomials. You can use the FOIL method (First, Outer, Inner, Last) to multiply two binomials.

Simplify each product. Write in standard form.

36. $(x + 3)(x + 5)$
37. $(5v + 2)(3v - 7)$
38. $(2b + 5)(3b - 2)$

39. $(k - 1)(-k + 4)$
40. $(p + 2)(p^2 + p + 1)$
41. $(4a - 1)(a - 5)$

42. $(y - 4)(y^2 - 5y - 2)$
43. $(3x + 4)(x + 2)$
44. $(-2h^2 + h - 1)(h - 5)$

45. $(q - 4)(q - 4)$
46. $(2k^3 + 5)^2$
47. $(8 - 3t^2)(8 + 3t^2)$

48. $(2m^2 + 5)(2m^2 - 5)$
49. $(w - 4)(w + 4)$
50. $(4g^2 - 5h^4)(4g^2 + 5h^4)$

51. Geometry A rectangle has dimensions $2x + 1$ and $x + 4$. Write an expression for the area of the rectangle as a product and as a polynomial in standard form.

52. Error Analysis Suppose a classmate claims that the difference between $(x^2 - y^2)$ and $(x - y)^2$ must be 0. Is your classmate correct? Explain your answer.

9-5 and 9-6 Objectives

▼ To factor trinomials (p. 519)

▼ To factor trinomials of the type $ax^2 + bx + c$ (p. 524)

Some quadratic trinomials are the product of two binomial factors. You can factor trinomials using tiles or by using FOIL. Factor any common monomial factors first.

Factor each expression.

53. $x^2 + 3x + 2$
54. $y^2 - 9y + 14$
55. $x^2 - 2x - 15$

56. $2w^2 - w - 3$
57. $b^2 - 7b + 12$
58. $2t^2 + 3t - 2$

59. $x^2 + 5x - 6$
60. $6x^2 + 10x + 4$
61. $21x^2 - 22x - 8$

62. $3x^2 + x - 2$
63. $15y^2 + 16y + 1$
64. $15y^2 - 16y + 1$

When you factor a **perfect-square trinomial,** the two binomial factors are the same.

$$a^2 + 2ab + b^2 = (a + b)(a + b) = (a + b)^2 \text{ and}$$
$$a^2 - 2ab + b^2 = (a - b)(a - b) = (a - b)^2$$

When you factor the difference of squares of two terms, the two binomial factors are the sum and the difference of the two terms.

$$a^2 - b^2 = (a + b)(a - b)$$

Factor each expression.

65. $q^2 + 2q + 1$ **66.** $b^2 - 16$ **67.** $x^2 - 4x + 4$

68. $4t^2 - 121$ **69.** $4d^2 - 20d + 25$ **70.** $9c^2 + 6c + 1$

71. $9k^2 - 25$ **72.** $x^2 + 6x + 9$ **73.** $24y^2 - 6$

74. Geometry Find an expression for the length of a side of the square with an area of $\frac{1}{4}d^2 + d + 1$.

75. Critical Thinking Suppose you are using tiles to factor a quadratic trinomial. What do you know about the factors of the trinomial if the tiles form a square?

76. The area of a rectangle is $25u^2 + 65u + 36$. If the dimensions of the rectangle are factors of $25u^2 + 65u + 36$, could the rectangle be a square? Explain.

To factor a polynomial, first see if you can factor out the GCF. If the polynomial has four or more terms, you can group the terms and look for a common binomial factor. Then you can use the Distributive Property to factor the polynomial. If you do not quickly recognize the binomial factors of a polynomial of the form $ax^2 + bx + c$, grouping the terms may help you factor the polynomial.

Find the GCF of the first two terms and the GCF of the last two terms for each polynomial.

77. $16x^3 + 12x^2 - 8x - 6$ **78.** $9k^3 + 15k^2 - 6k - 10$

79. $72y^3 + 24y^2 - 12y - 4$ **80.** $20n^4 - 10n^3 + 14n - 7$

Factor completely.

81. $6x^3 + 3x^2 + 8x + 4$ **82.** $20y^4 - 45y^2$

83. $9g^2 + 15g - 6$ **84.** $6c^2 - 5cd + d^2$

85. $11k^2 + 23k + 2$ **86.** $3u^2 + 21u + 18$

87. $15p^2 + 14p + 3$ **88.** $3u^2 - 21u + 18$

89. $15h^3 + 11h^2 - 45h - 33$ **90.** $30x^3 + 42x^2 - 5x - 7$

91. $12s^4t + 20s^3t - 8s^2t$ **92.** $2x^3 + 7x^2 + 4x + 14$

93. Geometry The volume of the rectangular prism is $6p^3 + 38p^2 + 40p$. Find expressions for the possible dimensions of the prism.

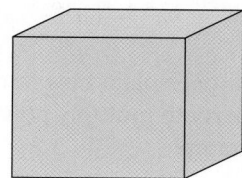

Chapter Test

Go Online
PHSchool.com
For: Chapter Test
Web Code: ata-0952

Write each polynomial in standard form. Then name each polynomial based on its degree and number of terms.

1. $y^2 + 2y + 5 - 3y^2 - 5y$

2. $4 - 5v - 12v - 6v^2 - 4 - 2v^3$

3. $-4x^4 + x^2 - 10 + 12x^4 - 7x^2$

4. $3k^5 + 4k^2 - 6k^5 - 5k^2$

Simplify. Write each answer in standard form.

5. $(4x^2 + 2x + 5) + (7x^2 - 5x + 2)$

6. $(9a^2 - 4 - 5a) - (12a - 6a^2 + 3)$

7. $(-4m^2 + m - 10) + (3m + 12 - 7m^2)$

8. $(3c - 4c^2 + c^3) - (5c^2 + 8c^3 - 6c)$

9. **Open-Ended** Write a trinomial with a degree of 6.

10. Write the standard form for the polynomial modeled at the right. Then factor the expression.

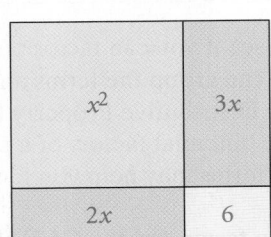

Simplify each product. Write in standard form.

11. $8b(3b + 7 - b^2)$

12. $-t(5t^2 + t)$

13. $3q(4 - q + 3q^3)$

14. $2c(c^5 + 4c^3)$

15. $(x + 6)(x + 1)$

16. $(d + 4)(d - 3)$

17. $(2h - 1)(h - 4)$

18. $(2m + 5)(3m - 7)$

19. $(p + 2)(2p^2 - 5p + 4)$

20. $(a - 4)(6a^2 + 10a - 3)$

21. $(3x + 5)(7x^2 - 2x + 1)$

Find the GCF of the terms of each polynomial.

22. $21x^4 + 18x^2 + 36x^3$

23. $3t^2 - 5t - 2t^4$

24. $-3a^{10} + 9a^5 + 6a^{15}$

25. $9m^3 - 7m^4 + 8m^2$

26. **Writing** Explain how to use the Distributive Property to multiply polynomials. Include an example.

Write an expression for each situation as a product. Then write each expression in standard form.

27. A plot of land has width x meters. The length of the plot of land is 5 meters more than 3 times its width. What is the area of the land?

28. The height of a box is 2 in. less than its width w. The length of the box is 3 in. more than 4 times its width. What is the volume of the box in terms of w?

Geometry Write an expression for the area of each shaded region. Write your answer in simplest form.

29.

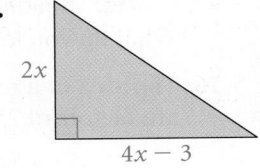

30.

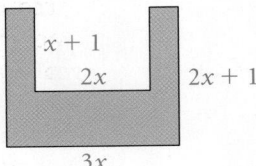

Factor each expression.

31. $w^2 - 5w - 14$

32. $g^2 + 10g + 25$

33. $9k^2 + 24k + 16$

34. $n^2 - 100$

35. $y^2 - 4y + 4$

36. $4x^2 - 49$

37. $4p^2 + 164p + 81$

38. $13c^2 - 52$

Write the missing value in each perfect-square trinomial.

39. $x^2 + \blacksquare x + 49$

40. $\blacksquare t^2 + 12t + 9$

41. $9x^2 - 30x + \blacksquare$

42. $4w^2 - \blacksquare w + 81$

Find the GCF of the first two terms and the GCF of the last two terms for each polynomial.

43. $6x^4 + 9x^3 - 8x + 12$

44. $16n^3 + 20n^2 - 4n - 5$

Factor completely.

45. $12n^3 + 15n^2 + 4n + 5$

46. $4x^2 - 10x + 6$

47. $x^3 - 5x^2 + 5x - 25$

48. $6r^3 - 9r^2 - 4r + 6$

49. $12y^3 + 28y^2 - 3y - 7$

50. $3n^3 - 4n^2 - 6n + 8$

51. **Open-Ended** Find three different values to complete the expression $x^2 + \blacksquare x + 30$ so that it can be factored into the product of two binomials. Show each factorization.

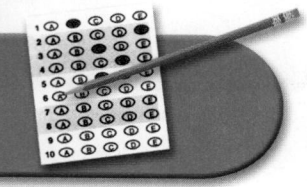
Regents Test Prep

Reading Comprehension Read the passage below, and then answer the questions on the basis of what is *stated* or *implied* in the passage.

> **Saving for College** For years, college costs have risen steadily. Although general inflation in our economy averages 3.0% to 3.5% annually, the rate of increase in college costs is about 5% a year. While many prospective students expect to benefit from financial aid, they also must prepare to pay some portion of the costs themselves.
>
> To help with these preparations, Congress has authorized special college-saver plans, called "529 Plans" because they are described in section 529 of the Internal Revenue Code. The plans allow the gains (interest from savings accounts or dividends from stocks) on college savings to grow without incurring federal income tax when the money is withdrawn for college expenses.
>
> If college savings are invested in stocks, you could expect an average increase of about 10.5% each year. This has been the historical rate of growth for stocks in the United States. Of course, it is impossible to predict what the growth will be in any particular year. But if you start saving for college early, with 5 or more years for your money to grow, you can expect good returns.

1. Suppose that total freshman year costs are $20,000. About what will the total costs be for senior year, assuming average increases?
   - **(1)** $20,000.00
   - **(2)** $23,000.00
   - **(3)** $23,152.50
   - **(4)** $24,310.13

2. Which expression represents annual college costs over time? Let t = the time in years and c = the starting cost in dollars.
   - **(1)** $c \cdot 0.05^t$
   - **(2)** $c \cdot 1.05^t$
   - **(3)** $t \cdot 1.05^c$
   - **(4)** $c \cdot 5$

3. Which equation represents the average value of stock investments over time? Let t = the time in years, x = the amount of the initial investment in dollars, and V = the value of the investment in dollars.
   - **(1)** $V = x \cdot 0.05^t$
   - **(2)** $V = x \cdot 1.105^t$
   - **(3)** $V = x \cdot 10.5^t$
   - **(4)** $V = x \cdot 210.5\%$

4. Suppose you have a scholarship that will pay 75% of your college costs each year. Total freshman year costs are $20,000. To prepare for your senior year costs, you invest $4,870 in bonds paying 4% a year at the beginning of your freshman year.

   Which answer gives the best estimates of the investment value and of your college costs at the beginning of your senior year?
   - **(1)** investment $5478, costs $5788
   - **(2)** investment $6571, costs $6500
   - **(3)** investment $5800, costs $23,000
   - **(4)** investment $5454, costs $5788

5. a. Suppose your grandparents put $1000 for your college costs in a savings account earning 6% interest compounded annually. In 10 years, how much money would be in the savings account?

   b. Suppose your grandparents did not know about the 529 plan. If you had to pay a 23% tax on the interest the account earned, how much would you have to pay?

Data Analysis

Activity Lab

Interpreting Statistical Results

NY A.S.2:
Determine whether the data to be analyzed is univariate or bivariate.

Univariate Data Statisticians often collect univariate, or one-variable, data. You see univariate data displayed in bar graphs, circle graphs, stem-and-leaf plots, histograms, and box-and-whisker plots.

Sample size affects how reliable data is. For example; suppose a survey of 3 students shows that 66.6% of ninth graders liked a certain movie. If one person changes his or her mind, the rating becomes 33.3% or 100%! This data is not reliable.

1 ACTIVITY

You will need a phone book for this activity.

1. **Open-Ended** Design an unbiased sampling method to select last names from the phone book. Describe your method.

2. **Data Collection** Use your method to take a sample of four last names. Find the average length of the four names.

3. Copy the table below. Enter your average under Trial 1 in the second row. Take 11 more samples, recording each average in the table.

Sample Size = 4	Trial Number	1	2	3	4	5	6	7	8	9	10	11	12
	Average Length	■	■	■	■	■	■	■	■	■	■	■	■
Sample Size = 10	Trial Number	1	2	3	4	5	6	7	8	9	10	11	12
	Average Length	■	■	■	■	■	■	■	■	■	■	■	■

4. Make a histogram using the axes like those shown at the right. (Graph how many samples had an average name length between 3 and 4, not including 4, between 4 and 5, not including 5, and so on.)

5. Repeat steps 2, 3, and 4 using a sample size of ten. Record your results in the fourth row of the table.

6. **Writing** Describe how increasing the sample size affected the results.

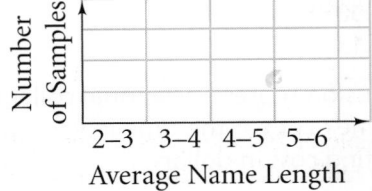

Statisticians sometimes use margin of error to help describe their level of confidence in statistical results. Margin of error depends on sample size n.

$$\text{margin of error} = \pm \frac{1}{\sqrt{n}}$$

If a sample of 120 residents shows that 68% favor building a new park, then the margin of error is $\pm \frac{1}{\sqrt{120}} \approx 0.09128$ or about $\pm 9\%$.

$$68 - 9 = 59 \quad \text{and} \quad 68 + 9 = 77$$

So, 59% to 77% of the entire population are probably in favor of the park.

Bivariate Data In Chapter 6, you used scatter plots to represent bivariate, or two-variable, data. Recall that predictions within the range of the data (interpolations) are more reliable than predictions beyond the data (extrapolations.)

2 ACTIVITY

The table shows the lengths and weights of several brook trout.

7. Use a graphing calculator to graph the data, find a line of best fit, and find the correlation coefficient.

8. a. Predict the weight of a 15-inch trout.
  b. Predict the weight of an 8-inch trout.
  c. Which prediction is more reliable? Explain.

9. Explain what the slope of the line of best fit represents.

Brook Trout			
Length	**Weight**	**Length**	**Weight**
10 in.	6 oz	22 in.	71 oz
13 in.	15 oz	17 in.	33 oz
16 in.	30 oz	20 in.	53 oz

The correlation coefficient r provides a measure of how reasonable it is to use the trend line to model the data.

3 ACTIVITY

The table shows how much oil and hydroelectric power (HEP) the top ten energy-consuming countries used in 2001. The unit for both oil and hydroelectric power is millions of tons of oil.

10. a. Use a graphing calculator to graph the data. Then find the correlation coefficient and the line of best fit.
  b. **Critical Thinking** Compare the r-value for this data with the r-value you found in Activity 2. What can you conclude about the usefulness of this line of best fit?

11. a. Is there an outlier? If so, which county is it?
  b. If you remove the outlier, does the data become more correlated or less correlated?

EXERCISES

12. A sample of 250 shoppers shows that 72% favor building a new mall. Interpret this survey result using the margin of error.

13. Data Collection You will need five circular objects and a tape measure.
  a. Measure the diameter and circumference of each circular object.
  b. Make a scatter plot of your data.
  c. Draw a trend line. What does the slope of the trend line represent?

14. a. Choose a yes-or-no question that you can ask your classmates.
  b. **Data Collection** Conduct the survey. Take ten samples of five classmates each.
  c. Describe your results. Do they represent the entire school population well? Explain.
  d. How could you make the results more reliable?

Country	Oil	HEP
United States	987	53
China	256	64
Russia	135	44
Japan	273	23
Germany	145	6
India	78	18
Canada	97	83
France	106	20
United Kingdom	84	2
South Korea	114	1

SOURCE: *BP Amoco Statistical Review of World Energy 2002*

What You've Learned

Learning Standards for Mathematics

In Chapter 5, you

- **NY A.G.3:** studied functions and their graphs.

In Chapter 8, you

- **NY A.G.4:** discovered that not all functions are linear as you explored exponential functions and fit data to them.

In Chapter 9, you

- **NY A.A.20:** factored trinomials of the form $ax^2 + bx + c$.

 **Check Your Readiness**   **GO for Help** to the Lesson in green.

Evaluating Expressions (Lesson 2-3)

Evaluate each expression for $a = -1, b = 3$, and $c = -2$.

1. $2a - b^2 + c$ **2.** $\frac{c^2 - ab}{2a}$ **3.** $bc - 3a^2$

4. $\frac{b^2 - 4ac}{2a}$ **5.** $5a + 2b(c - 1)$ **6.** $c^2 + 2ab - 1$

Evaluating Function Rules (Lesson 5-2)

Evaluate each function rule for $x = -6$.

7. $f(x) = -3x^2$ **8.** $y = x^2 - 10$ **9.** $h(x) = x^2 + 6x$ **10.** $y = (x - 1)^2$

11. $y = 5 - 2x^2$ **12.** $y = (1 + x)^2$ **13.** $g(x) = \frac{2}{3}x^2$ **14.** $y = (2x)^2$

Graphing Functions (Lesson 5-3)

Graph each function.

15. $y = x$ **16.** $y = -x^2$ **17.** $y = |x|$

Multiplying Binomials (Lesson 9-3)

Simplify each product using FOIL.

18. $(x + 2)(x - 3)$ **19.** $(2y + 1)(2y + 3)$ **20.** $(3x - 7)(x + 4)$

Factoring (Lessons 9-5 and 9-6)

Factor each expression.

21. $4x^2 + 4x + 1$ **22.** $5x^2 + 32x - 21$ **23.** $8x^2 - 10x + 3$

24. $m^2 - 7m - 18$ **25.** $12y^2 + 8y - 15$ **26.** $x^2 - 18x + 81$

Quadratic Equations and Functions

🔊 **Key Vocabulary**

- axis of symmetry (p. 551)
- completing the square (p. 579)
- discriminant (p. 592)
- maximum (p. 551)
- minimum (p. 551)
- parabola (p. 550)
- quadratic equation (p. 566)
- quadratic formula (p. 585)
- quadratic function (p. 550)
- quadratic parent function (p. 550)
- vertex (p. 551)
- Zero-Product Property (p. 572)

What You'll Learn Next

NY Learning Standards for Mathematics

- **NY A.G.4:** You will examine quadratic graphs and their equations.

- **NY A.A.8:** You will solve quadratic equations by various techniques such as factoring, finding square roots, completing the square, and applying the quadratic formula.

- **NY A.G.4:** You will determine an appropriate linear, quadratic, or exponential model for real-world data.

Exploring Quadratic Graphs

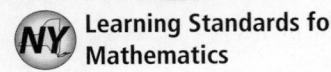

A.G.4 Identify and graph linear, quadratic (parabolic), absolute value, and exponential functions.

A.G.10 Determine the vertex and axis of symmetry of a parabola, given its graph.

✓ **Check Skills You'll Need**   **GO** for Help Lessons 1-2 and 5-3

Evaluate each expression for $h = 3, k = 2,$ and $j = -4.$
1. hkj **2.** kh^2 **3.** hk^2 **4.** $kj^2 + h$

Graph each equation.
5. $y = 2x - 1$ **6.** $y = |x| + 1$ **7.** $y = x^2 + 2$

◀)) **New Vocabulary** • **quadratic function** • **standard form of a quadratic function** • **quadratic parent function** • **parabola** • **axis of symmetry** • **vertex** • **minimum** • **maximum**

---

1 Graphing $y = ax^2$

Activity: Plotting Quadratic Curves

1. Graph the equations $y = x^2$ and $y = 3x^2$ on the same coordinate plane.

2. **a.** Describe how the graphs are alike.
   b. Describe how the graphs are different.

3. Predict how the graph of $y = \frac{1}{3}x^2$ will be similar to and different from the graph of $y = x^2$.

4. Graph $y = \frac{1}{3}x^2$. Were your predictions correct? Explain.

The functions shown above are quadratic functions.

 Key Concepts

Definition	**Standard Form of a Quadratic Function**

A **quadratic function** is a function that can be written in the form $y = ax^2 + bx + c$, where $a \neq 0$. This form is called the **standard form of a quadratic function.** **Examples** $y = 5x^2$ $y = x^2 + 7$ $y = x^2 - x - 3$

The simplest quadratic function, $f(x) = x^2$, or $y = x^2$, is the **quadratic parent function.**

The graph of a quadratic function is a U-shaped curve called a **parabola.** The graph of $y = x^2$, shown at the right, is a parabola.

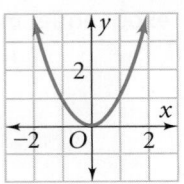

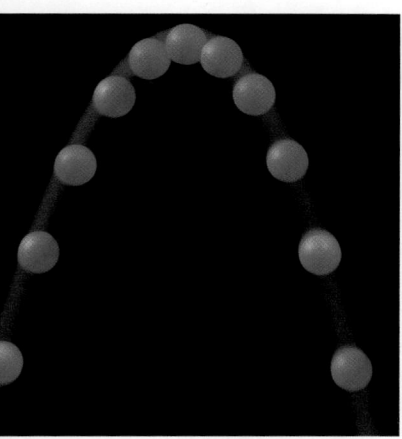

You can fold a parabola so that the two sides match exactly. This property is called *symmetry*. The fold or line that divides the parabola into two matching halves is called the **axis of symmetry.**

The highest or lowest point of a parabola is its **vertex,** which is on the axis of symmetry.

If $a > 0$ in $y = ax^2 + bx + c$
↓
the parabola opens upward.
↓
The vertex is the **minimum** point or lowest point of the parabola.

If $a < 0$ in $y = ax^2 + bx + c$
↓
the parabola opens downward.
↓
The vertex is the **maximum** point or highest point of the parabola.

1 EXAMPLE Identifying a Vertex

Identify the vertex of each graph. Tell whether it is a minimum or maximum.

a.
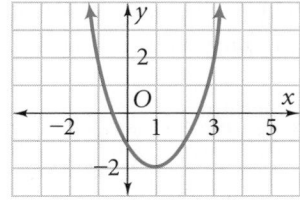

b.
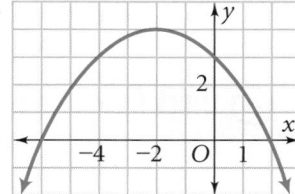

The vertex is $(1, -2)$. It is a minimum.   The vertex is $(-2, 4)$. It is a maximum.

 **Quick Check** ➊ Identify the vertex of each graph. Tell whether it is a minimum or maximum.

a.
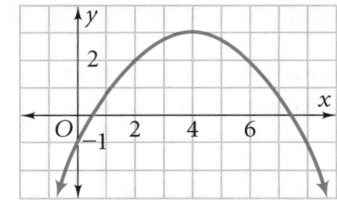

b.
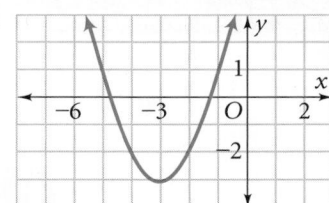

You can use the fact that a parabola is symmetric to graph it quickly. First find the coordinates of the vertex and several points on either side of the vertex. Then reflect the points across the axis of symmetry. For functions of the form $y = ax^2$, the vertex is at the origin.

2 EXAMPLE Graphing $y = ax^2$

Make a table of values and graph the quadratic function $y = \frac{1}{2}x^2$.

x	$y = \frac{1}{2}x^2$	(x, y)
0	$\frac{1}{2}(0)^2 = 0$	$(0, 0)$
2	$\frac{1}{2}(2)^2 = 2$	$(2, 2)$
4	$\frac{1}{2}(4)^2 = 8$	$(4, 8)$

Find the corresponding points on the other side of the axis of symmetry.

 **Quick Check** ➋ Make a table of values and graph the quadratic function $f(x) = -2x^2$.

The value of a, the coefficient of the x^2 term in a quadratic function, affects the width of a parabola as well as the direction in which it opens.

3 **EXAMPLE** **Comparing Widths of Parabolas**

Use the graphs below. Order the quadratic functions $f(x) = -4x^2$, $f(x) = \frac{1}{4}x^2$, and $f(x) = x^2$ from widest to narrowest graph.

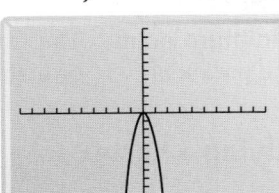

$$y = -4x^2$$

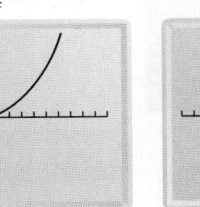

$$y = \frac{1}{4}x^2$$

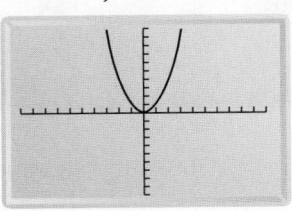

$$y = x^2$$

Of the three graphs, $f(x) = \frac{1}{4}x^2$ is the widest and $f(x) = -4x^2$ is the narrowest. So, the order from widest to narrowest is $f(x) = \frac{1}{4}x^2$, $f(x) = x^2$, and $f(x) = -4x^2$.

✓ **Quick Check** **3** Order the quadratic functions $y = x^2$, $y = \frac{1}{2}x^2$, and $y = -2x^2$ from widest to narrowest graph.

When $|m| < |n|$, the graph of $y = mx^2$ is wider than the graph of $y = nx^2$.

2 **Graphing $y = ax^2 + c$**

The y-axis is the axis of symmetry for functions in the form $y = ax^2 + c$. The value of c translates the graph up or down.

4 **EXAMPLE** **Graphing $y = ax^2 + c$**

Multiple Choice How is the graph of $y = 2x^2 + 3$ different from the graph of $y = 2x^2$?

Ⓐ It is shifted 3 units up.   Ⓑ It is shifted 3 units down.
Ⓒ It is shifted 3 units to the right.   Ⓓ It is shifted 3 units to the left.

x	$y = 2x^2$	$y = 2x^2 + 3$
−2	8	11
−1	2	5
0	0	3
−1	2	5
−2	8	11

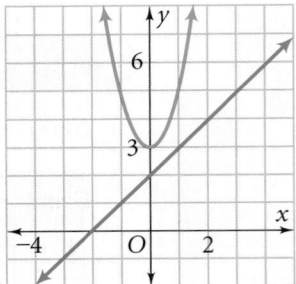

The graph of $y = 2x^2 + 3$ has the same shape as the graph of $y = 2x^2$, but it is shifted up 3 units. So A is the correct answer.

✓ **Quick Check** **4** **a.** Graph $y = x^2$ and $y = x^2 - 4$. Compare the graphs.
b. Critical Thinking Describe what positive and negative values of c do to the position of the vertex.

You can model the height of an object moving under the influence of gravity using a quadratic function. As an object falls, its speed continues to increase. Ignoring air resistance, you can find the approximate height of a falling object using the function $h = -16t^2 + c$. The height h is in feet, the time t is in seconds, and the initial height of the object c is in feet.

5 EXAMPLE **Real-World**  **Problem Solving**

Nature Suppose you see an eagle flying over a canyon. The eagle is 30 ft above the level of the canyon's edge when it drops a stick from its claws. The force of gravity causes the stick to fall toward Earth. The function $h = -16t^2 + 30$ gives the height of the stick h in feet after t seconds. Graph this quadratic function.

t	$h = -16t^2 + 30$
0	30
1	14
2	−34

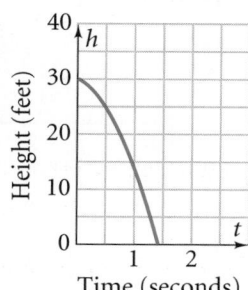

Height h is dependent on time t. Graph t on the x-axis and h on the y-axis. Use nonnegative values for t.

✓ Quick Check **5** **a.** Suppose a squirrel is in a tree 24 ft above the ground. She drops an acorn. The function $h = -16t^2 + 24$ gives the height of the acorn in feet after t seconds. Graph this function.

b. **Critical Thinking** Describe a reasonable domain and range for the function in Example 5.

EXERCISES

For more exercises, see *Extra Skill and Word Problem Practice*.

Practice and Problem Solving

A Practice by Example

 for Help

Example 1 (page 551)

Identify the vertex of each graph. Tell whether it is a minimum or maximum.

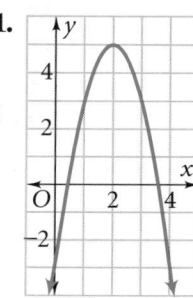

1.

2.

3.

Example 2 (page 551)

Graph each function.

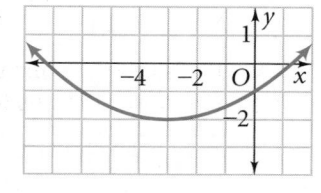

4. $y = -4x^2$ **5.** $f(x) = 1.5x^2$ **6.** $y = \frac{2}{3}x^2$

7. $f(x) = -\frac{1}{2}x^2$ **8.** $y = -\frac{1}{3}x^2$ **9.** $f(x) = 3x^2$

Example 3 (page 552)

Order each group of quadratic functions from widest to narrowest graph.

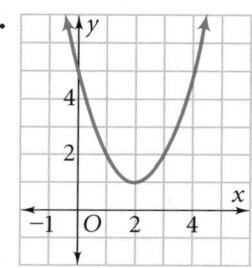

10. $y = 3x^2, y = \frac{1}{2}x^2, y = 4x^2$ **11.** $f(x) = 5x^2, f(x) = \frac{1}{3}x^2, f(x) = x^2$

12. $y = -\frac{1}{2}x^2, y = 5x^2, y = -\frac{1}{4}x^2$ **13.** $f(x) = -2x^2, f(x) = -\frac{2}{3}x^2, f(x) = -4x^2$

Example 4
(page 552)

Graph each function.

14. $f(x) = x^2 + 2$ **15.** $y = x^2 - 3$ **16.** $y = \frac{1}{2}x^2 + 4$

17. $f(x) = -x^2 - 1$ **18.** $y = -2x^2 + 2$ **19.** $f(x) = 4x^2 - 7$

Example 5
(page 553)

20. A gull drops a clam shell onto some rocks from a height of 50 ft. The function $h = -16t^2 + 50$ gives the shell's approximate height h in feet after t seconds. Graph the function.

B   **Apply Your Skills**

Match each graph with its function.

A. $f(x) = x^2 - 1$ **B.** $f(x) = x^2 + 4$ **C.** $f(x) = -x^2 + 2$

D. $f(x) = 3x^2 - 5$ **E.** $f(x) = -3x^2 + 8$ **F.** $f(x) = -0.2x^2 + 5$

21. 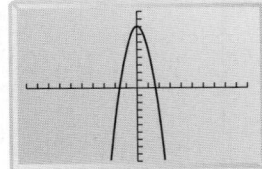   **22.** 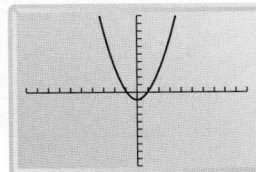   **23.**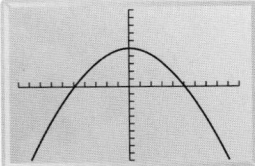

24. 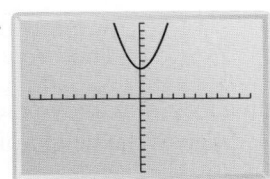   **25.** 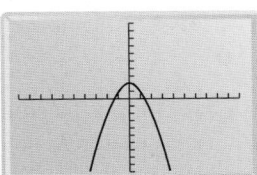   **26.**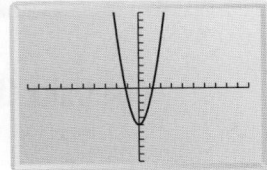

Writing   **Without graphing, describe how each graph differs from the graph of $y = x^2$.**

27. $y = 2x^2$ **28.** $y = -x^2$ **29.** $y = 1.5x^2$ **30.** $y = \frac{1}{2}x^2$

Graph each function.

31. $y = -\frac{1}{4}x^2 + 3$ **32.** $f(x) = -1.5x^2 + 5$ **33.** $y = 3x^2 - 6$

Trace each parabola on a sheet of paper and draw its axis of symmetry.

34. 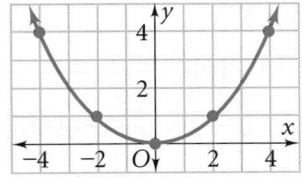   **35.**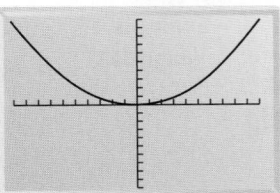

36.    **37.**

38. A bungee jumper dives from a platform. The function $h = -16t^2 + 200$ gives her approximate height h in feet after t seconds.
  a. Graph the function. Graph t on the x-axis and h on the y-axis.
  b. What will the jumper's height be after 1 second?
  c. What will the jumper's height be after 3 seconds?

39. Geometry Suppose that a pizza must fit into a box with a base that is 12 in. long and 12 in. wide. You can use the quadratic function $A = \pi r^2$ to find the area of a pizza in terms of its radius.
  a. What values of r make sense for the function?
  b. What values of A make sense for the function?
  c. Graph the function. Round values of A to the nearest tenth.

Three graphs are shown below. For Exercises 40–43, identify the graph(s) that fit each description.

40. $a > 0$

41. $a < 0$

42. $|a|$ has the greatest value.

43. $|a|$ has the least value.

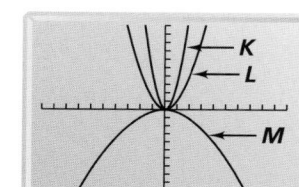

44. Gravity Suppose a person is riding in a hot-air balloon, 144 feet above the ground. He drops an apple. The height of the apple above the ground is given by the formula $h = -16t^2 + 144$, where h is height in feet and t is time in seconds.
  a. Graph the function.
  b. How far has the apple fallen from time $t = 0$ to $t = 1$?
  c. Critical Thinking Does the apple fall as far from time $t = 1$ to $t = 2$ as it does from time $t = 0$ to $t = 1$? Explain.

45. Multiple Choice Which function has a graph that is the same as the graph of $f(x) = x^2 + 1$ shifted 4 units down?
  Ⓐ $f(x) = x^2 + 5$ Ⓑ $f(x) = x^2 - 3$
  Ⓒ $f(x) = x^2 - 4$ Ⓓ $f(x) = x^2 + 4$

 **Challenge**

46. Critical Thinking Complete each statement. Assume $a \neq 0$.
  a. The graph of $y = ax^2 + c$ intersects the x-axis in two places when __?__.
  b. The graph of $y = ax^2 + c$ does not intersect the x-axis when __?__.

47. Landscaping The plan for a 20 ft-by-12 ft patio has a square garden in the middle of it. If each side of the garden is x ft, the function $y = 240 - x^2$ gives the area of the patio without the garden in square feet.
  a. Graph the function.
  b. What values make sense for the domain? Explain.
  c. What is the range of the function? Explain.
  d. Use the graph to estimate the side length of the garden if the area of the patio is 200 ft^2.

48. Consider the graphs of $y = ax^2$ and $y = (ax)^2$. Assume $a \neq 0$.
  a. For what values of a will both graphs lie in the same quadrants?
  b. For what values of a will the graph of $y = ax^2$ be wider than the graph of $y = (ax)^2$?

49. Architecture An architect wants to design an archway with the following requirements.
- The archway is 6 ft wide and has vertical sides 7 ft high.
- The top of the archway is modeled by the function $y = -\frac{1}{3}x^2 + 10$.

a. Sketch the archway by drawing vertical lines 7 units high at $x = -3$ and $x = 3$ and graphing the portion of the quadratic function that lies between $x = -3$ and $x = 3$.

b. The plan for the archway is then changed so that the top is modeled by the function $y = -0.5x^2 + 11.5$. Make a revised sketch of the archway.

NY REGENTS

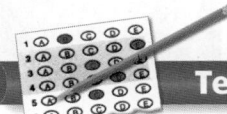 **Test Prep**

Multiple Choice

50. Which of the following equations has a graph that is narrower than the graph of $y = 2x^2 + 3$?
A. $y = 2x^2 - 3$ B. $y = -3x^2 + 2$
C. $y = 0.5x^2 + 3$ D. $y = -0.5x^2 - 2$

51. Which of the following equations has a graph that crosses the y-axis at a point lower than the graph of $y = -2x^2 - 1$?
F. $y = -3x^2 - 1$ G. $y = 3x^2 - 3$
H. $y = -3x^2 + 1$ J. $y = -3x^2 + 3$

52. The graph of $y = 4x^2 + 3$ does NOT pass through which of the following points?
A. $(0, 3)$ B. $(1, 7)$ C. $(-1, 7)$ D. $(3, 27)$

Extended Response

53. A construction worker drops a tool from the top of a building that is 200 ft high. The height of the tool above the ground can be modeled by $h = -16t^2 + 200$, where h is height in feet and t is time in seconds.
a. Make a table and graph this function.
b. Use your graph to estimate the amount of time it takes for the tool to hit the ground. Round to the nearest tenth of a second.

Mixed Review

 **for Help**

Lesson 9-8

Factor each expression.

54. $x^3 - 4x^2 + 2x - 8$ **55.** $15a^3 - 18a^2 - 10a + 12$

56. $7b^3 + 14b^2 + b + 2$ **57.** $y^3 + 3y^2 - 4y - 12$

58. $2n^3 - 2n^2 - 24n$ **59.** $30m^3 + 51m^2 + 9m$

Lesson 9-2

Simplify each expression.

60. $5x(3x - 4)$ **61.** $(n - 7)9n$ **62.** $-2t^2(6t - 11)$

63. $4m^2(3m^4 - m^3 + 5)$ **64.** $-5y(3y^5 + 2y^3 - 4)$ **65.** $3c^3(-4c^2 + 7c - 8)$

Lesson 7-4

66. Business The City Council invites your art club to sell helium balloons during a citywide celebration. The rental of the helium tank is $27.00 for the day. Each balloon costs $.20. If the balloons sell for $2.00 each, how many will your art club have to sell to break even?

Quadratic Functions

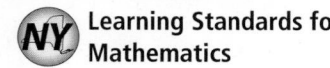

Learning Standards for Mathematics

A.A.8 Analyze and solve verbal problems that involve quadratic equations.

A.G.4 Identify and graph linear, quadratic (parabolic), absolute value, and exponential functions.

✓ **Check Skills You'll Need**

GO **for Help** Lessons 2-3 and 10-1

Evaluate the expression $\frac{-b}{2a}$ for the following values of a and b.

1. $a = -6, b = 4$
2. $a = 15, b = 20$
3. $a = -8, b = -56$
4. $a = -9, b = 108$

Graph each function.

5. $y = x^2$
6. $y = -x^2 + 2$
7. $y = \frac{1}{2}x^2 - 1$

---

1 ⟩ Graphing $y = ax^2 + bx + c$

In Lesson 10-1, you investigated the graphs of $y = ax^2$ and $y = ax^2 + c$. In the quadratic function $y = ax^2 + bx + c$, the value of b affects the position of the axis of symmetry.

Consider the graphs of the following functions.

$y = 2x^2 + 2x$
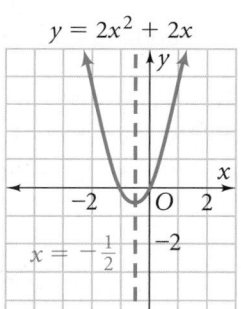

$y = 2x^2 + 4x$
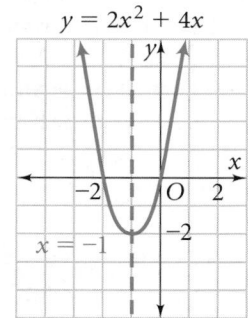

$y = 2x^2 + 6x$
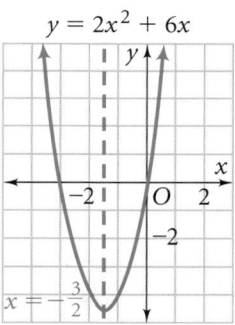

Notice that all three graphs have the same y-intercept. This is because in all three equations $c = 0$. The axis of symmetry changes with each change in the b value. The equation of the axis of symmetry is related to the ratio $\frac{b}{a}$.

equation:	$y = 2x^2 + 2x$	$y = 2x^2 + 4x$	$y = 2x^2 + 6x$
$\frac{b}{a}$:	$\frac{2}{2} = 1$	$\frac{4}{2} = 2$	$\frac{6}{2} = 3$
axis of symmetry:	$x = -\frac{1}{2}$	$x = -1, \text{ or } -\frac{2}{2}$	$x = -\frac{3}{2}$

The equation of the axis of symmetry is $x = -\frac{1}{2}\left(\frac{b}{a}\right)$, or $\frac{-b}{2a}$.

 Key Concepts

Property	**Graph of a Quadratic Function**

The graph of $y = ax^2 + bx + c$, where $a \neq 0$, has the line $x = \frac{-b}{2a}$ as its axis of symmetry. The x-coordinate of the vertex is $\frac{-b}{2a}$.

When you substitute $x = 0$ into the equation $y = ax^2 + bx + c$, $y = c$. So the y-intercept of a quadratic function is the value of c. You can use the axis of symmetry and the y-intercept to help you graph a quadratic function.

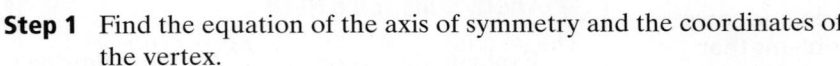
1 EXAMPLE Graphing $y = ax^2 + bx + c$

Graph the function $y = -3x^2 + 6x + 5$.

Step 1 Find the equation of the axis of symmetry and the coordinates of the vertex.

$x = \dfrac{-b}{2a} = \dfrac{-6}{2(-3)} = 1$ **Find the equation of the axis of symmetry.**

The axis of symmetry is $x = 1$.

$y = -3x^2 + 6x + 5$

$y = -3(1)^2 + 6(1) + 5$ **To find the y-coordinate of the vertex, substitute 1 for x.**

$\quad = 8$

The vertex is $(1, 8)$.

Step 2 Find two other points on the graph.

Use the y-intercept.

For $x = 0$, $y = 5$, so one point is $(0, 5)$.

Choose a value for x on the same side of the vertex as the y-intercept. Let $x = -1$.

$y = -3(-1)^2 + 6(-1) + 5$ **Find the y-coordinate for $x = -1$.**

$\quad = -4$

For $x = -1$, $y = -4$, so another point is $(-1, -4)$.

Step 3 Reflect $(0, 5)$ and $(-1, -4)$ across the axis of symmetry to get two more points. Then draw the parabola.

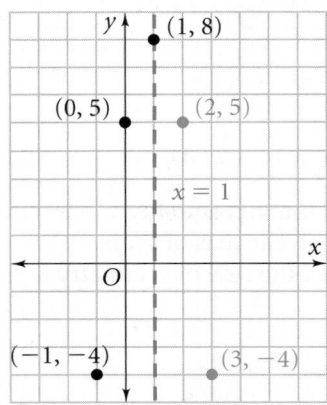
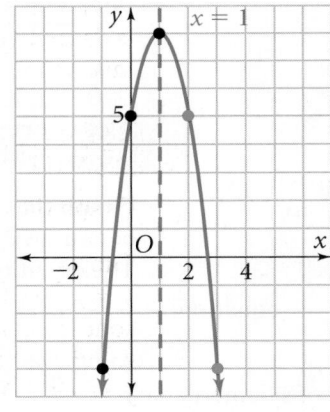

 **1** Graph $f(x) = x^2 - 6x + 9$. Label the axis of symmetry and the vertex.

You saw in the previous lesson that the formula $h = -16t^2 + c$ describes the height above the ground of an object falling from an initial height c, at time t. If an object is given an initial upward velocity v and continues with no additional force of its own, the formula $h = -16t^2 + vt + c$ describes its approximate height above the ground.

2 EXAMPLE Real-World  Problem Solving

Fireworks In professional fireworks displays, aerial fireworks carry "stars" upward, ignite them, and project them into the air.

Suppose a particular star is projected from an aerial firework at a starting height of 520 ft with an initial upward velocity of 72 ft/s. How long will it take for the star to reach its maximum height? How far above the ground will it be?

The equation $h = -16t^2 + 72t + 520$ gives the star's height h in feet at time t in seconds. Since the coefficient of t^2 is negative, the curve opens downward, and the vertex is the maximum point.

Step 1 Find the t-coordinate of the vertex.

$$\frac{-b}{2a} = \frac{-(72)}{2(-16)} = 2.25$$

After 2.25 seconds, the star will be at its greatest height.

Step 2 Find the h-coordinate of the vertex.

$h = -16(2.25)^2 + 72(2.25) + 520$ **Substitute 2.25 for t.**

$h = 601$ **Simplify using a calculator.**

The maximum height of the star will be 601 ft.

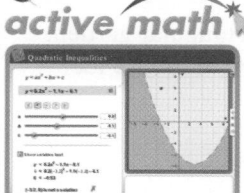 **Quick Check** **2** A ball is thrown into the air with an initial upward velocity of 48 ft/s. Its height h in feet after t seconds is given by the function $h = -16t^2 + 48t + 4$.
a. In how many seconds will the ball reach its maximum height?
b. What is the ball's maximum height?

---

2   **Graphing Quadratic Inequalities**

Graphing a quadratic inequality is similar to graphing a linear inequality. The curve is dashed if the inequality involves $<$ or $>$. The curve is solid if the inequality involves $\leq$ or $\geq$.

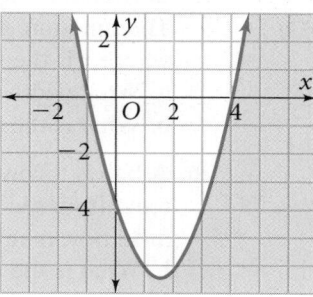

3 EXAMPLE Graphing Quadratic Inequalities

Graph the quadratic inequality $y \leq x^2 - 3x - 4$.

Graph the boundary curve, $y = x^2 - 3x - 4$. Use a solid line because the solution of the inequality $y \leq x^2 - 3x - 4$ includes the boundary. Shade below the curve.

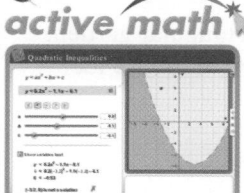 **Quick Check** **3** Graph each quadratic inequality.
a. $y \leq x^2 + 2x - 5$
b. $y > x^2 + x + 1$

A Practice by Example

Example 1
(page 558)

 GO for Help

Find the equation of the axis of symmetry and the coordinates of the vertex of the graph of each function.

1. $y = 2x^2 + 4$

2. $f(x) = 2x^2 + 4x - 5$

3. $y = x^2 - 8x - 9$

4. $y = 3x^2 - 9x + 5$

Match each graph with its function.

A. $y = x^2 - 6x$ **B.** $y = x^2 + 6x$ **C.** $y = -x^2 - 6x$

D. $y = -x^2 + 6x$ **E.** $y = -x^2 + 6$ **F.** $y = x^2 - 6$

5.
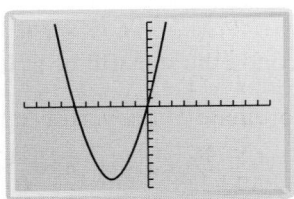

6.
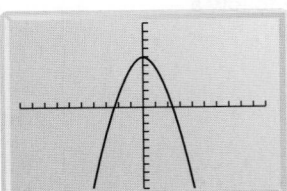

7.
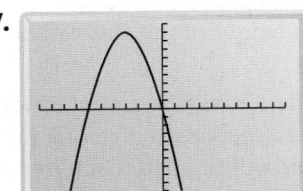

8.
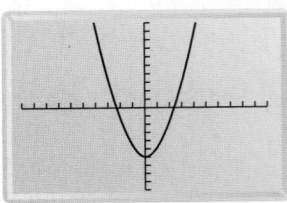

9.
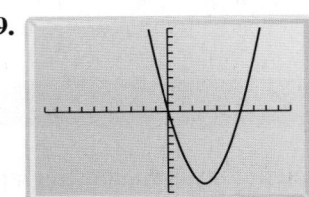

10.
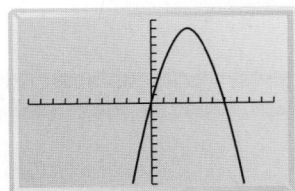

Graph each function. Label the axis of symmetry and the vertex.

11. $f(x) = x^2 + 4x + 3$

12. $y = 2x^2 - 6x$

13. $y = -x^2 + 4x - 4$

14. $y = 2x^2 + 3x + 1$

Example 2
(page 559)

15. Gardening Suppose you have 80 ft of fence to enclose a rectangular garden. The function $A = 40x - x^2$ gives you the area of the garden in square feet where x is the width in feet.
  a. What width gives you the maximum gardening area?
  b. What is the maximum area?

16. A ball is thrown into the air with an upward velocity of 40 ft/s. Its height h in feet after t seconds is given by the function $h = -16t^2 + 40t + 6$.
  a. In how many seconds does the ball reach its maximum height?
  b. What is the ball's maximum height?

Example 3
(page 559)

Graph each quadratic inequality.

17. $y > x^2$

18. $f(x) < -x^2$

19. $y \leq x^2 + 3$

20. $y < -x^2 + 4$

21. $y \geq -2x^2 + 6$

22. $f(x) > -x^2 + 4x - 4$

B **Apply Your Skills**

Graph each function. Label the axis of symmetry and the vertex.

23. $y = x^2 - 9x + 3$ **24.** $f(x) = -x^2 - 4x - 6$ **25.** $f(x) = x^2 - 2x + 1$

26. $y = 2x^2 + x - 3$ **27.** $y = x^2 + 3x + 2$ **28.** $y = -x^2 + 8x - 5$

29. $y = \frac{1}{2}x^2 + 2x + 1$ **30.** $y = \frac{1}{4}x^2 + 2x + 1$ **31.** $y = -\frac{1}{4}x^2 + 2x - 3$

Open-Ended For Exercises 32–34, give an example of a quadratic function for each description.

32. Its axis of symmetry is to the right of the y-axis.

33. Its graph opens downward and has its vertex at $(0, 0)$.

34. Its graph lies entirely above the x-axis.

35. Diving An athlete dives from the 3-meter springboard. Her height y, at horizontal distance x, can be approximated by the function $y = -1.2x^2 + 3.12x + 3$. Both the height and distance are in meters.
   a. How far has she traveled horizontally when she reaches her maximum height? Round to the nearest tenth of a meter.
   b. What is her maximum height? Round to the nearest tenth of a meter.

36. Road Construction An archway over a road is cut out of rock. Its shape is modeled by the quadratic function $y = -0.1x^2 + 12$ for $y \geq 0$.
   a. Write an inequality that describes the opening of the archway.
   b. Graph the inequality.
   c. **Critical Thinking** Can a camper 6 ft wide and 7 ft high fit under the arch without crossing the median line? Explain.

37. Multiple Choice A small company markets a new toy. The function $S = -64p^2 + 1600p$ predicts the total sales S in dollars as a function of the price p of the toy. What price will produce the highest total sales?
   (A) $64.00   (B) $25.00   (C) $12.50   (D) $8.00

Real-World **Connection**

After turning a somersault, the diver followed a parabolic path.

Estimation For each of the graphs below, estimate the area enclosed by the parabola, the x-axis, and the vertical lines $x = 1$ and $x = 7$. Follow the instructions below.

• Count the number of whole grid squares in the region.
• If half a square or more is included in the region, count it as one.
• If less than half a square is included in the region, do *not* count it.
• Add the counted squares to estimate the area.

38.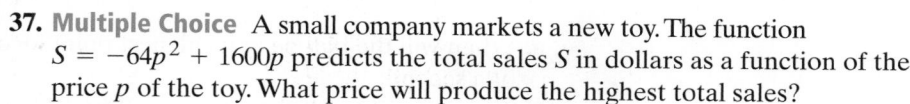

39.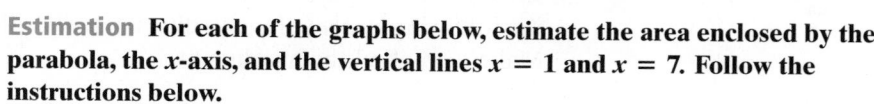

40. Writing Explain how changing the values of a, b, and c in a quadratic function affects the graph of the function.

41. Graphing Calculator Follow the steps below to find the vertex of the graph of $y = 2.3x^2 - 5.7x + 4.9$. Round to the nearest hundredth.

Step 1 Enter the function in the [Y=] screen. Graph the function.

Step 2 Use the **Calc** feature and select **minimum**.

Step 3 Move the cursor to the left side of the vertex. Press [ENTER].
Move the cursor to the right side of the vertex. Press [ENTER] twice.

42. Sports Suppose a volleyball player serves from 1 m behind the back line. If no other player touches the ball, it will land in bounds. The equation $h = -4.9t^2 + 3.82t + 1.7$ gives the ball's height h in meters in terms of time t in seconds.

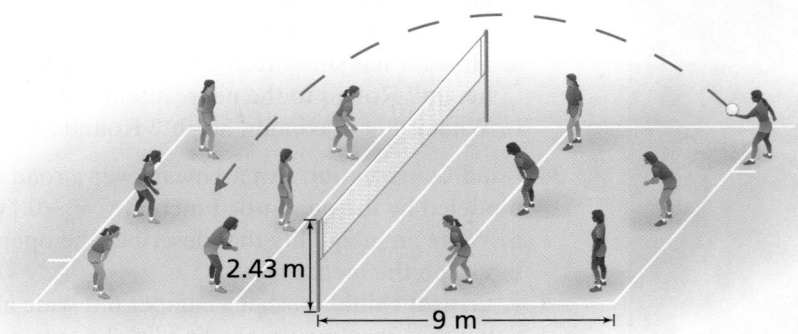

a. When will the ball be at its highest point? Round to the nearest tenth of a second.
b. The ball will reach the net at $t = 0.6$ s. Will it clear the net? Explain.

Challenge **43. Architecture** An architect designs a monument for a new park. The solution of the following system of inequalities describes the shape of the monument. Make a graph of the monument.

$$y \geq -x^2 + 6 \qquad y \leq -\tfrac{1}{2}x^2 + 8 \qquad y \geq 0$$

44. Sports Suppose a tennis player hits a ball over the net. The ball leaves his racket 0.5 m above the ground. The equation $h = -4.9t^2 + 3.8t + 0.5$ gives the ball's height h in meters at time t in seconds.
a. When will the ball be at the highest point in its path? Round to the nearest tenth of a second.
b. Critical Thinking If you double the answer from part (a), will you find the amount of time the ball is in the air before it hits the court? Explain.

Real-World **Connection**

A standard tennis court is 78 ft long and 36 ft wide.

45. The parabola shown at the right is of the form $y = x^2 + bx + c$.
a. Use the graph to find the y-intercept.
b. Find the equation of the axis of symmetry.
c. Use the vertex formula $x = \frac{-b}{2a}$ to find b.
d. Write the equation of the parabola.
e. Test one point using the equation from part (d).
f. Critical Thinking Would this method work if the value of a were not known? Explain.

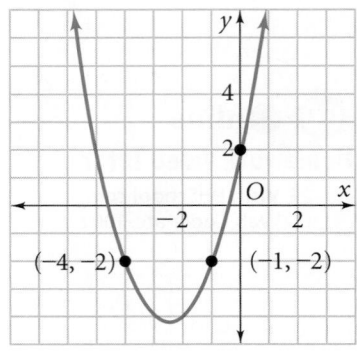

562 Chapter 10 Quadratic Equations and Functions

Multiple Choice

46. Which of the following are the coordinates of the vertex of $y = x^2 - 2x - 1$?

   A. $(1, -2)$ **B.** $(0, 3)$ **C.** $(1, -4)$ **D.** $(2, -3)$

47. Which parabola has a vertex at $(1.5, 2.25)$?

   F. $y = -x^2 - 2x$ **G.** $y = -x^2 - 3x$

   H. $y = -x^2 + 2x$ **J.** $y = -x^2 + 3x$

48. Which of the following is the graph of $y = 0.5x^2 - 2x + 1$?

  A.
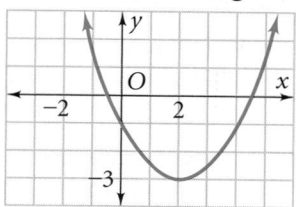

  B.
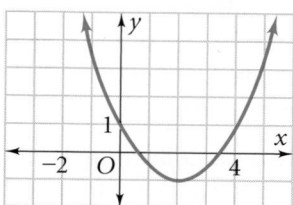

  C.
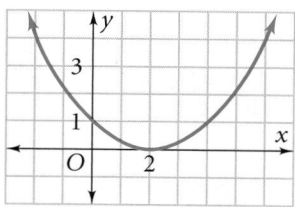

  D.
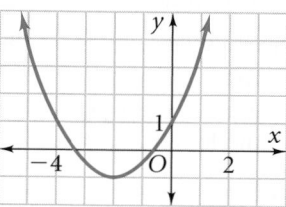

Short Response

49. An arrow is shot into the air. It follows a path given by the equation $y = -0.009x^2 + 0.3x + 4.5$, where x and y are in feet. Find its maximum height. Show your work.

Mixed Review

Lesson 10-1

Match each graph with its function.

  A. $y = \frac{1}{8}x^2 + 2$ **B.** $y = \frac{1}{2}x^2 + 2$ **C.** $y = -x^2 - 2$

  D. $y = x^2 + 2$ **E.** $y = -x^2 + 2$ **F.** $y = -\frac{1}{2}x^2 - 2$

50.
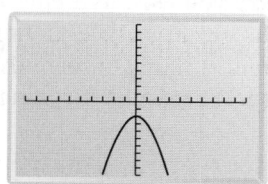

51.
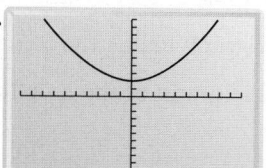

52.
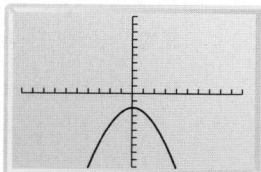

53.
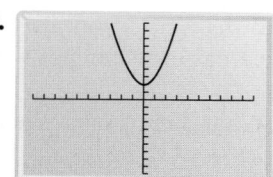

54.
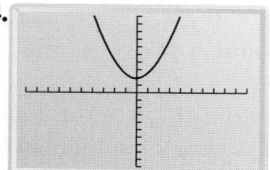

55.
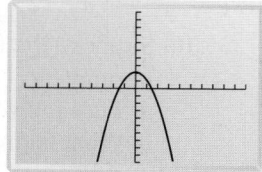

Lesson 9-3

Find each product.

56. $(c + 4)(c - 9)$ **57.** $(2x - 5)(x + 6)$ **58.** $(4t + 1)(5t + 3)$

59. $(7n^2 - 2)(3n^2 - 8)$ **60.** $(a + 5)(2a^2 - a + 4)$ **61.** $(3r^2 + 6r - 7)(2r - 1)$

Collecting Quadratic Data

In this activity, you will use a loop of string to make rectangles, record their dimensions, and explore the graph of length versus area.

NY **A.PS.2:** Recognize and understand equivalent representations of a problem situation or a mathematical concept.

1 ACTIVITY

Take a piece of string, no more than 40 cm long, and tie it into a loop. Use your thumbs and fingers to make a rectangle with the string. Hold it over a piece of graph paper to make right angles.

1. **Data Collection** Use the units on your graph paper to measure the length ℓ and width w of the rectangle. Record the measurements to the nearest tenth of a centimeter.

2. Copy and complete the first two columns in the table at the right. Be sure to include rectangles where the length is greater than the width and where the width is greater than the length.

Length (ℓ)	Width (w)	Area (A)
■	■	■
■	■	■
■	■	■
■	■	■
■	■	■

3. Record the area of each rectangle in the third column of the table.

4. Graph the length and area of each rectangle (length on the horizontal axis and area on the vertical axis).

5. **Writing** Based on your graph, explain why the data could NOT be modeled by a linear or exponential function.

2 ACTIVITY

6. **a.** Find the length of your loop of string.
   b. Write an expression for the width of any rectangle in terms of the length ℓ.
   c. Write a function for the area of any rectangle in terms of the length.

7. Graph the function on the same coordinate plane you used in Step 4.

8. **Writing** Does your graph match the function exactly? If not, explain why.

9. Find the vertex of the graph of the area function. What is the meaning of the vertex?

EXERCISES

10. Suppose the function $A = 50w - w^2$ models the areas of rectangles with a fixed perimeter.
    a. Find the vertex of the graph of this function. What dimensions give the greatest area?
    b. What is the fixed perimeter in this case?
    c. What is the greatest area for a rectangle with the given perimeter?

11. Suppose you repeated Activity 1 with a loop of string which was 140 cm long. Write a function to model the area A in terms of the length ℓ.

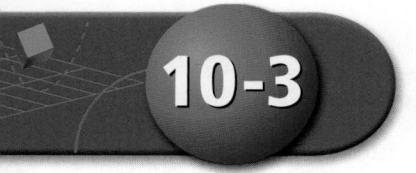

10-3

Solving Quadratic Equations

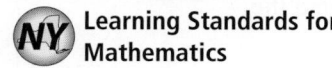
Learning Standards for Mathematics

A.A.8 Analyze and solve verbal problems that involve quadratic equations.

A.A.28 Understand the difference and connection between roots of a quadratic equation and factors of a quadratic expression.

A.G.8 Find the roots of a parabolic function graphically.

 **Check Skills You'll Need**

GO for Help   Lesson 3-8

Simplify each expression.

1. $\sqrt{36}$

2. $-\sqrt{81}$

3. $\pm\sqrt{121}$

4. $\sqrt{1.44}$

5. $\sqrt{0.25}$

6. $\pm\sqrt{1.21}$

7. $\sqrt{\frac{1}{4}}$

8. $\pm\sqrt{\frac{1}{9}}$

9. $\sqrt{\frac{49}{100}}$

New Vocabulary
• quadratic equation
• standard form of a quadratic equation
• zeros of the function
• roots of the equation

1 Solving Quadratic Equations by Graphing

Activity: Finding -intercepts

1. Find the x-intercepts of each graph.

a. $y = 2x - 3$

b. $y = x^2 + 3x - 4$

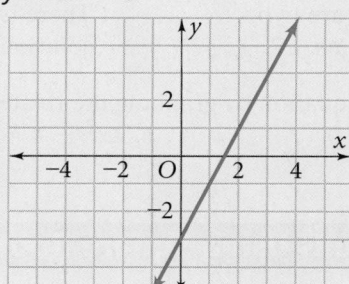 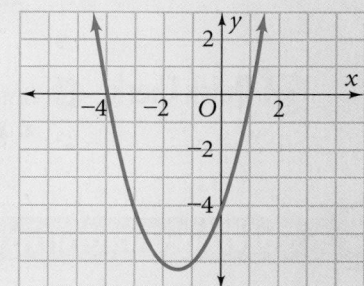

2. a. Solve $2x - 3 = 0$.

b. Is the solution of $2x - 3 = 0$ the same as the x-intercept of $y = 2x - 3$?

3. Do the x-intercepts that you found in Question 1b satisfy the equation $x^2 + 3x - 4 = 0$?

4. a. Graph $y = x^2 + x - 6$.

b. Find the x-intercepts of the graph of $y = x^2 + x - 6$.

c. Do the values you found in part (b) satisfy the equation $x^2 + x - 6 = 0$?

The equation $x^2 + 3x - 4 = 0$ is called a quadratic equation, and its related quadratic function is $y = x^2 + 3x - 4$. The solutions of a quadratic equation and the x-intercepts of its related quadratic function are the same.

Definition	Standard Form of a Quadratic Equation

A **quadratic equation** is an equation that can be written in the form $ax^2 + bx + c = 0$, where $a \neq 0$. This form is called the **standard form of a quadratic equation.**

A quadratic equation can have two, one, or no real-number solutions. In a future course you will learn about solutions of quadratic equations that are not real numbers. In this course *solutions* refers to real-number solutions.

The solutions of a quadratic equation and the related x-intercepts are often called **roots of the equation** or **zeros of the function.**

1 EXAMPLE Solving by Graphing

Solve each equation by graphing the related function.

a. $x^2 - 4 = 0$
Graph $y = x^2 - 4$.

b. $x^2 = 0$
Graph $y = x^2$.

c. $x^2 + 4 = 0$
Graph $y = x^2 + 4$.

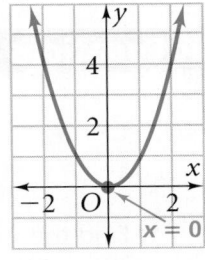

There are two solutions, ± 2.

There is one solution, 0.

There is no solution.

Quick Check ❶ Solve each equation by graphing the related function.
a. $x^2 - 1 = 0$
b. $2x^2 + 4 = 0$
c. $x^2 - 16 = -16$

2 Solving Quadratic Equations Using Square Roots

Recall from Lesson 3-9 that you can solve equations of the form $x^2 = a$ by finding the square roots of each side. We write the solution to $x^2 = 36$ as $\pm\sqrt{36}$ or ± 6.

2 EXAMPLE Using Square Roots

Solve $2x^2 - 98 = 0$.

$2x^2 - 98 + 98 = 0 + 98$ **Add 98 to each side.**

$2x^2 = 98$

$x^2 = 49$ **Divide each side by 2.**

$x = \pm\sqrt{49}$ **Find the square roots.**

$x = \pm 7$ **Simplify.**

Quick Check ❷ Solve each equation.
a. $t^2 - 25 = 0$
b. $3n^2 + 12 = 12$
c. $2g^2 + 32 = 0$

You can solve real-world problems by finding square roots. In many cases, the negative solution of a quadratic equation will not be a reasonable solution to the original problem.

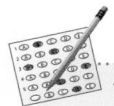

3 EXAMPLE   <u>Real-World</u> Problem Solving

Gridded Response A city is planning a circular duck pond for a new park. The depth of the pond will be 4 ft and the volume will be 20,000 ft³. Find the radius of the pond to the nearest tenth of a foot. Use the equation $V = \pi r^2 h$, where V is the volume, r is the radius, and h is the depth.

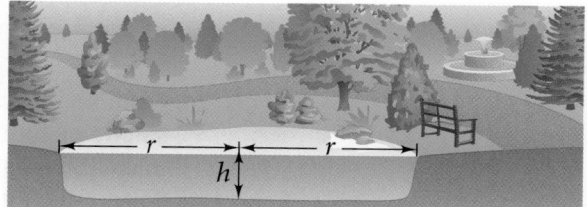

$$V = \pi r^2 h$$

$20,000 = \pi r^2(4)$ **Substitute 20,000 for V and 4 for h.**

$\dfrac{20,000}{(\pi \cdot 4)} = r^2$ **Put in calculator-ready form.**

$\sqrt{\dfrac{20,000}{(\pi \cdot 4)}} = r$ **Find the principal square root.**

$39.89422804 \approx r$ **Use a calculator.**

● The pond will have a radius of 39.9 ft.

 Quick Check **3** A city is planning a circular fountain. The depth of the fountain will be 3 ft and the volume will be 1800 ft³. Find the radius of the fountain.

EXERCISES

For more exercises, see *Extra Skill and Word Problem Practice.*

Practice and Problem Solving

 **Practice by Example**

Example 1
(page 566)

 **for Help**

Solve each equation by graphing the related function. If the equation has no solution, write *no solution*.

1. $x^2 - 9 = 0$ **2.** $x^2 + 5 = 0$ **3.** $4x^2 = 0$

4. $2x^2 - 8 = 0$ **5.** $x^2 + 16 = 0$ **6.** $\frac{1}{3}x^2 - 3 = 0$

7. $\frac{1}{2}x^2 + 1 = 0$ **8.** $x^2 + 7 = 7$ **9.** $\frac{1}{4}x^2 - 1 = 0$

Example 2
(page 566)

Solve each equation by finding square roots. If the equation has no solution, write *no solution*.

10. $k^2 = 49$ **11.** $b^2 = 441$ **12.** $m^2 - 225 = 0$

13. $c^2 + 25 = 25$ **14.** $x^2 - 9 = -16$ **15.** $4r^2 = 25$

16. $64p^2 = 4$ **17.** $6w^2 - 24 = 0$ **18.** $27 - y^2 = 0$

Example 3
(page 567)

Model each problem with a quadratic equation. Then solve. If necessary, round to the nearest tenth.

19. Find the side of a square with an area of 256 m².

20. Find the side of a square with an area of 90 ft².

21. Find the radius of a circle with an area of 80 cm².

22. Geometry Suppose a map company wants to produce a globe with a surface area of 450 in.^{2}. Use the formula $A = 4\pi r^2$, where A is the surface area and r is the radius of the sphere.

a. What should the radius be? Round to the nearest tenth of an inch.

b. Critical Thinking Why is the principal square root the only root that makes sense in this situation?

Mental Math Tell the number of solutions each equation has.

23. $y^2 = -36$ **24.** $a^2 - 12 = 6$ **25.** $n^2 - 15 = -15$

26. Framing Find dimensions for the square picture at the right that would make the area of the picture equal to 75% of the total area enclosed by the square frame. Round to the nearest tenth of an inch.

12 in

27. Suppose you have a can of paint that will cover 400 ft^2.

a. Find the radius of the largest circle you can paint. Round to the nearest tenth of a foot. (*Hint:* Use the formula $A = \pi r^2$.)

b. Suppose you have two cans of paint, which will cover a total of 800 ft^2. Find the radius of the largest circle you can paint. Round to the nearest tenth of a foot.

c. Critical Thinking Does the radius of the circle double when the amount of paint doubles? Explain.

Solve each equation by finding square roots. If the equation has no solution, write *no solution*. **If the value is irrational, round to the nearest tenth.**

28. $1.2q^2 - 7 = -34$ **29.** $49t^2 - 16 = -7$ **30.** $3d^2 - \frac{1}{12} = 0$

31. $\frac{1}{2}x^2 - 4 = 0$ **32.** $7h^2 + 0.12 = 1.24$ **33.** $-\frac{1}{4}x^2 + 3 = 0$

34. Physics The equation $d = \frac{1}{2}at^2$ gives the distance d an object starting at rest travels given acceleration a and time t. Suppose a ball rolls down the ramp shown at the right with acceleration $a = 2$ ft/s^2. Find the time it will take to roll from the top of the ramp to the bottom. Round to the nearest tenth of a second.

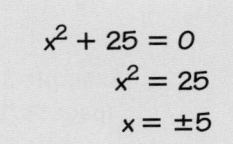

12 ft

35. Find a value for c such that the equation $x^2 - c = 0$ has 11 and -11 as solutions.

36. a. Critical Thinking For what values of n will $x^2 = n$ have two solutions?

b. For what value of n will $x^2 = n$ have exactly one solution?

c. For what values of n will $x^2 = n$ have no solution?

37. Error Analysis Michael's work is shown at the right. Explain the error that he made.

$x^2 + 25 = 0$
$x^2 = 25$
$x = \pm 5$

38. a. Solve $x^2 - 4 = 0$ and $2x^2 - 8 = 0$ by graphing their related functions.

b. Critical Thinking Why does it make sense that the graphs have the same x-intercepts?

39. Design Suppose your class wants to design a T-shirt logo similar to the one shown at the left. You want the orange region to have an area of 80 in.^{2}.
  a. Write expressions for the area of the square and of the circle.
  b. Write an equation for the area of the orange region.
  c. Solve the equation to find the radius of the circle and the side of the square. Round to the nearest tenth of an inch.

40. Open-Ended Write and solve equations in the form $ax^2 + c = 0$ for each of the following.
  a. The equation has no solutions.
  b. The equation has one solution.
  c. The equation has two solutions.

⬜ **Geometry** **Find the value of h for each triangle. If necessary, round to the nearest tenth.**

41.

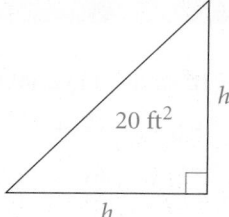

42.

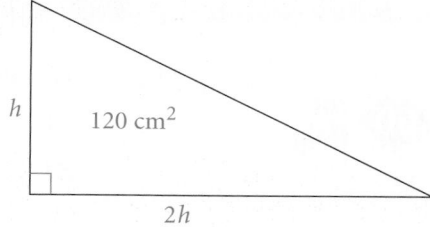

C **Challenge** 🌐 **43. Physics** The time t a pendulum takes to make a complete swing back and forth depends on the length of the pendulum. The formula $\ell = \frac{2.45t^2}{\pi^2}$ relates the length of a pendulum ℓ in meters to the time t in seconds.
  a. Find the length of the pendulum if $t = 1$ s. Round to the nearest tenth.
  b. Find t if $\ell = 1.6$ m. Round to the nearest tenth.
  c. Find t if $\ell = 2.2$ m. Round to the nearest tenth.
  ✏️ **d. Writing** You can adjust a clock that has a pendulum by making the pendulum longer or shorter. If a clock is running slowly, would you lengthen or shorten the pendulum to make the clock run faster? Explain.

44. a. Solve the equation $(x + 7)^2 = 0$.
  b. Find the vertex of the related function $y = (x + 7)^2$.
  c. Open-Ended Choose a value for h and repeat parts (a) and (b) using $(x + h)^2 = 0$ and $y = (x + h)^2$.
  d. Where would you expect to find the vertex of $y = (x - 4)^2$? Explain.

⬜ **45. Geometry** The trapezoid has an area of 1960 cm^2. Use the formula $A = \frac{1}{2}h(b_1 + b_2)$ to find the value of y.

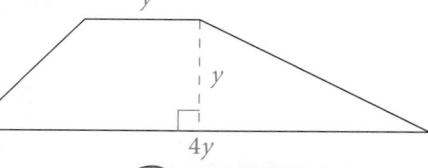

NY REGENTS

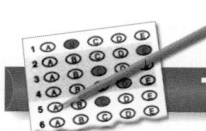

Test Prep

Multiple Choice

46. Which of the following shows all of the real solutions of $3x^2 - 48 = 0$?
  A. 4   **B.** -4, 4   **C.** -16, 16   **D.** no solution

47. For which of the following values of c does $5x^2 + c = 10$ have no real solutions?
  F. 1   **G.** 5   **H.** 9   **J.** 12

48. The x-intercepts of the graph at the right are solutions to which equation?

A. $-0.25x^2 = 4$ B. $-0.25x^2 + 4 = 0$

C. $-0.5x^2 + 2 = 0$ D. $-4x^2 + 1 = 0$

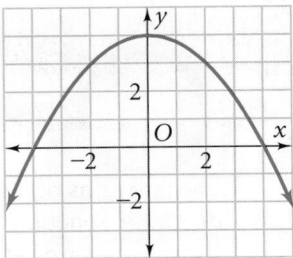

Short Response   **49.** Make a table for the function $y = 3x^2 - 7$. Graph the function. Estimate the value of the x-intercepts.

Extended Response   **50.** The surface area of a cube is 96 ft^2.

     a. Find the length of each edge. Show your work.

     b. If you double the length of each edge, what happens to the surface area of the cube? Show your work.

Mixed Review

Lesson 10-2

Graph each function. Label the axis of symmetry and the vertex.

51. $y = x^2 + 4x + 3$ **52.** $y = x^2 + 2$ **53.** $y = 2x^2 - 8x - 5$

54. $y = -x^2 + 6x - 1$ **55.** $y = 6x^2 - 12x + 1$ **56.** $y = -3x^2 + 18x$

Lesson 9-5

Factor each expression.

57. $x^2 + 5x + 4$ **58.** $y^2 - 15y + 26$ **59.** $a^2 + 3a - 10$

60. $z^2 - 6z - 72$ **61.** $c^2 - 14cd + 24d^2$ **62.** $t^2 + tu - 2u^2$

Lesson 8-2

Write each number in scientific notation.

63. 3,613,500   **64.** 0.0000348   **65.** 8.12

Write each number in standard notation.

66. 3.1×10^4 **67.** 7.01×10^5 **68.** 6.2×10^{-4}

✓ Checkpoint Quiz 1   Lessons 10-1 through 10-3

Graph each function. Label the axis of symmetry and the vertex.

1. $y = x^2 - 4$ **2.** $y = 8x^2 - 2x$ **3.** $f(x) = x^2 + 5x - 6$

4. Suppose you throw a ball up in the air. Its height h in feet after t seconds is given by the function $h = -16t^2 + 24t + 6$.

     a. When does the ball reach its maximum height?

     b. What is the ball's maximum height?

Solve each equation.

5. $2p^2 - 50 = 0$ **6.** $9h^2 = 49$

7. $t^2 - 64 = 0$ **8.** $6m^2 - 150 = 0$

Solve each equation by graphing the related function. If the equation has no solution, write *no solution*.

9. $3x^2 - 27 = 0$ **10.** $x^2 + 25 = 0$

The solutions of a quadratic equation are the *x*-intercepts of the related quadratic function. The solutions of a quadratic equation and the related *x*-intercepts are often called *roots* of the equation or *zeros* of the function.

NY A.CN.1: Understand and make connections among multiple representations of the same mathematical idea.

ACTIVITY

Use a graphing calculator to solve $x^2 - 6x + 3 = 0$.

Step 1

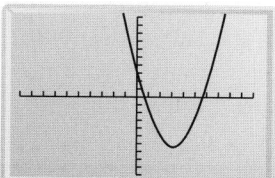

Enter $y = x^2 - 6x + 3$. Use the **CALC** feature. Select **2:ZERO**. The calculator will plot the graph.

Step 2

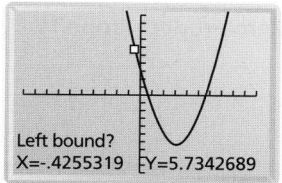

Left bound?
X=-.4255319 Y=5.7342689

Move the cursor to the left of the first *x*-intercept. Press ENTER to set the left bound.

Step 3

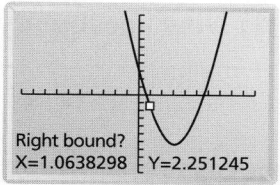

Right bound?
X=1.0638298 Y=2.251245

Move the cursor slightly to the right of the intercept. Press ENTER to set the right bound.

Step 4

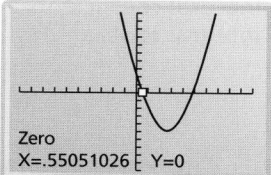

Zero
X=.55051026 Y=0

Press ENTER to display the first root, which is about 0.55.

Repeating the steps near the second intercept, you find that the second root is about 5.45. So the solutions are about 0.55 and 5.45.

Suppose you cannot see both of the *x*-intercepts on your graph. You can find the values of *y* that are close to zero by using the **TABLE** feature. You can use the **TBLSET** feature to control how the table behaves. Set ∆Tbl to 0.5. Set **Indpnt:** and **Depend:** to **Auto.** The calculator screen above shows part of the table for $y = 2x^2 - 48x + 285$.

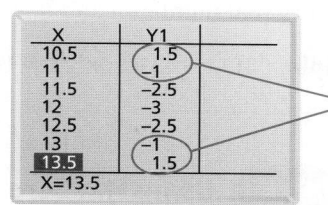

X	Y1
10.5	1.5
11	-1
11.5	-2.5
12	-3
12.5	-2.5
13	-1
13.5	1.5
X=13.5	

The graph crosses the *x*-axis when the values for *y* change signs. So the range of values of *x* should include 10.5 and 13.5.

EXERCISES

1. Find the *x*-intercepts of $y = 2x^2 - 48x + 285$. First use the WINDOW feature. Change **Xmin** to 10 and **Xmax** to 14. Round to the nearest hundredth.

Use a graphing calculator to solve each equation. Round to the nearest hundredth.

2. $x^2 - 6x - 16 = 0$
3. $2x^2 + x - 6 = 0$
4. $\frac{1}{3}x^2 + 8x - 3 = 0$
5. $x^2 - 18x + 5 = 0$
6. $0.25x^2 - 8x - 45 = 0$
7. $0.5x^2 + 3x - 36 = 0$

8. Geometry A rectangle has a length of *x* and a width of $2x + 3$. It has an area of 200 cm^2.
   a. Write an equation using the dimensions and the area.
   b. Graph the related function using a graphing calculator.
   c. Find the dimensions of the rectangle. Round to the nearest hundredth.

10-4

Factoring to Solve Quadratic Equations

NY Learning Standards for Mathematics

A.A.8 Analyze and solve verbal problems that involve quadratic equations.

A.A.27 Understand and apply the multiplication property of zero to solve quadratic equations with integral coefficients and integral roots.

 **Check Skills You'll Need**

GO for Help Lessons 3-1 and 9-6

Solve and check each equation.

1. $6 + 4n = 2$ 2. $\frac{a}{8} - 9 = 4$ 3. $7q + 16 = -3$

Factor each expression.

4. $2c^2 + 29c + 14$ 5. $3p^2 + 32p + 20$ 6. $4x^2 - 21x - 18$

🔊 **New Vocabulary** • Zero-Product Property

---

1   Solving Quadratic Equations

A.A.28 Understand the difference and connection between roots of a quadratic equation and factors of a quadratic expression.

In the previous lesson, you solved quadratic equations by finding square roots. This method works if $b = 0$. You can solve some quadratic equations when $b \neq 0$ by using the Zero-Product Property.

 Key Concepts

Property	**Zero-Product Property**

For every real number a and b, if $ab = 0$, then $a = 0$ or $b = 0$.

Example   If $(x + 3)(x + 2) = 0$, then $x + 3 = 0$ or $x + 2 = 0$.

---

1 EXAMPLE   **Using the Zero-Product Property**

Solve $(x + 5)(2x - 6) = 0$.

$(x + 5)(2x - 6) = 0$

$x + 5 = 0$ or $2x - 6 = 0$ **Use the Zero-Product Property.**

                         $2x = 6$ **Solve for x.**

$x = -5$ or $x = 3$

Check   Substitute -5 for x.   Substitute 3 for x.

           $(x + 5)(2x - 6) = 0$ $(x + 5)(2x - 6) = 0$

      $(-5 + 5)[2(-5) - 6] \stackrel{?}{=} 0$ $(3 + 5)[2(3) - 6] \stackrel{?}{=} 0$

                   $(0)(-16) = 0$ ✓ $(8)(0) = 0$ ✓

 **Quick Check** ❶ Solve each equation.

a. $(x + 7)(x - 4) = 0$ **b.** $(3y - 5)(y - 2) = 0$

c. $(6k + 9)(4k - 11) = 0$ **d.** $(5h + 1)(h + 6) = 0$

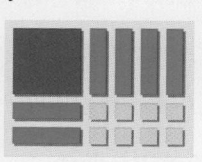
You can also use the Zero-Product Property to solve equations of the form $ax^2 + bx + c = 0$ if the quadratic expression $ax^2 + bx + c$ can be factored.

2 EXAMPLE Solving by Factoring

Solve $x^2 + 6x + 8 = 0$.

$$x^2 + 6x + 8 = 0$$
$$(x + 2)(x + 4) = 0 \qquad \text{Factor } x^2 + 6x + 8.$$
$$x + 2 = 0 \qquad \text{or} \qquad x + 4 = 0 \qquad \text{Use the Zero-Product Property.}$$
$$x = -2 \qquad \text{or} \qquad x = -4 \qquad \text{Solve for } x.$$

✓ **Quick Check** ➋ Solve $x^2 - 8x - 48 = 0$ by factoring.

Before solving a quadratic equation, you may need to add or subtract terms in order to write the equation in standard form. Then factor the quadratic expression.

3 EXAMPLE Solving by Factoring

Solve $2x^2 - 5x = 88$.

$$2x^2 - 5x - 88 = 0 \qquad \text{Subtract 88 from each side.}$$
$$(2x + 11)(x - 8) = 0 \qquad \text{Factor } 2x^2 - 5x - 88.$$
$$2x + 11 = 0 \qquad \text{or} \qquad x - 8 = 0 \qquad \text{Use the Zero-Product Property.}$$
$$2x = -11 \qquad \text{or} \qquad x = 8 \qquad \text{Solve for } x.$$
$$x = -5.5$$

✓ **Quick Check** ➌ Solve $x^2 - 12x = -36$.

4 EXAMPLE Real-World 🌐 Problem Solving

Manufacturing The diagram shows a pattern for an open-top box. The total area of the sheet of material used to manufacture the box is 288 in.^{2}. The height of the box is 3 in. Therefore, 3-in. $\times$ 3-in. squares are cut from each corner. Find the dimensions of the box.

Define Let x = width of a side of the box.
Then the width of the material = $x + 3 + 3 = x + 6$.
The length of the material = $x + 2 + 3 + 3 = x + 8$.

Relate length $\times$ width = area of the sheet

Write $(x + 8)(x + 6) = 288$

$$x^2 + 14x + 48 = 288 \qquad \text{Find the product } (x + 8)(x + 6).$$
$$x^2 + 14x - 240 = 0 \qquad \text{Subtract 288 from each side.}$$
$$(x + 24)(x - 10) = 0 \qquad \text{Factor } x^2 + 14x - 240.$$
$$x + 24 = 0 \qquad \text{or} \qquad x - 10 = 0 \qquad \text{Use the Zero-Product Property.}$$
$$x = -24 \qquad \text{or} \qquad x = 10 \qquad \text{Solve for } x.$$

The only reasonable solution is 10. So the dimensions of the box are 10 in. $\times$ 12 in. $\times$ 3 in.

waste material

✓ **Quick Check** ➍ Suppose that a box has a base with a width of x, a length of $x + 1$, and a height of 2 in. It is cut from a rectangular sheet of material with an area of 182 in.^{2}. Find the dimensions of the box.

EXERCISES

For more exercises, see *Extra Skill and Word Problem Practice*.

Practice and Problem Solving

A Practice by Example

Example 1
(page 572)

GO for Help

Use the Zero-Product Property to solve each equation.

1. $(x - 3)(x - 7) = 0$ **2.** $(x + 4)(2x - 9) = 0$ **3.** $t(t + 1) = 0$

4. $-3n(2n - 5) = 0$ **5.** $(7x + 2)(5x + 4) = 0$ **6.** $(4a - 7)(3a + 8) = 0$

Use algebra tiles to factor. Then solve using the Zero-Product Property.

7. $x^2 + 7x + 10 = 0$ **8.** $k^2 + 7k + 12 = 0$

Example 2
(page 573)

Solve by factoring.

9. $b^2 + 3b - 4 = 0$ **10.** $m^2 - 5m - 14 = 0$ **11.** $w^2 - 8w = 0$

12. $x^2 - 16x + 55 = 0$ **13.** $k^2 - 3k - 10 = 0$ **14.** $n^2 + n - 12 = 0$

Example 3
(page 573)

15. $x^2 + 8x = -15$ **16.** $t^2 - 3t = 28$ **17.** $n^2 = 6n$

18. $2c^2 - 7c = -5$ **19.** $3q^2 + 16q = -5$ **20.** $4y^2 = 25$

Example 4
(page 573)

21. Geometry The sides of a square are all increased by 3 cm. The area of the new square is 64 cm². Find the length of a side of the original square.

22. Geometry A rectangular box has volume 280 in.³. Its dimensions are 4 in. $\times$ $(n + 2)$ in. $\times$ $(n + 5)$ in. Find n. Use the formula $V = \ell w h$.

$2x + 2$
x

23. Construction You are building a rectangular wading pool. You want the area of the bottom to be 90 ft². You want the length of the pool to be 3 ft longer than twice its width. What will the dimensions of the pool be?

24. Sailing Suppose the area of the sail shown in the photo at the left is 110 ft². Find the dimensions of the sail.

25. The product of two consecutive numbers is 14 less than 10 times the smaller number. Find each number.

B Apply Your Skills

Write each equation in standard form. Then solve.

26. $2q^2 + 22q = -60$ **27.** $4 = -5n + 6n^2$

28. $6y^2 + 12y + 13 = 2y^2 + 4$ **29.** $3a^2 + 4a = 2a^2 - 2a - 9$

30. $3t^2 + 8t = t^2 - 3t - 12$ **31.** $4x^2 + 20 = 10x + 3x^2 - 4$

32. Manufacturing The length of an open box is 2 in. greater than its width. The box was made from an 80 in.² rectangular sheet of material. The height of the box is 1 in. Therefore 1-in. $\times$ 1-in. squares are cut from each corner. What were the dimensions of the original sheet of material? (*Hint:* Draw a diagram.)

33. Baseball Suppose you throw a baseball into the air with an initial upward velocity of 29 ft/s and an initial height of 6 ft. The formula $h = -16t^2 + 29t + 6$ gives the ball's height h in feet at time t in seconds.
   a. The ball's height h is 0 when it is on the ground. Find the number of seconds that pass before the ball lands by solving $0 = -16t^2 + 29t + 6$.
   b. Graphing Calculator Graph the related function for the equation in part (a). Use your graph to estimate the maximum height of the ball.

34. Writing Summarize the procedure for solving a quadratic equation by factoring. Include an example.

GO Online
Homework Video Tutor
Visit: PHSchool.com
Web Code: ate-1004

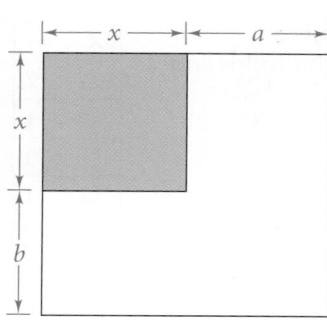

35. Geometry In the diagram at the left, a, b, and x represent positive integers. List several possible values for x, a, and b such that the large rectangle has an area of 56 square units.

36. Open-Ended Write and solve a quadratic equation using the Zero-Product Property.

37. a. Solve $x^2 = x$ and $x^2 = -x$ by factoring.
   b. What number is a solution to both equations?

Solve each cubic equation.

Sample
$$x^3 + 7x^2 + 12x = 0$$
$$x(x^2 + 7x + 12) = 0 \qquad \text{Factor out the GCF.}$$
$$x(x + 3)(x + 4) = 0 \qquad \text{Factor the quadratic trinomial.}$$
$$x = 0 \quad \text{or} \quad x + 3 = 0 \quad \text{or} \quad x + 4 = 0 \qquad \text{Use the Zero-Product Property.}$$
$$x = 0 \quad \text{or} \quad x = -3 \quad \text{or} \quad x = -4 \qquad \text{Solve for } x.$$

38. $x^3 - 10x^2 + 24x = 0$ **39.** $x^3 - 5x^2 + 4x = 0$ **40.** $3x^3 - 9x^2 = 0$

41. $x^3 + 3x^2 - 70x = 0$ **42.** $3x^3 - 30x^2 + 27x = 0$ **43.** $2x^3 = -2x^2 + 40x$

C Challenge **44. Construction** You are building a rectangular patio with two rectangular openings for gardens. You have 124 one-foot-square paving stones. Using the diagram below, what value of x would allow you to use all of the stones?

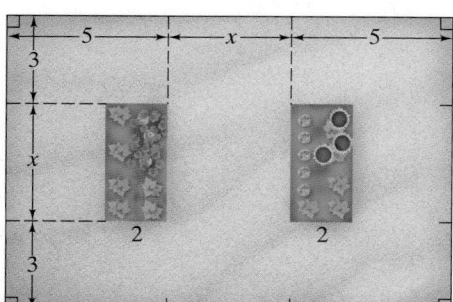

45. Find an equation that has the given numbers as solutions. For example, 4 and -3 are solutions to $x^2 - x - 12 = 0$.
   a. $-5, 8$ **b.** $3, -2$ **c.** $\frac{1}{2}, -10$ **d.** $\frac{2}{3}, -\frac{5}{7}$

Factor the expression on the left side of each equation by grouping. Then solve.

46. $x^3 + 5x^2 - x - 5 = 0$ **47.** $x^3 + x^2 - 4x - 4 = 0$

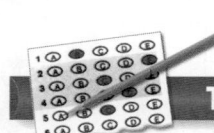

NY REGENTS

Test Prep

Multiple Choice

48. What are the solutions of $(x - 7)(2x + 8) = 0$?
   A. 7, 8   **B.** 7, 4   **C.** 7, -8 **D.** 7, -4

49. Which of the following is the sum of the solutions of $4x^2 - 35x - 9 = 0$?
   F. -8 **G.** 8   **H.** 8.75   **J.** 9.25

50. Which of the following is NOT a solution of $n(2n - 5)(10n + 3) = 0$?
   A. -3 **B.** -0.3 **C.** 0   **D.** 2.5

51. The table at the right shows x- and y-values for a given quadratic function. Based on the table, which of the following statements is NOT true?

x	y
-1	-7
0	-1
1	1
2	-1
3	-7

    F. The related equation has a solution between 0 and 1.
    G. The parabola opens upward.
    H. The related equation has more than one solution.
    J. The maximum y-value of the function is 1.

52. How many solutions does $(4x - 5)(x - 3)(x + 7) = 0$ have?
    A. 1   **B.** 2   **C.** 3   **D.** 4

Short Response   **53.** Find the solutions of the equation $3x^2 + 20x + 1 = 8$. Show your work.

Mixed Review

Lesson 10-3

Geometry **Model each situation with a quadratic equation. Then solve. Round answers to the nearest tenth.**

54. Find the side of a square with an area of 320 ft^2.

55. Find the radius of a circle with an area of 38 ft^2.

Lesson 9-6

Factor each expression.

56. $2x^2 + 13x + 15$ **57.** $3y^2 - 10y + 3$ **58.** $4t^2 + 5t - 6$

59. $6n^2 + 7n - 3$ **60.** $15a^3 - 50a^2 - 40a$ **61.** $-18b^3 + 42b^2 - 20b$

A P●int in Time

1500   **1600**   **1700**   **1800**   **1900**   **2000**

De la Cierva's work on problems of lift and gravity, like the work of aeronautical engineers of today, involved quadratic functions.

In 1923, Juan de la Cierva (1895–1936) designed the first successful autogyro, a rotor-based aircraft. The autogyro had rotating blades to give the aircraft lift, a propeller for forward thrust, and short, stubby wings for balance. Autogyros needed only short runways for takeoff and could descend almost vertically.

By hinging the rotor blades at the hub, de la Cierva allowed each blade to respond to aerodynamic forces. This was a significant contribution in the development of the modern helicopter.

PHSchool.com   **For:** More information about autogyros
                      Web Code: ate-2032

You can have roots other than square roots. The third root of a number x is written as $\sqrt[3]{x}$. For example, $\sqrt[3]{8} = 2$ because $2^3 = 8$. You can also express roots using exponents. The expression $a^{\frac{1}{n}}$ is defined as $\sqrt[n]{a}$. An example is $9^{\frac{1}{2}} = \sqrt[2]{9} = 3$.

Rational exponents follow the same rules as integer exponents, so $9^{\frac{1}{2}} \cdot 9^{\frac{1}{2}} = 9^{\left(\frac{1}{2}+\frac{1}{2}\right)} = 9^1 = 9$, just as $\sqrt{9} \cdot \sqrt{9} = 9$.

NY A.N.6 Evaluate expressions involving factorial(s), absolute value(s), and exponential expression(s).

1 EXAMPLE

Simplify $16^{\frac{1}{4}}$.

$$16^{\frac{1}{4}} = 2 \qquad 16 = 2 \cdot 2 \cdot 2 \cdot 2, \text{ so } \sqrt[4]{16} = 2$$

An exponent can be any rational number. You can simplify rational exponents using the property of raising a power to a power, $a^{mn} = (a^m)^n$.

2 EXAMPLE

Simplify $8^{\frac{2}{3}}$.

a.
$$8^{\frac{2}{3}} = 8^{\frac{1}{3} \cdot 2}$$ ⟵ Write $\frac{2}{3}$ as a product of 2 and $\frac{1}{3}$. ⟶ **b.** $8^{\frac{2}{3}} = 8^{2 \cdot \frac{1}{3}}$
$$= \left(8^{\frac{1}{3}}\right)^2$$ ⟵ Use the property of raising a power to a power. ⟶ $= \left(8^{\frac{1}{3}}\right)^2$
$$= (2)^2$$ ⟵ Simplify within the parentheses. ⟶ $= (64)^{\frac{1}{3}}$
$$= 4$$ ⟵ Simplify. ⟶ $= 4$

3 EXAMPLE

Simplify $\left(x^{\frac{3}{4}}\right)^5 \left(y^{\frac{1}{2}}\right) y^{\frac{2}{5}}$.

$$\left(x^{\frac{3}{4}}\right)^5 \left(y^{\frac{1}{2}}\right) y^{\frac{2}{5}} = \left(x^{\frac{15}{4}}\right)\left(y^{\frac{1}{2}} \cdot y^{\frac{2}{5}}\right)$$ Multiply exponents in $\left(x^{\frac{3}{4}}\right)^5$.

$$= x^{\frac{15}{4}}\left(y^{\frac{1}{2}+\frac{2}{5}}\right)$$ Add exponents of powers with the same base.

$$= x^{\frac{15}{4}} y^{\frac{9}{10}}$$ Simplify fractions. $\frac{1}{2} + \frac{2}{5} = \frac{9}{10}$. Leave $\frac{15}{4}$ as an improper fraction.

EXERCISES

Simplify each expression.

1. $100^{\frac{1}{2}}$ **2.** $25^{\frac{1}{2}}$ **3.** $8^{\frac{1}{3}}$ **4.** $\left(49^{\frac{1}{2}}\right)^3$

5. $\left(8^{\frac{1}{3}}\right)^2$ **6.** $8^{\frac{4}{3}}$ **7.** $25^{\frac{3}{2}}$ **8.** $64^{\frac{4}{3}}$

9. $\left(x^{\frac{1}{3}}\right)^6$ **10.** $\left(b^{\frac{1}{4}}\right)^4$ **11.** $\left(m^{\frac{2}{5}}\right)^{\frac{5}{3}}$ **12.** $\left(m^{\frac{2}{5}}\right)\left(m^{\frac{3}{5}}\right)$

Connecting Ideas With Concept Maps

You can show connections among ideas by making a diagram called a concept map.
The lines in a concept map connect related ideas.

EXAMPLE

Make a concept map using terms from
Chapter 10 for *quadratic relationships*.

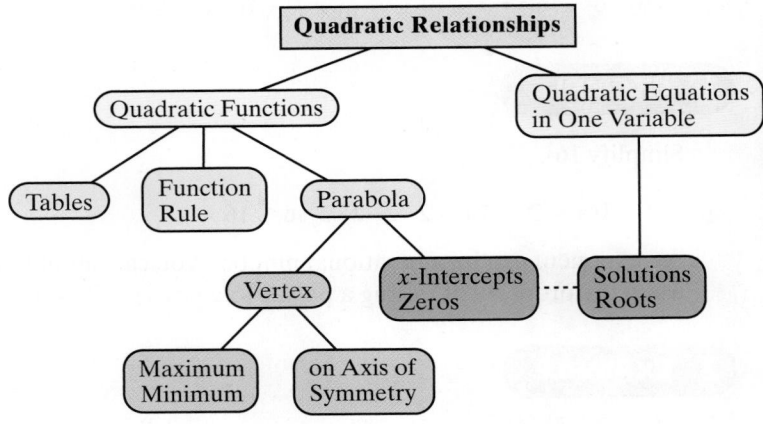

EXERCISES

1. Copy and complete the concept
map for *polynomial*. Fill in the
ovals using the appropriate terms
listed below.

binomial
constant
linear
monomial
trinomial
cubic
fourth power
quadratic

2. Copy and complete the concept map
for *power*. Give an example of each
term in the outer ovals.

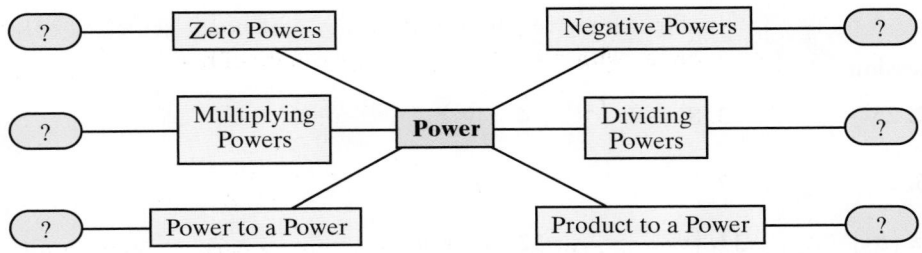

Completing the Square

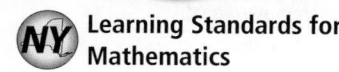
Learning Standards for Mathematics

A.A.8 Analyze and solve verbal problems that involve quadratic equations.

✓ **Check Skills You'll Need**

Find each square.

1. $(d - 4)^2$ **2.** $(x + 11)^2$ **3.** $(k - 8)^2$

Factor.

4. $b^2 + 10b + 25$ **5.** $t^2 + 14t + 49$ **6.** $n^2 - 18n + 81$

 **for Help** Lessons 9-4 and 9-7

🔊 **New Vocabulary** • completing the square

---

1   Solving by Completing the Square

Video Tutor Help
Visit: PHSchool.com
Web Code: ate-0775

In previous lessons, you solved quadratic equations by finding square roots and by factoring. These methods work in some cases. A third method, completing the square, works with every quadratic equation. Completing the square turns every quadratic equation into the form $m^2 = n$. You can model completing the square of a quadratic expression using algebra tiles.

The algebra tiles at the right represent the expression $x^2 + 8x$.

Here is the same expression rearranged to form part of a square. Notice that the x-tiles have been split evenly into two groups of four.

You can complete the square by adding 4^2, or 16, single tiles. The completed square is $x^2 + 8x + 16$ or $(x + 4)^2$.

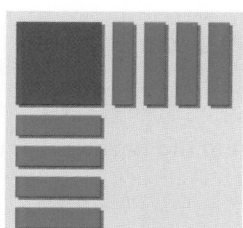

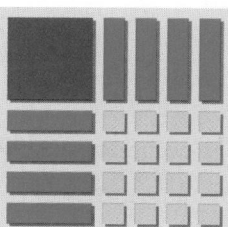

To explore this method algebraically, consider the equation below. In a perfect square trinomial, with $a = 1$, c must be the square of half of b.

$$(x + 4)^2 = x^2 + 2(4)x + 4^2$$
$$= x^2 + 8x + 16$$
$$\frac{8}{2} = 4 \rightarrow 4^2$$

You can change an expression like $x^2 + bx$ into a perfect square trinomial by adding $\left(\frac{b}{2}\right)^2$ to $x^2 + bx$. This process is called **completing the square.** The process is the same whether b is positive or negative.

Find the value of n such that $x^2 - 12x + n$ is a perfect square trinomial.

● The value of b is -12. The term to add to $x^2 - 12x$ is $\left(-\frac{12}{2}\right)^2$ or 36. So $n = 36$.

✅ **Quick Check** ① Find the value of n such that $x^2 + 22x + n$ is a perfect square trinomial.

The simplest equations in which to complete the square have the form $x^2 + bx = c$.

② **EXAMPLE** Solving $x^2 + bx = c$

Solve the equation $x^2 + 9x = 136$.

Step 1 Write the left side of $x^2 + 9x = 136$ as a perfect square.

$$x^2 + 9x = 136$$
$$x^2 + 9x + \left(\frac{9}{2}\right)^2 = 136 + \left(\frac{9}{2}\right)^2 \quad \text{Add } \left(\frac{9}{2}\right)^2, \text{ or } \frac{81}{4}, \text{ to each side of the equation.}$$
$$\left(x + \frac{9}{2}\right)^2 = \frac{544}{4} + \frac{81}{4} \quad \text{Write } x^2 + 9x + \left(\frac{9}{2}\right)^2 \text{ as a square.}$$
$$\text{Rewrite 136 as a fraction with denominator 4.}$$
$$\left(x + \frac{9}{2}\right)^2 = \frac{625}{4} \quad \text{Simplify the right side of the equation.}$$

Step 2 Solve the equation.

$$\left(x + \frac{9}{2}\right) = \pm\sqrt{\frac{625}{4}} \quad \text{Find the square root of each side.}$$
$$x + \frac{9}{2} = \pm\frac{25}{2} \quad \text{Simplify.}$$
$$x + \frac{9}{2} = \frac{25}{2} \quad \text{or} \quad x + \frac{9}{2} = -\frac{25}{2} \quad \text{Write as two equations.}$$
$$x = 8 \quad \text{or} \quad x = -17 \quad \text{Solve for } x \text{ by subtracting } \frac{9}{2} \text{ from each side of the equations and simplifying.}$$

✅ **Quick Check** ② Solve the equation $m^2 - 6m = 247$.

To solve an equation in the form $x^2 + bx + c = 0$, first subtract the constant term c from each side of the equation.

③ **EXAMPLE** Solving $x^2 + bx + c = 0$

Solve $x^2 - 20x + 32 = 0$.

Step 1 Rewrite the equation in the form $x^2 + bx = c$ and complete the square.

$$x^2 - 20x + 32 = 0$$
$$x^2 - 20x = -32 \quad \text{Subtract 32 from each side.}$$
$$x^2 - 20x + 100 = -32 + 100 \quad \text{Add } \left(-\frac{20}{2}\right)^2, \text{ or 100, to each side of the equation.}$$
$$(x - 10)^2 = 68 \quad \text{Write } x^2 - 20x + 100 \text{ as a square.}$$

Step 2 Solve the equation.

$$(x - 10) = \pm\sqrt{68} \quad \text{Find the square root of each side.}$$
$$x - 10 \approx \pm 8.25 \quad \text{Use a calculator to find } \sqrt{68}.$$
$$x - 10 \approx 8.25 \quad \text{or} \quad x - 10 \approx -8.25 \quad \text{Write as two equations.}$$
$$x \approx 8.25 + 10 \quad \text{or} \quad x \approx -8.25 + 10 \quad \text{Add 10 to each side.}$$
$$x \approx 18.25 \quad \text{or} \quad x \approx 1.75 \quad \text{Simplify.}$$

 ❸ Solve each equation. Round to the nearest hundredth.

a. $x^2 + 5x + 3 = 0$ **b.** $x^2 - 14x + 16 = 0$

The method of completing the square works when $a = 1$. To solve an equation like $3x^2 + 6x - 9 = 0$, you need to divide each side by 3 before completing the square.

$$3x^2 + 6x - 9 = 0 \quad \rightarrow \quad \frac{3x^2 + 6x - 9}{3} = \frac{0}{3} \quad \rightarrow \quad x^2 + 2x - 3 = 0$$

❹ EXAMPLE   **Real-World 🌐 Problem Solving**

Real-World 🌐 Connection

Careers Woodworkers use mathematical skills to design and create furniture.

Carpentry Suppose a woodworker wants to build a tabletop like the one shown at the right. If the surface area is 26 ft², what is the value of x?

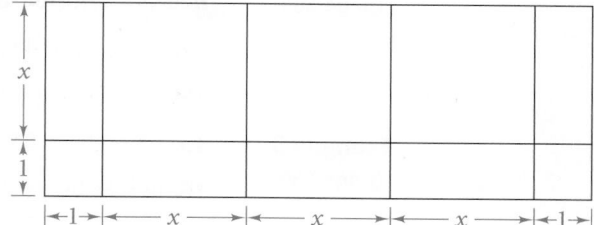

Define   width $= x + 1$
           length $= x + x + x + 1 + 1 = 3x + 2$

Relate   length $\times$ width $=$ area

Write $(3x + 2)(x + 1) = 26$
        $3x^2 + 5x + 2 = 26$

Step 1   Rewrite the equation in the form $x^2 + bx = c$.

$$3x^2 + 5x + 2 = 26$$
$$3x^2 + 5x = 24 \qquad \text{Subtract 2 from each side.}$$
$$x^2 + \tfrac{5}{3}x = 8 \qquad \text{Divide each side by 3.}$$

Step 2   Complete the square.

$$x^2 + \tfrac{5}{3}x + \tfrac{25}{36} = 8 + \tfrac{25}{36} \qquad \text{Add } \left(\tfrac{5}{6}\right)^2, \text{ or } \tfrac{25}{36}, \text{ to each side.}$$
$$\left(x + \tfrac{5}{6}\right)^2 = \tfrac{288}{36} + \tfrac{25}{36} \qquad \text{Write } x^2 + \tfrac{5}{3}x + \tfrac{25}{36} \text{ as a square. Rewrite 8 as a fraction with denominator 36.}$$
$$\left(x + \tfrac{5}{6}\right)^2 = \tfrac{313}{36} \qquad \text{Simplify.}$$

Step 3   Solve the equation.

$$\left(x + \tfrac{5}{6}\right) = \pm\sqrt{\tfrac{313}{36}} \qquad \text{Take the square root of each side.}$$
$$x + \tfrac{5}{6} \approx \pm 2.95 \qquad \text{Use a calculator. } \sqrt{\tfrac{313}{36}} \approx 2.95$$

$x + \tfrac{5}{6} \approx 2.95$ or $x + \tfrac{5}{6} \approx -2.95$	Write as two equations.
$x \approx 2.95 - \tfrac{5}{6}$ or $x \approx -2.95 - \tfrac{5}{6}$	Subtract $\tfrac{5}{6}$ from each side.
$x \approx 2.95 - 0.83$ or $x \approx -2.95 - 0.83$	$\tfrac{5}{6} \approx 0.83$, so substitute 0.83 for $\tfrac{5}{6}$.
$x \approx 2.12$ or $x \approx -3.78$	Use the positive answer for this problem.

● The value of x is about 2.12 ft.

 **Quick Check** ❹ Solve each equation. Round to the nearest hundredth.

a. $4a^2 - 8a = 24$ **b.** $5n^2 - 3n - 15 = 10$

EXERCISES

For more exercises, see *Extra Skill and Word Problem Practice*.

Practice and Problem Solving

A Practice by Example

Example 1
(page 580)

GO for Help

Find the value of *n* such that each expression is a perfect square trinomial.

1. $k^2 + 14k + n$ **2.** $m^2 - 8m + n$ **3.** $y^2 - 40y + n$

4. $p^2 - 6p + n$ **5.** $v^2 + 24v + n$ **6.** $w^2 - 36w + n$

Example 2
(page 580)

Solve each equation by completing the square. If necessary, round to the nearest hundredth.

7. $r^2 + 8r = 48$ **8.** $x^2 - 10x = 40$ **9.** $q^2 + 22q = -85$

10. $m^2 + 6m = 9$ **11.** $r^2 + 20r = 261$ **12.** $g^2 - 2g = 323$

Example 3
(page 580)

13. $r^2 - 2r - 35 = 0$ **14.** $x^2 + 10x + 17 = 0$ **15.** $p^2 - 12p + 11 = 0$

16. $w^2 + 3w - 5 = 0$ **17.** $m^2 + m - 28 = 0$ **18.** $a^2 + 9a - 682 = 0$

Example 4
(page 581)

What term do you need to add to each side to complete the square?

19. $2k^2 + 4k = 10$ **20.** $3x^2 + 12x = 24$ **21.** $5t^2 + 9t = 15$

Solve each equation by completing the square. If necessary, round to the nearest hundredth.

22. $4y^2 + 8y - 36 = 0$ **23.** $3q^2 - 12q = 15$ **24.** $2x^2 - 10x - 20 = 8$

25. a. Write an expression for the total area of the model below.

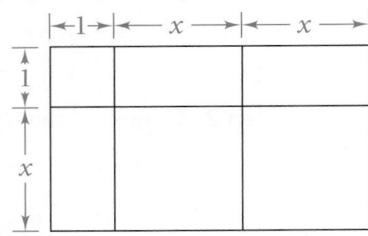

  b. The total area is 28 square units. Write an equation to find x.
  c. Solve by completing the square.

B Apply Your Skills

Solve each equation. If necessary, round to the nearest hundredth. If there is no solution, write *no solution*.

26. $b^2 + 4b + 1 = 0$ **27.** $c^2 + 7c = -12$ **28.** $h^2 + 6h - 40 = 0$

29. $y^2 - 8y = -12$ **30.** $4m^2 - 40m + 56 = 0$ **31.** $k^2 + 4k + 11 = -10$

32. $2x^2 - 15x + 6 = 41$ **33.** $3d^2 - 24d = 3$ **34.** $x^2 + 9x + 20 = 0$

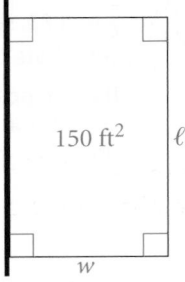

150 ft² ℓ

w

35. Gardening Suppose you want to enclose a rectangular garden plot against a house using fencing on three sides, as shown at the left. Assume you have 50 ft of fencing material and want to create a garden with an area of 150 ft².
  a. Let $w =$ the width. Write an expression for the length of the plot.
  b. Write and solve an equation for the area of the plot. Round to the nearest tenth of a foot.
  c. What dimensions should the garden have?
  d. Critical Thinking Find the area of the garden by using the dimensions you found in part (b). Does the area equal 150 ft²? Explain.

36. Error Analysis A classmate was completing the square to solve $4x^2 + 10x = 0$. For her first step she wrote $4x^2 + 10x + 25 = 25$. What was her error?

37. Writing Explain to a classmate how to solve $x^2 + 30x - 1 = 0$ by completing the square.

38. Open-Ended Write a quadratic equation and solve it by completing the square. Show your work.

Use each graph to estimate the values of x for which $f(x) = 5$. Write and solve an equation to find the values of x such that $f(x) = 5$. Round to the nearest hundredth.

39. $f(x) = x^2 - 4x - 1$

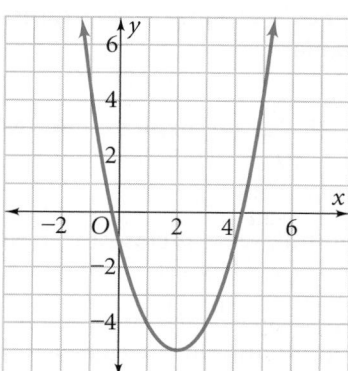

40. $f(x) = -\frac{1}{2}x^2 + 4x + 1$

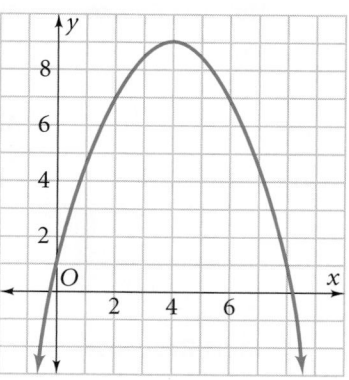

 41. Geometry A rectangle has a length of x. Its width is 3 feet longer than twice the length. Find the dimensions if its area is 80 ft². Round to the nearest tenth of a foot.

C **Challenge** 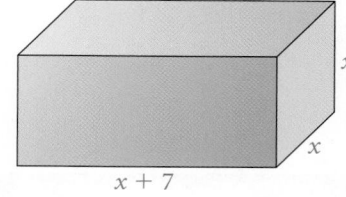 **42. Geometry** Suppose the prism shown at the right has the same surface area as an 8-in. cube.
  a. Write an expression for the surface area of the prism shown at the right.
  b. Write an equation that relates the surface area of the prism to the surface area of the 8-in. cube.
  c. Solve the equation you wrote in part (b) to find the dimensions of the prism.

43. Design Suppose you want to design a patio like the one shown below.

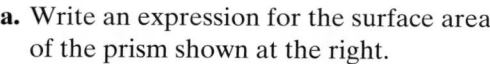

  a. Write an expression for the total area.
  b. If you want the total area to be 200 ft², what is the value of x?
  c. If you rounded the value of x to the nearest integer, what would the total area be?

44. a. Solve the equation $x^2 - 6x + 4 = 0$, but leave your answers in the form $p \pm \sqrt{q}$.

    b. Use the vertex formula $x = \frac{-b}{2a}$ to find the coordinates of the vertex of $y = x^2 - 6x + 4$.

    c. Critical Thinking Explain the relationship between your answers in part (a) and part (b).

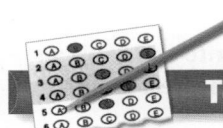

Test Prep

Multiple Choice

45. Which values of b will make the expression $x^2 + bx + 100$ a perfect square trinomial?

    A. $-10, 10$ **B.** $-20, 20$ **C.** $-25, 25$ **D.** $-50, 50$

46. Which of the following expressions is NOT a perfect square trinomial?

    F. $t^2 - 14t + 49$ **G.** $9b^2 + 66b + 121$

    H. $4m^2 - 24m + 36$ **J.** $81k^2 - 120k + 100$

47. Which of the following is closest to a solution of the equation $x^2 + 6x - 11 = 0$?

    A. -3 **B.** -1 **C.** 0 **D.** 2

Short Response

48. The area of the figure at the right is 200 cm². Find the value of x. Round to the nearest hundredth. Show your work.

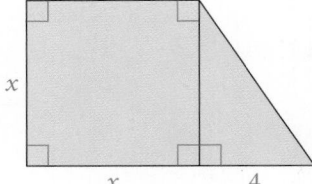

Extended Response

49. Kevin's office cubicle measures 8 ft by 12 ft.

    a. Write an equation to find the amount x that should be added to the current length and width to double the area.

    b. Solve the equation. Show your work.

    c. What will the new dimensions be?

Mixed Review

Lesson 10-4

Solve each equation.

50. $x^2 - 4x - 21 = 0$ **51.** $n^2 + 11n + 30 = 0$ **52.** $t^2 - 5t = 0$

53. $9v^2 - 64 = 0$ **54.** $4c^2 + 12c = -9$ **55.** $12w^2 = 28w + 5$

Lesson 9-7

Factor each expression.

56. $x^2 + 4x + 4$ **57.** $t^2 - 22t + 121$ **58.** $b^2 - 25$

59. $16c^2 + 24c + 9$ **60.** $49s^2 - 169$ **61.** $8m^3 - 18m$

62. $25m^2 + 120m + 144$ **63.** $400k^2 - 9$ **64.** $256g^2 - 121$

Lesson 8-4

Simplify.

65. $(r^3)^4$ **66.** $p(p^2)^6$ **67.** $-y^3(y^{-1})^2$

68. $(m^5)^{-8}$ **69.** $-w^7(w^8)^{-1}$ **70.** $t^8(t^{-7})^{-3}$

Using the Quadratic Formula

 **Learning Standards for Mathematics**

A.A.8: Analyze and solve verbal problems that involve quadratic equations.

✓ **Check Skills You'll Need**

GO for **Help**   Lesson 10-5

Find the value of c to complete the square for each expression.

1. $x^2 + 6x + c$ **2.** $x^2 + 7x + c$ **3.** $x^2 - 9x + c$

Solve each equation by completing the square.

4. $x^2 - 10x + 24 = 0$ **5.** $x^2 + 16x - 36 = 0$

6. $3x^2 + 12x - 15 = 0$ **7.** $2x^2 - 2x - 112 = 0$

🔊 **New Vocabulary** • **quadratic formula**

---

1 **Using the Quadratic Formula**

Real-World 🌐 Connection

The path of a golf ball can be modeled with a quadratic function.

In Lesson 10-5, you solved quadratic equations by completing the square. If you complete the square of the general equation $ax^2 + bx + c = 0$, you can derive the **quadratic formula,** which will solve any quadratic equation.

Step 1 Write $ax^2 + bx + c = 0$ so the coefficient of x^2 is 1.

$$ax^2 + bx + c = 0$$

$$x^2 + \frac{b}{a}x + \frac{c}{a} = 0 \quad \textbf{Divide each side by } a.$$

Step 2 Complete the square.

$$x^2 + \frac{b}{a}x = -\frac{c}{a} \qquad \textbf{Subtract } \frac{c}{a} \textbf{ from each side.}$$

$$x^2 + \frac{b}{a}x + \left(\frac{b}{2a}\right)^2 = -\frac{c}{a} + \left(\frac{b}{2a}\right)^2 \qquad \textbf{Add } \left(\frac{b}{2a}\right)^2 \textbf{ to each side.}$$

$$\left(x + \frac{b}{2a}\right)^2 = -\frac{c}{a} + \left(\frac{b}{2a}\right)^2 \qquad \textbf{Write the trinomial as a perfect square.}$$

$$= -\frac{4ac}{4a^2} + \frac{b^2}{4a^2} \qquad \textbf{Multiply } -\frac{c}{a} \textbf{ by } \frac{4a}{4a} \textbf{ to get like denominators, and simplify } \left(\frac{b}{2a}\right)^2.$$

$$= \frac{b^2 - 4ac}{4a^2} \qquad \textbf{Simplify the right side.}$$

Step 3 Solve the equation.

$$\sqrt{\left(x + \frac{b}{2a}\right)^2} = \sqrt{\frac{b^2 - 4ac}{4a^2}} \qquad \textbf{Take the square root of each side.}$$

$$x + \frac{b}{2a} = \pm\frac{\sqrt{b^2 - 4ac}}{2a} \qquad \textbf{Simplify the right side. } \frac{1}{\sqrt{4a^2}} = \pm\frac{1}{2a}$$

$$x = -\frac{b}{2a} \pm \frac{\sqrt{b^2 - 4ac}}{2a} \qquad \textbf{Subtract } \frac{b}{2a} \textbf{ from each side.}$$

$$x = \frac{-b \pm \sqrt{b^2 - 4ac}}{2a} \qquad \textbf{Simplify.}$$

Rule	Quadratic Formula

If $ax^2 + bx + c = 0$, and $a \neq 0$, then

$$x = \frac{-b \pm \sqrt{b^2 - 4ac}}{2a}$$

Be sure to write a quadratic equation in standard form before using the quadratic formula.

Video Tutor Help
Visit: PHSchool.com
Web Code: ate-0775

1 EXAMPLE Using the Quadratic Formula

Solve $x^2 + 6 = 5x$.

$x^2 - 5x + 6 = 0$ Subtract 5x from each side and write in standard form.

$x = \dfrac{-b \pm \sqrt{b^2 - 4ac}}{2a}$ Use the quadratic formula.

$x = \dfrac{-(-5) \pm \sqrt{(-5)^2 - (4)(1)(6)}}{2(1)}$ The coefficient of x^2 is 1. Substitute 1 for a, -5 for b, and 6 for c.

$x = \dfrac{5 \pm \sqrt{1}}{2}$ Simplify $-(-5)$ and the radicand.

$x = \dfrac{5 + 1}{2}$ or $x = \dfrac{5 - 1}{2}$ Write as two equations.

$x = 3$ or $x = 2$ Simplify.

Check   Substitute 3 for x.   Substitute 2 for x.

$(3)^2 + 6 \stackrel{?}{=} 5(3)$ $(2)^2 + 6 \stackrel{?}{=} 5(2)$

$9 + 6 \stackrel{?}{=} 15$ $4 + 6 \stackrel{?}{=} 10$

$15 = 15$ ✔ $10 = 10$ ✔

 **Quick Check**   **1** Use the quadratic formula to solve each equation.
     a. $x^2 - 2x - 8 = 0$ **b.** $x^2 - 4x = 117$

When the radicand in the quadratic formula is not a perfect square, you can use a calculator to approximate the solutions of an equation.

Vocabulary Tip

The radicand is the quantity inside the radical symbol. The radicand of $\sqrt{b^2 - 4ac}$ is $b^2 - 4ac$.

2 EXAMPLE Finding Approximate Solutions

Solve $2x^2 + 4x - 7 = 0$. Round the solutions to the nearest hundredth.

$x = \dfrac{-b \pm \sqrt{b^2 - 4ac}}{2a}$ Use the quadratic formula.

$x = \dfrac{-4 \pm \sqrt{4^2 - (4)(2)(-7)}}{2(2)}$ Substitute 2 for a, 4 for b, and -7 for c.

$x = \dfrac{-4 \pm \sqrt{72}}{4}$

$x = \dfrac{-4 + \sqrt{72}}{4}$ or $x = \dfrac{-4 - \sqrt{72}}{4}$ Write as two equations.

$x \approx \dfrac{-4 + 8.49}{4}$ or $x \approx \dfrac{-4 - 8.49}{4}$ Use a calculator. $\sqrt{72} \approx 8.49$

$x \approx 1.12$ or $x \approx -3.12$ Simplify. Round to the nearest hundredth.

 **Quick Check**   **2** Solve each equation. Round to the nearest hundredth.
     a. $-3x^2 + 5x - 2 = 0$ **b.** $7x^2 - 2x - 8 = 0$

You can use the quadratic formula to solve real-world problems. You must decide whether a solution makes sense in the real-world situation. For example, a negative value for time would not be a reasonable solution in most situations.

3 EXAMPLE Real-World Problem Solving

Sports Suppose a football player kicks a ball and gives it an initial upward velocity of 47 ft/s. The starting height of the football is 3 ft. If no one catches the football, how long will it be in the air?

Step 1 Use the vertical motion formula.

$$h = -16t^2 + vt + c$$ **The initial upward velocity is v, and the starting height is c.**

$$0 = -16t^2 + 47t + 3$$ **Substitute 0 for h, 47 for v, and 3 for c.**

Step 2 Use the quadratic formula.

$$x = \frac{-b \pm \sqrt{b^2 - 4ac}}{2a}$$

$$t = \frac{-47 \pm \sqrt{(47)^2 - (4)(-16)(3)}}{2(-16)}$$ **Substitute −16 for a, 47 for b, 3 for c, and t for x.**

$$t = \frac{-47 \pm \sqrt{2209 + 192}}{-32}$$ **Simplify.**

$$t = \frac{-47 \pm \sqrt{2401}}{-32}$$

$$t = \frac{-47 + 49}{-32} \quad \text{or} \quad t = \frac{-47 - 49}{-32}$$ **Write as two equations.**

$$t \approx -0.06 \quad \text{or} \quad t = 3$$ **Simplify. Use the positive answer because it is the only reasonable answer in this situation.**

● The football will be in the air for 3 seconds.

 **Quick Check** ❸ A football player kicks a ball with an initial upward velocity of 38.4 ft/s from a starting height of 3.5 ft.
a. Substitute the values into the vertical motion formula. Let $h = 0$.
b. Solve. If no one catches the ball, how long will it be in the air? Round to the nearest tenth of a second.

2 Choosing an Appropriate Method

There are many methods for solving a quadratic equation. You can always use the quadratic formula, but sometimes another method may be easier.

Method	When to Use
Graphing	Use if you have a graphing calculator handy.
Square Roots	Use if the equation has no x term.
Factoring	Use if you can factor the equation easily.
Completing the Square	Use if the x^2 term is 1, but you cannot factor the equation easily.
Quadratic Formula	Use if the equation cannot be factored easily or at all.

 **EXAMPLE** **Choosing an Appropriate Method**

Which method(s) would you choose to solve each equation? Justify your reasoning.

a. $2x^2 - 6 = 0$ Square roots; there is no x term.

b. $6x^2 + 13x - 17 = 0$ Quadratic formula; the equation cannot be factored easily.

c. $x^2 + 2x - 15 = 0$ Factoring; the equation is easily factorable.

d. $16x^2 - 96x + 45 = 0$ Quadratic formula; the equation cannot be factored easily, and the numbers are large.

e. $x^2 - 7x + 4 = 0$ Quadratic formula, completing the square, or graphing; the coefficient of the x^2 term is 1, but the equation is not factorable.

 Quick Check **4** Which method(s) would you choose to solve each equation? Justify your reasoning.

a. $13x^2 - 5x + 21 = 0$ **b.** $x^2 - x - 30 = 0$ **c.** $144x^2 = 25$

EXERCISES

For more exercises, see *Extra Skill and Word Problem Practice*.

Practice and Problem Solving

A **Practice by Example**

Example 1
(page 586)

 **for Help**

Use the quadratic formula to solve each equation. If necessary, round answers to the nearest hundredth.

1. $2x^2 + 5x + 3 = 0$ **2.** $5x^2 + 16x - 84 = 0$ **3.** $4x^2 - 12x + 9 = 0$

4. $3x^2 + 47x = -30$ **5.** $12x^2 - 77x - 20 = 0$ **6.** $3x^2 + 39x + 108 = 0$

7. $3x^2 + 40x - 128 = 0$ **8.** $2x^2 - 9x - 221 = 0$ **9.** $5x^2 - 68x = 192$

Example 2
(page 586)

10. $5x^2 + 13x - 1 = 0$ **11.** $2x^2 - 24x + 33 = 0$ **12.** $7x^2 + 100x - 4 = 0$

13. $8x^2 - 3x - 7 = 0$ **14.** $6x^2 + 5x - 40 = 0$ **15.** $3x^2 - 11x - 2 = 0$

Example 3
(page 587)

For Exercises 16 and 17, use the vertical motion formula $h = -16t^2 + vt + c$.

16. A child tosses a ball upward with a starting velocity of 10 ft/s from a height of 3 ft.
  a. Substitute the values into the vertical motion formula. Let $h = 0$.
  b. Solve. If it is not caught, how long will the ball be in the air? Round to the nearest tenth of a second.

17. A soccer ball is kicked with a starting upward velocity of 50 ft/s from a starting height of 3.5 ft.
  a. Substitute the values into the vertical motion formula. Let $h = 0$.
  b. Solve. If no one touches the ball, how long will the ball be in the air? Round to the nearest tenth of a second.

Example 4
(page 588)

Which method(s) would you choose to solve each equation? Justify your reasoning.

18. $x^2 + 2x - 13 = 0$ **19.** $4x^2 - 81 = 0$ **20.** $9x^2 - 31x = 51$

21. $3x^2 - 5x + 9 = 0$ **22.** $x^2 + 4x - 60 = 0$ **23.** $-4x^2 + 3x + 2 = 0$

 **Apply Your Skills**

Use any method you choose to solve each equation. If necessary, round to the nearest hundredth.

24. $2t^2 = 72$

25. $3x^2 + 2x - 4 = 0$

26. $5b^2 - 10 = 0$

27. $3x^2 + 4x = 10$

28. $m^2 - 4m = -4$

29. $13n^2 - 117 = 0$

30. $3s^2 - 4s = 2$

31. $5b^2 - 2b - 7 = 0$

32. $15x^2 - 12x - 48 = 0$

GO for Help

For a guide to solving
Exercise 33, see p. 591.

 33. Vertical Motion Suppose you throw a ball upward with a starting velocity of 30 ft/s. The ball is 6 ft high when it leaves your hand. After how many seconds will it hit the ground? Use the vertical motion formula $h = -16t^2 + vt + c$.

34. Geometry Suppose a rectangle has an area of 60 ft^2 and dimensions x and $(x + 1)$.
   a. Estimate each dimension of the rectangle to the nearest integer.
   b. Write a quadratic equation and use the quadratic formula to find each dimension to the nearest hundredth.

35. Writing Compare the way you solve the linear equation $mx + b = 0$ with the way you solve the quadratic equation $ax^2 + bx + c = 0$.

Geometry **Find the base and height of each triangle below. If necessary, round to the nearest hundredth.**

36.

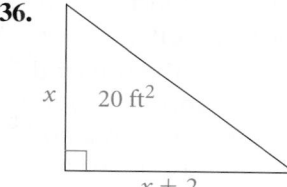

x 20 ft^2

$x + 2$

37.

Area = 50 cm^2

x

$x + 6$

38. Open-Ended Write a problem in which you find the area of a rectangle that you can solve using a quadratic equation. Draw a diagram and solve.

39. Critical Thinking How you can tell from the quadratic formula that a quadratic equation has one solution? Explain.

GO **nline**

Homework Video Tutor

Visit: PHSchool.com
Web Code: ate-1006

 **40. Multiple Choice** Refer to the cartoon. Suppose the man's starting upward velocity v is 5 ft/s. Use $0 = -16t^2 + vt + c$, where c is the starting height. Find the number of seconds t before he hits the water.

   Ⓐ about 1.6 s   Ⓑ about 1.9 s
   Ⓒ about 3.0 s   Ⓓ about 3.2 s

CLOSE TO HOME by John McPherson

LET'S SEE. 50 FEET UP. ACCELERATION OF GRAVITY IS 32 FEET/SEC²... WHICH MEANS... ..I'LL BE GOING 87 MILES AN HOUR WHEN I HIT THE POOL!!

THERE ARE TIMES WHEN BEING A WHIZ AT PHYSICS CAN BE A DEFINITE DRAWBACK.

C Challenge **41. Population** The function below models the United States population P in millions since 1900, where t is the number of years after 1900.

$$P = 0.0089t^2 + 1.1149t + 78.4491$$

   a. Use the function to estimate the United States population the year you graduate from high school.
   b. Estimate the United States population in 2025.
   c. Use the function to predict when the population will reach 300 million.

42. Critical Thinking The two solutions of any quadratic equation are

$$\frac{-b + \sqrt{b^2 - 4ac}}{2a} \text{ and } \frac{-b - \sqrt{b^2 - 4ac}}{2a}.$$

a. Find a formula for the sum of the solutions.
b. One solution of $2x^2 + 3x - 104 = 0$ is -8. Use the formula you found in part (a) to find the second solution.

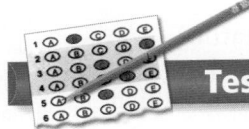

Test Prep

Short Response

The Gateway to the West

The Gateway Arch in St. Louis, Missouri, was completed in 1965. The arch spans 630 feet at its base and is 630 feet tall. More than 5100 tons of steel and 38,100 tons of concrete were used in its construction.

43. Use the data in the article above to answer the following questions.
   a. How many pounds of concrete were used in the construction of the Gateway Arch? Write your answer in scientific notation.
   b. How many more tons of concrete than steel were used? Write your answer in scientific notation.
   c. Suppose that a cleaner at the top of the Gateway Arch drops a cleaning brush. Use the vertical motion formula $h = -16t^2 + vt + c$. The starting upward velocity v is 0 and c is the starting height. How many seconds will the brush take to hit the ground?

44. Find the solutions to the equation $6x^2 - 40 = 11x$. Round to the nearest tenth.

Multiple Choice

45. The expression $\dfrac{9 \pm \sqrt{(-9)^2 - 4(5)(-7)}}{2(5)}$ gives the solutions to which of the following equations?

   A. $-9x^2 + 5x = -7$ **B.** $5x^2 + 7x = 9$
   C. $5x^2 - 9x = -7$ **D.** $5x^2 - 9x = 7$

46. The graph of $y = 15x^2 - 59x - 112$ crosses the x-axis closest to which of the following x-values?

   F. -1 **G.** 0   **H.** 3   **J.** 6

Mixed Review

Lesson 10-5

Solve each equation by completing the square. If necessary, round to the nearest hundredth.

47. $d^2 - 10d + 13 = 0$ **48.** $z^2 + 3z = -2$ **49.** $3x^2 + 18x - 1 = 0$

Lesson 9-8

Factor by grouping.

50. $2c^2 + 11c + 15$ **51.** $3z^2 + 10z - 8$ **52.** $5n^2 - 33n - 14$
53. $12v^2 + 32v - 35$ **54.** $6x^2 - 13x + 5$ **55.** $15t^2 + 19t + 6$

Understanding Word Problems   Read the exercise below and then follow along with what Hannah thinks and writes. Check your understanding with the exercise at the bottom of the page.

Vertical Motion Suppose you throw a ball in the air with a starting upward velocity of 30 ft/s. The ball is 6 ft high when it leaves your hand. After how many seconds will it hit the ground? Use the vertical motion formula $h = -16t^2 + vt + c$.

What Hannah Thinks

First, I'll write the values from the word problem and the vertical motion formula.

Next, I will substitute the starting upward velocity and the initial height into the formula.

When the ball hits the ground, the height will be zero. So *h* is zero.

I can use the quadratic formula to solve the equation. I'll write the quadratic formula and the values of *a*, *b*, and *c*.

I'll substitute *t* for *x* and substitute the values of *a*, *b*, and *c* into the quadratic formula. Then I simplify to find *t* when *h* equals zero.

Only the positive answer makes sense because I solved for time. So I'll write the positive answer with the units.

What Hannah Writes

starting upward velocity = 30 ft/s
starting height = 6 ft

$h = -16t^2 + vt + c$

$h = -16t^2 + 30t + 6$

$0 = -16t^2 + 30t + 6$

$x = \dfrac{-b \pm \sqrt{b^2 - 4ac}}{2a}$

$a = -16 \qquad b = 30 \qquad c = 6$

$t = \dfrac{-30 \pm \sqrt{30^2 - (4(-16)(6))}}{2(-16)}$

$ = \dfrac{-30 \pm \sqrt{900 - (-384)}}{-32}$

$t = \dfrac{-30 \pm \sqrt{1284}}{-32}$

$t \approx 2.1 \qquad \text{or} \qquad t \approx -0.2$

about 2.1 seconds

EXERCISE

Geometry Find the base and height of the triangle. If necessary, round to the nearest hundredth.

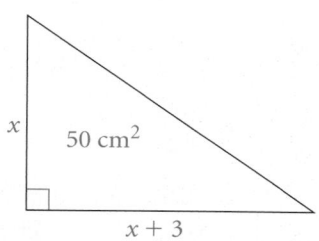

Using the Discriminant

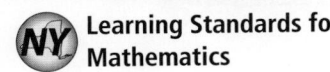

A.A.8 Analyze and solve verbal problems that involve quadratic equations.

✓ **Check Skills You'll Need**

GO for Help Lessons 2-3 and 10-6

Evaluate $b^2 - 4ac$ for the given values of a, b, and c.

1. $a = 3, b = 4, c = 8$
  −80

2. $a = -2, b = 0, c = 9$
  72

3. $a = 11, b = -5, c = 7$
  −283

Solve using the quadratic formula. If necessary, round to the nearest hundredth.

4. $3x^2 - 7x + 1 = 0$
  2.18, 0.15

5. $4x^2 + x - 1 = 0$
  0.39, −0.64

6. $x^2 - 12x + 35 = 0$
  7, 5

🔊 **New Vocabulary** • discriminant

---

1 Number of Real Solutions of a Quadratic Equation

Quadratic equations can have two, one, or no solutions. You can determine how many solutions a quadratic equation has, before you solve it, by using the discriminant. The **discriminant** is the expression under the radical in the quadratic formula.

$$x = \frac{-b \pm \sqrt{b^2 - 4ac}}{2a} \quad \longleftarrow \quad \text{the discriminant}$$

Consider the graphs of the functions below and the discriminant of each related equation.

$y = x^2 - 6x + 3$

$y = x^2 - 6x + 9$

$y = x^2 - 6x + 12$

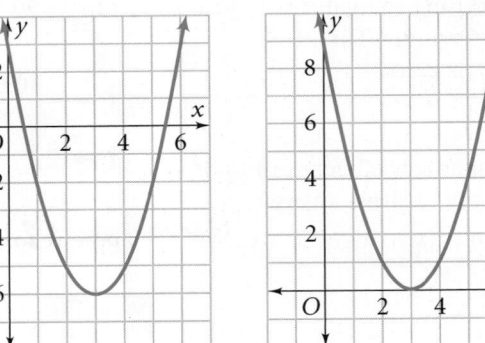

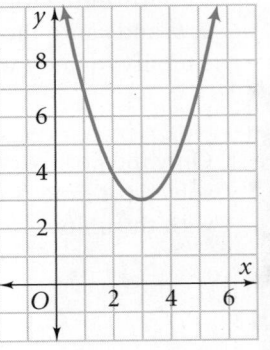

$x^2 - 6x + 3 = 0$	$x^2 - 6x + 9 = 0$	$x^2 - 6x + 12 = 0$
$b^2 - 4ac$: $(-6)^2 - 4(1)(3)$	$(-6)^2 - 4(1)(9)$	$(-6)^2 - 4(1)(12)$
$= 36 - 12$	$= 36 - 36$	$= 36 - 48$
$= 24$	$= 0$	$= -12$

The relationship you see between the graphs and discriminants above is true for all cases. If the discriminant is positive, there are two solutions. If the discriminant is zero, there is one solution. If the discriminant is negative, there are no solutions.

This graph shows the three cases together.

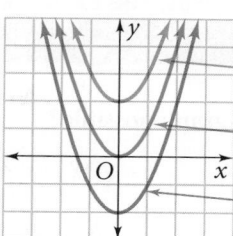

— Discriminant is negative.

— Discriminant is 0.

— Discriminant is positive.

 Key Concepts

Property	Property of the Discriminant

For the quadratic equation $ax^2 + bx + c = 0$, where $a \neq 0$, you can use the value of the discriminant to determine the number of solutions.

If $b^2 - 4ac > 0$, there are two solutions.

If $b^2 - 4ac = 0$, there is one solution.

If $b^2 - 4ac < 0$, there are no solutions.

1 EXAMPLE Using the Discriminant

Find the number of solutions of $3x^2 - 5x = 1$.

$3x^2 - 5x - 1 = 0$ **Write in standard form.**

$b^2 - 4ac = (-5)^2 - (4)(3)(-1)$ **Evaluate the discriminant. Substitute for *a*, *b*, and *c*.**

$\quad\quad\quad = 25 - (-12)$ **Use the order of operations.**

$\quad\quad\quad = 37$ **Simplify.**

● Since $37 > 0$, the equation has two solutions.

 **Quick Check** ❶ Find the number of solutions for each equation.

   a. $x^2 = 2x - 3$ **b.** $3x^2 - 4x = 7$ **c.** $5x^2 + 8 = 2x$

2 EXAMPLE <u>Real-World</u> 🌐 <u>Problem Solving</u>

Physics A construction worker on the ground tosses an apple to a fellow worker who is 20 ft above the ground. The starting height of the apple is 5 ft. Its initial upward velocity is 30 ft/s. Will the apple reach the second worker?

$h = -16t^2 + vt + c$ **Use the vertical motion formula.**

$20 = -16t^2 + 30t + 5$ **Substitute 20 for *h*, 30 for *v*, and 5 for *c*.**

$0 = -16t^2 + 30t - 15$ **Write in standard form.**

$b^2 - 4ac = (30)^2 - 4(-16)(-15)$ **Evaluate the discriminant.**

$\quad\quad\quad = 900 - 960$ **Use the order of operations.**

$\quad\quad\quad = -60$ **Simplify.**

● The discriminant is negative. The apple will not reach the second worker.

 **Quick Check** ❷ Suppose the same construction worker tosses an apple with an initial upward velocity of 32 ft/s. Will the apple reach the second worker?

EXERCISES

For more exercises, see *Extra Skill and Word Problem Practice.*

Practice and Problem Solving

A Practice by Example

Example 1
(page 593)

for Help

For which discriminant is each graph possible?

1.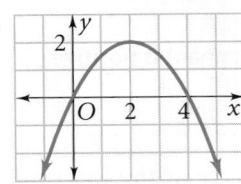

2.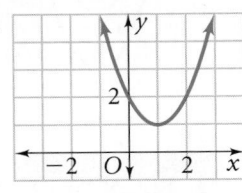

3.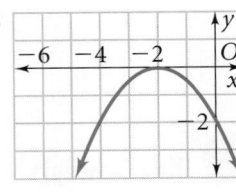

A. $b^2 - 4ac = 4$ **B.** $b^2 - 4ac = 0$ **C.** $b^2 - 4ac = -4$

Mental Math **Find the number of solutions of each equation.**

4. $x^2 - 3x + 4 = 0$ **5.** $x^2 - 6x + 9 = 0$ **6.** $x^2 + 4x - 2 = 0$

7. $x^2 - 1 = 0$ **8.** $x^2 - 2x - 3 = 0$ **9.** $x^2 + x = 0$

10. $2x^2 - 3x + 4 = 0$ **11.** $0 = x^2 - 6x + 5$ **12.** $x^2 - 7x + 6 = 0$

13. $x^2 + 2x + 1 = 0$ **14.** $0 = 2x^2 + 4x - 3$ **15.** $0 = x^2 + 2x + 9$

Example 2
(page 593)

16. Home Improvements The Reeves family's garden is 18 ft long and 15 ft wide. They want to decrease the length by x feet and increase the width by the same amount. The equation $A = (18 - x)(15 + x)$ models the new area of the garden. What value of x, if any, will give a new area of 280 ft^2?

17. Business An apartment rental agency uses the formula $I = 5400 + 300n - 50n^2$ to find its monthly income I based on renting n apartments. Will the agency's monthly income ever reach $7000? Explain.

18. Physics Suppose the equation $h = -16t^2 + 35t$ models the altitude a football will reach t seconds after it is kicked. Is the given altitude possible?
 a. $h = 16$ ft **b.** $h = 20$ ft **c.** $h = 30$ ft **d.** $h = 35$ ft

B Apply Your Skills

Find the number of x-intercepts of the related function of each equation.

19. $2x^2 + 4x = -15$ **20.** $4x^2 + 5x = -2$ **21.** $x^2 - 8x = -12$

22. $\frac{1}{2}x^2 + 4x = 7$ **23.** $0.25x^2 - 1.2x + 3.2 = 0$ **24.** $5x^2 = 3.5 + 4.7x$

25. Business A software company is producing a new computer application. The equation $S = p(54 - 0.75p)$ relates price p in dollars to total sales S in thousands of dollars.
 a. Write the equation in standard form.
 b. Use the discriminant to determine if it is possible for the company to earn $1,000,000 in sales.
 c. According to the model, what price would generate the greatest sales?
 d. Critical Thinking Total sales S decrease as p increases beyond the value in part (c). Why does this make sense in the given situation? Explain.

GO **O**nline
Homework Video Tutor
Visit: PHSchool.com
Web Code: ate-1007

26. Open-Ended For the equation $x^2 + 4x + k = 0$, find all values of k such that the equation has the given number of solutions.
 a. none **b.** one **c.** two

27. You can use a spreadsheet like the one at the right to find the discriminant for each value of *b* shown in column A.

	A	B	C	
1	b	x^2 + bx + 1 = 0	x^2 + bx + 2 = 0	
2	−3	▪	▪	
3	−2	▪	▪	
4	−1	▪	▪	
5	0	▪	▪	
6	1	▪	▪	
7	2	▪	▪	
8	3	▪	▪	

  a. What spreadsheet formula would you use to find the value in cell B2? What value would you use for cell C2?

  b. Find the integer values of *b* for which $x^2 + bx + 1 = 0$ has no solutions.

28. Electrical Engineering The function $P = 3i^2 - 2i + 450$ models the power *P* in an electric circuit with a current *i*. Can the power in this circuit ever be zero? If so, what is the value of *i*?

29. Error Analysis Kenji claimed that the discriminant of $2x^2 + 5x - 1 = 0$ was 17. What error did he make?

30. Find the value of the discriminant and the solutions of each equation. If necessary, round to the nearest hundredth.

  a. $x^2 - 6x + 5 = 0$ **b.** $x^2 + x - 20 = 0$ **c.** $2x^2 - 7x - 3 = 0$

  d. Reasoning When the discriminant is a perfect square, are the solutions rational or irrational? Explain.

Real-World Connection

Careers An electrical engineer designs circuits that are used in a wide variety of devices.

Does the graph of each function cross the *x*-axis? If so, find the *x*-intercepts.

31. $y = x^2 - 2x + 5$ **32.** $y = 2x^2 - 4x + 3$ **33.** $y = 4x^2 + x - 5$

34. $y = -3x^2 - x + 2$ **35.** $y = x^2 - 5x + 7$ **36.** $y = 2x^2 - 3x - 5$

37. Writing How can you use the discriminant to write an equation that has two solutions?

Challenge

Reasoning For each condition given, tell whether $ax^2 + bx + c = 0$ will have two solutions *sometimes*, *always*, or *never*.

38. $b^2 < 4ac$ **39.** $b^2 = 0$ **40.** $ac < 0$

41. Critical Thinking The graph of a quadratic equation includes the points $(2, -1)$ and $(3, 2)$. How many solutions does the related equation have? Explain.

42. Critical Thinking The discriminant of $0 = 2x^2 + 6x + 7$ is −20. The discriminant of $0 = 2x^2 + 8x + 10$ is −16. Without graphing, determine which related function has a vertex closer to the *x*-axis. Explain.

NY REGENTS

Test Prep

Multiple Choice

43. Which of the following is equal to $b^2 - 4ac$, if $a = 1$, $b = 7$, and $c = -4$?
  A. −111   **B.** 33   **C.** 65   **D.** 113

44. Which of the following equations has NO real-number solutions?
  F. $3x^2 - 5x + 1 = 0$ **G.** $3x^2 - 5x + 4 = 0$
  H. $-3x^2 - 11x + 4 = 0$ **J.** $-2x^2 - 3x + 1 = 0$

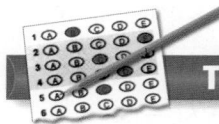

45. In a quadratic equation, $a = 1.1$, $b = 5$, and $c = 5$. How many real-number solutions does the equation have?
  A. 0   **B.** 1   **C.** 2   **D.** 3

46. Which does NOT show the number of real-number solutions a quadratic equation has?
  F. quadratic formula   **G.** graph of the equation
  H. discriminant   **J.** degree of the polynomial

47. The graph of a quadratic function is shown at the right. Based on the graph, what is the most likely value for the discriminant?
  A. 0   **B.** -2
  C. 2   **D.** $(-4, 1)$

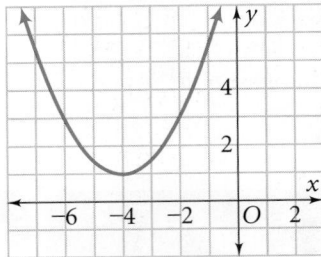

Short Response   **48.** A rectangle has a perimeter of 50 cm. Is it possible for it to have an area of 136 cm^2? If so, what are the dimensions? Show your work.

Mixed Review

Lesson 10-6   **Use the quadratic formula to solve each equation. If necessary, round to the nearest hundredth.**

  49. $4x^2 + 4x - 3 = 0$ **50.** $x^2 + 2x - 7 = 0$ **51.** $6x^2 - 2x - 1 = 0$

  52. $x^2 + x = 5$ **53.** $3x^2 - 8x + 1 = 0$ **54.** $2x^2 - 7x = -6$

Lesson 8-8   **Find the balance in each account.**

  55. $1000 principal earning 3% compounded quarterly; after 3 years

  56. $200 principal earning 4.5% compounded quarterly; after 10 years

  57. $5000 principal earning 5% annual interest compounded monthly; after 4 years

Lesson 8-6   **Determine whether each sequence is *arithmetic* or *geometric*.**

  58. $5, 9, 13, \ldots$ **59.** $-11, -16, -21, \ldots$ **60.** $10, 20, 40, \ldots$ **61.** $3, 6, 9, \ldots$

✓ Checkpoint Quiz 2   Lessons 10-4 through 10-7

Solve each equation. If necessary, round to the nearest tenth. If there is no solution, write *no solution*.

  1. $(x + 3)(x - 7) = 0$ **2.** $x^2 + 12x + 27 = 0$ **3.** $x^2 - 5x = 50$

  4. $x^2 + 2x - 1 = 0$ **5.** $x^2 - 5x - 4 = 0$ **6.** $x^2 - 8x - 33 = 0$

  7. $4x^2 - x - 3 = 0$ **8.** $4x^2 - x + 3 = 0$ **9.** $2x^2 - 3x + 1 = 0$

  10. Use the discriminant to determine the number of solutions of the equation $4x^2 - 3x + 5 = 0$.

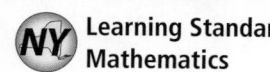

Choosing a Linear, Quadratic, or Exponential Model

NY Learning Standards for Mathematics

A.R.2 Recognize, compare, and use an array of representational forms.

✓ Check Skills You'll Need

Graph each function.

1. $y = 3x - 1$

2. $y = \frac{1}{4}x + 2$

3. $y = 2^x$

4. $y = \left(\frac{1}{3}\right)^x$

5. $y = x^2 + 5$

6. $y = 2x^2 - 1$

GO for Help Lessons 6-2, 8-7, and 10-1

1 Choosing a Linear, Quadratic, or Exponential Model

You can use the linear, exponential, or quadratic functions you have studied to model some sets of data. Recall the general appearance of each type of function.

 Key Concepts

Summary	Linear, Quadratic, and Exponential Functions

Linear
$y = mx + b$

Quadratic
$y = ax^2 + bx + c$

Exponential
$y = a \cdot b^x$

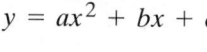

You may be able to use the graph of data points to determine a model for the data.

1 EXAMPLE Choosing a Model by Graphing

Graph each set of points. Which model is most appropriate for each set?

a. $(-3, 6), (-2, 2), (0, -2),$
$(3, 6), (1, -1), (2, 2)$

b. $(-2, 5), (0, 3),$
$(1, 2.5), (2, 2)$

c. $(-2, 4), (-1, 2),$
$(1, -2), (2, -4)$

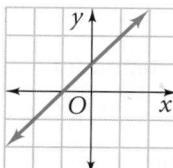

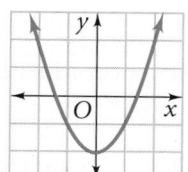

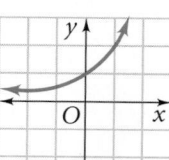

Quadratic model

Exponential model

Linear model

✓ **Quick Check** ① Graph each set of points. Which model is most appropriate for each set?
    a. $(-1.5, -2), (0, 2), (1, 4), (2, 6)$ **b.** $(-1, 1), (0, 0), (1, 1), (2, 4)$
    c. $(-1, 0.5), (0, 1), (1, 2), (2, 4)$

You can also analyze data numerically to find the best model.

In Lesson 5-7, you learned that the terms of an arithmetic sequence have a common difference. You can model an arithmetic sequence with a linear function.

	x	y	
+1	−2	−2	+4
+1	−1	2	+4
+1	0	6	+4
+1	1	10	+4
	2	14	

The *y*-coordinates have a common difference of 4. A linear model fits the data.

In Lesson 8-6, you learned that the terms of a geometric sequence have a common ratio. You can model a geometric sequence with an exponential function.

	x	y	
+1	−2	$\frac{1}{9}$	×3
+1	−1	$\frac{1}{3}$	×3
+1	0	1	×3
+1	1	3	×3
	2	9	

The *y*-coordinates have a common ratio of 3. An exponential model fits the data.

Data from quadratic functions show a different pattern. For linear data, the first differences are the same. For quadratic data, the second differences are the same. If data have a common second difference, then you can model them with a quadratic function.

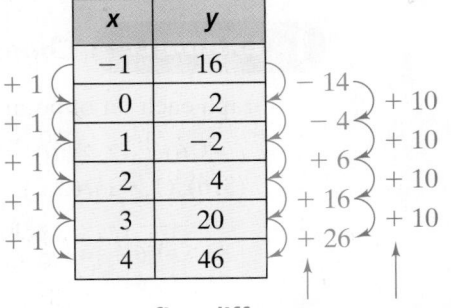

first differences   second differences

The *y*-coordinates have a common second difference of 10. A quadratic model fits the data.

② **EXAMPLE** **Modeling Data**

x	y
−1	20
0	8
1	3.2
2	1.28
3	0.512

a. Which kind of function best models the data at the left? Write an equation to model the data.

Step 1 Graph the data.

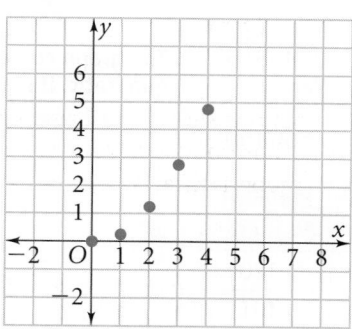

Step 2 The data appear to suggest an exponential model. Test for a common ratio.

x	y	
−1	20	$8 \div 20 = 0.4$
0	8	$3.2 \div 8 = 0.4$
1	3.2	$1.28 \div 3.2 = 0.4$
2	1.28	$0.512 \div 1.28 = 0.4$
3	0.512	

(+1 between each x value)

There is a common ratio, 0.4.

Step 3 Write an exponential model.

Relate $y = a \cdot b^x$

Define Let a = the initial value, 8.
Let b = the decay factor, 0.4.

Write $y = 8 \cdot 0.4^x$

Step 4 Test two points other than $(0, 8)$.

$y = 8 \cdot 0.4^1$ $y = 8 \cdot 0.4^3$

$y = 8 \cdot 0.4$ $y = 8 \cdot 0.064$

$y = 3.2$ $y = 0.512$

$(1, 3.2)$ and $(3, 0.512)$ are both data points.

The equation $y = 8 \cdot 0.4^x$ models the data.

x	y
0	0
1	0.3
2	1.2
3	2.7
4	4.8

b. Which kind of function best models the data at the left? Write an equation to model the data.

Step 1 Graph the data.

Step 2 The data appear to be quadratic. Test for a common second difference.

x	y		
0	0	+ 0.3	
1	0.3		+ 0.6
2	1.2	+ 0.9	+ 0.6
3	2.7	+ 1.5	+ 0.6
4	4.8	+ 2.1	

(+1 between each x value)

There is a common second difference, 0.6.

Step 3 Write a quadratic model.

$y = ax^2$

$1.2 = a(2)^2$ **Use a point other than (0, 0) to find a.**

$1.2 = 4a$ **Simplify.**

$0.3 = a$ **Divide each side by 4.**

$y = 0.3x^2$ **Write a quadratic function.**

Step 4 Test two points other than $(2, 1.2)$ and $(0, 0)$.

$y = 0.3(3)^2$ $y = 0.3(4)^2$

$y = 0.3 \cdot 9$ $y = 0.3 \cdot 16$

$y = 2.7$ $y = 4.8$

$(3, 2.7)$ and $(4, 4.8)$ are both data points.

The equation $y = 0.3x^2$ models the data.

2 Which kind of function best models the data in each table? Write an equation to model the data.

a.

x	y
0	4
1	4.4
2	4.84
3	5.324
4	5.8564

b.

x	y
0	$\frac{1}{2}$
1	1
2	$1\frac{1}{2}$
3	2

c.

x	y
0	0
1	−0.5
2	−2
3	−4.5
4	−8

While real-world data seldom fall exactly into linear, exponential, or quadratic patterns, you can find a best-possible model.

3 EXAMPLE Real-World 🌐 Problem Solving

Zoology Suppose you are studying frogs that live in a nearby wetland area. The data at the right were collected by a local conservation organization. They indicate the number of frogs estimated to be living in the wetland area over a five-year period. Determine which kind of function best models the data. Write an equation to model the data.

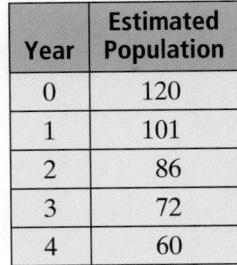

Year	Estimated Population
0	120
1	101
2	86
3	72
4	60

Step 1 Graph the data to decide which model is most appropriate.

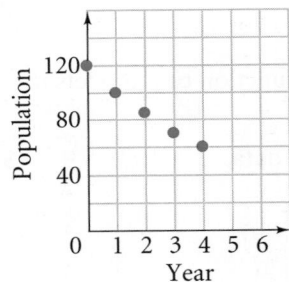

The graph curves, and it does not look quadratic. It may be exponential.

Real-World 🌐 Connection

Frog populations around the world have been declining over the past 20 years.

Step 2 Test for a common ratio.

	Year	Estimated Population	
+1	0	120	101 ÷ 120 ≈ 0.84
+1	1	101	86 ÷ 101 ≈ 0.85
+1	2	86	72 ÷ 86 ≈ 0.84
+1	3	72	60 ÷ 72 ≈ 0.83
	4	60	

The common ratio is roughly 0.84.

The population of frogs is roughly 0.84 times its value the previous year.

Step 3 Write an exponential model.

Relate $y = a \cdot b^x$

Define   Let a = the initial value, 120.
              Let b = the decay factor, 0.84.

Write $y = 120 \cdot 0.84^x$

Step 4 Test two points other than $(0, 120)$.

$$y = 120 \cdot 0.84^3 \qquad\qquad y = 120 \cdot 0.84^4$$
$$y \approx 71 \qquad\qquad\qquad\quad y \approx 60$$

The point $(3, 71)$ is close to the data point $(3, 72)$. The predicted value $(4, 60)$ matches the corresponding data point. The equation $y = 120 \cdot 0.84^x$ models the data.

 Quick Check ❸ The profits of a small company are shown in the table at the right. Let $x = 0$ correspond to the year 2000.
  a. Determine which kind of function best models the data.
  b. Write an equation to model the data.

Year	Profit (dollars)
0	0
1	5,000
2	20,000
3	44,000
4	79,000

EXERCISES

For more exercises, see *Extra Skill and Word Problem Practice.*

Practice and Problem Solving

 **Practice by Example**

Example 1
(page 597)

GO for Help

Graph each set of points. Which model is most appropriate for each set?

1. $(-2, -3), (-1, 0), (0, 1), (1, 0), (2, -3)$ **2.** $(-2, -8), (0, -4), (3, 2), (5, 6)$

3. $(-3, 6), (-1, 0), (0, -1), (1, -1.5)$ **4.** $(-2, 5), (-1, -1), (0, -3), (1, -1), (2, 5)$

5. $\left(-2, -5\frac{8}{9}\right), \left(-1, -5\frac{2}{3}\right), (0, -5), (2, 3)$ **6.** $(-3, 8), (-1, 6), (0, 5), (2, 3), (3, 2)$

Example 2
(page 599)

Which kind of function best models the data in each table? Write an equation to model the data.

7.

x	y
0	0
1	1.5
2	6
3	13.5
4	24

8.

x	y
0	-5
1	-3
2	-1
3	1
4	3

9.

x	y
0	0
1	2.8
2	11.2
3	25.2
4	44.8

10.

x	y
0	1
1	1.2
2	1.44
3	1.728
4	2.0736

11.

x	y
0	5
1	2
2	0.8
3	0.32
4	0.128

12.

x	y
0	2
1	1.5
2	1
3	0.5
4	0

Example 3
(page 600)

13. The table at the right shows the end-of-the-month balance in a checking account.
  a. Graph the data. Does the graph suggest a linear, exponential, or quadratic model?
  b. Find the differences of consecutive terms. Are they roughly the same?
  c. Estimate a common difference based on your answer to part (b).
  d. Write an equation to model the data.

Month	Balance (dollars)
1	58
2	123
3	187
4	251

Age (years)	Value (dollars)
0	16,500
1	14,500
2	12,750
3	11,200
4	9900

 14. Car Value The value of a car over several years is shown in the table at the left.
 a. Determine which model is most appropriate for the data.
 b. Write an equation to model the data.

 15. Physics Your class collected the data in the table at the right by rolling a ball down a ramp. The ramp had the same angle throughout.
 a. Find the differences of consecutive terms.
 b. Find the second differences.
 c. Write an equation to model the data. Let t be the time in seconds and d be the distance in centimeters.
 d. Based on your equation, how far would the ball have rolled down the ramp in 2.5 seconds?

Time (seconds)	Distance (centimeters)
0	0
1	41
2	164
3	370

B Apply Your Skills

16. The table below shows the projected population of a small town. Let $t = 0$ correspond to the year 2020.
 a. Graph the data. Does the graph suggest a linear, exponential, or quadratic model?
 b. What is the difference in years?
 c. Find the differences of consecutive terms. Divide by the difference in years to find possible common differences.
 d. Write a linear equation to model the data based on your answer to part (c).

Year	Population
0	5100
5	5700
10	6300
15	6900

Year	Population (millions)
0	4457
5	4855
10	5284
15	5691
20	6080

SOURCES: U. S. Census Bureau. Go to **www.PHSchool.com** for a data update.

Web Code: atg-9041

 17. Population The table at the left shows the world population in millions from 1980 to 2000. The year $t = 0$ corresponds to 1980.
 a. What is the difference in years?
 b. Find the differences of consecutive terms. Divide by the difference in years to find possible common differences.
 c. Find the average of the common differences you found in part (b).
 d. Write a linear equation to model the data based on your answer to part (c).
 e. Use your equation to predict the world population in 2010.

18. Writing Explain in writing to a classmate how to decide whether a linear, exponential, or quadratic function is the most appropriate equation to model a set of data.

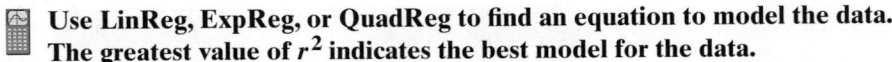 Use LinReg, ExpReg, or QuadReg to find an equation to model the data. The greatest value of r^2 indicates the best model for the data.

19.
x	y
0	1.7
1	1.4
2	4.7
3	7.9

20.
x	y
0	2.0
1	1.5
2	1.2
3	0.9

21.
x	y
0	2.8
1	1.4
2	2.7
3	9.8

22.
x	y
−1	4.3
0	5.1
1	4.3
2	2.2
3	1.3

23.
x	y
−1	4.6
0	3.5
1	2.4
2	1.3
3	0.2

24.
x	y
−1	0.04
0	0.10
1	0.26
2	0.68
3	1.76

GO Online
Homework Video Tutor
Visit: PHSchool.com
Web Code: ate-1008

25. a. Make a table of five points using consecutive x-values for each function. Find the common second difference.

   i. $f(x) = x^2 - 3$ **ii.** $f(x) = 3x^2$ **iii.** $f(x) = 4x^2 - 5x$

   b. What is the relationship between the common second difference and the coefficient of x^2?

   c. Critical Thinking Explain how you could use this relationship to help you model data if the function were not given.

26. Open-Ended Write a set of data you could model with a quadratic function.

27. Physics A group of students dropped ping-pong balls and found the distance the balls traveled, given a certain amount of time. The diagram below shows their results.

   a. Determine which model is most appropriate for the data.

   b. Write an equation to model the data. If necessary, round to the nearest tenth.

   c. Use your equation to predict the distance a ping-pong ball would fall in 2 s.

0.2 s	0.4 s	0.6 s	0.8 s	1.0 s
0.5 ft	2.2 ft	4.9 ft	8.7 ft	13.6 ft

C **Challenge**

28. Data Collection Complete the chart at the right for your town.

   a. Use ExpReg or QuadReg to find an equation to model the data.

   b. Use the equation you found in part (a) to predict the current population of your town.

   c. Use the equation you found in part (a) to predict the population of your town in 2010.

Year	Population (thousands)
1970	■
1980	■
1990	■
2000	■

29. Data Analysis There are times when data show trends but not consistent ratios or differences. The data in the table below show the total U.S. retail auto sales for the years 1980 through 2000. The value $t = 0$ corresponds to the year 1980.

   a. Find the ratios of consecutive entries in the sales columns. Round to the nearest hundredth.

   b. Find the differences of consecutive entries in the sales columns.

   c. Find the second differences of consecutive entries in the sales column.

   d. Which ratio is most different from the others? Explain.

   e. Critical Thinking If sales data increase every year, can a second difference be negative? Explain.

   f. Graph the data. Draw a curve to show the trend.

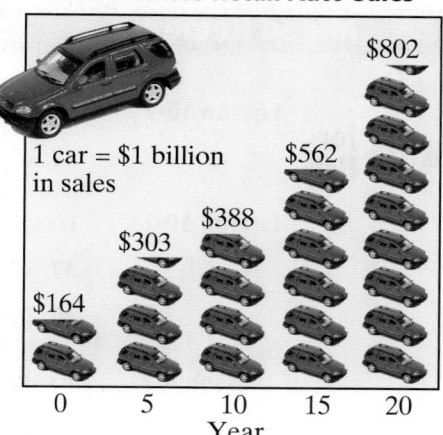

United States Retail Auto Sales

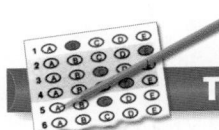
Multiple Choice

30. Which equation best models the data in the table at the right?
   A. $y = 4x$
   B. $y = 2x + 2$
   C. $y = 2^x$
   D. $y = 2x^2$

x	y
0	2
1	4
2	6
3	8
4	10

31. Which of the following sets of data is best described by an exponential model?
   F. $(-1, 16)$, $\left(-\frac{1}{2}, 4\right)$ $(0, 2)$, $\left(\frac{1}{2}, -2\right)$, $(1, 4)$
   G. $(1, -7)$, $(2, -4)$, $(3, -1)$, $(4, 2)$, $(5, 5)$
   H. $(1, 1)$, $(3, 3)$, $(0, 0.5)$, $(5, 8)$, $(7, 14)$
   J. $(-2, -4)$, $(-1, 3)$, $(0, 8)$, $(1, 14)$, $(2, 12)$

Short Response

32. The population of a town was 33,500 in 2000. The population is increasing by about 1.4% each year. Write an equation that will predict the population n years after 2000. Let 2000 correspond to $n = 0$. Predict the town's population in 2010.

Extended Response

33. Suppose you put marbles into a cup hanging from an elastic band (spring). You measure the distance d from the floor in centimeters as the number n of marbles is increased.

n	0	1	2	3	4	5
d	43.5	41	38.5	36	33.5	31

   a. Which type of model best fits this data set?
   b. Write an equation for the data.
   c. Suppose the pattern shown above continues. Find the least number of marbles you need to make the cup rest on the floor.

Mixed Review

GO for Help

Lesson 10-7

Find the number of x-intercepts of each function.

34. $y = -x^2$ **35.** $y = x^2 + 3x + 4$ **36.** $y = 4x^2 - 10x + 3$

Lesson 10-1

Graph each function.

37. $y = -2x^2$ **38.** $f(x) = \frac{1}{4}x^2$ **39.** $f(x) = x^2 + 4$

40. $y = -x^2 - 2$ **41.** $y = 3x^2 + 1$ **42.** $y = -\frac{1}{2}x^2 + 1$

Lesson 8-7

Evaluate each function rule for the given value.

43. $y = 2^x$ for $x = -3$ **44.** $f(x) = -2^x$ for $x = 5$

45. $g(t) = 2 \cdot 3^t$ for $t = -3$ **46.** $f(t) = 10 \cdot 5^t$ for $t = 2$

47. $y = \left(\frac{1}{2}\right)^t$ for $t = -4$ **48.** $y = 9 \cdot \left(\frac{3}{2}\right)^x$ for $x = 3$

Cubic Functions

In this chapter you learned about quadratic functions. You can also construct and explore functions in which the highest power is greater than two. Functions of the form $y = ax^3 + bx^2 + cx + d$, where $a \neq 0$, are called cubic functions.

EXAMPLE

Make a table of values and graph $y = x^3$ and $y = \frac{1}{3}x^3$.

$$y = x^3 \qquad\qquad\qquad\qquad y = \frac{1}{3}x^3$$

x	$y = x^3$	$y = \frac{1}{3}x^3$
-2	-8	$-2\frac{2}{3}$
-1	-1	$-\frac{1}{3}$
0	0	0
1	1	$\frac{1}{3}$
2	8	$2\frac{2}{3}$

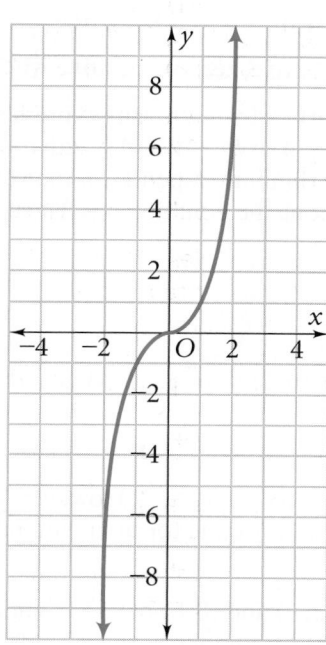
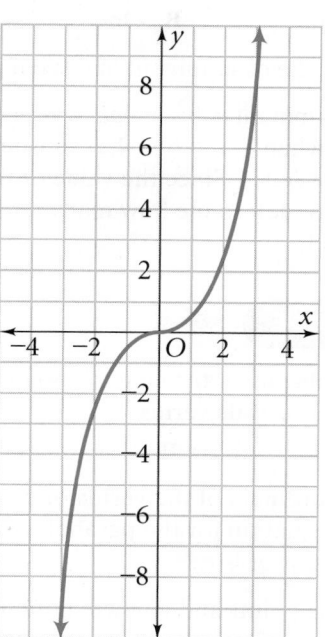

EXERCISES

1. **Critical Thinking** Describe the similarities and differences in the graphs in the Example.

2. **a.** Graph each of the following functions.
   i. $y = -x^3$ **ii.** $y = 2x^3$ **iii.** $y = -\frac{1}{4}x^3$
   b. Describe the similarities and differences in the graphs in part (a).

3. **Critical Thinking** Does a play the same role in cubic functions of the form $y = ax^3$ as it does in quadratic functions? Explain.

4. **a.** Do cubic graphs of the form $y = ax^3$ have an axis of symmetry?
   b. **Critical Thinking** Do cubic graphs have any form of symmetry? Explain.

5. The volume V of a sphere with radius r is given by the function $V = \frac{4}{3}\pi r^3$.
   a. Graph $V = \frac{4}{3}\pi r^3$ from $r = 0$ to $r = 2$.
   b. Use the graph to estimate the radius of a sphere with a volume of 10 ft^3.

Choosing "Cannot Be Determined"

Some multiple-choice questions do not contain enough information. One of the answer choices will then be "Cannot be determined." In such cases, you will not be able to find a specific answer. However, just because the answer choice "Cannot be determined" appears, do not assume that you cannot answer the question; the choice may have been put there as a distraction.

1 EXAMPLE

How many real solutions does the equation $x^2 + bx + 4 = 0$ have if $b > 0$?

A. None **B.** One **C.** Two **D.** Three **E.** Cannot be determined

Calculate the discriminant and see if it is negative, zero, or positive when $b > 0$.

The discriminant is $b^2 - 4ac = b^2 - 4(1)(4) = b^2 - 16$. This expression is negative when $-4 < b < 4$, zero when $b = -4$ or $b = 4$, and positive when $b < -4$ or $b > 4$. Since the value of the discriminant depends on the value of b, you cannot determine how many real solutions the equation has. The correct answer is E.

2 EXAMPLE

A parabola passes through the points $(2, 8)$, $(6, 2)$, and $(8, 8)$. What is the x-coordinate of its vertex?

A. 4 **B.** 5 **C.** 7 **D.** 8 **E.** Cannot be determined

The x-coordinate of the vertex of $y = ax^2 + bx + c$ is $-\frac{b}{2a}$. However, you don't have an equation for the parabola, so you can't use the formula. You might think that the answer is E.

Graph the points and you will notice that $(2, 8)$ and $(8, 8)$ have the same y value, so they are symmetrically placed about the axis of symmetry. The axis of symmetry will be $x = \frac{2 + 8}{2} = 5$. Since the x-coordinate of the vertex is on the axis of symmetry, the correct answer is B. Choice E is a distraction; it was put there in case you thought that the problem could not be done.

EXERCISES

1. Use the question below to answer parts a–c.
How many real solutions does the equation $x^2 + 4x - n^2 = 0$ have?
  A. None **B.** One **C.** Two **D.** Three **E.** Cannot be determined

  a. What is the discriminant of the equation?
  b. Explain why the discriminant is always positive for all values of n.
  c. What is the correct answer to the question?

2. If $x \neq 0$ and $y \neq 0$, and $x^2 + 2xy + y^2 = 0$, which of the following could be the ratio of x to y?
  A. $1 : 1$ **B.** $2 : 1$ **C.** $-1 : 1$ **D.** $1 : 2$ **E.** Cannot be determined

Chapter Review

Vocabulary Review

axis of symmetry (p. 551)
completing the square (p. 579)
discriminant (p. 592)
maximum (p. 551)
minimum (p. 551)
parabola (p. 550)

quadratic equation (p. 566)
quadratic formula (p. 585)
quadratic function (p. 550)
quadratic parent function (p. 550)
roots of the equation (p. 566)
standard form of a quadratic equation
(p. 566)

standard form of a quadratic function
(p. 550)
vertex (p. 551)
Zero-Product Property (p. 572)
zeros of the function (p. 566)

Choose the term that correctly completes each sentence.

1. The U-shaped graph of a quadratic function is a *(parabola, perfect square)*.

2. If the quadratic expression $ax^2 + bx + c$ *cannot* be factored, one good way to solve the equation $ax^2 + bx + c = 0$ is to use *(completing the square, the Zero-Product Property)*.

3. If $a^2 = b$ and $a > 0$, then a is the *(negative square root, principal square root)* of b.

4. The *(radicand, vertex)* of a parabola is the point at which the parabola intersects the axis of symmetry.

5. The *(discriminant, axis of symmetry)* can be used to determine the number of solutions of a quadratic equation.

Go Online
PHSchool.com

For: Vocabulary quiz
Web Code: atj-1051

Skills and Concepts

10-1 and 10-2 Objectives

▼ To graph quadratic functions of the form $y = ax^2$ (p. 550)

▼ To graph quadratic functions of the form $y = ax^2 + c$ (p. 552)

▼ To graph quadratic functions of the form $y = ax^2 + bx + c$ (p. 557)

▼ To graph quadratic inequalities (p. 559)

A function of the form $y = ax^2 + bx + c$, where $a \neq 0$, is a **quadratic function.** The shape of its graph is a **parabola.** The **axis of symmetry** of a parabola divides it into two congruent halves. The **vertex** of a parabola is the point at which the parabola intersects the axis of symmetry. The axis of symmetry is the line with the equation $x = \frac{-b}{2a}$. The x-coordinate of the vertex of the parabola is $\frac{-b}{2a}$.

The value of a in a quadratic function $y = ax^2 + bx + c$ determines the width of the parabola and whether it opens upward or downward. The value of c is the y-intercept of the graph. Changing the value of c shifts the parabola up or down.

When the parabola opens downward, the y-coordinate of the vertex is a **maximum** point of the function. When the parabola opens upward, the y-coordinate of the vertex is a **minimum** point of the function.

Open-Ended **Give an example of a quadratic function for each description.**

6. Its graph opens downward.

7. Its vertex is at the origin.

8. Its graph opens upward.

9. Its graph is wider than $y = x^2$.

Graph each function.

10. $y = \frac{2}{3}x^2$

11. $y = -x^2 + 1$

12. $y = x^2 - 4$

13. $y = 5x^2 + 8$

State whether each function has a *maximum* or *minimum* point.

14. $y = 4x^2 + 1$

15. $y = -3x^2 - 7$

16. $y = \frac{1}{2}x^2 + 9$

17. $y = -x^2 + 6$

Graph each function. Label the axis of symmetry and the vertex.

18. $y = -\frac{1}{2}x^2 + 4x + 1$

19. $y = -2x^2 - 3x + 10$

20. $y = x^2 + 6x - 2$

21. $y = \frac{1}{2}x^2 + 2x - 3$

Graph each quadratic inequality.

22. $y \leq 3x^2 + x - 5$

23. $y > 3x^2 + x - 5$

24. $y \geq -x^2 - x - 8$

25. $y < x^2 - 4x$

10-3 and 10-4 Objectives

▼ To solve quadratic equations by graphing (p. 565)

▼ To solve quadratic equations using square roots (p. 566)

▼ To solve quadratic equations by factoring (p. 572)

The **standard form of a quadratic equation** is $ax^2 + bx + c = 0$, where $a \neq 0$. Quadratic equations can have two, one, or no solutions. You can solve some quadratic equations by graphing the related function and finding the x-intercepts. If the quadratic expression $ax^2 + bx + c$ can be factored, you can use the **Zero-Product Property** to find the solutions of the equation $ax^2 + bx + c = 0$. This property states that for all real numbers a and b, if $ab = 0$, then $a = 0$ or $b = 0$.

Solve each equation. If the equation has no solution, write *no solution*.

26. $6(x^2 - 2) = 12$

27. $-5m^2 = -125$

28. $9(w^2 + 1) = 9$

29. $3r^2 + 27 = 0$

30. $4 = 9k^2$

31. $4n^2 = 64$

Write each equation in standard form. Then solve by factoring.

32. $x^2 + 7x + 12 = 0$

33. $5x^2 - 10x = 0$

34. $2x^2 - 9x = x^2 - 20$

35. $2x^2 + 5x = 3$

36. $3x^2 - 5x = -3x^2 + 6$

37. $x^2 - 5x + 4 = 0$

38. Geometry The area of a circle is given by the formula $A = \pi r^2$. Find the radius of a circle with area 16 in.^{2}. Round to the nearest tenth of an inch.

10-5 Objective

▼ To solve quadratic equations by completing the square (p. 579)

You can solve any quadratic equation by writing it in the form $x^2 + bx = c$, **completing the square,** and finding the square roots of each side of the equation.

Solve each equation by completing the square. If necessary, round to the nearest hundredth.

39. $x^2 + 6x - 5 = 0$

40. $x^2 = 3x - 1$

41. $2x^2 + 7x = -6$

42. Gardening Alice is planning a rectangular garden. Its length is 3 ft less than twice its width. Its area is 170 ft^2. Find the dimensions of the garden.

You can solve the quadratic equation $ax^2 + bx + c = 0$ when $a \neq 0$ by using the **quadratic formula** $x = \frac{-b \pm \sqrt{b^2 - 4ac}}{2a}$. When a quadratic equation is in the form $ax^2 + bx + c = 0$ $(a \neq 0)$, the **discriminant** is $b^2 - 4ac$.

If $b^2 - 4ac > 0$, there are two solutions.
If $b^2 - 4ac = 0$, there is one solution.
If $b^2 - 4ac < 0$, there is no solution.

Find the number of solutions of each equation.

43. $x^2 - 10 = 3$

44. $3x^2 = 27$

45. $x^2 + 3 = 2x$

46. $x^2 + 10x = -25$

Solve each equation using the quadratic formula. Round to the nearest hundredth.

47. $4x^2 + 3x - 8 = 0$

48. $2x^2 - 7x = -3$

49. $-x^2 + 8x + 4 = 5$

50. $9x^2 - 270 = 0$

Writing Solve each equation. Explain why you chose the method you used.

51. $5x^2 - 10 = x^2 + 90$

52. $9x^2 + 30x - 29 = 0$

53. $2x^2 - 9x = x^2 - 20$

54. $x^2 - 6x + 9 = 0$

55. $x^2 + 3x - 225 = 3x$

56. $x^2 + 8x = 4$

57. **Geometry** A square pool has side length p. The border of the pool is 1 ft wide. The combined area of the border and the pool is 400 ft^2. Find the length and the area of the pool.

58. **Vertical Motion** Suppose you throw a ball in the air. The ball is 6 ft high when it leaves your hand. Use the equation $0 = -16t^2 + 20t + 6$ to find the number of seconds t that the ball is in the air.

Graphing data points or analyzing data numerically can help you find the best model. Linear data have a common difference. Exponential data have a common ratio. Quadratic data have a common second difference.

Graph each set of points. Which model is most appropriate for each set?

59. $(-3, 0), (1, 4), (-1, 6), (2, 0)$

60. $(0, 5), (1, 3), (3, -1), (-1, 7)$

61. $(0, 6), (5, 2), (1, 4), (8, 1.5), (2, 3)$

62. $(1, 4), (4, 2), (2, 3), (5, 3.5), (6, 5)$

Write an equation to model the data.

63.

x	y
−1	2.5
0	5
1	10
2	20
3	40

64.

x	y
−1	−5
0	−2
1	1
2	4
3	7

65.

x	y
−3	4
−2	1
−1	0
0	1
1	4

66.

x	y
0	0.5
1	5
2	50
3	500
4	5000

Chapter Test

Go Online
PHSchool.com

For: Chapter Test
Web Code: ata-1052

Match each graph with its function.

   A. $y = 3x^2$ **B.** $y = -3x^2 + 1$

   C. $y = -2x^2$ **D.** $y = x^2 - 3$

1.   **2.**

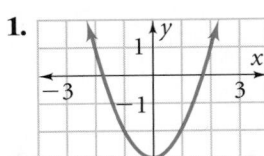   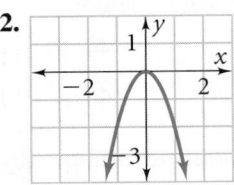

3.   **4.**

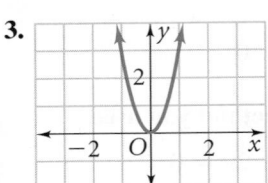   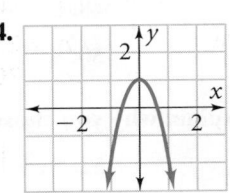

Find the equation of the axis of symmetry and the coordinates of the vertex of the graph of each function. Is the vertex a maximum or a minimum?

  5. $y = 3x^2 - 7$ **6.** $y = x^2 - 3x + 2$

  7. $y = -2x^2 + 10x - 1$ **8.** $y = \frac{1}{2}x^2 + 6x$

Graph each function.

  9. $y = x^2 - 4$ **10.** $y = -x^2 + 1$

 11. $y = 5x^2$ **12.** $y = \frac{1}{2}x^2 + 3x$

 13. $y = x^2 - 3x + 5$ **14.** $y = 2x^2 - 5$

15. Writing Explain what you can determine about the shape of a parabola from its equation alone.

Find the number of x-intercepts of each function.

 16. $y = 5x^2$ **17.** $y = 3x^2 + 10$

 18. $y = -2x^2 + x + 7$ **19.** $y = x^2 - 4x$

Graph each inequality.

 20. $y \le 2x^2 - 1$ **21.** $y > \frac{1}{2}x^2$

Solve each equation.

 22. $x^2 + 11x - 26 = 0$ **23.** $x^2 - 2x = 35$

 24. $x^2 - 19x + 80 = -8$ **25.** $x^2 - 5x = -4x$

 26. $2x^2 = 7 + 13x$ **27.** $5x^2 - 8x = 8 - 5x$

Find the number of solutions of each equation.

 28. $x^2 + 4x = -4$ **29.** $x^2 + 8 = 0$

 30. $2x^2 + x = 0$ **31.** $3x^2 - 9x = -5$

32. Find a nonzero value of k such that $kx^2 - 10x + 25 = 0$ has one solution.

Solve each equation. If necessary, round to the nearest hundredth.

 33. $2x^2 = 50$ **34.** $-3x^2 + 7x = -10$

 35. $x^2 + 6x + 9 = 25$ **36.** $-x^2 - x + 2 = 0$

 37. $x^2 + 4x = 1$ **38.** $12x^2 + 16x - 28 = 0$

39. Open-Ended Write an equation of a parabola that has two x-intercepts and a maximum value. Include a graph of your parabola.

Model each problem with a quadratic equation. Then solve.

40. Geometry The volume V of a cylinder is given by the formula $V = \pi r^2 h$, where r is the radius of the cylinder and h is the height. A cylinder with height 10 ft has volume 140 ft^3. To the nearest tenth of a foot, what is the radius of the cylinder?

41. Landscaping The area of a rectangular patio is 800 ft^2. The patio's length is twice its width. Find the dimensions of the patio.

Identify each graph as *linear*, *quadratic*, or *exponential*. Write an equation that models the data shown in each graph.

42.   **43.**

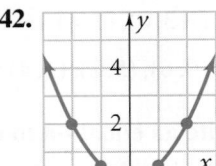   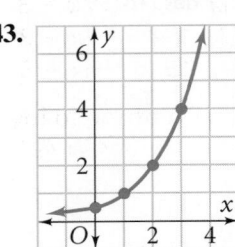

44.   **45.**

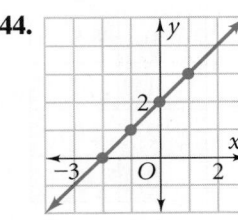   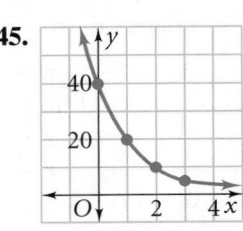

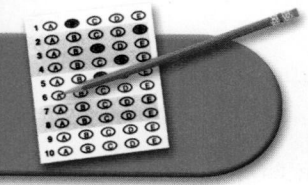

Regents Test Prep

Multiple Choice

For Exercises 1–11, choose the correct letter.

1. Choose the best approximation of the solutions of $3x^2 - 5x + 1 = 0$.
   (1) 2 and −3   (2) 1.5 and 1.75
   (3) −3 and 2   (4) 1.5 and 0.25

2. Which is the equation of the axis of symmetry of the graph of $y = 5x^2 - 2x + 3$?
   (1) $x = \frac{1}{5}$ (2) $x = -\frac{1}{5}$
   (3) $y = \frac{1}{5}$ (4) $y = -\frac{1}{5}$

3. Between which two consecutive integers is $\sqrt{52}$?
   (1) 5 and 6   (2) 6 and 7
   (3) 7 and 8   (4) 8 and 9

4. A line perpendicular to $y = 3x - 2$ passes through the point (0, 6). Which other point lies on the line?
   (1) (9, 3)   (2) (−9, 3)
   (3) (−9, −3)   (4) (9, −3)

5. What is the probability of *not* rolling a 1 or 2 on a number cube?
   (1) $\frac{1}{6}$ (2) $\frac{1}{3}$ (3) $\frac{1}{2}$ (4) $\frac{2}{3}$

6. If $b = 2a - 16$ and $b = a + 2$, then what is the value of a?
   (1) 5   (2) 6   (3) 14   (4) 18

7. What is the standard form of the product $(3x - 1)(5x + 3)$?
   (1) $15x^2 + 2x - 3$
   (2) $15x^2 + 2x + 3$
   (3) $15x^2 - 4x - 3$
   (4) $15x^2 + 4x - 3$

8. Which of the following solve(s) $x^2 + 4x + 4 = 49$?
   I. 5   II. −9
   III. $\sqrt{47}$ IV. $-\sqrt{47}$
   (1) I only   (2) III only
   (3) III and IV   (4) I and II

9. How many solutions are there for the following system?
$$y = x - 3$$
$$6y - x = 24$$
   (1) 0   (2) 1   (3) 2   (4) 3

10. Mandy has $2.40 in coins, all in dimes and nickels. The total number of coins is 30. If she has d dimes and n nickels, which system models the situation?
    (1) $d + n = 30$
           $10d + 5n = 240$
    (2) $d + n = 30$
           $10d + 5n = 2.40$
    (3) $d + n = 240$
           $10d + 5n = 30$
    (4) $d + n = 2.40$
           $10d + 5n = 30$

11. Which function best models the data in the table at the right?
    (1) $y = 0.2(3^x)$
    (2) $y = 0.4x^2 + 0.2$
    (3) $y = 0.4x + 0.2$
    (4) $y = 0.4(0.2^x)$

x	y
0	0.2
1	0.6
2	1.8
3	5.4
4	16.2

Short Response

12. Find the slope of the line that passes through (−1, 3) and (4, 6).

13. A new company employed 12 people. Two years later, it employed a total of 20 people. What was the percent of increase rounded to the nearest percent?

14. **Geometry** The length of a rectangle is 6 m less than twice its width. The area of the rectangle is 140 m^2. Find the dimensions of the rectangle. Show your work.

15. Describe the shape of the graph for each type of function: linear, quadratic, and absolute value.

16. Simplify $9a + 3b - 3 - 4a + 7 - 8b$. Show your work.

Extended Response

17. Graph $y = 4x^2 - 3x$. Show the vertex, axis of symmetry, and x-intercepts of the equation.

Surface Areas and Volumes

You can use formulas to find the surface area and volume of a rectangular prism. To find the surface area of a prism you use the lateral area and the base areas. The lateral area is product of the perimeter of the base and the height. To find the volume you find the product of the base area and the height.

NY **A.G.2:** Use formulas to calculate volume and surface area of rectangular prisms and cylinders.

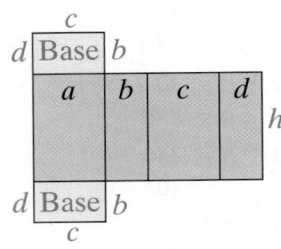

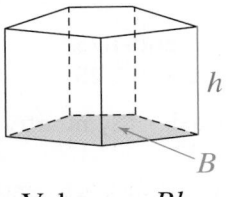

Volume = Bh

L.A.= Lateral Area = ph

perimeter ↗ ↖ heigl

Area of a base ↓

Surface Area = L.A. + 2B

1 EXAMPLE

Use formulas to calculate the surface area and the volume of the rectangular prism.

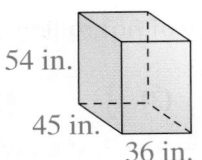

54 in.

45 in.

36 in.

Step 1 Find the lateral area.

$$L.A. = ph$$
$$= (45 + 36 + 45 + 36)54$$
$$= 8{,}748$$

Step 2 Find the surface area.

$$S.A. = L.A. + 2B$$
$$= 8{,}748 + 2(45 \cdot 36)$$
$$= 11{,}988$$

The surface area of the rectangular prism is 11,988 in.^{2}.

Use the formula $V = Bh$ to find the volume of the prism.

$$V = Bh$$
$$= 45 \cdot 36 \cdot 54$$
$$= 87{,}480$$

The volume of the rectangular prism is 87,480 in.^{3}.

You can also use formulas to calculate the surface area and volume of cylinders.

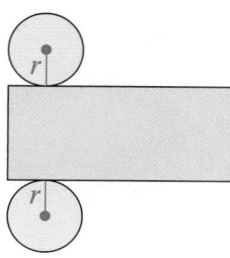

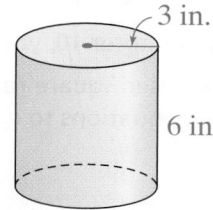

Volume = Bh

L.A. = $2\pi rh$

S.A. = L.A. + $2B$

2 EXAMPLE

Use formulas to calculate the surface area and the volume of the cylinder to the nearest inch.

Step 1 Find the lateral area.

$$\text{L.A.} = 2\pi rh$$
$$= 2(3.14(3)(6)$$
$$= 113.04$$

Step 2 Find the surface area.

$$\text{S.A.} = \text{L.A.} + 2B$$
$$= 113.04 + 2(3.14)3^2$$
$$= 169.56$$

The surface area of the cylinder is about 170 in.

Use the formula $V = Bh$ to find the volume of the cylinder.

$$V = Bh$$
$$V = \pi r^2 h$$
$$= 3.14 \cdot 3^2 \cdot 6$$
$$= 169.56$$

The volume of the cylinder is about 170 in.^{2}.

EXERCISES

Use formulas to calculate the surface area and volume of each solid.

1.
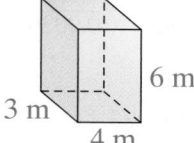
6 m
3 m
4 m

2.
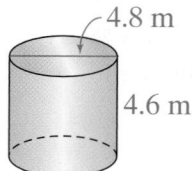
4.8 m
4.6 m

3.
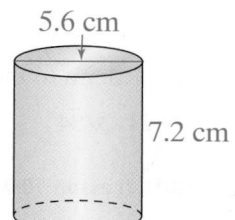
5.6 cm
7.2 cm

4.
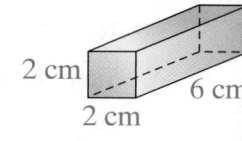
2 cm
2 cm
6 cm

What You've Learned

NY **Learning Standards for Mathematics**

In Chapter 3, you
- **NY A.A.26:** studied ratios and how to solve a proportion for a given variable.

In Chapter 8, you
- **NY A.A.12:** simplified expressions containing exponents.

In Chapter 10, you
- Used square roots and evaluated the discriminants of quadratic equations to determine the number of solutions.

 **Check Your Readiness**  **for Help** to the Lesson in green.

Calculating the Mean (Lesson 1-6)

Find the mean of each set of data.

1. $1, 4, 2, 5, 3, 2$

2. $40, 55, 60, 52$

3. $1.6, 2.1, 1.8, 1.8$

4. $-4, 2, 0, -1$

5. $214, 198, 202$

6. $3, 2, 2, 3, 4, 4$

Solving Proportions (Lesson 3-4)

Solve each proportion.

7. $\frac{8}{x} = \frac{24}{9}$

8. $\frac{k-4}{27} = \frac{1}{3}$

9. $\frac{5}{6} = \frac{25}{c}$

10. $\frac{42}{x+8} = \frac{7}{2}$

11. $\frac{9}{13} = \frac{y}{65}$

12. $\frac{6}{x-5} = \frac{1}{2}$

13. $\frac{3n-1}{14} = \frac{4}{7}$

14. $\frac{4}{33} = \frac{8}{w}$

15. $\frac{y-1}{5} = \frac{y+3}{7}$

Finding Square Roots (Lesson 3-8)

Simplify each expression.

16. $\sqrt{4}$

17. $\sqrt{225}$

18. $\sqrt{\frac{9}{25}}$

19. $-\sqrt{0.0036}$

Simplify each expression. Round to the nearest hundredth.

20. $\sqrt{40}$

21. $\sqrt{84}$

22. $\sqrt{104}$

23. $\sqrt{3.2}$

Using the Discriminant (Lesson 10-7)

Find the number of real solutions of each equation.

24. $x^2 + 6x + 1 = 0$

25. $x^2 - 5x - 6 = 0$

26. $x^2 - 2x + 9 = 0$

27. $4x^2 - 4x = -1$

28. $6x^2 + 5x - 2 = -3$

29. $(2x - 5)^2 = 121$

Radical Expressions and Equations

🔊 **Key Vocabulary**

- angle of depression (p. 651)
- angle of elevation (p. 650)
- conjugates (p. 623)
- cosine (p. 646)
- extraneous solution (p. 631)
- radical equation (p. 629)
- radical expression (p. 616)
- rationalize (p. 619)
- sine (p. 646)
- square root function (p. 638)
- tangent (p. 646)
- trigonometric ratios (p. 646)

What You'll Learn Next

NY Learning Standards for Mathematics

- **NY A.N.2:** You will simplify expressions containing radicals.

- **NY A.N.3:** You will solve radical equations.

- **NY A.A.44:** You will use trigonometry to find the lengths of sides of right triangles.

Data Analysis   **Activity Lab** Applying what you learn, you will explore conditional probability, on pages 660–661.

615

Simplifying Radicals

 Learning Standards for Mathematics

A.N.2 Simplify radical terms (no variable in the radicand).

A.N.3 Perform the four arithmetic operations using like and unlike radical terms and express the results in simplest form.

✓ **Check Skills You'll Need**

Complete each equation.

1. $a^3 = a^2 \cdot a^\blacksquare$ **2.** $b^7 = b^6 \cdot b^\blacksquare$ **3.** $c^6 = c^3 \cdot c^\blacksquare$ **4.** $d^8 = d^4 \cdot d^\blacksquare$

Find the value of each expression.

5. $\sqrt{4}$ **6.** $\sqrt{169}$ **7.** $\sqrt{25}$ **8.** $\sqrt{49}$

GO for Help   Lessons 8-3 and 3-8

◀))) **New Vocabulary**   • radical expression   • rationalize

---

1   Simplifying Radical Expressions Involving Products

Radical expressions like $2\sqrt{3}$ and $\sqrt{x + 3}$ contain a radical. You read $\sqrt{x + 3}$ as "the square root of the quantity x plus three." You can simplify a radical expression by removing perfect-square factors from the radicand. Recall that a radicand is the quantity or expression under the radical sign.

 Key Concepts

Property	**Multiplication Property of Square Roots**
For every number $a \geq 0$ and $b \geq 0$, $\sqrt{ab} = \sqrt{a} \cdot \sqrt{b}$.	
Example $\sqrt{54} = \sqrt{9} \cdot \sqrt{6} = 3 \cdot \sqrt{6} = 3\sqrt{6}$	

You can use the Multiplication Property of Square Roots to simplify radical expressions by rewriting the radicand as a product of the perfect-square factors times the remaining factors.

1 EXAMPLE   **Removing Perfect-Square Factors**

Simplify $\sqrt{192}$.

$\sqrt{192} = \sqrt{64 \cdot 3}$ **64 is a perfect square and a factor of 192.**

$\quad\quad = \sqrt{64} \cdot \sqrt{3}$ **Use the Multiplication Property of Square Roots.**

$\quad\quad = 8\sqrt{3}$ **Simplify $\sqrt{64}$.**

✓ **Quick Check** ❶ Simplify each radical expression.

a. $\sqrt{50}$ **b.** $-5\sqrt{300}$ **c.** $\sqrt{18}$

You can simplify radical expressions that contain variables. A variable with a nonzero, even exponent is a perfect square. Variables with odd exponents (other than 1 and −1) are the product of a perfect square and the variable. For example, $n^3 = n^2 \cdot n$, so $\sqrt{n^3} = \sqrt{n^2 \cdot n}$. Assume that all variables of all radicands represent nonnegative numbers.

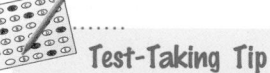 **EXAMPLE** Removing Variable Factors

Multiple Choice Simplify $\sqrt{45a^5}$.

   Ⓐ $a^2\sqrt{45}$ Ⓑ $3a^2\sqrt{5a}$ Ⓒ $5\sqrt{3a^5}$ Ⓓ $9a^4\sqrt{5a}$

$$\sqrt{45a^5} = \sqrt{9a^4 \cdot 5a} \quad \textbf{9a^4 is a perfect square and a factor of 45a^5.}$$
$$= \sqrt{9a^4} \cdot \sqrt{5a} \quad \textbf{Use the Multiplication Property of Square Roots.}$$
$$= 3a^2\sqrt{5a} \quad \textbf{Simplify } \sqrt{9a^4}.$$

So B is the correct answer.

✓ Quick Check ② Simplify each radical expression.

  a. $\sqrt{27n^2}$ **b.** $-a\sqrt{60a^7}$ **c.** $\sqrt{x^2y^5}$

You can use the Multiplication Property of Square Roots to write $\sqrt{a} \cdot \sqrt{b} = \sqrt{ab}$. Sometimes the product of two radical expressions has a perfect-square factor.

 **EXAMPLE** Multiplying Two Radicals

Simplify each radical expression.

a. $\sqrt{8} \cdot \sqrt{12} = \sqrt{8 \cdot 12}$ **Use the Multiplication Property of Square Roots.**
$$= \sqrt{96} \quad \textbf{Simplify under the radical.}$$
$$= \sqrt{16 \cdot 6} \quad \textbf{16 is a perfect square and a factor of 96.}$$
$$= \sqrt{16} \cdot \sqrt{6} \quad \textbf{Use the Multiplication Property of Square Roots.}$$
$$= 4\sqrt{6} \quad \textbf{Simplify } \sqrt{16}.$$

b. $3\sqrt{2b} \cdot 4\sqrt{10b} = 12\sqrt{20b^2}$ **Multiply the whole numbers and use the Multiplication Property of Square Roots.**
$$= 12\sqrt{4b^2 \cdot 5} \quad \textbf{4b^2 is a perfect square and a factor of 20b^2.}$$
$$= 12\sqrt{4b^2} \cdot \sqrt{5} \quad \textbf{Use the Multiplication Property of Square Roots.}$$
$$= 12 \cdot 2b\sqrt{5} \quad \textbf{Simplify } \sqrt{4b^2}.$$
$$= 24b\sqrt{5} \quad \textbf{Simplify.}$$

Test-Taking Tip

Another method of simplifying is using prime factors.
$$\sqrt{8} \cdot \sqrt{12}$$
$$= \sqrt{2 \cdot 2 \cdot 2 \cdot 2 \cdot 2 \cdot 3}$$
$$= \sqrt{2^2 \cdot 2^2 \cdot 2 \cdot 3}$$
$$= 2 \cdot 2\sqrt{6}$$
$$= 4\sqrt{6}$$

✓ Quick Check ③ Simplify each radical expression.

  a. $\sqrt{13} \cdot \sqrt{52}$ **b.** $5\sqrt{3c} \cdot \sqrt{6c}$ **c.** $2\sqrt{5a^2} \cdot 6\sqrt{10a^3}$

④ **EXAMPLE** **Real-World 🌐 Problem Solving**

Sightseeing You can use the formula $d = \sqrt{1.5h}$ to estimate the distance d in miles to a horizon when h is the height of the viewer's eyes above the ground in feet. Estimate the distance a visitor at the Washington Monument can see to the horizon from the observation windows. Round your answer to the nearest mile.

$$d = \sqrt{1.5h}$$
$$= \sqrt{1.5 \cdot 500} \quad \textbf{Substitute 500 for h.}$$
$$= \sqrt{750} \quad \textbf{Multiply.}$$
$$\approx 27 \quad \textbf{Use a calculator.}$$

500 ft

The distance a visitor can see is about 27 miles.

✓ Quick Check ④ Suppose you are looking out a second floor window 25 ft above the ground. Find the distance you can see to the horizon. Round your answer to the nearest mile.

You can use the Division Property of Square Roots to simplify expressions.

Key Concepts

Property	Division Property of Square Roots
	For every number $a \geq 0$ and $b > 0$, $\sqrt{\dfrac{a}{b}} = \dfrac{\sqrt{a}}{\sqrt{b}}$.
	Example $\sqrt{\dfrac{16}{25}} = \dfrac{\sqrt{16}}{\sqrt{25}} = \dfrac{4}{5}$

When the denominator of the radicand is a perfect square, it is easier to simplify the numerator and denominator separately.

5 EXAMPLE Simplifying Fractions Within Radicals

Vocabulary Tip

Read $\dfrac{\sqrt{2}}{3}$ as "the square root of 2 over 3," and $\sqrt{\dfrac{2}{3}}$ as "the square root of two thirds."

Simplify each radical expression.

a. $\sqrt{\dfrac{11}{49}} = \dfrac{\sqrt{11}}{\sqrt{49}}$ Use the Division Property of Square Roots.

$= \dfrac{\sqrt{11}}{7}$ Simplify $\sqrt{49}$.

b. $\sqrt{\dfrac{25}{b^4}} = \dfrac{\sqrt{25}}{\sqrt{b^4}}$ Use the Division Property of Square Roots.

$= \dfrac{5}{b^2}$ Simplify $\sqrt{25}$ and $\sqrt{b^4}$.

 **5** Simplify each radical expression.

a. $\sqrt{\dfrac{144}{9}}$ **b.** $\sqrt{\dfrac{25p^3}{q^2}}$ **c.** $\sqrt{\dfrac{75}{16t^2}}$

When the denominator of the radicand is not a perfect square, it may be easier to divide first and then simplify the radical expression.

6 EXAMPLE Simplifying Radicals by Dividing

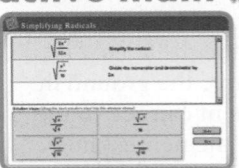

For: Radical Activity
Use: Interactive Textbook, 11-1

Simplify each radical expression.

a. $\sqrt{\dfrac{88}{11}} = \sqrt{8}$ Divide.

$= \sqrt{4 \cdot 2}$ 4 is a perfect square and a factor of 8.

$= \sqrt{4} \cdot \sqrt{2}$ Use the Multiplication Property of Square Roots.

$= 2\sqrt{2}$ Simplify $\sqrt{4}$.

b. $\sqrt{\dfrac{12a^3}{27a}} = \sqrt{\dfrac{4a^2}{9}}$ Divide the numerator and denominator by $3a$.

$= \dfrac{\sqrt{4a^2}}{\sqrt{9}}$ Use the Division Property of Square Roots.

$= \dfrac{\sqrt{4} \cdot \sqrt{a^2}}{\sqrt{9}}$ Use the Multiplication Property of Square Roots.

$= \dfrac{2a}{3}$ Simplify $\sqrt{4}$, $\sqrt{a^2}$ and $\sqrt{9}$.

 **6** Simplify each radical expression.

a. $\sqrt{\dfrac{90}{5}}$ **b.** $\sqrt{\dfrac{48}{75}}$ **c.** $\sqrt{\dfrac{27x^3}{3x}}$

A radicand in the denominator of a radical expression may not be a perfect square. To simplify, you may need to **rationalize** the denominator. To do this, you multiply the numerator and the denominator by the same radical expression. You choose a radical expression that will make the denominator a perfect square.

 **EXAMPLE** **Rationalizing a Denominator**

Simplify by rationalizing the denominator.

a. $\dfrac{2}{\sqrt{5}} = \dfrac{2}{\sqrt{5}} \cdot \dfrac{\sqrt{5}}{\sqrt{5}}$ **Multiply by $\dfrac{\sqrt{5}}{\sqrt{5}}$ to make the denominator a perfect square.**

$= \dfrac{2\sqrt{5}}{\sqrt{25}}$ **Use the Multiplication Property of Square Roots.**

$= \dfrac{2\sqrt{5}}{5}$ **Simplify $\sqrt{25}$.**

b. $\dfrac{\sqrt{7}}{\sqrt{8n}} = \dfrac{\sqrt{7}}{\sqrt{8n}} \cdot \dfrac{\sqrt{2n}}{\sqrt{2n}}$ **Multiply by $\dfrac{\sqrt{2n}}{\sqrt{2n}}$ to make the denominator a perfect square.**

$= \dfrac{\sqrt{14n}}{\sqrt{16n^2}}$ **Use the Multiplication Property of Square Roots.**

$= \dfrac{\sqrt{14n}}{4n}$ **Simplify $\sqrt{16n^2}$.**

  Simplify by rationalizing the denominator.

a. $\dfrac{3}{\sqrt{3}}$ **b.** $\dfrac{\sqrt{5}}{\sqrt{18t}}$ **c.** $\sqrt{\dfrac{7m}{10}}$

The summary below can help you determine whether a radical expression is in simplest radical form.

 **Key Concepts**

Summary	Simplest Radical Form

A radical expression is in simplest radical form when all three statements are true.

- The radicand has no perfect-square factors other than 1.
- The radicand has no fractions.
- The denominator of a fraction has no radical.

EXERCISES

For more exercises, see *Extra Skill and Word Problem Practice.*

Practice and Problem Solving

A   **Practice by Example**

Simplify each radical expression.

Examples 1, 2
(pages 616, 617)

1. $\sqrt{200}$	**2.** $\sqrt{98}$	**3.** $\sqrt{75}$	**4.** $-\sqrt{80}$
5. $-3\sqrt{120}$	**6.** $5\sqrt{320}$	**7.** $\sqrt{28n^2}$	**8.** $\sqrt{108b^4}$
9. $3\sqrt{12x^2}$	**10.** $\sqrt{4n^3}$	**11.** $\sqrt{20a^5}$	**12.** $-\sqrt{48b^4}$

Example 3
(page 617)

13. $\sqrt{10} \cdot \sqrt{40}$	**14.** $3\sqrt{6} \cdot \sqrt{6}$	**15.** $\sqrt{22} \cdot \sqrt{11}$	**16.** $2\sqrt{18} \cdot 7\sqrt{6}$
17. $\sqrt{7} \cdot \sqrt{21}$	**18.** $-3\sqrt{20} \cdot \sqrt{15}$	**19.** $\sqrt{3n} \cdot \sqrt{24n}$	**20.** $2\sqrt{7t} \cdot \sqrt{14t}$
21. $\sqrt{3x} \cdot \sqrt{51x^3}$	**22.** $5\sqrt{8t} \cdot \sqrt{32t^5}$	**23.** $\sqrt{2a^2} \cdot \sqrt{9a^4}$	**24.** $-2\sqrt{6a^3} \cdot \sqrt{3a}$

Example 4
(page 617)

For Exercises 25–27, use the formula $d = \sqrt{1.5h}$ to approximate distance d in miles to a horizon when h is the height in feet of the viewer's eyes above the ground. Round your answer to the nearest mile.

25. Find the distance you can see to the horizon from a height of 6 feet.

26. Find the distance you can see to the horizon from a height of 100 feet.

27. Find the distance you can see to the horizon from a height of 200 feet.

Example 5
(page 618)

Simplify each radical expression.

28. $\sqrt{\dfrac{21}{49}}$

29. $3\sqrt{\dfrac{3}{4}}$

30. $\sqrt{\dfrac{625}{100}}$

31. $\sqrt{\dfrac{120}{121}}$

32. $\sqrt{\dfrac{5}{9a^2}}$

33. $\sqrt{\dfrac{7}{16c^2}}$

34. $\sqrt{\dfrac{75a}{49}}$

35. $\sqrt{\dfrac{8n^3}{81}}$

Example 6
(page 618)

36. $\sqrt{\dfrac{15}{5}}$

37. $\sqrt{\dfrac{54}{24}}$

38. $\sqrt{\dfrac{60}{5}}$

39. $-\sqrt{\dfrac{160}{8}}$

40. $\sqrt{\dfrac{140x^3}{5x}}$

41. $\sqrt{\dfrac{3s^3}{27s}}$

42. $\sqrt{\dfrac{30a^5}{40a}}$

43. $\sqrt{\dfrac{63y}{7y^3}}$

Example 7
(page 619)

Simplify each radical expression by rationalizing the denominator.

44. $\dfrac{3}{\sqrt{2}}$

45. $\dfrac{5}{\sqrt{5}}$

46. $\dfrac{\sqrt{3}}{\sqrt{7x}}$

47. $\dfrac{2\sqrt{2}}{\sqrt{5n}}$

48. $\dfrac{9}{\sqrt{8}}$

49. $\dfrac{12}{\sqrt{12}}$

50. $\dfrac{3\sqrt{2}}{\sqrt{9b}}$

51. $\dfrac{5\sqrt{11}}{\sqrt{20y}}$

B **Apply Your Skills**

Writing **Explain why each radical expression is or is not in simplest radical form.**

52. $\dfrac{13}{\sqrt{4}}$

53. $\dfrac{3}{\sqrt{3}}$

54. $4\sqrt{3}$

55. $5\sqrt{30}$

56. Suppose a and b are positive integers.
  a. Verify that if $a = 18$ and $b = 10$, then $\sqrt{a} \cdot \sqrt{b} = 6\sqrt{5}$.
  b. **Open-Ended** Find two other pairs of positive integers a and b such that $\sqrt{a} \cdot \sqrt{b} = 6\sqrt{5}$.

Simplify each radical expression.

57. $\sqrt{12} \cdot \sqrt{75}$

58. $\sqrt{26 \cdot 2}$

59. $\dfrac{\sqrt{72}}{\sqrt{64}}$

60. $\dfrac{-2}{\sqrt{a^3}}$

61. $\dfrac{\sqrt{180}}{\sqrt{3}}$

62. $\dfrac{\sqrt{x^2}}{\sqrt{y^3}}$

63. $\dfrac{-3\sqrt{2}}{\sqrt{6}}$

64. $\sqrt{8} \cdot \sqrt{10}$

65. $\sqrt{20a^2b^3}$

66. $\sqrt{a^3b^5c^3}$

67. $\sqrt{\dfrac{3m}{16m^2}}$

68. $\dfrac{16a}{\sqrt{6a^3}}$

Solve each equation. Leave your answer in simplest radical form.

69. $x^2 + 6x - 9 = 0$

70. $n^2 - 2n + 1 = 5$

71. $3y^2 - 4y - 2 = 0$

72. **a.** Show work to verify that $\sqrt{50}$ equals $5\sqrt{2}$.
  b. **Writing** Explain why $5\sqrt{2}$ is in simplest radical form.

73. **Open-Ended** What are three numbers whose square roots can be written in the form $a\sqrt{3}$ for some integer value of a?

74. **Newspaper Layout** A square picture on the front page of a newspaper occupies an area of 24 in.^{2}.
  a. Find the length of each side in simplest radical form.
  b. Calculate the length of each side to the nearest hundredth of an inch.

GO nline
Homework Video Tutor
Visit: PHSchool.com
Web Code: ate-1101

Simplify each radical expression.

75. $\sqrt{24} \cdot \sqrt{2x} \cdot \sqrt{3x}$ **76.** $2b(\sqrt{5b})^2$ **77.** $\sqrt{45a^7} \cdot \sqrt{20a}$

78. Physics The time that a pendulum of a grandfather clock takes to swing back and forth one cycle is the period of the pendulum. The formula for finding the period T in seconds is $T = 2\pi\sqrt{\frac{L}{32}}$, where L is the length of the pendulum in feet. Find the period of a pendulum that is 8 feet long. Write your answer in terms of π.

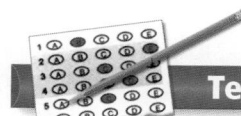

Test Prep

Multiple Choice

79. Simplify $\sqrt{80}$.

    A. $10\sqrt{8}$ **B.** $8\sqrt{10}$ **C.** $4\sqrt{5}$ **D.** 40

80. Simplify $5\sqrt{3x^2} \cdot \sqrt{6x}$.

    F. $15x\sqrt{2x}$ **G.** $5x\sqrt{18x}$ **H.** $3x\sqrt{10x}$ **J.** $6x\sqrt{5x}$

81. Which of the following equals $\frac{2}{3}$?

    A. $\sqrt{\frac{9}{25}}$ **B.** $\sqrt{\frac{20}{45}}$ **C.** $2\sqrt{\frac{4}{27}}$ **D.** $\sqrt{\frac{6}{9}}$

82. Which of the following equals $1.5\sqrt{0.038}$?

    F. $150\sqrt{3.8}$ **G.** $15\sqrt{3.8}$ **H.** $15\sqrt{0.38}$ **J.** $15\sqrt{0.00038}$

Short Response

83. A square window has an area of 96 ft². What is the length of each side of the window in simplest radical form? Show your work.

Mixed Review

Lesson 10-8

Which kind of function best models the data in each table? Write an equation to model the data.

84.

x	y
−1	0.2
0	0
1	0.2
2	0.8
3	1.8
4	3.2

85.

x	y
−1	1.6
0	4
1	10
2	25
3	62.5
4	156.25

86.

x	y
−1	11.2
0	7
1	2.8
2	−1.4
3	−5.6
4	−9.8

Lesson 10-2

Graph each function. Label the axis of symmetry and the vertex.

87. $f(x) = x^2 + 8x - 4$ **88.** $y = x^2 - 10x + 7$ **89.** $y = 3x^2 + 12x - 5$

Lesson 9-1

Simplify. Write each answer in standard form.

90. $(n^2 + 5n - 1) + (2n^2 + 6)$ **91.** $(4v^2 + 8v - 2) - (v^2 + 9v + 7)$

92. $(5t^3 - 14t) + (8t^2 - 11)$ **93.** $(2b^2 - 12b - 8) - (5b^2 + 11b + 13)$

Operations With Radical Expressions

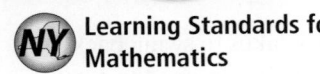
Learning Standards for Mathematics

A.N.3 Perform the four arithmetic operations using like and unlike radical terms and express the results in simplest form.

✓ **Check Skills You'll Need**

GO **for Help** Lesson 11-1

Simplify each radical expression.

1. $\sqrt{52}$ **2.** $\sqrt{200}$ **3.** $4\sqrt{54}$ **4.** $\sqrt{125x^2}$

Rationalize each denominator.

5. $\dfrac{\sqrt{3}}{\sqrt{11}}$ **6.** $\dfrac{\sqrt{5}}{\sqrt{8}}$ **7.** $\dfrac{\sqrt{15}}{\sqrt{2x}}$

🔊 **New Vocabulary** • like radicals • unlike radicals • conjugates

---

1   Simplifying Sums and Differences

For radical expressions, **like radicals** have the same radicand. **Unlike radicals** do not have the same radicand. For example, $4\sqrt{7}$ and $-12\sqrt{7}$ are like radicals, but $3\sqrt{11}$ and $2\sqrt{5}$ are unlike radicals. To simplify sums and differences, you use the Distributive Property to combine like radicals.

1 EXAMPLE   **Combining Like Radicals**

Simplify $\sqrt{2} + 3\sqrt{2}$.

$\sqrt{2} + 3\sqrt{2} = 1\sqrt{2} + 3\sqrt{2}$ **Both terms contain $\sqrt{2}$.**

$\qquad\qquad\quad = (1 + 3)\sqrt{2}$ **Use the Distributive Property to combine like radicals.**

$\qquad\qquad\quad = 4\sqrt{2}$ **Simplify.**

✓ **Quick Check** ❶ Simplify each expression.
a. $-3\sqrt{5} - 4\sqrt{5}$ **b.** $\sqrt{10} - 5\sqrt{10}$

You may need to simplify a radical expression to determine if you have like radicals.

Online
active math

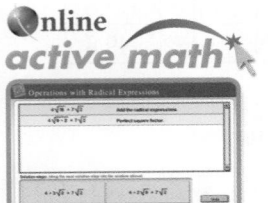

For: Radical Activity
Use: Interactive Textbook, 11-2

2 EXAMPLE   **Simplifying to Combine Like Radicals**

Simplify $7\sqrt{3} - \sqrt{12}$.

$7\sqrt{3} - \sqrt{12} = 7\sqrt{3} - \sqrt{4 \cdot 3}$ **4 is a perfect square and a factor of 12.**

$\qquad\qquad\quad = 7\sqrt{3} - \sqrt{4} \cdot \sqrt{3}$ **Use the Multiplication Property of Square Roots.**

$\qquad\qquad\quad = 7\sqrt{3} - 2\sqrt{3}$ **Simplify $\sqrt{4}$.**

$\qquad\qquad\quad = (7 - 2)\sqrt{3}$ **Use the Distributive Property to combine like radicals.**

$\qquad\qquad\quad = 5\sqrt{3}$ **Simplify.**

✓ **Quick Check** ❷ Simplify each expression.
a. $3\sqrt{20} + 2\sqrt{5}$ **b.** $3\sqrt{3} - 2\sqrt{27}$

When simplifying a radical expression like $\sqrt{3}(\sqrt{6} + 7)$, use the Distributive Property to multiply $\sqrt{3}$ times $(\sqrt{6} + 7)$.

3 EXAMPLE   Using the Distributive Property

Simplify $\sqrt{3}(\sqrt{6} + 7)$.

$\sqrt{3}(\sqrt{6} + 7) = \sqrt{18} + 7\sqrt{3}$ **Use the Distributive Property.**

$\qquad\qquad = \sqrt{9} \cdot \sqrt{2} + 7\sqrt{3}$ **Use the Multiplication Property of Square Roots.**

$\qquad\qquad = 3\sqrt{2} + 7\sqrt{3}$ **Simplify.**

 **Quick Check** ❸ Simplify each radical expression.

a. $\sqrt{5}(2 + \sqrt{10})$ **b.** $\sqrt{2x}(\sqrt{6x} - 11)$ **c.** $\sqrt{5a}(\sqrt{5a} + 3)$

If both radical expressions have two terms, you can multiply the same way you find the product of two binomials, by using FOIL.

4 EXAMPLE   Simplifying Using FOIL

Simplify $(\sqrt{5} - 2\sqrt{15})(\sqrt{5} + \sqrt{15})$.

$(\sqrt{5} - 2\sqrt{15})(\sqrt{5} + \sqrt{15})$

$\quad = \sqrt{25} + \sqrt{75} - 2\sqrt{75} - 2\sqrt{225}$ **Use FOIL.**

$\quad = 5 - \sqrt{75} - 2(15)$ **Combine like radicals and simplify $\sqrt{25}$ and $\sqrt{225}$.**

$\quad = 5 - \sqrt{25 \cdot 3} - 30$ **25 is a perfect square factor of 75.**

$\quad = 5 - \sqrt{25} \cdot \sqrt{3} - 30$ **Use the Multiplication Property of Square Roots.**

$\quad = 5 - 5\sqrt{3} - 30$ **Simplify $\sqrt{25}$.**

$\quad = -25 - 5\sqrt{3}$ **Simplify.**

 **Quick Check** ❹ Simplify each radical expression.

a. $(2\sqrt{6} + 3\sqrt{3})(\sqrt{6} - 5\sqrt{3})$ **b.** $(\sqrt{7} + 4)^2$

Video Tutor Help

Visit: PHSchool.com
Web Code: ate-0775

Conjugates are the sum and the difference of the same two terms. The radical expressions $\sqrt{5} + \sqrt{2}$ and $\sqrt{5} - \sqrt{2}$ are conjugates. The product of two conjugates results in a difference of two squares.

$(\sqrt{5} + \sqrt{2})(\sqrt{5} - \sqrt{2}) = (\sqrt{5})^2 - (\sqrt{2})^2$

$\qquad\qquad\qquad\qquad = 5 - 2$

$\qquad\qquad\qquad\qquad = 3$

Notice that the product of these conjugates has no radical.

You recall that a simplified radical expression has no radical in the denominator. When a denominator contains a sum or a difference including radical expressions, you can rationalize the denominator by multiplying the numerator and the denominator by the conjugate of the denominator. For example, to simplify a radical expression like $\frac{6}{\sqrt{5} - \sqrt{2}}$, you multiply by $\frac{\sqrt{5} + \sqrt{2}}{\sqrt{5} + \sqrt{2}}$.

**EXAMPLE** Rationalizing a Denominator Using Conjugates

Simplify $\dfrac{6}{\sqrt{5} - \sqrt{2}}$.

$\dfrac{6}{\sqrt{5} - \sqrt{2}} = \dfrac{6}{\sqrt{5} - \sqrt{2}} \cdot \dfrac{\sqrt{5} + \sqrt{2}}{\sqrt{5} + \sqrt{2}}$ **Multiply the numerator and the denominator by the conjugate of the denominator.**

$= \dfrac{6(\sqrt{5} + \sqrt{2})}{5 - 2}$ **Multiply in the denominator.**

$= \dfrac{6(\sqrt{5} + \sqrt{2})}{3}$ **Simplify the denominator.**

$= 2(\sqrt{5} + \sqrt{2})$ **Divide 6 and 3 by the common factor 3.**

$= 2\sqrt{5} + 2\sqrt{2}$ **Simplify the expression.**

✔ Quick Check ⑤ Simplify each expression.

a. $\dfrac{4}{\sqrt{7} + \sqrt{5}}$

b. $\dfrac{-4}{\sqrt{10} + \sqrt{8}}$

c. $\dfrac{-5}{\sqrt{11} - \sqrt{3}}$

You can solve a ratio involving radical expressions.

6 EXAMPLE Real-World 🌐 Problem Solving

Art The ratio length : width of this painting by Mondrian is approximately equal to the *golden ratio* $(1 + \sqrt{5}) : 2$. The length of the painting is 81 inches. Find the width of the painting in simplest radical form. Then find the approximate width to the nearest inch.

Define 81 = length of painting
x = width of painting

Relate $(1 + \sqrt{5}) : 2 =$ length : width

Write $\dfrac{1 + \sqrt{5}}{2} = \dfrac{81}{x}$

$x(1 + \sqrt{5}) = 162$ **Cross multiply.**

$\dfrac{x(1 + \sqrt{5})}{(1 + \sqrt{5})} = \dfrac{162}{(1 + \sqrt{5})}$ **Divide both sides by $(1 + \sqrt{5})$.**

$x = \dfrac{162}{(1 + \sqrt{5})} \cdot \dfrac{(1 - \sqrt{5})}{(1 - \sqrt{5})}$ **Multiply the numerator and the denominator by the conjugate of the denominator.**

$x = \dfrac{162(1 - \sqrt{5})}{1 - 5}$ **Multiply in the denominator.**

$x = \dfrac{162(1 - \sqrt{5})}{-4}$ **Simplify the denominator.**

$x = \dfrac{-81(1 - \sqrt{5})}{2}$ **Divide 162 and −4 by the common factor −2.**

$x = 50.06075309$ **Use a calculator.**

$x \approx 50$

The exact width of the painting is $\dfrac{-81(1 - \sqrt{5})}{2}$ inches. The approximate width of the painting is 50 inches.

Real-World 🌐 Connection

Mondrian painted straight lines at right angles because he felt this to be the angle of complete equilibrium.

**Quick Check** ⑥ Another painting has a length : width ratio approximately equal to the golden ratio $(1 + \sqrt{5}) : 2$. Find the length of a painting if the width is 34 inches.

EXERCISES

For more exercises, see *Extra Skill and Word Problem Practice.*

Practice and Problem Solving

A Practice by Example

Example 1
(page 622)

Simplify each expression.

1. $-3\sqrt{6} + 8\sqrt{6}$

2. $16\sqrt{10} + 2\sqrt{10}$

3. $\sqrt{5} - 3\sqrt{5}$

4. $6\sqrt{7} - 4\sqrt{7}$

5. $15\sqrt{2} - \sqrt{2}$

6. $-5\sqrt{3} - 3\sqrt{3}$

Example 2
(page 622)

Tell whether each pair of expressions can be simplified to like radicals.

7. $\sqrt{2}, \sqrt{32}$

8. $\sqrt{3}, \sqrt{75}$

9. $\sqrt{5}, \sqrt{50}$

Simplify each expression.

10. $\sqrt{18} + \sqrt{2}$

11. $2\sqrt{12} - 7\sqrt{3}$

12. $\sqrt{8} + 2\sqrt{2}$

13. $4\sqrt{5} - 2\sqrt{45}$

14. $3\sqrt{7} - \sqrt{28}$

15. $-4\sqrt{10} + 6\sqrt{40}$

Example 3
(page 623)

16. $\sqrt{2}(\sqrt{8} - 4)$

17. $\sqrt{3}(\sqrt{27} + 1)$

18. $2\sqrt{3}(\sqrt{3} - 1)$

19. $\sqrt{3}(\sqrt{15} + 2)$

20. $\sqrt{2}(3 + 3\sqrt{2})$

21. $\sqrt{6}(\sqrt{6} - 5)$

Example 4
(page 623)

22. $(3\sqrt{2} + \sqrt{3})(\sqrt{2} - 5\sqrt{3})$

23. $(2\sqrt{5} - \sqrt{6})(4\sqrt{5} - 3\sqrt{6})$

24. $(\sqrt{7} - 2)^2$

25. $(2\sqrt{10} + \sqrt{3})^2$

26. $(2\sqrt{11} + 5)(\sqrt{11} + 2)$

27. $(4 - \sqrt{13})(9 + \sqrt{13})$

Example 5
(page 624)

28. $\dfrac{8}{\sqrt{7} - \sqrt{3}}$

29. $\dfrac{-12}{\sqrt{8} - \sqrt{2}}$

30. $\dfrac{48}{\sqrt{6} - \sqrt{18}}$

31. $\dfrac{3}{\sqrt{10} - \sqrt{5}}$

32. $\dfrac{-40}{\sqrt{11} - \sqrt{3}}$

33. $\dfrac{9}{\sqrt{12} - \sqrt{11}}$

Example 6
(page 624)

Find an exact solution for each equation. Find the approximate solution to the nearest tenth.

34. $\dfrac{5\sqrt{2}}{\sqrt{2} - 1} = \dfrac{x}{\sqrt{2}}$

35. $\dfrac{3}{1 + \sqrt{5}} = \dfrac{1 - \sqrt{5}}{x}$

36. $\dfrac{\sqrt{2} - 1}{\sqrt{2} + 1} = \dfrac{x}{2}$

37. The ratio of the length to the width of a painting is $(1 + \sqrt{5}) : 2$. The length is 12 ft. What is the width?

B Apply Your Skills

Simplify each expression.

38. $\sqrt{40} + \sqrt{90}$

39. $3\sqrt{2}(2 + \sqrt{6})$

40. $\sqrt{12} + 4\sqrt{75} - \sqrt{36}$

41. $(\sqrt{3} + \sqrt{5})^2$

42. $\dfrac{\sqrt{13} + \sqrt{10}}{\sqrt{13} - \sqrt{5}}$

43. $(\sqrt{7} + \sqrt{8})(\sqrt{7} + \sqrt{8})$

44. $2\sqrt{2}(-2\sqrt{32} + \sqrt{8})$

45. $4\sqrt{50} - 7\sqrt{18}$

46. $\dfrac{2\sqrt{12} + 3\sqrt{6}}{\sqrt{9} - \sqrt{6}}$

47. Chemistry The ratio of the rates of diffusion of two gases is given by the formula $\dfrac{r_1}{r_2} = \dfrac{\sqrt{m_2}}{\sqrt{m_1}}$, where m_1 and m_2 are the masses of the molecules of the gases. Find $\dfrac{r_1}{r_2}$ if $m_1 = 12$ units and $m_2 = 30$ units.

Geometry Find the exact perimeter of each figure below.

48.

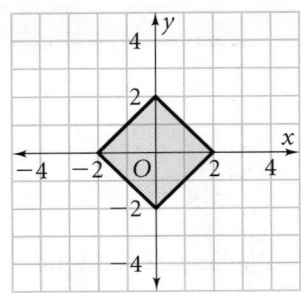

49.

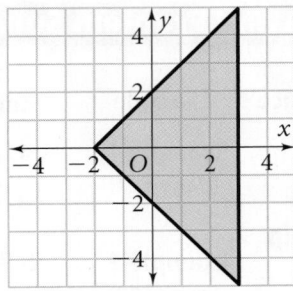

50.

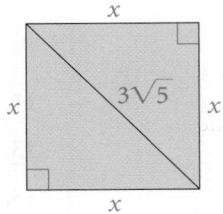

51.

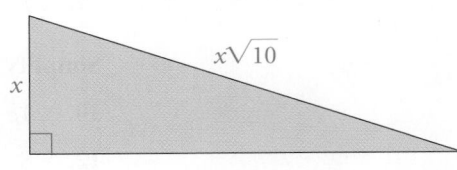

52. Open-Ended Make up three sums that are less than or equal to 50. Use the square roots of 2, 3, 5, or 7, and the whole numbers less than 10. For example, $8\sqrt{5} + 9\sqrt{7} \le 50$.

53. Error Analysis When simplifying $\sqrt{24} + \sqrt{48}$, a student wrote $2\sqrt{6} + 2\sqrt{24}$.
 a. What error did the student make?
 b. Simplify $\sqrt{24} + \sqrt{48}$ correctly.

54. You can make a box kite like the one at the right in the shape of a rectangular solid. The opening at each end of the kite is a square.
 a. Suppose the sides of the square are 2 ft long. How long are the diagonal struts used for bracing?
 b. Suppose each side of the square has length s. Find the length of the diagonal struts in terms of s. Write your answer in simplest form.

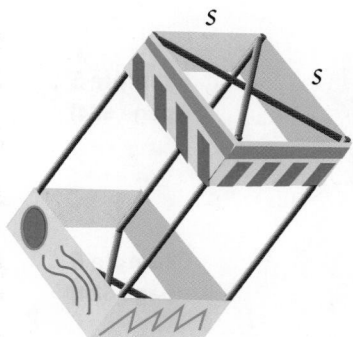

🌐 **Investments** For Exercises 55–57, the formula $r = \sqrt{\dfrac{A}{P}} - 1$ gives the interest rate r that will allow principal P to grow into amount A in two years, if the interest is compounded annually. Use the formula to find the interest rate you would need to meet each goal.

55. Suppose you have $500 to deposit into an account. Your goal is to have $595 in that account at the end of the second year.

56. Suppose you have $550 to deposit into an account. Your goal is to have $700 in that account at the end of two years.

57. Suppose you have $600 to deposit into an account. Your goal is to have $800 in that account at the end of two years.

GO **O**nline
Homework Video Tutor
Visit: PHSchool.com
Web Code: ate-1102

58. a. Suppose n is an even number. Simplify $\sqrt{x^n}$.
 b. Suppose n is an odd number greater than 1. Simplify $\sqrt{x^n}$.

59. Critical Thinking Simplify $\dfrac{a\sqrt{b}}{b\sqrt{a}}$.

60. Find the value of the numerical expression for Professor Hinkle's age in the cartoon.

 61. Writing Explain why $\sqrt{3} + \sqrt{6}$ cannot be simplified.

62. a. Copy and complete the table.

a	b	$\sqrt{a}$	$\sqrt{b}$	$\sqrt{a} + \sqrt{b}$	$\sqrt{a+b}$
1	0	■	■	■	■
16	1	■	■	■	■
25	9	■	■	■	■
64	36	■	■	■	■
100	81	■	■	■	■

b. Does $\sqrt{a} + \sqrt{b}$ always equal $\sqrt{a+b}$? Explain.

63. Error Analysis Explain the error in the work below.

$$\sqrt{41} = \sqrt{16+25} = \sqrt{16} + \sqrt{25} = 4 + 5 = 9$$

 **Challenge**

Simplify each expression.

64. $\sqrt{18} + \dfrac{3}{\sqrt{2}}$

65. $\dfrac{\sqrt{28}}{3} + \dfrac{3}{\sqrt{7}}$

66. $\sqrt{\dfrac{3}{5}} + \sqrt{\dfrac{5}{3}}$

67. $\dfrac{\sqrt{27} + \sqrt{48} - \sqrt{75}}{\sqrt{3}}$

68. $\sqrt{288} + \sqrt{50} - \sqrt{98}$

69. $\left(\sqrt{2} + \sqrt{32}\right)\left(\sqrt{2} + \sqrt{8} + \sqrt{32}\right)$

70. $\dfrac{\sqrt{5} + \sqrt{10} - \sqrt{15}}{\sqrt{10} - \sqrt{5}}$

71. Find the length of each hypotenuse. Write your answers in simplified radical form.

a.

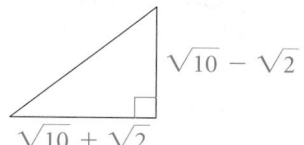

$\sqrt{10} - \sqrt{2}$

$\sqrt{10} + \sqrt{2}$

b.

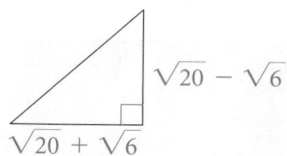

$\sqrt{20} - \sqrt{6}$

$\sqrt{20} + \sqrt{6}$

c. If the length of the legs of a right triangle are $\sqrt{p} + \sqrt{q}$ and $\sqrt{p} - \sqrt{q}$, write an expression for the length of the hypotenuse.

 **REGENTS**

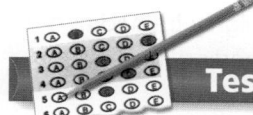 **Test Prep**

Multiple Choice

72. Simplify $4\sqrt{75} + \sqrt{27}$.

A. $12\sqrt{3}$ **B.** $23\sqrt{3}$ **C.** $4\sqrt{102}$ **D.** $5\sqrt{102}$

73. Which radical expression is NOT equal to $5\sqrt{2}$?

   F. $\sqrt{8} + \sqrt{18}$ **G.** $\sqrt{98} - \sqrt{8}$

   H. $-\sqrt{32} + \sqrt{162}$ **J.** $\sqrt{48} + \sqrt{2}$

Short Response   **74.** Simplify $(3\sqrt{5} - \sqrt{2})(\sqrt{5} + 5\sqrt{2})$. Show your work.

Extended Response   **75.** Explain the steps needed to simplify $\dfrac{5}{\sqrt{7} + \sqrt{21}}$.

Mixed Review

Lesson 11-1   **Simplify each radical expression.**

GO for Help

76. $\sqrt{8} \cdot \sqrt{6}$ **77.** $\dfrac{\sqrt{12}}{\sqrt{18}}$ **78.** $\sqrt{5 \cdot 10}$

79. $\sqrt{40b^5}$ **80.** $\dfrac{\sqrt{2x^2}}{\sqrt{4x^6}}$ **81.** $\sqrt{\dfrac{24v}{v^8}}$

82. $\sqrt{\dfrac{300}{3}}$ **83.** $\sqrt{11n} \cdot \sqrt{22n^3}$ **84.** $-\sqrt{\dfrac{28b^5}{7b}}$

Lesson 10-4   **Solve each equation by factoring.**

85. $5t^2 - 35t = 0$ **86.** $p^2 - 7p - 18 = 0$ **87.** $k^2 + 12k + 27 = 0$

88. $y^2 - 2y = 24$ **89.** $m^2 + 30 = -17m$ **90.** $2a^2 = -7a - 3$

Lesson 9-4   **Find each product.**

91. $(b + 11)(b + 11)$ **92.** $(2p + 7)(2p + 7)$ **93.** $(5g - 7)(5g + 7)$

94. $(3x + 1)(3x - 1)$ **95.** $\left(\frac{1}{3}k - 9\right)\left(\frac{1}{3}k + 9\right)$ **96.** $(d - 1.1)(d - 1.1)$

Algebra at Work

·········· Auto Mechanic

Auto mechanics work to see that car engines get the most out of every gallon of gasoline. Formulas used by mechanics often involve radicals. For example, a car gets its power when gas and air in each cylinder are compressed and ignited by a spark plug. An engine's efficiency e is given by the formula $e = \dfrac{c - \sqrt{c}}{c}$, where c is the compression ratio.

Because of the complexity of such formulas and of modern high-performance engines, today's auto mechanic must be a highly trained and educated professional who understands algebra, graph reading, and the operation of computerized equipment.

Go Online
PHSchool.com

For: Information about a career as an auto mechanic
Web Code: atb-2031

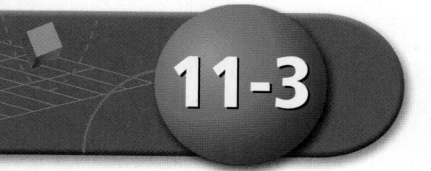

Solving Radical Equations

NY Learning Standards for Mathematics

A.PS.9 Interpret solutions within the given constraints of a problem.

 Check Skills You'll Need

GO for Help   Lesson 3-8

Evaluate each expression for the given value.

1. $\sqrt{x} - 3$ for $x = 16$ **2.** $\sqrt{x + 7}$ for $x = 9$ **3.** $2\sqrt{x + 3}$ for $x = 1$

Simplify each expression.

4. $(\sqrt{3})^2$ **5.** $(\sqrt{x + 1})^2$ **6.** $(\sqrt{2x - 5})^2$

 **New Vocabulary** • radical equation • extraneous solution

---

1 Solving Radical Equations

GO Online

Video Tutor Help
Visit: PHSchool.com
Web Code: ate-0775

A **radical equation** is an equation that has a variable in a radicand. You can often solve a radical equation by getting the radical by itself on one side of the equation. Then you square both sides. Remember that the expression under a radical must be nonnegative.

$$\text{When } x \geq 0, (\sqrt{x})^2 = x.$$

1 EXAMPLE **Solving by Isolating the Radical**

Solve each equation. Check your solution.

a. $\sqrt{x} - 3 = 4$

$\sqrt{x} = 7$ Get the radical on the left side of the equation.

$(\sqrt{x})^2 = 7^2$ Square both sides.

$x = 49$

Check $\sqrt{x} - 3 = 4$

$\sqrt{49} - 3 \stackrel{?}{=} 4$ Substitute 49 for x.

$7 - 3 = 4$ ✓

b. $\sqrt{x - 3} = 4$

$(\sqrt{x - 3})^2 = 4^2$ Square both sides.

$x - 3 = 16$ Solve for x.

$x = 19$

Check $\sqrt{x - 3} = 4$

$\sqrt{19 - 3} \stackrel{?}{=} 4$ Substitute 19 for x.

$\sqrt{16} = 4$ ✓

 **Quick Check** **1** Solve each equation. Check your solution.

a. $\sqrt{x} + 7 = 12$ **b.** $\sqrt{a} - 4 = 5$ **c.** $\sqrt{c - 2} = 6$

For an equation like $2\sqrt{x} = 8$, you could square both sides first, or you could divide by 2 to get $\sqrt{x}$ alone on one side of the equation.

2 EXAMPLE Real-World Problem Solving

Designing a Ride On a roller coaster ride, your speed in a loop depends on the height of the hill you have just come down and the radius of the loop in feet. The equation $v = 8\sqrt{h - 2r}$ gives the velocity v in feet per second of a car at the top of the loop.

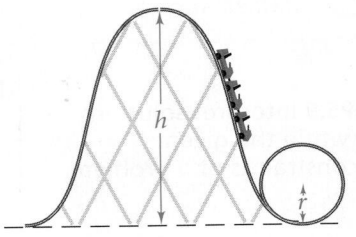

Suppose the loop has a radius of 18 ft. You want the car to have a velocity of 30 ft/s at the top of the loop. How high should the hill be?

Solve $v = 8\sqrt{h - 2r}$ for h when $v = 30$ and $r = 18$.

$$30 = 8\sqrt{h - 2(18)}$$ **Substitute 30 for v and 18 for r.**

$$\frac{30}{8} = \frac{8\sqrt{h - 2(18)}}{8}$$ **Divide each side by 8 to isolate the radical.**

$$3.75 = \sqrt{h - 36}$$ **Simplify.**

$$(3.75)^2 = \left(\sqrt{h - 36}\right)^2$$ **Square both sides.**

$$14.0625 = h - 36$$

$$50.0625 = h$$

The hill should be about 50 ft high.

Quick Check **2 a.** Find the height of the hill when the velocity at the top of the loop is 35 ft/s, and the radius of the loop is 24 ft.

      b. Critical Thinking Would you expect the velocity of the car to increase as the radius of the loop increases? Explain.

You can square both sides of an equation to solve an equation involving radical expressions.

3 EXAMPLE Solving With Radical Expressions on Both Sides

Gridded Response Solve $\sqrt{3n - 2} = \sqrt{n + 6}$.

$$\left(\sqrt{3n - 2}\right)^2 = \left(\sqrt{n + 6}\right)^2$$ **Square both sides.**

$$3n - 2 = n + 6$$ **Simplify.**

$$3n = n + 8$$ **Add 2 to each side.**

$$2n = 8$$ **Subtract n from each side.**

$$n = 4$$ **Divide each side by 2.**

Check $\sqrt{3n - 2} = \sqrt{n + 6}$

$$\sqrt{3(4) - 2} \stackrel{?}{=} \sqrt{4 + 6}$$ **Substitute 4 for n.**

$$\sqrt{10} = \sqrt{10} \checkmark$$

The solution is 4. Grid 4.

Quick Check **3** Solve $\sqrt{3t + 4} = \sqrt{5t - 6}$. Check your answer.

When you solve an equation by squaring each side, you create a new equation. This new equation may have solutions that do not solve the original equation.

Original equation	Square of each side	New equation	Solutions of new equation
$x = 2$ $\longrightarrow$	$(x)^2 = (2)^2$ $\longrightarrow$	$x^2 = 4$ $\longrightarrow$	$2, -2$

In the example above, -2 does not satisfy the original equation. It is an extraneous solution. An **extraneous solution** is a solution that does not satisfy the original equation. Be sure to check all solutions in the original equation to determine whether a solution is extraneous.

4 EXAMPLE Identifying Extraneous Solutions

Solve $x = \sqrt{x + 6}$.

$\begin{aligned} (x)^2 = (\sqrt{x + 6})^2 && \text{Square both sides.} \\ x^2 = x + 6 && \text{Simplify.} \\ x^2 - x - 6 = 0 && \text{Subtract } x \text{ and 6 from both sides.} \\ (x - 3)(x + 2) = 0 && \text{Solve the quadratic equation by factoring.} \\ (x - 3) = 0 \ \text{ or } \ (x + 2) = 0 && \text{Use the Zero-Product Property.} \\ x = 3 \ \text{ or } \ \ \ \ \ \ \ \ x = -2 && \text{Solve for } x. \end{aligned}$

Check $\qquad\qquad x = \sqrt{x + 6}$

$\begin{aligned} 3 \overset{?}{=} \sqrt{3 + 6} \qquad\qquad -2 \overset{?}{=} \sqrt{-2 + 6} && \text{Substitute 3 and 2 for } x. \\ 3 = 3 \ \checkmark \qquad\qquad\quad -2 \neq 2 \end{aligned}$

● The solution to the original equation is 3. The value -2 is an extraneous solution.

Problem Solving Hint

For Quick Check 4b, you can use any method to solve a quadratic equation: graphing, factoring, or the quadratic formula.

✓ **Quick Check** **4 a. Critical Thinking** How could you determine that -2 was not a solution of $x = \sqrt{x + 6}$ without going through all the steps of the check?
b. Solve $y = \sqrt{y + 2}$. Check your solutions.

It is possible that the only solution you get after squaring both sides of an equation is extraneous. In that case, the original equation has no solution.

5 EXAMPLE No Solution

Solve $\sqrt{2x} + 6 = 4$.

$\begin{aligned} \sqrt{2x} = -2 \\ (\sqrt{2x})^2 = (-2)^2 && \text{Square both sides.} \\ 2x = 4 \\ x = 2 \end{aligned}$

Check $\sqrt{2x} + 6 = 4$

$\begin{aligned} \sqrt{2(2)} + 6 \overset{?}{=} 4 && \text{Substitute 2 for } x. \\ \sqrt{4} + 6 \overset{?}{=} 4 \\ 2 + 6 \neq 4 && x = 2 \text{ does not solve the original equation.} \end{aligned}$

● $\sqrt{2x} + 6 = 4$ has no solution.

✓ **Quick Check** **5** Solve $8 - \sqrt{2n} = 20$. Check your solution.

EXERCISES

For more exercises, see *Extra Skill and Word Problem Practice*.

Practice and Problem Solving

A Practice by Example

Example 1
(page 629)

GO for Help

Example 2
(page 630)

Solve each radical equation. Check your solution.

1. $\sqrt{x} + 3 = 5$

2. $\sqrt{t} + 2 = 9$

3. $\sqrt{s} - 1 = 5$

4. $\sqrt{n + 7} = 12$

5. $\sqrt{a - 6} = 3$

6. $\sqrt{z} - 7 = -3$

7. Distance The time t in seconds it takes an object to fall d feet is given by $t = \sqrt{\frac{d}{16}}$. Find the distance an object falls after 6 seconds.

8. Power The current of an electrical circuit I in amps is related to the power P in watts and the resistance R in ohms by the formula $I = \sqrt{\frac{P}{R}}$. Find the power (to the nearest watt) when the current is 8 amps and the resistance is 9.4 ohms.

Example 3
(page 630)

Solve each radical equation. Check your solution.

9. $\sqrt{3x + 1} = \sqrt{5x - 8}$

10. $\sqrt{2y} = \sqrt{9 - y}$

11. $\sqrt{7v - 4} = \sqrt{5v + 10}$

12. $\sqrt{s + 10} = \sqrt{6 - s}$

13. $\sqrt{n + 5} = \sqrt{5n - 11}$

14. $\sqrt{3m + 1} = \sqrt{7m - 9}$

Examples 4, 5
(page 631)

Tell which solutions, if any, are extraneous for each equation.

15. $-z = \sqrt{-z + 6}; z = -3, z = 2$

16. $\sqrt{12 - n} = n; n = -4, n = 3$

17. $y = \sqrt{2y}; y = 0, y = 2$

18. $2a = \sqrt{4a + 3}; a = \frac{3}{2}, a = -\frac{1}{2}$

19. $x = \sqrt{28 - 3x}; x = 4, x = -7$

20. $-t = \sqrt{-6t - 5}; t = -5, t = -1$

Solve each radical equation. Check your solution. If there is no solution, write *no solution*.

21. $x = \sqrt{2x + 3}$

22. $n = \sqrt{4n + 5}$

23. $\sqrt{3b} = -3$

24. $2y = \sqrt{5y + 6}$

25. $-2\sqrt{2r + 5} = 6$

26. $\sqrt{d + 12} = d$

27. $\sqrt{z + 5} = 2z$

28. $2t = \sqrt{5t - 1}$

B Apply Your Skills

29. Geometry In the right triangle $\triangle ABC$, the altitude $\overline{CD}$ is at a right angle to the hypotenuse. You can use $CD = \sqrt{(AD)(DB)}$ to find missing lengths.
  a. Find AD if $CD = 10$ and $DB = 4$.
  b. Find DB if $AD = 20$ and $CD = 15$.

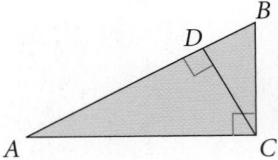

30. Packaging The volume V in cubic units for a cylindrical can is given by the formula $V = \pi r^2 h$, where r is the radius of the can and h is the height. The volume of the can is 98 in.^{3}, and the height of the can is 5 in. Find the radius of the can.

31. Writing Explain what is meant by an extraneous solution.

32. Open-Ended Write two radical equations that have 3 for a solution.

33. The formula $t = \sqrt{\frac{n}{16}}$ gives the time t in seconds for an object that is initially at rest to fall n feet. Find the distance an object falls in the first 10 seconds.

GO Online
Homework Video Tutor
Visit: PHSchool.com
Web Code: ate-1103

Solve each radical equation. Check your solution. If there is no solution, write *no solution*.

34. $\sqrt{5x + 10} = 5$

35. $-6 - \sqrt{3y} = -3$

36. $\sqrt{7p + 5} = \sqrt{p - 3}$

37. $a = \sqrt{7a - 6}$

38. $\sqrt{y + 12} = 3\sqrt{y}$

39. $\sqrt{x - 10} = 1$

40. $\frac{x}{2} = \sqrt{3x}$

41. $\frac{c}{3} = \sqrt{c - 2}$

42. $7 = \sqrt{x + 5}$

43. $3 - \sqrt{4a + 1} = 12$

44. a. The equation $v = 8\sqrt{h - 2r}$ gives the velocity v in feet per second of a car at the top of the loop of a roller coaster. Find the radius of the loop when the hill is 150 ft high and the velocity of the car is 30 ft/s.
   b. Find the approximate speed in mi/h for 30 ft/s. (*Hint:* 1 mi = 5280 ft)
   c. Critical Thinking Would you expect the velocity of the car to increase or decrease as the radius of the loop increases? As the height of the hill decreases?
   d. Explain your reasoning in your answer for part (c).

 45. a. Graphing Calculator Graph the equations $y = \sqrt{2x + 1}$ and $y = \sqrt{3x - 5}$.
   b. What is the solution to the system of equations?
   c. Solve $\sqrt{2x + 1} = \sqrt{3x - 5}$. Compare this solution with your answer to part (b).

 46. Packaging The diagram at the right shows a piece of cardboard that makes a box when sections of it are folded and taped. The ends of the box are x inches by x inches and the body of the box is 10 inches long.
   a. Write an equation for the volume V of the box.
   b. Solve the equation in part (a) for x.
   c. Find the integer values of x that would give the box a volume between 40 in.^{3} and 490 in.^{3}, inclusive.

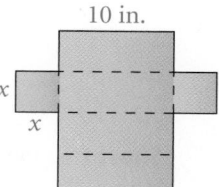

47. a. Solve $x^2 - 3 = 4$.
   b. Solve $\sqrt{x} - 3 = 4$.
   c. How are solving part (a) and solving part (b) alike? How are they different?

For a guide to solving Exercise 46, see p. 635.

Challenge

Solve each radical equation. Check your solution. If there is no solution, write *no solution*.

48. $\sqrt{x^2 - 6x} = 4$

49. $\sqrt{2x^2 + 8x} = x$

50. $\sqrt{x^2 + 4x + 5} = x$

51. $x + 3 = \sqrt{x^2 - 4x - 1}$

52. Writing Explain how you would solve the equation $\sqrt{2x} + \sqrt{x + 2} = 0$.

53. Critical Thinking Explain the difference between squaring $\sqrt{x - 1}$ and squaring $\sqrt{x} - 1$.

 54. Physics The equation $T = \sqrt{\frac{2\pi^2 r}{F}}$ gives the time T in seconds it takes a satellite with mass 0.5 kilograms to complete one orbit of radius r meters. The force F in newtons pulls the body toward the center of the orbit.
   a. It takes 2 seconds for an object to make one revolution with a force of 10 newtons. Find the radius of the orbit.
   b. Find the radius of the orbit if the force is 160 newtons and $T = 2$.

Real-World Connection

In the yo-yo trick called "Around the World," the yo-yo is a satellite circling a person's hand.

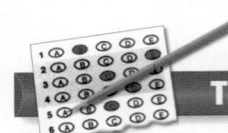
Multiple Choice

55. Which choice shows the solution(s) of $k = \sqrt{5k + 6}$?
   A. 6, −1   **B.** −6, 1   **C.** 6   **D.** −1

56. Which choice shows the solution(s) of $\sqrt{2x + 1} = \sqrt{3x - 5}$?
   F. −1   **G.** 6   **H.** $\frac{4}{5}$, 6   **J.** no solution

57. Which radical equation has no solution?
   A. $-\sqrt{x} = -25x$ **B.** $\sqrt{3x + 1} = -10$
   C. $-3\sqrt{3x} = -5$ **D.** $\frac{x}{2} = \sqrt{x - 1}$

58. Which shows the most appropriate first step in solving $\sqrt{2x} + 1 = 5$?
   F. $(\sqrt{2x} + 1)^2 = 5^2$ **G.** $(\sqrt{2x})^2 + 1^2 = 5^2$
   H. $\sqrt{2x} = 4$ **J.** $\sqrt{2x} - 4 = 0$

59. What is the solution of $\sqrt{n + 8} = \sqrt{3n}$?
   I. −8   **II.** 2   **III.** 4   **IV.** −2
   A. I only
   B. III only
   C. II and III
   D. III and IV

60. For which values of a does $\sqrt{x^2 - a} = 3$ have no solution?
   F. $a < -9$
   G. $a > -9$
   H. $a = x$ and $a = 3$
   J. The equation has a solution for all real values of a.

Short Response

61. Solve $\sqrt{15 - 5x} = \sqrt{4x - 3}$. Show your work.

Mixed Review

GO for Help

Lesson 11-2

Simplify each expression.

62. $\sqrt{20} + \sqrt{45}$ **63.** $\sqrt{3}(\sqrt{6} + 4)$

64. $\sqrt{72} - \sqrt{50}$ **65.** $\sqrt{2}(2\sqrt{8} + 4\sqrt{18})$

66. $\dfrac{8}{\sqrt{5} + \sqrt{3}}$ **67.** $\sqrt{12}(7\sqrt{6} + \sqrt{24})$

Lesson 10-5

Solve by completing the square. If necessary, round to the nearest tenth.

68. $w^2 + 2w - 11 = 0$ **69.** $k^2 - 8k = 3$

70. $d^2 + 5d + 1 = 0$ **71.** $2a^2 + 20a = 16$

72. $\frac{1}{5}x^2 + 2x = 4$ **73.** $6g^2 - 9g - 30 = 0$

Lessons 9-5, 9-6

Factor each expression.

74. $x^2 + 10x - 24$ **75.** $m^2 - 14m + 13$

76. $b^2 + 16b - 36$ **77.** $2p^2 + 15p + 7$

78. $3d^2 + 12d - 15$ **79.** $4v^2 - 25v + 25$

Guided Problem Solving

Understanding Word Problems Read the problem below, and then follow along with what Lily thinks as she solves the problem. Check your understanding by solving the exercise at the bottom of the page.

Packaging The diagram at the right shows a piece of cardboard that makes a box when sections of it are folded and taped. The ends of the box are x inches by x inches and the box is 10 inches long.

a. Write an equation for the volume V of the box.
b. Solve the equation in part (a) for x.
c. Find the integer values of x that give the box a volume between 40 in.^{3} and 490 in.^{3}, inclusive.

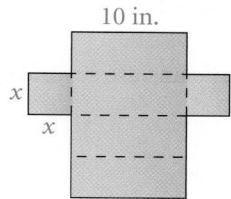

What Lily Thinks

What will this box look like? I'll draw a sketch.

Okay, I can see now that the box has length 10 in., width x in., and height x in. I can substitute these into the formula for the volume of a rectangular prism.

Part (b) asks me to solve the equation for x. Since x is a dimension of the box, it can't be negative. So the solution is the positive square root only.

Part (c) asks me to write a sentence showing that the volume is between 40 in.^{3} and 490 in.^{3}. Oh, I can write an inequality! Then I can solve for x.

Wait! The problem asks for *integer values* only.

What Lily Writes

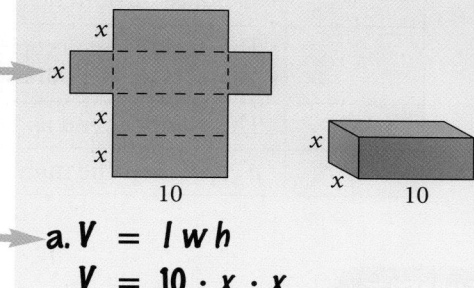

a. $V = lwh$
$V = 10 \cdot x \cdot x$
$V = 10x^2$

b. $V = 10x^2$
$\dfrac{V}{10} = x^2$
$\pm\sqrt{\dfrac{V}{10}} = x$
$x = \sqrt{\dfrac{V}{10}} = \dfrac{\sqrt{10V}}{10}$

c. $40 \le$ Volume ≤ 490
$40 \le 10x^2 \le 490$
$4 \le x^2 \le 49$
$2 \le x \le 7$

The possible integer values of x are 2, 3, 4, 5, 6, and 7.

EXERCISE

Geometry Each edge of a cube has length e.
a. Write an equation for the surface area, S, of the cube in terms of e.
b. Find the integer values of e that will give the cube a surface area between 726 cm^2 and 1176 cm^2, inclusive.

Standard Deviation

Statisticians use measures of variation to describe how data are distributed. Range and interquartile range (the difference between the third and first quartiles) are two measures of variation. Another measure of variation is *standard deviation*. **Standard deviation** is a measure of how the data vary, or deviate, from the mean.

Statisticians use several special symbols in the formula for standard deviation:

$$\sigma = \sqrt{\frac{\Sigma(x - \bar{x})^2}{n}}.$$

Symbol	Description
σ	The Greek letter sigma represents standard deviation.
Σ	The capital sigma represents the sum of a series of numbers or of a data set.
$\bar{x}$	This symbol, read as "*x*-bar," represents the mean.
n	n represents the number of values in a data set.

1 EXAMPLE

Find the mean and the standard deviation for the data: 12.6, 15.1, 11.2, 17.9, 18.2. Use a table to help organize your work.

Step 1 Find the mean of the data: $\bar{x}$.

$$\frac{12.6 + 15.1 + 11.2 + 17.9 + 18.2}{5} = 15$$

Step 2 Find the difference between each value and mean: $x - \bar{x}$.

Step 3 Square each difference: $(x - \bar{x})^2$.

Step 4 Find the average (mean) of these squares: $\frac{\Sigma(x - \bar{x})^2}{n}$.

$$\frac{5.76 + 0.01 + 14.44 + 8.41 + 10.24}{5} = 7.772$$

x	$\bar{x}$	$(x-\bar{x})$	$(x-\bar{x})^2$
12.6	15	−2.4	5.76
15.1	15	0.1	0.01
11.2	15	−3.8	14.44
17.9	15	2.9	8.41
18.2	15	3.2	10.24

Step 5 Take the square root to find the standard deviation: $\sqrt{\frac{\Sigma(x - \bar{x})^2}{n}}$.

$$\sqrt{7.772} \approx 2.79$$

The mean is 15, and the standard deviation is about 2.79.

A small standard deviation (compared to the data values) means that the data are clustered tightly around the mean. As the data become more widely distributed, the standard deviation increases.

EXAMPLE

Baseball The table at the right shows the batting averages of the American League Batting Champions. Use a graphing calculator to find the mean and the standard deviation to the nearest thousandth.

Batting Averages of American League Batting Champions From 1995 to 2004				
.356	.358	.347	.339	.357
.372	.350	.349	.326	.372

Step 1 Press [STAT] [ENTER] and then enter the data as **L1**.
Step 2 Press [STAT] and then select **CALC** and **1; 1-Var Stats**.
Step 3 Press [ENTER]. The calculator will display several values, including the mean and the standard deviation.

1–Var Stats
x̄=.3526 ⟵ **mean**
Σx=3.526
Σx²=1.245024
Sx=.0139698087
σx=.0132529242 ⟵ **standard deviation**
↓n=10

● The mean batting average is about .353, and the standard deviation is about 0.013.

In a data list, every value falls within some number of standard deviations of the mean. When a value falls within one standard deviation of the mean, it is in the range of values from one standard deviation below the mean to one standard deviation above.

3 **EXAMPLE**

Use the data from Example 2. Within how many standard deviations of the mean do all of the values fall?

Step 1 Draw a number line. Plot the data values and the mean. **Step 2** Mark off intervals of 0.013 on either side of the mean.

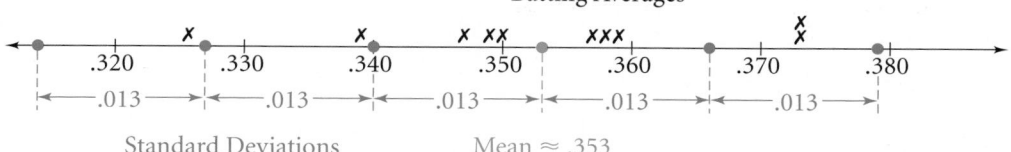

Batting Averages

.320   .330   .340   .350   .360   .370   .380

◄—.013—►◄—.013—►◄—.013—►◄—.013—►◄—.013—►

Standard Deviations   Mean ≈ .353

● All of the values fall within three standard deviations of the mean.

EXERCISES

For each data set, find the mean, the standard deviation, and the number of standard deviations that includes all the data values.

1. 4, 8, 5, 12, 3, 9, 5, 2

2. 102, 98, 103, 86, 101, 110

3. Explain why the value .326 in Example 3 can be described as an outlier.

4. Critical Thinking Suppose two sets of data have the same mean and range but different standard deviations. What does this indicate about how the data in each set are distributed?

5. a. Data Collection Gather numerical data on a topic of your choosing.
   b. Find the mean and the standard deviation.
   c. Make a line plot of the data showing the mean, the data, and the standard deviations. How many standard deviations are needed to include all the data?

11-4

Graphing Square Root Functions

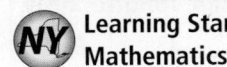
A.R.8 Use mathematics to show and understand mathematical phenomena (e.g., compare the graphs of the functions represented by the equations $y = x^2$ and $y = -x^2$).

 Check Skills You'll Need

GO for Help Lessons 3-8 and 10-1

Graph each pair of quadratic functions on the same graph.

1. $y = x^2, y = x^2 + 3$

2. $y = x^2, y = x^2 - 4$

Evaluate each expression for the given value of x.

3. $\sqrt{x}$ for $x = 4$

4. $\sqrt{x + 7} - 3$ for $x = 2$

5. $3\sqrt{x} + 2$ for $x = 9$

New Vocabulary • square root function

---

1 Graphing Square Root Functions

Vocabulary Tip

The square root function is the simplest example of a radical function.

A **square root function** is a function that contains the independent variable in the radicand. The function $y = \sqrt{x}$ is the simplest square root function. For x-values that are not perfect squares, you can approximate the y-values to the nearest tenth.

You can graph a square root function by plotting points. Plot the least value in the domain and several other points. Then join the points using a curve.

x	y
0	0
1	1
2	1.4
4	2
6	2.4
9	3

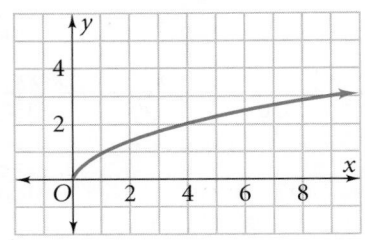

For real numbers, the value of the radicand cannot be negative. So the domain is limited to those values of x that make the radicand greater than or equal to 0.

1 EXAMPLE Finding the Domain of a Square Root Function

Find the domain of each function.

a. $y = \sqrt{x + 3}$

$x + 3 \geq 0$ ←Make the radicand $\geq$ 0.→

$x \geq -3$

The domain is the set of all numbers greater than or equal to -3.

b. $y = 3\sqrt{2x - 8}$

$2x - 8 \geq 0$

$2x \geq 8$

$x \geq 4$

The domain is the set of all numbers greater than or equal to 4.

 **Quick Check** Find the domain of $y = \sqrt{x - 7}$.

 **EXAMPLE** <u>Real-World</u> Problem Solving

Measurement For good weather conditions, police can use the formula $r = 2\sqrt{5L}$ to find the approximate speed r of a car that leaves a skid mark of length L in feet. Graph the function.

Length of Skid Mark (ft)	Speed (mi/h)
0	0
10	14.1
20	20
30	24.5
40	28.3

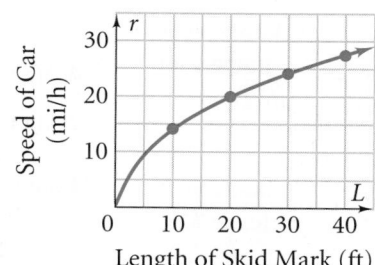

 **2** **a.** Copy and extend the graph in Example 2 for skid mark lengths from 60 ft to 100 ft.
b. **Critical Thinking** How long is the skid mark when a car's speed before putting on the brakes is 85 mi/h?

2 Translating Graphs of Square Root Functions

For any positive number k, $y = \sqrt{x} + k$ translates the graph of $y = \sqrt{x}$ up k units, while $y = \sqrt{x} - k$ translates the graph of $y = \sqrt{x}$ down k units.

3 **EXAMPLE** Graphing a Vertical Translation

Graph $y = \sqrt{x} + 3$ by translating the graph of $y = \sqrt{x}$.

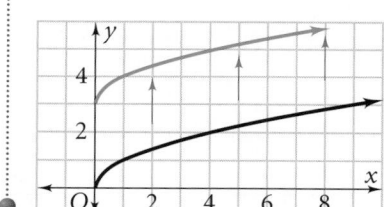

For the graph of $y = \sqrt{x} + 3$, the graph of $y = \sqrt{x}$ is shifted 3 units up.

 **3** Graph $f(x) = \sqrt{x} - 4$ by translating the graph of $f(x) = \sqrt{x}$.

For any positive number h, $y = \sqrt{x + h}$ translates the graph of $y = \sqrt{x}$ to the left h units, while $y = \sqrt{x - h}$ translates the graph of $y = \sqrt{x}$ to the right h units.

4 **EXAMPLE** Graphing a Horizontal Translation

Graph $f(x) = \sqrt{x + 4}$ by translating the graph of $f(x) = \sqrt{x}$.

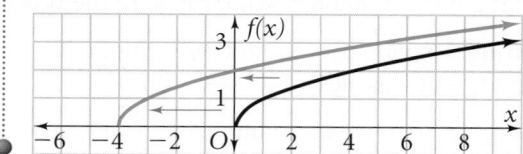

For the graph of $f(x) = \sqrt{x + 4}$, the graph of $f(x) = \sqrt{x}$ is shifted 4 units to the left.

 **4** Graph $f(x) = \sqrt{x - 3}$ by translating the graph of $f(x) = \sqrt{x}$.

EXERCISES

For more exercises, see *Extra Skill and Word Problem Practice.*

Practice and Problem Solving

 Practice by Example

Example 1
(page 638)

 GO for Help

Find the domain of each function.

1. $y = \sqrt{x - 2}$ **2.** $f(x) = \sqrt{4x - 3}$ **3.** $y = \sqrt{1.5x}$

4. $f(x) = \sqrt{7 + x}$ **5.** $y = \sqrt{x + 3} - 1$ **6.** $f(x) = \sqrt{x - 5} + 1$

7. $f(x) = \sqrt{3x + 5}$ **8.** $f(x) = \sqrt{2 + x}$ **9.** $f(x) = \sqrt{6x - 8} + 1$

Example 2
(page 639)

Make a table of values and graph each function.

10. $y = \sqrt{2x}$ **11.** $f(x) = 2\sqrt{x}$ **12.** $y = \sqrt{4x - 8}$

13. $y = \sqrt{3x}$ **14.** $f(x) = 3\sqrt{x}$ **15.** $y = -3\sqrt{x}$

16. Physics You can use the function $v = \sqrt{64h}$ to find the velocity v of an object, ignoring air resistance, after it has fallen h feet. Make a table of values and graph the function.

Examples 3, 4
(page 639)

Match each graph with its function.

17. $y = \sqrt{x} + 4$ **18.** $y = \sqrt{x} - 2$ **19.** $y = \sqrt{x + 4}$ **20.** $y = \sqrt{x - 2}$

A.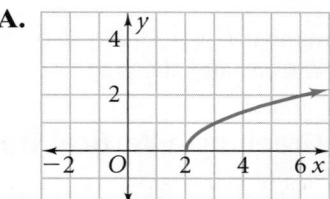

B.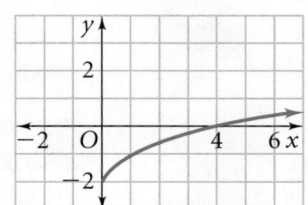

C.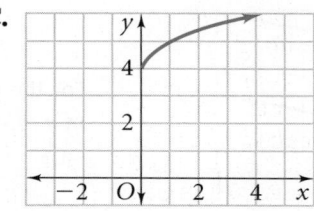

D.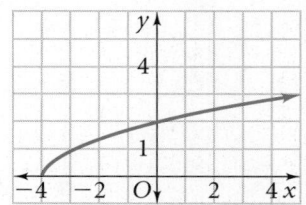

Graph each function by translating the graph of $y = \sqrt{x}$.

21. $y = \sqrt{x} + 5$ **22.** $y = \sqrt{x} - 5$ **23.** $y = \sqrt{x} - 3$

24. $y = \sqrt{x + 2}$ **25.** $f(x) = \sqrt{x - 2}$ **26.** $f(x) = \sqrt{x - 4}$

27. $y = \sqrt{x} + 1$ **28.** $y = \sqrt{x + 1}$ **29.** $y = \sqrt{x - 1}$

 Apply Your Skills

30. What are the domain and the range of the function $y = \sqrt{2x - 8}$?

31. What are the domain and the range of the function $y = \sqrt{8 - 2x}$?

32. Writing Explain how to find the domain of a square root function. Include an example.

33. Open-Ended Give an example of a square root function in each form. Choose $n \neq 0$.
    a. $y = \sqrt{x} + n$ **b.** $y = \sqrt{x + n}$ **c.** $y = n\sqrt{x}$
    d. Graph each function in parts (a)–(c).

 Writing Describe how to translate the graph of $y = \sqrt{x}$ to obtain the graph of each function.

34. $y = \sqrt{x + 8}$ **35.** $f(x) = \sqrt{x} - 10$

36. $f(x) = \sqrt{x} + 12$ **37.** $y = \sqrt{x - 9}$

Make a table of values and graph each function.

38. $y = \sqrt{x - 2.5}$ **39.** $f(x) = 4\sqrt{x}$ **40.** $y = \sqrt{x + 6}$

41. $y = \sqrt{0.5x}$ **42.** $y = \sqrt{x - 2} + 3$ **43.** $f(x) = \sqrt{x + 2} - 4$

44. $y = \sqrt{2x} + 3$ **45.** $y = \sqrt{2x + 6} + 1$ **46.** $y = \sqrt{3x - 3} - 2$

47. Multiple Choice Which graph best models the function $y = \sqrt{x - 1} + 3$?

 A
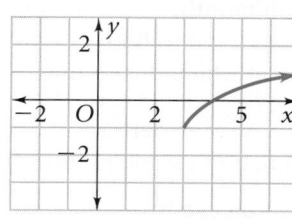

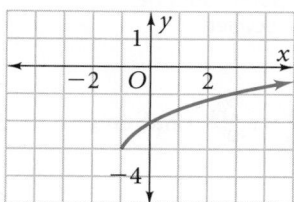 B

 C
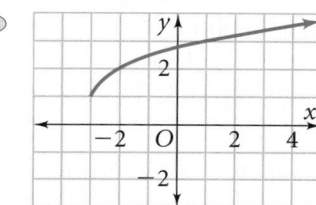

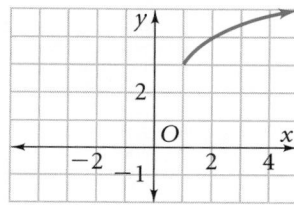 D

Real-World 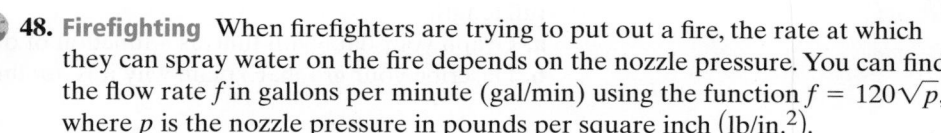 **Connection**

Careers There are more than 300,000 paid firefighters in the United States, with thousands more volunteer firefighters.

48. Firefighting When firefighters are trying to put out a fire, the rate at which they can spray water on the fire depends on the nozzle pressure. You can find the flow rate f in gallons per minute (gal/min) using the function $f = 120\sqrt{p}$, where p is the nozzle pressure in pounds per square inch (lb/in.²).
 a. What is the domain of the function?
 b. Graph the function.
 c. Use the graph to estimate the pressure when the flow rate is 800 gal/min.

49. The graph of $x = y^2$ is shown at the right.
 a. Is this the graph of a function?
 b. How does $x = y^2$ relate to the square root function $y = \sqrt{x}$?
 c. Critical Thinking What is a function for the part of the graph that is shown in Quadrant IV?

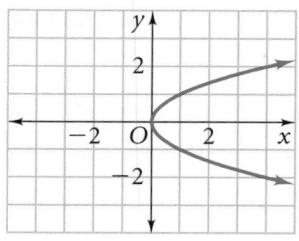

50. Without graphing, determine which graph rises more steeply, $y = \sqrt{3x}$ or $y = 3\sqrt{x}$. Explain your answer.

Determine whether each statement is true or false. If it is false, explain why.

51. If $\sqrt{x} = 9$, then $x = 3$.

52. The expression $3 + 4\sqrt{x}$ is equivalent to $7\sqrt{x}$.

53. For $\sqrt{x + 5} = 12$, $x = 139$.

54. $\sqrt{x + 5} = 2$ has no solution.

Real-World  **Connection**

Single-use cameras were introduced in 1986. By 1992, manufacturers had redesigned the cameras so that their parts could be reused or recycled.

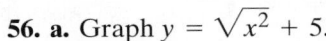 **55. Business** Last year a store had an advertising campaign. The graph shows the sales for single-use cameras. The function $n = 27\sqrt{5t} + 53$ models the sales volume n for the cameras as a function of time t, the number of months after the start of the advertising campaign.

a. Evaluate the function to find how many disposable cameras the store sold in the seventh month.

b. Solve an equation to find the month in which the number of single-use cameras sold was about 175.

Single-Use Camera Sales

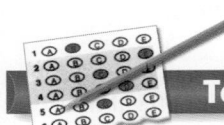 **Challenge**

56. a. Graph $y = \sqrt{x^2} + 5$.

b. Write a function for the graph you drew that does not require a radical.

57. Explain how you can translate $y = \sqrt{x}$ to graph $y = \sqrt{x - 2} + 3$.

58. a. Graph each function.
    i. $y = \sqrt{4x}$ **ii.** $y = \sqrt{5x}$ **iii.** $y = \sqrt{6x}$ **iv.** $y = \sqrt{-6x}$

b. Critical Thinking Describe how the graph of $y = \sqrt{nx}$ changes as the value of n varies.

59. Data Collection Roll a ball down a ramp which is at least 6 ft long. Record the time the ball takes to roll several different distances down the ramp, up to its full length.

a. Graph your data with time as a function of distance (d, t).

b. Describe your graph. Explain why it is *not* linear.

NY REGENTS

Test Prep

Multiple Choice

60. Which of the following is the graph of $y = \sqrt{1 - x}$?

A.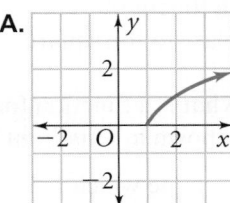

B.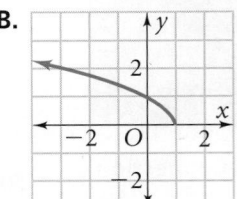

C.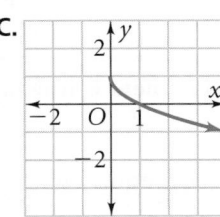

D.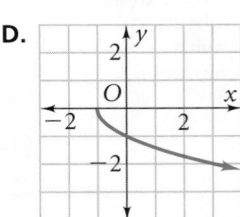

61. What is the least possible value for x for the graph of $y = \sqrt{2x - 44} + 3$?
    F. 3   **G.** 19   **H.** 22   **J.** 25

62. What is the greatest possible value of y for the graph of $y = -\sqrt{5x - 10}$?

    A. -10 **B.** -2 **C.** 0 **D.** 2

63. If $x = -2$, which function has the least value?

    F. $y = \sqrt{7 - x}$ **G.** $y = \sqrt{7 + x}$

    H. $y = \sqrt{-x} - 7$ **J.** $y = \sqrt{-x} + 7$

64. If $a = \sqrt{24}$, and $\frac{a}{b} = \frac{\sqrt{6}}{c}$, then what does $\frac{c}{b}$ equal?

    A. $\frac{1}{4}$ **B.** $\frac{1}{2}$ **C.** 2 **D.** 4

65. In which quadrant(s) is the graph of $y = \sqrt{x} + 7$?

    F. I, II, and III   **G.** I and IV   **H.** I   **J.** I and II

Short Response   **66.** Graph $y = \sqrt{x - 6}$.

Mixed Review

Lesson 11-3   **Solve each equation. Check your solutions.**

    67. $\sqrt{x} + 7 = 11$ **68.** $\sqrt{c + 1} = \sqrt{2c - 6}$

    69. $\sqrt{x} - 4 = 9$ **70.** $13 = 5\sqrt{m - 8}$

    71. $\sqrt{k + 3} + 12 = 6$ **72.** $\sqrt{5h - 2} = \sqrt{2h}$

Lesson 10-6   **Use the quadratic formula to solve each equation.**

    73. $2x^2 + 4x - 7 = 0$ **74.** $x^2 - 8x - 23 = 0$

    75. $5x^2 - x + 11 = 0$ **76.** $9x^2 + 6x - 10 = 0$

    77. $1.2x^2 + x + 6 = 0$ **78.** $9x^2 + 13x - 7 = 0$

Lesson 9-6   **Factor completely.**

    79. $2x^2 - 7x - 4$ **80.** $3x^2 + x - 10$

    81. $4x^2 + 20x + 9$ **82.** $2x^2 - 10x - 48$

    83. $4x^2 - 4x - 60$ **84.** $x^3 - 12x^2 - 13x$

✓ Checkpoint Quiz 1   Lessons 11-1 through 11-4

Simplify each expression.

    1. $-10\sqrt{7} + 2\sqrt{7}$ **2.** $\sqrt{16} - 5\sqrt{2}$

    3. $2\sqrt{5} + 3\sqrt{25}$ **4.** $\sqrt{6}\left(\sqrt{12} - \sqrt{3}\right)$

    5. $\left(\sqrt{3} + \sqrt{2}\right)^2$ **6.** $\dfrac{\sqrt{8} - \sqrt{27}}{\sqrt{6} - \sqrt{5}}$

Solve each equation.

    7. $4 - \sqrt{m} = -12$ **8.** $\sqrt{t + 5} = \sqrt{2t - 3}$ **9.** $r = \sqrt{4r + 5}$

10. Writing Describe how to translate the graph of $y = \sqrt{x}$ in order to graph $y = \sqrt{x} + 2$. Then graph the function.

Throughout history, surveyors, navigators and astronomers have used right triangles to find distances they could not measure directly. Early mathematicians measured and recorded ratios in right triangles to help solve these measurement problems.

In this activity you will explore the ratios of sides of similar right triangles.

1 ACTIVITY

1. On graph paper, draw a right triangle ABC with the right angle at C. Use a protractor to make $m\angle A = 55°$. Draw a second right triangle, GHK, with the right angle at K. Make $m\angle G = 55°$ and $GK \neq AC$.

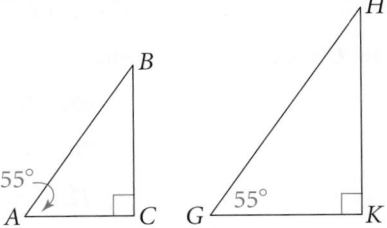

2. **a.** Copy the table at the right. Measure and record the lengths of the hypotenuses, $\overline{AB}$ and $\overline{GH}$, to the nearest tenth of a centimeter.
   b. Find the lengths of legs $\overline{AC}$ and $\overline{GK}$.
   c. Calculate and record the ratios $\dfrac{\text{leg}}{\text{hypotenuse}}$ for each triangle. Round to the nearest hundredth.

Right Triangle	Leg	Hypotenuse	Ratio
$\triangle ABC$	$\overline{AC}$	$\overline{AB}$	$\dfrac{AC}{AB}$
	■	■	■
$\triangle GHK$	$\overline{GK}$	$\overline{GH}$	$\dfrac{GK}{GH}$
	■	■	■

3. How do corresponding ratios in the two triangles compare?

4. Sketch $\triangle PQR$ with a right angle at R and a 55° angle at P. Suppose the length of the hypotenuse PQ is 20 cm. Estimate the length of $\overline{PR}$.

In this activity, you will explore how angles in right triangles relate to ratios of side lengths.

2 ACTIVITY

At the bottom of a piece of graph paper, draw right triangle ABC with the right angle at C. Make $AC = 12$ cm and $m\angle A = 20°$.

Extend $\overline{CB}$ to the top of the paper.

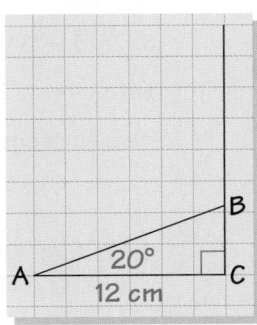

5. Copy the table at the right. Measure the hypotenuse and record its length to the nearest tenth of a centimeter.

6. On the same piece of paper, draw right triangles with $m\angle A = 30°, 40°, 50°$, and $60°$. Use the same points A and C for all of the triangles. Record the length of each hypotenuse in your table.

7. Calculate $\dfrac{AC}{\text{hypotenuse}}$ for each triangle. Round to the nearest hundredth.

Angle	Hypotenuse	$\dfrac{AC}{\text{Hypotenuse}}$
20°	▪	▪
30°	▪	▪
40°	▪	▪
50°	▪	▪
60°	▪	▪

8. What happens to the ratio as $m\angle A$ increases? Why?

9. What happens to the ratio $\dfrac{AC}{\text{hypotenuse}}$ as $m\angle A$ gets very close to 0?

From Activity 1, you know that for a given angle, corresponding ratios of sides in similar triangles are the same. From Activity 2 you found leg-hypotenuse ratios for certain angles. In this activity, you will apply those ratios to estimate measures in right triangles.

3 ACTIVITY

10. a. Estimation Use the ratio you found in Activity 2 to estimate the length of hypotenuse $\overline{ST}$.
b. Measure $\overline{ST}$ and compare it with your estimate.

11. Sketch a right triangle XYZ with a right angle at Z and a $40°$ angle at X. Suppose the hypotenuse $\overline{XY}$ is 19 cm. Estimate the length of leg $\overline{XZ}$.

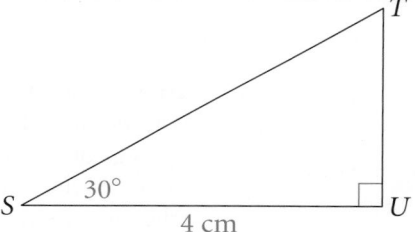

EXERCISES

12. Repeat Activity 1 using a pair of larger right triangles.
a. How do your results compare?
b. Critical Thinking Explain why using a larger right triangle might lead to more accurate results.

Use your data from Activity 2 to estimate the value of x in each triangle.

13.

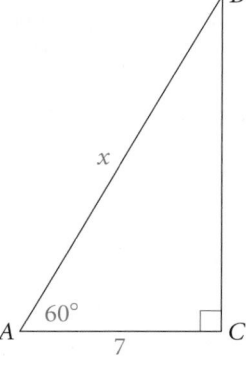

14.

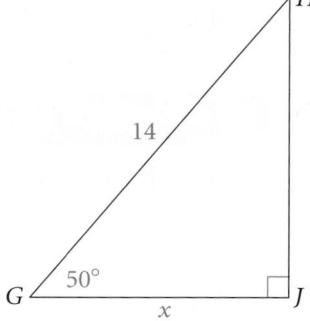

15. Critical Thinking Explain why $\dfrac{\text{leg}}{\text{hypotenuse}}$ cannot be greater than 1.

Trigonometric Ratios

 **Learning Standards for Mathematics**

A.A.42 Find the sine, cosine, and tangent ratios of an angle of a right triangle, given the lengths of the sides.

A.A.44 Find the measure of a side of a right triangle, given an acute angle and the length of another side.

✓ **Check Skills You'll Need**

GO for Help   Lessons 1-6 and 3-4

Let $c = \frac{A}{H}, s = \frac{O}{H}, t = \frac{O}{A}$. Calculate c, s, and t for the given values.

1. $A = 3, O = 4, H = 5$

2. $A = 5, O = 12, H = 13$

Solve each equation.

3. $\frac{15}{x} = \frac{0.75}{1}$ **4.** $\frac{x}{20} = \frac{0.34}{1}$ **5.** $\frac{0.84}{1} = \frac{21}{x}$ **6.** $\frac{x}{0.52} = \frac{14}{1}$

◀)) **New Vocabulary** • trigonometric ratios • sine • cosine • tangent

---

1 Finding Trigonometric Ratios

Ratios of the sides of a right triangle are called **trigonometric ratios.** In $\triangle ABC$, at the right, you see the relationships between an angle and the legs of a triangle. The same letter indicates the length of a side of the triangle and the angle opposite that side.

You can use these relationships to express trigonometric ratios.

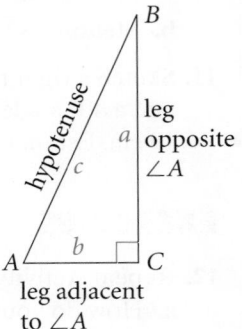

$$\textbf{sine of } \angle A = \frac{\text{length of leg opposite } \angle A}{\text{length of hypotenuse}} \quad \text{or} \quad \sin A = \frac{a}{c} = \frac{\text{opposite leg}}{\text{hypotenuse}}$$

$$\textbf{cosine of } \angle A = \frac{\text{length of leg adjacent to } \angle A}{\text{length of hypotenuse}} \quad \text{or} \quad \cos A = \frac{b}{c} = \frac{\text{adjacent leg}}{\text{hypotenuse}}$$

$$\textbf{tangent of } \angle A = \frac{\text{length of leg opposite } \angle A}{\text{length of leg adjacent to } \angle A} \quad \text{or} \quad \tan A = \frac{a}{b} = \frac{\text{opposite leg}}{\text{adjacent leg}}$$

1 EXAMPLE Finding Trigonometric Ratios

Use the triangle below. Find $\sin A$, $\cos A$, and $\tan A$.

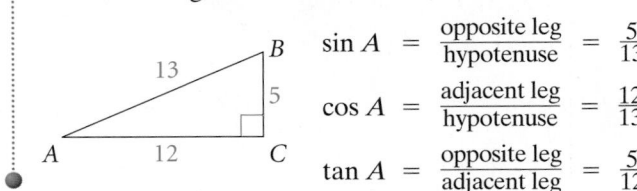

$\sin A = \dfrac{\text{opposite leg}}{\text{hypotenuse}} = \dfrac{5}{13}$

$\cos A = \dfrac{\text{adjacent leg}}{\text{hypotenuse}} = \dfrac{12}{13}$

$\tan A = \dfrac{\text{opposite leg}}{\text{adjacent leg}} = \dfrac{5}{12}$

✓ **Quick Check**  **a.** Use the triangle in Example 1. Find $\sin B$, $\cos B$, and $\tan B$.

b. Critical Thinking What is the relationship between $\sin A$ and $\cos B$?

You can use a calculator to find the values of the trigonometric ratios when you know the measure of an angle.

2 EXAMPLE Finding a Trigonometric Ratio With a Calculator

Find sin 50°.

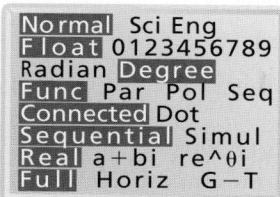

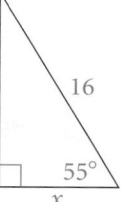

Use degree mode when finding trigonometric ratios.

To find sin 50°, press SIN **50** ENTER.

• Rounded to the nearest ten-thousandth, sin 50° is 0.7660.

✓ Quick Check ❷ Find the value of each expression. Round to the nearest ten-thousandth.
  a. sin 70°   **b.** cos 70°
  c. tan 70°   **d.** sin 5°

You can use trigonometry to find missing lengths in a right triangle when you know the length of one side and the measure of one of the angles.

3 EXAMPLE Finding Missing Side Lengths

Find the value of x in the triangle at the right.

Step 1 Decide which trigonometric ratio to use.

  You know the angle and the length of the hypotenuse.
  You are trying to find the adjacent side. Use the cosine.

Step 2 Write an equation and solve.

$$\cos 55° = \frac{\text{adjacent leg}}{\text{hypotenuse}}$$

$$\cos 55° = \frac{x}{16}$$ **Substitute x for adjacent leg and 16 for hypotenuse.**

$$x = 16(\cos 55°)$$ **Solve for x.**

16 × COS 55 ENTER = *9.177222982*   **Use a calculator.**

$$x \approx 9.2$$ **Round to the nearest tenth.**

• The value of x is about 9.2.

✓ Quick Check ❸ Find the value of x in each triangle. Round to the nearest tenth.
  a. 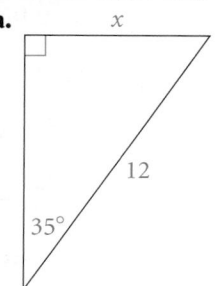   **b.**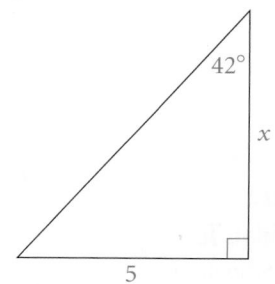

EXERCISES

For more exercises, see *Extra Skill and Word Problem Practice.*

Practice and Problem Solving

 A **Practice by Example**

 for Help

Example 1
(page 646)

Use △ *RST* at the right. Find the value of each expression.

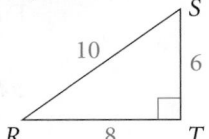

1. sin *R*　　　　**2.** cos *R*
3. tan *R*　　　　**4.** sin *S*
5. cos *S*　　　　**6.** tan *S*

Example 2
(page 647)

Find the value of each expression.
Round to the nearest ten-thousandth.

7. sin 32°　　　**8.** cos 55°　　　**9.** tan 52°　　　**10.** sin 85°

11. cos 15°　　**12.** tan 87°　　**13.** cos 30°　　**14.** sin 5°

Example 3
(page 647)

Find the value of *x* to the nearest tenth.

15. 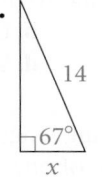　　**16.** 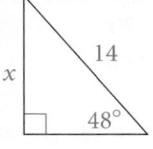　　**17.**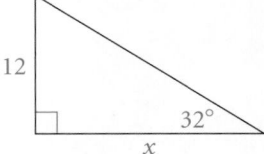

18. 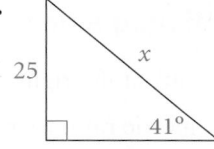　　**19.** 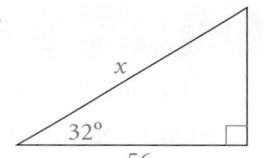　　**20.**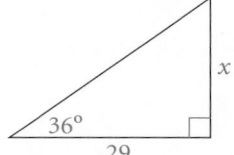

B **Apply Your Skills**

Suppose △ *ABC* has right angle *C*. Find the measures of the other sides to the nearest whole number.

21. *m∠A* = 40°, *BC* = 5　　　　　**22.** *m∠A* = 32°, *AB* = 42

23. *m∠B* = 71°, *AC* = 17　　　　**24.** *m∠B* = 5°, *BC* = 50

25. a. Find the values of each pair of expressions.
　　　i. sin 80°, cos 10°　　　　　　**ii.** cos 25°, sin 65°
　　b. What do you notice about your values and the angles in each pair?
　　c. Critical Thinking Explain why your results make sense.

For each triangle, find the value of the variable. Then find sin *A*, cos *A*, and tan *A*.

26. 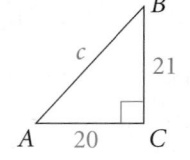　　　　　　　**27.**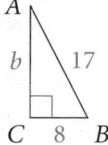

 online
Homework Video Tutor
Visit: PHSchool.com
Web Code: ate-1105

28. 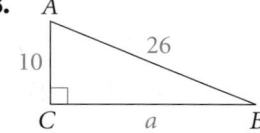　　　　　　　**29.**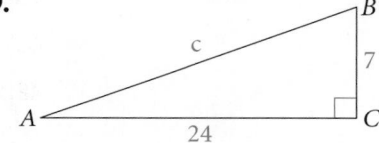

Find the value of the variable(s) in each figure to the nearest tenth.

30.
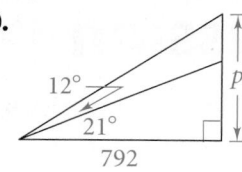

31.
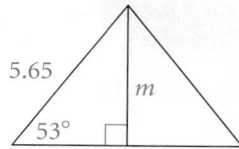

32.
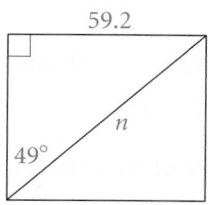

33.
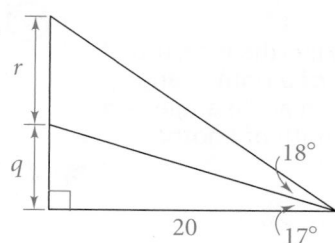

C Challenge

34. Open-Ended Draw a right triangle. Measure one acute angle and one leg. Use a trigonometric ratio to find the length of the other leg.

35. A line passes through the origin and forms an angle of 14° with the positive x-axis. Find the slope of the line. Round to the nearest hundreth.

Multiple Choice

36. Find the value of x to the nearest tenth.
  A. 7.2 **B.** 9.1
  C. 10.9 **D.** 11.9

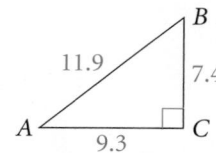

37. $\triangle KLM$ is a right triangle with a right angle at M. Which of the given statements is false?

  F. $\sin K = \dfrac{LM}{KL}$ **G.** $\cos K = \dfrac{KL}{KM}$ **H.** $\tan K = \dfrac{LM}{KM}$ **J.** $\cos L = \dfrac{LM}{KL}$

Short Response

38. Using the triangle at the right, find $\sin A$, $\cos A$, and $\tan A$.

B
11.9 7.4
A
9.3 C

Mixed Review

Lesson 11-4

Graph each function.

39. $y = \sqrt{5x}$ **40.** $y = \sqrt{x + 3}$ **41.** $y = \sqrt{x - 7}$

Lesson 10-7

Find the number of solutions of each equation.

42. $2x^2 - x - 4 = 0$ **43.** $7x^2 + x + 20 = 0$ **44.** $9x^2 + 6x + 1 = 0$

Lesson 9-7

Factor each expression.

45. $n^2 - 400$ **46.** $x^2 - 30x + 225$ **47.** $100p^2 - 49$
48. $\frac{1}{16}d^2 - \frac{9}{4}$ **49.** $98w^2 - 128$ **50.** $x^2 + 26x + 169$

Angles of Elevation and Depression

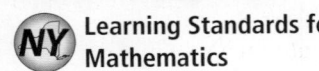

Learning Standards for Mathematics

A.A.44 Find the measure of a side of a right triangle, given an acute angle and the length of another side.

✓ **Check Skills You'll Need**

GO ▶ **for Help** Lesson 11-5

Find the value of each expression. Round to the nearest ten-thousandth.

1. $\sin 36°$ **2.** $\tan 56°$ **3.** $\cos 78°$ **4.** $\sin 73°$

 **New Vocabulary** • angle of elevation • angle of depression

---

1 **Solving Problems Using Trigonometric Ratios**

You can use trigonometric ratios to measure distances indirectly when you know an angle of elevation. An **angle of elevation** is an angle from the horizontal up to a line of sight.

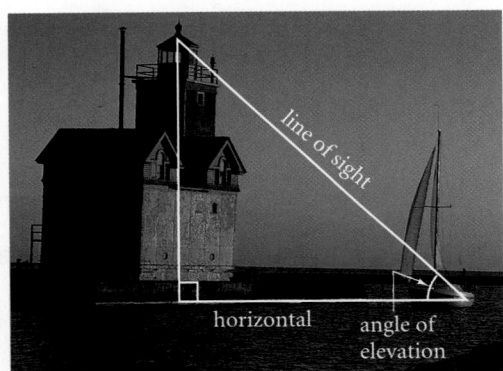

horizontal   angle of elevation

1 **EXAMPLE** **Using Angle of Elevation**

Navigation Suppose the angle of elevation from a rowboat to the light of a lighthouse is 35°. You know that the lighthouse is 96 ft tall. How far from the lighthouse is the rowboat? Round your answer to the nearest foot.

Draw a diagram.

96 ft

35°

x

Define Let x = the distance from the boat to the lighthouse.

Relate You know the angle of elevation and the opposite leg. You are trying to find the adjacent leg. Use the tangent.

Write $\tan A = \dfrac{\text{opposite leg}}{\text{adjacent leg}}$

$\tan 35° = \dfrac{96}{x}$ **Substitute for the angle and the legs.**

$x(\tan 35°) = 96$ **Multiply each side by x.**

$x = \dfrac{96}{\tan 35°}$ **Divide each side by tan 35°.**

$x \approx 137.1022086$ **Use a calculator.**

$x \approx 137$ **Round to the nearest unit.**

● The rowboat is about 137 ft from the lighthouse.

① The angle of elevation from a point on the ground 300 ft from a tower is 42°. How tall is the tower?

An **angle of depression** is measured from the horizontal down to a line of sight. In the picture below, a ranger is looking down from the tower to the distant fire.

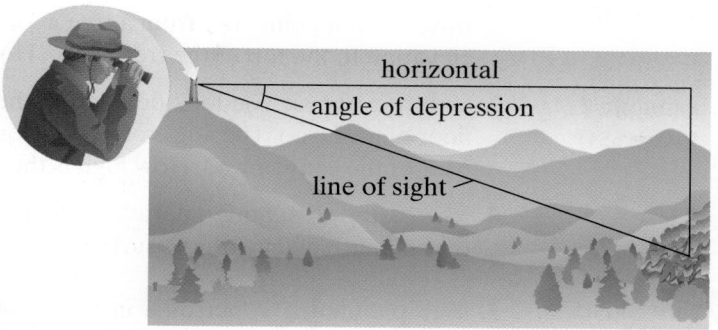

horizontal

angle of depression

line of sight

When you solve real-world problems involving trigonometry, you frequently have to round your answers. Round your answers to the measurements used in the problem. If the problem has measurements to the nearest foot, round your answer to the nearest foot. If the problem has measurements to the nearest 10,000 feet, round your answer to the nearest 10,000 feet.

2 EXAMPLE **Using Angle of Depression**

Aviation A pilot is flying a plane 20,000 ft above the ground. The pilot begins a 2° descent to an airport runway. How far is the airplane from the start of the runway (in ground distance)?

Draw a diagram.

Define Let x = the ground distance from the start of the runway.

Relate You know the angle of depression and the opposite leg. You are trying to find the adjacent leg. Use the tangent.

Write $\tan B = \dfrac{\text{opposite leg}}{\text{adjacent leg}}$

$\tan 2° = \dfrac{20{,}000}{x}$ **Substitute for the angle and the legs.**

$x(\tan 2°) = 20{,}000$ **Multiply each side by x.**

$x = \dfrac{20{,}000}{\tan 2°}$ **Divide each side by tan 2°.**

$x \approx 572{,}725.0657$ **Use a calculator.**

$x \approx 570{,}000$ **Round to the nearest 10,000 feet.**

The airplane is about 570,000 feet (or about 108 miles) from the start of the runway.

✓ **Quick Check** ② Suppose the pilot in Example 2 is flying at an altitude of 26,000 ft when the airplane begins a 2° descent. How far is the airplane from the start of the runway?

EXERCISES

For more exercises, see *Extra Skill and Word Problem Practice*.

Practice and Problem Solving

 Practice by Example

Example 1
(page 650)

 **for Help**

Example 2
(page 651)

1. **Aviation** Suppose you live about 5.1 miles from a tower. From your home, you see a plane directly above the tower. Your angle of elevation to the plane is 21°. What is the plane's altitude?

2. **Nature** At a point 80 ft from a tree a forest ranger measures the angle of elevation to the top of the tree as 65°. How tall is the tree?

3. **Nature** Suppose you look down from the top of a vertical cliff to a cabin on the floor of a canyon. The angle of depression is 58°. The cabin is 510 ft from the base of the canyon wall. How high is the canyon wall?

4. **Navigation** A submarine travels 2.6 miles diving at an angle of 8°. How deep is the submarine beneath the surface of the water?

 **Apply Your Skills**

5. **Multiple Choice** A person is in a lighthouse 225 ft above sea level. She sees a ship in the harbor. The angle of depression from her position to the ship is 48°. Which equation can she use to find the distance from the ship to the shore?

   A. $\tan 48° = \frac{225}{x}$

   B. $\tan 48° = \frac{x}{225}$

   C. $\tan 225° = \frac{48}{x}$

   D. $\tan 225° = \frac{x}{48}$

6. **Engineering** To support a pole, a cable is drawn tight between the pole and the ground. The cable is 4.1 m from the base of the pole, and the angle of elevation from the bottom of the cable to the top of the pole is 47°. How tall is the pole?

7. **Recreation** Use the Ferris wheel photo at the right.
   a. Find the height at the top of the Ferris wheel.
   b. If the hub or center of the wheel is about 91 feet off the ground, what is the radius of the wheel?

8. **Hobbies** Suppose you are flying a kite. The kite string is 60 m long, and the angle of elevation of the string is 65° from your hand. Your hand is 1 m above the ground. How high above the ground is the kite?

9. An architect is designing an access ramp. The angle of the ramp with the ground will be 5°. The top of the ramp will be 5 ft above the ground. How long will the ramp be?

 **Online**
Homework Video Tutor
Visit: PHSchool.com
Web Code: ate-1106

10. a. **Aviation** A pilot is flying a plane at an altitude of 30,000 ft. The pilot begins a 1° descent to an airport runway. How far is the airplane from the start of the runway (in ground distance)?
    b. What is your answer to part (a) in miles?

 **Challenge**

11. At a certain point in a large, level park, the angle of elevation to the top of an office building is 30°. If you move 400 ft closer to the building, the angle of elevation is 45°. To the nearest 10 feet, how tall is the building?

12. A truck travels out of a valley up a long hill which climbs steadily at a 5° angle. The top of the hill is 800 vertical feet above the valley floor. How many miles long is the road?

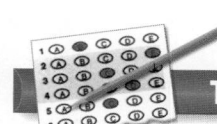
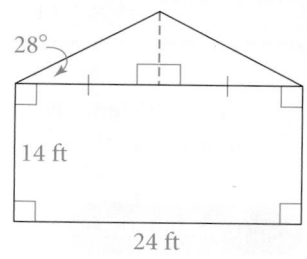
Multiple Choice

13. A carpenter makes a rooftop like the one at the right. How high above the ground is the peak of the roof?
  A. about 19 ft   **B.** about 19.2 ft
  C. about 20.4 ft   **D.** about 20.7 ft

28°

14 ft

24 ft

14. A right triangle *GHK* has a hypotenuse of 17 ft. ∠*G* is 78°. What is the perimeter to the nearest foot?

  F. 37 ft   **G.** 38 ft   **H.** 40 ft   **J.** 42 ft

Extended Response

15. A mouse is at the edge of a cliff 850 ft above the base of the cliff. He sees a fox in a canyon. The angle of depression from his position to the fox is 56°. Draw and label a diagram for the situation. Then write and solve a trigonometric equation that will determine how far the fox is from the base of the cliff. Round to the nearest foot.

Mixed Review

GO **for Help**

Lesson 11-5

Use △*ABC* at the right. **Find the value of each expression.**

16. sin *A*   **17.** cos *B*

18. tan *A*   **19.** tan *B*

20. sin *B*   **21.** cos *A*

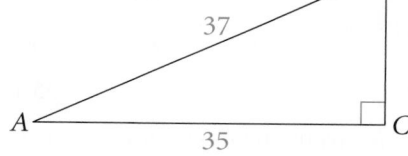

B

37

A

35

C

Lesson 6-6

Determine whether each pair of lines is *parallel*, *perpendicular* or *neither*.

22. $y = 4x + 1$
    $-8x + 2y = 12$

23. $y + x = 5$
    $y - x = 9$

24. $-0.5x - 6 = y$
    $y + 7 = -2x$

Checkpoint Quiz 2   **Lessons 11-5 through 11-6**

Find the value of *x* to the nearest tenth.

1.

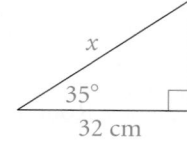

x

35°

32 cm

2.

60 in.

x

33°

3.

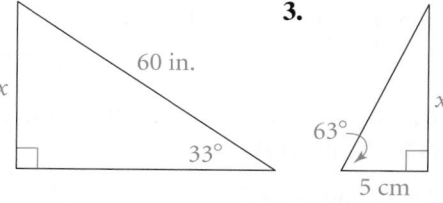

x

63°

5 cm

4. The length of a tree's shadow is 15 m. The angle of elevation from the top of the tree to the sun is 38°. What is the height of the tree?

5. A 7-meter ladder rests against the side of a house. The angle of elevation of the ladder is 75°. How high is the top of the ladder?

Finding Angles in Right Triangles

Given the lengths of two sides in a right triangle, you can find the measure of an acute angle by using the *inverse of a trigonometric function*.

> **NY** **A.A.43:** Determine the measure of an angle of a right triangle, given the length of any two sides of the triangle.

EXAMPLE   **Finding Angle Measures**

In ABC, $\angle C$ is a right angle, $c = 11$, and $b = 9$. Find $m\angle A$ to the nearest tenth of a degree.

Step 1 Draw a diagram.

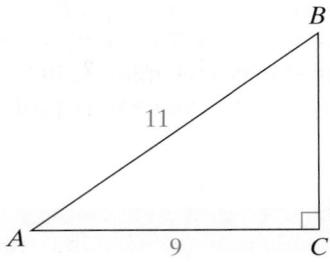

Step 2 Use a cosine ratio.

$$\cos A = \frac{9}{11}$$

$$m\angle A = \cos^{-1}\frac{9}{11} \quad \text{The inverse of cosine is written } \cos^{-1}.$$

$$= 35.108 \quad \text{Use a calculator.}$$

To the nearest tenth of a degree, $m\angle A$ is $35.1°$.

EXERCISES

In ABC, $\angle C$ is a right angle. Find $m\angle A$ to the nearest tenth of a degree.

1.

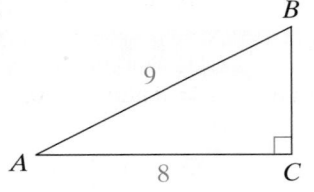

2.

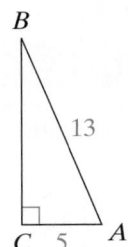

3.

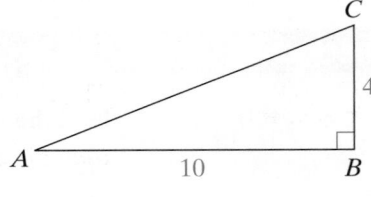

In ABC, $\angle C$ is a right angle. Find the remaining sides and angles. Round your answers to the nearest tenth.

4. $AB = 12$ and $AC = 7$ **5.** $AB = 20$ and $BC = 13$ **6.** $AC = 15$ and $BC = 13$

Chapter Review

Vocabulary Review

📢 angle of depression (p. 651)
angle of elevation (p. 650)
conjugates (p. 623)
cosine (p. 646)
extraneous solution (p. 631)

like radicals (p. 622)
radical equation (p. 629)
radical expression (p. 616)
rationalize (p. 619)
sine (p. 646)

square root function (p. 638)
standard deviation (p. 636)
tangent (p. 646)
trigonometric ratios (p. 646)
unlike radicals (p. 622)

Choose the vocabulary term that correctly completes each sentence.

1. Two radical expressions that are the sum and the difference of the same two terms are __?__.

2. One method of simplifying a radical expression is to __?__ the denominator.

3. A(n) __?__ is a value that satisfies the new equation but not the original equation.

4. Radicals with the same radicand are __?__.

5. In a right triangle, the __?__ is the trigonometric ratio of the length of the leg opposite an angle to the length of the hypotenuse of the triangle.

6. A horizontal and the line of sight to an object above the horizontal form a(n) __?__.

7. A(n) __?__ is a function that contains the independent variable in the radicand.

8. Ratios of two sides of a right triangle are called __?__.

Go Online
PHSchool.com
For: Vocabulary quiz
Web Code: atj-1151

Skills and Concepts

11-1 Objectives

▼ To simplify radicals involving products (p. 616)

▼ To simplify radicals involving quotients (p. 618)

You can simplify some radical expressions by using products or quotients.

The Multiplication Property of Square Roots states that for $a \geq 0$ and $b \geq 0$, $\sqrt{ab} = \sqrt{a} \cdot \sqrt{b}$.

The Division Property of Square Roots states that for $a \geq 0$ and $b > 0$, $\sqrt{\frac{a}{b}} = \frac{\sqrt{a}}{\sqrt{b}}$.

Simplify each radical expression.

9. $\sqrt{32} \cdot \sqrt{144}$

10. $\sqrt{\frac{84}{121}}$

11. $\sqrt{96c^3} \cdot \sqrt{25c}$

12. $\frac{10}{\sqrt{13}}$

13. $\sqrt{5h^3} \cdot \sqrt{50h}$

14. $3\sqrt{20t^6}$

15. $\frac{\sqrt{8}}{\sqrt{32}}$

16. $\frac{2\sqrt{6k^2}}{\sqrt{9k^4}}$

17. $\sqrt{\frac{225}{100}}$

18. A rectangle is 7 times as long as it is wide. Its area is 1400 cm². Find the dimensions of the rectangle in simplest radical form.

11-2 Objectives

▼ To simplify sums and differences (p. 622)

▼ To simplify products and quotients (p. 623)

You can use the Distributive Property to simplify expressions with sums and differences of radicals. First, simplify the radicals and check for **like radicals**.

When a denominator contains a sum or a difference including radical expressions, you can **rationalize** the denominator by multiplying the numerator and the denominator by the **conjugate** of the denominator.

Simplify each radical expression.

19. $6\sqrt{7} - 2\sqrt{28}$

20. $5(\sqrt{20} + \sqrt{80})$

21. $\sqrt{54} - 2\sqrt{6}$

22. $\sqrt{125} - 3\sqrt{5}$

23. $\sqrt{10}(\sqrt{10} - \sqrt{20})$

24. $(\sqrt{2} + \sqrt{7})(3\sqrt{2} - \sqrt{7})$

25. $(\sqrt{5} + 4\sqrt{3})^2$

26. $\dfrac{3}{\sqrt{6} - \sqrt{3}}$

27. $\dfrac{\sqrt{7} + \sqrt{28}}{\sqrt{7} - \sqrt{14}}$

28. $\sqrt{28} + 5\sqrt{63}$

29. $\dfrac{2}{\sqrt{8} + \sqrt{7}}$

30. $3\sqrt{5}(\sqrt{10} - 2\sqrt{15})$

11-3 Objectives

▼ To solve equations containing radicals (p. 629)

▼ To identify extraneous solutions (p. 631)

A **radical equation** has a variable in the radicand. Sometimes you can solve such an equation by squaring both sides. You may also square both sides of the equation when each side is a square root.

Squaring both sides of a radical equation may produce an **extraneous solution.** It is not a solution of the original equation.

Solve each radical equation.

31. $\sqrt{x + 7} = 3$

32. $\sqrt{x} + 3\sqrt{x} = 16$

33. $4 = \sqrt{n - 1}$

34. $\sqrt{p} - 5 = 8$

35. $\sqrt{x + 7} = \sqrt{2x - 1}$

36. $\sqrt{x} - 5 = 4$

37. $\sqrt{n} + 6 = 16$

38. $8\sqrt{b} - \sqrt{b} = 14$

Tell which of the given solutions is extraneous for each equation.

39. $\sqrt{4x} = x - 3$
    $x = 1, x = 9$

40. $\sqrt{d - 3} = 5 - d$
    $d = 4, d = 7$

Solve each radical equation. Check your solution. If there is no solution, write *no solution*.

41. $\sqrt{2x} = \sqrt{5x + 3}$

42. $\sqrt{2d + 1} = \sqrt{3d - 2}$

43. $5\sqrt{x} = \sqrt{10x + 60}$

44. $\sqrt{-5b + 2} = \sqrt{-4b - 4}$

45. $3\sqrt{x} + 4\sqrt{x} = 0$

46. $\sqrt{k + 1} = \sqrt{7 + 2k}$

47. A rectangle has a width of $2\sqrt{5}$ cm and an area of 50 cm². Find the length of the rectangle.

48. The volume V of a cylinder is given by $V = \pi r^2 h$, where r is the radius of a cylinder and h is its height. If the volume of the cylinder is 54 in.³, and its height is 2 in., what is its radius to the nearest 0.01 in.?

11-4 Objectives

▼ To graph square root functions (p. 638)

▼ To translate graphs of square root functions (p. 639)

The simplest square root function is $y = \sqrt{x}$. The graphs of $y = \sqrt{x} + k$ and $y = \sqrt{x} - k$ are vertical translations of $y = \sqrt{x}$. The graphs of $y = \sqrt{x - h}$ and $y = \sqrt{x + h}$ are horizontal translations of $y = \sqrt{x}$. To find the domain of a square root function, solve the inequality where the radicand is greater than or equal to zero.

Make a table of values and graph each function.

49. $y = \sqrt{\dfrac{x}{2}}$ **50.** $y = \dfrac{\sqrt{x}}{2}$ **51.** $y = \sqrt{2x}$ **52.** $y = 1 + \sqrt{x}$

Find the domain of each function. Then graph each function.

53. $y = \sqrt{x} + 5$ **54.** $y = \sqrt{x - 2}$ **55.** $y = \sqrt{x + 1}$ **56.** $f(x) = 2\sqrt{x}$

11-5 Objective

▼ To find trigonometric ratios (p. 646)

Trigonometric ratios are triangle measurement ratios. For a right triangle of a given shape, the ratios do not change no matter how large or small the triangle is. Three trigonometric ratios—**sine** (sin), **cosine** (cos), and **tangent** (tan)—are shown below.

sine of $\angle A = \dfrac{\text{length of leg opposite } \angle A}{\text{length of hypotenuse}}$

cosine of $\angle A = \dfrac{\text{length of leg adjacent to } \angle A}{\text{length of hypotenuse}}$

tangent of $\angle A = \dfrac{\text{length of leg opposite } \angle A}{\text{length of leg adjacent to } \angle A}$

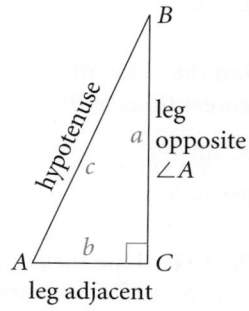

Suppose △ABC has right angle C. Find the measures of the other sides to the nearest whole number.

57. $AB = 12, m\angle A = 34°$ **58.** $BC = 9, m\angle B = 72°$

59. $AC = 8, m\angle B = 52°$ **60.** $AB = 25, m\angle A = 12°$

61. $BC = 18, m\angle A = 42°$ **62.** $AC = 20, m\angle B = 80°$

11-6 Objective

▼ To solve problems using trigonometric ratios (p. 650)

You can use trigonometric ratios to measure distances indirectly. You can use an **angle of elevation** or **angle of depression** to measure heights indirectly.

63. During a violent thunderstorm, a tree near John's house was broken by the wind. The top part of the tree was bent so that it touched the ground 21 ft from the base of the tree. If the broken top part of the tree made a 48° angle with the ground, how tall was the tree before the storm?

64. An architect is designing an access ramp. The angle of the ramp with the ground will be 5°. The top of the ramp will be 5 ft above the ground. How long will the ramp be?

65. Suppose you are lying on the ground looking up at a California redwood tree. Your angle of elevation to the top of the tree is 42°. You are 280 feet from the base of the tree. How tall is the tree?

Chapter Test

Simplify each radical expression.

1. $\sqrt{300}$

2. $-\sqrt{90}$

3. $\sqrt{12} \cdot \sqrt{18}$

4. $2\sqrt{7} \cdot 4\sqrt{7}$

5. $\sqrt{5h^3} \cdot \sqrt{2h^5}$

6. $-\sqrt{12b^2} \cdot 2\sqrt{8b}$

7. $\sqrt{\dfrac{20t^5}{5t}}$

8. $\sqrt{\dfrac{98m}{2m^9}}$

Solve each radical equation. Check your solution. If there is no solution, write *no solution*.

9. $\sqrt{7n + 3} = \sqrt{n + 15}$

10. $2\sqrt{k - 11} = \sqrt{1 - 5k}$

11. $-8 = \sqrt{4x}$

12. $\sqrt{45 + 3c} = 3$

Find the value of each expression. Round to the nearest thousandth.

13. $\sin 49°$

14. $\tan 83°$

15. $\cos 12°$

16. $\sin 75°$

17. A ski slope drops steadily from 1000 feet above sea level to 200 feet above sea level. The slope forms an angle of 23°. How long is the ski slope? Round your answer to the nearest foot.

Find each side length to the nearest tenth.

18. AB

19. AC

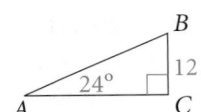

Simplify each radical expression.

20. $\sqrt{\dfrac{128}{64}}$

21. $\sqrt{\dfrac{27}{75}}$

22. $\sqrt{48}$

23. $\sqrt{12} \cdot \sqrt{8}$

24. $3\sqrt{32} + 5\sqrt{2}$

25. $2\sqrt{27} + 5\sqrt{3}$

26. $7\sqrt{125} - 3\sqrt{175}$

27. $\sqrt{128} - \sqrt{192}$

28. $\dfrac{15}{\sqrt{3}}$

29. $\dfrac{8}{\sqrt{10} + \sqrt{6}}$

30. Which expression shows $\sqrt{24x^2y^3}$ written in simplest radical form?

A. $2xy\sqrt{12xy^2}$

B. $2xy\sqrt{6y}$

C. $xy\sqrt{24y}$

D. $4xy\sqrt{3y}$

31. Open-Ended Write an expression involving addition of two like radicals. Simplify the sum.

Solve each radical equation.

32. $3\sqrt{x} + 2\sqrt{x} = 10$

33. $8 = \sqrt{5x - 1}$

34. $5\sqrt{x} = \sqrt{15x + 60}$

35. $\sqrt{x} = \sqrt{2x - 7}$

36. $3\sqrt{x + 3} = 2\sqrt{x + 9}$

37. $\sqrt{3x} = x - 6$

38. A rectangle is 5 times as long as it is wide. The area of the rectangle is 100 ft². How wide is the rectangle? Express your answer in simplest radical form.

Find the domain of each function. Graph the function.

39. $y = 3\sqrt{x}$

40. $y = \sqrt{x} + 4$

41. $y = \sqrt{x - 4}$

42. $y = \sqrt{x + 9}$

43. The hypotenuse of a right triangle is 26 cm. The length of one leg is 10 cm. Find the length of the other leg.

44. Writing Explain how to graph $y = \sqrt{x} - 3$ by translating the graph of $y = \sqrt{x}$.

45. Geometry The formula for the volume V of a cylinder with height h and radius r is $V = \pi r^2 h$. Solve for r in terms of V and h.

46. From ground level you can see a satellite dish on the roof of a building 60 ft high. The angle of elevation is 62°. How far away from you is the building?

Find the values for $\triangle RST$.

47. $\overline{RT}$

48. $\overline{ST}$

49. $\tan T$

50. $\sin T$

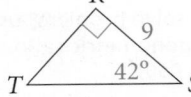

51. Geometry In $\triangle ABC$, $\angle C$ is a right angle, $AB = 7$, and $m\angle B = 28°$. What are the lengths of $\overline{BC}$ and $\overline{AC}$ to the nearest tenth?

52. Ken has a 20-ft ladder to use for washing windows. When leaned against a building, the ladder forms an angle of 75° with the ground. How far from the side of the building is the base of the ladder?

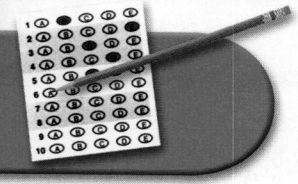

Regents Test Prep

Reading Comprehension Read the passage below and then answer the questions on the basis of what is *stated* or *implied* in the passage.

Pricing Products Carlos and Anna have created a new electronic game that they think will be a big hit. But they can't decide how much to sell it for. They have manufactured 2500 to start out, at a cost to them of $18 for each game, but they doubt they can sell all of them right away.

They have completed a market study. By interviewing potential buyers, they have learned that if they set a price of $40, they should be able to sell 1000 games during the first six months. They have also discovered that for every $5 they increase the price they will lose 50 sales.

Carlos and Anna realize that the data are important. They need help in determining the price that will maximize their sales revenue, which is the amount of total sales.

1. What is the total cost to Carlos and Anna for manufacturing 2500 electronic games?
   (1) $36,000   (2) $45,000
   (3) $50,000   (4) $60,000

2. At a selling price of $40 each, how much revenue does the market research indicate they could expect in the first six months?
   (1) $100,000   (2) $55,000
   (3) $40,000   (4) $22,000

3. Suppose Carlos and Anna raise the price to $45. From their market analysis, how many sales and how much revenue can they expect?
   (1) 1050, $57,750
   (2) 1050, $47,250
   (3) 1000, $45,000
   (4) 950, $42,750

4. Their accountant tells them that they can use the function $r = 40{,}000 + 3000n - 250n^2$ to find the price for the game that will produce the greatest revenue r. The variable n is the number of times they increase the price $5. What type of function is this?
   (1) linear
   (2) quadratic
   (3) exponential
   (4) absolute value

5. The accountant found the function in Exercise 4 by writing functions for the price p and the number of sales s, given the number of times n the price is increased. Which functions did he use?
   (1) $s = 1000 - 5n$, $p = 40 - 5n$
   (2) $s = 1000 + 50n$, $p = 40 - 5n$
   (3) $s = 1000 - 50n$, $p = 40n$
   (4) $s = 1000 - 50n$, $p = 40 + 5n$

6. Carlos and Anna want to find the maximum of the function $r = 40{,}000 + 3000n - 250n^2$, where r is the revenue and n is the number of times the price is increased.
   a. How many times can they increase the price to obtain the maximum?
   b. What is the maximum revenue possible using this model?
   c. What would be the number of sales?
   d. Based on this information, what price should Carlos and Anna set for the electronic game?

7. Suppose the market analysis also indicated that with each $5 decrease in price, Carlos and Anna could expect an increase of 50 sales. Would you suggest they decrease the price to increase the sales? Explain.

Data Analysis

Activity Lab

Conditional Probability

NY **A.S.18:** Know the definition of conditional probability and use it to solve for probabilities in finite sample spaces.

A conditional probability contains a condition that may limit the sample space for an event. You can write a conditional probability using the notation $P(B\,|\,A)$, read "the probability of event B, given event A."

1 ACTIVITY

The table shows sales at several car dealerships.

	$15,000 or less	more than $15,000
Domestic	15	8
Foreign	11	12

1. a. How many domestic cars were sold?
  b. How many foreign cars were sold?
2. a. How many cars were sold for $15,000 or less?
  b. How many cars were sold for more than $15,000?

To find $P(\text{domestic}\,|\,\$15,000\text{ or less})$ first determine the sample space. The condition "$15,000 or less" limits the sample space to 26 possible outcomes. Of those outcomes, 15 cars are domestic. So, $P(\text{domestic}\,|\,\$15,000\text{ or less})$ is $\frac{15}{26}$.

3. a. Find $P(\text{more than }\$15,000\,|\,\text{foreign})$ and $P(\text{foreign}\,|\,\text{more than }\$15,000)$.
  b. Critical Thinking Are the probabilities the same? Why or why not?

You can use tree diagrams to solve problems involving conditional probability and events which are not equally likely.

2 ACTIVITY

Consider the following situation.

- When the owner throws a flying disk 20 ft or less, her dog catches it 40% of the time.
- When the owner throws a flying disk more than 20 ft, her dog catches it 75% of the time.
- The owner throws the flying disk more than 20 ft 80% of the time.

To organize these probabilities in a tree diagram, use L (≤ 20 ft), M (> 20 ft), C (caught), and N (not caught).

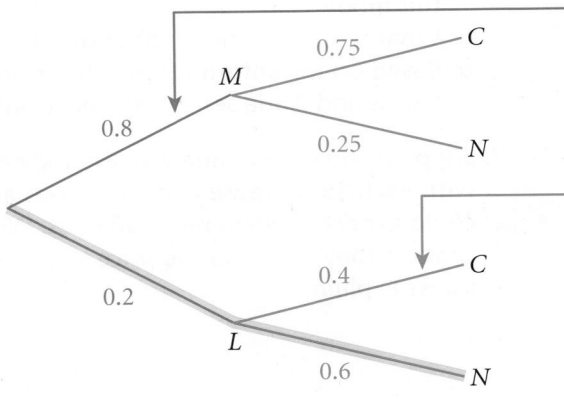

The first branch represents simple probability. $P(M) = 0.8$

The second branch represents a conditional probability. $P(C\,|\,L) = 0.4$

The highlighted path represents $P(L \text{ and } N)$. $P(L \text{ and } N) = P(L) \cdot P(N\,|\,L) = 0.12$

Use the tree diagram at the bottom of page 660 to answer the following questions.

4. Find $P(\text{caught} \mid > 20 \text{ ft})$.

5. Find $P(\text{not caught} \mid \leq 20 \text{ ft})$.

6. a. Suppose the dog returns the flying disk 90% of the time after the dog catches it, and 15% of the time when the dog doesn't catch it. Copy the tree diagram and extend it to include this information.

   b. On your tree diagram, highlight the branch that shows the probability that the flying disk was thrown more than 20 ft, caught and returned. Find this probability.

EXERCISES

7. Data Collection Conduct a survey of at least ten people. Find out if they are right- or left-handed and if they have at least one left-handed parent.
   a. Copy the table below and record your data.

	Left-handed Parents	Right-handed Parents
Left-handed	■	■
Right-handed	■	■

   b. Find $P(\text{left-handed} \mid \text{left-handed parent})$.
   c. Find $P(\text{right-handed parents} \mid \text{left-handed})$.
   d. Interpret your results.

8. Critical Thinking Consider the theoretical probabilities for a pair of number cubes. Does $P(\text{even on cube 1} \mid \text{odd on cube 2}) = P(\text{odd on cube 2} \mid \text{even on cube 1})$? Explain your answer.

For Exercises 9–12, use the table at the right. Find each probability.

9. $P(\text{female recipient} \mid \text{associate's degree})$

10. $P(\text{male recipient} \mid \text{bachelor's degree})$

11. $P(\text{advanced degree} \mid \text{female recipient})$

12. $P(\text{female recipient} \mid \text{advanced degree})$

Projected Number of Degree Recipients in 2010 (thousands)

Degree	Male	Female
Associate's	224	387
Bachelor's	547	776
Advanced	245	322

Source: *U.S. National Center for Education Statistics*

13. a. Data Collection Conduct a survey of at least 20 people. Record their gender and whether or not their first name ends with a vowel (including *y*).
   b. Make a tree diagram using gender as the first branch and last letter of first name as the second branch.
   c. Find $P(\text{female} \mid \text{vowel})$ and $P(\text{vowel} \mid \text{male})$.
   d. Interpret your results.
   e. Critical Thinking Do you think it makes more sense to have gender as the first branch or the second branch? Explain.

What You've Learned

NY Learning Standards for Mathematics

In Chapter 2, you
● **NY A.S.20:** used ratios to express probability.

In Chapter 5, you
● **NY A.S.22:** modeled some situations with a direct variation.

In Chapter 8, you
● **NY A.S.12:** used the properties of exponents to simplify expressions containing exponents that are zero or negative.

In Chapters 10 and 11, you
● Solved quadratic and radical equations and checked for extraneous solutions.

 **Check Your Readiness**

 for Help to the Lesson in green.

Adding and Subtracting Fractions (Skills Handbook page 760)

Add or subtract. Write each answer in simplest form.

1. $\frac{2}{3} + \frac{1}{2}$ **2.** $\frac{3}{13} + \frac{6}{13}$ **3.** $\frac{16}{25} + \frac{3}{10}$ **4.** $\frac{5}{9} - \frac{5}{36}$

Finding Probabilities (Lesson 2-6)

You have three $1 bills and two $5 bills in your pocket. You choose two bills without looking. Find each probability.

5. P(two $1 bills) **6.** P(two $5 bills)

7. P(two bills of the same kind) **8.** P(two different kinds of bills)

Simplifying Expressions (Lesson 8-5)

Simplify each expression.

9. $\frac{6w^3x^2}{2wx}$ **10.** $\frac{81r^{10}s^6}{(3r^2s)^4}$ **11.** $\frac{(5k^5)(2k^3)}{(2k^2)^2}$

Solving Radical Equations (Lesson 11-3)

Solve each radical equation. If there is no solution, write *no solution*.

12. $\sqrt{x} - 4 = 6$ **13.** $\sqrt{3x} + 5 = 2$ **14.** $2x = \sqrt{3x + 1}$

Finding the Domain (Lesson 11-4)

Find the domain of each function.

15. $f(x) = 5 - \sqrt{x}$ **16.** $y = -2 + \sqrt{3x}$ **17.** $y = \sqrt{10 - 3x}$

Rational Expressions and Functions

◀)) **Key Vocabulary**

- asymptote (p. 664)
- combination (p. 706)
- multiplication counting principle (p. 700)
- permutation (p. 701)
- rational equation (p. 692)
- rational expression (p. 672)
- rational function (p. 664)

What You'll Learn Next

NY **Learning Standards for Mathematics**

- **NY A.A.16:** You will combine rational expressions using addition, subtraction, multiplication, and division.

- **NY A.N.7:** You will use permutations and combinations to find the number of outcomes of real-world situations.

Activity Lab Applying what you learn, you will solve inverse variation equations for vibrating strings in order to construct a guitar "neck," on pages 722–723.

12-1

Graphing Rational Functions

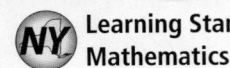
A.R.2 Recognize, compare, and use an array of representational forms.

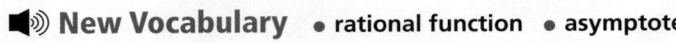

✓ **Check Skills You'll Need**

GO for Help   Lessons 5-3, 8-7, and 10-1

Evaluate each function for $x = -2, 0, 3$.

1. $f(x) = x - 8$ **2.** $g(x) = x^2 + 4$ **3.** $y = 3^x$

Graph each function.

4. $f(x) = 2x + 1$ **5.** $g(x) = -x^2$ **6.** $y = 2^x$

◀))) **New Vocabulary**   • rational function   • asymptote

---

1 Graphing Rational Functions

You can write the inverse variation $xy = 3$ as $y = \frac{3}{x}$. This is a rational function.

A **rational function** can be written in the form $f(x) = \frac{\text{polynomial}}{\text{polynomial}}$. The value of the variable cannot make the denominator equal to 0.

1 EXAMPLE   Graphing a Rational Function

Travel On any trip, the time you travel in a car varies inversely with your average speed. The function $t = \frac{60}{r}$ models the time it will take you to travel 60 miles at different rates of speed. Graph this function.

Step 1 Make a table of values.   **Step 2** Plot the points.

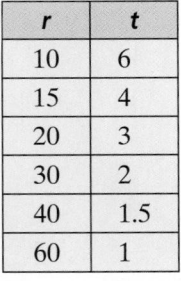

r	t
10	6
15	4
20	3
30	2
40	1.5
60	1

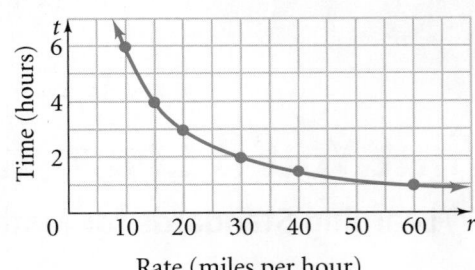

✓ **Quick Check** ❶ **a.** The function $t = \frac{40}{r}$ models the time it will take you to travel 40 miles at different rates of speed. Graph this function.

b. Critical Thinking Why is it reasonable in this situation to graph the function in Quadrant I only?

As you can see in the graph in Example 1, the graph approaches both axes but does not cross either axis. A line is an **asymptote** of a graph if the graph of the function gets closer to the line as x or y gets larger in absolute value. In the graph in Example 1, the horizontal and vertical axes are asymptotes.

The graphs of many rational functions are related to each other. Look at the graphs of the functions below.

$$y = \frac{1}{x}$$

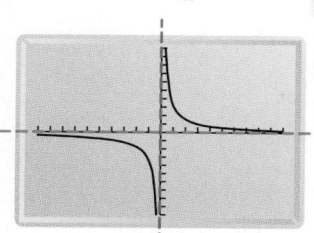

$$y = \frac{1}{x - 2}$$

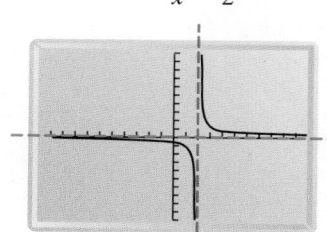

vertical asymptote at $x = 0$

horizontal asymptote at $y = 0$

vertical asymptote at $x = 2$

horizontal asymptote at $y = 0$

The graphs are identical in shape, but the second graph is translated two units to the right.

When the numerator and denominator of a rational function have no common factors other than 1, there is a vertical asymptote at the x-value that makes the denominator equal zero. This is because division by zero is undefined. The domain of a function does not include that x-value where there is a vertical asymptote.

2 EXAMPLE **Using a Vertical Asymptote**

Identify the vertical asymptote of $y = \frac{4}{x + 3}$. Then graph the function.

Step 1 Find the vertical asymptote.

$x + 3 = 0$ **The numerator and denominator have no common factors. Find any value(s) where the denominator equals zero.**

$x = -3$ **This is the equation of the vertical asymptote.**

Step 2 Make a table of values. Use values of x near -3, the asymptote.

Step 3 Graph the function.

Use a dashed line for the asymptote $x = -3$.

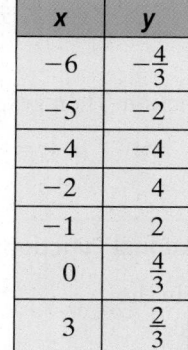

x	y
-6	$-\frac{4}{3}$
-5	-2
-4	-4
-2	4
-1	2
0	$\frac{4}{3}$
3	$\frac{2}{3}$

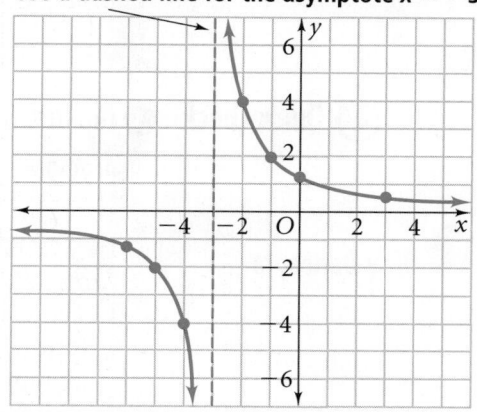

✓ Quick Check **2** Identify the vertical asymptote of each function. Then graph the function.

a. $f(x) = \frac{1}{x + 2}$

b. $h(x) = \frac{2}{x - 3}$

You can also see how to shift the graph of a rational function vertically.

$$y = \frac{1}{x}$$

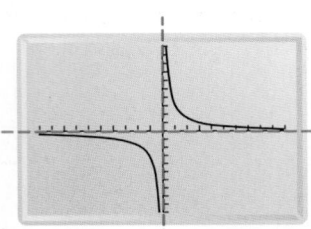

$$y = \frac{1}{x} - 2$$

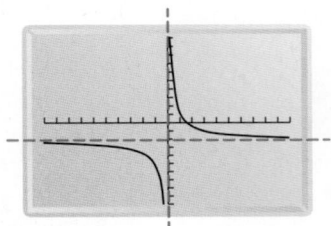

vertical asymptote at $x = 0$
horizontal asymptote at $y = 0$

vertical asymptote at $x = 0$
horizontal asymptote at $y = -2$

The graphs are identical in shape, but the second graph is translated two units down.

Online
active math

For: Rational Function Activity
Use: Interactive Textbook, 12-1

3 **EXAMPLE** **Using Vertical and Horizontal Asymptotes**

Identify the asymptotes of $y = \frac{4}{x + 3} + 2$. Then graph the function.

Step 1 From the form of the function, you can see that there is a vertical asymptote at $x = -3$ and a horizontal asymptote at $y = 2$. Sketch the asymptotes.

Step 2 Make a table of values using values of x near -3.

Step 3 Graph the function.

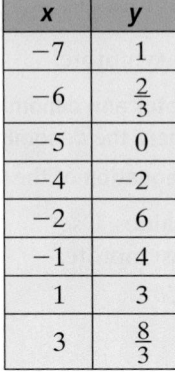

x	y
-7	1
-6	$\frac{2}{3}$
-5	0
-4	-2
-2	6
-1	4
1	3
3	$\frac{8}{3}$

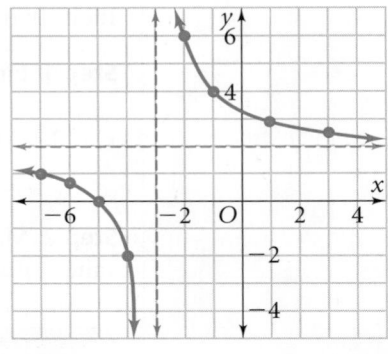

 **Quick Check** **3** Identify the asymptotes of each function. Then graph the function.

a. $f(x) = \frac{1}{x + 2} - 3$ **b.** $y = \frac{1}{x - 4} + 1$

Key Concepts

Summary	**Graphs of Rational Functions**

The graph of a rational function in the form $y = \frac{a}{x - b} + c$ has a vertical asymptote at $x = b$ and a horizontal asymptote at $y = c$. The graph is a translation of $y = \frac{a}{x}$, b units right or left (for b positive or negative) and c units up or down (for c positive or negative).

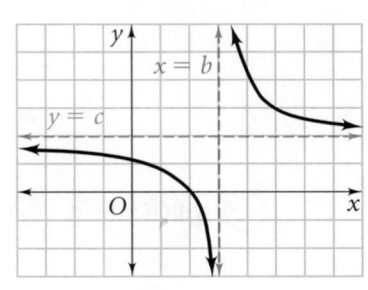

Types of Functions

You can think of graphs of functions with similar features as families of functions. You have studied six families of functions this year. Their properties and graphs are shown in this summary.

Key Concepts | **Summary** | **Families of Functions**

Linear function
$y = mx + b$

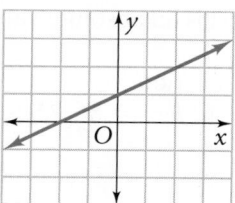

parent function: $f(x) = x$
slope $= m$
y-intercept $= b$
The greatest exponent is 1.

Quadratic function
$y = ax^2 + bx + c$

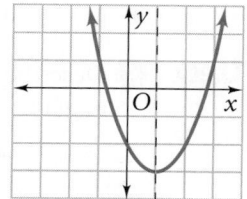

parent function: $f(x) = x^2$
parabola with axis of
symmetry at $x = -\frac{b}{2a}$
The greatest exponent is 2.

Absolute value function
$y = |x - a| + b$

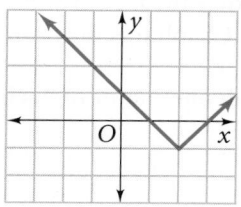

parent function: $f(x) = |x|$
shift $y = |x|$ horizontally a units
shift $y = |x|$ vertically b units
vertex at (a, b)
The greatest exponent is 1.

Exponential function
$y = ab^x$

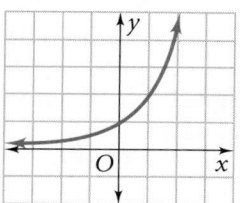

growth for $b > 1$
decay for $0 < b < 1$
The variable is the exponent.

Square root function
$y = \sqrt{x - b} + c$

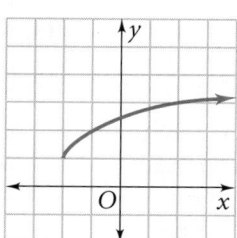

shift $y = \sqrt{x}$ horizontally b units
shift $y = \sqrt{x}$ vertically c units
The variable is under the radical.

Rational function
$y = \frac{a}{x - b} + c$

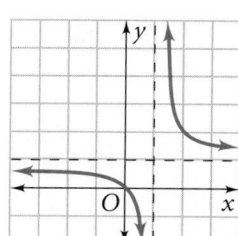

vertical asymptote at $x = b$
horizontal asymptote at $y = c$
The variable is in the denominator.

4 EXAMPLE Identifying Functions

Describe the graph of each function.

a. $y = 5^x$ — The graph is of exponential growth.

b. $y = 5x$ — The graph is a line with slope 5 and y-intercept 0.

c. $y = \frac{x}{5}$ — The graph is a line with slope $\frac{1}{5}$ and y-intercept 0.

d. $y = \frac{5}{x}$ — The graph is a rational function with vertical asymptote at $x = 0$ and horizontal asymptote at $y = 0$.

e. $y = 5x^2$ — The graph is a parabola with an axis of symmetry at $x = 0$.

f. $y = \sqrt{x - 5}$ — The graph is the radical function $y = \sqrt{x}$ shifted right 5 units.

g. $y = |x - 5|$ — The graph is an absolute value function with a vertex at $(5, 0)$.

✓ Quick Check **④** Describe the graph of each function.

a. $g(x) = |x + 4|$ **b.** $f(x) = 8 \cdot 2^x$ **c.** $h(x) = \frac{2}{x + 1}$

EXERCISES

For more exercises, see *Extra Skill and Word Problem Practice.*

A Practice by Example

Example 1
(page 664)

Example 2
(page 665)

Graph each function.

1. $y = \frac{3}{x}$ **2.** $y = \frac{4}{x}$ **3.** $f(x) = \frac{5}{x}$ **4.** $h(x) = \frac{6}{x}$

5. The function $t = \frac{24}{r}$ models the time it will take you to travel 24 miles at different rates of speed. Graph this function.

What value of x makes the denominator of each function equal zero?

6. $f(x) = \frac{3}{x}$ **7.** $y = \frac{1}{x - 2}$ **8.** $y = \frac{x}{x + 2}$ **9.** $h(x) = \frac{3}{2x - 4}$

Identify the asymptotes of each graph.

10. 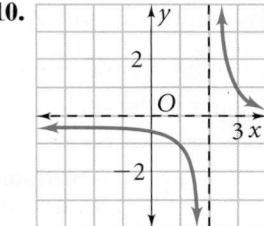 **11.**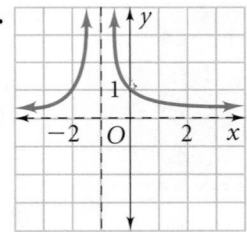

12. 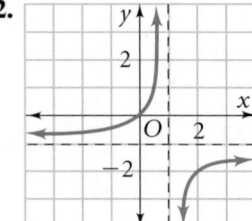 **13.**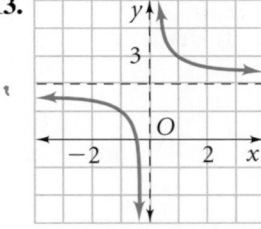

Identify the vertical asymptote of each function. Then graph the function.

14. $g(x) = \frac{10}{x}$ **15.** $f(x) = \frac{12}{x}$ **16.** $y = \frac{1}{x + 1}$

17. $f(x) = \frac{1}{x - 5}$ **18.** $g(x) = \frac{4}{x + 4}$ **19.** $y = \frac{2}{x + 4}$

Example 3
(page 666)

Identify the asymptotes of each function. Then graph the function.

20. $y = \frac{1}{x} - 5$ **21.** $y = \frac{1}{x} + 5$ **22.** $y = \frac{1}{x} - 6$

23. $h(x) = \frac{2}{x + 1} + 4$ **24.** $f(x) = \frac{1}{x - 3} - 5$ **25.** $h(x) = \frac{1}{x - 1} - 2$

Example 4
(page 668)

Describe the graph of each function.

26. $y = 4x + 1$ **27.** $h(x) = |x - 4|$ **28.** $y = 0.4^x$

29. $f(x) = \frac{x}{4}$ **30.** $y = \frac{4}{x} + 1$ **31.** $h(x) = \sqrt{x - 4} + 1$

32. $g(x) = x^2 - 4$ **33.** $f(x) = \frac{4}{x + 4} - 1$ **34.** $g(x) = 4x^2 + 2x + 1$

B   **Apply Your Skills**

Describe how the graphs of each function are translations of the graph of $f(x) = \frac{7}{x}$.

35. $g(x) = \frac{7}{x + 1}$ **36.** $y = \frac{7}{x - 3}$ **37.** $y = \frac{7}{x} - 15$ **38.** $f(x) = \frac{7}{x + 12}$

39. $g(x) = \frac{7}{x} + 12$ **40.** $h(x) = \frac{7}{x + 3}$ **41.** $g(x) = \frac{7}{x} - 2$ **42.** $y = \frac{7}{x + 3} - 2$

GO **Online**

Homework Video Tutor

Visit: PHSchool.com
Web Code: ate-1201

Identify the asymptotes of each function. Then graph the function.

43. $f(x) = \frac{-1}{x}$ **44.** $y = \frac{-4}{x}$ **45.** $g(x) = \frac{-2}{x + 4}$

46. $y = \frac{-1}{x} + 1$ **47.** $y = \frac{1}{x + 1} + 4$ **48.** $g(x) = \frac{1}{x + 1} - 3$

49. $f(x) = \frac{1}{x - 1} + 3$ **50.** $g(x) = \frac{2}{x + 5} + 1$ **51.** $h(x) = \frac{-4}{x - 3} - 2$

52. Open-Ended Write two rational functions whose graphs are identical except that one has been shifted vertically 3 units.

53. Light In the formula $I = \frac{445}{x^2}$, I is the intensity of light in lumens at a distance x feet from a light bulb with 445 watts. What is the intensity of light 5 ft from the light bulb? 15 ft from the light bulb?

54. a. Graph $y = \frac{1}{x}$ and $y = \frac{1}{x^2}$.
    b. What are the vertical and horizontal asymptotes of the graph of each function?
    c. What is the range of $y = \frac{1}{x}$? Of $y = \frac{1}{x^2}$?

55. Physics As radio signals move away from a transmitter, they become weaker. The function $s = \frac{1600}{d^2}$ relates the strength s of a signal at a distance d miles from a transmitter.
    a. Graphing Calculator Graph the function. For what distances is $s \leq 1$?
    b. Find the signal strength at 10 mi, 1 mi, and 0.1 mi.
    c. Critical Thinking Suppose you drive by the transmitter for one radio station while your car radio is tuned to a second station. The signal from the transmitter can interfere and come through your radio. Use your results from part (b) to explain why.

Real-World **Connection**

Careers Photographers use high-intensity lights to get dramatic photos.

 56. Writing Describe the similarities and differences between the graphs of $y = \frac{3}{x}$ and $y = \frac{-3}{x}$.

C   **Challenge**

Graph each function. Include a dashed line for each asymptote.

57. $g(x) = \frac{x}{x - 1}$ **58.** $y = \frac{1}{(x - 1)^2}$

59. $y = \frac{2}{(x - 2)(x + 2)}$ **60.** $y = \frac{1}{x^2 - 2x}$

61. Use the graph at the right. It is a translation of the graph of $y = \frac{1}{x}$.
  a. What are the asymptotes of the graph?
  b. Reasoning Write a function rule for the graph.

62. Graph $f(x) = \frac{(x + 2)(x + 1)}{x + 2}$ and $g(x) = x + 1$. Are the graphs the same? Explain.

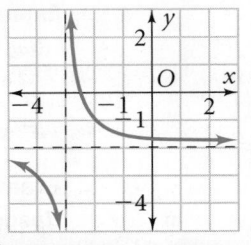

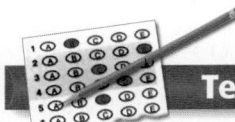

Test Prep

Multiple Choice

63. Which describes the graph of the function $y = \frac{3}{x} + 2$?
  A. a line with slope 3 and y-intercept 2
  B. a line with slope $\frac{1}{3}$ and y-intercept 2
  C. a rational function with asymptotes at $x = 0$ and $y = 2$
  D. a rational function with asymptotes at $x = -3$ and $y = -2$

64. The graph of $y = \frac{2}{x}$ is translated down 3 units. What is the equation of the new graph?
  F. $y = \frac{2}{x + 3}$ **G.** $y = \frac{2}{x - 3}$
  H. $y = \frac{2}{x} + 3$ **J.** $y = \frac{2}{x} - 3$

65. Which equation represents an exponential function?
  A. $y = x^2 - 1$ **B.** $y = 6^x$
  C. $y = \frac{3}{x - 7} + 1$ **D.** $y = 2x + 5$

66. Which statement does NOT describe the graph of $y = \frac{2}{x - 3}$.
  F. The function is undefined when $x = 3$ and so it has a vertical asymptote at $x = 3$.
  G. The graph has an axis of symmetry.
  H. The graph has a horizontal asymptote at $y = 2$.
  J. The graph crosses the y-axis at $\left(0, -\frac{2}{3}\right)$.

Short Response

67. Describe the graph of $y = 1 + \sqrt{x + 3}$.

Mixed Review

Lesson 10-7

Find the number of solutions of each equation.

68. $x^2 + x + 1 = 0$ **69.** $x^2 + 2x + 1 = 0$ **70.** $x^2 - 8x = 7$

71. $-3x^2 + 4x = -5$ **72.** $2x^2 + 5 = 0$ **73.** $9x^2 + 144 = 7x$

Lesson 9-8

Factor completely.

74. $3d^2 - 108$ **75.** $2m^2 - 14m - 120$ **76.** $t^3 - t^2 + 3t - 3$

77. $3h^3 + 3h^2 - 5h - 5$ **78.** $8x^2 - 72$ **79.** $28t^3 - 8t^2 + 63t - 18$

Lesson 5-6

Write the equation of an inverse variation that includes the given point.

80. $(3, 7)$ **81.** $(8, 2)$ **82.** $(4, 5.5)$ **83.** $(6.2, 3.4)$

Graphing Rational Functions

FOR USE WITH LESSON 12-1

Functions like $y = \frac{1}{x}$, $y = \frac{1}{x+2}$, and $y = \frac{1}{x} - 4$ are rational functions. When you use a graphing calculator to graph a rational function, sometimes false connections appear on the screen. When this happens, you need to make adjustments to see the true shape of the graph.

Graph the function $y = \frac{1}{x+2} - 4$. You can enter this as $y = 1 \div (x + 2) - 4$. The graph of the function may look like the graph at the right on your screen. The highest point and lowest point on the graph that appear on the screen are not supposed to connect. If you use the trace feature on the calculator, no point on the graph lies on this connecting line. So this is a false connection.

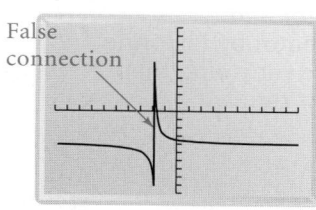

Here's how you can graph a rational function and avoid false connections.

Step 1 Press the MODE key. Then scroll down and right to highlight the word **Dot**. Then press ENTER.

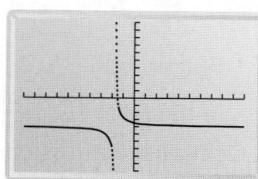

Step 2 Graph again. Now the false connection is gone!

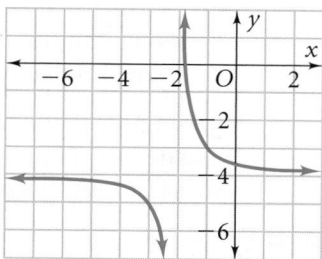

Step 3 Use the TRACE key or TABLE key to find points on the graph. Sketch the graph.

EXERCISES

Use a graphing calculator to graph each function. Then sketch the graph.

1. $y = \frac{4}{x}$

2. $y = \frac{-2}{x}$

3. $y = \frac{1}{x+3}$

4. $y = \frac{1}{x-4}$

5. $y = \frac{1}{x} + 3$

6. $y = \frac{1}{x} - 4$

7. $y = \frac{1}{x-2} - 3$

8. $y = \frac{4}{x+1} - 3$

9. $y = \frac{-4}{x+1} - 3$

10. a. Graph $y = \frac{1}{x}$, $y = \frac{1}{x-3}$, and $y = \frac{1}{x+4}$.

   b. Make a Conjecture How does adding or subtracting a number in the denominator translate the graph?

11. a. Graph $y = \frac{1}{x}$, $y = \frac{1}{x} - 3$, and $y = \frac{1}{x} + 4$.

   b. Make a Conjecture How does adding or subtracting a number to the expression translate the graph?

Simplifying Rational Expressions

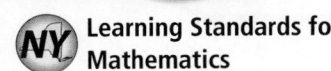

Learning Standards for Mathematics

A.A.16 Simplify fractions with polynomials in the numerator and denominator by factoring both and renaming them to lowest terms.

Check Skills You'll Need

GO for Help Skills Handbook page 758 and Lesson 9-5

Write each fraction in simplest form.

1. $\frac{8}{2}$ **2.** $-\frac{15}{24}$ **3.** $\frac{25}{35}$

Factor each quadratic expression.

4. $x^2 + x - 12$ **5.** $x^2 + 6x + 8$ **6.** $x^2 - 2x - 15$

7. $x^2 + 8x + 16$ **8.** $x^2 - x - 12$ **9.** $x^2 - 7x + 12$

New Vocabulary • rational expression

---

1 Simplifying Rational Expressions

Fractions like $\frac{5}{9}$, $\frac{7}{12}$, and $\frac{1}{2}$ are rational numbers. An expression which can be written in the form $\frac{\text{polynomial}}{\text{polynomial}}$ is a **rational expression.** Here are some examples of rational expressions.

$$\frac{1}{x} \qquad \frac{x + 2}{x - 3} \qquad \frac{x^2 - 5}{x^2 - 10x + 25}$$

Of course, the value of the expression in the denominator cannot be zero, since division by zero is undefined. For the rest of this chapter, assume that the values of the variables that make the denominator zero are excluded from the domain.

Like rational numbers, a rational expression is in simplest form if the numerator and denominator have no common factors except 1. For example, $\frac{z + 5}{10z}$ is in simplest form since no factor of $10z$ is a factor of $z + 5$.

1 EXAMPLE Simplifying a Rational Expression

Simplify $\frac{6x + 12}{x + 2}$.

$$\frac{6x + 12}{x + 2} = \frac{6(x + 2)}{x + 2}$$ **Factor the numerator. The denominator cannot be factored.**

$$= \frac{6(x + 2)^{\,1}}{x + 2_{\,1}}$$ **Divide out the common factor $x + 2$.**

$$= 6$$ **Simplify.**

Quick Check ① Simplify each expression.

a. $\frac{15b}{25b^2}$ **b.** $\frac{12c^2}{3c + 6}$ **c.** $\frac{4m - 2}{2m - 1}$ **d.** $\frac{20 + 4t}{t + 5}$

Recall that you learned to factor quadratic expressions in Lessons 9-5 and 9-6. You may need to factor a quadratic expression to simplify a rational expression.

2 EXAMPLE Simplifying a Rational Expression

Simplify $\dfrac{2x - 12}{x^2 - 7x + 6}$.

$$\dfrac{2x - 12}{x^2 - 7x + 6} = \dfrac{2(x - 6)}{(x - 6)(x - 1)} \qquad \text{Factor the numerator and the denominator.}$$

$$= \dfrac{2(x - 6)^{1}}{{}_{1}(x - 6)(x - 1)} \qquad \text{Divide out the common factor } x - 6.$$

$$= \dfrac{2}{x - 1} \qquad \text{Simplify.}$$

✓ Quick Check ➋ Simplify each expression.

a. $\dfrac{3x + 12}{x^2 - x - 20}$ **b.** $\dfrac{2z - 2}{z^2 - 4z + 3}$ **c.** $\dfrac{8a + 16}{2a^2 + 5a + 2}$ **d.** $\dfrac{c^2 - c - 6}{c^2 + 5c + 6}$

The numerator and denominator of $\dfrac{x - 3}{3 - x}$ are opposites. To simplify the expression, you can factor -1 from $3 - x$ to get $-1(-3 + x)$, which you can rewrite as $-1(x - 3)$. Then simplify $\dfrac{x - 3}{-1(x - 3)}$.

3 EXAMPLE Recognizing Opposite Factors

Simplify $\dfrac{5x - 15}{9 - x^2}$.

$$\dfrac{5x - 15}{9 - x^2} = \dfrac{5(x - 3)}{(3 - x)(3 + x)} \qquad \text{Factor the numerator and the denominator.}$$

$$= \dfrac{5(x - 3)}{-1(x - 3)(3 + x)} \qquad \text{Factor } -1 \text{ from } 3 - x.$$

$$= \dfrac{5(x - 3)^{1}}{-1_{1}(x - 3)(x + 3)} \qquad \text{Divide out the common factor } x - 3.$$

$$= -\dfrac{5}{x + 3} \qquad \text{Simplify.}$$

✓ Quick Check ➌ Simplify each expression.

a. $\dfrac{x - 4}{4 - x}$ **b.** $\dfrac{8 - m}{m^2 - 64}$ **c.** $\dfrac{8 - 4r}{r^2 + 2r - 8}$ **d.** $\dfrac{2c^2 - 2}{3 - 3c^2}$

You can use a rational expression to model some real-world situations.

4 EXAMPLE Evaluating a Rational Expression

Baking The baking time for bread depends, in part, on its size and shape. A good approximation for the baking time, in minutes, of a cylindrical loaf is $\dfrac{60 \cdot \text{volume}}{\text{surface area}}$, or $\dfrac{30rh}{r + h}$, where the radius r and the length h of the baked loaf are in inches. Find the baking time for a loaf that is 5 inches long and has a radius of 4 inches. Round your answer to the nearest minute.

$$\dfrac{30rh}{r + h} = \dfrac{30(4)(5)}{4 + 5} \qquad \text{Substitute 4 for } r \text{ and 5 for } h.$$

$$= \dfrac{600}{9} \qquad \text{Simplify.}$$

$$\approx 67 \qquad \text{Round to the nearest whole number.}$$

The baking time is approximately 67 minutes.

Real-World 🌐 Connection

For a given volume of dough, the greater the surface area is, the shorter the baking time.

a. Find the baking time for a loaf that is 4 inches long and has a radius of 3 inches. Round your answer to the nearest minute.

b. Critical Thinking The ratio $\frac{60 \cdot \text{volume}}{\text{surface area}}$ for a cylinder is $\frac{60\pi r^2 h}{2\pi r^2 + 2\pi rh}$. Simplify this expression to show that it is the same as the expression evaluated in Example 4.

EXERCISES

For more exercises, see *Extra Skill and Word Problem Practice.*

Practice and Problem Solving

 **A** **Practice by Example**

Example 1
(page 672)

Simplify each expression.

1. $\frac{6a + 9}{12}$

2. $\frac{4x^3}{28x^4}$

3. $\frac{2m - 5}{6m - 15}$

4. $\frac{2p - 24}{4p - 48}$

5. $\frac{3x^2 - 9x}{x - 3}$

6. $\frac{3x + 6}{3x^2}$

Example 2
(page 673)

7. $\frac{2x^2 + 2x}{3x^2 + 3x}$

8. $\frac{2b - 8}{b^2 - 16}$

9. $\frac{m + 6}{m^2 - m - 42}$

10. $\frac{w^2 + 7w}{w^2 - 49}$

11. $\frac{a^2 + 2a + 1}{5a + 5}$

12. $\frac{m^2 + 7m + 12}{m^2 + 6m + 8}$

13. $\frac{c^2 - 6c + 8}{c^2 + c - 6}$

14. $\frac{b^2 + 8b + 15}{b + 5}$

15. $\frac{m + 4}{m^2 + 2m - 8}$

Example 3
(page 673)

16. $\frac{5 - 4n}{4n - 5}$

17. $\frac{12 - 4t}{t^2 - 2t - 3}$

18. $\frac{4m - 8}{4 - 2m}$

19. $\frac{m - 2}{4 - 2m}$

20. $\frac{v - 5}{25 - v^2}$

21. $\frac{4 - w}{w^2 - 8w + 16}$

Example 4
(page 673)

Baking Use the expression $\frac{30rh}{r + h}$ to estimate the baking time in minutes for each type of bread. Round your answer to the nearest minute.

22. baguette: $r = 1.25$ in., $h = 26$ in.

23. pita: $r = 3.5$ in., $h = 0.5$ in.

24. biscuit: $r = 1$ in., $h = 0.75$ in.

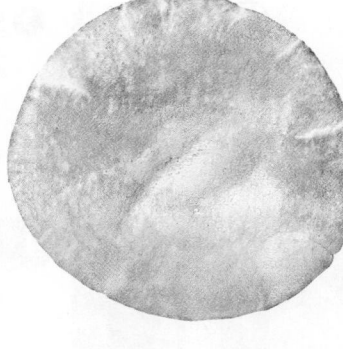

B **Apply Your Skills**

Simplify each expression.

25. $\frac{2r^2 + 9r - 5}{r^2 + 10r + 25}$

26. $\frac{7z^2 + 23z + 6}{z^2 + 2z - 3}$

27. $\frac{5t^2 + 6t - 8}{3t^2 + 5t - 2}$

28. $\frac{32a^3}{16a^2 - 8a}$

29. $\frac{3z^2 + 12z}{z^4}$

30. $\frac{2s^2 + s}{s^3}$

31. $\frac{4a^2 - 8a - 5}{15 - a - 2a^2}$

32. $\frac{16 + 16m + 3m^2}{m^2 - 3m - 28}$

33. $\frac{10c + c^2 - 3c^3}{5c^2 - 6c - 8}$

34. **Open-Ended** Write an expression that has 2 and −3 excluded from the domain.

35. a. Construction To keep heating costs down for a structure, architects want the ratio of surface area to volume as small as possible. Find an expression for the ratio of the surface area to volume for each shape.

   i. square prism   **ii.** cylinder

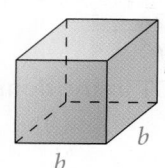 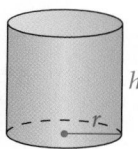

b. Find the ratio for each figure when $b = 12$ ft, $h = 18$ ft, and $r = 6$ ft.

36. Error Analysis Explain what error the student made in simplifying the rational expression at the right.

$$\frac{x^2 + 2x}{2x} = \frac{x^2 + \cancel{2x}}{\cancel{2x}}$$
$$= x^2$$

37. Writing Explain why $\frac{x^2 - 9}{x + 3}$ is not the same as $x - 3$.

Probability If a point is selected at random from a figure and is equally likely to be any point in the figure, then the probability that the point is in a shaded part of the figure is $\frac{\text{area of shaded part}}{\text{area of whole figure}}$. Find the probability that a point will be in the shaded part of each figure.

38.

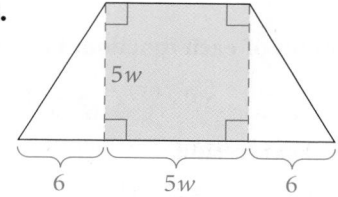

39.

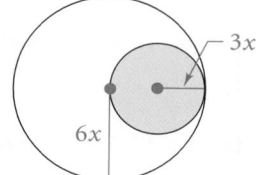

40.

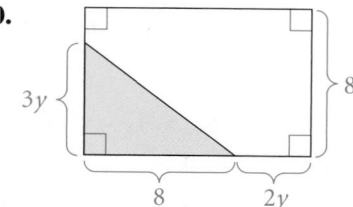

41.

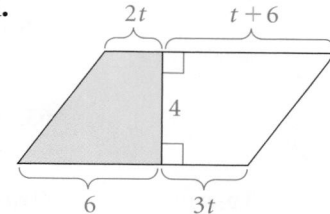

C **Challenge** **Simplify each expression.**

42. $\dfrac{m^2 - n^2}{m^2 + 11mn + 10n^2}$ **43.** $\dfrac{a^2 - 5ab + 6b^2}{a^2 + 2ab - 8b^2}$ **44.** $\dfrac{36v^2 - 49w^2}{18v^2 + 9vw - 14w^2}$

Reasoning Determine whether each statement is *sometimes*, *always*, or *never* true for real numbers a and b.

45. $\dfrac{2b}{b} = 2$ **46.** $\dfrac{ab^3}{b^4} = ab$ **47.** $\dfrac{a^2 + 6a - 5}{2a + 2} = \dfrac{a + 5}{2}$

NY REGENTS

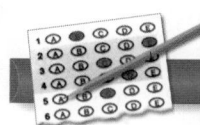

Test Prep

Multiple Choice

48. Which expression simplifies to -1?

   A. $\dfrac{x + 1}{x - 1}$ **B.** $\dfrac{r + 3}{3 - r}$ **C.** $\dfrac{n - 2}{2 - n}$ **D.** $\dfrac{4 - p}{4 + p}$

49. Simplify $\dfrac{y^2 + 8y - 9}{y^2 - 81}$.

   F. $\dfrac{8y}{9}$ **G.** $\dfrac{y + 9}{y + 81}$ **H.** $\dfrac{1}{9}$ **J.** $\dfrac{y - 1}{y - 9}$

50. Which expression is in simplest form?

   A. $\dfrac{t + 1}{t^2 - 1}$ **B.** $\dfrac{2n - 1}{n^2 + 4}$ **C.** $\dfrac{c - 7}{7 - c}$ **D.** $\dfrac{2r - 4}{8 + 6r}$

51. What values of x are NOT in the domain of $\dfrac{(x + 1)(x - 3)}{(x - 2)(x - 5)}$?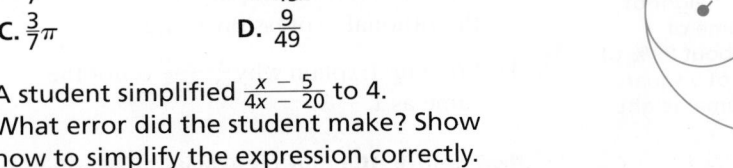

   F. -1 and 3 **G.** 1 and -3
   H. 2 and 5 **J.** -2 and -5

52. What is the ratio of the area of the small circle to the area of the large circle?

   A. $\dfrac{3}{7}$ **B.** $\dfrac{9\pi}{49x}$

   C. $\dfrac{3}{7}\pi$ **D.** $\dfrac{9}{49}$

Short Response

53. A student simplified $\dfrac{x - 5}{4x - 20}$ to 4. What error did the student make? Show how to simplify the expression correctly.

Mixed Review

Lesson 12-2

Identify the asymptotes of each function. Then graph the function.

54. $h(x) = \dfrac{8}{x} + 2$ **55.** $f(x) = \dfrac{8}{x - 4}$ **56.** $g(x) = \dfrac{8}{x} - 4$

Lesson 11-1

Simplify each radical expression.

57. $\sqrt{20} \cdot \sqrt{10}$ **58.** $\sqrt{a^4 b^7 c^8}$ **59.** $\dfrac{\sqrt{80}}{\sqrt{10}}$

60. $\sqrt{\dfrac{2m}{25m^5}}$ **61.** $\sqrt{90h^2 k^4}$ **62.** $\sqrt{6} \cdot \sqrt{8}$

63. $\sqrt{\dfrac{72}{2a^2}}$ **64.** $\sqrt{9x} \cdot \sqrt{11x}$ **65.** $\sqrt{\dfrac{28y^5}{7y^2}}$

Lesson 10-1

Order each group of quadratic functions from widest to narrowest graph.

66. $y = x^2, y = 3x^2, y = -2x^2$ **67.** $y = \dfrac{1}{3}x^2, y = \dfrac{1}{4}x^2, y = \dfrac{2}{5}x^2$
68. $y = 2x^2, y = 0.5x^2, y = -4x^2$ **69.** $y = -x^2, y = 2.3x^2, y = -3.8x^2$

Checkpoint Quiz 1   Lessons 12-1 through 12-2

Identify the asymptotes of each function. Then graph the function.

   1. $f(x) = \dfrac{2}{x - 4}$ **2.** $y = \dfrac{1}{x} + 4$

Simplify each expression.

   3. $\dfrac{6x^2 - 24}{x + 2}$ **4.** $\dfrac{3c + 9}{3c - 9}$

   5. $\dfrac{k - 2}{k^2 + 2k - 8}$

12-3

Multiplying and Dividing Rational Expressions

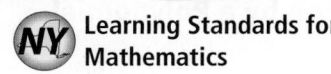

Learning Standards for Mathematics

A.A.18 Multiply and divide algebraic fractions and express the product or quotient in simplest form.

✓ **Check Skills You'll Need**

GO for **Help** Lessons 8-3 and 9-6

Simplify each expression.

1. $r^2 \cdot r^8$

2. $b^3 \cdot b^4$

3. $c^7 \div c^2$

4. $3x^4 \cdot 2x^5$

5. $5n^2 \cdot n^2$

6. $15a^3(-3a^2)$

Factor each polynomial.

7. $2c^2 + 15c + 7$

8. $15t^2 - 26t + 11$

9. $2q^2 + 11q + 5$

---

1 Multiplying Rational Expressions

Multiplying rational expressions is similar to multiplying rational numbers. If a, b, c, and d represent polynomials (with $b \neq 0$ and $d \neq 0$), then $\frac{a}{b} \cdot \frac{c}{d} = \frac{ac}{bd}$.

1 EXAMPLE Multiplying Rational Expressions

Multiply.

a. $\dfrac{3}{x} \cdot \dfrac{4}{x^2}$

$\dfrac{3}{x} \cdot \dfrac{4}{x^2} = \dfrac{12}{x^3}$ **Multiply the numerators and multiply the denominators.**

b. $\dfrac{x}{x+4} \cdot \dfrac{x-3}{x-2}$

$\dfrac{x}{x+4} \cdot \dfrac{x-3}{x-2} = \dfrac{x(x-3)}{(x+4)(x-2)}$ **Multiply the numerators and multiply the denominators. Leave the answer in factored form.**

✓ **Quick Check** ❶ Multiply.

a. $\dfrac{6}{a^2} \cdot \dfrac{-2}{a^3}$

b. $\dfrac{x-5}{x+3} \cdot \dfrac{x-7}{x}$

As with rational numbers, the product $\frac{ac}{bd}$ may not be in simplest form. Look for factors common to the numerator and the denominator to divide out.

2 EXAMPLE Using Factoring

Multiply $\dfrac{2x+1}{3}$ and $\dfrac{6x}{4x^2-1}$.

$\dfrac{2x+1}{3} \cdot \dfrac{6x}{4x^2-1} = \dfrac{2x+1}{3} \cdot \dfrac{6x}{(2x+1)(2x-1)}$ **Factor the denominator.**

$= \dfrac{2x+1}{1}\dfrac{^1}{3} \cdot \dfrac{6^2x}{1(2x+1)(2x-1)}$ **Divide out the common factors 3 and (2x + 1).**

$= \dfrac{2x}{2x-1}$ **Simplify.**

✓ **Quick Check** ❷ Multiply $\dfrac{x-2}{8x}$ and $\dfrac{-8x-16}{x^2-4}$.

You can also multiply a rational expression by a polynomial. Leave the product in factored form.

3 EXAMPLE Multiplying a Rational Expression by a Polynomial

Multiply $\frac{3s + 2}{2s + 4}$ and $s^2 + 5s + 6$.

$$\frac{3s + 2}{2s + 4} \cdot (s^2 + 5s + 6) = \frac{3s + 2}{2(s + 2)} \cdot \frac{(s + 2)(s + 3)}{1}$$ **Factor.**

$$= \frac{3s + 2}{2_1(s + 2)} \cdot \frac{(s + 2)^1(s + 3)}{1}$$ **Divide out the common factor $s + 2$.**

$$= \frac{(3s + 2)(s + 3)}{2}$$ **Leave in factored form.**

✓ **Quick Check** ③ Multiply.

a. $\frac{3}{c} \cdot (c^3 - c)$ **b.** $\frac{2v}{v + 3} \cdot (v^2 - 2v - 15)$ **c.** $(m - 1) \cdot \frac{4m + 8}{m^2 - 1}$

---

2   Dividing Rational Expressions

Recall that $\frac{a}{b} \div \frac{c}{d} = \frac{a}{b} \cdot \frac{d}{c}$, where $b \neq 0, c \neq 0$, and $d \neq 0$.

When you divide rational expressions that can be factored, first rewrite the expression using the reciprocal before dividing out common factors.

4 EXAMPLE Dividing Rational Expressions

Vocabulary Tip

The <u>vinculum</u> or fraction bar is a grouping symbol.

Divide $\frac{a^2 + 7a + 10}{a - 6}$ by $\frac{a + 5}{a^2 - 36}$.

$$\frac{a^2 + 7a + 10}{a - 6} \div \frac{a + 5}{a^2 - 36} = \frac{a^2 + 7a + 10}{a - 6} \cdot \frac{a^2 - 36}{a + 5}$$ **Multiply by $\frac{a^2 - 36}{a + 5}$,**

                                            the reciprocal of $\frac{a + 5}{a^2 - 36}$.

$$= \frac{(a + 2)(a + 5)}{(a - 6)} \cdot \frac{(a - 6)(a + 6)}{a + 5}$$ **Factor.**

$$= \frac{(a + 2)(a + 5)^1}{1(a - 6)} \cdot \frac{(a - 6)^1(a + 6)}{1 a + 5}$$ **Divide out the common factors $a + 5$ and $a - 6$.**

$$= (a + 2)(a + 6)$$ **Leave in factored form.**

✓ **Quick Check** ④ Divide.

a. $\frac{a - 2}{ab} \div \frac{a - 2}{a}$ **b.** $\frac{5m + 10}{2m - 20} \div \frac{7m + 14}{14m - 20}$ **c.** $\frac{6n^2 - 5n - 6}{2n^2 - n - 3} \div \frac{2n - 3}{n + 1}$

The reciprocal of a polynomial such as $5x^2 + 5x$ is $\frac{1}{5x^2 + 5x}$.

5 EXAMPLE Dividing a Rational Expression by a Polynomial

Divide $\frac{x^2 + 3x + 2}{4x}$ by $(5x^2 + 5x)$.

$$\frac{x^2 + 3x + 2}{4x} \div \frac{5x^2 + 5x}{1} = \frac{x^2 + 3x + 2}{4x} \cdot \frac{1}{5x^2 + 5x}$$ **Multiply by the reciprocal of $5x^2 + 5x$.**

$$= \frac{(x + 1)(x + 2)}{4x} \cdot \frac{1}{5x(x + 1)}$$ **Factor.**

$$= \frac{(x + 1)^1(x + 2)}{4x} \cdot \frac{1}{5x_1(x + 1)}$$ **Divide out the common factor $x + 1$.**

$$= \frac{x + 2}{20x^2}$$ **Simplify.**

a. $\frac{3x^3}{2} \div \left(-15x^5\right)$ **b.** $\frac{y+3}{y+2} \div (y+2)$ **c.** $\frac{z^2 + 2z - 15}{z^2 + 9z + 20} \div (z - 3)$

EXERCISES

For more exercises, see *Extra Skill and Word Problem Practice.*

Practice and Problem Solving

Ⓐ Practice by Example

Example 1
(page 677)

Example 2
(page 677)

Example 3
(page 678)

Example 4
(page 678)

Multiply.

1. $\frac{7}{3} \cdot \frac{5x}{12}$ **2.** $\frac{3}{t} \cdot \frac{4}{t}$ **3.** $\frac{5}{3a^2} \cdot \frac{8}{a^3}$

4. $\frac{m-2}{m+2} \cdot \frac{m}{m-1}$ **5.** $\frac{2x}{x+1} \cdot \frac{x-1}{3}$ **6.** $\frac{6x^2}{5} \cdot \frac{2}{x+1}$

7. $\frac{4c}{2c+2} \cdot \frac{c+1}{c-1}$ **8.** $\frac{5x^3}{x^2} \cdot \frac{3x^4}{6x}$ **9.** $\frac{3t}{t-2} \cdot \frac{3t-6}{t^2}$

10. $\frac{m-2}{3m+9} \cdot \frac{2m+6}{2m-4}$ **11.** $\frac{x-5}{4x+6} \cdot \frac{6x+9}{3x-15}$ **12.** $\frac{4x+1}{5x+10} \cdot \frac{30x+60}{2x-2}$

13. $\frac{4t+4}{t-3} \cdot \left(t^2 - t - 6\right)$ **14.** $\frac{2m+1}{3m-6} \cdot \left(9m^2 - 36\right)$ **15.** $\left(x^2 - 1\right) \cdot \frac{x-2}{3x+3}$

Find the reciprocal of each expression.

16. $\frac{2}{x+1}$ **17.** $\frac{-6d^2}{2d-5}$ **18.** $c^2 - 1$ **19.** $s + 4$

Divide.

20. $\frac{x-1}{x+4} \div \frac{x+3}{x+4}$ **21.** $\frac{3t+12}{5t} \div \frac{t+4}{10t}$ **22.** $\frac{y-4}{10} \div \frac{4-y}{5}$

23. $\frac{x-3}{6} \div \frac{3-x}{2}$ **24.** $\frac{x^2 + 6x + 8}{x^2 + x - 2} \div \frac{x+4}{2x+4}$ **25.** $\frac{2n^2 - 5n - 3}{4n^2 - 12n - 7} \div \frac{4n+5}{2n-7}$

Example 5
(page 678)

26. $\frac{3x+9}{x} \div (x+3)$ **27.** $\frac{11k+121}{7k-15} \div (k+11)$ **28.** $\frac{x^2 + 10x - 11}{x^2 + 12x + 11} \div (x-1)$

Ⓑ Apply Your Skills

Multiply or divide.

29. $\frac{t^2 + 5t + 6}{t-3} \cdot \frac{t^2 - 2t - 3}{t^2 + 3t + 2}$ **30.** $\frac{c^2 + 3c + 2}{c^2 - 4c + 3} \div \frac{c+2}{c-3}$

31. $\frac{7t^2 - 28t}{2t^2 - 5t - 12} \cdot \frac{6t^2 - t - 15}{49t^3}$ **32.** $\frac{5x^2 + 10x - 15}{5 - 6x + x^2} \div \frac{2x^2 + 7x + 3}{4x^2 - 8x - 5}$

33. $\frac{x^2 + x - 6}{x^2 - x - 6} \div \frac{x^2 + 5x + 6}{x^2 + 4x + 4}$ **34.** $\left(\frac{x^2 - 25}{x^2 - 4x}\right)\left(\frac{x^2 + x - 20}{x^2 + 10x + 25}\right)$

35. Error Analysis In the work shown at the right, what error did the student make in dividing the rational expressions?

$$\frac{3a}{a+2} \div \frac{(a+2)^2}{a-4} = \frac{3a}{\cancel{a+2}} \div \frac{\cancel{(a+2)^2}}{a-4}$$

$$= 3a \div \frac{a+2}{a-4}$$

$$= 3a \cdot \frac{a-4}{a+2}$$

$$= \frac{3a(a-4)}{a+2}$$

36. Open-Ended Write two rational expressions. Find the product.

37. Critical Thinking For what values of x is the expression $\frac{2x^2 - 5x - 12}{6x} \div \frac{-3x - 12}{x^2 - 16}$ undefined?

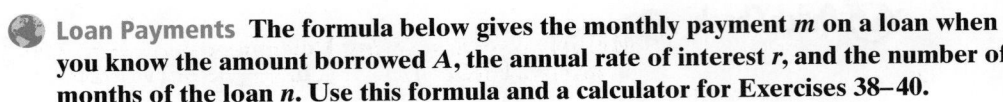

Loan Payments The formula below gives the monthly payment m on a loan when you know the amount borrowed A, the annual rate of interest r, and the number of months of the loan n. Use this formula and a calculator for Exercises 38–40.

$$m = \frac{A\left(\frac{r}{12}\right)\left(1 + \frac{r}{12}\right)^n}{\left(1 + \frac{r}{12}\right)^n - 1}$$

38. What is the monthly payment on a loan of $1500 at 8% annual interest for 18 months?

39. What is the monthly payment on a loan of $3000 at 6% annual interest for 24 months?

40. Suppose your parents want to buy the house shown at the left. They have $20,000 for a down payment. Their mortgage will have an annual interest rate of 6%. The loan is to be repaid over a 30-year period.
  a. How much will your parents have to borrow?
  b. How many monthly payments will there be?
  c. What will the monthly payment be?
  d. How much will it cost your parents to repay this mortgage over the 30-year period?

Geometry Find the volume of each rectangular solid.

41.

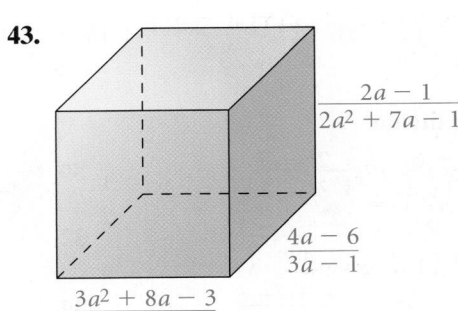

$\frac{x - 5}{3x + 2}$ $\frac{x - 2}{x^2 + 2x - 35}$ $\frac{3x + 2}{4}$

42.

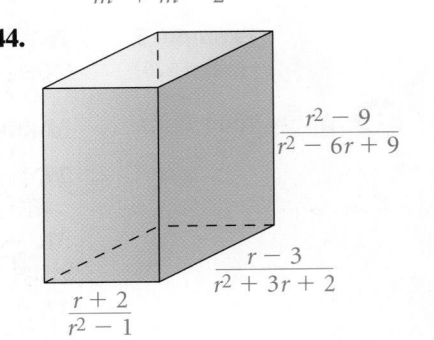

$\frac{2m + 4}{m}$ $\frac{m^2 - m - 6}{m^2 + m - 2}$ $\frac{m^3}{m^2 + m - 12}$

43.

$\frac{2a - 1}{2a^2 + 7a - 15}$ $\frac{4a - 6}{3a - 1}$ $\frac{3a^2 + 8a - 3}{2a^2 + 5a - 3}$

44.

$\frac{r^2 - 9}{r^2 - 6r + 9}$ $\frac{r - 3}{r^2 + 3r + 2}$ $\frac{r + 2}{r^2 - 1}$

45. Writing Robin's first step in finding the product $\frac{2}{w} \cdot w^5$ was to rewrite the expression as $\frac{2}{w} \cdot \frac{w^5}{1}$. Why do you think Robin did this?

46. Probability If a point is selected at random from a figure and is equally likely to be any point in the figure, then the probability that the point is in a shaded part of the figure is $\frac{\text{area of shaded part}}{\text{area of whole figure}}$. Suppose two points are randomly selected.
  a. What is the probability that both points will be in the shaded part?
  b. What is the probability that one point will be in the shaded part and the other point will *not* be in the shaded part?

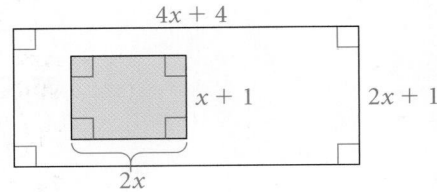

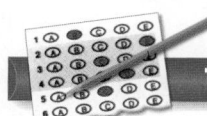

Multiply or divide. (*Hint:* Remember that $\dfrac{\frac{a}{b}}{\frac{c}{d}} = \dfrac{a}{b} \div \dfrac{c}{d}$.)

47. $\dfrac{3m^3 - 3m}{4m^2 + 4m - 8} \cdot (6m^2 + 12m)$

48. $\dfrac{t^2 - r^2}{t^2 + tr - 2r^2} \cdot \dfrac{t^2 + 3tr + 2r^2}{t^2 + 2tr + r^2}$

49. $\dfrac{5x^2}{y^2 - 25} \div \dfrac{5xy - 25x}{y^2 - 10y + 25}$

50. $\dfrac{2a^2 - ab - 6b^2}{2b^2 + 9ab - 5a^2} \div \dfrac{2a^2 - 7ab + 6b^2}{a^2 - 4b^2}$

51. $\dfrac{\frac{3m}{m-1}}{\frac{6m^2}{m-2}}$

52. $\dfrac{\frac{3x}{x^2-1}}{\frac{6}{x^2-x-2}}$

53. $\dfrac{\frac{w-3}{w^2-4}}{\frac{w^2-9}{w-2}}$

NY REGENTS

Test Prep

Multiple Choice

54. Simplify $(2x - 5) \cdot \dfrac{2x}{2x^2 - 9x + 10}$.

    A. 1
    B. $\dfrac{2x}{x - 2}$
    C. $\dfrac{x - 5}{-4x + 5}$
    D. $\dfrac{2x - 5}{-8x - 10}$

55. Simplify $\dfrac{a^2 - 9}{a^2} \div \dfrac{a + 3}{a}$.

    F. -3
    G. $\dfrac{a - 3}{a}$
    H. $\dfrac{a(a - 3)}{a^2}$
    J. $\dfrac{a(a^2 - 9)}{a^2(a + 3)}$

56. Which expression is equivalent to $\dfrac{r^2 - 1}{r} \div (2r^2 - 2)$?

    A. $\dfrac{r^2 - 1}{r} \cdot \dfrac{1}{2}(r^2 - 1)$
    B. $\dfrac{r^2 - 1}{r} \cdot \dfrac{2}{r^2 - 1}$

    C. $\dfrac{r^2 - 1}{r} \cdot \left(\dfrac{1}{2r^2} - 2\right)$
    D. $\dfrac{r^2 - 1}{r} \cdot \dfrac{1}{2r^2 - 2}$

57. Which CANNOT be the first step in multiplying $\dfrac{x^2 - 2x - 3}{x + 3}$ by $\dfrac{2x + 6}{2x + 2}$?

    F. Multiply the numerators.
    G. Find the reciprocal of $\dfrac{2x + 6}{2x + 2}$.
    H. Factor each polynomial.
    J. Multiply the denominators.

Short Response

58. Simplify $\dfrac{x^2 - 1}{x} \cdot \dfrac{3x}{x - 1}$. Write the product in factored form. Show your work.

Mixed Review

Lesson 12-2

Simplify each expression.

59. $\dfrac{5b - 25}{10}$
    60. $\dfrac{36k^3}{48k^4}$
    61. $\dfrac{7m - 14}{3m - 6}$

62. $\dfrac{7q^5}{28q}$
    63. $\dfrac{15t^2 - 27}{24}$
    64. $\dfrac{6m^3}{12m - 18m^2}$

65. $\dfrac{5a^2}{10a^4 - 15a^2}$
    66. $\dfrac{2z^2 - 11z - 21}{z^2 - 6z - 7}$
    67. $\dfrac{4c^2 - 36c + 81}{4c^2 - 2c - 72}$

Lesson 10-2

Graph each function. Label the axis of symmetry and the vertex.

68. $y = x^2 + 10x - 2$
    69. $y = x^2 - 10x - 2$
    70. $y = 2x^2 + x + 5$

Lesson 3-9

Assume a and b are legs of a right triangle, and c is the hypotenuse. Find the length of the missing side of each right triangle. If necessary, round to the nearest tenth.

71. $a = 2, b = 8$
    72. $a = 3.1, b = 4.3$
    73. $a = \sqrt{7}, c = \sqrt{32}$

74. $a = \sqrt{10}, b = \sqrt{111}$
    75. $a = \frac{1}{5}, b = \frac{1}{12}$
    76. $a = 2\frac{1}{3}, b = 6\frac{2}{3}$

Dividing Polynomials

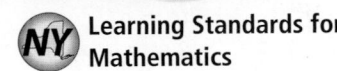
Learning Standards for
Mathematics

A.A.14 Divide a polynomial
by a monomial or
binomial, where the
quotient has no
remainder.

✓ Check Skills You'll Need

GO ➤ for Help Lessons 9-1 and 9-3

Write each polynomial in standard form.

1. $9a - 4a^2 + 1$

2. $3x^2 - 6 + 5x - x^3$

3. $-2 + 8t$

Find each product.

4. $(2x + 4)(x + 3)$

5. $(-3n - 4)(n - 5)$

6. $(3a^2 + 1)(2a - 7)$

---

1   Dividing Polynomials

To divide a polynomial by a monomial, divide each term of the polynomial by the monomial divisor.

1 EXAMPLE   Dividing a Polynomial by a Monomial

Divide $8x^3 + 4x^2 - 12x$ by $2x^2$.

$$(8x^3 + 4x^2 - 12x) \div 2x^2 = (8x^3 + 4x^2 - 12x)\frac{1}{2x^2}$$ **Multiply by the reciprocal of $2x^2$.**

$$= \frac{8x^3}{2x^2} + \frac{4x^2}{2x^2} - \frac{12x}{2x^2}$$ **Use the Distributive Property.**

$$= 4x^1 + 2x^0 - \frac{6}{x}$$ **Use the division rules for exponents.**

$$= 4x + 2 - \frac{6}{x}$$ **Simplify.**

✓ Quick Check

1 Divide.

a. $(3m^3 - 6m^2 + m) \div 3m^2$

b. $(8t^5 + 16t^3 - 4t^2 + 2t) \div 4t^2$

The process of dividing a polynomial by a binomial is similar to long division. For example, consider dividing 737 by 21.

$$
\begin{array}{r}
35 \\
21\overline{)737} \\
63 \\
\hline
107 \\
105 \\
\hline
2
\end{array}
$$

1. **Divide:** 21 can go into 73 about 3 times.

2. **Multiply** 3 × 21, then **subtract** from 73.

3. **Bring down** the 7. Divide: 107 ÷ 21 ≈ 5.

4. Multiply 5 × 21, and then subtract from 107.

5. The remainder is 2.

$737 \div 21 = 35\frac{2}{21}$

You can summarize the process for long division as

"Divide, multiply, subtract, bring down, and repeat as necessary."

In the division above, the answer is written as a mixed number: $35\frac{2}{21}$ means $35 + \frac{2}{21}$. In dividing polynomials, write the answer as quotient $+ \frac{\text{remainder}}{\text{divisor}}$.

When the divisor and dividend are in standard form, divide the first term of the dividend by the first term of the divisor to find the first term of the quotient.

② EXAMPLE Dividing a Polynomial by a Binomial

Divide $2y^2 + 3y - 40$ by $y + 5$.

Step 1 Begin the long division process.

$\downarrow$ **Align terms by their degrees.**
So put 2y above 3y of the dividend.

$$
\begin{array}{r}
2y \\
y + 5 \overline{)2y^2 + 3y - 40} \\
\underline{2y^2 + 10y} \\
-7y - 40
\end{array}
$$

Divide: Think $2y^2 \div y = 2y$.
Multiply: $2y(y + 5) = 2y^2 + 10y$. Then subtract.
Bring down −40.

Step 2 Repeat the process: divide, multiply, subtract, and bring down.

$$
\begin{array}{r}
2y - 7 \\
y + 5 \overline{)2y^2 + 3y - 40} \\
\underline{2y^2 + 10y} \\
-7y - 40 \\
\underline{-7y - 35} \\
-5
\end{array}
$$

Divide: $-7y \div y = -7$.
Multiply: $-7(y + 5) = -7y - 35$. Then subtract.
The remainder is −5.

The answer is $2y - 7 + \frac{-5}{y + 5}$, or $2y - 7 - \frac{5}{y + 5}$.

Quick Check ② Divide.

    a. $(2b^2 - b - 3) \div (b + 1)$ **b.** $(6m^2 - 5m - 7) \div (2m + 1)$

When the dividend is in standard form and a power is missing, add a term of that power with 0 as its coefficient. For example, rewrite $4b^3 + 5b - 3$, as $4b^3 + 0b^2 + 5b - 3$.

③ EXAMPLE Dividing Polynomials With a Zero Coefficient

Geometry The width and area of a rectangle are shown in the figure. What is an expression for the length?

$w = (2b - 1)$ in.

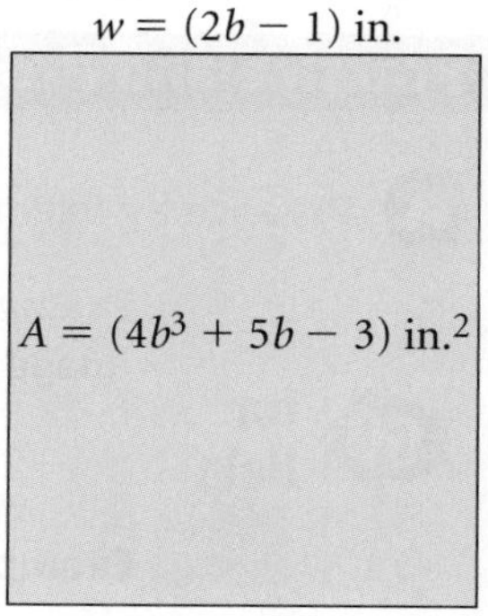

Since $A = \ell w$, divide the area by the width to find the length.

$$
\begin{array}{r}
2b^2 + b + 3 \\
2b - 1 \overline{)4b^3 + 0b^2 + 5b - 3} \\
\underline{4b^3 - 2b^2} \\
2b^2 + 5b \\
\underline{2b^2 - b} \\
6b - 3 \\
\underline{6b - 3} \\
0
\end{array}
$$

Rewrite the dividend with $0b^2$.

$A = (4b^3 + 5b - 3)$ in.²

An expression for the length of the rectangle is $(2b^2 + b + 3)$ in.

Quick Check ③ Divide.

    a. $(t^4 + t^2 + t - 3) \div (t - 1)$ **b.** $(c^3 - 4c + 12) \div (c + 3)$

To use the process for long division, write any divisor or dividend in standard form before you begin to divide.

4 EXAMPLE Reordering Terms and Dividing Polynomials

Divide $-3x + 4 + 9x^2$ by $1 + 3x$.

Rewrite $-3x + 4 + 9x^2$ as $9x^2 - 3x + 4$ and $1 + 3x$ as $3x + 1$. Then divide.

$$
\begin{array}{r}
3x - 2 \\
3x + 1{\overline{\smash{\big)}\,9x^2 - 3x + 4}} \\
\underline{9x^2 + 3x} \\
-6x + 4 \\
\underline{-6x - 2} \\
6
\end{array}
$$

The answer is $3x - 2 + \dfrac{6}{3x + 1}$.

 Quick Check ④ Divide.

a. $\left(10x - 1 + 8x^2\right) \div (1 + 2x)$ b. $\left(9 - 6a^2 - 11a\right) \div (3a - 2)$

 Key Concepts

Summary	Dividing a Polynomial by a Polynomial

Step 1   Arrange the terms of the dividend and divisor in standard form.

Step 2   Divide the first term of the dividend by the first term of the divisor. This is the first term of the quotient.

Step 3   Multiply the first term of the quotient by the divisor and place the product under the dividend.

Step 4   Subtract this product from the dividend.

Step 5   Bring down the next term.

Repeat Steps 2–5 as necessary until the degree of the remainder is less than the degree of the divisor.

EXERCISES

For more exercises, see *Extra Skill and Word Problem Practice.*

Practice and Problem Solving

A Practice by Example

Divide.

Example 1
(page 682)

 for Help

1. $\left(x^6 - x^5 + x^4\right) \div x^2$ 2. $\left(12x^8 - 8x^3\right) \div 4x^4$

3. $\left(9c^4 + 6c^3 - c^2\right) \div 3c^2$ 4. $\left(n^5 - 18n^4 + 3n^3\right) \div n^3$

5. $\left(8q^2 - 32q\right) \div 2q^2$ 6. $\left(-7t^5 + 14t^4 - 28t^3 + 35t^2\right) \div 7t^2$

Example 2
(page 683)

7. $\left(x^2 - 5x + 6\right) \div (x - 2)$ 8. $\left(2t^2 + 3t - 11\right) \div (t - 3)$

9. $\left(n^2 - 5n + 4\right) \div (n - 4)$ 10. $\left(y^2 - y + 2\right) \div (y + 2)$

11. $\left(3x^2 - 10x + 3\right) \div (x - 3)$ 12. $\left(-4q^2 - 22q + 12\right) \div (2q + 1)$

Example 3
(page 683)

13. $\left(5t^2 - 500\right) \div (t + 10)$ 14. $\left(2w^3 + 3w - 15\right) \div (w - 1)$

15. $\left(3b^3 - 10b^2 + 4\right) \div (3b - 1)$ 16. $\left(c^3 - c^2 - 1\right) \div (c - 1)$

17. $\left(t^3 - 6t - 4\right) \div (t + 2)$ 18. $\left(n^3 - 25n - 50\right) \div (n + 2)$

19. Geometry The width of a rectangle is $(r - 5)$ cm and the area is $(r^3 - 24r - 5)$ cm². What is the length?

20. Geometry The base of a triangle is $(c + 2)$ ft and the area is $(2c^3 + 16)$ ft². What is the height? ($Hint:$ The formula for the area of a triangle is $A = \frac{1}{2}bh$.)

Example 4
(page 684)

Divide.

21. $(49 + 16b + b^2) \div (b + 4)$

22. $(a^2 - 6 + 3a) \div (4 + a)$

23. $(39w + 14 + 10w^2) \div (72 + w)$

24. $(4t + t^2 - 9) \div (4 + t)$

25. $(-13x + 2x^3 - 6 - x^2) \div (x - 3)$

26. $(6 - q + 3q^3 - 4q^2) \div (q - 2)$

B Apply Your Skills

27. $(6x^4 + 4x^3 - x^2) \div 2x^3$

28. $(c^3 + 11c^2 - 15c + 8) \div c$

29. $(8b + 2b^3) \div (b - 1)$

30. $(4y + y^3 - 7) \div (y - 5)$

31. $(56a^2 + 4a - 12) \div (2a + 1)$

32. $(5t^4 - 10t^2 + 6) \div (t + 5)$

33. $(3k^3 - 0.9k^2 - 1.2k) \div 3k$

34. $(-7s + 6s^2 + 5) \div (2s + 3)$

35. $(-2z^3 - z + z^2 + 1) \div (z + 1)$

36. $(6m^3 + 3m + 70) \div (m + 4)$

37. $(64c^3 - 125) \div (5 - 4c)$

38. $(21 - 5r^4 - 10r^2 + 2r^6) \div (r^2 - 3)$

39. $(2t^4 - 2t^3 + 3t - 1) \div (2t^3 + 1)$

40. $(z^4 + z^2 - 2) \div (z + 3)$

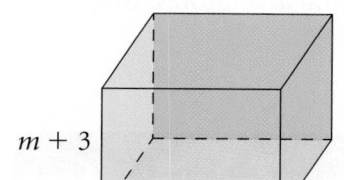

GO Online
Homework Video Tutor
Visit: PHSchool.com
Web Code: ate-1204

41. a. Open-Ended Write a binomial and a trinomial using the same variable.
   b. Divide the trinomial by the binomial.

42. Use the function rule $y = \frac{2x + 5}{x + 3}$.
   a. Rewrite the function rule as a quotient plus a remainder.
   b. Make a table of values and graph the function.
   c. What are the vertical and horizontal asymptotes?

43. Writing Suppose you divide a polynomial by a binomial. Explain how you know if the binomial is a factor of the polynomial.

44. Geometry The volume of the rectangular prism shown at the left is $m^3 + 8m^2 + 19m + 12$. Find the area of the base of the prism.

$m + 3$

45. a. Find $(d^2 - d + 1) \div (d + 1)$.
   b. Find $(d^3 - d^2 + d - 1) \div (d + 1)$.
   c. Find $(d^4 - d^3 + d^2 - d + 1) \div (d + 1)$.
   d. Patterns Predict the result of dividing $d^5 - d^4 + d^3 - d^2 + d - 1$ by $d + 1$.
   e. Verify your prediction by dividing the polynomials.

46. Critical Thinking Find the value of k if $x + 3$ is a factor of $x^2 - x - k$.

47. a. Solve $d = rt$ for t.
   b. Use your answer from part (a) to find an expression for the time it takes to travel a distance of $t^3 - 6t^2 + 5t + 12$ miles at a rate of $t + 1$ miles per hour.

C Challenge

Divide.

48. $(4a^3b^4 - 6a^2b^5 + 10a^2b^4) \div 2ab^2$

49. $(15x^2 + 7xy - 2y^2) \div (5x - y)$

50. $(90r^6 + 28r^5 + 45r^3 + 2r^4 + 5r^2) \div (9r + 1)$

51. $(2b^6 + 2b^5 - 4b^4 + b^3 + 8b^2 - 3) \div (b^3 + 2b^2 - 1)$

52. In Lesson 12-1, you saw horizontal and vertical asymptotes of rational functions. Some rational functions have asymptotes that are neither horizontal nor vertical. Consider the function $f(x) = \frac{x^2 + 2x - 5}{x + 5}$.
  a. Divide $x^2 + 2x - 5$ by $x + 5$.
  b. Rewrite the function using the quotient.
  c. As $|x|$ increases, the remainder will become smaller, approaching zero, though never reaching it. So an asymptote is defined by the quotient without the remainder. Write an equation of that asymptote.
  d. Graph the original equation and the asymptote.

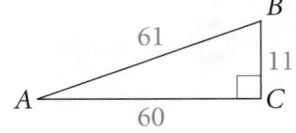

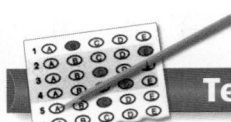

Multiple Choice

53. Which of the following expressions equals $(3x^3 - 4x - 1) \div (x + 1)$?
  A. $3x^2 - 3x - 7 + \frac{6}{x + 1}$ **B.** $3x^2 - 3x - 7 - \frac{8}{x + 1}$
  C. $3x^2 - 3x - 1$ **D.** $3x^2 - 7x + 8$

54. What is the remainder when $x^2 - 4$ is divided by $x - 3$?
  F. -13 **G.** 5 **H.** $\frac{-13}{x - 3}$ **J.** $\frac{5}{x + 3}$

55. Which of the following must be true for $(x^2 + 2x + 1) \div (x + 3)$?
  I. The remainder is negative.
  II. The dividend is in standard form.
  III. The quotient is larger than the divisor for positive values of x.
  A. I only **B.** II only **C.** I and II **D.** II and III

Short Response

56. The volume of a rectangular prism is $2x^3 + 5x^2 + x - 2$. The height of the prism is $2x - 1$, and the length of the prism is $x + 2$. Find the width of the prism. Show your work.

Extended Response

57. a. Write the function $y = \frac{3x + 10}{x + 4}$ as a quotient plus a remainder.
  b. Make a table and find the vertical and horizontal asymptotes. Graph the function.

Mixed Review

Lesson 12-3

Multiply or divide.

58. $\frac{n^2 + 7n - 8}{n - 1} \cdot \frac{n^2 - 4}{n^2 + 6n - 16}$

59. $\frac{6t^2 - 30t}{2t^2 - 53t - 55} \cdot \frac{6t^2 + 35t + 11}{18t^2}$

60. $\frac{3c^2 - 4c - 32}{2c^2 + 17c + 35} \div \frac{c - 4}{c + 5}$

61. $\frac{x^2 + 9x + 20}{x^2 + 5x - 24} \div \frac{x^2 + 15x + 56}{x^2 + x - 12}$

Lesson 11-5

Use $\triangle ABC$ at the right. Find the value of each expression.

62. $\sin A$ **63.** $\cos B$ **64.** $\tan B$

65. $\cos A$ **66.** $\tan A$ **67.** $\sin B$

Lesson 3-8

Find the value of each expression. If the value is irrational, round to the nearest hundredth.

68. $\sqrt{28.9}$ **69.** $\sqrt{289}$ **70.** $-\sqrt{161.29}$ **71.** $\sqrt{4000}$ **72.** $\sqrt{40}$

12-5

Adding and Subtracting Rational Expressions

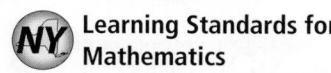

Learning Standards for Mathematics

A.A.17 Add or subtract fractional expressions with monomial or like binomial denominators.

GO for Help Lessons 2-2 and 9-5

✓ **Check Skills You'll Need**

Simplify each expression.

1. $\frac{4}{9} + \frac{2}{9}$

2. $\frac{3}{7} - \frac{5}{7}$

3. $\frac{1}{2} + \left(-\frac{5}{2}\right)$

4. $\frac{5}{6} + \frac{2}{9}$

5. $\frac{1}{4} - \frac{1}{3}$

6. $\frac{5}{12} - \frac{3}{4}$

7. $\frac{4x}{9} + \frac{2x}{9}$

8. $\frac{7x}{12} - \frac{x}{12}$

9. $\frac{7}{12y} - \frac{1}{12y}$

Factor each quadratic expression.

10. $x^2 + 3x + 2$

11. $y^2 + 7y + 12$

12. $t^2 - 4t + 4$

---

1 Adding and Subtracting Rational Expressions With Like Denominators

Adding rational expressions with like denominators is like adding rational numbers with like denominators. If a, b, and c represent polynomials (with $c \neq 0$), then $\frac{a}{c} + \frac{b}{c} = \frac{a + b}{c}$.

1 EXAMPLE Adding Expressions With Like Denominators

Add $\frac{2}{x + 3}$ and $\frac{5}{x + 3}$.

$$\frac{2}{x + 3} + \frac{5}{x + 3} = \frac{2 + 5}{x + 3} \quad \textbf{Add the numerators.}$$

$$= \frac{7}{x + 3} \quad \textbf{Simplify the numerator.}$$

 **① Add.**

a. $\frac{3}{x + 2} + \frac{2}{x + 2}$

b. $\frac{y}{y - 5} + \frac{3y}{y - 5}$

c. $\frac{5n}{n + 1} + \frac{2n}{n + 1}$

Similarly, you can subtract rational expressions with like denominators.

2 EXAMPLE Subtracting Expressions With Like Denominators

Subtract $\frac{2n + 1}{2n^2 + 5n - 3}$ from $\frac{3n + 4}{2n^2 + 5n - 3}$.

$$\frac{3n + 4}{2n^2 + 5n - 3} - \frac{2n + 1}{2n^2 + 5n - 3} = \frac{3n + 4 - (2n + 1)}{2n^2 + 5n - 3} \quad \textbf{Subtract the numerators.}$$

$$= \frac{3n + 4 - 2n - 1}{2n^2 + 5n - 3} \quad \textbf{Use the Distributive Property.}$$

$$= \frac{n + 3}{2n^2 + 5n - 3} \quad \textbf{Simplify the numerator.}$$

$$= \frac{n + 3^{1}}{(2n - 1)_1 (n + 3)} \quad \textbf{Factor the denominator. Divide out the common factor } n + 3.$$

$$= \frac{1}{2n - 1} \quad \textbf{Simplify.}$$

 **Quick Check** ② Subtract.

a. $\dfrac{4}{t-2} - \dfrac{5}{t-2}$ **b.** $\dfrac{7b-2}{3b+6} - \dfrac{b+7}{3b+6}$ **c.** $\dfrac{2c+1}{5m+2} - \dfrac{3c-4}{5m+2}$

2   Adding and Subtracting Rational Expressions With Unlike Denominators

To add or subtract rational expressions with different denominators, you can write the expressions with the least common denominator (LCD), which is the least common multiple (LCM) of the denominators.

<table>
<tr><td>LCM of Whole Numbers</td><td>LCM of Variable Expressions</td></tr>
<tr><td>$4 = 2 \cdot 2$</td><td>$4x = 2 \cdot 2 \cdot x$</td></tr>
<tr><td>$6 = 2 \cdot 3$</td><td>$6x^2 = 2 \cdot 3 \cdot x \cdot x$</td></tr>
<tr><td>$LCM = 2 \cdot 2 \cdot 3 = 12$</td><td>$LCM = 2 \cdot 2 \cdot 3 \cdot x \cdot x = 12x^2$</td></tr>
</table>

3   EXAMPLE   Adding Expressions With Monomial Denominators

Add $\dfrac{2}{3x} + \dfrac{1}{6}$.

Step 1 Find the LCD of $\dfrac{2}{3x}$ and $\dfrac{1}{6}$.

$3x = 3 \cdot x$ **Factor each denominator.**

$6 = 2 \cdot 3$

$LCD = 2 \cdot 3 \cdot x = 6x$.

Step 2 Rewrite using the LCD and add.

$\dfrac{2}{3x} + \dfrac{1}{6} = \dfrac{2 \cdot 2}{2 \cdot 3x} + \dfrac{1 \cdot x}{6 \cdot x}$ **Rewrite each fraction using the LCD.**

$\phantom{\dfrac{2}{3x} + \dfrac{1}{6}} = \dfrac{4}{6x} + \dfrac{x}{6x}$ **Simplify numerators and denominators.**

$\phantom{\dfrac{2}{3x} + \dfrac{1}{6}} = \dfrac{4+x}{6x}$ **Add the numerators.**

 **Quick Check** ③ Add or subtract.

a. $\dfrac{3}{7y^4} + \dfrac{2}{3y^2}$ **b.** $\dfrac{4}{25x} - \dfrac{49}{100}$ **c.** $\dfrac{5}{12b} + \dfrac{15}{36b^2}$

You can also find the LCD of rational expressions that have polynomials with two or more terms in the denominator.

4   EXAMPLE   Adding Expressions With Polynomial Denominators

Add $\dfrac{5}{c+2}$ and $\dfrac{6}{c-3}$.

Step 1 Find the LCD of $c + 2$ and $c - 3$.
Since there are no common factors, the LCD is $(c+2)(c-3)$.

Step 2 Rewrite using the LCD and add.

$\dfrac{5}{c+2} + \dfrac{6}{c-3} = \dfrac{5(c-3)}{(c+2)(c-3)} + \dfrac{6(c+2)}{(c+2)(c-3)}$ **Rewrite the fractions using the LCD.**

$\phantom{\dfrac{5}{c+2} + \dfrac{6}{c-3}} = \dfrac{5c-15}{(c+2)(c-3)} + \dfrac{6c+12}{(c+2)(c-3)}$ **Simplify each numerator.**

$\phantom{\dfrac{5}{c+2} + \dfrac{6}{c-3}} = \dfrac{5c-15+6c+12}{(c+2)(c-3)}$ **Add the numerators.**

$\phantom{\dfrac{5}{c+2} + \dfrac{6}{c-3}} = \dfrac{11c-3}{(c+2)(c-3)}$ **Simplify the numerator.**

GO Online

Video Tutor Help
Visit: PHSchool.com
Web Code: ate-0775

✓ **Quick Check** ④ Add.

   a. $\dfrac{5}{t+4}+\dfrac{3}{t-1}$ **b.** $\dfrac{m}{2m+1}+\dfrac{3}{m-1}$ **c.** $\dfrac{-2}{a+2}+\dfrac{3a}{2a-1}$

You can combine rational expressions to investigate real-world situations.

5 EXAMPLE   Real-World 🌐 Problem Solving

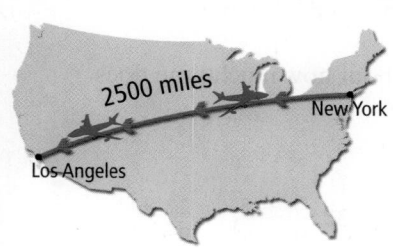

2500 miles

New York

Los Angeles

Air Travel The ground speed for jet traffic from Los Angeles to New York City can be about 15% faster than the ground speed from New York City to Los Angeles. This difference is due to a strong westerly wind at high altitudes. If r is a jet's ground speed from New York City to Los Angeles, write and simplify an expression for the round-trip air time. The two cities are about 2500 miles apart.

NYC to LA time: $\dfrac{2500}{r}$ time $=\dfrac{\text{distance}}{\text{rate}}$

LA to NYC time: $\dfrac{2500}{1.15r}$ time $=\dfrac{\text{distance}}{\text{rate}}$ **15% more than a number is 115% of the number.**

An expression for the total time is $\dfrac{2500}{r}+\dfrac{2500}{1.15r}$.

$$\dfrac{2500}{r}+\dfrac{2500}{1.15r}=\dfrac{2875}{1.15r}+\dfrac{2500}{1.15r}$$ **Rewrite using the LCD, 1.15r.**

$$=\dfrac{5375}{1.15r}$$ **Add the numerators.**

$$\approx\dfrac{4674}{r}$$ **Simplify.**

✓ **Quick Check** ⑤ **Air Travel** The distance between Atlanta, Georgia, and Albuquerque, New Mexico, is about 1270 miles. The ground speed for jet traffic from Atlanta to Albuquerque can be about 12% faster than the ground speed from Albuquerque to Atlanta. Use r for a jet's ground speed. Write and simplify an expression for the round-trip air time.

EXERCISES

For more exercises, see *Extra Skill and Word Problem Practice.*

Practice and Problem Solving

A Practice by Example

Example 1
(page 687)

GO for Help

Example 2
(page 687)

Add or subtract.

1. $\dfrac{5}{2m}+\dfrac{4}{2m}$ **2.** $\dfrac{4}{6t-1}+\dfrac{3}{6t-1}$ **3.** $\dfrac{n}{n+3}+\dfrac{2}{n+3}$

4. $\dfrac{5}{c-5}+\dfrac{9}{c-5}$ **5.** $\dfrac{s^2+3}{4s^2+2}+\dfrac{s^2-2}{4s^2+2}$ **6.** $\dfrac{5c}{2c+7}+\dfrac{c-28}{2c+7}$

7. $\dfrac{1}{2-b}-\dfrac{4}{2-b}$ **8.** $\dfrac{5}{t^2+1}-\dfrac{6}{t^2+1}$ **9.** $\dfrac{3t}{2t-3}-\dfrac{5t}{2t-3}$

10. $\dfrac{2y+1}{y-1}-\dfrac{y+2}{y-1}$ **11.** $\dfrac{3n+2}{n+4}-\dfrac{n-6}{n+4}$ **12.** $\dfrac{3}{b-3}-\dfrac{b}{b-3}$

Example 3
(page 688)

Find the LCD of each pair of expressions.

13. $\dfrac{1}{2},\dfrac{4}{x^2}$ **14.** $\dfrac{b}{6},\dfrac{2b}{9}$ **15.** $\dfrac{1}{z},\dfrac{3}{7z}$ **16.** $\dfrac{8}{5b},\dfrac{12}{7b^3c}$

Add or subtract.

17. $\dfrac{7}{3a}+\dfrac{2}{5}$ **18.** $\dfrac{4}{x}-\dfrac{2}{3}$ **19.** $\dfrac{6}{5x^8}+\dfrac{4}{3x^6}$

20. $\dfrac{3}{8m^3}+\dfrac{1}{12m^2}$ **21.** $\dfrac{27}{n^3}-\dfrac{9}{7n^2}$ **22.** $\dfrac{9}{4x^2}+\dfrac{9}{5}$

Example 4
(page 688)

Add.

23. $\dfrac{9}{m + 2} + \dfrac{8}{m - 7}$ **24.** $\dfrac{a}{a + 3} + \dfrac{4}{a + 5}$ **25.** $\dfrac{a}{a + 3} + \dfrac{a + 5}{4}$

26. $\dfrac{c}{c + 5} + \dfrac{4}{c + 3}$ **27.** $\dfrac{5}{t^2} + \dfrac{4}{t + 1}$ **28.** $\dfrac{3}{2a + 1} + \dfrac{6}{2a - 1}$

Example 5
(page 689)

29. Exercise Jane walks one mile from her house to her grandparents' house. Then she returns home, walking with her grandfather. Her return rate is 70% of her normal walking rate. Let r represent her normal walking rate.
  a. Write an expression for the amount of time Jane spends walking.
  b. Simplify your expression.
  c. Suppose Jane's normal walking rate is 3 mi/h. About how much time does she spend walking?

B **Apply Your Skills**

Add or subtract.

30. $\dfrac{y^2 + 2y - 1}{3y + 1} - \dfrac{2y^2 - 3}{3y + 1}$ **31.** $\dfrac{h^2 + 1}{2t^2 - 7} + \dfrac{h}{2t^2 - 7}$ **32.** $\dfrac{r - 5}{9 + p^3} - \dfrac{2k + 1}{9 + p^3}$

33. $\dfrac{2 - x}{xy^2z} - \dfrac{5 + z}{xy^2z}$ **34.** $\dfrac{k}{2m^2} + \dfrac{3k}{2m}$ **35.** $\dfrac{12}{ab} - \dfrac{15}{bc}$

36. $\dfrac{c^2}{ab} - \dfrac{a^2}{bc}$ **37.** $9 + \dfrac{x - 3}{x + 2}$ **38.** $\dfrac{t}{2t - 3} - 11$

39. $\dfrac{x}{x^2 - 9} - \dfrac{x}{x^2 + 6x + 9}$ **40.** $\dfrac{k - 24}{k^2 - 3k - 18} - \dfrac{3}{k + 3} + \dfrac{k + 1}{k - 6}$

41. Error Analysis A student wrote that $\dfrac{2}{x + 3} + \dfrac{3}{x + 1} = \dfrac{5}{2x + 4}$. What error did the student make?

42. Rowing A rowing team practices rowing 2 mi upstream and 2 mi downstream. The team can row downstream 25% faster than they can row upstream.
  a. Let r represent their rate upstream. Write and simplify an expression for the amount of time they spend rowing.
  b. Let d represent their rate downstream. Write and simplify an expression for the amount of time they spend rowing.
  c. Critical Thinking Do the expressions you wrote in parts (a) and (b) represent the same time? Explain.

43. Writing When adding or subtracting rational expressions, will the answer be in simplest form if you use the LCD? Explain.

Real-World **Connection**

Dragon boat racing has become popular in North American cities in recent years.

44. Open-Ended Write two rational expressions with different denominators. Find the LCD and add the two expressions.

For $f(x) = 8x, g(x) = \dfrac{1}{x}$, **and** $h(x) = \dfrac{4}{x - 5}$, **perform the indicated operation.**

Samples $f(x) + g(x) = 8x + \dfrac{1}{x}$ $f(x) \div g(x) = 8x \div \dfrac{1}{x}$

$\qquad\qquad\qquad = \dfrac{8x}{1}\left(\dfrac{x}{x}\right) + \dfrac{1}{x} \qquad\qquad = 8x\left(\dfrac{x}{1}\right)$

$\qquad\qquad\qquad = \dfrac{8x^2}{x} + \dfrac{1}{x} \qquad\qquad\quad = 8x^2$

$\qquad\qquad\qquad = \dfrac{8x^2 + 1}{x}$

nline
Homework Video Tutor

Visit: PHSchool.com
Web Code: ate-1205

45. $f(x) - g(x)$ **46.** $f(x) \cdot g(x)$ **47.** $g(x) - h(x)$

48. $f(x) \cdot h(x)$ **49.** $g(x) \div h(x)$ **50.** $h(x) \div f(x)$

C **Challenge**

Simplify each expression.

51. $\dfrac{7d - 2}{d^2 + 2d - 8} - \dfrac{4}{d + 4} - \dfrac{d}{d - 2}$ **52.** $\dfrac{7x - 10}{x^3 + x^2 - 10x} - \dfrac{1}{x + 5}$

53. $\dfrac{x^2}{x^2 + x - 12} - \dfrac{x}{x + 4} \cdot \dfrac{3}{x - 3}$ **54.** $\dfrac{2}{a - 5} \cdot \dfrac{a - 1}{a + 2} - \dfrac{a - 2}{2 + a} \cdot \dfrac{3}{5 - a}$

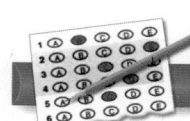
Multiple Choice

55. Subtract $\frac{2x}{3x-2}$ from $\frac{5x}{3x-2}$.

A. -2 B. $\frac{-3x}{3x-2}$ C. $\frac{7x}{3x-2}$ D. $\frac{3x}{3x-2}$

56. What is the least common denominator of $\frac{x}{x^2-1}$ and $\frac{-2}{x-1}$?

F. $x+1$ G. $x-1$ H. x^2-1 J. $(x^2-1)(x-1)$

57. A band director found that he could line up the musicians in the brass section in rows of 4, 5, or 8. What is the least number of brass players in the band?

A. 16 players   B. 24 players   C. 32 players   D. 40 players

Short Response

58. The members of a bicycle club rode a 20-mile round-trip route. On the way back, they had a tail wind and averaged 3 mi/h faster than on the first 10 miles of the trip.

a. Use r for the rate. Write an expression for the total ride time. Simplify the expression.

b. Suppose the bicyclists averaged a rate of 12 mi/h for the first half of the ride. How long did the round trip take? Show your work.

Mixed Review

Go for Help

Lesson 12-4

Divide.

59. $(2x^4+8x^3-4x^2) \div 4x^2$

60. $(10b+5b^3) \div (b+2)$

Lesson 11-3

Solve each radical equation. Check your answers. If there is no solution, write *no solution*.

61. $x = \sqrt{5x+6}$ **62.** $n = \sqrt{24-5n}$ **63.** $\sqrt{16y} = -8$

Lesson 10-3

Solve each equation by finding square roots. Round to the nearest tenth. If the equation has no solution, write *no solution*.

64. $a^2 - 48 = 0$ **65.** $2n^2 = 30$ **66.** $3p^2 + 60 = 0$

Algebra at Work

· Electrician

More than half a million men and women work as electricians. All are highly skilled technicians licensed by the states in which they work. Electricians use formulas containing rational expressions. For example, when a circuit connected in parallel contains two resistors with resistances R_1 and R_2 ohms, the total resistance R_T (in ohms) of the circuit can be found using the formula $\frac{1}{R_T} = \frac{1}{R_1} + \frac{1}{R_2}$.

Go Online
PHSchool.com

For: Information about a career as an electrician
Web Code: atb-2031

Solving Rational Equations

NY Learning Standards for Mathematics

A.A.20 Factor algebraic expressions completely, including trinomials with a lead coefficient of one (after factoring a GCF).

A.A.26 Solve algebraic proportions in one variable which result in linear or quadratic equations.

✓ **Check Skills You'll Need**

GO for Help Lessons 3-4 and 12-5

Solve each proportion.

1. $\frac{1}{x} = \frac{3}{5}$

2. $\frac{3}{t} = \frac{5}{2}$

3. $\frac{m}{3} = \frac{27}{m}$

Find the LCD of each group of expressions.

4. $\frac{3}{4n}; \frac{1}{2}; \frac{2}{n}$

5. $\frac{1}{3x}; \frac{2}{5}; \frac{4}{3x}$

6. $\frac{1}{8y}; \frac{1}{y^2}; \frac{5}{6}$

🔊 **New Vocabulary** • rational equation

---

1 Solving Rational Equations

A **rational equation** contains one or more rational expressions. One method for solving rational equations is similar to the method you learned in Chapter 2 for solving equations with rational numbers.

$$\frac{1}{2}x + \frac{3}{10} = \frac{1}{5} \qquad \text{The denominators are 2, 10, and 5. The LCD is 10.}$$

$$10\left(\frac{1}{2}x + \frac{3}{10}\right) = 10\left(\frac{1}{5}\right) \qquad \text{Multiply each side by 10.}$$

$$^{5}10\left(\frac{1}{2_1}x\right) + {}^{1}10\left(\frac{3}{10_1}\right) = {}^{2}10\left(\frac{1}{5_1}\right) \qquad \text{Use the Distributive Property.}$$

$$5x + 3 = 2 \qquad \text{No fractions! This equation is easier to solve.}$$

GO Online

Video Tutor Help

Visit: PHSchool.com
Web Code: ate-0775

1 EXAMPLE Solving Equations With Rational Expressions

Solve $\frac{1}{2x} + \frac{3}{10} = \frac{1}{5x}$. Check the solution.

$$\frac{1}{2x} + \frac{3}{10} = \frac{1}{5x} \qquad \begin{array}{l}\text{The denominators are } 2x, 10, \text{ and } 5x.\\ \text{The LCD is } 10x.\end{array}$$

$$10x\left(\frac{1}{2x} + \frac{3}{10}\right) = 10x\left(\frac{1}{5x}\right) \qquad \text{Multiply each side by } 10x.$$

$$^{5}10x\left(\frac{1}{2x_1}\right) + {}^{1}10x\left(\frac{3}{10_1}\right) = {}^{2}10x\left(\frac{1}{5x_1}\right) \qquad \text{Use the Distributive Property.}$$

$$5 + 3x = 2 \qquad \text{No rational expressions! Now you can solve.}$$

$$3x = -3 \qquad \text{Subtract 5 from each side.}$$

$$x = -1 \qquad \text{Divide each side by 3, and then simplify.}$$

Check See if -1 makes $\frac{1}{2x} + \frac{3}{10} = \frac{1}{5x}$ true.

$$\frac{1}{2(-1)} + \frac{3}{10} \stackrel{?}{=} \frac{1}{5(-1)}$$

$$-\frac{5}{10} + \frac{3}{10} = -\frac{1}{5} \checkmark$$

✓ **Quick Check**  **1** Solve each equation. Check your solution.

a. $\frac{1}{3} + \frac{1}{3x} = \frac{1}{6}$

b. $\frac{4}{c} = \frac{3}{2c} - \frac{1}{5}$

To solve some rational equations, you need to factor a quadratic expression.

2 EXAMPLE **Solving by Factoring**

Solve $\frac{5}{x^2} = \frac{6}{x} - 1$. Check the solution.

$$x^2\left(\frac{5}{x^2}\right) = x^2\left(\frac{6}{x} - 1\right)$$ **Multiply each side by the LCD, x^2.**

$$^1x^2\left(\frac{5}{_1x^2}\right) = {}^xx^2\left(\frac{6}{x_1}\right) - x^2(1)$$ **Use the Distributive Property.**

$$5 = 6x - x^2$$ **Simplify.**

$$x^2 - 6x + 5 = 0$$ **Collect terms on one side.**

$$(x - 5)(x - 1) = 0$$ **Factor the quadratic expression.**

$$(x - 5) = 0 \quad \text{or} \quad (x - 1) = 0$$ **Use the Zero-Product Property.**

$$x = 5 \quad \text{or} \quad x = 1$$ **Solve.**

Check $\frac{5}{5^2} \stackrel{?}{=} \frac{6}{5} - 1 \qquad \frac{5}{1^2} \stackrel{?}{=} \frac{6}{1} - 1$

$\frac{1}{5} = \frac{1}{5}$ ✓ $\qquad\qquad 5 = 5$ ✓

✓ Quick Check **2** Solve each equation. Check your solution.

a. $\frac{5}{m} = \frac{2}{m^2} + 2$ **b.** $t - 2 = \frac{8 - 2t}{t - 1}$

You can solve a work problem by finding the part of the job each person does in one unit of time (hour or minute). Find the part of the job each person does, and then write an equation.

3 EXAMPLE **Work-Rate Problem**

Volunteerism Max can wash and wax a car in 60 min. His older sister Kayla can do the same job in 45 min. How long will the car take them if they work together?

Define Let n = the time to complete the job if they work together (in minutes).

Relate

| fraction of job Max can do in 1 minute | + | fraction of job Kayla can do in 1 minute | = | fraction of job completed in 1 minute |

Write $\qquad \frac{1}{60} \qquad + \qquad \frac{1}{45} \qquad = \qquad \frac{1}{n}$

$180n\left(\frac{1}{60} + \frac{1}{45}\right) = 180n\left(\frac{1}{n}\right)$ **Multiply each side by the LCD, $180n$.**

$3n + 4n = 180$ **Use the Distributive Property.**

$7n = 180$ **Simplify.**

$n = \frac{180}{7}$, or $25\frac{5}{7}$ **Simplify.**

It will take the two of them about 26 minutes to wash the car working together.

Check Max will do $\frac{180}{7} \cdot \frac{1}{60} = \frac{3}{7}$ of the job, and Kayla will do $\frac{180}{7} \cdot \frac{1}{45} = \frac{4}{7}$ of the job. Together, they will do $\frac{3}{7} + \frac{4}{7} = 1$, or the whole job.

Real-World Connection

Many student groups hold car washes to raise money.

✓ Quick Check **3** Peggy can pick a bushel of apples in 45 min. Peter can pick a bushel of apples in 75 min. How long will it take them to pick a bushel if they work together?

Some rational equations are proportions. You can solve them by using cross products.

4 EXAMPLE **Solving a Rational Proportion**

Solve $\frac{2}{x} = \frac{1}{x + 4}$. Check the solution.

$$\frac{2}{x} = \frac{1}{x + 4}$$

$$2(x + 4) = x(1) \qquad \textbf{Write cross products.}$$

$$2x + 8 = x \qquad \textbf{Use the Distributive Property.}$$

$$x = -8 \qquad \textbf{Solve for } x.$$

Check $\quad \frac{2}{x} = \frac{1}{x + 4}$

$$\frac{2}{-8} \overset{?}{=} \frac{1}{-8 + 4}$$

$$-\frac{1}{4} = -\frac{1}{4} \checkmark$$

Quick Check **4** Solve each equation. Check your solution.

a. $\frac{3}{a} = \frac{5}{a - 2}$

b. $\frac{n}{5} = \frac{4}{n + 1}$

The process of cross multiplying or multiplying by the LCD may give an extraneous solution to an equation, or a solution that makes a denominator in the original equation equal zero. An extraneous solution solves a new equation but not the original one. So you must check your solutions.

5 EXAMPLE **Checking to Find an Extraneous Solution**

Solve $\frac{-2}{x - 2} = \frac{x - 4}{x^2 - 4}$.

$$\frac{-2}{x - 2} = \frac{x - 4}{x^2 - 4}$$

$$-2(x^2 - 4) = (x - 2)(x - 4) \qquad \textbf{Write cross products.}$$

$$-2x^2 + 8 = x^2 - 6x + 8 \qquad \textbf{Simplify each side of the equation.}$$

$$0 = 3x^2 - 6x \qquad \textbf{Collect terms on one side.}$$

$$0 = 3x(x - 2) \qquad \textbf{Factor.}$$

$$0 = 3x \text{ or } 0 = x - 2 \qquad \textbf{Use the Zero-Product Property.}$$

$$x = 0 \qquad x = 2 \qquad \textbf{Solve.}$$

Check $\frac{-2}{x - 2} = \frac{x - 4}{x^2 - 4}$

$$\frac{-2}{0 - 2} \overset{?}{=} \frac{0 - 4}{0^2 - 4} \qquad\qquad \frac{-2}{2 - 2} \overset{?}{=} \frac{2 - 4}{2^2 - 4}$$

$$\frac{-2}{-2} \overset{?}{=} \frac{-4}{-4} \qquad\qquad \frac{-2}{0} = \frac{-2}{0} \qquad \textbf{Undefined!}$$

$$1 = 1 \checkmark$$

The equation has one solution, 0.

Quick Check **5** Solve each equation. Check your solutions.

a. $\frac{2}{c^2} = \frac{2}{c^2 + 1}$

b. $\frac{w^2}{w - 1} = \frac{1}{w - 1}$

 **Practice by Example**

Examples 1, 2
(pages 692, 693)

 for Help

Solve each equation. Check your solutions.

1. $\frac{1}{2} + \frac{2}{x} = \frac{1}{x}$

2. $5 + \frac{2}{p} = \frac{17}{p}$

3. $\frac{3}{a} - \frac{5}{a} = 2$

4. $y - \frac{6}{y} = 5$

5. $\frac{5}{2s} + \frac{3}{4} = \frac{9}{4s}$

6. $\frac{1}{t-2} = \frac{t}{8}$

7. $\frac{2}{c-2} = 2 - \frac{4}{c}$

8. $\frac{5}{3p} + \frac{2}{3} = \frac{5+p}{2p}$

9. $\frac{8}{x+3} = \frac{1}{x} + 1$

10. $\frac{v+2}{v} + \frac{4}{3v} = 11$

11. $\frac{4}{3(c+4)} + 1 = \frac{2c}{c+4}$

12. $\frac{3+a}{2a} = \frac{1}{3} + \frac{5}{6a}$

13. $7 + \frac{3}{x} = \frac{7}{x} + 9$

14. $\frac{a}{a+3} = \frac{2a}{a-3} - 1$

15. $\frac{z}{z+2} - \frac{1}{z} = 1$

Example 3
(page 693)

16. Gardening Marian can weed a garden in 3 hours. Robin can weed the same garden in 4 hours. How long will the weeding take them if they work together?

17. David can unload a delivery truck in 20 min. Allie can unload a delivery truck in 35 min. How long will the unloading take them if they work together?

Examples 4, 5
(page 694)

Solve each equation. Check your solutions. If there is no solution, write *no solution*.

18. $\frac{5}{x+1} = \frac{x+2}{x+1}$

19. $\frac{4}{c+4} = \frac{c}{c+25}$

20. $\frac{3}{m-1} = \frac{2m}{m+4}$

21. $\frac{2x+4}{x-3} = \frac{3x}{x-3}$

22. $\frac{30}{x+3} = \frac{30}{x-3}$

23. $\frac{x+2}{x+4} = \frac{x-2}{x-1}$

 **Apply Your Skills**

Solve each equation. Check your solutions.

24. $\frac{2r}{r-4} - 2 = \frac{4}{r+5}$

25. $6 - \frac{2}{b} = \frac{-5}{b-3}$

26. $\frac{r+1}{r-1} = \frac{r}{3} + \frac{2}{r-1}$

27. $\frac{3}{s-1} + 1 = \frac{12}{s^2-1}$

28. $\frac{d}{d+2} - \frac{2}{2-d} = \frac{d+6}{d^2-4}$

29. $\frac{u+1}{u} + \frac{1}{2u} = 4$

30. $\frac{s}{3s+2} + \frac{s+3}{2s-4} = \frac{-2s}{3s^2-4s-4}$

31. $\frac{u+1}{u+2} = \frac{-1}{u-3} + \frac{u-1}{u^2-u-6}$

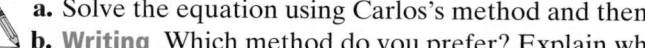 **GO for Help**

For a guide to solving
Exercise 32, see p. 698.

32. Two pipes fill a storage tank in 9 hours. The smaller pipe takes 3 times as long to fill the tank as the larger pipe. How long would it take the larger pipe to fill the tank alone?

33. A teacher assigned the equation $\frac{40}{x} = \frac{15}{x-20}$. Carlos studied the equation and said, "I'll start by finding the LCD." Ingrid studied the equation and said, "I'll start by cross multiplying."
   a. Solve the equation using Carlos's method and then Ingrid's method.
   b. Writing Which method do you prefer? Explain why.
   c. Critical Thinking Will Ingrid's method work for all rational equations? Explain.

34. Find the value of each variable. $\begin{bmatrix} \frac{5a}{3} & \frac{7}{3b} \\ 2c-15 & \frac{5}{2d} + \frac{3}{4} \end{bmatrix} = \begin{bmatrix} 2 + \frac{7a}{6} & 9 \\ \frac{1}{5c} & \frac{9}{4d} \end{bmatrix}$

35. a. Write two functions using the expressions on the two sides of the equation $\frac{6}{x^2} + 1 = \frac{(x+7)^2}{6}$. Graph the functions.
   b. Find the coordinates of the points of intersection.
   c. Are the x-values of the points of intersection solutions to the equation? Explain.

 **GO Online**
Homework Video Tutor
Visit: PHSchool.com
Web Code: ate-1206

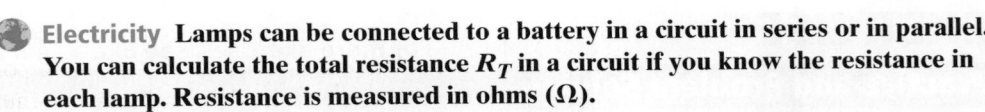 **Electricity** Lamps can be connected to a battery in a circuit in series or in parallel. You can calculate the total resistance R_T in a circuit if you know the resistance in each lamp. Resistance is measured in ohms (Ω).

Circuit connected in series.

Circuit connected in parallel.

Real-World Connection

The resistance of a conductor is the opposition it gives to the flow of electrical current through it.

Circuit Connected in Series

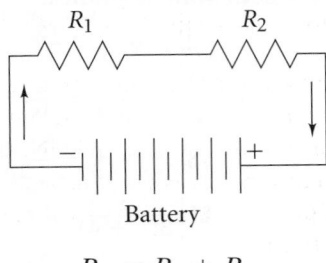

Battery

$$R_T = R_1 + R_2$$

36. Find R_T.

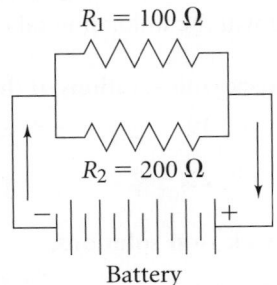

$R_1 = 100\ \Omega$

$R_2 = 200\ \Omega$

Battery

38. Find R_T. (*Hint:* The lamps in this circuit are connected in series *and* in parallel.)

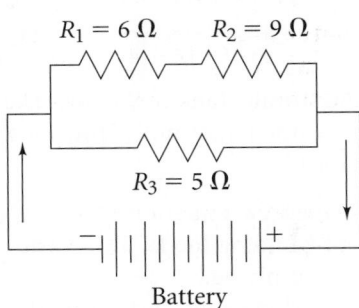

$R_1 = 6\ \Omega \quad R_2 = 9\ \Omega$

$R_3 = 5\ \Omega$

Battery

Circuit Connected in Parallel

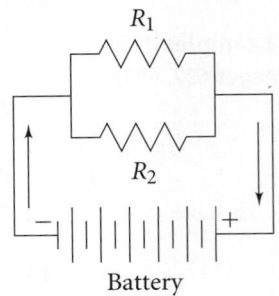

R_1

R_2

Battery

$$\frac{1}{R_T} = \frac{1}{R_1} + \frac{1}{R_2}$$

37. $R_T = 12\ \Omega$; find R_2.

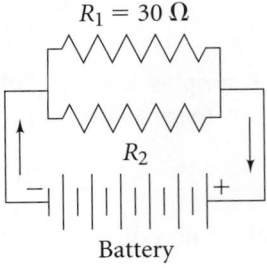

$R_1 = 30\ \Omega$

R_2

Battery

39. $R_T = 10\ \Omega$; find R_2.

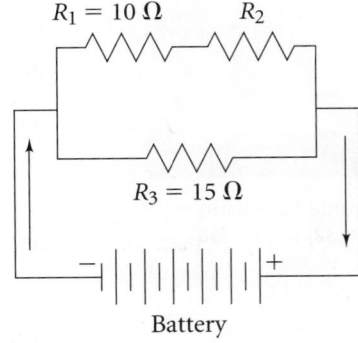

$R_1 = 10\ \Omega \quad R_2$

$R_3 = 15\ \Omega$

Battery

40. Open-Ended Write a rational equation that has 3 as a solution.

41. Travel A plane flies 450 mi/h. It can travel 980 miles with a wind in the same amount of time as it travels 820 mi against the wind. Solve the equation $\frac{980}{450 + s} = \frac{820}{450 - s}$ to find the speed s of the wind.

 **Challenge** **Solve each equation. Be sure to check your answers.**

42. $\dfrac{x - 6}{x + 3} + \dfrac{2x}{x - 3} = \dfrac{4x + 3}{x + 3}$

43. $\dfrac{n}{n - 2} + \dfrac{n}{n + 2} = \dfrac{n}{n^2 - 4}$

44. $\dfrac{2}{r} + \dfrac{1}{r^2} + \dfrac{r^2 + r}{r^3} = \dfrac{1}{r}$

45. $\dfrac{3}{t} - \dfrac{t^2 - 2t}{t^3} = \dfrac{4}{t^2}$

46. It takes Jon 75 min to paint a room. It takes Jeff 60 min and Jackie 80 min each to paint the same room. How long will the painting take if all three work together?

47. Chemistry A chemist has one solution that is 80% acid and a second solution that is 30% acid. The chemist needs to mix some of both solutions to make 50 liters of a solution that is 62% acid. Let s = the number of liters of the 80% solution used in the mixture.

  a. Write an expression for the amount of acid in s liters of the 80% solution.
  b. Write an expression for the number of liters of the 30% acid used in the mixture.
  c. Write an expression for the amount of acid in a 30% acid solution.
  d. Write an equation that combines the amount of acid in each solution to make the total amount of acid in 50 liters of 62% acid solution.
  e. Solve the equation you wrote in part (d).
  f. How many liters of each solution will the chemist need to make 50 liters of 62% acid solution?

48. Sumi can wash the windows of an office building in $\frac{3}{4}$ the time it takes her apprentice. One day they worked on a building together for 2 h 16 min, and then Sumi continued alone. It took 4 h 32 min more to complete the job. How long would it take her apprentice to wash all the windows alone?

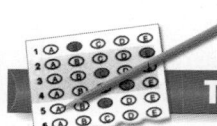

NY REGENTS

Test Prep

Multiple Choice

49. Which is a solution of $\frac{2}{n} + \frac{1}{2} = \frac{1}{n}$?
  A. -4 **B.** -2 **C.** 2   **D.** 4

50. Which inequality contains both solutions of $x = \frac{1}{2} + \frac{3}{x}$?
  F. $-1 < x < 3$ **G.** $-2 < x \le 2$ **H.** $-2 \le x < 0$ **J.** $-3 \le x \le -1$

51. What is the least common denominator of $\frac{1}{x}$, $\frac{x}{3}$, and $\frac{3}{2x}$?
  A. $2x$ **B.** $3x$ **C.** $6x$ **D.** $6x^2$

Short Response

52. You are trying to find the number you would add to both the numerator and denominator of $\frac{3}{16}$ to make a fraction equal to $\frac{1}{2}$. Write a rational equation that can be used to find this number. Then solve the equation to find the number and check your solution.

Mixed Review

Lesson 12-5

Add or subtract.

53. $\dfrac{5}{x^2 y^2 z} - \dfrac{8}{x^2 y^2 z}$ **54.** $\dfrac{3h^2}{2t^2 - 8} + \dfrac{h}{t - 2}$ **55.** $\dfrac{k - 11}{k^2 + 6k - 40} - \dfrac{5}{k - 4}$

Lesson 11-4

Graph each function either by translating the graph of $y = \sqrt{x}$ or by making a table of values.

56. $f(x) = -2\sqrt{x}$ **57.** $y = \sqrt{x + 7}$ **58.** $f(x) = \sqrt{x - 2} - 8$
59. $y = \sqrt{0.25x}$ **60.** $y = \sqrt{2x} + 3$ **61.** $y = \sqrt{4x - 2} - 2$

Lesson 10-4

Solve each equation by factoring.

62. $x^2 + 23x + 90 = 0$ **63.** $x^2 - 19x + 88 = 0$ **64.** $x^2 + 22x - 23 = 0$
65. $x^2 + 2x - 48 = 0$ **66.** $x^2 + 52 = -17x$ **67.** $x^2 + 92x - 9 = -100$

Understanding Word Problems Read the exercise below and then follow along with what Deon thinks and writes. Check your understanding with the exercise at the bottom of the page.

Two pipes fill a storage tank in 9 hours. The smaller pipe takes 3 times as long to fill the tank as the larger pipe. How long would it take the larger pipe to fill the tank alone?

What Deon Thinks

First I'll write down the important information from the problem.

I'll define a variable. I'll use the time it takes the larger pipe to fill the tank alone as the variable.

It would take the small pipe three times longer to fill the tank.

Next I'll write an equation for the amount of work done in one hour.

I'll multiply both sides of the equation by the LCD of 9, n, and $3n$, which is $9n$. Then I'll simplify and solve the equation.

Finally, I write my answer.

What Deon Writes

time to fill is 9 hours

large pipe fills the tank 3 times faster than small one

n = time in hours for large pipe to fill the tank alone.

$3n$ = time in hours for small pipe to fill the tank alone.

$$\frac{1}{9} = \frac{1}{n} + \frac{1}{3n}$$

$$9n\left(\frac{1}{9}\right) = 9n\left(\frac{1}{n} + \frac{1}{3n}\right)$$

$$n = 9 + 3$$

$$n = 12$$

It takes 12 hours for the larger pipe to fill the tank alone.

EXERCISES

1. Sam can shovel the driveway in 30 minutes. His cousin Dawn can shovel the same driveway in 50 minutes. How long will it take them if they work together?

2. A homeowner uses two pumps to pump out water from a flooded basement. Together they take 160 minutes to complete the job. The larger pump works twice as fast as the smaller pump. How long would it take the smaller pump to complete the job alone?

12-7

Counting Methods and Permutations

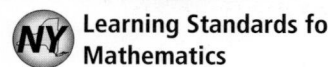
Learning Standards for Mathematics

A.N.7 Determine the number of possible events, using counting techniques or the Fundamental Principle of Counting.

A.N.8 Determine the number of possible arrangements (permutations) of a list of items.

 Check Skills You'll Need

GO for Help Lessons 2-6 and 2-7

You roll a number cube. Find each probability.

1. P(even number)

2. P(prime number)

3. P(a number greater than 5)

4. P(a negative number)

You roll a blue number cube and a yellow number cube. Find each probability.

5. P(blue 1 and yellow 2)

6. P(blue even and yellow odd)

🔊 **New Vocabulary** • **Multiplication Counting Principle** • **permutation**

---

1 | **Using the Multiplication Counting Principle**

Activity: Determining Order

Suppose you have to read the following works over the summer for English class next year.

1. In how many different orders can you read the books?

2. Describe how you determined the number of different ways you could order the books.

3. Suppose you choose one of the orders at random. What is the probability that you will read the books in alphabetical order by title? By author?

You can find the possible orders of objects by making an organized list. Another way is to make a tree diagram. Both methods help you see if you have thought of all of the possibilities.

Suppose you have three shirts and two pair of pants that coordinate well. Make a tree diagram to find the number of possible outfits you have.

Shirts	Pants	Outfits
Shirt 1 →	Pants 1 →	Shirt 1, Pants 1
	Pants 2 →	Shirt 1, Pants 2
Shirt 2 →	Pants 1 →	Shirt 2, Pants 1
	Pants 2 →	Shirt 2, Pants 2
Shirt 3 →	Pants 1 →	Shirt 3, Pants 1
	Pants 2 →	Shirt 3, Pants 2

● There are six possible outfits.

Quick Check ① **a.** Suppose you have two T-shirts and four pairs of shorts you could bring for gym class. Make a tree diagram to find the number of possible outfits for gym.

b. Critical Thinking Would you want to use a tree diagram to find the number of outfits for five T-shirts and eight pairs of shorts? Explain.

Recall that when one event does not affect the result of a second event, the events are *independent*. When events are independent, you can find the number of outcomes using the Multiplication Counting Principle.

Key Concepts

Rule	Multiplication Counting Principle

If there are m ways to make a first selection and n ways to make a second selection, there are $m \times n$ ways to make the two selections.

Example For five shirts and eight pairs of shorts, the number of possible outfits is $5 \cdot 8 = 40$.

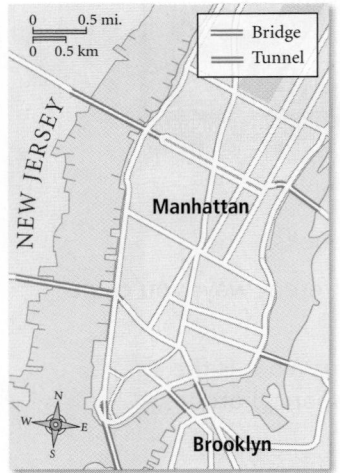

0   0.5 mi.
0   0.5 km

═══ Bridge
═══ Tunnel

NEW JERSEY

Manhattan

Brooklyn

② **EXAMPLE** Using the Multiplication Counting Principle

Travel On the map there are two tunnels for cars going from New Jersey to Manhattan, New York, and three bridges from Manhattan to Brooklyn, New York. How many routes using a tunnel and then a bridge are there from New Jersey to Brooklyn through Manhattan?

$$2 \cdot 3 = 6 \leftarrow \text{routes from New Jersey to Brooklyn through Manhattan}$$

routes by tunnel, from New Jersey to Manhattan   **routes by bridge, from Manhattan to Brooklyn**

● There are six possible routes from New Jersey to Brooklyn.

Quick Check ② Pizza At the neighborhood pizza shop, there are five vegetable toppings and three meat toppings for a pizza. How many different pizzas can you order with one meat and one vegetable topping?

One kind of counting problem is to find the number of possible arrangements of the objects in a set. Here are the possible arrangements for the letters A, B, and C without repeating any letters.

ABC   BAC   CAB   ACB   BCA   CBA

Each of the arrangements is a permutation. A **permutation** is an arrangement of objects in a specific order.

Vocabulary Tip

A short way to write the product in Example 3 is 9!, read "nine factorial." For positive integers, <u>factorial</u> means the product of integers from the given number to 1.

3 EXAMPLE Counting Permutations

Baseball How many different batting orders can you have with 9 baseball players?

There are 9 choices for the first batter, 8 for the second, 7 for the third, and so on.

$9 \cdot 8 \cdot 7 \cdot 6 \cdot 5 \cdot 4 \cdot 3 \cdot 2 \cdot 1 = 362{,}880$ **Use a calculator.**

There are 362,880 possible batting orders.

 **Quick Check** **3** **Swimming** A swimming pool has eight lanes. In how many ways can eight swimmers be assigned lanes for a race?

In how many ways can you select a right, center, and left fielder from eight players on a baseball team? To answer this question, you need to find the number of permutations of 8 objects (players) arranged 3 at a time.

Key Concepts

Definition	Permutation Notation

The expression $_nP_r$ represents the number of permutations of n objects arranged r at a time.

n will be the first factor.

$$_nP_r = \underbrace{n(n-1)(n-2) \ldots}$$

Stop when you have r factors.

Example $_8P_3$ represents 8 objects (players) chosen 3 at a time, or $8 \cdot 7 \cdot 6 = 336$.

4 EXAMPLE Using Permutation Notation

Go Online
PHSchool.com

For: Graphing calculator procedures
Web Code: ate-2116

Simplify $_7P_4$.

Method 1 Use pencil and paper.

$_7P_4 = 7 \cdot 6 \cdot 5 \cdot 4$ **The first factor is 7, and there are 4 factors.**

$= 840$ **Simplify.**

Method 2 Use a graphing calculator. Use MATH to select nPr in the PRB screen.

$_7P_4 = 840$

```
7 nPr 4
              840
```

 **Quick Check** **4** Simplify each expression.

a. $_9P_3$ **b.** $_7P_3$ **c.** $_5P_2$

In some situations, order matters, but repetition is allowed. In that case, use the Multiplication Counting Principle. However, when repetition is not allowed, use permutations to find the number of arrangements.

5 **EXAMPLE**  **Real-World** Problem Solving

Computers Suppose you use six different letters to make a computer password. Find the number of possible six-letter passwords.

There are 26 letters in the alphabet. You are finding the number of permutations of 26 different letters arranged 6 at a time.

$$_{26}P_6 = 26 \cdot 25 \cdot 24 \cdot 23 \cdot 22 \cdot 21 \qquad \textbf{Use a calculator.}$$
$$= 165,765,600$$

● There are 165,765,600 six-letter passwords in which letters do not repeat.

Real-World **Connection**

Using 100,000 encryption operations per second, a 6-letter password can be cracked in about 50 min.

 **Quick Check**

5 **a.** Suppose your cousin needs to choose a four-digit number to use with a new debit card. Find the number of possible four-digit numbers without repeating a digit.

b. Critical Thinking Is a six-letter password or a six-digit number harder for someone to guess? Explain.

EXERCISES

For more exercises, see *Extra Skill and Word Problem Practice*.

Practice and Problem Solving

A Practice by Example

Example 1
(page 700)

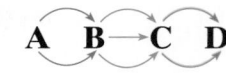

1. Job Interview James must wear a shirt and tie for a job interview. He has two dress shirts and five ties. Use a tree diagram to find the number of shirt-tie choices he has.

2. Catering Suppose your grandparents are planning an anniversary party. The caterer offers the following choices for a menu. Guests can have either spinach salad or chef's salad, and either vegetable soup or chicken soup. The main course is chicken, beef, or salmon. Use a tree diagram to find the number of possible menus.

Example 2
(page 700)

3. Telephones A seven-digit telephone number can begin with any digit except 0 or 1.
   a. How many possible choices are there for the first digit? The second digit? The third digit? The seventh digit?
   b. How many different seven-digit telephone numbers are possible?

4. Use the diagram and the Multiplication Counting Principle to find each of the following:
   a. the number of routes from A to C
   b. the number of routes from A to D

A B C D

Example 3
(page 701)

5. Sports In ice-skating competitions, the order in which competitors skate is determined by a drawing. Suppose there are ten skaters in the finals. How many different orders are possible for the final program?

6. Suppose you are lining up with four cousins for a photo. How many different arrangements are possible?

Example 4
(page 701)

Simplify each expression.

7. $_8P_4$ **8.** $_9P_4$ **9.** $_6P_4$ **10.** $_5P_4$

11. $_7P_7$ **12.** $_7P_6$ **13.** $_7P_5$ **14.** $_7P_2$

Example 5
(page 702)

15. On a bookshelf there are ten books: five novels, two volumes of short stories, and three biographies. In how many ways can you select four books to read in order?

16. A student council has 24 members. A 3-person committee must arrange a car wash. Each person on the committee will have a task: one person will find a location, another person will organize publicity, and the third person will schedule workers. In how many different ways can three students be chosen and given a job?

 Apply Your Skills

Which is greater?

17. $_8P_6$ or $_6P_2$ **18.** $_9P_7$ or $_9P_2$ **19.** $_{10}P_3$ or $_8P_4$

Use the tiles at the left for Exercises 20 and 21.

20. a. How many vowels are there? How many consonants are there?
b. In how many ways can you choose at random a vowel and then a consonant?
c. **Critical Thinking** Would your answer to part (b) be different if you chose a consonant first and then a vowel? Explain.

21. a. What is the number of possible arrangements in which you can select three letters?
b. Find the probability that you select C, A, and then R.

22. a. Lena wants a password that uses the four letters of her name. How many permutations are possible using each letter only once?
b. A four-letter password with no repeated letters is assigned randomly from the alphabet. What is the probability that it uses the letters L, E, N, and A?
 c. **Writing** Is creating a password based on your name a good idea? Explain your reasoning.

Simplify each expression.

23. $\dfrac{_5P_3}{_5P_2}$ **24.** $\dfrac{_4P_3}{_4P_2}$ **25.** $\dfrac{_7P_3}{_7P_2}$

26. Multiple Choice Suppose your cousin needs to choose a six-digit number to use with a new debit card, and repetition is allowed. How many six-digit passwords are possible?

Ⓐ 1,000,000 Ⓑ 151,200 Ⓒ 720 Ⓓ 60

27. License Plates In Indiana, a regular license plate has two numbers that are fixed by county, then one letter, and then four numbers.
a. How many different license plates are possible in each county?
b. There are 92 counties in Indiana. How many license plates are possible in the entire state?

28. a. **Open-Ended** Use the letters of your last name. If any letters are repeated, use only one of them. For example, for the last name Bell, use the letters B, E, and L. In how many ways can the resulting letters be arranged?
b. In how many ways can two different letters be selected and arranged from your last name?

Homework Video Tutor
Visit: PHSchool.com
Web Code: ate-1207

29. Travel The International Airline Transportation Association (IATA) assigns three-letter codes to each airport. For example, LAX is the code for the Los Angeles International Airport. Letters can be repeated within a code.
 a. How many possible codes are there?
 b. Of the possible codes, the IATA has kept 50 codes for internal purposes. How many codes are available for airports?

 30. Radio Stations The call letters of radio and television stations in the United States generally begin with the letter W east of the Mississippi River and the letter K west of the Mississippi. Repetition of letters is allowed.
 a. How many different call letters are possible if each station uses a W or K followed by three letters?
 b. How many different call letters are possible if each station uses a W or K followed by four letters?

 **Real-World Connection**

NASDAQ (NAZ dak) started as an acronym for National Association of Securities Dealers Automated Quotations. Today its volume of trade is larger than the New York Stock Exchange.

 31. Stock Exchange The companies listed on the New York Stock Exchange have a one- to three-letter ticker symbol. The companies listed on the NASDAQ exchange have a four- or five-letter symbol. Letters can be repeated in a ticker symbol.
 a. How many companies can be listed on the New York Stock Exchange?
 b. How many companies can be listed on the NASDAQ Exchange?
 c. Which stock exchange can list more companies, and how many more?

 **Challenge**

32. You know there are 3!, or 6, arrangements of 3 objects. Consider the number of clockwise arrangements possible for objects placed in a loop, without a beginning or end.

ABC, BCA, and CAB are all part of one possible clockwise loop arrangement of the letters A, B, and C.

 a. Find the number of clockwise loop arrangements possible for letters A, B, and C.
 b. Use the diagram at the right to help find the number of loop arrangements possible for A, B, C, and D.
 c. Write an expression for the number of clockwise loop arrangements for n objects.

33. Reasoning Suppose $_nP_r = 210$ and $r = 3$. What is the value of n?

34. A three-digit number is formed by randomly selecting from the digits 1, 2, 3, 4, and 5 without replacement.
 a. How many different three-digit numbers can be formed?
 b. How many different numbers with three odd digits can be formed?
 c. Find the probability that the number formed has three odd digits.

35. Critical Thinking Assume a and b are positive integers. Is the statement $(a - b)! = a! - b!$ true or false? If it is true, explain why. If it is false, give a counterexample.

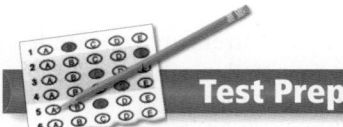 **Test Prep**

Gridded Response

36. The school cafeteria has lunches of one entrée, one vegetable, one salad, and one dessert. The menu has choices of 2 entrées, 3 vegetables, 3 salads, and 4 desserts. How many different lunches are there?

37. You are planning your schedule for next year. You already know your classes for the last three periods. There are four choices for period 1, two choices for period 2, and two choices for period 3. How many combinations are there for the first three periods?

38. Southside High School T-shirts come in two styles, three colors, and four sizes. A student wants to buy a large or extra large black T-shirt. How many choices does the student have?

39. Simplify $_4P_3$.

Mixed Review

Lesson 12-6

Solve each equation. Check your solution.

40. $\frac{1}{x} + \frac{1}{2x} = 5$

41. $\frac{2}{n} + \frac{1}{n+1} = \frac{11}{n^2+n}$

42. $\frac{m}{2} = \frac{24-m}{m}$

43. $\frac{10}{3v+6} = \frac{3v}{v+2} + \frac{v^2}{3v+6}$

44. $\frac{3w}{w-1} - \frac{2w}{w+3} = \frac{8w+40}{w^2+2w-3}$

45. $\frac{h-5}{h+4} + \frac{h+1}{h+3} = \frac{-6h-6}{h^2+7h+12}$

Lesson 11-5

Suppose $\triangle ABC$ has right angle C. Find the lengths of the other sides to the nearest whole number.

46. $m\angle A = 45°, BC = 5$

47. $m\angle A = 32°, AB = 64$

48. $m\angle B = 75°, AC = 20$

49. $m\angle B = 8°, BC = 48$

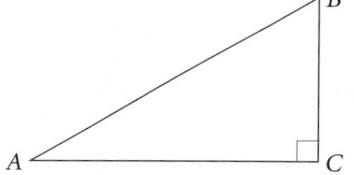

Lesson 10-5

Solve by completing the square. If there is no solution, write *no solution*.

50. $x^2 + 12x + 1 = 0$

51. $x^2 + 6x + 20 = 0$

52. $x^2 + 8x = 11$

53. $x^2 + 7x + 1 = 13$

54. $2x^2 + 8x + 7 = 9$

55. $x^2 - 18x + 65 = 0$

✓ Checkpoint Quiz 2   Lessons 12-3 through 12-7

Find each product or quotient.

1. $\frac{x^2-4}{x+3} \cdot \frac{x^2+7x+12}{x-2}$

2. $\frac{z+5}{z} \div \frac{3z+15}{4z}$

3. $\frac{a^3+2a^2-1}{a-3}$

Find each sum or difference.

4. $\frac{9}{x-3} - \frac{4}{x-3}$

5. $\frac{8}{m+2} - \frac{6}{3-m}$

6. $\frac{6}{t} + \frac{3}{t^2}$

Solve each equation.

7. $\frac{9}{t} + \frac{3}{2} = 12$

8. $\frac{10}{z+4} = \frac{30}{2z+3}$

9. $c - \frac{8}{c} = -7$

10. Hiking Suppose there are four different trails up a mountain.
  a. In how many ways can someone climb up and down the mountain?
  b. In how many ways can someone climb up and down the mountain if that person does not want to take the same trail in both directions?

Combinations

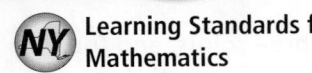
Learning Standards for Mathematics

A.N.8 Determine the number of possible arrangements (permutations) of a list of items.

 **Check Skills You'll Need**

Evaluate each expression.

1. $_5P_3$ **2.** $_6P_3$ **3.** $_7P_3$ **4.** $_7P_4$

GO for Help   Lessons 12-7 and 2-7

A **and** *B* **are independent events. Find** *P(A* **and** *B)* **for the given probabilities.**

5. $P(A) = \frac{1}{3}, P(B) = \frac{3}{4}$ **6.** $P(A) = \frac{1}{8}, P(B) = \frac{5}{9}$

7. $P(A) = \frac{9}{10}, P(B) = \frac{5}{6}$ **8.** $P(A) = 0.35, P(B) = 0.2$

🔊 **New Vocabulary**   • combination

---

1   Combinations

Suppose you are making a sandwich with three of these ingredients: turkey, cheese, tomato, and lettuce. Below are the permutations of the four ingredients chosen three at a time.

turkey cheese **tomato**
turkey tomato cheese
cheese tomato turkey
cheese turkey tomato
tomato turkey cheese
tomato cheese turkey

turkey cheese lettuce
turkey lettuce cheese
cheese lettuce turkey
cheese turkey lettuce
lettuce turkey cheese
lettuce cheese turkey

turkey tomato lettuce
turkey lettuce tomato
tomato lettuce turkey
tomato turkey lettuce
lettuce turkey tomato
lettuce tomato turkey

cheese **tomato lettuce**
cheese lettuce tomato
tomato lettuce cheese
tomato cheese lettuce
lettuce cheese tomato
lettuce tomato cheese

For most people, the *order* of the ingredients within the sandwich does not matter. Ignoring order, there are only four different sandwiches. Each sandwich type is a **combination,** a collection of objects without regard to order.

Every combination of three ingredients has $_3P_3$ or 6 permutations, which are considered the same. Dividing the total number of permutations, $_4P_3$ or 24, by 6 gives the number of different sandwiches without regard to order.

number of different sandwiches =

$$= \frac{\text{number of permutations of ingredients } (_4P_3)}{\text{number of permutations of ingredients in one sandwich } (_3P_3)}$$

$$= \frac{4 \cdot 3 \cdot 2}{3 \cdot 2 \cdot 1}$$

$$= 4$$

Key Concepts

Definition	Combination Notation

The expression $_nC_r$ represents the number of combinations of n objects arranged r at a time.

$$_nC_r = \frac{_nP_r}{_rP_r} = \frac{n(n-1)(n-2)\ldots}{r(r-1)(r-2)\ldots} \quad \begin{array}{l} \leftarrow r \text{ factors, starting with } n \\ \leftarrow r \text{ factors, starting with } r \end{array}$$

Example $_4C_3 = \frac{_4P_3}{_3P_3} = \frac{4 \cdot 3 \cdot 2}{3 \cdot 2 \cdot 1} = 4$

1 EXAMPLE   Counting Combinations

Simplify $_8C_5$.

$$_8C_5 = \frac{_8P_5}{_5P_5} \qquad \text{Write using permutation notation.}$$

$$= \frac{8 \cdot 7 \cdot 6 \cdot 5 \cdot 4}{5 \cdot 4 \cdot 3 \cdot 2 \cdot 1} \qquad \text{Write the product represented by the notation.}$$

$$= 56 \qquad \text{Simplify.}$$

Quick Check ❶ Simplify each expression.
a. $_4C_2$ **b.** $_7C_3$ **c.** $_{10}C_4$

You can use combinations to model real-world situations.

2 EXAMPLE   Real-World Problem Solving

Juries Twenty people report for jury duty. How many different twelve-person juries can be chosen?

The order in which jury members are listed once the jury members are chosen does not distinguish one jury from another. You need the number of combinations of 20 potential jurors chosen 12 at a time. Evaluate $_{20}C_{12}$.

Method 1 Use pencil and paper.

$$_{20}C_{12} = \frac{_{20}P_{12}}{_{12}P_{12}}$$

$$= \frac{20 \cdot 19 \cdot 18 \cdot 17 \cdot 16 \cdot 15 \cdot 14 \cdot 13 \cdot 12 \cdot 11 \cdot 10 \cdot 9}{12 \cdot 11 \cdot 10 \cdot 9 \cdot 8 \cdot 7 \cdot 6 \cdot 5 \cdot 4 \cdot 3 \cdot 2 \cdot 1}$$

$$= 125{,}970 \qquad \text{Use a calculator.}$$

Method 2 Use a graphing calculator.
Use MATH to select **nCr** in the **PRB** screen.
$$_{20}C_{12} = 125{,}970$$

20 nCr 12
125970

There are 125,970 different twelve-person juries that can be chosen from a group of 20 people.

Real-World Connection

Most civil and criminal cases in the United States are tried by a judge with a jury of 12 people.

Quick Check ❷ **a.** For your history report, you can choose to write about two of a list of five presidents of the United States. Use $_nC_r$ notation to write the number of combinations possible for your report.
     b. Calculate the number of combinations of presidents on whom you could report.
     c. Critical Thinking Explain why you should model this situation with combinations, not permutations.

Some probability problems can be solved using permutations.

3 EXAMPLE Using Permutations in Probability

Music Suppose you have six new CDs (two rock, two rhythm and blues, and two jazz) to put on a rack. If you choose the CDs at random, what is the probability that the first one is jazz and the second one is rhythm and blues?

number of ways to select jazz then rhythm and blues $= 2 \cdot 2$ **Use the Multiplication Counting Principle.**

number of ways to order 6 CDs arranged 2 at a time $= {}_6P_2$

$P(\text{jazz, then rhythm and blues}) = \dfrac{\text{number of ways to select jazz, then rhythm and blues}}{\text{number of ways to order all 6 CDs}}$

$= \dfrac{2 \cdot 2}{{}_6P_2}$ **Substitute.**

$= \dfrac{4}{30}$ **Simplify the numerator. Find ${}_6P_2$.**

$= \dfrac{2}{15}$ **Simplify.**

● The probability that you choose jazz, and then rhythm and blues, is $\frac{2}{15}$.

✓ **Quick Check** ❸ Suppose you have three novels and two history books to read over the summer. If you choose books at random, what is the probability that you first choose a novel, and then a history book?

You can rewrite the probability formula using combination notation, where f is the number of favorable items, t is the total number of items, and r is the number of items being chosen.

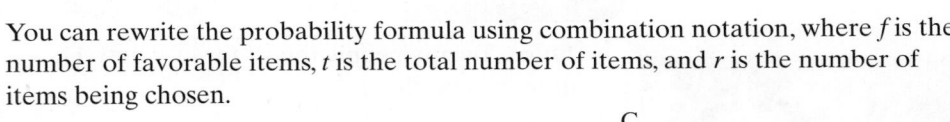

$$P(\text{event}) = \frac{\text{number of favorable outcomes}}{\text{total number of outcomes}} = \frac{{}_fC_r}{{}_tC_r}$$

4 EXAMPLE Using Combinations in Probability

Money Suppose you have six Virginia quarters and four Connecticut quarters in a change purse. You choose three quarters without looking. What is the probability that all the quarters you choose have a boat on them?

There are 10 quarters in all. The six Virginia quarters have boats on them.

number of favorable outcomes $= {}_6C_3$ **number of ways to choose 3 quarters from the 6 quarters with boats on them**

number of possible outcomes $= {}_{10}C_3$ **number of ways to choose 3 quarters from 10 possible quarters**

$P(3 \text{ quarters with boats})$
$= \dfrac{\text{number of favorable outcomes}}{\text{total number of outcomes}}$ **Use the definition of probability.**

$= \dfrac{{}_6C_3}{{}_{10}C_3}$ **Substitute.**

$= \dfrac{20}{120} = \dfrac{1}{6}$ **Simplify each expression. Simplify fraction.**

● The probability that you choose three quarters with boats is $\frac{1}{6}$, or about 17%.

Real-World Connection

The United States Mint began issuing quarters for each state in 1999.

Quick Check 4 **a.** Suppose you have the quarters in Example 4. You choose four quarters at random. How many combinations are possible?
   b. How many combinations have only trees?
   c. What is the probability that all of the quarters have trees?

EXERCISES

For more exercises, see *Extra Skill and Word Problem Practice.*

Practice and Problem Solving

 **A** **Practice by Example**

Example 1
(page 707)

Simplify each expression.

1. $_6C_6$ **2.** $_6C_5$ **3.** $_6C_4$ **4.** $_6C_3$ **5.** $_6C_2$

6. $_6C_1$ **7.** $_8C_6$ **8.** $_8C_2$ **9.** $_7C_5$ **10.** $_7C_2$

Example 2
(page 707)

11. Law For some civil cases, at least nine of twelve jurors must agree on a verdict. How many combinations of nine jurors are possible on a twelve-person jury?

12. For your birthday you received a gift certificate from a music store for three CDs. There are eight you would like to have. If you select the CDs at random, how many different groups of three CDs could you select?

Example 3
(page 708)

13. The colors red, orange, yellow, green, blue, indigo, and violet are written on slips of paper and placed in a hat. What is the probability that the slips will be chosen in the order of the colors of a rainbow (the order listed above)?

Example 4
(page 708)

14. The letters A, B, C, D, E, F, G, H, I, and J are written on slips of paper and placed in a hat. Two letters are then drawn from the hat.
   a. What is the number of possible combinations of two letters?
   b. How many combinations consist only of the vowels A, E, or I?
   c. What is the probability that the letters chosen would consist only of vowels?
   d. What is the probability that the letters chosen would be only B, C, D, or F?

15. a. The class president plans to randomly select a committee of three people from three boys and five girls. How many committees are possible?
   b. How many possible committees have all boys?
   c. What is the probability that the committee will have all boys?
   d. Critical Thinking What is the probability that the committee will have no boys?

B **Apply Your Skills**

Find the number of combinations of letters taken three at a time that can be formed from each set of cards.

16.    **17.**

18.    **19.**

Classify each situation as a permutation or a combination problem. Explain.

20. A locker contains eight books. You select three books at random. How many different sets of books could you select?

21. You take four books out of the library to read during spring vacation. In how many different orders can you read the four books?

Visit: PHSchool.com
Web Code: ate-1208

22. a. Geometry Draw four points on your paper like those in Figure 1. Draw line segments so that every point is joined to every other point.
   b. How many segments did you draw?
   c. Find the number of segments using combinations. You are joining four points, two at a time.
   d. How many segments would you need to join each point to all the others in Figure 2?

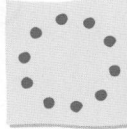

Figure 1   Figure 2

23. A famous problem known as the handshake problem asks, "If ten people in a room shake hands with everyone else in the room, how many different handshakes occur?"
   a. Is this a combination or a permutation problem? Justify your answer.
   b. Solve the problem.
   c. Critical Thinking Is the handshake problem similar to the problem in Exercise 22 part (d)? Explain.

 24. Locks A lock like the one at the left is called a combination lock. However, mathematically speaking, it should be called a permutation lock! This is because the order of the numbers *is* important. Suppose a three-number sequence opens the lock, and no numbers are repeated.
   a. How many different sequences are possible?
   b. How many sequences use 32 as the first number?
   c. What is the probability that the sequence of numbers that opens this lock uses 32 as the first number?
   d. Explain why the lock is unlikely to be opened by someone who does not know the correct sequence.

 25. Writing Explain the difference between a permutation and a combination.

Simplify each expression.

26. $_2C_2 + {_2C_1} + {_2C_0}$

27. $_3C_3 + {_3C_2} + {_3C_1} + {_3C_0}$

28. a. How many different 3-digit numbers are possible using 2, 3, 5, and 7 if you do not repeat any digits?
   b. What is the probability that a number selected at random is even?

29. Open-Ended Write and solve two problems: one that can be solved using permutations and one that can be solved using combinations.

30. Twelve computer monitors are stored in a warehouse. The warehouse manager knows that three monitors are defective, but the report telling him which ones are defective is missing. He selects five monitors at random to begin testing.
   a. How many different choices of five monitors does the manager have?
   b. In how many ways could he happen to select five monitors that include the defective ones?
   c. What is the probability that he will find the three defective monitors when he tests the first five monitors?

Challenge   **Determine whether each statement is *sometimes*, *always*, or *never* true.**

31. $_nC_1 = n$

32. $_3C_x > x$

33. $_nC_{(n-1)} = n$

34. The numerals 3, 4, 5, 6, 7, 10, 12, and 13 are written on slips of paper and placed in a hat. Three slips of paper are then drawn from the hat. What is the probability that the numbers will form a Pythagorean triple?

35. On a particular test, you must answer seven out of ten questions. You also must answer at least four of the first six questions.

  a. In how many ways can you choose four of the first six questions?

  b. How many questions are left after you have answered four of the first six questions? How many must you still answer?

  c. In how many ways can you choose questions to finish the test?

  d. How many different ways are there of completing the test (meeting all its requirements)?

36. a. Graph the function $f(x) = {}_xC_2$ using the replacement set $\{2, 3, 4, 5\}$.

  b. **Critical Thinking** What is $f(x)$? Explain.

  c. **Critical Thinking** Explain why you should not connect the points you graphed in part (a).

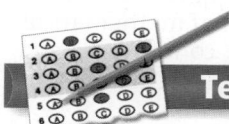

Multiple Choice

37. A florist has twelve different flowers to choose from to make a bouquet. In how many different ways can the florist choose three of these flowers?

  A. 36   **B.** 220   **C.** 1320   **D.** 1728

38. What is ${}_5C_3 - {}_8C_2$?

  F. -18 **G.** -2 **H.** -1 **J.** 4

39. Suppose a teacher gives a four-question true-or-false quiz. The correct answers are F, T, T, and F. If a student guesses the answers, how many of the outcomes have exactly 2 correct answers?

  A. 1   **B.** 4   **C.** 6   **D.** 16

Short Response

40. The Pizza Palace offers five different toppings, but you have just enough money for a two-topping pizza. How many different two-topping pizzas can you choose? Is this a combination or a permutation? Explain.

Mixed Review

Lesson 12-7

Simplify each expression.

41. ${}_4P_2$ **42.** ${}_7P_6$ **43.** ${}_7P_4$ **44.** ${}_9P_1$ **45.** ${}_{23}P_1$

46. License Plates In North Carolina, a regular license plate for cars and light trucks has seven characters: three letters followed by four digits. The letters G, I, O, Q, and U are not used. How many different license plates are possible?

Lessons 12-6 and 11-3

Solve each equation. Check your solution. If there is no solution, write *no solution*.

47. $\dfrac{4}{x} + \dfrac{x}{x-4} = 1$ **48.** $\sqrt{r-7} = \sqrt{2r+4}$ **49.** $\dfrac{2}{z} - \dfrac{3}{2z} = 5$

50. $\sqrt{3v+10} = v$ **51.** $\dfrac{1}{a+2} + \dfrac{1}{a-2} = \dfrac{10}{a-2}$ **52.** $\sqrt{16m} = \sqrt{m^2}$

Lesson 10-6

Use the Quadratic Formula to solve each equation. If necessary, round answers to the nearest hundredth.

53. $2x^2 + 12x - 11 = 0$ **54.** $x^2 - 7x + 2 = 0$ **55.** $x^2 + 4x - 8 = 0$

56. $-9x^2 + x + 15 = 0$ **57.** $-x^2 - 3x + 18 = 0$ **58.** $1.5x^2 - 20x - 1 = 0$

Networks

A special kind of graph called a network is helpful in planning things like flight schedules, teams in a tournament, or class assignments for teachers. In a network, endpoints are *vertices*, and the segments connecting vertices are *edges*. A *path* is a sequence of edges from one vertex to another.

NY A.R.1: Use physical objects, diagrams, charts, tables, graphs, symbols, equations, or objects created using technology as representations of mathematical concepts.

1 ACTIVITY

The network below shows the possible routes for an air carrier between several cities.

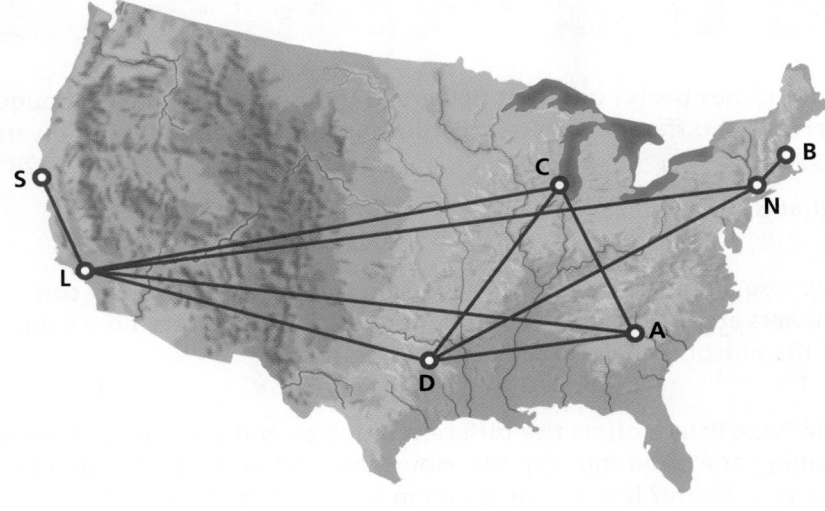

1. a. How many west-to-east paths are there from Los Angeles to Boston?
   b. Describe each path.

2. Which path represents the most direct flight?

3. If the air carrier wants to develop another Los Angeles-to-Boston route, how could they do this without making a direct flight?

4. A circuit is a path that starts and ends at the same vertex. describe a circuit that starts in Atlanta and has four edges.

You can draw networks for given situations.

2 ACTIVITY

Basketball teams from five different schools have participated in a tournament.

   Bakerstown has played Allensville, Carterberg, and David City.
   Carterberg has played Evans Crossing, Allensville, and Bakertown.

5. Draw a network for the information given.

6. Based on the information given, how many more games must be played for each team to have played each other?

The number of edges that meet at a vertex determines the *degree* of the vertex. An Euler circuit is a path that begins and ends at the same vertex, passing through each vertex and edge, but not going through any edge more than once.

3 ACTIVITY

The network below shows the possible routes for an air carrier between several cities.

I.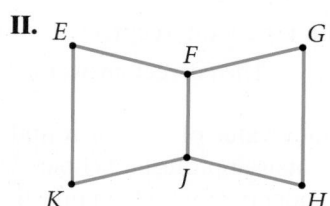

II.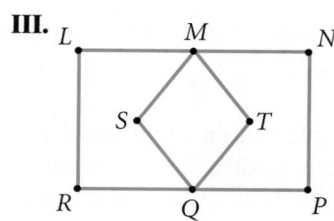

III.

For each graph, start at any point. Trace along the edges to get back to the original vertex without tracing along an edge more than once, and without lifting your pencil.

7. For each graph, find the degree for each vertex.

8. Make a conjecture about how to use the degree of the vertices in a network to determine if the network is an Euler circuit.

EXERCISES

9. Buses travel on main roads between several locations in a downtown area. Direct routes connect each of the locations below. Draw the network.

- the library to city hall, the stadium, the high school, and the mall
- city hall to the library, the airport, and the mall
- the high school to the library and the mall

a. Draw the network.
b. Suppose a traveler takes a bus from the airport. Which stops would they have to make before they got the high school?
c. What is the degree of the vertex that represents the library?

The network represents the streets between three friends' houses and their school.

10. Can Sarah travel to school without passing one of her friends' houses?

11. Which of the three friends is unlikely to visit one of the others on the way to school? Explain.

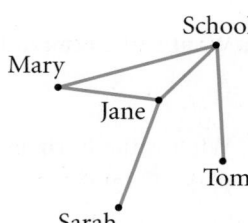

Draw a network for each description. Determine if it is an Euler circuit.

12. two vertices of degree 2 and two vertices of degree 3

13. one vertex of degree 1, three vertices of degree 2, and one vertex of degree 3

Answering the Question Asked

When answering a question, be sure to answer the question that is asked. Read the question carefully and identify the quantity that you are asked to find.

1 EXAMPLE

What is the y-intercept of the graph of $y = \frac{6}{x+2} - 4$?

A. -4 **B.** -2 **C.** -1 **D.** -0.5 **E.** 0

The question is asking for the y-intercept. To find the y-intercept, let $x = 0$. Therefore $y = \frac{6}{0+2} - 4 = \frac{6}{2} - 4 = 3 - 4 = -1$. The correct answer is C.

In Example 1, notice that the choice for A is the y-value of the horizontal asymptote. Choice B is the x-value of the vertical asymptote, and choice D is the x-intercept. Usually answer choices only have meaning for a given question, so you must be careful to answer the question asked.

2 EXAMPLE

If $5x^2 - x + 4$ is divided by $x + 1$, what is the remainder?

F. $5x - 6$ **G.** $5x + 1$ **H.** 10 **J.** 0 **K.** -6

Use long division.

$$
\begin{array}{r}
5x - 6 \\
x + 1 \overline{)5x^2 - x + 4} \\
\underline{5x^2 + 5x} \\
-6x + 4 \\
\underline{-6x - 6} \\
10
\end{array}
$$

Choice F is the quotient, and you might be tempted to choose it. The question, however, is asking for the remainder. Choice H is the remainder. That is the correct answer to the question.

EXERCISES

1. What is the remainder when $3x^3 - 2x^2 - 4$ is divided by $x^2 - 2$?

     A. $6x - 8$ **B.** $6x + 8$ **C.** $3x + 2$ **D.** $3x - 2$ **E.** $3x$

2. What is the horizontal asymptote of $y = \frac{3}{x+2} + 2$?

     F. $y = -2$ **G.** $y = 2$ **H.** $y = 0.5$ **J.** $y = 1.5$ **K.** $y = 3$

3. If $\frac{4x}{3} + \frac{4x}{5} = 1$, what is the value of $4x$?

     A. $\frac{15}{32}$ **B.** $\frac{15}{8}$ **C.** 4 **D.** 8 **E.** 15

Chapter Review

Vocabulary Review

🔊 **asymptote** (p. 664)
combination (p. 706)
multiplication counting principle
   (p. 700)

permutation (p. 701)
rational equation (p. 692)

rational expression (p. 672)
rational function (p. 664)

Go Online
PHSchool.com
For: Vocabulary quiz
Web Code: atj-1251

Choose the correct vocabulary term to complete each sentence.

1. A(n) __?__ is in simplest form if the numerator and the denominator have no common factors other than 1.

2. The y-axis is a vertical __?__ of the function $y = \frac{1}{x}$.

3. A(n) __?__ is an arrangement of some or all of a set of objects in a specific order.

4. The first step in solving a(n) __?__ is to find the least common denominator.

Skills and Concepts

12-1 Objectives

▼ To graph rational functions (p. 664)

▼ To identify types of functions (p. 667)

A **rational function** has a polynomial of at least degree 1 in the denominator. The graph of a rational function may have asymptotes. A line is an **asymptote** of a graph if the graph of the function gets closer to the line as x or y gets larger in absolute value.

What value of x makes the denominator of each function equal zero?

5. $y = \frac{7}{x + 2}$ 6. $f(x) = \frac{9}{x}$ 7. $y = \frac{x}{4 - x}$ 8. $f(x) = \frac{6}{5x + 3}$

Identify the asymptotes of each function. Then graph the function.

9. $y = \frac{8}{x}$ 10. $xy = 20$ 11. $y = \frac{6}{x - 5}$ 12. $y = \frac{3}{x} + 2$

13. **Open-Ended** Write the equation of a rational function with a graph that is in three quadrants only.

14. **Writing** Explain why the function $f(x) = \frac{5}{x + 3}$ has asymptotes.

12-2 Objective

▼ To simplify rational expressions (p. 672)

A **rational expression** is an expression that can be written in the form $\frac{\text{polynomial}}{\text{polynomial}}$. The **domain** of a variable is all real numbers excluding the values for which the denominator is zero. A rational expression is in simplest form when the numerator and denominator have no common factors other than 1.

Simplify each expression.

15. $\frac{6t^2}{16t^3}$ 16. $\frac{h - 9}{3h - 27}$ 17. $\frac{k - 10}{20 - 2k}$

18. $\frac{x^2 - 4}{x + 2}$ 19. $\frac{5x}{20x + 15}$ 20. $\frac{6x - 18}{x - 3}$

21. $\frac{-3t}{t^3 - t^2}$ 22. $\frac{z + 2}{2z^2 + z - 6}$ 23. $\frac{x^2 - 3x - 10}{x^2 - x - 20}$

12-3 Objectives

▼ To multiply rational expressions (p. 677)

▼ To divide rational expressions (p. 678)

You can multiply and divide rational expressions.

$$\frac{a}{b} \cdot \frac{c}{d} = \frac{ac}{bd}$$ where b and d are nonzero.

$$\frac{a}{b} \div \frac{c}{d} = \frac{a}{b} \cdot \frac{d}{c}$$ where b, c, and d are nonzero.

Multiply or divide.

24. $\frac{8}{m-3} \cdot \frac{3m}{m+1}$

25. $\frac{4t-12}{t^2-9} \cdot (3t+9)$

26. $\frac{4n+8}{3n} \div \frac{4}{9n}$

27. $\frac{2e+1}{8e-4} \div \frac{4e^2+4e+1}{4e-2}$

12-4 Objective

▼ To divide polynomials (p. 682)

To divide a polynomial by a monomial, divide each term by the monomial divisor. To divide a polynomial by another polynomial, use long division. When dividing polynomials, write the answer as quotient $+ \frac{\text{remainder}}{\text{divisor}}$.

Divide.

28. $(14x^2 - 28x) \div 7x$

29. $(24x^6 + 32x^5 - 8x^2) \div 8x^2$

30. $(50x^5 - 7x^4 + x^2) \div x^3$

31. $(x^2 + 8x + 2) \div (x + 1)$

32. $(x^2 + 8x - 16) \div (x + 4)$

33. $(8x^3 - 22x^2 - 5x + 12) \div (4x + 3)$

12-5 Objectives

▼ To add and subtract rational expressions with like denominators (p. 687)

▼ To add and subtract rational expressions with unlike denominators (p. 688)

You can add and subtract rational expressions. Restate each expression with the LCD as the denominator, and then add or subtract the numerators.

If a, b, and c represent polynomials (with $c \neq 0$),

then $\frac{a}{c} + \frac{b}{c} = \frac{a+b}{c}$.

Add or subtract.

34. $\frac{8x}{x-7} - \frac{4}{x-7}$

35. $\frac{6}{7x} + \frac{1}{4}$

36. $\frac{x}{4+x} - \frac{5}{x-2}$

37. $\frac{10x}{6+5x} + \frac{12}{5x+6}$

38. $\frac{9}{3x-1} + \frac{5x}{2x+3}$

39. $\frac{7m}{m^2-1} - \frac{10}{m+1}$

40. What is the LCD of $\frac{1}{4}$, $\frac{2}{x}$, $\frac{5x}{3x-2}$, and $\frac{3}{8x}$?

  A. $96x^3 - 64x^2$

  B. $12x + 2$

  C. $24x^2 - 16x$

  D. $27x^2 - 6x - 8$

12-6 Objectives

▼ To solve rational equations (p. 692)

▼ To solve proportions (p. 694)

You can use the least common denominator (LCD) to solve **rational equations.** Check possible solutions to make sure each answer satisfies the original equation.

Solve each equation. Check your solution.

41. $\frac{1}{2} + \frac{3}{t} = \frac{5}{8}$

42. $9 + \frac{1}{t} = \frac{1}{4}$

43. $\frac{3}{m-4} + \frac{1}{3(m-4)} = \frac{6}{m}$

44. $\frac{2c}{c-4} - 2 = \frac{4}{c+5}$

45. $\dfrac{5}{2x - 3} = \dfrac{7}{3x}$ **46.** $\dfrac{2}{x} = \dfrac{2}{x^2} + \dfrac{1}{2}$

47. $\dfrac{1}{x} + \dfrac{10}{x^2} = \dfrac{3}{x}$ **48.** $\dfrac{4}{10x} = \dfrac{2}{3x - 14}$

49. Business A new photocopier can make 72 copies in 2 min. When an older photocopier is working, the two photocopiers can make 72 copies in 1.5 min. How long will it take the older photocopier working alone to make 70 copies?

12-7 Objectives

▼ To use the multiplication counting principle (p. 699)

▼ To find permutations (p. 701)

You can find the number of different outcomes using the **multiplication counting principle.** If there are m ways to make a first selection and n ways to make a second selection, there are $m \times n$ ways to make the two selections.

A **permutation** is an arrangement of objects in a definite order. To calculate $_nP_r$, the number of permutations of n objects taken r at a time, use the following formula.

$$_nP_r = n(n - 1)(n - 2)\ldots \leftarrow r \text{ factors starting with } n$$

Simplify each expression.

50. $_5P_3$ **51.** $_8P_4$ **52.** $_6P_4$ **53.** $_5P_2$

54. $_9P_4$ **55.** $_7P_4$ **56.** $_7P_3$ **57.** $_9P_5$

58. a. Telephones Before 1995, three-digit area codes could begin with any number except 0 or 1. The middle number was either 0 or 1, and the last number could be any digit. How many possible area codes were there?

 b. Beginning in 1995, area codes were not limited to having 0 or 1 as the middle number. How many new area codes became available?

12-8 Objectives

▼ To find combinations (p. 706)

▼ To find probability with counting techniques (p. 708)

A **combination** is an arrangement of objects without regard to order.

To find $_nC_r$, the combinations of n objects taken r at a time, use the following formula:

$$_nC_r = \dfrac{_nP_r}{_rP_r} = \dfrac{n(n - 1)(n - 2)\ldots}{r(r - 1)(r - 2)\ldots} \quad \begin{matrix} \leftarrow r \text{ factors starting with } n \\ \leftarrow r \text{ factors starting with } r \end{matrix}$$

Simplify each expression.

59. $_6C_5$ **60.** $_{10}C_4$ **61.** $_9C_2$ **62.** $_{11}C_3$

63. $_5C_4$ **64.** $_5C_1$ **65.** $_{12}C_8$ **66.** $_{12}C_4$

67. Nutrition You want to have three servings of dairy products without having the same food more than once. Milk, yogurt, cottage cheese, and cheddar cheese are in the refrigerator. How many different combinations can you have?

68. A group of 15 friends wants to go to an amusement park. Unfortunately, there are only 10 seats available on the bus they would take. How many combinations of 10 friends could take the bus to the amusement park?

Chapter Test

What value of x makes the denominator of each function equal zero?

1. $f(x) = \dfrac{19 + x}{x - 5}$

2. $y = \dfrac{x}{7x + 1}$

3. $y = \dfrac{3}{8 - x}$

4. $f(x) = \dfrac{2x}{8x - 12}$

Simplify each expression.

5. $\dfrac{6p - 30}{3p - 15}$

6. $\dfrac{7p - 8}{8 - 7p}$

7. $\dfrac{n^2 + 4n - 5}{n + 5}$

8. $\dfrac{2x + 18}{x^2 + 11x + 18}$

Find the LCD of each pair of expressions.

9. $\dfrac{5}{h}, \dfrac{6}{3h}$

10. $\dfrac{8}{x^2}, \dfrac{7}{3x}$

11. $\dfrac{x}{12}, \dfrac{11}{8x}$

12. $\dfrac{4}{a^2 b^3}, \dfrac{3}{9ab^4}$

Identify the asymptotes of each function. Then graph the function.

13. $y = \dfrac{6}{x}$

14. $xy = 20$

15. $y = \dfrac{1}{x} + 3$

16. $y = \dfrac{4}{x} + 3$

Multiply or divide.

17. $\dfrac{3}{x - 2} \cdot \dfrac{x^2 - 4}{12}$

18. $\dfrac{5x}{x^2 + 2x} \div \dfrac{30x^2}{x + 2}$

19. $\dfrac{4w}{3w - 5} \cdot \dfrac{7}{2w}$

20. $\dfrac{6c - 2}{c + 5} \div \dfrac{3c - 9}{c}$

21. Open-Ended Write a rational expression for which 6 and 3 are excluded from the domain.

Divide.

22. $(12x^4 + 9x^3 - 10x^2) \div 3x^3$

23. $(x^4 - 16) \div (x + 2)$

24. $(4x^4 - 6x^3 + x + 7) \div (2x - 1)$

25. $(6x^3 - 11x^2 - 16x + 13) \div (3x + 2)$

26. If three people working together can clean an office suite in two hours, how long will it take a crew of four people to clean the office?

Solve each equation. Check your solution.

27. $\dfrac{v}{3} + \dfrac{v}{v + 5} = \dfrac{-4}{v + 5}$

28. $\dfrac{16}{x + 10} = \dfrac{8}{2x - 1}$

29. $\dfrac{2}{3} + \dfrac{t + 6}{t - 3} = \dfrac{18}{2(t - 3)}$

Add or subtract.

30. $\dfrac{5}{t} + \dfrac{t}{t + 1}$

31. $\dfrac{9}{n} - \dfrac{8}{n + 1}$

32. $\dfrac{2y}{y^2 - 9} - \dfrac{1}{y - 3}$

33. $\dfrac{4b - 2}{3b} + \dfrac{b}{b + 2}$

Classify each situation as a permutation or a combination problem. Explain your choice. Then solve the problem.

34. The 30-member debate club needs a president and treasurer. How many different pairs are possible?

35. How many different ways can you choose two books from the six books on your shelf?

36. You have enough money for two extra pizza toppings. If there are six possible toppings, how many different pairs of toppings can you choose?

Find the number of combinations of letters taken four at a time that can be formed from each set of letters.

37. A E I O U Y

38. E Q U A T I O N

39. L E A R N

40. A B C D E F G

Simplify each expression.

41. $_4C_3$

42. $_8P_6$

43. $_{10}P_7$

44. $_5C_2$

45. You have 5 kinds of wrapping paper and 4 different bows. How many different combinations of paper and a bow can you have?

46. Probability There are 15 books on your summer reading list. Three of them are plays, one is poetry, and the rest are novels. What is the probability that the first three books you read are two novels and a play if you choose the books at random?

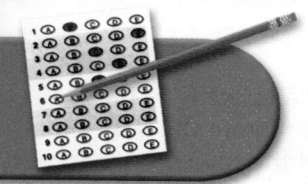

Regents Test Prep

Go **O**nline
PHSchool.com
For: End-of-Course Test
Web Code: ata-1254

Multiple Choice

For Exercises 1–49, choose the correct letter.

1. If $\frac{2x}{3} = 5$, $\frac{2y - 2}{4} = 3$, and $\frac{z}{2} + \frac{z}{3} = 5$, which of the following is true?
(1) $x > y$ (2) $y < z$
(3) $x = z$ (4) $z > x$

2. Business A manufacturing company spends $1200 each day on plant costs plus $7 per item for labor and materials. The items sell for $23 each. How many items must the company sell in one day to equal its daily costs?
(1) 52   (2) 75   (3) 150   (4) 200

3. A rectangle has a perimeter of 72 in. The length is 3 in. more than twice the width. What is the length of the rectangle in inches?
(1) 11   (2) 22   (3) 25   (4) 36

4. The test scores of one student are 79, 82, 83, 87, and 94. Find the mean of these scores.
(1) 15   (2) 83   (3) 85   (4) 94

5. Consumer Russell wants to purchase a computer that costs $1575. With a better disk drive, the cost of the computer will go up 8%. How much will an upgraded computer cost?
(1) $1701   (2) $1449
(3) $1458.33   (4) $1712

6. What is true of the graphs of the two lines $3y - 8 = -5x$ and $3x = 2y - 18$?
(1) no intersection   (2) intersect at $(2, -6)$
(3) intersect at $(-2, 6)$ (4) identical

7. Which relations are functions?

I.

x	1	−1	2	1
y	3	4	5	7

II.

x	1	2	3	4
y	1	1	3	5

III.

x	0	1	2	3
y	0	1	3	2

(1) I   (2) II
(3) II and III   (4) I, II, and III

8. What is the value of $f(x) = \frac{-3}{x - 1}$ for $x = -1$?
(1) 2   (2) $\frac{3}{2}$ (3) undefined   (4) $-\frac{3}{2}$

9. Which function is modeled by the table?

x	−1	1	3	5
y	−5	−1	3	7

(1) $y = 2x$ (2) $y = 2x - 3$
(3) $y = \frac{1}{2}x$ (4) $y = \frac{1}{2}x + 3$

10. Which expression is equivalent to $(j^2 k^3)(jk^2)$?
(1) $j^2 k^2$ (2) $j^3 k^5$ (3) $j^2 k^6$ (4) $j^3 k^6$

11. A parachutist opens her parachute at 800 ft. Her rate of change in altitude is -30 ft/s. Which equation represents her altitude a in feet t seconds after she opens her parachute?
(1) $a = 30t$ (2) $a = 800 - 30t$
(3) $a = 800 + 30t$ (4) $a = -30t$

12. What is the slope of a line perpendicular to $3x + 2y = 7$?
(1) $-\frac{3}{2}$ (2) $-\frac{2}{3}$ (3) $\frac{2}{3}$ (4) $\frac{3}{2}$

13. A and B are independent events. If $P(A) = \frac{5}{6}$ and $P(A \text{ and } B) = \frac{1}{8}$, what is $P(B)$?
(1) $\frac{1}{10}$ (2) $\frac{3}{20}$ (3) $\frac{1}{5}$ (4) $\frac{1}{4}$

14. What is the solution of the system $-2x - 3y = -15$ and $3x + 2y = 0$?
(1) $(-6, 9)$ (2) $(-6, -9)$ (3) $(6, -9)$ (4) $(6, 9)$

15. Which of the following points are solutions of $4y - 3x \le 8$?
I. $(0, 2)$ II. $\left(-3, \frac{1}{4}\right)$
III. $(5, -17.6)$ IV. $\left(-4, \frac{2}{5}\right)$
(1) I only   (2) IV only
(3) II and IV   (4) I and III

16. Which statement is true for every solution of the following system?
$$2y > x + 4$$
$$3y + 3x > 13$$
(1) $x \le -3$ (2) $y < 5$
(3) $x > 4$ (4) $y > 1$

17. Simplify $(2.5 \times 10^4)(3.0 \times 10^{-15})$.
(1) 7.5×10^{19} (2) 7.5×10^{-19}
(3) 7.5×10^{11} (4) 7.5×10^{-11}

18. Evaluate $a^2b^3c^{-1}$ for $a = 2$, $b = -1$, and $c = -2$.
 (1) -2 **(2)** 2 **(3)** 4 **(4)** 8

19. Simplify $\frac{20x^2y^4}{30x^5y^2}$.

 (1) $\frac{4x^7y^6}{6}$ **(2)** $\frac{2y^2}{3x^3}$

 (3) $\frac{3y^6}{2x^7}$ **(4)** $\frac{2}{3}x^{10}y^8$

20. Interest Suppose you deposit \$1000 in an account paying 5.5% interest, compounded annually. Which expression represents the value of the investment after 10 years?
 (1) $1000 \cdot 1.55^{10}$ **(2)** $1000 \cdot 1.055^{10}$
 (3) $1000 \cdot 0.055^{10}$ **(4)** $1000 \cdot 10^{1.055}$

21. Simplify $(3x^2 - 7x - 2) - (8x - 3)$.
 (1) $3x^2 + x - 5$ **(2)** $-5x^3 + 1$
 (3) $3x^2 - 5x - 5$ **(4)** $3x^2 - 15x + 1$

22. Find the greatest common factor (GCF) of the terms of $12x^5 + 4x^3 - 16x^2$.
 (1) $4x^3$ **(2)** x^3 **(3)** $4x^2$ **(4)** $-2x$

23. What is the standard form of the product $(3x - 1)(5x + 3)$?
 (1) $15x^2 + 2x - 3$ **(2)** $15x^2 + 2x + 3$
 (3) $15x^2 + 4x + 3$ **(4)** $15x^2 + 4x - 3$

24. Simplify $(2x - 3)(5x + 4)$.
 (1) $10x^2 - 12$ **(2)** $10x^2 - 7x - 12$
 (3) $10x^2 + 23x - 12$ **(4)** $10x^2 - 7x + 12$

25. Factor $x^2 + 3x - 10$.
 (1) $(x - 2)(x + 5)$ **(2)** $(x + 2)(x - 5)$
 (3) $(x - 2)(x - 5)$ **(4)** $-(x + 2)(x + 5)$

26. What is the maximum value of y in $y = -3x^2 - 6x - 1$?
 (1) -2 **(2)** -1 **(3)** 1 **(4)** 2

27. Find the solutions of $2x^2 + 5x + 3 = 0$.
 (1) $-3, -1$ **(2)** $-3, -2$
 (3) $\frac{3}{2}, 1$ **(4)** $-\frac{3}{2}, -1$

28. Which are the solutions of $x^2 + 4x - 5 = 0$?
 (1) $1, 5$ **(2)** $-1, 5$
 (3) $1, -5$ **(4)** $-1, -5$

29. How many solutions are there to the quadratic equation $2x^2 + 5x + 1 = 0$?
 (1) 0 **(2)** 1 **(3)** 2 **(4)** many

30. Each leg of a right isosceles triangle is 8 cm long. What is the length of the hypotenuse to the nearest tenth?
 (1) 27.7 cm **(2)** 16 cm
 (3) 13.9 cm **(4)** 11.3 cm

31. A support wire from the top of a tower is 100 ft long. It is anchored at a spot 60 ft from the base of the tower. Find the height of the tower.
 (1) 160 ft **(2)** 80 ft
 (3) 40 ft **(4)** $4\sqrt{10}$ ft

32. What are the solutions of the equation $x^2 - 6x - 11 = 0$?
 (1) -8 and 3 **(2)** $3 \pm 4\sqrt{5}$
 (3) 8 and -3 **(4)** $3 \pm 2\sqrt{5}$

33. Which expression is equivalent to $6a^2 \times 6a^5 \div 2a$?
 (1) $18a^{11}$ **(2)** $12a^8$ **(3)** $18a^6$ **(4)** $12a^{13}$

34. Triangle DEF is a right triangle with a right angle at F. Which of the following is false?
 (1) $\sin D = \frac{EF}{DE}$ **(2)** $\cos D = \frac{DE}{DF}$
 (3) $\sin E = \frac{DF}{DE}$ **(4)** $\cos E = \frac{EF}{DE}$

35. Max has \$510 in his savings account and Rose has \$360 in her savings acount. Starting today, Max will add \$40 to his account each month, and Rose will add \$55 to her account each month. Which system of equations can be used to find the number of months m it will take for Max and Rose to have the same account balance b?
 (1) $b = 40m + 510$ **(2)** $m = 40b + 510$
      $b = 55m + 360$ $m = 55b + 360$
 (3) $b = 40m + 360$ **(4)** $m = 40b + 360$
      $b = 55m + 510$ $m = 55b + 510$

36. Bakery A baker can shape 2 loaves of bread in 5 minutes. How many loaves can the baker shape in an hour?
 (1) 120 **(2)** 12 **(3)** 60 **(4)** 24

37. Which expression is equal to $\frac{3x - 9}{x^2 - 6x + 9}$?
 (1) $\frac{1}{3}x - \frac{1}{3}$ **(2)** $\frac{3}{x - 3}$
 (3) $\frac{1}{x + 3}$ **(4)** $x^2 + \frac{1}{3}x - \frac{1}{3}$

38. Divide $\frac{x - 4}{x^2 + 2x} \div \frac{x^2 - 16}{x^2 - x}$.
 (1) $\frac{x - 1}{x^2 + 6x + 8}$ **(2)** $\frac{1}{x^2 + x - 4}$
 (3) $\frac{x^2 - 6x + 4}{x^2 + x - 4}$ **(4)** 1

39. Divide $\frac{5}{m^3}$ by $\frac{10}{m^2}$.
 (1) $\frac{m}{2}$ **(2)** $\frac{2}{m}$ **(3)** $\frac{50}{m^5}$ **(4)** $\frac{1}{2m}$

40. Multiply $\frac{x - 1}{x + 3} \cdot \frac{x - 3}{x^2 - 1}$.
 (1) $\frac{x - 1}{x^2 - 1}$ **(2)** $\frac{x - 3}{x^2 + 4x + 3}$
 (3) $\frac{x - 3}{x + 3}$ **(4)** $\frac{x - 3}{x^2 + 2x - 3}$

41. What is the slope of the line?

(1) $-\frac{2}{3}$ **(2)** $\frac{2}{3}$

(3) $-\frac{3}{2}$ **(4)** $\frac{3}{2}$

42. Simplify $|{-5.2}| - |2.1|$.
     (1) 3.1   **(2)** -3.1 **(3)** 7.3   **(4)** -7.3

43. Solve $5(1 + d) \geq 3(2d + 9)$.
     (1) $d \leq 22$ **(2)** $d \leq -22$
     (3) $d \geq -22$ **(4)** $d \geq -2$

44. Simplify the expression $2b - 3a + b + a$.
     (1) -1 **(2)** ab **(3)** $b - a$ **(4)** $3b - 2a$

45. What is the function rule for the table at the right?

     (1) $y = -\frac{1}{4}x + 2$

     (2) $y = x + 13$

     (3) $y = \frac{1}{4}x + 10$

     (4) $y = 9.9(1.024^x)$

x	y
4	11
8	12
12	13
16	14

46. What is the solution of the system?
$$3x + 2y = 10$$
$$y = 2x - 2$$
     (1) $(1, 2)$ **(2)** $(3, 2)$ **(3)** $(-1, 3)$ **(4)** $(2, 2)$

47. Which set of data has a median of 4.5?
     (1) $3, 11, 4, 1, 15, 4, 7\frac{1}{2}, 1$
     (2) $2, 6, 10, 1, 3, 4, 9, 2, 5$
     (3) $8, 4, 4, 1, 3\frac{1}{4}, 3, 10, 2$
     (4) $4, 2, 10, 1, 3, 5, 7, 13$

Short Response

48. Find the value of $_{10}P_3$.

49. Simplify $\dfrac{4x + 10}{2x^2 + 7x + 5}$.

50. Factor $4x^2 - 12x + 9$.

51. Geometry What is the length of the diagonal of a rectangle with sides 6 cm and 10 cm?

52. Writing What are extraneous solutions? In what type of equations do they occur?

53. Probability A bag contains 5 green cubes and 7 yellow cubes. You pick two cubes without replacing the first one.
     a. What is the probability of choosing a yellow cube and then a green cube?
     b. What is the probability of choosing two yellow cubes?

54. Evaluate the expression $a(bc)^4$ for $a = -2$, $b = -1$, and $c = 4$.

55. Solve $\frac{x}{6} + \frac{1}{2} = \frac{3}{x}$.

56. The product of two positive integers is 45. The first is 4 less than the second. Find the integers.

57. The solution of the system $ax - 3y = 13$ and $x - by = 8$ is $(2, -3)$. Find a and b.

58. Transportation In 1996, the City Council of New York City voted to increase the number of taxis in the city from 11,787 to 12,187. What was the percent of increase?

59. Find the asymptotes, the x-intercept, and the y-intercept of the graph of $y = \frac{5}{x - 2} + 1$.

60. Serena bought a sweatshirt on sale for \$32. The regular price was \$42. What was the percent of decrease, to the nearest tenth of a percent?

61. Do the equations $x - 2 = 5$ and $\frac{x}{x - 7} - \frac{2}{x - 7} = \frac{5}{x - 7}$ have the same solution(s)?

62. Suppose $x - 472 = 1634$. Find the value of $x + 472$.

Extended Response

63. There is a linear relationship between the total length ℓ of a certain species of snake and the tail length (t) of this snake. Here are the measurements for two snakes of this species. Snake 1: $\ell = 150$ mm and $t = 19$ mm; Snake 2: $\ell = 300$ mm and $t = 40$ mm. Use the ordered pairs (ℓ, t).
     a. Find a linear equation for these data points.
     b. Use this linear equation to estimate the tail length of a snake with a total length of 200 mm.
     c. Use this linear equation to estimate the total length of a snake with a tail length of 61 mm.

64. Geometry In $\triangle ABC$, $\angle C$ is a right angle, $AB = 7$, and $m\angle B = 28°$. What are the lengths of $\overline{BC}$ and $\overline{AC}$ to the nearest hundredth?

65. The table shows the closing prices of a stock over a period of 5 days. Graph this relation and determine whether it is a linear function. Explain why or why not.

Day	1	2	3	4	5
Price	12	$12\frac{1}{2}$	$12\frac{1}{4}$	13	$13\frac{1}{4}$

Activity Lab

Good Vibrations

Applying Functions When you pluck a guitar string, it vibrates at a frequency corresponding to the note being played. In an acoustic guitar the hollow chamber and vibration of the wood shell amplify the sound produced by the strings. Electric guitars rely on electromagnetic pickups to detect the string's vibration and relay a signal to an amplifier. You can use the frequencies of musical notes to calculate the length of a guitar string.

NY A.CN.6: Recognize and apply mathematics to situations in the outside world.

Bridge

Electric Plucking

In 1952 Gibson introduced a solid-body electric guitar incorporating design elements by guitarist Les Paul.

Activity 1

Materials: pencil, paper or poster board

Cut a 70-cm length of 6-cm wide paper or poster board. This paper will represent the length of a string from nut to bridge. (See diagram and photo.)

a. Use the frequency 220.0 cycles/s and length 70 cm to find the constant k for the A string on page 723.

b. Use the value of k and the frequencies in the diagram to calculate the positions of each of the frets shown. Draw the frets on your model.

Building a Guitar

A master guitar maker invests about 120 hours of intensive work to build a guitar. The tonal properties of the wood used for the soundboard are crucial to the sound the guitar produces.

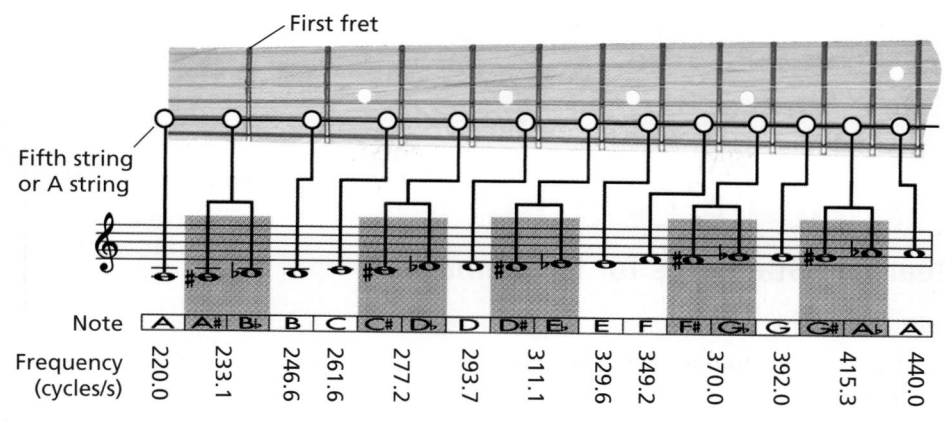

First fret

Fifth string
or A string

Musical Interlude

When you pluck the A string, it vibrates according to the equation below. Pressing a string against a fret changes its length and consequently its frequency.

$f = \dfrac{k}{L}$, where

f = frequency of string vibration (cycles/s)

k = constant

L = length of string (cm)

Note	A	A#	B♭	B	C	C#	D♭	D	D#	E♭	E	F	F#	G♭	G	G#	A♭	A
Frequency (cycles/s)	220.0	233.1	246.6	261.6	277.2	293.7	311.1	329.6	349.2	370.0	392.0	415.3	440.0

Activity 2

a. The distance from C to G is 7 frets, which corresponds to 7 half-steps or 3.5 full steps. The C and G notes form a musical interval called a perfect fifth. Calculate the ratio of the frequency of G to the frequency of C.

b. Locate other perfect fifths and calculate each frequency ratio. What do you notice?

c. Calculate the frequency ratio of an octave (6 steps) and of a major third (2 steps).

d. **Make a Conjecture** Predict the ratio of a perfect fourth (2.5 steps). Check your prediction by calculating the ratio.

Activity 3

Materials: graphing calculator or graphing utility

Use your model from Activity 1.

a. Plot the points (Frequency, String Length).

b. Plot the points (Fret Number, Frequency).

c. **Reasoning** Which of the sets of points plotted above lie on the graph of an inverse function? Explain.

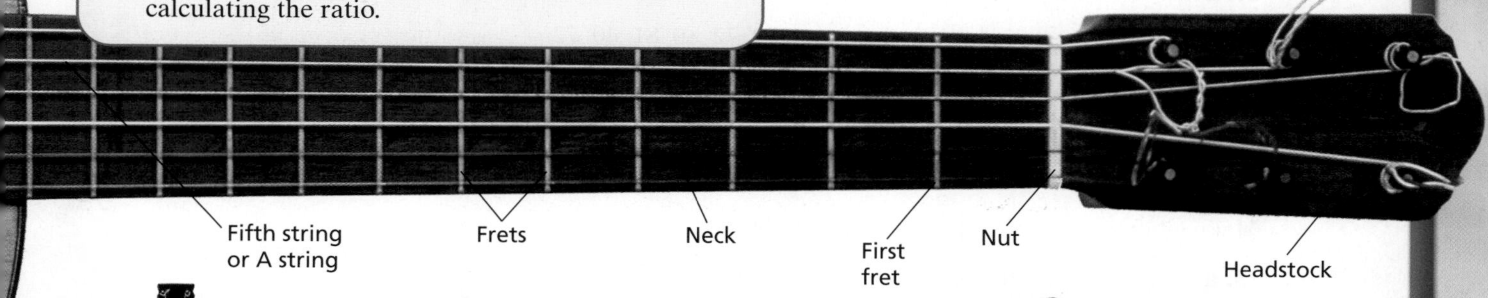

Fifth string
or A string   Frets   Neck   First fret   Nut   Headstock

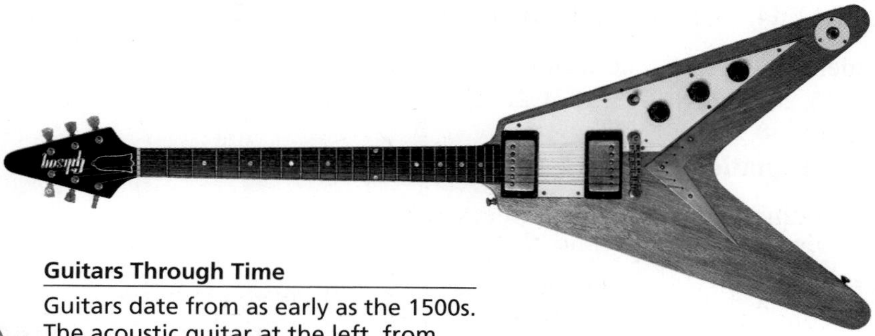

Guitars Through Time

Guitars date from as early as the 1500s. The acoustic guitar at the left, from 1627, has wire strings, which produce greater volume. The electric guitar above, from 1958, has two volume controls and one tone control.

Go Online
PHSchool.com

For: More information about guitars
Web Code: ate-1253

What You've Learned

 Learning Standards for Mathematics

In Chapter 1, you
- **NY A.S.12:** showed the relationship between two sets of real world data using a scatter plot.

In Chapter 5, you
- **NY A.G.4:** studied function rules, and modeled data using equations, tables, and graphs.

In Chapter 7, you
- **NY A.A.7:** analyzed and solved problems whose solution required solving systems of linear equations in two variables.

 **Check Your Readiness**

GO for Help to the Lesson in green.

Using Measures of Central Tendency (Lesson 1-6)

Find the mean, median, and mode. Which measure of central tendency best describes the data?

1. resting heart rate in beats per minute: 79 72 80 81 40 72

2. time spent playing video games per day:
   30 min, 0 min, 1 h, 45 min, 0 min, 55 min, 0 min, 6 h, 50 min, 1.5 h, 0 min

Displaying Data (Skills Handbook page 774)

Make a box-and-whisker plot for each set of data.

3. ages of members in the drama club:
   13, 18, 16, 13, 14, 14, 14, 15, 18, 17, 15, 16, 16, 15, 16, 16, 14, 14, 17, 17, 19

4. number of students in various school clubs:
   5, 20, 30, 12, 10, 15, 20, 6, 10, 8, 9, 17, 18, 4, 22, 27

Solving Systems of Equations (Lesson 7-1)

Solve each system of equations by graphing. Check each solution.

5. $y = x - 1$
   $y = 3x - 7$

6. $y = 2x + 5$
   $x + y = 8$

7. $y - x = -4$
   $3x - y = 0$

Solving Quadratic Equations (Lesson 10-3)

Solve each equation by graphing the related function.

8. $a^2 - 36 = 0$

9. $8x^2 - 32 = 0$

10. $x^2 - x - 6 = 0$

New York Additional Topics

◄)) **Key Vocabulary**

- bias (p. NY 728)
- bivariate data (p. NY 727)
- box-and-whisker plot (p. NY 734)
- causation (p. NY 749)
- complement of a set (p. NY 739)
- disjoint sets (p. NY 740)
- extrapolation (p. NY 748)
- interpolation (p. NY 748)
- intersection (p. NY 744)
- interval notation (p. NY 740)
- maximum (p. NY 733)
- minimum (p. NY 733)
- null set (p. NY 739)
- percentile rank (p. NY 733)
- qualitative data (p. NY 726)
- quantitative data (p. NY 726)
- roster form (p. NY 738)
- set (p. NY 738)
- set-builder notation (p. NY 738)
- subset (p. NY 739)
- union (p. NY 743)
- univariate (p. NY 727)

What You'll Learn Next

 **Learning Standards for Mathematics**

- **NY A.S.15:** You will identify and describe sources of bias and its effect, drawing conclusions from data.
- **NY A.S.6:** You will explain how to use the five statistical summaries (minimum, maximum, and the three quartiles) to construct a box-and-whisker plot.
- **NY A.A.29:** You will use set-builder notation and/or interval notation to illustrate the elements of a set, given the elements in roster form.
- **NY A.A.31:** Find the intersection of sets (no more than three sets) and/or union of sets (no more than three sets).
- **NY A.S.13:** You will explain the difference between correlation and causation.
- **NY A.A.11:** You will solve a system of one linear and one quadratic equation in two variables.

 **Online**

New York Additional Topics online
Visit: PHSchool.com
Web Code: azq-5500

Analyzing Data and Identifying Bias

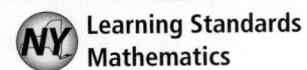
Learning Standards for Mathematics

A.S.1 Categorize data as qualitative or quantitative.

A.S.2 Determine whether data to be analyzed is univariate or bivariate.

A.S.3 Determine when collected data or display of data may be biased.

✓ **Check Skills You'll Need**   **GO** for Help   Lesson 1-5

1. Make a scatter plot of the data below.

Television Hours	2	1.5	1	2.5	0	1
Exercise Hours	4	1.5	3	0	4	3.5

🔊 **New Vocabulary** • **quantitative** • **qualitative** • **univariate** • **bivariate** • **bias**

---

1   Identifying Types of Data

You can collect data using measurements or by using categories. **Quantitative** data measures quantity and can be described numerically, such as test scores and elapsed time. **Qualitative** data name qualities and can be words or numbers, such as sports and zip codes.

Type of Data	Quantitative	Qualitative
Description	Has units, can be measured and numerically compared	Describes a category, cannot be measured or numerically compared
Examples	**Age:** 13 years **Weight:** 214 grams **Commute:** 23 minutes	**Hair color:** brown **Attitude:** optimistic **Zip code:** 02865

Even though zip code is a number, it is qualitative because it is not a measurement.

1 EXAMPLE   Classifying Data

Determine whether each data set is qualitative or quantitative.

a. favorite movies

The data are not numerical quantities. This is qualitative data.

b. number of classmates who are taking Spanish.

The data are numerical quantities. This is quantitative data.

✓ **Quick Check** ❶ Determine whether the data sets are qualitative or quantitative.
    a. cost of CDs   **b.** eye color

The kind of data you are working with determines the type of graph you use to display the data. Some data involve two variables, and some involve only one. A set of data that uses only one variable is **univariate.** A set of data that uses two variables is **bivariate.**

2 EXAMPLE Identifying Data

Determine whether each data set is univariate or bivariate.

a. The atomic weights of calcium and copper.

There is only one variable, atomic weight. The data set is univariate.

b. The volume of a cube and the length of an edge of the cube.

There are two variables, edge length and volume. The data set is bivariate.

 **Quick Check** **2** Determine whether the data sets are univariate or bivariate.

    a. height of a mammal and the mammal's weight

    b. the cost of high-speed Internet service from two different suppliers

2   Sampling and Surveys

Statisticians collect information about specific groups of objects or people, called a population. When a population is too large to survey, statisticians sample a part of it to find characteristics of the whole. Three sampling methods are shown below.

Name	Sampling Method	Example
Random	Survey a population at random.	Survey people whose names are drawn out of a hat.
Systematic	Select a number n at random. Then survey every nth person.	Select the number 5 at random. Survey every fifth person.
Stratified	Separate a population into smaller groups, each with a certain characteristic. Then survey at random within each group.	Separate a high school into four groups by grade level. Survey a random sample of students from each grade.

When designing a survey, you should choose a sample that reflects the population.

3 EXAMPLE Choosing a Sample

DVD Rentals You want to find out how many DVDs students at your school rent in a month. You decide to interview every tenth teenager you see at a mall. Describe this sampling method. State whether this plan will give a good sample.

Because you are interviewing every tenth teenager, this method is systematic. This is not a good sample because it will likely include students who do not attend your school.

 **Quick Check** **3** You revise your plan and decide to interview all students leaving a school assembly who are wearing the school colors. Describe this sampling method. State whether this plan will give a good sample.

A survey question has **bias** when it makes assumptions that may or may not be true. Bias can influence opinion and can make one answer seem better than another. When statisticians survey a sample of a population, they try to write questions that are free from bias.

4 EXAMPLE Determining Bias in a Survey Question

A reporter wanted to find out what kinds of movies were most popular with local residents. The survey question the reporter used was "Do you prefer exciting action movies or studious documentaries?" Is this question biased? Explain.

The question is biased because the words *exciting* and *studious* make action films sound more interesting than documentaries.

Quick Check 4 A survey asked the question "What is your favorite food?" Is this question biased? Explain.

The method of collecting data is important. All voluntary response samples are biased because they do not include the reactions of non-respondents. Another issue that affects the validity of a survey is whether the sample accurately represents the population.

5 EXAMPLE Determining Bias in Survey Results

Sports You want to determine what percent of teens ages 14 to 18 watch wrestling on TV. You go to a high school wrestling match and ask every third teenager whether he or she watches wrestling on TV. Explain how this might cause bias in the results of your survey.

The sample chosen is not representative of the population. People who attend a high school wrestling match may be more likely to watch wrestling on TV.

Quick Check 5 You want to know how many of your classmates have cell phones. To determine this, you send every classmate an e-mail asking, "Do you own a cell phone?" Explain how this method of gathering data might affect the results of your survey.

When you read the results of a survey, you need to be critical not only of the sampling techniques, but also of the manner in which the results are displayed.

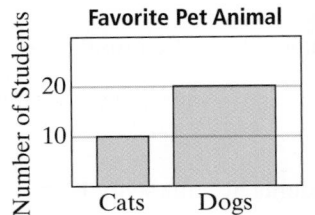
Favorite Pet Animal

6 EXAMPLE Determining Bias in a Display of Data

What makes the graph at the left misleading?

The bar on the right has increased not only in height, but also in width. The area of the second bar is more than two times the area of the first bar. This gives the impression that the difference is much greater than it actually is.

Quick Check 6 What makes the graph below misleading?

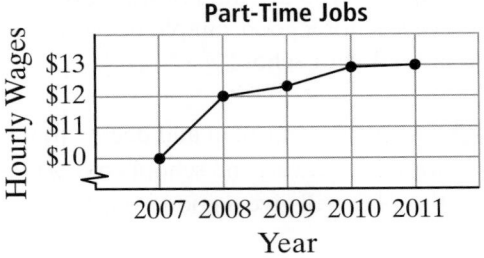

EXERCISES

For more exercises, see *Extra Skill and Word Problem Practice.*

Practice and Problem Solving

A Practice by Example

Example 1
(page NY 726)

Determine whether each data set is qualitative or quantitative.

1. favorite recording stars
2. best-selling DVDs
3. number of gigabytes in a memory card
4. prices of HDTVs

Example 2
(page NY 727)

Determine whether each data set is univariate or bivariate.

5. number of CDs you own
6. a person's age and height
7. your zip code
8. circumference and radius
9. amount of fuel used and number of miles driven

Example 3
(page NY 727)

For each survey plan, determine whether the plan is random, systematic, or stratified. State whether the plan will give a good sample.

10. A candidate calls every fiftieth name in the phone book to find out whether the person likes the candidate.

11. A factory tests the quality of the last 25 out of 1000 shirts made.

12. A printer selects 10 of the 450 packages of inserts at random to see if all the inserts were printed properly.

Example 4
(page NY 728)

Determine whether each question is biased. Explain your answer.

13. What kind of healthy toppings do you like on pizza?

14. Where would you most like to go for a vacation?

Example 5
(page NY 728)

15. You want to find out how much time people in your town spend doing volunteer work. You call 100 homes in the community during the day. Of those surveyed, 85% are over the age of 60. Explain how this might create bias in your survey results.

16. A perfume manufacturer sends out a sample of a new scent to 500 homes. Included is a response card asking recipients how much they like the scent. Each responder will have a chance to win a bottle of perfume. How will this affect the results?

Example 6
(page NY 728)

Explain how each graph is misleading.

17.

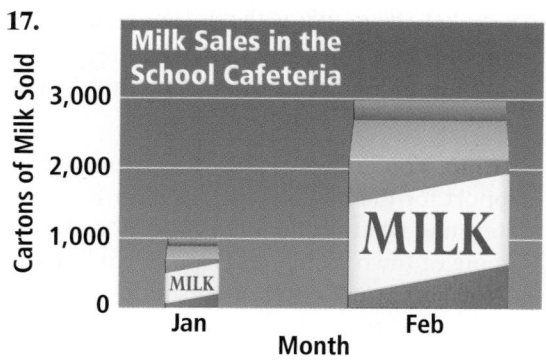

18.

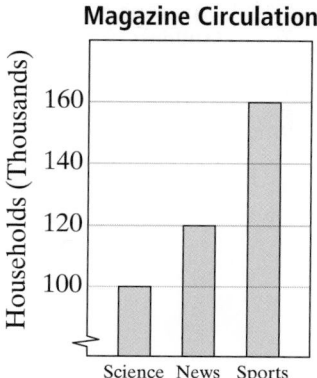

B **Apply Your Skills**

Real-World Connection

The Eiffel Tower was named for its designer, Gustave Eiffel, who also designed the Statue of Liberty framework.

19. Travel A travel agent wants to determine whether a trip to France is a popular vacation for young adults. Explain how each factor could create a bias in the survey results.
  a. The agent interviews people at an international airport.
  b. The agent asks, "Would you prefer to vacation in France or Italy?"
  c. Of the people interviewed, 86% took a French class in high school.

Classify the data as qualitative or quantitative and univariate or bivariate.

20. average number of visitors per day at each of six different theme parks

21. record low temperature in Rochester, New York

22. names of U.S. presidents born in each state

23. favorite color and a person's gender

24. Reasoning You review the results of survey questions given to two random samples of students from your school. The results are shown in the table below. Why are the results not the same?

Favorite Color

Color	Red	Blue	Green	Yellow	Pink	Purple	Black
Group A	8	6	4	2	5	4	1
Group B	7	7	3	0	6	5	2

25. Sports A student posts a survey on her Web site asking readers to choose their favorite sport to play from a list of five sports. The results are shown below.

Students' Favorite Sports

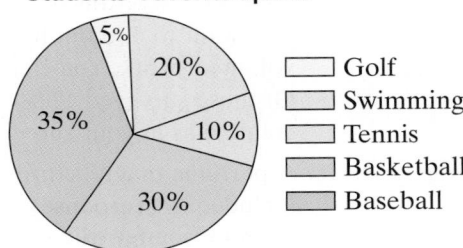

☐ Golf
☐ Swimming
☐ Tennis
☐ Basketball
☐ Baseball

  a. What biases might exist as a result of the design of this survey?
  b. Do you believe the results of this survey are valid? Explain your answer.

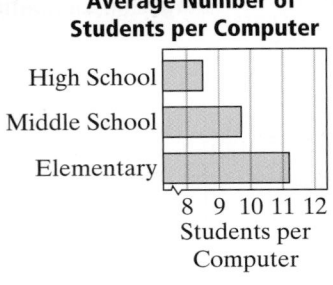

Average Number of Students per Computer

26. The graph at the left gives the impression that the number of students per computer in elementary school is three times the number of students per computer in high school. Draw a graph that accurately displays the data.

27. Writing You are going to write an article for the school newspaper about support for the mayor's proposal for bike paths. For each situation below, determine whether the data collection method will result in an unbiased sample of town residents. Explain your answer.
  a. You survey every tenth person leaving a bicycle repair store.
  b. You conduct a telephone poll every morning Monday through Friday for one week.
  c. You send an e-mail to 100 classmates chosen at random.
  d. You poll every fifth person at a popular local sandwich shop.

C **Challenge**

28. **Data Collection** You want to find out what kinds of pets the families of students attending your school own.
    a. Write an unbiased question for your survey. Will you be collecting quantitative or qualitative data?
    b. Choose a population and sampling method. Describe them both.
    c. Collect the data as you described and display the results in a graph.

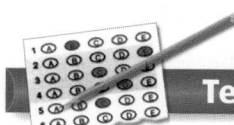

NY REGENTS

Test Prep

Multiple Choice

For Exercises 29 and 30, use the graph below.

29. The graph makes it appear that there are how many times as many students per teacher at North HS as there are at West HS?
    A. 2 times   B. 3 times
    C. 4 times   D. 5 times

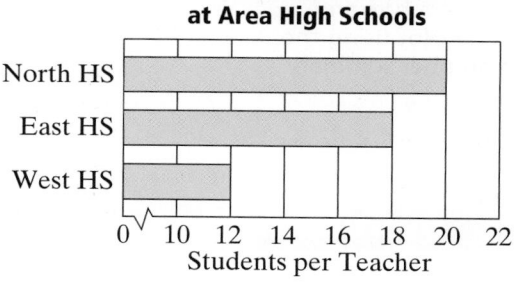

Student-Teacher Ratios at Area High Schools

30. Why might this graph give a distorted picture of the data?
    F. The longest horizontal bar is on top.
    G. The vertical lines are evenly spaced.
    H. There is a break in the horizontal axis.
    J. The horizontal bars are different lengths.

Short Response

31. Use the table at the right to make two different line graphs.
    a. Draw one graph to suggest that sales more than doubled from 1996 to 1999.
    b. Draw the other graph to suggest that sales increased only slightly during the same period.

Annual Sales

Year	Sales
1996	$87 million
1997	$87 million
1998	$88 million
1999	$90 million

Mixed Review

Lesson 1-1

Write a variable expression for each word phrase.

32. twenty-five less than x
33. the product of n and 3
34. ten decreased by t
35. a number x divided by 4
36. a number n increased by 5
37. two more than y

Lesson 1-2

Simplify.

38. $3(4 + 2)^2$
39. $49 - (4 \cdot 2)^2$
40. $-3^2 + 5 \cdot 2^3$
41. $2(9 - 4)^2$
42. $25 - (3 \cdot 2)^2$
43. $2 \cdot (-2)^4 + 10^1$
44. $(-4)(-6)^2(2)$
45. $(4 + 8)^2 \div 4^2$
46. $(12 - 3)^2 \div (2^2 - 1^2)$

Quartiles and Box-and-Whisker Plots

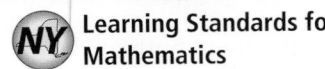 Learning Standards for Mathematics

A.S.11 Find the percentile rank of an item in a data set and identify the point values for first, second, and third quartiles.

A.S.5, A.S.9, A.S.6 Construct, analyze, and interpret a box-and-whisker plot using the five number summary.

✓ **Check Skills You'll Need**

Find the range and the median.

1. 3 8 12 15 20

2. 2.1 3.3 −5.4 0.8 3.5

3. 0 2 7 10 −1 −4 −11

4. 64 16 23 57 14 22

GO for Help Lesson 1-6

◀)) **New Vocabulary**
- first quartile
- third quartile
- five number summary
- minimum
- maximum
- percentiles
- percentile rank
- quartile
- box-and-whisker plot
- interquartile range

---

1 Quartiles and Percentiles

In Lesson 1-6, you used measures of central tendency to describe data sets. In this lesson, you will use **quartiles** to separate data in four equal parts. The median separates the data into upper and lower halves. The **first quartile** (Q_1) is the median of the lower half of the data. The **third quartile** (Q_3) is the median of the upper half of the data.

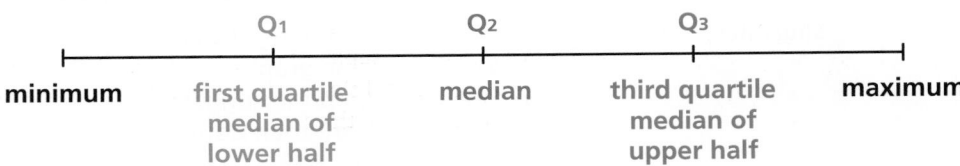

minimum | Q_1 first quartile / median of lower half | Q_2 median | Q_3 third quartile / median of upper half | maximum

1 EXAMPLE **Finding Median, First, and Third Quartiles**

For the data below, find the median (Q_2), first quartile (Q_1), and third quartile (Q_3).

39 27.5 26 31 42 26 37 30 22

Step 1 Arrange the data in order from least to greatest. Find the median.

22 26 26 27.5 30 31 37 39 42 ◀─**The median (Q_2) is the middle value 30.**

Step 2 Find the first quartile and third quartile, which are the medians of the lower and upper halves.

22 26 | 26 27.5 30 31 37 | 39 42

first quartile (Q_1) $= \dfrac{26 + 26}{2} = 26$ third quartile (Q_3) $= \dfrac{37 + 39}{2} = 38$

 **Quick Check** **1** For the following data, find the median, the first quartile, and the third quartile.
480 500 600 250 660 570 490 610 530

The **five number summary** of a set of data includes the minimum, the first quartile, the median, the third quartile, and the maximum. The **minimum** of a data set is the least value. The **maximum** is the greatest value.

2 **EXAMPLE** **Finding the Five Number Summary**

Find the five number summary for the data below.

125 80 140 135 126 140 350

Step 1 Arrange the data in order from least to greatest.
80 125 126 135 140 140 350

Step 2 Find the minimum, maximum, and median.
80 125 126 135 140 140 350

The minimum is 80. The maximum is 350. The median is 135.

Step 3 Find the first quartile and third quartile.

$$\text{first quartile } (Q_1) = \frac{125 + 126}{2} = 125.5$$

$$\text{third quartile } (Q_3) = \frac{140 + 140}{2} = 140$$

The five number summary is minimum = 80, Q_1 = 125.5, median = 135, Q_3 = 140, and maximum = 350.

 **Quick Check** **2** Find the five number summary for each set of data.

a. 95 85 75 85 65 60 100 105 75 85 75

b. 11 19 7 5 21 53

Data sets can be separated into smaller equal parts. **Percentiles** separate data sets into 100 equal parts. The **percentile rank** of a score is the percentage of scores that are less than or equal to that score.

3 **EXAMPLE** **Finding a Percentile Rank**

Of 25 test scores, eight are less than or equal to 75. What is the percentile rank of a test score of 75?

Write a ratio of the number of scores less than or equal to 75 compared to the total number of test scores.

$\dfrac{8}{25}$ ←———— **Number of test scores less than or equal to 75**
←———— **Total number of test scores**

$\dfrac{8}{25} = 0.32$ **Rewrite $\frac{8}{25}$ as a percent.**

$\quad\;\; = 32\%$

The percentile rank of 75 is 32.

 **Quick Check** **3** **a.** Of the 25 test scores, 15 scores are less than or equal to 85. What is the percentile rank of 85?

b. **Critical Thinking** Is it possible to have a percentile rank of 0? Explain.

A **box-and-whisker plot** is a graph that summarizes a data set along a number line. The box represents the first quartile, the median, and the third quartile of the data. The whiskers on the left and right sides represent the minimum and maximum values. Use the five number summary to construct a box-and-whisker plot.

Box-and-Whisker Plot

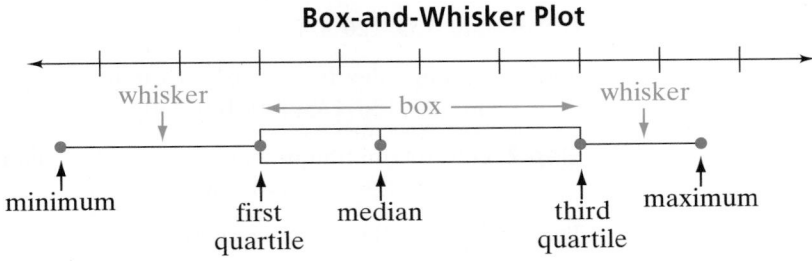

4 **EXAMPLE** Real-World Problem Solving

Agriculture The table below shows United States crops harvested from 1992 to 2004. Find the five number summary and construct a box-and-whisker plot of the data.

Step 1 Arrange the data in order from least to greatest. Find the minimum, maximum, and the median.

308 314 316 317 321 321 321
324 325 326 326 327 332

Crops Harvested

Year	Acres (millions)	Year	Acres (millions)
1992	317	1999	327
1993	308	2000	325
1994	321	2001	321
1995	314	2002	316
1996	326	2003	324
1997	332	2004	321
1998	326		

SOURCE: *Statistical Abstract of the United States.*

Step 2 Find the first quartile and third quartile.

$$\text{first quartile} = \frac{316 + 317}{2} = \frac{633}{2} = 316.5$$

$$\text{third quartile} = \frac{326 + 326}{2} = \frac{652}{2} = 326$$

Step 3 Draw a number line. Construct the box-and-whisker plot. Mark the values of the minimum, first quartile, median, third quartile, and maximum.

Crops Harvested (millions of acres)

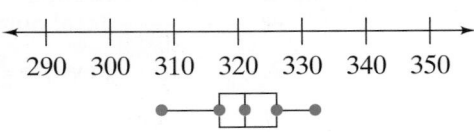

Quick Check **4** Construct a box-and-whisker plot for the following distances of bird migrations in thousands of miles.

5 2.5 6 8 9 2 1 4 6.2 18

5 EXAMPLE Using a Graphing Calculator

Use a graphing calculator to make a box-and-whisker plot of the data below. Find the five number summary.

10 26 18 35 14 11 17 29 31 25 27 20 19 12 13 26

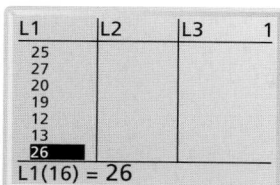

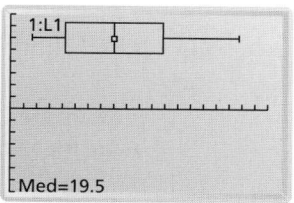

Step 1
Press STAT
1: Edit ... ENTER.
Enter data under
L1. (To clear L1,
move the cursor
over L1 and press
CLEAR, ENTER.)

Step 2
Press 2nd Y= for
STAT PLOTS.
Enter 1.
Highlight **ON** and
also Type:
box-and-whisker
plot. Press ENTER.

Step 3
Press ZOOM and
enter 9. Press
TRACE and move the
cursor across the
box-and-whisker
plot to find the five
number summary.

The five number summary is minimum = 10, Q_1 = 13.5, median = 19.5, Q_3 = 26.5, maximum = 35.

✓ **Quick Check** ⑤ Use a graphing calculator to find the five number summary for the following set of student test grades shown below.

98 92 76 84 93 82 74 68 85 91 77 83 94 97 72 88 70 84 87 82

The **interquartile range** is the difference between the first and third quartiles. In a box-and-whisker plot, this is the width of the box. You can use a parallel box-and-whisker plot to compare two sets of data.

6 EXAMPLE Comparing Data Sets

The box-and-whisker plot below shows average monthly rainfall for Miami and New Orleans. Which city shows the greater range in average monthly rainfall?

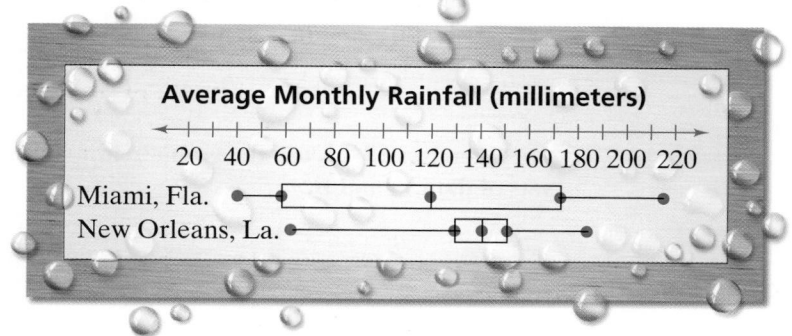

Miami has the longer box-and-whisker plot, so it has the greater range.

✓ **Quick Check** ⑥ What do the interquartile ranges tell you about the average monthly rainfall for Miami and New Orleans?

EXERCISES

For more exercises, see *Extra Skill and Word Problem Practice*.

Practice and Problem Solving

A Practice by Example

Example 1
(page NY 732)

Example 2
(page NY 733)

Example 3
(page NY 733)

Example 4
(page NY 734)

Example 5
(page NY 735)

Example 6
(page NY 735)

Find the median, the first quartile, and the third quartile.

1. 12 10 11 7 9 10 5

2. 4.5 3.2 6.3 5.2 5 4.8 6 3.9 12

3. 55 53 67 52 50 49 51 52 52

4. 101 100 100 105 101 102 104

5. Find the five number summary for the set of data.

   1 1 3 3 4 5 7 7 8 9 9 32 31 29 35 30 32 11

6. Of 20 scores, 19 are less than or equal to 10. Find the percentile rank of 10.

7. Gerry ranks 48th in a class of 120 students. What is his percentile rank?

8 Construct a box-and-whisker plot for the ages of the members of the art club.

   13 18 16 14 15 17 15 16 16 15 16 14 14 17 19

9. Use a graphing calculator to draw a box-and-whisker plot for the data below. Then find the interquartile range.

   16 18 59 75 30 34 25 49 27 16 21 58 71 19 50

10. Compare the box-and-whisker plots below. What can you conclude?

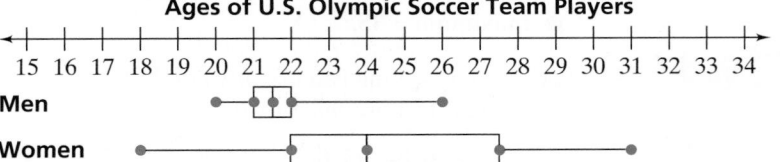

Ages of U.S. Olympic Soccer Team Players

B Apply Your Skills

For Exercises 11–16, use the data below. Find each measure.

   1.8 2.5 3.9 4.6 4.7 4.8 4.8 4.9

11. median

12. first quartile

13. minimum

14. maximum

15. third quartile

16. percentile rank of 4.6

17. Use the box-and-whisker plot below. What can you conclude about the areas of state parks?

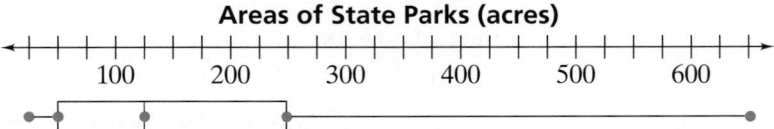

Areas of State Parks (acres)

For Exercises 18–21, use the graph below showing box-and-whisker plots for two sets of data, *A* and *B*.

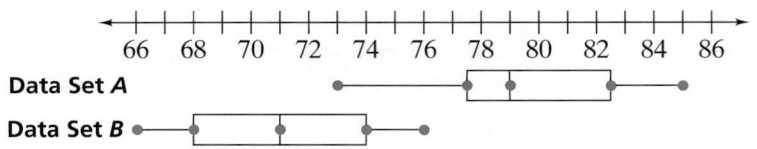

18. Which set of data has the greater range?

19. Which set of data has the lesser median?

20. Which set of data has the greater interquartile range?

21. Which set of data has the lesser minimum?

For Exercises 22–24, use the five number summary given below.

Minimum = 10 $Q_1 = 20$ Median = 30 $Q_3 = 42$ Maximum = 75

22. Find the interquartile range.   **23.** Find the range of the data.

24. Use the data to construct a box-and-whisker plot.

25. Social Studies The plots below compare the percents of the voting-age population who said they registered to vote in U.S. elections to the percents who said they voted. Which conclusion best reflects the data collected?

Percents of Population Who Registered and Voted, 1990–2000

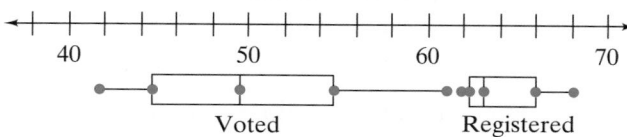

A. The percent who voted was about 15% less than the percent who registered.
B. The percent who voted was about half the percent who registered.
C. The percent who voted was equal to the percent who registered.
D. The percent who voted was about 15% more than the percent who registered.

26. Error Analysis In a class of 250 students, Emily had the tenth highest grade average. She computed her percentile rank as 4. What was her error?

C Challenge   **27. Reasoning** Can you find the mean, median, and mode of a set of data by looking at a box-and-whisker plot? Explain.

NY REGENTS

Test Prep

Multiple Choice   **28.** Find the first quartile and third quartile in the following data set.

17 20 30 19 20 18 25 28 31 23 17 29 31 33 28

A. 19 and 30   **B.** 17 and 25   **C.** 19.5 and 30.5   **D.** 17 and 33

29. Of 20 test scores, sixteen are less than or equal to 80. What is the percentile rank of a test score of 80?

F. 16th   **G.** 85th   **H.** 25th   **J.** 80th

Short Response   **30.** Describe how you could find the 75th percentile score in a set of 20 scores.

Mixed Review

Lesson 1-1   **Define variables and write an equation to model each situation.**

31. The total cost of the number of CDs times $5.00.

32. The perimeter of a square equals 4 times the length of the side.

Lesson 1-3   **Decide whether each statement is *true* or *false*. If the statement is false, give the counterexample.**

33. All whole numbers are rational numbers.

34. The square root of a number is always smaller than the number.

NY-3 Working With Sets

NY Learning Standards for Mathematics

A.A.29 Use set-builder notation and/or interval notation to illustrate the elements of a set, given the elements in roster form.

A.A.30 Find the complement of a subset of a given set, within a given universe.

✓ Check Skills You'll Need

GO ▶ for Help   Lesson 1-3

1. List the factors of 12.
2. List the first five prime numbers.

For Exercises 3–5, match each number in the first column with the corresponding letter in the second column.

3. natural numbers
4. whole numbers
5. integers

A. whole numbers and their opposites
B. the nonnegative integers
C. the counting numbers

🔊 **New Vocabulary** • set • roster form • set-builder notation • subset • null set • universal set • complement of a set • interval notation

---

1 | Working with Sets

Vocabulary Tip

The word *set* is often used to describe a group of things such as a set of dishes or a set of luggage.

A **set** is a collection of elements or members. There are various ways to describe sets. Use braces, { }, to denote a set.

Roster form lists elements of a set within braces. For example, you can write the set consisting of 1, 2, and 3 as {1, 2, 3}. Use " … " to describe an infinite set, such as the set of natural numbers {1, 2, 3, …}.

Set-builder notation describes elements of a set. It uses a variable and limits, or conditions, on the variable. For example, you read the set-builder notation $\{x \mid x \text{ is a factor of } 12\}$ as "the set of all values of x such that x is a factor of 12."

1 EXAMPLE Using Roster Form and Set-Builder Notation

a. Write "P is the set of whole numbers less than 5" in roster form.

"P is" can be written as "$P =$."   **List all whole numbers less than 5.**

$$P = \{0, 1, 2, 3, 4\}$$

b. Write "P is the set of whole numbers less than 5" using set-builder notation.

Use a variable.   **Describe limits on the variable.**

$$P = \{x \mid x \text{ is a whole number and } x < 5\}$$

Read "$\{x \mid$" as "the set of all values of x such that."

✓ **Quick Check** ❶ M is the set of odd whole numbers less than 12.
  a. Write set M in roster form.
  b. Write set M in set-builder notation.

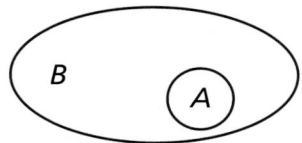

A **subset** consists of elements from any given set. If $B = \{1, 2, 3, 4, 5, 6, 7\}$ and $A = \{1, 2, 5\}$, then A is a subset of B. That means A is contained in B and is indicated by $A \subseteq B$.

The **null set,** or empty set, is a set that contains no elements. The null set is a subset of every set. Use { } or Ø to represent the null set.

2 EXAMPLE Subsets

List all possible subsets of the set $\{1, 2, 3\}$.

{ }   **Start with the null set.**

$\{1\}, \{2\}, \{3\}$ **List the subsets with one element.**

$\{1, 2\}, \{1, 3\}, \{2, 3\}$ **List the subsets with two elements.**

$\{1, 2, 3\}$ **List the original set. It is always considered a subset.**

● The 8 subsets of $\{1, 2, 3\}$ are { }, $\{1\}$, $\{2\}$, $\{3\}$, $\{1, 2\}$, $\{1, 3\}$, $\{2, 3\}$, and $\{1, 2, 3\}$.

✓ Quick Check **2** List the subsets of the set $U = \{x, y\}$.

When working with sets, the largest set you are using is sometimes called the **universal set,** or universe. The **complement of a set** contains the elements of a universal set not contained in a given set. The complement of set A is indicated by A'.

In the figures below, set U represents the universal set. Notice that $A \subseteq U$ and $A' \subseteq U$.

Set *A* is shaded.   **The complement of set *A* is shaded.**

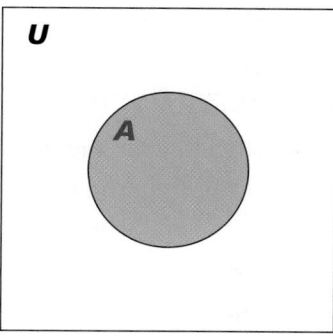   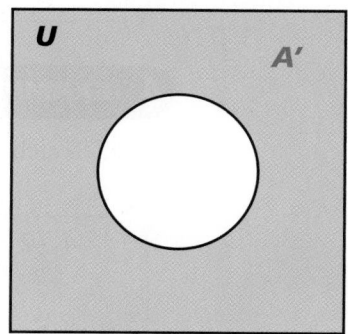

3 EXAMPLE Finding the Complement

Given $A \subseteq W$, $W = \{$days of the week$\}$, and $A = \{$Saturday, Sunday$\}$. Find the complement of set A.

Elements of W not included in A are Monday, Tuesday, Wednesday, Thursday, and Friday.

● So, $A' = \{$Monday, Tuesday, Wednesday, Thursday, Friday$\}$.

✓ Quick Check **3** Given $P \subseteq T$, $T = \{1, 2, 3, 4, 5, 6, 7, 8\}$, and $P = \{2, 4, 6, 8\}$. Find P'.

An inequality such as $x \leq -2$ describes a portion of a number line called an interval. You can also describe intervals using interval notation. **Interval notation** includes three special symbols:

parentheses: Use (or) when endpoints of the interval are not included ($<$ or $>$).

brackets: Use [or] when endpoints of the interval are included ($\leq$ and $\geq$).

infinity: Use ∞ when the interval continues forever in a positive direction.

         Use $-\infty$ when the interval continues forever in a negative direction.

Inequality	Graph	Interval Notation
$2 \leq x \leq 6$		$[2, 6]$
$2 < x < 6$		$(2, 6)$
$2 < x \leq 6$		$(2, 6]$
$2 \leq x < 6$		$[2, 6)$
$x \geq 3$		$[3, \infty)$
$x < 3$		$(-\infty, 3)$

4 **EXAMPLE**   **Using Interval Notation**

a. Sketch the graph of $[-2, 8)$ and write the interval as an inequality.

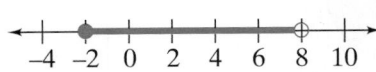

The closed dot shows that –2 is included.
The open dot shows that 8 is not included.
Shade between –2 and 8.

The interval notation $[-2, 8)$ represents the inequality $-2 \leq x < 8$.

b. Sketch the graph of $x \leq 8$ and write the inequality as an interval.

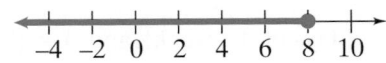

The closed dot shows that 8 is included.
Shade to the left of 8.

The interval notation $(-\infty, 8]$ represents the inequality $x \leq 8$.

 **Quick Check**   **4**   **a.** Sketch the graph of $x < 5$ and write the inequality as an interval.
                 b. Sketch the graph of $(-2, 5]$ and write the interval as an inequality.

EXERCISES

For more exercises, see *Extra Skill and Word Problem Practice.*

Practice and Problem Solving

A Practice by Example

Example 1
(page NY 738)

GO for Help

Write in roster form.

1. $M = \{x \mid x$ is a positive multiple of 2 and $x < 18\}$

2. $T = \{x \mid x$ is an integer and $x \geq 12\}$

3. $N = \{n \mid n$ is an odd integer, $n > 0\}$

4. $X = \{x \mid x$ is an even prime$\}$

Write in set-builder notation.

5. $B = \{11, 12, 13, 14, \ldots\}$

6. $M = \{2, 3, 5, 7, 11, 13, 17, 19\}$

7. $S = \{1, 2, 3, 4, 6, 12\}$

8. $G = \{\ldots, -2, -1, 0, 1, 2, \ldots\}$

Example 2
(page NY 739)

List all subsets for the each set.

9. $\{a, b\}$ **10.** $\{0, 1, 2\}$ **11.** $\{1, 2, z\}$

Example 3
(page NY 739)

12. Given $A \subseteq B$, $B = \{1, 2, 3, 4, 5\}$, and $A = \{2, 3\}$. Find A'.

13. Given $P \subseteq U$, $U = \{1, 2, 3, 4, 5, 6, 7, 8, 9\}$, and $P = \{2, 4, 6, 8\}$. Find P'.

Example 4
(page NY 740)

Write interval notation for each graph.

14.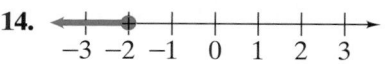

15.

16.

17.

B Apply Your Skills

Write each situation in interval notation.

18. Let t be truck weight in tons.

19. Let s be speed in mi/h.

Write the following inequalities in interval notation.

20. $0 \leq x \leq 7$ **21.** $-4 \leq z < -2$ **22.** $-6 \leq a < 5$

23. $-3 \leq x < 2$ **24.** $-3 \leq y \leq 5$ **25.** $-4 < a < 2$

26. $25 < z \leq 30$ **27.** $15 < x < 55$ **28.** $-75 \leq p \leq 150$

Suppose $U = \{0, 1, 2, 3, 4, 5, 6\}$, $A = \{2, 4, 6\}$, and $B = \{1, 2, 3\}$. Identify each statement as *true* or *false*. Justify your answer.

29. $A \subseteq U$ **30.** $U \subseteq B$ **31.** $B \subseteq A$ **32.** $\emptyset \subseteq B$

Nutrition For Exercises 33–35, use the following data. Answer in roster form. Rosalie's breakfast consists of peanut butter, toast, grapefruit, and orange juice. Miguel's breakfast consists of orange juice, cereal, milk, and toast.

33. Rosalie and Miguel are eating breakfast at the same table. Write the set of their breakfast foods.

34. Write the set of foods Rosalie and Miguel have in common.

35. Open-Ended Write the set of breakfast foods you eat that are not in Rosalie's set or Miguel's set.

For Exercises 36–39, refer to universal set $U = \{a, b, c, d, e, f, g, 1, 2, 3, 4, 5, 6, 7\}$.

36. Let $P = \{numbers\}$. Find P'.

37. Let $V = \{letters\}$. Find V'.

38. Let $W = \{even\ numbers\}$. Find W'.

39. Let $M = \{odd\ numbers\ and\ letters\}$. Find M'.

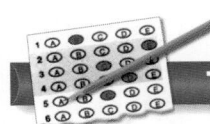 **Challenge** If the given set A has n elements, then set A has 2^n subsets. How many subsets are there for each finite set?

40. $R = \{even\ numbers\ less\ than\ 20\}$ **41.** $Q = \{\ \}$

Test Prep

Multiple Choice

42. Set $U = \{natural\ numbers\}$ and set $E = \{2, 4, 6, 8, \ldots\}$. What is E'?
   A. $\{1, 3, 5, 7, \ldots\}$ **B.** $\{0, 2, 4, 6, 8, \ldots\}$
   C. $\{all\ positive\ integers\}$ **D.** $\{2, 4, 6, 8, \ldots\}$

43. Which of the following is interval notation for "all numbers less than 10"?
   F. $\{\ldots, 10\}$ **G.** $(\infty, 10)$ **H.** $(-\infty, 10)$ **J.** $(-\infty, 10]$

Short Response

44. Write $-8 \leq x < 4$ in interval notation and sketch the graph.

Mixed Review

Lesson 4-5

For Exercises 45–48, solve each compound inequality.

45. $-2b - 1 < 7$ or $3b + 2 < 26$ **46.** $4 + q \geq 3$ or $5q < -30$

47. $2k + 6 \leq -1$ or $-2k + 6 \leq 6$ **48.** $3c - 5 > 10$ or $\frac{-3c - 2}{5} \leq -3$

49. Write the compound inequality that represents "all real numbers that are less than -4 or greater than 8."

Lesson NY-2

For Exercises 50–53 use the data below.
55, 50, 60, 65, 65, 35, 40, 40, 45, 65, 55, 75, 30, 35, 55, 60, 45, 55, 35, 55

50. Make a box-and-whisker plot.

51. Label the median, first quartile, third quartile, and outliers.

52. What are the values for the five number summary?

53. What is the value at the fiftieth percentile?

Union and Intersection of Sets

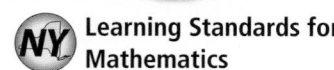 **Learning Standards for Mathematics**

A.A.31 Find the intersection of sets (no more than three sets) and/or union of sets (no more than three sets).

✓ Check Skills You'll Need

GO for Help Lesson NY-3

Write each set in set-builder notation.
1. $A = \{0, 1, 2, 3, 4, 5, 6, 7, 8, 9\}$ **2.** $B = \{1, 3, 5, 7\}$

Write each set in roster form.
3. $C = \{n \mid n$ is an even number between -15 and $-5\}$
4. $D = \{k \mid k$ is a composite number between 7 and 17$\}$

◀)) **New Vocabulary** • union • intersection • disjoint sets

---

1 Operations on Sets

Vocabulary Tip

$A \cup B$ is read "A union B," or "the union of sets A and B."

In Lesson NY-3, you examined different ways to express sets. In this lesson, you will perform operations on these sets.

The **union** of two or more sets is the set that contains all elements of the sets. The symbol for union is ∪. To find the union of two sets, list the elements that are in either set, or in both sets. In the Venn diagram below, $A \cup B$ is shaded.

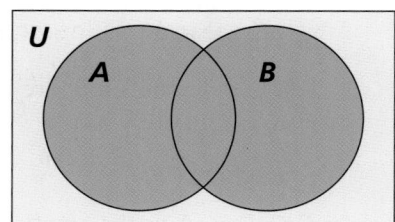

1 EXAMPLE Union of Sets

A bouquet of flowers contains roses, carnations, and baby's breath. A second bouquet has roses, lilies, and daisies. Both bouquets are put in the same vase. Use union of sets to find the set of flowers in the vase.

first bouquet: $B = \{$roses, carnations, baby's breath$\}$ ◀─┐ **Write each**
second bouquet: $S = \{$roses, lilies, daisies$\}$ ◀─┘ **bouquet as a set.**

List the flowers that are in either bouquet, or in both bouquets.

● $B \cup W = \{$roses, carnations, baby's breath, lilies, daisies$\}$

✓ **Quick Check** **❶** $P = \{5, 10, 15, 20\}$ and $Q = \{8, 10, 18, 20\}$. Find $P \cup Q$.

Vocabulary Tip

$A \cap B$ is read "A intersection B," or "the intersection of sets A and B."

The **intersection** of sets is the set of elements that are common to two or more sets. The symbol for intersection is $\cap$. When you find the intersection of two sets, list only the elements that are in both sets. The shaded area below shows $A \cap B$.

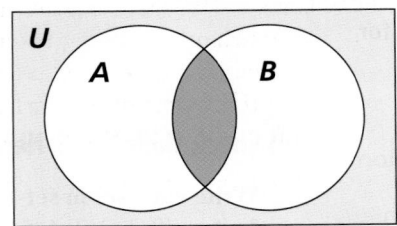

Disjoint sets have no elements in common. The intersection of disjoint sets is the empty set. The diagram below shows two disjoint sets.

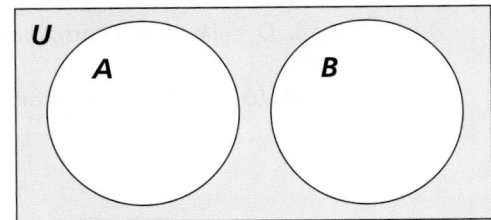

2 EXAMPLE Intersection of Sets

Set $C = \{x \mid x$ is a natural number less than 20$\}$, set $D = \{y \mid y$ is an odd integer$\}$, and set $E = \{z \mid z$ is a multiple of 4$\}$.

a. Find $C \cap D$.

List the elements that are both odd integers and natural numbers less than 20.
$C \cap D = \{1, 3, 5, 7, 9, 11, 13, 15, 17, 19\}$

b. Find $C \cap E$.

List the elements that are both multiples of 4 and natural numbers less than 20.
$C \cap E = \{4, 8, 12, 16\}$

c. Find $D \cap E$.

$D \cap E = \varnothing$, or the empty set. There are no multiples of 4 that are also odd, so these are disjoint sets. They have no elements in common.

 **Quick Check** **2** $X = \{2, 4, 6, 8, 10\}$, $Y = \{0, 2, 5, 7, 8\}$, and $Z = \{n \mid n$ is an odd integer$\}$.
 a. Find $X \cap Y$. **b.** Find $X \cap Z$. **c.** Find $Z \cap Y$.

2 Solving Problems With Venn Diagrams

You can use Venn diagrams to show the relationships between sets. You can write the elements inside the appropriate section of the diagram. Elements in the intersection of sets appear in the overlapping sections of the Venn diagram.

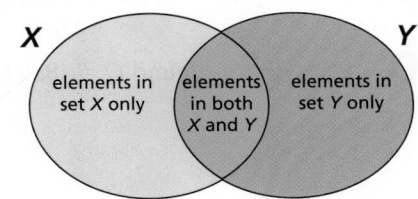

Mt. Rushmore is
1,745 m tall.

3 EXAMPLE   Venn Diagrams: Three Sets

Presidents The set of presidents on Mount Rushmore can be represented by the set R = {Washington, Jefferson, Lincoln, T. Roosevelt}. The set of presidents born in February can be represented by the set F = {Washington, Harrison, Lincoln, Reagan}. Presidents who had beards can be represented by the set B = {Lincoln, Grant, Garfield, Harrison}. Draw a Venn diagram to represent the union and intersection of these sets.

Draw and label three intersecting circles to represent the sets.

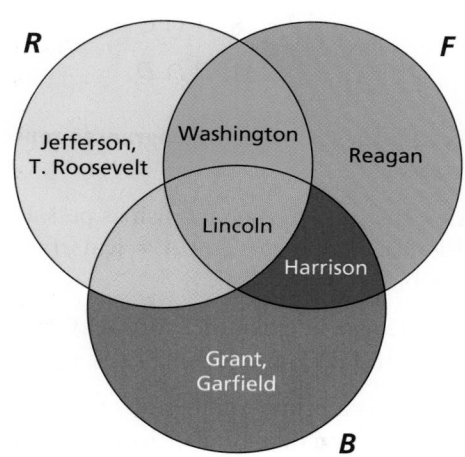

Step 1 Lincoln is in all three sets. Put Lincoln in the center area.

Step 2 $R \cap F$ = {Washington, Lincoln} Put Washington in the green area.

Step 3 $F \cap B$ = {Lincoln, Harrison} Put Harrison in the purple area.

Step 4 $R \cap F$ = {Lincoln} The orange area remains empty.

Step 5 Place the remaining elements of each set into the corresponding yellow, blue, and red sections.

Note that each element only appears in the Venn diagram in one location.

 **③** Let X = {a, e, i, o, u}, Y = {a, b, c, d, e, f}, and Z = {a, c, e}. Draw a Venn diagram to represent the intersection and union of the sets.

You can also use Venn diagrams to show the number of elements in the union or intersection of a set.

4 EXAMPLE   Real-World 🌐 Connection

Pets Of 50 cat and dog owners surveyed, 25 have a cat. Ten owners have a dog and a cat. How many owners have a dog?

Step 1 Draw a Venn diagram. Let C = owners with cats and D = owners with dogs.

Step 2 The intersection of C and D represents the owners with both cats and dogs. $C \cap D$ = 10.

Vocabulary Tip

The total number is the union of the two sets, $C \cup D$.

Step 3 Find the number of owners with only a cat. $25 - 10 = 15$. Enter 15 into the Venn diagram.

Step 4 The total number of owners is 50. Subtract to find the number of owners with only a dog. $50 - 15 - 10 = 25$.

The total number of dog owners is $25 + 10 = 35$.

 **④** Of 240 college freshmen, 152 are taking history and 81 are taking science and history. How many freshmen are taking history but not science?

EXERCISES

For more exercises, see *Extra Skill and Word Problem Practice*.

Practice and Problem Solving

 Practice by Example

 **GO for Help**

Example 1
(page NY 743)

Example 2
(page NY 744)

**For Exercises 1–12, find each union or intersection. Let $A = \{2, 5\}$, $B = \{5, 7, 9\}$,
$C = \{x \mid x$ is an odd number less than 9$\}$, and $D = \{x \mid x$ is an even number less than 9$\}$.**

1. $A \cup B$ 2. $A \cup C$ 3. $A \cup D$

4. $B \cup C$ 5. $B \cup D$ 6. $C \cup D$

7. $A \cap B$ 8. $A \cap C$ 9. $A \cap D$

10. $B \cap C$ 11. $B \cap D$ 12. $C \cap D$

Example 3
(page NY 745)

For Exercises 13–16, draw a Venn diagram to represent the union and intersection of the given sets.

13. **Pets** Alex has cats, rabbits, and fish as pets. Becky has cats and dogs. Cory has cats, birds, fish, and turtles. Let $A = \{$cats, rabbits, fish$\}$, $B = \{$cats, dogs$\}$, and $C = \{$cats, birds, fish, turtles$\}$.

14. Let $X = \{x \mid x$ is a letter in the word ALGEBRA$\}$, $Y = \{y \mid y$ is a letter in the word GEOMETRY$\}$, and $Z = \{z \mid z$ is a letter in the word CALCULUS$\}$.

15. Let $P = \{x \mid x$ is a prime number less than 10$\}$, $C = \{1, 2, 3, 4, 5, 6, 7, 8, 9, 10\}$, and $O = \{a \mid a$ is an odd number less than 10$\}$.

16. Let $L = \{a, b, c, 1, 2, 3,$ horse, cow, pig$\}$, $M = \{-1, 0, 1, b, y,$ pig, duck, $\triangle\}$, and $N = \{c, 3,$ duck, $\triangle\}$.

Example 4
(page NY 745)

17. Of 75 people with cell phones, 42 take pictures with their cell phone and 36 use their cell phone to take pictures and send text messages. How many use their phones to send text messages?

18. Of 100 people in a band, 70 members said they play a sport and 50 members play a sport and take music lessons. How many take music lessons?

19. An ice cream shop owner surveys 200 people who eat chocolate and vanilla ice cream. If 154 people like both flavors, and 196 people like vanilla, how many people like chocolate?

 Apply Your Skills

20. **Reasoning** Find two sets A and B such that $A \cup B = \{1, 2, 3, 4, 5\}$ and $A \cap B = \{2\}$.

21. **Writing** Let $M = \{x \mid x$ is a multiple of 3$\}$ and $N = \{x \mid x$ is a multiple of 4$\}$. Describe the intersection of M and N.

22. **Critical Thinking** Set X has 10 elements, set Y has 15 elements, and $X \cap Y$ has 5 elements. How many elements are in $X \cup Y$? Explain.

For Exercises 23–25, identify each statement as true or false. Use a Venn diagram or give a counterexample to justify your answer.

23. The intersection of two sets is always a subset of their union.

24. Two sets that contain no elements in common are disjoint sets.

25. The intersection of the set of even numbers and the set of prime numbers is the empty set.

The planetarium at the American Museum of Natural History in New York City is spherical.

26. Of students surveyed, 80 had visited the planetarium, 75 had visited Ellis Island, and 65 had been to Niagara Falls. 30 students went to both the planetarium and Ellis Island, 25 went to the planetarium and Niagara Falls, and 10 went to Ellis Island and Niagara Falls. Only 5 went to all three.
   a. How many students went only to the planetarium?
   b. How many students went only to Niagara Falls?
   c. How many students did not go to Ellis Island but went to the planetarium or Niagra Falls?
   d. How many total students were surveyed?

27. Open-Ended Find two sets A and B such that $A \cup B = \{1, 2, 3, 4, 5\}$ and $A \cap B = \varnothing$.

28. Let $P = \{x \mid x$ is a prime number$\}$ and $Q = \{x \mid x$ is a multiple of 6$\}$. Describe the intersection of P and Q.

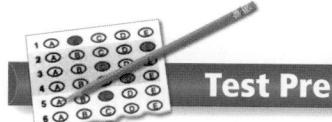 **Challenge**

For Exercises 29–32, write each set in roster notation. Let $X = \{a, c, e, g, i\}$, $Y = \{b, c, e, f\}$, and $Z = \{b, d, f\}$.

29. $Z \cap (X \cup Y)$

30. $Z \cup (X \cap Y)$

31. $Y \cup (Z \cap X)$

32. $(Y \cup Z) \cap X$

33. Critical Thinking Is the equality $(A \cup B) \cap C = A \cup (B \cap C)$ *always*, *sometimes*, or *never* true? Justify your answer.

 **REGENTS**

Test Prep

Multiple Choice

34. Consider the set D of all numbers in the interval $(-\infty, 5)$. What is the complement of D?
   A. $(-\infty, -5)$ **B.** $[5, +\infty)$ **C.** $(-\infty, 5)$ **D.** $(5, +\infty)$

35. Set $X = \{x \mid x$ is a factor of 12$\}$ and set $Y = \{y \mid y$ is a factor of 16$\}$. Which set represents $X \cap Y$?
   F. $\varnothing$ **G.** $\{0, 1, 2, 4\}$
   H. $\{1, 2, 4\}$ **J.** $\{1, 2, 3, 4, 6, 8, 12, 16\}$

Short Response

36. Let the universal set U represent the integers from 1 to 10. Use the subsets of $M = \{x \mid x$ is a factor of 8$\}$, and $N = \{1, 4, 6, 7, 10\}$ to find the complement of $M \cup N$.

Mixed Review

Lesson NY-1

Classify each as qualitative or quantitative data.

37. number of movies watched

38. favorite pizza toppings

39. colors of passing cars

40. SAT math scores

Lesson NY-2

Find the median, the first quartile, and the third quartile of each data set.

41. 3   5   8   2   11   15   13

42. 17   20   11   34   91   12   29

43. 105   121   151   89   93   141

NY-5 Related Data Sets

A.S.13 Understand the difference between correlation and causation.

A.S.14 Identify the variables that may have a correlation but not a causal relationship.

A.S.17 Use a reasonable line of best fit to make a prediction involving interpolation and extrapolation.

☑ **Check Skills You'll Need**

GO **for Help** Lesson 1-5

Is there a positive, negative, or no correlation for each scatter plot?

1.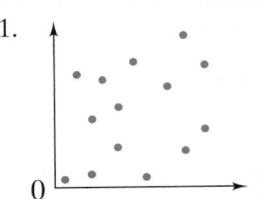
2.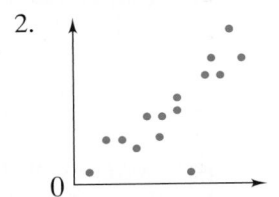
3.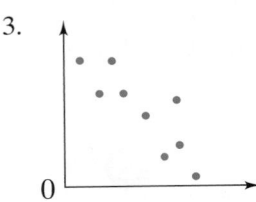

◀)) **New Vocabulary** • interpolation • extrapolation • causation

1 Predictions

The scatter plot below shows education and income data with a line of best fit. **Interpolation** is estimating a value among existing data. **Extrapolation** is predicting a value outside the existing data.

1 EXAMPLE Interpolation

Use the scatter plot to estimate the income of a person with 13 years of education.

Since $x = 13$ is visible in the graph, use interpolation to read the graph of the line.

The y-value is about 63.

● The estimated income for a person with 13 years of education is $63,000.

☑ **Quick Check** ❶ Use the scatter plot to estimate income of a person with 9 years of education.

Learn and Earn

(Scatter plot: y-axis labeled "Annual Incomes ($1,000)" from 0 to 120; x-axis labeled "Years of Education Completed" from 8 to 16.)

2 EXAMPLE Extrapolation

Use the scatter plot above. Estimate the income of a person with 18 years of education.

Since $x = 18$ is not visible in the graph, use extrapolation to extend the graph of the line. When $x = 18$, the y-value is about 116.

● The estimated income for the average person with 18 years of education is $116,000.

☑ **Quick Check** ❷ Use the scatter plot above. Estimate the income of a person with 7.5 years of education.

For help with using a graphing calculator to find a line of best fit, go to Lesson 6-7.

Average Ticket Price	Numbers of Admissions (millions)
$4.18	1292
$4.42	1339
$4.69	1481
$5.39	1421
$5.81	1639

You can use a graphing calculator to draw a scatter plot and find the line of best fit.

3 EXAMPLE Find the Line of Best Fit and Make Predictions

Use the data in the table at the left and a graphing calculator to estimate the number of admissions for ticket prices of $5.00 and $6.00.

Step 1 Use a graphing calculator to plot the data and find the line of best fit.

Step 2 Use TRACE to find the y-value for an x-value of $5.00.

Step 3 Use TRACE to find the y-value for an x-value of $6.00.

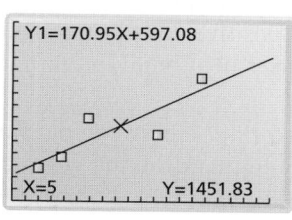

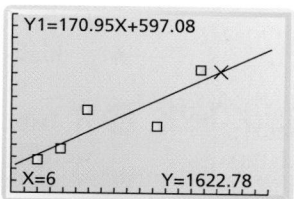

The y-value is about 1452.

The y-value is about 1623.

An interpolated estimate for $5.00 is about 1452 million admissions.
An extrapolated estimate for $6.00 is about 1623 million admissions.

✓ Quick Check **3 a.** Find an estimated number of admissions for a ticket price of $4.60.
b. Is your estimate an interpolation or an extrapolation? Explain.

---

2 Relationships Between Data Sets

Problem Solving Hint
Causal relationships always have a correlation. However, quantities that have a correlation may not have causation.

Causation is when a change in one quantity causes a change in a second quantity. For example, as study time increases, test scores increase.

4 EXAMPLE Identifying Whether Relationships are Causal

Error Analysis Mr. Singh made a scatter plot of the number of sunflower seeds he ate and the number of runs the New York Yankees scored. He concluded that eating more seeds caused the Yankees to score more runs. Explain Mr. Singh's reasoning error.

The graph shows a positive correlation between the number of seeds and the number of runs.

There is no causal relationship between the number of seeds and the number of runs.

✓ Quick Check **4** Is there a causal relationship between the cost of a gallon of gas and the color of cars that people choose? Explain.

Seeds and Runs Scored

(scatter plot: Runs Scored vs. Seeds Eaten)

5 EXAMPLE Understanding Correlation and Causation

In each of the following situations, is there likely to be a correlation? Does that correlation reflect a causal relationship? Explain.

a. the price of gas and the number of vacation miles driven
b. the price of a car and the price of gas
c. the speed of a car and the wear on its tires

a. Yes, there is likely to be a negative correlation and also a causal relationship. As the price of gas goes up, people tend to drive less on vacation.
b. Some data sets may show a correlation. However, there is no causal relationship, because a rise in the car price does not cause a rise in the gas price.
c. There is likely to be a positive correlation. There is also causation, because high-speed driving causes heat buildup that increases tire wear.

✓ Quick Check ⑤ A scatter plot shows a correlation between daily amount of water drunk and weight gain. Does this mean that drinking water causes weight gain? Explain.

EXERCISES

For more exercises, see *Extra Skill and Word Problem Practice*.

Practice and Problem Solving

Ⓐ Practice by Example

Examples 1 and 2
(page NY 748)

Use the line of best fit at the right to estimate the number of fat grams for each given amount of Calories. Identify each estimate as interpolation or extrapolation.

1. 359 Calories   **2.** 114 Calories
3. 605 Calories   **4.** 514 Calories

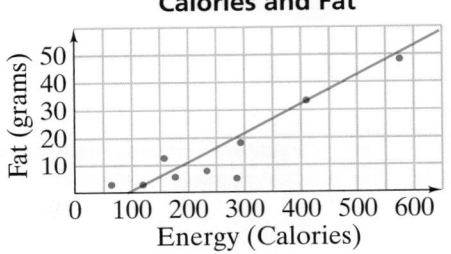

Example 3
(page NY 749)

Use the table below for Exercises 5–8. Use a graphing calculator to estimate the number of television sets in homes based on the following values for newspaper circulation (in millions).

5. 58
6. 62
7. 60.5
8. 55

Year	91	92	93	94	95	96
Daily Newspaper Circulation (millions)	61	60	60	59	57	57
Television Sets in Homes (millions)	193	192	201	211	217	233

Examples 4 and 5
(pages NY 749 and NY 750)

Tell whether each correlation is a causal relationship. Justify your answer.

9. Weight and height have a positive correlation. Is this a causation? (Does gaining weight cause you to grow taller?)

10. Age and time spent playing video games is a negative correlation. Is this a causation? (Does getting older cause you to play fewer video games?)

11. Number of cavities and size of vocabulary have a positive correlation. Is this a causation? (Does getting more cavities cause the size of your vocabulary to increase?)

12. Time spent studying for a test and the test score have a positive correlation. Is this a causation?

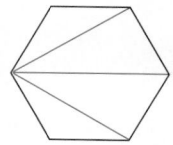

B **Apply Your Skills**

13. Geometry From one vertex of a hexagon, you can draw three diagonals. Is there a causal relationship between the number of sides of a polygon and the number of diagonals? Explain.

14. Writing Look at the scatter plot for Example 1 showing education and income data. Is there a correlation? Is there a causal relationship? Explain.

15. Critical Thinking Is there a correlation between the age of a used car and the asking price for selling the car? Is there a causal relationship? Explain.

16. Open-Ended Describe two data sets that are correlated but do not have a causal relationship.

C **Challenge**

17. The table shows the number of hours of daylight on the first day of each month. For example, (3, 11) represents 11 hours of daylight on March 1. Use a graphing calculator to find the line of best fit and make the following estimates.
  a. the number of hours of daylight on February 16 ($x = 2.5$)
  b. the number of hours of daylight on July 1 ($x = 7$)
  c. Critical Thinking Explain why the answer to part(b) is less accurate than the answer to part (a).
  d. Is an interpolated or an extrapolated estimate more likely to be accurate? Why?

Month	Hours of Daylight
1	9.21
2	9.92
3	11.02
4	12.41
5	13.68
6	14.63

NY REGENTS

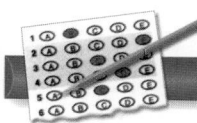

Test Prep

Multiple Choice

18. Which correlation is most likely to be a causation as well?
  A. the number of ice cream trucks and the number of shark attacks
  B. the age of a car and the number of miles on the car
  C. the latitude of a city and its population
  D. a baseball player's batting average and his weight

Short Response

19. Use a graphing calculator to estimate the amount the student has in her bank account after 6 weeks.

Week	1	3	4	7	9
Account Balance	$35	$68	$85	$105	$136

Mixed Review

Lesson NY-4

20. Draw a Venn diagram to represent the union and intersection of the following sets. $A = \{1, 2, 3, 4, 6\}$, $B = \{1, 5, 6\}$, and $C = \{2, 4, 6, 7\}$.

Find each union or intersection. Let $A = \{x \mid x$ is prime number greater than 2 and less than 20\}, $B = \{1, 5, 9, 13, 19\}$, and $C = \{x \mid x$ is multiple of 3 less than 21\}.

21. $A \cap B$ **22.** $A \cap C$ **23.** $A \cup B$ **24.** $C \cap B$

Lesson NY-3

List all subsets for each set.

25. $\{x, y\}$ **26.** $\{1, 3, 5\}$ **27.** $\{a, b, c, d\}$

28. Given $A \subseteq B$, $B = \{1, 3, 6, 9, 12\}$, and $A = \{3, 9\}$. Find A'.

Systems of Linear and Quadratic Equations

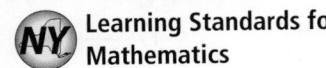
A.A.11 Solve a system of
one linear and one
quadratic equation in
two variables, where only
factoring is required.

A.G.9 Solve systems of linear
and quadratic equations
graphically.

✓ Check Skills You'll Need

1. Solve the system using substitution.
$$x = y + 2$$
$$4x + y = 8$$

3. Solve $x^2 - 5x + 6 = 0$ by factoring.

 **for Help**   Lessons 7-1, 7-2, and 10-4

2. Solve the system by graphing.
$$y = 2x + 3$$
$$x = y$$

---

1   Solving Systems Using Graphing

In Lesson 7-1, you solved systems of linear equations graphically and algebraically.
A system of linear equations can have either one solution, no solutions, or infinitely
many solutions. In Chapter 10, you solved quadratic equations graphically and
algebraically.

In this lesson, you will study systems of linear and quadratic equations. This type of
system can have one solution, two solutions, or no solutions.

$$y = x^2 - 4$$
$$y = -3$$

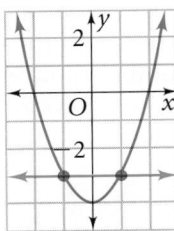

two solutions

$$y = x^2$$
$$y = 0$$

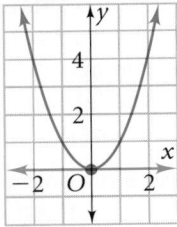

one solution

$$y = x^2 + 4$$
$$y = x + 1$$

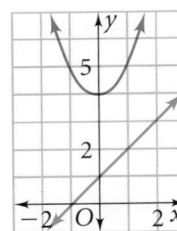

no solutions

1 EXAMPLE   Solve by Graphing

Solve the following system by graphing. $y = x^2 + x - 2$
$$y = -x + 1$$

Graph both equations on the same coordinate plane.
Identify the point(s) of intersection, if any.

The points $(-3, 4)$ and $(1, 0)$ are the solutions of
the system.

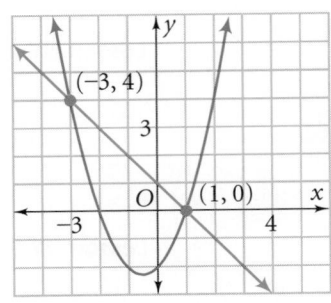

✓ Quick Check   **①** Solve the system by graphing. $y = 2x + 2$
$$y = -x^2 - x + 2$$

2 **EXAMPLE** 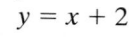 **Graphing to Count Solutions**

Find the number of solutions for the system. $y = 2x^2 + 3$
$$y = x + 2$$

Step 1 Graph both equations on the same coordinate plane.

Step 2 Identify the point(s) of intersection, if any.

There are no points of intersection, so there is no solution to the system of equations.

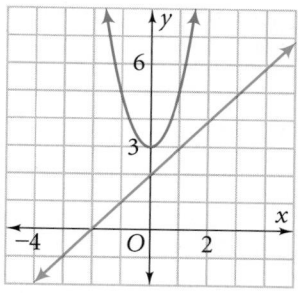

Quick Check **2** Find the number of solutions for each system.

a. $y = x - 4$
$y = 2x^2 + x$

b. $y = x^2 - 6x + 10$
$y = 1$

---

2 **Solving Systems Using Algebraic Methods**

In Lesson 7-3, you solved linear systems using elimination. The same technique can be applied to systems of linear and quadratic equations.

3 **EXAMPLE** **Using Elimination**

Solve the following system of equations: $y = x^2 - 11x - 36$
$$y = -12x + 36$$

Step 1 Eliminate y.

$$\begin{aligned} y &= x^2 - 11x - 36 \\ -(y &= \quad\; -12x + 36) \\ \hline 0 &= x^2 + x - 72 \end{aligned}$$

Subtract the two equations.
Subtraction Property of Equality

Step 2 Factor and solve for x.

$$0 = x^2 + x - 72$$
$$0 = (x + 9)(x - 8)$$ Factor.
$$x + 9 = 0 \quad \text{or} \quad x - 8 = 0$$ Zero-Product Property
$$x = -9 \quad \text{or} \quad\quad x = 8$$

Step 3 Find the corresponding y values. Use either equation.

$$y = x^2 - 11x - 36 \qquad\qquad y = x^2 - 11x - 36$$
$$y = (-9)^2 - 11(-9) - 36 \qquad y = (8)^2 - 11(8) - 36$$
$$y = 81 + 99 - 36 \qquad\qquad y = 64 - 88 - 36$$
$$y = 144 \qquad\qquad\qquad\quad y = -60$$

The solutions are $(-9, 144)$ and $(8, -60)$.

Quick Check **3** Solve the system using elimination. $y = x^2 + 4x - 1$
$$y = 3x + 1$$

4 EXAMPLE **Using Substitution**

Solve the following system of equations: $y = x^2 - 6x + 9$ and $y + x = 5$.

Step 1 Solve $y + x = 5$ for y.

$$y + x - x = 5 - x$$ — Subtract x from both sides.
$$y = 5 - x$$

Step 2 Write a single equation containing only one variable.

$$y = x^2 - 6x + 9$$
$$5 - x = x^2 - 6x + 9$$ — Substitute $5 - x$ for y.
$$5 - x - (5 - x) = x^2 - 6x + 9 - (5 - x)$$ — Subtract $5 - x$ from both sides.
$$0 = x^2 - 5x + 4$$

Step 3 Factor and solve for x.

$$0 = (x - 4)(x - 1)$$ — Factor.
$$x - 4 = 0 \quad \text{or} \quad x - 1 = 0$$ — Zero-Product Property
$$x = 4 \quad \text{or} \quad x = 1$$

Step 4 Find the corresponding y-values. Use either equation.

$$y = -x^2 + 4x + 1 \qquad\qquad y = -x^2 + 4x + 1$$
$$= -(4^2) + 4(4) + 1 \qquad\qquad = -(1^2) + 4(1) + 1$$
$$= 1 \qquad\qquad\qquad\qquad\quad = 4$$

● The solutions of the system are $(4, 1)$ and $(1, 4)$.

✓ **Quick Check** ④ Solve the system using substitution. $y - 30 = 12x$
$$y = x^2 + 11x - 12$$

In Lesson 10-7, you used the discriminant to find the number of solutions of a quadratic equation. With systems of linear and quadratic equations you can also use the discriminant once you eliminate a variable.

5 EXAMPLE **Using the Discriminant to Count Solutions**

At how many points do the graphs of $y = 2$ and $y = x^2 + 4x + 7$ intersect?

Step 1 Eliminate y from the system. Write the resulting equation in standard form.

$$y = x^2 + 4x + 7$$
$$-\underline{(y = \qquad\qquad 2)}$$ — Subtract the two equations.
$$0 = x^2 + 4x + 5$$ — Subtraction Property of Equality

Step 2 Determine whether the discriminant, $b^2 - 4ac$, is positive, 0, or negative.

$$b^2 - 4ac = 4^2 - 4(1)(5)$$ — Evaluate the discriminant.
$$= 16 - 20$$ — Use $a = 1$, $b = 4$, and $c = 5$.
$$= -4$$

● Since the discriminant is -4, there are no solutions. The graphs do not intersect.

✓ **Quick Check** ⑤ At how many points do the graphs of $y = x^2 - 2$ and $y = x + 5$ intersect?

6 EXAMPLE **Solve Using a Graphing Calculator**

Solve the system of equations $y = -x^2 + 4x + 1$ and $y = -x + 5$ using a graphing calculator.

Step 1

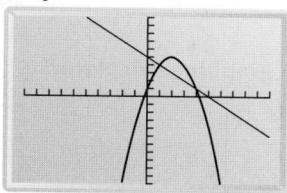

Enter $y = -x^2 + 4x + 1$ and $y = -x + 5$ into Y1 and Y2. Press **GRAPH** to display the system.

Step 2

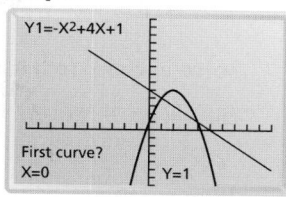

Use the **CALC** feature. Select 5: Intersect.

Step 3

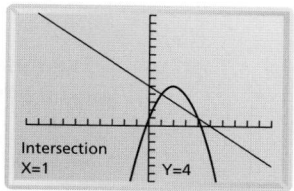

Move the cursor close to a point of intersection. Press **ENTER** three times to find the point of intersection.

Step 4 Repeat Steps 2 and 3 to find the second intersection point.

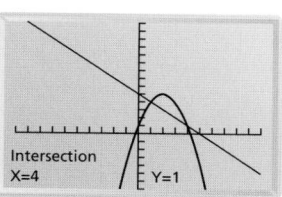

● The solutions of the system are $(1, 4)$ and $(4, 1)$.

✓ Quick Check **6** Solve the system using a graphing calculator. $y = x^2 - 2$
$y = -x$

EXERCISES

For more exercises, see *Extra Skill and Word Problem Practice.*

Practice and Problem Solving

A Practice by Example

Examples 1 and 2
(pages NY 752 and NY 753)

Solve each system by graphing. Find the number of solutions for each system.

1. $y = x^2 + 1$
$y = x + 1$

2. $y = x^2 + 4$
$y = 4x$

3. $y = x^2 - 5x - 4$
$y = -2x$

4. $y = x^2 + 2x + 4$
$y = x + 1$

5. $y = x^2 + 2x + 5$
$y = -2x + 1$

6. $y = 3x + 4$
$y = -x^2$

Example 3
(page NY 753)

Solve each system using elimination.

7. $y = -x + 3$
$y = x^2 + 1$

8. $y = x^2$
$y = x + 2$

9. $y = -x - 7$
$y = x^2 - 4x - 5$

10. $y = x^2 + 11$
$y = -12x$

11. $y = 5x - 20$
$y = x^2 - 5x + 5$

12. $y = x^2 - x - 90$
$y = x + 30$

Example 4
(page NY 754)

Solve each system using substitution.

13. $y = x^2 - 2x - 6$
$y = 4x + 10$

14. $y = 3x - 20$
$y = -x^2 + 34$

15. $y = x^2 + 7x + 100$
$y + 10x = 30$

16. $-x^2 - x + 19 = y$
$x = y + 80$

17. $3x - y = -2$
$2x^2 = y$

18. $y = 3x^2 + 21x - 5$
$-10x + y = -1$

Example 5
(page NY 754)

Use the discriminant to find the number of solutions for each system.

19. $y = x^2 - 5x - 8$
$y = x$

20. $y = -x^2 - 3$
$y = 9 + 2x$

21. $y = -3x - 6$
$y = 2x^2 - 7x$

22. $y = 25x^2 - 9x + 2$
$y + 2 = 11x$

23. $y = -x^2 - 4x + 9$
$y + 5x = -7$

24. $4x^2 + 20x + 29 = y$
$8x + y + 20 = 0$

Example 6
(page NY 755)

Solve each system using a graphing calculator.

25. $y = x^2 - 2x - 2$
$y = -2x + 2$

26. $y = -x^2 + 2$
$y = 4 - 0.5x$

27. $y = x - 5$
$y = x^2 - 6x + 5$

28. $y = -0.5x^2 - 2x + 1$
$y + 3 = -x$

29. $y = 2x^2 - 24x + 76$
$y + 7 = 11$

30. $-x^2 - 8x - 15 = y$
$-x + y = 3$

B **Apply Your Skills**

31. Critical Thinking The graph at the right shows a quadratic function and the linear function $y = d$.
　a. If the linear function were changed to $y = d + 3$, how many solutions would the system have?
　b. If the linear function were changed to $y = d - 5$, how many solutions would the system have?

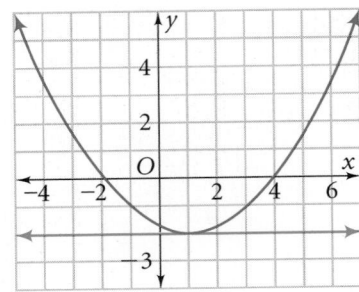

Solve each system using either elimination or substitution.

32. $y = 2x^2 + 13x$
$y = -9 - 6x$

33. $y = -8x$
$y = 1 + 16x^2$

34. $y = x^2 + 9x - 91$
$x = \frac{y}{3}$

35. $y + 20x = 39$
$15 + 4x^2 + 9x = y$

36. $y = x^2 - 12x - 20$
$y = 25(4 - x)$

37. $5x^2 + 14x + 1 = y$
$-12 + y + 40x = 0$

38. Graphing Calculator The screen at the right shows the y- and x-values for the system $y = x^2 - 6x + 8$ and $y = x - 1$. Use the table to find the ·solutions of the system.

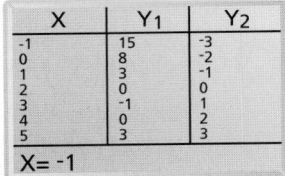

39. Writing Explain why a system of linear and quadratic equations cannot have an infinite number of solutions.

Use substitution and the quadratic formula to find the solutions of each system. Round your answers to the nearest hundredth.

40. $y = 2x^2 + 4x - 1$
$-5x + y = 5$

41. $2y + 4 = x$
$y + x^2 = 4$

42. $3x = y - 7$
$y = 6x^2 - 4x + 1$

43. The graph at the right shows the system $y = x^2 - 5$ and $y = x$. Find the values of x such that the y-values on the parabola are 10 units greater than the corresponding y-values on the line. Round your answers to the nearest hundredths.

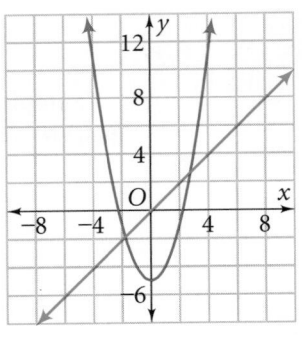

44. Critical Thinking Solve the system $y = x^2 + x + 25$ and $y = x$ using substitution. How can you tell that the system has no solutions *without* using graphing, the discriminant, or the quadratic formula?

C Challenge

45. Geometry The figures below show rectangles that are centered on the *y*-axis with bases on the *x*-axis and upper vertices defined by the function $y = -0.3x^2 + 4$. Find the area of each rectangle. Round to the nearest hundredth.

a.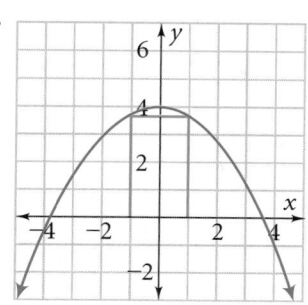

b.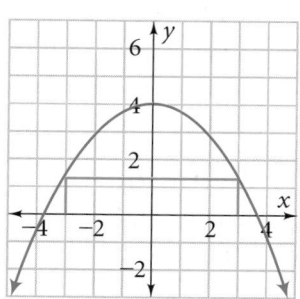

c. Find the *x*- and *y*-coordinates of the vertices of the square constructed in the same manner.

d. Find the area of the square. Round to the nearest hundredth.

NY REGENTS

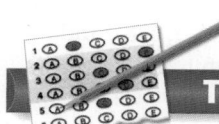
Test Prep

Multiple Choice

46. Which coordinate pair is a solution to the following system?
$$y = x^2 + 2x - 2$$
$$y = x + 10$$

A. (4, 14) **B.** (3, 13) **C.** (2, 12) **D.** (−3, 7)

47. The graph at the right shows the system $y = x + 4$ and $y = -x^2 + x$. How many solutions does the system have?

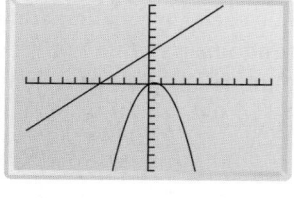

  F. one solution
  G. two solutions
  H. no solutions
  J. cannot be determined

Short Response

48. Use the discriminant to determine the number of solutions of the system.

$$y = 49x^2 - 2x + 34$$
$$y + 30 = 100x$$

Extended Response

49. Solve the system using substitution and factoring. Show your work.

$$x^2 + 3x - 23 = y$$
$$\frac{y}{5} - 5 = x$$

Mixed Review

Lesson NY-5

Tell whether each correlation is a causal relationship. Justify your answer.

50. Hours of computer use and television viewing have a negative correlation. Is this a causation?

51. The number of ice cream trucks in a town on a given day and the high temperature have a positive correlation. Is this a causation?

Lesson NY-4

Find each union or intersection. Let $A = \{1, 3, 6\}$, $B = \{2, 6, 8\}$, $C = \{x \mid x \text{ is an even number less than 6}\}$, and $D = \{x \mid x \text{ is a multiple of 3 less than 12}\}$.

52. $A \cup B$ **53.** $A \cap B$ **54.** $A \cap D$ **55.** $C \cap D$

56. $B \cap C$ **57.** $D \cup C$ **58.** $D \cup A$ **59.** $A \cap C$

Chapter Test

Determine whether each data set is qualitative or quantitative.

1. the time it takes to run a marathon

2. your favorite ice cream flavors

3. an automobile license plate number

4. the cost of a movie ticket

State whether each data set is univariate or bivariate.

5. the length and width of a rectangle

6. the ages of your family members

7. Find the five number summary for the test scores below.

   95 80 75 85 55 65 93 105 75 85 75

8. Of the 20 test scores, 12 scores are less than or equal to 73. What is the percentile rank of 73?

For each survey plan, determine whether the plan is random, systematic, or stratified. State whether the plan will give a good sample.

9. A government agency tests the quality of every 200th tire out of 30,000 tires made.

10. A candy company checks 100 of the 5500 packages of candy at random to see if all the candy boxes were packaged properly.

Determine whether each survey is biased. Explain your answer.

11. You ask, "What kind of healthy fast foods do you like?"

12. You want to find out how much time people in your neighborhood spend reading for leisure. You go door to door after school. Of those surveyed, 85% are in the age ranges of 8–16 and 60 and older.

Use this information for Exercises 13–14: A is the set of even whole numbers less than 18.

13. Write set A in roster form.

14. Write set A in set-builder notation.

15. List the subsets of the set $U = \{a, b, c\}$.

16. Given $B \subseteq A$, $A = \{1, 2, 3, 4, 5, 6, 7, 8\}$, and $B = \{2, 4, 6, 8\}$. Find B'.

17. Sketch the graph of $x > 7$ and write the inequality as an interval.

18. Sketch the graph of $[-3, 8]$ and write the interval as an inequality.

19. Given $A = \{1, 2, 3, 4, 5, 6, 7, 8, 9\}$ and $B = \{2, 4, 6, 8\}$, find $A \cup B$.

20. Let $P = \{1, 5, 7, 9, 13\}$, $R = \{1, 2, 3, 4, 5, 6, 8\}$, and $Q = \{1, 3, 5\}$. Draw a Venn diagram to represent the intersection and union of the sets.

21. Let $N = \{\text{multiples of 2}\}$ and $P = \{\text{multiples of 6}\}$. Describe the intersection of N and P.

22. Is there a correlation between the age of a coin and the face value of a coin? Is there a causal relationship? Explain.

Climate Data

City	Location Latitude (°N)	Daily Mean Temperature (°F)
Atlanta, GA	34	61
Duluth, MN	47	39
Honolulu, HI	21	77
Houston, TX	30	68
Juneau, AK	58	41
Portland, ME	44	45
San Diego, CA	33	64

Use the table above for Exercises 23–26. Use a graphing calculator to estimate the daily mean temperature for the following cities and their latitudes.

23. New York, NY (41° N)   24. Key West, FL (24° N)

25. Anchorage, AK (61° N)   26. San Juan, P.R. (18° N)

Solve each system of equations.

27. $y = x^2 - 2$
    $y = x$

28. $y = x^2 + 2$
    $y = 2x - 2$

29. $y = x^2 - 3x - 2$
    $y = -2x$

Regents Test Prep

Multiple Choice

For Exercises 1–6, choose the correct answer.

1. Which data set is qualitative?
   (1) highway mileage for hybrid cars
   (2) cost of cars
   (3) best-selling compact cars
   (4) annual operating cost for automobiles

2. Which of the following describes the set
   $A = \{2, 4, 8, 10\}$?
   (1) $A = \{n \mid n$ is an even number and $n < 11\}$
   (2) $A = \{x \mid x$ is a factor of 20$\}$
   (3) $A = \{m \mid m$ is a multiple of 2$\}$
   (4) $A = \{b \mid b$ is an even factor of 40 and $b < 20\}$

3. The box-and-whisker plot shows the prices of jeans in a shop.

   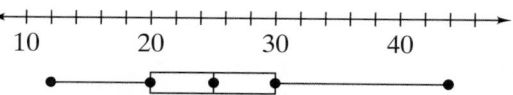

   Which of the following statements is NOT true?
   (1) The range is $32.
   (2) The first quartile is $20.
   (3) The interquartile range is $14.
   (4) The maximum price is $44.

4. Which set of data is bivariate?
   (1) the number of Calories in several different sandwiches
   (2) the capacity of memory cards and the cost of each card
   (3) the weight of several different types of snowboards
   (4) the cost of 100 minutes of mobile phone use from five different companies

5. Which of the following is the best example of correlation *without* causation?
   (1) the number of trees in a park and the number of squirrels living in the park
   (2) ocean temperature and number of hurricanes in a season
   (3) the number of hours exercising per week and resting heart rate
   (4) the number of pairs of shoes owned and average annual income

6. Let $W = \{1, 2, 4, 9\}$ and $Y = \{2, 6, 9\}$. Find $W \cap Y$.
   (1) $\{1, 2, 4, 6, 9\}$ (2) $\{2, 9\}$
   (3) $\{2\}$ (4) $\{1, 4\}$

Short Response

Find each answer.

7. Determine if the following question is biased. Explain your answer.

   Do you enjoy peaceful walks in the woods or shopping in a noisy mall?

8. Write interval notation for the graph below.

   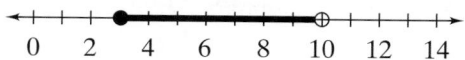

9. Find the five number summary for the following data.

   2 5 4 12 10 8 25 17 13

10. Of 25 test scores, 20 scores are less than or equal to 91. What is the percentile rank of 91?

11. Given $A = \{C, A, T\}$ and $B = \{P, U, M, A\}$. Find $A \cup B$.

12. The table shows an infant's height during her first four months.

Age (months)	0	1	2	3	4
Height (cm)	50	53	57	59	62

    Make a graph to estimate her height at six months.

13. Given $N \subseteq M$, $M = \{11, 13, 15, 17, 19, 21\}$, and $N = \{11, 13, 17, 19\}$, find N'.

Show all of your work.

14. Graph the system and find its solution(s).
    $$y = x - 2$$
    $$y = x^2 - 4$$

15. Out of 52 students who had at least one sibling, 29 had at least one sister and 14 had at least one brother and one sister. How many students had at least one brother but no sisters?

16. Solve the system. $y + 6x = 3$
    $$y = x^2 + 2x - 30$$

Chapter 1

Extra Practice: Skills and Word Problems

● **Lesson 1-1** Define variables and write an equation to model each situation.

1. The total length of the edges of a cube is 12 times the length of an edge.

2. The total cost of lunch is $5.50 times the number of people at the table.

3. The area of a rectangle is 12 cm times the length of the rectangle.

4. The cost of a telephone call is 75 cents plus 25 cents times the number of minutes.

● **Lesson 1-2** Simplify each expression.

5. $4 + 3 \cdot 8$　　**6.** $2 \cdot 3^2 - 7$　　**7.** $6 \cdot (5 - 2) - 9$　　**8.** $2 - 12 \div 3$

9. $4^2 + 8 \div 2$　　**10.** $\frac{1}{2} \div \frac{4}{3}$　　**11.** $-6 \cdot 4.2 - 5 \div 2$　　**12.** $9 - (3 + 1)^2$

13. $2 + 6 \cdot 8 \div 4$　　**14.** $6 + 8 \div 2 - 3$　　**15.** $10 \div 5 \cdot 2 + 6$　　**16.** $5 + 4 \cdot (8 - 6)^2$

● **Lesson 1-3** Use <, =, or > to compare.

17. $0.45 \blacksquare 0.54$　　**18.** $-1.08 \blacksquare -1.008$　　**19.** $\frac{3}{7} \blacksquare \frac{11}{25}$　　**20.** $1.4 \blacksquare \frac{18}{11}$

21. $0.444\ldots \blacksquare \frac{4}{9}$　　**22.** $\frac{4}{13} \blacksquare \frac{4}{15}$　　**23.** $0.101101110\ldots \blacksquare \frac{1}{9}$　　**24.** $\pi \blacksquare \frac{22}{7}$

● **Lesson 1-4** The relationships in the tables are functions. Write a function rule for each.

25.

Ears of Corn	Total Cost
1	$0.20
2	$0.40
3	$0.60
4	$0.80

26.

Time (hours)	Cost of Canoe Rental
1	$12
2	$19
3	$26
4	$33

27.

Number of Tickets	Total Cost
1	$9
2	$18
3	$27
4	$36

● **Lesson 1-5**

28. Make a scatter plot of the data. Describe the trend of the data.

Cover Price and Number of Pages of Some Magazines

Cover Price	$2.25	$2.50	$3.75	$3.00	$4.95	$1.95	$2.95	$2.50
Number of Pages	208	68	122	124	234	72	90	90

● **Lesson 1-6** Find the mean, median, and mode for each set of data.

29. $36, 42, 35, 40, 35, 51, 41, 35$　　**30.** $1.2, 0.9, 0.7, 1.1, 0.8, 1.3, 0.6$　　**31.** $5, 8, 6, 8, 3, 5, 8, 6, 5, 9$

32. A student surveyed the members of the drama club. She included a question about age. Her results are shown at the right.
　a. Make a stem-and-leaf plot for the data.
　b. What is the median age of the drama club members?

Ages of Drama Club Members
14 18 16 15 17 14 15 18 15
13 14 15 18 14 17 16 14 16

● **Lesson 1-1** **Define variables and write an equation for each phrase.**

33. The total cost of gas is the number of gallons times $2.25.

34. The length of the trip is 35 minutes plus extra time spent in traffic.

35. The number of tickets available is 4200 minus the number of tickets sold.

● **Lesson 1-2** **Use the formula for the area of a trapezoid $A = h\left(\dfrac{b_1 + b_2}{2}\right)$, where A is area, b_1 and b_2 are the length of the bases and h is the height, to answer each question.**

36. What is the area of a trapezoidal pool with and a height of 15 yd and bases of 14 yd and 26 yd?

37. How many square feet of grass are there on a trapezoidal field with a height of 75 ft and bases of 125 ft and 81 ft?

● **Lesson 1-3** **Is each statement *true* or *false*? If the statement is false, give a counterexample.**

38. The product of a rational number and an integer is not an integer.

39. The quotient of two integers is an integer.

40. The sum of two rational numbers is a rational number.

● **Lesson 1-4** **Identify the independent and dependent quantity in each situation.**

41. A museum charges $10 admission for each person.

42. Marta is reimbursed $.20 for each mile that she drives on company business.

43. Each acre of cropland takes a farmer 90 minutes to plow.

44. Tom earns $12 for each lawn he cuts. He can cut up to seven lawns working on Saturday and Sunday. Identify the independent and dependent quantities for this situation. Then find reasonable domain and range values.

● **Lesson 1-5** **Would you expect a *positive correlation*, a *negative correlation*, or *no correlation* between the two data sets? Explain why.**

45. the population of a town and the length of its name

46. the temperature during a lacrosse game and the amount of water the players need to drink

47. the number of computers sold and the price of the computer

● **Lesson 1-6** **Make a stem-and-leaf plot for each set of data. Then find the mean, median, and mode.**

48. Students' grades on a test:

95, 62, 68, 72, 88, 75, 78, 81, 88, 91, 72, 68, 75, 78, 82, 88, 88, 76, 93

49. Amount of precipitation for the past 12 months in inches:

0.4, 0.6, 1.2, 1.0, 0.9, 1.4, 1.6, 1.1, 0.9, 1.1, 0.2, 0.3, 0.5

● **Lessons 2-1 to 2-3** Simplify each expression.

1. $22 + (-33)$

2. $45 + (-54)$

3. $-\frac{4}{3} - \frac{4}{5}$

4. $\frac{4}{13} - \frac{4}{13}$

5. $|12 - 21|$

6. $|12| - |-21|$

7. $-(-(11 - 22))$

8. $\left|\frac{2}{3} + \frac{4}{5}\right|$

9. $(-2)(44)$

10. $(-3)^2$

11. -3^2

12. $\left(\frac{3}{2}\right)\left(-\frac{22}{33}\right)$

13. $\frac{3^2}{2^3}$

14. $\frac{-5^2}{(-5)^2}$

15. $81 \div (-9)$

16. $\frac{4^2}{5^2}$

17. $\frac{2 \cdot 3 + 4}{2(3 + 4)}$

18. $1 + \dfrac{1}{2 + \frac{1}{3}}$

19. $\left(\frac{5}{7}\right)^2$

20. $\frac{2}{3} \div \frac{4}{9}$

21. $\begin{bmatrix} -3 & 0 \\ 11 & -5 \end{bmatrix} + \begin{bmatrix} -4 & 6 \\ -8 & 13 \end{bmatrix}$

22. $\begin{bmatrix} 6 & 12 \\ -9 & 7 \end{bmatrix} - \begin{bmatrix} 8 & -6 \\ 15 & 0 \end{bmatrix}$

23. $\begin{bmatrix} 4.2 & 0.6 \\ 1.7 & 9.5 \end{bmatrix} + \begin{bmatrix} 5.8 & -3.5 \\ 0.2 & 4.9 \end{bmatrix}$

● **Lessons 2-4 and 2-5** Simplify each expression.

24. $-4(a + 3)$

25. $-12\left(\frac{4}{3}x - 1\right)$

26. $5 + 6(m + 1)$

27. $\frac{4}{9}(18 - 9t)$

28. $1 + 3 + 5 + 7$

29. $1 - 3 + 5 - 7$

30. $-3(7w) + 7(3w)$

31. $2(1 - d) - (2d + 1)$

32. $6c + 2(4c - 3)$

33 $5(2 - j) + (2j - 3)$

34. $\frac{1}{3}(12 - 6r)$

35. $6\left(\frac{1}{2} - \frac{2}{3}y\right)$

● **Lesson 2-6** The results of rolling a number cube 54 times are shown below. Use the results to find each probability.

```
6 3 4 5 1 1 5 5 3 6 3 2 1 3 3 3 2 1
2 3 6 3 3 4 5 1 2 2 6 3 3 6 5 4 5 3
2 5 1 4 5 2 6 2 5 2 1 2 5 3 2 4 6 3
```

36. $P(3)$

37. $P(4)$

38. $P(\text{not } 5)$

39. $P(7)$

40. $P(\text{even number})$

41. $P(\text{not } 1)$

42. $P(1)$

43. $P(\text{odd number})$

● **Lesson 2-7** You roll a blue number cube and a red number cube. Find each probability.

44. $P(\text{blue 3 and red 2})$

45. $P(\text{blue odd and red 6})$

46. $P(\text{blue 5 and red less than 4})$

47. $P(\text{same number})$

48. $P(\text{both numbers less than three})$

49. $P(\text{both numbers greater than 5})$

You have 3 green marbles, 5 red marbles, and 1 yellow marble in a bag. You pick two marbles from the bag. You pick the second one without replacing the first one. Find each probability.

50. $P(\text{red then green})$

51. $P(\text{yellow then red})$

52. $P(\text{two greens})$

● **Lessons 2-1 to 2-3 Use addition, subtraction, multiplication, or division to answer each question.**

53. A pot of water has a temperature of 25°C. How many degrees should you raise the temperature to boil the water at 100°C?

54. Romana is 68 in. tall and Sophie is 73 in. tall. How much taller is Sophie than Romana?

55. Your goal is to save $50. So far, you have saved $34. How much more do you need to save?

56. A company buys 1500 small items from a manufacturer for $.02 apiece. What is the total cost of the items?

57. If 36 people are at a pizza party, how many eight-piece pizzas need to be ordered so each person can get two pieces of pizza?

58. Your parents drove the family car 462 miles on 14 gallons of gas. On average, how many miles did the car travel on each gallon of gas?

● **Lesson 2-4 Use the Distributive Property to find each price.**

59. Cereal is on sale for $3.95 per box. What is the cost of seven boxes?

60. The school drama club is putting on a play. If there will be five shows and the auditorium holds 480 people, how many people can see the show?

61. What is the total cost of four CDs on sale for $12.15 each?

● **Lesson 2-5 Identify the property of real numbers shown in each situation.**

62. The cost of one item sold for $14.50 is $14.50.

63. You can find the cost of fish by multiplying the price per pound by the amount or by multiplying the amount by the price per pound.

64. To find total time spent doing homework in a week, you add the amount from each day. You find that the total is the same no matter what order you use.

● **Lesson 2-6 Find the theoretical probability of each event.**

65. What is the probability of rolling an even number on a number cube?

66. Tiles with the letters from the word PROBABILITY are placed in a bag. Include the letter *y* as a vowel. What is the probability of selecting a vowel at random?

67. A spinner has four equal sections. Two are red, one is blue, and one is yellow. What is the probability of spinning yellow?

● **Lesson 2-7 A bag contains 5 green marbles, 15 purple marbles, and 10 yellow marbles.**

68. What is the probability of selecting a green marble, replacing it, and then selecting another green marble?

69. What is the probability of selecting a yellow marble, not replacing it, and then selecting a green marble?

Extra Practice: Skills and Word Problems

● **Lessons 3-1 to 3-2** Solve each equation.

1. $8p - 3 = 13$ **2.** $8j - 5 + j = 67$ **3.** $-n + 8.5 = 14.2$ **4.** $6(t + 5) = -36$

5. $m - 9 = 11$ **6.** $\frac{1}{2}(s + 5) = 7.5$ **7.** $7h + 2h - 3 = 15$ **8.** $\frac{7}{12}x = \frac{3}{14}$

9. $3r - 8 = -32$ **10.** $8g - 10g = 4$ **11.** $-3(5 - t) = 18$ **12.** $3(c - 4) = -9$

Define a variable and write an equation for each situation. Then solve.

13. Your test scores for the semester are 87, 84, and 85. Can you raise your test average to 90 with your next test?

14. You spend $\frac{1}{2}$ of your allowance each week on school lunches. Each lunch costs $1.25. How much is your weekly allowance?

● **Lesson 3-3** Solve each equation. If the equation is an identity, write *identity*. If it has no solution, write *no solution*.

15. $4h + 5 = 9h$ **16.** $2(3x - 6) = 3(2x - 4)$ **17.** $7t = 80 + 9t$

18. $m + 3m = 4$ **19.** $-b + 4b = 8b - b$ **20.** $6p + 1 = 3(2p + 1)$

21. $10z - 5 + 3z = 8 - z$ **22.** $3(g - 1) + 7 = 3g + 4$ **23.** $17 - 20q = (-13 - 5q)4$

● **Lessons 3-4 and 3-5** Solve each proportion.

24. $\frac{3}{4} = \frac{-6}{m}$ **25.** $\frac{t}{7} = \frac{3}{21}$ **26.** $\frac{9}{j} = \frac{3}{16}$ **27.** $\frac{2}{5} = \frac{w}{65}$

28. $\frac{s}{15} = \frac{4}{45}$ **29.** $\frac{9}{4} = \frac{x}{10}$ **30.** $\frac{10}{q} = \frac{8}{62}$ **31.** $\frac{3}{2} = \frac{18}{y}$

32. The scale on a map is 1 in. : 15 mi. The distance between two cities is 25 mi. Find the distance in inches between the cities on the map.

● **Lesson 3-6**

33. **Transportation** A bus traveling 40 mi/h and a car traveling 50 mi/h cover the same distance. The bus travels 1 h more than the car. How many hours did each travel?

● **Lesson 3-7** Find each percent of change. Describe each as a percent of increase or decrease. Round to the nearest percent.

34. $4.50 to $5.00   **35.** 56 in. to 65 in.   **36.** 18 oz to 12 oz

37. 1 s to 3 s   **38.** 8 lb to 5 lb   **39.** 6 km to 6.5 km

● **Lesson 3-8** Find the square roots of each number.

40. 25   **41.** $\frac{4}{9}$ **42.** 64   **43.** $\frac{25}{36}$ **44.** 0.81   **45.** 900

● **Lesson 3-9** Determine whether the given lengths are sides of a right triangle.

46. 15, 36, 39   **47.** 3, 7, 10   **48.** 8, 15, 17   **49.** $\sqrt{3}, \sqrt{4}, \sqrt{5}$

For the values given, *a* and *b* are legs of a right triangle. Find the length of the hypotenuse. If necessary, round to the nearest tenth.

50. $a = 6, b = 8$ **51.** $a = 5, b = 9$ **52.** $a = 4, b = 10$ **53.** $a = 9, b = 1$

● **Lessons 3-1 to 3-3** **Write an equation to model each situation. Then solve.**

54. A DVD club charges a monthly membership fee of $4.95 and $11.95 for each DVD purchased. If a customer's bill for the month was $64.70, how many DVDs did the customer purchase?

55. A lawyer charges $100 per month to be put on retainer for a client. The lawyer also charges an hourly rate of $75 for work done. How many hours does the lawyer have to work for a client, in one month, to charge $625?

56. A rectangular pool is twice as long as it is wide. What are the dimensions of the pool if the perimeter is 42 yd?

57. Two friends rent an apartment together. They agree that one person will pay 1.5 times what the other person pays. If the rent is $850, how much will each friend pay?

58. A shopper's discount club charges a monthly fee of $15 and sells gasoline for $2.05 per gallon. The gas station across the street sells gasoline for $2.35 per gallon and charges no fee. How many gallons of gasoline would you have to buy in one month to spend the same amount at either store?

59. Michael and Kevin are running. Kevin gets a 3-mile head start and runs at a rate of 5.5 mi/h. Michael runs at a rate of 7 mi/h. How many hours will it take Michael to catch up with Kevin?

● **Lessons 3-4 and 3-5**

60. A 12-ounce can of green beans is sold for $1.45. What is the price per pound?

61. A sailboat is traveling at a speed of 10 nautical miles per hour. If 1 nautical mile is 6076 ft, what is the speed of the sailboat in feet per second?

62. A 40 : 1 scale model of an airplane is being used to conduct wind-tunnel tests. If the model is 4.5 feet long, how long is the actual airplane?

● **Lesson 3-6**

63. Joe and Gayna live 3 miles apart. They both leave their houses at the same time and walk to meet each other. Joe walks at 2.5 mi/h and Gayna walks at 3.5 mi/h. How far will Joe walk before he meets Gayna?

● **Lesson 3-7** **Find the percent of change. Describe the percent of change as an increase or decrease.**

64. A $1500 computer is on sale for $1275.

65. The value of a stamp collection increases from $160 to $180 in one year.

● **Lesson 3-8**

66. What is the length of each side of a square garden with an area of 70 ft^2?

● **Lesson 3-9** **Use the Pythagorean Theorem to answer each question.**

67. A 20-ft ladder is placed 5 ft from the base of a building. How high on the building will the ladder reach?

68. A soccer field is 80 yd long and 35 yd wide. What is the diagonal distance across the field?

Chapter 4

Extra Practice: Skills and Word Problems

● **Lessons 4-1 to 4-4** Solve each inequality. Graph and check your solution.

1. $-8w < 24$

2. $9 + p \le 17$

3. $\frac{r}{4} > -1$

4. $7y + 2 \le -8$

5. $t - 5 \ge -13$

6. $9h > -108$

7. $8w + 7 > 5$

8. $\frac{s}{6} \le 3$

9. $\frac{6c}{5} \ge -12$

10. $-8\ell + 3.7 \le 31.7$

11. $9 - t \le 4$

12. $m + 4 \ge 8$

13. $y + 3 < 16$

14. $n - 6 \le 8.5$

15. $12b - 5 > -29$

16. $4 - a > 15$

17. $4 - x \le 3$

18. $1 - 4d \ge 4 - d$

19. $n + 7 \le 3n - 1$

20. $\frac{s}{2} + 1 < s + 2$

21. $3 - \frac{2x}{3} > 5$

22. $8r - \frac{r}{6} > \frac{1}{6} - 8$

23. $1.4 + 2.4x < 0.6$

24. $x - 2 < 3x - 4$

25. The booster club raised \$102 in their car wash. They want to buy \$18 soccer balls for the soccer team. Write and solve an inequality to find how many soccer balls they can buy.

26. You earn \$7.50 per hour and need to earn \$35. Write and solve an inequality to find how many hours you must work.

● **Lesson 4-5** Solve each compound inequality.

27. $8 < w + 3 < 10$

28. $-6 < t - 1 < 6$

29. $6m - 15 \le 9$ or $10m > 84$

30. $9j - 5j \ge 20$ and $8j > -36$

31. $37 < 3c + 7 < 43$

32. $3 < 5 + 6h < 10$

33. $1 + t < 4 < 2 + t$

34. $2 + 3w < -1 < 3w + 5$

35. $2x - 3 \le x$ and $2x + 1 \ge x + 3$

36. $3n - 7 > n + 1$ or $4n - 5 < 3n - 3$

● **Lesson 4-6** Choose a variable and write an absolute value inequality that represents each set of numbers.

37. all real numbers less than 2 units from 0

38. all real numbers more than 0.5 units from 4.5

39. all real numbers less than 1 unit from -4

40. all real numbers 3 or more units from -1

41. all real numbers less than or equal to 5 units from 3

Solve each inequality. Graph and check your solution.

42. $|x| < 5$

43. $|t| > 1$

44. $|t| - 5 \le 3$

45. $|-6m + 2| > 20$

46. $|3c| - 1 \ge 11$

47. $|8 - w| \le 8$

48. $|2b + 3| < 7$

49. $|c - 5| \le 6$

50. $|n| + 4 \le 5$

51. Write an absolute value inequality that has numbers between 2 and 3 as the solutions.

52. Holes with radius 3 cm must be drilled in sheets of metal. The radius must have an error no more than 0.01 cm. Write an absolute value inequality whose solutions are acceptable radii.

● **Lesson 4-1 Define a variable and write an inequality for each situation.**

53. A car dealership sells at least 35 cars each week.

54. No more than 425 tickets to a musical will be sold.

55. You must be at least 18 years old to vote.

56. The party store sold more than 720 balloons in July.

● **Lessons 4-2 to 4-4 Write and solve an inequality for each situation.**

57. Suppose you are trying to increase your coin collection to at least 500 coins. How many more coins do you need if you already have a collection of 375 coins?

58. Janet has a balance of $125 on a credit card. On her next statement, she wants to reduce her balance to no more than $60. How much does she need to pay off?

59. A homeroom class with 25 students is holding a fund-raiser to support school sports. Their goal is to raise at least $200. On average, how much money does each student need to contribute to meet or exceed the goal?

60. You are reading a book with 19 chapters. How many chapters should you read each week if you want to finish the book in 5 weeks or less?

61. The sophomore class is putting on a variety show to raise money. It costs $700 to rent the banquet hall they are going to use. If they charge $15 for each ticket, how many tickets do they need to sell in order to raise at least $1000?

62. A technical-support company charges $10 per month plus $35 per hour of phone support. If you need to spend less than $100 per month on support, how many hours can you get?

● **Lesson 4-5 Write a compound inequality for each situation. Graph your solution.**

63. Water will not be in liquid form when it is colder than 32°F or warmer than 212°F.

64. The width of a parking space needs to be at least 8 feet and no more than 11 feet.

65. A car salesman has been told to sell a particular car for more than $14,500 and up to the sticker price of $15,755.

● **Lesson 4-6 Write and solve an absolute value inequality for each situation.**

66. The ideal diameter of an aircraft tire is 105 inches. The acceptable error for each tire is 0.175 inches. Find the range of acceptable tire diameters.

67. A tractor crankshaft is designed to have a radius of 4.25 cm. The acceptable error for the radius is 0.005 cm. Find the range of acceptable radii for the crankshaft.

68. The ideal weight of an exercise ball is 175 ounces. Each ball can have an error of 0.35 ounces. What is the range of acceptable weights?

Extra Practice: Skills and Word Problems

● **Lesson 5-1** Sketch a graph to describe each situation. Label each section of the graph.

1. the number of apples on a tree over one year

2. the amount of milk in your bowl as you eat cereal

3. the energy you use in a 24-h period

4. your distance from home plate after your home run

● **Lesson 5-2** Find the range of each function when the domain is $\{-4, -1, 0, 3\}$.

5. $y = 6x - 5$

6. $y = |x| - 2$

7. $y = x^2 + 3x + 1$

8. $y = \frac{1}{2}x + 8$

9. $y = -x^2 - x$

10. $y = \frac{2}{3}x$

Use a mapping diagram to determine whether each relation is a function.

11. $\{(1, 2), (2, 3), (3, 4), (4, 5), (5, 6)\}$

12. $\{(5, 2), (1, 3), (4, 7), (5, 6), (0, 4)\}$

13. $\{(3.4, 2), (5.6, 2), (0.1, 2), (2.8, 2)\}$

14. $\{(6, 7), (5, 2), (7, 7), (4, 3), (0, 0)\}$

● **Lesson 5-3** Graph each function.

15. $y = 2x + 1$

16. $y = 4 - x$

17. $y = |x| - 3$

● **Lesson 5-4** Write a function rule for each table.

18.

x	f(x)
−3	−1
−1	1
1	3
3	5

19.

x	f(x)
0	0
3	6
6	12
9	18

20.

x	f(x)
21	14
25	18
29	'22
33	26

21.

x	f(x)
−8	−4
−6	−3
−4	−2
−2	−1

● **Lesson 5-5** Graph the direct variation that includes the given point. Write the equation of the line.

22. $(5, 4)$

23. $(7, 7)$

24. $(-3, -10)$

25. $(4, -8)$

26. $(-2, 9)$

● **Lesson 5-6** Find the constant of variation k for each inverse variation.

27. $y = 10$ when $x = 7$

28. $y = -8$ when $x = 12$

29. $y = 0.2$ when $x = 4$

Each pair of points is on the graph of an inverse variation. Find the missing value.

30. $(5.4, 3)$ and $(2, y)$

31. $(x, 4)$ and $(5, 6)$

32. $(3, 6)$ and $(9, y)$

33. $(100, 2)$ and $(x, 25)$

34. $(6, 1)$ and $(x, -2)$

35. $(8, y)$ and $(-2, 4)$

● **Lesson 5-7** Find the second and fifth terms of each sequence.

36. $A(n) = 22 + (n - 1)11$

37. $A(n) = -2 + (n - 1)(-2)$

38. $A(n) = -2 + (n - 1)$

39. $A(n) = 1 + 4(n - 1)$

● Lesson 5-1 **Sketch a graph to represent each situation.**

40. the speed of a bus as it makes its last stop to pick up students and then continues until it arrives at school

41. the temperature of a lake from January through December

42. the height of a roller coaster during a ride

● Lesson 5-2 **Evaluate each function rule to find the range for the domain {1, 4, 9}.**

43. The function $f(x) = 20 - x$ represents the amount of change you receive after paying for an item that costs x dollars with a \$20 bill.

44. The function $f(x) = x^2$ represents the area of a square with a side length of x.

45. The function $f(x) = 8x$ represents the number of pieces of pizza from x eight-piece pizzas.

● Lesson 5-3 **Make a table of values and graph each function.**

46. The function $f(x) = 175 + x$ represents the amount of money in a savings account that started with \$175 after a deposit of x dollars.

47. The function $f(x) = 4x$ represents the perimeter of a square with side length x.

● Lesson 5-4 **Write a function rule for each situation.**

48. the circumference of a circle $C(r)$ when you know the radius r

49. the area of a 100-yard-long field $A(w)$ when you know the width w

50. the distance run in feet $D(m)$ when you know the distance in miles m

51. the number of pounds $P(n)$ when you know the number of ounces n

● Lessons 5-5 and 5-6 **Write a direct or inverse variation to model each situation. Then answer the question.**

52. After 30 minutes a car moving at a constant speed has traveled 25 miles. Moving at the same speed, how far will it travel in 140 minutes?

53. The perimeter of a square depends on the length of a side of the square. What is the perimeter of a square with side length 13.4 in.?

54. Two rectangular fields have the same area. One measures 75 yd by 60 yd. If the other has a length of 72 yd, what is its width?

55. Kevin is training to run in a half marathon. Initially, he could run 6 miles per hour for 2 hours. Two months later he ran the same distance in 1 hour and 45 minutes. What is his new speed?

● Lesson 5-7 **Use inductive reasoning to answer the question.**

56. A train arrives at a subway station at 9:05 A.M. Another train arrives at 9:12 A.M. and a third arrives at 9:19 A.M. If the trains keep running on the same schedule, at what times will each of the next three trains arrive?

57. The balance of a car loan starts at \$9,200 and decreases by \$180 each month. What will the balance be at the end of the next three months?

Extra Practice: Skills and Word Problems

● **Lesson 6-1 Find the rate of change for each situation.**

1. growing from 1.4 m to 1.6 m in one year

2. bicycling 3 mi in 15 min and 7 mi in 55 min

3. growing 22.4 mm in 14 s

4. reading 8 pages in 9 min and 22 pages in 30 min

● **Lessons 6-2 and 6-3 Find the slope and y-intercept.**

5. $y = 6x + 8$ **6.** $3x + 4y = -24$ **7.** $2y = 8$ **8.** $y = \frac{-3}{4}x - 8$

Graph each equation.

9. $y = 2x - 3$ **10.** $y = \frac{2}{3}x - 4$ **11.** $y = -\frac{1}{2}x + 4$ **12.** $y = -\frac{5}{4}x$

● **Lessons 6-4 and 6-5 Find the x- and y-intercepts for each equation.**

13. $6x + y = 12$ **14.** $y = -7x$ **15.** $y = \frac{1}{2}x + 3$ **16.** $-2y = 5x - 12$

Write the equation in point-slope form for the line through the given point with the given slope.

17. $(4, 6); m = -5$ **18.** $(3, -1); m = 1$ **19.** $(8, 5); m = \frac{1}{2}$ **20.** $(0, -6); m = \frac{4}{3}$

Graph each equation.

21. $x + 4y = 8$ **22.** $y - 5 = -2(x + 1)$ **23.** $x + 3 = 0$

24. $4x - 3y = 12$ **25.** $y = -1$ **26.** $y + 1 = -\frac{1}{2}(x + 2)$

A line passes through the given points. Write an equation for the line in slope-intercept form.

27. $(2, 5)$ and $(4, 8)$ **28.** $(1, 6)$ and $(7, 3)$ **29.** $(-2, 4)$ and $(3, 9)$ **30.** $(1, 6)$ and $(9, -4)$

31. $(0, -7)$ and $(-1, 0)$ **32.** $(7, 0)$ and $(3, -4)$ **33.** $(0, 0)$ and $(-7, 1)$ **34.** $(10, 0)$ and $(0, 7)$

● **Lesson 6-5 Write an equation in standard form that satisfies the given conditions.**

35. parallel to $y = 4x + 1$, through $(-3, 5)$ **36.** perpendicular to $y = -x - 3$, through $(0, 0)$

37. perpendicular to $3x + 4y = 12$, through $(7, 1)$ **38.** parallel to $2x - y = 6$, through $(-6, -9)$

39. parallel to the x-axis and through $(4, -1)$ **40.** through $(4, 44)$ and parallel to the y-axis

● **Lesson 6-6**

41. a. Graph the (ages, grades) data of some students in a school at the right.
   b. Draw a trend line.
   c. Find the equation of the line of best fit.

> $(10, 6), (16, 10), (15, 10), (18, 12), (17, 11),$
> $(17, 12), (19, 12), (16, 11), (11, 7), (15, 9), (13, 8)$

● **Lesson 6-7 Graph each equation by translating $y = |x|$ or $y = -|x|$.**

42. $y = |x| + 1$ **43.** $y = |x + 2|$ **44.** $y = |x - 2|$ **45.** $y = -|x - 1|$

46. $y = -|x + 1|$ **47.** $y = -|x| + 1$ **48.** $y = |x + 0.5|$ **49.** $y = |x| - 4$

● **Lesson 6-1 Find the rate of change for each situation.**

50. The cost of four movie tickets is $30 and the cost of seven tickets is $52.50.

51. Five seconds after jumping out of the plane, a sky diver is 10,000 ft above the ground. After 30 seconds, the sky diver is 3,750 ft above the ground.

● **Lesson 6-2 Write an equation in slope-intercept form for each situation.**

52. A skateboard ramp is 5 ft high and 12 ft long from end to end.

53. An airplane with no fuel weighs 2575 lbs. Each gallon of gasoline added to the fuel tanks weighs 6 lbs.

● **Lesson 6-3 Write a function for each situation and then determine a reasonable range for the function.**

54. You plan to buy no more than three gallons of milk. Each gallon costs $3.75.

55. The amount of time spent in line at a grocery store depends on how many people are in line ahead of you. Each person takes about two minutes to check out at the cash register.

● **Lesson 6-4 Write an equation in standard form for each situation.**

56. Juan can ride his bike at 12 mi/h and walk at 4 mi/h. Write an equation that relates the amount of time he can spend riding or walking combined, to travel 20 miles.

57. You have $25 to buy supplies for a class party. Juice costs $3 per bottle and chips cost $2 per bag. Write an equation that relates the amount of juice and chips you can buy using $25.

● **Lesson 6-5 Write an equation in point-slope form for each situation.**

58. A train travels at a rate of 70 mi/h. Two hours after leaving the station it was 210 miles from its destination.

59. An escalator has a slope of $\frac{3}{4}$. After traveling forward 32 feet, the escalator is 24 feet above the floor.

● **Lesson 6-6 Tell whether each statement is *true* or *false*. Explain your choice.**

60. Two airplanes traveling at the same rate leave an airport 1 hour apart. The graphs of the distance each plane travels will be parallel.

61. Two lines with negative slopes can be perpendicular.

● **Lesson 6-7**

62. Use a calculator to find a line of best fit for the data. Find the value of the correlation coefficient r. Let $x = 0$ correspond to 1960.

Total U.S. Vehicle Production (millions)

1960	1970	1980	1990	2000
7.9	8.8	8.0	9.8	12.8

● **Lesson 6-8**

63. A car traveling at a rate of 50 mi/h passes a rest area 30 minutes after the beginning of the trip. Write an absolute value equation that represents the car's distance from the rest area.

Extra Practice: Skills and Word Problems

● **Lesson 7-1** Solve each system by graphing.

1. $x - y = 7$

$3x + 2y = 6$

2. $y = 2x + 3$

$y = -\frac{3}{2}x - 4$

3. $y = -2x + 6$

$3x + 4y = 24$

● **Lesson 7-2** Solve each system by using substitution.

4. $x - y = 13$

$y - x = -13$

5. $3x - y = 4$

$x + 5y = -4$

6. $x + y = 4$

$y = 7x + 4$

● **Lesson 7-3** Solve each system by elimination.

7. $x + y = 19$

$x - y = -7$

8. $-3x + 4y = 29$

$3x + 2y = -17$

9. $3x + y = 3$

$-3x + 2y = -30$

10. $6x + y = 13$

$y - x = -8$

11. $4x - 9y = 61$

$10x + 3y = 25$

12. $4x - y = 105$

$x + 7y = -10$

● **Lesson 7-4** Write a system of equations to model each problem and solve.

13. Suppose you have 12 coins that total 32 cents. Some of the coins are nickels and the rest are pennies. How many of each coin do you have?

14. Claire bought three bars of soap and five sponges for $2.31. Steve bought five bars of soap and three sponges for $3.05. Find the cost of each item.

15. The perimeter of a rectangular lot is 74 feet. The cost of fencing along the two lengths is $1 per foot, and the cost of fencing along the two widths is $3.50 per foot. Find the dimensions of the lot if the total cost of the fencing is $159.

16. A chemist wants to make a 10% solution of fertilizer. How much water and how much of a 30% solution should the chemist mix to get 30 L of a 10% solution?

17. Fruit drink A consists of 6% pure fruit juice and drink B consists of 15% pure fruit juice. How much of each kind of drink should you mix together to get 4 L of a 10% concentration of fruit juice?

18. A motor boat traveled 12 miles with the current, turned around, and returned 12 miles against the current to its starting point. The trip with the current took 2 hours and the trip against the current took 3 hours. Find the speed of the boat and the speed of the current.

● **Lesson 7-5** Graph each linear inequality.

19. $y < x$

20. $y < x - 4$

21. $y > -6x + 5$

22. $y \leq 14 - x$

23. $y \geq \frac{1}{4}x - 3$

24. $2x + 3y \leq 6$

● **Lesson 7-6** Solve each system by graphing.

25. $y \leq 5x + 1$

$y > x - 3$

26. $y > 4x + 3$

$y \geq -2x - 1$

27. $y > -x + 2$

$y > x - 4$

28. $y < -2x + 1$

$y > -2x - 3$

29. $y \leq 5$

$y \geq -x + 1$

30. $y \leq 5x - 2$

$y > 3$

Lesson 7-1 **Write and solve a system of equations by graphing.**

31. One calling card has a $.50 connection fee and charges $.02 per minute. Another card has a $.25 connection fee and charges $.03 per minute. After how many minutes would a call cost the same amount using either card?

32. Suppose that you have $75 in your savings account and you save an additional $5 per week. Your friend has $30 in his savings account and saves an additional $10 per week. In how many weeks will you both have the same amount of money in your accounts?

Lesson 7-2 **Write and solve a system of equations by substitution.**

33. A farmer grows corn and soybeans on her 300-acre farm. She wants to plant 110 more acres of soybeans than corn. How many acres of each crop does she need to plant?

34. The perimeter of a rectangle is 34 cm. The length is 1 cm longer than the width. What are the dimensions of the rectangle?

Lesson 7-3 **Write and solve a system of equations using elimination.**

35. Two groups of people order food at a restaurant. One group orders 4 hamburgers and 7 chicken sandwiches for $34.50. The other group orders 8 hamburgers and 3 chicken sandwiches for $30.50. Find the cost of each item.

36. The sum of two numbers is 25. Their difference is 9. What are the two numbers?

Lesson 7-4 **Write and solve a system of equations for each situation by any method. Explain why you chose the method.**

37. The ratio of boys to girls at a college is 4 : 5. How many boys and girls are there if the total number of students is 3321?

38. A boat travels 18 miles downstream in 1.5 hours. It then takes the boat 3 hours to travel upstream the same distance. Find the speed of the boat in still water and the speed of the current.

Lesson 7-5 **Write and graph a linear inequality for each situation.**

39. Suppose you can spend up to $10 on bananas and apples. Apples cost $3 per pound and bananas cost $1 per pound. List three possible combinations of apples and bananas you can buy.

40. Trenton is going to make a rectangular garden in his yard. He wants the perimeter to be no larger than 40 ft. What are three possible sets of dimensions that the garden can have?

Lesson 7-6

41. Hideo plans to spend no more than $60 at an entertainment store on DVDs and CDs. DVDs cost $17 each and CDs cost $14 each. He wants to buy at least two items. Write and graph a system of linear inequalities that describes the situation. What are three possible combinations of CDs and DVDs that he can buy? Write and graph a system of inequalities that describes the situation.

● **Lessons 8-1 to 8-5** Simplify each expression. Use only positive exponents.

1. $(2t)^{-6}$ **2.** $5m^5m^{-8}$ **3.** $(4.5)^4(4.5)^{-2}$ **4.** $(m^7t^{-5})^2$

5. $(x^2n^4)(n^{-8})$ **6.** $(w^{-2}j^{-4})^{-3}(j^7j^3)$ **7.** $(t^6)^3(m)^2$ **8.** $(3n^4)^2$

9. $\dfrac{r^5}{g^{-3}}$ **10.** $\dfrac{1}{a^{-4}}$ **11.** $\dfrac{w^7}{w^{-6}}$ **12.** $\dfrac{6}{t^{-4}}$

13. $\dfrac{a^2b^{-7}c^4}{a^5b^3c^{-2}}$ **14.** $\dfrac{(2t^5)^3}{4t^8t^{-1}}$ **15.** $\left(\dfrac{a^6}{a^7}\right)^{-3}$ **16.** $\left(\dfrac{c^5c^{-3}}{c^{-4}}\right)^{-2}$

Evaluate each expression for $m = 2$, $t = -3$, $w = 4$, and $z = 0$.

17. t^m **18.** t^{-m} **19.** $(w \cdot t)^m$ **20.** $w^m \cdot t^m$

21. $(w^z)^m$ **22.** $w^m w^z$ **23.** $z^{-t}(m^t)^z$ **24.** $w^{-t}t^t$

Write each number in scientific notation.

25. 34,000,000   **26.** 0.00063   **27.** 1500   **28.** 0.0002

29. 360,000   **30.** 6,200,000,000   **31.** 0.05   **32.** 0.000000000891

Write each number in standard notation.

33. 8.05×10^6 **34.** 3.2×10^{-7} **35.** 9.0×10^8 **36.** 4.25×10^{-4}

37. 2.35×10^2 **38.** 6.3×10^4 **39.** 2.001×10^{-5} **40.** 5.2956×10^3

● **Lesson 8-6** Find the common ratio of each sequence. Then find the next two terms.

41. $12, 18, 27, \ldots$ **42.** $2, 1, 0.5, \ldots$ **43.** $-1, -0.2, -0.04, \ldots$ **44.** $-2, -4, -8, \ldots$

45. $2, 6, 18, \ldots$ **46.** $1.2, -0.6, 0.3, \ldots$ **47.** $30, 10, \frac{10}{3}, \ldots$ **48.** $-2.25, -9, -36, \ldots$

● **Lesson 8-7** Evaluate each function for the domain $\{-1, 0, 1, 2\}$. As the values of the domain increase, do the values of the function *increase* or *decrease*?

49. $y = 3^x$ **50.** $y = \left(\frac{3}{4}\right)^x$ **51.** $y = 1.5^x$ **52.** $y = \frac{1}{2} \cdot 3^x$

53. $y = -3 \cdot 7^x$ **54.** $y = -(4)^x$ **55.** $y = 3 \cdot \left(\frac{1}{5}\right)^x$ **56.** $y = 2^x$

57. $y = 2 \cdot 3^x$ **58.** $y = (0.8)^x$ **59.** $y = 2.5^x$ **60.** $y = -4 \cdot (0.2)^x$

● **Lesson 8-8** Identify each function as *exponential growth* or *exponential decay*. Then identify the growth factor or decay factor.

61. $y = 8^x$ **62.** $y = \frac{3}{4} \cdot 2^x$ **63.** $y = 9 \cdot \left(\frac{1}{2}\right)^x$ **64.** $y = 4 \cdot 9^x$

65. $y = 0.65^x$ **66.** $y = 3 \cdot 1.5^x$ **67.** $y = \frac{2}{5} \cdot \left(\frac{1}{4}\right)^x$ **68.** $y = 0.1 \cdot 0.9^x$

Write an exponential function to model each situation. Find each amount after the specified time.

69. $200 principal, 4% compounded annually for 5 years

70. $1000 principal, 3.6% compounded monthly for 10 years

71. $3000 investment, 8% loss each year for 3 years

● **Lesson 8-1**

72. Suppose an investment doubles in value every 5 years. This year the investment is worth $12,480. How much will it be worth 10 years from now? How much was it worth 5 years ago?

● **Lesson 8-2** **Write each number in scientific notation.**

73. A bacteria culture has a population of approximately 7,500,000,000.

74. The diameter of a blood cell is about 0.0000082 m.

● **Lessons 8-3 and 8-4** **Write each answer in scientific notation.**

75. A light-year is the distance light travels in one year. If the speed of light is about 3×10^5 km/s, how long is a light-year in kilometers? (Use 365 days for the length of a year).

76. The radius of Earth is approximately 6.4×10^6 m. Use the formula $V = \frac{4}{3}\pi r^3$ to find the volume of Earth.

77. A spherical cell has a radius of 2.75×10^{-6} m. Use the formula for the surface area of a sphere $S.A. = 4\pi r^2$ to find the surface area of a cell.

● **Lesson 8-5**

78. What is the volume of a cube with a side length of $\frac{4}{5}$ m?

79. The speed of sound is approximately 1.2×10^3 km/h. How long does it take for sound to travel 7.2×10^2 km? Write your answer in minutes.

● **Lesson 8-6** **Write a rule and find the given term in each geometric sequence described below.**

80. What is the ninth term when the first term is 4 and the common ratio is -6?

81. What is the sixth term when the first term is 60 and the common ratio is 0.4?

● **Lesson 8-7** **Write and solve an exponential equation to answer each question.**

82. Suppose an investment of $5,000 doubles every 12 years. How much is the investment worth after 36 years? After 48 years?

83. Suppose 15 animals are taken to an island, and then their population triples every 8 months. How many animals will there be in 4 years?

● **Lesson 8-8** **Find the balance in each account.**

84. You deposit $2500 in a savings account with 3% interest compounded annually. What is the balance in the account after 6 years?

85. You deposit $750 in an account with 7% interest compounded semiannually. What is the balance in the account after 4 years?

86. You deposit $520 in an account with 4% interest compounded monthly. What is the balance in the account after 5 years?

● **Lesson 9-1** Simplify. Write each answer in standard form.

1. $(5x^3 + 3x^2 - 7x + 10) - (3x^3 - x^2 + 4x - 1)$

2. $(x^2 + 3x - 2) + (4x^2 - 5x + 2)$

3. $(4m^3 + 7m - 4) + (2m^3 - 6m + 8)$

4. $(8t^2 + t + 10) - (9t^2 - 9t - 1)$

5. $(-7c^3 + c^2 - 8c - 11) - (3c^3 + 2c^2 + c - 4)$

6. $(6v + 3v^2 - 9v^3) + (7v - 4v^2 - 10v^3)$

7. $(s^4 - s^3 - 5s^2 + 3s) - (5s^4 + s^3 - 7s^2 - s)$

8. $(9w - 4w^2 + 10) + (8w^2 + 7 + 5w)$

9. The sides of a rectangle are $4t - 1$ and $5t + 9$. Write an expression for the perimeter of the rectangle.

10. Three consecutive integers are $n - 1, n,$ and $n + 1$. Write an expression for the sum of the three integers.

● **Lesson 9-2** Simplify each product.

11. $4b(b^2 + 3)$

12. $9c(c^2 - 3c + 5)$

13. $8m(4m - 5)$

14. $5k(k^2 + 8k)$

15. $5r^2(r^2 + 4r - 2)$

16. $2m^2(m^3 + m - 2)$

17. $-3x(x^2 + 3x - 1)$

18. $-x(1 + x + x^2)$

Find the GCF of the terms of each polynomial. Factor.

19. $t^6 + t^4 - t^5 + t^2$

20. $3m^2 - 6 + 9m$

21. $16c^2 - 4c^3 + 12c^5$

22. $8v^6 + 2v^5 - 10v^9$

23. $6n^2 - 3n^3 + 2n^4$

24. $5r + 20r^3 + 15r^2$

25. $9x^6 + 5x^5 + 4x^7$

26. $4d^8 - 2d^{10} + 7d^4$

● **Lessons 9-3 and 9-4** Simplify each product. Write in standard form.

27. $(5c + 3)(-c + 2)$

28. $(3t - 1)(2t + 1)$

29. $(w + 2)(w^2 + 2w - 1)$

30. $(3t + 5)(t + 1)$

31. $(2n - 3)(2n + 4)$

32. $(b + 3)(b + 7)$

33. $(3x + 1)^2$

34. $(5t + 4)^2$

35. $(w - 1)(w^2 + w + 1)$

36. $(a + 4)(a - 4)$

37. $(3y - 2)(3y + 2)$

38. $(w^2 + 2)(w^2 - 2)$

39. **Geometry** A rectangle has dimensions $3x - 1$ and $2x + 5$. Write an expression for the area of the rectangle as a product and in standard form.

40. Write an expression for the product of the two consecutive odd integers $n - 1$ and $n + 1$.

● **Lessons 9-5 to 9-7** Factor each expression.

41. $x^2 - 4x + 3$

42. $3x^2 - 4x + 1$

43. $v^2 + v - 2$

44. $5t^2 - t - 18$

45. $m^2 + 9m - 22$

46. $x^2 - 2x - 15$

47. $2n^2 + n - 3$

48. $2h^2 - 5h - 3$

49. $m^2 - 25$

50. $9y^2 - 1$

51. $9y^2 + 6y + 1$

52. $p^2 + 2p + 1$

53. $x^2 + 6x + 9$

54. $25x^2 - 9$

55. $4t^2 + t - 3$

56. $9c^2 - 169$

57. $4m^2 - 121$

58. $3v^2 + 10v - 8$

59. $4g^2 + 4g + 1$

60. $-w^2 + 5w - 4$

61. $9t^2 + 12t + 4$

62. $12m^2 - 5m - 2$

63. $36s^2 - 1$

64. $c^2 - 10c + 25$

● **Lesson 9-8** Factor each expression.

65. $3y^3 + 9y^2 - y - 3$

66. $3u^3 + u^2 - 6u - 2$

67. $w^3 - 3w^2 + 3w - 9$

68. $4z^3 + 2z^2 - 2z - 1$

69. $3x^3 + 8x^2 - 3x$

70. $y^5 - 9y$

71. $2p^3 - 4p^2 + 2p - 4$

72. $3y^3 - 3y^2 - 6y$

● **Lesson 9-1 Find an expression for the perimeter of each figure.**

73. A rectangle has side lengths $4k - 3$ and $2k + 2$.

74. A triangle has side lengths $2t^2$, $4t - 3$, and $10 - 2t$.

75. A rhombus has two sides $3d^3 - 2d$ long and two sides $d^2 + 5$ long.

● **Lesson 9-2**

76. A rectangular roof has a length of $13g$ and a width of $4g + 7$. Write an expression for the area of the roof.

77. A cylinder has a base area of $3w^2 + 5$ and a height of $4w$. Find an expression for the volume.

● **Lesson 9-3**

78. A circular pool has a radius of $5p - 3$ m. Write an expression for area the pool.

79. An office building has a rectangular base with side lengths of $12y - 7$ and $22y + 4$. Write an expression for the area of a floor in the office building.

● **Lesson 9-4**

80. Suppose you play a game with two number cubes. Let A represent rolling a number less than 4 and B represent rolling a number greater than 4. The probability of A is $\frac{1}{2}$. The probability of B is $\frac{1}{3}$.
  a. Find $\left(\frac{1}{2}A + \frac{1}{3}B\right)^2$
  b. What is the probability that both cubes show a number less than 4?
  c. What is the probability that one cube shows a number less than 4 and the other cube shows a number greater than 4?

81. Suppose there are two squares with side lengths of $4x - 3$ and $2x + 4$. Write an expression for the area of each square. Find the area of each square if $x = 6$ cm.

● **Lesson 9-5**

82. Write an expression for the side length of a square that has an area of $h^2 - 12h + 36$.

83. Write an expression for the radius of a circular flower garden with an area of $\pi m^2 + 14\pi m + 49\pi$.

● **Lessons 9-6 to 9-8 Use factoring to find expressions for possible dimensions of each figure.**

84. A rectangular parking lot has an area of $10w^2 - 9w - 40$.

85. A rectangular door has an area of $12d^2 - 31d + 14$

86. A circular window has an area of $49\pi v^2 + 84\pi v + 36\pi$.

87. A rectangular field has an area of $64m^2 - 169n^2$.

88. A rectangular prism has a volume of $6t^3 + 44t^2 + 70t$.

89. A rectangular field has an area of $10k^3 + 25k^2 - 6k - 15$.

● **Lesson 10-1** **Without graphing, describe how each graph differs from the graph of**
$y = x^2$.

1. $y = 3x^2$

2. $y = -4x^2$

3. $y = -0.5x^2$

4. $y = 0.2x^2$

5. $y = x^2 - 4$

6. $y = x^2 + 1$

7. $y = 2x^2 + 5$

8. $y = -0.3x^2 - 7$

● **Lesson 10-2** **Identify the axis of symmetry and the vertex of each function.**

9. $y = 3x^2$

10. $y = -2x^2 + 1$

11. $y = 0.5x^2 - 3$

12. $y = -x^2 + 2x + 1$

13. $y = 3x^2 + 6x$

14. $y = \frac{3}{4}x^2$

15. $y = 2x^2 - 9$

16. $y = -5x^2 + x + 4$

17. $y = x^2 - 8x$

Graph each quadratic inequality.

18. $y > x^2 - 4$

19. $y < 2x^2 + x$

20. $y \leq x^2 + x - 2$

● **Lessons 10-3 and 10-4** **Solve each equation. If the equation has no solution, write *no solution*.**

21. $x^2 = 36$

22. $x^2 + x - 2 = 0$

23. $c^2 - 100 = 0$

24. $9d^2 = 25$

25. $(x - 4)^2 = 100$

26. $3x^2 = 27$

27. $2x^2 - 54 = 284$

28. $7n^2 = 63$

29. $h^2 + 4 = 0$

● **Lessons 10-5 and 10-6** **Solve each equation. If the equation has no solution, write *no solution*.**

30. $x^2 + 6x - 2 = 0$

31. $x^2 - 5x = 7$

32. $x^2 - 10x + 3 = 0$

33. $2x^2 - 4x + 1 = 0$

34. $3x^2 + x + 5 = 0$

35. $\frac{1}{2}x^2 - 3x - 8 = 0$

36. $x^2 + 8x + 4 = 0$

37. $x^2 - 2x - 6 = 0$

38. $-3x^2 + x - 7 = 0$

39. $x^2 + 5x + 6 = 0$

40. $d^2 - 144 = 0$

41. $c^2 + 6 = 2 - 4c$

42. $x^2 + 4x = 2x^2 - x + 6$

43. $3x^2 + 2x - 12 = x^2$

44. $r^2 + 4r + 1 = r$

45. $d^2 + 2d + 10 = 2d + 100$

46. $3c^2 + c - 10 = c^2 - 5$

47. $t^2 - 3t - 10 = 0$

48. Agriculture You are planting a rectangular garden. It is 5 feet longer than 3 times its width. The area of the garden is 250 ft^2. Find the dimensions of the garden.

● **Lesson 10-7** **Find the number of solutions of each equation.**

49. $3x^2 + 4x - 7 = 0$

50. $5x^2 - 4x = -6$

51. $x^2 - 20x + 101 = 1$

52. $2x^2 - 8x + 9 = 4$

53. $4x^2 - 5x + 6 = 0$

54. $x^2 - 2x + 7 = 0$

● **Lesson 10-8** **Graph each set of data. Which model is most appropriate for each set?**

55. $(2, 4), (4, 4), (1, 2), (5, 1.5)$

56. $(3, 8), (4, 6), (5, 5), (6, 4), (7, 3)$

57. $(0, 7), (1, 3), (3, 0.5), (2, 1)$

● **Lesson 10-1**

58. Water from melting snow drips from a roof at a height of 40 ft. The function $h = -16t^2 + 40$ gives the approximate height h in feet of a drop of water t seconds after it falls. Graph the function.

● **Lesson 10-2** **The formula $h = -16t^2 + vt + c$ describes the height of an object thrown into the air, where h is the height, t is the time in seconds, v is the initial velocity, and c is the initial height. Use the formula to answer each question.**

59. A football is thrown with an upward velocity of 15 ft/s from an initial height of 5 feet. How long will it take for the football to reach its maximum height?

60. A ball is thrown from the top of a 50-ft building with an upward velocity of 24 ft/s. When will it reach its maximum height? How far above the ground will it be?

● **Lesson 10-3** **Model each problem with a quadratic equation. Then solve. If necessary, round to the nearest tenth.**

61. Find the radius of a circular lid with an area of 12 in.^{2}.

62. Find the side length of a square sandbox with an area of 150 ft^2.

63. Find the diameter of a circular pond with an area of 300 m^2.

● **Lesson 10-4** **Answer each question by factoring a quadratic equation.**

64. The length of an open-top box is 4 cm longer than its width. The box was made from a 480-cm^2 rectangular sheet of material with 6 cm-by-6 cm squares cut from each corner. The height of the box is 6 cm. Find the dimensions of the box.

65. Suppose you throw a rugby ball into the air with an initial upward velocity of 29 ft/s and an initial height of 6 ft. The formula $h = -16t^2 + 29t + 6$ gives the ball's height h in feet at time t seconds. Solve the equation for $h = 0$ to find when the ball will hit the ground.

● **Lesson 10-5** **Solve by completing the square.**

66. A rectangular patio has a length of $x + 6$ m, a width of $x + 8$ m, and a total area of 400 m^2. Find the dimensions to the nearest tenth.

● **Lessons 10-6 and 10-7**

67. A tennis ball is hit with a vertical velocity of 40 ft/s from an initial height of 7 ft. In how many seconds will the ball hit the ground?

68. A ball is thrown from an initial height of 6 feet at a rate of 42 ft/s to someone standing on a roof 30 feet above the ground. Use the discriminant to determine if the ball will reach the person on the roof.

● **Lesson 10-8**

69. Use a graphing calculator to determine what kind of function best models the data. Let $t = 0$ correspond to the year 2000. Write an equation that models the data.

Bird Population in the Town Park				
2000	2001	2002	2003	2004
225	207	185	168	160

● **Lesson 11-1** Simplify each radical expression.

1. $\dfrac{\sqrt{27}}{\sqrt{81}}$

2. $\sqrt{\dfrac{25}{4}}$

3. $\sqrt{\dfrac{50}{9}}$

4. $\dfrac{\sqrt{72}}{\sqrt{50}}$

5. $\sqrt{25} \cdot \sqrt{4}$

6. $\sqrt{27} \cdot \sqrt{3}$

7. $\sqrt{\dfrac{44x^4}{11}}$

8. $\dfrac{\sqrt{3c^2}}{\sqrt{27}}$

9. $\sqrt{45} \cdot \sqrt{18}$

● **Lesson 11-2** Simplify each radical expression.

10. $\sqrt{75} - 4\sqrt{75}$

11. $\sqrt{5}(\sqrt{20} - \sqrt{80})$

12. $\sqrt{6}(\sqrt{6} - 3)$

13. $3\sqrt{300} + 2\sqrt{27}$

14. $5\sqrt{2} \cdot 3\sqrt{50}$

15. $\sqrt{8} - 4\sqrt{2}$

16. $\dfrac{\sqrt{z^3}}{\sqrt{5z}}$

17. $(\sqrt{5} + 1)(\sqrt{5} - 1)$

18. $(\sqrt{3} + \sqrt{2})^2$

19. $\dfrac{1}{\sqrt{2} + 1}$

20. $\dfrac{2}{\sqrt{2} - 2}$

21. $\dfrac{\sqrt{3} + 1}{\sqrt{2} + 1}$

● **Lesson 11-3** Solve each radical equation. Check your solution.

22. $\sqrt{3x + 4} = 1$

23. $6 = \sqrt{8x - 4}$

24. $2x = \sqrt{14x - 6}$

25. $\sqrt{2x + 5} = \sqrt{3x + 1}$

26. $2x = \sqrt{6x + 4}$

27. $\sqrt{5x + 11} = \sqrt{7x - 1}$

28. $\sqrt{3x - 2} = x$

29. $\sqrt{x + 7} = x + 1$

30. $\sqrt{x + 3} = \dfrac{x + 9}{5}$

● **Lesson 11-4** Find the domain of each function. Then graph the function.

31. $y = \sqrt{x + 5}$

32. $y = \sqrt{x} - 2$

33. $y = \sqrt{x + 1}$

34. $y = \sqrt{x} - 4$

35. $y = \sqrt{x - 3}$

36. $y = \sqrt{x} + 6$

● **Lesson 11-5** Use $\triangle ABC$ to find the value of each expression.

37. $\sin A$

38. $\cos A$

39. $\tan A$

40. $\sin B$

41. $\cos B$

42. $\tan B$

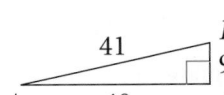

Find the value of each expression. Round to the nearest ten-thousandth.

43. $\sin 72°$

44. $\cos 29°$

45. $\tan 48°$

46. $\tan 52°$

47. $\sin 65°$

48. $\cos 43°$

● **Lesson 11-6**

49. A forester is 50 m from the base of a tree and measures the angle between the ground and the top of the tree. If the angle is 76°, find the height of the tree. Round your answer to the nearest meter.

50. A pilot is flying at an altitude of 25,000 ft. The angle of depression from her position to the start of runway is 2°. How far is the airplane from the start of the runway (in ground distance)?

● **Lesson 11-1** Use the formula $d = \sqrt{1.5h}$ to estimate the distance d in miles to a horizon when h is the height in feet of the viewer's eyes above the ground. **Round your answer to the nearest mile.**

51. Find the distance you can see to the horizon from the top of a building that is 355 ft high.

52. Find the distance you can see to the horizon from an airplane that is 36,000 ft above the ground.

53. Find the distance you can see to the horizon from 15 ft above the ground.

● **Lesson 11-2** **Find an exact solution. Then find an approximate solution to the nearest tenth.**

54. The ratio of the length of a painting to its width is $(1 + \sqrt{5}) : 1$. The length is 25 in. What is the width?

55. The ratio of the length of a painting to its width is $(1 + \sqrt{3}) : 1$. The length is 40 cm. What is the width?

● **Lesson 11-3** **Write and solve a radical equation for each situation.**

56. What is the diameter of a circular livestock pen if the area is 450 m^2?

57. On a roller coaster ride, your speed in a loop depends on the height h of the preceding hill and the radius of the loop in feet. The equation $v = 8\sqrt{h - 2r}$ gives the velocity v in feet per second of a car at the top of the loop. Suppose a loop has a radius of 25 ft and you want the car to have a velocity of 25 ft/s at the top of the loop. How high should the preceding hill be?

● **Lesson 11-4** **Write and graph a radical expression to answer each question.**

58. Suppose a motorcycle company uses the function $n = 30\sqrt{4t} + 90$ to model the sales volume n, as a function of time t, the number of months after the start of the advertising camping.
a. Graph the function.
b. Use the graph to estimate when 200 motorcycles will be sold.

● **Lessons 11-5 and 11-6** **Use trigonometric ratios to answer each question.**

59. A painter leans a 20-ft ladder up against the side of a house making a right triangle. How far up the side of the house will the top of the ladder rest if the ladder makes a 75° angle with the ground?

60. The length of a rectangular field is 100 yd and the diagonal across the field forms an angle of 25° with the sideline. What is the length of the diagonal line?

61. A pilot is flying a plane 6500 ft above the ground. The pilot begins a 3° descent to an airport runway. Find the horizontal distance between the airplane and the start of the runway?

62. While hiking you see an 100-ft observation tower on top of a 1100-ft hill. The angle of elevation to the top of the tower is 6°? How far away is the tower?

● **Lesson 12-1** Identify the asymptotes of each function. Then graph the function.

1. $y = \dfrac{6}{x}$ **2.** $y = \dfrac{8}{x + 2}$ **3.** $y = \dfrac{4}{x} - 3$ **4.** $y = \dfrac{5}{x + 1} + 3$

5. $y = \dfrac{5}{x} - 1$ **6.** $y = \dfrac{2}{x - 1}$ **7.** $y = \dfrac{3}{x} + 4$ **8.** $y = \dfrac{2}{x + 1} - 1$

● **Lesson 12-2** Simplify each expression.

9. $\dfrac{4t^2}{16t}$ **10.** $\dfrac{c - 5}{c^2 - 25}$ **11.** $\dfrac{4m - 12}{m - 3}$ **12.** $\dfrac{a^2 + 2a - 3}{a + 3}$

13. $\dfrac{x + 7}{x^2 + 8x + 7}$ **14.** $\dfrac{t - 4}{4 - t}$ **15.** $\dfrac{m^2 + 7m + 10}{m^2 + 8m + 15}$ **16.** $\dfrac{6b^2 + 42b}{b^3}$

● **Lessons 12-3 to 12-5** Simplify each expression.

17. $\dfrac{4}{x} - \dfrac{3}{x}$ **18.** $\dfrac{6t}{5} + \dfrac{4t}{5}$ **19.** $\dfrac{6}{c} + \dfrac{4}{c^2}$

20. $\dfrac{6}{3d} - \dfrac{4}{3d}$ **21.** $\dfrac{5s^4}{10s^3}$ **22.** $\dfrac{4n^2}{7} \cdot \dfrac{14}{2n^3}$

23. $\dfrac{8b^2 - 4b}{3b^2} \div \dfrac{2b - 1}{9b}$ **24.** $\dfrac{v^5}{v^3} \cdot \dfrac{4v^{-1}}{v^2}$ **25.** $\dfrac{5}{t + 4} + \dfrac{3}{t - 4}$

26. $\dfrac{8}{m^2 + 6m + 5} + \dfrac{4}{m + 1}$ **27.** $\dfrac{3y}{4y - 8} \div \dfrac{9y}{2y^2 - 4y}$ **28.** $\dfrac{4}{d^2} - \dfrac{3}{d^3}$

Divide.

29. $(2x^3 - x^2 - 13x - 6) \div (x - 3)$ **30.** $(3x^3 - 3) \div (x + 1)$

31. $(3x^3 + 5x^2 - 22x + 24) \div (x + 4)$ **32.** $(3x^3 - 3) \div (x - 1)$

● **Lesson 12-6** Solve each equation. Check your answer.

33. $\dfrac{1}{4} + \dfrac{1}{x} = \dfrac{3}{8}$ **34.** $\dfrac{4}{m} - 3 = \dfrac{2}{m}$ **35.** $\dfrac{1}{b - 3} = \dfrac{1}{4b}$

36. $\dfrac{4}{x - 1} = \dfrac{3}{x}$ **37.** $\dfrac{4}{n} + \dfrac{5}{9} = 1$ **38.** $\dfrac{x}{x + 2} = \dfrac{x - 3}{x + 1}$

39. $t - \dfrac{8}{t} = \dfrac{17}{t}$ **40.** $\dfrac{x + 2}{x + 5} = \dfrac{x - 4}{x + 4}$ **41.** $\dfrac{1}{5} - \dfrac{1}{x} = \dfrac{1}{2}$

42. $\dfrac{4}{c + 1} - \dfrac{2}{c - 1} = \dfrac{3c + 6}{c^2 - 1}$ **43.** $\dfrac{4}{m + 3} = \dfrac{6}{m - 3}$ **44.** $\dfrac{4}{t + 5} + 1 = \dfrac{15}{t^2 - 25}$

● **Lessons 12-7 and 12-8** Simplify each expression.

45. $_6C_4$ **46.** $_7P_2$ **47.** $_{10}C_5$ **48.** $_8C_7$

49. $_{12}P_6$ **50.** $_9C_7$ **51.** $_{10}C_6$ **52.** $_{10}P_6$

53. How many four-letter groups can be made with the letters A, B, C, and D if no letter can be repeated?

54. Two people are running for president, three are running for vice-president, and two are running for speaker. How many election results are possible?

55. a. Suppose your bank assigns a four-digit personal identification number for your bank card. The first digit cannot be zero. How many different numbers are possible?
 b. What is the probability that the number you are given is 4861?

● **Lesson 12-1** **Graph a rational function to answer each question.**

56. The function $I = \frac{350}{x^2}$ gives the intensity I in lumens at a distance of x feet from a light bulb with 350 watts. What is the intensity of light when you are 40 feet away?

57. The function $s = \frac{1500}{d^2}$ gives the signal strength s of a radio signal at a distance d miles from a transmitter. What is the signal strength when you are 25 miles away?

● **Lesson 12-2** **The expression $\frac{30rh}{r + h}$ gives the approximate baking time for bread in minutes based on the radius r and the height h of the baked loaf in inches. Use this expression to find the baking time for each loaf.**

58. a roll that will have a radius of 2 in. and a height of 1.5 in.

59. a loaf of Italian bread that will have a length of 18 in. and a radius of 4 in.

● **Lesson 12-3**

60. Find an expression for the area of a rectangular field with a length of $\frac{4y + 10}{3y^2 - 4}$ m and a width of $\frac{y^3 + 6}{2y + 5}$ m.

● **Lesson 12-4**

61. What is the area of the base of a hexagonal prism with a volume of $15x^3 - 26x^2 + 43x - 14$ m³ and a height of $5x - 2$ m?

● **Lesson 12-5**

62. A canoe travels 5 miles downstream at rate r, and then 5 miles back upstream at 70% of its downstream rate. Write and simplify a rational expression for the total time of the trip.

63. Louis walks to get to his friend's house 2 miles away. When he returns home he jogs at a rate that is 160% of his walking rate r. Write and simplify a rational expression for the total time of his trip.

● **Lesson 12-6**

64. Suppose you can clean your kitchen in 30 minutes. Your friend can clean your kitchen in 45 minutes. How long will it take you to clean the kitchen together?

65. One employee of a landscaping company can mow a lawn in 20 minutes. Another employee can do it in 15 minutes. How long will it take them to mow the lawn together?

● **Lessons 12-7 and 12-8** **Find the number of possibilities for each situation.**

66. Suppose you have nine different shirts. How many ways can you select five shirts to wear in order?

67. How many different ways can four different officer positions be filled from a school club with 35 members?

68. Suppose you have a summer reading list with 15 books and you have to read three. How many different combinations of books can you read?

Skills Handbook

Problem Solving Strategies

You may find one or more of these strategies helpful in solving a word problem.

STRATEGY	WHEN TO USE IT
Draw a Diagram	The problem describes a picture or diagram.
Try, Check, Revise	Solving the problem directly is too complicated.
Look for a Pattern	The problem describes a relationship.
Make a Table	The problem has data that need to be organized.
Solve a Simpler Problem	The problem is complex or has numbers that are too cumbersome to use at first.
Use Logical Reasoning	You need to reach a conclusion using given information.
Work Backward	You need to find the number that led to the result in the problem.

Problem Solving: Draw a Diagram

> **EXAMPLE**
>
> Two cars started from the same point. One traveled east at 45 mi/h and the other west at 50 mi/h. How far apart were the cars after 5 hours?
>
> Draw a diagram.
>
> West ←⎯⎯⎯ 50 mi/h · 5 ⎯⎯⎯ Start ⎯⎯⎯ 45 mi/h · 5 ⎯⎯⎯→ East
>
> The first car traveled 45 · 5 or 225 mi. The second car traveled 50 · 5 or 250 mi.
>
> The diagram shows that the two distances should be added: $225 + 250 = 475$ mi.
>
> • After 5 hours, the cars were 475 mi apart.

EXERCISES

1. Jason, Lee, Melda, Aaron, and Bonnie want to play one another in tennis. How many games will be played?

2. A playground, a zoo, a picnic area, and a flower garden will be in four corners of a new park. Straight paths will connect each of these areas to all the other areas. How many pathways will be built?

3. Pedro wants to tack 4 posters on a bulletin board. He will tack the four corners of each poster, overlapping the sides of each poster a little bit. What is the least number of tacks that Pedro can use?

Problem Solving: Try, Check, Revise

When you are not sure how to start, guess an answer and then test it. In the process of testing a guess, you may see a way of revising your guess to get closer to the answer or to get the exact answer.

EXAMPLE

Maria bought books and CDs as gifts. Altogether she bought 12 gifts and spent $84. The books cost $6 each and the CDs cost $9 each. How many of each gift did she buy?

Trial 6 books   **Test** $6 \cdot \$6 =$ $36
        6 CDs $6 \cdot \$9 =$ $\underline{+\$54}$
                                          $90

Revise your guess. You need fewer CDs to bring the total cost down.

Trial 7 books   **Test** $7 \cdot \$6 =$ $42
        5 CDs $5 \cdot \$9 =$ $\underline{+\$45}$
                                          $87

The cost is still too high.

Trial 8 books   **Test** $8 \cdot \$6 =$ $48
        4 CDs $4 \cdot \$9 =$ $\underline{+\$36}$
                                          $84

Maria bought 8 books and 4 CDs.

EXERCISES

1. Find two consecutive odd integers whose product is 323.

2. Find three consecutive integers whose sum is 81.

3. Find four consecutive integers whose sum is 138.

4. Mika bought 9 rolls of film to take 180 pictures on a field trip. Some rolls had 36 exposures and the rest had 12 exposures. How many of each type did Mika buy?

5. Tanya is 18 years old. Her brother Shawn is 16 years younger. How old will Tanya be when she is 3 times as old as Shawn?

6. Steven has 100 ft of fencing and wants to build a fence in the shape of a rectangle to enclose the largest possible area. What should be the dimensions of the rectangle?

7. The combined ages of a mother, her son, and her daughter are 61 years. The mother is 22 years older than her son and 31 years older than her daughter. How old is each person?

8. Kenji traveled 210 mi in a two-day bicycle race. He biked 20 mi farther on the first day than he did on the second day. How many miles did Kenji travel each day?

9. Darren is selling tickets for the school play. Regular tickets are $4 each, and student tickets are $3 each. Darren sells a total of 190 tickets and collects $650. How many of each kind of ticket did he sell?

Problem Solving: Look for a Pattern and Make a Table

Some problems describe relationships that involve regular sequences of numbers or other things. To solve the problem you need to be able to recognize and describe the pattern that gives the relationship for the numbers or things. One way to organize the information given is to make a table.

EXAMPLE

A tree farm is planted as shown at the right. The dots represent trees. The lot will be enlarged by adding larger squares. How many trees will be in the fifth square?

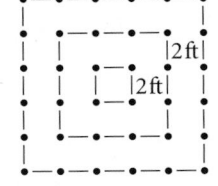

Make a table to help find a pattern.

Square position	1st	2nd	3rd	4th	5th
Number of trees	4	12	20	■	■

Pattern: 8 more trees are planted in each larger square.

The fourth square will have 28 trees. The fifth square will have 36 trees.

EXERCISES

1. Kareem made a display of books at a book fair. One book was in the first row, and each of the other rows had two more books than the row before it. How many books does Kareem have if he has nine rows?

2. Chris is using green and white tiles to cover her floor. If she uses tiles in the pattern G, W, G, G, W, G, G, W, G, G, what will be the color of the twentieth tile?

3. Jay read one story the first week of summer vacation, 3 stories the second week, 6 stories the third week, and 10 stories the fourth week. He kept to this pattern for eight weeks. How many stories did he read the eighth week?

4. Jan has 6 coins, none of which is a half dollar. The coins have a value of $.85. What coins does she have?

5. Sam is covering a wall with rows of red, white, and blue siding. The red siding is cut in 1.8-m strips, the white in 2.4-m strips, and the blue in 1.2-m strips. What is the shortest length that Sam can cover with uncut strips to form equal rows of each color?

6. A train leaves a station at 8:00 A.M. and averages 40 mi/h. Another train leaves the same station one hour later and averages 50 mi/h traveling in the same direction on a parallel track. At what time will the second train catch up with the first train? How many miles would each train have traveled by that time?

7. The soccer team held a car wash and earned $200. They charged $7 per truck and $5 per car. In how many different ways could the team have earned the $200?

8. Students are going to march in a parade. There will be one first-grader, two second-graders, three third-graders, and so on through the twelfth grade. How many students will march in the parade?

Problem Solving: Solve a Simpler Problem

By solving one or more simpler problems you can often find a pattern that will help solve a more complicated problem.

Skills Handbook

EXAMPLE

How many different rectangles are in a strip with 10 squares?

Begin with one square, and then add one square at a time. Determine whether there is a pattern.

Squares in the strip:	1	2	3	4	5
Number of rectangles:	1	3	6	10	15
Pattern:		$1 + 2 = 3$	$3 + 3 = 6$	$6 + 4 = 10$	$10 + 5 = 15$

Now continue the pattern.

Squares in the strip:	6	7	8	9	10
Number of rectangles:	21	28	36	45	55
Pattern:	$15 + 6 = 21$	$21 + 7 = 28$	$28 + 8 = 36$	$36 + 9 = 45$	$45 + 10 = 55$

● There are 55 rectangles in a strip with 10 squares.

EXERCISES

1. Lockers in the east wing of Hastings High School are numbered 1–120. How many contain the digit 8?

2. What is the sum of all the numbers from 1 to 100? (*Hint:* What is $1 + 100$? What is $2 + 99$?)

3. Suppose your heart beats 70 times per minute. At this rate, how many times had it beaten by the time you were 10?

4. For a community project you have to create the numbers 1 through 148 using large cardboard digits covered with glitter, which you make by hand. How many cardboard digits will you make?

5. There are 64 teams competing in the state soccer championship. If a team loses a game it is eliminated. How many games have to be played in order to get a single champion team?

6. You work in a supermarket. Your boss asks you to arrange oranges in a pyramid for a display. The pyramid's base should be a square with 25 oranges. How many layers of oranges will be in your pyramid? How many oranges will you need?

7. Kesi has her math book open. The product of the two page numbers on the facing pages is 1056. What are the two page numbers?

8. There are 12 girls and 11 boys at a school party. A photographer wants to take a picture of each boy with each girl. How many photographs must the photographer set up?

Problem Solving: Use Logical Reasoning

Some problems can be solved without the use of numbers. They can be solved by the use of logical reasoning, given some information.

 **EXAMPLE**

Joe, Melissa, Liz and Greg play different sports. Their sports are running, basketball, baseball, and tennis. Liz's sport does not use a ball. Joe hit a home run in his sport. Melissa is the sister of the tennis player. Which sport does each play?

Make a table to organize what you know.

	Running	Basketball	Baseball	Tennis
Joe	✗	✗	✓	✗
Melissa				✗
Liz	✓	✗	✗	✗
Greg				

←— A home run means Joe plays baseball.
←— Melissa cannot be the tennis player.
←— Liz must run, since running does not involve a ball.

Use logical reasoning to complete the table.

	Running	Basketball	Baseball	Tennis
Joe	✗	✗	✓	✗
Melissa	✗	✓	✗	✗
Liz	✓	✗	✗	✗
Greg	✗	✗	✗	✓

←— The only option for Greg is tennis.

● Greg plays tennis, Melissa plays basketball, Liz runs, and Joe plays baseball.

EXERCISES

1. Juan has a dog, a horse, a bird, and a cat. Their names are Bo, Cricket, K.C., and Tuffy. Tuffy and K.C. cannot fly or be ridden. The bird talks to Bo. Tuffy runs from the dog. What is each pet's name?

2. A math class has 25 students. There are 13 students who are only in the band, 4 students who are only on the swimming team, and 5 students who are in both groups. How many students are not in either group?

3. Annette is taller than Heather but shorter than Garo. Tanya's height is between Garo's and Annette's. Karin would be the shortest if it weren't for Alexa. List the names in order from shortest to tallest.

4. The Robins, Wrens, and Sparrows teams played one another twice in basketball. The Robins won 3 of their games. The Sparrows won 2 of their games. How many games did each team win and lose?

5. The girls' basketball league uses a telephone tree when it needs to cancel its games. The leader takes 1 min to call 2 players. These 2 players take 1 min to call 2 more players, and so on. How many players will be called in 6 min?

6. Miss White, Miss Gray, and Miss Black are wearing single-colored dresses that are white, gray, and black. Miss Gray remarks to the woman wearing a black dress that no woman's dress color matches her last name. What color dress is each woman wearing?

Problem Solving: Work Backward

To solve some problems, you need to start with the end result and work backward to the beginning.

EXAMPLE

On Monday, Rita withdrew $150 from her savings account. On Wednesday, she deposited $400 into her account. She now has $1000. How much was in her account on Monday before she withdrew the money?

money in account now	$1000
Undo the deposit.	$-\$400$
	$\$600$
Undo the withdrawal.	$+\$150$
	$\$750$

● Rita had $750 in her account on Monday before withdrawing money.

EXERCISES

1. Ned gave Connie the following puzzle: I am thinking of a number. I doubled it, then tripled the result. The final result was 36. What is my number?

2. Fernando gave Maria the following puzzle: I am thinking of a number. I divide it by 3. Then I divide the result by 5. The final result is 8. What is my number?

3. A teacher lends pencils to students. She gave out 7 pencils in the morning, collected 5 before lunch, and gave out 3 after lunch. At the end of the day she had 16 pencils. How many pencils did the teacher have at the start of the day?

4. This week Sandy withdrew $350 from her savings account. She made a deposit of $125, wrote a check for $ 275, and made a deposit of $150. She now has $225 in her account. How much did she have in her account at the beginning of the week?

5. Jeff paid $12.50, including a $1.60 tip, for a taxi ride from his home to the airport. City Cab charges $1.90 for the first mile plus $.15 for each additional $\frac{1}{6}$ mile. How many miles is Jeff's home from the airport?

6. Ben sold $\frac{1}{4}$ as many tickets to the fund-raiser as Charles. Charles sold 3 times as many as Susan. Susan sold 4 fewer than Tom. Tom sold 12 tickets. How many did Ben sell?

7. Two cars start traveling towards each other. One car averages 30 mi/h and the other 40 mi/h. After 4 h the cars are 10 mi apart. How far apart were the cars when they started?

8. Nina has a dentist appointment at 8:45 A.M. She wants to arrive 10 min early. Nina needs to allow 25 min to travel to the appointment and 45 min to dress and have breakfast. What is the latest time Nina should get up?

9. Jordan spent the day exploring her neighborhood and ended up at the park 2 mi east and 1 mi north of the center of town. She started from her house and walked $\frac{1}{2}$ mi north. Next she walked 3 mi west and then $1\frac{1}{2}$ mi south. Finally, she walked $\frac{1}{4}$ mi east to the park. Where is Jordan's house in relation to the center of town?

Prime Numbers and Composite Numbers

A prime number is a whole number greater than 1 that has exactly two factors, the number 1 and itself.

Prime number	2	5	17	29
Factors	1, 2	1, 5	1, 17	1, 29

A composite number is a number that has more than two factors. The number 1 is neither prime nor composite.

Composite number	6	15	48
Factors	1, 2, 3, 6	1, 3, 5, 15	1, 2, 3, 4, 6, 8, 12, 16, 24, 48

1 EXAMPLE

Is 51 prime or composite?

$51 = 3 \cdot 17$ **Try to find factors other than 1 and 51.**

51 is a composite number.

You can use a factor tree to find the prime factors of a number. When all the factors are prime numbers, it is called the prime factorization of the number.

2 EXAMPLE

Use a factor tree to write the prime factorization of 28.

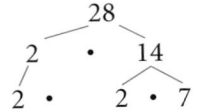   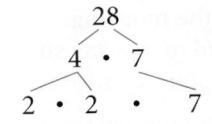   **The order of listing the factors may be different, but the end result is the same.**

The prime factorization of 28 is $2 \cdot 2 \cdot 7$.

EXERCISES

Is each number prime or composite?

1. 9	**2.** 16	**3.** 34	**4.** 61	**5.** 7	**6.** 13
7. 12	**8.** 40	**9.** 57	**10.** 64	**11.** 120	**12.** 700
13. 39	**14.** 23	**15.** 63	**16.** 19	**17.** 522	**18.** 101

List all the factors of each number.

19. 46	**20.** 32	**21.** 11	**22.** 65	**23.** 27	**24.** 29
25. 205	**26.** 123	**27.** 24	**28.** 162	**29.** 88	**30.** 204
31. 183	**32.** 6	**33.** 98	**34.** 92	**35.** 59	**36.** 47

Use a factor tree to write the prime factorization of each number.

37. 18	**38.** 20	**39.** 27	**40.** 54	**41.** 64	**42.** 96
43. 100	**44.** 125	**45.** 84	**46.** 150	**47.** 121	**48.** 226

Factors and Multiples

A common factor is a number that is a factor of two or more numbers. The greatest common factor (GCF) is the greatest number that is a common factor of two or more numbers.

1 EXAMPLE

Find the GCF of 24 and 64.

Method 1 List all the factors of each number.

Factors of 24   1, 2, 3, 4, 6, 8, 12, 24   **Find the common factors: 1, 2, 4, 8.**

Factors of 64   1, 2, 4, 8, 16, 32, 64   **The greatest common factor is 8.**

GCF (24, 64) = 8

Method 2 Use the prime factorization of each number.

$24 = 2 \cdot 2 \cdot 2 \cdot 3$ **Find the prime factorization of each number.**

$64 = 2 \cdot 2 \cdot 2 \cdot 2 \cdot 2 \cdot 2$

$GCF = 2 \cdot 2 \cdot 2 = 8$ **Use each factor the number of times it appears as a common factor.**

A common multiple is a number that is a multiple of two or more numbers. The least common multiple (LCM) is the least number that is a common multiple of two or more numbers.

2 EXAMPLE

Find the LCM of 12 and 18.

Method 1 List the multiples of each number.

Multiples of 12   12, 24, 36, . . .   **List a number of multiples until you find**

Multiples of 18   18, 36, . . .   **the first common multiple.**

LCM (12, 18) = 36

Method 2 Use the prime factorization of each number.

$12 = 2 \cdot 2 \cdot 3$

$18 = 2 \cdot 3 \cdot 3$

$LCM = 2 \cdot 2 \cdot 3 \cdot 3 = 36$ **Use each prime factor the greatest number of times it appears in either number.**

EXERCISES

Find the GCF of each set of numbers.

1. 12 and 22   **2.** 7 and 21   **3.** 24 and 48   **4.** 17 and 51

5. 9 and 12   **6.** 10 and 25   **7.** 21 and 49   **8.** 27 and 36

9. 10, 30, and 25   **10.** 56, 84, and 140 **11.** 42, 63, and 105   **12.** 20, 28, and 40

Find the LCM of each set of numbers.

13. 16 and 20   **14.** 14 and 21   **15.** 11 and 33   **16.** 8 and 9

17. 5 and 12   **18.** 54 and 84   **19.** 48 and 80   **20.** 25 and 36

21. 10, 15, and 25   **22.** 6, 7, and 12   **23.** 5, 8, and 20   **24.** 18, 21, and 36

Divisibility

An integer is divisible by another integer if the remainder is zero. You can use the following tests to determine whether a number is divisible by the numbers below.

Number	Divisibility Test
2	The ones digit is 0, 2, 4, 6, or 8.
3	The sum of the digits is divisible by 3.
4	The number formed by the last two digits is divisible by 4.
5	The ones digit is 0 or 5.
6	The number is divisible by 2 and by 3.
8	The number formed by the last three digits is divisible by 8.
9	The sum of the digits is divisible by 9.
10	The ones digit is 0.

EXAMPLE

Use the divisibility tests to determine the numbers by which 2116 is divisible.

2: Yes; the ones digit is 6.
3: No; the sum of the digits is $2 + 1 + 1 + 6 = 10$, which is *not* divisible by 3.
4: Yes; the number formed by the last two digits is 16, which is divisible by 4.
5: No; the ones digit is 6, *not* 0 or 5.
6: No; 2116 is *not* divisible by 3.
8: No; the number formed by the last three digits is 116, which is *not* divisible by 8.
9: No; the sum of the digits is $2 + 1 + 1 + 6 = 10$, which is *not* divisible by 9.
10: No; the ones digit is 6, *not* 0.

● 2116 is divisible by 2 and 4.

EXERCISES

Determine whether each number is divisible by 2, 3, 4, 5, 6, 8, 9, or 10.

1. 236	**2.** 72	**3.** 105	**4.** 108	**5.** 225	**6.** 364
7. 1234	**8.** 4321	**9.** 7848	**10.** 3366	**11.** 1421	**12.** 1071
13. 78,765	**14.** 30,303	**15.** 4104	**16.** 700	**17.** 868	**18.** 1155

19. Reasoning Since 435 is divisible by both 3 and 5, it is also divisible by what number?

20. Find a number greater than 1000 that is divisible by 4, 5, and 9.

21. Critical Thinking If a is divisible by 2, what can you conclude about $a + 1$? Justify your answer.

Using Estimation

To make sure the answer to a problem is reasonable, you can estimate before you calculate. If the answer is close to your estimate, the answer is probably correct.

1 EXAMPLE

Estimate to find whether each answer is reasonable.

a. Calculation   Estimate

$126.91	$\approx$	$130
$14.05	$\approx$	$10
+$25.14	$\approx$	+$30
$266.10		$170

The answer is not close to the estimate. It is *not* reasonable. The calculation is *incorrect*.

b. Calculation   Estimate

372.85	$\approx$	370
−227.31	$\approx$	−230
145.54		140

The answer is close to the estimate. It *is* reasonable. The calculation is *correct*.

For some situations, like estimating a grocery bill, you may not need an exact answer. A *front-end estimate* will give you a good estimate that is usually closer to the exact answer than an estimate you would get by rounding. Add the front-end digits, estimate the sum of the remaining digits by rounding, and then combine sums.

2 EXAMPLE

Tomatoes cost $3.54, squash costs $2.75, and lemons cost $1.20. Estimate the total cost of the produce.

Add the front-end digits.

3.54	$\rightarrow$	0.50	**Estimate by rounding.**
2.75	$\rightarrow$	0.80	
+1.20	$\rightarrow$	+ 0.20	
6	+	1.50 = 7.50	

The total cost is about $7.50.

EXERCISES

Estimate by rounding.

1. the sum of $15.70, $49.62, and $278.01

2. 563 − 125

3. the sum of $163.90, $107.21, and $33.56

4. 824 − 467

Use front-end estimation.

5. $1.65 + $5.42 + $9.89

6. 1.369 + 7.421 + 2.700

7. 9.563 − 2.480

8. 1.17 + 3.92 + 2.26

9. 8.611 − 1.584

10. $2.52 + $3.04 + $5.25

Estimate using a method of your choice.

11. Ticket prices at an amusement park cost $11.25 for adults and $6.50 for children under 12. Estimate the cost for three children and one adult.

12. Esmeralda has a new checking account. So far, she has deposited $177, $250, and $193. She has also written a check for $26.89. Estimate her current balance.

Simplifying Fractions

A fraction can name a part of a group or region. The region below is divided into 10 equal parts and 6 of the equal parts are shaded.

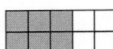 $\dfrac{6}{10}$ ← Numerator ← Denominator   **Read: six tenths**

A fraction can have many names. Different names for the same fraction are called equivalent fractions. You can find an equivalent fraction for any given fraction by multiplying the numerator and denominator of the given fraction by the same number.

1 EXAMPLE

Write five equivalent fractions for $\dfrac{3}{5}$.

$\dfrac{3}{5} = \dfrac{3 \cdot 2}{5 \cdot 2} = \dfrac{6}{10}$ $\dfrac{3}{5} = \dfrac{3 \cdot 3}{5 \cdot 3} = \dfrac{9}{15}$ $\dfrac{3}{5} = \dfrac{3 \cdot 4}{5 \cdot 4} = \dfrac{12}{20}$ $\dfrac{3}{5} = \dfrac{3 \cdot 5}{5 \cdot 5} = \dfrac{15}{25}$ $\dfrac{3}{5} = \dfrac{3 \cdot 6}{5 \cdot 6} = \dfrac{18}{30}$

The fraction $\dfrac{3}{5}$ is in simplest form because its numerator and denominator are relatively prime, that is, their only common factor is the number 1. To write a fraction in simplest form, divide its numerator and denominator by their greatest common factor (GCF).

2 EXAMPLE

Write $\dfrac{6}{24}$ in simplest form.

Step 1 Find the GCF of 6 and 24.

$6 = 2 \cdot 3$ **Multiply the common prime factors.**

$24 = 2 \cdot 2 \cdot 2 \cdot 3$ **GCF = 2 · 3 = 6.**

Step 2 Divide the numerator and denominator of $\dfrac{6}{24}$ by the GCF, 6.

$\dfrac{6}{24} = \dfrac{6 \div 6}{24 \div 6} = \dfrac{1}{4}$ **simplest form**

EXERCISES

Write five equivalent fractions for each fraction.

1. $\dfrac{4}{7}$ **2.** $\dfrac{9}{16}$ **3.** $\dfrac{3}{8}$ **4.** $\dfrac{8}{17}$ **5.** $\dfrac{5}{6}$ **6.** $\dfrac{7}{10}$

Complete each statement.

7. $\dfrac{3}{7} = \dfrac{\blacksquare}{21}$ **8.** $\dfrac{5}{8} = \dfrac{20}{\blacksquare}$ **9.** $\dfrac{11}{12} = \dfrac{44}{\blacksquare}$ **10.** $\dfrac{12}{16} = \dfrac{\blacksquare}{4}$ **11.** $\dfrac{50}{100} = \dfrac{1}{\blacksquare}$

12. $\dfrac{5}{9} = \dfrac{\blacksquare}{27}$ **13.** $\dfrac{3}{8} = \dfrac{\blacksquare}{24}$ **14.** $\dfrac{5}{6} = \dfrac{20}{\blacksquare}$ **15.** $\dfrac{12}{20} = \dfrac{\blacksquare}{5}$ **16.** $\dfrac{75}{150} = \dfrac{1}{\blacksquare}$

Which fractions are in simplest form?

17. $\dfrac{4}{12}$ **18.** $\dfrac{3}{16}$ **19.** $\dfrac{5}{30}$ **20.** $\dfrac{9}{72}$ **21.** $\dfrac{11}{22}$ **22.** $\dfrac{24}{25}$

Write in simplest form.

23. $\dfrac{8}{16}$ **24.** $\dfrac{7}{14}$ **25.** $\dfrac{6}{9}$ **26.** $\dfrac{20}{30}$ **27.** $\dfrac{8}{20}$ **28.** $\dfrac{12}{40}$

29. $\dfrac{15}{45}$ **30.** $\dfrac{14}{56}$ **31.** $\dfrac{10}{25}$ **32.** $\dfrac{9}{27}$ **33.** $\dfrac{45}{60}$ **34.** $\dfrac{20}{35}$

Fractions and Decimals

You can write a fraction as a decimal.

1 EXAMPLE

Write $\frac{3}{5}$ as a decimal.

$$5\overline{)3.0} \quad \begin{array}{c} 0.6 \end{array}$$

$$-3.0 \quad \textbf{Divide the numerator by the denominator.}$$

The decimal for $\frac{3}{5}$ is 0.6.

You can write a decimal as a fraction.

2 EXAMPLE

Write 0.38 as a fraction.

$0.38 = 38$ hundredths $= \frac{38}{100} = \frac{19}{50}$

Some fractions have decimal forms that do not end, but do repeat.

3 EXAMPLE

Write $\frac{3}{11}$ as a decimal.

Divide the numerator by the denominator. The remainders 8 and 3 keep repeating. Therefore 2 and 7 will keep repeating in the quotient.

$\frac{3}{11} = 0.2727 \ldots = 0.\overline{27}$

$$\frac{3}{11} = 11\overline{)3.0000 \ldots} \quad \begin{array}{c} 0.2727 \end{array}$$

$$\begin{array}{r} 22 \\ \hline 80 \\ 77 \\ \hline 30 \\ 22 \\ \hline 80 \\ 77 \\ \hline 3 \end{array}$$

You can write a repeating decimal as a fraction.

4 EXAMPLE

Write 0.363636 . . . as a fraction.

Let $x = 0.363636 \ldots$

Then $100x = 36.36363636 \ldots$ **When 2 digits repeat, multiply by 100.**

$99x = 36$ **Subtract the first equation from the second.**

$x = \frac{36}{99}$ or $\frac{4}{11}$ **Divide each side by 99.**

EXERCISES

Write as a decimal.

1. $\frac{3}{10}$ **2.** $\frac{13}{12}$ **3.** $\frac{4}{20}$ **4.** $\frac{25}{75}$ **5.** $\frac{5}{7}$ **6.** $4\frac{3}{25}$

7. $\frac{5}{9}$ **8.** $5\frac{7}{8}$ **9.** $\frac{2}{7}$ **10.** $\frac{3}{15}$ **11.** $\frac{16}{100}$ **12.** $2\frac{2}{5}$

Write as a fraction in simplest form.

13. 0.07   **14.** 0.25   **15.** 0.875   **16.** 0.4545 . . .   **17.** 6.333 . . .   **18.** 7.2626 . . .

19. 0.77 . . .   **20.** 3.1313 . . .   **21.** 0.375   **22.** 0.8333 . . .   **23.** 6.48   **24.** 0.8

Adding and Subtracting Fractions

You can add and subtract fractions when they have the same denominator.
Fractions with the same denominator are called like fractions.

1 EXAMPLE

a. Add $\frac{4}{5} + \frac{3}{5}$.

$$\frac{4}{5} + \frac{3}{5} = \frac{4 + 3}{5} = \frac{7}{5} = 1\frac{2}{5} \leftarrow$$

Add or subtract the numerators and keep the same denominator.

b. Subtract $\frac{5}{9} - \frac{2}{9}$.

$$\rightarrow \frac{5}{9} - \frac{2}{9} = \frac{5 - 2}{9} = \frac{3}{9} = \frac{1}{3}$$

Fractions with unlike denominators are called unlike fractions. To add or subtract fractions with unlike denominators, find the least common denominator (LCD) and write equivalent fractions with the same denominator. Then add or subtract the like fractions.

2 EXAMPLE

Add $\frac{3}{4} + \frac{5}{6}$.

$$\frac{3}{4} + \frac{5}{6} = \frac{9}{12} + \frac{10}{12}$$

Find the LCD. The LCD is the same as the least common multiple (LCM).
The LCD of 4 and 6 is 12.

$$= \frac{9 + 10}{12} = \frac{19}{12} \text{ or } 1\frac{7}{12}$$

Write equivalent fractions with the same denominator.

To add or subtract mixed numbers, add or subtract the fractions. Then add or subtract the whole numbers. Sometimes when subtracting mixed numbers you may have to regroup.

3 EXAMPLE

Subtract $5\frac{1}{4} - 3\frac{2}{3}$.

$$5\frac{1}{4} - 3\frac{2}{3} = 5\frac{3}{12} - 3\frac{8}{12}$$

Write equivalent fractions with the same denominator.

$$= 4\frac{15}{12} - 3\frac{8}{12}$$

Write $5\frac{3}{12}$ as $4\frac{15}{12}$ so you can subtract the fractions.

$$= 1\frac{7}{12}$$

Subtract the fractions. Then subtract the whole numbers.

EXERCISES

Add. Write each answer in simplest form.

1. $\frac{2}{7} + \frac{3}{7}$

2. $\frac{3}{8} + \frac{7}{8}$

3. $\frac{6}{5} + \frac{9}{5}$

4. $\frac{4}{9} + \frac{8}{9}$

5. $6\frac{2}{3} + 3\frac{4}{5}$

6. $1\frac{4}{7} + 2\frac{3}{14}$

7. $4\frac{5}{6} + 1\frac{7}{18}$

8. $2\frac{4}{5} + 3\frac{6}{7}$

9. $4\frac{2}{3} + 1\frac{6}{11}$

10. $3\frac{7}{9} + 5\frac{4}{11}$

11. $8 + 1\frac{2}{3}$

12. $8\frac{1}{5} + 3\frac{3}{4}$

13. $11\frac{3}{8} + 2\frac{1}{16}$

14. $9\frac{1}{12} + 8\frac{3}{4}$

15. $33\frac{1}{3} + 23\frac{2}{5}$

Subtract. Write each answer in simplest form.

16. $\frac{7}{8} - \frac{3}{8}$

17. $\frac{9}{10} - \frac{3}{10}$

18. $\frac{17}{5} - \frac{2}{5}$

19. $\frac{11}{7} - \frac{2}{7}$

20. $\frac{5}{11} - \frac{4}{11}$

21. $8\frac{5}{8} - 6\frac{1}{4}$

22. $3\frac{2}{3} - 1\frac{8}{9}$

23. $8\frac{5}{6} - 5\frac{1}{2}$

24. $12\frac{3}{4} - 4\frac{5}{6}$

25. $17\frac{2}{7} - 8\frac{2}{9}$

26. $7\frac{3}{4} - 3\frac{3}{8}$

27. $4\frac{1}{12} - 1\frac{11}{12}$

28. $5\frac{5}{8} - 2\frac{7}{16}$

29. $11\frac{2}{3} - 3\frac{5}{6}$

30. $25\frac{5}{8} - 17\frac{15}{16}$

Multiplying and Dividing Fractions

To multiply two or more fractions, multiply the numerators, multiply the denominators, and simplify the product, if necessary.

1 EXAMPLE

Multiply $\frac{3}{7} \cdot \frac{5}{6}$.

$\frac{3}{7} \cdot \frac{5}{6} = \frac{3 \cdot 5}{7 \cdot 6} = \frac{15}{42} = \frac{15 \div 3}{42 \div 3} = \frac{5}{14}$

Sometimes you can simplify before multiplying.

$\frac{3^1}{7} \cdot \frac{5}{6_2} = \frac{5}{14}$ **Divide a numerator and a denominator by a common factor.**

To multiply mixed numbers, change the mixed numbers to improper fractions and multiply the fractions. Write the product as a mixed number.

2 EXAMPLE

Multiply $2\frac{4}{5} \cdot 1\frac{2}{3}$.

$2\frac{4}{5} \cdot 1\frac{2}{3} = \frac{14}{{}_1 5} \cdot \frac{5^1}{3} = \frac{14}{3} = 4\frac{2}{3}$

To divide fractions, change the division problem to a multiplication problem. Remember that $8 \div \frac{1}{4}$ is the same as $8 \cdot 4$.

To divide mixed numbers, change the mixed numbers to improper fractions and divide the fractions.

3 EXAMPLE

a. Divide $\frac{4}{5} \div \frac{3}{7}$.

$\frac{4}{5} \div \frac{3}{7} = \frac{4}{5} \cdot \frac{7}{3}$ ⟵ **Multiply by the reciprocal of the divisor.** ⟶

$= \frac{28}{15}$ ⟵ **Simplify the answer.** ⟶

$= 1\frac{13}{15}$ ⟵ **Write as a mixed number.**

b. Divide $4\frac{2}{3} \div 7\frac{3}{5}$.

$4\frac{2}{3} \div 7\frac{3}{5} = \frac{14}{3} \div \frac{38}{5}$

$= \frac{14^7}{3} \cdot \frac{5}{38_{19}}$

$= \frac{35}{57}$

EXERCISES

Multiply. Write your answers in simplest form.

1. $\frac{2}{5} \cdot \frac{3}{4}$
2. $\frac{3}{7} \cdot \frac{4}{3}$
3. $1\frac{1}{2} \cdot 5\frac{3}{4}$
4. $3\frac{4}{5} \cdot 10$
5. $5\frac{1}{4} \cdot \frac{2}{3}$

6. $4\frac{1}{2} \cdot 7\frac{1}{2}$
7. $3\frac{2}{3} \cdot 6\frac{9}{10}$
8. $6\frac{1}{2} \cdot 7\frac{2}{3}$
9. $2\frac{2}{5} \cdot 1\frac{1}{6}$
10. $4\frac{1}{9} \cdot 3\frac{3}{8}$

11. $3\frac{1}{5} \cdot 1\frac{7}{8}$
12. $7\frac{5}{6} \cdot 4\frac{1}{2}$
13. $1\frac{2}{3} \cdot 5\frac{9}{10}$
14. $3\frac{3}{4} \cdot 5\frac{1}{3}$
15. $1\frac{2}{3} \cdot 3\frac{9}{16}$

Divide. Write your answers in simplest form.

16. $\frac{3}{5} \div \frac{1}{2}$
17. $\frac{4}{5} \div \frac{9}{10}$
18. $2\frac{1}{2} \div 3\frac{1}{2}$
19. $1\frac{4}{5} \div 2\frac{1}{2}$
20. $3\frac{1}{6} \div 1\frac{3}{4}$

21. $5 \div \frac{3}{8}$
22. $\frac{4}{9} \div \frac{3}{5}$
23. $\frac{5}{8} \div \frac{3}{4}$
24. $2\frac{1}{5} \div 2\frac{1}{2}$
25. $6\frac{1}{2} \div \frac{1}{4}$

26. $1\frac{3}{4} \div 4\frac{3}{8}$
27. $\frac{8}{9} \div \frac{2}{3}$
28. $\frac{1}{5} \div \frac{1}{3}$
29. $2\frac{2}{5} \div 7\frac{1}{5}$
30. $7\frac{2}{3} \div \frac{2}{9}$

Fractions, Decimals, and Percents

Percent means per hundred. 50% means 50 per hundred. $50\% = \frac{50}{100} = 0.50$

You can write fractions as percents by writing the fractions as decimals first. Then move the decimal point two places to the right and write a percent sign.

1 EXAMPLE

Write each number as a percent.

a. $\frac{3}{5}$ **b.** $\frac{7}{20}$ **c.** $\frac{2}{3}$

$\frac{3}{5} = 0.6$ $\frac{7}{20} = 0.35$ $\frac{2}{3} = 0.66\overline{6}$

Move the decimal point two places to the right and write a percent sign.

$0.6 = 60\%$ $0.35 = 35\%$ $0.66\overline{6} = 66.\overline{6}\% \approx 66.7\%$

You can write percents as decimals by moving the decimal point two places to the left and removing the percent sign.

You can write a percent as a fraction with the denominator of 100. You then simplify it, if possible.

2 EXAMPLE

Write each number as a decimal and as a fraction or mixed number.

a. 25%   **b.** $\frac{1}{2}\%$ **c.** 360%

$25\% = 0.25$ $\frac{1}{2}\% = 0.5\% = 0.005$ $360\% = 3.6$

$25\% = \frac{25}{100} = \frac{1}{4}$ $\frac{1}{2}\% = \frac{\frac{1}{2}}{100} = \frac{1}{2} \div 100$ $360\% = \frac{360}{100} = \frac{18}{5} = 3\frac{3}{5}$

$= \frac{1}{2} \cdot \frac{1}{100} = \frac{1}{200}$

EXERCISES

Write each number as a percent. If necessary, round to the nearest tenth.

1. 0.56   **2.** 0.09   **3.** 6.02   **4.** 5.245

5. 8.2   **6.** 0.14   **7.** $\frac{1}{7}$ **8.** $\frac{9}{20}$

9. $\frac{1}{9}$ **10.** $\frac{5}{6}$ **11.** $\frac{3}{4}$ **12.** $\frac{7}{8}$

Write each number as a decimal.

13. 7%   **14.** 8.5%   **15.** 0.9%   **16.** 250%   **17.** 83%   **18.** 110%

19. 15%   **20.** 72%   **21.** 0.03%   **22.** 36.2%   **23.** 365%   **24.** 101%

Write each number as a fraction or mixed number in simplest form.

25. 19%   **26.** $\frac{3}{4}\%$ **27.** 450%   **28.** $\frac{4}{5}\%$ **29.** 64%   **30.** $\frac{2}{3}\%$

31. 24%   **32.** 845%   **33.** $\frac{3}{8}\%$ **34.** 480%   **35.** 60%   **36.** 350%

37. 2%   **38.** 16%   **39.** 66%   **40.** $\frac{4}{7}\%$ **41.** 125%   **42.** 84%

Exponents

You can express $2 \cdot 2 \cdot 2 \cdot 2 \cdot 2$ as 2^5. The raised number 5 shows the number of times 2 is used as a factor. The number 2 is the base. The number 5 is the exponent.

$2^5 \leftarrow$ **exponent**

$\uparrow$ **base**

Factored Form	Exponential Form	Standard Form
$2 \cdot 2 \cdot 2 \cdot 2 \cdot 2$	2^5	32

A number with an exponent of 1 is the number itself: $8^1 = 8$.
Any number, except 0, with an exponent of 0 is 1: $5^0 = 1$.

1 EXAMPLE

Write using exponents.

a. $8 \cdot 8 \cdot 8 \cdot 8 \cdot 8$
b. $2 \cdot 9 \cdot 9 \cdot 9 \cdot 9 \cdot 9 \cdot 9$
c. $6 \cdot 6 \cdot 10 \cdot 10 \cdot 10 \cdot 6 \cdot 6$

Count the number of times the number is used as a factor.

$= 8^5$
$= 2 \cdot 9^6$
$= 6^4 \cdot 10^3$

2 EXAMPLE

Write in standard form.

a. 2^3
b. $8^2 \cdot 3^4$
c. $10^3 \cdot 15^2$

Write in factored form and multiply.

$2 \cdot 2 \cdot 2 = 8$
$8 \cdot 8 \cdot 3 \cdot 3 \cdot 3 \cdot 3 = 5184$
$10 \cdot 10 \cdot 10 \cdot 15 \cdot 15 = 225,000$

In powers of 10, an exponent tells how many zeros are in the equivalent standard form.

$10^1 = 10$
$10^2 = 10 \cdot 10 = 100$
$10^3 = 10 \cdot 10 \cdot 10 = 1000$

$10^4 = 10 \cdot 10 \cdot 10 \cdot 10 = 10,000$
$10^5 = 10 \cdot 10 \cdot 10 \cdot 10 \cdot 10 = 100,000$
$10^6 = 10 \cdot 10 \cdot 10 \cdot 10 \cdot 10 \cdot 10 = 1,000,000$

You can use exponents to write numbers in expanded form.

3 EXAMPLE

Write 739 in expanded form using exponents.

$739 = 700 + 30 + 9 = (7 \cdot 100) + (3 \cdot 10) + (9 \cdot 1) = (7 \cdot 10^2) + (3 \cdot 10^1) + (9 \cdot 10^0)$

EXERCISES

Write using exponents.

1. $6 \cdot 6 \cdot 6 \cdot 6$
2. $7 \cdot 7 \cdot 7 \cdot 7 \cdot 7$
3. $5 \cdot 2 \cdot 2 \cdot 2 \cdot 2$

4. $3 \cdot 3 \cdot 3 \cdot 3 \cdot 3 \cdot 14 \cdot 14$
5. $4 \cdot 4 \cdot 3 \cdot 3 \cdot 2$
6. $3 \cdot 5 \cdot 5 \cdot 7 \cdot 7 \cdot 7$

Write in standard form.

7. 4^3
8. 9^4
9. 12^2
10. $6^2 \cdot 7^1$
11. $11^2 \cdot 3^3$

Write in expanded form using exponents.

12. 658
13. 1254
14. 7125
15. 83,401
16. 294,863

Measuring and Classifying Angles

An angle is a geometric figure formed by two rays with a common endpoint. The rays are sides of the angle and the endpoint is the vertex of the angle. An angle is measured in degrees. The symbol for an angle is ∠.

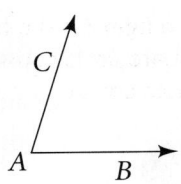

The angle picture at the right can be named in three different ways: ∠A, ∠BAC, or ∠CAB.

Angles can be classified by their measures.

Acute angle
less than 90°

Right angle
90°

Obtuse angle
greater than 90°
but less than 180°

Straight angle
180°

EXAMPLE

Measure the angle. Is it *acute, right, obtuse* or *straight*?

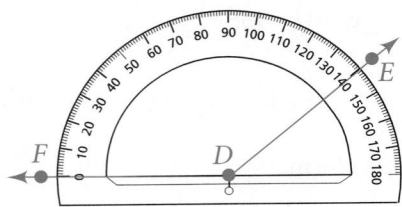

Line up side *DF* through 0° with the vertex at the center of the protractor. Read the scale number through which side *DE* passes.

● The measure of the angle is 140°. The angle is obtuse.

EXERCISES

Measure each angle. Is the angle *acute, right, obtuse*, or *straight*?

1.

     A

2.

     B

3.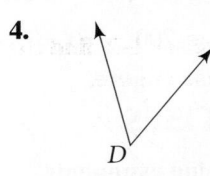

4.

     D

Draw an angle with the given measure.

5. 45°   **6.** 95°   **7.** 120°   **8.** 170°

9. 20°   **10.** 85°   **11.** 150°   **12.** 160°

13. Open-Ended Draw a triangle. Use a protractor to find the measure of each angle of your triangle.

Perimeter, Area, and Volume

The perimeter of a figure is the distance around the figure. The area of a figure is the number of square units contained in the figure. The volume of a space figure is the number of cubic units contained in the space figure.

1 EXAMPLE

Find the perimeter of each figure.

a.

5 in.
3 in.
4 in.

Add the measures of the sides.
$3 + 4 + 5 = 12$
The perimeter is 12 in.

b.

3 cm
4 cm

Use the formula $P = 2\ell + 2w$.
$P = 2(3) + 2(4)$
$= 6 + 8 = 14$
The perimeter is 14 cm.

2 EXAMPLE

Find the area of each figure.

a.

5 in.
6 in.

Use the formula $A = bh$.
$A = 6 \cdot 5 = 30$
The area is 30 in.^{2}.

b.

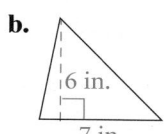
6 in.
7 in.

Use the formula $A = \frac{1}{2}(bh)$.
$A = \frac{1}{2}(7 \cdot 6) = 21$
The area is 21 in.^{2}.

3 EXAMPLE

Find the volume of each figure.

a.

6 in.
3 in.
5 in.

Use the formula $V = Bh$.
(B = area of the base
$= 3 \cdot 5 = 15$).
$V = 15 \cdot 6 = 90$ in.^{3}
The volume is 90 in.^{3}.

b.

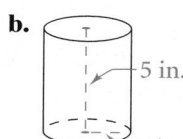
5 in.
2 in.

Use the formula $V = \pi r^2 h$.
$V = 3.14 \cdot 2^2 \cdot 5$
$= 3.14 \cdot 4 \cdot 5 = 62.8$ in.^{3}
The volume is 62.8 in.^{3}.

EXERCISES

For Exercises 1–2, find the perimeter of each figure. For Exercises 3–4, find the area of each figure.

1.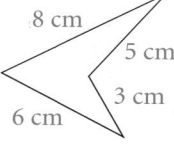
8 cm
5 cm
3 cm
6 cm

2.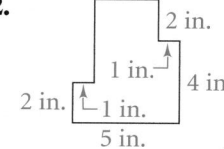
2 in.
1 in.
4 in
2 in.
1 in.
5 in.

3.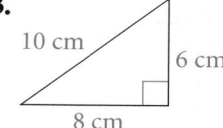
10 cm
6 cm
8 cm

4.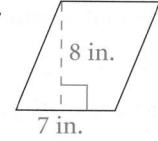
8 in.
7 in.

Find the volume of each figure.

5.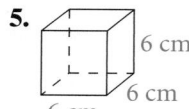
6 cm
6 cm
6 cm

6.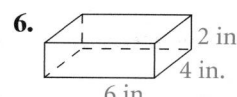
2 in.
4 in.
6 in.

7.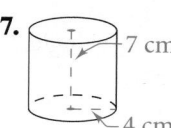
7 cm
4 cm

Translations

A translation is a transformation that moves a figure so that every point in the figure moves the same direction and the same distance. Each translated figure is an image of the original figure. If point A is on the original figure, the corresponding point on the image is A'.

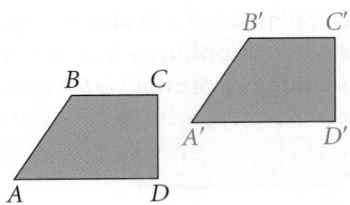

1 EXAMPLE

Graph the image of $\triangle ABC$ after a translation of 4 units right and 2 units down.

To get from $A(-6, 1)$ to $A'(-2, -1)$ by a translation, move
4 units to the right: add 4 to the x-value, $-6 + 4 = -2$,
and 2 units down: add -2 to the y-value, $1 + (-2) = -1$.

● Similarly, the coordinates of B' are $(3, 3)$ and of C' are $(2, 0)$.

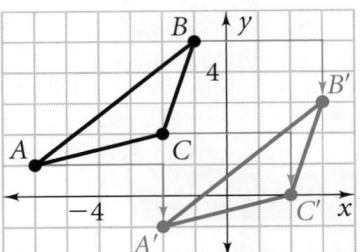

You can describe a translation using arrow ($\rightarrow$) notation. The translation from A to A' in Example 1 can be written $A(-6, 1) \rightarrow A'(-2, -1)$.

2 EXAMPLE

Write a rule to describe the translation of $\triangle PQR$ to $\triangle P'Q'R'$.

Use any point on the figure and its image to find the horizontal and vertical translations. Try $P(3, 2)$ and its image $P'(-2, 5)$.

horizontal translation: $-2 - 3 = -5$ 5 units left
vertical translation: $5 - 2 = 3$ 3 units up

● The rule for the translation is $(x, y) \rightarrow (x - 5, y + 3)$.

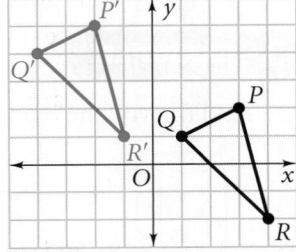

EXERCISES

The vertices of a triangle are given. Graph the triangle and its image after a translation of the specified number of units in each direction.

1. $P(2, 5)$, $Q(5, 6)$, $R(5, 0)$; 4 units right, 5 units down

2. $U(0, 0)$, $V(-5, -3)$, $W(-6, 0)$; 2 units left, 2 units up

Write a rule to describe each translation.

3. $A(4, -4) \rightarrow A'(2, -5)$

4. $G(-1, 9) \rightarrow G'(1, -9)$

5. $M(-2, 0) \rightarrow M'\left(2, \frac{1}{2}\right)$

6. $D(2, 4) \rightarrow D'(-3, 5)$

7. $Q(7, -3) \rightarrow Q'(9, 2)$

8. $W(6, -4) \rightarrow W'(2, 2)$

9.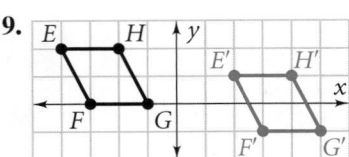

10.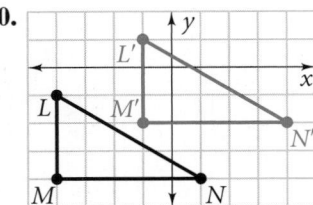

11.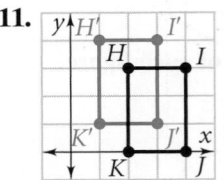

Reflections

A reflection is a transformation that flips a figure over a line, called a line of reflection. The picture at the right shows $\triangle DEF$ and its reflection over the line of reflection ℓ. The image is $\triangle D'E'F'$.

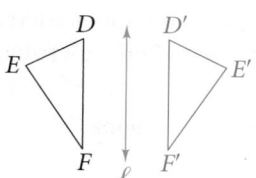

1 EXAMPLE

Reflect the point $P(2, 5)$ over the line $y = 2$. What are the coordinates of its image P'?

● The coordinates of P' are $(2, -1)$.

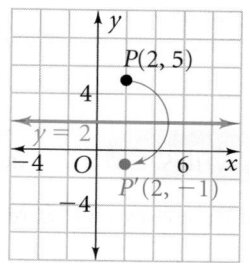

2 EXAMPLE

The vertices of $\triangle ABC$ are $A(-7, 0)$, $B(-5, 6)$, and $C(-3, 4)$. Graph the image of the triangle after a reflection over the line $x = 1$.

To get from $A(-7, 0)$ to $A'(9, 0)$, point A is reflected the same distance on the other side of the line $x = 1$. The coordinates of the other two vertices in the reflected image are $B'(7, 6)$ and $C'(5, 4)$.

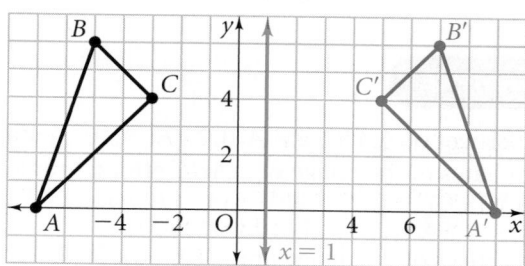

EXERCISES

Graph each point and its image after a reflection over the specified line.

1. $K(2, 3)$; $x = 1$

2. $D(-3, 1)$; $y = -2$

3. $Z(2, 5)$; x-axis

The vertices of various figures are given. Graph both the figure and its image after a reflection over the specified line.

4. $R(2, 4)$, $S(-1, 3)$, $T(2, 0)$; $y = -2$

5. $C(1, 4)$, $D(1, 7)$, $E(-2, 7)$; $x = 3$

6. $H(-5, 5)$, $I(6, 0)$, $J(0, 0)$; y-axis

7. $K(-2, -3)$, $L(-7, -3)$, $M(-7, -5)$, $N(-2, -5)$; x-axis

The reflected image of each figure is shown in red. Identify the line of reflection.

8.
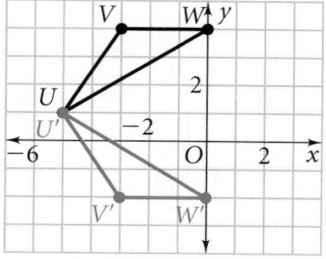

9.
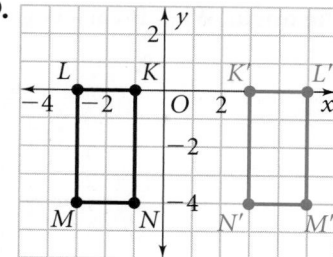

Rotations

A rotation is a transformation that turns a figure about a fixed point, called the *center of rotation*. You can rotate a figure up to 360°. All rotations shown on this page are counterclockwise.

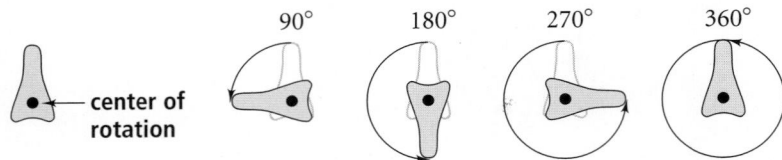

1 EXAMPLE

Find the image of $P(1, 2)$ after a rotation of 90° about the origin.

● The coordinates of P' are $(-2, 1)$.

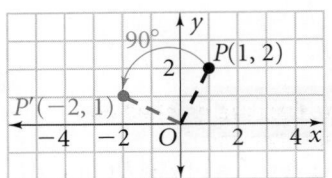

2 EXAMPLE

The vertices of $\triangle ABC$ are $A(0, 2)$, $B(-2, 0)$, and $C(2, 2)$. Find the coordinates of the image of $\triangle ABC$ after a rotation of 180° about the origin.

● The vertices of the image are $A'(0, -2)$, $B'(2, 0)$, and $C'(-2, -2)$.

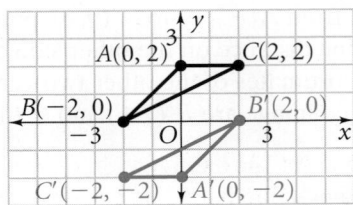

EXERCISES

Graph each point. Then rotate it the given number of degrees counterclockwise about the given center of rotation and graph the new point.

1. $D(2, 4)$; 90° about the origin

2. $G(-3, -1)$; 180° about $(0, 0)$

3. $Z(4, -2)$; 90° about $(2, 0)$

4. $M(-3, 2)$; 180° about $(2, 3)$

The vertices of a triangle are given. On separate coordinate planes, graph each triangle and its image after a rotation of (a) 90° and (b) 180° about the origin.

5. $A(2, 0), B(7, 2), C(7, 0)$

6. $E(0, 3), F(-5, 4), G(-3, 0)$

7. $L(-2, -1), M(-5, -1), N(-2, -5)$

8. $Q(0, -2), R(0, 0), S(2, 0)$

The triangles in the exercises below were formed by rotating the triangle at the left counterclockwise about the origin. What is each angle of rotation?

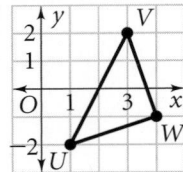

9.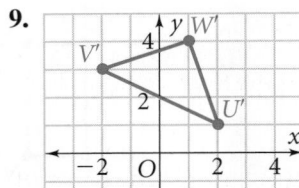

10.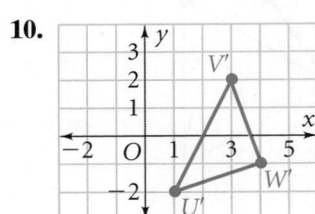

Line Plots

A line plot is created by placing a mark above a number line corresponding to the location of each data item. Line plots have two main advantages:

- You can see the frequency of data items.
- You can see how the data items compare.

EXAMPLE

The table at the right gives the heights (in inches) of a group of twenty-five adults. Display the data in a line plot. Describe the data shown in the line plot.

Height of Adults (inches)

59	60	63	63	64
64	64	65	65	65
67	67	67	67	68
68	68	69	70	70
71	72	73	73	77

The data are graphed on a number line.

The title describes the data.

An *X* represents one element of the data set.

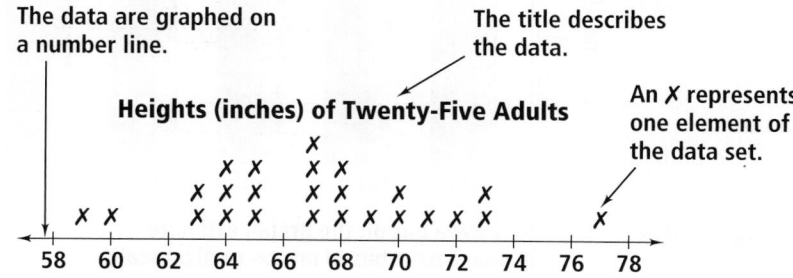

Heights (inches) of Twenty-Five Adults

The line plot shows that most of the heights are concentrated around 67 inches, the maximum value is 77, and the minimum value is 59.

EXERCISES

Display each set of data in a line plot.

1. 3, 6, 4, 3, 6, 0, 4, 5, 0, 4, 6, 1, 5, 1, 0, 5, 5, 6, 5, 3

2. 19, 18, 18, 18, 19, 20, 19, 18, 18, 17, 18, 20, 19, 17

Draw a line plot for each frequency table.

3.

Number	1	2	3	4	5	6
Frequency	4	1	0	5	7	2

4.

Number	12	13	15	16	18	19
Frequency	2	5	1	3	6	3

5. Olympics Here are the numbers of gold medals won by different countries during the 1998 Winter Olympics (Bulgaria had the least with 1 gold medal and Germany had the most with 12 gold medals).
1, 1, 2, 2, 2, 2, 3, 3, 5, 5, 6, 6, 9, 10, 12
Display the data in a line plot. Describe the data shown in the line plot.

Bar Graphs

Bar graphs are used to compare amounts. The horizontal axis shows the categories and the vertical axis shows the amounts. A multiple bar graph includes a key.

EXAMPLE

Draw a bar graph for the data in the table below.

Median Household Income

State	1995	1997	1999
Calif.	$40,457	$41,203	$43,744
Conn.	$43,993	$45,657	$50,798
Ind.	$36,496	$40,367	$40,929
Tex.	$35,024	$36,408	$38,978
Utah	$39,879	$44,401	$46,094

SOURCE: U.S. Census Bureau

The categories (in the first column) are placed on the horizontal scale. The amounts (in the second, third, and fourth columns) are placed on the vertical scale.

Graph the data for each state. Use the values in the top row to create the key.

The highest projected income is $50,798. So a reasonable range for the vertical scale is 0 to $55,000.

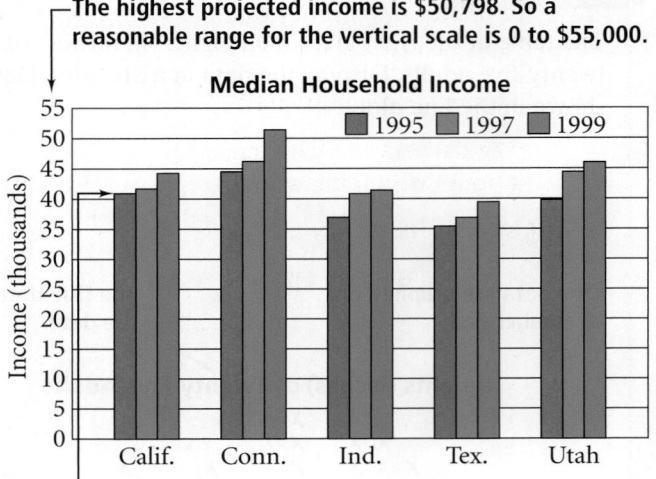

To draw a bar on the graph, estimate its placement based on the vertical scale.

EXERCISES

1. Draw a bar graph for the data in the table below.

Highest Temperatures (°F)

City	March	June	August
Juneau, AK	61	86	83
Denver, CO	84	104	101
Atlanta, GA	89	101	102
Honolulu, HI	88	92	93
Detroit, MI	81	104	100
Buffalo, NY	81	96	99
Houston, TX	91	103	107

2. a. Critical Thinking If one more column of data were added to the table in the example, how would the bar graph be different?

   b. If one more row of data were added to the table in the example, how would the bar graph be different?

Histograms

A histogram is a bar graph that shows the frequency, or number of times, a data item occurs. Histograms often combine data into intervals of equal size. The intervals do not overlap.

 **EXAMPLE**

The data at the right show the number of hours of battery life for different brands of batteries used in portable CD players. Use the data to make a histogram.

Hours of Battery Life

12 9 10 14 10 11
10 18 21 10 14 22

Step 1 Decide on an interval size.

The data start at 9 hours and go to 22 hours. Use equal-sized intervals of 4 hours, beginning with 8 hours. So the first interval will be 8–11.

Step 2 Make a frequency table.

Battery Life

Hours	Tally	Frequency
8–11	ℍℍ I	6
12–15	|||	3
16–19			1
20–23	||	2

Step 3 Make a histogram.

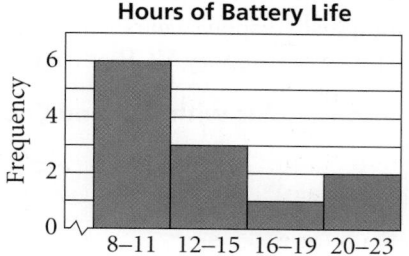

EXERCISES

1. Students answered a survey question about how long it takes to get ready in the morning. The histogram at the right shows the survey results.

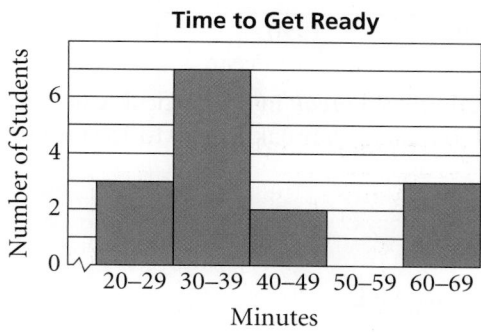

  a. Which interval indicates the answers most students gave?
  b. How many students answered the survey question?
  c. Why might no students have given an answer in the interval 50–59?
  d. Critical Thinking With the information you have, could you redraw the histogram with intervals half their current size? Explain why or why not.

2. a. An Internet company surveyed its users. The first 25 people who responded gave the ages shown at the right. What intervals would you use to make a histogram?
  b. Make a frequency table for the data.
  c. Make a histogram.

Age of Internet Users

25, 43, 65, 12, 8, 30, 44, 68, 18, 21, 25, 33, 37, 54, 61, 29, 31, 38, 22, 48, 19, 34, 55, 14, 21

3. a. Data Collection Survey your class to find out what day of the month they were born. For example, 12 if a student's birthday is August 12th.
  b. What intervals would you use to make a histogram?
  c. Make a frequency table for the data.
  d. Make a histogram.

Line Graphs

Line graphs are used to display the change in a set of data over a period of time. A multiple-line graph shows change in more than one category of data over time. You can use a line graph to look for trends and make predictions.

Graph the data in the table below.

Households with VCR and Cable TV (millions)

Year	1985	1990	1995	1996	1997	1998
VCR	18	63	77	79	82	83
Cable TV	36	52	60	63	64	66

SOURCE: Television Bureau of Advertising, Inc., *Trends in Television*

Since the data show changes over time for two sets of data, use a double line graph. The horizontal scale displays years. The vertical scale displays the number of households for each category, VCR and cable TV.

Households with VCR and Cable TV

Notice that there is a *break* in the vertical scale, which goes from 0 to 85. A zigzag line is used to indicate a break from 0 to 15 since there is no data to graph in this part of the *y*-axis.

EXERCISES

Graph the following data.

1.
Market Shares (percent)

Year	1994	1995	1996	1997	1998	1999	2000
Rap/Hip Hop	7.9	6.7	8.9	10.1	9.7	10.8	12.9
Pop	10.3	10.1	9.3	9.4	10.0	10.3	11.0

SOURCE: The Recording Industry of America

2.
Percents of Schools with Internet Access

Year	1995	1996	1997	1998	1999
Elementary	46	61	75	88	94
Secondary	65	77	89	94	98

SOURCE: U.S. National Center for Education Statistics

Circle Graphs

A circle graph is an efficient way to present certain types of data. The graphs show data as percents or fractions of a whole. The total must be 100% or 1. Circle graphs are used to show the parts of the whole. The angles at the center are central angles, and each angle is proportional to the percent or fraction of the total.

What Do You Think Is the Number One Problem in the World Today?

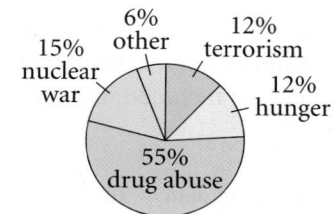

SOURCE: *The Second Kids' World Almanac*

EXAMPLE

The table below shows the number of people in the United States who have at least one grandchild under the age of 18. Draw a circle graph for the data.

Ages of U.S. Grandparents

Age	People (millions)
44 and under	3.6
45–54	10.3
55–64	15.0
65 and over	18.2

Step 1 Add to find the total number.

$3.6 + 10.3 + 15.0 + 18.2 = 47.1$ (million)

Step 2 For each central angle, set up a proportion to find the measure. Use a calculator to solve each proportion.

$\frac{3.6}{47.1} = \frac{a}{360°}$ $\frac{10.3}{47.1} = \frac{b}{360°}$ $\frac{15.0}{47.1} = \frac{c}{360°}$ $\frac{18.2}{47.1} = \frac{d}{360°}$

$a \approx 27.5°$ $b \approx 78.7°$ $c \approx 114.6°$ $d \approx 139.1°$

Step 3 Use a compass to draw a circle. Draw the approximate central angles with a protractor.

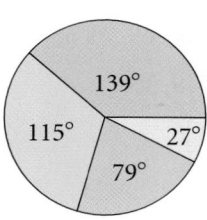

Step 4 Label each sector. Add any necessary information.

Ages of U.S. Grandparents

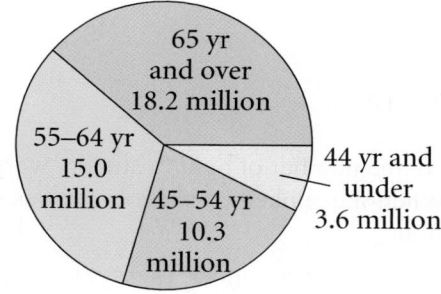

EXERCISES

1. a. Use the data in the table to draw a circle graph.
  b. Approximately what percent of students ride the bus?
  c. Approximately how many times more students walk than ride in a car?

Transportation Mode	Walk	Bicycle	Bus	Car
Number of Students	252	135	432	81

2. Data Collection Survey your class to find out how they get to school. Use the data to draw a circle graph.

Box-and-Whisker Plots

To show how data items are spread out, you can arrange a set of data in order from least to greatest. The maximum, minimum, and median give you some information about the data. You can better describe the data by dividing it into fourths.

The lower quartile is the median of the lower half of the data. The upper quartile is the median of the upper half of the data. If the data set has an odd number of items, the median is not included in either the upper half or the lower half.

The data below describe the highway gas mileage (mi/gal) for several cars.

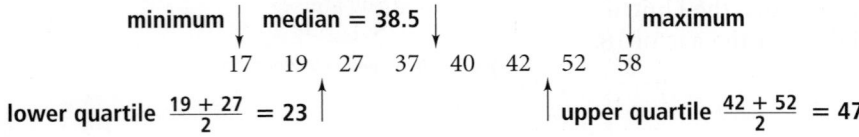

minimum | median = 38.5 | maximum

17   19   27   37   40   42   52   58

lower quartile $\frac{19 + 27}{2} = 23$ upper quartile $\frac{42 + 52}{2} = 47$

A box-and-whisker plot is a visual representation of data. The box-and-whisker plot below displays the gas mileage information.

Highway Gas Mileage (mi/gal)

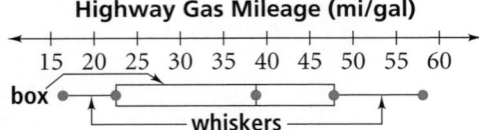

15 20 25 30 35 40 45 50 55 60

box   whiskers

The box represents the data from the lower quartile to the upper quartile. The vertical line segment represents the median. Horizontal line segments called whiskers show the spread of the data to the minimum and to the maximum.

EXERCISES

Create a box-and-whisker plot for each data set.

1. {3, 2, 3, 4, 6, 6, 7}

2. {1, 1.5, 1.7, 2, 6.1, 6.2, 7}

3. {1, 2, 5, 6, 9, 12, 7, 10}

4. {65, 66, 59, 61, 67, 70, 67, 66, 69, 70, 63}

5. {29, 32, 40, 31, 33, 39, 27, 42}

6. {3, 3, 5, 7, 1, 10, 10, 4, 4, 7, 9, 8, 6}

7. {1, 1.2, 1.3, 4, 4.1, 4.2, 7}

8. {1, 3.8, 3.9, 4, 4.3, 4.4, 7, 5}

9. **Jobs** Below are the number of hours a student worked each week at her summer job. When she applied for the job, she was told that the typical work week was 29 hours.

29, 25, 21, 20, 17, 16, 15, 33, 33, 30, 15

   a. Make a box-and-whisker plot for the data.
   b. How many weeks are above the upper quartile? What are the numbers of hours worked?
   c. What is the median number of hours she worked? What is the mean? Compare them to the typical work week.

10. **Writing** In what ways are histograms and box-and-whisker plots alike, and in what ways are they different?

Choosing an Appropriate Graph

The type of data you want to display can suggest an appropriate graph. You can have data by categories (qualitative data), such as states (page 770), years (page 773), or mode of transportation (page 773). You may also have measurement data (quantitative data), such as height (page 769), time to get ready in the morning (page 771), or gas mileage (page 774).

The table below lists some common types of graphs and how they are frequently used.

Graph	Use
Bar Graph	To display frequency of categories
Circle Graph	To show categories as part of a whole
Line Graph	To show trends over time
Line Plot, Histogram, Stem-and-Leaf Plot	To display frequency distribution of measurement data
Box-and-Whisker Plot	To summarize the distribution of measurement data
Scatter Plot	To display possible relationships in data pairs

EXAMPLE

Would you use a line graph or a circle graph to display the percent of fiction books published each year for the last ten years?

A circle graph shows percents, but it would not allow you to show the change over time. A line graph would be more appropriate.

EXERCISES

Choose the appropriate graph to display each set of data. Explain your choice.

1. circle graph or bar graph
   how much the average family spends on rent, food, transportation, utilities, and entertainment in October

2. bar graph or line graph
   the number of runners in the Olympic marathon for each of the last five Olympic games

3. scatter plot or double bar graph
   the ages of twelve cars and their levels of emissions

4. double box-and-whisker plot or scatter plot
   the heights of men and women playing professional basketball in 2004

Open-Ended **For each type of graph, describe a set of data that would be appropriate.**

5. stem-and-leaf plot

6. double line graph

7. circle graph

Misleading Graphs

There are many ways to graph data that show the data accurately. There are also ways to graph data that are misleading. One way that is misleading is graphing data that has less than the whole vertical axis, but doesn't point this out with the use of a break symbol on the axis.

1 EXAMPLE

What impression does the graph at the right give? What is actually true about the data?

Company Revenues

The graph implies that company revenues are growing quickly. Actually, company revenues are increasing more slowly.

Another way a graph can mislead is by using shapes that increase in both height and width, which implies a much bigger increase has happened, since the area increases much more than the height alone.

2 EXAMPLE

What makes the graph at the right misleading? Explain.

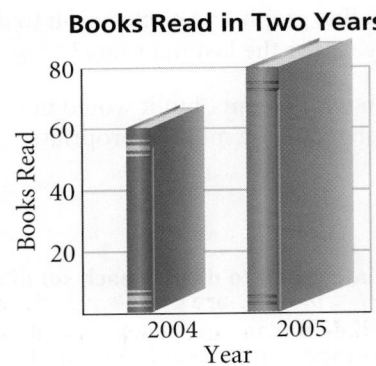

Books Read in Two Years

Looking at the vertical axis, you can see that the number of books read increased by one third. However, the bar on the right increased not only in height, but width. The area of the second bar is more than two times the area of the first bar. This gives the impression that the increase was much greater than it really was.

EXERCISES

For each graph below, (a) explain how the graph is misleading, and (b) explain how to redraw the graph so it is not misleading.

1.

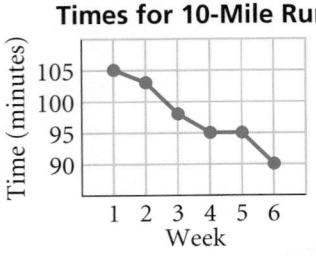

Times for 10-Mile Run

2.

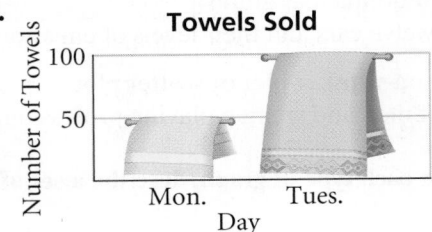

Towels Sold

3. Choose one of the graphs on this page. Redraw the graph to display the data accurately.

Spreadsheets

You can use a spreadsheet to evaluate formulas. Spreadsheets use the symbols $+$ for addition and $-$ for subtraction, but different symbols for other operations.

Multiplication: $* \rightarrow$ $10 * 2 = 10 \cdot 2 = 20$

Division: $/ \rightarrow$ $10 / 2 = 10 \div 2 = 5$

Exponent: $\wedge \rightarrow$ $10 \wedge 2 = 10^2 = 100$

EXAMPLE

Evaluate the formula $P = 2\ell + 2w$ for $\ell = 3$ and for whole-number values of w from 8 to 11.

Enter the values of ℓ and w into the first two columns. Cell A2 has the value of ℓ. Cell B2 has the first value of w. In cell C2, enter the expression $= 2*A2 + 2*B2$ to find the perimeter of a figure with length 3 and width 8.

Column names

	A	B	C
1	L	W	2L + 2W
2	3	8	= 2*A2 + 2* B2
3	3	9	
4	3	10	
5	3	11	

Row numbers

Cell B5

The spreadsheet evaluates the expression automatically.

Copy the expression in cell C2 into cells C3, C4, and C5. The spreadsheet automatically updates for the values of ℓ and w in rows 3, 4, and 5.

✗	✓	= 2*A2 + 2* B2

	A	B	C
1	L	W	2L + 2W
2	3	8	22
3	3	9	24
4	3	10	26
5	3	11	28

EXERCISES

Suppose the values of a, b, and c are in cells A2, B2, and C2 of a spreadsheet. Write the expression you would use to enter each formula in the spreadsheet.

1. $P = a + b + c$

2. $T = \dfrac{3a + 5b}{8}$

3. $R = \frac{1}{2}bc$

4. $A = c^2$

5. You deposit $200 in an account that earns 6% compounded annually for three years. The spreadsheet below shows the balance at the end of each year.

	A	B	C	D	E
1	Year	Start of Year	Rate	Interest	End of Year
2	1st	$200.00	0.06	$12.00	$212.00
3	2nd	$212.00	0.06	$12.72	$224.72
4	3rd	$224.72	0.06	$13.48	$238.20

a. In which cell of the spreadsheet would you find the formula $= B3 * C3$?

b. In which cell of the spreadsheet would you find the formula $= B4 + D4$?

Tables

Table 1 Measures

United States Customary	Metric

Length

12 inches (in.) = 1 foot (ft)	10 millimeters (mm) = 1 centimeter (cm)
36 in. = 1 yard (yd)	100 cm = 1 meter (m)
3 ft = 1 yard	1000 mm = 1 meter
5280 ft = 1 mile (mi)	1000 m = 1 kilometer (km)
1760 yd = 1 mile	

Area

144 square inches (in.^2) = 1 square foot (ft^2)	100 square millimeters (mm^2) = 1 square centimeter (cm^2)
9 ft^2 = 1 square yard (yd^2)	10,000 cm^2 = 1 square meter (m^2)
43,560 ft^2 = 1 acre (a)	10,000 m^2 = 1 hectare (ha)
4840 yd^2 = 1 acre	

Volume

1728 cubic inches (in.^3) = 1 cubic foot (ft^3)	1000 cubic millimeters (mm^3) = 1 cubic centimeter (cm^3)
27 ft^3 = 1 cubic yard (yd^3)	1,000,000 cm^3 = 1 cubic meter (m^3)

Liquid Capacity

8 fluid ounces (fl oz) = 1 cup (c)	1000 milliliters (mL) = 1 liter (L)
2 c = 1 pint (pt)	1000 L = 1 kiloliter (kL)
2 pt = 1 quart (qt)	
4 qt = 1 gallon (gal)	

Weight or Mass

16 ounces (oz) = 1 pound (lb)	1000 milligrams (mg) = 1 gram (g)
2000 pounds = 1 ton (t)	1000 g = 1 kilogram (kg)
	1000 kg = 1 metric ton

Temperature

32°F = freezing point of water	0°C = freezing point of water
98.6°F = normal body temperature	37°C = normal body temperature
212°F = boiling point of water	100°C = boiling point of water

Time

60 seconds (s) = 1 minute (min)	365 days = 1 year (yr)
60 minutes = 1 hour (h)	52 weeks (approx.) = 1 year
24 hours = 1 day (d)	12 months = 1 year
7 days = 1 week (wk)	10 years = 1 decade
4 weeks (approx.) = 1 month (mo)	100 years = 1 century

Table 2 Reading Math Symbols

Symbol	Meaning	Page
·	multiplication sign, times ($\times$)	p. 4
=	equals	p. 5
( )	parentheses for grouping	p. 9
a^n	nth power of a	p. 9
%	percent	p. 11
[ ]	brackets for grouping	p. 12
...	and so on	p. 17
$-a$	opposite of a	p. 17
$\neq$	is not equal to	p. 17
π	pi, an irrational number, approximately equal to 3.14	p. 18
$<$	is less than	p. 19
$>$	is greater than	p. 19
$\leq$	is less than or equal to	p. 19
$\geq$	is greater than or equal to	p. 19
$|a|$	absolute value of a	p. 20
(x, y)	ordered pair	p. 24
$\overline{AB}$	segment with endpoints A and B	p. 25
°	degree(s)	p. 58
$\begin{bmatrix} 1 & 2 \\ 3 & 4 \end{bmatrix}$	matrix	p. 59
$\frac{1}{a}, a \neq 0$	reciprocal of a	p. 72
$P(\text{event})$	probability of the event	p. 93
$\overset{?}{=}$	Is the statement true?	p. 120
$\angle A$	angle A	p. 123
$a : b$	ratio of a to b	p. 142
$\approx$	is approximately equal to	p. 143
$\triangle ABC$	triangle ABC	p. 149
AB	length of $\overline{AB}$; distance between points A and B	p. 149
$\cong$	is congruent to	p. 150
$\sqrt{x}$	nonnegative square root of x	p. 176
$\pm$	plus or minus	p. 176
{ }	set braces	p. 257
$f(x)$	f of x; the function value at x	p. 258
x_1, x_2, etc.	specific values of the variable x	p. 310
y_1, y_2, etc.	specific values of the variable y	p. 310
$\overleftrightarrow{AB}$	line through points A and B	p. 310
m	slope of a linear function	p. 318
b	y-intercept of a linear function	p. 318
a^{-n}	$\frac{1}{a^n}, a \neq 0$	p. 431
$\overline{x}$	mean of data values of x	p. 636
$\sin A$	sine of $\angle A$	p. 646
$\cos A$	cosine of $\angle A$	p. 646
$\tan A$	tangent of $\angle A$	p. 646
$m\angle A$	measure of angle A	p. 648
$_nP_r$	permutations of n things taken r at a time	p. 701
$_nC_r$	combinations of n things taken r at a time	p. 707
$\wedge$	raised to a power (in a spreadsheet formula)	p. 777
$*$	multiply (in a spreadsheet formula)	p. 777
$/$	divide (in a spreadsheet formula)	p. 777

Table 3 Squares and Square Roots

Number n	Square n^2	Positive Square Root $\sqrt{n}$	Number n	Square n^2	Positive Square Root $\sqrt{n}$	Number n	Square n^2	Positive Square Root $\sqrt{n}$
1	1	1.000	51	2601	7.141	101	10,201	10.050
2	4	1.414	52	2704	7.211	102	10,404	10.100
3	9	1.732	53	2809	7.280	103	10,609	10.149
4	16	2.000	54	2916	7.348	104	10,816	10.198
5	25	2.236	55	3025	7.416	105	11,025	10.247
6	36	2.449	56	3136	7.483	106	11,236	10.296
7	49	2.646	57	3249	7.550	107	11,449	10.344
8	64	2.828	58	3364	7.616	108	11,664	10.392
9	81	3.000	59	3481	7.681	109	11,881	10.440
10	100	3.162	60	3600	7.746	110	12,100	10.488
11	121	3.317	61	3721	7.810	111	12,321	10.536
12	144	3.464	62	3844	7.874	112	12,544	10.583
13	169	3.606	63	3969	7.937	113	12,769	10.630
14	196	3.742	64	4096	8.000	114	12,996	10.677
15	225	3.873	65	4225	8.062	115	13,225	10.724
16	256	4.000	66	4356	8.124	116	13,456	10.770
17	289	4.123	67	4489	8.185	117	13,689	10.817
18	324	4.243	68	4624	8.246	118	13,924	10.863
19	361	4.359	69	4761	8.307	119	14,161	10.909
20	400	4.472	70	4900	8.367	120	14,400	10.954
21	441	4.583	71	5041	8.426	121	14,641	11.000
22	484	4.690	72	5184	8.485	122	14,884	11.045
23	529	4.796	73	5329	8.544	123	15,129	11.091
24	576	4.899	74	5476	8.602	124	15,376	11.136
25	625	5.000	75	5625	8.660	125	15,625	11.180
26	676	5.099	76	5776	8.718	126	15,876	11.225
27	729	5.196	77	5929	8.775	127	16,129	11.269
28	784	5.292	78	6084	8.832	128	16,384	11.314
29	841	5.385	79	6241	8.888	129	16,641	11.358
30	900	5.477	80	6400	8.944	130	16,900	11.402
31	961	5.568	81	6561	9.000	131	17,161	11.446
32	1024	5.657	82	6724	9.055	132	17,424	11.489
33	1089	5.745	83	6889	9.110	133	17,689	11.533
34	1156	5.831	84	7056	9.165	134	17,956	11.576
35	1225	5.916	85	7225	9.220	135	18,225	11.619
36	1296	6.000	86	7396	9.274	136	18,496	11.662
37	1369	6.083	87	7569	9.327	137	18,769	11.705
38	1444	6.164	88	7744	9.381	138	19,044	11.747
39	1521	6.245	89	7921	9.434	139	19,321	11.790
40	1600	6.325	90	8100	9.487	140	19,600	11.832
41	1681	6.403	91	8281	9.539	141	19,881	11.874
42	1764	6.481	92	8464	9.592	142	20,164	11.916
43	1849	6.557	93	8649	9.644	143	20,449	11.958
44	1936	6.633	94	8836	9.695	144	20,736	12.000
45	2025	6.708	95	9025	9.747	145	21,025	12.042
46	2116	6.782	96	9216	9.798	146	21,316	12.083
47	2209	6.856	97	9409	9.849	147	21,609	12.124
48	2304	6.928	98	9604	9.899	148	21,904	12.166
49	2401	7.000	99	9801	9.950	149	22,201	12.207
50	2500	7.071	100	10,000	10.000	150	22,500	12.247

Table 4 Trigonometric Ratios

Angle	Sine	Cosine	Tangent		Angle	Sine	Cosine	Tangent
1°	0.0175	0.9998	0.0175		46°	0.7193	0.6947	1.0355
2°	0.0349	0.9994	0.0349		47°	0.7314	0.6820	1.0724
3°	0.0523	0.9986	0.0524		48°	0.7431	0.6691	1.1106
4°	0.0698	0.9976	0.0699		49°	0.7547	0.6561	1.1504
5°	0.0872	0.9962	0.0875		50°	0.7660	0.6428	1.1918
6°	0.1045	0.9945	0.1051		51°	0.7771	0.6293	1.2349
7°	0.1219	0.9925	0.1228		52°	0.7880	0.6157	1.2799
8°	0.1392	0.9903	0.1405		53°	0.7986	0.6018	1.3270
9°	0.1564	0.9877	0.1584		54°	0.8090	0.5878	1.3764
10°	0.1736	0.9848	0.1763		55°	0.8192	0.5736	1.4281
11°	0.1908	0.9816	0.1944		56°	0.8290	0.5592	1.4826
12°	0.2079	0.9781	0.2126		57°	0.8387	0.5446	1.5399
13°	0.2250	0.9744	0.2309		58°	0.8480	0.5299	1.6003
14°	0.2419	0.9703	0.2493		59°	0.8572	0.5150	1.6643
15°	0.2588	0.9659	0.2679		60°	0.8660	0.5000	1.7321
16°	0.2756	0.9613	0.2867		61°	0.8746	0.4848	1.8040
17°	0.2924	0.9563	0.3057		62°	0.8829	0.4695	1.8807
18°	0.3090	0.9511	0.3249		63°	0.8910	0.4540	1.9626
19°	0.3256	0.9455	0.3443		64°	0.8988	0.4384	2.0503
20°	0.3420	0.9397	0.3640		65°	0.9063	0.4226	2.1445
21°	0.3584	0.9336	0.3839		66°	0.9135	0.4067	2.2460
22°	0.3746	0.9272	0.4040		67°	0.9205	0.3907	2.3559
23°	0.3907	0.9205	0.4245		68°	0.9272	0.3746	2.4751
24°	0.4067	0.9135	0.4452		69°	0.9336	0.3584	2.6051
25°	0.4226	0.9063	0.4663		70°	0.9397	0.3420	2.7475
26°	0.4384	0.8988	0.4877		71°	0.9455	0.3256	2.9042
27°	0.4540	0.8910	0.5095		72°	0.9511	0.3090	3.0777
28°	0.4695	0.8829	0.5317		73°	0.9563	0.2924	3.2709
29°	0.4848	0.8746	0.5543		74°	0.9613	0.2756	3.4874
30°	0.5000	0.8660	0.5774		75°	0.9659	0.2588	3.7321
31°	0.5150	0.8572	0.6009		76°	0.9703	0.2419	4.0108
32°	0.5299	0.8480	0.6249		77°	0.9744	0.2250	4.3315
33°	0.5446	0.8387	0.6494		78°	0.9781	0.2079	4.7046
34°	0.5592	0.8290	0.6745		79°	0.9816	0.1908	5.1446
35°	0.5736	0.8192	0.7002		80°	0.9848	0.1736	5.6713
36°	0.5878	0.8090	0.7265		81°	0.9877	0.1564	6.3138
37°	0.6018	0.7986	0.7536		82°	0.9903	0.1392	7.1154
38°	0.6157	0.7880	0.7813		83°	0.9925	0.1219	8.1443
39°	0.6293	0.7771	0.8098		84°	0.9945	0.1045	9.5144
40°	0.6428	0.7660	0.8391		85°	0.9962	0.0872	11.4301
41°	0.6561	0.7547	0.8693		86°	0.9976	0.0698	14.3007
42°	0.6691	0.7431	0.9004		87°	0.9986	0.0523	19.0811
43°	0.6820	0.7314	0.9325		88°	0.9994	0.0349	28.6363
44°	0.6947	0.7193	0.9657		89°	0.9998	0.0175	57.2900
45°	0.7071	0.7071	1.0000		90°	1.0000	0.0000	

Properties and Formulas

Chapter 1

Order of Operations
1. Perform any operation(s) inside grouping symbols.
2. Simplify powers.
3. Multiply and divide in order from left to right.
4. Add and subtract in order from left to right.

The Midpoint Formula
The midpoint M of a line segment with endpoints $A(x_1, y_1)$ and $B(x_2, y_2)$ is $\left(\dfrac{x_1 + x_2}{2}, \dfrac{y_1 + y_2}{2}\right)$.

Chapter 2

Identity Property of Addition
For every rational number $n, n + 0 = n$.

Inverse Property of Addition
For every rational number n, there is an additive inverse $-n$ such that $n + (-n) = 0$.

Identity Property of Multiplication
For every real number $n, 1 \cdot n = n$.

Multiplication Property of Zero
For every real number $n, n \cdot 0 = 0$.

Multiplication Property of -1
For every real number $n, -1 \cdot n = -n$.

Inverse Property of Multiplication
For every nonzero real number a, there is a multiplicative inverse $\frac{1}{a}$ such that $a\left(\frac{1}{a}\right) = 1$.

Distributive Property
For every real number a, b, and c:
$a(b + c) = ab + ac$
$(b + c)a = ba + ca$
$a(b - c) = ab - ac$
$(b - c)a = ba - ca$

Commutative Property of Addition
For every real number a and $b, a + b = b + a$.

Commutative Property of Multiplication
For every real number a and $b, a \cdot b = b \cdot a$.

Associative Property of Addition
For every real number a, b, and c,
$(a + b) + c = a + (b + c)$.

Associative Property of Multiplication
For every real number a, b, and c,
$(a \cdot b) \cdot c = a \cdot (b \cdot c)$.

Probability Formula
$P(\text{event}) = \dfrac{\text{number of favorable outcomes}}{\text{number of possible outcomes}}$

Probability of Complement Formula
$P(\text{event}) + P(\text{not event}) = 1$;
$P(\text{not event}) = 1 - P(\text{event})$

Probability of Two Independent Events
If A and B are independent events,
$P(A \text{ and } B) = P(A) \cdot P(B)$.

Probability of Two Dependent Events
If A and B are dependent events,
$P(A \text{ then } B) = P(A) \cdot P(B \text{ after } A)$.

Chapter 3

Addition Property of Equality
For every real number a, b, and c, if $a = b$, then $a + c = b + c$.

Subtraction Property of Equality
For every real number a, b, and c, if $a = b$, then $a - c = b - c$.

Multiplication Property of Equality
For every real number a, b, and c, if $a = b$, then $a \cdot c = b \cdot c$.

Division Property of Equality
For every real number a, b, and c, with $c \neq 0$, if $a = b$, then $\frac{a}{c} = \frac{b}{c}$.

Cross Products of a Proportion
If $\frac{a}{b} = \frac{c}{d}$, then $ad = bc$.

Percent Error Formula
$\text{percent error} = \dfrac{\text{greatest possible error}}{\text{measurement}}$

The Pythagorean Theorem
In a right triangle, the sum of the squares of the lengths of the legs is equal to the square of the length of the hypotenuse. $a^2 + b^2 = c^2$

The Converse of the Pythagorean Theorem
If a triangle has sides of lengths a, b, and c, and $a^2 + b^2 = c^2$, then the triangle is a right triangle with hypotenuse of length c.

The Distance Formula
The distance d between any two points (x_1, y_1) and (x_2, y_2) is $d = \sqrt{(x_2 - x_1)^2 + (y_2 - y_1)^2}$.

Chapter 4
The following properties of inequality are also true for $\geq$ and $\leq$.

Addition Property of Inequality
For every real number a, b, and c,
if $a > b$, then $a + c > b + c$;
if $a < b$, then $a + c < b + c$.

Subtraction Property of Inequality
For every real number a, b, and c,
if $a > b$, then $a - c > b - c$;
if $a < b$, then $a - c < b - c$.

Multiplication Property of Inequality
For every real number a and b, and for $c > 0$,
if $a > b$, then $ac > bc$;
if $a < b$, then $ac < bc$.

For every real number a and b, and for $c < 0$,
if $a > b$, then $ac < bc$;
if $a < b$, then $ac > bc$.

Division Property of Inequality
For every real number a and b, and for $c > 0$,
if $a > b$, then $\frac{a}{c} > \frac{b}{c}$;
if $a < b$, then $\frac{a}{c} < \frac{b}{c}$.

For every real number a and b, and for $c < 0$,
if $a > b$, then $\frac{a}{c} < \frac{b}{c}$;
if $a < b$, then $\frac{a}{c} > \frac{b}{c}$.

Reflexive Property of Equality
For every real number a, $a = a$.

Symmetric Property of Equality
For every real number a and b,
if $a = b$, then $b = a$.

Transitive Property of Equality
For every real number a, b, and c,
if $a = b$ and $b = c$, then $a = c$.

Transitive Property of Inequality
For every real number a, b, and c,
if $a < b$ and $b < c$, then $a < c$.

Chapter 5

Arithmetic Sequence
The form for the rule of an arithmetic sequence is $A(n) = a + (n - 1)d$, where $A(n)$ is the nth term, a is the first term, $n - 1$ is the term number, and d is the common difference.

Chapter 6

Slope
$$\text{slope} = \frac{\text{vertical change}}{\text{horizontal change}} = \frac{\text{rise}}{\text{run}}$$

Slope-Intercept Form of a Linear Equation
The slope-intercept form of a linear equation is $y = mx + b$, where m is the slope and b is the y-intercept.

Standard Form of a Linear Equation
The standard form of a linear equation is $Ax + By = C$, where A, B, and C are real numbers and A and B are not both zero.

Point-Slope Form of a Linear Equation
The point-slope form of the equation of a nonvertical line that passes through the point (x_1, y_1) with slope m is $y - y_1 = m(x - x_1)$.

Slopes of Parallel Lines
Nonvertical lines are parallel if they have the same slope and different y-intercepts. Any two vertical lines are parallel.

Slopes of Perpendicular Lines
Two lines are perpendicular if the product of their slopes is -1. A vertical and a horizontal line are perpendicular.

Chapter 7

Solutions of Systems of Linear Equations
A system of linear equations can have one solution, no solution, or infinitely many solutions:
- If the lines have different slopes, the lines intersect, so there is one solution.
- If the lines have the same slopes and different y-intercepts, the lines are parallel, so there are no solutions.
- If the lines have the same slopes and the same y-intercepts, the lines are the same, so there are infinitely many solutions.

Chapter 8

Zero as an Exponent
For every nonzero number a, $a^0 = 1$.

Negative Exponent
For every nonzero number a and integer n, $a^{-n} = \frac{1}{a^n}$.

Scientific Notation
A number in scientific notation is written as the product of two factors in the form $a \times 10^n$, where n is an integer and $1 \le a < 10$.

Multiplying Powers with the Same Base
For every nonzero number a and integers m and n, $a^m \cdot a^n = a^{m+n}$.

Dividing Powers with the Same Base
For every nonzero number a and integers m and n, $\frac{a^m}{a^n} = a^{m-n}$.

Raising a Power to a Power
For every nonzero number a and integers m and n, $\left(a^m\right)^n = a^{mn}$.

Raising a Product to a Power
For every nonzero number a and b and integer n, $(ab)^n = a^n b^n$.

Raising a Quotient to a Power
For every nonzero number a and b and integer n, $\left(\frac{a}{b}\right)^n = \frac{a^n}{b^n}$.

Geometric Sequence
The form for the rule of a geometric sequence is $A(n) = a \cdot r^{n-1}$, where $A(n)$ is the nth term, a is the first term, $n - 1$ is the term number, and r is the common ratio.

Exponential Growth and Decay
An exponential function has the form $y = a \cdot b^x$, where a is a nonzero constant, b is greater than 0 and not equal to 1, and x is a real number.
- The function $y = a \cdot b^x$ models exponential growth for $a > 0$ and $b > 1$. b is the growth factor.
- The function $y = a \cdot b^x$ models exponential decay for $a > 0$ and $0 < b < 1$. b is the decay factor.

Chapter 9

Factoring Special Cases
For every nonzero number a and b:
$a^2 - b^2 = (a + b)(a - b)$
$a^2 + 2ab + b^2 = (a + b)(a + b) = (a + b)^2$
$a^2 - 2ab + b^2 = (a - b)(a - b) = (a - b)^2$

Chapter 10

Graph of a Quadratic Function
The graph of $y = ax^2 + bx + c$, where $a \ne 0$, has the line $x = \frac{-b}{2a}$ as its axis of symmetry. The x-coordinate of the vertex is $\frac{-b}{2a}$.

Zero-Product Property
For every real number a and b, if $ab = 0$, then $a = 0$ or $b = 0$.

Quadratic Formula
If $ax^2 + bx + c = 0$ and $a \ne 0$, then $x = \frac{-b \pm \sqrt{b^2 - 4ac}}{2a}$.

Property of the Discriminant
For the quadratic equation $ax^2 + bx + c = 0$, where $a \ne 0$, the value of the discriminant $b^2 - 4ac$ tells you the number of solutions.
- If $b^2 - 4ac > 0$, there are two real solutions.
- If $b^2 - 4ac = 0$, there is one real solution.
- If $b^2 - 4ac < 0$, there are no real solutions.

Chapter 11

Multiplication Property of Square Roots
For every number $a \ge 0$ and $b \ge 0$, $\sqrt{ab} = \sqrt{a} \cdot \sqrt{b}$.

Division Property of Square Roots
For every number $a \ge 0$ and $b > 0$, $\sqrt{\frac{a}{b}} = \frac{\sqrt{a}}{\sqrt{b}}$.

Trigonometric Ratios

sine of $\angle A = \dfrac{\text{length of leg opposite } \angle A}{\text{length of hypotenuse}}$

cosine of $\angle A = \dfrac{\text{length of leg adjacent to } \angle A}{\text{length of hypotenuse}}$

tangent of $\angle A = \dfrac{\text{length of leg opposite } \angle A}{\text{length of leg adjacent to } \angle A}$

Chapter 12

Multiplication Counting Principle

If there are m ways to make a first selection and n ways to make a second selection, there are $m \times n$ ways to make the two selections.

Permutation Notation

The expression $_nP_r$ stands for the number of permutations of n objects chosen r at a time.
$$_nP_r = n(n-1)(n-2)\ldots \} \; r \text{ factors}$$

Combination Notation

The expression $_nC_r$ stands for the number of combinations of n objects chosen r at a time.
$$_nC_r = \frac{_nP_r}{_rP_r} = \frac{n(n-1)(n-2)\ldots}{r(r-1)(r-2)\ldots}$$

Formulas of Geometry

You will use a number of geometric formulas as you work through your algebra book. Here are some perimeter, area, and volume formulas.

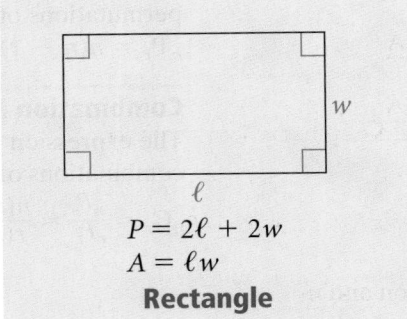

$$P = 2\ell + 2w$$
$$A = \ell w$$

Rectangle

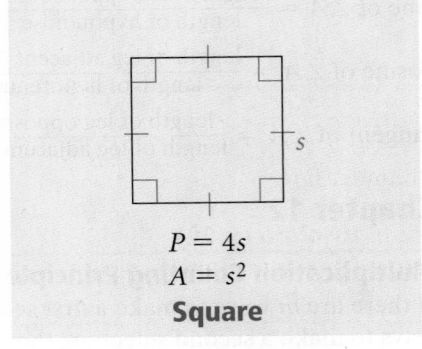

$$P = 4s$$
$$A = s^2$$

Square

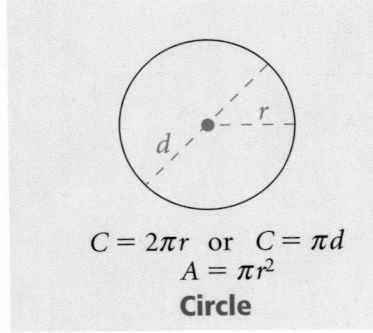

$$C = 2\pi r \quad \text{or} \quad C = \pi d$$
$$A = \pi r^2$$

Circle

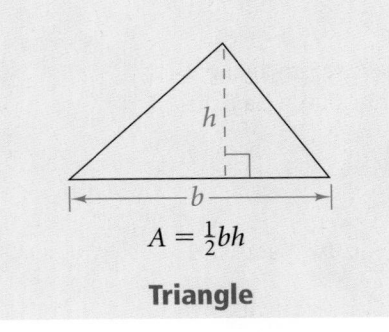

$$A = \tfrac{1}{2}bh$$

Triangle

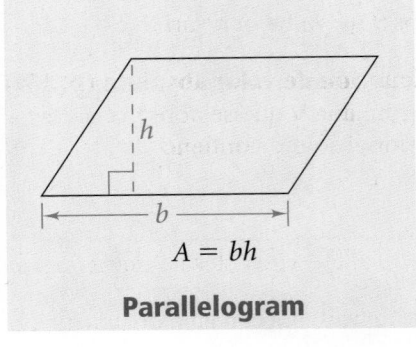

$$A = bh$$

Parallelogram

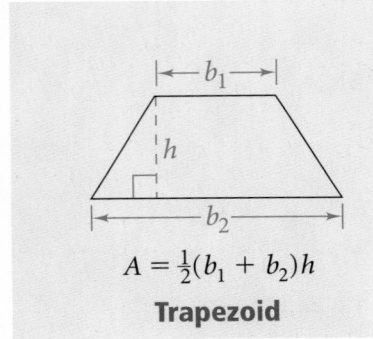

$$A = \tfrac{1}{2}(b_1 + b_2)h$$

Trapezoid

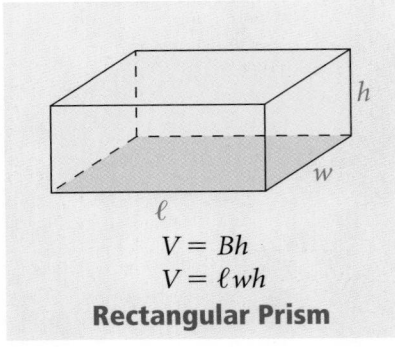

$$V = Bh$$
$$V = \ell wh$$

Rectangular Prism

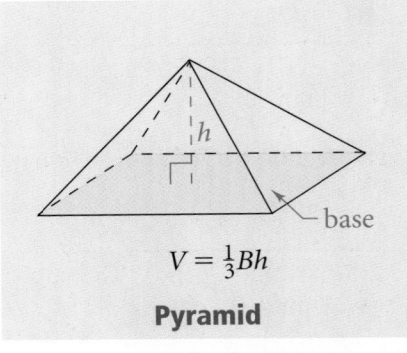

$$V = \tfrac{1}{3}Bh$$

Pyramid

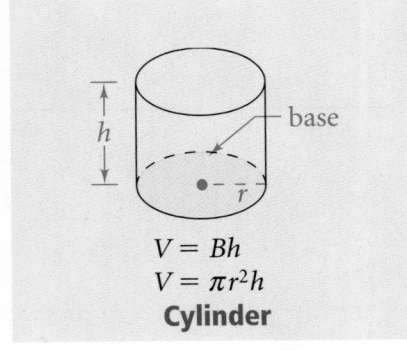

$$V = Bh$$
$$V = \pi r^2 h$$

Cylinder

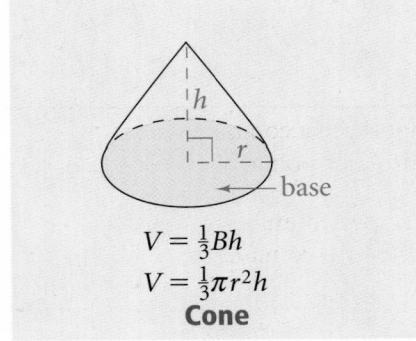

$$V = \tfrac{1}{3}Bh$$
$$V = \tfrac{1}{3}\pi r^2 h$$

Cone

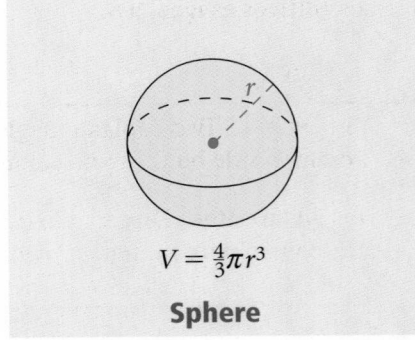

$$V = \tfrac{4}{3}\pi r^3$$

Sphere

English/Spanish Illustrated Glossary

EXAMPLES

Absolute value (p. 20)   The distance that a number is from zero on a number line.

-7 is 7 units from 0, so $|-7| = 7$.

Valor absoluto (p. 20)   La distancia a la que un número está del cero en una recta numérica.

Absolute value equation (p. 359)   Equation whose graph forms a V that opens up or down. An absolute value function contains the absolute value of a variable.

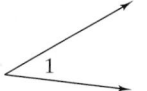

$y = |3 - x|$

Ecuación de valor absoluto (p. 359)   La ecuación cuya gráfica forma una V que se abre hacia arriba o hacia abajo. La función del valor absoluto contiene el valor de una variable.

Accuracy   How close a numerical measure is to its actual value.

Exactitud   La cercanía de una medida numérica a su valor verdadero.

Acute angle (p. 764)   An angle whose measure is greater than $0°$ and less than $90°$.

Ángulo agudo (p. 764) Un ángulo que mide más de $0°$ y menos de $90°$.

Additive inverse (p. 56)   The opposite of a number. Additive inverses sum to 0.

-5 and 5 are additive inverses because $-5 + 5 = 0$.

Inverso aditivo (p. 56)   El opuesto de un número. La suma de inversos aditivos es igual a 0.

Adjacent angles   Two coplanar angles that share a common vertex and a common side but have no common interior points.

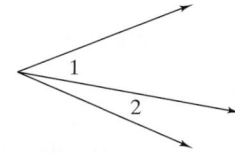

$\angle 1$ and $\angle 2$ are adjacent angles.

Ángulos adyacentes   Dos ángulos que comparten un mismo vértice y un lado común pero no tienen puntos interiores comunes.

Adjacent side of an acute angle in a right triangle   *See* **Sine**.

Lado adyacente de un ángulo agudo en un triánglo rectángulo *Ver* **Sine.**

English/Spanish Glossary

Algebra The branch of mathematics that uses letters and/or symbols to represent numbers and express mathematical relationships.

Álgebra Rama de las matemáticas que usa letras y/o símbolos para representar números y expresar relaciones matemáticas.

Algebraic equation A mathematical statement that contains an equal sign is written using one or more variables and constants.

Ecuación algebraica Una declaración matemática con signo de igualdad. Se escribe usando una o más variables y constantes.

$$3y + 5 = 1$$

$$\sqrt{2x - 5} = 11$$

Algebraic expression (p. 4) A mathematical phrase that can include numbers, variables, and operation symbols.

Expresión algebraica (p. 4) Proposición matemática que incluye números, variables y símbolos de operaciones.

$7 + x$ is an algebraic expression.

Algebraic fraction A fraction that contains an algebraic expression in its numerator and/or denominator.

Fracción algebraica Una expresión que contiene una expresión algebraica en el numerador y/o en el denominador.

$$\frac{2x + 4}{x + 2}$$

Algebraic representation The use of an equation or algebraic expression to model a mathematical relationship.

Representación algebraica El uso de una ecuación o expresión algebraica para representar una relación matemática.

Algorithm A defined series of steps for carrying out a computation or process.

Algoritmo Un conjunto de pasos para llevar a cabo un cálculo o proceso.

Angle (p. 764) A geometric figure formed by two rays that have a common endpoint.

Ángulo (p. 764) Figura geométrica formada por dos rayos que tienen el mismo extremo.

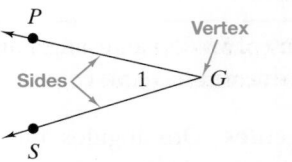

Angle of depression (p. 651) An angle from the horizontal down to a line of sight. It is used to measure heights indirectly.

Ángulo de depresión (p. 651) Un ángulo de la horizontal hacia la línea de vista. Ángulo con que se miden indirectamente las alturas.

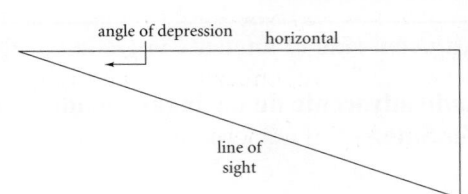

Angle of elevation (p. 650) An angle from the horizontal up to a line of sight. It is used to measure heights indirectly.

Ángulo de elevación (p. 650) Ángulo de la horizontal hacia la línea de vista. Ángulo con que se miden las alturas indirectamente.

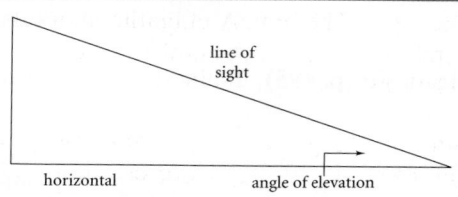

Approximate value A value for some quantity, accurate to a specified degree.

Valor aproximado Valor de una cantidad, cercano a un valor dado.

A board that measures 4 feet 2 inches has an approximate length to the nearest foot of 4 feet.

Arithmetic sequence (p. 293) A number sequence formed by adding a fixed number to each previous term.

Progresión aritmética (p. 293) Sucesión numérica que se obtiene al sumar un número constante a cada término consecutivo.

$4, 7, 10, 13, \ldots$ is an arithmetic sequence.

Associative Properties of Addition and Multiplication (p. 86) For any numbers a, b, and c, $(a + b) + c = a + (b + c)$ and $(ab)c = a(bc)$.

Propiedad asociativa de la suma y de la multiplicación (p. 86) Para cualesquiera números a, b y c, $(a + b) + c = a + (b + c)$ y $(ab)c = a(bc)$.

$(2 + 7) + 3 = 2 + (7 + 3)$
$(9 \cdot 4)5 = 9(4 \cdot 5)$

Asymptote (p. 664) A line the graph of a function gets closer to as x or y gets larger in absolute value.

Asíntota (p. 664) Línea recta a la que la gráfica de una función se acerca indefinidamente, mientras el valor absoluto de x o y aumenta.

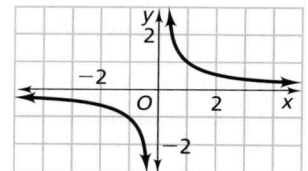

The y-axis is a vertical asymptote for $y = \frac{1}{x}$. The x-axis is a horizontal asymptote for $y = \frac{1}{x}$

Axis of symmetry (p. 551) The line that divides a parabola into two matching halves.

Eje de simetría (p. 551) Línea recta que divide una parábola en dos mitades exactamente iguales.

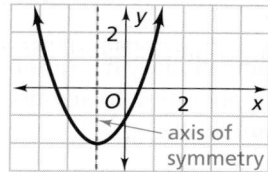

B

Base (p. 9) A number that is multiplied repeatedly.

Base (p. 9) El número que se multiplica repetidas veces.

$4^5 = 4 \cdot 4 \cdot 4 \cdot 4 \cdot 4$. The base 4 is used as a factor 5 times.

English/Spanish Glossary

EXAMPLES

Binomial (p. 495)   A polynomial of two terms.

$3x + 7$ is a binomial.

Binomio (p. 495)   Polinomio compuesto de dos términos.

Bivariate data (p. NY 727)   Data involving two variables.

Datos bivariantes (p. NY 727)   Datos que incluyen dos variables.

Box-and-whisker plot (p. NY 734)   A box-and-whisker plot is a graph that shows the distribution of data along a number line. Quartiles divide the data into four equal parts. A visual display of a set of data showing the five number summary: minimum, first quartile, median, third quartile, and maximum. This plot shows the range of scores within each quarter of the data.

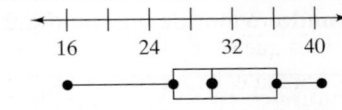

EXAMPLE   The box-and-whisker plot at the right is for the data
16 19 26 27 27 29 30 31 34 35 37 39 40.

The first quartile (Q_1) is 26.5. The median (Q_2) is 30. The third quartile (Q_3) is 36.

Gráfica de caja y brazos (p. NY 734)   Una gráfica de caja y brazos presenta la distribución de datos en una recta numérica. Los cuartiles dividen los datos en cuatro partes iguales. Una demostración visual de un conjunto de datos que muestra el resumen de cinco números—dos extremos, cuartil inferior (Q_1), mediana (Q_2) y cuartil superior (Q_3). Estos puntos muestran su distribución dentro de cada cuarto de los datos.

Circle   A circle is the set of all points in a plane that are equidistant from a given point, called the center.

EXAMPLE   Circle O

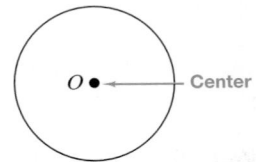

Círculo   Un círculo es un conjunto de puntos en un plano que se hallan a la misma distancia de un punto dado llamado el centro.

Closure   A set S and a binary operation are said to exhibit closure if applying the binary operation to any two elements in S produces a value that is a member of S.

Cerradura   Se dice que un conjunto S y una operación binaria expresan cerradura al aplicar una operación binaria donde el resultado de esta operación realizada con cualesquiera dos elementos del conjunto S produce un valor del mismo conjunto S.

Coefficient (p. 81)   The numerical factor when a term has a variable.

In the expression $2x + 3y + 16$, 2 and 3 are coefficients.

Coeficiente (p. 81)   Factor numérico de un término que contiene una variable.

Combination (pp. 706–707) An arrangement of some or all of a set of objects without regard to order. The number of combinations

$$= \frac{\text{total number of permutations}}{\text{number of times the objects in each group are repeated}}.$$ You can use the notation $_nC_r$ to write the number of combinations of n objects chosen r at a time.

Combinación (pp. 706–707) Grupo que contiene algunos o todos los objetos de un conjunto, cualquiera que sea su orden. El número de combinaciones

$$= \frac{\text{número total de permutaciones}}{\text{número de veces que se repiten los objetos de cada grupo}}.$$ La notación $_nC_r$ expresa el número de combinaciones de n objetos escogidos r en un momento dado.

The number of combinations 10 things can be taken 4 at a time:

$$_{10}C_4 = \frac{_{10}P_4}{_4P_4} = \frac{10 \cdot 9 \cdot 8 \cdot 7}{4 \cdot 3 \cdot 2 \cdot 1} = 210$$

Common base(s) Exponential expressions or equations that have the same or equivalent bases.

Base común Expresión o ecuación exponencial que tiene bases equivalentes o del mismo valor.

a) 2 is the common base in 2^3 and 2^4.

b) In the equation $3 \times 3 = 3^2$, the common base is 3.

Common difference (p. 293) The fixed number added to each term of an arithmetic sequence.

Diferencia común (p. 293) Número constante que se usa a cada término para formar una progresión aritmética.

The common difference is 3 in the arithmetic sequence 4, 7, 10, 13, . . .

Common factor (p. 530) A number, polynomial, or quantity that divides two or more numbers or algebraic expressions evenly.

Factor común (p. 530) Número, polinomial o cantidad que divide dos o más números o expresiones algebraicas equitativamente.

1, 3, 5, 15, are common factors of 15 and 30.

$2x$ is a common factor of $4xy$ and $6x^2$.

Common ratio (p. 460) The fixed number used to find terms in a geometric sequence.

Razón común (p. 460) Número constante que se usa para hallar los términos en una progresión geométrica.

The common ratio is $\frac{1}{3}$ in the geometric sequence 9, 3, 1, $\frac{1}{3}$, . . .

Commutative Properties of Addition and Multiplication (p. 86) For any numbers a and b, $a + b = b + a$ and $ab = ba$.

Propiedad conmutativa de la suma (p. 86) Para cualquier número a y b, $a + b = b + a$, y $ab = ba$.

$6 + 4 = 4 + 6$
$9 \cdot 5 = 5 \cdot 9$

Complement of an event (p. 94) All possible outcomes that are not in the event.

$P(\text{complement of event}) = 1 - P(\text{event})$

Complemento de un suceso (p. 94) Todos los resultados posibles que no se dan en el suceso.

$P(\text{complemento de un suceso}) = 1 - P(\text{suceso})$

The complement of rolling a 1 or a 2 on a number cube is rolling a 3, 4, 5, or 6.

Complement of a set (p. NY 739) The elements of a universe not contained in a given set; the subset that must be added to any given subset to yield the original set. The complement of set A is indicated by A'.

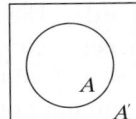

Complemento de un conjunto (p. NY 739) Elementos de un conjunto universal que no están contenidos en un conjunto dado; es el subconjunto que debe ser sumado al conjunto dado para que dé el conjunto original. El complemento del conjunto A se indica con A'.

Completing the square (p. 579) A method of solving quadratic equations. Completing the square turns every quadratic equation into the form $x^2 = c$.

$x^2 + 6x - 7 = 11$ is rewritten as $(x + 3)^2 = 25$ by completing the square.

Completar el cuadrado (p. 579) Método para solucionar ecuaciones cuadráticas. Cuando se completa el cuadrado se transforma la ecuación cuadrática a la fórmula $x^2 = c$.

Compound inequalities (p. 227) Two inequalities that are joined by *and* or *or*.

$5 < x$ and $x < 10$

$14 < x$ or $x \geq -3$

Desigualdades compuestas (p. 227) Dos desigualdades que están enlazadas por medio de una *y* o una *o*.

Compound interest (p. 476) Interest paid on both the principal and the interest that has already been paid.

For an initial deposit of $1000 at a 6% interest rate with interest compounded quarterly, the function $y = 1000\left(\frac{0.06}{4}\right)^{4x}$ gives the account balance y after x years.

Interés compuesto (p. 476) Interés calculado tanto sobre el capital como sobre los intereses ya pagados.

Conclusion (p. 183) In a conditional, the part following *then*. *See* **Conditional**.

In the conditional "If an animal has four legs, then it is a horse," the conclusion is "it is a horse."

Conclusión (p. 183) En un enunciado condicional, la parte que sigue a *entonces*. *Ver* **Conditional**.

Conditional (p. 183) An *if-then* statement.

If an animal has four legs, then it is a horse.

Condicional (p. 183) Un enunciado de la forma *si-entonces*.

Conjecture (p. 292) Conclusion reached by inductive reasoning.

Conjetura (p. 292) Conclusión a la que se llega mediante el razonamiento inductivo.

Conjugates (p. 623) The sum and the difference of the same two terms.

$\left(\sqrt{3} + 2\right)$ and $\left(\sqrt{3} - 2\right)$ are conjugates.

Valores conjugados (p. 623) La suma y resta de los mismos dos términos.

Consecutive integers (p. 159) Integers that differ by one.

Números enteros consecutivos (p. 159) Números enteros cuya diferencia es 1.

$-5, -4$, and -3 are three consecutive integers.

Constant (p. 81) A term that has no variable factor.

Constante (p. 81) Término que tiene un valor fijo.

In the expression $4x + 13y + 17$, 17 is a constant term.

Constant of variation for direct variation (p. 278) The nonzero constant k in the function $y = kx$.

Constante de variación en variaciones directas (p. 278) La constante k cuyo valor no es cero en la función $y = kx$.

For the function $y = 24x$, 24 is the constant of variation.

Constant of variation for inverse variation (p. 285) The nonzero constant k in the function $y = \frac{k}{x}$.

Constante de variación en variaciones inversas (p. 285) La constante k cuyo valor no es cero en la función $y = \frac{k}{x}$.

For the equation $y = \frac{8}{x}$, 8 is the constant of variation.

Continuous Data (p. 264) Data where numbers between any two data values have meaning.

Datos continuos (p. 264) Datos en los cuales los números entre dos valores cualesquiera tienen un significado.

temperature, length, or weight

Converse (p. 183) The statement obtained by reversing the *if* and *then* parts of an *if-then* statement.

Expresión recíproca (p. 183) La que se obtiene al invertir los componentes *si* y *entonces* de un enunciado condicional.

The converse of "If I was born in Houston, then I am a Texan," would be "If I am a Texan, then I was born in Houston."

Coordinate plane (p. 24) A plane formed by two number lines that intersect at right angles.

Plano de coordenadas (p. 24) Se forma cuando dos rectas numéricas se cortan formando ángulos rectos.

Coordinates (p. 24) The numbers that make an ordered pair and identify the location of a point.

Coordenadas (p. 24) Números ordenados por pares que determinan la posición de un punto sobre un plano.

The coordinates of R are $(-4, 1)$.

English/Spanish Glossary

Correlation (p. 34) A statistical measure that quantifies how pairs of variables are related.

Correlación (p. 34) Medida estadística que cuantifica la relación entre dos variables.

Positive correlation

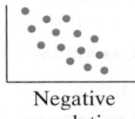
Negative correlation

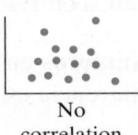
No correlation

Correlation coefficient (p. 351) A number that tells how closely the equation of best fit models the data. The value r of the correlation coefficient is in the range $-1 \le r \le 1$.

Coeficiente de correlación (p. 351) Valor que indica qué tan bien la ecuación más adecuada expresa los datos. El valor r del coeficiente de correlación está dentro del rango $-1 \le r \le 1$.

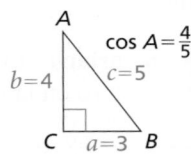
```
LinReg
  y=ax+b
  a=.0134039132
  b=-.3622031627
  r²=.886327776
  r=.9414498267
  ■
```

The correlation coefficient for the data points $(68, 0.5)$, $(85, 0.9)$, $(100, 0.9)$, and $(108, 1.1)$ is approximately 0.94.

Cosine (p. 646) In a right triangle, such as $\triangle ABC$ with right $\angle C$,

cosine of $\angle A = \dfrac{\text{length of side adjacent to } \angle A}{\text{length of hypotenuse}}$ or $\cos A = \dfrac{b}{c}$.

Coseno (p. 646) En un triángulo rectángulo tal que $\triangle ABC$ con $\angle C$ recto, el coseno de $\angle A = \dfrac{\text{longitud del lado adyacente a } \angle A}{\text{longitud de la hipotenusa}}$, o $\cos A = \dfrac{b}{c}$.

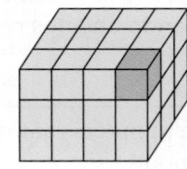
$\cos A = \dfrac{4}{5}$, $b = 4$, $c = 5$, $a = 3$

Counterexample (p. 18) Any example that proves a statement false.

Contraejemplo (p. 18) Todo ejemplo que pruebe la falsedad de un enunciado.

Statement: All apples are red.

Counterexample: A Granny Smith apple is green.

Cross products (p. 144) In a proportion, $\frac{a}{b} = \frac{c}{d}$, the products ad and bc. These products are equal.

Productos cruzados (p. 144) En una proporción, $\frac{a}{b} = \frac{c}{d}$, los productos ad y bc. Estos productos son iguales.

$\dfrac{3}{4} = \dfrac{6}{8}$

The cross products are $3 \cdot 8$ and $4 \cdot 6$.
$3 \cdot 8 = 24$ and $4 \cdot 6 = 24$.

Cubic unit A cubic unit is the amount of space occupied by a cube with edges one unit long.

Unidad cúbica Una unidad cúbica es el espacio que ocupa un cubo cuyos lados miden una unidad de longitud.

Cumulative frequency table A table that shows how often each item, number, or range of numbers occurs in a set of data. This table displays the total number of scores that fall into each of several cumulative intervals. The cumulative intervals are created by adding the preceding tallies (of lower scores) to the new tallies for each interval.

Tabla de frecuencia acumulada Tabla que demuestra la frecuencia en que cada número o rango de números ocurre en un conjunto. Esta tabla demuestra el número total de puntajes que caen entre varios intervalos acumulados. Los intervalos acumulados se dan al sumar los puntajes anteriores (de puntajes bajos) con los puntajes nuevos para cada intervalo.

Data: 5, 7, 6, 8, 9, 5, 13, 2, 1, 6, 5, 14, 10, 5, 9

Interval	Cumulative Frequency
1-5	6
1-10	13
1-15	15

Cumulative frequency histogram A histogram where each bar contains all the data up to and including the data in that bar's interval.

Histograma de frecuencia acumulada Histograma donde cada barra contiene información que alcanza e incluye la información en el intervalo de la barra.

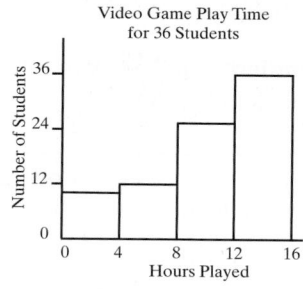

D

Decay factor (p. 478) The base b in the exponential function $y = ab^x$, where $0 < b < 1$.

The decay factor of the function $y = 5(0.3)^x$ is 0.3.

Factor de decremento (p. 478) La base b en la función exponencial $y = ab^x$ donde $0 < b < 1$.

Deductive reasoning (p. 88) A process of reasoning logically from given facts to a conclusion.

Razonamiento deductivo (p. 88) Proceso de razonar lógicamente para llegar a una conclusión a partir de datos dados.

Degree of a monomial (p. 495) The sum of the exponents of the variables of a monomial.

$-4x^3y^2$ is a monomial of degree 5.

Grado de un monomio (p. 495) La suma de los exponentes de las variables de un monomio.

Degree of a polynomial (p. 495) The degree of the term with the greatest exponent for a polynomial in one variable.

The degree of $3x^2 + x - 9$ is 2.

Grado de un polinomio (p. 495) El grado del término con el mayor exponente en un polinario de una variable.

Degree of a vertex (p. 713) The number of edges that meet at a vertex.

Grado de un vértice (p. 713) El número de aristas que concurren en un vértice.

Dependent events (p. 103)   Two events in which the occurrence of one event affects the probability of the second event.

Sucesos dependientes (p. 103)   Cuando el resultado de un suceso influye en la probabilidad de que ocurra el segundo suceso, los sucesos son dependientes.

You have a bag with marbles of different colors. If you pick a marble from the bag and pick another without replacing the first, the events are dependent events.

Dependent variable (p. 28)   A variable that provides the output values of a function.

Variable dependiente (p. 28)   Variable de la que dependen los valores de salida de una función.

In the equation $y = 3x$ the value of y depends upon the value of x.

Difference of squares (p. 514)   A difference of squares is an expression of the form $a^2 - b^2$. It can be factored as $(a + b)(a - b)$.

Diferencia de cuadrados (p. 514)   La diferencia de cuadrados es una expresión de la forma $a^2 - b^2$. Se puede factorizar como $(a + b)(a - b)$.

$25a^2 - 4 = (5a + 2)(5a - 2)$
$m^6 - 1 = (m^3 + 1)(m^3 - 1)$

Dilation (p. 150)   A transformation in which a figure and its image are similar.

Dilatación (p. 150)   Transformación en la cual una figura y su imagen son semejantes.

Quadrilateral $PQRS$ is dilated by a scale factor of $\frac{1}{2}$.

Direct variation (p. 278)   A linear function that can be expressed in the form $y = kx$, where $k \neq 0$.

Variación directa (p. 278)   Función lineal que puede expresarse como $y = kx$, donde $k \neq 0$.

$y = 18x$ is a direct variation.

Discrete data (p. 264)   Data that involve a count of items.

Datos discretos (p. 264)   Datos que implican un conteo de objetos.

number of people or number of cars

Discriminant (p. 592)   The quantity $b^2 - 4ac$ is the discriminant of $ax^2 + bx + c = 0$.

Discriminante (p. 592)   La cantidad $b^2 - 4ac$ es el discriminante de $ax^2 + bx + c = 0$.

The discriminant of $2x^2 + 9x - 2 = 0$ is 65.

Disjoint sets (p. NY 744)   Sets with no elements in common.

Conjuntos desunidos (p. NY 744)   Conjuntos sin elementos en común.

$\{1, 2, 3\}$ and $\{4, 5, 6\}$ are disjoint sets.

Distance Formula (p. 186)   The distance d between any two points (x_1, y_1) and (x_2, y_2) is

$$d = \sqrt{(x_2 - x_1)^2 + (y_2 - y_1)^2}.$$

Fórmula de distancia (p. 186)   La distancia d entre dos puntos cualesquiera (x_1, y_1) y (x_2, y_2) es

$$d = \sqrt{(x_2 - x_1)^2 + (y_2 - y_1)^2}.$$

The distance between $(-2, 4)$ and $(4, 5)$ is

$$d = \sqrt{(4 - (-2))^2 + (5 - 4)^2}$$
$$= \sqrt{(6)^2 + (1)^2}$$
$$= \sqrt{37}$$

Distributive Property (p. 79)   For every real number a, b, and c:

$a(b + c) = ab + ac \quad (b + c)a = ba + ca$

$a(b - c) = ab - ac \quad (b - c)a = ba - ca$

Propiedad Distributiva (p. 79)   Para cada número real a, b y c:

$a(b + c) = ab + ac \quad (b + c)a = ba + ca$

$a(b - c) = ab - ac \quad (b - c)a = ba - ca$

$3(19 + 4) = 3(19) + 3(4)$

$(19 + 4)3 = 19(3) + 4(3)$

$7(11 - 2) = 7(11) - 7(2)$

$(11 - 2)7 = 11(7) - 2(7)$

Domain (of a function) (p. 29)   The possible values for the input, or the independent variable, of a function.

Dominio (de una función) (p. 29)   Todos los valores posibles para la entrada, o variable independiente, de una función.

In the function $f(x) = x + 22$, the domain is all real numbers.

Edges (p. 712)   Segments connecting vertices in a network.

Aristas (p. 712)   Segmentos que unen los vértices de una red.

Element (p. 59)   An item in a matrix.

Elemento (p. 59)   Componente de una matriz.

$$\begin{bmatrix} 5 & -2 \\ 7 & 3 \end{bmatrix}$$

$5, 7, -2$, and 3 are the four elements of the matrix.

Elements (of a set) (p. 233)   Members of a set.

Elementos (p. 233)   Partes integrantes de un conjunto.

Cats and dogs are elements of the set of mammals.

Elimination method (p. 387)   A method for solving a system of linear equations. You add or subtract the equations to eliminate a variable.

Método de eliminación (p. 387)   Método para resolver un sistema de ecuaciones lineales. Se suman o se restan las ecuaciones para eliminar una variable.

$3x + y = 19$

$2x - y = 1$

$x + 0 = 18 \qquad x = 18$

$2(18) - y = 1 \rightarrow$ Substitute 18 for x

$36 - y = 1 \qquad$ in the second equation.

$y = 35 \rightarrow$ Solve for y.

Empirical probability   An estimate of the probability of an event based on the results of repeated trials of the event.

Probabilidad empírica   Una estimación de la probabilidad que ocurra un evento basandose en repetidos ensayos del evento.

Equation (p. 5)   A mathematical sentence that uses an equal sign.

Ecuación (p. 5)   Enunciado matemático que tiene el signo de igual.

$x + 5 = 3x - 7$

Equivalent equations (p. 118)   Equations that have the same solution.

Ecuaciones equivalentes (p. 118)   Ecuaciones que tienen la misma solución.

$\frac{9}{3} = 3$ and $\frac{9}{3} + a = 3 + a$ are equivalent equations.

Equivalent forms   Different ways of writing numbers or expressions that have equal values.

Formas equivalentes   Diferentes maneras de escribir números o expresiones de mismo valor.

$\frac{2}{3}$ is equivalent to $\frac{8}{12}$.

$3 \times 3 \times 5 \times 5 \times 5 \times 5$ is equivalent to $3^2 5^4$.

Equivalent inequalities (p. 206)   Equivalent inequalities have the same set of solutions.

Desigualdades equivalentes (p. 206)   Las desigualdades equivalentes tienen el mismo conjunto de soluciones.

$x + 4 < 7$ and $x < 3$ are equivalent inequalities.

Evaluate (p. 10)   Substitute a given number for each variable, and then simplify.

Evaluar (p. 10)   Método de sustituir cada variable por un número dado para luego simplificar la expresión.

To evaluate $3x + 4$ for $x = 2$, substitute 2 for x and simplify.
$3(2) + 4$
$6 + 4$
10

Event (p. 93)   Any group of outcomes in a situation involving probability.

Suceso (p. 93)   En la probabilidad, cualquier grupo de resultados.

When rolling a number cube, there are six possible outcomes. Rolling an even number is an event with three possible outcomes, 2, 4, and 6.

Experimental probability (p. 94)   The ratio of the number of times an event actually happens to the number of times the experiment is done.

$P(\text{event}) = \frac{\text{number of times an event happens}}{\text{number of times the experiment is done}}$

Probabilidad experimental (p. 94)   La razón entre el número de veces que un suceso sucede en la realidad y el número de veces que se hace el experimento.

$P(\text{suceso}) = \frac{\text{número de veces que sucede un suceso}}{\text{número de veces que se hace el experimento}}$

A baseball player's batting average shows how likely it is that a player will get a hit, based on previous times at bat.

Exponent (p. 9)   A number that shows repeated multiplication.

Exponente (p. 9)   Denota el número de veces que debe multiplicarse.

$$3^4 = 3 \cdot 3 \cdot 3 \cdot 3$$

The exponent 4 indicates that 3 is used as a factor four times.

Exponential decay (p. 478)   A situation modeled with a function of the form $y = ab^x$, where $a > 0$ and $0 < b < 1$.

Decremento exponencial (p. 478)   Para $a > 0$ y $0 < b < 1$, la función $y = ab^x$ representa el decremento exponencial.

$$y = 5(0.1)^x$$

Exponential form   An expression or equation containing exponents.

Forma exponencial   Expresión o ecuación que contiene exponentes.

In exponential form, $32 = 2^5$

Exponential function (p. 468)   A function that repeatedly multiplies an initial amount by the same positive number. You can model all exponential functions using $y = ab^x$, where a is a nonzero constant, $b > 0$, $b \neq 1$.

Función exponencial (p. 468)   Función que multiplica repetidas veces una cantidad inicial por el mismo número positivo. Todas las funciones exponenciales se pueden representar mediante $y = ab^x$, donde a es una constante con valor distinto de cero, $b > 0$ y $b \neq 1$.

$$y = 4.8(1.1)^x$$

Exponential growth (p. 475)   A situation modeled with a function of the form $y = ab^x$, where $a > 0$ and $b > 1$.

Incremento exponencial (p. 475)   Para $a > 0$ y $b > 1$, la función $y = ab^x$ representa el incremento exponencial.

$$y = 100(2)^x$$

Expression   *See* **Algebraic expression.**

Expresión   *Ver* **Algebraic expression.**

Extraneous solution (p. 631)   An apparent solution of the equation that does not satisfy the original equation.

Solución extraña (p. 631)   Solución aparente de una ecuación que no satisface la ecuación original.

$$\frac{b}{b + 4} = 3 - \frac{4}{b + 4}$$

Multiply by $(b + 4)$.

$$b = 3(b + 4) - 4$$
$$b = 3b + 12 - 4$$
$$-2b = 8$$
$$b = -4$$

Replace b with -4 in the original equation. The denominator is 0, and so -4 is an extraneous solution.

English/Spanish Glossary

Extrapolate The process of using a given data set to estimate the value of a function or measurement beyond the values already known.

Extrapolar El proceso en el que se usa la información de un conjunto para estimar el valor de una función o medida no incluidos en dichos valores.

Extremes of a proportion (p. 143) In the proportion, $\frac{a}{b} = \frac{c}{d}$, a and d are the extremes.

The product of the extremes of $\frac{x}{4} = \frac{x+3}{2}$ is $2x$.

Valores extremos de una proporción (p. 143) En la proporción $\frac{a}{b} = \frac{c}{d}$, , a y d son los valores extremos.

Factor (noun) A whole number that is a divisor of another number; an algebraic expression that is a divisor of another algebraic expression.

3 is a factor of 12.
$(x + 2)$ is a factor of $x^2 - x - 6$.

Factor (sustantivo) Un número cardinal por el que se divide otro número; expresión algebraica por la que se divide otra expresión algebraica.

Factor (verb) To find the numbers or algebraic expressions that give an indicated product.

To factor $x^2 - x - 6$,
write $(x - 3)(x + 2)$.

Descomponer en factores (verbo) Encontrar los números o expresiones algebraicas que producen cierto resultado.

Factor by grouping (p. 534) A method of factoring that uses the distributive property to remove a common binomial factor of two pairs of terms.

The expression $7x(x - 1) + 4(x - 1)$ can be factored as $(7x + 4)(x - 1)$.

Factor común por agrupación de términos (p. 534) Método de factorización que aplica la propiedad distributiva para sacar un factor común de dos pares de términos en un binomio.

Factorial The product of all positive integers less than or equal to a whole number. Note: $0! = 1$ and $1! = 1$.

$6! = 6 \cdot 5 \cdot 4 \cdot 3 \cdot 2 \cdot 1 = 720$

$n! = n \cdot (n - 1) \cdot (n - 2) \cdots 3 \cdot 2 \cdot 1$

Factorial El producto de todos los números cardinales inferiores a o iguales a un número entero. Nota: $0! = 1$ y $1! = 1$.

Five number summary (p. NY 733) *See* **Box-and-whisker plot.**

Resumen de cinco números (p. NY 733) *Ver* **Box-and-whisker plot.**

Frequency table   A frequency table is a list of items that shows the number of times, or frequency, with which they occur.

EXAMPLE   This frequency table shows the number of household telephones for the students in one school class.

Tabla de frecuencia   Una tabla de frecuencia registra el número de veces, o frecuencia, con que se ha producido un determinado tipo de resultado.

Household Telephones

Phones	Tally	Frequency
1	卌 |||	8
2	卌		6
3						4

Function (p. 27)   A relation that assigns exactly one value in the range to each value of the domain.

Función (p. 27)   La relación que asigna exactamente un valor del rango a cada valor del dominio.

Earned income is a function of the number of hours worked. If you earn \$4.50/h, then your income is expressed by the function $f(h) = 4.5h$.

Function notation (p. 258)   To write a rule in function notation, you use the symbol $f(x)$ in place of y.

Notación de una función (p. 258)   Para expresar una regla en notación de función se usa el símbolo $f(x)$ en lugar de y.

$f(x) = 3x - 8$ is in function notation.

Function rule (p. 27)   An equation that describes a function.

Regla de una función (p. 27)   Ecuación que describe una función.

$y = 4x + 1$ is a function rule.

Geometric sequence (p. 460)   A number sequence formed by multiplying a term in a sequence by a fixed number to find the next term.

Progresión geométrica (p. 460)   Tipo de sucesión numérica formada al multiplicar un término de la secuencia por un número constante, para hallar el siguiente término.

$9, 3, 1, \frac{1}{3}, \ldots$ is an example of a geometric sequence.

Graphical representation   A graph or graphs used to model a mathematical relationship.

Representación gráfica   Gráfica o gráficas que se usan para representar una relación matemática.

Graphical solution of a system of equations   The set of points in the plane whose coordinates are solutions to a system of equations.

Gráfica de soluciones en un sistema de ecuaciones   Conjunto de puntos en un plano cuyas coordenadas son soluciones a un sistema de ecuaciones.

Greatest common factor (GCF) (p. 501) The greatest number or expression that is a factor of two or more numbers or expressions.

12 is the GCF of 24 and 36.

$5xy$ is the GCF of $25x^2y$ and $10xy^2$

Máximo común divisor (MCD) (p. 501) El mayor divisor o expresión que tienen en común dos o más números o expresiones.

Greatest possible error (p. 169) One half of the measuring unit, for any measurement.

The mass of a rock is measured as 3.8 g. The greatest possible error is one half 0.1 g, or 0.05 g.

Máximo error posible (p. 169) Para una medición dada, la mitad de la unidad de medida.

Growth factor (p. 475) The number b in an exponential growth function of the form $y = ab^x$, where $b > 1$.

The growth factor of $y = 7(1.3)^x$ is 1.3.

Factor incremental (p. 475) Para la función exponencial incremental $y = ab^x$, donde $b > 1$, b es el factor incremental.

Histogram (p. 304) A histogram is a bar graph in which the heights of the bars give the frequencies of the data. There are no spaces between bars.

EXAMPLE This histogram gives the frequencies of board-game purchases at a local toy store.

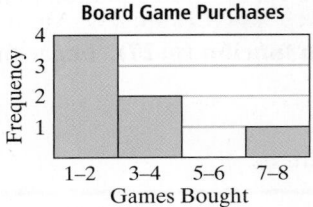

Histograma (p. 304) Un histograma es una gráfica de barras en la cual la altura de las barras representa la frecuencia de los datos. No hay espacio entre las barras.

Hypotenuse (p. 181) The side opposite the right angle in a right triangle. It is the longest side in the triangle.

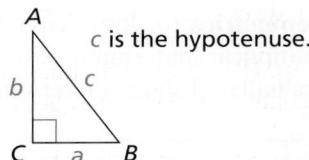

c is the hypotenuse.

Hipotenusa (p. 181) En un triángulo rectángulo, el lado opuesto al ángulo recto. Es el lado más largo del triángulo.

Hypothesis (p. 183) The part following *if* in a conditional. *See* **Conditional.**

In the conditional "If an animal has four legs, then it is a horse," the hypothesis is "an animal has four legs."

Hipótesis (p. 183) En un enunciado condicional, la parte que sigue a *si. Ver* **Conditional.**

Identity (p. 136) An equation that is true for every value.

$5 - 14x = 5\left(1 - \frac{14}{5}x\right)$ is an identity because it is true for any value of x.

Identidad (p. 136) Una ecuación que es verdadera para todos los valores.

Image   The resulting point or set of points under a given transformation.

Imagen   El punto o conjunto de puntos que resultan de una transformación.

$A'B'C'D'$ is the image of $ABCD$.

Impossible event/outcome (p. 93)   An event that cannot occur. The probability of an impossible event equals zero.

Suceso/resultado imposible (p. 93)   Un evento que no puede ocurrir. La probabilidad de un suceso imposible es cero.

rolling a sum of 13 when tossing two number cubes labeled 1 to 6

Independent events (p. 102)   Two events for which the outcome of one does not affect the other.

Sucesos independientes (p. 102)   Dos sucesos son independientes si el resultado de uno de ellos no influye en el resultado del otro.

Picking a colored marble from a bag, and then replacing it and picking another marble, are two independent events.

Independent variable (p. 28)   A variable that provides the input values of a function.

Variable independiente (p. 28)   Variable de la que dependen los valores de entrada de una función.

In the equation $y = 3x$, x is the independent variable.

Inductive reasoning (p. 292)   Making conclusions based on observed patterns.

Razonamiento inductivo (p. 292)   Sacar conclusiones a partir de patrones observados.

Inequality (p. 19)   A mathematical sentence that compares the values of two expressions using an inequality symbol.

Desigualdad (p. 19)   Expresión matemática que compara el valor de dos expresiones con el símbolo de desigualdad.

$3 < 7$

Infinitely many solutions (p. 376)   The number of solutions of a system of equations in which the graphs of the equations are the same line.

Infinitamente muchas soluciones (p. 376)   El número de soluciones de un sistema de ecuaciones cuyas gráficas son la misma recta.

The system $2x + 4y = 8$ and $y = -\frac{1}{2}x + 2$ has infinitely many solutions.

Integers (p. 17)   Whole numbers and their opposites.

Números enteros (p. 17)   Números que constan exclusivamente de una o más unidades, y sus opuestos.

$\ldots -3, -2, -1, 0, 1, 2, 3, \ldots$

Interest period (p. 476)   The length of time over which interest is calculated.

Período de interés (p. 476)   Plazo para el cual se calcula el interés a pagar.

Interpolate (p. NY 748)   The process of using a given data set to estimate the value of a function or measurement between the values already known.

Interpolar (p. NY 748)   Proceso en el que se usa un conjunto dado para estimar el valor de la función o medida entre valores ya conocidos.

Intersection (p. NY 744)   The set of elements that are common to two or more sets.

Intersección (p. NY 744)   El conjunto de elementos que son comunes a dos o más conjuntos.

If $C = \{1, 2, 3, 4\}$ and $D = \{2, 4, 6, 8\}$, then the intersection of C and D, or $C \cap D$, equals $\{2, 4\}$.

Inverse operations (p. 118)   Operations that undo one another.

Operaciones inversas (p. 118)   Las operaciones que se cancelan una a la otra.

Addition and subtraction are inverse operations. Multiplication and division are inverse operations.

Inverse variation (p. 285)   A function that can be written in the form $xy = k$ or $y = \frac{k}{x}$. The product of the quantities remains constant, so as one quantity increases, the other decreases.

Variación inversa (p. 285)   Función que puede expresarse como $xy = k$ ó $y = \frac{k}{x}$. El producto de las cantidades permanece constante, de modo que al aumentar una cantidad, disminuye la otra.

The length x and the width y of a rectangle with a fixed area vary inversely. If the area is 40, $xy = 40$.

Irrational number (p. 18)   A number that cannot be written as a ratio of two integers. Irrational numbers in decimal form are nonterminating and nonrepeating.

Número irracional (p. 18)   Número que no puede expresarse como razón de dos números enteros. Los números irracionales en forma decimal no tienen término y no se repiten.

$\sqrt{11}$ and π are irrational numbers.

L

Leading coefficient   The coefficient of the first term of a polynomial when the polynomial is in standard form.

Coeficiente principal   El coeficiente del primer término de un polinomial cuando el polinomial está en forma general.

5 is the leading coefficient of $5x^2 - 9x + 7$.

-4 is the leading coefficient of $1 - 4n^2 + 7n$.

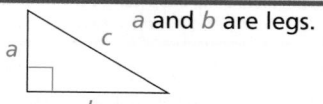

Leg (p. 181)   Each of the sides that form the right angle of a right triangle.

Cateto (p. 181)   Cada uno de los dos lados que forman el ángulo recto en un triángulo rectángulo.

a and *b* are legs.

Like radicals (p. 622)   Radical expressions with the same radicands.

Radicales semejantes (p. 622)   Expresiones radicales con los mismos radicandos.

$3\sqrt{7}$ and $-5\sqrt{7}$ are like radicals.

Like terms (p. 81)   Terms with exactly the same variable factors in a variable expression.

Términos semejantes (p. 81)   Términos con los mismos factores variables en una expresión variable.

$4y$ and $16y$ are like terms.

Line of best fit (p. 351)   The most accurate trend line on a scatter plot showing the relationship between two sets of data.

Recta de mayor aproximación (p. 351)   La línea de tendencia en un diagrama de puntos que más se acerca a los puntos que representan la relación entre dos conjuntos de datos.

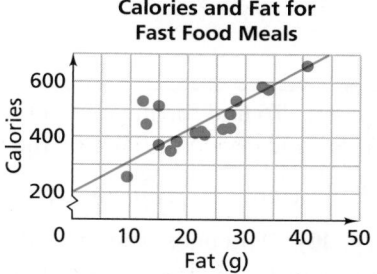

Linear equation (p. 317)   An equation whose graph forms a straight line.

Ecuación lineal (p. 317)   Ecuación cuya gráfica es una línea recta.

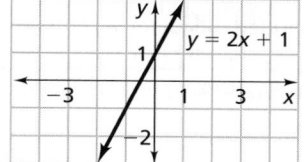

Linear inequality (p. 405)   A mathematical sentence that describes a region of the coordinate plane having a boundary line. Each point in the region is a solution of the inequality.

Desigualdad lineal (p. 405)   Expresión matemática que describe una región del plano de coordenadas que tiene una recta límite. Cada punto de la región es una solución de la desigualdad.

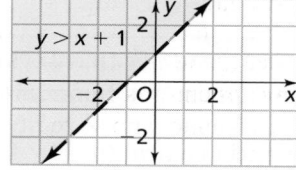

Linear parent function (p. 317)   The simplest form of a linear function.

Función lineal elemental (p. 317)   La forma más simple de una función lineal.

$y = x$

Literal equation (p. 140)   An equation involving two or more variables.

Ecuación literal (p. 140)   Ecuación que incluye dos o más variables.

$4x + 2y = 18$ is a literal equation.

English/Spanish Glossary

M

Matrix (p. 59) A rectangular arrangement of numbers. The number of rows and columns of a matrix determines its size. Each item in a matrix is an element.

Matriz (p. 59) Conjunto de números dispuestos en forma de rectángulo. La cantidad de filas y columnas de una matriz determina su tamaño. Cada cifra de la matriz es un elemento.

$\begin{bmatrix} 2 & 5 & 6.3 \\ -8 & 0 & -1 \end{bmatrix}$ is a 2×3 matrix.

Maximum (p. 551) The y-coordinate of the vertex of a parabola that opens downward.

Valor máximo (p. 551) La coordenada y del vértice en una parábola que se abre hacia abajo.

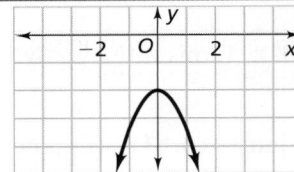

Since the parabola opens downward, the y-coordinate of the vertex is the function's maximum value.

Mean (p. 40) To find the mean of a set of numbers, find the sum of the numbers and divide the sum by the number of items.

The mean is $\dfrac{\text{sum of data items}}{\text{number of data items}}$.

Media (p. 40) Para encontrar la media de un conjunto de números, se suman todos los números y el resultado se divide por la cantidad de números sumados.

La media es $\dfrac{\text{suma de los datos}}{\text{cantidad de números sumados}}$.

In the data set 12, 11, 12, 10, 13, 12, and 7, the mean is

$\dfrac{12 + 11 + 12 + 10 + 13 + 12 + 7}{7} = 11$.

Means of a proportion (p. 143) In the proportion $\frac{a}{b} = \frac{c}{d}$, b and c are the means.

Valores medios de una proporción (p. 143) En la proporción $\frac{a}{b} = \frac{c}{d}$, b y c son los valores medios.

The product of the means of $\frac{x}{4} = \frac{x + 3}{2}$ is $4(x + 3)$ or $4x + 12$.

Measure of central tendency (p. 40) Mean, median, and mode. They are used to organize and summarize a set of data.

Medida de tendencia central (p. 40) La media, la mediana y la moda. Se usan para organizar y resumir un conjunto de datos.

For examples, see mean, median, and mode.

Median (p. 40) The middle value in an ordered set of numbers.

Mediana (p. 40) El valor del medio en un conjunto ordenado de números.

In the data set 7, 10, 11, 12, 12, 12, and 13, the median is 12.

Midpoint (p. 25) The point M that divides a segment $\overline{AB}$ into two equal segments, $\overline{AM}$ and $\overline{MB}$.

Punto medio (p. 25) El punto M que divide un segmento $\overline{AB}$ en dos segmentos iguales, $\overline{AM}$ y $\overline{MB}$.

M is the midpoint of $\overline{XY}$.

X   M   Y

Midpoint Formula (p. 25) The midpoint M of a line segment with endpoints $A(x_1, y_1)$ and $B(x_2, y_2)$ is $\left(\dfrac{x_1 + x_2}{2}, \dfrac{y_1 + y_2}{2}\right)$.

Fórmula del punto medio (p. 25) El punto medio M de un segmento con puntos extremos $A(x_1, y_1)$ y $B(x_2, y_2)$ es $\left(\dfrac{x_1 + x_2}{2}, \dfrac{y_1 + y_2}{2}\right)$.

The midpoint of a segment with endpoints $A(3, 5)$ and $B(7, 1)$ is $(5,3)$.

Minimum (p. 551) The y-coordinate of the vertex of a parabola that opens upward.

Valor mínimo (p. 551) La coordenada y del vértice en una parábola que se abre hacia arriba.

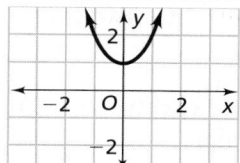

Since the parabola opens upward, the y-coordinate of the vertex is the function's minimum value.

Mode (p. 40) The data item that occurs the greatest number of times in a data set. A data set may have no mode, one mode, or more than one mode.

Moda (p. 40) El dato que ocurre el mayor número de veces en un conjunto de datos. El conjunto de datos puede no tener moda, o tener una o más modas.

In the data set 7, 7, 9, 10, 11, and 13, the mode is 7.

Monomial (p. 494) An expression that is a number, a variable, or a product of a number and one or more variables.

Monomio (p. 494) Expresión algebraica que puede ser un número, una variable o el producto de un número y una o más variables.

$9, n$, and $-5xy^2$ are examples of monomials.

Multiple representations Various ways to present, interpret, communicate, and connect mathematical information and relationships.

Representación múltiple Varias maneras de presentar, interpretar, comunicar y conectar información y relaciones matemáticas.

algebraically, geometrically, graphically, numerically, and verbally

---

Multiplication counting principle (p. 700) If there are m ways to make the first selection and n ways to make the second selection, there are $m \times n$ ways to make the two selections.

Principio de conteo en la multiplicación (p. 700) Si hay m maneras de hacer la primera selección y n maneras de hacer la segunda selección, quiere decir que hay $m \times n$ maneras de hacer las dos selecciones.

For 5 shirts and 8 pairs of shorts, the number of possible outfits is $5 \cdot 8 = 40$.

EXAMPLES

Multiplicative inverse (p. 72)   Given a nonzero rational number $\frac{a}{b}$, the multiplicative inverse, or reciprocal, is $\frac{b}{a}$. The product of a nonzero number and its multiplicative inverse is 1.

$\frac{3}{4}$ is a multiplicative inverse of $\frac{4}{3}$ because $\frac{3}{4} \times \frac{4}{3} = 1$.

Inverso multiplicativo (p. 72)   Dado un número racional $\frac{a}{b}$ distinto de cero, el inverso multiplicativo, o recíproco, es $\frac{b}{a}$. El producto de un número distinto de cero y su inverso multiplicativo es 1.

Mutually exclusive events   Two events that cannot occur at the same time.

Eventos mutuamente excluyentes   Dos eventos que no pueden ocurrir al mismo tiempo.

Natural numbers (p. 17)   The counting numbers.

1, 2, 3, . . .

Números naturales (p. 17)   Los números que se emplean para contar.

Negative correlation (p. 34)   The relationship between two sets of data, in which one set of data decreases as the other set of data increases.

Correlación negativa (p. 34)   Relación entre dos conjuntos de datos en la que uno de los conjuntos disminuye a medida que el otro aumenta.

Negative reciprocal (p. 344)   A number of the form $-\frac{b}{a}$, where $\frac{a}{b}$ is a nonzero rational number. The product of a number and its negative reciprocal is -1.

$\frac{2}{5}$ and $-\frac{5}{2}$ are negative reciprocals because $\left(\frac{2}{5}\right)\left(-\frac{5}{2}\right) = -1$.

Recíproco negativo (p. 344)   El recíproco negativo de un número racional $\frac{a}{b}$ cuyo valor no es cero es $-\frac{b}{a}$. El producto de un número y su recíproco negativo es -1.

Negative square root (p. 176)   A number of the form $-\sqrt{b}$, which is the negative square root of b.

-7 is the negative square root of $\sqrt{49}$.

Raíz cuadrada negativa (p. 176) $-\sqrt{b}$ es la raíz cuadrada negativa de b.

Network (p. 712)   A graph composed of endpoints and segments.

Red (p. 712)   Gráfica compuesta por puntos extremos y segmentos.

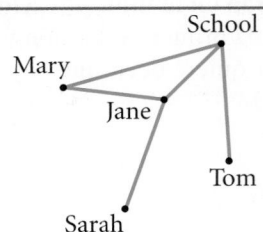

No correlation (p. 34)   There does not appear to be a relationship between two sets of data.

Sin correlación (p. 34)   No hay relación entre dos conjuntos de datos.

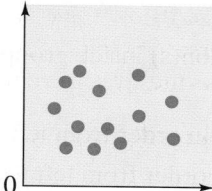

No solution (p. 376)   When the graphs of the equations in a system are parallel with no point of intersection.

Sin solución (p. 376)   Cuando las gráficas de las ecuaciones de un sistema son paralelas y no existe entre ellas ningún punto de intersección.

There is no solution to the system of equations $x + y = 5$ and $x + y = -3$.

Null set (p. NY 739)   A set that has no elements.

Conjunto vacío (p. NY 739)   Conjunto que no tiene elementos.

$\{\}$ or $\varnothing$

Odds (p. 94)   A ratio that describes the likelihood of an event. Odds in favor of an event are number of favorable outcomes : number of unfavorable outcomes.

Probabilidad a favor (p. 94)   Razón que describe la posibilidad de que se produzca un suceso. La probabilidad a favor de que se produzca determinado resultado es la razón del número de resultados favorables : número de resultados no favorables.

You have 3 red marbles and 5 blue marbles. The odds in favor of selecting red are 3 : 5.

Open sentence (p. 5)   An equation that contains one or more variables.

Ecuación abierta (p. 5)   Ecuación que contiene una o más variables.

$5 + x = 12$ is an open sentence.

Opposite (p. 20)   A number that is the same distance from zero on the number line as a given number, but lies in the opposite direction.

Opuestos (p. 20)   Dos números son opuestos si están a la misma distancia del cero en la recta numérica, pero en sentido opuesto.

-3 and 3 are opposites.

Opposite side in a right triangle   *See* **Cosine.**

Lado opuesto de un triángulo rectángulo *Ver* **Cosine.**

Order of operations (p. 10)

1. Perform any operation(s) inside grouping symbols.

2. Simplify powers.

3. Multiply and divide in order from left to right.

4. Add and subtract in order from left to right.

$$6 - (4^2 - [2 \cdot 5]) \div 3$$
$$= 6 - (16 - 10) \div 3$$
$$= 6 - 6 \div 3$$
$$= 6 - 2$$
$$= 4$$

Orden de las operaciones (p. 10)

1. Se hacen las operaciones que están dentro de símbolos de agrupación.

2. Se simplifican todos los términos que tengan exponentes.

3. Se hacen las multiplicaciones y divisiones en orden de izquierda a derecha.

4. Se hacen las sumas y restas en orden de izquierda a derecha.

Ordered pair (p. 24) Two numbers that identify the location of a point.

Par ordenado (p. 24) Un par ordenado de números que denota la ubicación de un punto.

The ordered pair $(4, -1)$ identifies the point 4 units to the right on the x-axis and 1 unit down on the y-axis.

Origin (p. 24) The point at which the axes of the coordinate plane intersect.

Origen (p. 24) Punto de intersección de los ejes del plano de coordenadas.

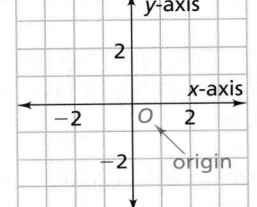

Outcome (p. 93) The result of a single trial in a probability experiment.

Resultado (p. 93) Lo que se obtiene al hacer una sola prueba en un experimento de probabilidad.

The outcomes of rolling a number cube are 1, 2, 3, 4, 5, and 6.

Outlier (p. 40) A data value that is much higher or lower than the other data values in the set.

Valor extremo (p. 40) Valor que es mucho mayor o mucho menor que los demás valores de un conjunto.

For the set of values 2, 5, 3, 7, 12, the data value 12 is an outlier.

P

Parabola (p. 550) The graph of a quadratic function.

Parábola (p. 550) La gráfica de una función cuadrática.

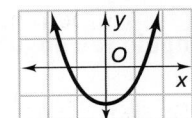

Parallel lines (p. 343) Two lines in the same plane that never intersect. Parallel lines have the same slope.

Rectas paralelas (p. 343) Dos rectas situadas en el mismo plano que nunca se cortan. Las rectas paralelas tienen la misma pendiente.

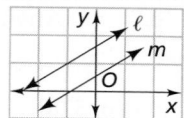

Lines ℓ and m are parallel.

Parameter A quantity or constant whose value varies with the circumstances of its application.

Parámetro Una cantidad o constante cuya valor varía dependiendo de las circunstancias en las que se aplica.

In $y = ax^2$, a is a parameter.

Parent function (p. 317) A family of functions is a group of functions with common characteristics. A parent function is the simplest function with these characteristics.

Función elemental (p. 317) Una familia de funciones es un grupo de funciones con características en común. La función elemental es la función más simple que reúne esas características.

$y = x^2$ is the parent function for the family of quadractic equations of the form $y = ax^2 + bx + c$.

Path (p. 712) A sequence of edges from one vertex to another in a network.

Recorrido (p. 712) Secuencia de aristas desde un vértice hasta otro en una red.

Percent error (p. 170) The ratio of the greatest possible error to the measurement.

Error porcentual (p. 170) La razón del máximo error posible a la medida obtenida.

The diameter of a CD is measured as 12.1 cm. The greatest possible error is 0.05 cm. The percent error is $\frac{0.05}{12.1} \approx 0.4\%$.

Percent of change (p. 168) The ratio of the amount of change to the original amount expressed as a percent.

Porcentaje de cambio (p. 168) La razón de la cantidad de cambio a la cantidad original, expresada como un porcentaje.

The price of a sweater was $20. The price increases $2. The percent of change is $\frac{2}{20} = 10\%$.

Percent of decrease (p. 168) The percent of change found when the original amount decreases.

Porcentaje de disminución (p. 168) El porcentaje de cambio que resulta cuando la cantidad original disminuye.

The price of a sweater was $22. The price decreases $2. The percent of decrease is $\frac{2}{22} \approx 9\%$.

Percent of increase (p. 168) The percent of change found when the original amount increases.

Porcentaje de aumento (p. 168) El porcentaje de cambio que resulta cuando la cantidad original aumenta.

See example for percent of change.

English/Spanish Glossary

Perfect square trinomial (p. 528)   Any trinomial of the form $a^2 + 2ab + b^2$ or $a^2 - 2ab + b^2$.

$(x + 3)^2 = x^2 + 6x + 9$

Trinomio cuadrado perfecto (p. 528)   Todo trinomio de la forma $a^2 + 2ab + b^2$ ó $a^2 - 2ab + b^2$.

Perfect squares (p. 177)   Numbers whose square roots are integers.

The numbers $1, 4, 9, 16, 25, 36, \ldots$ are perfect squares because they are the squares of integers.

Cuadrado perfecto (p. 177)   Número cuya raíz cuadrada es un número entero.

Perimeter (p. 765)   The perimeter of a figure is the distance around the figure. To find the perimeter of a rectangle, find the sum of the lengths of all its sides, or use the formula $P = 2\ell + 2w$.
EXAMPLE   The perimeter of $ABCD$ is 12 ft.

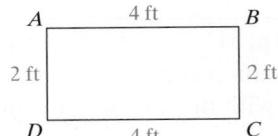

Perímetro (p. 765)   El perímetro de una figura es la suma de las longitudes de sus lados. Para hallar el perímetro de un rectángulo, halla la suma de los largos de todos los lados o usa la fórmula $P = 2\ell + 2w$.

Permutation (p. 701)   An arrangement of some or all of a set of objects in a specific order. You can use the notation $_nP_r$ to express the number of permutations, where n equals the number of objects available and r equals the number of selections to make.

How many ways can 5 children be arranged three at a time?
$_5P_3 = 5 \cdot 4 \cdot 3 = 60$ arrangements

Permutación (p. 701)   Disposición de algunos o de todos los objetos de un conjunto en un orden determinado. El número de permutaciones se puede expresar con la notación $_nP_r$, donde n es igual al número total de objetos y r es igual al número de selecciones que han de hacerse.

Perpendicular lines (p. 344)   Lines that intersect to form right angles. Two lines are perpendicular if the product of their slopes is -1.

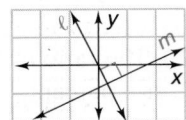

Rectas perpendiculares (p. 344)   Rectas que forman ángulos rectos en su intersección . Dos rectas son perpendiculares si el producto de sus pendientes es -1.

Lines ℓ and m are perpendicular.

Point-slope form (p. 336)   A linear equation of a nonvertical line written as $y - y_1 = m(x - x_1)$. The line passes through the point (x_1, y_1) with slope m.

An equation with a slope of $-\frac{1}{2}$ passing through $(2, -1)$ would be written $y + 1 = -\frac{1}{2}(x - 2)$ in point-slope form.

Forma punto-pendiente (p. 336)   La ecuación lineal de una recta no vertical que pasa por el punto (x_1, y_1) con pendiente m está dada por $y - y_1 = m(x - x_1)$.

Polygon   A polygon is a closed plane figure with at least three sides.

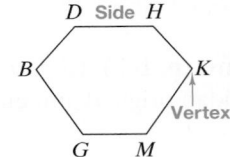

Polígono   Un polígono es una figura plana cerrada formada por tres o más lados.

Polynomial (p. 495) A monomial or the sum or difference of two or more monomials. A quotient with a variable in the denominator is not a polynomial.

$2x^2, 3x + 7, 28,$ and $-7x^3 - 2x^2 + 9$ are all polynomials.

Polinomio (p. 495) Un monomio o la suma o diferencia de dos o más monomios. Un cociente con una variable en el denominador no es un polinomio.

Positive correlation (p. 34) The relationship between two sets of data in which both sets of data increase together.

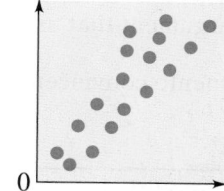

Correlación positiva (p. 34) La relación entre dos conjuntos de datos en la que ambos conjuntos incrementan a la vez.

Power (p. 9) The base and the exponent of an expression of the form a^n.

5^4

Potencia (p. 9) La base y el exponente de una expresión de la forma a^n.

Premise A proposition upon which an argument is based or from which a conclusion is drawn.

Premisa Una afirmación en la cual se basa un argumento o se saca una conclusión.

Prime factorization (p. 754) The prime factorization of a number is the expression of the number as the product of its prime factors.

The prime factorization of 30 is $2 \cdot 3 \cdot 5$.

Descomposición en factores primos (p. 754) La descomposición en factores primos de un número es la expresión de dicho número como el producto de sus factores primos.

Principal square root (p. 176) A number of the form $\sqrt{b}$. The expression $\sqrt{b}$ is called the principal (or positive) square root of b.

5 is the principal square root of $\sqrt{25}$.

Raíz cuadrada principal (p. 176) La expresión $\sqrt{b}$ se llama raíz cuadrada principal (o positiva) de b.

Probability (p. 93) How likely it is that an event will occur (written formally as P(event)).

You have 4 red marbles and 3 white marbles. The probability that you select one red marble, and then, without replacing it, randomly select another red marble is $P(\text{red}) = \frac{4}{7} \cdot \frac{3}{6} = \frac{2}{7}$.

Probabilidad (p. 93) La posibilidad de que un suceso ocurra, escrita formalmente P(suceso).

Product property of proportions In a proportion $\frac{a}{b} = \frac{c}{d}$, the product of the means (b and c) equals the product of the extremes (a and d), or $b \cdot c = a \cdot d$.

Producto de la propiedad de proporciones En la proporción $\frac{a}{b} = \frac{c}{d}$, el producto de los medios (b y c) equivale al producto de los extremos (a y d). Es decir: $b \cdot c = a \cdot d$.

Proof A convincing argument that uses deductive reasoning.

Demostración Argumento convincente que usa razonamiento deductivo.

Properties of equality (p. 118) For all real numbers a, b, and c:

Addition: If $a = b$, then $a + c = b + c$.

Subtraction: If $a = b$, then $a - c = b - c$.

Multiplication: If $a = b$, then $a \cdot c = b \cdot c$.

Division: If $a = b$, and $c \neq 0$, then $\frac{a}{c} = \frac{b}{c}$.

Since $\frac{2}{4} = \frac{1}{2}$, $\frac{2}{4} + 5 = \frac{1}{2} + 5$.

Since $\frac{9}{3} = 3$, $\frac{9}{3} - 6 = 3 - 6$.

Propiedades de la igualdad (p. 118) Para todos los números reales a, b y c:

Suma: Si $a = b$, entonces $a + c = b + c$.

Resta: Si $a = b$, entonces $a - c = b - c$.

Multiplicación: Si $a = b$, entonces $a \cdot c = b \cdot c$.

División: Si $a = b$, y $c \neq 0$, entonces $\frac{a}{c} = \frac{b}{c}$.

Properties of the real numbers Rules that apply to operations with real numbers.

Propiedades de los números reales Reglas que se aplican a operaciones matemáticas con números reales.

Commutative Property
$a + b = b + a$ $\qquad$ $ab = ba$

Associative Property
$a + (b + c) = (a + b) + c$ $\quad$ $a(bc) = (ab)c$

Distributive Property
$a(b + c) = ab + ac$

Identity
$a + 0 = a$ $\qquad$ $a \cdot 1 = a$

Inverse
$a + (-a) = 0$ $\qquad$ $a \cdot \frac{1}{a} = 1$

Zero Property
$a \cdot 0 = 0$

Proportion (p. 143) An equation that states that two ratios are equal.

$\frac{a}{b} = \frac{c}{d}$ where $b \neq 0$ and $d \neq 0$

$\frac{7.5}{9} = \frac{5}{6}$

Proporción (p. 143) Es una ecuación que establece que dos razones son iguales.

$\frac{a}{b} = \frac{c}{d}$ por $b \neq 0$ y $d \neq 0$

Proportional Two variables are proportional if they maintain a constant ratio.

Proporcional Dos variables son proporcionales si mantienen una razón constante.

The perimeter of any square is proportional to the length of one of its sides because the ratio of the length of one side to the perimeter is always 1:4.

Pythagorean Theorem (p. 181) In any right triangle, the sum of the squares of the lengths of the legs is equal to the square of the length of the hypotenuse: $a^2 + b^2 = c^2$.

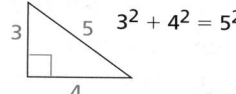

$3^2 + 4^2 = 5^2$

Teorema de Pitágoras (p. 181) En un triángulo rectángulo, la suma de los cuadrados de los catetos es igual al cuadrado de la hipotenusa: $a^2 + b^2 = c^2$.

Quadrants (p. 24) The four parts into which the coordinate plane is divided by its axes.

Cuadrantes (p. 24) El plano de coordenadas está dividido por sus ejes en cuatro regiones llamadas cuadrantes.

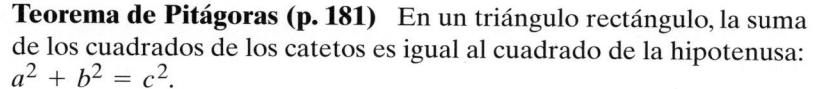

Quadratic equation (p. 566) An equation you can write in the standard form $ax^2 + bx + c = 0$, where $a \neq 0$.

$4x^2 + 9x - 5 = 0$

Ecuación cuadrática (p. 566) Ecuación que puede expresarse de la forma normal como $ax^2 + bx + c = 0$, en la que $a \neq 0$.

Quadratic formula (p. 585) If $ax^2 + bx + c = 0$ and $a \neq 0$, then

$x = \frac{-b \pm \sqrt{b^2 - 4ac}}{2a}$.

Fórmula cuadrática (p. 585) Si $ax^2 + bx + c = 0$ y $a \neq 0$, entonces

$x = \frac{-b \pm \sqrt{b^2 - 4ac}}{2a}$.

$2x^2 + 10x + 12 = 0$

$x = \frac{-b \pm \sqrt{b^2 - 4ac}}{2a}$

$x = \frac{-10 \pm \sqrt{10^2 - 4(2)(12)}}{2(2)}$

$x = \frac{-10 \pm \sqrt{4}}{4}$

$x = \frac{-10 + 2}{4}$ or $\frac{-10 - 2}{4}$

$x = -2$ or -3

Quadratic function (p. 550)   A function of the form $y = ax^2 + bx + c$, where $a \neq 0$. The graph of a quadratic function is a parabola, a U-shaped curve that opens up or down.

$y = 5x^2 - 2x + 1$ is a quadratic function.

Función cuadrática (p. 550)   La función $y = ax^2 + bx + c$, en la que $a \neq 0$. La gráfica de una función cuadrática es una parábola, o curva en forma de U que se abre hacia arriba o hacia abajo.

Quartiles (p. NY 733)   Quartiles are numbers that divide a data set into four equal parts.

See **Box-and-whisker plot**.

Cuartiles (p. NY 733)   Los cuartiles son números que dividen un conjunto de datos en cuatro partes iguales.

R

Radical   The root of a quantity as indicated by the radical sign.

Radical   La raíz de un número así como lo indica el signo radical.

Radical equation (p. 629)   An equation that has a variable in a radicand.

$\sqrt{x} - 2 = 12$
$\sqrt{x} = 14$

Ecuación radical (p. 629)   Ecuación que tiene una variable en un radicando.

Radical expression (p. 616)   Expression that contains a radical.

$\sqrt{3}$, $\sqrt{5x}$, and $\sqrt{x - 10}$ are examples of radical expressions.

Expresión radical (p. 616)   Expresiones que contienen radicales.

Radicand (p. 176)   The expression under the radical sign.

The radicand of the radical expression $\sqrt{x + 2}$ is $x + 2$.

Radicando (p. 176)   La expresión que aparece debajo del signo radical.

Random event (p. 107)   An event in which you cannot predict individual outcomes.

Suceso aleatorio (p. 107)   Suceso para el cual no se pueden predecir resultados individuales.

Range (p. 42)   The difference between the greatest and the least data values for a set of data.

For the set 2, 5, 8, 12, the range is $12 - 2 = 10$.

Rango (p. 42)   Diferencia entre el valor mayor y el menor en un conjunto de datos.

Range (p. 29) The possible values of the output, or dependent variable, of a function.

Rango (p. 29) El conjunto de todos los valores posibles de salida, o variable dependiente, de una función.

In the function $y = |x|$, the range is the set of all nonnegative numbers.

Rate (p. 142) A ratio of a to b where a and b represent quantities measured in different units.

Razón (p. 142) La relación que existe entre a y b cuando a y b son cantidades medidas con distintas unidades.

Traveling 125 miles in 2 hours results in the rate $\frac{125 \text{ miles}}{2 \text{ hours}}$ or 62.5 mi/h.

Rate of change (p. 308) The relationship between two quantities that are changing. The rate of change is also called slope.

$$\text{rate of change} = \frac{\text{change in the dependent variable}}{\text{change in the independent variable}}$$

Tasa de cambio (p. 308) La relación entre dos cantidades que cambian. La tasa de cambio se llama también pendiente.

$$\text{tasa de cambio} = \frac{\text{cambio en la variable dependiente}}{\text{cambio en la variable independiente}}$$

Video rental for 1 day is $1.99. Video rental for 2 days is $2.99.

$$\text{rate of change} = \frac{2.99 - 1.99}{2 - 1}$$
$$= \frac{1.00}{1}$$
$$= 1$$

Ratio (p. 142) A comparison of two numbers by division.

Razón (p. 142) Comparación de dos números por división.

$\frac{5}{7}$ and $7:3$ are ratios.

Rational coefficient A coefficient that is a rational number.

Coeficiente racional Un coeficiente que es un número racional.

Rational equation (p. 692) An equation containing rational expressions.

Ecuación racional (p. 692) Ecuación que contiene expresiones racionales.

$\frac{1}{x} = \frac{3}{2x - 1}$ is a rational equation.

Rational expression (p. 672) A ratio of two polynomials. The value of the variable cannot make the denominator equal to 0.

Expresión racional (p. 672) Una razón de dos polinomios. El valor de la variable no puede hacer el denominador igual a 0.

$\frac{3}{x^3 + x}$ when $x \neq 0$

Rational function (p. 664) A function that can be written in the form $f(x) = \frac{\text{polynomial}}{\text{polynomial}}$. The value of the variable cannot make the denominator equal to 0.

Función racional (p. 664) Función que puede expresarse de forma $f(x) = \frac{\text{polinomio}}{\text{polinomio}}$. El valor de la variable no puede hacer el denominador igual a 0.

$y = \frac{x}{x^2 + 2}$

Rational number (p. 17) A real number that can be written as a ratio of two integers. Rational numbers in decimal form are terminating or repeating.

$\frac{2}{3}$, 1.548, and 2.292929 . . . are all rational numbers.

Número racional (p. 17) Número real que puede expresarse como la razón de dos números enteros. Los números racionales en forma decimal son exactos o periódicos.

Rationalize (p. 619) Rewrite as a rational number. Rationalizing the denominator of a radical expression may be necessary to obtain the simplest radical form.

$$\frac{2}{\sqrt{5}} = \frac{2}{\sqrt{5}} \cdot \frac{\sqrt{5}}{\sqrt{5}} = \frac{2\sqrt{5}}{\sqrt{25}} = \frac{2\sqrt{5}}{5}$$

Racionalizar (p. 619) Escribir una expresión matemática en forma de número racional. A veces es necesario racionalizar el denominador de una expresión radical a fin de obtener la forma radical más simple.

Real number (p. 18) A number that is either rational or irrational.

5, -3, $\sqrt{11}$, 0.666 . . . , $5\frac{4}{11}$, 0, and π are all real numbers.

Número real (p. 18) Un número que no es racional ni irracional.

Reciprocal (p. 73) Given a nonzero rational number $\frac{a}{b}$, the reciprocal, or multiplicative inverse, is $\frac{b}{a}$. The product of a nonzero number and its reciprocal is 1.

$\frac{2}{5}$ and $\frac{5}{2}$ are reciprocals because $\frac{2}{5} \times \frac{5}{2} = 1$.

Recíproco (p. 73) El recíproco, o inverso multiplicativo, de un número racional $\frac{a}{b}$ cuyo valor no es cero es $\frac{b}{a}$. El producto de un número que no es cero y su valor recíproco es 1.

Rectangular coordinates An ordered pair of real numbers that establishes the location of a point in a coordinate plane using the distances from two perpendicular intersecting lines called the coordinate axes.

$(5, -3)$

Coordenadas rectangulares Un par de ordenados de números reales que establece la ubicación de un punto en un plano coordenado usando la distancia a rectas perpendiculares, llamadas ejes.

Regular polygon A polygon which is both equilateral and equiangular.

Polígono regular Un polígono que es equilateral y equiangular.

Relation (p. 257) Any set of ordered pairs.

$\{(0, 0), (2, 3), (2, -7)\}$ is a relation.

Relación (p. 257) Cualquier conjunto de pares ordenados.

Relative error   The ratio of the absolute error in a measurement to the size of the measurement; often written as a percent and called the percent of error.

Error relativo   La razón de un error absoluto en una medida a el tamaño de la medida; a menudo se escribe como un porcentaje y se denomina el porcentaje de error.

Right angle   A right angle is an angle with a measure of 90°.
EXAMPLE $\angle CDE$ is a right angle.

Ángulo recto   Un ángulo recto es un ángulo que mide 90°.

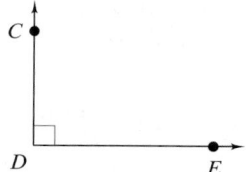

Right triangle   A right triangle is a triangle with one right angle.
EXAMPLE $\triangle ABC$ is a right triangle, since $\angle B$ is a right angle.

Triángulo rectángulo   Un triángulo rectángulo es aquél que posee un ángulo recto.

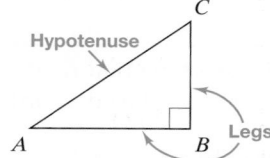

Roots of the equation (p. 566)   The solutions of an equation.

Raíces de la ecuación (p. 566)   Las soluciones de una ecuación.

Roster form (p. NY 738)   A notation for listing all of the elements in a set using set braces and commas.

Lista (p. NY 738)   Una notación en la que se enlistan todos los elementos en un conjunto usando llaves y comma.

The set of prime numbers less than 10, expressed in roster form is {2, 3, 5, 7}.

Sample space (p. 93)   All possible outcomes of an event.

Espacio de muestra (p. 93)   Todos los resultados posibles de un suceso.

When tossing two coins one at a time, the sample space is (H, H), (T, T), (H, T), (T, H).

Scale (p. 151)   The ratio of a distance in a drawing to the actual distance.

Escala (p. 151)   La razón entre la distancia expresada en un dibujo y la distancia real.

For a drawing in which a 2-in. length represents an actual length of 18 ft, the scale is 1 in. : 9 ft.

Scale drawing (p. 151) An enlarged or reduced drawing similar to an actual object or place.

Dibujo a escala (p. 151) Dibujo que muestra de mayor o menor tamaño un objeto o lugar dado.

Scale factor (p. 150) The ratio of the dimensions of a similar image of a figure to its original dimensions.

Factor de escala (p. 150) La razón de las dimensiones de una imagen semejante de una figura a sus dimensiones originales.

Scatter plot (p. 33) A graph that relates data of two different sets. The two sets of data are displayed as ordered pairs.

Diagrama de puntos (p. 33) Gráfica que muestra la relación entre dos conjuntos. Los datos de ambos conjuntos se presentan como pares ordenados.

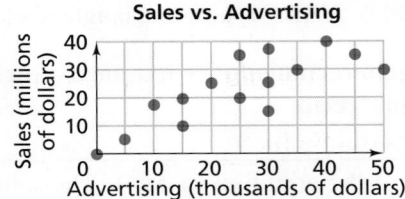

The scatter plot displays the amount spent on advertising (in thousands of dollars) versus product sales (in millions of dollars).

Scientific notation (p. 436) A number expressed in the form $a \times 10^n$, where n is an integer and $1 < a < 10$.

Notación científica (p. 436) Un número expresado en forma de $a \times 10^n$, donde n es un número entero y $1 < a < 10$.

3.4×10^6

Sector of a circle A region bounded by an arc of the circle and the two radii to the endpoints of the arc.

Sector circular Región de un círculo abarcada por dos rayos y el arco entre ellos.

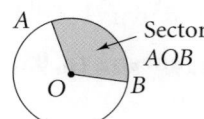

Semicircle Either of the arcs of a circle determined by the endpoints of a diameter.

Semicírculo Cualquiera de los arcos de un círculo determinado por el extremo del diámetro.

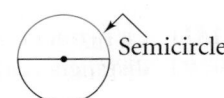

Sequence (p. 293) A number pattern.

Progresión (p. 293) Sucesión de números.

$-4, 5, 14, 23$ is a sequence.

Set (p. NY 738)   A well-defined collection of elements.

Conjunto (p. NY 738)   Un grupo bien definido de elementos.

The set of integers:

$Z = \{\ldots, -3, -2, -1, 0, 1, 2, 3, \ldots\}$

Set-builder notation (p. NY 738)   A notation used to describe the elements of a set.

Notación conjuntista (p. NY 738)   Notación que se usa para describir los elementos de un conjunto.

The set of all positive real numbers in set builder notation is $\{x : x \in R$ and $x > 0\}$. This is read as "the set of all values of x such that x is a real number and x is greater than 0."

Similar figures (p. 150)   Figures that have the same shape, but not necessarily the same size.

Figuras semejantes (p. 150)   Figuras que tienen la misma forma pero no necesariamente el mismo tamaño.

$\triangle DEF$ and $\triangle GHI$ are similar.

Simple interest (p. 167)   Interest paid only on the principal.

Interés simple (p. 167)   Intéres basado en el capital solamente.

The interest on $1000 at 6% for 5 years is $1000(0.06)5 = $300.

Simplest form   An expression that has been rewritten as simply as possible using the rules of arithmetic and algebra.

Mínima expresión   Expresión escrita de la manera más sencilla posible usando reglas aritméticas y algebraicas.

Simplify (p. 9)   Replace an expression with its simplest name or form.

Simplificar (p. 9)   Reemplazar una expresión por su versión o forma más simple.

$\dfrac{3 + 5}{8}$

Simulation (p. 100)   A simulation is a model of a real-world situation used to find probability.

Simulación (p. 100)   Una simulación es un modelo de una situación real que se usa para hallar la probabilidad.

A baseball team has an equal chance of winning or losing its next game. You can toss a coin to simulate the situation.

Sine (p. 646)   In a right triangle, such as $\triangle ABC$, with right $\angle C$, sine of $\angle A = \dfrac{\text{length of side opposite } \angle A}{\text{length of hypotenuse}}$ or $\sin A = \dfrac{a}{c}$.

Seno (p. 646)   En un triángulo rectángulo tal que $\triangle ABC$ con $\angle C$ recto, el seno de $\angle A = \dfrac{\text{longitud del lado opuesto a } \angle A}{\text{longitud de la hipotenusa}}$, o sen $A = \dfrac{a}{c}$.

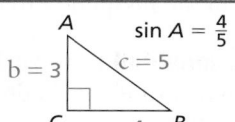

Slope (p. 310)   The ratio of the vertical change to the horizontal change.

$$\text{slope} = \frac{\text{vertical change}}{\text{horizontal change}} = \frac{y_2 - y_1}{x_2 - x_1}, \text{where } x_2 - x_1 \neq 0$$

Pendiente (p. 310)   La razón del cambio vertical al cambio horizontal.

$$\text{pendiente} = \frac{\text{cambio vertical}}{\text{cambio horizontal}} = \frac{y_2 - y_1}{x_2 - x_1}, \text{donde } x_2 - x_1 \neq 0$$

The slope of the line below is $\frac{2}{4} = \frac{1}{2}$.

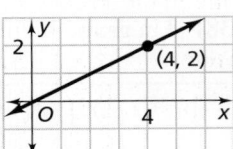

Slope-intercept form (p. 318)   A linear equation of a nonvertical line written as $y = mx + b$, where m is the slope and b is the y-intercept.

Forma pendiente-intercepto (p. 318)   La ecuación lineal de una recta no vertical expresada como $y = mx + b$, donde m es la pendiente y b es el intercepto en y.

$y = 8x + 2$

Solution of a system of linear equations (p. 374)   Any ordered pair in a system that makes all the equations of that system true.

Solución de un sistema de ecuaciones lineales (p. 374)   Todo par ordenado de un sistema que hace verdaderas todas las ecuaciones de ese sistema.

$(2, 1)$ is a solution of the system
$y = 2x - 3$
$y = x - 1$
because the ordered pair makes each equation true.

Solution of a system of linear inequalities (p. 411)   Any ordered pair that makes all of the inequalities in the system true.

Solución de un sistema de desigualdades lineales (p. 411)   Todo par ordenado que hace verdaderas todas las desigualdades del sistema.

The shaded purple area shows the solution of the system
$y > 2x - 5$
$3x + 4y < 12$.

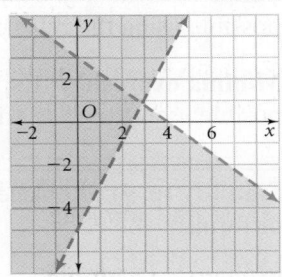

Solution of an equation (p. 118)   Any value or values that make an equation true.

Solución de una ecuación (p. 118)   Cualquier valor o valores que hagan verdadera una ecuación.

In the equation $y + 22 = 11, -11$ is the solution.

Solution of an inequality (one variable) (p. 200)   Any value or values of a variable in the inequality that makes an inequality true.

Solución de una desigualdad (una variable) (p. 200)   Cualquier valor o valores de una variable de la desigualdad que hagan verdadera la desigualdad.

The solution of the inequality $x < 9$ is all numbers less than 9.

Solution of an inequality (two variables) (p. 405)   Any ordered pair that makes the inequality true.

Solución de una desigualdad (dos variables) (p. 405)   Cualquier par ordenado que haga verdadera la desigualdad.

Each ordered pair in the pink area and on the solid pink line is a solution of $3x - 5y \leq 10$.

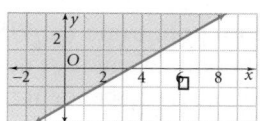

Solution set   Any and all value(s) of the variable(s) that satisfy an equation, inequality, system of equations, or system of inequalities.

Conjunto solución   Conjunto de todos los valores de las variables que satisfacen determinada ecuación, desigualdad, sistema de ecuaciones o sistema de desigualdades.

Square root (p. 176)   A number b such that $a^2 = b$. $\sqrt{b}$ is the principal square root. $-\sqrt{b}$ is the negative square root.

Raíz cuadrada (p. 176)   Si $a^2 = b$, entonces a es la raíz cuadrada de b. $\sqrt{b}$ es la raíz cuadrada principal. $-\sqrt{b}$ es la raíz cuadrada negativa.

-3 and 3 are square roots of 9.

Square root function (p. 638)   A function that contains the independent variable in the radicand.

Función de raíz cuadrada (p. 638)   Una función que contiene la variable independiente en el radicando.

$y = \sqrt{2x}$ is a square root function.

Square unit   The amount of area covered by a square with side length one unit long.

Unidades cuadradas   La cantidad de área que abarca un cuadrado cuyo lado mide una unidad.

Standard deviation (p. 636)   A measure of how data varies, or deviates, from the mean.

Desviación típica (p. 636)   Medida de cómo los datos varían, o se desvían, de la media.

Use the following formula to find the standard deviation.

$$\sigma = \sqrt{\frac{\Sigma(x - \bar{x})^2}{n}}$$

Standard form of a linear equation (p. 330)   The form of a linear equation $Ax + By = C$, where A, B, and C are real numbers and A and B are not both zero.

Forma normal de una ecuación lineal (p. 330)   La forma normal de una ecuación lineal es $Ax + By = C$, donde A, B y C son números reales, y donde A y B no son iguales a cero.

$6x - y = 12$

English/Spanish Glossary

Standard form of a polynomial (p. 495)   The form of a polynomial in which the degree of the terms decreases from left to right (also *descending order*).

$$15x^3 + x^2 + 3x - 9$$

Forma normal de un polinomio (p. 495)   Cuando el grado de los términos de un polinomio disminuye de izquierda a derecha, está en forma normal, o en orden descendente.

Standard form of a quadratic equation (p. 566)   The form of a quadratic equation written $ax^2 + bx + c = 0$.

$$-x^2 + 2x + 9 = 0$$

Forma normal de una ecuación cuadrática (p. 566)   Cuando una ecuación cuadrática se expresa de forma $ax^2 + bx + c = 0$.

Standard form of a quadratic function (p. 550)   The form of a quadratic function written $y = ax^2 + bx + c$, where $a \neq 0$.

$$= 2x^2 - 5x + 2$$

Forma normal de una función cuadrática (p. 550)   Cuando una ecuación cuadrática se expresa como $y = ax^2 + bx + c$, donde $a \neq 0$.

Stem-and-leaf plot (p. 42)   A display of data made by using the digits of the values.

Diagrama de tallo y hojas (p. 42)   Un arreglo de los datos que use los dígitos de los valores.

Subset (p. NY 739)   A set consisting of elements from a given set; it may be the empty set.

If $B = \{1, 2, 3, 4, 5, 6, 7\}$ and $A = \{1, 2, 5\}$, then A is a subset of B.

Subconjunto (p. NY 739)   Un conjunto que consiste en los elementos de un conjunto dado; puede ser un conjunto vacío.

Substitution method (p. 382)   A method of solving a system of equations by replacing one variable with an equivalent expression containing the other variable.

$$y = 2x + 5$$
$$x + 3y = 7$$
$$x + 3(2x + 5) = 7$$

Método de sustitución (p. 382)   Método para resolver un sistema de ecuaciones en el que se reemplaza una variable por una expresión equivalente que contenga la otra variable.

Substitution property   Any quantity can be replaced by an equal quantity.

If $a \times b = x$ and $x = c$, then $a \times b = c$.

Propiedad de sustitución   Cualquier cantidad que puede ser reemplazada por una cantidad equivalente.

Surface area Surface area is the sum of the areas of the base(s) and lateral faces of a space figure.

EXAMPLE The surface area of the prism is the sum of the areas of its faces.

$$(12 + 12 + 12 + 12 + 9 + 9) \text{ in.}^2 = 66 \text{ in.}^2$$

Each square = 1 in.^{2}

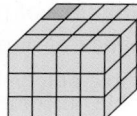

Área total El área total es la suma de las áreas de la o las bases y de las caras laterales de una figura tridimensional.

Survey (p. 105) A gathering of facts or opinions by asking people questions through an interview or questionnaire.

Encuesta (p. 105) Preguntas hechas a personas con el fin de recopilar hechos u opiniones por medio de entrevistas o cuestionarios.

System of linear equations (p. 374) Two or more linear equations using the same variables.

$$y = 5x + 7, y = \frac{1}{2}x - 3$$

Sistema de ecuaciones lineales (p. 374) Dos o más ecuaciones lineales que usen las mismas variables.

System of linear inequalities (p. 411) Two or more linear inequalities using the same variables.

$$y \leq x + 11, y < 5x$$

Sistema de desigualdades lineales (p. 411) Dos o más desigualdades lineales que usen las mismas variables.

Tangent (p. 646) In a right triangle, such as $\triangle ABC$ with right $\angle C$, tangent of $\angle A = \dfrac{\text{length of side opposite } \angle A}{\text{length of side adjacent to } \angle A}$, or $\tan A = \dfrac{a}{b}$.

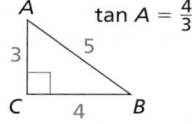

Tangente (p. 646) En un triángulo rectángulo tal que $\triangle ABC$, con $\angle C$ recto, la tangente de $\angle A = \dfrac{\text{longitud del lado opuesto a } \angle A}{\text{longitud del lado adyacente a } \angle A}$, o la $\tan A = \dfrac{a}{b}$.

Term (p. 81) A number, variable, or the product or quotient of a number and one or more variables.

The expression $5x + \frac{y}{2} - 8$ has three terms: $5x, \frac{y}{2}$, and -8.

Término (p. 81) Un número, una variable o el producto o cociente de un número y una o más variables.

Term of a sequence (p. 293) Any number in a sequence.

-4 is the first term of the sequence $-4, 5, 14, 23$.

Término de una progresión (p. 293) Todos los números de una progresión.

English/Spanish Glossary

EXAMPLES

Theoretical probability (p. 93)   The ratio of the number of favorable outcomes to the number of possible outcomes if all outcomes have the same chance of happening.

$$P(\text{event}) = \frac{\text{number of favorable outcomes}}{\text{number of possible outcomes}}$$

Probabilidad teórica (p. 93)   Si cada resultado tiene la misma probabilidad de darse, la probabilidad teórica de un suceso se calcula como la razón del número de resultados favorables al número de resultados posibles.

$$P(\text{suceso}) = \frac{\text{número de resultados favorables}}{\text{número de resultados posibles}}$$

In tossing a coin, the probabilities of getting a head or tail are equally likely. The likelihood of getting a head is $P(\text{head}) = \frac{1}{2}$.

Translation (p. 359)   A transformation that shifts a graph horizontally, vertically, or both.

Traslación (p. 359)   Proceso de mover una gráfica horizontalmente, verticalmente o en ambos sentidos.

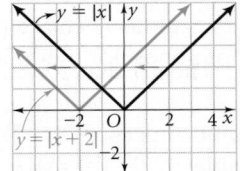

$y = |x + 2|$ is a translation of $y = |x|$.

Trend line (p. 341)   A line on a scatter plot drawn near the points. It shows a correlation.

Línea de tendencia (p. 341)   Línea de un diagrama de puntos que se traza cerca de los puntos para mostrar una correlación.

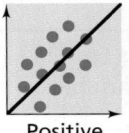 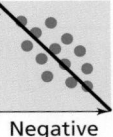
Positive   Negative

Trigonometric ratios (p. 646)   The ratios of the sides of a right triangle. *See* **Cosine**, **Sine**, and **Tangent**.

Razones trigonométricas (p. 646)   Las razones de los lados de un triángulo rectángulo. *Ver* **Cosine**, **Sine,** and **Tangent**.

Trigonometry   Trigonometry is a branch of mathematics involving triangle measures.

Trigonometría   La trigonometría es la rama de las matemáticas que se relaciona con la medida de triángulos.

Trinomial (p. 495)   A polynomial of three terms.

Trinomio (p. 495)   Polinomio compuesto de tres términos.

$3x^2 + 2x - 5$

Undefined   An expression in mathematics which does not have meaning and therefore is not assigned a value.

Indefinido   Expresión matemática que no tiene sentido y por ende no tiene un valor asignado.

When $x = 4$, the expression $\dfrac{x + 3}{x - 4}$ is undefined.

When $x < 2$, the expression $\sqrt{x - 2}$ is undefined in the Real numbers.

Uniform motion (p. 159)   The motion of an object moving at a constant rate.

Movimiento uniforme (p. 159)   El movimiento de un objeto que se mueve a una velocidad constante.

Union (p. NY 743)   The set that contains all of the elements of two or more sets.

Unión (p. NY 743)   El conjunto que contiene todos los elementos de dos o más conjuntos.

If $A = \{1, 3, 6, 9\}$ and $B = \{1, 5, 10\}$, then the union of A and B, or $A \cup B$, equals $\{1, 3, 5, 6, 9, 10\}$.

Unit analysis (p. 142)   The process of selecting conversion factors to produce the appropriate units.

Análisis de unidades (p. 142)   Proceso de seleccionar factores de conversión para producir las unidades apropiadas.

To change ten feet to yards, multiply by the conversion factor $\dfrac{1 \text{ yd}}{3 \text{ ft}}$.
$10 \text{ ft}\left(\dfrac{1 \text{ yd}}{3 \text{ ft}}\right) = 3\frac{1}{3} \text{ yd}$

Unit rate (p. 142)   A rate with a denominator of 1.

Razón en unidades (p. 142)   Razón cuyo denominador es 1.

The unit rate for 120 miles driven in 2 hours is 60 mi/h.

Univariate (p. NY 727)   A set of data involving only one variable.

Univariante (p. NY 727)   Conjunto que involucra sólo una variable.

Universe   The set of all possible specified elements from which subsets are formed. Also known as the universal set.

Universo   Conjunto de todos los posibles elementos específicos del cual se forma un subconjunto. También se conoce como conjunto universal.

Unlike radicals (p. 622)   Radical expressions that do not have the same radicands.

Radicales no semejantes (p. 622)   Expresiones radicales que no tienen radicandos semejantes.

$\sqrt{2}$ and $\sqrt{3}$ are unlike radicals.

English/Spanish Glossary

Valid argument A logical argument supported by known facts or assumed axioms; an argument in which the premise leads to a conclusion.

Argumento válido Argumento lógico apoyado por hechos o axiomas; sun argumento en el que la premisa lleva a una conclusión.

Variable (p. 4) A symbol, usually a letter, that represents one or more numbers.

Variable (p. 4) Símbolo, generalmente una letra, que representa uno o más valores.

x is a variable in the equation $9 - x = 3$.

Venn diagram (p. NY 744) A Venn diagram is a diagram that uses regions, usually circles, to show how sets of numbers or objects are related.

Diagrama de Venn (p. NY 744) Un diagrama de Venn es un diagrama que usa regiones, generalmente círculos, para mostrar cómo se relacionan conjuntos de números u objetos.

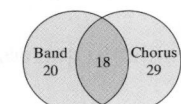

Vertex (p. 551) The highest or lowest point on a parabola. The axis of symmetry intersects the parabola at the vertex.

Vértice (p. 551) El punto más alto o más bajo de una parábola. El punto de intersección del eje de simetría y la parábola.

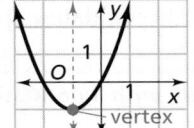

Vertex of an angle (p. 135) The vertex of an angle is the point of intersection of two sides of an angle or figure.

Vértice de un ángulo (p. 135) El vértice de un ángulo es el punto de intersección de dos lados de un ángulo o de una figura.

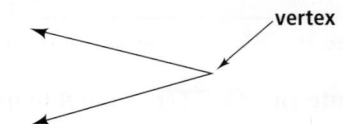

Vertex of a polygon A point where the edges of a polygon intersect.

Vértice de un polígono El vértice de un polígono es cualquier punto donde se encuentran dos lados de un polígono.

Vertical-line test (p. 258) A method used to determine if a relation is a function or not. If a vertical line passes through a graph more than once, the graph is not the graph of a function.

Prueba de la recta vertical (p. 258) Método que permite determinar si una relación es o no es una función. Si una recta vertical corta la gráfica más de una vez, la gráfica no es de función.

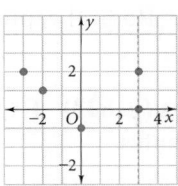

A line would pass through $(3, 0)$ and $(3, 2)$, so the relation is not a function.

Visualization   A mental image based on a given description.

Visualización   Imagen mental basada en una descripción dada.

Volume (p. 765)   The volume of a space figure is the number of cubic units needed to fill it.
EXAMPLE The volume of the rectangular prism is 36 in.^{3}.

Volumen (p. 765)   El volumen de una figura tridimensional es el número de unidades cúbicas necesarias para llenar el espacio interior de dicha figura.

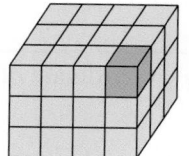

Each cube is 1 in.^{3}.

Whole numbers (p. 17)   The nonnegative integers.

$0, 1, 2, 3, \ldots$

Números cardinales (p. 17)   Todos los números enteros que no son negativos.

x-axis (p. 24)   The horizontal axis of the coordinate plane.

Eje x (p. 24)   El eje horizontal del plano de coordenadas.

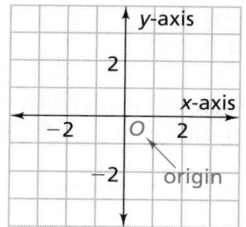

x-coordinate (p. 24)   The location on the x-axis of a point in the coordinate plane.

Coordenada x (p. 24)   La ubicación de un punto sobre el eje x en el plano de coordenadas.

In the ordered pair $(4, -1)$, 4 is the x-coordinate.

x-intercept (p. 330)   The x-coordinate of the point where a line crosses the x-axis.

Intercepto en x (p. 330)   La coordenada x del punto donde una recta corta el eje x.

The x-intercept of $3x + 4y = 12$ is 3.

y-axis (p. 24)   The vertical axis of the coordinate plane.

Eje y (p. 24)   El eje vertical del plano de coordenadas.

English/Spanish Glossary

y-coordinate (p. 24) The location on the y-axis of a point in the coordinate plane.

In the ordered pair $(4, -1)$, -1 is the y-coordinate.

Coordenada y (p. 24) La ubicación de un punto sobre el eje y en el plano de coordenadas.

y-intercept (p. 317) The y-coordinate of the point where a line crosses the y-axis.

The y-intercept of $y = 5x + 2$ is 2.

Intercepto en y (p. 317) La coordenada y del punto donde una recta corta el eje y.

z-coordinate The third coordinate in any (x, y, z) ordered triple; the number represents how many units the point is located above or below the xy-plane.

Coordenada z La tercera coordenada en cualquier ordenado triple (x, y, z); el número representa cuantas unidades abajo o arriba del plano-xy se ubica el punto.

Zero-product property (p. 572) For all real numbers a and b, if $ab = 0$, then $a = 0$ or $b = 0$.

$$x(x + 3) = 0$$
$$x = 0 \text{ or } x + 3 = 0$$
$$x = 0 \text{ or } x = -3$$

Propiedad del producto cero (p. 572) Para todos los números reales a y b, si $ab = 0$, entonces $a = 0$ ó $b = 0$.

Zeros of the function (p. 566) The x-intercepts of the graph of a function.

The zeros of $y = x^2 - 4$ are ± 2.

Ceros de la función (p. 566) En la gráfica de una función, cada intercepto en x representa un cero.

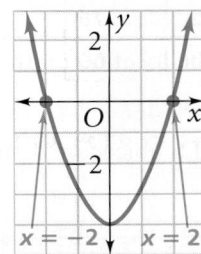

Chapter 1

Check Your Readiness   p. 2

1. $\frac{4}{5}$ **2.** $\frac{5}{7}$ **3.** $\frac{3}{7}$ **4.** $\frac{1}{7}$ **5.** $\frac{12}{13}$ **6.** $\frac{7}{24}$ **7.** $\frac{6}{11}$ **8.** $1\frac{9}{20}$ **9.** $\frac{23}{39}$
10. $12\frac{23}{40}$ **11.** $4\frac{7}{12}$ **12.** $1\frac{5}{6}$ **13.** $13\frac{13}{28}$ **14.** 9^5 **15.** $5 \cdot 7^6$
16. $2^2 \cdot 3^6$ **17.** $\approx$2,650,000 voters **18.** Texas; $\approx$1,000,000 voters

Lesson 1-1   pp. 4–6

Check Skills You'll Need 1. Reasonable **2.** Incorrect **3.** Incorrect **4.** Reasonable

Quick Check 1a. $\frac{4 \cdot 2}{c}$ **b.** $t - 15$ **2.** Let n be the number. **a.** $n - 9$ **b.** $2n + 31$
3a. $c = 15n$ **b.** Each CD costs $10.99.
4. Answers may vary. Sample: $e =$ money earned, $s =$ money saved, $s = \frac{1}{2}e$

Lesson 1-2   pp. 9–12

Check Skills You'll Need 1. 4 **2.** 3 **3.** 1 **4.** 4 **5.** 7 **6.** 4 **7.** 8 **8.** 60 **9.** 35 **10.** 9 **11.** 18 **12.** 36

Quick Check 1a. 4 **b.** 10 **c.** 29 **d.** 134 **2a.** 3 **b.** 8 **c.** 6 **d.** 45
3.

Price p	$p + 0.05p$	Cost c
$5	5 + 0.05(5)	$5.25
$10	10 + 0.05(10)	$10.50
$15	15 + 0.05(15)	$15.75
$20	20 + 0.05(20)	$21.00

$5.25, $10.50, $15.75, $21.00
4a. 26 **b.** 10.5 **5a.** 1764 **b.** 1134 **c.** 15,876 **6a.** 95 **b.** 21 **c.** 29 **7.** 63,000 ft^2

Lesson 1-3   pp. 17–20, 23

Check Skills You'll Need 1. $\frac{1}{2}$ **2.** $\frac{1}{20}$ **3.** $\frac{13}{4}$ **4.** $\frac{13}{40}$ **5.** 0.4 **6.** 0.375 **7.** $0.\overline{6}$ **8.** $3.\overline{5}$

Quick Check 1a. integers, rational numbers **b.** rational numbers **c.** rational numbers **d.** natural numbers, whole numbers, integers, rational numbers **2.** rational numbers **3a.** true **b.** False; answers may vary. Sample: $\frac{3}{1} = 3$ is a whole number. **4.** $-\frac{2}{3}, -\frac{5}{8}, \frac{1}{12}$ **5a.** 5 **b.** 4 **c.** 3.7 **d.** $\frac{5}{7}$

Checkpoint Quiz 1 1. $6 + 4$ **2.** $\frac{c}{2}$ **3.** $4.3a$ **4.** $b + c + 2a$ **5.** 28 **6.** 58 **7.** 60 **8.** 252 **9.** False; answers may vary. Sample: The opposite of -2 is 2, but $-2 > 2$. **10.** natural numbers or whole numbers

Lesson 1-4   pp. 27–29

Check Skills You'll Need
1. $2x + 10$ **2.** $\frac{t}{4}$ **3.** $8 - 6n$ **4.** $7 - 2p$

Quick Check 1. $m = 60h$ **2a.** 36 **b.** No; you need 11 toothpicks to make 2 houses and 16 toothpicks to make 3 houses. **c.** $y = 2 + 3x$ **3.** independent: weight of turkey; dependent: cooking time
4. independent: the amount of money he can spend; dependent: number of songs he can download. domain: $3.00 to $6.00; range: 4 to 8 songs.

Lesson 1-5   pp. 33–34, 37

Check Skills You'll Need

1–4.

Quick Check 1.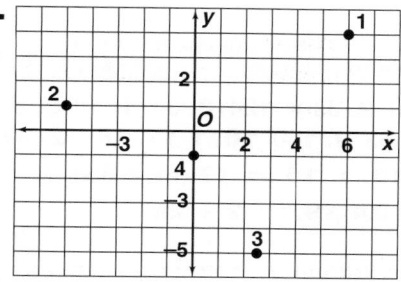

2a. 4-year-old car with an asking price of $14,900 **b.** $5000

Checkpoint Quiz 2 1. $w = 35m$ **2.** depth below sea level; pressure **3.** money in the account; time in account **4.** number of papers delivered; money earned
5a. 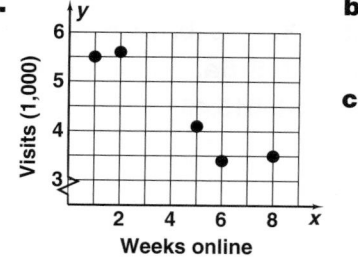 **b.** negative correlation **c.** {1, 2, 5, 6, 8}

Check Skills You'll Need 1. 1.9, 2.4, 3.6, 7.5, 9.8
2. 58, 72, 98, 144, 195, 235 **3.** $-12, -8, -3, 0, 7, 14$
4. $-4\frac{3}{8}, -3\frac{2}{3}, -2\frac{5}{8}, 2\frac{1}{2}, 4\frac{1}{2}, 6\frac{1}{4}$ **5.** 5 **6.** 7

Quick Check 1a. about \$6.53/h; about \$6.38/h;
\$6.25/h **b.** Median; the mean is still larger than 8
of 10 wages. **2.** No; you would need at least a 104
to have a 92 average. **3.** Maine: mean $= 0.4°F$,
range $= 13°F$; Michigan: mean $= 4.4°F$,
range $= 34°F$. On average, Maine was colder. The
temperatures for Michigan were more spread out.

4.
0	2 8 8
1	4
2	6
3	5
4	3 3 5
6	0

4 | 3 means 4.3

5a. city: 28 mi/gal;
highway: 32 mi/gal
b. city: 23 mi/gal; 31 mi/gal;
highway: 32 mi/gal; 38 mi/gal
c. city: 15 mi/gal;
highway: 14 mi/gal

Chapter 2

Check Your Readiness p. 54

1. $0.45n = 3.60$ **2.** $3s = 124$ **3.** 34 **4.** 10 **5.** 5
6. 30 **7.** 33 **8.** 63 **9.** 140 **10.** 1 **11.** 18 **12.** 39
13. $<$ **14.** $<$ **15.** $>$ **16.** $=$ **17.** $m = 55h$
18. $t = 0.8c + 0.4$ **19.** $l = 2 + 2w$

Check Skills You'll Need 1. 6 **2.** 17 **3.** 14 **4.** 59 **5.** 1.3
6. 5.2 **7.** 10.9 **8.** 17.1 **9.** $\frac{4}{5}$ **10.** $1\frac{2}{9}$ **11.** $1\frac{1}{4}$ **12.** $\frac{5}{8}$

Quick Check 1a. -2 **b.** -2 **c.** -11 **d.** 6 **2a.** -11
b. -17.4 **c.** $-1\frac{1}{4}$ **d.** $\frac{1}{18}$ **3.** $-15 + 18 = 3$, rise of 3°
4a. -11.4 **b.** -9.1 **c.** 15.6 **d.** 14.59 **5.** Choices of
variable may vary. Sample: $c =$ change in temp.,
$-14 + c$; $-25°F$

6a. $\begin{bmatrix} -4 \\ 1.5 \\ -16 \end{bmatrix}$ **b.** $\begin{bmatrix} -9 & \frac{1}{8} \\ 1\frac{1}{4} & -1 \end{bmatrix}$

Check Skills You'll Need 1. -6 **2.** 7 **3.** -3.79 **4.** $\frac{7}{19}$
5. 1 **6.** 6 **7.** 5 **8.** $\frac{1}{2}$

Quick Check 1a. -4 **b.** -5 **c.** -11 **d.** 5
2a. -1 **b.** 14 **c.** 4 **d.** 3 **3a.** -8 **b.** 12 **c.** 8.0
d. $-\frac{1}{18}$ **4a.** 1 **b.** 1 **c.** 6 **d.** 6 **5a.** -5 **b.** 5 **c.** 9
d. 5 **6.** ABC: \$32.47; PQR: \$15.46

Check Skills You'll Need 1. -8 **2.** -25 **3.** -24 **4.** -72
5. 8, 10, 12 **6.** 0, -2, -4 **7.** 3, 0, -3 **8.** 0, 6, 12

Quick Check 1a. -24 **b.** 50 **c.** 39.2 **d.** $-\frac{1}{2}$
2a. -56 **b.** 336 **c.** -56 **3a.** $-24.75°F$ **b.** 15.25°F
4a. -64 **b.** 16 **c.** 0.09 **d.** $-\frac{9}{16}$ **5a.** -6 **b.** 4
c. -1 **d.** 13 **6a.** $-4\frac{1}{2}$ **b.** $-\frac{1}{5}$ **c.** $29\frac{1}{2}$ **7.** -10

Check Skills You'll Need 1. 33 **2.** -22 **3.** 1 **4.** -1 **5.** $3t$
6. $-4m$

Quick Check 1a. 1339 **b.** 2121 **c.** 2352
d. 1485 **2.** \$17.70 **3a.** $6 - 14t$
b. $1.2 + 3.3c$ **4a.** $-2x - 1$ **b.** $-3 + 8a$ **5a.** $13y$
b. $2t$ **c.** $-12w^3$ **d.** $9d$
6a. $-2(t + 7)$ **b.** $14(8 + w)$

Checkpoint Quiz 1 1. $-\frac{1}{9}$ **2.** -20 **3.** 5 **4.** $\begin{bmatrix} 0 & -5 \\ -2.1 & 8 \\ -23 & 11 \end{bmatrix}$

5. 120 **6.** 1 **7.** -1.6 **8.** $t =$ temperature change;
$-5 + t$; 8°F **9.** $8x - 12$ **10.** $49t$

Check Skills You'll Need 1. 19 **2.** -30 **3.** 26 **4.** 140
5. -1 **6.** 3 **7.** $1 + x$ **8.** $5t - 8$ **9.** $-7m$

Quick Check 1a. Ident. Prop. of Mult.; m is
mult. by the mult. identity, 1. **b.** Assoc. Prop. of
Add.; the grouping of the terms changes.
c. Assoc. Prop. of Mult.; the grouping of the
factors changes. **d.** Ident. Prop. of Add.; the
ident. for add., 0, is added. **e.** Comm. Prop. of
Mult.; the order of the factors changes.
f. Comm. Prop. of Add.; the order of the terms
changes. **2.** \$8.80
3a. $5a + 6 + a$

$= 5a + a + 6$	Comm. Prop. of Add.
$= (5a + a) + 6$	Assoc. Prop. of Add.
$= (5a + 1a) + 6$	Ident. Prop. of Mult.
$= (5 + 1)a + 6$	Dist. Prop.
$= 6a + 6$	addition

b. $2(3t - 1) + 2$

$= 6t - 2 + 2$	Dist. Prop.
$= 6t + (-2) + 2$	def. of subtr.
$= 6t + [(-2) + 2]$	Assoc. Prop. of Add.
$= 6t + 0$	Inv. Prop. of Add.
$= 6t$	Ident. Prop. of Add.

Lesson 2-6
pp. 93–95, 99

Check Skills You'll Need **1.** 32% **2.** 9% **3.** 22.5% **4.** 18%

Quick Check **1.** $\frac{2}{7}$ **2.** increases **3.** 6 : 2 or 3 : 1 **4.** 98%
5. about 35,260 light bulbs

Checkpoint Quiz 2 **1.** Dist. Prop. **2.** Comm. Prop. of Add. **3.** Assoc. Prop. of Add. **4.** Simplify. **5.** Dist. Prop. **6.** Simplify. **9.** $\frac{2}{7}$ **10.** 2352 bicycles

Lesson 2-7
pp. 101–103

Check Skills You'll Need **1.** $\frac{1}{3}$ **2.** $\frac{1}{3}$ **3.** $\frac{1}{6}$ **4.** 0 **5.** $\frac{1}{6}$ **6.** $\frac{1}{4}$
7. $1\frac{3}{5}$

Quick Check **1.** $\frac{1}{18}$ **2.** $\frac{4}{225}$ **3.** $\frac{4}{105}$ **4a.** $\frac{2}{39}$ **b.** $\frac{2}{39}$
c. No; according to the Comm. Prop. of Mult., the order of the terms does not change the result.

Chapter 3

Check Your Readiness
p. 116

1. $m = 7.5n$ **2.** $t = 200 - 13w$ **3.** 3 **4.** −10 **5.** 8
6. −8 **7.** 7.14 **8.** 16.4 **9.** $-\frac{9}{20}$ **10.** $-\frac{7}{15}$ **11.** 17
12. −3 **13.** 576 **14.** −2.75 **15.** $16k^2$ **16.** $13xy$
17. $2t + 2$ **18.** $12x - 4$

Lesson 3-1
pp. 119–121

Check Skills You'll Need **1.** 19; Add. Prop. of Eq.
2. 5.2; Subtr. Prop. of Eq. **3.** 14; Add. Prop. of Eq.
4. 33; Add. Prop. of Eq. **5.** 32; Mult. Prop. of Eq.
6. 3; Div. Prop. of Eq. **7.** $\frac{1}{5}$; **8.** −9; **9.** $3\frac{3}{4}$

Quick Check **1a.** 5 **b.** 243 **c.** 3 **2.** 2

3a. 29 **b.** Let t = total weekly salary and
s = weekly sales; $t = 125 + \frac{1}{12}s$; $1560

4.

$-9 - 4m = 3$	Original equation
$-9 + 9 - 4m = 3 + 9$	Add. Prop. of Eq.
$-4m = 12$	Simplify.
$\frac{-4m}{-4} = \frac{12}{-4}$	Div. Prop. of Eq.
$m = -3$	Simplify.

Lesson 3-2
pp. 126–128

Check Skills You'll Need **1.** $-n$ **2.** $8b - 2$ **3.** $9w - 45$
4. $-10b + 120$ **5.** $-3x + 12$ **6.** $30 - 5w$ **7.** 43
8. 26 **9.** −45 **10.** −35

Quick Check **1a.** 8 **b.** 3 **c.** 3 **d.** 4 **2.** 135 ft
3a. −1 **b.** −1 **4a.** $\frac{5}{6}$ **b.** 624 **5a.** 28 **b.** 4

Lesson 3-3
pp. 134–136

Check Skills You'll Need **1.** $4x$ **2.** $-4x$ **3.** 0 **4.** 0 **5.** −2
6. −5 **7.** −7 **8.** $\frac{1}{3}$

Quick Check **1a.** $-\frac{4}{7}$ **b.** −2 **c.** $-\frac{5}{2}$ **d.** 10
2. at least 15 bottles **3.** 10 **4a.** no solution
b. identity

Lesson 3-4
pp. 142–145

Check Skills You'll Need **1.** $\frac{7}{12}$ **2.** $\frac{4}{7}$ **3.** $\frac{3}{4}$ **4.** 4 **5.** $\frac{3}{4}$ **6.** $\frac{1}{2}$

Quick Check **1.** $1.025/rose; $1.25 rose; Main Street Florist **2a.** about 3.4 mi/h; $d = 3.4t$
b. about 3.4 mi **3.** 13.2 ft/min **4a.** $6\frac{2}{3}$ **b.** $6\frac{6}{7}$ **c.** 5.4
5a. $8\frac{1}{3}$ **b.** 33.6 **c.** 48 **6a.** 5 **b.** −8.75 **c.** −21

Lesson 3-5
pp. 149–151, 155

Check Skills You'll Need **1.** $\frac{6}{7}$ **2.** $\frac{3}{4}$ **3.** $\frac{1}{2}$ **4.** $2\frac{4}{5}$ **5.** $2\frac{2}{15}$
6. $6\frac{2}{3}$ **7.** $1\frac{1}{9}$ **8.** 10 **9.** $\frac{3}{5}$

Quick Check **1.** 10.5 cm

2.

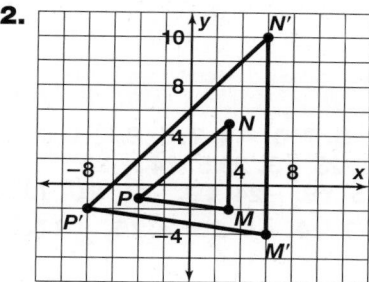

3a. 9.75 ft **b.** 42 ft **4a.** about 21 mi **b.** 3.5 in.

Checkpoint Quiz 1 **1.** 2.5 **2.** 2 **3.** no solution
4. $6\frac{3}{4}$ **5.** −20 **6.** 4.4 **7.** 33.6 min **8.** 3.125 cm
9. 4.5 ft **10.** 35 ft

Lesson 3-6
pp. 158–162

Check Skills You'll Need **1.** $25q$ **2.** 2ℓ **3.** $34h$ **4.** $5x$
5. $3.99n$

Quick Check **1.** 5 cm **2a.** Let x = the first integer.
b. $x + 1$ is the second integer and $x + 2$ is the third integer. **c.** $3x + 3 = 48$; 15, 16, 17
3a. $3\frac{2}{3}$ h **b.** $1\frac{2}{3}$ h **4.** 1 h **5.** John: 38 mi/h; Sarah: 50 mi/h

Lesson 3-7
pp. 168–170

Check Skills You'll Need **1.** $\frac{x}{20} = \frac{20}{100}$, 4 **2.** $\frac{8}{20} = \frac{x}{100}$, 40%
3. $\frac{18}{x} = \frac{90}{100}$, 20 **4.** $\frac{27}{x} = \frac{90}{100}$, 30 **5.** 16 **6.** 32

Quick Check **1a.** 8% **b.** 7% **2.** 1899%
3. 0.5 cm **4.** 86.25 ft^2, 106.25 ft^2 **5a.** about 0.3%
b. about 0.03% **6.** about 66%

Lesson 3-8   pp. 176–178, 180

Check Skills You'll Need 1. 121 **2.** 144 **3.** −144 **4.** 2.25
5. 0.36 **6.** $\frac{1}{4}$ **7.** $\frac{4}{9}$ **8.** $\frac{16}{25}$

Quick Check 1a. 7 **b.** ±6 **c.** −11 **d.** $\frac{1}{5}$
2a. irrational **b.** rational **c.** irrational **d.** rational
3. −11 and −10 **4.** 4.22 **5.** 156.5 ft

Checkpoint Quiz 2 1. 14 cm **2.** 153, 155, 157
3. Beth: 63 mi/h; Marcus: 50 mi/h **4.** about 0.2%
5. 2 **6.** 303.75 cm^2; 340.75 cm^2 **7.** −16 **8.** $\frac{8}{11}$
9. ±0.05 **10.** 7 and 8

Lesson 3-9   pp. 181–184

Check Skills You'll Need 1. 61 **2.** 65 **3.** 25t^2 **4.** 14
5. $\frac{5}{7}$ **6.** 1.2

Quick Check 1. 25 cm **2.** 6.9 mi **3.** yes **4.** no

Chapter 4

Check Your Readiness   p. 198

1. > **2.** = **3.** > **4.** < **5.** 7 **6.** −4 **7.** 1 **8.** 2 **9.** 3
10. −12 **11.** 32.4 **12.** 23 **13.** 29.5 **14.** −28
15. −12 **16.** 48 **17.** 5 **18.** −24 **19.** −10 **20.** 1.85
21. −24 **22.** −2 **23.** 3 **24.** −4 **25.** 3 **26.** $\frac{1}{2}$ **27.** $\frac{5}{2}$
28. 4.1 **29.** 48

Lesson 4-1   pp. 200–202

Check Skills You'll Need 1–5.
6. > **7.** < **8.** = **9.** = **10.** > **11.** <

Quick Check 1a. no **b.** yes **c.** yes **d.** yes **2a.** no **b.**
no **c.** yes **d.** yes
3a.   **b.**

c.   **4a–b.** Choice of
variable may vary. **a.** $x \geq 2$ **b.** $x < 0$ **5a.** No;
speeds cannot be negative, so you can't use all
real numbers. **b.** No; answers may vary.
Sample: Hourly wages are not likely to be in
hundreds of dollars.

Lesson 4-2   pp. 206–208

Check Skills You'll Need 1. > **2.** < **3.** > **4.** 9 **5.** −2
6. −9 **7.** $\frac{1}{6}$

Quick Check **1.** $m > 2$;
2. $n \leq 5$;
3. $t \geq 5$;
4. $b \geq 53$

Lesson 4-3   pp. 212–215, 217

Check Skills You'll Need 1. 16 **2.** $-\frac{2}{3}$ **3.** −6 **4.** 6.4
5. −18 **6.** 18 **7.** $x \leq -1$ **8.** $x > 3$

Quick Check 1a. $b > 2$;
b. $d \geq 2\frac{1}{2}$;
c. $y \leq -1.5$;
2a. $k < 4$;
b. $t > -\frac{1}{2}$;
c. $w \leq -10$;
3a. $t > 4$;
b. $w \leq -4$;
c. $n > -3$;
4. $0.4c > 327$; $c > 317.5$; 818 calendars

Checkpoint Quiz 1 1. $c > 5$;
2. $x < -6$;
3. $p \leq -6$;
4. $y \geq 2$;
5. $g < -8$;
6. $b \leq -5$;
7a. yes **b.** no **c.** yes **d.** no **8a.** no **b.** no **c.** no
d. yes **9a.** $m + 38 + 50 \geq 180$ **b.** $m \geq 92$
10a. $1.50p \leq 20$ **b.** 13 plants

Lesson 4-4   pp. 219–221

Check Skills You'll Need 1. −2 **2.** no solution **3.** $-3\frac{1}{3}$
4. identity **5.** $1\frac{2}{9}$ **6.** −11 **7.** 40 cm **8.** 13 in.

Quick Check 1a. $x \geq -6$ **b.** $t < 1$ **c.** $n > 3$
2. $2(12) + 2w \leq 40$, so the banner's
width must be 8 feet or less. **3a.** $p < -1$
b. $m \leq -3$ **c.** $b > 3$ **4.** $b > 3$ **5.** $x \leq 2\frac{1}{4}$

Lesson 4-5 — pp. 227–229, 232

Check Skills You'll Need 1.
(number line: 6 7 8 9 10 11 12, open circle at 8, closed dot at 10)

2. (number line: −6 −5 −4 −3 −2 −1, closed dots)

3. (number line: 5 6 7 8 9 10 11 12 13 14, closed dot at 7, open circle at 12)

4. 6 **5.** −3 **6.** 14 **7.** 3

Quick Check 1a. $n > -2$ and $n < 9$ or $-2 < n < 9$; (number line: −4 −2 0 2 4 6 8 10)

b. $3.50 \leq b \leq 6$; (number line: 0 1 2 3 4 5 6 7)

2a. $-2 \leq x < 5$; (number line: −4 −2 0 2 4 6)

b. $-1 < x < 4$; (number line: −2 −1 0 1 2 3 4 5)

c. $-4 \leq n < -2$; (number line: −5 −4 −3 −2 −1 0 1)

3a. $6.7 \leq p \leq 8.5$ **b.** $5.2 \leq p \leq 7$. No; readings in this range are unlikely if the first readings are high.
4. $n \leq -5$ or $n \geq 3$; (number line: −6 −4 −2 0 2 4 6)

5. $x < 2$ or $x \geq 3$; (number line: −1 0 1 2 3 4)

Checkpoint Quiz 2

1. $d < -3$ (number line: −5 −4 −3 −2 −1 0 1)

2. $n \geq -2$ (number line: −4 −3 −2 −1 0 1 2 3 4)

3. $-2 \leq m \leq 1$ (number line: −3 −2 −1 0 1 2)

4. $s < 2$ (number line: −1 0 1 2 3)

5. $p > 4$ (number line: −2 0 2 4 6 8)

6. $x \leq -4$ or $x > 4$ (number line: −6 −4 −2 0 2 4 6)

7. $c < 8$ **8.** $65 \leq t \leq 75$
9. $2(15) + 2(w) \leq 48,\ w \leq 9$ **10.** $x \geq -19$

Lesson 4-6 — pp. 235–237

Check Skills You'll Need 1. 15 **2.** 3 **3.** 6 **4.** −7 **5.** 24
6. 2 **7.** = **8.** > **9.** < **10.** > **11.** > **12.** =

Quick Check 1a. −1, 1 **b.** −5, 5 **c.** −2, 2
d. No; an absolute value cannot be negative.
2a. −4, 8 **b.** no solution **c.** −2, 2
3a. $w < -7$ or $w > 3$, (number line: −8 −6 −4 −2 0 2 4 6)

b. all real numbers **4.** 33.81 oz to 33.91 oz, inclusive

Chapter 5

Check Your Readiness — p. 250

1. Let n = number of pens and t = total price; $t = 0.59n$. **2.** Let h = height of house and t = height of tower; $t = h + 200$. **3.** Let s = length of a side and p = perimeter; $p = 3s$. **4.** −7 **5.** −18 **6.** 2 **7.** −1

8–11.
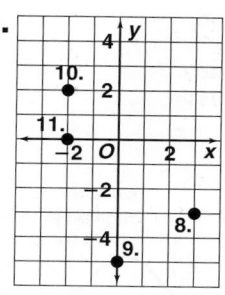

12. $6\frac{2}{5}$ **13.** 0.6 **14.** 4.5
15. −58 **16.** −4, 0
17. 3, 7 **18.** no solution

Lesson 5-1 — pp. 252–253

Check Skills You'll Need 1. C **2.** D **3.** E **4.** A **5.** (0, 0)
6. (−4, −2) **7.** (−3, 3)

Quick Check 1–2. Labels may vary. Samples are given.

1.
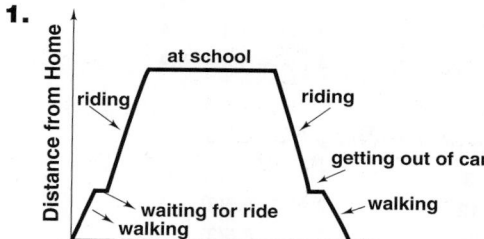

2.
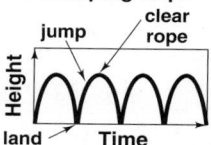
Height While Jumping Rope
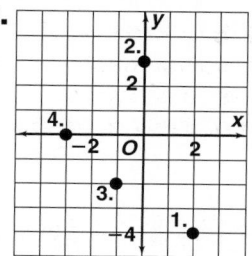

3. F

Lesson 5-2 — pp. 257–259, 262

Check Skills You'll Need

1–4.
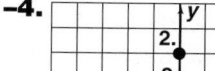

5. −17 **6.** −54 **7.** 108

Quick Check 1a. not a function **b.** function
2a. function **b.** not a function

3.

x	8 − 3x	y
1	8 − 3(1)	5
2	8 − 3(2)	2
3	8 − 3(3)	−1
4	8 − 3(4)	−10

4a. {−8, −6, −1} **b.** {−20, 0, 8}
c. {1, 5, 26}

Checkpoint Quiz 1 **1–3.** Graphs may vary. Samples are given. **1.**

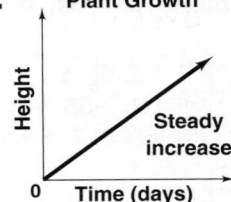
Plant Growth
Steady increase

2.

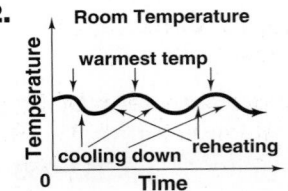
Room Temperature
warmest temp
cooling down   reheating

3.

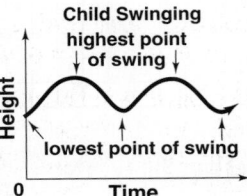
Child Swinging
highest point of swing
lowest point of swing

4. Yes; it passes the vertical line test.

5.

x	−5x	f(x)
1	−5(1)	−5
2	−5(2)	−10
3	−5(3)	−15
4	−5(4)	−20

6.

x	x + 1.4	g(x)
1	1 + 1.4	2.4
2	2 + 1.4	3.4
3	3 + 1.4	4.4
4	4 + 1.4	5.4

7.

n	$3n^2$	f(n)
1	$3(1)^2$	3
2	$3(2)^2$	12
3	$3(3)^2$	27
4	$3(4)^2$	48

8.

x	2 − 0.5x	y
1	2 − 0.5(1)	1.5
2	2 − 0.5(2)	1
3	2 − 0.5(3)	0.5
4	2 − 0.5(4)	0

9. function **10.** not a function

Lesson 5-3   pp. 263–265

Check Skills You'll Need

1. ind: time; dep: distance
2. ind: number of apples; dep: price

Quick Check **1a.** They would be the same.

b. Tables may vary. Sample:

x	f(x)
0	4
1	7
−1	1
−2	−2

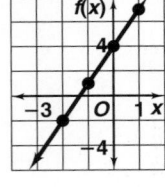

2a. Tables may vary. Sample:

c	P(c)
0	300
100	550
200	800
300	1050
500	1550

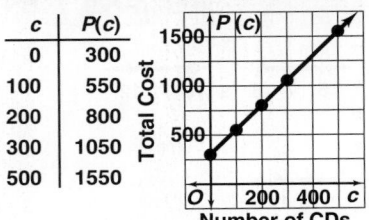

b. $550 to $1050

3a. Tables may vary. Sample: continuous;

t	4000 + 500t	h
0	4000 + 500(0)	4000
1	4000 + 500(1)	4500
2	4000 + 500(2)	5000
3	4000 + 500(3)	5500

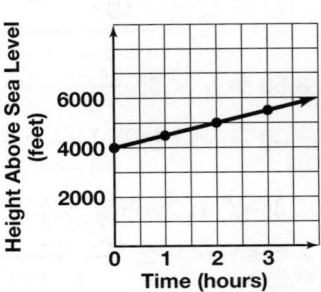

b. Tables may vary. Sample: discrete;

t	6.50t	c
1	6.50(1)	6.50
2	6.50(2)	13.00
3	6.50(3)	19.50
4	6.50(4)	26.00

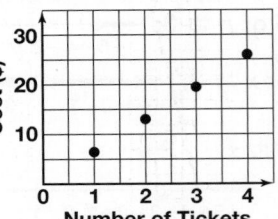

4a. Tables may vary. Sample:

x	y
−2	1
−1	0
0	−1
1	0
2	1

b. Tables may vary. Sample:

x	f(x)
−2	3
−1	0
0	−1
1	0
2	3

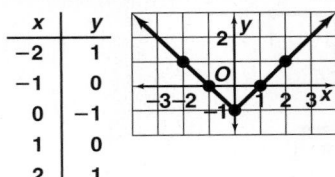

Lesson 5-4   pp. 270–271

Check Skills You'll Need **1–6.** Tables may vary. Samples are given.

1.

x	f(x)
−1	−6
0	−1
1	4
2	9
3	14

2.

x	y
−2	10
−1	7
0	4
1	1
2	−2

3.

t	g(t)
−2	−7.4
−1	−7.2
0	−7
1	−6.8
2	−6.6

4.

x	y
−2	−7
−1	−3
0	1
1	5
2	9

5.

x	f(x)
−2	8
−1	7
0	6
1	5
2	4

6.

d	c(d)
−2	−1.1
−1	−0.1
0	0.9
1	1.9
2	2.9

7. 3 **8.** −2 **9.** 4

Quick Check 1a. $f(x) = x - 2$ **b.** $y = 2x$
c. $y = x + 2$ **2a.** $C(x) = 1.19x$ **b.** \$14.28
3. $p(n) = 15n - 199$

Lesson 5-5   pp. 277–280, 283

Check Skills You'll Need 1. 7.5 **2.** 20 **3.** 5 **4.** 10 **5.** 14.4
6. 81

Quick Check 1a. yes; $\frac{2}{7}$ **b.** no **c.** yes; 7.5
2. $y = 2x$ **3.** $y = 12x$ **4a.** no **b.** yes; $y = 1.5x$
5. 2.5 lb

Checkpoint Quiz 2

1.

x	y
−1	−3
0	1
1	5

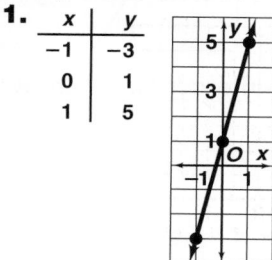

2.

x	y
−2	−1
0	0
2	1

$\frac{1}{2}$

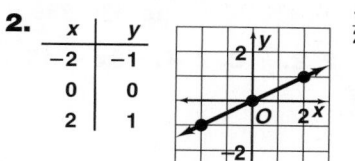

3.

x	f(x)
−1	3
0	0
1	−3

−3

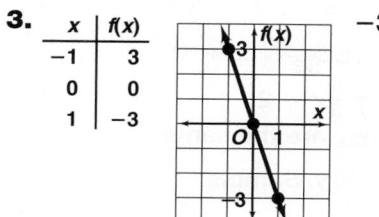

4.

x	y
−1	5
0	2
1	−1

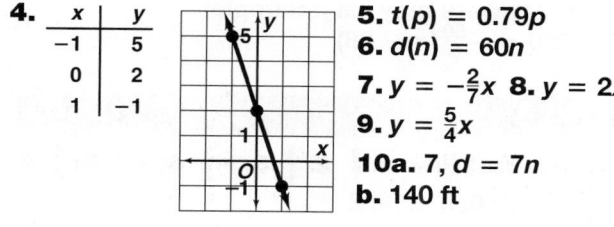

5. $t(p) = 0.79p$
6. $d(n) = 60n$
7. $y = -\frac{2}{7}x$ **8.** $y = 2x$
9. $y = \frac{5}{4}x$
10a. 7, $d = 7n$
b. 140 ft

Lesson 5-6   pp. 284–287

Check Skills You'll Need 1. 5 **2.** −7 **3.** $\frac{1}{3}$ **4.** 4 **5.** $y = 2x$
6. $y = 0.5x$ **7.** $y = -0.25x$ **8.** $y = 0.4x$

Quick Check 1. $xy = 18$ **2a.** 15 **b.** 5 **3a.** $5.\overline{3}$ ft
b. 120 lb **4a.** inverse variation; $xy = 36$ **b.** direct
variation; $y = 4x$ **5a.** Direct variation, since the
ratio $\frac{\text{cost}}{\text{sweater}}$ is constant at \$15 each. **b.** Inverse
variation, since the total number of miles walked
each day is a constant product of 5.

Lesson 5-7   pp. 292–294

Check Skills You'll Need 1. 12, 15, 18 **2.** 15, 22, 29
3. −2.6, −5.6, −8.6 **4.** 14 **5.** −17 **6.** −1.9

Quick Check 1a. "Multiply the previous term by 3";
243, 729. **b.** "Add 6 to the previous term"; 33, 39.
c. "Multiply the previous term by −2"; 32, −64.
2a. 12 **b.** −5 **3a.** −5, 10, 28 **b.** 6.3, 31.3, 61.3

Chapter 6

Check Your Readiness   p. 306

1. 2 **2.** 5 **3.** $\frac{1}{12}$ **4.** 7 **5.** 7

6.

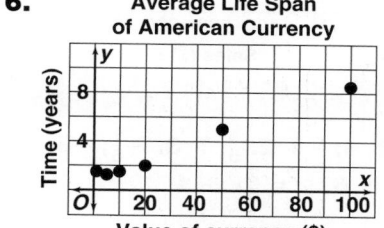

Average Life Span of American Currency

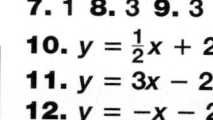

7. 1 **8.** 3 **9.** 3
10. $y = \frac{1}{2}x + 2$
11. $y = 3x - 2$
12. $y = -x - 2$

13.

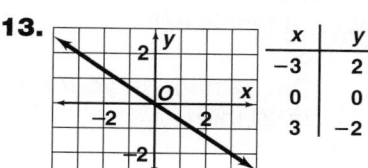

x	y
−3	2
0	0
3	−2

14.

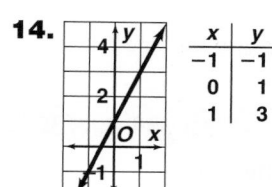

x	y
−1	−1
0	1
1	3

15.

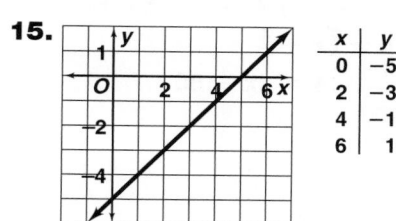

x	y
0	−5
2	−3
4	−1
6	1

Instant Check System™ Answers

Lesson 6-1 — pp. 308–311

Check Skills You'll Need **1.** -12 **2.** 12 **3.** -5 **4.** 5 **5.** 2
6. $-\frac{1}{3}$ **7.** -3 **8.** $\frac{3}{5}$ **9.** $\frac{1}{4}$ **10.** -1

Quick Check **1a.** 15 **b.** No; the rate of change for each consecutive pair of days does
not have to be the same. **2.** 50 mi/h **3a.** $\frac{3}{5}$ **b.** $-\frac{1}{6}$
4a. 1 **b.** $-\frac{3}{2}$ **c.** $\frac{d-b}{c-a}$ **5a.** 0 **b.** undefined

Lesson 6-2 — pp. 317–319

Check Skills You'll Need **1.** 15 **2.** -11 **3.** 6 **4.** 5
5. -7 **6.** 1

Quick Check **1a.** $m = \frac{7}{6}$; $b = -\frac{3}{4}$ **b.** The line shifts
3 units down; the equation becomes $y = 3x - 8$.
2. $y = \frac{2}{5}x - 1$ **3a.** $y = \frac{1}{2}x + 1$ **b.** No; any two
points will give the same slope.

4. 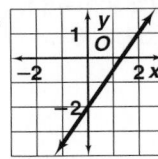 **5.**

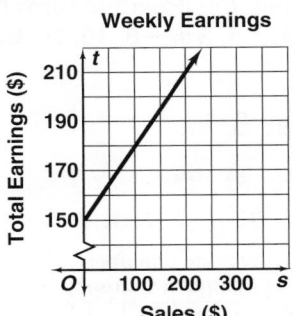

Weekly Earnings

Lesson 6-3 — pp. 324–325

Check Skills You'll Need **1.** 4; -3 **2.** $\frac{4}{5}$; 7 **3.** $-\frac{1}{3}$; -8
4. 6; 0 **5.** $y = 5x - 1$ **6.** $y = \frac{2}{5}x + 4$
7. $y = 6x + \frac{1}{4}$

Quick Check **1.** t = number of tennis balls,
n = number of cans; $t = 3n$

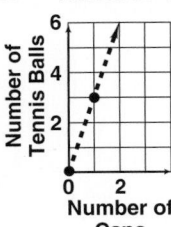

2a. 12 mi/h **b.** your rate
is 10 mi/h

Lesson 6-4 — pp. 330–332, 335

Check Skills You'll Need **1.** $y = -3x + 5$
2. $y = 2x + 10$ **3.** $y = x - 6$ **4.** $y = -5x + 2$
5. $y = -\frac{1}{3}x + \frac{1}{9}$ **6.** $y = \frac{2}{5}x + \frac{4}{5}$
7. $625x + 850 = 775$ **8.** $4 = 2x - 50$
9. $900 - 222x = 1000$

Quick Check **1.** -3; $\frac{4}{3}$

2. 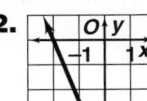 **3a.** 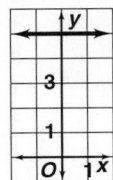 **b.**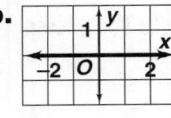

c. **d.**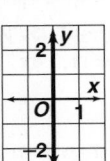

4. $2x + 5y = 5$
5. $4x + 5y = 250$

Checkpoint Quiz 1 **1.** $-\frac{5}{7}$ **2.** $\frac{3}{4}$ **3.** -4 **4.** -1
5. \$121.75 billion

6. 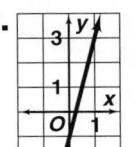 **7.** 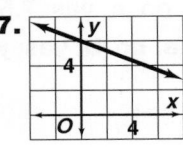 **8.**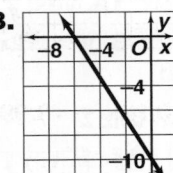

9. 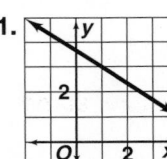 **10.** same y-intercepts; different slopes

Lesson 6-5 — pp. 336–339

Check Skills You'll Need **1.** -2 **2.** $\frac{1}{2}$ **3.** 2 **4.** $-3x + 15$
5. $5x + 10$ **6.** $-\frac{4}{9}x + \frac{8}{3}$

Quick Check

1.

2. $y + 8 = \frac{2}{5}(x - 10)$
3a. $y + 5 = \frac{8}{3}(x + 1)$
b. $y = \frac{8}{3}x - 2\frac{1}{3}$
c. They are the same.
4. Yes; answers may vary. Sample:
$y - 5 = \frac{2}{5}(x - 19)$
5. Yes; answers may vary. Sample:
$y - 3030 = -\frac{50}{3}(x - 68)$

Lesson 6-6 — pp. 343–345

Check Skills You'll Need **1.** $\frac{2}{1}$ **2.** $\frac{3}{4}$ **3.** $-\frac{5}{2}$ **4.** $-\frac{5}{7}$ **5.** $\frac{5}{3}$; 4
6. $\frac{5}{3}$; -8 **7.** 6; 0 **8.** 6; 2

Quick Check **1.** Yes; same slope, different
y-intercepts **2.** $y = 3x - 12$ **3.** $y = -\frac{4}{3}x + 9\frac{1}{3}$
4. $y = 2x + 4$

Lesson 6-7
pp. 350–352, 356

Check Skills You'll Need

1.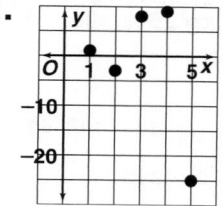
2.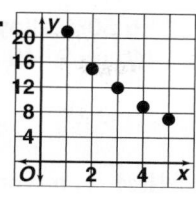

Quick Check

1.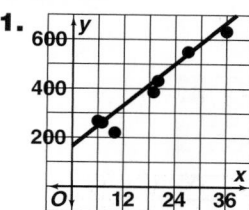
Answers may vary. Sample: $y = 14x + 166$; 362 calories

2. $y = 0.52x - 43.99$; 0.990

Checkpoint Quiz 2 1. $y - 4 = -\frac{1}{4}(x - 3)$
2. $y + 3 = 18x$ **3.** $y = -5$ **4.** $y + 6 = -\frac{2}{3}(x - 2)$
5. $y - 4 = -(x - 5)$
6. $y - 6 = -\frac{3}{2}(x + 2)$ **7.** $y - 2 = \frac{1}{4}x$
8. $y - 2 = -\frac{3}{2}(x + 6)$ **9.** Answers may vary.
Sample: $y = 5.33x + 1.34$ **10.** $y = -6.07x + 62.71$

Lesson 6-8
pp. 359–361

Check Skills You'll Need **1.** 5 **2.** 5 **3.** 18 **4.** 12

5.
x	y
0	6
1	5
2	4
3	3

6.
x	y
0	1
1	2
2	3
3	4

7.
x	y
0	1
1	2
-1	0
-2	1

Quick Check 1a. Answers may vary. Sample: same shape, different y-intercepts 0 and 3
b. Answers may vary. Sample: same shape, different y-intercepts 0 and -3

2a. 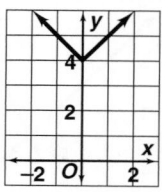 **b.**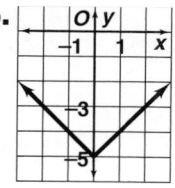

3a. $y = |x| + 2$ **b.** $y = |x| - 5$

4a. 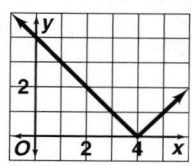 **b.**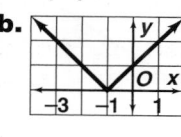

5a. $y = |x - 5|$ **b.** $y = |x + 7|$

Chapter 7

Check Your Readiness
p. 372

1. identity **2.** 1 **3.** no solution **4.** 3 **5.** $\frac{3}{2}$
6. no solution **7.** $y = \frac{3}{2}x + 1$ **8.** $y = -\frac{1}{5}x + 2$
9. $y = -x - 4$
10. $-10 < x < 3$;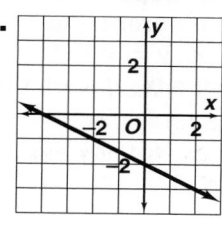
11. $x < 12$ or $x > 60$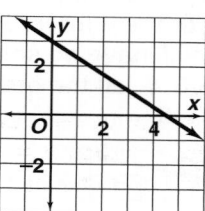

12a. $C(t) = 39.50t$ **b.** $118.50 **c.** 6

13. 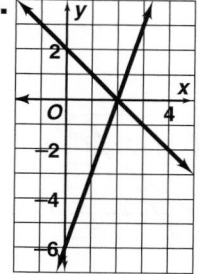 **14.**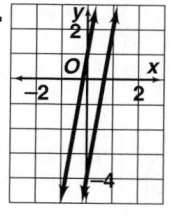

15.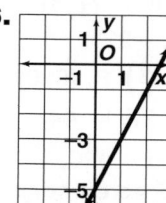

Lesson 7-1
pp. 374–376

Check Skills You'll Need **1.** $1\frac{2}{3}$ **2.** $3\frac{1}{2}$ **3.** 7

4. 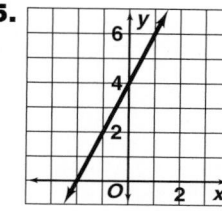 **5.**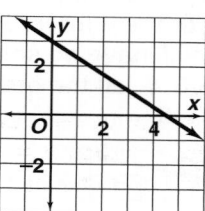

6. 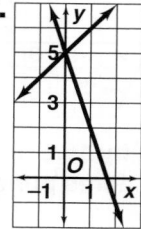 **7.**

Quick Check

1a. (−1, 4); 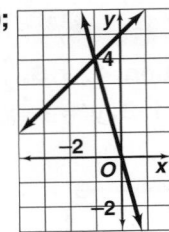 **b.** (−2, 3);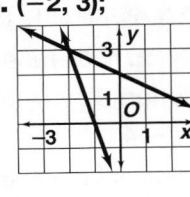

2. (H)d = 3d; + 5 (H)d = 4d + 1
3a. (1.5, 6) **b.** 1.5h after the second walker starts, both have walked 6 mi. **4.** If the slopes are the same, but the y-intercepts are different, the system will have no solution. Ex.: $y = x − 1$, $y = x + 3$ **5.** all the ordered pairs (x, y) such that $y = \frac{1}{5}x + 9$

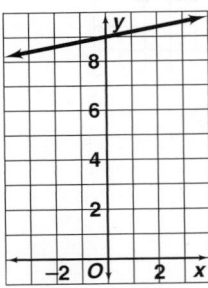

Lesson 7-2   pp. 382–383, 386

Check Skills You'll Need 1. $−4\frac{2}{3}$ **2.** $1\frac{1}{2}$ **3.** 15 **4.** no **5.** no

Quick Check 1. (3, 6) **2.** (2.3, 1.6)
3. 3 cm by 12 cm

Checkpoint Quiz 1

1. (1, −1); 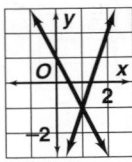 **2.** (3, 2);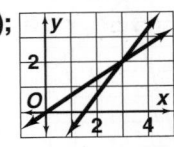

3. (−4, −2); 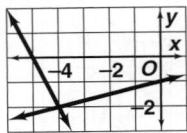 **4.** (2, −8) **5.** (1, 7)
6. (−15, −22)
7. $\left(\frac{28}{5}, −\frac{6}{5}\right)$ **8.** (6, 0)
9. 2L + 2W = 44,

3W = L; 5.5 cm by 16.5 cm **10.** p + c = 420, c = p + 250; 85 acres pumpkins, 335 acres corn

Lesson 7-3   pp. 387–390

Check Skills You'll Need 1. (8, 29) **2.** (2, −6) **3.** (3, −4)

Quick Check 1. (2, 3) **2.** 30 adult; 34 student **3.** (1, −2) **4.** 85 cards, 135 gift wrap **5.** (1, −2)

Lesson 7-4   pp. 396–399

Check Skills You'll Need 1. 3.25 h **2.** 275 mi

Quick Check 1. 32 kg 50% alloy; 8 kg 25% alloy **2.** 3850 copies **3.** 440 mi/h; 40 mi/h

Lesson 7-5   pp. 404–406, 410

Check Skills You'll Need 1. never **2.** always
3. sometimes **4.** $y = \frac{2}{3}x − 3$ **5.** $y = −3x + 6$
6. $y = \frac{3}{4}x + \frac{1}{4}$

Quick Check

1. 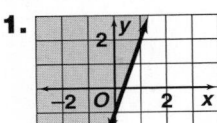 **2.**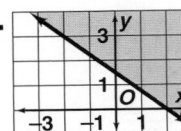

3. Answers may vary. Sample: 6 lb hamburger and 2 lb chicken, 3 lb hamburger and 6 lb chicken, 5 lb hamburger and 3 lb chicken

Checkpoint Quiz 2 1. (11, −4) **2.** (−9, 1) **3.** (3.5, 10)
4. no solution **5.** $\left(15, −\frac{1}{2}\right)$ **6.** n + d = 21, 0.05n + 0.10d = 1.70; 8 nickels, 13 dimes
7. y = 200 + 0.35x, y = 1.20x; about 236 ice cream cones
8. x + y = 4, x − y = 3; 3.5 mi/h, 0.5 mi/h

9. 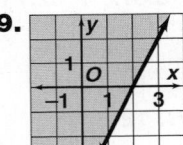 **10.**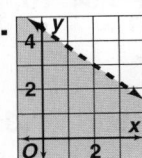

Lesson 7-6   pp. 411–414

Check Skills You'll Need 1. (2, 0);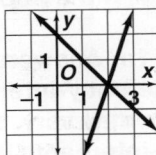

2. no solution; 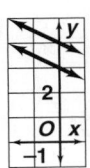 **3.** (4, 0);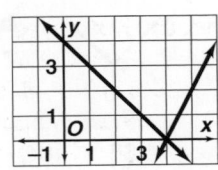

4. 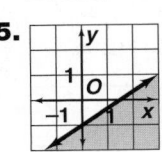 **5.** 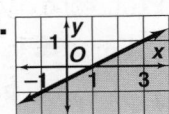 **6.**

Quick Check

1.

2a. $y \geq \frac{1}{2}x - 1$ and $y < 2$
b. $y \leq -\frac{2}{3}x$ and $x > -3$

3. Garden Dimensions

4a. Answers may vary. Sample: 2 20¢ stamps and 6 34¢ stamps; 4 20¢ stamps and 5 34¢ stamps **b.** no **c.** You cannot have a negative or fractional number of stamps.

Chapter 8

Check Your Readiness

1. 0.7 **2.** 6.4 **3.** 0.008 **4.** 3.5 **5.** $0.\overline{27}$ **6.** 49
7. 5.027 **8.** 0.75 **9.** 4 **10.** 100 **11.** 49 **12.** 17
13. -31 **14.** $33\frac{1}{3}\%$ **15.** 25% **16.** $\{-24.5, -8, 0\}$
17. $\{-32.875, 10, 18\}$ **18.** $\{-11, -1, 16.5\}$
19. 9, 11 **20.** 14, 20 **21.** 31, 37

Lesson 8-1
pp. 430–432

Check Skills You'll Need **1.** 8 **2.** $\frac{1}{16}$ **3.** 4 **4.** -27 **5.** -27
6. 3 **7.** $\frac{1}{2}$ **8.** -1 **9.** 4

Quick Check **1a.** $\frac{1}{81}$ **b.** 1 **c.** $-\frac{1}{64}$ **d.** $\frac{1}{7}$ **e.** $-\frac{1}{9}$
2a. $\frac{11}{m^5}$ **b.** $\frac{7t^2}{s^4}$ **c.** $2a^3$ **d.** $\frac{1}{n^5v^2}$ **3a.** $-\frac{1}{8}$ **b.** $-\frac{1}{50}$ **c.** $\frac{1}{16}$
d. $-12\frac{1}{2}$ **4.** 600; 5400; for $x = -2$, the population is 600, 2 months before the population is 5400. For $x = 0$, it is the population when time is 0.

Lesson 8-2
pp. 436–438

Check Skills You'll Need **1.** 60,000 **2.** 0.07 **3.** 820,000
4. 0.003 **5.** 34 **6.** 524 **7.** 367.8

Quick Check **1a.** yes **b.** No; $52 > 10$. **c.** No; $0.04 < 1$. **2a.** 2.67×10^5 **b.** 4.6205×10^7
c. 3.25×10^{-5} **d.** 9.0×10^{-9} **e.** 436 is 436 times greater than 1, and $436 = 4.36 \times 10^2$.
Then $(4.36 \times 10^2) \cdot 10^9 = 4.36 \times 10^{11}$.
3a. 3,200,000,000,000 **b.** 50,700 **c.** 0.00056
d. 0.083 **4.** electron, proton, neutron
5. 60.2×10^{-5}, 61×10^{-2}, 0.067×10^3, 63×10^4
6a. 1.5×10^4 **b.** 8×10^{-10}

Lesson 8-3
pp. 441–443

Check Skills You'll Need **1.** t^7 **2.** $(6 - m)^3$ **3.** $(r + 5)^5$
4. 5^3s^3 **5.** -625 **6.** 625 **7.** 1 **8.** $\frac{1}{625}$

Quick Check **1a.** 5^9 **b.** 2^1 **c.** 7^5 **2a.** $7n^6$ **b.** $28x^2y^7$
c. $\frac{7m^3}{n^2}$
3a. 1.5×10^{12} **b.** 4.5×10^2 **c.** 6.3×10^{-14}
4. about 2.56×10^{13} red blood cells

Lesson 8-4
pp. 447–449, 452

Check Skills You'll Need **1.** 3^6 **2.** 2^{12} **3.** 5^{28} **4.** 7^3 **5.** x^6
6. a^6 **7.** $\frac{1}{y^6}$ **8.** $\frac{1}{n^6}$
Quick Check **1.** a^{28}; $\frac{1}{a^{28}}$ **2a.** $\frac{1}{t^{12}}$ **b.** a^{18}
3a. $16z^4$ **b.** $\frac{1}{16g^{10}}$ **4a.** $81c^{26}$ **b.** $864a^{18}b^6$
c. $\frac{5400n^3}{m^3}$ **5a.** 5.22×10^5 joules
b. about 1.7×10^8 h

Checkpoint Quiz 1 **1.** $\frac{1}{45}$ **2.** r^{20} **3.** $6x^{17}$ **4.** $\frac{mq^2}{n^4}$ **5.** $\frac{1}{a}$
6. $\frac{64m^6}{9}$ **7.** $6m^9$ **8.** $\frac{27t^6}{8}$ **9.** 500; 2000; 16,000
10a. 6.8×10^3 km **b.** about 1.45×10^8 km^2
c. 145,000,000 km^2

Lesson 8-5
pp. 453–455

Check Skills You'll Need **1.** $\frac{1}{4}$ **2.** 5 **3.** $\frac{3}{5}$ **4.** 31 **5.** $\frac{2}{5}$ **6.** $\frac{4}{15}$
7. $\frac{2}{7}$ **8.** $\frac{2}{7}$ **9.** $\frac{y}{3}$ **10.** $\frac{2y^2}{x}$ **11.** $\frac{c}{4}$ **12.** $\frac{4}{n^2}$
Quick Check **1a.** $\frac{1}{b^5}$ **b.** z^5 **c.** $\frac{1}{a^2b^2}$ **d.** $\frac{n}{m^4}$
e. $\frac{xz^7}{y^5}$ **2a.** 2.5×10^{-6} **b.** 3.0×10^{16} **c.** $3.\overline{3} \times 10^2$
d. about 9.59×10^{-3} tons **3a.** $\frac{9}{x^4}$ **b.** $\frac{x^3}{y^6}$
4a. $\frac{64}{27}$ **b.** -32 **c.** $\frac{s}{2r}$ **d.** $\frac{m^2}{49a^2}$

Lesson 8-6
pp. 460–463

Check Skills You'll Need **1.** 2 **2.** -2 **3.** -1.2 **4.** 3.5
5. 32, 64 **6.** 108, 324 **7.** 3.2, 6.4 **8.** 12.5, 6.25

Quick Check **1a.** $\frac{1}{5}$ **b.** 2 **c.** $\frac{3}{2}$ **2a.** 81, 243, 729 **b.** 7.5, -3.75, 1.875 **c.** 17.6, 35.2, 70.4
3a. arithmetic **b.** geometric **c.** arithmetic
4a. 4; 972; 708,588 **b.** -2; -6250; $-97,656,250$
5. $A(n) = 200 \cdot 0.56^{n-1}$; 35.1 cm

Lesson 8-7
pp. 468–470, 473

Check Skills You'll Need

1. 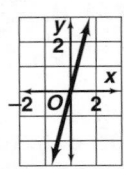 **2.** 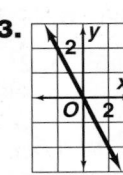 **3.**

4. 9 **5.** $\frac{1}{125}$ **6.** 162 **7.** $\frac{2}{9}$ **8.** $\frac{3}{2}$ **9.** 90

Quick Check 1a. $\frac{1}{16}$, 1, 64 **b.** $\frac{2}{5}$, 10, 1250
c. $-\frac{2}{9}$, -2, -54 **2.** 40,960 animals

3a. 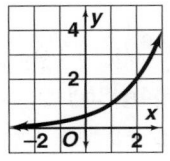 **b.**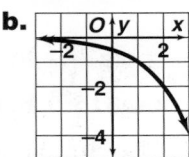

4a.

b. $f(0) = 1$, so copies are made at the same size of the original, or at 100%.

CheckPoint Quiz 2 1. 3^{12} **2.** $\frac{y^{15}}{x^{10}}$ **3.** $\frac{4n^{12}}{25m^6}$ **4.** $\frac{1}{t^{10}}$
5. geometric **6.** geometric **7.** arithmetic
8a. -100 **b.** $-\frac{1}{5}$ or -0.2
c. $A(n) = -100 \cdot (-0.2)^{n-1}$ **d.** -0.16; -0.0064
9a. $A(n) = 40 \cdot (0.85)^{n-1}$ **b.** 209 mm
10a. 1.08×10^8 **b.** 2.49×10^8 **c.** about $2.31

Lesson 8-8
pp. 475–479

Check Skills You'll Need 1. 12% decr. **2.** 20% incr.
3. 6% incr. **4.** 31% decr. **5.** 36% incr.

Quick Check 1a. $y = 4512 \cdot 1.025^x$ **b.** about 4859
students **2a.** $3038.72 **b.** $3038.72
c. $(1 + r)$ is the same as 100% + 100r% written
as a decimal. **3a.** $2813.83 **b.** $210.23; $220.99;
$256.67 **4a.** 4 half-lives **b.** 25 mCi; 12.5 mCi
c. 15 mCi; 3.75 mCi **5a.** 604,000 **b.** 0.982
c. $y = 604,000 \cdot (0.982)^x$ **d.** about 420,017 people

Chapter 9

Check Your Readiness
p. 492

1. 1, 2, 3, 4, 6, 12 **2.** 1, 2, 4, 7, 8, 14, 28, 56 **3.** 1, 31
4. 1, 3, 9, 27 **5.** 1, 2, 5, 10, 11, 22, 55, 110 **6.** 1, 5,
13, 65 **7.** 1, 2, 5, 10, 25, 50 **8.** 1, 2, 4, 5, 8, 10, 20,
25, 40, 50, 100, 200 **9.** 1, 11 **10.** 1, 2, 3, 6, 7, 14,
21, 42 **11.** 1, 2, 3, 6, 11, 22, 33, 66 **12.** 1, 73

13. $3x^2 - 4x$ **14.** $2b + 6$ **15.** $3y^2 - 11y$
16. $-\frac{2}{3}w^2 + \frac{7}{6}w$ **17.** $7z - 66$ **18.** $4x^2 - 2x$
19. $-6t^2 + 10t$ **20.** $-p^2 - 2p$ **21.** $49w^2$ **22.** $25n^3$
23. $9z^4$ **24.** $10t^7$ **25.** $16y^6$ **26.** $81a^2b^2$ **27.** $4x^4$
28. $36p^8$ **29.** x^2y^4 **30.** $3c$ **31.** $-\frac{1}{4t^3}$ **32.** $2a^2$

Lesson 9-1
pp. 494–496

Check Skills You'll Need 1. $19t$ **2.** $39g$ **3.** $-8k$ **4.** $11b - 6$
5. $-3n^2$ **6.** $7x^2$

Quick Check 1. 0; the degree of a nonzero
constant is 0. **2a.** $-9x^4 + 6x^2 + 7$; fourth
degree trinomial **b.** $-y^3 + 3y - 4$; cubic trinomial
c. $-4v + 8$; linear binomial **3a.** $20m^2 + 9$
b. $4t^2 + 5$ **c.** $16w^3 + 8w^2 + 4$
d. $11p^3 + 17p^2 + 13p$ **4a.** $-8v^3 + 13v^2 - 4v$
b. $28d^3 - 30d^2 - 3d$ **c.** $-2x^2 + 4x - 7$

Lesson 9-2
pp. 500–501

Check Skills You'll Need 1. 906 **2.** 287 **3.** 4536
4. $24 + 20x$ **5.** $-16y - 8$ **6.** $25v - 5$ **7.** $7p - 14$
8. $54 - 9x$ **9.** $-8x + 2$

Quick Check 1a. $20b^3 + 4b^2 + 24b$
b. $-21h^3 + 56h^2 + 7h$ **c.** $2x^3 - 12x^2 + 10x$
2a. $5v^3$ **b.** 3 **c.** $2b$ **3a.** $4x(2x - 3)$ **b.** $5d(d^2 + 2)$
c. $6m(m^2 - 2m - 4)$

Lesson 9-3
pp. 505–507, 510

Check Skills You'll Need 1. $4r^2 - 4r$ **2.** $6h^3 + 48h^2 - 18h$
3. $2y^5 - 7y^2$ **4.** $x^3 + 8x^2 + 2x + 1$ **5.** $8t^3 + t + 6$
6. $5w^2 - 27w$ **7.** $-2b^2 - 15b$
8. $7m^3 + 27m^2 - 6m$ **9.** $d^5 - 4d^3 - 18d^2$

Quick Check 1a. $12h^2 + 4h - 21$ **b.** $40m^2 + 11m - 2$ **c.** $63a^2 - 20a - 32$ **2a.** $6x^2 + 23x + 20$
b. $6x^2 + 7x - 20$ **c.** $6x^2 - 7x - 20$
d. $6x^2 - 23x + 20$ **3a.** $25x^2 + 28x + 16$
b. $x^2 + 2x - 2$ **4.** $12n^3 - 10n^2 + 34n - 56$

Checkpoint Quiz 1 1. $9x^2 + 10x + 1$ **2.** $6b^2 - 13b + 9$
3. $36w^2 - 11w$ **4.** $-48k^3 + 42k^2$ **5.** $x^2 - 2x - 15$
6. $12n^5 + 2n^4 - 30n^2 - 5n$
7. $4g^4 + 8g^3 + 7g^2 + 32g - 36$ **8.** $2(6y^2 - 5)$
9. $5t(t^5 + 5t^2 - 2)$ **10.** $9v^2(2v^2 + 3v + 4)$

Lesson 9-4
pp. 512–515

Check Skills You'll Need 1. $49x^2$ **2.** $9v^2$ **3.** $16c^2$ **4.** $25g^6$
5. $j^2 + 12j + 35$ **6.** $6b^2 - 34b + 48$
7. $20y^2 - 3y - 2$ **8.** $x^2 - x - 12$
9. $8c^4 - 78c^2 - 20$ **10.** $54y^4 - 21y^2 - 3$

842 Answers to Instant Check System

Quick Check 1a. $t^2 + 12t + 36$
b. $25y^2 + 10y + 1$ **c.** $49m^2 - 28mp + 4p^2$
d. $81c^2 - 144c + 64$ **2a.** $\frac{1}{9}A^2 + \frac{4}{9}AB + \frac{4}{9}B^2$
b. $\frac{1}{9}$ **c.** $\frac{4}{9}$ **d.** $\frac{4}{9}$ **3a.** 961 **b.** 841 **c.** 9604
d. 41,209 **4a.** $d^2 - 121$ **b.** $c^4 - 64$ **c.** $81v^6 - w^8$
5a. 396 **b.** 399 **c.** 3599 **d.** 8091

Lesson 9-5   pp. 519–521

Check Skills You'll Need 1. 1, 2, 3, 4, 6, 8, 12, 24 **2.** 1, 2, 3, 4, 6, 12 **3.** 1, 2, 3, 6, 9, 18, 27, 54 **4.** 1, 3, 5, 15 **5.** 1, 2, 3, 4, 6, 9, 12, 18, 36 **6.** 1, 2, 4, 7, 8, 14, 28, 56 **7.** 1, 2, 4, 8, 16, 32, 64 **8.** 1, 2, 3, 4, 6, 8, 12, 16, 24, 32, 48, 96

Quick Check 1a. $(g + 5)(g + 2)$ **b.** $(v + 20)\cdot$
$(v + 1)$ **c.** $(a + 10)(a + 3)$ **2a.** $(k - 5)(k - 5)$
b. $(x - 2)(x - 9)$ **c.** $(q - 12)(q - 3)$ **3a.** $(m + 10)\cdot$
$(m - 2)$ **b.** $(p - 8)(p + 5)$ **c.** $(y + 7)(y - 8)$
4a. $(x + 8y)(x + 3y)$ **b.** $(v + 8w)(v - 6w)$
c. $(m - 20n)(m + 3n)$

Lesson 9-6   pp. 524–525

Check Skills You'll Need 1. $6x$ **2.** 7 **3.** 2 **4.** $(x + 1)(x + 4)$
5. $(y - 7)(y + 4)$ **6.** $(t - 5)(t - 6)$

Quick Check 1a. $(2y + 1)(y + 2)$
b. $(3n - 1)(2n - 7)$ **c.** $(2y - 1)(y - 2)$
2a. $(5d + 1)(d - 3)$ **b.** $(2n + 3)(n - 1)$
c. $(5p - 9)(4p + 1)$ **3a.** $2(v - 1)(v - 5)$
b. $2(2y + 1)(y + 3)$ **c.** $6(3k + 1)(k - 1)$

Lesson 9-7   pp. 528–530, 533

Check Skills You'll Need 1. $9x^2$ **2.** $25y^2$ **3.** $225h^4$
4. $4a^2b^4$ **5.** $c^2 - 36$ **6.** $p^2 - 22p + 121$
7. $16d^2 + 56d + 49$

Quick Check 1a. $(x + 4)^2$ **b.** $(n + 8)^2$
c. $(n - 8)^2$ **2a.** $(3g - 2)^2$ **b.** $(2t + 9)^2$
c. $(2t - 9)^2$ **3a.** $(x + 6)(x - 6)$
b. $(m + 10)(m - 10)$ **c.** $(p + 7)(p - 7)$
4a. $(3v + 2)(3v - 2)$ **b.** $(5x + 8)(5x - 8)$
c. $(2w + 7)(2w - 7)$ **5a.** $2(2y + 5)(2y - 5)$
b. $3(c + 5)(c - 5)$ **c.** $7(2k + 1)(2k - 1)$

Checkpoint Quiz 2 1. $k^2 - 14k + 49$ **2.** $25t^2 +$
$90t + 81$ **3.** $h^2 - 22h + 121$ **4.** $(v + 10)^2$
5. $(p - 10)(p + 4)$ **6.** $(k - 12)(k - 5)$
7. $(2x + 11)(x + 1)$ **8.** $(5m + 7)(2m + 1)$
9. $3(w + 2)(w - 4)$ **10.** $(3t + 5)(3t - 5)$

Lesson 9-8   pp. 534–536

Check Skills You'll Need 1. 2 **2.** $3r$ **3.** $5h$ **4.** $4m$
5. $v^3 + 3v^2 + 5v + 15$ **6.** $2q^3 - 10q^2 - 4q + 20$
7. $6t^2 - 7t - 20$ **8.** $4x^3 + 7x^2 + 10x - 3$

Quick Check 1a. $(5t^3 + 6)(t + 4)$
b. $(w^2 - 7)(2w + 1)$ **2.** $3m(3m^2 + 2)(5m - 1)$
3a. $(9d + 5)(7d + 1)$ **b.** $(11k + 5)(k + 4)$
c. $(4y - 7)(y + 10)$ **4a.** Answers may vary.
Sample: $2g$, $(3g + 4)$, and $(g + 2)$ **b.** m, $(3m + 1)$,
and $(m + 3)$

Chapter 10

Check Your Readiness   p. 548

1. -13 **2.** $-\frac{7}{2}$ **3.** -9 **4.** $-\frac{1}{2}$ **5.** -23 **6.** -3 **7.** -108
8. 26 **9.** 0 **10.** 49 **11.** -67 **12.** 25 **13.** 24 **14.** 144

15. 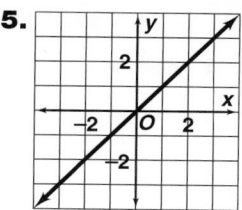 **16.**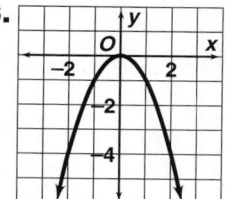

17.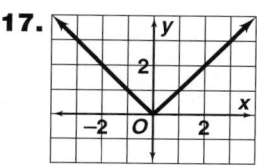

18. $x^2 - x - 6$ **19.** $4y^2 + 8y + 3$ **20.** $3x^2 +$
$5x - 28$ **21.** $(2x + 1)^2$ **22.** $(5x - 3)(x + 7)$
23. $(4x - 3)(2x - 1)$ **24.** $(m - 9)(m + 2)$
25. $(6y - 5)(2y + 3)$ **26.** $(x - 9)^2$

Lesson 10-1   pp. 550–553

Check Skills You'll Need 1. -24 **2.** 18 **3.** 12 **4.** 35

5. 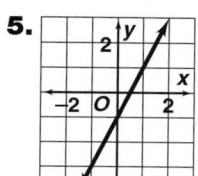 **6.** 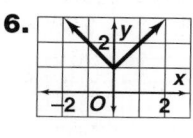 **7.**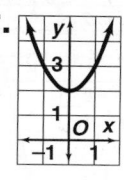

Quick Check 1a. $(4, 3)$; max. **b.** $(-3, -3)$; min.

2.

x	$f(x) = -2x^2$	(x, y)
0	$-2(0)^2 = 0$	$(0, 0)$
1	$-2(1)^2 = -2$	$(1, -2)$
2	$-2(2)^2 = -8$	$(2, -8)$

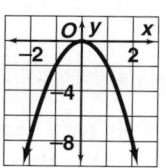

3. $y = \frac{1}{2}x^2$, $y = x^2$, $y = -2x^2$

4a. 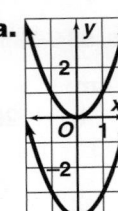 The graph of $y = x^2 - 4$ has the same shape as the graph of $y = x^2$, and it is shifted down 4 units.

b. Positive values of c shift the vertex up. Negative values of c shift the vertex down.

5a. 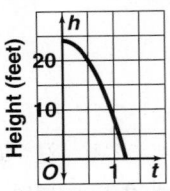 **b.** Domain: 0 to about 1.5 seconds; Range: 0 to 30 feet.

Lesson 10-2   pp. 557–559

Check Skills You'll Need 1. $\frac{1}{3}$ **2.** $-\frac{2}{3}$ **3.** $-3\frac{1}{2}$ **4.** 6

5. 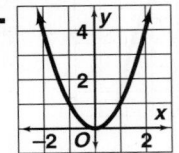 **6.**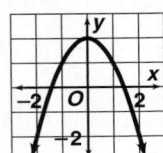

7.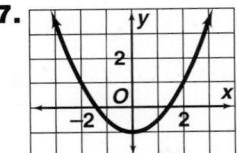

Quick Check

1. 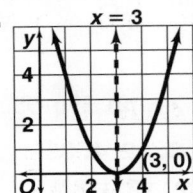 **2a.** 1.5 s **b.** 40 ft

3a. 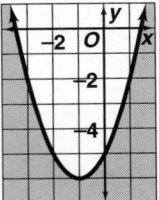 **b.**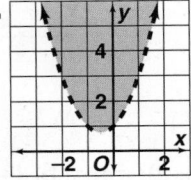

Lesson 10-3   pp. 565–567, 570

Check Skills You'll Need 1. 6 **2.** −9 **3.** ±11 **4.** 1.2
5. 0.5 **6.** ±1.1 **7.** $\frac{1}{2}$ **8.** $\pm\frac{1}{3}$ **9.** $\frac{7}{10}$

Quick Check

1a. 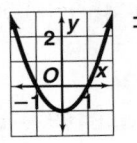 ±1 **b.** 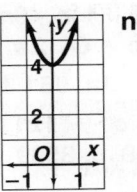 no solution

c. 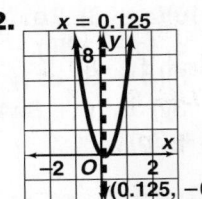 $x = 0$

2a. ±5 **b.** 0 **c.** no solution **3.** about 14 ft

Checkpoint Quiz 1

1. 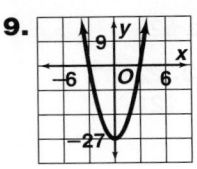 **2.**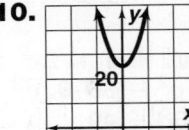

3.

$x = -2.5$

$(-2.5, -12.25)$

4a. 0.75 s **b.** 15 ft **5.** ±5 **6.** $\pm\frac{7}{3}$ **7.** ±8 **8.** ±5

9. ±3 **10.** no solution

Lesson 10-4   pp. 572–573

Check Skills You'll Need 1. −1 **2.** 104 **3.** $-2\frac{5}{7}$
4. $(2c + 1)(c + 14)$ **5.** $(3p + 2)(p + 10)$
6. $(4x + 3)(x - 6)$

Quick Check 1a. −7, 4 **b.** $\frac{5}{3}$, 2 **c.** $-\frac{3}{2}$, $\frac{11}{4}$ **d.** $-\frac{1}{5}$, −6
2. −4, 12 **3.** 6 **4.** 9 in. × 10 in. × 2 in.

Lesson 10-5   pp. 579–581

Check Skills You'll Need 1. $d^2 - 8d + 16$
2. $x^2 + 22x + 121$ **3.** $k^2 - 16k + 64$ **4.** $(b + 5)^2$
5. $(t + 7)^2$ **6.** $(n - 9)^2$

Quick Check 1. 121 **2.** 19, −13
3a. −0.70, −4.30 **b.** 12.74, 1.26 **4a.** −1.65, 3.65
b. 2.56, −1.96

Lesson 10-6 pp. 585–588

Check Skills You'll Need 1. 9 **2.** $\frac{49}{4}$ **3.** $\frac{81}{4}$ **4.** 6, 4
5. 2, −18 **6.** 1, −5 **7.** −7, 8

Quick Check 1a. 4, −2 **b.** 13, −9 **2a.** 0.67, 1 **b.** 1.22,
−0.94 **3a.** $0 = -16t^2 + 38.4t + 3.5$
b. $t \approx 2.5$; 2.5 s **4a.** Quadratic formula; the
equation cannot be factored easily. **b.** Factoring;
the equation is easily factorable. **c.** Square roots;
there is no x term.

Lesson 10-7 pp. 592–593, 596

Check Skills You'll Need 1. −80 **2.** 72 **3.** −283 **4.** 2.18,
0.15 **5.** 0.39, −0.64 **6.** 7, 5

Quick Check 1a. 0 **b.** 2 **c.** 0 **2.** yes

Checkpoint Quiz 2 1. −3, 7 **2.** −3, −9 **3.** −5, 10
4. 0.4, −2.4 **5.** 5.7, −0.7 **6.** 11, −3 **7.** 1, $-\frac{3}{4}$
8. no solution **9.** 1, 0.5 **10.** 0

Lesson 10-8 pp. 597–601

Check Skills You'll Need

1. 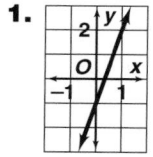 **2.**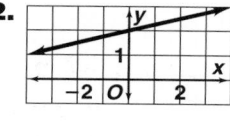

3. 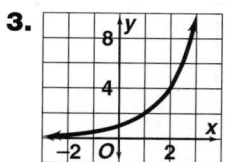 **4.**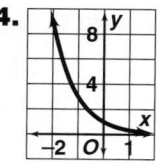

5. 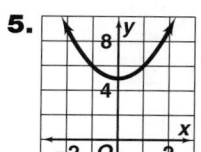 **6.**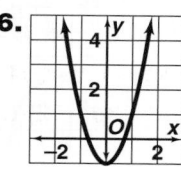

Quick Check

1a. 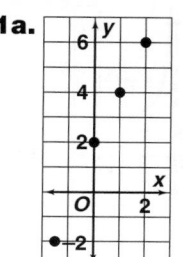 **b.**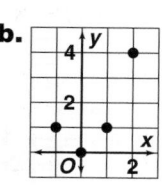

quadratic

linear

c.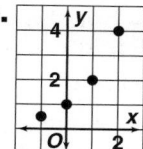

exponential

2a. exponential; $y = 4 \cdot 1.1^x$ **b.** linear; $y = \frac{1}{2}x + \frac{1}{2}$
c. quadratic; $y = -\frac{1}{2}x^2$ **3a.** quadratic
b. $y = 5000x^2$

Chapter 11

Check Your Readiness p. 614

1. $2.8\overline{3}$ or $2\frac{5}{6}$ **2.** 51.75 **3.** 1.825 **4.** −0.75 **5.** $204.\overline{6}$
6. 3 **7.** 3 **8.** 13 **9.** 30 **10.** 4 **11.** 45 **12.** 17 **13.** 3
14. 66 **15.** 11 **16.** 2 **17.** 15 **18.** $\frac{3}{5}$ **19.** −0.06
20. 6.32 **21.** 9.17 **22.** 10.20 **23.** 1.79 **24.** 2 **25.** 2
26. 0 **27.** 1 **28.** 2 **29.** 2

Lesson 11-1 pp. 616–619

Check Skills You'll Need 1. 1 **2.** 1 **3.** 3 **4.** 4 **5.** 2 **6.** 13
7. 5 **8.** 7

Quick Check 1a. $5\sqrt{2}$ **b.** $-50\sqrt{3}$ **c.** $3\sqrt{2}$
2a. $3n\sqrt{3}$ **b.** $-2a^4\sqrt{15a}$ **c.** $xy^2\sqrt{y}$ **3a.** 26
b. $15c\sqrt{2}$ **c.** $60a^2\sqrt{2a}$ **4.** 6 mi **5a.** 4 **b.** $\frac{5p\sqrt{p}}{q}$
c. $\frac{5\sqrt{3}}{4t}$ **6a.** $3\sqrt{2}$ **b.** $\frac{4}{5}$ **c.** $3x$ **7a.** $\sqrt{3}$ **b.** $\frac{\sqrt{10t}}{6t}$
c. $\frac{\sqrt{70m}}{10}$

Lesson 11-2 pp. 622–624

Check Skills You'll Need 1. $2\sqrt{13}$ **2.** $10\sqrt{2}$ **3.** $12\sqrt{6}$
4. $5x\sqrt{5}$ **5.** $\frac{\sqrt{33}}{11}$ **6.** $\frac{\sqrt{10}}{4}$ **7.** $\frac{\sqrt{30x}}{2x}$

Quick Check 1a. $-7\sqrt{5}$ **b.** $-4\sqrt{10}$ **2a.** $8\sqrt{5}$
b. $-3\sqrt{3}$ **3a.** $2\sqrt{5} + 5\sqrt{2}$ **b.** $2x\sqrt{3} - 11\sqrt{2x}$
c. $5a + 3\sqrt{5a}$ **4a.** $-33 - 21\sqrt{2}$ **b.** $23 + 8\sqrt{7}$
5a. $2(\sqrt{7} - \sqrt{5})$ **b.** $-2(\sqrt{10} - 2\sqrt{2})$
c. $\frac{-5(\sqrt{11} + \sqrt{3})}{8}$ **6.** 55 in.

Lesson 11-3 pp. 629–631

Check Skills You'll Need 1. 1 **2.** 4 **3.** 4 **4.** 3 **5.** $x + 1$
6. $2x - 5$

Quick Check 1a. 25 **b.** 81 **c.** 38 **2a.** about
67 ft **b.** No; $h - 2r$ decreases as r increases.
3. 5 **4a.** A principal square root must be a
nonnegative number. **b.** 2 **5.** no solution

Lesson 11-4
pp. 638–639, 643

Check Skills You'll Need

1. 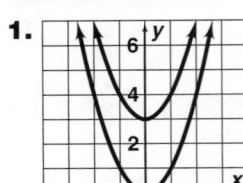 **2.**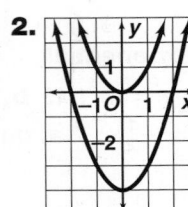

3. 2 **4.** 0 **5.** 11

Quick Check **1.** $x \geq 7$

2a.

b. 361.25 ft

3. 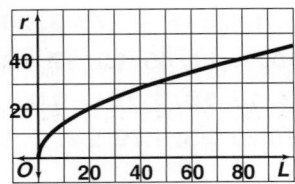 **4.**

Checkpoint Quiz 2 **1.** $-8\sqrt{7}$ **2.** $4 - 5\sqrt{2}$
3. $2\sqrt{5} + 15$ **4.** $3\sqrt{2}$ **5.** $5 + 2\sqrt{6}$
6. $4\sqrt{3} + 2\sqrt{10} - 9\sqrt{2} - 3\sqrt{15}$ **7.** 256
8. 8 **9.** 5 **10.** It is translated 2 units left.

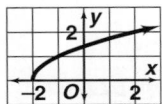

Lesson 11-5
pp. 646–647

Check Skills You'll Need **1.** $\frac{3}{5}; \frac{4}{5}; \frac{4}{3}$ **2.** $\frac{5}{13}; \frac{12}{13}; \frac{12}{5}$ **3.** 20
4. 6.8 **5.** 25 **6.** 7.28

Quick Check **1a.** $\frac{12}{13}; \frac{5}{13}; \frac{12}{5}$ **b.** They are equal.
2a. 0.9397 **b.** 0.3420 **c.** 2.7475 **d.** 0.0872
3a. 6.9 **b.** 5.6

Lesson 11-6
pp. 650–651, 653

Check Skills You'll Need **1.** 0.5878 **2.** 1.4826 **3.** 0.2079
4. 0.9563

Quick Check **1.** about 270 ft **2.** about 745,000 ft

Checkpoint Quiz 2 **1.** 39.1 cm + 4 **2.** 32.7 in.
3. 9.8 cm **4.** about 11.7 m **5.** 2. about 6.8 m

Chapter 12

Check Your Readiness
p. 662

1. $1\frac{1}{6}$ **2.** $\frac{9}{13}$ **3.** $\frac{47}{50}$ **4.** $\frac{5}{12}$ **5.** $\frac{3}{10}$ **6.** $\frac{1}{10}$ **7.** $\frac{2}{5}$ **8.** $\frac{3}{5}$
9. $3w^2x$ **10.** r^2s^2 **11.** $\frac{5}{2}k^4$ **12.** 100 **13.** no
solution **14.** 1 **15.** $x \geq 0$ **16.** $x \geq 0$ **17.** $x \leq \frac{10}{3}$

Lesson 12-1
pp. 664–668

Check Skills You'll Need **1.** $-10; -8; -5$ **2.** 8, 4, 13
3. $\frac{1}{9}$, 1, 27

4. 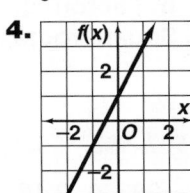 **5.**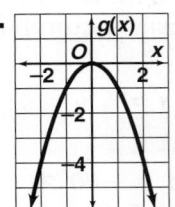

6.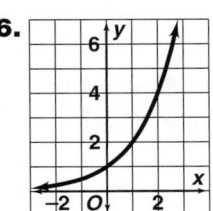

Quick Check

1a.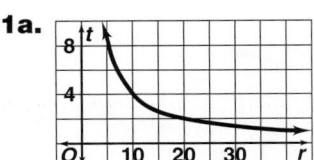

b. Speed and time are both positive values, so graphing it in the first quadrant only makes sense.

2a. asymptote $x = -2$; **b.** asymptote $x = 3$;

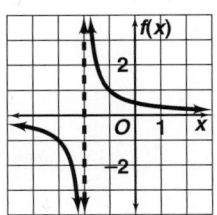 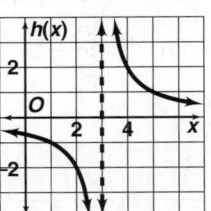

3a. $x = -2, y = -3$; **b.** $x = 4, y = 1$;

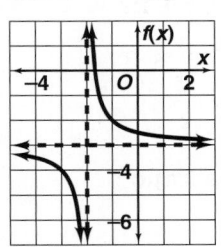 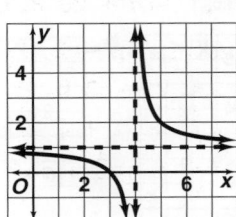

4a. an absolute value function with vertex $(-4, 0)$
b. a graph of exponential growth **c.** a rational function with asymptotes $x = -1$ and $y = 0$

Lesson 12-2 pp. 672–674, 676

Check Skills You'll Need 1. 4 **2.** $-\frac{5}{8}$ **3.** $\frac{5}{7}$
4. $(x + 4)(x - 3)$ **5.** $(x + 4)(x + 2)$
6. $(x - 5)(x + 3)$ **7.** $(x + 4)^2$ **8.** $(x + 3)(x - 4)$
9. $(x - 3)(x - 4)$

Quick Check 1a. $\frac{3}{5b}$ **b.** $\frac{4c^2}{c + 2}$ **c.** 2 **d.** 4
2a. $\frac{3}{x - 5}$ **b.** $\frac{2}{z - 3}$ **c.** $\frac{8}{2a + 1}$ **d.** $\frac{c - 3}{c + 3}$ **3a.** -1
b. $-\frac{1}{m + 8}$ **c.** $-\frac{4}{r + 4}$ **d.** $-\frac{2}{3}$ **4a.** 51 min **b.** $\frac{30rh}{r + h}$

Checkpoint Quiz 1 1. $x = 4, y = 0;$

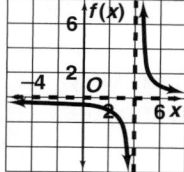

2. $x = 0, y = 4;$ **3.** $6(x - 2)$ **4.** $\frac{c + 3}{c - 3}$ **5.** $\frac{1}{k + 4}$

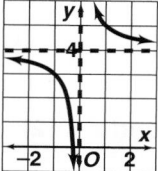

Lesson 12-3 pp. 677–679

Check Skills You'll Need 1. r^{10} **2.** b^7 **3.** c^5 **4.** $6x^9$
5. $5n^4$ **6.** $-45a^5$ **7.** $(2c + 1)(c + 7)$
8. $(15t - 11)(t - 1)$ **9.** $(2q + 1)(q + 5)$

Quick Check 1a. $\frac{-12}{a^5}$ **b.** $\frac{(x - 5)(x - 7)}{x(x + 3)}$ **2.** $-\frac{1}{x}$
3a. $3(c - 1)(c + 1)$ **b.** $2v(v - 5)$ **c.** $\frac{4(m + 2)}{m + 1}$ **4a.** $\frac{1}{b}$
b. $\frac{5(7m - 10)}{7(m - 10)}$ **c.** $\frac{3n + 2}{2n - 3}$ **5a.** $-\frac{1}{10x^2}$ **b.** $\frac{y + 3}{(y + 2)^2}$
c. $\frac{1}{z + 4}$

Lesson 12-4 pp. 682–684

Check Skills You'll Need 1. $-4a^2 + 9a + 1$ **2.** $-x^3 + 3x^2 + 5x - 6$ **3.** $8t - 2$ **4.** $2x^2 + 10x + 12$
5. $-3n^2 + 11n + 20$ **6.** $6a^3 - 21a^2 + 2a - 7$

Quick Check 1a. $m - 2 + \frac{1}{3m}$ **b.** $2t^3 + 4t - 1 + \frac{1}{2t}$ **2a.** $2b - 3$ **b.** $3m - 4 - \frac{3}{2m + 1}$
3a. $t^3 + t^2 + 2t + 3$ **b.** $c^2 - 3c + 5 - \frac{3}{c + 3}$
4a. $4x + 3 - \frac{4}{2x + 1}$ **b.** $-2a - 5 - \frac{1}{3a - 2}$

Lesson 12-5 pp. 687–689

Check Skills You'll Need 1. $\frac{2}{3}$ **2.** $-\frac{2}{7}$ **3.** -2 **4.** $1\frac{1}{18}$
5. $-\frac{1}{12}$ **6.** $-\frac{1}{3}$ **7.** $\frac{2x}{3}$ **8.** $\frac{x}{2}$ **9.** $\frac{1}{2y}$ **10.** $(x + 2)(x + 1)$
11. $(y + 3)(y + 4)$ **12.** $(t - 2)^2$

Quick Check 1a. $\frac{5}{x + 2}$ **b.** $\frac{4y}{y - 5}$ **c.** $\frac{7n}{n + 1}$
2a. $-\frac{1}{t - 2}$ **b.** $\frac{2b - 3}{b + 2}$ **c.** $\frac{-c + 5}{5m + 2}$ **3a.** $\frac{9 + 14y^2}{21y^4}$
b. $\frac{16 - 49x}{100x}$ **c.** $\frac{5(b + 1)}{12b^2}$ **4a.** $\frac{8t + 7}{(t + 4)(t - 1)}$
b. $\frac{m^2 + 5m + 3}{(2m + 1)(m - 1)}$ **c.** $\frac{3a^2 + 2a + 2}{(a + 2)(2a - 1)}$
5. $\frac{1270}{r} + \frac{1270}{1.12r} \approx \frac{2404}{r}$

Lesson 12-6 pp. 692–694

Check Skills You'll Need 1. $1\frac{2}{3}$ **2.** $1\frac{1}{5}$ **3.** $-9, 9$ **4.** $4n$
5. $15x$ **6.** $24y^2$

Quick Check 1a. -2 **b.** $-\frac{25}{2}$ **2a.** $\frac{1}{2}, 2$ **b.** $-2, 3$
3. $28\frac{1}{8}$ min **4a.** -3 **b.** $-5, 4$ **5a.** no solution
b. -1

Lesson 12-7 pp. 699–702, 705

Check Skills You'll Need 1. $\frac{1}{2}$ **2.** $\frac{1}{2}$ **3.** $\frac{1}{6}$ **4.** 0 **5.** $\frac{1}{36}$ **6.** $\frac{1}{4}$

Quick Check
1a. Shirt 1 → Shorts 1 → Shirt 1, Shorts 1
        → Shorts 2 → Shirt 1, Shorts 2
        → Shorts 3 → Shirt 1, Shorts 3
        → Shorts 4 → Shirt 1, Shorts 4
  Shirt 2 → Shorts 1 → Shirt 2, Shorts 1
        → Shorts 2 → Shirt 2, Shorts 2
        → Shorts 3 → Shirt 2, Shorts 3
        → Shorts 4 → Shirt 2, Shorts 4
There are eight possible outfits.
b. Answers may vary. Sample: No, because it would take a lot of room to make such a diagram.
2. 15 pizzas **3.** 40,320 ways **4a.** 504 **b.** 210 **c.** 20
5a. 5040 **b.** A six-letter password; there are more possible passwords, since $26 > 10$.

Checkpoint Quiz 2 1. $(x + 2)(x + 4)$ **2.** $\frac{4}{3}$
3. $a^2 + 5a + 15 + \frac{44}{a - 3}$ **4.** $\frac{5}{x - 3}$ **5.** $\frac{2(7m - 6)}{(m + 2)(m - 3)}$
6. $\frac{3(2t + 1)}{t^2}$ **7.** $\frac{6}{7}$ **8.** -9 **9.** $-8, 1$ **10a.** 16 **b.** 12

Lesson 12-8 pp. 706–709

Check Skills You'll Need 1. 60 **2.** 120 **3.** 210 **4.** 840
5. $\frac{1}{4}$ **6.** $\frac{5}{72}$ **7.** $\frac{3}{4}$ **8.** 0.07

Quick Check 1a. 6 **b.** 35 **c.** 210 **2a.** $_5C_2$ **b.** 10
c. Order does not matter. 3. $\frac{3}{10}$ **4a.** 210 **b.** 1
c. $\frac{1}{210}$

Chapter NY

Check Your Readiness
p. NY 724

1. about 70.7, 75.5, 72; median **2.** about 62.7, 45, 0; median

3.

Ages of Members of the Drama Club

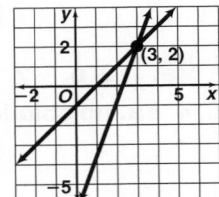

4.

Number of Students in Various Clubs

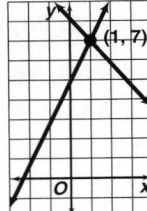

5. (3, 2);

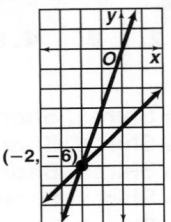

6. (1, 7);

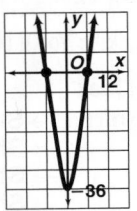

7. (−2, −6);

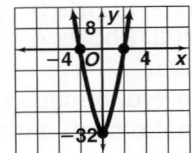

8. ±6;

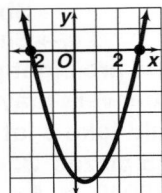

9. ±2;

10. −2, 3;

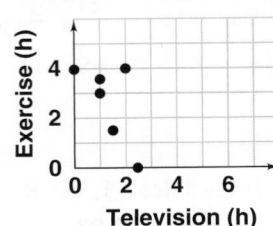

Lesson NY-1
pp. NY 726–NY 728

Check Skills You'll Need 1.

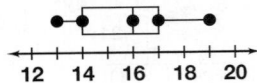

Quick Check 1a. quantitative **b.** qualitative **2a.** bivariate **b.** univariate **3.** stratified sample; not a good sample **4.** Unbiased; there are no

influencing words included. **5.** People who have e-mail might be more inclined to use technology. You are excluding people from the survey who do not have e-mail capability. **6.** The vertical axis does not start at zero.

Lesson NY-2
pp. NY 732–NY 735

Check Skills You'll Need 1. 17, 12 **2.** 8.9, 2.1 **3.** 21, 0 **4.** 50, 22.5

Quick Check 1. 530, 485, 605 **2a.** 60, 75, 85, 95, 105 **b.** 5, 7, 15, 21 **3a.** 60 **b.** No; since the score is equal to itself, the numerator is always greater than zero. **4.**

Distances of Migration of Birds (thousands of miles)

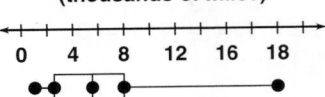

5. 68, 76.5, 84, 91.5, 98 **6.** Answers may vary. Sample: New Orleans has more rainy months than Miami.

Lesson NY-3
pp. NY 738–NY 740

Check Skills You'll Need 1. 1, 2, 3, 4, 6, 12 **2.** 2, 3, 5, 7, 11 **3.** c **4.** b **5.** a

Quick Check 1a. $M = \{1, 3, 5, 7, 9, 11\}$ **b.** $M = \{x \mid x$ is an odd whole number and $x < 12\}$ **2.** $\{\ \}, \{x\}, \{y\}, \{x, y\}$ **3.** $P = \{1, 3, 5, 7\}$

4a. $(-\infty, 5)$;

b. $-2 < x \le 5$;

Lesson NY-4
pp. NY 743–NY 745

Check Skills You'll Need
1. $A = \{x \mid x$ is a whole number less than 10\}
2. $B = \{x \mid x$ is an odd whole number less than 9\}
3. $C = \{-14, -12, -10, -8, -6\}$
4. $D = \{8, 9, 10, 12, 14, 15, 16\}$

Quick Check 1. $\{5, 8, 10, 15, 18, 20\}$
2a. $\{2, 8\}$ **b.** Ø **c.** $\{5, 7\}$

3.

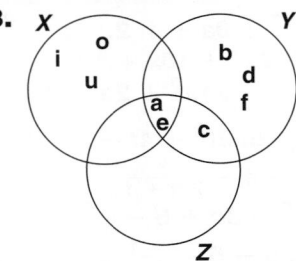

4. 71

Check Skills You'll Need
1. no correlation **2.** positive correlation
3. negative correlation

Quick Check 1. about $20,000
2. about $5000 **3a.** about 1383
b. interpolation; it is among existing data
4. No; there is no causal relationship between the cost of gas and the color of cars that people choose. **5.** No; drinking water does not cause weight gain.

Check Skills You'll Need 1. (2, 0)
2. (−3, −3) **3.** 2, 3

Quick Check 1. (0, 2), (−3, −4);
2a. no solutions **b.** one
solution **3.** (−2, −5), (1, 4)
4. (−6, 42), (7, 114) **5.** two
points **6.** (−2, 2), (1, −1)

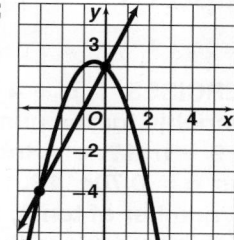

Instant Check System™ Answers

Selected Answers

Chapter 1

Lesson 1-1 pp. 6–8

EXERCISES 1. $p + 4$ **3.** $12 - m$ **9–11.** Choice of variable for the number may vary. **9.** $2n + 2$ **11.** $9 - n$ **17.** c = total cost, n = number of cans, $c = 0.70n$ **19.** ℓ = total length in feet, n = number of tents, $\ell = 60n$ **21–23.** Choices of variables may vary. Samples are given. **21.** w = number of workers, r = number of radios, $r = 13w$ **23.** n = number of sales, t = total earnings, $t = 0.4n$ **25.** $9 + k - 17$ **27.** $37t - 9.85$ **35–37.** Answers may vary. Samples are given. **35.** the difference of 3 and t **37.** the quotient of y and 5 **39.** Choices of variables may vary. Sample is given. n = number of days, c = change in height (m), $c = 0.165n$

Lesson 1-2 pp. 12–15

EXERCISES 1. 59 **3.** 7 **7.** 21 **9.** 124

13.

Original Price P	$P - 0.15P$	Sale Price S
$ 12	12–0.15(12)	$ 10.20
$ 16	16–0.15(16)	$ 13.60
$ 20	20–0.15(20)	$ 17.00
$ 25	25–0.15(25)	$ 21.25

15. 22 **17.** 44 **21.** 704 **23.** 185 **29.** 18 **31.** 0 **35.** 8 cm^3 **37.** 21 ft^3 **41.** 15 **43.** 111 **51.** 17 **53.** 143 **59.** $.16 **63a.** 23.89 in.^{3} **b.** 2.0 in.^{3} **c.** 47.38 in.^{2}

Lesson 1-3 pp. 20–23

EXERCISES 1. integers, rational numbers **3.** rational numbers **11.** Answers may vary. Sample: −17 **13.** Answers may vary. Sample: 0.3 **15.** whole numbers **17.** whole numbers **19.** true **21.** False; answers may vary. Sample: 6 **25.** < **27.** = **29.** $-9\frac{3}{4}, -9\frac{2}{3}, -9\frac{7}{12}$ **31.** −1.01, −1.001, −1.0009 **35.** 9 **37.** 0.5 **43.** Answers may vary. Sample: $\frac{5}{1}$ **45.** Answers may vary. Sample: $\frac{1034}{1000}$ **47.** natural numbers, whole numbers, integers, rational numbers **49.** rational numbers **51.** = **53.** < **57.** 6 **59.** a **67.** sometimes **69.** always

Lesson 1-4 pp. 29–32

EXERCISES 1. $s = 4c$ **3.** $c = 40 + 25h$ **5.** number of minutes; cost **7.** number of gallons of gas; distance traveled; 10 to 12 gallons; 250 to 300 miles **9.** $w = 125m$ **11.a.** Yes; there is one range value for each domain value **b.** year, number of students

Lesson 1-5 pp. 35–37

EXERCISES 3. negative correlation **5.** no correlation **17.** Answers may vary. Sample: ages of adults and the number of hours they sleep.

Lesson 1-6 pp. 43–45

EXERCISES 1. 12; 11; 10; median **3.** 63; 52; none; median **5.** $\frac{3.8 + 4.2 + 5.3 + x}{4} = 4.8$; 5.9 **7.** $\frac{100 + 121 + 105 + 113 + 108 + x}{6} = 112$; 125 **9.** 18 **11.** 20 **21.** −3.1, −2, −1, and −2, 15 **25.a.** A—5.7875, 5.75, 5.4, 1.2
    B—5.5625, 5.45, none, 2.9
b. A—mean. There are no outliers.
    B—median. The mean is thrown off by high outliers
c. Plant A has better quality control because there is a smaller range.

Chapter Review pp. 47–49

1. median **2.** evaluate **3.** algebraic expression **4.** rational **5.** absolute value **6.** rational number **7.** x-coordinate **8.** scatter plot **9.** negative correlation **10.** power **11.** origin **12.** function **13.** Let n = the number; $5 + 3n$. **14.** Let n = the number; $30 - n$. **15.** Let n = the number; $\frac{7}{n}$. **16.** Let n = the number; $n(12)$. **17.** 9 **18.** 64 **19.** 4 **20.** 8 **21.** real numbers, rational numbers **22.** real numbers, irrational numbers **23.** real numbers, rational numbers **24.** real numbers, rational numbers, natural numbers, whole numbers, integers **25.** real numbers, rational numbers, natural numbers, whole numbers, integers **26.** money spent; number of oranges bought; $0 to $8.00; 0 to 10 oranges **27.** hours worked; money earned; 0 to 18 hours; $0 to $117 **28.** $y = 7x$ **29.** $y = 3x + 3$ **30.** $y = 24 - 4x$

31.a.

b. positive correlation

32. 85, 85, 87 **33.** 30.6, 27, 24 **34.** 2.3, 2.3, 2.3
35. 42, 42, 37 **36.** 18, 18.4, 19.9, 3.8 **37.** 6 **38.** 7

Chapter 2

Lesson 2-1   pp. 59–63

EXERCISES 1. $6 + (-3)$; 3 **3.** $-5 + 7$; 2 **5.** 15
7. -19 **25.** $-47 + 12 = -35$, 35 ft
27. $-6 + 13 = 7$, 7°F **29.** -1.7 **31.** -8.7
37. Choices of variable may vary. $c =$ change in
amount of money, $74 + c$ **a.** \$92 **b.** \$45 **c.** \$27
d. \$1.94 **39.** $\begin{bmatrix} -18.2 \\ 11.6 \\ 19.1 \end{bmatrix}$ **41.** $\begin{bmatrix} 1.8 & 22 \\ -\frac{1}{2} & 7 \end{bmatrix}$
43. $-5, -3, -1\frac{1}{2}, 9$ **45.** 13.8 million people
47. Weaving; add the numbers in each column.
49. 0 **51.** 1 **57.** The sum of -227 and 319; the
sum of -227 and 319 is positive, while the sum
of 227 and -319 is negative. **59.** -0.3 **61.** -0.6
65. D **67.** The matrices are not the same size, so
they can't be added.

Lesson 2-2   pp. 66–68

EXERCISES 1. -1

3. -6

9. -4 **11.** -10 **21.** 3 **23.** 6
29. -10 **31.** 1 **37.** \$50.64 **39.** -1.5 **41.** 5.5
53. false; $2 - (-1) = 3$, $3 \not< 2$ or -1 **57.** $\begin{bmatrix} -\frac{1}{4} & 0 & -3 \end{bmatrix}$

Lesson 2-3   pp. 73–76

EXERCISES 1. -15 **3.** 15 **13.** -12 **15.** -15
25. -64 **27.** 4 **33.** 8 **35.** 81 **41.** -4 **43.** 6 **49.** -7
51. 0 **55.** -15 **57.** $-\frac{6}{25}$ **59.** $-8, -7\frac{1}{4}, -3\frac{1}{2}, 4$
63. -6 **65.** $-\frac{1}{10}$ **75.** 0.1 is the multiplicative
inverse of 10 because $0.1(10) = 1$. (-10 is the
opposite of 10.)
77. $\begin{bmatrix} -22 & 10 \\ 18 & -12 \\ 8 & -6 \end{bmatrix}$

Lesson 2-4   pp. 82–85

EXERCISES 1. 2412 **3.** 5489 **9.** \$3.96 **11.** \$29.55
15. $7t - 28$ **17.** $3m + 12$ **27.** $-x - 3$ **29.** $-3 - x$
35. $-3t$ **37.** $7x$ **43.** $3(m - 7)$ **45.** $2(b + 9)$
47. 44,982 **49.** 14.021 **53.** $-28; -18; -3; 4$
55. $-18; -6; 0; 5$ **57.** $6\frac{7}{100}$ $\left(8 + \frac{4}{3}p\right)$ **59.** $\frac{17}{z - 34}$
65. $4.78d$ **67.** $-6.1t^2 + 13.7t$ **75.** terms: $-7t, 6v$,
$7, -19y$, coefficients: $-7, 6, -19$, constant: 7
77. 10 pennies, 7 nickels, 4 quarters, 1 dime

Lesson 2-5   pp. 88–90

EXERCISES 1. Ident. Prop. of Add.; 0, the identity
for addition, is added. **3.** Ident. Prop. of Mult.;
1, the identity for multiplication, is multiplied.
11. 7400 **13.** 4200 **17a.** def. of subtr. **b.** Dist.
Prop. **c.** addition
19. $25 \cdot 1.7 \cdot 4$
     $= 25 \cdot 4 \cdot 1.7$ Comm. Prop. of Mult.
     $= (25 \cdot 4) \cdot 1.7$ Assoc. Prop. of Mult.
     $= 100 \cdot 1.7$ mult.
     $= 170$ mult.
21. $8 + 9m + 7$
     $= 9m + 8 + 7$ Comm. Prop. of Add.
     $= 9m + (8 + 7)$ Assoc. Prop. of Add.
     $= 9m + 15$ add.
25. $2 + g\left(\frac{1}{g}\right) = 2 + 1$ Inv. Prop. of Mult.
          $= 3$ add.
27. $(3^2 - 2^3)(8759) = (9 - 8)(8759)$ mult.
          $= [9 + (-8)](8759)$ def. of subtr.
          $= 1(8759)$ add.
          $= 8759$ Ident. Prop. of Mult.
33. no **35.** yes **41.** No; $(5 - 3) - 1 = 2 - 1 = 1$,
while $5 - (3 - 1) = 5 - 2 = 3$. **43.** No;
$16 \div (4 \div 2) = 16 \div 2 = 8$, while $(16 \div 4) \div 2 =$
$4 \div 2 = 2$.

Lesson 2-6   pp. 96–99

EXERCISES 1. $\frac{1}{2}$ **3.** $\frac{1}{6}$ **11.** $\frac{5}{6}$ **13.** 1 **15.** $2 : 4$ or $1 : 2$
17. $5 : 1$ **21.** 24% **23.** 15% **29.** $\frac{1}{6}$ **31.** $\frac{1}{3}$ **33.** $\frac{1}{2}$
35. $\frac{1}{450}$ **43.** C **45.a.** 14% **b.** 15% **47.** $\frac{3}{16}$ **49.** $\frac{7}{16}$
51. $1 : 3$ **53.** $4 : 4$ or $1 : 1$

Lesson 2-7   pp. 104–106

EXERCISES 1. $\frac{1}{36}$ **3.** $\frac{1}{18}$ **9.** $\frac{4}{81}$ **11.** $\frac{1}{9}$ **15.** $\frac{2}{11}$ **17.** $\frac{1}{55}$
21. $\frac{2}{7}$ **23.** $\frac{1}{6}$ **25.** $\frac{1}{9}$ **31.** Indep.; the data set hasn't
changed. **33.a.** 0.58 **b.** 0.003248 **35.** D
43.a. 8% **b.** 4 customers

1. term **2.** matrix **3.** reciprocal **4.** Distributive Property **5.** Commutative **6.** an outcome **7.** complement of an event **8.** independent **9.** sample space **10.** additive inverse **11.** -17 **12.** -5 **13.** 9.9 **14.** 24.9 **15.** -12 **16.** 0 **17.** 10 **18.** -40 **19.** $-\frac{5}{9}$ **20.** 2 **21.** $-\frac{5}{4}$ **22.** $\frac{15}{8}$ **23.** $4m + 3$ **24.** $b + 10$ **25.** $-5w + 20$ **26.** $36 - 27j$ **27.** $-3 + 30y$ **28.** $-2r + 1$ **29.** $35b + 5$ **30.** $7 - 25v$ **31.** $9t - \frac{6}{5}$ **32.** $-18 + 9m$ **33.** $-4 + x$ **34.** $10g + 1.5$ **35.** Assoc. Prop. of Add. **36.** Indent. Prop. of Add. **37.** Comm. Prop. of Mult. **38.** Dist. Prop.

39. $19 + 56\left(\frac{1}{56}\right)$

$\quad = 19 + 1 \quad$ Inv. Prop. of Mult.

$\quad = 20$ add.

40. $-12p + 45 - 7p$

$= -12p + 45 + (-7p)$	def. of subtr.
$= -12p + (-7p) + 45$	Comm. Prop. of Add.
$= [-12 + (-7)]p$	Dist. Prop.
$= -19p + 45$	add.

41. $2(7 - v) - 3v$

$= (14 - 2v) - 3v$	Dist. Prop.
$= 14 + (-2v) + -(3v)$	def. of. subtr.
$= 14 + (-2v + -3v)$	Assoc. Prop. of Add.
$= 14 + [(-2) + (-3)]v$	Dist. Prop.
$= 14 + (-5v)$	add.
$= 14 - 5v$	def. of subtr.

42. $24\,abc - 24\,bac$

$= 24\,abc - 24\,abc$	Comm. Prop of Mult.
$= 24\,abc + (-24\,abc)$	def. of subtr.
$= 0$	Inv. Prop. of Add.

43. $4 \cdot 13 \cdot 25 \cdot 1$

$= 4 \cdot 13 \cdot (25 \cdot 1)$	Assoc. Prop. of Mult.
$= 4 \cdot 13 \cdot 25$	Ident. Prop. of Mult.
$= 4 \cdot 25 \cdot 13$	Comm. Prop. of Mult.
$= (4 \cdot 25) \cdot 13$	Assoc. Prop. of Mult.
$= 100 \cdot 13$	Mult.
$= 1300$	Mult.

44. $4[m - 2(2m + 3)]$

$= 4(m - 4m - 6)$	Dist. Prop.
$= 4[m + (-4m) + (-6)]$	def. of subtr.
$= 4[1m + (-4m + (-6)]$	Ident. Prop. of Mult.
$= 4[m(1 + (-4)) - 6]$	Dist. Prop. of Mult.
$= 4[(-3m) - 6]$	add
$= -12m - 24$	Dist. Prop.

45. 0 **46.** $\frac{5}{6}$ **47.** $\frac{1}{3}$ **48.** $\frac{1}{6}$ **49.** $\frac{5}{6}$ **50.** $\frac{1}{2}$ **51.** $1:7$ **52.** $1:1$ **53.** $3:5$ **54.** $3:1$ **55.** $1:3$ **56.** $1:1$ **57a.** $\frac{1}{4}$ **b.** Sample: P(not 3 heads) means the probability of getting 0, 1, 2, or 4 heads. **58.** $\frac{11}{14}$ **59.** dependent; $\frac{2}{45}$ **60.** independent; $\frac{1}{20}$ **61.** Dep.; $\frac{5}{18}$ **62.** Indep.; $\frac{1}{4}$ **63.** Dep.; $\frac{1}{18}$ **64.** Answers may vary. Sample: In probability, two events are dependent if the outcome of one influences the outcome of the other. In everyday language, if one person is dependent on another, the first person relies on the second for support. **65.** Indep.; the result of one number cube does not affect the other. **66.** Dep.; once you select one sock, there are fewer socks when you make the second selection.

Chapter 3

EXERCISES 1. -10 **3.** -1 **21.** $2n + 4028 = 51{,}514$; 23,743 books **23.** $c = 39.95 + 0.35m$; 85 min **25.** Add. Prop. of Eq., Simplify., Mult. Prop. of Eq., Simplify. **27.** Subtr. Prop. of Eq., Simplify., Div. Prop. of Eq., Simplify. **29.** 75 **31.** 1

41.

$7 - 3k - 7 = -14 - 7$	Subtr. Prop. of Eq.
$-3k = -21$	Simplify.
$\frac{-3k}{-3} = \frac{-21}{-3}$	Div. Prop. of Eq.
$k = 7$	Simplify.

43.

$\frac{-y}{2} + 14 - 14 = -1 - 14$	Subtr. Prop. of Eq.
$\frac{-y}{2} = -15$	Simplify.
$\frac{-y}{2}(-2) = -15(-2)$	Mult. Prop. of Eq.
$y = 30$	Simplify.

45. $P = 0.8c - 500$; \$6437.50 **47.** 43 **49.** 2 **55.** 15.5 **57.** 31.5 **59.** The neg. sign was dropped in the term $-3y$; -1. **63.** $100°$; $98.6°$; $20°$; $32°$; $-40°$

EXERCISES 1. 9 **3.** $5\frac{4}{7}$ **11.** $x + 9 + x = 25$; 8 ft by 9 ft **13.** 8 **21.** 11 **23.** 46 **31.** 21 **33.** 3.5 **39.** 2 **41.** 5 **55.** $4\frac{2}{3}$ **57.** 92 mi

EXERCISES 1. 3 **11.** $16.95 + 0.05m = 22.95 + 0.02m$ 200 min **13.** 4.25

17a. Answers may vary. Sample:

$\quad$ 0: $\quad 9 = 9$

$\quad$ 3: $\quad -9 = -9$

-4: $\quad 33 = 33$

-6: $\quad 45 = 45$ **b.** identity

19. no solution **21.** identity **25.** 0 **27.** 10 **33.** $1200 + 9b = 25b$; 75 bags **35.** $a = 3$, $b = 6$, $c = 5$, $d = \frac{1}{3}$ **37.** The student subtracted y from both sides instead of adding y to both sides; 5.3.

EXERCISES 1. \$9.50/h **3.** 131 cars/week **9.** A **11.** B **13.** 480 **15.** 10,800 **17.** 11.25 **19.** 25.2 **25.** 105.6 km **27.** $8\frac{11}{12}$ **29.** $-3\frac{1}{2}$ **33.** 18.75 **35.** 18.25 **39.** 15 mi/h **41.** 1 mi/h **45.** $5.\overline{3}$ **49.** about 646 students **57.** 48 V

Lesson 3-5 pp. 152–155

EXERCISES 1. $\overline{AB} \cong \overline{PQ}$, $\overline{BC} \cong \overline{QR}$, $\overline{CA} \cong \overline{RP}$; $\angle A \cong \angle P$, $\angle B \cong \angle Q$, $\angle C \cong \angle R$ **3.** 3.125 ft
5. 80 in. **11.** 10.8 in. **13.** 145.25 mi
15. 350 mi **19.** $2\frac{2}{3}$ in. by 4 in. **27.** 3 ft **33.** C

Lesson 3-6 pp. 162–165

EXERCISES 1a. Let w = width **b.** $\ell = w + 3$
c. $2w + 2(w + 3) = 30$; 6 **d.** 9 in. **3.** 9 cm; 18 cm
5. 304, 305, 306 **7.** −148, −150
11a. x; $2\frac{1}{4} - x$ **b.** $22x = 72 - 3x$, $1\frac{1}{3}$ h **15.** −31,
−29, −27 **17.** 12:30 P.M. **21.** 175 mi/h; 375 mi/h
23. 1993, 1994, 1995 **27.** 6 6-V; 4 12-V

Lesson 3-7 pp. 171–173

EXERCISES 1. 50%; increase **3.** 25%; increase
13. 39% **15.** 0.5 ft **17.** 0.005 g **19.** 19.25 cm^2,
29.25 cm^2 **21.** 46.75 in.^{2}, 61.75 in.^{2} **25.** 25%
27. 12.5% **29a.** 48 cm^3 **b.** 74.375 cm^3
c. 28.125 cm^3 **d.** 26.375 cm^3 **e.** 55%
31. 22%; decrease **33.** 175%; increase **39.** 2%
41. 1 mm **45.** 24.5 cm^2, 25.5 cm^2 **47.** 54.1 in.^{2},
54.3 in.^{2} **51.** 11%

Lesson 3-8 pp. 178–180

EXERCISES 1. 13 **3.** $\frac{1}{3}$ **13.** irrational
15. irrational **17.** 5 and 6 **19.** −12 and −11
21. 3.46 **23.** 107.47 **25.** 0.93 **27.** 0 **29.** $\pm\frac{3}{7}$
41. $-\frac{2}{5}$ **43.** 1.26

Lesson 3-9 pp. 184–187

EXERCISES 1. 10 **3.** 17 **17.** no
19. yes **23.** no **25.** yes **27.** $\frac{4}{15}$ **29.** 1.25
33. C **35.** 559.9 **37.** 9.7 **41.** 12.8 ft **45.** A figure
is a square; the figure is a rectangle; if a figure is
a rectangle then the figure is a square; false.
47. An angle is a right angle; its measure is 90°;
if the measure of an angle is 90°, then it is a right
angle; true. **49.** 6 in.

Chapter Review pp. 191–193

1. inverse operations **2.** Solutions of equivalent
equations **3.** rate **4.** cross products **5.** greatest
possible error **6.** 4 **7.** 3 **8.** 5 **9.** 3 **10.** 2
11. 11 **12.** Let p = number of people; $6p + 3 =$
27; 4 people

13.
$$314 = -n + 576$$
$$314 - 576 = -n + 576 - 576 \quad \text{Subtr. Prop. of Eq.}$$
$$-262 = -n \quad \text{Simplify.}$$
$$-1(-262) = -1(-n) \quad \text{Mult. Prop. of Eq.}$$
$$262 = n \quad \text{Simplify.}$$
14. $-\frac{1}{4}w - 1 = 6$
$$-\frac{1}{4}w - 1 + 1 = 6 + 1 \quad \text{Add. Prop. of Eq.}$$
$$-\frac{1}{4}w = 7 \quad \text{Simplify.}$$
$$-4(-\frac{1}{4}w) = -4(7) \quad \text{Mult. Prop. of Eq.}$$
$$w = -28 \quad \text{Simplify.}$$
15. $3h - 4 = 5$
$$3h - 4 + 4 = 5 + 4 \quad \text{Add. Prop. of Eq.}$$
$$3h = 9 \quad \text{Simplify.}$$
$$\frac{3h}{3} = \frac{9}{3} \quad \text{Div. Prop. of Eq.}$$
$$h = 3 \quad \text{Simplify.}$$
16. −18 **17.** 4 **18.** $-\frac{1}{2}$
19. $\frac{3}{2}$ **20.** 2 **21.** 0 **22.** no solution **23.** identity
24. 20 **25.** 10 **26.** $2x + 2(x - 6) = 72$; width =
15 cm, length = 21 cm **27.** 150 mi/h **28.** 3.4 mi/h
29. 0.2 mi/h **30.** 2 **31.** 2.3 **32.** −6 **33.** 20
34. 6 **35.** 5 **36.** 7.5 **37.** 19.5 **38.** 36 ft
39. $4r = 7.6$; 1.9 h or 1 h 54 min **40.** $n + (n + 1) +$
$(n + 2) = 582$; 193, 194, 195 **41.** $10t + 18(t - 2) =$
209; 6.75 h or 6 h 45 min **42.** 13%; increase
43. 25%; decrease **44.** 33.3%; decrease
45. 5.4% **46.** irr.; 9.27 **47.** rat.; −11
48. irr.; ±0.71 **49.** irr.; 1.60 **50.** rat.; $-\frac{2}{5}$ **51.** 5.8
52. 17.8 **53.** 14.8 **54.** 9.8 **55.** yes **56.** yes

Chapter 4

Lesson 4-1 pp. 202–205

EXERCISES 1. yes **3.** yes **5.** no **9a.** no **b.** no **c.**
yes **11a.** no **b.** no **c.** no **15.** C **17.** D
19. (number line from −1 to 3, open circle) **21.** (number line from −7 to −2, closed dot)
27–35. Choice of variable may vary. **27.** $x > -3$
29. $x \geq 1$ **33.** Let s = number of students. $s \leq 48$
35. Let w = safe number of watts. $w \leq 60$
39. b is greater than 0. **41.** z is greater than or
equal to −5.6.
53. (number line −1 to 3, open circle) **55.** (number line −2 to 2, closed dot)
57. (number line −2 to 2, closed dot) **59.** (number line 0 to 5, closed dot)
65. Since 1231 < 1513, Option A < Option B.
67. Put an open dot at 3 and color the rest of the
number line.

Lesson 4-2 — pp. 208–211

EXERCISES 1. 5 **3.** 4.3 **5.** $t < 1$

7. $d \geq 10$

21. $\frac{5}{3}$ **23.** $w \leq 5$;

25. $b > -7$;

39. $s + 637 \leq 2000$, $1363 **41.** $r + 17 + 12 \geq 50$, 21 reflectors **43.** Subtract 9 from each side.
45. $w \geq 11$ **47.** $y < 3.1$ **63.** nearly 51.2 MB
67. $x \geq 1$ **69.** $t \leq -3$

Lesson 4-3 — pp. 215–217

EXERCISES 1. $t \geq -4$;

3. $w \leq -2$;

17. $t < -3$;

19. $w \leq -5$;

29. $4.5c \geq 300$, 67 cars **31–33.** Answers may vary. Samples are given. **31.** $-2, -3, -4, -5$
33. $-3, -4, -5, -6$ **39.** Multiply each side by -4 and reverse the inequality symbol. **41.** Divide each side by 5. **45.** -2 **47.** 4 **51.** x and y are equal. **53–55.** Estimates may vary. **53.** $j > -6$
55. $s \geq 28$ **59.** $d \leq -7$ **61.** $s < -\frac{1}{4}$ **75.** Yes; in each case, y is greater than 6.

Lesson 4-4 — pp. 222–225

EXERCISES 1. $d \leq 4$ **3.** $x > -2\frac{1}{2}$

11. $27 \geq 2s + 8$ and $s \leq 9.5$, so the two equal sides must be no longer than 9.5 cm. **13.** $j \geq 1$
15. $h > 5$ **23.** $t \leq -1$ **25.** $n \geq 2$ **31.** $v \geq -4$
33. $r \geq -1\frac{4}{5}$ **37.** Subtract 8 from each side.
39. Add 5 and subtract y from each side. **43.** for x = the number of guests, $(0.75)200 + 1.25x \geq 250$, $x \geq 80$, so at least 80, guests must attend.
45. B **47.** F **53.** $m \geq 2$ **55.** $n \geq -2\frac{5}{8}$
73. Distribute 4 to 2 as well as n, so $n > -7$.

Lesson 4-5 — pp. 229–232

EXERCISES 1. $-4 < x$ and $x < 6$ or $-4 < x < 6$;

3. $23 < c < 23.5$;

5. $-5 < j < 5$;

7. $2 < n \leq 6$;

17. $n \leq -3$ or $n \geq 5$;

19. $h < 1$ or $h > 3$;

21. $b < -2$ or $b > 2$;

23. $c < 2$ or $c \geq 3$;

29. $-2 < x < 3$ **31.** $x \leq 0$ or $x > 2$ **33.** $q \leq -2\frac{2}{3}$ or $q > 4$ **35.** $4 \leq t \leq 14$ **39.** D **43.** $6 < x < 30$
45. $11 < x < 21$ **47.** $15 \leq D \leq 30$

Extension — p. 233

1. $A \cup B = \{2, 5, 7, 8, 9\}$, $A \cap B = \{5\}$

3. $C \cup D = \{2, 4, 6, 7, 8\}$, $C \cap D = \{2,8\}$

5. $B \cup D = \{2, 4, 5, 6, 7, 8, 9\}$, $B \cap D = \{\}$ or $\emptyset$

7a. $P \cap Q = \{x | x,$ is a multiple of 12$\}$,

b. Yes; since $40 \in Q$, $40 \in P \cup Q$

Lesson 4-6 — pp. 237–240

EXERCISES 1. $-2, 2$ **3.** $-\frac{1}{2}, \frac{1}{2}$ **13.** 3, 13 **15.** $-3, 1$
23. $k < -2.5$ or $k > 2.5$;

25. $-8 < x < 2$;

35. between 12.18 mm and 12.30 mm, inclusive
37. $-9, 9$ **39.** $-1\frac{1}{2}, 1\frac{1}{2}$ **53.** $|n| > 7.5$
55. $|n + 1| \geq 3$ **57.** 39%, 45% **63.** $|x - 2| = 4$
65. $|x - 12\frac{1}{2}| = 3\frac{1}{2}$

Extension — p. 241

1. Symmetric Prop. of Equality **7.** Add. Prop. of Ineq., Add. Prop. of Ineq., Transitive Prop. of Ineq.

Chapter Review — pp. 243–245

1. C **2.** B **3.** A **4.** C **5.** D **6.**

7.

8.

9.

10. $n < -2$ **11.** $n \geq -3.5$ **12.** $n > -6$ **13.** $n \geq 2$
14. Let p = number of people, $p \geq 600$.
15. Let n = number of people, $n \leq 15$. **16.** Let t = temperature in degrees Fahrenheit, $t < 32$.

17. $h > -1$; 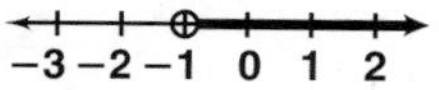
−3 −2 −1 0 1 2

18. $t < -5$; 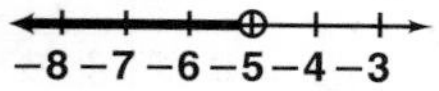
−8 −7 −6 −5 −4 −3

19. $m \geq -3$; 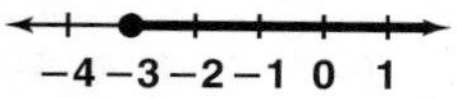
−4 −3 −2 −1 0 1

20. $w \geq -2$; 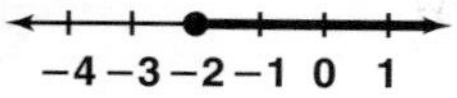
−4 −3 −2 −1 0 1

21. $q > -2.5$; 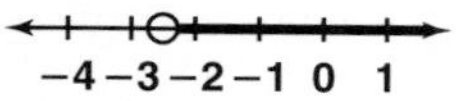
−4 −3 −2 −1 0 1

22. $y > -14$; 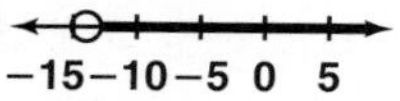
−15 −10 −5 0 5

23. $n \leq 15$; 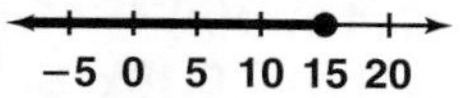
−5 0 5 10 15 20

24. $d \geq 4$; 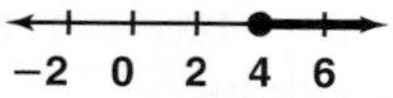
−2 0 2 4 6

25. $-2 \leq t$; 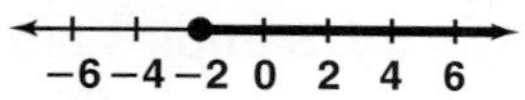
−6 −4 −2 0 2 4 6

26. $0 < c$; 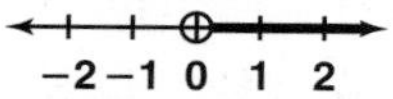
−2 −1 0 1 2

27. $2.5 \geq u$; 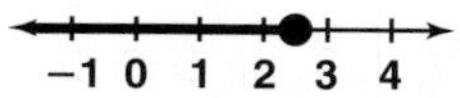
−1 0 1 2 3 4

28. $-9 < p$; 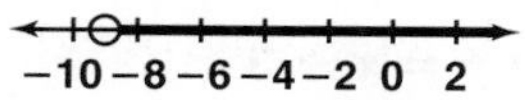
−10 −8 −6 −4 −2 0 2

29. $3.50 + 2.75 + x \leq 12.00$,
$x \leq 5.75$

30. $7.25h \geq 200$, $h \geq 27.586$
You must work at least 28 h.

31. $n > -2$ **32.** $k \leq -\frac{1}{2}$ **33.** $b < 40$ **34.** $c \leq -2$
35. $m < -6$ **36.** $t > 3$ **37.** $x \geq 2$ **38.** $y \leq -56$
39. $x < \frac{8}{3}$ **40.** $190 + 0.04x \geq 500$, $x \geq 7750$

41. 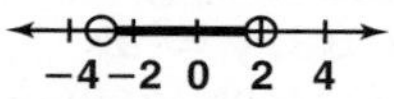
−4 −2 0 2 4

42. 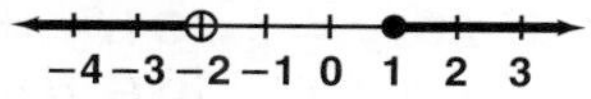
−4 −3 −2 −1 0 1 2 3

43. 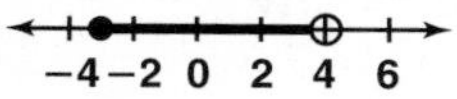
−4 −2 0 2 4 6

44. $-2 \leq z < 4$, 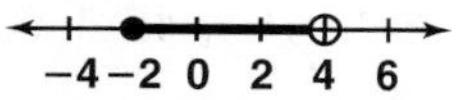
−4 −2 0 2 4 6

45. $-\frac{5}{2} \leq d < 4$, 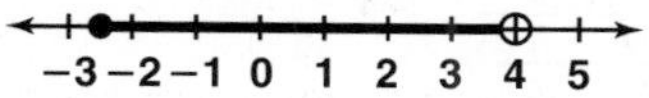
−3 −2 −1 0 1 2 3 4 5

46. $-\frac{3}{2} \leq b < 0$, 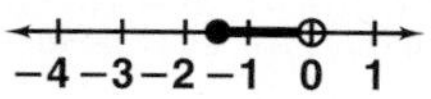
−4 −3 −2 −1 0 1

47. $t \leq -2$ or $t \geq 7$, 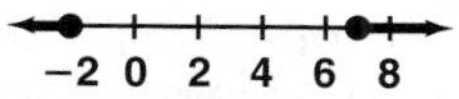
−2 0 2 4 6 8

48. $2 \leq a < 5$, 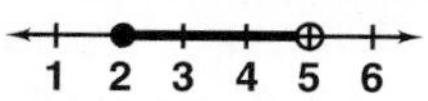
1 2 3 4 5 6

49. $2 \leq a < 4$, 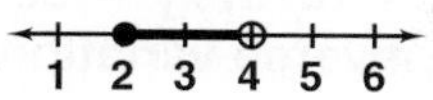
1 2 3 4 5 6

50. $75 \leq t \leq 89$

51. $|n + 2| > 3$
52. $|n - 12| \leq 5$ **53.** 5 or -5
54. $n \leq -6$ or $n \geq 2$ **55.** $-3 \leq x \leq 3$
56. $-9.6 < m < 9.6$ **57.** $x < 3$ or $x > 4$
58. 6.5 or -12.5 **59.** 8 **60.** all real numbers
61. no solution **62.** $k < -7$ or $k > -3$
63. -5 or 1 **64.** $z \leq -0.25$ or $z \geq 0.25$
65. $2.74 \leq d \leq 2.86$ **66.** $19.6 \leq \ell \leq 20.4$

Chapter 5

Lesson 5-1 pp. 254–256

EXERCISES 1. Labels may vary. Sample is given.

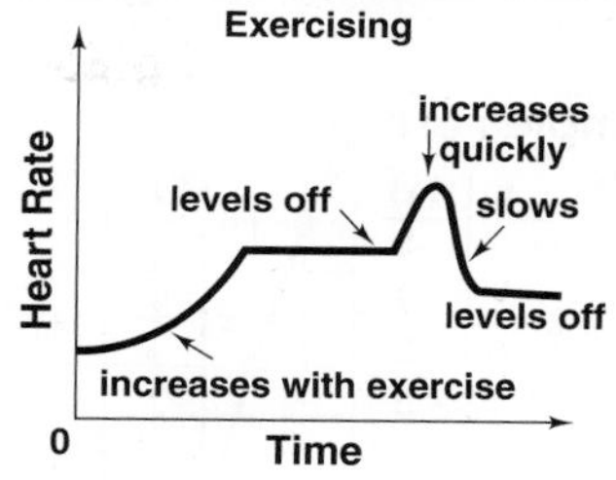
Exercising
Heart Rate
increases
quickly
levels off
slows
levels off
increases with exercise
0
Time

5.

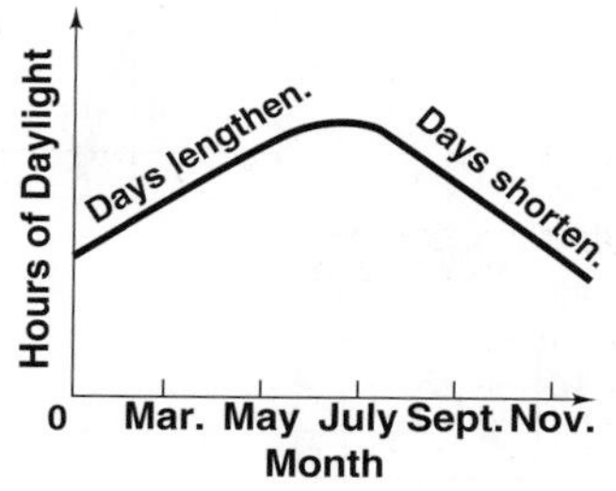
Hours of Daylight
Days lengthen.
Days shorten.
0
Mar. May July Sept. Nov.
Month

9. C; the temperature increases steadily and then alternates cooling and warming as the oven turns off and on during a cooking cycle.

11a.

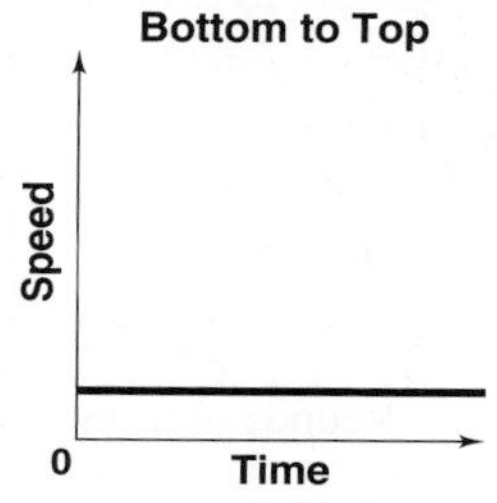
Bottom to Top
Speed
0
Time

b.

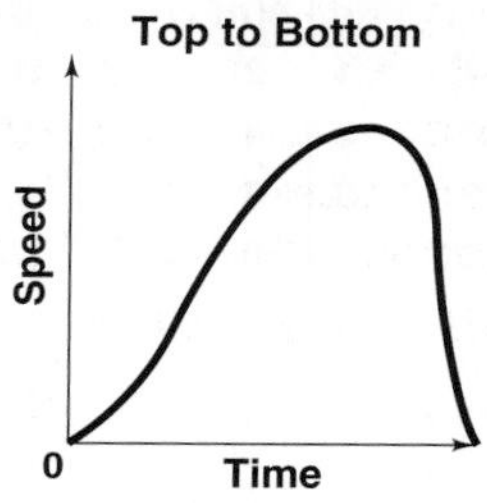
Top to Bottom
Speed
0
Time

No; the graphs are different because you have a constant speed traveling up but not down.

15a.

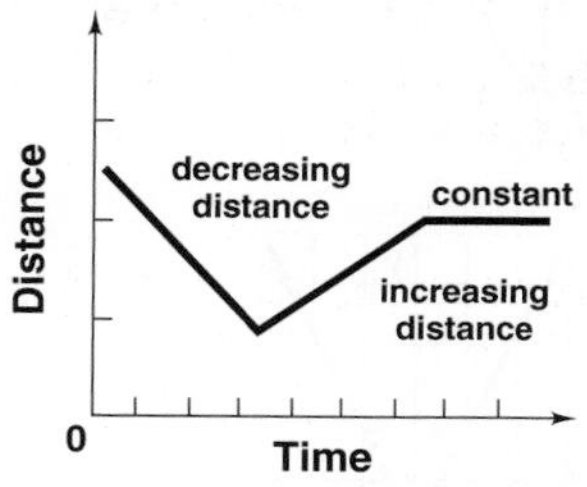
Distance
decreasing
distance
constant
increasing
distance
0
Time

b. section showing the distance decreasing
c. first 2 sections

Lesson 5-2 pp. 259–261

EXERCISES 1. no **3.** yes **5.** yes **7.** no

9.

x	$x + 7$	$f(x)$
1	$1 + 7$	8
2	$2 + 7$	9
3	$3 + 7$	10
4	$4 + 7$	11

11.

x	x^2	$f(x)$
1	1^2	1
2	2^2	4
3	3^2	9
4	4^2	16

17. {0.5, 53} **19.** {−27, −7, −2, 8, 48}
21. no **23.** yes; {−4, −1, 0, 3}; {−4}

27. {−3, 3, 15.8} **29.** {−0.5, 0, 2.7}
33. no **35.** yes **37.** yes **39.** no

Lesson 5-3   pp. 266–267

EXERCISES
1–3. Tables may vary. Samples are given.

1.

x	f(x)
−1	3
0	0
1	−3

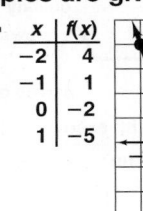

3.

x	f(x)
−2	4
−1	1
0	−2
1	−5

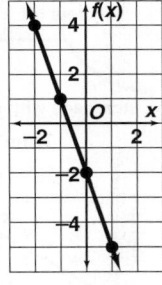

11a.

ℓ	P(ℓ)
1	5
2	10
3	15
4	20

11b.

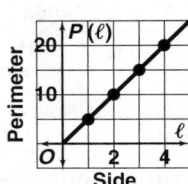

13a. continuous **13b.**

n	A(n)
1	2
2	4
3	6
4	8

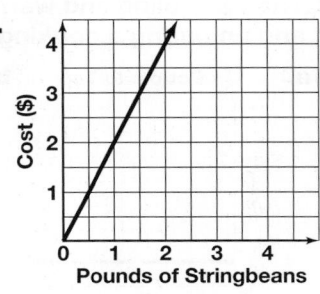

15. 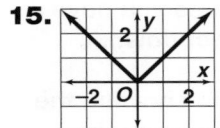 **17.**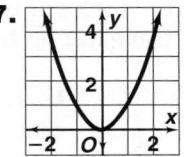

27. 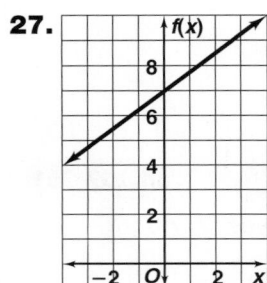 **29.**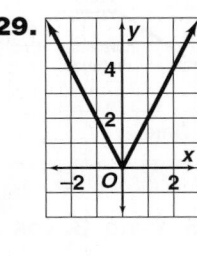

37a. 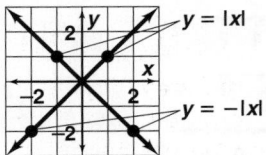 **39a.**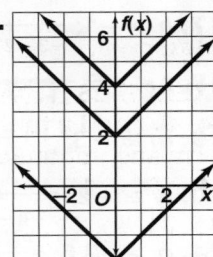

Lesson 5-4   pp. 272–275

EXERCISES 1. B **3.** C **11.** $d(n) = 45n$
13. $e(n) = 6.37n$ **17a.** $f(x) = 0.19x$ **b.** \$1.52
19. $f(x) = 1000x$ **21a.** $C(a) = 10a + 1$ **b.** \$31
c. 61; the total cost of 12 books
27–29. Tables may vary. Samples are given.

27.

x	y
−1	3
0	2
1	1
2	0
3	−1

$y = −x + 2$

29.

x	y
−4	−1
−2	1
0	3

$y = x + 3$

Lesson 5-5   pp. 280–283

EXERCISES 1. no **3.** yes; −2 **11.** $y = \frac{1}{5}x$
13. $y = \frac{9}{5}x$ **23.** Choices of variables may vary.
$E(h) = 7.10h$ **25.** no **27a.** $\frac{20}{50}$ or 0.4 **b.** $f = 0.4w$,
52 lb **29.** $y = \frac{1}{6}x$ **31.** $y = −\frac{36}{25}x$ **33.** $y = 9x$
41. 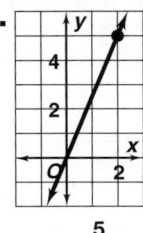 **45a.** $\frac{1}{32}$ **b.** $b = \frac{1}{32}w$

$y = \frac{5}{2}x$

Lesson 5-6   pp. 288–290

EXERCISES 1. $xy = 18$ **3.** $xy = 56$ **11.** 15 **13.** 7
23. $13.\overline{3}$ mi/h **25.** inverse variation; $xy = 60$
27. Direct variation; the ratio $\frac{\text{cost}}{\text{pound}}$ is constant at
\$1.79. **29.** Inverse variation; the product of the
length and width remains constant with an area
of 24 square units. **31.** 1.1; $rt = 1.1$ **33.** 1; $ab = 1$
37. Inverse variation; the product of the rate and
the time is always 150. **39.** 121 ft **41.** direct
variation; $y = 0.4x$; 8 **43.** inverse variation;
$xy = 48$; 0.5 **47.** A

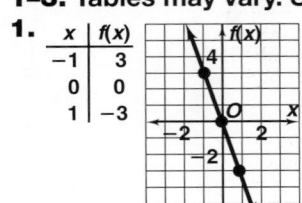
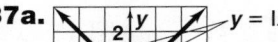

Lesson 5-7

pp. 294–297

EXERCISES 1. "Add 2 to the previous term"; 12, 14. **3.** "Add 2 to the first term, 3 to the second term and continue, adding 1 more each time"; 18, 24. **13.** 3 **15.** −11 **23.** −3, 15, 39 **25.** 17, 44, 80 **35.** $3\frac{1}{4}$, $3\frac{1}{2}$ **37.** $\frac{4}{27}$, $\frac{4}{81}$ **47.** $4500, $4350, $4200, $4050, $3900; the balance after 4 payments **49.** No; there is no common difference. **51.** No; there is no common difference. **57.** 4.5, −4.5, −22.5 **59.** 1, $2\frac{3}{5}$, $5\frac{4}{5}$ **63.** value of new term = value of previous term + 6 **65.** value of new term = value of previous term − 2.5

Chapter Review

pp. 299–301

1. C **2.** D **3.** A **4.** E **5.** B **6.** G **7.** F
8. Answers may vary. Sample: A computer rental costs $2.50/h. If you start with a fixed amount of money, the longer you work on the computer, the less money you will have left.
9. Answers may vary. Sample: A residential thermostat senses when the temperature in the room falls below the set level. The heater is turned on until the temperature is 3°F above the set level. The heater is then turned off. The graph shows the air temperature rising while the heater is working, and falling after the heater is turned off.
10. Answers may vary. Sample: An elevator is on the second floor. Someone gets in, goes to the 11th floor, and gets off.
11–14. Answers may vary. Samples are given.

11.

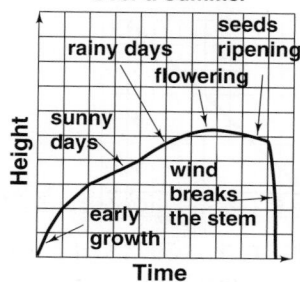

Height of a Sunflower Over a Summer

12.

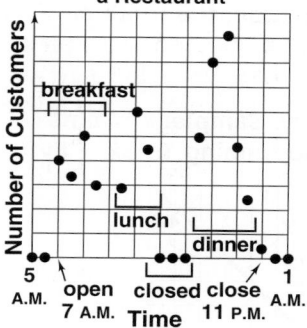

Number of People in a Restaurant

13.

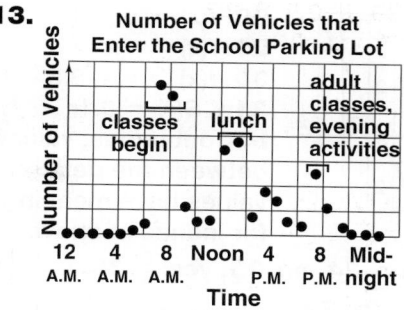

Number of Vehicles that Enter the School Parking Lot

14.

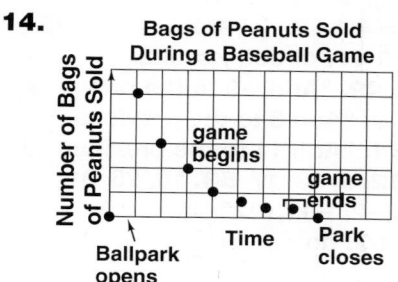

Bags of Peanuts Sold During a Baseball Game

15. {−23, −7, −3, 13} **16.** {1, 3, 3.5, 5.5} **17.** {1, 2, 17, 26} **18.** {−10, 2, 5, 17} **19.** no **20.** yes **21.** yes **22.** no **23.** A relation is a function when each value of the domain corresponds to exactly one value of the range.
24–27. Tables may vary. Samples are given.

24.

x	f(x)
−1	−2
0	−3
1	−2

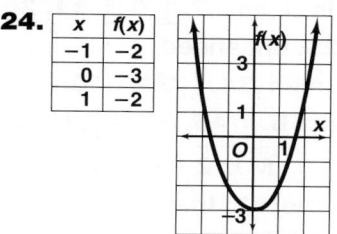

25.

x	f(x)
−2	−2
0	−3
2	−4

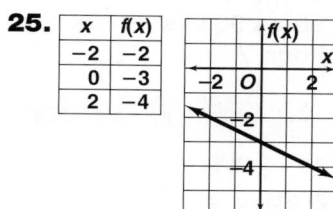

26.

x	y
−2	−5
−1	−6
0	−7
1	−6
2	−5

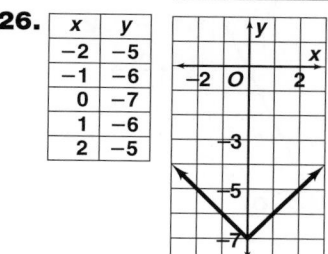

27.

x	y
-2	-3
-1	-1
0	1
1	3

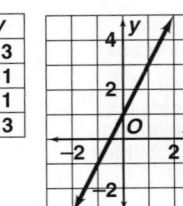

28. $f(x) = x + 1$
29. $f(x) = -x$
30. $f(x) = x + 3.5$
31a. $S(r) = 0.1r$
b. Continuous; values between the data values have meaning for inches of snow.

32. yes; −3 **33.** no **34.** no **35.** yes; $\frac{2}{5}$ **36.** $y = \frac{1}{5}x$
37. $y = x$ **38.** $y = 2x$ **39.** $y = -3x$ **40.** $xy = 6$
41. $xy = 9$ **42.** $xy = 4.4$ **43.** 4 **44.** 2.5 **45.** 18.75
46. inverse; $xy = 70$ **47.** direct; $y = 8.2x$
48. inverse; $xy = 3$ **49.** "Add −9 to the previous term"; 63, 54, 45. **50.** "Add 3 to the previous term"; 17, 20, 23. **51.** "Add 11 to the previous term"; 56, 67, 78. **52.** 3, 13, 17 **53.** 10, 25, 31 **54.** 4.5, 12, 15 **55.** −2, −17, −23
56. yes; 42, 49, 56 **57.** no

Chapter 6

Lesson 6-1   pp. 312–315

EXERCISES 1. 3; the temperature increases 3°F each hour. **3.** $-\frac{1}{15}$ gal/mi; the amount of fuel consumed each month is $\frac{1}{15}$ gal. **7.** $\frac{1}{2}$ **9.** $\frac{2}{3}$ **11.** 2
13. $-\frac{3}{2}$ **23.** undefined **25.** undefined
27. $\frac{9}{10}$ in./month **29.** 30 mi/h **31.** $\frac{1}{6}$ **33.** −20
37.

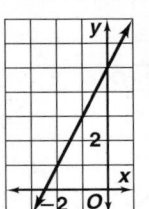

41a. $\frac{2}{3}$ **b.** $\frac{2}{3}$ **45.** $\overline{JK}$: $-\frac{1}{2}$; $\overline{KL}$: 2; $\overline{ML}$: $-\frac{1}{2}$; $\overline{MJ}$:: 2 **49.** −6 **51.** 4
55. true **57.** true
61a. 111; $111/h
b. about 56 customers per hour

Lesson 6-2   pp. 320–323

EXERCISES 1. −2; 1 **3.** 1; $-\frac{5}{4}$ **11.** $y = 3x + \frac{2}{9}$
13. $y = 1$ **23.** $y = \frac{3}{4}x + 2$ **25.** $y = \frac{1}{2}x + \frac{1}{2}$
29.

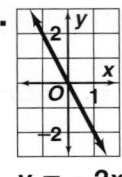

$y = \frac{2}{3}x - 1$

31.

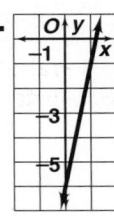

$y = 2x + 5$

41. −3; 2
43. 9; $\frac{1}{2}$

51.

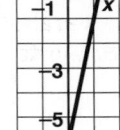

$y = -2x$

53.

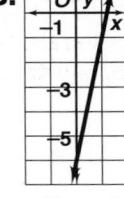

$y = 5x - 6$

59. no **61.** no
67. $y = -4x + 7$
69. $y = -\frac{1}{4}x + 5$

Lesson 6-3   pp. 325–328

1. $y = 1.2x$

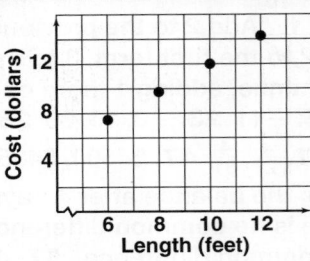

3. $y = 25 - 3w$

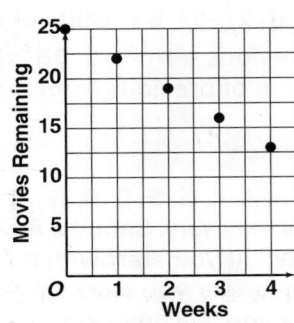

5a. Yes; the candle burns continuously.
b. h = height of candle in inches and t = time in hours the candle has burned; $h = 8 - 2t$
c. The graph of the line would be steeper but the y-intercept would be the same. **9a.** yes; both time and distance are continuous. **b.** The line will be less steep because the kayakers are going more slowly. Since the kayakers are heading back to camp, the slope is negative. **11.** C

Lesson 6-4   pp. 333–335

EXERCISES 1. 18; 9 **3.** −6; 30 **11.** C
13.   **15.**

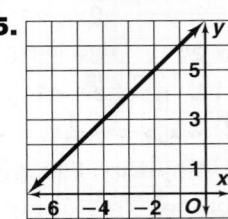

19. horizontal **21.** horizontal
23.   **25.**   **27.** $-3x + y = 1$
                                                29. $x - 2y = 6$

37a. Answers may vary. Sample: x = time walking; y = time running **b.** $3x + 8y = 15$
39.   **41.**

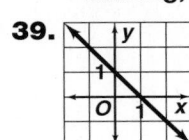

47a. $3x + 7y = 28$
49. $y = \frac{4}{5}x + 10$

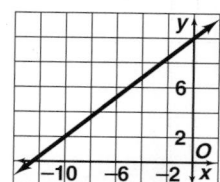

51. $y = -\frac{4}{5}x - 3$
57. $-3x$ instead of $3x$
59. $y = -2$ **61.** $x = -2$

Lesson 6-5   pp. 339–341

EXERCISES 1. 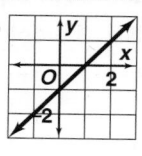 **3.**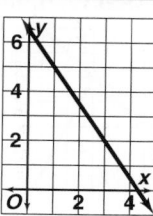

11. $y - 2 = -\frac{5}{3}(x - 4)$ **13.** $y + 7 = -\frac{3}{2}(x + 2)$
19–21. Answers may vary for the point indicated by the equation. **19.** $y = 1(x + 1); y = x + 1$
21. $y + 2 = -\frac{6}{5}(x - 4); y = -\frac{6}{5}x + \frac{14}{5}$ **31.** Yes;
answers may vary. Sample: $y - 9 = -2(x + 4)$
33. no **37–39.** Answers may vary for point indicated by the equation. **37.** $y + 3 = \frac{2}{5}(x - 1)$ **39.** $y - 4 = \frac{3}{2}(x - 1); -3x + 2y = 5$
55. $y = -2.6x + 315.6$ **57.** y-intercept changes

Lesson 6-6   pp. 346–349

EXERCISES 1. $\frac{1}{2}$ **3.** 1 **7.** no, different slopes
9. yes, same slopes and different y-intercepts
13. $y = 6x$ **15.** $y = -2x - 1$ **19.** $-\frac{1}{2}$ **21.** $-\frac{5}{7}$
25. $y = -\frac{1}{2}x$ **27.** $y = 3x - 10$ **31.** $y = \frac{5}{4}x + 1$
33. parallel **35.** neither **41.** $y = -\frac{4}{5}x - \frac{19}{5}$;
$y = -\frac{4}{5}x + \frac{3}{5}$ **43.** $y = -\frac{1}{2}x; y = 2x$ **47.** about $\frac{5}{4}$
49. Answers may vary. Sample: $\frac{5}{4} \cdot \left(-\frac{1}{2}\right) \neq -1$
53. No; the slopes are not neg. reciprocals.
55. False; the product of two positive numbers can't be -1. **59.** The slope of $\overleftrightarrow{JK}$ is $\frac{1}{5}$. The slope of $\overleftrightarrow{KL}$ is -2. The slope of $\overleftrightarrow{LM}$ is $\frac{1}{6}$. The slope of $\overleftrightarrow{JM}$ is -4. The quadrilateral is not a parallelogram.

Lesson 6-7   pp. 352–356

EXERCISES 1–3. Trend lines may vary. Samples are given. **1.** $y - 52.5 = 2(x - 91)$
3. $y - 16.4 = 0.64(x - 69.9)$
7. $y = -1.06x + 92.31; -0.970$
9. $y = -1.63x + 556.76; -0.725$

13a.

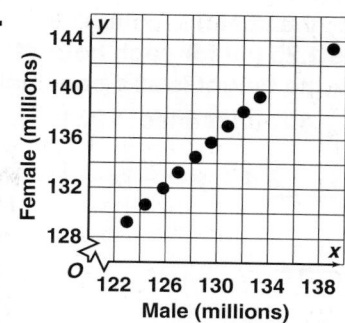

b. Answers may vary. Sample: $y = 0.939x + 13.8$
c. 154,650,000 **d.** Answers may vary. Sample: No, the year is too far in the future. **17.** $y = 0.37x - 28.66$; $12.04 billion **19a.** (2, 3) and (6, 6); $y = 0.75x + 1.5$ **b.** $y = 0.75x + 1.21$

Lesson 6-8   pp. 361–363

EXERCISES 1. Answers may vary. Sample: same shape, shifted 3 units up **3.** Answers may vary. Sample: same shape, shifted 7 units down

5. 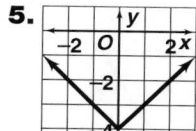 **7.**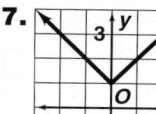

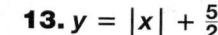

11. $y = |x| - 6$ **13.** $y = |x| + \frac{5}{2}$

17. 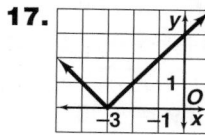 **19.**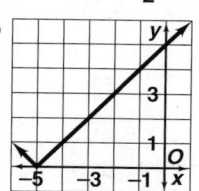

23. $y = |x - 9|$ **25.** $y = |x + \frac{3}{2}|$

29. 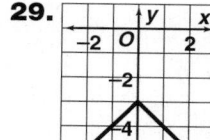 **33.** $y = -|x + 2.25|$
35. $y = -|x - 4|$

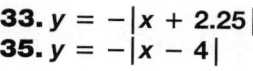

37. 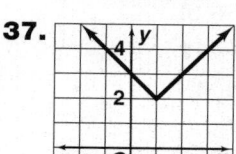 **39.**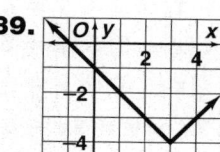

Chapter Review   pp. 365–367

1. perpendicular lines **2.** parallel lines
3. translation **4.** slope **5.** y-intercept **6.** 8 oz/mo

Selected Answers

Selected Answers   **859**

7. 3.375 in./wk **8.** 5; the speed is 5 mi/h. **9.** −1.25; gasoline decreases 1.25 gal for each hour of driving time. **10.** 0; the height is at a constant level of 150 ft. **11.** $\frac{1}{4}$ **12.** undefined **13.** 1

14. $y = -3$ 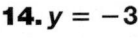 **15.** $y = -7x + \frac{1}{2}$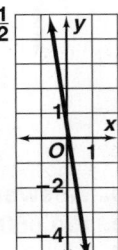

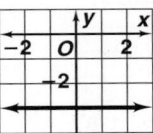

16. $y = \frac{2}{5}x$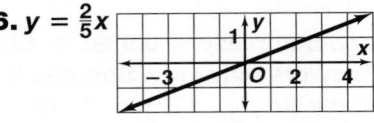

17. $y = -\frac{1}{2}x - \frac{1}{2}$ **18.** $y = \frac{1}{4}x - 3$

19a. $p = 0.25s + 75$

b. 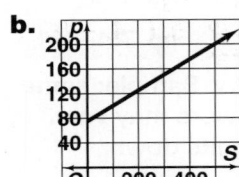 **c.** \$275 **d.** 75; weekly salary when no sales are made

20. 2; 5 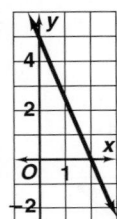 **21.** 8; −13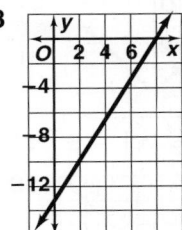

22. −1; $-\frac{1}{3}$ 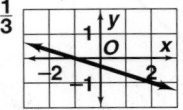 **23.** $3x - 5y = -35$ **24.** $x + 3y = -6$

25. $4x - 3y = 15$
26. $y + 2 = 2(x - 1)$
27. $y + 2 = \frac{3}{4}(x - 1)$
28. $y + 2 = -3(x - 1)$ **29.** $y + 2 = 0$
30. $y - 3 = \frac{1}{3}(x - 4)$ or $y - 1 = \frac{1}{3}(x + 2)$
31. $y + 4 = -\frac{6}{5}(x - 5)$ or $y - 2 = -\frac{6}{5}x$
32. $y = \frac{1}{2}(x + 1)$ or $y + 1 = \frac{1}{2}(x + 3)$
33. $y + 1 = 5(x - 2)$ or $y = 5x - 11$
34. $y - 5 = \frac{1}{3}(x - 3)$ or $y = \frac{1}{3}x + 4$
35. $y + 5 = 9x$ or $y = 9x - 5$
36. $y - 10 = -\frac{1}{8}(x - 4)$ or $y = -\frac{1}{8}x + 10\frac{1}{2}$
37. $l = 3.5w$
38a. Answers may vary. Sample: $y = 1.28x - 60.2$
b. For sample in (a): 80.6 lb/person

39. 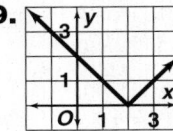 **40.**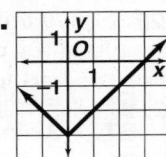

Chapter 7

| Lesson 7-1 | pp. 377–379 |

EXERCISES 1. Yes, (−1, 5) makes both equations true. **3.** Yes, (−1, 5) makes both equations true.

5. (0, 2); 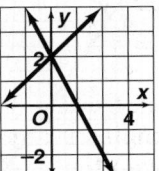 **13a.** 3 weeks **b.** \$35

15. no solution;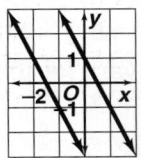

19. no solution; same slope, different y-int.
21. one solution; different slopes **23.** A

25. (20, 60); 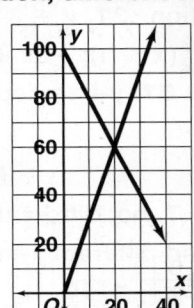 **35.** (−2, 10)
37. (−0.9, 1.6)

| Lesson 7-2 | pp. 384–386 |

EXERCISES 1. D **3.** B **5–13.** Coordinates given in alphabetical order. **5.** (9, 28) **7.** $\left(6\frac{1}{3}, -\frac{1}{3}\right)$

11. (2, 0) **13.** (6, −2) **17.** 4 cm by 13 cm
19. (15, 15) **21.** (−4, 4) **23.** 15 video rentals
25. D **27.** estimate: (2,3); (−2, 3) **29.** estimate:
(−3.5, −3.5); $\left(-\frac{10}{3}, -\frac{11}{3}\right)$ **35.** $\left(-\frac{1}{2}, -\frac{1}{2}\right)$ **37.** $\left(2, -\frac{1}{2}\right)$

| Lesson 7-3 | pp. 390–393 |

EXERCISES 1. (1, 3) **3.** (5, −17) **7a.** $x + y = 20$, $x - y = 4$ **b.** 12 and 8 **9.** (−5, 1) **11.** $\left(-2, -\frac{5}{2}\right)$
15a. $30w + \ell = 17.65$, $20w + 3\ell = 25.65$
b. \$.39 for a wallet size, \$5.95 for an 8 × 10
17. (−1, −3) **19.** (2, −2) **23–25.** Choice of

method may vary. Samples are given. **23.** (−1, −2); substitution; both solved for *y* **25.** (10, 2); substitution; one eq. solved for *x* **33.** (10, −6) **35.** (−15, −1) **39.** 9

Lesson 7-4   pp. 399–402

EXERCISES 1a. $4a + 5b = 6.71$
b. $5a + 3b = 7.12$ **c.** pen: $1.19, pencil: $.39

3a.

a	b	24
0.04a	0.08b	0.05(24)

b. $a + b = 24$; $0.04a + 0.08b = 1.2$ **c.** 18 kg A, 6 kg B **5.** 600 games **7a.** $s + c = 2.75$
b. $s − c = 1.5$ **c.** 2.125 mi/h **d.** 0.625 mi/h
9–11. Answers may vary. Samples are given.
9. Substitution; one eq. is solved for *t*.
11. Elimination; subtract to eliminate *m*.
15a. $t = 99 − 3.5m$; $t = 0 + 2.5m$; $t = 41.25°C$, $m = 16.5$ min **b.** After 16.5 min, the temp. of either piece will be 41.25°C. **19a.** 42 mi/h
b. 12mi/h

Lesson 7-5   pp. 407–410

EXERCISES 1. no **3.** yes **7.** A **9.** B

11.
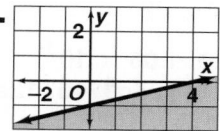

19. $y \leq \frac{2}{3}x − \frac{7}{3}$;   **21.** $y \leq \frac{2}{3}x − \frac{8}{3}$;
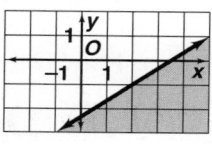   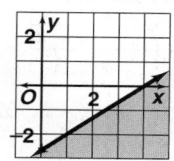

23a. $3x + 5y \leq 48$
b.
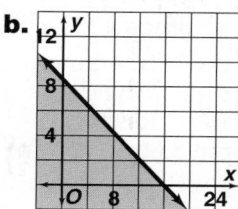
c. Answers may vary. Sample: 8 blue and 4 gold, 2 blue and 8 gold, 12 blue and 2 gold **d.** No; you cannot buy −2 rolls of paper.

25.   **35.** $x \leq −3$
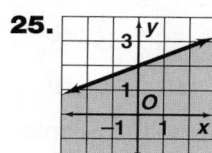

39. $y < 0$;   **43.** $y < x + 2$
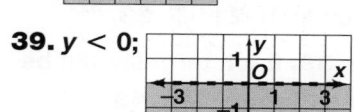

Lesson 7-6   pp. 414–418

EXERCISES 1. no **3.** no **5.**
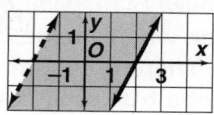

17. $y \geq −\frac{1}{2}x − 2$ and $y \leq \frac{1}{2}x + 2$

19. $y \leq −\frac{2}{3}x − 4$ and $y \geq \frac{1}{5}x − 3$
21.
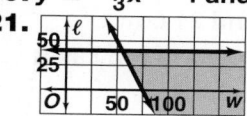

23. $x + y \geq 50$, $4x + 3y \leq 180$
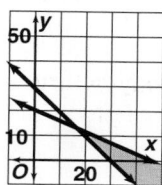
25. $x \leq 3$, $x \geq −3$, $y \leq 3$, $y \geq −3$
27. $y \geq \frac{2}{3}x − 2$, $y < \frac{2}{3}x + 2$
31a. triangle **b.** (2, 2), (−4, −1), (−4, 2) **c.** 9 units^2

35a. $x \geq 1$, $10.99x + 4.99y \leq 45$
b. (3, 0), (3, 1), (3, 2), (4, 0)
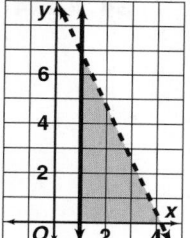

Chapter Review   pp. 421–423

1. elimination **2.** solution of the system of linear equations **3.** system of linear inequalities
4. solution of the inequality **5.** substitution
6. A **7.** No; (2, 5) only satisfies one equation.
8. Infinitely many; the equations are equivalent.
9. Answers may vary. Sample: systems with noninteger solutions

10.
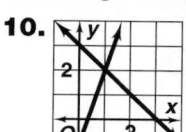
(1, 2) **11.**
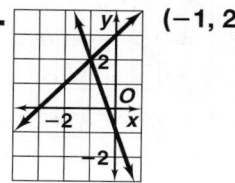
(−1, 2)

12.   **13.**
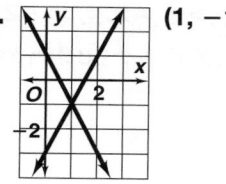
(1, −1)
no solution

14. $(-2, 5)$ **15.** $\left(-4\frac{1}{2}, -6\right)$ **16.** $\left(-1\frac{1}{9}, -\frac{5}{9}\right)$ **17.** $(2, 2)$

18. Answers may vary. Sample: There is no solution when you get a false equation such as $0 = 2$. There are infinitely many solutions when you get a true equation such as $5 = 5$.

19a. $x + y = 24, 4x + 5y = 100$ **b.** $(20, 4)$
c. 20 4-point, 4 5-point **20.** $(-6, 23)$ **21.** $(1, -1)$
22. $(6, 4)$ **23.** $\left(5\frac{5}{11}, 1\frac{7}{11}\right)$ **24.** $x + y = 34$,
$2x + 4y = 110$; 13 chickens and 21 cows
25. $10\frac{2}{3}$ fl oz **26.** 63° and 27° **27.** 18 ft by 39 ft
28. \$1.29 **29.** 154 km/h

30. 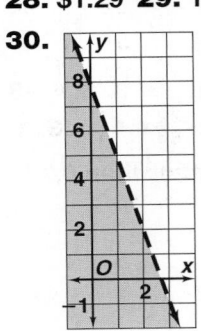 **31.**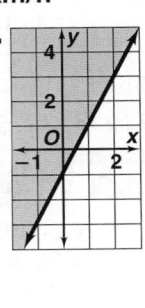

32. 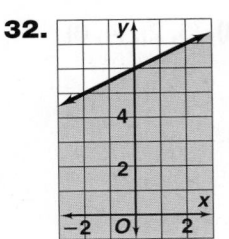 **33.** 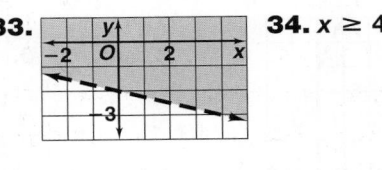 **34.** $x \geq 4$

35. $y \leq 3x + 3$ **36.** $2x + 3y \geq -6$

37. 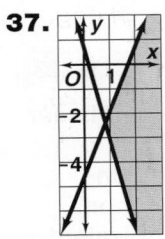 **38.**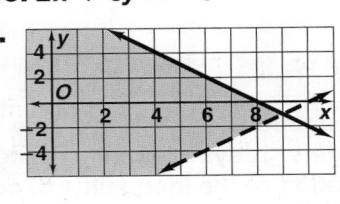

39. 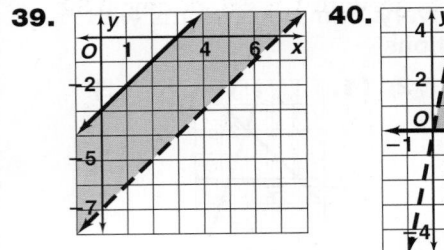 **40.**

41. $y \leq 3, y > x$ **42.** $y > -2x + 2, y > \frac{4}{5}x - 4$
43. $x > -1, y \leq x + 5$ **44.** $y \leq -\frac{3}{2}x + 3$,
$y \geq -\frac{1}{2}x - 1$

Chapter 8

Lesson 8-1 p. 433–435

EXERCISES 1. -1 **3.** $\frac{1}{25}$ **13.** -2 **15.** $0; -3$
17. $3a$ **19.** x^7 **33.** $\frac{1}{25}$ **35.** $-\frac{1}{9}$ **45a.** \$20.48; \$.32
b. No; the value of the allowance rapidly becomes very great. **47.** pos. **51.** 10^{-1} **53.** 10^{-3}
57. 0.000001 **59.** 0.03 **61a.** $5^{-2}, 5^{-1}, 5^0, 5^1, 5^2$
b. 5^4 **c.** $\frac{a^n}{1}$ **63.** 45 **65.** 40 **69.** $\frac{2}{9}$ **71.** $\frac{1}{16}$

73.

a	4	$\frac{1}{3}$	6	$\frac{7}{8}$	2
a^{-1}	$\frac{1}{4}$	3	$\frac{1}{6}$	$\frac{8}{7}$	0.5

75. A, B, D

77. No; $3x^{-2} \cdot 3x^2 = 9 \cdot x^0 = 9$. The product of reciprocals should be 1. **79a.** 1 correct, 0.4096; 2 correct, 0.1536; 3 correct, 0.0256; 4 correct, 0.0016 **b.** 0 or 1

Lesson 8-2 pp. 438–440

EXERCISES 1. No; $55 > 10$. **3.** No; $0.9 < 1$.
7. 9.04×10^9 **9.** 9.3×10^6 **15.** 500 **17.** 2040
27. C,A,B, **29.** 2.4×10^{15} **31.** 3.18×10^{-3}
35. 7×10^1 **37.** 4.6×10^{-2}

Lesson 8-3 pp. 443–446

EXERCISES 1. 2^{10} **3.** 1 **7.** c^5 **9.** $\frac{10}{t^7}$
23. 6×10^9 **25.** 3.4×10^{-5} **29.** 1.08×10^{21}
dollars **31.** 9 **33.** -3 **41.** $4x^4$ **43.** $4c^4$ **45.** $12a^7$
47. $3^4 \cdot 2^2$ **49.** 8.0×10^5 **51.** 1.2×10^{-4}
59. 7.65×10^{14} **61.** 7.039305×10^{-7} **63.** about
6.7×10^{33} molecules **65.** x^3 **67.** $5c^3$

Lesson 8-4 pp. 449–452

EXERCISES 1. c^{10} **7.** $\frac{1}{t^{14}}$ **9.** $625y^4$ **11.** $49a^2$
23. 1.6×10^{11} **25.** 8×10^{-30} **31.** 8.57375×10^{-10} m³ **33.** -4 **35.** -3 **41.** The student who
wrote $x^5 + x^5 = 2x^5$ is correct; x^5 times x^5 is x^{10}.
43. $243x^3$ **45.** $30x^2$ **51a.** $24x^2$; $96x^2$ **b.** 4 times
c. $8x^3$; $64x^3$ **d.** 8 times **53.** $(ab)^5$ **55.** $\left(\frac{2x}{y}\right)^2$
59a. about 5.15×10^{14} m² **b.** about 3.60×10^{14} m²
c. about 1.37×10^{18} m³

Lesson 8-5 pp. 456–459

EXERCISES 1. 7 **3.** -3 **5.** $\frac{1}{4}$ **7.** $\frac{1}{c^3}$ **13.** 5×10^7
15. 6×10^2 **19a.** 3.86×10^{11} h; 2.65×10^8 people
b. about 1457 h **c.** about 4.0 h **21.** $\frac{9}{25}$ **23.** $\frac{32x^5}{y^5}$
37. 5^3 simplifies to 125. **39.** Each term should be
raised to the 4th power and simplified. **43.** $\frac{1}{16m^{12}}$

45. a^6 **53.** $\frac{a^5c^5}{b^3}$ **55.** 5 **63a.** The student treated
$\frac{5^4}{5}$ as $\left(\frac{5}{5}\right)^4$. **b.** 125 **65.** $\left(\frac{m}{n}\right)^7$ **67.** 10^{10} **75.** dividing
powers with the same base, def. of neg. exponent
77. mult. powers with the same base

EXERCISES 1. 4 **3.** 0.1 **7.** 40, 80, 160 **9.** 20.25,
30. 375, 45.5625 **13.** geometric **15.** geometric
19. 5; 135; 10,935 **21.** 5; −135; −10,935
25. $A(n) = 6 \cdot 0.5^{n-1}$; 0.375 **27.** $A(n) = 7 \cdot (1.1)^{n-1}$;
9.317 **29a.** $A(n) = 100 \cdot (0.64)^{n-1}$
b. about 10.74 cm **31.** 1, 0.2, 0.04;
$A(n) = 625 \cdot (0.2)^{n-1}$ **33.** 1, −0.5, 0.25;
$A(n) = 16 \cdot (-0.5)^{n-1}$ **37.** arithmetic; 3, 1, −1
39. geometric; 1.125, 0.5625, 0.28125
41a. $A(n) = 36 \cdot (0.9)^{n-1}$ **b.** 6; $n = 1$ corresponds
to the first swing, because $A(1) = 36$. **c.** 21.3 cm

EXERCISES 1. 216 **3.** 2.5 **9.** $160,000; $320,000
11. $16,000, $32,000 **13.** C **15.** B **17.** C

19. 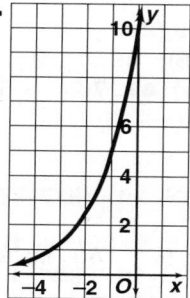   **21.**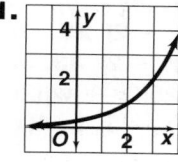

25. 0.04, 0.2, 1, 5, 25, 125;
increase **27.** 100, 10, 1, 0.1,
0.01, 0.001; decrease

37a.

x	y
1	−2
2	4
3	−8
4	16
5	−32

39. $f(t) = 200 \cdot t^2$
41. $f(x) = 100x^2$

EXERCISES 1. 20; 2 **3.** 10,000; 1.01 **7.** 1.05
9. 1.0875 **11.** 0.75%, 0.25% **13.** 1.125%; 0.375%
17. $16,661.35 **19.** $28,338.18 **21a.** 3 half-lives
b. 3.125 mCi **23.** 0.1 **25.** 0.9 **27.** exp. decay
29. exp. decay **31.** $y = 130,000 \cdot (1.01)^x$; about
142,179 people **33.** $y = 2400 \cdot (1.07)^x$; $4721.16
35a. $y = 584 \cdot (1.065)^x$; $3862.79 **37.** Neither; it is
not just one straight line. **39.** Neither; it
decreases and and then increases, unlike an
exponential function. **41.** exponential function

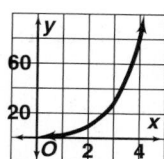

45. 4 half-lives
47a. $y = 6,284,000 \cdot (1.01)^x$
b. 7,667,674 people **49.** 88%
51. 46.1%

1. exponential growth **2.** growth factor
3. Scientific notation **4.** exponential decay
5. decay factor **6.** Compound interest
7. common ratio **8.** interest period **9.** geometric
sequence **10.** exponential function **11.** $\frac{d^6}{b^4}$ **12.** $\frac{y^8}{x^2}$
13. $\frac{7h^3}{k^8}$ **14.** $\frac{q^4}{p^2}$ **15.** $\frac{625}{16}$ or $39\frac{1}{16}$ **16.** $-\frac{1}{8}$ **17.** $-\frac{1}{8}$
18. $\frac{1}{49y^4}$ **19.** $\frac{9x^2}{w^4y^7}$ **20.** 36 **21.** $\frac{4}{9}$ **22.** $\frac{9}{8}$ or $1\frac{1}{8}$ **23.** 1
24. 108 **25.** C **26.** No; for values other than 0,
$(-3b)^4 = 81b^4 \neq -12b^4$. **27.** No; 950 > 10.
28. No; 72.35 > 10. **29.** yes **30.** No; 0.84 < 1.
31. 2.793×10^6 mi **32.** 1.89×10^8 cars and
trucks **33.** $2d^5$ **34.** $q^{12}r^4$ **35.** $-20c^4m^2$ **36.** 1.34^2
or 1.7956 **37.** $\frac{243x^2y^{14}}{64}$ **38.** $-\frac{4}{3r^{10}z^8}$

39. about 7.8×10^3 pores **40.** Answers may
vary. Sample: Simplify $(2a^{-2})^{-2}(-3a)^2$; $\frac{9a^6}{4}$.
41. $\frac{1}{w^3}$ **42.** $\frac{1}{64}$ **43.** $7x^2$ **44.** $\frac{n^{35}}{v^{21}}$ **45.** $\frac{c^3}{e^{11}}$
46. 2×10^{-3} **47.** 2.5×10^1 **48.** 5×10^{-5}
49. 3×10^3 **50.** Answers may vary. Sample:
Simplify and use div. prop: $\left(\frac{a^2}{2}\right)^{-3}$; use raising a
quot. to a power prop.: $\frac{a^{-6}}{2^{-3}}$; use the def. of neg.
exp.: $\frac{2^3}{a^6}$ or $\frac{8}{a^6}$ **51.** 0.1 **52.** 3 **53.** $-\frac{1}{2}$
54. geometric; $\frac{25}{4}, \frac{25}{16}, \frac{25}{64}$ **55.** neither; −30, −25,
−19 **56.** arithmetic; 42, 49, 56 **57.** 6, 12, 24, 48
58. 7.5, 5.625, 4.21875 **59a.** 2430 bacteria
b. about 180 min **60.** $a = 100, b = 1.025$
61. $a = 32, b = 0.75$ **62.** $a = 0.4, b = 2$
63. growth; 3 **64.** growth; 1.5 **65.** decay; 0.32
66. decay; $\frac{1}{4}$

67. 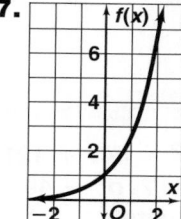   **68.**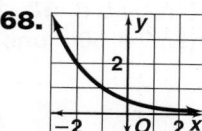

69. 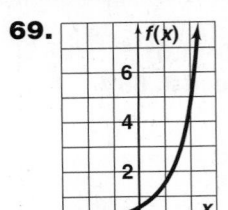 **70.**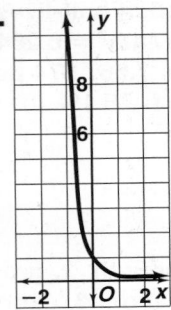

71. about 8.2 mg

Chapter 9

EXERCISES 1. 1 **3.** 0 **9.** quadratic trinomial
11. cubic trinomial **15.** $-3x^2 + 4x$; quadratic
binomial **17.** $c^2 + 4c - 2$; quadratic trinomial
21. $8m^2 + 15$ **23.** $8w^2 - 3w + 4$ **29.** $b + 1$
31. $7n^4 + n^3$ **35.** $18y^2 + 5y$ **37.** $-7z^3 + 6z^2 + 2z - 5$ **39.** $28c - 16$ **43.** $-x^4 + x^3 + 15x$
45. $-h^{10} - 5h^9 + 8h^5 + 2h^4$ **49.** $5x + 18$
51. No; both terms of a binomial cannot be
constants.

EXERCISES 1. $8m^2 + 48m$ **3.** $63k^2 + 36k$ **13.** 3
15. 12 **19.** $2(3x - 2)$ **21.** $5(2x^3 - 5x^2 + 4)$
25. Karla; Kevin multiplied $-2x$ by 3 instead of -3.
27. $-12a^3 + 15a^2 - 27a$ **29.** $-60c^3 + 36c^2 - 48c$
33a. $A = 16\pi x^2 - 4x^2$ **b.** $A = 4x^2(4\pi - 1)$
35. $24x(x^2 - 4x + 2)$ **37.** $x^2(5x^2 + 4x + 3)$

EXERCISES 1. 30 **3.** 7 **5.** $x^2 + 7x + 10$
7. $k^2 + k - 42$ **11.** $r^2 + 2r - 24$ **13.** $x^2 - x - 42$
21. $-2x^2 + 5x + 48$ units^2 **23.** $a^3 - 6a^2 + 9a - 4$
25. $3k^3 + 19k^2 - 33k + 56$ **27.** $2t^3 - 17t^2 + 36t - 15$ **29.** $48w^3 - 28w^2 - 2w + 2$ **31.** $p^2 + p - 56$ **33.** $25c^2 - 40c - 9$ **39a.** $2x^2 + 12x + 16$
b. $12x + 16$ **c.** 10 ft by 5 ft **43.** $1.5x^2 + 2.5x - 1$
units^2 **45.** $n^3 + 15n^2 + 56n$ units^3

EXERCISES 1. $c^2 + 2c + 1$ **3.** $4v^2 + 44v + 121$
11. 9801 **13.** 91,204 **15.** $x^2 - 16$ **17.** $d^2 - 49$
21. 899 **23.** 2496 **27.** $(10x + 15)$ units^2
29. $25p^2 - 10pq + q^2$ **31.** $x^2 - 14xy + 49y^2$
45. $p^2 - 81q^2$ **47.** $49b^2 - 64c^2$ **65.** $2x^2 - 23x + 66$ **67.** $3y^2 + 4y + 1$ **73.** 8.713×10^3
75. 6.8952×10^4

EXERCISES 1. 5 **3.** 7 **5.** $(r + 3)(r + 1)$ **7.** $(k + 3)$
$(k + 2)$ **17.** 5 **19.** 9 **21.** $(x + 4)(x - 1)$ **23.** $(y + 5) \cdot$
$(y - 4)$ **31.** B **33.** $(t + 9v)(t - 2v)$ **35.** $(p - 8q) \cdot$
$(p - 2q)$ **39–41.** Answers may vary. Samples are
given. **39.** 18; $(x - 6)(x + 3)$, 28; $(x - 7)(x + 4)$,10;
$(x - 5)(x + 2)$ **41.** 7; $(x + 4)(x + 3)$, 8; $(x + 6)(x + 2)$,
13; $(x + 12)(x + 1)$ **43.** $(k + 2)(k + 8)$ **45.** $(n - 4) \cdot$
$(n + 14)$ **55.** $4x^2 + 12x + 5$; $(2x + 1)(2x + 5)$

EXERCISES 1. $(2n + 1)(n + 7)$ **3.** $(11w - 3)(w - 1)$
13. $(2t - 3)(t + 1)$ **15.** $(2q + 3)(q - 7)$
29. Answers may vary. Sample: 18; $(5m - 4)$
$(3m + 6)$, 54; $(5m - 2)(3m + 12)$, 117; $(5m - 1)$
$(3m + 24)$ **33.** $(9p + 4)(6p + 7)$ **35.** $(7x - 2)$
$(2x - 7)$

EXERCISES 1. $(c + 5)^2$ **3.** $(h + 6)^2$ **7.** $(2m + 5)$
9. $(5g - 4)$ **11.** $(8r - 9)^2$ **13.** $(x + 2)(x - 2)$
23. $(7y + 2)(7y - 2)$ **25.** $(2m + 9)(2m - 9)$
31. $3(m + 2)(m - 2)$ **33.** $3(x + 8)^2$ **39.** 11, 9
41. 15, 5 **45.** $25(2v + w)(2v - w)$ **47.** $7(2c + 5d)^2$
55a. $4(x + 5)(x - 5)$ **b.** $4(x + 5)(x - 5)$

EXERCISES 1. $2m^2$; 3 **3.** $2z^2$; -5 **5.** $(2n^2 + 1) \cdot$
$(3n + 4)$ **7.** $(3t + 1)(3t - 1)(3t + 5)$ **11.** $2(2v^2 + 1) \cdot$
$(3v - 8)$ **13.** $2(m^2 + 2)(10m - 9)$ **17.** $(6p + 5) \cdot$
$(2p + 1)$ **19.** $(6n - 1)(3n + 10)$ **27.** $5k$, $(k + 2)$, and
$(k + 4)$ **29.** $2(10t^2 - 11)(3t - 10)$ **31.** $4(3x - 7y) \cdot$
$(x + 2y)$ **35.** $(7w^2 - 4)(2w + 7)$ **37.** $2(2t^2 + 3) \cdot$
$(11t - 1)$ **39.** $2w$, $(6w + 5)$, and $(7w + 1)$

1. A **2.** D **3.** E **4.** C **5.** B **6.** $-6y^2 + 8y + 2$;
quadratic trinomial **7.** $9h^2 + 1$; quadratic
binomial **8.** $3k^5 + k$; fifth degree binomial
9. $7t^2 + 8t + 9$; quadratic trinomial **10.** x^2y^2;
fourth degree monomial **11.** $x^3 + x^2 + 5$; cubic
trinomial **12.** Answers may vary.
Sample: $3z^4 - 5z^2 + 1$; 4 **13.** $-b^5 + 2b^3 + 6$
14. $8g^4 - 5g^2 + 11g + 5$
15. $7x^3 + 8x^2 - 3x + 12$ **16.** $t^3 - 5t^2 + 12t - 8$
17. $4y^2 + 3y + 4$
18. $7w^5 - 5w^4 - 7w^3 + w^2 + 3w - 3$
19. $-40x^2 + 16x$ **20.** $35g^3 + 15g^2 - 45g$
21. $-40t^4 + 24t^3 - 32t^2$ **22.** $5m^3 + 15m^2$
23. $-6w^4 - 8w^3 + 20w^2$ **24.** $-3b^3 + 5b^2 + 10b$
25. $3x$; $3x(3x^3 + 4x^2 + 2)$

26. $4t^2$; $4t^2(t^3 - 3t + 2)$
27. $10n^3$; $10n^3(4n^2 + 7n - 3)$
28. 2; $2(k^4 + 2k^3 - 3k - 4)$ **29.** $3d$; $3d(d - 2)$
30. $2m^2$; $2m^2(5m^2 - 6m + 2)$ **31.** 5; $5(2v - 1)$
32. $4w$; $4w(3w^2 + 2w + 5)$
33. $3d^3$; $3d^3(6d^2 + 2d + 3)$ **34.** 12; if the GCF of x and y is 3, the GCF of $4x$ and $4y$ is $4 \cdot 3$ or 12.
35. Kim; 4, m, and n are factors of both monomials. The GCF is their product.
36. $x^2 + 8x + 15$ **37.** $15v^2 - 29v - 14$
38. $6b^2 + 11b - 10$ **39.** $-k^2 + 5k - 4$
40. $p^3 + 3p^2 + 3p + 2$ **41.** $4a^2 - 21a + 5$
42. $y^3 - 9y^2 + 18y + 8$ **43.** $3x^2 + 10x + 8$
44. $-2h^3 + 11h^2 - 6h + 5$ **45.** $q^2 - 8q + 16$
46. $4k^6 + 20k^3 + 25$ **47.** $64 - 9t^4$ **48.** $4m^4 - 25$
49. $w^2 - 16$ **50.** $16g^4 - 25h^8$ **51.** $(2x + 1)(x + 4)$; $2x^2 + 9x + 4$ **52.** No; $(x - y)^2 = x^2 - 2xy + y^2 \neq x^2 - y^2$. **53.** $(x + 2)(x + 1)$ **54.** $(y - 7)(y - 2)$
55. $(x - 5)(x + 3)$ **56.** $(2w - 3)(w + 1)$
57. $(b - 3)(b - 4)$ **58.** $(2t - 1)(t + 2)$
59. $(x + 6)(x - 1)$ **60.** $2(3x + 2)(x + 1)$
61. $(7x + 2)(3x - 4)$ **62.** $(3x - 2)(x + 1)$
63. $(15y + 1)(y + 1)$ **64.** $(15y - 1)(y - 1)$
65. $(q + 1)^2$ **66.** $(b + 4)(b - 4)$ **67.** $(x - 2)^2$
68. $(2t + 11)(2t - 11)$ **69.** $(2d - 5)^2$ **70.** $(3c + 1)^2$
71. $(3k + 5)(3k - 5)$ **72.** $(x + 3)^2$ **73.** $6(2y + 1) \cdot$
$(2y - 1)$ **74.** $\frac{1}{2}d + 1$ **75.** The factors are the same.
76. No; only the square $(5u + 6)^2$ would have $25u^2$ and 36 as the first and last terms, however $2(5u)(6) \neq 65u$. **77.** $4x^2$; 2 **78.** $3k^2$; 2
79. $24y^2$; 4 **80.** $10n^3$; 7 **81.** $(3x^2 + 4)(2x + 1)$
82. $5y^2(2y + 3)(2y - 3)$ **83.** $3(3g - 1)(g + 2)$
84. $(3c - d)(2c - d)$ **85.** $(11k + 1)(k + 2)$
86. $3(u + 6)(u + 1)$ **87.** $(5p + 3)(3p + 1)$
88. $3(u - 6)(u - 1)$ **89.** $(h^2 - 3)(15h + 11)$
90. $(5x + 7)(6x^2 - 1)$ **91.** $4s^2t(3s - 1)(s + 2)$
92. $(x^2 + 2)(2x + 7)$ **93.** $2p$, $(p + 5)$, and $(3p + 4)$

Chapter 10

Lesson 10-1   pp. 553–556

EXERCISES 1. (2, 5); max. **3.** (2, 1); min.

5.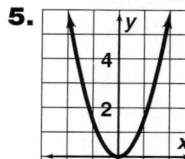
11. $f(x) = \frac{1}{3}x^2$, $f(x) = x^2$, $f(x) = 5x^2$
13. $f(x) = -\frac{2}{3}x^2$, $f(x) = -2x^2$, $f(x) = -4x^2$

15.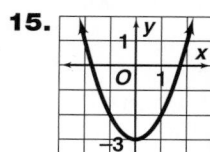
21. E **23.** F **27.** The graph of $y = 2x^2$ is narrower.
29. The graph of $y = 1.5x^2$ is narrower.

31. 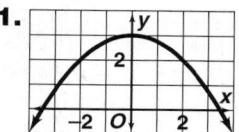 **35.**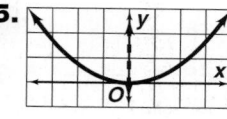

41. M **43.** M

Lesson 10-2   pp. 560–563

EXERCISES 1. $x = 0$, (0, 4) **3.** $x = 4$, (4, −25)
5. B **7.** C

11. 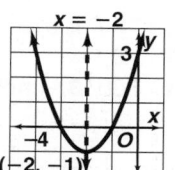 **15a.** 20 ft **b.** 400 ft^2

17. 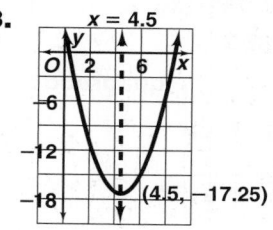 **23.**

33. Answers may vary. Sample: $y = -3x^2$
37. C **39.** 26 units^2

Lesson 10-3   pp. 567–570

EXERCISES

1. 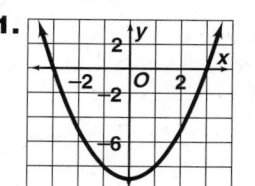 ±3

11. ±21 **13.** 0 **19.** $x^2 = 256$; 16 m **21.** $\pi r^2 = 80$;
5.0 cm **23.** none **25.** one **29.** $\pm\frac{3}{7}$ **31.** ±2.8
35. 121 **41.** 6.3 ft

Lesson 10-4   pp. 574–576

EXERCISES 1. 3, 7 **3.** 0, −1 **7.** −2, −5 **9.** 1, −4
11. 0, 8 **21.** 5 cm **25.** 2 and 3 or 7 and 8
27. $6n^2 - 5n - 4 = 0$; $\frac{4}{3}$, $-\frac{1}{2}$
29. $a^2 + 6a + 9 = 0$; −3

Lesson 10-5   pp. 582–584

EXERCISES 1. 49 **3.** 400 **7.** 4, −12 **9.** −5, −17
19. 1 **21.** $\frac{81}{100}$ **23.** 5, −1 **25a.** $(2x + 1)(x + 1)$ **b.**
$2x^2 + 3x + 1 = 28$ **c.** 3 **27.** −3, −4 **29.** 6, 2
39. 5.16, −1.16 **41.** 5.6 ft by 14.2 ft

Lesson 10-6   pp. 588–590

EXERCISES 1. $-1, -1.5$ **3.** 1.5 **17a.** $0 = -16t^2 + 50t + 3.5$ **b.** $t \approx 3.2$; 3.2 s **19.** Factoring or square roots; the equation is easily factorable and there is no x term. **21.** Quadratic formula; the equation cannot be factored. **25.** $0.87, -1.54$ **27.** $1.28, -2.61$ **33.** about 2.1 s **37.** 13.44 cm and 7.44 cm

Lesson 10-7   pp. 594–596

EXERCISES 1. A **3.** B **5.** 1 **7.** 2 **17.** No; the discriminant is negative. **19.** 0 **21.** 2 **25a.** $S = -0.75p^2 + 54p$ **b.** no **c.** \$36 **31.** no **33.** yes; $1, -1.25$

Lesson 10-8   pp. 601–604

EXERCISES

1. 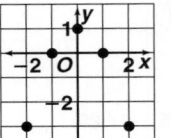   quadratic

7. quadratic; $y = 1.5x^2$ **9.** quadratic; $y = 2.8x^2$

13a. 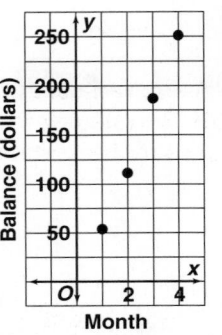   linear

b. 65, 64, 64; yes **c.** 64 **d.** $y = 64x - 5$ **15a.** 41, 123, 206 **b.** 82, 83 **c.** $d = 41t^2$ **d.** 256.25 cm **19.** $y = 0.875x^2 - 0.435x + 1.515$ **21.** $y = 2.125x^2 - 4.145x + 2.955$ **27a.** quadratic **b.** $d = 13.6t^2$ **c.** 54.4 ft **35.** 0

Extension   p. 605

1. Both graphs have the same shape, go through the origin, and lie in Quadrants I and III. The graph of $y = x^3$ is narrower than the graph of $y = \frac{1}{3}x^3$. **3.** Yes; the sign of a changes which quadrants the graphs are in, and the larger $|a|$, the narrower the graph.

5a.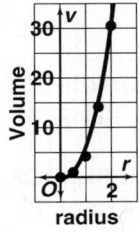

b. about 1.3 ft

Chapter Review   pp. 607–609

1. parabola **2.** completing the square **3.** principal square root **4.** vertex **5.** discriminant **6–9.** Answers may vary. Samples are given. **6.** $y = -2x^2$ **7.** $y = 2x^2$ **8.** $y = x^2$ **9.** $y = \frac{1}{2}x^2$

10. 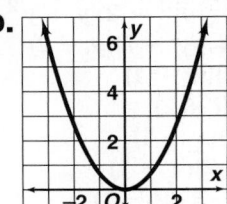   **11.**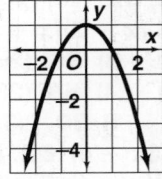

12. 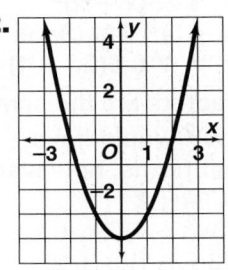   **13.**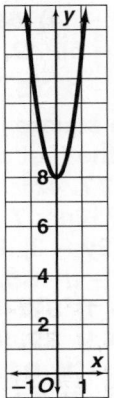

14. min. **15.** max. **16.** min. **17.** max.

18. 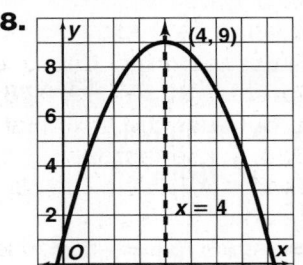   **19.**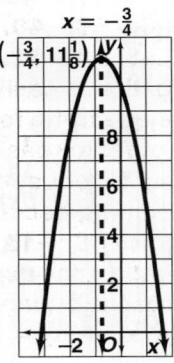

20.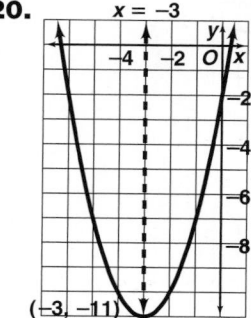

21.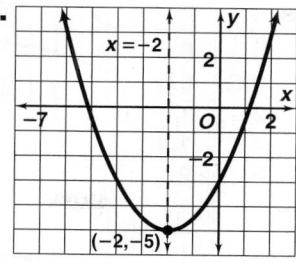

22. 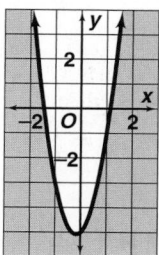 **23.**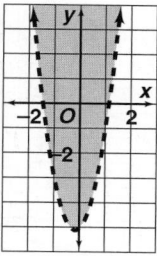

24. 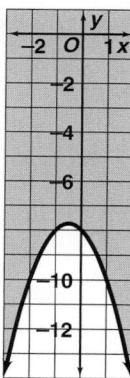 **25.**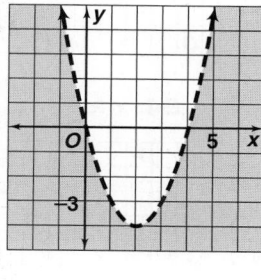

26. 2, −2 **27.** 5, −5 **28.** 0 **29.** no solution
30. $-\frac{2}{3}, \frac{2}{3}$ **31.** −4, 4 **32.** −4, −3 **33.** 0, 2 **34.** 4, 5
35. $-3, \frac{1}{2}$ **36.** $-\frac{2}{3}, 1\frac{1}{2}$ **37.** 1, 4 **38.** 2.3 in.
39. −6.74, 0.74 **40.** 0.38, 2.62 **41.** −2, $-1\frac{1}{2}$
42. 10 ft × 17 ft **43.** 2 **44.** 2 **45.** 0 **46.** 1
47. −1.84, 1.09 **48.** 0.5, 3 **49.** 0.13, 7.87
50. −5.48, 5.48 **51.** −5, 5; use factoring, because
the equation is easily factorable. **52.** −4.12, 0.78;
use the quadratic formula, because the trinomial
does not factor easily. **53.** 4, 5; use factoring,
because the equation is easily factorable. **54.** 3;
use factoring, because the equation is easily
factorable. **55.** −15, 15; use square roots, because
the equation has no x term. **56.** −8.47, 0.47;
complete the square, because the equation is in
the form $x^2 + bx = c$. **57.** 18 ft; 324 ft^2 **58.** 1.5 s

59. 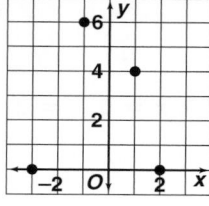 quadratic

60. 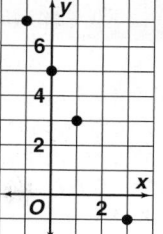 linear

61. exponential

62. quadratic

63. $y = 5(2)^x$ **64.** $y = 3x − 2$ **65.** $y = (x + 1)^2$
66. $y = \frac{1}{2}(10^x)$

Chapter 11

Lesson 11-1 pp. 619–621

EXERCISES 1. $10\sqrt{2}$ **3.** $5\sqrt{3}$

25. 3 mi **27.** 17 mi **29.** $\frac{3\sqrt{3}}{2}$ **31.** $\frac{2\sqrt{30}}{11}$ **45.** $\sqrt{5}$
47. $\frac{2\sqrt{10n}}{5n}$ **53.** not simplest form; radical in the
denominator of a fraction **55.** Simplest form;
radicand has no perfect-square factors other
than 1. **57.** 30 **59.** $\frac{3\sqrt{2}}{4}$ **69.** $-3 \pm 3\sqrt{2}$
71. $\frac{2 \pm \sqrt{10}}{3}$

Lesson 11-2 pp. 625–628

EXERCISES 1. $5\sqrt{6}$ **3.** $-2\sqrt{5}$ **7.** yes **9.** no
11. $-3\sqrt{3}$ **13.** $-2\sqrt{5}$ **35.** $-\frac{4}{3}$; −1.3
37. about 7.4 ft **39.** $6\sqrt{2} + 6\sqrt{3}$ or $6(\sqrt{2} + \sqrt{3})$
41. $8 + 2\sqrt{15}$ **47.** $\frac{\sqrt{10}}{2}$ **49.** $(10 + 10\sqrt{2})$ units
55. 9.1% **57.** 15.5%

Lesson 11-3 pp. 632–634

EXERCISES 1. 4 **3.** 36 **7.** 576 ft **9.** 4.5 **11.** 7
15. 2 **17.** none **21.** 3 **23.** no solution **29a.** 25
b. 11.25 **33.** 1600 ft **35.** no solution **37.** 1, 6

Lesson 11-4   pp. 640–643

EXERCISES 1. $x \geq 2$ **3.** $x \geq 0$

11.

x	$f(x)$
0	0
1	2
4	4

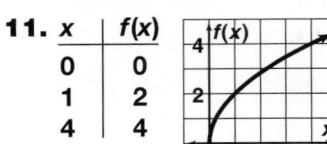

17. D **19.** C

21. 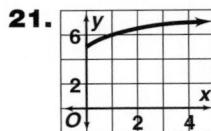   **31.** $x \leq 4$; $y \geq 0$ **35.** Translate the graph of $y = \sqrt{x}$ 10 units down. **37.** Translate the graph of $y = \sqrt{x}$ 9 units right.

39.

x	$f(x)$
0	0
1	4
2	5.7
4	8

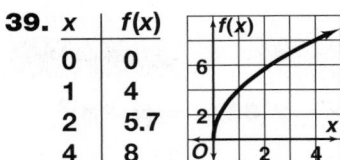

47. D **51.** False; x must equal 81.
53. True

Lesson 11-5   pp. 648–649

EXERCISES 1. $\frac{3}{5}$ **3.** $\frac{3}{4}$ **7.** 0.5299 **9.** 1.2799

15. 5.5 **17.** 19.2 **21.** $AC \approx 6$; $AB \approx 8$ **23.** $BC \approx$ 6; $AB \approx 18$ **27.** $b = 15$; $\sin A = \frac{8}{17}$; $\cos A = \frac{15}{17}$; $\tan A = \frac{8}{15}$ **31.** 4.5 **33.** $q \approx 6.1$; $r \approx 7.9$

Lesson 11-6   pp. 652–653

1. about 2 mi **3.** about 816 ft **5.** A

9. about 57 ft

Chapter Review   pp. 655–657

1. conjugates **2.** rationalize **3.** extraneous solution **4.** like radicals **5.** sine **6.** angle of elevation **7.** square root function **8.** trig. ratios
9. $48\sqrt{2}$ **10.** $\frac{2\sqrt{21}}{11}$ **11.** $20c^2\sqrt{6}$ **12.** $\frac{10\sqrt{13}}{13}$
13. $5h^2\sqrt{10}$ **14.** $6t^3\sqrt{5}$ **15.** $\frac{1}{2}$ **16.** $\frac{2\sqrt{6}}{3k^2}$ **17.** $\frac{3}{2}$
18. $10\sqrt{2}$ cm by $70\sqrt{2}$ cm **19.** $2\sqrt{7}$ **20.** $30\sqrt{5}$
21. $\sqrt{6}$ **22.** $2\sqrt{5}$ **23.** $10 - 10\sqrt{2}$
24. $-1 + 2\sqrt{14}$ **25.** $53 + 8\sqrt{15}$ **26.** $\sqrt{6} + \sqrt{3}$
27. $-3 - 3\sqrt{2}$ **28.** $17\sqrt{7}$ **29.** $4\sqrt{2} - 2\sqrt{7}$
30. $15\sqrt{2} - 30\sqrt{3}$ **31.** 2 **32.** 16 **33.** 17 **34.** 169
35. 8 **36.** 81 **37.** 100 **38.** 4 **39.** 1 **40.** 7
41. no solution **42.** 3 **43.** 4 **44.** no solution
45. 0 **46.** no solution **47.** $5\sqrt{5}$ cm **48.** 2.93 in.

49.

x	y
0	0
0.5	0.5
2	1
4	1.4
8	2

50.

x	y
0	0
1	$\frac{1}{2}$
4	1
9	$1\frac{1}{2}$

51.

x	y
0	0
$\frac{1}{2}$	1
2	2
8	4

52.

x	y
0	1
1	2
4	3
9	4

53. $x \geq 0$;   **54.** $x \geq 2$;

55. $x \geq -1$;   **56.** $x \geq 0$;

57. $AC \approx 10$; $BC \approx 7$ **58.** $AB \approx 29$; $AC \approx 28$
59. $AB \approx 10$; $BC \approx 6$ **60.** $AC \approx 24$; $BC \approx 5$
61. $AB \approx 27$; $AC \approx 20$ **62.** $AB \approx 20$; $BC \approx 4$
63. about 55 ft **64.** about 57 ft **65.** about 252 ft

Chapter 12

Lesson 12-1   pp. 668–670

EXERCISES

1.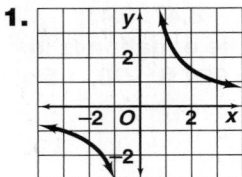

7. 2 **9.** 2 **11.** $x = -1$, $y = 0$ **13.** $x = 0$, $y = 2$

15. $x = 0$;

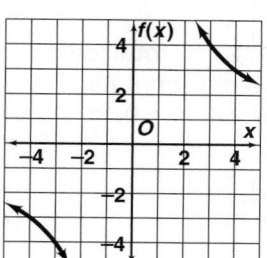

21. $x = 0, y = 5$;

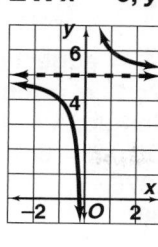

27. absolute value function with vertex (4, 0)
29. line with slope $\frac{1}{4}$, y-int. 0 **35.** moves graph
1 unit to the left **37.** lowers graph 15 units
43. $x = 0, y = 0$;

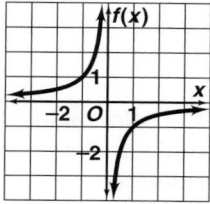

53. 17.8 lumens; $1.9\overline{7}$ lumens

Technology   p. 671

1.

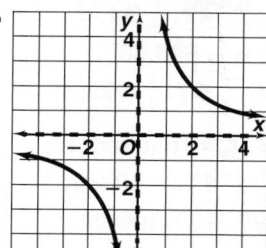

11a.

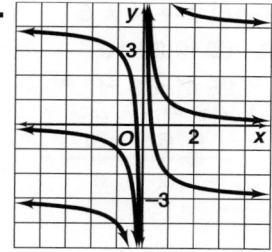

b. Adding translates the graph up, subtracting translates the graph down.

Lesson 12-2   pp. 674–676

EXERCISES 1. $\frac{2a + 3}{4}$ **3.** $\frac{1}{3}$ **23.** 13 min **25.** $\frac{2r - 1}{r + 5}$
27. $\frac{5t - 4}{3t - 1}$ **35a. i.** $\frac{2b + 4h}{bh}$ **ii.** $\frac{2h + 2r}{rh}$ **b.** $\frac{4}{9}$; $\frac{4}{9}$ **39.**
$\frac{1}{4}$ **41.** $\frac{t + 3}{3(t + 2)}$

Lesson 12-3   pp. 679–681

EXERCISES 1. $\frac{35x}{36}$ **3.** $\frac{40}{3a^5}$ **17.** $-\frac{2d - 5}{6d^2}$
19. $\frac{1}{s + 4}$ **21.** 6 **23.** $-\frac{1}{3}$ **29.** $t + 3$
31. $\frac{3t - 5}{7t^2}$ **39.** \$132.96 **41.** $\frac{x - 2}{4(x + 7)}$ **43.** $\frac{2}{a + 5}$

Lesson 12-4   pp. 684–686

EXERCISES 1. $x^4 - x^3 + x^2$ **3.** $3c^2 + 2c - \frac{1}{3}$
19. $(r^2 + 5r + 1)$ cm **21.** $b + 12 + \frac{1}{b + 4}$ **23.**
$10w - 681 + \frac{49,046}{w + 72}$
27. $3x + 2 - \frac{1}{2x}$ **29.** $2b^2 + 2b + 10 + \frac{10}{b - 1}$
47a. $t = \frac{d}{r}$ **b.** $t^2 - 7t + 12$

Lesson 12-5   pp. 689–691

EXERCISES 1. $\frac{9}{2m}$ **3.** $\frac{n + 2}{n + 3}$ **13.** $2x^2$ **15.** $7z$
17. $\frac{35 + 6a}{15a}$ **19.** $\frac{18 + 20x^2}{15x^8}$ **23.** $\frac{17m - 47}{(m + 2)(m - 7)}$
25. $\frac{a^2 + 12a + 15}{4(a + 3)}$ **31.** $\frac{h^2 + h + 1}{2t^2 - 7}$ **33.** $\frac{-3 - x - z}{xy^2z}$
45. $\frac{8x^2 - 1}{x}$ **47.** $\frac{-3x - 5}{x(x - 5)}$

Lesson 12-6   pp. 695–697

EXERCISES 1. -2 **3.** -1 **17.** ≈ 12.7 min
19. 10, -10 **21.** 4 **25.** $\frac{1}{2}$, 2 **27.** -5, 2 **37.** 20 Ω
39. 20 Ω

Lesson 12-7   pp. 702–705

EXERCISES

1. 10 choices
Shirt 1 → Tie 1 → S1, T1
      → Tie 2 → S1, T2
      → Tie 3 → S1, T3
      → Tie 4 → S1, T4
      → Tie 5 → S1, T5
Shirt 2 → Tie 1 → S2, T1
      → Tie 2 → S2, T2
      → Tie 3 → S2, T3
      → Tie 4 → S2, T4
      → Tie 5 → S2, T5
3a. 8, 10, 10, 10 **b.** 8,000,000 telephone numbers
5. 3,628,800 orders **7.** 1680 **9.** 360 **15.** 5040
17. $_8P_6$ **19.** $_8P_4$ **21a.** 24 **b.** $\frac{1}{24}$ **23.** 3 **25.** 5
27a. 260,000 license plates **b.** 23,920,000 license plates **29a.** 17,576 codes **b.** 17,526 codes

EXERCISES **1.** 1 **3.** 15 **11.** 220 **13.** $\frac{1}{5040}$ **15a.** 56 **b.** 1 **c.** $\frac{1}{56}$ **d.** $\frac{5}{28}$ **17.** 1 **19.** 4 **21.** permutation, since order *is* important **27.** 8

Chapter Review   pp. 715–717

1. rational expression **2.** asymptote
3. permutation **4.** rational equation **5.** −2 **6.** 0
7. 4 **8.** $-\frac{3}{5}$

9. $y = 0, x = 0$;   **10.** $y = 0, x = 0$;

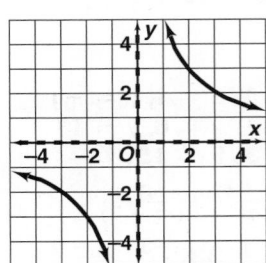 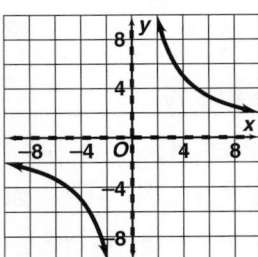

11. $y = 0, x = 5$;   **12.** $y = 2, x = 0$;

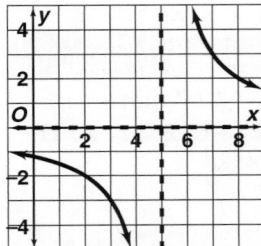 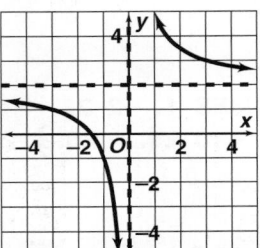

13. Answers may vary. Sample: $y = \frac{1}{x + 1}$

14. The graph of $f(x) = \frac{5}{x + 3}$ gets closer and closer to the lines $x = -3$ and $y = 0$. **15.** $\frac{3}{8t}$
16. $\frac{1}{3}$ **17.** $-\frac{1}{2}$ **18.** $x - 2$ **19.** $\frac{x}{4x + 3}$ **20.** 6
21. $\frac{-3}{t(t - 1)}$ **22.** $\frac{1}{2z - 3}$ **23.** $\frac{x + 2}{x + 4}$
24. $\frac{24m}{(m - 3)(m + 1)}$ **25.** 12 **26.** 3(n + 2)
27. $\frac{1}{2(2e + 1)}$ **28.** $2x - 4$ **29.** $3x^4 + 4x^3 - 1$
30. $50x^2 - 7x + \frac{1}{x}$ **31.** $x + 7 - \frac{5}{x + 1}$
32. $x + 4 - \frac{32}{x + 4}$ **33.** $2x^2 - 7x + 4$ **34.** $\frac{8x - 4}{x - 7}$
35. $\frac{7x + 24}{28x}$ **36.** $\frac{x^2 - 7x - 20}{x^2 + 2x - 8}$ **37.** 2
38. $\frac{15x^2 + 13x + 27}{(3x - 1)(2x + 3)}$ **39.** $\frac{-3m + 10}{(m + 1)(m - 1)}$
40. C **41.** 24 **42.** $-\frac{4}{35}$ **43.** 9 **44.** −14 **45.** −21
46. 2 **47.** 5 **48.** −7 **49.** ≈ 6 min **50.** 60
51. 1680 **52.** 360 **53.** 20 **54.** 3024 **55.** 840
56. 210 **57.** 15,120 **58a.** 160 possible area codes **b.** 640 new area codes **59.** 6 **60.** 210
61. 36 **62.** 165 **63.** 5 **64.** 5 **65.** 495 **66.** 495
67. 4 combinations **68.** 3003 ways

Extra Practice

CHAPTER 1 **1.** S = edge length; ℓ = total length of edges; $\ell = 12s$ **3.** e = length of rectangle, A = area of rectangle; $A = 12e$ **17.** < **19.** <
25. $C = 0.20E$ **27.** $C = 9n$
29. mean = 39.375
median = 38
mode = 35
31. mean = 6.3
median = 6
mode = 5,8
33. c = cost of gas
g = gallons
$c = 2.25g$
35. a = available tickets
s = tickets sold
$a = 4200 - s$

37. 7725 ft^2 **39.** Answer may vary Sample: false; $1 \div 2 = \frac{1}{2}$ **41.** number of people; total charge **43.** acres plowed; time **45.** no correlation; The length of a town's name is not likely to affect its population. **47.** negative; The lower the price, the more likely people are to buy it.

CHAPTER 2 **1.** −11 **3.** $\frac{-32}{15}$ **25.** $-16x + 12$
27. $8 - 4t$ **37.** $\frac{5}{54}$ **39.** 0 **45.** $\frac{1}{12}$ **47.** $\frac{1}{6}$ **51.** $\frac{5}{72}$
53. 75°C **55.** $16 **59.** $27.65 **61.** $48.60
63. Comm. Prop. of Mult. **65.** $\frac{1}{2}$ **67.** $\frac{1}{4}$ **69.** $\frac{5}{87}$

CHAPTER 3 **1.** 2 **3.** −5.7 **13.** t = test score; $\frac{87 + 84 + 85 + t}{4} = 90$; no **15.** 1 **17.** −40 **25.** 1
27. 26 **33.** bus: 5h; car: 4h **35.** 16% increase
37. 200% increase **41.** $\pm\frac{2}{3}$ **43.** $\pm\frac{5}{6}$
47. no; $3^2 + 7^2 \neq 10^2$ **49.** no; $3 + 4 \neq 5$
51. 10.3 **53.** 9.1 **55.** 7h **57.** $340; $510
61. about 16.9 ft/sec **63.** 1.25 mi **65.** 12.5% increase **67.** about 19.4 ft

CHAPTER 4 **1.** $w > -3$; ⊕ at −3 on number line −4−3−2−1 0 1
3. $r > -4$; ⊕ at −4 on number line −8−4 0 4 8 12 **25.** $18x \leq 102$, 5 balls **27.** $5 < w < 7$
29. $m \leq 4$ or $m > 8.4$ **37.** $|x| < 2$ **39.** $|x + 4| < 1$
43. $t > 1$ or $t < -1$; −4−2 0 2 4
45. $m > 3\frac{2}{3}$ or $m < -3$; −8−4 0 4 8
51. $|x - 2.5| < 0.5$
53. c is the number of cars sold; $c \geq 35$
55. y is age in years; $y \geq 18$
57. $375 + c \geq 500$, $c \geq 125$ **59.** $25c \geq 200$, $c \geq 8$
63. $32 \leq t \leq 212$ −100 0 100 200 300
67. $|r - 4.25| \leq 0.005$; $4.245 \leq r \leq 4.255$

CHAPTER 5 1.

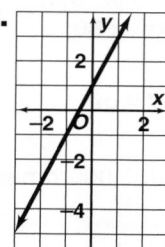

5. {−29, −11, −5, 13} **7.** {5, −1, 1, 19} **11.** yes
13. yes

15. 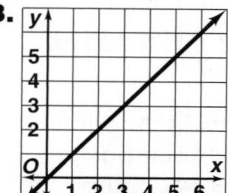 **17.**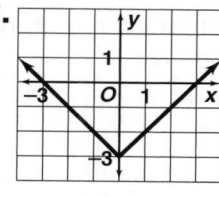

19. f(x) = 2x **21.** f(x) = $\frac{1}{2}$x

23. 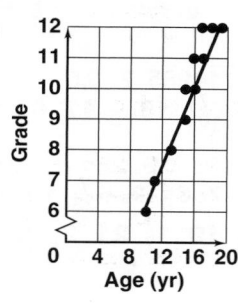 y = x

27. 70 **29.** 0.8 **31.** 7.5
33. 8 **37.** −4, −10 **39.** 5, 17 **43.** {19, 16, 11}
45. {8, 32, 72} **49.** A(w) = 100w
51. P(n) = $\frac{n}{16}$ **53.** P = 4s; 53.6 in.
55. rt = 12; about 6.9 mi/h
57. $9020, $8840, $8660

CHAPTER 6 1. 0.2 m/yr **3.** 1.6 mm/s
5. slope = 6, y-intercept = 8
7. slope = 0, y-intercept = 4
13. x-intercept = 2, y-intercept = 12
15. x-intercept = −6, y-intercept = 3
17. y − 6 = −5(x − 4) **19.** y − 5 = $\frac{1}{2}$(x − 8)

21.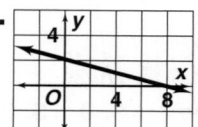

27. y = $\frac{3}{2}$x + 2
29. y = x + 6
35. 4x − y = −17
37. 4x − 3y = 25

41a–b.

c. grade = 0.720 · age − 1.118

43.

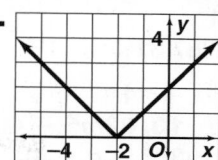

51. −250 ft/s **53.** y = 6x + 2575
55. t = 2n; even whole numbers
57. 3j + 2c = 25 **59.** y − 24 = $\frac{3}{4}$(x − 32)
61. False; the slopes of perpendicular lines have a product of −1, so one must be positive and the other must be negative. **63.** y = 50|x − 0.5|

CHAPTER 7 1. (4, −3);

1.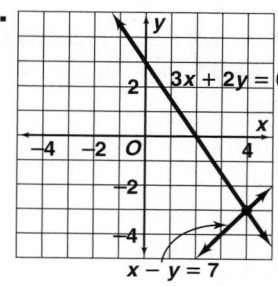

5. x = 1, y = −1
7. x = 6, y = 13
9. x = 4, y = −9
13. x + y = 12,
5x + y = 32, 5
nickels,
7 pennies
15. 2x + 2y = 74,
7x + 2y = 159,
length:
20 ft, width: 17 ft

19. 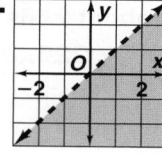 **25.**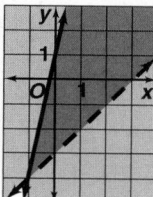

31. y = 0.5 + 0.02x
y = 0.25 + 0.03x
25 minutes
33. c + s = 300, s = c + 110; corn: 95 acres;
soybeans: 205 acres
35. 4h + 7c = 34.50, 8h + 3c = 30.50; h: $2.50,
c: $3.50 **37.** b = $\frac{4}{5}$g, b + g = 3321; 1476 boys,
1845 girls

CHAPTER 8 1. $\frac{1}{64t^6}$ **3.** 20.25 **17.** 9 **19.** 144
25. 3.4 × 10⁷ **27.** 1.5 × 10³ **33.** 8,050,000
35. 900,000,000 **41.** 1.5; 40.5, 60.75
43. 0.2; −0.008, −0.0016
49. {$\frac{1}{3}$, 1, 3, 9}; increase **51.** {$\frac{2}{3}$, 1, $\frac{3}{2}$, $\frac{9}{4}$}; increase
61. exponential growth; growth factor = 8
63. exponential decay; decay factor = $\frac{1}{2}$
69. y = 200(1.04)ˣ; $243.33
71. y = 3000(0.92)ˣ; $2336.06
73. 7.5 × 10⁹ **75.** about 9.5 × 10¹² km

77. about 9.5×10^{-11} m² **79.** 36 min
81. $A(n) = 60 \cdot (0.4)^{n-1}$; 0.6144
83. $f(x) = 15 \cdot 3^x$; 10,935 **85.** $987.61

CHAPTER 9 **1.** $2x^3 + 4x^2 - 11x + 11$
3. $6m^3 + m + 4$ **11.** $4b^3 + 12b$ **13.** $32m^2 - 40m$
19. $t^2(t^4 - t^3 + t^2 + 1)$ **21.** $4c^2(3c^3 - c + 4)$
27. $-5c^2 + 7c + 6$ **29.** $w^3 + 4w^2 + 3w - 2$
39. $(3x - 1)(2x + 5)$, $6x^2 + 13x - 5$
41. $(x - 3)(x - 1)$ **43.** $(v - 1)(v + 2)$
65. $(y + 3)(3y^2 - 1)$ **67.** $(w - 3)(w^2 + 3)$
73. $12k - 2$ **75.** $6d^3 + 2d^2 - 4d + 10$
77. $12w^3 + 20w$ **79.** $264y^2 - 106y - 28$
81. $16x^2 - 24x + 9$ cm², $4x^2 + 16x + 16$cm²;
441 cm², 256 cm² **83.** $m + 7$
85. $(12d - 7)$ by $(d - 2)$
87. $(8m - 13n)$ by $(8m + 13n)$

CHAPTER 10 **1.** narrower **3.** wider and reflected
over the x-axis **9.** $x = 0$, $(0, 0)$ **11.** $x = 0$, $(0, -3)$

19.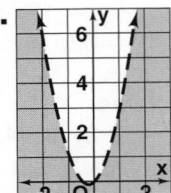
21. 6, −6
23. 10, −10
31. $\frac{-5 \pm \sqrt{53}}{2}$
33. $\frac{2 \pm \sqrt{2}}{2}$
49. 2 **51.** 1

59. 0.46875s **61.** $12 = \pi r^2$; 2.0 in.
63. $\left(\frac{D}{2}\right)^2 \pi = 300$; 19.6m **65.** 2s **67.** about 2.66s
69. quadratic; $y = 1.786x^2 - 24.043x + 226.371$

CHAPTER 11 **1.** $\frac{\sqrt{3}}{3}$ **3.** $\frac{5\sqrt{2}}{3}$ **11.** −10
13. $36\sqrt{3}$ **23.** 5 **25.** 4

31. $x \geq -5$;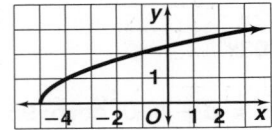

37. $\frac{9}{41}$ **39.** $\frac{9}{40}$ **43.** 0.9511 **45.** 1.1106 **49.** 201 m
51. 23 mi **53.** 5 mi **55.** $-20(1 - \sqrt{3})$; 14.6 cm
57. about 60 **59.** about 19 ft
61. 124,027 ft (about 23 mi)

CHAPTER 12
1. $x = 0$, $y = 0$;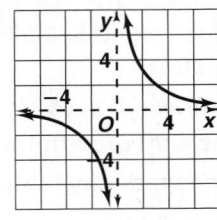

9. $\frac{t}{4}$ **11.** 4 **17.** $\frac{1}{x}$ **19.** $\frac{6c + 4}{c^2}$ **29.** $2x^2 + 5x + 2$

31. $3x^2 - 7x + 6$ **33.** 8
35. −1 **45.** 15 **47.** 252
53. 24 **55a.** 9000 **b.** $\frac{1}{9000}$
59. about 98 min **61.** $3x^2 - 4x + 7$ m²
63. $\frac{2}{r} + \frac{2}{1.6r} = \frac{3.25}{r}$ **65.** $8\frac{4}{7}$ min **67.** 52,360

Algebra Skills Handbook

p. 748 **1.** 10 games **3.** 9 tacks

p. 749 **1.** 17 and 19, or −19 and −17 **3.** 33, 34, 35,
and 36 **5.** 24 years old **7.** mother: 38 yr; son:
16 yr; daughter: 7 yr **9.** regular tickets: 80;
student tickets: 110

p. 750 **1.** 81 books **3.** 36 stories **5.** 7.2 m
7. 6 ways

p. 751 **1.** 21 lockers **3.** about 370,000,000 times
5. 63 games **7.** 32 and 33

p. 752 **1.** dog: K. C.; horse: Bo; bird: Cricket; cat:
Tuffy **3.** Alexa, Karin, Heather, Annette, Tanya,
Garo **5.** 126 players

p. 753 **1.** 6 **3.** 21 pencils **5.** 11 mi **7.** 290 mi
9. $4\frac{3}{4}$ mi east, 2 mi north

p. 754 **1.** composite **3.** composite **5.** prime
7. composite **9.** composite **11.** composite
13. composite **15.** composite **17.** composite
19. 1, 2, 23, 46 **21.** 1, 11 **23.** 1, 3, 9, 27 **25.** 1, 5,
41, 205 **27.** 1, 2, 3, 4, 6, 8, 12, 24 **29.** 1, 2, 4, 8, 11,
22, 44, 88 **31.** 1, 3, 61, 183 **33.** 1, 2, 7, 14, 49, 98
35. 1, 59 **37.** $2 \cdot 3 \cdot 3$ **39.** $3 \cdot 3 \cdot 3$ **41.** $2 \cdot 2 \cdot 2 \cdot 2 \cdot 2 \cdot 2$
43. $2 \cdot 2 \cdot 5 \cdot 5$ **45.** $2 \cdot 2 \cdot 3 \cdot 7$ **47.** $11 \cdot 11$

p. 755 **1.** 2 **3.** 24 **5.** 3 **7.** 7 **9.** 5 **11.** 21 **13.** 80
15. 33 **17.** 60 **19.** 240 **21.** 150 **23.** 40

p. 756 **1.** 2, 4 **3.** 3, 5 **5.** 3, 5, 9 **7.** 2 **9.** 2, 3, 4, 6, 8,
9 **11.** none **13.** 3, 5 **15.** 2, 3, 4, 6, 8, 9 **17.** 2, 4
19. 15 **21.** Answers may vary. Sample: $a + 1$ is
not divisible by 2. Dividing by 2 will leave a
remainder of 1.

p. 757 **1.** $350 **3.** $300 **5.** $17.00 **7.** 7.10 **9.** 7.00
11. $30.80

p. 758 **1.** $\frac{8}{14}$, $\frac{12}{21}$, $\frac{16}{28}$, $\frac{20}{35}$, $\frac{24}{42}$ **3.** $\frac{6}{16}$, $\frac{9}{24}$, $\frac{12}{32}$, $\frac{15}{40}$, $\frac{18}{48}$
5. $\frac{10}{12}$, $\frac{15}{18}$, $\frac{20}{24}$, $\frac{25}{30}$, $\frac{30}{36}$ **7.** 9 **9.** 48 **11.** 2 **13.** 9 **15.** 3
17. no **19.** no **21.** no **23.** $\frac{1}{2}$ **25.** $\frac{2}{3}$ **27.** $\frac{2}{5}$ **29.** $\frac{1}{3}$
31. $\frac{2}{5}$ **33.** $\frac{3}{4}$

p. 759 **1.** $0.\overline{3}$ **3.** $0.\overline{2}$ **5.** $0.\overline{714285}$ **7.** $0.\overline{5}$
9. $0.\overline{285714}$ **11.** 0.16 **13.** $\frac{7}{100}$ **15.** $\frac{7}{8}$ **17.** $6\frac{1}{3}$ **19.** $\frac{7}{9}$
21. $\frac{3}{8}$ **23.** $6\frac{12}{25}$

p. 760 **1.** $\frac{5}{7}$ **3.** 3 **5.** $10\frac{7}{15}$ **7.** $6\frac{2}{9}$ **9.** $6\frac{7}{33}$ **11.** $9\frac{2}{3}$
13. $13\frac{7}{16}$ **15.** $56\frac{11}{15}$ **17.** $\frac{3}{5}$ **19.** $1\frac{2}{7}$ **21.** $2\frac{3}{8}$ **23.** $3\frac{1}{3}$
25. $9\frac{4}{63}$ **27.** $2\frac{1}{6}$ **29.** $7\frac{5}{6}$

p. 761 **1.** $\frac{3}{10}$ **3.** $8\frac{5}{8}$ **5.** $3\frac{1}{2}$ **7.** $25\frac{3}{10}$ **9.** $2\frac{4}{5}$ **11.** 6
13. $9\frac{5}{6}$ **15.** $5\frac{15}{16}$ **17.** $\frac{8}{9}$ **19.** $\frac{18}{25}$ **21.** $13\frac{1}{3}$ **23.** $\frac{5}{6}$ **25.** 26
27. $1\frac{1}{3}$ **29.** $\frac{1}{3}$

p. 762 **1.** 56% **3.** 602% **5.** 820% **7.** 14.3%
9. 11.1% **11.** 75% **13.** 0.07 **15.** 0.009 **17.** 0.83
19. 0.15 **21.** 0.0003 **23.** 3.65 **25.** $\frac{19}{100}$ **27.** $4\frac{1}{2}$
29. $\frac{16}{25}$ **31.** $\frac{6}{25}$ **33.** $\frac{3}{800}$ **35.** $\frac{3}{5}$ **37.** $\frac{1}{50}$ **39.** $\frac{33}{50}$ **41.** $1\frac{1}{4}$

p. 763 **1.** 6^4 **3.** $5 \cdot 2^4$ **5.** $4^2 \cdot 3^2 \cdot 2$ **7.** 64 **9.** 144
11. 3267 **13.** $(1 \cdot 10^3) + (2 \cdot 10^2) + (5 \cdot 10^1) + (4 \cdot 10^0)$
15. $(8 \cdot 10^4) + (3 \cdot 10^3) + (4 \cdot 10^2) + (1 \cdot 10^0)$

p. 764 **1.** 100°, obtuse **3.** 180°, straight
5–12. Check students' work. **13.** Check students' work.

p. 765 **1.** 22 cm **3.** 24 cm^2 **5.** 216 cm^3
7. 351.68 cm^3

p. 766 **1.**

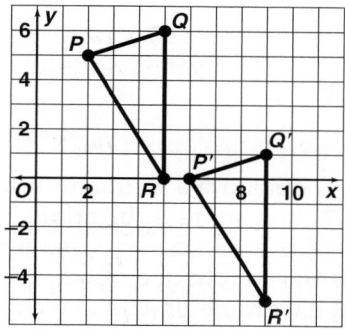

3. $(x, y) \rightarrow (x - 2, y - 1)$ **5.** $(x, y) \rightarrow (x + 4, y + \frac{1}{2})$
7. $(x, y) \rightarrow (x + 2, y + 5)$ **9.** $(x, y) \rightarrow (x + 6, y - 1)$
11. $(x, y) \rightarrow (x - 1, y + 1)$

p. 767 **1.**

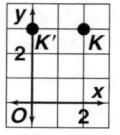

3.

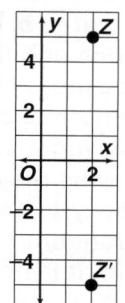

5.

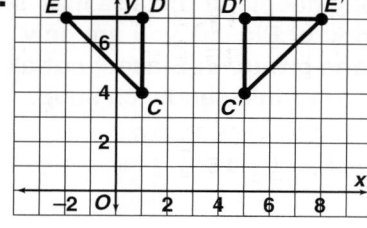

7.

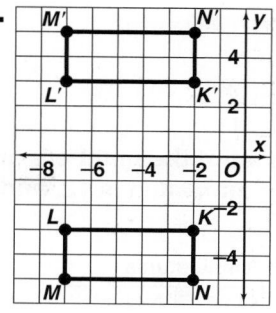

9. $x = 1$

p. 768 **1.**

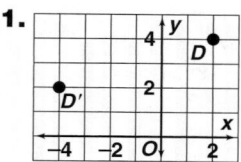

3.

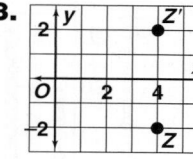

5a.

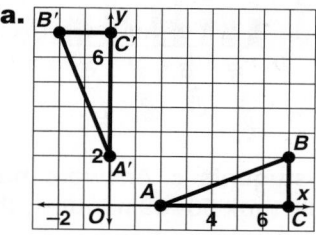

b.

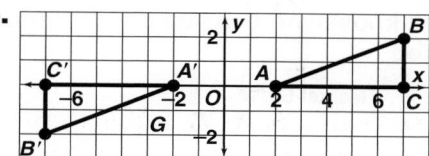

7a.

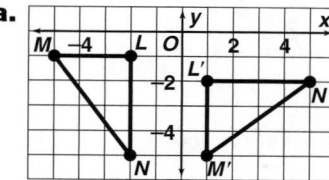

b.

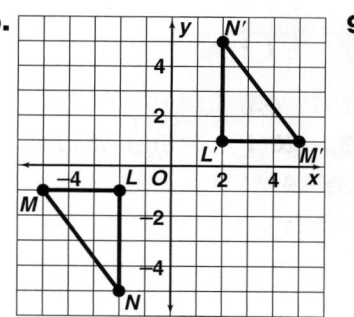

9. 90°

p. 769 **1.**

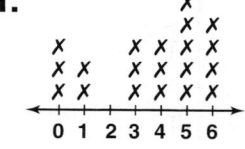

3.

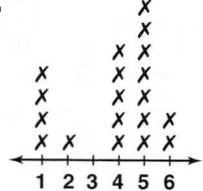

5.

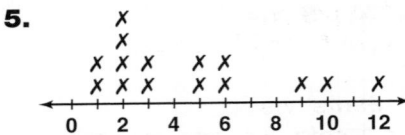

The line plot shows that most of the numbers are concentrated around 2, the maximum is 12, and the minimum is 1.

p. 770 1.

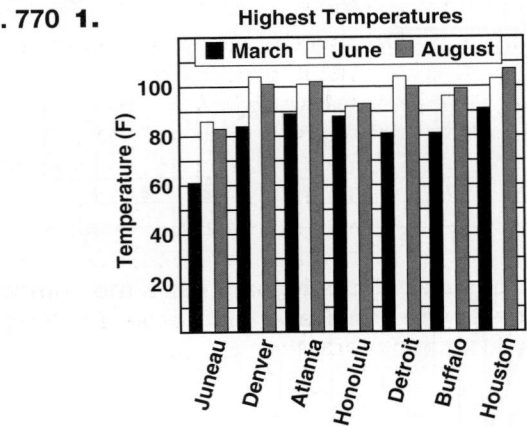

p. 771 1a. 30–39 **b.** 15 students **c.** Answers may vary. Sample: If it actually took 50–59 minutes, the student might estimate by saying 1 hour.
d. No; you don't know where inside each interval the answers are.

p. 772 1.

Market Share

p. 773 1a.

Transportation Mode

135 Bicycle
252 Walk
81 Car
432 Bus

b. 50% **c.** 3 times

p. 774 1.

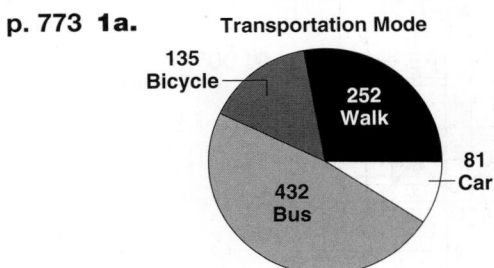

3.

9a. **b.** 2; 33, 33

c. 21 h; 23 h; both are much less than the 29-h typical week.

p. 775 1. Circle graph; the categories seem to cover how an average family spends all of its money in October. A circle graph also shows the percentage of a category more easily than a bar graph. **3.** Scatter plot; the levels of emissions most likely increase as the age of a car increases. A double bar graph would only be used if the data had an additional subject.

p. 776 1a. The graph implies that the runner's time on the 10-mile run was more than cut in half in six weeks, while actually the runner's time decreased by about 14%. **b.** Add a break on the vertical axis, or redraw the graph with the vertical axis starting at zero.

p. 777 1. = A2 + B2 + C2 **3.** = 1/2 * B2 * C2
5.a. D3 **b.** E4

Chapter NY

EXERCISES 1. qualitative **3.** quantitative
5. univariate **7.** univariate **9.** bivariate
11. Stratified; it is a bad sample because every shirt does not have a chance of being chosen.
13. Biased; the word *healthy* is biased. **15.** During the day many people are working and therefore not answering the phone. **17.** The milk carton increases in height as well as width. This gives the impression that the increase was much greater than it actually was. **19a.** People also travel by land and water. **b.** The question limits the choices to France and Italy. **c.** People who took French may be more likely to want to travel to France than the normal population.
21. quantitative; univariate **23.** qualitative; bivariate **25a.** Answers may vary. Sample: Her friends are more likely to see her website than people chosen at random. There are sports that are not listed. **b.** No, it is not a random survey and the list of sports is biased. **27a.** Biased; people leaving a bicycle shop probably like to bike. **b.** Biased; many people of the age that like to bicycle are at work. **c.** Biased; you are only surveying a select age group. **d.** Biased; not all town residents will go to this sandwich shop.
29. B

Lesson NY-2   pp. NY 736–NY 737

EXERCISES 1. 10, 7, 11 **3.** 52, 50.5, 54
5. 1, 4, 8.5, 30, 35 **7.** 61 **9.** 39 **11.** 4.65 **13.** 1.8
15. 4.8 **17.** Answers may vary. Sample: The
acreages vary considerably, from about 25 to
650 acres. However, about half of the parks are
between 50 and 250 acres, with a median of
125 acres. **19.** data set B **21.** data set B **23.** 65
25. A **27.** You can only find the median. **29.** J

Lesson NY-3   pp. NY 741–NY 742

EXERCISES 1. $M = \{2, 4, 6, 8, 10, 12, 14, 16\}$
3. $N = \{1, 3, 5, 7, 9, 11, \ldots\}$ **5.** Answers may vary.
Sample: $B = \{x \mid x$ is an integer and $x \geq 11\}$
7. Answers may vary. Sample: $S = \{x \mid x$ is a
factor of 12$\}$ **9.** $\{ \}, \{a\}, \{b\}, \{a, b\}$ **11.** $\{ \}, \{1\}, \{2\},$
$\{Z\}, \{1, 2\}, \{1, Z\}, \{2, Z\}, \{1, 2, Z\}$ **13.** $P' = \{1, 3, 5, 7,$
$9\}$ **15.** $[2, \infty)$ **17.** $(-4, \infty)$ **19.** $(0, 25]$ **21.** $[-4, -2)$
23. $[-3, 2)$ **25.** $(-4, 2)$ **27.** $(15, 55)$ **29.** True;
every element of A is an element of U. **31.** False;
not every element of B is in A. **33.** {peanut butter,
toast, grapefruit, orange juice, cereal, milk}
35. Check students' work.
37. $V' = \{1, 2, 3, 4, 5, 6, 7\}$ **39.** $M' = \{2, 4, 6\}$

Lesson NY-4   pp. NY 746–NY 747

EXERCISES 1. $\{2, 5, 7, 9\}$ **3.** $\{2, 4, 5, 6, 8\}$
5. $\{2, 4, 5, 6, 7, 8, 9\}$ **7.** $\{5\}$ **9.** $\{2\}$ **11.** ø
13.

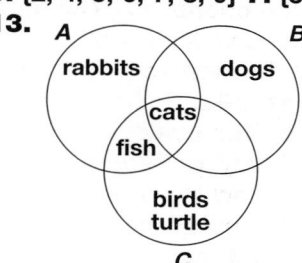

15.

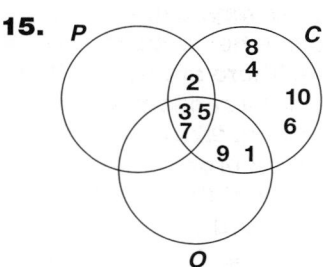

17. 69 people **19.** 158 people
21. The intersection of M and N will be
the multiples of 12.
23. True; check students' work.
25. False; 2 is in both sets. **27.** Answers may
vary. Sample: $A = \{1, 2, 3\}$ $B = \{4, 5\}$

Lesson NY-5   pp. NY 750–NY 751

EXERCISES 1–8. Answers may vary. Samples
are given. **1.** 25 grams; interpolation
3. about 50 grams; extrapolation
5. about 216 million television sets **7.** about
195 million television sets **9.** No; gaining weight
does not cause you to grow taller. **11.** No;
getting more cavities does not cause the size of
your vocabulary to increase. **13.** Yes; the
number of sides of a polygon causes the number
of diagonals you can draw to increase. **15.** Yes;
the age of a car causes the value of the car to
decrease.

Lesson NY-6   pp. NY 755–NY 757

EXERCISES 1. $(0, 1), (1, 2)$; 2;

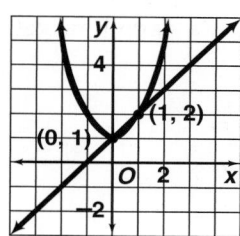

3. $(-1, 2), (4, -8)$; 2;

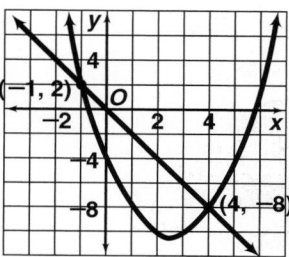

5. (−2, 5); 1;

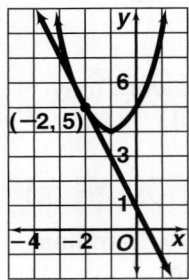

7. (−2, 5), (1, 2) **9.** (1, −8), (2, −9) **11.** (5, 5)
13. (−2, 2), (8, 42) **15.** (−10, 130), (−7, 100)

17. (−0.5, 0.5), (2, 8) **19.** 2 **21.** no solutions
23. 2 **25.** (−2, 6), (2, −2)
27. (2, −3), (5, 0) **29.** (6, 4)
31a. 2 solutions **b.** no solutions
33. (−0.25, 2) **35.** (−8, 199), (0.75, 24)
37. (−11, 452), (0.2, 4) **39.** A system of linear equations can have an infinite number of solutions because the graphs can be identical lines. However, a linear and quadratic function cannot have the same graph so a linear and quadratic system cannot have an infinite number of solutions. **41.** (−2.71, −3.36), (2.21, −0.89)
43. −3.41, 4.41

Index

Index

Index

Index

Rectangular solid, volume of, 680
Recursive formula, 296
Reflections, 767
Reflexive Property of Equality, 241
Regression, linear, 351. *See also* Best fit, line of
Reifenstuhl, Rocky, 85
Relations, 257–258, 300
Relationships
  causal, 36
  among function rule, table, and graph, 269
  modeling with equations, 5–6
  modeling with variables, 4–5
Repeating decimals, 759
Resnik, Judith A., 323
Review. *See also* Assessment; Extra Practice; Mixed Review
  Graphing on the Coordinate Plane, 24
  Proportions and Percents, 166–167
  Solving One-Step Equations, 118
  Surface Area and Volume, 612–613
  Transforming Formulas, 140–141
Rhombus, 348
Right angle, 764
Right triangles
  finding missing side lengths in, 647
  45-45-90, 188
  hypotenuse of, 181, 193
  identifying, 183–184
  isosceles, 188
  legs of, 181, 193
  Pythagorean Theorem and, 181–184, 188–189, 193
  ratios in, 644–645
  special, 188–189, 193
  30-60-90, 189
  trigonometric ratios in, 646–647, 657
Roots, of equation, 566, 571. *See also* Square root(s); Square root functions
Roster form, NY 738
Rotations, 768
Rubric-based exercises. *See* Extended Response exercises; Short Response exercises
Rule(s)
  for arithmetic sequences, 293–294
  for difference of squares, 530
  of exponential decay, 478
  of exponential growth, 475
  function. *See* Function rules
  Multiplication Counting Principle, 700
  Quadratic Formula, 586
  solving absolute value equations, 236
  solving absolute value inequalities, 236
Ruler, 114, 153

S

Sample, NY 727
Sample size, 546

Sample space, 93, 110
SAT preparation. *See* Test Prep
Scalar, 75
Scalar multiplication, 75
Scale, 151, 192
Scale drawings, 151, 192
Scale factor, 150, 156–157
Scale model, 271
Scatter plots, 33–34
  defined, 33, 49
  making, 33, 350–351
  trends in, 34, 350–351
Scientific calculator. *See* Calculator; Graphing calculator
Scientific notation, 436–438
  defined, 436, 486
  multiplying numbers in, 438, 442–443
  recognizing, 436
  using, 437–438
  writing numbers in, 436–437
Segment, midpoint of, 25
Sequences
  arithmetic, 293–294, 301, 461, 598
  defined, 293, 301
  Fibonacci, 296
  function rules for, 461–463
  geometric, 460–463, 487, 598
  terms of, 293–294, 301, 461, 462
  writing rules for, 293–294
Set, NY 738
Set-builder notation, NY 738
Short Response exercises, 108, 113, 148, 155, 173, 187, 205, 211, 225, 232, 240, 247, 256, 261, 268, 275, 282, 290, 297, 315, 323, 328, 335, 349, 369, 379, 386, 393, 402, 409, 418, 440, 446, 451, 465, 473, 482, 489, 499, 503, 510, 517, 523, 527, 538, 563, 570, 576, 584, 590, 596, 604, 611, 621, 628, 634, 643, 649, 670, 676, 681, 686, 691, 697, 711, 721
Sierpinski's Triangle, 464
Sigma, 636
Signs
  addition and, 57, 109
  division and, 72, 110
  multiplication and, 70, 110
Signum function, 267
Similar figures, 149–151, 192
Simpler problem, solving, 751
Simplifying
  absolute value, 65
  algebraic expressions, 80–81, 453–454
  defined, 9, 48
  exponential expressions, 71, 431, 455, 485
  expressions, 10–12, 48, 79, 80
  fractions, 618, 758
  like radicals, 622, 656
  numerical expressions, 10, 79
  order of operations and, 10
  powers, 12, 431, 448–449
  product raised to a power, 448–449
  products and quotients, 616–619, 623–624, 655

  radical expressions, 616–619, 623–624, 655, 656
  rational expressions, 672–674
  square root expressions, 176–177
  using prime factors, 617
Simulations, 100, 107. *See also* Mathematical modeling
Sine, 646, 657, 781
Skewed histogram, 305
Skill maintenance. *See* Mixed Review
Skills Handbook, 748–777
  Adding and Subtracting Fractions, 760
  Bar Graphs, 770
  Box-and-Whisker Plots, 774
  Choosing an Appropriate Graph, 775
  Circle Graphs, 773
  Divisibility, 756
  Draw a Diagram, 748
  Exponents, 763
  Factors and Multiples, 755
  Fractions, Decimals, and Percents, 762
  Fractions and Decimals, 759
  Histograms, 771
  Line Graphs, 772
  Line Plots, 769
  Look for a Pattern, 750
  Make a Table, 750
  Measuring and Classifying Angles, 764
  Misleading Graphs, 776
  Multiplying and Dividing Fractions, 761
  Perimeter, Area, and Volume, 765
  Prime Numbers and Composite Numbers, 754
  Reflections, 767
  Rotations, 768
  Simplifying Fractions, 758
  Solve a Simpler Problem, 751
  Spreadsheets, 777
  Translations, 766
  Try, Check, Revise, 749
  Use Estimation, 757
  Use Logical Reasoning, 752
  Work Backward, 753
Slope
  defined, 310, 365
  finding, 310–312
  formula for, 310
  of horizontal and vertical lines, 311
  identifying, 318
  negative, 312
  of parallel lines, 343, 366
  of perpendicular lines, 344, 366
Slope-intercept form of linear equations, 318, 339, 342, 366
Solid, rectangular, 680
Solution(s)
  of absolute value equation, 235–236, 245
  of absolute value inequality, 236–237, 245
  "Cannot be determined," 606
  of compound inequality, 227–229, 245
  defined, 118
  of equation, 118. *See also* Solving equations
  extraneous, 631, 656, 694
  with grouping symbols, 127

Index

solution of, 411–413, 423
writing and using, 413–414
Systems of quadratic equations, NY 752

T

Table(s), 269
  direct variation and, 279–280, 287
  finding rate of change using, 309
  frequency, 304, 305
  input-output, 258
  inverse variation and, 287
  linear functions and, 263, 277, 279–280, 309
  making from a function rule, 259
  of math symbols, 779
  modeling functions with, 263–265
  patterns and, 270
  solving equations using, 125, 134, 160
  solving inequalities using, 218, 224
  solving problems using, 160, 750
  solving systems of linear equations using, 380
  of squares and square roots, 780
  of units of measure, 778
  writing equations using, 338
  writing function rule from, 259, 270
TABLE key, of graphing calculator, 671
Takahashi, Naoko, 148
Take It to the Net. *See* Internet Links
Tangent, 646, 657, 781
TBLSET feature, of graphing calculator, 269, 380, 571
Technology. *See also* Calculator; Computers; Graphing calculator; Internet Links; Spreadsheets
  Activity Lab, 77, 100, 107, 125, 218, 269, 316, 358, 380, 394–395, 419, 474, 571, 671
  exercise, 445
  fulcrum, 286
  International Space Station, 612–613
  telecommunications, 457
  telecommunications device for the deaf (TDD), 261
  windlass, 280
Temperature, formulas for, 51, 71, 75, 124, 141
Terms
  defined, 81
  in geometric sequences, 461, 462
  like, 81, 110, 126, 192
  in sequence, 293–294, 301, 461, 462
Test(s), preparing for
  Gridded Response examples, 80, 221, 294, 311, 389, 469, 567, 630
  Gridded Response exercises, 68, 113, 124, 180, 216–217, 247, 261, 341, 369, 386, 435, 489, 532–533, 611, 704–705
  Multiple Choice examples, 5, 34, 71, 102, 127, 150, 208, 213, 253, 271, 318, 345, 375, 448, 455, 484, 500, 506, 552, 617
  Multiple Choice exercises, 7, 8, 13, 15, 22–23, 31–32, 36–37, 38, 39, 45, 51, 61, 62–63, 67, 75, 76, 83, 84–85, 89, 90, 97, 98–99, 105, 106, 113, 123,

131–132, 137, 139, 147, 148, 153, 154, 165, 173, 179, 185, 187, 203, 205, 210–211, 222, 225, 230, 232, 239–240, 247, 255, 256, 268, 273, 275, 282, 289, 290, 296–297, 315, 321, 323, 327, 328, 335, 346, 349, 355, 357, 363, 369, 379, 385, 393, 401–402, 409, 417, 418, 433, 439, 440, 445–446, 451, 457, 459, 465, 471, 472–473, 481, 482, 489, 499, 503, 509, 510, 517, 523, 527, 538, 555, 556, 561, 563, 569–570, 575–576, 584, 589, 590, 595–596, 604, 611, 621, 627–628, 634, 641, 642–643, 649, 652, 653, 670, 675–676, 681, 686, 691, 697, 703, 711, 719–721, NY 731, NY 737, NY 742, NY 747, NY 750, NY 757
  Test Prep, 8, 15, 22–23, 31–32, 36–37, 45, 51, 62–63, 68, 76, 84–85, 90, 98–99, 106, 113, 124, 131–132, 139, 148, 154–155, 165, 173, 180, 187, 195, 205, 210–211, 216–217, 225, 232, 239–240, 247, 256, 261, 268, 275, 282, 290, 296–297, 303, 315, 323, 328, 335, 341, 349, 355–356, 363, 369, 379, 386, 393, 401–402, 409, 418, 425, 435, 440, 445–446, 451, 459, 465, 472–473, 482, 489, 499, 503, 510, 517, 523, 527, 532–533, 538–539, 545, 556, 563, 569–570, 575–576, 584, 590, 595–596, 604, 611, 621, 627–628, 634, 642–643, 649, 653, 659, 670, 675–676, 681, 686, 691, 697, 704–705, 711, 719–721, NY 731, NY 737, NY 742, NY 747, NY 750, NY 757
  Test-Taking Strategies, 108, 190, 242, 298, 364, 420, 484, 540, 606, 654, 714, 748–753
  Test-Taking Tips, 5, 34, 71, 80, 94, 103, 127, 150, 162, 208, 213, 221, 253, 271, 294, 311, 318, 345, 375, 389, 448, 455, 469, 500, 506, 552, 567, 617, 630
Test Prep
  Cumulative Review, 113, 247, 369, 489, 611, 719–721
  exercises, 8, 15, 22–23, 31–32, 36–37, 45, 62–63, 68, 76, 84–85, 90, 98–99, 106, 124, 131–132, 139, 148, 154–155, 165, 173, 180, 187, 205, 210–211, 216–217, 225, 232, 239–240, 256, 261, 268, 275, 282, 290, 296–297, 315, 323, 328, 335, 341, 349, 355–356, 363, 379, 386, 393, 401–402, 409, 418, 435, 440, 445–446, 451, 459, 465, 472–473, 482, 499, 503, 510, 517, 523, 527, 532–533, 538–539, 556, 563, 569–570, 575–576, 584, 590, 595–596, 604, 621, 627–628, 634, 642–643, 649, 653, 670, 675–676, 681, 686, 691, 697, 704–705, 711, NY 731, NY 737, NY 742, NY 747, NY 750, NY 757
  Reading Comprehension, 51, 195, 303, 425, 545, 659
Testing. *See* Assessment
Test-Taking Strategies
  Answering the Question Asked, 714
  Choosing "Cannot Be Determined," 606
  Drawing a Diagram, 364
  Eliminating Answers, 540

Finding Multiple Correct Answers, 420
Interpreting Data, 242
Testing Multiple Choices, 484
Using a Variable, 298
Using Estimation, 654
Writing Extended Responses, 190
Writing Short Responses, 108
Test-Taking Tips, 5, 34, 71, 80, 94, 103, 127, 150, 162, 208, 213, 221, 253, 271, 294, 311, 318, 345, 375, 389, 448, 455, 469, 500, 506, 552, 567, 617, 630
Theorems
  45-45-90 Triangle, 188
  Pythagorean, 181–184, 193
  30-60-90 Triangle, 189
  Triangle Inequality, 210
Theoretical probability, 93–95
  defined, 93, 110
  finding, 94
  odds, 94–95, 111, 197
30-60-90 triangles, 189. *See also* Right triangles
Third quartile, NY 732–NY 737
Tiles. *See* Algebra tiles
Time. *See* Distance-rate-time problems
TRACE feature, of graphing calculator, 269, 671
Transformations, 766, 767, 768. *See also* Translations
Transforming formulas, 140–141
Transitive Property
  of Equality, 241
  of Inequality, 241
Translations, 766
  defined, 359, 367
  graphing, 360, 361, 367, 639
  horizontal, 361, 367, 639, 766
  vertical, 359–360, 367, 639, 766
Trapezoids, area of, 12, 15, 16, 508, 569
Tree diagrams, 660–661, 699–700
Trend. *See* Correlation(s)
Trend line, 34, 350–351, 352, 354, 355, 356, 357
Triangle(s)
  area of, 13, 113, 267, 685, 765
  legs of, 181, 193
  length of side of, 150
  Pascal's, 539
  perimeter of, 154, 165, 231, 288
  proportions in, 149
  right. *See* Right triangles
  Sierpinski's, 464
  similar, 150
  sum of angles of, 130
Triangle Inequality Theorem, 210
Trigonometric ratios
  cosine, 646, 657, 781
  defined, 646, 657
  finding, 646–647
  in right triangles, 646–647, 657
  sine, 646, 657, 781
  solving problems using, 650–651
  table of, 781
  tangent, 646, 657, 781

Trinomials
    factoring, 519–521, 524–525, 528–530, 535–536, 543
    multiplying a binomial by, 507
    multiplying a monomial by, 500
    perfect-square, 528–529, 543
Try, Check, Revise, 749
Two-step equations, 119–121, 191

U

Uniform histogram, 305
Uniform motion problems, 159–162
Union, of two sets, NY 743
Unit analysis, 143
Unit fractions, 63
Unit rate, 142
Univariate, NY 727
Univariate data, 546
Universal set, NY 738–NY 739
Unlike denominators, 688–689, 716
Unlike fractions, 760
Unlike radicals, 622, 656
Use Logical Reasoning, 752

V

Value, absolute. *See* Absolute value
Variable(s)
    on both sides of equation, 134–136
    on both sides of inequality, 221
    defined, 4, 48
    defining in terms of another, 158–159
    dependent, 29–30, 48
    domain of, 672, 715
    eliminating, 390, 422
    independent, 29–30, 48
    modeling relationships with, 4–5
    on one side of inequality, 219–221
    using, 298
Variable expressions, 26
Variation
    constant of, 278, 285, 301
    direct, 277–280, 286–287, 301
    linear functions and direct, 317
    inverse, 284–287, 291, 301
Venn diagrams, 39, NY 744–NY 745
Vermeer, Johannes, 113
Vertex
    of angle, 764
    degree of, 713
    identifying, 551
    in network, 712, 713
    of parabola, 551, 607
Vertical angles, 135
Vertical asymptote, 665–666, 714
Vertical lines, 311, 331
Vertical-line test, 258, 259, 260, 300
Vertical motion formula, 587
Vertical translations, 359–360, 367, 639, 766

Video Tutor Help Online, 10, 65, 102, 159, 201, 221, 263, 278, 310, 344, 390, 412, 448, 453, 506, 535, 579, 586, 623, 629, 688, 692
Video Tutor, Homework, 7, 13, 21, 30, 36, 44, 60, 67, 74, 84, 89, 97, 105, 123, 131, 138, 147, 154, 164, 172, 179, 186, 204, 210, 215, 223, 231, 239, 255, 260, 267, 273, 289, 296, 314, 321, 326, 334, 341, 347, 355, 362, 378, 385, 392, 401, 409, 416, 434, 440, 444, 450, 457, 464, 472, 480, 498, 502, 508, 516, 522, 526, 532, 538, 555, 561, 568, 574, 583, 589, 594, 602, 620, 626, 632, 641, 648, 652, 669, 675, 680, 685, 690, 695, 703, 709
Vinculum, 678
Vocabulary
    exercise, 256
    New Vocabulary, 4, 9, 17, 27, 33, 40, 56, 69, 79, 86, 93, 101, 134, 142, 149, 158, 168, 176, 181, 200, 206, 227, 257, 263, 277, 284, 292, 308, 317, 330, 336, 343, 350, 359, 374, 382, 387, 404, 411, 436, 460, 468, 475, 494, 528, 534, 550, 565, 572, 579, 585, 592, 616, 622, 629, 638, 646, 650, 664, 672, 692, 699, 706, NY 726, NY 732, NY 738, NY 743, NY 758
    Vocabulary Builder, 175, 234, 342, 578
    Vocabulary Quiz Online, 47, 109, 191, 243, 299, 365, 421, 485, 541, 607, 655, 715
    Vocabulary Review, 47, 109, 191, 243, 299, 365, 421, 485, 541, 607, 655, 715
    Vocabulary Tips, 4, 18, 19, 41, 72, 93, 94, 128, 135, 144, 176, 177, 181, 200, 201, 207, 223, 227, 259, 278, 285, 292, 308, 317, 337, 359, 376, 382, 387, 393, 431, 460, 495, 501, 586, 618, 638, 678, 701
Vocabulary Builder
    Connecting Ideas with Concept Maps, 578
    High-Use Academic Words, 234
    Learning Vocabulary, 175
    Understanding Vocabulary, 342
Vocabulary Quiz Online, 47, 109, 191, 243, 299, 365, 421, 485, 541, 607, 655, 715
Vocabulary Review, 47, 109, 191, 243, 299, 365, 421, 485, 541, 607, 655, 715
Vocabulary Tips, 4, 18, 19, 41, 72, 93, 94, 128, 135, 144, 176, 177, 181, 200, 201, 207, 223, 227, 259, 278, 285, 292, 308, 317, 337, 359, 376, 382, 387, 393, 431, 460, 495, 501, 586, 618, 638, 678, 701
Volume
    of cube, 450
    of cylinder, 14, 503, 538, 610, 632, 656
    defined, 765
    finding percent error in calculating, 170, 174
    formulas for, 16, 436, 516, 605, 765
    of rectangular prism, 174, 445, 536, 538, 543, 685
    of rectangular solid, 680
    scale factor and, 156–157
    of sphere, 14, 16, 436, 516, 605

W

Whole numbers, 17, 18, 48
WINDOW feature, of graphing calculator, 136, 269, 316
Word problems, 16, 329, 403, 591, 635, 698
Working backward, 753
Writing
    absolute value equations, 360, 361
    algebraic expressions, 4
    compound inequalities, 227, 229
    equation in standard form, 332
    equation of direct variation, 277–279
    exercises, 7, 13, 14, 21, 36, 38, 39, 44, 50, 61, 67, 74, 83, 89, 97, 104, 111, 112, 123, 130, 138, 147, 153, 164, 172, 179, 185, 203, 210, 216, 222, 230, 239, 246, 255, 261, 273, 281, 289, 294, 295, 300, 302, 304, 305, 314, 322, 327, 334, 335, 348, 354, 362, 370, 378, 392, 400, 408, 417, 420, 422, 424, 427, 439, 445, 451, 457, 463, 472, 481, 486, 488, 498, 502, 508, 516, 522, 526, 531, 538, 544, 546, 554, 562, 564, 569, 574, 583, 589, 595, 602, 609, 610, 613, 620, 627, 632, 633, 635, 640, 641, 643, 658, 669, 675, 680, 685, 690, 695, 703, 710, 715, 721, 774
    expressions, 81
    Extended Responses, 190
    function rules, 27–28, 48, 259, 270–271
    functions, 121
    inequalities, 201–202, 227, 229
    linear equations, 317–318, 338–339, 350–352
    numbers in scientific notation, 436–437
    rules for sequences, 293–294
    Short Responses, 108
    systems of linear equations, 396–399
    systems of linear inequalities, 413–414

X

x-axis, 24
x-bar symbol, 636
x-coordinate, 24
x-intercept, 330–331, 366, 565, 714

Y

Y= feature, of graphing calculator, 218
y-axis, 24
y-coordinate, 24, 598
y-intercept, 317, 318, 331, 366, 714

Z

Zero, 17
    division by, 73
    as exponent, 431, 485
    Multiplication Property of, 70, 87
Zero coefficient, 683
Zero-Product Property, 572–573, 608
Zeros of the function, 566, 571
ZOOM feature, of graphing calculator, 125, 136, 269, 316, 358

Acknowledgments

Staff Credits

The people who made up the High School Mathematics team—representing design services, editorial, editorial services, education technology, image services, marketing, market research, production services, publishing processes, and strategic markets—are listed below. Bold type denotes the core team members.

Leora Adler, **Scott Andrews,** Carolyn Artin, Stephanie Bradley, Amy D. Breaux, Peter Brooks, Judith D. Buice, Ronit Carter, Lisa J. Clark, Bob Cornell, Sheila DeFazio, Marian DeLollis, Kathleen Dempsey, Jo DiGiustini, Delphine Dupee, Emily Ellen, Janet Fauser, Debby Faust, Suzanne Feliciello, Frederick Fellows, Jonathan Fisher, Paula Foye, Paul Frisoli, **Patti Fromkin,** Paul Gagnon, Melissa Garcia, Jonathan Gorey, Jennifer Graham, **Ellen Granter,** Barbara Hardt, Daniel R. Hartjes, Richard Heater, Kerri Hoar, Jayne Holman, Karen Holtzman, Angela Husband, Kevin Jackson-Mead, Al Jacobson, Misty-Lynn Jenese, **Elizabeth Lehnertz,** Carolyn Lock, Diahanne Lucas, Catherine Maglio, **Cheryl Mahan,** Barry Maloney, Meredith Mascola, Ann McSweeney, Eve Melnechuk, Terri Mitchell, Sandy Morris, Cindy Noftle, Marsha Novak, Marie Opera, Jill Ort, **Michael Oster,** Steve Ouellette, Dorothy M. Preston, Rashid Ross, Donna Russo, Suzanne Schineller, Siri Schwartzman, Malti Sharma, Dennis Slattery, Emily Soltanoff, Deborah Sommer, Kathryn Smith, Lisa Smith-Ruvalcaba, Kara Stokes, Mark Tricca, Paula Vergith, Nate Walker, Diane Walsh, Merce Wilczek, **Joe Will,** Amy Winchester, Mary Jane Wolfe, Helen Young, Carol Zacny

Additional Credits: J. J. Andrews, Sarah Aubry, Jonathan Kier, Lucinda O'Neill, Sara Shelton, Ted Smykal, Michael Torocsik

Cover Design

Brainworx Studio

Cover Photos

Fern, Corbis; Rollercoaster, Robin Smith/Stone/ Getty Images, Inc.

Technical Illustration

Nesbitt Graphics, Inc.

Photo Research

Sue McDermott, Magellan Visual Research

Illustration

Argosy Illustration: 159, 161t, 161b, 222, 308, 321
Ken Batelman: 415, 562, 567, 651
John Edwards and Associates: 260, 280, 340, 432, 437, 451
Annette LeBlanc: 239
Ron Magnes: 164
Steve McEntee: 398
Morgan Cain & Associates: 59, 33, 36, 202, 142, 154, 575, 583
Matthew Pippin: 83
Brucie Rosch: 8, 17, 22; 66, 206, 42, 206, 153b, 391, 482, 494, 569
Phil Scheuer: 61
Neil Stewart: 476, 477
J/B Woolsey and Associates: 151t, 152, 114, 115
XNR Productions: 204, 224, 147, 151, 153, 700

Photography

Front matter: Page vi fl, Gary Holscher/Getty Images, Inc.; **vi l,** Chris Stevens/Dorling Kindersley; **vi ct,** Hans Sautter/Aurora Photos; **vi cb,** Michael Moran/Dorling Kindersley; **vi r,** PhotoDisc/Getty Images, Inc.; **vi fr,** Carr Clifton/Minden Pictures; **viii,** Jon Riley/Getty Images, Inc.; **x,** Frank Lane/Parfitt/Getty Images, Inc.; **xi,** Bob Daemmrich/The Image Works; **xii,** Rob Atkins/Getty Images, Inc.; **xiii,** Joseph McBride/Getty Images, Inc.; **xiv,** Wilfried Krecichwost/Getty Images, Inc.; **xix,** Jeff Greenberg/Photo Edit; **xv,** John Lund/Getty Images, Inc.; **xvi,** Ron Kimball/Ron Kimball Stock; xvii, R.D. Rubic/Precision Chromes, Inc.; **xviii,** Zefa/London/ Corbis Stock Market

Chapter 1: Pages 2, 3, Chris Noble/Getty Images, Inc.; **5,** Jon Riley/Getty Images, Inc.; **7,** Bob Daemmrich/Stock Boston; **12,** Steve Bronstein/The Image Bank/Getty Images, Inc.; **13,** PhotoEdit; **14 both,** The Granger Collection; **17,** Arnulf Husmo/Getty Images, Inc.; **20,** Dugald Bremner/Getty Images, Inc.; **21,** © Bill Amend/Universal Press Syndicate; 28, Richard Haynes; **30,** ImageState/Alamy Images; **32,** Bob Daemmrich Photography, Inc.; **34,** Russ Lappa; **41,** R. Krubner/Robertstock; **44,** Fred Whitehouse/Animals/Earth Scenes; **52 tr,** Corbis; **52 bl,** Dorling Kindersley/Charlestown Shipwreck Heritage Centre; **52 br,** EyeWire/Getty Images, Inc.; **53 tl,** AP/Wide World Photos; **53 r,** Stephen Frink/Getty Images, Inc.

Chapter 2: Pages 54, 55, David Madison/Getty Images, Inc.; **58,** John Elk, III/Stock Boston; **61,** Russ Lappa; **62,** Copyright © 1980 by Thaves. Distributed from www.thecomics.com.; **63,** British Museum/Michael Holford; **67,** Hal Beral/Visuals Unlimited; **71,** Nicholas Devore III/Bruce Coleman, Inc.; **75,** Courtesy of Cedar Point/Photo by Dan Feicht; **83,** Alan Thornton/Getty Images, Inc.; **87,** Russ Lappa; **92 all,** Russ Lappa; **94,** Russ Lappa; **95,** Lawrence Migdale/Stock Boston; **97,** David Young-Wolff/PhotoEdit; **98,** Duomo Photography, Inc.; **102,** Russ Lappa; **104,** Mark Thayer; **105,** Peter Timmermans/ Getty Images, Inc.; **114,** Jim Mone/AP/Wide World Photos; **114-115,** The Brearley Collection; **115,** ©1992 Ron Vesely/Major League Baseball Photos

Chapter 3: Pages 116, 117, V.C.L./Getty Images, Inc.; **120,** Redferns Music Picture Library/Alamy Images; **123,** SuperStock, Inc.; **127,** Kevin R. Morris/Getty Images, Inc.; **130,** VCL/Alistair Berg//Getty Images, Inc.; **132,** David Frazier//Getty Images, Inc.; **135,** Michelle Bridwell/PhotoEdit; **137,** Russ Lappa; **143 t,** Reuters New Media, Inc./Corbis; **143 b,** Frank Lane/Parfitt//Getty Images, inc.; **147,** Charles Gupton/Corbis; **149 both,** Russ Lappa; **160,** The Image Works; **169,** Don Mason/Corbis; **170,** Russ Lappa; **172,** NOAA; **173,** Mulvehill/The Image Works; **178,** David Austen/Getty Images, Inc.; **179 t,** NASA; **179 b,** Reprinted by permission: Tribune Media Services; **182,** Spencer Jones/Getty Images, Inc.; **185,** AFP/Corbis

Chapter 4: Pages 198, 199, Bohemian Nomad Picturemakers/Corbis; **202,** Tony Freeman/PhotoEdit; **204,** Poulides/Thatcher/Getty Images, Inc.; **209,** Clive Brunskill/ Allsport/Getty Images, Inc.; **210,** Corbis; **211,** Mug Shots/Corbis;

896 Acknowledgments